A Dictionary of
British
Place-Names

FIRST EDITION REVISED

A. D. MILLS

OXFORD

UNIVERSITY PRESS

OXFORD
UNIVERSITY PRESS

Great Clarendon Street, Oxford OX2 6DP

Oxford University Press is a department of the University of Oxford.
It furthers the University's objective of excellence in research, scholarship,
and education by publishing worldwide in

Oxford New York

Auckland Cape Town Dar es Salaam Hong Kong Karachi
Kuala Lumpur Madrid Melbourne Mexico City Nairobi
New Delhi Shanghai Taipei Toronto

With offices in

Argentina Austria Brazil Chile Czech Republic France Greece
Guatemala Hungary Italy Japan Poland Portugal Singapore
South Korea Switzerland Thailand Turkey Ukraine Vietnam

Oxford is a registered trade mark of Oxford University Press
in the UK and in certain other countries

Published in the United States
by Oxford University Press Inc., New York

British Library Cataloguing in Publication Data
Data available

Library of Congress Cataloging in Publication Data
Data available

Typeset by
SPI Publisher Services, Pondicherry, India
Printed in Great Britain on acid-free paper by
Clays Ltd, St Ives plc

ISBN 978-0-19-960908-6

A Dictionary of

British Place-Names

A. D. Mills is Emeritus Reader in English, University of London, and is a member of the Council of the English Place-Name Society and of the Society for Name Studies in Britain and Ireland. His previous books include *A Dictionary of English Place-Names* (Oxford University Press), *The Place-Names of Dorset* (English Place-Name Society), *The Place-Names of the Isle of Wight* (Paul Watkins), and *A Dictionary of London Place Names* (Oxford University Press).

(⊕) SEE WEB LINKS

This is a web-linked dictionary. There is a list of recommended web links in the Appendix of Web Links, on page 531. To access the websites, go to the dictionary's web page at www.oup.com/uk/reference/resources/ britishplacenames, click on **Web links** in the Resources section and click straight through to the relevant websites.

FOR
SOLVEJG

Preface

This new dictionary includes a representative selection of some 17,000 major place-names from the whole of the British Isles: England, Scotland and the Scottish islands, Wales, Northern Ireland and the Republic of Ireland, the Channel Islands, and the Isle of Man. It incorporates a fully revised, updated and slightly expanded version of my *Dictionary of English Place-Names* (Oxford University Press 1991, second edition 1998), together with wholly new additional entries from the other countries and the islands.

The revisions and refinements in the etymologies of a number of the English place-names included reflect the results of the most recent published research, partly that produced in the latest county volumes and other monographs published by the English Place-Name Society, partly that made available in various other major studies that have appeared in the last few years. The most important of these publications are listed in the updated Select Bibliography at the end of the book.

New entries covering a good selection of place-names from Ireland (both Northern Ireland and the Republic), Scotland and Wales have been compiled by Mr Adrian Room (well known to many readers as a prolific author of numerous books of reference over a wide range of subjects, including names), originally for *The Oxford Names Companion* (2002). The Publishers and I are particularly grateful to Mr Room for most diligently and efficiently providing the initial draft for these additional entries, as well as for drafting new entries for some thirty English district council names not included in my *Dictionary of English Place-Names*. Mr Room is also responsible for the useful and informative sections on Irish, Scottish and Welsh place-names in the enlarged Introduction.

We have been most fortunate in being able to call on the specialized knowledge of three eminent place-name experts to review the Irish, Scottish and Welsh entries respectively. Those for Northern Ireland and the Republic have been reviewed by Dr Kay Muhr, Senior Research Fellow of the Northern Ireland Place-Name Project in the Department of Irish and Celtic Studies at Queen's University, Belfast, a former President of the Society for Name Studies in Britain and Ireland: the Northern Ireland entries are particularly indebted to her colleague Pat McKay's *Dictionary of Ulster Place-Names* (Belfast 1999). Those for Scotland have been reviewed by Professor W. F. H. Nicolaisen, Honorary Research Professor at the University of Aberdeen and former President of the International Council of Onomastic Sciences. The entries for Wales have been reviewed by Professor Hywel Wyn Owen, Director of the Place-Name Research Centre at the University of Wales, Bangor and Director of the Board of Celtic Studies Place-Name Survey of Wales: indeed the Welsh place-name entries are heavily indebted to Professor Owen's *The Place-Name of Wales* (University of Wales Press, Pocket Guide series, 1998, reprint 2000). The Publishers and I would like to thank these scholars for all their specialist help and advice and for giving their time and expertise to the respective entries in the final list of names. Without their valuable and unique contribution, this volume would not have been possible.

Finally I would like to recall acknowledgements made in the first and second editions of my *Dictionary of English Place-Names* (1991 and 1998) to the late Professor Kenneth Cameron and the late Mr Victor Watts who in their capacity as successive Honorary Directors of the English Place-Name Survey kindly gave their permission to quote from the detailed county volumes published by the English Place-Name Society. The untimely deaths of these two distinguished and respected scholars is a sad loss to place-name studies. To them, and to the many others with whom I have discussed names or corresponded, or whose work has been published on various aspects of the subject either recently or over the last few decades, I remain deeply indebted. However I do of course take sole responsibility for the views finally proposed in this book, as well as for any errors or deficiencies remaining.

David Mills

West Wickham
February 2003

Preface to the Revised Edition

For this revised edition a few light corrections have been made to existing entries and a number of etymologies have been revised or refined in the light of recent research (examples include Blore, Corscombe, Grasby, Hameringham, Penruddock, Rattlesden, Shoreditch and Wambrook). Some 200 new entries have been added. These include several old names dating back to the Anglo-Saxon period (Baulking, Deanshanger, Evesbatch, Tibenham, Worgret), some of them of Scandinavian Viking origin (Antrobus, Galby, Minsmere) and some Celtic (Conock, Dunchideock). Also added are a number of more recent (and often somewhat exotic) transferred names (California, Botany Bay, Lilliput, Quebec), a handful of names that are particularly iconic or at least rich in association (Anfield, Maiden Castle, Piccadilly, Sutton Hoo), and many names (both older and more recent) that are unusual or even downright curious (Boot, Crackpot, Flash, Indian Queens, Old Wives Lees, Pant, Pennycomequick and Wham).

David Mills

Monks Eleigh
July 2011

Contents

Introduction

Place-Names and their Meanings

Place-names, those familiar but curious labels for places that feature in all their splendid variety on map and signpost, fulfil such an essential function in our daily lives that we take them very much for granted. Yet the place-names of the British Isles are as much part of our cultural heritage as the various languages, historical events and landscapes from which they spring, and almost every place-name has an older original meaning behind its modern form.

Indeed the place-names of the countries and regions that make up the British Isles present us with the most astonishing linguistic richness and diversity. The picture is a complicated one. Names of English (Old English) origin predominate in much of England, but there are significant numbers of Scandinavian (Old Danish and Old Norse) names in the north-west, north, and east of the country, names of Cornish origin in Cornwall, and an overall sprinkling of Norman-French, Latin (rare), and ancient Celtic (Brittonic) names add spice to the mixture. In Ireland, (much of) Scotland, and in Wales, names of Celtic (respectively Irish and Scots Gaelic and Welsh) origin predominate, but again there are significant contributions from English, Scandinavian and Norman-French.

It is probably the case that most people will have wondered at some time or other about the original meaning of a place-name—the name of their home town or of the other familiar places encountered en route to work by road or rail, the names of stations and destinations and those seen on roadsigns and signposts, and the more unusual names discovered on trips into the countryside or on holiday. Why Eccles, Stoke Poges, Great Snoring, or Leighton Buzzard? What is the meaning of names like Bangor, Banff, Bootle, or Ballynabrackey? How on earth did Croydon, Cricieth, and Crewe get their names, not to mention Billericay, Tipperary, and Drumnadrochit?

In fact all these names, like the vast majority of the names included in this dictionary, have original meanings that are not always apparent from their modern forms. That is because most place-names today are what could be termed 'linguistic fossils'. Although they originated as living units of speech, coined by our distant ancestors as descriptions of places in terms of their topography, appearance, situation, use, ownership, or other association, most have become, in the course of time, mere labels, no longer possessing a clear linguistic meaning. This is perhaps not surprising when one considers that most place-names are a thousand years old or more, and are expressed in vocabulary that may have evolved differently from the equivalent words in the ordinary language, or that may now be completely extinct or obscure.

Of course some place-names, even very old ones, have apparently changed very little through the many centuries of their existence, and may still convey something of their original meaning when the words from which they are composed have survived in the ordinary language (even though the features to which they refer may have changed or disappeared). Thus English names such as Claybrooke, Horseheath, Marshwood, Nettlebed, Oxford, Saltmarshe, Sandford, and Woodbridge are shown by their early spellings to be virtually self-explanatory, having undergone little or no change in form or spelling over a very long period.

But even a casual glance at the alphabetical list of place-names will show that such instant etymologies are usually a delusion. The modern form of a name can never be assumed to convey its original meaning without early spellings to confirm it, and indeed many names that look equally obvious and easy to interpret prove to have quite unexpected meanings in the light of the evidence of early records. Thus in England the name Easter is 'the sheep-fold', Slaughter 'the muddy place', Swine 'the creek or channel', and Wool 'the spring or springs'—the inevitable association of such names with well-known words in the ordinary vocabulary is understandable but quite misleading, for they all derive from old words which survive in fossilized form in place-names but which are no longer found in the language.

Names then can never be taken at their face value, but can only be correctly interpreted after the careful scrutiny of the earliest attested spellings in the light of the dialectal development of the sounds of the appropriate language, after wide comparisons have been made with similar or identical names, and after other linguistic, historical, and geographical factors have been taken into account. These fundamental principles of place-name etymology are most clearly illustrated by the names which now have identical forms but which prove to have quite distinct origins: for example, the English name Broughton (found also in Wales) occurs several times but has no less than three different origins ('brook farmstead', 'hill farmstead', and 'fortified farmstead'), the various places called Hinton fall into two distinct groups ('high farmstead' or 'farmstead belonging to a religious community'), and even a place-name like Ashford can be deceptive and means something other than 'ash-tree ford' in two instances. On the other hand, English names now with different spellings can turn out to have identical origins: thus Aldermaston and Alderminster are both 'nobleman's farmstead', Chiswick and Keswick are both 'cheese farm', Hatfield and Heathfield are both 'heathy open land', and Naunton, Newington, Newnton, Newton, and Niton are all 'new farmstead'. Even place-names from quite different linguistic backgrounds can turn out to have identical meanings. Like Blackpool in Lancashire, the name Dublin means 'the black pool' (referring no doubt to the dark waters of the River Liffey), and the Cornish name Penzance means 'holy headland' just like Holyhead in Wales. It goes without saying that guesswork on the basis of a modern form is of little use, and that each name must be the subject of individual scrutiny. For the same reason it should be remembered that the interpretation offered for a particular name in the list may not apply to another name with identical modern spelling occurring elsewhere, which might well have a quite different origin and meaning on the evidence of its early spellings and of other information.

Scope and Arrangement of the Dictionary

The main object of this dictionary is simple—to explain the most likely meanings and origins of some 17,000 British place-names in a clear, concise, and easily accessible form, based on the evidence and information so far available. The names included have been selected because they appear in all or several of the popular touring atlases, containing maps on a scale of three or four miles to the inch, produced by the Ordnance Survey and by the motoring organizations and other publishers. Thus the names of all the better-known places in the British Isles have been included: the names of towns and cities, of a good number of villages and hamlets and city suburbs, together with the names of counties and districts (old and new) and of many rivers and coastal features.

The entries are strictly alphabetical, each name being referred to the county or unitary authority in which the place is located. Priority in the entries is given to what the individual name 'means'. Thus wherever possible the suggested original meaning, that thought most likely as deduced from the evidence of early spellings and other information and from the fuller discussions of the name available in more detailed studies, is presented as a 'translation' into a modern English phrase of the old words or 'elements' that make up the name. The elements themselves are usually then cited in their original spelling and language, Celtic, Old English, Old Scandinavian, Gaelic, Welsh, or other as the case may be (a Glossary of some of the most common elements being provided at the end of the book).

Most names can be satisfactorily explained with respect to the elements from which they are derived, although the precise shades of meaning of the individual elements or of a particular compound may not always be easy to ascertain. For some names the evidence so far available is not decisive, and explanations may be somewhat provisional. A few remain doubtful or obscure or partly so. It is of course possible that earlier or better evidence may still come to light for some names, especially for places in those English counties like Durham, Hampshire, Kent, Lancashire, Somerset, and Suffolk for which there is as yet no English Place-Name Society survey, or for places in parts of Ireland, Scotland and Wales where detailed and systematic surveys have still to be completed.

Alternative explanations have often been given for names where two or more interpretations seem possible. For instance it is often difficult to say whether the qualifying element of a compound name is a personal name or a significant word, as in English names like Eversden, Hauxley, Hinxhill, Ranskill, and Yearsley. However for reasons of space some alternative explanations considered rather unlikely, problematical, or controversial have been omitted from the entries, in favour of those judged most plausible. Alternative interpretations of this kind are of course more fully rehearsed and discussed in the detailed surveys and monographs.

It should perhaps be pointed out that although the explanations suggested are considered to be the most likely, and are as accurate and reliable as possible within the limitations of scope and space imposed, final certainty in establishing the original meanings of many older place-names is unlikely to be achieved because of the nature of the materials. Given the archaic character of many place-names, and the fact that we can rarely know precisely when and by whom they were originally coined or came into use (as opposed to when they first appear in written records), there will always be an element of conjecture in their interpretation. However the study of place-names is a continually developing and evolving field, as the last few decades have shown, and further revision and refinement of etymologies is bound to come out of current and future research.

Inevitably the rather concise explanations of meaning and origin attempted in this dictionary, although based on the latest research, have meant leaving aside other important considerations. It has not been possible to enter into the complexities of philological argument, or to explore questions as to the precise nature or location of a topographical or habitative feature, or to examine the identity and status of a person associated with a place and the precise significance of that association. Such matters as these, and many other considerations bearing on the significance of a place-name in its historical, archaeological, and geographical context, are of course explored more fully in the various county surveys and studies of name-groups listed in the Select Bibliography, and they should be consulted by the interested reader wanting further information.

Although the scope of the present work does not allow for the presentation of a full range of early attested spellings such as would be required to provide visible support for the etymologies proposed in many cases, at least one early spelling (usually the earliest known) has been cited for most names, together with its date, to give some idea of the age of the name in question and of its original form. The sources of such spellings are not usually given, except where the source is of particular importance or interest, such as the Domesday Book of 1086 (abbreviated as 'DB'). Where spellings from Domesday Book or other early sources are followed by '[*sic*]'—Latin for 'thus'—this indicates that the spelling is cited exactly as it appears in that source even though it is apparently rather erratic or corrupt (the Norman scribes clearly had difficulty with the pronunciation and spelling of many British names!). In entries from the Celtic countries, since most Irish place-names and many Scottish and Welsh place-names have alternative Celtic and English or anglicized forms (not always corresponding in meaning), these have usually been noted. As before, readers needing further information about these alternative forms, or about the dates and sources of early spellings, or wishing to refer to fuller displays of spellings, should consult the detailed regional surveys and monographs (where these exist) listed in the Select Bibliography.

Elements and personal names cited with an asterisk are postulated or hypothetical forms, that is although there may be good evidence for their assumed existence in the early languages in question, they are either not recorded in independent use or are only found in use at a later date. To avoid unnecessary complication, the terminology for the provenance of elements and personal names has been somewhat generalized: for instance Old English (OE) stands for all dialects, Anglian, West Saxon, etc.; Old Scandinavian (OScand.) embraces Old Norse and Old Danish as well as forms more correctly labelled Anglo-Scandinavian; Old French (OFrench) includes Norman-French, Anglo-Norman, etc.; and the term Celtic is used for British, Primitive Welsh, and the other early related Brittonic languages. Similarly the term 'personal name' is used of personal names proper as well as of bynames formed in the early period.

In compound names, where both elements are from the same language the term or abbreviation for that language appears only once: e.g. OE *sand* + *wīc* for Sandwich. Where two elements are from different languages in a so-called hybrid name, each element is separately labelled, e.g. OE **wilig* + OScand. *toft* for Willitoft. Cross references to other place-names in the alphabetical list are given in small capitals. Place-names and river-names no longer in current use are printed in italics (e.g. *Ashwell* under Ashwellthorpe, *Ravenser* under Spurn Head, and River *Ann* under Amport and Andover).

Elliptical place-names of various kinds have been given the fuller meanings that seemed appropriate. Thus English names of the type Byfleet and Underbarrow, consisting of preposition + noun, literally 'by stream' and 'under hill', have been translated '(place) by the stream', '(place) under the hill' to bring out the implicit meaning. So-called folk-names (originally the names of family or tribal groups rather than of places) have been similarly treated, for example Barking has been rendered '(settlement of) the family or followers of a man called **Berica*'.

The Chronology and Languages of English Place-Names

Place-names show an astonishing capacity for survival, as the dates alone of most of the earliest spellings testify, even though it should be remembered that every name will of course be older than its earliest occurrence in the records, often a good deal older. In general it might be claimed that most of the English names included in this book are about a thousand years old, and that a good many are older than that. The various strata of English place-names reflect all the great historical migrations, conquests, and settlements of the past and the different languages spoken by successive waves of inhabitants.

Some river-names, few in number but the most ancient of all, seem to belong to an unknown early Indo-European language which is neither Celtic nor Germanic. Such pre-Celtic names, sometimes termed 'Old European', may have been in use among the very early inhabitants of these islands in Neolithic times, and it is assumed that they were passed on to Celtic settlers arriving from the Continent about the 4th century BC. Among the ancient names that possibly belong to this small but important group are Colne, Humber, Itchen, and Wey.

During the last four centuries BC there took place the invasions and settlements of the Iron-Age Celts, peoples speaking various Celtic dialects which can be divided into two main groups, Goidelic or Gaelic (later differentiated into Irish, Scots, and Manx) and Brittonic or British (later differentiated into Welsh and Cornish). Celtic place-names coined in British (really the language of the ancient Britons) were in use for several centuries and some have survived from the period when this Celtic language was spoken over the whole of what is now England as well as further west. These early place-names of Celtic or British origin were borrowed by the Anglo-Saxons when they came to Britain from the 5th century AD onwards and are found all over England, only sporadically in the east but increasing in numbers further west towards Cornwall and Wales where they are of course still predominant. Celtic place-names belong for the most part to several well-defined categories: names of tribes or territories like Devon and Leeds, names of important towns and cities like Carlisle, York, and Dover, names of hills and forests (now often transferred to places) like Crick, Mellor, Penge, and Lytchett, and most frequent of all, river-names like Avon, Exe, Frome, Peover, and Trent. There are also a good many hybrid names, consisting of a Celtic name to which an Old English element has been added, like Lichfield, Chatham, Bredon, and Manchester. Some places, important at a very early date, had Celtic names in Romano-British times which were later replaced, for instance Cambridge was *Duroliponte*, Canterbury was *Durovernum*, and Leicester was *Ratae*: for these, reference should be made to the fuller treatments of individual names in the county surveys or to the specialized study by Rivet and Smith (*see* Bibliography).

The Roman occupation of Britain during the first four centuries AD left little mark on place-names, for it is clear that Latin was mainly the official written language of government and administration rather than the spoken language of the countryside. Thus Celtic names, though usually Latinized in written sources, continued to be used throughout this period and were not replaced. However a few early names like Catterick and Lincoln contain Latin elements, and others like Eccles and (probably) Caterham were coined from Celtic elements that were borrowed from Latin during this early period. The small part played by Latin in place-name formation during the Romano-British period should be distinguished from the later influence of Latin on English place-names during the medieval period. In the Middle Ages, Latin was again the language of the church and administration, and this Medieval Latin was widely used in affixes like *Forum* 'market', *Magna* 'great', and *Regis* 'of the king' to distinguish places with identical names, as well as occasionally in the formation of names like Bruera, Dacorum, and Pontefract.

The Anglo-Saxon conquest and settlement of Britain began in the 5th century AD, spreading from east to west and culminating in the occupation of the whole of what is now England (except for Cornwall and some areas along the Welsh border), as well as south-east Scotland, by the 9th century. These new settlers were the Angles, Saxons, and Jutes, Germanic tribes from Northern Europe whose language was Anglo-Saxon, now usually called Old English to emphasize its continuity with Middle and Modern English. It is in this language, Old English, that the great majority of the place-names now in use in England were coined. This dominant stratum in English place-names (apart from those of Cornwall) is a result of the political domination by the Anglo-Saxons of the Celtic-speaking Britons and the gradual imposition of the Old English language on them. Many Celtic names were borrowed by the incomers as already mentioned (important evidence for the survival of a British population and for continuity and contact between the two peoples), but thousands of new names were coined in Old English during the Anglo-Saxon period between the 5th and 11th centuries. Thus the majority of English towns and villages, and a good many hamlets and landscape features, have names of Old English origin that predate the Norman Conquest. These names vary in age, and it is not always easy to tell which names belong to the earlier phases of the settlement and which to the later part of the Anglo-Saxon period, although detailed studies have shown that many of the names containing the elements *hām*, *-ingas*, *-inga-*, *ēg*, *feld*, *ford*, and *dūn* are among the earliest. It should in any case be remembered that all names are older than their earliest recorded spelling, so that names first mentioned in, for example, Domesday Book (1086) or even in a 12th-century source usually have their origins in this period.

The Scandinavian invasions and settlements took place during the 9th, 10th, and 11th centuries and resulted in many place-names of Scandinavian origin in the north, north-west and east of England (as well of course as in many other parts of the British Isles). The Vikings came to Britain from two Scandinavian countries, Denmark and Norway, the Danes settling principally in East Anglia, the East Midlands, and a large part of Yorkshire, whilst the Norwegians were mainly concentrated in the north-west, especially Lancashire and Cumbria (as well as in areas outside England, particularly northern and western Scotland, the Isle of Man and the Scottish islands, and coastal districts of Ireland and Wales). The Germanic languages spoken by these Vikings, Old Danish and Old Norse (in this book referred to jointly as Old Scandinavian), were similar in many ways to Old English, but there were also striking differences in sound system and vocabulary which reveal themselves in the early spellings of many place-names from the areas mentioned. Although names of Scandinavian origin are rare to the south of Watling Street (because that formed the boundary of the Danelaw, which was the area subject to Danish law, established in the late 9th century), the distribution of Scandinavian names in the north and east varies greatly, parts of Norfolk, Leicestershire, Lincolnshire, and Yorkshire being among the areas with the thickest concentration. To explain such large numbers of Scandinavian place-names in these areas, recent scholarship has suggested that in addition to settlements made by Viking warriors and their descendants there was probably a large-scale migration and colonization from the Scandinavian homelands in the wake of the invasion. Many hundreds of names in the areas mentioned are completely of Scandinavian origin (Kirkby, Lowestoft, Scunthorpe, Braithwaite), others are hybrids, a mixture of Scandinavian and English (Grimston, Durham, Welby), and some (on account of the similarity of some Old English and Old Scandinavian words) could be from either language (Crook, Kettleburgh, Lytham, Snape). In addition many place-names of Old English origin were modified by Scandinavian speech in these areas, for example by the substitution of *sk* and *k* sounds for *sh* and *ch* in names like Skidbrooke, Skipton, Keswick, and Kippax.

The number of English place-names of French origin is relatively small, in spite of the far-reaching effects of the Norman Conquest on English social and political life and on the English language in general. It is clear that by 1066, most settlements and landscape features already had established names, but the new French-speaking aristocracy and ecclesiastical hierarchy often gave distinctively French names to their castles, estates, and monasteries (Battle, Belvoir, Grosmont, Montacute, Richmond), some of them transferred directly from France, and there are a few names of French origin referring to landscape and other features (Devizes, Malpas). However the French influence on English place-names is perhaps most evident in the way the names of the great French-speaking feudal families were affixed to the names of the manors they possessed. These manorial additions result in a great many hybrid 'double-barrelled' names which contribute considerable variety and richness to the map of England. Most of them serve to distinguish one manor from another with an identical name, and of course the surnames of the more powerful land-owning families occur in a good many different place-names (Kingston Lacy, Stanton Lacy, Sutton Courtenay, Hirst Courtney, Drayton Bassett, Wootton Bassett, and so on). Some place-names of this type are not easily recognizable from their modern spellings, since the manorial affixes are now compounded with the original elements (Herstmonceux, Owermoigne, Stogursey). A further important aspect of the French influence on English place-names is the way it affected their spelling and pronunciation. Norman scribes had difficulty with some English sounds, often substituting their own (as seen for instance in the spellings of Domesday Book and other early medieval sources). Some of these Norman spellings have had a permanent effect on the names in question and have remained in use, disguising the original forms (Cambridge, Cannock, Diss, Durham, Nottingham, Salop, Trafford).

Of course not all of the names on the modern map, even names of sizeable settlements or well-known features, are as old as most of those so far mentioned. Other names besides the French names already noted originated in the Middle English period, that is between the 12th and 15th centuries inclusive. These include settlement names incorporating post-Conquest personal names and surnames like Bassenthwaite, Forston, and Vauxhall, names containing old elements but not on early record like Bournemouth and Paddock Wood, and various other names such as Broadstairs, Forest Row, Poplar, and Sacriston.

Finally there are some place-names, perhaps surprisingly few, which originate in the post-medieval period or even in quite recent times. Many of course are names of new industrial towns or of suburban developments, others are names of coastal resorts or ports or of new administrative districts. Most of these 'modern' names seem rather artificial creations compared with the earlier place-names that began life as actual descriptions of habitations or natural features. Some are in fact simply straight transfers of older names without any change of form (like the London borough-name Waltham Forest, or the 'revived' district names Bassetlaw and Dacorum), some are based on rather fanciful identifications of ancient names made by early antiquarians (like Adur and Morecambe), and others are new adaptations of existing old names with some sort of addition (like Devonport, Thamesmead, and New Brighton). Of the newly formed modern names, some are straightforwardly descriptive of a local feature whether natural (Highcliffe) or man-made (Ironbridge), others are named from a building around which the settlement developed (the pub in Nelson and Queensbury, the chapel in St Helens and Chapel St Leonards), some are named from fields (Hassocks and Whyteleafe), others refer to local products (Coalville, Port Sunlight), commemorate a famous historical event (Peacehaven, Vigo, Waterloo) or even a famous novel (Westward Ho!). In addition a good number of the names coined in more recent times commemorate entrepreneurs or other notable individuals, some consisting simply of their names (Fleetwood, Peterlee, Telford), others incorporating these into a sort of spurious form that looks older than it is (Carterton, Maryport, Stewartby), others referring to landowners (Camden Town) or local families (Burgess Hill, Gerrards Cross).

Some Different Place-Name Types and Structures

All English place-names, whether of Celtic, Old English, or Scandinavian origin, can be divided into three main groups: folk-names, habitative names, and topographical names.

Of the three, **folk-names** form the smallest group though nevertheless a very important and interesting one. Place-names in this category were originally the names of the inhabitants of a place or district. Thus tribal names came to denote the district occupied by the tribe, as with Essex and Sussex (both old Anglo-Saxon kingdoms), and Norfolk and Suffolk (divisions of the Anglo-Saxon kingdom of East Anglia). The names Jarrow, Hitchin, and Ripon also represent tribes (later their territories) from Anglo-Saxon times, and names like Clewer and Ridware must represent the settlements of smaller groups. Of particular interest, because they are to be associated with the early phases of the Anglo-Saxon settlement, are the names formed with the suffix -*ingas* ('people of', 'dwellers at') like Hastings, Reading, and Spalding, all of them originally denoting family or tribal groups, later their settlements.

Habitative names form a much larger group. They denoted inhabited places from the start, whether homesteads, farms or enclosures, villages or hamlets, strongholds, cottages, or other kinds of building or settlement. In names of this type the second element describes the kind of habitation, and among others the Old English elements *hām* 'homestead', *tūn* 'farm', *worth* 'enclosure', *wīc* 'dwelling', *cot* 'cottage', *burh* 'stronghold', and the Old Scandinavian elements *bý* 'farmstead' and *thorp* 'outlying farmstead' are particularly common, as in names like Streatham, Middleton, Lulworth, Ipswich, Didcot, Aylesbury, Grimsby, and Woodthorpe. Detailed studies of the various habitative elements have shown that they had a wide range of meanings which varied according to their use at different periods or in different parts of the country or in combination with other elements. For example Old English *tūn* may have its original meaning 'enclosure' in some names, whereas in others 'farmstead', 'village', 'manor', or 'estate' may be more appropriate. The reader is recommended to consult the Glossary at the end of the book to discover some of the alternative meanings evidenced for other habitative elements like *beretūn, burh, thorp, wīc, *wīc-hām*, and so on.

Topographical names also form a very large and diverse group. Some may have consisted originally of a description of some topographical or physical feature, either natural or man-made, which was then transferred to the settlement near the feature named, probably at a very early date. Others may have been applied as settlement-names to already established (pre-English) settlement sites characterized by the topographical feature. Thus names for rivers and streams, springs and lakes, fords and roads, marshes and moors, hills and valleys, woods and clearings, and various other landscape features are also the names of inhabited places. Typical examples of the type are Sherborne, Fulbrook, Bakewell, Tranmere, Oxford, Breamore, Stodmarsh, Swindon, Goodwood, Bromsgrove, Bexley, and Hatfield—all have second elements that denote topographical features. Indeed our early ancestors made use of a vast topographical vocabulary, applied with precision and subtlety in any one period or locality to the natural and artificial features they depended upon for their subsistence and survival. However, the meanings of topographical terms can vary a good deal from name to name, for some elements used over a long period in the formation of English place-names underwent considerable changes of meaning during medieval times, for instance Old English *feld* originally 'open land' developed a later sense 'enclosed plot', Old English *wald* 'forest' came to mean 'open upland', and Old English *lēah* 'wood' became 'woodland clearing' and then 'meadow'. The choice of the most likely meaning for one of these elements in an individual name is therefore a matter of judgement, based among other things on locality, the nature of the compound, and assumptions about the age of the name. Moreover recent research has increasingly shown that what seem to be similar terms for hills or valleys, woodland or marshland, or agricultural land, had fine distinctions of meaning in early times. For instance the different Old English terms for 'hill' like *dūn, hyll, hrycg, hōh, hēafod*, and **ofer*, far from being synonymous, seem to have had their own specialized meanings. In addition these and other common topographical elements like *ēg* 'island', *hamm* 'enclosure', and *halh* 'nook' were each capable of a wide range of extended meanings according to date, region, and the character of the landscape itself. Indeed the meanings suggested for names containing these elements can often be checked and refined by those with a close knowledge of the local topography of the places in question. The Glossary at the end of the book provides a selection of the meanings found for some of these topographical elements and gives an idea of the great range and variety of this vocabulary.

From the structural point of view, most English place-names are compounds, that is they consist of two elements, the first of which usually qualifies the second. The first element in such compounds may be a noun, an adjective, a river-name, a personal name, or a tribal name. The names mentioned in the last paragraph are typical examples of compound place-names formed during the Old English period. However some place-names, known as simplex, consist of one element only, at least to begin with: examples include names like Combe ('the valley'), Hale, Lea, Stoke, Stowe, Thorpe, Worth, and Wyke. Less common are names consisting of three elements such as Claverton ('burdock ford farmstead'), Redmarley, Woodmansterne, and Wotherton; in most of these the third element has probably been added later to an already existing compound. There are also other kinds of place-name composition, one of the most frequent being the use of the medial connective particle *-ing-* in place-names like Paddington, probably best explained as 'estate associated with a man called Padda'. In addition some compound place-names in the western parts of England (especially in Cornwall, counties bordering Wales, and in Cumbria) have a different formation. They are so-called 'name-phrases' in which the usual order of elements as found in English place-names is reversed following Celtic practice. In this group are names like Aspatria ('Patrick's ash-tree'), Bewaldeth, Brigsteer, Landulph, and Tremaine. Of course names with this characteristic Celtic word-order are also predominant throughout Ireland, Wales, and much of Scotland.

So-called 'double-barrelled' names, usually originating as ordinary simplex or compound names but later having an affix added to distinguish them from similar or identical names, are often of the manorial type already mentioned in which the affix is the name of a land-owning individual or family (for example Langton Matravers or Leighton Buzzard). But many other kinds of affix occur, most of them dating from the 13th or 14th centuries. Some refer to the size or shape of the place as in Much Wenlock or Long Buckby, others to geographical position relative to neighbouring manors as in High Barnet or Nether and Over Haddon, others to soil conditions as in Black Callerton or Dry Doddington, or to a local product as in

Iron Acton and Saffron Walden. Some affixes indicate the presence of a castle or other building or the existence of a market, as in Castle Rising, Steeple Bumpstead, Market Harborough, and Chipping Sodbury. A good number of the most notable affixes are Latin, as already noted, among them such resounding examples as Barton in Fabis, Ryme Intrinseca, Toller Porcorum, and Whitchurch Canonicorum (the last three from Dorset). But even the 'double-barrelled' names are not always what they seem: names like East Garston and Tur Langton which at face value seem to belong to this category turn out to have unexpected origins as ordinary compounds that are now completely disguised.

Many old place-names, especially compounds, have undergone some degree of reduction or contraction in the long period since they were first coined. Some names originally consisting of several syllables like Brighton or York have been considerably reduced by the centuries of use in speech. A common characteristic of compound place-names, and one which often helps to disguise their origin, is the shortening of original long vowels and diphthongs, as in compound words in the ordinary vocabulary. Just as *holi-* and *bon-* in the compounds *holiday* and *bonfire* represent *holy* and *bone* with their historically long vowels (*hālig* and *bān* in Old English), so in compound place-names Old English elements like *brād* 'broad', *brōm* 'broom', *hām* 'homestead', *stān* 'stone', and *strēt* 'street' occur with shortened vowels in names like Bradford, Bromley, Hampstead, Stanley, Stondon, Stratford, and Stretton. The same tendency, together with weakening of stress, also affected the second elements of compound names, resulting in some originally distinct elements coinciding in form and pronunciation. Once shortened, the important Old English habitative element *hām* 'homestead, village' came to sound like the quite separate topographical element *hamm* 'enclosure, river-meadow'. As a result, without definite evidence of one kind or another, it is not possible to be sure whether a number of place-names originally contained *hām* or *hamm* (in such cases both elements will usually have been cited as possible alternatives). The same combination of shortening of vowel and weakening of stress leads to the confusion of other elements that were originally quite distinct, among them Old English *dūn* 'down, hill', *denu* 'valley', and *tūn* 'farmstead': thus the modern forms of Croydon (from *denu*), Morden (from *dūn*), and Islington (also from *dūn*) belie their origins.

Some of the archaic features of English place-names are grammatical in origin. Old English was a highly inflected language, and although certain grammatical endings of Old English nouns, adjectives, and personal names disappeared from the ordinary language by the 11th or 12th centuries, they have left their permanent mark on a good number of place-names. Thus the genitive (i.e. possessive) singular of so-called 'weak' nouns and personal names (Old English *-n*) often survives as *-n-* in names like Dagenham (Old English *Dæccan hām* '*Dæcca's homestead'), Graveney, Putney, Tottenham, and Watnall. There are also many fossilized remains of the Old English dative endings of nouns and adjectives in place-names, since place-names would often naturally occur in adverbial or prepositional contexts requiring the dative case in Old English. Thus the old dative singular ending of the 'weak' adjective (*-an*) is often preserved in the middle of a modern name, as for instance in Bradnop (Old English *brādan hope* '(at) the broad valley'), Bradenham, Henley, and Stapenhill. Even more common are modern place-names that reflect the old dative case ending of an Old English noun. Thus the names Cleeve, Hale, and Sale derive from old dative forms of the words *clif, halh,* and *salh,* and most examples of the name Barrow represent Old English *bearwe* '(at) the wood or grove'. The common element *burh* 'fortified place' (the word *borough* in modern English) often appears in place-names as *Bury, -bury* from the Old English dative singular form *byrig,* but as *Burgh, -borough* from the nominative case of the same word. The distinctive dative plural ending *-um* of the Old English noun has also left its trace in the modern forms of many place-names, especially in the Midlands and North. Instances include Coatham, Cotham, Coton, Cottam, and Cotton (all probably from Old English *cotum* '(at) the cottages'), Laneham (from Old English *lanum* '(at) the lanes'), and other similar names like Downholm, Newsham, and Oaken. Occasionally too, in the place-names of the old Danelaw area of the North and East, old grammatical endings from the early Scandinavian languages spoken by the Vikings have been preserved. Thus there are traces of an old genitive (possessive) ending *-ar* in names like Helperthorpe, Osmotherley, and Windermere, an Old Scandinavian plural ending *-ar* is reflected in the modern forms of Sawrey and Burton upon Stather, and the Old Scandinavian dative plural *-um* is found in names like Arram and Kelham.

It should perhaps be noted here that many of the shorter Old English and Scandinavian men's names in use in the Anglo-Saxon period and incorporated into place-names, especially those ending in *-a* or *-i*, actually resemble names used for women in modern times. For this reason particular care has been taken in the explanations of place-names to indicate the gender of the person involved. Examples of such masculine personal names liable to be misinterpreted by the modern reader include Anna (in Amble and Ancaster), Betti (in Beachley and Bettiscombe), Emma (in Emley), Hilda (in Hillingdon), Káti (in Cadeby), Lill (in Lilleshall), and Sali (in Saleby). To a more limited extent the opposite may also be true, that some Old English and Scandinavian women's names may have rather a masculine look to the modern reader, and place-names incorporating these are explained with this in mind, examples being Helperby, Kenilworth, and Wilbraham containing the feminine personal names Hjalp, Cynehild, and Wilburh respectively.

The phenomenon known as 'back-formation' accounts for a good many modern river-names. Once the original meaning of a place-name was forgotten, there was sometimes a tendency for antiquarians and others to try to reinterpret it as if it contained the name of the river or stream on which the place was situated. Thus Plym came to be the name of the river at Plympton because the village name (historically 'farmstead of the plum-tree') came to be understood as 'farmstead on the stream called Plym'. Other examples of back-formation include Arun from Arundel, Chelmer from Chelmsford, Len from Lenham, Mole from Molesey, Roch from Rochdale, Rom from Romford, Stort from Stortford, and Wandle from Wandsworth. Many of these rivers and streams are known to have had genuine earlier names which were replaced by the new back-formations, usually from about the 16th century onwards.

Irish Place-Names

A glance at a map of Ireland will show a preponderance of anglicized Irish place-names, both in the Republic and in Northern Ireland. Most of them are descriptive of some kind of settlement, building, or natural feature and include such frequently found elements as *Bally-* (Irish *baile*, 'farmstead, townland') *Carrick-* or *Carrig-* (*carraig*, 'rock'), *Derry-* (*doire*, 'oak grove'), *Drum-* (*droim*, 'ridge'), *Inch-* or *Inish-* (*inis*, 'island'), *Kil-* or *Kill-* (*cill*, 'church', but sometimes *coill*, 'wood'), *Knock-* (*cnoc*, 'hill'), *Letter-* (*leitir*, 'hillside'), and *Slieve-* (*sliabh*, 'mountain'). Both *Lis-* (*lios*) and *Rath-* (*ráth*) are usually translated 'ring fort', the former word denoting the fort as a whole and the latter usually implying the presence of a church or monastery. *Dun-* (*dún*), on the other hand, usually rendered 'fort' or 'fortress', denotes the dwelling of a king or chieftain. The commonly occurring *Ballin-* or *Ballina-* may represent either *baile na*, 'homestead of the ...', or *béal átha na*, 'ford-mouth of the ...'. Examples of such names and their respective Irish forms are *Ballyshrule* (*Baile Sruthail*), 'townland of the stream', *Carrickfergus* (*Carraig Fhearghais*), 'Fergus's rock', *Derryboy* (*Doire Buí*), 'yellow oak grove', *Drumahoe* (*Droim na hUamha*), 'ridge of the cave', *Kilteely* (*Cill Tíle*), 'Tíl's church', *Killadangan* (*Coill an Daingin*), 'wood of the fortress', *Knockmoyle* (*Cnoc Maol*), 'bald hill', *Letterbrick* (*Leitir Bruic*), 'hillside of the badger', *Slieveardagh* (*Sliabh Ardach*), 'mountain of the high field', *Lismore* (*Lios Mór*), 'big fort', *Rathfeigh* (*Ráth Faiche*), 'fort of the green', *Dungannon* (*Dún Geanainn*), 'Geanann's fort', *Ballinderreen* (*Baile an Doirín*), 'townland of the little oak grove', and *Ballingar* (*Béal Átha na gCarr*), 'ford-mouth of the carts'.

Other common Irish place-name words include *achadh*, 'field', *aird*, 'promontory', *beag*, 'small', *beann* or *binn*, 'peak', *bóthar*, 'road', *caol*, 'narrow (place)', *cloch*, 'stone', *cluain*, 'meadow', *craobh*, 'tree', *domhnach*, 'church' (from Latin *dominicum*), *glas*, 'grey-green', *gleann*, 'valley', *loch*, 'lake', *mainistir*, 'monastery' (from Latin *monasterium*), *ros*, 'promontory', or sometimes 'grove', *teach*, 'house', often implying a saint's house, *teampall*, 'church' (from Latin *templum*), *tobar*, 'well', and *tóchar*, 'causeway'. Examples of names incorporating these are *Aghaboe* (*Achadh Bó*), 'field of the cows', *Ardmore* (*Aird Mhór*), 'great promontory', *Beginish* (*Beag Inis*), 'small island', *Benbeg* (*Beann Beag*), 'little peak', *Boherard* (*Bóthar Ard*), 'high road', *Kealkill* (*An Caolchoill*), 'the narrow wood', *Clonakilty* (*Cloich na Coillte*), 'stone of the woods', *Clonmore* (*Cluain Mhór*), 'large pasture', *Creevelea* (*Craobh Liath*), 'grey sacred tree', *Donaghmore* (*Domhnach Mór*), 'big church', *Glaslough* (*Glasloch*), 'grey-green lake', *Glendowan* (*Gleann Domhain*), 'deep valley', *Loughrea* (*Loch Riach*), 'grey lake', *Monasterevin* (*Mainistir Eimhín*), 'Eimhín's monastery', *Rosslare* (*Ros Láir*), 'middle promontory', *Timolin* (*Tigh Moling*), 'Moling's house', *Templemichael* (*Teampall Mhichil*), 'Michael's church', *Tobercurry* (*Tobar an Choire*), 'well of the cauldron', and *Ballintogher* (*Baile an Tóchair*), 'homestead of the causeway'.

Caiseal and *caisleán* both mean 'castle' (from Latin *castellum*), but the former word is normally used of a ring fort with stone walls, while the latter is used specifically of a medieval or post-medieval castle. *Caiseal* is mostly found in the north-west of the country. Examples are *Cashelgarran* (*Caiseal an Ghearráin*), 'fort of the horse', and *Castlederg* (*Caisleán na Deirge*), 'castle of the (river) Derg'.

Viking settlers introduced Scandinavian names to the east and south coasts of Ireland from the 9th century. Many end in *-ford*, which is not 'ford' but 'sea inlet' (OScand. *fjǫrthr*, English *fjord*), as for *Carlingford*, 'inlet of the hag', *Strangford*, 'inlet with a strong current', and *Waterford*, 'inlet where wethers are loaded'. *Wicklow* is 'Vikings' meadow' (*víkingr + ló*), while Dublin's *Howth* is *hofuth*, 'headland'. *Wexford* combines OIrish *escir*, 'sandbank', and OScand. *fjǫrthr*. Inland, *Leixlip* is 'salmon leap' (*leax + hlaup*). Most places with Scandinavian names have unrelated Irish names, so Waterford is *Port Láirge*, 'bank of the haunch', and Wexford *Loch Garman*, 'lake of the (river) Garma'. The province names *Leinster*, *Munster*, and *Ulster* have an OScand. genitive *-s* before Irish *tír*, 'territory'. The name of *Dublin* is Irish (*dubh + linn*, 'black pool'), but was used by the Vikings for the town they built by the River Liffey. The city's official Irish name is *Baile Átha Cliath*, 'town of the hurdle ford'.

Anglo-Norman place-names appeared in the 12th century. Few now remain, but examples are *Mitchelstown*, 'Mitchel's homestead', and *Pomeroy*, 'apple orchard'. Some Anglo-Norman names were subsequently gallicized, such as *Ballylanders* (Irish *Baile an Londraigh*), 'de Londra's homestead', presumably denoting an Anglo-Norman family from London.

The Plantations of the 16th and 17th centuries brought an influx of more directly English names, such as *Cookstown* and *Draperstown*. The Irish names are often unrelated, as *An Chorr Chríochach*, 'the boundary hill', for Cookstown, and *Baile na Croise*, 'town of the cross', for Draperstown. *Maryborough*, named from Queen Mary I of England, is now known by its Irish name of *Port Laoise*, 'port of the tribe of Laeighis', while *Queen's County*, after the same monarch, is now again *Laois*, earlier anglicized as *Leix*. *Offaly* (*Uibh Fhailí*), '(place of the) descendants of Failge', was similarly *King's County*, after Mary's husband, King Philip II of Spain, but *Kingstown*, now again *Dun Laoghaire*, 'Laoghaire's fort', was a later renaming, after King George IV.

Scottish Place-Names

Scotland's earliest Celtic names are British (Cumbric), in a language akin to Welsh spoken by the 'Ancient' Britons, or Pictish, in a similar language spoken by the Picts. Names such as

Glasgow, 'green hollow', and *Melrose*, 'bare moor', are British, while Pictish produced the distinctive *Pit-* names found in the northeast of Scotland, Fife, and Angus, such as *Pitcairn*, *Pitlochry*, and *Pittenweem*. The element represents Pictish **pett*, 'portion (of land)', a word ultimately related to English *piece*.

The majority of Scotland's names are Gaelic, however, in the language of the original Scots from Ireland. They incorporate such common words as *ard*, 'height, point' (*Ardnamurchan*, 'point of the otters'), *baile*, 'homestead, village' (*Ballantrae*, 'village on the shore'), *beinn*, 'mountain' (*Benbrack*, 'speckled mountain'), *ceann*, 'head, end' (*Kintyre*, 'end of the land'), *cill*, 'church' (*Kilbride*, 'church of St Bridget'), *druim*, 'ridge' (*Drumnadrochit*, 'ridge of the bridge'), *dùn*, 'fort' (*Dunoon*, 'river fort'), *inis*, 'island' (*Inchcolm*, 'St Columba's island'), *inbhir*, 'river mouth' (*Inverness*, 'mouth of the River Ness'), *loch*, 'lake' (*Lochinvar*, 'lake of the height'), *ros*, 'promontory' (*Rosemarkie*, 'point of land by the horse stream'), and *srath*, 'valley' (*Strathallan*, 'valley of the Allan Water').

South-eastern Scotland has a number of English names, introduced by the Northumbrian Angles from the 7th century. Examples are *Haddington*, 'farm associated with Hada', *Prestwick*, 'outlying farm of the priests', and *Whithorn*, 'white building'. From the 8th century, Scandinavian names began to appear, especially in northern and north-western Scotland. Among them are *Dingwall*, 'field of the assembly', *Kirkwall*, 'church bay', *Lerwick*, 'mud inlet', *Scalloway*, 'bay by the shielings', and *Stornoway*, 'steering bay'. OScand. *bólstathr*, 'homestead', lies behind a number of names in the north, and especially in Orkney and Shetland, such as *Isbister*, 'east dwelling', *Kirkabister*, 'church homestead', *Lybster*, 'dwelling place by the slope', and *Scrabster*, 'homestead of the young seagull'. The word is greatly reduced in *Skibo*, 'Skíthi's farm'. OScand. *kirkja*, 'church', the source of Scottish *kirk*, gave such names as *Kirkcudbright*, 'St Cuthbert's church', *Kirkoswald*, 'St Oswald's church', and the common *Kirkton*, 'village with a church'. In the first two of these, the word order is Celtic, not Scandinavian.

More recent names include Norman French names such as *Beauly* ('beautiful place'), 'military' names of the 17th and 18th centuries, such as *Fort Augustus, Fort George*, and *Fort William*, and names dating from this same period that commemorate a settlement's founder or his wife or daughter, such as *Bettyhill, Campbeltown, Fraserburgh, Grantown, Helensburgh*, and *Jemimaville*. *Edinburgh*, popularly derived from St Edwin, 7th-century king of Northumbria, in fact dates from before his time.

Welsh Place-Names

The majority of place-names in Wales are Welsh and descriptive of a natural feature or location, such as *Moelfre*, 'bare hill', or *Penmaenmawr*, 'headland of the great rock'. Common Welsh words in place-names include *aber*, 'river-mouth' (*Aberystwyth*, 'mouth of the Ystwyth'), *caer*, 'fort' (*Caerphilly*, 'Ffili's fort'), *cwm*, 'valley' (*Cwmafan*, 'valley of the Afan'), *llan*, 'church' (*Llanfair*, 'St Mary's church'), *llyn*, 'lake' (*Llyn Tegid*, 'Tegid's lake'), *nant*, 'stream, valley' (*Nantyglo*, 'valley of the coal'), *tre* or *tref*, 'farm' (*Trefeglwys*, 'church farm'), *pont*, 'bridge' (*Pontypridd*, 'bridge by the earthen house'), and *porth*, 'harbour' (*Porthcawl*, 'harbour of the sea kale').

Viking raids on the coasts of Wales from the 9th century have left their imprint in distinctive Scandinavian names, especially those of islands such as *Anglesey*, 'Ongull's island', *Caldy*, 'cold island', *Ramsey*, 'wild garlic island', and *Skomer*, 'cloven island'. Coastal towns with Viking names include *Fishguard*, 'fish yard', *Milford Haven*, 'harbour at the sandy inlet', and *Swansea*, 'Sveinn's island'. Norman names followed from the 11th century, such as *Beaumaris*, 'beautiful marsh', *Grosmont*, 'great hill', and *Malpas*, 'difficult passage'. English names became established from the 12th century, such as *Chepstow*, 'market place', *Haverfordwest*, 'western ford of the goats', *Holyhead*, 'holy headland', and *Wrexham*, 'Wryhtel's river meadow'. *Snowdon*, 'snow hill', is recorded in 1095.

Distinctive names of the 18th and 19th centuries are those adopted from biblical places, such as *Bethesda* and *Carmel*. They are mainly names of small villages that arose around a Nonconformist chapel. The slate quarries of *Bethesda* made it a town with a name of this type. Mine owners, ironmasters, or other industrial or commercial entrepreneurs gave their names to such places as *Griffithstown, Morriston, Port Talbot,* and *Tredegar*. Such names are mostly found in South Wales.

A number of places in Wales have a name that is an anglicized form of the Welsh original, such as *Cardiff* (*Caerdydd*) and *Denbigh* (*Dinbych*), while several places with Scandinavian or English names have an unrelated Welsh name, such as *Abergwaun*, 'mouth of the (river) Gwaun', for Fishguard, *Abertawe*, 'mouth of the (river) Tawe', for Swansea, and *Caergybi*, 'Cybi's fort', for Holyhead. *Montgomery*, named from its Norman lord, is Welsh *Trefaldwyn*, 'town of Baldwin', another Norman.

The Wider Significance of Place-Names

It will of course already be apparent that the interest of a place-name does not stop at its etymological meaning and derivation: rather these provide the basic and essential starting point for the fuller appreciation of a place-name's significance in its wider linguistic, historical, archaeological, or geographical context. Although the scope of the present book does not allow this fuller exploration and appraisal of individual names, a few other points will be touched upon here in addition to those already mentioned, and readers are recommended to follow up such aspects as may interest them in the various studies listed in the Select Bibliography.

Place-names can tell us a great deal about tribal migrations, invasions and settlements. In England, older names of Celtic origin (supported by place-names containing such elements as Old English *walh* 'a Briton') testify to direct communication between the Celtic Britons and the English-speaking Anglo-Saxon invaders, and indeed indicate the survival of a British population in some districts. The vast majority of English place-names reflect the steady progress from east to west and the overwhelming success of the Anglo-Saxon invasions and settlements of the 5th century onwards, certain name-types being particularly associated with the early phases of immigration and colonization and others reflecting the gradual establishment of a new administrative and manorial system and the continued exploitation of the land for agriculture. Names of French origin, and those with manorial affixes consisting of Norman-French family names, are reminders of the Norman Conquest and its widespread political, social and linguistic consequences, including the imposition of the feudal system. In the north, north-west and east of England, as well as in other parts of the British Isles such as northern and western Scotland, the Isle of Man and the Scottish islands, and coastal districts of Ireland and Wales, the distribution and considerable numbers of Scandinavian place-names suggest the extent and relative density of the Viking settlements in those areas. In addition the distribution of particular place-name elements can often be significant: for instance the 300 or so names in north-east Scotland containing the word **pett* 'share of land' (such as Pitlochry and Pitlurg) can be used to indicate the area settled by the Picts, and this clear linguistic evidence is strongly supported by archaeological finds.

A very small number of place-names have pagan associations, some for instance providing evidence for the worship of the heathen deities Woden, Thunor and Tiw in early Anglo-Saxon England before the conversion of the English to Christianity in the 7th century. Names like Wednesfield, Thundersley and Tysoe are among the names referring to these gods, and Wye and Harrow contain words for a heathen temple. In Ireland place-names like Armagh and Maynooth refer to pagan Irish deities, and Celtic river-names like Brent in England, Dee in England and Scotland, and Bann and Shannon in Ireland, all with meanings like 'goddess, holy one', suggest a cult of river worship in ancient times. On the other hand place-names with Christian associations are extremely common in England as well as in Ireland, Scotland,

and Wales. Many refer to churches and other holy places (such as Hawkchurch, Ormskirk, Kidderminster, Templeoran, Kilmarnock and Llandudno), some to crosses and holy springs (as in Crosby, Holywell, Ruthwell, Crosspatrick and Tobermory), and others to ownership by priests or other ecclesiastics (for example Monkton, Fryerning, Abbotsbury, Prestatyn and Kilnamanagh).

Many place-names from all parts of the British Isles provide information of archaeological interest. Some places are named from their situation on Roman roads or on ancient routes and tracks. Others contain elements referring to earthworks and fortifications, ranging from Iron-Age hill forts and Roman camps to medieval strongholds and castles. Particularly common are names referring to ancient burial sites, burial mounds and tumuli.

Numerous place-names illustrate the socal structures and legal customs of early times in the various countries and regions of the British Isles. All ranks of society are represented in place-names, from kings and queens and others of noble birth to the humble peasant. Some names reflect the early divisions of the social hierarchy, others indicate various aspects of land tenure, others inform us about sites where important meetings and assemblies were held, others reveal details of ancient boundaries, lookout places, old land disputes, and even leisure activities like sport and hunting.

Many persons and families from many different periods of history and from many different linguistic and cultural backgrounds are in a sense commemorated in the place-names of the British Isles. Some of these can of course be identified with particular men and women or families known from the historical record, but about the vast majority of them nothing more is known other than what the place-names themselves tell us. Some may have been important overlords or chieftains, many must have been thegns or noblemen granted their estates by kings or bishops, others may have been farmers or relatively humble peasants. A small but significant minority of the people named are women. These are probably unlikely to have been secular leaders, but a few seem to have been religious persons or founders of churches and the rest were no doubt the widows or daughters of manorial lords who had been granted their estates in earlier times.

British place-names provide abundant evidence for early personal names of all kinds, some of them well known from the surviving historical records, others rare and more hypothetical. In those regions and countries where the various Celtic languages continued to flourish—Cornwall, the Isle of Man, Ireland, Scotland, and Wales—the personal names, like the place-names in which they appear, are overwhelmingly Celtic in form and origin. In England, many of the Old English personal names for both men and women fell into disuse after 1066, being largely replaced by Christian names introduced by the Normans. Because of these drastic changes in name-giving fashions, our knowledge of the personal nomenclature of the Old English period is incomplete, for it is quite clear that only a proportion of the numerous names in use during the six centuries after the Anglo-Saxon settlement have survived in the ordinary written records. Here place-names provide good evidence for the existence of many personal names of both men and women that are otherwise unrecorded. Such personal names, inferred from comparative evidence and postulated to occur with varying degrees of certainty in particular place-names, are customarily asterisked (like the *Dæcca in Dagenham or the *Berica in Barking) to indicate that they are not found in independent use.

The way place-names from all parts of the British Isles reflect the face of the landscape, utilizing a rich and diverse vocabulary to describe every undulation and type of terrain, has already been touched upon. The natural history of this varying landscape is also abundantly represented, as is clear from the many different species of trees and plants, wild animals and birds, fish and even insects, that are evidenced among the elements found in names. Moreover, place-names reflect every aspect of human activity in the different regions over a long period, from the utilization and development of the land by our forefathers for agricultural purposes, to their exploitation of the environment for communications, trade, and industry.

The numerous British place-names containing woodland terms provide good evidence for the distribution, use and management of woods, copses and groves in early times. Those containing words for types of woodland clearing indicate areas of former forest and the particular purposes for which clearings were used. Names derived from elements denoting various kinds of field, pasture, meadow, arable and enclosure suggest different aspects of land-use in the subsistence agriculture of our ancestors, in which arable land had to be broken in to produce crops, meadow-land provided hay, and pastureland and enclosures were needed for animals. Place-names give information too about the kind or quality of the soil, the crops grown and harvested, the domestic creatures reared, the practice of transhumance, and the goods produced.

The importance of river valleys for early settlement, providing fertile soils, ease of access and a good water supply, is reflected in the number of places named from the rivers or streams on which they are situated. Many other place-names refer to the roads or routes on which they stand, also essential for communications and trade. The large group of names containing elements meaning 'ford' or 'bridge' show the vital part played by river crossings, whilst those derived from words for 'landing-place' or 'harbour' suggest the early importance of trade and transport by water. Many elements indicate the local industries and occupations of particular places and regions, as for example milling, fishing, salt-making, charcoal-burning, coal-mining, pot-making, iron-working, quarrying, timber production, bee-keeping, and many others.

The study of place-names has made important contributions to our knowledge of the original vocabulary of Old English as well as that of the early Celtic and Scandinavian languages. Dozens of words once used in living speech may never have found their way into literary or historical writings before they went out of use, but such words often occur in place-names formed in the early period. It will be apparent from the alphabetical list of names that this archaic vocabulary (customarily asterisked to show that it is only evidenced in place-names and not otherwise recorded) is well represented among the entries. Moreover many words recorded in independent use are evidenced much earlier in place-names than in the ordinary languages, and these too are by convention asterisked. Many other old words, once part of the living languages but now lost from the general vocabulary, survive in fossilized form in place-names. A good selection of these old words, those most frequently found in the place-names of the British Isles, are listed for convenience in the Glossary of Some Common Elements at the end of the book.

It is of course the case that the current local pronunciation of place-names, especially in the Celtic language areas of the British Isles but also quite often in England, differs from what the modern spelling might lead us to expect (English examples might include Beaulieu, Bicester, Chiswick, Cholmondeley, Stiffkey and Towcester). Although such matters are outside the scope of this book, the historical and linguistic reasons for these characteristics and disparities are often of some interest and are dealt with in the detailed regional or county surveys, whilst the current pronunciations of many names can be found in the specialized pronouncing dictionaries. Indeed further information of any kind about any of the names included in this dictionary, as well as information about names not included for reasons of space, should be sought in the regional surveys or other monographs and studies dealing with particular name-types and groups of names, a selection of which is listed in the Bibliography.

List of Counties and Unitary Authorities

England

B. & NE. Som.	Bath & North East Somerset
Barns.	Barnsley
Beds.	Bedfordshire
Birm.	Birmingham
Black. w. Darw.	Blackburn with Darwen
Bmouth.	Bournemouth
Bolton	
Bpool.	Blackpool
Brack. For.	Bracknell Forest
Brad.	Bradford
Bright. & Hove	Brighton & Hove
Brist.	City of Bristol
Bucks.	Buckinghamshire
Bury	
Calder.	Calderdale
Cambs.	Cambridgeshire
Ches.	Cheshire
Cornwall	
Covtry.	Coventry
Cumbria	
Darltn.	Darlington
Derby	City of Derby
Derbys.	Derbyshire
Devon	
Donc.	Doncaster
Dorset	
Dudley	
Durham	
E. R. Yorks.	East Riding of Yorkshire
E. Sussex	East Sussex
Essex	
Gatesd.	Gateshead
Glos.	Gloucestershire
Gtr. London	Greater London
Halton	
Hants.	Hampshire
Hartlepl.	Hartlepool
Herefs.	Herefordshire
Herts.	Hertfordshire
I. of Scilly	Isles of Scilly
I. of Wight	Isle of Wight
Kent	
Kirkl.	Kirklees
Knows.	Knowsley
K. upon Hull	City of Kingston upon Hull
Lancs.	Lancashire
Leeds	
Leic.	City of Leicester
Leics.	Leicestershire
Lincs.	Lincolnshire
Lpool.	Liverpool
Luton	
Manch.	Manchester
Medway	
Middlesbr.	Middlesbrough
Milt. K.	Milton Keynes
Newc. upon T.	Newcastle upon Tyne
Norfolk	
NE. Lincs.	North East Lincolnshire
N. Lincs.	North Lincolnshire
N. Som.	North Somerset
N. Tyne.	North Tyneside
N. Yorks.	North Yorkshire
Northants.	Northamptonshire
Northum.	Northumberland
Nott.	City of Nottingham
Notts.	Nottinghamshire
Oldham	
Oxon.	Oxfordshire
Peterb.	City of Peterborough
Plym.	City of Plymouth
Poole	
Portsm.	City of Portsmouth
Readg.	Reading
Red. & Cleve.	Redcar & Cleveland
Rochdl.	Rochdale
Rothm.	Rotherham
Rutland	
Salford	

Sandw.	Sandwell
Sefton	
Sheff.	Sheffield
Shrops.	Shropshire
Slough	
Solhll.	Solihull
Somerset	
S. Glos.	South
Gloucestershire	
S. Tyne.	South Tyneside
Sotn.	City of Southampton
St Hel.	St Helens
Staffs.	Staffordshire
Sthend.	Southend-on-Sea
Stock. on T.	Stockton on Tees
Stockp.	Stockport
Stoke	City of Stoke
Suffolk	
Sundld.	Sunderland
Surrey	
Swindn.	Swindon
Tamesd.	Tameside
Tel. & Wrek.	Telford & Wrekin
Thurr.	Thurrock
Torbay	
Traffd.	Trafford
Wakefd.	Wakefield
Warrtn.	Warrington
Warwicks.	Warwickshire
W. Berks.	West Berkshire
Wigan	
Wilts.	Wiltshire
Winds. & Maid.	Windsor & Maidenhead
Wirral	
W. Sussex	West Sussex
Wokhm.	Wokingham
Wolverh.	Wolverhampton
Worcs.	Worcestershire
Wsall.	Walsall
York	City & County of York

Northern Ireland

Antrim
Armagh
Derry
Down
Fermanagh
Tyrone

Republic of Ireland

Carlow
Cavan
Clare
Cork
Donegal
Dublin
Galway
Kerry
Kildare
Kilkenny
Laois
Leitrim
Limerick
Longford
Louth
Mayo
Meath
Monaghan
Offaly
Roscommon
Sligo
Tipperary
Waterford
Westmeath
Wexford
Wicklow

Scotland

Abdn.	Aberdeen City
Aber.	Aberdeenshire
Ang.	Angus
Arg.	Argyll and Bute
Edin.	City of Edinburgh
Clac.	Clackmannanshire
Dumf.	Dumfries and Galloway
Dund.	Dundee City
E. Ayr.	East Ayrshire
E. Dunb.	East Dunbartonshire
E. Loth.	East Lothian
E. Renf.	East Renfrewshire
Falk.	Falkirk
Fife	
Glas.	Glasgow City
Highland	
Invclyd.	Inverclyde
Midloth.	Midlothian

Moray	
N. Ayr.	North Ayrshire
N. Lan.	North Lanarkshire
Orkn.	Orkney
Perth.	Perth and Kinross
Renf.	Renfrewshire
Sc. Bord.	Scottish Borders
Shet.	Shetland
S. Ayr.	South Ayrshire
S. Lan.	South Lanarkshire
Stir.	Stirling
W. Dunb.	West Dunbartonshire
W. Isles	Western Isles (Eilean Siar)
W. Loth.	West Lothian

Wales

Angl.	Isle of Anglesey
Blae.	Blaenau Gwent
Bri.	Bridgend
Cphy.	Caerphilly
Card.	Cardiff

Carm.	Carmarthenshire
Cergn.	Ceredigion
Conwy	
Denb.	Denbighshire
Flin.	Flintshire
Gwyd.	Gwynedd
Mer. T.	Merthyr Tydfil
Mon.	Monmouthshire
Neat.	Neath Port Talbot
Newpt.	Newport
Pemb.	Pembrokeshire
Powys	
Rhon.	Rhondda Cynon Taf
Swan.	Swansea
Torf.	Torfaen
Vale Glam.	Vale of Glamorgan
Wrex.	Wrexham

The Channel Islands and Isle of Man

Guernsey
Jersey
Isle of Man

Abbreviations

Anglo-Scand.	Anglo-Scandinavian	ModE	Modern English
c.	*circa*	OCornish	Old Cornish
	('approximately')	OE	Old English (the
cent.	century		English language
DB	Domesday Book		*c.*450–*c.*1100)
	(includes *Great*	OFrench	Old French
	Domesday, *Little*	OGaelic	Old Gaelic
	Domesday, and *Exon*	OGerman	Old German
	Domesday)	OIrish	Old Irish
eModE	early Modern English	OScand.	Old Scandinavian (the
	(the English language		language of the
	*c.*1500–*c.*1650)		Vikings, comprising
EPNS	English Place-Name		Old Danish and Old
	Society		Norse)
ME	Middle English (the	OWelsh	Old Welsh
	English language	pers. name	personal name
	*c.*1100–*c.*1500)	Scand.	Scandinavian
MIrish	Middle Irish	St	Saint

Maps

ORKNEY

SHETLAND

1 KINCARDINESHIRE
2 DUNBARTONSHIRE
3 STIRLINGSHIRE
4 CLACKMANNANSHIRE
5 KINROSS-SHIRE
6 RENFREWSHIRE
7 WEST LOTHIAN
8 MIDLOTHIAN
9 PEEBLES-SHIRE
10 SELKIRKSHIRE

CAITH-NESS

SUTHERLAND

ROSS
AND
CROMARTY

NAIRN
MORAY-SHIRE
BANFFSHIRE

INVERNESS-SHIRE

ABERDEENSHIRE

ANGUS

PERTH

FIFE

EAST
LOTHIAN

ARGYLLSHIRE

BUTE

LANARK-SHIRE

BERWICK-SHIRE

AYRSHIRE

ROXBURGH-SHIRE

DUMFRIES-SHIRE

NORTHUMBERLAND

KIRKUD-BRIGHTSHIRE

WIGTOWNSHIRE

CUMBERLAND

DURHAM

WESTMORLAND

ISLE OF MAN

North Riding

YORKSHIRE

East
Riding

West
Riding

LANCASHIRE

FLINTSHIRE

ANGELSEY

CHESHIRE

DERBY.

NOTTS.

LINCOLNSHIRE

CAERNARVONSHIRE

DENBIGH-SHIRE

MERIONETHSHIRE

STAFFS.

LEICS.

RUTLAND

NORFOLK

MONTGOMERYSHIRE

SHROPSHIRE

CARDIGANSHIRE

PEMBROKESHIRE

RADNOR-SHIRE

HEREFORD-SHIRE

WORCS.

WARKS.

NORTHANTS.

HUNTS.

CAMBS.

SUFFOLK

CARMARTHEN-SHIRE

BRECKNOCK-SHIRE

BEDS.

GLOUCESTER-SHIRE

OXON.

BUCKS.

HERTS.

ESSEX

GLAMORGAN

MONMOUTHSHIRE

BERKSHIRE

G. LONDON

WILTSHIRE

SURREY

KENT

SOMERSET

HAMPSHIRE

SUSSEX

DEVON

DORSET

ISLE OF WIGHT

CORNWALL

ISLES OF
SCILLY

0   50   100   150 km

England, Scotland, and Wales: Historical Counties

England: Counties and Unitary Authorities

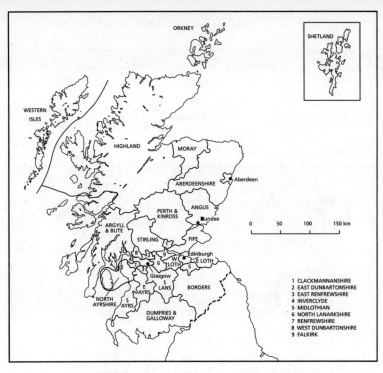

Scotland: Counties and Unitary Authorities

1 CLACKMANNANSHIRE
2 EAST DUNBARTONSHIRE
3 EAST RENFREWSHIRE
4 INVERCLYDE
5 MIDLOTHIAN
6 NORTH LANARKSHIRE
7 RENFREWSHIRE
8 WEST DUNBARTONSHIRE
9 FALKIRK

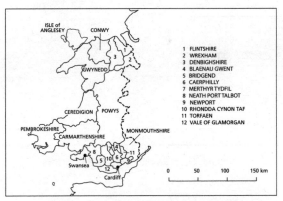

Wales: Counties and Unitary Authorities

1 FLINTSHIRE
2 WREXHAM
3 DENBIGHSHIRE
4 BLAENAU GWENT
5 BRIDGEND
6 CAERPHILLY
7 MERTHYR TYDFIL
8 NEATH PORT TALBOT
9 NEWPORT
10 RHONDDA CYNON TAF
11 TORFAEN
12 VALE OF GLAMORGAN

Ireland: Counties and Unitary Authorities

Ab Kettleby Leics. *See* KETTLEBY.

Abbas Combe Somerset. *See* COMBE.

Abberley Worcs. *Edboldelege* 1086 (DB). 'Woodland clearing of a man called Ēadbeald'. OE pers. name + *lēah*.

Abberton Essex. *Edburgetuna* 1086 (DB). 'Farmstead or estate of a woman called Ēadburh'. OE pers. name + *tūn*.

Abberton Worcs. *Eadbrihtincgtun* 972, *Edbretintune* 1086 (DB). 'Estate associated with a man called Ēadbeorht'. OE pers. name + *-ing-* + *tūn*.

Abberwick Northum. *Alburwic* 1170. 'Dwelling or (dairy) farm of a woman called Aluburh or Alhburh'. OE pers. name + *wīc*.

Abbess Roding Essex. *See* RODING.

Abbey Dore Herefs. *See* DORE.

Abbey Hulton Staffs. *See* HULTON.

Abbey Town Cumbria. *Abbey Towne* 1649. 'Estate by the abbey', with reference to the former abbey of Holme Cultram.

Abbeydorney (*Mainistir Ó dTorna*) Kerry. 'Abbey of Uí Thorna'.

Abbeyfeale (*Mainistir na Féile*) Limerick. 'Abbey of the River Feale'.

Abbeylara (*Mainistir Leathrátha*) Longford. 'Abbey of the half ring-fort'.

Abbeyleix (*Mainistir Laoise*) Laois. 'Abbey of Laois'.

Abbeyshrule (*Mainistir Shruthla*) Longford. 'Abbey of the stream'.

Abbeystead Lancs. *Abbey* 1323. '(Deserted) site of the abbey', with reference to the abbey of Wyresdale. ME *abbeye* + *stede*.

Abbots, Abbotts as affix. See main name, e.g. for **Abbots Bickington** (Devon) *see* BICKINGTON.

Abbotsbury Dorset. *Abbedesburie* 946, *Abedesberie* 1086 (DB). 'Fortified house or manor of the abbot'. OE *abbod* + *burh* (dative *byrig*). With reference to early possession by the abbot of Glastonbury.

Abbotsford Sc. Bord. '(Place of the) abbot's ford'. The mansion was built by Sir Walter Scott in 1816 on land owned by the Abbot of Melrose by a ford over the Tweed.

Abbotsham Devon. *Hama* 1086 (DB), *Abbudesham* 1238. OE *hamm* 'enclosure' with the later addition of *abbod* 'abbot' (referring to early possession by the abbot of Tavistock).

Abbotskerswell Devon. *Cærswylle* 956, *Carsuella* 1086 (DB), *Karswill Abbatis* 1285. 'Spring or stream where water-cress grows'. OE *cærse* + *wella*. Affix from early possession by the abbot of Horton.

Abbotsley Cambs. *Adboldesle* 12th cent. 'Woodland clearing of a man called Ealdbeald'. OE pers. name + *lēah*.

Abdon Shrops. *Abetune* 1086 (DB). 'Farmstead or estate of a man called Abba'. OE pers. name + *tūn*.

Aberaeron Cergn. *ad ostium Ayron* 1184. 'Mouth of the River Aeron'. Welsh *aber*. The Celtic river name means 'battle'.

Aberafan. *See* ABERAVON.

Aberavon (*Aberafan*) Neat. *Abberauyn* c.1400, *Aberavan* 1548. 'Mouth of the River Afan'. Welsh *aber*. The river-name is probably derived from a pers. name.

Abercraf Powys. *Abercraven* 1680. 'Mouth of the River Craf'. Welsh *aber*. The river-name means 'garlic' (Welsh *craf*).

Aberdâr. *See* ABERDARE.

Aberdare (*Aberdâr*) Rhon. *Aberdar* 1203. 'Mouth of the River Dâr'. Welsh *aber*. The Celtic river name means 'oak river'.

Aberdaron Gwyd. 'Mouth of the River Daron'. Welsh *aber*. The Celtic river name means 'oak river'.

Aberdaugleddau. *See* MILFORD HAVEN.

Aberdeen Abdn. *Aberdon* c.1187, *Aberden* c.1214. 'Mouth of the River Don'. Pictish *aber*. The river is named after *Devona*, a Celtic goddess. Modern Aberdeen is at the mouth of the DEE, but the name relates to Old Aberdeen, at the mouth of the Don.

Aberdour Fife. *Abirdaur* 1226. 'Mouth of the River Dour'. Pictish *aber*. The river-name means 'water'.

Aberdovey. *See* ABERDYFI.

Aberdyfi Gwyd. *Aberdewi* 12th cent., *aber dyfi* 14th cent. 'Mouth of the River Dyfi'. Welsh *aber*. The river-name probably means 'dark one'.

Aberfeldy Perth. 'Confluence of Peallaidh'. Pictish *aber*. *Peallaidh* is the name of a water sprite said to haunt the place where the Moness Burn enters the Tay.

Aberford Leeds. *Ædburford* 1176. 'Ford of a woman called Êadburh'. OE pers. name + *ford*.

Aberfoyle Stir. *Abirfull* 1481. 'Confluence of the pool'. Pictish *aber* + Gaelic *poll* (genitive *phuill*). The two headstreams of the River Forth unite near here, and are joined by the River Foyle.

Abergafenni. *See* ABERGAVENNY.

Abergavenny (*Abergafenni* or *Y Fenni*) Mon. *Gobannio* 4th cent., *Abergavenni* 1175. 'Mouth of the River Gafenni'. Welsh *aber*. The river-name probably means 'the smith', referring to the ironworks exploited here by the Romans, whose fort was *Gobannum*, from the same Celtic source.

Abergele Conwy. *Opergelei* 9th cent., *Abergele* 1257. 'Mouth of the River Gele'. Welsh *aber*. The river-name means 'blade' (OWelsh *gelau*).

Abergwaun. *See* FISHGUARD.

Aberhonddu. *See* BRECON.

Aberlour Moray. 'Confluence of the Lour Burn'. Pictish *aber*. The river-name means 'babbling brook'. The town's formal name is *Charlestown of Aberlour*, after *Charles* Grant, who laid the original village out in 1812.

Abermo. *See* BARMOUTH.

Abernethy Perth. *Aburnethige* c.970. 'Confluence of the River Nethy'. Pictish *aber*. The river-name means 'pure' (Pictish *nectona*). Nectonos is also the name of a Celtic water divinity.

Aberpennar. *See* MOUNTAIN ASH.

Aber-porth Cergn. *Aberporth* 1284. 'Estuary in the bay'. Welsh *aber* + *porth*.

Aber-soch Gwyd. *Absogh* 1350, *Avon Soch* 1598. 'Mouth of the River Soch'. Welsh *aber*. The river-name meaning 'nosing one', referring to the way the river 'roots' its course through the land.

Abersychan Torf. *Aber Sychan* c.1850. 'Mouth of the River Sychan'. Welsh *aber*. The river-name is based on Welsh *sych*, 'dry', implying a river that dries up in summer.

Abertawe. *See* SWANSEA.

Aberteifi. *See* CARDIGAN.

Aberteleri. *See* ABERTILLERY.

Abertillery (*Abertyleri* or *Aberteleri*) Blae. *Teleri* 1332, *Aber-Tilery* 1779. 'Confluence of the River Teleri'. Welsh *aber*. The river-name derives from a pers. name.

Abertyleri. *See* ABERTILLERY.

Aberystwyth Cergn. *Aberestuuth* 1232, *aber ystwyth* 14th cent., *Aberystwith, or Aberrheidol* 1868. 'Mouth of the River Ystwyth'. Welsh *aber*. The river-name means 'winding one' (Welsh *ystwyth*). Aberystwyth is now at the mouth of the *Rheidol*, but the name relates to the *Ystwyth*, to the south, where a Norman castle was built in 1110.

Abingdon Oxon. *Abbandune* 968, *Ab(b)endone* 1086 (DB). 'Hill of a man called Æbba or of a woman called Æbbe'. OE pers. name (genitive *-n*) + *dūn*.

Abinger Surrey. *Abinceborne* [sic] 1086 (DB), *Abingewurd* 1191. 'Enclosure of the family or followers of a man called Abba', or 'enclosure at Abba's place'. OE pers. name + *-inga-* or *-ing* + *worth*. **Abinger Hammer** is named from the former iron foundry here, called *The Hammer Mill* 1600.

Abington, 'estate associated with a man called Abba', OE pers. name + *-ing-* + *tūn*: **Abington, Great & Abington, Little** Cambs. *Abintone* 1086 (DB). **Abington Pigotts** Cambs. *Abintone* 1086 (DB), *Abington Pigots* 1635. Manorial affix from the *Pykot* family, here from the 15th cent.

Abington (*Mainistir Uaithne*) Limerick. 'Abbey of (the district of) Uaithne'.

Ablington Glos. *Eadbaldingtun* 855. 'Estate associated with a man called Eadbeald'. OE pers. name + *-ing-* + *tūn*.

Ablington Wilts. *Alboldintone* 1086 (DB). 'Estate associated with a man called Ealdbeald'. OE pers. name + *-ing-* + *tūn*.

Aboyne Aber. *Obyne* 1260. Obscure. The full formal name is *Charleston of Aboyne*, after *Charles* Gordon, 1st Earl of *Aboyne* (d.1681), who erected a burgh of barony here in 1670.

Abram Wigan. *Adburgham* late 12th cent. 'Homestead or enclosure of a woman called Ēadburh'. OE pers. name + *hām* or *hamm*.

Abridge Essex. *Affebrigg* 1203. 'Bridge of a man called Æffa'. OE pers. name + *brycg*.

Abthorpe Northants. *Abetrop* 1190. 'Outlying farmstead or hamlet of a man called Abba'. OE pers. name + OE *throp* or OScand. *thorp*.

Aby Lincs. *Abi* 1086 (DB). 'Farmstead or village on the stream'. OScand. *á* + *bý*.

Acaster Malbis (York) **& Acaster Selby** (N. Yorks). *Acastre* 1086 (DB), *Acaster Malebisse* 1252, *Acastre Seleby* 1285. 'Fortification on the river'. OScand. *á* (perhaps replacing OE *ēa*) + OE *ceaster*. Manorial affixes from lands here held by the *Malbis* family and by Selby Abbey.

Accrington Lancs. *Akarinton* 12th cent. 'Farmstead or village where acorns are found or stored'. OE *æcern* + *tūn*.

Achabog (*Achadh Bog*) Monaghan. *Aghabog* 1665. 'Soft field'.

Acharacle Highland. 'Torquil's ford'. OScand. pers. name + Gaelic *àth*.

Achnashellach Forest Highland. *Auchnashellicht* 1543. 'Field of the willows'. Gaelic *achadh* + *na* + *seileach*, with addition of English *forest*.

Achonry (*Achadh Conaire*) Sligo. 'Field of Conaire'.

Achray Forest Stir. *Achray* 1791. Probably 'shaking ford'. Gaelic *àth* + *chrathaidh*. A 'shaking ford' is a quagmire. *Forest* was added by the Forestry Commission.

Achurch Northants. *Asencircan c.*980, *Asechirce* 1086 (DB). 'Church of a man called *Asa or Ási'. OE or OScand. pers. name + OE *cirice*.

Acklam, '(place at) the oak woods or clearings', OE *āc* + *lēah* (in a dative plural form *lēagum*): **Acklam** Middlesbr. *Aclum* 1086 (DB). **Acklam** N. Yorks. *Aclun* 1086 (DB).

Acklington Northum. *Eclinton* 1177. Probably 'estate associated with a man called Ēadlāc'. OE pers. name + *-ing-* + *tūn*.

Ackton Wakefd. *Aitone* [sic] 1086 (DB), *Aicton c.*1166. 'Oak-tree farmstead'. OScand. *eik* + OE *tūn*.

Ackworth, High & Ackworth, Low Wakefd. *Aceuurde* 1086 (DB). 'Enclosure of a man called Acca'. OE pers. name + *worth*.

Aclare (*Áth an Chláir*) Sligo. 'Ford of the plain'.

Acle Norfolk. *Acle* 1086 (DB). 'Oak wood or clearing'. OE *āc* + *lēah*.

Acol Kent. *Acholt* 1270. 'Oak wood'. OE *āc* + *holt*.

Acomb, '(place at) the oak-trees', OE *āc* in a dative plural form *ācum*: **Acomb** Northum. *Akum* 1268. **Acomb** York. *Akum* 1222.

Aconbury Herefs. *Akornebir* 1213. 'Old fort inhabited by squirrels'. OE *ācweorna* + *burh* (dative *byrig*).

Acre, Castle, Acre, South & Acre, West Norfolk. *Acre* 1086 (DB), *Castelacr* 1235, *Sutacra* 1242, *Westacre* 1203. 'Newly cultivated land'. OE *æcer*. Distinguishing affixes from OFrench *castel* (with reference to the Norman castle here), OE *sūth* and *west*.

Acton, a common name, usually 'farmstead or village by the oak-tree(s)' or 'specialized farm where oak timber is worked', OE *āc* + *tūn*; examples include: **Acton** Gtr. London. *Acton* 1181. **Acton Beauchamp** Herefs. *Aactune* 727. Manorial affix from the *Beauchamp* family, here from the 12th cent. **Acton Burnell & Acton Pigott** Shrops. *Actune, Æctune* 1086

(DB), *Akton Burnell* 1198, *Acton Picot* 1242. Manorial affixes from the *Burnell* and *Picot* families, here in the 12th cent. **Acton, Iron** S. Glos. *Actune* 1086 (DB), *Irenacton* 1248. Affix is OE *īren* 'iron', referring to old iron-workings here. **Acton Round** Shrops. *Achetune* 1086 (DB), *Acton la Runde* 1284. Affix is ME *ro(u)nd* 'round in shape' or from its early possession by the Earls of *Arundel* (perhaps falsely interpreted as containing the word *ro(u)nd*). **Acton Scott** Shrops. *Actune* 1086 (DB), *Scottes Acton* 1289. Manorial affix from the *Scot* family, here in the 13th cent. **Acton Trussell** Staffs. *Actone* 1086 (DB), *Acton Trussel* 1481. Manorial affix from the *Trussell* family, here in the 14th cent.

However some Actons have a different origin: **Acton** Dorset. *Tacatone* 1086 (DB). Probably 'farmstead or village where young sheep are reared'. OE **tacca + tūn*. Initial *T-* was dropped in the 16th cent. due to confusion with the preposition *at*. **Acton** Suffolk. *Acantun c.*995, *Achetuna* 1086 (DB). 'Farmstead or village of a man called Ac(c)a'. OE pers. name + *tūn*. **Acton Turville** S. Glos. *Achetone* 1086 (DB), *Acton Torvile* 1284. Identical in origin with the previous name. Manorial affix from the *Turville* family, here from the 13th cent.

Acton Armagh. *Acton* 1619. The village was founded by Charles Poyntz in 1600 and named after his native *Iron Acton*, Glos.

Adare (*Áth Dara*) Limerick. 'Ford of the oak'.

Adbaston Staffs. *Edboldestone* 1086 (DB). 'Farmstead or village of a man called Ēadbald'. OE pers. name + *tūn*.

Adber Dorset. *Eatan beares* 956, *Ateberie* 1086 (DB). 'Grove of a man called Ēata'. OE pers. name + *bearu*.

Adderbury, East & Adderbury, West Oxon. *Eadburggebyrig c.*950, *Edburgberie* 1086 (DB). 'Stronghold of a woman called Ēadburh'. OE pers. name + *burh* (dative *byrig*).

Adderley Shrops. *Eldredelei* 1086 (DB). 'Woodland clearing of a woman called Althrȳth'. OE pers. name + *lēah*.

Adderstone Northum. *Edredeston* 1233. 'Farmstead or village of a man called Ēadrēd'. OE pers. name + *tūn*.

Addingham Brad. *Haddincham c.*972, *Odingehem* 1086 (DB). Probably 'homestead associated with a man called Adda'. OE pers. name + *-ing-* + *hām*.

Addington, 'estate associated with a man called Eadda or Ǣddi', OE pers. name + *-ing-* + *tūn*: **Addington** Bucks. *Edintone* 1086 (DB). **Addington** Gtr. London. *Eddintone* 1086 (DB). **Addington** Kent. *Eddintune* 1086 (DB). **Addington, Great & Addington, Little** Northants. *Edintone* 1086 (DB).

Addiscombe Gtr. London. *Edescamp* 1229. 'Enclosed land of a man called Ǣddi'. OE pers. name + *camp*.

Addlestone Surrey. *Attelesdene* 1241. 'Valley of a man called **Ættel*'. OE pers. name + *denu*.

Addlethorpe Lincs. *Arduluetorp* 1086 (DB). 'Outlying farmstead or hamlet of a man called Eardwulf'. OE pers. name + OScand. *thorp*.

Adel Leeds. *Adele* 1086 (DB). From OE *adela* 'dirty, muddy place'.

Adeney Tel. & Wrek. *Eduney* 1212. 'Island, or dry ground in marsh, of a woman called Ēadwynn'. OE pers. name + *ēg*.

Aderavoher (*Eadar dhá Bhóthair*) Sligo. 'Place between two roads'.

Adisham Kent. *Adesham* 616, *Edesham* 1086 (DB). 'Homestead of a man called **Ēadi* or Ǣddi'. OE pers. name + *hām*.

Adlestrop Glos. *Titlestrop* [*sic*] 714, *Tedestrop* 1086 (DB). 'Outlying farmstead or hamlet of a man called **Tætel*'. OE pers. name + *throp*. Initial *T-* disappeared from the 14th cent. due to confusion with the preposition *at*.

Adlingfleet E. R. Yorks. *Adelingesfluet* 1086 (DB). 'Water-channel or stream of the prince or nobleman'. OE *ætheling* + *flēot*.

Adlington, 'estate associated with a man called Ēadwulf', OE pers. name + *-ing-* + *tūn*; **Adlington** Ches. *Eduluintune* 1086 (DB). **Adlington** Lancs. *Edeluinton c.*1190.

Admaston Staffs. *Ædmundeston* 1176. 'Farmstead or village of a man called Ēadmund'. OE pers. name + *tūn*.

Admington Warwicks. *Edelmintone* 1086 (DB). 'Estate associated with a man called Æthelhelm'. OE pers. name + *-ing-* + *tūn*.

Adrigole (*Eadargóil*) Cork. 'Place between forks'.

Adrivale (*Eadargóil*) Cork. 'Place between forks'.

Adstock Bucks. *Edestoche* 1086 (DB). 'Outlying farmstead or hamlet of a man called Æddi or Eadda'. OE pers. name + *stoc*.

Adstone Northants. *Atenestone* 1086 (DB). 'Farmstead or village of a man called **Ættīn*'. OE pers. name + *tūn*.

Adur W. Sussex, district name from the River **Adur**, a late back-formation from *Portus Adurni* 'Adurnos's harbour', said to be at the mouth of this river by the 17th cent. antiquarian poet Drayton.

Adwell Oxon. *Advelle* 1086 (DB). 'Spring or stream of a man called Ead(d)a'. OE pers. name + *wella*.

Adwick le Street, Adwick upon Dearne Donc. *Adeuuic* 1086 (DB). 'Dwelling or (dairy) farm of a man called Adda'. OE pers. name + *wīc*. Distinguishing affixes from the situation of one Adwick on a Roman road (OE *strǣt*) and of the other on the River Dearne (*see* BOLTON UPON DEARNE).

Affpuddle Dorset. *Affapidele* 1086 (DB). 'Estate on the River Piddle of a man called Æffa'. OE pers. name + river-name (*see* PIDDLEHINTON).

Agangarrive Hill (*Aigeán Garbh*) Antrim. 'Rough hill'.

Agglethorpe N. Yorks. *Aculestorp* 1086 (DB). 'Outlying farmstead or hamlet of a man called Ācwulf'. OE pers. name + OScand. *thorp*.

Agha (*Achadh*) Carlow. 'Field'.

Aghaboe (*Achadh Bó*) Laois. 'Field of the cows'.

Aghabrack (*Achadh Breac*) Westmeath. 'Speckled field'.

Aghabullogue (*Achadh Bolg*) Cork. 'Field of the bulges'.

Aghacashel (*Achadh an Chaisil*) Limerick. 'Field of the stone fort'.

Aghacommon (*Achadh Camán*) Armagh. *Aghcamon* 1617. 'Field of the little bends'.

Aghada (*Achadh Fada*) Cork. 'Long field'.

Aghadarragh (*Achadh Darach*) Tyrone. 'Field of the oak-tree'.

Aghadaugh (*Achadh Damh*) Westmeath. 'Field of the oxen'.

Aghadoe (*Achadh dá Eo*) Kerry. 'Field of the two yews'.

Aghadowey (*Achadh Dubhthaigh*) Derry. *Achad Dubthaig c.*1170. 'Dubhthach's field'.

Aghadown (*Achadh Dúin*) Cork. 'Field of the fort'.

Aghafatten (*Achadh Pheatáin*) Antrim. *Aghafatten* 1780. Possibly 'Peatán's field'.

Aghagallon (*Achadh Gallan*) Tyrone, Antrim. 'Field of the standing stone'.

Aghagower (*Achadh Ghobair*) Mayo. 'Field of the spring'.

Aghalane (*Achadh Leathan*) Fermanagh. *Aghalane* 1622. 'Broad field'.

Aghalee (*Achadh Lí*) Antrim. *Acheli* 1306. 'Field'. The second element is obscure.

Aghamore (*Achadh Mór*) Mayo. 'Big field'.

Aghanacliff (*Achadh na Cloiche*) Louth. 'Field of the stones'.

Aghanloo (*Áth Lú*) Derry. *Athlouge* 1397. 'Lú's ford'.

Aghatubrid (*Achadh Tiobraid*) Kerry. 'Field of the well'.

Aghavannagh (*Achadh Bheannach*) Wicklow. 'Hilly field'.

Aghavea (*Achadh beithe*) Fermanagh. 'Birch field'.

Aghawoney (*Achadh Mhóna*) Donegal. 'Field of the bog'.

Agher (*Achair*) Meath. 'Space'.

Aghern (*Áth Chairn*) Cork. 'Ford of the cairn'.

Aghery Lough (*Loch Eachraí*) Down. 'Lake of the horses'.

Aghintain (*Achadh an tSéin*) Tyrone. *Aghityan* 1613. 'Field of the good luck'.

Aghinver (*Achadh Inbhir*) Fermanagh. 'Field of the river-mouth'.

Aghleam (*Eachléim*) Mayo. 'Horse leap'.

Aghlish (*Eaglais*) Kerry. 'Church'.

Aghnabohy (*Achadh na Boithe*) Westmeath. 'Field of the huts'.

Aghnahily (*Achadh na hAille*) Laois. 'Field of the cliff'.

Aghnamullen (*Achadh na Muileann*) Monaghan. *Aughamullen* 1530. 'Field of the mills'.

Aghnaskeagh (*Achadh na Scéithe*) Louth. 'Field of the shields'.

Aghory (*Áth Óraí*) Armagh. *Aghoorier* 1610. 'Ford of the boundary'.

Aghowla (*Achadh Abhla*) Limerick. 'Field of the apple-tree'.

Aghowle (*Achadh Abhla*) Limerick, Wicklow. 'Field of the apple-tree'.

Aghyaran (*Achadh Uí Áráin*) Tyrone. *Agharan* 1666. 'Field of Uí Árán'.

Aghyowle (*Achadh Abhla*) Fermanagh. 'Field of the apple-tree'.

Agivey (*Áth Géibhe*) Derry. *Athgeybi* 1492. Possibly 'ford of the fetter'.

Aglish (*An Eaglais*) Waterford. 'The church'.

Agola (*Áth Gobhlach*) Antrim. 'Ford of the fork'.

Ahafona (*Áth an Phóna*) Kerry. 'Ford of the pound'.

Ahakista (*Áth an Chiste*) Kerry. 'Ford of the box'.

Ahalia, Lough (*Loch an tSáile*) Galway. 'Lake of the salt water'.

Ahascragh (*Áth Eascrach*) Galway. 'Ford of the gravel ridge'.

Ahenny (*Áth Eine*) Tipperary. 'Ford of the fire'.

Ahoghill (*Áth Eochaille*) Antrim. *Achochill* 1306. 'Ford of the yew wood'.

Aikton Cumbria. *Aictun c.*1200. 'Oak-tree farmstead'. OScand. *eik* + OE *tūn*.

Aille (*Aille*) Clare, Mayo. 'Cliffs'.

Aillenaveagh (*Aill na bhFiach*) Galway. 'Cliff of the ravens'.

Ailsa Craig (island) S. Ayr. Gaelic *allasa* + *creag* 'rock'. The first element is obscure.

Ailsworth Peterb. *Ægeleswurth* 948, *Eglesworde* 1086 (DB). 'Enclosure of a man called *Ægel*. OE pers. name + *worth*.

Ainderby Steeple N. Yorks. *Eindrebi* 1086 (DB), *Aynderby wyth Stepil* 1316. 'Farmstead or village of a man called Eindrithi'. OScand. pers. name + *bý*. Affix is OE *stēpel* 'church steeple, tower'.

Ainsdale Sefton. *Einuluesdel* 1086 (DB). 'Valley of a man called *Einulfr*. OScand. pers. name + *dalr*.

Ainstable Cumbria. *Ainstapillith c.*1210. 'Slope where bracken grows'. OScand. *einstapi* + *hlíth*.

Ainsworth Bury. *Haineswrthe c.*1200. 'Enclosure of a man called *Ægen*. OE pers. name + *worth*.

Aintree Sefton. *Ayntre* 1220. 'Solitary tree'. OScand. *einn* + *tré*.

Airdrie N. Lan. *Airdrie* 1584. 'High slope'. Gaelic *ard* + *ruigh*.

Airmyn E. R. Yorks. *Ermenie* 1086 (DB). 'Mouth of the River Aire'. River-name + OScand. *mynni*. The river-name **Aire** is possibly from OScand. *eyjar* 'islands', but may be of Celtic or pre-Celtic origin with a meaning 'strongly flowing'.

Airton N. Yorks. *Airtone* 1086 (DB). 'Farmstead on the River Aire'. Old river-name (*see* AIRMYN) + OE *tūn*.

Aisby, 'farmstead or village of a man called Ási', OScand. pers. name + *bý*: **Aisby** Lincs., near Blyton. *Asebi* 1086 (DB). **Aisby** Lincs., near Sleaford. *Asebi* 1086 (DB).

Aisholt, Lower Somerset. *Æscholt* 854. 'Ash-tree wood'. OE *æsc* + *holt*.

Aiskew N. Yorks. *Echescol* [*sic*] 1086 (DB), *Aykescogh* 1235. 'Oak wood'. OScand. *eik* + *skógr*.

Aislaby N. Yorks., near Pickering. *Aslache(s)bi* 1086 (DB). 'Farmstead or village of a man called Áslákr'. OScand. pers. name + *bý*.

Aislaby N. Yorks., near Whitby. *Asulue(s)bi* 1086 (DB). 'Farmstead or village of a man called Ásulfr'. OScand. pers. name + *bý*.

Aisthorpe Lincs. *Estorp* 1086 (DB). 'East outlying farmstead or hamlet'. OE *ēast* + OScand. *thorp*.

Akeld Northum. *Achelda* 1169. 'Oak-tree slope'. OE *āc* + *helde*.

Akeley Bucks. *Achelei* 1086 (DB). 'Oak wood or clearing'. OE *ācen* + *lēah*.

Akeman Street (Roman road from Bath to St Albans). *Accemannestrete* 12th cent. OE *strǣt* 'Roman road' with an uncertain first element also found in an early alternative name for BATH, *Acemannes ceastre* 10th cent., from OE *ceaster* 'Roman town or city'. *Ac(c)emannes-* may reflect a British name for BATH, **Acumannā* 'Aquae-place', with reference to the Roman name *Aquae Sulis* 'waters of Sulis, a pagan goddess'.

Akenham Suffolk. *Acheham* 1086 (DB). 'Homestead of a man called Aca'. OE pers. name (genitive -*n*) + *hām*.

Alberbury Shrops. *Alberberie* 1086 (DB). 'Stronghold or manor of a woman called Aluburh'. OE pers. name + *burh* (dative *byrig*).

Albourne W. Sussex. *Aleburn* 1177. 'Stream where alders grow'. OE *alor* + *burna*.

Albrighton Shrops., near Shifnal. *Albricstone* 1086 (DB). 'Farmstead or village of a man called Æthelbeorht'. OE pers. name + *tūn*.

Albrighton Shrops., near Shrewsbury. *Etbritone* 1086 (DB). 'Farmstead or village of a man called Ēadbeorht'. OE pers. name + *tūn*.

Alburgh Norfolk. *Aldeberga* 1086 (DB). 'Old mound or hill', or 'mound or hill of a man called Alda'. OE (*e*)*ald* or OE pers. name + *beorg*.

Albury, 'old or disused stronghold', OE (*e*)*ald* + *burh* (dative *byrig*): **Albury** Herts. *Eldeberie* 1086 (DB). **Albury** Surrey. *Ealdeburi* 1062, *Eldeberie* 1086 (DB).

Alby Hill Norfolk. *Alebei* 1086 (DB). 'Farmstead or village of a man called Áli'. OScand. pers. name + *bý*.

Alcaston Shrops. *Ælmundestune* 1086 (DB). 'Farmstead or village of a man called Ealhmund'. OE pers. name + *tūn*.

Alcester Warwicks. *Alencestre* 1138. 'Roman town on the River Alne'. Celtic river-name (*see* ALNE) + OE *ceaster*.

Alciston E. Sussex. *Alsistone* 1086 (DB). 'Farmstead or village of a man called Ælfsige or Ealhsige'. OE pers. name + *tūn*.

Alconbury Cambs. *Acumesberie* [*sic*] 1086 (DB), *Alcmundesberia* 12th cent. 'Stronghold of a man called Ealhmund'. OE pers. name + *burh* (dative *byrig*).

Aldborough, 'old or disused stronghold', OE (*e*)*ald* + *burh*: **Aldborough** Norfolk. *Aldeburg* 1086 (DB). **Aldborough** N. Yorks. *Burg* 1086 (DB), *Aldeburg* 1145. Here referring to a Roman fort.

Aldbourne Wilts. *Ealdincburnan c.*970, *Aldeborne* 1086 (DB). 'Stream associated with a man called Ealda'. OE pers. name + -*ing-* + *burna*.

Aldbrough, 'old or disused stronghold', OE (*e*)*ald* + *burh*: **Aldbrough** E. R. Yorks. *Aldenburg* 1086 (DB). **Aldbrough** N. Yorks. *Aldeburne* [*sic*] 1086 (DB), *Aldeburg* 1247.

Aldbury Herts. *Aldeberie* 1086 (DB). 'Old or disused stronghold'. OE (*e*)*ald* + *burh* (dative *byrig*).

Aldeburgh Suffolk. *Aldeburc* 1086 (DB). 'Old or disused stronghold'. OE (*e*)*ald* + *burh*. The river-name **Alde** is a back-formation from the place name.

Aldeby Norfolk. *Aldebury* 1086 (DB), *Aldeby c.*1180. 'Old or disused stronghold'. OE (*e*)*ald* + *burh* (dative *byrig*) replaced by OScand. *bý* 'farmstead'.

Aldenham Herts. *Ældenham* 785, *Eldeham* 1086 (DB). 'Old homestead', or 'homestead of a man called Ealda'. OE (*e*)*ald* (dative -*an*) or OE pers. name (genitive -*n*) + *hām*.

Alderbury Wilts. *Æthelware byrig* 972, *Alwarberie* 1086 (DB). 'Stronghold of a woman called **Æthelwaru*'. OE pers. name + *burh* (dative *byrig*).

Alderford Norfolk. *Alraforda* 1163. 'Ford where alders grow'. OE *alor* + *ford*.

Alderholt Dorset. *Alreholt* 1285. 'Alder wood'. OE *alor* + *holt*.

Alderley Glos. *Alrelie* 1086 (DB). 'Woodland clearing where alders grow'. OE *alor* + *lēah*.

Alderley Edge Ches. *Aldredelie* 1086 (DB). 'Woodland clearing of a woman called Althrȳth'. OE pers. name + *lēah*. The 19th-cent. addition *Edge* is taken from the abrupt escarpment here, itself called Alderley Edge (from OE *ecg*).

Aldermaston W. Berks. *Ældremanestone* 1086 (DB). 'Farmstead of the chief or nobleman'. OE *(e)aldormann + tūn*.

Alderminster Warwicks. *Aldermanneston* 1167. Identical in origin with the previous name.

Alderney (island) Channel Islands. *Aurene c.*1042, *Aureneie* 1238. 'Gravel or mudflat island'. OScand. **aurinn* (an adjectival derivative of *aurr* 'gravel, mud') + *ey*. In the 4th cent. *Maritime Itinerary*, Alderney is recorded as *Riduna*, possibly a Celtic name meaning 'before the hill' and perhaps originally a name for the port dating from the Roman period.

Aldershot Hants. *Halreshet* 1171. 'Projecting piece of land where alders grow'. OE *alor + scēat*.

Alderton, usually 'estate associated with a man called Ealdhere', OE pers. name + *-ing- + tūn*: **Alderton** Glos. *Aldritone* 1086 (DB). **Alderton** Northants. *Aldritone* 1086 (DB). **Alderton** Wilts. *Aldrintone* 1086 (DB).
   However two Aldertons have a different origin, 'farmstead where alders grow', OE *alor + tūn*: **Alderton** Shrops. *Olreton* 1309. **Alderton** Suffolk. *Alretuna* 1086 (DB).

Alderwasley Derbys. *Alrewaseleg* 1251. 'Clearing by the alluvial land where alders grow'. OE *alor + *wæsse + lēah*.

Aldfield N. Yorks. *Aldefeld* 1086 (DB). 'Old (i.e. long used) stretch of open country'. OE *ald + feld*.

Aldford Ches. *Aldefordia* 1153. 'The old (i.e. formerly used) ford'. OE *ald + ford*.

Aldham, 'the old homestead', or 'homestead of a man called Ealda', OE *eald* or OE pers. name + *hām*: **Aldham** Essex. *Aldeham* 1086 (DB). **Aldham** Suffolk. *Aldeham* 1086 (DB).

Aldingbourne W. Sussex. *Ealdingburnan c.*880, *Aldingeborne* 1086 (DB). 'Stream associated with a man called Ealda'. OE pers. name + *-ing- + burna*.

Aldingham Cumbria. *Aldingham* 1086 (DB). Probably 'homestead of the family or followers of a man called Alda'. OE pers. name + *-inga- + hām*.

Aldington, 'estate associated with a man called Ealda', OE pers. name + *-ing- + tūn*: **Aldington** Kent. *Aldintone* 1086 (DB).

Aldington Worcs. *Aldintona* 709, *Aldintone* 1086 (DB).

Aldreth Cambs. *Alrehetha* 1170. 'Landing-place by the alders'. OE *alor + hȳth*.

Aldridge Wsall. *Alrewic* 1086 (DB). 'Dwelling or farm among alders'. OE *alor + wīc*.

Aldringham Suffolk. *Alrincham* 1086 (DB). 'Homestead of the family or followers of a man called Aldhere'. OE pers. name + *-inga- + hām*.

Aldsworth Glos. *Ealdeswyrthe* 1004, *Aldeswrde* 1086 (DB). 'Enclosure of a man called **Ald*. OE pers. name + *worth*.

Aldwark, 'old fortification', OE *(e)ald + weorc*: **Aldwark** Derbys. *Aldwerk* 1140. **Aldwark** N. Yorks. *Aldeuuerc* 1086 (DB).

Aldwick W. Sussex. *Aldewyc* 1235. 'Old dwelling', or 'dwelling of a man called Ealda'. OE *eald* or OE pers. name + *wīc*.

Aldwincle Northants. *Eldewincle* 1086 (DB). 'River-bend of a man called Ealda'. OE pers. name + **wincel*.

Aldworth W. Berks. *Elleorde* [*sic*] 1086 (DB), *Aldewurth* 1167. 'Old enclosure', or 'enclosure of a man called Ealda'. OE *eald* or OE pers. name + *worth*.

Aldwych Gtr London. Street named after the medieval settlement recorded as *Aldewich* 1211. 'The old trading place'. OE *ald + wīc*.

Alexandria W. Dunb. '(Place of) Alexander'. The town arose in the mid-18th cent. and the name was given in *c.*1760 for *Alexander* Smollett (d.1799), MP for Bonhill.

Alfington Devon. *Alfinton* 1244. 'Estate associated with a man called Ælf'. OE pers. name + *-ing- + tūn*.

Alfold Surrey. *Alfold* 1227. 'Old fold or enclosure'. OE *eald + fald*.

Alford Aber. *Afford c.*1200, *Afurd* 1654. Origin and meaning obscure.

Alford Lincs. *Alforde* 1086 (DB). Probably 'old ford'. OE *ald + ford*. The road carried by the ford may be Roman.

Alford Somerset. *Aldedeford* 1086 (DB). 'Ford of a woman called Ealdgȳth'. OE pers. name + *ford*.

Alfreton Derbys. *Elstretune* [*sic*] 1086 (DB), *Alferton* 12th cent. 'Farmstead or village of a man called Ælfhere'. OE pers. name + *tūn*.

Alfrick Worcs. *Alcredeswike* early 13th cent. 'Dwelling or farm of a man called Ealhrǣd'. OE pers. name + *wīc*.

Alfriston E. Sussex. *Alvricestone* 1086 (DB). 'Farmstead or village of a man called Ælfrīc'. OE pers. name + *tūn*.

Alhampton Somerset. *Alentona* 1086 (DB). 'Estate on the River Alham'. Celtic river-name (of uncertain meaning) + OE *tūn*.

Alkborough N. Lincs. *Alchebarge* 1086 (DB), *Alchebarua* 12th cent. Probably 'hill of a man called Al(u)ca'. OE pers. name + *beorg*.

Alkerton Oxon. *Alcrintone* 1086 (DB). 'Estate associated with a man called Ealhhere'. OE pers. name + *-ing-* + *tūn*.

Alkham Kent. *Ealhham c.*1100. 'Homestead in a sheltered place, or used as a sanctuary'. OE *ealh* + *hām*.

Alkington Shrops. *Alchetune* 1086 (DB), *Alkinton* 1256. 'Estate associated with a man called Ealha'. OE pers. name + *-ing-* + *tūn*.

Alkmonton Derbys. *Alchementune* 1086 (DB). 'Farmstead or village of a man called Ealhmund'. OE pers. name + *tūn*.

All Cannings Wilts. *See* CANNINGS.

All Stretton Shrops. *See* STRETTON.

Allendale Town Northum. *Alewenton* 1245. 'Settlement by (in the valley of) the River Allen'. Celtic or pre-Celtic river-name (of uncertain meaning) + OE *tūn*, with the later insertion of OScand. *dalr* 'valley'.

Allenheads Northum. '(Place by) the source of the River Allen'. Celtic or pre-Celtic river-name + OE *hēafod*.

Allensmore Herefs. *More* 1086 (DB), *Aleinesmor* 1220. 'Marshy ground of a man called Ala(i)n'. OFrench pers. name + OE *mōr*.

Aller Somerset. *Alre* late 9th cent., 1086 (DB). '(Place at) the alder-tree'. OE *alor*.

Allerby Cumbria. *Aylwardcrosseby* 1260, *Aylewardby c.*1275. 'Farmstead (with crosses) of a man called Ailward (Æthelweard)'. OE pers. name + OScand. *bý* (earlier *krossa-bý*).

Allerdale (district) Cumbria. *Alnerdall* 11th cent., *Aldersdale* 1268. 'Valley of the River Ellen'. A revival of an old ward-name. Celtic river-name (of uncertain meaning) + OScand. *dalr*.

Allerford Somerset, near Minehead. *Alresford* 1086 (DB). 'Alder-tree ford'. OE *alor* + *ford*.

Allerston N. Yorks. *Alurestan* 1086 (DB). 'Boundary stone of a man called Ælfhere'. OE pers. name + *stān*.

Allerthorpe E. R. Yorks. *Aluuarestorp* 1086 (DB). 'Outlying farmstead or hamlet of a man called Ælfweard or *Alfvarthr*'. OE or OScand. pers. name + OScand. *thorp*.

Allerton, usually 'farmstead or village where alder-trees grow', OE *alor* + *tūn*: **Allerton** Brad. *Alretune* 1086 (DB). **Allerton** Lpool. *Alretune* 1086 (DB). **Allerton Bywater** Leeds. *Alretune* 1086 (DB), *Allerton by ye water* 1430. Affix 'by the water' (OE *wæter*) refers to its situation on the River Aire. **Allerton, Chapel** Leeds. *Alretun* 1086 (DB), *Chapel Allerton* 1360. Affix is ME *chapele* 'a chapel'.

However some Allertons have a different origin: **Allerton, Chapel** Somerset. *Alwarditone* 1086 (DB). 'Farmstead of a man called Ælfweard'. OE pers. name + *tūn*. Affix as in previous name. **Allerton Mauleverer** N. Yorks. *Aluertone* 1086 (DB), *Aluerton Mauleuerer* 1231. 'Farmstead of a man called Ælfhere'. OE pers. name + *tūn*. Manorial affix from the *Mauleverer* family, here from the 12th cent.

Allesley Covtry. *Alleslega* 1176. 'Woodland clearing of a man called Ælle'. OE pers. name + *lēah*.

Allestree Derby. *Adelardestre* 1086 (DB). 'Tree of a man called Æthelheard'. OE pers. name + *trēow*.

Allexton Leics. *Adelachestone* 1086 (DB). 'Farmstead or village of a man called *Æthellāc*'. OE pers. name + *tūn*.

Allhallows Medway. *Ho All Hallows* 1285. Named from the 12th-cent. church of All Saints here. For *Ho* in the early form, *see* HOO.

Allihies (*Na hAilichí*) Cork. 'The cliff fields'.

Allington, a common name, has a number of different origins: **Allington** Kent, near Lenham. *Alnoitone* 1086 (DB), *Eilnothinton* 1242. 'Farmstead associated with a man called Æthelnōth'. OE pers. name + *-ing-* + *tūn*.

Allington Kent, near Maidstone. *Elentun* 1086 (DB). 'Farmstead associated with a man called Ælla or Ælle'. OE pers. name + -*ing*- + *tūn*.

Allington Lincs. *Adelingetone* 1086 (DB). 'Farmstead of the princes'. OE *ætheling* + *tūn*.

Allington Wilts., near Amesbury. *Aldintona* 1178. 'Farmstead associated with a man called Ealda'. OE pers. name + -*ing*- + *tūn*. **Allington** Wilts., near Devizes. *Adelingtone* 1086 (DB). 'Farmstead of the princes'. OE *ætheling* + *tūn*.

Allington, East Devon. *Alintone* 1086 (DB). 'Farmstead associated with a man called Ælla or Ælle'. OE pers. name + -*ing*- + *tūn*.

Allistragh (*An tAileastrach*) Armagh. *Tallastagh* 1609. 'Place of the wild irises'.

Allithwaite Cumbria. *Hailiuethait c.*1170. 'Clearing of a man called Eilífr'. OScand. pers. name + *thveit*.

Alloa Clac. *Alveth* 1357. 'Rocky plain'. Gaelic *allmhagh*.

Allonby Cumbria. *Alayneby* 1262. 'Farmstead or village of a man called Alein'. OFrench pers. name + OScand. *bý*.

Allow (*Abhainn Alla*) (river) Cork. 'River of (the district of) Ealla'.

Alloway S. Ayr. *Auleway* 1324. 'Rocky plain'. Gaelic *allmhagh*.

Allt Melyd. See MELIDEN.

Allweston Dorset. *Alfeston* 1214, *Alfletheston* 1244, *Alueueston* 1268. 'Farmstead or village of a woman called Ælfflǣd or Ælfgifu'. OE pers. name + *tūn*.

Almeley Herefs. *Elmelie* 1086 (DB). 'Elm wood or clearing'. OE *elm* + *lēah*.

Almer Dorset. *Elmere* 943. 'Eel pool'. OE *ǣl* + *mere*.

Almington Staffs. *Almentone* 1086 (DB). 'Farmstead or village of a man called Alhmund'. OE pers. name + *tūn*.

Almondbury Kirkl. *Almaneberie* 1086 (DB). 'Stronghold of the whole community'. OScand. *almenn* (genitive plural *almanna*) + OE *burh* (dative *byrig*).

Almondsbury S. Glos. *Almodesberie* 1086 (DB). 'Stronghold of a man called Æthelmōd or Æthelmund'. OE pers. name + *burh* (dative *byrig*).

Alne N. Yorks. *Alna c.*1050, *Alne* 1086 (DB). A Celtic name of uncertain meaning.

Alne, Great & Alne, Little Warwicks. *Alne* 1086 (DB). Named from the River Alne, a Celtic river-name probably identical in origin with the River Aln (*see* ALNHAM).

Alness Highland. *Alenes* 1226. 'Allan station, level place on the River Allan'. Gaelic *Alanais*. Pre-Celtic river-name ('flowing water') + Gaelic *fas*.

Alnham Northum. *Alneham* 1228. 'Homestead on the River Aln'. Celtic river-name (earlier *Alaunos*, of uncertain meaning) + OE *hām*.

Alnmouth Northum. *Alnemuth* 1201. 'Mouth of the River Aln'. Celtic river-name (*see* ALNHAM) + OE *mūtha*.

Alnwick Northum. *Alnewich* 1178. 'Dwelling or farm on the River Aln'. Celtic river-name (*see* ALNHAM) + OE *wīc*.

Alperton Gtr. London. *Alprinton* 1199. 'Estate associated with a man called Ealhbeorht'. OE pers. name + -*ing*- + *tūn*.

Alphamstone Essex. *Alfelmestuna* 1086 (DB). 'Farmstead or village of a man called Ælfhelm'. OE pers. name + *tūn*.

Alpheton Suffolk. *Alflede(s)ton* 1204. 'Farmstead or village of a woman called Ælfflǣd or Æthelflǣd'. OE pers. name + *tūn*.

Alphington Devon. *Alfintune c.*1060, *Alfintone* 1086 (DB). 'Estate associated with a man called Ælf'. OE pers. name + -*ing*- + *tūn*.

Alport Derbys. *Aldeport* 12th cent. 'Old town'. OE (*e*)*ald* + *port*.

Alpraham Ches. *Alburgham* 1086 (DB). 'Homestead of a woman called Alhburh'. OE pers. name + *hām*.

Alresford Essex. *Ælesford c.*1000, *Eilesforda* 1086 (DB). 'Ford of the eel, or of a man called *Ægel*. OE *ǣl* or OE pers. name + *ford*.

Alresford, New & Alresford, Old Hants. *Alresforda* 701, *Alresforde* 1086 (DB). 'Alder-tree ford'. OE *alor* (genitive *alres*) + *ford*. The river-name **Alre** is a back-formation from the place name.

Alrewas Staffs. *Alrewasse* 942, *Alrewas* 1086 (DB). 'Alluvial land where alders grow'. OE *alor* + *wæsse*.

Alsager Ches. *Eleacier* [*sic*] 1086 (DB), *Allesacher* 13th cent. 'Cultivated land of a man called Ælle'. OE pers. name + *æcer*.

Alsh, Loch. *See* KYLE OF LOCHALSH.

Alsop en le Dale Derbys. *Elleshope* 1086 (DB), *Alsope in le dale* 1535. 'Valley of a man called Ælle'. OE pers. name + *hop*. Later affix means 'in the valley'.

Alston Cumbria. *Aldeneby* 1164–71, *Aldeneston* 1209. 'Farmstead or village of a man called Halfdan'. OScand. pers. name + OE *tūn* (earlier OScand. *bý*).

Alstone Glos. *Ælfsigestun* 969. 'Farmstead or village of a man called Ælfsige'. OE pers. name + *tūn*.

Alstonefield Staffs. *Ænestanefelt* [*sic*] 1086 (DB), *Alfstanesfeld* 1179. 'Open country of a man called Ælfstān'. OE pers. name + *feld*.

Altagowlan (*Alt an Ghabhláin*) Roscommon. 'Hillside of the fork'.

Altamuskin (*Alt na Múscán*) Tyrone. *Altmuskan* 1611. 'Hillside of the loose clay'.

Altan (*Altán*) Donegal. 'Little height'.

Altan, Lough (*Loch Allltáin*) Donegal. 'Lake of the flocks'.

Altarnun Cornwall. *Altrenune* c.1100. 'Altar of St Nonn'. Cornish *alter* 'altar of a church' + female saint's name.

Altavilla (*Alta a' Bhile*) Limerick. 'Hillside of the sacred tree'.

Altcar, Great (Lancs.) **& Altcar, Little** (Sefton). *Acrer* [*sic*] 1086 (DB), *Altekar* 1251. 'Marsh by the River Alt'. Celtic river-name (meaning 'muddy river') + OScand. *kjarr*.

Altham Lancs. *Elvetham* c.1150. 'Enclosure or river-meadow where there are swans'. OE *elfitu* + *hamm*.

Althorne Essex. *Aledhorn* 1198. '(Place at) the burnt thorn-tree'. OE *æled* + *thorn*.

Althorp Northants. *Olletorp* 1086 (DB), *Olethorp* 1208. 'Outlying farmstead or hamlet of a man called *Olla*'. OE pers. name + OScand. *thorp*.

Althorpe N. Lincs. *Aletorp* 1086 (DB). 'Outlying farmstead or hamlet of a man called Áli'. OScand. pers. name + *thorp*.

Altinure (*Alt an Iúir*) Cavan, Derry. 'Height of the yew-tree'.

Altishahane (*Alt Inse Uí Chatháin*) Tyrone. *Altonisechan* c.1655. 'Hillside of the island of Uí Chatháin'.

Altnamachin (*Alt na Meacan*) Armagh. 'Hillside of the root vegetables'.

Altnaveagh (*Alt na bhFiach*) Tyrone. 'Hillside of the ravens'.

Altofts Wakefd. *Altoftes* c.1090. Probably 'the old homesteads'. OE *ald* + OScand. *toft*.

Alton, usually 'farmstead at the source of a river', OE *æwiell* + *tūn*: **Alton** Hants. *Aultone* 1086 (DB). **Alton Pancras** Dorset. *Awultune* 1012, *Altone* 1086 (DB), *Aweltone Pancratii* 1226. Affix from the dedication of the church to St Pancras. **Alton Priors** Wilts. *Aweltun* 825, *Auuiltone* 1086 (DB), *Aulton Prioris* 1199. Affix from its early possession by the Priory of St Swithin at Winchester.
   However other Altons have a different origin: **Alton** Derbys. *Alton* 1296. 'Old farmstead'. OE *ald* + *tūn*. **Alton** Staffs. *Elvetone* 1086 (DB). 'Farmstead of a man called *Ælfa*'. OE pers. name + *tūn*.

Altore (*Altóir*) Roscommon. 'Altar'.

Altrincham Traffd. *Aldringeham* 1290. 'Homestead of the family or followers of a man called Aldhere', or 'homestead at the place associated with Aldhere'. OE pers. name + *-inga-* or *-ing* + *hām*.

Alum Bay I. of Wight, first recorded in 1720 and so called from the large quantities of alum mined here as early as the 16th cent. **Alum Chine** Dorset, also on record from the 18th cent., alludes to mining of the same mineral, used in paper-making and leather-tanning.

Alva Clac. *Alweth* 1489. 'Rocky plain'.

Alvah Aber. *Alveth* 1308. 'Rocky plain'. The name applies to both *Bridge of Alvah* and *Kirktown of Alvah*.

Alvanley Ches. *Elveldelie* 1086 (DB). 'Woodland clearing of a man called Ælfweald'. OE pers. name + *lēah*.

Alvaston Derby. *Alewaldestune* c.1002, *Alewoldestune* 1086 (DB). 'Farmstead or village of a man called Æthelwald or Ælfwald'. OE pers. name + *tūn*.

Alvechurch Worcs. *Ælfgythe cyrcan* 10th cent., *Alvievecherche* 1086 (DB). 'Church of a woman called Ælfgȳth'. OE pers. name + *cirice*.

Alvecote Warwicks. *Avecote c.*1160. 'Cottage(s) of a man called Afa'. OE pers. name + *cot.*

Alvediston Wilts. *Alfwieteston* 1165. Probably 'farmstead or village of a man called Ælfgeat'. OE pers. name + *tūn*.

Alveley Shrops. *Alvidelege* 1086 (DB). 'Woodland clearing of a woman called Ælfgȳth'. OE pers. name + *lēah*.

Alverdiscott Devon. *Alveredescota* 1086 (DB). 'Cottage of a man called Ælfrēd'. OE pers. name + *cot*.

Alverstoke Hants. *Stoce* 948, *Alwarestoch* 1086 (DB). 'Outlying farmstead or hamlet of a woman called Ælfwaru or *Æthelwaru'. OE pers. name + *stoc*.

Alverstone I. of Wight. *Alvrestone* 1086 (DB). 'Farmstead or village of a man called Ælfrēd'. OE pers. name + *tūn*.

Alverton Notts. *Aluriton* 1086 (DB). 'Estate associated with a man called Ælfhere'. OE pers. name + -*ing*- + *tūn*.

Alvescot Oxon. *Elfegescote* 1086 (DB). 'Cottage of a man called Ælfhēah'. OE pers. name + *cot*.

Alveston S. Glos. *Alwestan* 1086 (DB). 'Boundary stone of a man called Ælfwīg'. OE pers. name + *stān*.

Alveston Warwicks. *Eanulfestun* 966, *Alvestone* 1086 (DB). 'Farmstead or village of a man called Ēanwulf'. OE pers. name + *tūn*.

Alvingham Lincs. *Aluingeham* 1086 (DB). 'Homestead of the family or followers of a man called Ælf'. OE pers. name + -*inga*- + *hām*.

Alvington Glos. *Eluinton* 1220. Probably identical in origin with the next name.

Alvington, West Devon. *Alvintone* 1086 (DB). 'Estate associated with a man called Ælf'. OE pers. name + -*ing*- + *tūn*.

Alwalton Cambs. *Æthelwoldingtun* 955, *Alwoltune* 1086 (DB). 'Estate associated with a man called Æthelwald'. OE pers. name + -*ing*- + *tūn*.

Alwinton Northum. *Alwenton* 1242. 'Farmstead or village on the River Alwin'.

Celtic or pre-Celtic river-name (of uncertain meaning) + OE *tūn*.

Alwoodley Leeds. *Aluuoldelei* 1086 (DB), *Adelwaldesleia* 1166. 'Woodland clearing of a man called Æthelwald'. OE pers. name + *lēah*.

Alyth Perth. *Alicht c.*1249, *Alyth* 1327. Perhaps 'rocky place'. Gaelic *eileach*.

Ambergate Derbys. , a recent name, first recorded in 1836, referring to a toll-gate near the River **Amber** (a pre-Celtic river-name of uncertain meaning) which also gives name to the district of **Amber Valley**.

Amberley, probably 'woodland clearing frequented by a bird such as the bunting or yellow-hammer', OE *amer* + *lēah*: **Amberley** Glos. *Unberleia* [*sic*] 1166, *Omberleia, Amberley c.*1240. **Amberley** W. Sussex. *Amberle* 957, *Ambrelie* 1086 (DB).

Ambersham, South W. Sussex. *Æmbresham* 963. 'Homestead or river-bend land of a man called *Æmbre'. OE pers. name + *hām* or *hamm*.

Amble Northum. *Ambell* 1204, *Anebell* 1256. Probably 'promontory of a man called *Amma or Anna'. OE pers. name + *bile*.

Amblecote Dudley. *Elmelecote* 1086 (DB). Probably 'cottage of a man called *Æmela'. OE pers. name + *cot*.

Ambleside Cumbria. *Ameleseta c.*1095. 'Shieling or summer pasture by the river sandbank'. OScand. *á* + *melr* + *sǽtr*.

Ambleston (*Treamlod*) Pemb. *Amleston* 1230. 'Amelot's farm'. OFrench pers. name + OE *tūn* (Welsh *tref*).

Ambrosden Oxon. *Ambresdone* 1086 (DB). Possibly 'hill of a man called *Ambre', OE pers. name + *dūn*. Alternatively 'hill of the bunting' if the first element is rather OE *amer*.

Amcotts N. Lincs. *Amecotes* 1086 (DB). 'Cottages of a man called *Amma'. OE pers. name + *cot*.

Amersham Bucks. *Agmodesham* 1066, *Elmodesham* 1086 (DB). 'Homestead or village of a man called Ealhmund'. OE pers. name + *hām*.

Amesbury Wilts. *Ambresbyrig c.*880, *Ambresberie* 1086 (DB). Possibly 'stronghold of a man called *Ambre', OE pers. name + *burh* (dative *byrig*). Alternatively '(disused) stronghold frequented by buntings' if the first element is rather OE *amer*.

Amington Staffs. *Ermendone* [*sic*] 1086 (DB), *Aminton* 1150. Probably 'estate associated with a man called *Earma'. OE pers. name + *-ing-* + *tūn*.

Amlwch Angl. *Anulc* 1254, *Amelogh* 1352. '(Place) near the swamp'. Welsh *am* + *llwch*.

Ammanford (*Rhydaman*) Carm. *Amman* 1541. 'Ford over the River Aman'. OE *ford* (Welsh *rhyd*). The river-name means 'pig' (Welsh *banw*), for a river that 'roots' its way through the ground.

Amotherby N. Yorks. *Aimundrebi* 1086 (DB). 'Farmstead or village of a man called Eymundr'. OScand. pers. name + *bý*.

Ampleforth N. Yorks. *Ampreforde* 1086 (DB). 'Ford where dock or sorrel grows'. OE *ampre* + *ford*.

Ampney Crucis, Ampney St Mary, Ampney St Peter & Down
Ampney Glos. *Omenie* 1086 (DB), *Ameneye Sancte Crucis* 1287, *Ammeneye Beate Marie* 1291, *Amenel Sancti Petri* c.1275, *Dunamenell* 1205. Possibly named from **Ampney Brook**, 'stream of a man called *Amma'. OE pers. name (genitive *-n*) + *ēa*. Alternatively the second element may be OE *ēg* 'island, area of raised ground in marsh'. Distinguishing affixes from the dedication of the churches to the Holy Rood (Latin *crucis* 'of the cross'), St Mary and St Peter, and from OE *dūne* 'lower downstream'.

Amport Hants. *Anna de Port* c.1270. 'Estate on the River *Ann* held by a family called *de Port*'. Celtic river-name (meaning 'ash-tree stream') + manorial affix (from its Domesday owner), *see* ANDOVER.

Ampthill Beds. *Ammetelle* 1086 (DB). 'Anthill, hill infested with ants'. OE *ǣmette* + *hyll*.

Ampton Suffolk. *Hametuna* 1086 (DB). 'Farmstead or village of a man called *Amma'. OE pers. name + *tūn*.

Amwell, Great & Amwell, Little Herts. *Emmewelle* 1086 (DB). 'Spring or stream of a man called *Æmma'. OE pers. name + *wella*.

Anascaul (*Abhainn an Scáil*) (river) Kerry. 'River of the phantom'.

Ancaster Lincs. *Anecastre* 12th cent. Probably 'Roman fort or town associated with a man called An(n)a'. OE pers. name + *cæster*.

Ancroft Northum. *Anecroft* 1195. 'Lonely or isolated enclosure'. OE *āna* + *croft*.

Anderby Lincs. *Andreby* c.1135. Possibly 'farmstead or village of a man called Arnthórr'. OScand. pers. name + *bý*. Alternatively the first element may be OScand. *andri* 'snow-shoe' perhaps in the sense 'billet of wood'.

Andersonstown (*Baile Andarsan*) Antrim. *Anderson's Town Village* 1832. 'Anderson's town'.

Anderton Ches. *Anderton* 1184. 'Farmstead or village of a man called Ēanrēd or Eindrithi'. OE or OScand. pers. name + *tūn*.

Andover Hants. *Andeferas* 955, *Andovere* 1086 (DB). '(Place by) the ash-tree waters'. Celtic river-name *Ann* (an earlier name for the River Anton and Pillhill Brook) with the Celtic word also found in DOVER.

Andoversford Glos. *Onnan ford* 759, *Anneford* c.1243, *Annesford* 1327. 'Ford of a man called Anna'. OE pers. name + *ford*.

Anerley Gtr. London, so named from a solitary house built here by a Scotsman in the mid-19th cent. which he called by the dialect word *anerly* 'lonely'.

Anfield Lpool. *Hongfield* 1642. Possibly 'field on a slope'. ME *hange* + *feld*.

Angersleigh Somerset. *Lega* 1086 (DB), *Aungerlegh* 1354. OE *lēah* 'woodland clearing' with manorial addition from the *Aunger* family, here in the 13th cent.

Anglesey (*Môn*) (island) Angl. *ynys uon* 815, *Anglesege* 1098, *Ongulsey* 13th cent., *Anglesey, or Anglesea* 1868. 'Ongull's island'. OScand. pers. name + *ey*. The Welsh name cannot satisfactorily be explained. Hence the Roman name, *Mona*.

Angmering W. Sussex. *Angemæringum* c.880, *Angemare* 1086 (DB). '(Settlement of) the family or followers of a man called *Angenmǣr'. OE pers. name + *-ingas* (dative *-ingum*).

Angram '(place at) the pastures or grasslands', OE **anger* in a dative plural form **ang(e)rum*: **Angram** N. Yorks., near Keld. *Angram* late 12th cent. **Angram** N. Yorks., near York. *Angrum* 13th cent.

Angus (the unitary authority). *Enegus* 12th cent. '(Place of) Angus'. With reference to *Angus*, 8th-cent. king of the Picts.

Ankail (*Eing Caol*) Kerry. 'Narrow strip'.

Anlaby E. R. Yorks. *Unlouebi* [*sic*] 1086 (DB), *Anlauebi* 1203. 'Farmstead or village of a man called Óláfr'. OScand. pers. name + *bý*.

Anmer Norfolk. *Anemere* [*sic*] 1086 (DB), *Anedemere* 1291. 'Duck pool'. OE *æned* + *mere*.

Ann, Abbotts Hants. *Anne* 901, 1086 (DB), *Anne Abbatis* c.1270. 'Estate on the River *Ann* belonging to the abbot'. Celtic river-name (meaning 'ash-tree stream') with manorial affix referring to early possession by Hyde Abbey at Winchester.

Anna Valley Hants., a recent name, coined from the old river-name *Ann* as in previous name, *see* ANDOVER.

Annabella (*Eanach Bile*) Cork. 'Marsh of the sacred tree'.

Annacarty (*Áth na Cairte*) Tipperary. 'Ford of the cart'.

Annaclone (*Eanach Cluana*) Down. *Enaghluan* 1422. 'Marsh of the haunch-like hill'.

Annacloy (*Áth na Cloiche*) Down. *Annacloy* 1621. 'Ford of the stone'.

Annacotty (*Áth an Choite*) Limerick. 'Ford of the boat'.

Annacurragh (*Eanach Churraigh*) Wicklow. 'Marsh of the bog'.

Annadorn (*Áth na nDorn*) Down. *Annaghdorney* 1627. 'Ford of the fists'.

Annaduff (*Eanach Dubh*) Leitrim. 'Black marsh'.

Annagassan (*Áth na gCasán*) Louth. 'Ford of the paths'.

Annagh (*Eanach*) Mayo. 'Marsh'.

Annaghdown (*Eanach Dúin*) Galway. 'Marsh of the fortress'.

Annaghmore (*Eanach Mór*) Armagh, Laois, Offaly. 'Big marsh'.

Annahilt (*Eanach Eilte*) Down. (*Molibae*) *Enaig Elti* c.830. 'Marsh of the doe'.

Annakisha (*Áth na Cise*) Cork. 'Ford of the wicker causeway'.

Annalee (*Abhainn Eanach Lao*) (river) Cavan. 'River of the marsh of the calf'.

Annalong (*Áth na Long*) Down. *Analong* c.1655. 'Ford of the ships'.

Annalore (*Áth na Lobhar*) Monaghan. 'Ford of the leper'.

Annamoe (*Áth na mBó*) Wicklow. 'Ford of the cows'.

Annan Dumf. *Anava* 7th cent., *Estrahanent* 1124, *Stratanant* 1152, *Annandesdale* 1179. '(Place by the) River Annan'. The Celtic (or pre-Celtic) river-name means 'water'. The last three forms of the name above have added an element meaning 'valley' (Cumbric *ystrad*, Gaelic *srath*, OScand. *dalr* or OE *dæl*).

Annaveagh (*Áth na bhFiada*) Monaghan. 'Ford of the deer'.

Annesley Woodhouse Notts. *Aneslei* 1086 (DB), *Anseleia* c.1190, *Annesley Wodehouse* 13th cent. Possibly 'woodland clearing of a man called *Ān*'. OE pers. name + *lēah*. Alternatively perhaps identical with ANSLEY Warwicks. The 13th-cent. addition denotes 'woodland hamlet'.

Annfield Plain Durham. A modern name, first recorded 1857, perhaps identical with ANFIELD Lpool.

Annsborough (*Baile Anna*) Down. *Anne-borough* 1823. After *Annsborough House*, itself perhaps named after the *Annesley* family.

Ansford Somerset. *Almundesford* 1086 (DB). 'Ford associated with a man called Ealhmund'. OE pers. name + *ford*.

Ansley Warwicks. *Hanslei* 1086 (DB), *Anesteleye* 1235. Probably 'woodland clearing with a hermitage'. OE *ānsetl* + *lēah*.

Anslow Staffs. *Eansythelege* 1012. 'Woodland clearing of a woman called Ēanswīth'. OE pers. name + *lēah*.

Anstey, Ansty, a name found in various counties, from OE *ānstīg* 'single track, track linking other routes' or *anstig* 'steep track'; examples include: **Anstey** Leics. *Anstige* 1086 (DB). **Anstey, East & Anstey, West** Devon. *Anesti(n)ga* 1086 (DB). **Ansty** Warwicks. *Anestie* 1086 (DB). **Ansty** Wilts. *Anestige* 1086 (DB). **Ansty Cross, Higher Ansty** Dorset. *Anesty* 1219.

Anston, North & Anston, South Rothm. *Anestan, Litelanstan* 1086 (DB), *Northanstan, Suthanstan* 1297. 'The single or solitary stone'. OE *āna* + *stān*.

Anstruther Fife. *Ainestrooder* 1178–88, *Anestrothir c.*1205. Perhaps 'stream of Ethernan'. Gaelic *sruthair*.

Anthorn Cumbria. *Eynthorn* 1279. 'Solitary thorn-tree'. OScand. *einn* + *thorn*.

Antingham Norfolk. *Antingham* 1044–7, 1086 (DB). 'Homestead of the family or followers of a man called *Anta'. OE pers. name + *-inga-* + *hām*.

Antonine Wall Dumf, Falk. The wall from the Forth to the Clyde was built in AD 142 for the Roman emperor *Antoninus* Pius.

Antony Cornwall. *Antone* 1086 (DB). 'Farmstead of a man called Anna or *Anta'. OE pers. name + *tūn*.

Antrim (*Aontroim*, earlier *Aontreibh*) Antrim. (*Fiontan*) *Oentreibh* 612. 'Single house'.

Antrobus Ches. *Entrebus* 1086 (DB), *Anderbusk* 1295. Possibly 'bush of a man called Endrithi or *Andrithi'. OScand. pers. name + *buskr*.

Anwick Lincs. *Amuinc* [*sic*], *Haniwic* 1086 (DB), *Amewic* 1218, *Anewic c.*1221. 'Dwelling or farm of a man called Amma or Anna'. OE pers. name + *wīc*.

Anyalla (*Eanaigh Gheala*) Monaghan. *Anyalle* 1591. 'White marshes'.

Aperfield Gtr. London. *Apeldrefeld* 1242. 'Open land where apple-trees grow'. OE *apuldor* + *feld*.

Apethorpe Northants. *Patorp* [*sic*] 1086 (DB), *Apetorp* 1162. 'Outlying farmstead or hamlet of a man called Api'. OScand. pers. name + *thorp*.

Apley Lincs. *Apeleia* 1086 (DB). 'Apple wood'. OE *æppel* + *lēah*.

Apperknowle Derbys. *Apelknol* 1317. 'Apple-tree hillock'. OE *æppel* + *cnoll*.

Apperley Glos. *Apperleg* 1210. 'Wood or clearing where apple-trees grow'. OE *apuldor* + *lēah*.

Appin (district) Highland. 'Abbey land'. Gaelic *apainn*.

Appleby, 'farmstead or village where apple-trees grow', OE *æppel* (perhaps replacing OScand. *epli*) + OScand. *bý*: **Appleby** Cumbria. *Aplebi* 1130. **Appleby** N. Lincs. *Aplebi* 1086 (DB). **Appleby Magna & Appleby Parva** Leics. *Æppelby* 1002, *Apelbi* 1086 (DB). Distinguishing affixes are Latin *magna* 'great' and *parva* 'little'.

Applecross Highland. *Aporcrosan c.*1080. 'Mouth of the River Crosan'. Pictish *aber*. The river-name means 'little cross' (Gaelic *cros* + diminutive suffix *-an*).

Appledore, '(place at) the apple-tree', OE *apuldor*: **Appledore** Devon. *le Apildore* 1335. **Appledore** Kent. *Apuldre* 10th cent., *Apeldres* 1086 (DB).

Appleford Oxon. *Æppelford c.*895, *Apleford* 1086 (DB). 'Ford where apple-trees grow'. OE *æppel* + *ford*.

Appleshaw Hants. *Appelsag* 1200. 'Small wood where apple-trees grow'. OE *æppel* + *sceaga*.

Appleton, 'farmstead where apples grow, apple orchard', OE *æppel-tūn*; examples include: **Appleton** Oxon. *Æppeltune* 942, *Apletune* 1086 (DB). **Appleton** Warrtn. *Epletune* 1086 (DB). **Appleton, East** N. Yorks. *Apelton* 1086 (DB). **Appleton-le-Moors** N. Yorks. *Apeltun* 1086 (DB). Affix means 'near the moors'. **Appleton-le-Street** N. Yorks. *Apletun* 1086 (DB). Affix means 'on the main road'. **Appleton Roebuck** N. Yorks. *Æppeltune c.*972, *Apleton* 1086 (DB), *Appleton Roebucke* 1664. Manorial affix from the *Rabuk* family, here in the 14th cent. **Appleton Wiske** N. Yorks. *Apeltona* 1086 (DB). Affix refers to its situation on the River Wiske (from OE *wisc* 'marshy meadow').

Appletreewick N. Yorks. *Apletrewic* 1086 (DB). 'Dwelling or farm by the apple-trees'. OE *æppel-trēow* + *wīc*.

Appley Bridge Lancs. *Appelleie* 13th cent. 'Apple-tree wood or clearing'. OE *æppel* + *lēah*.

Apsley End Beds. *Aspele* 1230. 'Aspen-tree wood'. OE *æspe* + *lēah*.

Apuldram W. Sussex. *Apeldreham* 12th cent. 'Homestead or enclosure where apple-trees grow'. OE *apuldor* + *hām* or *hamm*.

Aran Islands (*Árainn*) Galway. 'Islands of the ridge'.

Aranmore (*Árainn Mhór*) Donegal. 'Big ridge'.

Arberth. *See* NARBERTH.

Arboe (*Ard Bó*) Tyrone. (*Colman*) *Airdi Bó* c.830. 'Height of the cows'.

Arborfield Wokhm. *Edburgefeld* c.1190, *Erburgefeld* 1222. Probably 'open land of a woman called Hereburh'. OE pers. name + *feld*.

Arbroath Ang. *Aberbrothok* 1178, *Arbroath, or Aberbrothwick* 1868. 'Mouth of the River Brothock'. Pictish *aber*. The river-name means 'seething one' (Gaelic *brothach*).

Ardagh (*Ardach*) Longford. 'High field'.

Ardaghy (*Ardachadh*) Monaghan. 'High field'.

Ardakillen (*Ard an Choillín*) Roscommon. 'Height of the little wood'.

Ardanleagh (*Ardán Liath*) Limerick. 'Little grey height'.

Ardara (*Ard an Rátha*) Donegal. *Árd an Rátha* c.1854. 'Height of the fort'.

Ardaragh (*Ard Darach*) Down. *Ardarre* 1549. 'Height of the oak-tree'.

Ardattin (*Ard Aitinn*) Carlow. 'Height of the gorse'.

Ardavagga (*Ard a' Mhagaidh*) Offaly. 'Height of merriment'.

Ardballymore (*Ardvhaile Mór*) Westmeath. 'Big high homestead'.

Ardcath (*Ard an Chatha*) Meath. 'Height of the battle'.

Ardcolm (*Ard Coilm*) Wexford. 'Height of Colm'.

Ardcrony (*Ard Cróine*) Tipperary, Wexford. 'Cróne's height'.

Ardee (*Baile Átha Fhirdhia*) Louth. 'Ferdia's ford'.

Ardeen (*Ardín*) Cork, Kerry. 'Little height'.

Ardeley Herts. *Eardeleage* 939, *Erdelei* 1086 (DB). 'Woodland clearing of a man called *Earda'. OE pers. name + *lēah*.

Arden (old forest) Warwicks. *See* HENLEY-IN-ARDEN.

Arderin (*Ard Éireann*) Laois, Offaly. 'Height of Ireland'.

Ardersier Highland. *Ardrosser* 1227, *Arderosseir* 1257. 'Promontory of the artisan'. Gaelic *ard-na-saor*.

Ardfert (*Ard Fhearta*) Kerry. 'Height of the grave'.

Ardfield (*Ard Ó bhFicheallaigh*) Cork. 'Height of Uí Fhicheallaigh'.

Ardfinnan (*Ard Fhíonáin*) Tipperary. 'Fíonán's height'.

Ardgarvan (*Ard an Garbháin*) Derry. *Ardagarnen* 1616. 'Height of the gravel'.

Ardgivna (*Ard Goibhne*) Sligo. 'Height of the smith'.

Ardglass (*Aird Ghlais*) Down. (*go*)*hAird Glais* 1433. 'Grey point'.

Ardglass (*Ard Glas*) Cork. 'green height'.

Ardgroom (*Dhá Dhrom*) Cork. 'Two ridges'.

Ardingary (*Ard an Gháire*) Donegal. 'Height of shouting'.

Ardingly W. Sussex. *Erdingelega* early 12th cent. 'Woodland clearing of the family or followers of a man called *Earda'. OE pers. name + *-inga-* + *lēah*.

Ardington Oxon. *Ardintone* 1086 (DB). Probably 'estate associated with a man called *Earda'. OE pers. name + *-ing-* + *tūn*.

Ardivaghan (*Ard Uí Mhocháin*) Westmeath. 'Ó Mocháin's height'.

Ardkearagh (*Ard Caorach*) Kerry. 'Height of sheep'.

Ardkeen (*Ard Caoin*) Down. *Ardkene* 1306. 'Pleasant height'.

Ardlea (*Ard Liath*) Laois. 'Grey height'.

Ardleigh Essex. *Erleiam* [*sic*] 1086 (DB), *Ardlega* 12th cent. Probably 'woodland clearing with a dwelling place'. OE *eard* + *lēah*.

Ardley Oxon. *Eardulfes lea* 995, *Ardulveslie* 1086 (DB). 'Woodland clearing of a man called Eardwulf'. OE pers. name + *lēah*.

Ardlougher (*Ard Luachra*) Cavan. *Ardloagher* 1611. 'Height of the rushes'.

Ardmillan (*Ard an Mhuilinn*) Down. 'Height of the mill'.

Ardmore (*Aird Mhór*) Armagh, Derry, Galway, Mayo, Waterford. 'Big height'.

Ardmore Point Highland (Islay). 'Large point'. Gaelic *ard* + *mór*, and English *point*.

Ardmorney (*Ard Murnaigh*) Westmeath. 'Morna's height'.

Ardnacrohy (*Ard na Croiche*) Limerick. 'Height of the gallows'.

Ardnacrusha (*Ard na Croise*) Clare. 'Height of the cross'.

Ardnaglug (*Ard na gClog*) Westmeath. 'Height of the bell'.

Ardnagroghery (*Ard na gCrochaire*) Cork. 'Height of the hangmen'.

Ardnamoghill (*Ard na mBuachaill*) Donegal. 'Height of the boys'.

Ardnamurchan (peninsula) Highland. *Art Muirchol c.*700, *Ardnamurchin* 1309. 'Point of the otters (literally "sea dogs")'. Gaelic *ard* + *na* + *muir* + *cù* (genitive plural *chon*). The first form of the name above seems to suggest a final element *chol*, 'sin', with 'sea sins' implying piracy.

Ardnapreaghaun (*Ard na bPréachán*) Limerick. 'Height of the crows'.

Ardnaree (*Ard na Ria*) Mayo. 'Height of the executions'.

Ardnurcher (*Áth an Urchair*) Westmeath. 'Ford of the cast'.

Ardpatrick (*Ard Pádraig*) Limerick. 'Patrick's height'.

Ardrahan (*Ard Raithin*) Galway. 'Height of ferns'.

Ardress (*An tArdriasc*) Armagh. *Tardresk* 1609. 'High bog'.

Ardrishaig Arg. 'Height of the brambles'. Gaelic *ard* + *dris*.

Ardroe (*Aird Rua*) Galway. 'Red point'.

Ardross Highland. 'Height of the moorland'. Gaelic *ard* + *ros*. Alternatively, 'height of the headland', from the same elements.

Ardrossan N. Ayr. *Ardrossene c.*1320. 'Height of the little headland'. Gaelic *ard* + *ros* + diminutive suffix -*an*.

Ards Peninsula (*Aird Uladh*) Down. *(i)nAird Ulad c.*830. 'Peninsula of the Ulstermen'.

Ardscull (*Ard Scol*) Kildare. 'Height of schools'.

Ardsheelane (*Ard Síoláin*) Kerry. 'Height of Síolán'.

Ardsley Barns. *Erdeslaia* 12th cent. 'Woodland clearing of a man called Eorēd or Ēanrēd'. OE pers. name + *lēah*.

Ardsley East Leeds. *Erdeslawe* 1086 (DB). 'Mound of a man called Eorēd or Ēanrēd'. OE pers. name + *hlāw*.

Ardstraw (*Ard Sratha*) Tyrone. (*muintir*) *Aird Sratha c.*900. 'Height of the river island'.

Ardtole (*Ard Tuathail*) Down. 'Tuathal's height'.

Ardwick Manch. *Atherdwic, Atheriswyke, Aderwyk* 1282. From OE *wīc* 'specialized farm or building' with an uncertain male pers. name, possibly OE *Æthelrǣd* or *Ēadrǣd*.

Areley Kings Worcs. *Erneleia c.*1138, *Kyngges Arley* 1405. 'Wood or clearing frequented by eagles'. OE *earn* + *lēah*. Affix *Kings* because it was part of a royal manor.

Argideen (*Airgidín*) (river) Cork. 'Silver (river)'.

Argos Hill E. Sussex. First recorded as *Argarshill, Ergershill* 1547, probably so called from the name of a local family.

Argyll (district) Arg. *Arregaithel c.*970, *Argail* 1292. 'Coastland of the Gaels'. Gaelic *oirthir Ghaideal*.

Arkendale N. Yorks. *Arghendene* 1086 (DB). Possibly 'valley of a man called *Eorcna*. OE pers. name + *denu* (replaced from the 14th cent. by OScand. *dalr*). Alternatively the first element may be OE *arce* (genitive *arcan*) 'ark, chest', perhaps used in a transferred topographical sense.

Arkesden Essex. *Archesdana* 1086 (DB). Possibly 'valley of a man called Arnkel'. OScand. pers. name + OE *denu*.

Arkholme Lancs. *Ergune* 1086 (DB). '(Place at) the shielings or hill-pastures'. OScand. *erg* in a dative plural form *ergum*.

Arkley Gtr. London. *Arkeleyslond* 1332, *Arkeley* 1547. Possibly 'woodland clearing by the ark, or where arks are made'. OE *(e)arc* 'ark, chest, bin or other receptacle' + *lēah*.

Arklow (*An tInbhear Mór*) Wicklow. *Herketelou* 1177. 'Arnkell's meadow'. OScand. pers. name + *ló*. The Irish name means 'the big estuary'.

Arksey Donc. *Archeseia* 1086 (DB). Possibly 'island, or dry ground in marsh, of a man called Arnkel'. OScand. pers. name + OE *ēg*.

Arlecdon Cumbria. *Arlauchdene c.*1130. Possibly 'valley of the stream frequented by eagles'. OE *earn* + *lacu* + *denu*.

Arlescote Warwicks. *Orlavescote* 1086 (DB). 'Cottage(s) of a man called Ordlāf'. OE pers. name + *cot*.

Arlesey Beds. *Alricheseia* 1062, 1086 (DB). 'Island or well-watered land of a man called Ælfrīc'. OE pers. name + *ēg*.

Arless (*Ardlios*) Laois. 'High fort'.

Arleston Tel. & Wrek. *Erdelveston* 1180. 'Farmstead or village of a man called Eardwulf'. OE pers. name + *tūn*.

Arley, usually 'wood or clearing frequented by eagles', OE *earn* + *lēah*: **Arley** Warwicks. *Earnlege* 1001, *Arlei* 1086 (DB). **Arley, Upper** Worcs. *Earnleie* 996, *Ernlege* 1086 (DB).
  However the following may have a different origin: **Arley** Ches. *Arlegh* 1340. Possibly 'grey wood', or 'wood on a boundary'. OE *hār* + *lēah*.

Arlingham Glos. *Erlingeham* 1086 (DB). 'Homestead or enclosure of the family or followers of a man called *Eorl(a)'. OE pers. name + *-inga-* + *hām* or *hamm*.

Arlington Devon. *Alferdintona* 1086 (DB). 'Estate associated with a man called Ælffrith'. OE pers. name + *-ing-* + *tūn*.

Arlington E. Sussex. *Erlington* 1086 (DB). 'Estate associated with a man called Eorl(a)'. OE pers. name + *-ing-* + *tūn*.

Arlington Glos. *Ælfredincgtune* 1004, *Alvredintone* 1086 (DB). 'Estate associated with a man called Ælfrēd'. OE pers. name + *-ing-* + *tūn*.

Armadale Highland (Skye). *Armidill* 1723. 'Arm-shaped valley'. OScand. *armr* + *dalr*.

Armadale W. Loth. The town arose in the mid-19th cent. and was named after William Honeyman, Lord Armadale, who took his title from *Armadale*, Highland, near Melvich. Its own name has the same origin as ARMADALE, Skye.

Armagh (*Ard Mhacha*) Armagh. *Ard Macha* 444. 'Macha's height' or 'height of the plain'.

Armathwaite Cumbria. *Ermitethwait* 1212. 'Clearing of the hermit'. ME *ermite* + OScand. *thveit*.

Arminghall Norfolk. *Hameringahala* 1086 (DB). Possibly 'nook of land of the family or followers of a man called *Ambre or Ēanmǣr'. OE pers. name + *-inga-* + *halh*.

Armitage Staffs. *Armytage* 1520. '(Place at) the hermitage'. ME *ermitage*.

Armoy (*Oirthear Maí*) Antrim. *Airther Maigi c.*900. 'East of the plain'.

Armscote Warwicks. *Eadmundescote* 1042. 'Cottage(s) of a man called Ēadmund'. OE pers. name + *cot*.

Armthorpe Donc. *Ernulfestorp* 1086 (DB). 'Outlying farmstead or hamlet of a man called Earnwulf or Arnulfr'. OE or OScand. pers. name + OScand. *thorp*.

Arncliffe N. Yorks. *Arneclif* 1086 (DB). 'Cliff of the eagles'. OE *earn* + *clif*.

Arncott, Upper & Arncott, Lower Oxon. *Earnigcote* 983, *Ernicote* 1086 (DB). 'Cottage(s) associated with a man called *Earn'. OE pers. name + *-ing-* + *cot*.

Arne Dorset. *Arne* 1268. Probably OE *ærn* 'house or building'. Alternatively '(place at) the heaps of stones or tumuli', from OE *hær* in a dative plural form *harum*.

Arnesby Leics. *Erendesbi* 1086 (DB). 'Farmstead or village of a man called Iarund or *Erendi'. OScand. pers. name + *bý*.

Arnold, 'nook of land frequented by eagles', OE *earn* + *halh*: **Arnold** E. R. Yorks. *Ærnhale* 1190. **Arnold** Notts. *Ernehale* 1086 (DB).

Arnside Cumbria. *Harnolvesheuet* 1184–90. 'Hill or headland of a man called Earnwulf or Arnulfr'. OE or OScand. pers. name + OE *hēafod*.

Arra Mountain (*Sliabh Ára*) Tipperary. 'Mountain of (the district of) Ára'.

Arram E. R. Yorks. *Argun* 1086 (DB). '(Place at) the shielings or hill-pastures'. OScand. *erg* in a dative plural form *ergum*.

Arran (island) N. Ayr. Meaning uncertain. Ancient name.

Arranmore (*Árainn Mhór*) Donegal. *hAruinn Uí Dhomhnuill* c.1600. 'Great ridge'.

Arrathorne N. Yorks. *Ergthorn* 13th cent. 'Thorn-tree by the shieling or hill-pasture'. OScand. *erg* + *thorn*.

Arreton I. of Wight. *Eaderingtune* c.880, *Adrintone* 1086 (DB). 'Estate associated with a man called Ēadhere'. OE pers. name + *-ing-* + *tūn*.

Arrington Cambs. *Earnningtone* c.950, *Erningtune* 1086 (DB). Probably 'farmstead of the family or followers of a man called *Earn(a)'. OE pers. name + *-inga-* + *tūn*.

Arrow (river) Herefs. *See* STAUNTON ON ARROW.

Arrow Warwicks. *Arne* [sic] 710, *Arue* 1086 (DB). Named from the River **Arrow**, an ancient Celtic or pre-Celtic river-name probably meaning 'swift one'.

Arryheernabin (*Áirí Thír na Binne*) Donegal. 'Shieling of the country of the peak'.

Arthington Leeds. *Hardinctone* 1086 (DB). 'Estate associated with a man called *Earda'. OE pers. name + *-ing-* + *tūn*.

Arthingworth Northants. *Arningvorde* 1086 (DB). 'Enclosure associated with a man called *Earn(a)'. OE pers. name + *-ing-* + *worth*.

Articlave (*Ard an Chléibh*) Derry. *Ard Cleibh* c.1680. 'Height of the basket'.

Artiferrall (*Ard Tighe Fearghail*) Antrim. 'Height of Fearghal's house'.

Artigarvan (*Ard Tí Garbháin*) Tyrone. *Ordogarvan* c.1655. 'Height of Garbhán's house'.

Artnagross (*Ard na gCros*) Antrim. *Artnagross* 1780, 'Height of the crosses'.

Artrea (*Ard Tré*) Tyrone. (*airchindeach*) *Arda Trea* 1127. 'Tré's height'.

Arundel W. Sussex. *Harundel* 1086 (DB). 'Valley where the plant horehound grows'. OE *hārhūne* + *dell*. The river-name **Arun** is a back-formation from the place name.

Arvagh (*Ármhach*) Cavan. *Arvaghbeg, Arvaghmore* 1630. 'Battlefield'.

Asby, 'farmstead or village where ash-trees grow', OScand. *askr* + *bý*: **Asby** Cumbria, near Arlecdon. *Asbie* 1654. **Asby, Great & Asby, Little** Cumbria. *Aschaby* c.1160.

Ascot, Ascott, 'eastern cottage(s)', OE *ēast* + *cot*: **Ascot** Winds. **&** Maid. *Estcota* 1177. **Ascott under Wychwood** Oxon. *Estcot* 1220. Affix means 'near the forest of Wychwood' (an OE name, *Huiccewudu* 840, meaning 'wood of a tribe called the *Hwicce*').

Asdee (*Eas Daoi*) Kerry. 'Dark waterfall'.

Asenby N. Yorks. *Æstanesbi* 1086 (DB), 'farmstead or village of a man called Eysteinn'. OScand. pers. name + *bý*.

Asfordby Leics. *Osferdebie* 1086 (DB), *Asfordebi*, 1184. Possibly 'farmstead of a man called Ásfrøthr', OScand. pers. name + *bý*. Or an OE *Æscford* 'ash-tree ford' + *bý*.

Asgarby, 'farmstead or village of a man called Ásgeirr', OScand. pers. name + *bý*: **Asgarby** Lincs., near Sleaford. *Asegarby* 1201. **Asgarby** Lincs., near Spilsby. *Asgerebi* 1086 (DB).

Ash, '(place at) the ash-tree(s)', OE *æsc*; examples include: **Ash** Kent, near Sandwich. *Æsce* c.1100. **Ash** Surrey. *Essa* 1170. **Ash Magna** Shrops. *Magna Asche* 1285. Affix is Latin *magna* 'great'. **Ash Priors** Somerset. *Æsce* 1065, *Esse Prior* 1263. Affix from its early possession by the Prior of Taunton.

Ashampstead W. Berks. *Essamestede* 1155–8. 'Homestead by the ash-tree(s)'. OE *æsc* + *hām-stede*.

Ashbocking Suffolk. *Assa* 1086 (DB), *Bokkynge Assh* 1411. '(Place at) the ash-tree(s)'. OE *æsc* + manorial affix from the *de Bocking* family, here in the 14th cent.

Ashbourne Derbys. *Esseburne* 1086 (DB). 'Stream where ash-trees grow'. OE *æsc* + *burna*.

Ashbrittle Somerset. *Aisse* 1086 (DB), *Esse Britel* 1212. '(Place at) the ash-tree(s)'. OE *æsc* + manorial affix from its possession by a man called *Bretel* in 1086.

Ashburnham E. Sussex. *Esseborne* 1086 (DB), *Esburneham* 12th cent. 'Meadow by the stream where ash-trees grow'. OE *æsc* + *burna* with the later addition of *hamm*. The river here is still called **Ashburn**.

Ashburton Devon. *Æscburnan lande* 1008–12, *Essebretone* 1086 (DB). 'Farmstead or village by the stream where ash-trees grow'. OE *æsc* + *burna* + *tūn* (in the earliest spelling alternating with OE *land* 'cultivated land, estate').

Ashbury, 'stronghold where ash-trees grow', OE *æsc* + *burh* (dative *byrig*): **Ashbury** Devon. *Esseberie* 1086 (DB). **Ashbury** Oxon. *Eissesberie* [*sic*] 1086 (DB), *Æsseberia* 1187.

Ashby, a common name in the North and Midlands, usually 'farmstead or village where ash-trees grow', OE *æsc* or OScand. *askr* + OScand. *bý*; however 'farmstead of a man called Aski', OScand. pers. name + *bý*, is a possible alternative for some names; examples include: **Ashby** N. Lincs. *Aschebi* 1086 (DB). **Ashby by Partney** Lincs. *Aschebi* 1086 (DB). *See* PARTNEY. **Ashby, Canons** Northants. *Ascebi* 1086 (DB), *Essheby Canons* 13th cent. Affix from the priory here, founded in the 12th cent. **Ashby, Castle** Northants. *Asebi* 1086 (DB), *Castel Assheby* 1361. Affix from the former castle here. **Ashby, Cold** Northants. *Essebi* 1086 (DB), *Caldessebi* c.1150. Affix is OE *cald* 'cold, exposed'. **Ashby cum Fenby** NE. Lincs. *Aschebi* 1086 (DB). Fenby is *Fen(de)bi* 1086 (DB), 'farmstead in a fen or marsh', OE *fenn* + OScand. *bý*; Latin *cum* is 'with'. **Ashby de la Launde** Lincs. *Aschebi* 1086 (DB). Manorial affix from the *de la Launde* family, here in the 14th cent. **Ashby de la Zouch** Leics. *Ascebi* 1086 (DB), *Esseby la Zuche* 1205. Manorial affix from the *(de) la Zuche* family, here in the 12th cent. **Ashby Folville** Leics. *Ascebi* 1086 (DB). Manorial affix from the *de Foleuilla* family, here in the 12th cent. **Ashby Magna & Ashby Parva** Leics. *Essebi* 1086 (DB). Affixes are Latin *magna* 'great', *parva* 'little'. **Ashby, Mears** Northants. *Asbi* 1086 (DB), *Esseby Mares* 1281. Manorial affix from the *de Mares* family, here in the 12th cent. **Ashby Puerorum** Lincs. *Aschebi* 1086 (DB). Latin affix means 'of the boys', in allusion to a bequest for the support of the choir-boys of Lincoln Cathedral. **Ashby St Ledgers** Northants. *Ascebi* 1086 (DB), *Esseby Sancti Leodegarii* c.1230. Affix from the dedication of the church to St Leger. **Ashby St Mary** Norfolk. *Ascebei* 1086 (DB). Affix from

the dedication of the church. **Ashby, West** Lincs. *Aschebi* 1086 (DB).

Ashcombe Devon. *Aissecome* 1086 (DB). 'Valley where ash-trees grow'. OE *æsc* + *cumb*.

Ashcott Somerset. *Aissecote* 1086 (DB). 'Cottage(s) where ash-trees grow'. OE *æsc* + *cot*.

Ashdon Essex. *Æstchendune* c.1036, *Ascenduna* 1086 (DB). 'Hill overgrown with ash-trees'. OE *æscen* + *dūn*.

Ashdown Forest Sussex. *Essendon* 1207. Identical in origin with the previous name.

Asheldham Essex. *Assildeham* c.1130. 'Homestead of a woman called *Æschild'. OE pers. name + *hām*.

Ashen Essex. *Asce* 1086 (DB), *Asshen* 1344. '(Place at) the ash-trees'. OE *æsc* in a dative plural form *æscum*.

Ashendon Bucks. *Assedune* 1086 (DB). 'Hill overgrown with ash-trees'. OE *æscen* + *dūn*.

Ashfield, 'open land where ash-trees grow', OE *æsc* + *feld*: **Ashfield** Notts. *Esfeld* 1216. An old name now revived as a district name. **Ashfield cum Thorpe** Suffolk. *Assefelda* 1086 (DB). The preposition *cum* is Latin for 'with'; Thorpe is from OScand. *thorp* 'secondary settlement, dependent outlying farmstead or hamlet'. **Ashfield, Great** Suffolk. *Eascefelda* 1086 (DB).

Ashford, usually 'ford where ash-trees grow', OE *æsc* + *ford*: **Ashford** Devon. *Aiseforda* 1086 (DB). **Ashford Bowdler & Ashford Carbonel** Shrops. *Esseford* 1086 (DB), *Asford Budlers, Aysford Carbonel* 1255. Manorial affixes from the *de Boulers* and *Carbunel* families, here at an early date. **Ashford in the Water** Derbys. *Æscforda* 926, *Aisseford* 1086 (DB). Affix 'in the Water' occurs from the late 17th cent., no doubt with reference to the meandering course of the River Wye here.

However two Ashfords have a different origin: **Ashford** Kent. *Essetesford* 1086 (DB). 'Ford by a clump of ash-trees'. OE **æscet* + *ford*. **Ashford** Surrey. *Ecelesford* 969, *Exeforde* 1086 (DB). Probably 'ford of a man called *Eccel'. Celtic pers. name + OE ford.

Ashill Norfolk. *Assclea* 1086 (DB). 'Ash-tree wood'. OE *æsc* + *lēah*.

Ashill Somerset. *Aisselle* 1086 (DB). 'Hill where ash-trees grow'. OE *æsc* + *hyll*.

Ashingdon Essex. *Assandun* 1016, *Nesenduna* [sic] 1086 (DB). 'Hill of the ass, or of a man called *Assa'. OE *assa* or OE pers. name (genitive -*n*) + *dūn*.

Ashington Northum. *Essenden* 1205. 'Valley where ash-trees grow'. OE *æscen* + *denu*.

Ashington W. Sussex. *Essingetona* 1073. 'Farmstead of the family or followers of a man called Æsc'. OE pers. name + -*inga*- + *tūn*.

Ashleworth Glos. *Escelesuuorde* 1086 (DB). 'Enclosure of a man called *Æscel'. OE pers. name + *worth*.

Ashley, a common name, 'ash-tree wood or clearing', OE *æsc* + *lēah*; examples include: **Ashley** Cambs. *Esselie* 1086 (DB). **Ashley** Ches. *Ascelie* 1086 (DB). **Ashley** Devon. *Esshelegh* 1238. **Ashley** Dorset. *Asseleghe* 1246. **Ashley** Hants., near Lymington. *Esselie* 1086 (DB). **Ashley** Hants., near Winchester. *Asselegh* 1275. **Ashley** Northants. *Ascele* 1086 (DB). **Ashley** Staffs. *Esselie* 1086 (DB). **Ashley Green** Bucks. *Essleie* 1227, *Assheley grene* 1468.

Ashling, East & Ashling, West W. Sussex. *Estlinges* 1185. Probably '(settlement of) the family or followers of a man called *Æscla'. OE pers. name + -*ingas*.

Ashmansworth Hants. *Æscmæreswierthe* 909. 'Enclosure by the ash-tree pool'. OE *æsc* + *mere* + *worth*.

Ashmore Dorset. *Aisemare* 1086 (DB). 'Pool where ash-trees grow'. OE *æsc* + *mere*.

Ashorne Warwicks. *Hassorne* 1196. 'Horn-shaped hill where ash-trees grow'. OE *æsc* + *horn*.

Ashover Derbys. *Essovre* 1086 (DB). 'Ridge or slope where ash-trees grow'. OE *æsc* + *ofer*.

Ashow Warwicks. *Asceshot* [sic] 1086 (DB), *Essesho* 12th cent. 'Hill-spur of the ash-tree or of a man called Æsc'. OE *æsc* or OE pers. name + *hōh*.

Ashperton Herefs. *Spertune* [sic] 1086 (DB), *Aspretonia* 1144. Probably 'farmstead or village of a man called Æscbeorht or Æscbeorn'. OE pers. name + *tūn*.

Ashprington Devon. *Aisbertone* 1086 (DB). 'Estate associated with a man called Æscbeorht or Æscbeorn'. OE pers. name + -*ing*- + *tūn*.

Ashreigney Devon. *Aissa* 1086 (DB), *Esshereingni* 1238. '(Place at) the ash-tree(s)'.

OE *æsc* + manorial affix from the *de Regny* family, here in the 13th cent.

Ashtead Surrey. *Stede* 1086 (DB), *Estede* c.1150. 'Place where ash-trees grow'. OE *æsc* + *stede*.

Ashton, a common name, usually 'farmstead where ash-trees grow', OE *æsc* + *tūn*; examples include: **Ashton** Ches. *Estone* 1086 (DB). **Ashton** Herefs. *Estune* 1086 (DB). **Ashton** Northants., near Oundle. *Ascetone* 1086 (DB). **Ashton-in-Makerfield** Wigan. *Eston* 1212. Affix is an old district name (*Macrefeld* 1121), from a Celtic word meaning 'wall, ruin' + OE *feld* 'open land'. **Ashton Keynes** Wilts. *Æsctun* 880–5, *Essitone* 1086 (DB), *Aysheton Keynes* 1572. Manorial affix from the *de Keynes* family, here from the 13th cent. **Ashton, Long** N. Som. *Estune* 1086 (DB), *Longe Asshton* 1467. Affix from the length of the village. **Ashton, Steeple & Ashton, West** Wilts. *Æystone* 964, *Aistone* 1086 (DB), *Westaston* 1248, *Stepelaston* 1268. Distinguishing affixes from OE *stīepel* 'church steeple' and *west*. **Ashton under Hill** Worcs. *Æsctun* 991, *Essetone* 1086 (DB), *Assheton Underhill* 1544. Affix 'under the hill' refers to BREDON Hill. **Ashton-under-Lyne** Tamesd. *Haistune* c.1160, *Asshton under Lyme* 1305. Affix is from an old district name *Lyme*, possibly 'escarpment', *see* BURSLEM. **Ashton upon Mersey** Traffd. *Asshton* 1408. On the River Mersey, 'boundary river' from OE *mære* (genitive -*s*) + *ēa*.
      However some Ashtons have a different origin: **Ashton** Northants., near Northampton. *Asce* 1086 (DB), *Asshen* 1296. '(Place at) the ash-trees'. OE *æsc* in a dative plural form *æscum*. **Ashton, Higher & Ashton, Lower** Devon. *Aiserstone* 1086 (DB). 'Farmstead of a man called Æschere'. OE pers. name + *tūn*.

Ashurst, 'wooded hill growing with ash-trees', OE *æsc* + *hyrst*: **Ashurst** Kent. *Aeischerste* c.1100. **Ashurst** W. Sussex. *Essehurst* 1164.

Ashurstwood W. Sussex. *Foresta de Esseherst* 1164. Identical in origin with the previous names.

Ashwater Devon. *Aissa* 1086 (DB), *Esse Valteri* 1270. '(Place at) the ash-tree(s)'. OE *æsc* + manorial affix from its possession by a man called *Walter* in the 13th cent.

Ashwell, 'spring or stream where ash-trees grow', OE *æsc* + *wella*: **Ashwell** Herts.

Asceuuelle 1086 (DB). **Ashwell** Rutland. *Exewelle* 1086 (DB).

Ashwellthorpe Norfolk. *Aescewelle, Thorp* c.1066. 'Hamlet belonging to a place called *Ashwell* ("ash-tree spring or stream")'. OE *æsc* + *wella* + OScand. *thorp*.

Ashwick Somerset. *Escewiche* 1086 (DB). 'Dwelling or farmstead where ash-trees grow'. OE *æsc* + *wīc*.

Ashwicken Norfolk. *Wiche* 1086 (DB), *Askiwiken* 1275. '(Place at) the dwellings or buildings'. OE *wīc* in a dative plural form *wīcum* or a ME plural form *wiken*. Later addition may be OE *æsc* 'ash-tree' or a pers. name.

Askam in Furness Cumbria. *Askeham* 1535. Possibly '(place at) the ash-trees'. OScand. *askr* in a dative plural form *askum*. For the affix, *see* BARROW IN FURNESS.

Askamore (*An Easca Mhór*) Wexford. 'The big bog'.

Askanagap (*Easca na gCeap*) Wicklow. 'Bog of the stumps'.

Askeaton (*Eas Géitine*) Limerick. *Eas-Gephtine* n.d. 'Géitine's waterfall'.

Askern Donc. *Askern* c.1170. 'House near the ash-tree'. OScand. *askr* + OE *ærn*.

Askerswell Dorset. *Oscherwille* 1086 (DB). 'Spring or stream of a man called Ōsgār'. OE pers. name + *wella*.

Askham, 'homestead or enclosure where ash-trees grow', OE *æsc* (replaced by OScand. *askr*) + OE *hām* or *hamm*: **Askham** Notts. *Ascam* 1086 (DB). **Askham Bryan & Askham Richard** York. *Ascham* 1086 (DB), *Ascam Bryan* 1285, *Askham Ricardi* 1291. Manorial additions from early possession by men called *Brian* and *Richard*.

Askham Cumbria. *Askum* 1232. '(Place at) the ash-trees'. OScand. *askr* in the dative plural form *askum*.

Askrigg N. Yorks. *Ascric* 1086 (DB). Probably 'ash-tree ridge'. OScand. *askr* + OE **ric*.

Askwith N. Yorks. *Ascvid* 1086 (DB). 'Ash-tree wood'. OScand. *askr* + *vithr*.

Aslackby Lincs. *Aslachebi* 1086 (DB). 'Farmstead or village of a man called Áslákr'. OScand. pers. name + *bý*.

Aslacton Norfolk. *Aslactuna* 1086 (DB). 'Farmstead or village of a man called Áslákr'. OScand. pers. name + OE *tūn*.

Aslockton Notts. *Aslachetune* 1086 (DB). Identical in origin with the previous name.

Aspatria Cumbria. *Aspatric* c.1160. 'Ash-tree of St Patrick'. OScand. *askr* + Celtic pers. name. The order of elements is Celtic.

Aspenden Herts. *Absesdene* 1086 (DB). 'Valley where aspen-trees grow'. OE *æspe* + *denu*.

Aspley Guise Beds. *Æpslea* 969, *Aspeleia* 1086 (DB), *Aspeleye Gyse* 1363. 'Aspen-tree wood or glade'. OE *æspe* + *lēah*, with manorial affix from the *de Gyse* family, here in the 13th cent.

Aspull Wigan. *Aspul* 1212. 'Hill where aspen-trees grow'. OE *æspe* + *hyll*.

Assaroe (*Easa Rua*) Donegal. 'Red waterfall'.

Asselby E. R. Yorks. *Aschilebi* 1086 (DB). 'Farmstead or village of a man called Áskell'. OScand. pers. name + *bý*.

Assendon, Lower & Assendon, Middle Oxon. *Assundene* late 10th cent. 'Valley of the ass, or of a man called *Assa'. OE *assa* or OE pers. name (genitive *-n*) + *denu*.

Assington Suffolk. *Asetona* 1086 (DB), *Assintona* 1175. 'Estate associated with a man called *As(s)a'. OE pers. name + *-ing-* + *tūn*.

Assolus (*Áth Solais*) Cork. 'Ford of light'.

Astbury Ches. *Astbury* 1093. 'East manor or stronghold'. OE *ēast* + *burh* (dative *byrig*).

Astcote Northants. *Aviescote* 1086 (DB). 'Cottage(s) of a man called Æfic'. OE pers. name + *cot*.

Asterley Shrops. *Estrelega* 1208. 'More easterly woodland clearing'. OE *ēasterra* + *lēah*.

Asterton Shrops. *Esthampton* 1255. 'Eastern home farm'. OE *ēast* + *hām-tūn*.

Asthall Oxon. *Esthale* 1086 (DB). 'East nook(s) of land'. OE *ēast* + *h(e)alh*.

Asthall Leigh Oxon. *Estallingeleye* 1272. 'Woodland clearing of the people of Asthall'. ASTHALL + OE *-inga-* + *lēah*.

Astley, 'east wood or clearing', OE *ēast* + *lēah*: **Astley** Shrops. *Hesleie* 1086 (DB). **Astley**

Warwicks. *Estleia* 1086 (DB). **Astley** Wigan.
Asteleghe c.1210. **Astley** Worcs. *Eslei* 1086
(DB). **Astley Abbots** Shrops. *Estleia* c.1090,
Astleye Abbatis late 13th cent. Affix alludes to
early possession by Shrewsbury Abbey.

Aston, a common name, usually 'eastern
farmstead or estate', OE *ēast* + *tūn*; examples
include: **Aston** Birm. *Estone* 1086 (DB). **Aston**
Flin. *Estone* 1086 (DB). **Aston** Rothm. *Estone*
1086 (DB). **Aston Blank** or **Cold Aston**
Glos. *Eastunæ* 716–43, *Estone* 1086 (DB). Affix
may be OFrench *blanc* 'white, bare'. **Aston
Cantlow** Warwicks. *Estone* 1086 (DB), *Aston
Cantelou* 1273. Manorial affix from the *de
Cantilupe* family, here in the 13th cent. **Aston,
Chetwynd** Tel. & Wrek. *Estona* 1155, *Greate
Aston alias Chetwynde Aston* 1619. For affix, *see*
CHETWYND. **Aston Clinton** Bucks. *Estone*
1086 (DB), *Aston Clinton* 1237–40. Manorial affix
from the *de Clinton* family, here in the late 12th
cent. **Aston Fields** Worcs. *Eastun* 767, *Estone*
1086 (DB), *Aston Fields* 1649. **Aston Ingham**
Herefs. *Estune* 1086 (DB), *Estun Ingan* 1242.
Manorial affix from the *Ingan* family, here in the
13th cent. **Aston Rowant** Oxon. *Estone* 1086
(DB), *Aston Roaud* 1318. Manorial affix from
Rowald de Eston, here in 1236. **Aston,
Steeple** Oxon. *Estone* 1086 (DB), *Stipelestun*
1220. Affix is OE *stīepel* 'church steeple'. **Aston
upon Trent** Derbys. *Estone* 1086 (DB).
For the river-name, *see* TRENTHAM. **Aston,
Wheaton** Staffs. *Estone* 1086 (DB), *Wetenaston*
1248. Affix is OE *hwǣten* 'growing with wheat'.
Aston, White Ladies Worcs. *Eastune* 977,
Estun 1086 (DB), *Whitladyaston* 1481. Affix from
its possession by the Cistercian nuns of
Whitstones.
    However the following has a different
origin: **Aston on Clun** Shrops. *Assheston*
1291. 'Ash-tree farmstead'. OE *æsc* + *tūn*.
For the river-name, *see* CLUN.

Astrop Northants. *Estrop* 1200. 'East hamlet'.
OE *ēast* + *throp*.

Astwood, 'east wood', OE *ēast* + *wudu*:
Astwood Milt. K. *Estwode* 1151–4. **Astwood**
Worcs. *Estwode* 1182.

Aswarby Lincs. *Asuuardebi* 1086 (DB).
'Farmstead or village of a man called Ásvarthr'.
OScand. pers. name + *bý*.

Aswardby Lincs. *Asewrdeby* c.1155. Identical
in origin with the previous name.

Atch Lench Worcs. *See* LENCH.

Atcham Shrops. *Atingeham* 1086 (DB).
'Homestead of the family or followers of a
man called Ætti or Ēata', or 'homestead at
the place associated with Ætti or Ēata'. OE pers.
name + *-inga-* or *-ing-* + *hām*. Alternatively
the final element may be *hamm* 'land in a
river-bend'.

Athboy (*Baile Átha Buí*) Meath. 'Town of the
yellow ford'.

Athcarne (*Áth Chairn*) Meath. 'Ford of the
cairn'.

Athea (*Áth an tSléibhe*) Limerick. 'Ford of the
mountain'.

Athelhampton Dorset. *Pidele* 1086 (DB),
Pidele Athelamston 1285. Originally named from
the River Piddle on which it stands (*see*
PIDDLEHINTON), later 'farmstead of a man
called Æthelhelm', OE pers. name + *tūn*.

Athelington Suffolk. *Alinggeton* 1219.
'Farmstead or village of the princes'. OE
ætheling + *tūn*.

Athelney Somerset. *Æthelingaeigge* 878,
Adelingi 1086 (DB). 'Island, or dry ground in
marsh, of the princes'. OE *ætheling* + *ēg*.

Athenboy (*Aiteann Buí*) Westmeath. 'Yellow
gorse'.

Athenry (*Baile Átha an Rí*) Galway. 'Town of
the ford of the kings'.

Atherfield, Little I. of Wight. *Aderingefelda*
959, *Avrefel* 1086 (DB). 'Open land of the family
or followers of a man called Ēadhere or
Æthelhere'. OE pers. name + *-inga-* + *feld*.

Atherington Devon. *Hadrintone* 1272.
'Estate associated with a man called Ēadhere or
Æthelhere'. OE pers. name + *-ing-* + *tūn*.

Atherstone Warwicks. *Aderestone* 1086 (DB),
Atheredestone 1221. 'Farmstead or village of a
man called Æthelrēd'. OE pers. name + *tūn*.

Atherstone on Stour Warwicks.
Eadrichestone 710, *Edricestone* 1086 (DB).
'Farmstead or village of a man called Ēadrīc'.
OE pers. name + *tūn*. Affix from its situation on
the River Stour, a Celtic or OE river-name
probably meaning 'the strong one'.

Atherton Wigan. *Aderton* 1212. 'Farmstead
or village of a man called Æthelhere'. OE pers.
name + *tūn*.

Athgarvan (*Áth Garbháin*) Kildare. 'Garbhán's ford'.

Athgreany (*Áth Gréine*) Wicklow. 'Grian's ford'.

Athlacca (*An tÁth Leacach*) Limerick. 'The ford with flagstones'.

Athleague (*Áth Liag*) Roscommon. 'Ford of the boulders'.

Athlone (*Baile Áth Luain*) Westmeath. 'Town of Luan's ford'.

Athnid (*Áth Nid*) Tipperary. 'Ford of the nest'.

Atholl. *See* BLAIR ATHOLL.

Athy (*Baile Áth Í*) Kildare. 'Town of the yew ford'.

Atlow Derbys. *Etelawe* 1086 (DB). 'Burial-mound of a man called Eatta'. OE pers. name + *hlāw*.

Attanagh (*Áth Tanaí*) Laois. 'Shallow ford'.

Attatantee (*Áit a' tSean Tighe*) Donegal. 'Place of the old house'.

Attenborough Notts. *Adinburcha* 12th cent. 'Stronghold associated with a man called Adda or Æddi'. OE pers. name + *-ing-* + *burh*.

Attical (*Áit Tí Chathail*) Down. *Atty Caell* c.1659. 'Place of Cathal's house'.

Atticonor (*Áit Tí Chonuir*) Westmeath. 'Place of Conor's house'.

Attleborough Norfolk. *Ateburc* 1086 (DB). 'Stronghold of a man called Ætla'. OE pers. name + *burh*.

Attleborough Warwicks. *Atteleberga* 12th cent. 'Hill or mound of a man called Ætla'. OE pers. name + *beorg*.

Attlebridge Norfolk. *Atlebruge* 1086 (DB). 'Bridge of a man called Ætla'. OE pers. name + *brycg*.

Attymachugh (*Áit Tí Mhic Aodha*) Mayo. 'Place of Mac Aodha's house'.

Attymass (*Áit Tí an Mheasaigh*) Mayo. 'Place of the church of the Measach'.

Attymon (*Áth Tíomáin*) Galway. 'Tíomán's ford'.

Atwick E. R. Yorks. *Attingwik* 12th cent. 'Dwelling or dairy-farm of a man called Atta'. OE pers. name + *-ing-* + *wīc*.

Atworth Wilts. *Attenwrthe* 1001. 'Enclosure of a man called Atta'. OE pers. name + *worth*.

Aubourn Lincs. *Aburne* 1086 (DB), *Alburn* 1275. 'Stream where alder-trees grow'. OE *alor* + *burna*.

Auchinleck E. Ayr. *Auechinlec* 1239. 'Field of the flat stones'. Gaelic *achadh nan leac*.

Auchterarder Perth. *Vchterardouere* c.1200. 'Upland of high water'. Gaelic *uachdar* + *ard* + *dobhar*.

Auchtermuchty Fife. *Vchtermuckethin* c.1210. 'Upland of the pig place'. Gaelic *uachdar* + *muccatu*.

Auckland, Bishop, Auckland, St Helen & Auckland, West Durham. *Alclit* c.1040. Probably a Celtic name meaning 'rock or hill on a river called Clyde ("the cleansing one")'. Clyde was probably the original name of the River Gaunless (from OScand. **gagnlauss* 'unprofitable one'). Later distinguishing affixes from possession by the bishop of Durham and from the church dedication to St Helen.

Auckley Donc. *Alchelie* 1086 (DB). 'Woodland clearing of a man called Alca or *Alha'. OE pers. name + *lēah*.

Aucloggeen (*Áth Cloigín*) Galway. 'Ford of the little bell'.

Audenshaw Tamesd. *Aldwynshawe* c.1200. 'Copse of a man called Aldwine'. OE pers. name + *sceaga*.

Audlem Ches. *Aldelime* 1086 (DB). 'Old *Lyme*', or 'the part of *Lyme* belonging to a man called Alda'. OE *ald* or OE pers. name + old district name *Lyme*, probably 'escarpment', *see* BURSLEM.

Audley Staffs. *Aldidelege* 1086 (DB). 'Woodland clearing of a woman called Aldgȳth'. OE pers. name + *lēah*.

Aughall (*Eochaill*) Tipperary. 'Yew wood'.

Aughamullen (*Achadh Ó Maoláin*) Tyrone. *Aghmoylan* 1609. 'Ó Maoláin's field'.

Augher (*Eochair*) Tyrone. *Ogher* c.1655. 'Border'.

Aughil (*Eochaill*) Derry. 'Yew wood'.

Aughinish Island (*Eachinis*) Galway, Limerick. 'Horse island'.

Aughnacloy (*Achadh na Cloiche*) Tyrone. *Aghenecloy c.*1655. 'Field of the stone'.

Aughnagomaun (*Achadh na gComán*) Tipperary. 'Hurling field'.

Aughnahoy (*Achadh na hÁithe*) Antrim. 'Field of the kiln'.

Aughnanure (*Achadh an Iúr*) Galway. 'Field of the yew'.

Aughnasheelan (*Achadh na Síleann*) Leitrim. 'Field of the withies'.

Aughrim (*Eachroim*) Derry, Galway, Wexford. 'Horse ridge'.

Aughris Head (*Ceann Eachrois*) Sligo. 'Horse promontory'.

Aughton, 'farmstead where oak-trees grow', OE *āc* + *tūn*: **Aughton** E. R. Yorks. *Actun* 1086 (DB). **Aughton** Lancs., near Lancaster. *Acheton* 1212. **Aughton** Lancs., near Ormskirk. *Achetun* 1086 (DB). **Aughton** Rothm. *Actone* 1086 (DB).

Aughvolyshane (*Áth Bhuaile Shéain*) Tipperary. 'Ford of Seán's milking place'.

Ault Hucknall Derbys. *See* HUCKNALL.

Aultbea Highland. 'Stream where birch-trees grow'. Gaelic *allt* + *beithe*.

Aunsby Lincs. *Ounesbi* 1086 (DB), *Outhenby* 1281. Probably 'farmstead or village of a man called Authunn'. OScand. pers. name + *bý*. Alternatively the first element may be OScand. *authn* 'uncultivated land, deserted farm'.

Aust S. Glos. *Austan* 794. Possibly from Latin *Augusta*, perhaps alluding to the crossing of the River Severn here used by the Roman Second Legion, the *Legio Augusta*. Alternatively from the Latin pers. name *Augustinus*.

Austerfield Donc. *Eostrefeld c.*715, *Oustrefeld* 1086 (DB). 'Open land with a sheepfold'. OE *eowestre* + *feld*.

Austrey Warwicks. *Alduluestreow* 958, *Aldulvestreu* 1086 (DB). 'Tree of a man called Ealdwulf'. OE pers. name + *trēow*.

Austwick N. Yorks. *Ousteuuic* 1086 (DB). 'East dwelling or dairy-farm'. OScand. *austr* + OE *wīc*.

Authorpe, 'outlying farmstead or hamlet of a man called Aghi', OScand. pers. name + *thorp*: **Authorpe** Lincs. *Agetorp* 1086 (DB). **Authorpe Row** Lincs. *Aghetorp c.*1115.

Avaghon, Lough (*Loch an Meatháin*) Monaghan. 'Lake of the saplings'.

Avalbane (*Abhall Bán*) Monaghan. 'White orchard'.

Avebury Wilts. *Aureberie* 1086 (DB), *Aveberia c.*1180. Probably 'stronghold of a man called Afa'. OE pers. name + *burh* (dative *byrig*).

Aveley Thurr. *Aluitheleam* 1086 (DB). 'Woodland clearing of a woman called Ælfgȳth'. OE pers. name + *lēah*.

Avening Glos. *Æfeningum* 896, *Aveninge* 1086 (DB). '(Settlement of) the people living by the River *Avon*'. Celtic river-name (perhaps the original name of the stream here, meaning simply 'river') + OE *-ingas*.

Averham Notts. *Aigrun* 1086 (DB). Probably '(place by) the floods or high tides' (with reference to the Trent bore). OE *ēgor* in a dative plural form *ēgrum*.

Aveton Gifford Devon. *Avetone* 1086 (DB), *Aveton Giffard* 1276. 'Farmstead on the River Avon'. Celtic river-name (meaning simply 'river') + OE *tūn* + manorial affix from the *Giffard* family, here in the 13th cent.

Aviemore Highland. 'Big hill face'. Gaelic *aghaid* + *mór*.

Avington W. Berks. *Avintone* 1086 (DB). 'Estate associated with a man called Afa'. OE pers. name + *-ing-* + *tūn*.

Avoca (*Abhóca*) (river) Wicklow. The name was adopted from Ptolemy's *Oboka* (2nd cent.) for a length of the AVONMORE.

Avon, old river-name found several times in England, from a Celtic word meaning simply 'river'. Of these, the Bristol or Lower **Avon** gives name to AVONMOUTH; the Wilts./Hants. **Avon** to AVON, NETHERAVON, and UPAVON; the Devon **Avon** to AVETON GIFFORD; the Glos./Warwicks. **Avon** to STRATFORD UPON AVON.

Avon Hants. *Avere* 1086 (DB). Named from the River Avon, *see* previous entry.

Avon Dassett Warwicks. *See* DASSETT.

Avonbeg (*An Abhainn Bheag*) (river) Wicklow. 'The little river'. The name contrasts with the AVONMORE.

Avonmore (*An Abhainn Mhór*) Cork, Sligo, Wicklow. 'The big river'.

Avonmouth Brist., a modern port at the mouth of the River Avon (*Afenemuthan* 10th cent.). Celtic river-name (*see* AVON) + OE *mūtha*.

Awbeg (*Abha Bheag*) Cork, Limerick. 'Little river'.

Awbridge Hants. *Abedric* 1086 (DB). 'Ridge of the abbot'. OE *abbod* + *hrycg*.

Awe, Loch Arg. *Aba* 700. 'Loch of (the River) Awe'. Gaelic *loch*. The Celtic river-name means 'river'.

Awliscombe Devon. *Aulescome* 1086 (DB). Probably 'valley near the fork of a river'. OE *āwel* + *cumb*.

Awre Glos. *Avre* 1086 (DB). Probably '(place at) the alder-tree'. OE *alor*.

Awsworth Notts. *Ealdeswyrthe* 1002, *Eldesvorde* 1086 (DB). 'Enclosure of a man called *Eald*. OE pers. name + *worth*.

Axbridge Somerset. *Axanbrycg* 10th cent. 'Bridge over the River Axe'. Celtic river-name (meaning uncertain) + OE *brycg*.

Axford Wilts. *Axeford* 1184. 'Ford by the ash-trees'. OE *æsc* + *ford*.

Axminster Devon. *Ascanmynster* late 9th cent., *Aixeministra* 1086 (DB). 'Monastery or large church by the River Axe'. Celtic river-name + OE *mynster*.

Axmouth Devon. *Axanmuthan* c.880, *Alsemuda* 1086 (DB). 'Mouth of the River Axe'. Celtic river-name + OE *mūtha*.

Aycliffe Durham. *Heaclif* 1109, *Acleia* c.1190. Early spellings suggest alternation between 'high oak-tree bank', OE *hēah* + *āc* + *clif*, and 'oak-tree wood or clearing', OE *āc* + *lēah*.

Aylburton Glos. *Ailbricton* 12th cent. 'Farmstead or village of a man called *Æthelbeorht*'. OE pers. name + *tūn*.

Ayle Northum., named from the river on which it stands, Ayle Burn (*Alne* 1347), a Celtic river-name of uncertain meaning.

Aylesbeare Devon. *Ailesberga* 1086 (DB). 'Grove of a man called *Ægel*'. OE pers. name + *bearu*.

Aylesbury Bucks. *Ægelesburg* late 9th cent., *Eilesberia* 1086 (DB). 'Stronghold of a man called *Ægel*'. OE pers. name + *burh* (dative *byrig*).

Aylesby NE. Lincs. *Alesbi* 1086 (DB). 'Farmstead or village of a man called Áli'. OScand. pers. name + *bý*.

Aylesford Kent. *Æglesforda* 10th cent., *Ailesford* 1086 (DB). 'Ford of a man called *Ægel*'. OE pers. name + *ford*.

Aylesham Kent. *Elisham* 1367. Possibly 'homestead or enclosure of a man called *Ægel*'. OE pers. name + *hām* or *hamm*.

Aylestone Leic. *Aileston* 1086 (DB). 'Farmstead or village of a man called *Ægel*'. OE pers. name + *tūn*.

Aylmerton Norfolk. *Almartune* 1086 (DB). 'Farmstead or village of a man called *Æthelmǣr*'. OE pers. name + *tūn*.

Aylsham Norfolk. *Ailesham* 1086 (DB). 'Homestead of a man called *Ægel*'. OE pers. name + *hām*.

Aylton Herefs. *Aileuetona* 1138. 'Farmstead or village of a woman called *Æthelgifu*'. OE pers. name + *tūn*.

Aymestrey Herefs. *Elmodestreu* 1086 (DB). 'Tree of a man called *Æthelmund*'. OE pers. name + *trēow*.

Aynho Northants. *Aienho* 1086 (DB). 'Hillspur of a man called *Æga*'. OE pers. name (genitive *-n*) + *hōh*.

Ayot St Lawrence & Ayot St Peter Herts. *Aiegete* c.1060, *Aiete* 1086 (DB). 'Gap or pass of a man called *Æga*'. OE pers. name + *geat*. Distinguishing affixes from the dedications of the churches at the two places.

Ayr S. Ayr. *Ar* 1177. '(Mouth of the river) Ayr'. The Celtic (or pre-Celtic) river-name means simply 'river'. Ayr was formerly known as *Inverayr*, but the 'mouth' element (Gaelic *inbhir*) was dropped.

Aysgarth N. Yorks. *Echescard* 1086 (DB). 'Gap or open place where oak-trees grow'. OScand. *eiki* + *skarth*.

Ayston Rutland. *Æthelstanestun* 1046.
'Farmstead or village of a man called
Æthelstān'. OE pers. name + *tūn*.

Aythorpe Roding Essex. *See* RODING.

Ayton, 'farmstead or estate on the river'
(probably denoting a settlement
which performed a special local function
in relation to the river), OE *ēa* (modified
by OScand. *á*) + *tūn*: **Ayton, East & Ayton,
West** N. Yorks. *Atune* 1086 (DB).
Ayton, Great & Ayton, Little N. Yorks.
Atun 1086 (DB).

Azerley N. Yorks. *Asserle* 1086
(DB). 'Woodland clearing of a man
called Atsurr'. OScand. pers. name +
OE *lēah*.

B

Babbacombe Torbay. *Babbecumbe* c.1200. 'Valley of a man called Babba'. OE pers. name + *cumb*.

Babcary Somerset *Babba Cari* 1086 (DB). 'Estate on the River Cary held by a man called Babba'. OE pers. name + Celtic or pre-Celtic river-name (*see* CARY FITZPAINE)

Babergh (district) Suffolk. *Baberga* 1086 (DB). 'Mound of a man called Babba'. A revival of an old hundred-name. OE pers. name + *beorg*.

Babraham Cambs. *Badburgham* 1086 (DB). 'Homestead or village of a woman called *Beaduburh'. OE pers. name + *hām*.

Babworth Notts. *Baburde* 1086 (DB). 'Enclosure of a man called Babba'. OE pers. name + *worth*.

Babylon. *See* BRABLING GREEN.

Backbarrow Cumbria. *Bakbarowe* 1537. 'Hill with a ridge'. OE *bæc* + *beorg*.

Backford Ches. *Bacfort* 1150. 'Ford by a ridge'. OE *bæc* + *ford*.

Backwell N. Som. *Bacoile* [sic] 1086 (DB), *Bacwell* 1202. 'Spring or stream near a ridge'. OE *bæc* + *wella*.

Backworth N. Tyne. *Bacwrth* 12th cent. 'Enclosure of a man called Bacca'. OE pers. name + *worth*.

Baconsthorpe Norfolk. *Torp, Baconstorp* 1086 (DB). 'Outlying farmstead or hamlet of a family called Bacon'. Norman surname + OScand. *thorp*.

Bacton, 'farmstead of a man called Bacca', OE pers. name + *tūn*: **Bacton** Herefs. *Bachetone* 1086 (DB). **Bacton** Norfolk. *Baketuna* 1086 (DB). **Bacton** Suffolk. *Bachetuna* 1086 (DB).

Bacup Lancs. *Fulebachope* c.1200, *Bacop* 1324. 'Valley by a ridge'. OE *bæc* + *hop* (prefixed by OE *fūl* 'foul, muddy' in the early spelling).

Badbury Swindn. *Baddeburi* 955, *Badeberie* 1086 (DB). Probably 'stronghold of a man called Badda'. OE pers. name + *burh* (dative *byrig*).

Badby Northants. *Baddanbyrig* 944, *Badebi* 1086 (DB). Identical in origin with the previous name, though the second element was replaced at an early date by OScand. *bý* 'farmstead, village'.

Baddeley Green Stoke. *Baddilige* 1227. 'Woodland clearing of a man called Badda'. OE pers. name + *lēah*.

Baddesley, 'woodland clearing of a man called *Bæddi', OE pers. name + *lēah*: **Baddesley Ensor** Warwicks. *Bedeslei* 1086 (DB), *Baddesley Endeshouer* 1327. Manorial affix from the *de Edneshoure* family (from EDENSOR), here in the 13th cent. **Baddesley, North** Hants. *Bedeslei* 1086 (DB).

Baddow, Great & Baddow, Little Essex. *Beadewan* c.975, *Baduuen* 1086 (DB). Probably a Celtic river-name (an old name for the River Chelmer), of uncertain origin and meaning.

Badenoch (district) Highland. *Badenach* 1229. 'Submerged land'. Gaelic *bàithteanach*. The region lies to the south of the river Spey, which is liable to flood.

Badger Shrops. *Beghesovre* 1086 (DB). 'Hillspur of a man called *Bæcg'. OE pers. name + *ofer*.

Badgeworth, Badgworth, 'enclosure of a man called *Bæcga', OE pers. name + *worth*: **Badgeworth** Glos. *Beganwurthan* 862, *Beiewrda* 1086 (DB). **Badgworth** Somerset. *Bagewerre* 1086 (DB).

Badingham Suffolk. *Badincham* 1086 (DB). 'Homestead or village associated with a man called *Bēada'. OE pers. name + *-ing-* + *hām*.

Badlesmere Kent. *Badelesmere* 1086 (DB). Probably 'pool of a man called *Bæddel'. OE pers. name + *mere*.

Badminton, Great & Badminton, Little S. Glos. *Badimyncgtun* 972, *Madmintune* [*sic*] 1086 (DB). 'Estate associated with a man called Baduhelm'. OE pers. name + *-ing-* + *tūn*.

Badsey Worcs. *Baddeseia* 709, *Badesei* 1086 (DB). 'Island, or dry ground in marsh, of a man called *Bæddi*. OE pers. name + *ēg*.

Badsworth Wakefd. *Badesuuorde* 1086 (DB). 'Enclosure of a man called *Bæddi*. OE pers. name + *worth*.

Badwell Ash Suffolk. *Badewell* 1254, *Badewelle Asfelde* 13th cent. 'Spring or stream of a man called Bada'. OE pers. name + *wella* + later affix from shortened form of GREAT ASHFIELD.

Bae Cinmel. *See* KINMEL BAY.

Bae Colwyn. *See* COLWYN BAY.

Bag Enderby Lincs. *See* ENDERBY.

Bagborough, West Somerset. *Bacganbeorg* 904, *Bageberge* 1086 (DB). Probably 'hill of a man called Bacga'. OE pers. name + *beorg*.

Bagby N. Yorks. *Baghebi* 1086 (DB). 'Farmstead or village of a man called Baggi'. OScand. pers. name + *bý*.

Bagendon Glos. *Benwedene* [*sic*] 1086 (DB), *Baggingeden* 1220. 'Valley of the family or followers of a man called *Bæcga*. OE pers. name + *-inga-* + *denu*.

Baginton Warwicks. *Badechitone* 1086 (DB). 'Estate associated with a man called Badeca'. OE pers. name + *-ing-* + *tūn*.

Baglan Neat. *Bagelan* 1199. '(Church of) Baglan'. From the dedication of the church to St Baglan.

Bagley Shrops. *Bageleia c.*1090. Probably 'wood or clearing frequented by badgers'. OE **bagga* + *lēah*.

Bagnall Staffs. *Badegenhall* 1273. Probably 'nook of land of a man called Badeca'. OE pers. name (genitive *-n*) + *halh*.

Bagshot Surrey. *Bagsheta* 1164. 'Projecting piece of land frequented by badgers'. OE **bagga* + *scēat*.

Bagshot Wilts. *Bechesgete* 1086 (DB). 'Gate or gap of a man called *Beocc*. OE pers. name + *geat*.

Bagthorpe Norfolk. *Bachestorp* 1086 (DB). Probably 'outlying farmstead or hamlet of a man called Bakki or Bacca'. OScand. or OE pers. name + OScand. *thorp*.

Bagworth Leics. *Bageworde* 1086 (DB). 'Enclosure of a man called Bacga'. OE pers. name + *worth*.

Baildon Brad. *Bægeltune, Bældune c.*1030, *Beldune* 1086 (DB). 'Circle hill'. OE **bægel* + *dūn*.

Baile Átha Cliath. *See* DUBLIN.

Bailehaise (*Béal Átha hÉis*) Cavan. (*go*) *Béal Átha Haeis* 1644. 'Ford-mouth of the track'.

Baileysmill (*Muileann Bháille*) Down. *Bailey's Mill* 1814. 'Bailey's mill'.

Bailieborough (*Coill an Chollaigh*) Cavan. *Kilcothie al. Bailie-Borrow* 1629. 'Bailie's town'. William *Bailie* built Bailieborough Castle here in 1610. The Irish name means 'wood of the boar'.

Bainbridge N. Yorks. *Bainebrigg* 1218. 'Bridge over the River Bain'. OScand. river-name ('the short or helpful one') + OE *brycg*.

Bainton Peterb. *Badingtun c.*980. 'Estate associated with a man called Bada'. OE pers. name + *-ing-* + *tūn*.

Bainton E. R. Yorks. *Bagentone* 1086 (DB). 'Estate associated with a man called Bǣga'. OE pers. name + *-ing-* + *tūn*.

Bakewell Derbys. *Badecanwelle* 949, *Badequella* 1086 (DB). 'Spring or stream of a man called Badeca'. OE pers. name + *wella*.

Bala Gwyd. 'Outlet'. *la Bala* 1331, *the Bala* 1582. Welsh *bala*. The town is at the point where the Dee leaves Bala Lake, whose Welsh name is *Llyn Tegid*, 'Tegid's lake' (pers. name + Welsh *llyn*).

Balbane (*Baile Bán*) Donegal. 'White townland'.

Balboru (*Béal Boromha*) Clare. 'Mouth of the (river) Borumha'.

Balbriggan (*Baile Brigín*) Dublin. 'Brigín's townland'.

Balcombe W. Sussex. *Balecumba* late 11th cent., *Baldecombe* 1279. Possibly 'valley of a man called Bealda'. OE pers. name + *cumb*. Alternatively the first element may be OE *bealu* 'evil, calamity'.

Balcomie Fife. *Balcolmie* 1253. 'Estate of (Gille) Colm'. Gaelic *baile*.

Baldersby N. Yorks. *Baldrebi* 1086 (DB). 'Farmstead or village of a man called Baldhere'. OE pers. name + OScand. *bý*.

Balderstone Lancs. *Baldreston* 1323. 'Farmstead of a man called Baldhere'. OE pers. name + *tūn*.

Balderton Notts. *Baldretune* 1086 (DB). Identical in origin with the previous name.

Baldock Herts. *Baldoce* c.1140. This place was founded in the 12th cent. by the Knights Templar, who called it *Baldac*, the OFrench form for the Arabian city of Baghdad.

Baldon, Marsh & Baldon, Toot Oxon. *Balde(n)done* 1086 (DB), *Mersse Baldindon* 1241, *Totbaldindon* 1316. 'Hill of a man called Bealda'. OE pers. name + *dūn*. Distinguishing affixes from OE *mersc* 'marsh' and *tōt(e)* 'look-out hill'.

Baldoyle (*Baile Dúill*) Dublin. 'Dúghall's townland'.

Baldwinholme Cumbria. *Baldewinholme* 1278. 'Island or water-meadow of a man called Baldwin'. OGerman pers. name + OScand. *holmr*.

Bale Norfolk. *Bathele* 1086 (DB). 'Woodland clearing where there are springs used for bathing'. OE *bæth* + *lēah*.

Balemartine Arg. (Tiree). *Balmartin* 1654. 'Martin's village'. Gaelic *baile*.

Balerno Edin. *Belhernoch* 1280. 'Townland where sloe-trees grow'. Gaelic *baile* + *airneach*.

Balfeddock (*Baile Feadóg*) Louth. 'Townland of the plover'.

Balfour Orkn. (Shapinsay). 'Farm with pasture'. Gaelic *baile* + *pór*.

Balgowan Highland. 'Farm of the smith'. Gaelic *baile* + *gobhan*.

Balgown Highland. (Skye). 'Farm of the smith'. Gaelic *baile* + *gobhan*.

Balham Gtr. London. *Bælgenham* 957, *Belgeham* 1086 (DB). Probably 'smooth or rounded enclosure'. OE **bealg* + *hamm*.

Balintore Highland. 'Village where bleaching is done'. Gaelic *baile* + *todhair*.

Balkholme E. R. Yorks. *Balcholm* 1199. 'Island with a low ridge', or 'island of a man called Balki'. OE *balca* or OScand. pers. name + OScand. *holmr*.

Balla (*Balla*) Mayo. 'Spring'.

Ballachulish Highland. 'Townland by the strait'. Gaelic *baile* + *caolas* (genitive *chaolais*).

Ballacolla (*Baile Cholla*) Laois. 'Colla's townland'.

Balladian (*Bealach an dá Éan*) Monaghan. 'Pass of the two birds'.

Ballagan Point (*Gob Bhaile Uí Ágáin*) Louth. 'Point of the townland of Ó Ágán'.

Ballagh (*Bealach*) Fermanagh, Galway, Limerick, Tipperary. 'Pass'.

Ballaghaderreen (*Bealach an Doirín*) Roscommon. 'Pass of the little oak grove'.

Ballaghanery (*Bealach an Aoire*) Down. 'Pass of the shepherd'.

Ballaghkeen (*Bealach Caoin*) Wexford. 'Smooth pass'.

Ballaghmoon (*Bealach Mughna*) Carlow. 'Mughan's pass'.

Ballanagare (*Béal Átha na gCarr*) Roscommon. 'Ford-mouth of the carts'.

Ballanruan (*Baile an Ruáin*) Clare. 'Townland of the red area'.

Ballantrae S. Ayr. 'Village on the shore'. Gaelic *baile* + *traigh*.

Ballard (*Baile Ard*) Offaly. 'High townland'.

Ballater Aber. *Balader* 1704, *Ballader* 1716. Origin obscure. The name originally applied to the pass where the Ballater Burn flows through the mountains.

Balleen (*Bailín*) Kerry. 'Little townland'.

Ballickmoyler (*Baile Mhic Mhaoilir*) Laois. 'Townland of Mac Maoilir'.

Balliggan (*Baile Uí Uiginn*) Down. *Ballyhiggin* 1623. 'Townland of Ó hUiginn'.

Ballina (*Béal an Átha*) Mayo, Tipperary. 'Ford-mouth'.

Ballinaboy (*Béal Átha na Bá[ighe]*) Galway. 'Ford-mouth of the bay'.

Ballinabrackey (*Buaile na Bréachmhaí*) Meath. 'Milking place of the wolf plain'.

Ballinabranagh (*Baile na mBreatnach*) Carlow. 'Townland of the Breatnachs'.

Ballinacarriga (*Béal na Carraige*) Cork. 'Mouth of the rocks'.

Ballinacarrow (*Baile na Cora*) Sligo. 'Townland of the weir'.

Ballinaclash (*Baile na Claise*) Wicklow. 'Townland of the ravine'.

Ballinaclashet (*Baile na Claise*) Cork. 'Townland of the ravine'.

Ballinaclough (*Baile na Cloiche*) Tipperary. 'Townland of the stone'.

Ballinacor (*Baile na Cora*) Wicklow. 'Townland of the weir'.

Ballinacurra (*Baile na Cora*) Cork, Limerick. 'Townland of the weir'.

Ballinadee (*Baile na Daibhche*) Cork. 'Townland of the well'.

Ballinafad (*Béal an Átha Fada*) Sligo. 'Mouth of the long ford'.

Ballinafid (*Baile na Feide*) Westmeath. 'Townland of the end'.

Ballinagar (*Béal Átha na gCarr*) Offaly. 'Ford-mouth of the carts'.

Ballinaglerach (*Baile na gCléireach*) Leitrim. 'Townland of the clerics'.

Ballinagree (*Baile na Graí*) Cork. 'Townland of the stud'.

Ballinahinch (*Baile na hInse*) Tipperary. 'Townland of the water-meadow'.

Ballinakill (*Baile na Coille*) Laois. 'Townland of the wood'.

Ballinalack (*Béal Átha na Leac*) Westmeath. 'Ford-mouth of the flagstones'.

Ballinalea (*Buaile na Lao*) Wicklow. 'Milking place of the calves'.

Ballinalee (*Béal Átha na Lao*) Longford. 'Ford-mouth of the calves'.

Ballinamallard (*Béal Átha na Mallacht*) Fermanagh. 'Ford-mouth of the curses'.

Ballinamara (*Baile na Marbh*) Kilkenny. 'Townland of the dead'.

Ballinameen (*Béal an Átha Mín*) Roscommon. 'Mouth of the smooth ford'.

Ballinamuck (*Béal Átha na Muc*) Longford. 'Ford-mouth of the pigs'.

Ballinard (*Baile an Aird*) Tipperary. 'Townland of the height'.

Ballinascarty (*Béal na Scairte*) Cork. 'Mouth of the thicket'.

Ballinascorny (*Baile na Scórnad*) Dublin. 'Townland of the gullet'.

Ballinaskeagh (*Baile na Sceach*) Down. 'Townland of the thorns'.

Ballinasloe (*Béal Átha na Sluaigheadh*) Galway. 'Ford-mouth of the military expeditions'.

Ballinaspick (*Baile an Easpaig*) Waterford. 'Townland of the bishop'.

Ballinattin (*Baile na Aiteann*) Tipperary, Waterford. 'Townland of the gorse'.

Ballinchalla (*Baile an Chaladh*) Mayo. 'Townland of the landing-place'.

Ballinclashet (*Baile na Claise*) Cork. 'Townland of the ravine'.

Ballincloher (*Baile an Cloichir*) Kerry. 'Townland of the stony place'.

Ballincollig (*Baile an Chollaigh*) Cork. 'Townland of the boar'.

Ballincrea (*Baile an Chraoibh*) Kilkenny. 'Townland of the sacred tree'.

Ballincurrig (*Baile an Churraigh*) Cork. 'Townland of the marsh'.

Ballindaggan (*Baile an Daingin*) Waterford. 'Townland of the fortress'.

Ballindangan (*Baile an Daingin*) Cork. 'Townland of the fortress'.

Ballindarragh (*Baile na Dara*) Fermanagh. 'Townland of the oak'.

Ballinderreen (*Baile an Doirín*) Galway. 'Townland of the little oak grove'.

Ballinderry (*Baile an Doire*) Antrim, Tipperary. 'Townland of the oak wood'.

Ballindrait (*Baile an Droichid*) Donegal. *Ballendraite c.*1655. 'Townland of the bridge'.

Ballinea (*Béal an Átha*) Westmeath. 'Ford-mouth'.

Ballineddan (*Baile an Fheadáin*) Wicklow. 'Townland of the little stream'.

Ballineen (*Béal Átha Fhinín*) Cork. 'Ford-mouth of Finín'.

Ballinenagh (*Baile an Aonaigh*) Limerick. 'Townland of the assembly'.

Ballinfull (*Baile an Phoill*) Sligo. 'Townland of the pool'.

Ballingaddy (*Baile an Ghadaí*) Limerick. 'Townland of the thief'.

Ballingar (*Béal Átha na gCarr*) Offaly. 'Ford-mouth of the carts'.

Ballingarrane (*Baile an Gharráin*) Limerick. 'Townland of the grove'.

Ballingarry (*Baile an Gharraí*) Limerick, Tipperary. 'Townland of the garden'.

Ballingeary (*Béal Átha an Ghaorthaidh*) Cork. 'Ford-mouth of the wooded valley'.

Ballingham Herefs. *Badelingeham* 1215. 'Homestead of the family or followers of a man called *Badela', or 'homestead at the place associated with *Badela'. OE pers. name + *-inga-* or *-ing* + *hām*. Alternatively the second element may be OE *hamm* 'land in a river-bend'.

Ballinglen (*Baile an Ghleanna*) Wicklow. 'Townland of the valley'.

Ballingurteen (*Baile an Ghoirtín*) Cork. 'Townland of the little tilled field'.

Ballinhassig (*Béal Átha an Cheasaigh*) Cork. 'Ford-mouth of the wicker causeway'.

Ballinkelleen (*Baile an Chillín*) Carlow. 'Townland of the little church'.

Ballinleeny (*Baile an Laighnigh*) Limerick. 'Townland of the Laighneach'.

Ballinlough (*Baile an Locha*) Meath, Roscommon. 'Townland of the lake'.

Ballinlug (*Baile an Loig*) Galway. 'Townland of the hollow'.

Ballinluska (*Baile an Loiscthe*) Cork. 'Townland of the burnt ground'.

Ballinran (*Baile an Raithin*) Down. *Ballynynranny* 1540. 'Townland of the ferns'.

Ballinridderra (*Baile an Ridire*) Westmeath. 'Townland of the knight'.

Ballinrobe (*Baile an Róba*) Mayo. 'Town of the (river) Róba'.

Ballinskelligs (*Baile an Sceilg*) Kerry. 'Townland of the rocks'.

Ballinspittle (*Béal Átha an Spidéil*) Cork. 'Mouth of the fort of the hospital'.

Ballintaggart (*Baile an tSagairt*) Armagh, Down, Kerry. 'Townland of the priest'.

Ballintannig (*Baile an tSeanaigh*) Cork. 'Townland of the fox'.

Ballinteean (*Baile a' tSiodháin*) Mayo. 'Townland of the fairy hill'.

Ballinteer (*Baile an tSaoir*) Derry, Dublin. 'Townland of the craftsman'.

Ballintemple (*Baile an Teampaill*) Cork. 'Townland of the church'.

Ballintlieve (*Baile an tSléibhe*) Down, Meath. 'Townland of the mountain'.

Ballintober (*Baile an Tobair*) Mayo, Roscommon. 'Townland of the well'.

Ballintogher (*Baile an Tóchair*) Sligo. 'Townland of the causeway'.

Ballintoy (*Baile an Tuaighe*) Antrim. *Ballenatoy* 1603. Possibly 'townland of the ruler of the tuath'. A *tuath* was a petty Irish kingdom.

Ballintra (*Baile an tSratha*) Donegal. *Baile an tSrátha* 1934. 'Townland of the river-meadow'.

Ballintrillick (*Béal Átha an Trí Liag*) Sligo. 'Ford-mouth of the three flagstones'.

Ballintubbert (*Baile na Tiobrad*) Laois. 'Townland of the well'.

Ballinturly (*Baile an Turlaigh*) Roscommon. 'Townland of the fen'.

Ballinunty (*Baile an Fhantaigh*) Tipperary. 'Fant's townland'.

Ballinure (*Baile an Iúir*) Galway, Tipperary. 'Townland of the yew'.

Ballinvana (*Baile an Bhána*) Limerick. 'Townland of the green field'.

Ballinvinny (*Baile an Mhuine*) Cork. 'Townland of the thicket'.

Ballinvonear (*Baile an Mhóinéir*) Cork. 'Townland of the meadow'.

Ballinwully (*Baile an Mhullaigh*) Roscommon. 'Townland of the summit'.

Ballisk (*Baile Uisce*) Down. 'Townland of the water'.

Ballitore (*Béal Átha an Tuair*) Kildare. 'Ford-mouth of the bleach green'.

Ballivor (*Baile Íomhair*) Meath. 'Íomhar's townland'.

Ballmacoda (*Baile Mhac Óda*) Cork. 'Mac Códa's townland'.

Ballnaclogh (*Baile na Cloiche*) Tipperary. 'Townland of the stone'.

Balloo (*Baile Aodha*) Down. *Ballow* 1605. 'Aodh's townland'.

Balloughmore (*Bealach Mór*) Laois. 'Big pass'.

Balloughter (*Baile Uachtair*) Wexford. 'Upper townland'.

Ballougry (*Baile Dhúdhoire*) Derry. *Ballidowgry* 1637. 'Townland of the black oak wood'.

Ballsmill (*Baile na gCléireach*) Armagh. *Ballinaglera* 1838. 'Ball's mill'. Thomas *Ball* was granted land here in the 17th cent. The Irish name means 'townland of the clerics'.

Ballyaghlis (*Baile na hEachlaisce*) Down. *Ballyhaghliske* 1625. 'Townland of the horse enclosure'.

Ballyagran (*Béal Átha Grean*) Limerick. 'Ford-mouth of the gravel'.

Ballyallaght (*Baile Uí Allachta*) Antrim. 'Townland of Ó Allacht'.

Ballyallinan (*Baile Uí Áilleanáin*) Limerick. 'Townland of Ó hÁilleanáin'.

Ballyalton (*Baile Altúin*) Down. *Ballyawlton* 1547. 'Alton's townland'.

Ballyandreen (*Baile Aindrín*) Cork. 'Aindrín's townland'.

Ballyanne (*Baile Anna*) Wexford. 'Anna's townland'.

Ballyardle (*Baile Ardghail*) Down. *Bally Ardell* 1661. 'Ardel's townland'.

Ballybay (*Béal Átha Beithe*) Monaghan. *Balloghnebegh* 1591. 'Ford-mouth of the birch-tree'.

Ballybeen (*Baile Bín*) Down. *Ballibeine* 1605. 'Bean's townland'.

Ballybeg (*Baile Beag*) Tipperary. 'Small townland'.

Ballyboden (*Baile Baodáin*) Dublin. 'Baodán's townland'.

Ballybofey (*Bealach Féich*) Donegal. *Srath Bó Diaich* 1548. 'Fiach's pass'.

Ballybogey (*Baile an Bhogaigh*) Antrim. *Ballyboggy* 1669. 'Townland of the swamp'.

Ballyboghal (*Baile Bachaille*) Dublin. 'Townland of the crozier'.

Ballybought (*Baile Bocht*) Antrim. 'Poor townland'.

Ballyboy (*Baile Átha Buí*) Offaly. 'Townland of the yellow ford'.

Ballyboyland (*Baile Uí Bhaolláin*) Antrim. *Bellebolan c.*1659. 'Townland of Ó Baollán'.

Ballybrack (*Baile Breac*) Dublin, Kerry, Tyrone. 'Speckled townland'.

Ballybrittas (*Baile Briotáis*) Laois. 'Townland of the wooden palisade'.

Ballybrood (*Baile Bhrúid*) Limerick. 'Townland of the ashes'.

Ballybrophy (*Baile Uí Bhróithe*) Laois. 'Townland of Ó Bróithe'.

Ballybunion (*Baile an Bhuinneánaigh*) Kerry. 'Buinnéan's townland'.

Ballycahill (*Bealach Achaille*) Tipperary. 'Pass of Achall'.

Ballycallan (*Baile Uí Challáin*) Kilkenny. 'Townland of Ó Calláin'.

Ballycanew (*Baile Uí Chonnmhaí*) Wexford. 'Townland of Ó Connmhaí'.

Ballycarnahan (*Baile Uí Chearnacháin*) Kerry. 'Ó Cearnacháin's townland'.

Ballycarny (*Baile Uí Chearnaigh*) Wexford. 'Townland of Ó Cearnaigh'.

Ballycarra (*Baile na Cora*) Mayo. 'Townland of the weir'.

Ballycarry (*Baile Cora*) Antrim. *Ballycarry* 1669. 'Townland of the weir'.

Ballycashin (*Baile Uí Chaisín*) Waterford. 'Ó Caisín's townland'.

Ballycassidy (*Baile Uí Chaiside*) Fermanagh. *Ballicashedy* 1659. 'Ó Caiside's townland'.

Ballycastle (*Baile an Chaisil*) Mayo. 'Townland of the stone fort'.

Ballycastle (*Baile Chaisleáin*) Antrim. *Baile Caislein* 1565. 'Townland of the castle'.

Ballyclare (*Bealach Cláir*) Antrim. *Balleclare* 1620. 'Pass of the plain'.

Ballyclerahan (*Baile Uí Chléireacháin*) Tipperary. 'Townland of Ó Cléireacháin'.

Ballyclery (*Baile Uí Chléirigh*) Galway. 'Townland of Ó Cléirigh'.

Ballyclogh (*Baile Cloch*) Cork. 'Townland of the stones'.

Ballycommon (*Baile Uí Chomáin*) Offaly, Tipperary. 'Townland of Ó Comán'.

Ballyconneely (*Baile Conaola*) Galway. 'Townland of Conaola'.

Ballyconnell (*Béal Átha Conaill*) Cavan. *Beallaconnell* 1630. 'Ford-mouth of Conall'.

Ballyconree (*Baile Con Raoi*) Galway. 'Townland of Cú Ruí'.

Ballycooge (*Baile Chuag*) Wicklow. 'Townland of the cuckoo'.

Ballycopeland (*Baile Chóplainn*) Down. *Ballicoppland* 1605. 'Copeland's townland'.

Ballycorick (*Béal Átha Chomhraic*) Clare. 'Ford-mouth of the confluence'.

Ballycotton (*Baile Choitín*) Cork. 'Coitín's townland'.

Ballycraddock (*Baile Chreadóig*) Waterford. 'Townland of the clay'.

Ballycrissane (*Baile Crosáin*) Galway. 'Townland of the cross'.

Ballycroghan (*Baile Cruacháin*) Down. 'Townland of the little rick'.

Ballycrossaun (*Baile Crosáin*) Galway. 'Townland of the cross'.

Ballycroy (*Baile Chruaiche*) Mayo. 'Townland of the rick'.

Ballycullane (*Baile Uí Choileáin*) Wexford. 'Ó Coileáin's townland'.

Ballyculter (*Baile Uí Choltair*) Down. *Balinculter* 1183. 'Ó Coltair's townland'.

Ballycumber (*Béal Átha Chomair*) Offaly. 'Ford-mouth of the confluence'.

Ballydaheen (*Baile Dáithín*) Cork. 'Dáthín's townland'.

Ballydangan (*Baile Daighean*) Roscommon. 'Townland of the stronghold'.

Ballydavid (*Baile Dháibhí*) Galway. 'Dóibhí's townland'.

Ballydavid (*Baile Dáibhí*) Kerry. 'Dáith's townland'.

Ballydehob (*Béal an dá Chab*) Cork. 'Mouth of the two openings'.

Ballydesmond (*Baile Deasumhan*) Cork. 'Town of south Munster'.

Ballydonegan (*Baile Uí Dhonnagáin*) Cork. 'Ó Donnagáin's townland'.

Ballydonohoe (*Baile Uí Dhonnchadha*) Kerry. 'Ó Donnchadha's townland'.

Ballydooley (*Baile Uí Dhúlaoich*) Roscommon. 'Townland of Ó Dúlaoich'.

Ballyduff (*An Baile Dubh*) Kerry, Waterford. 'The black townland'.

Ballydugan (*Baile Uí Dhúgáin*) Down. *Boile Í Dhubhagán* 1646. 'Ó Dúgáin's townland'.

Ballyeaston (*Baile Uistín*) Antrim. *Austin's town* 1306. 'Austin's townland'.

Ballyeighter (*Baile Íochtair*) Clare. 'Lower townland'.

Ballyengland (*Baile an Aingleontaigh*) Limerick. 'Townland of the Aingleontach'.

Ballyfarna (*Bealach Fearna*) Mayo. 'Pass of the alder'.

Ballyfarnon (*Béal Átha Fearnáin*) Roscommon. 'Ford-mouth of the alders'.

Ballyfeard (*Baile Feá Aird*) Cork. 'Townland of the high wood'.

Ballyfermot (*Baile Formaid*) Dublin. 'Townland of Formad'.

Ballyferriter (*Baile an Fheirtéaraigh*) Kerry. 'Ferriter's townland'.

Ballyfin (*An Baile Fionn*) Laois. 'The white townland'.

Ballyforan (*Béal Átha Feorainne*) Roscommon. 'Ford-mouth of the brink'.

Ballyfore (*Baile Fuar*) Offaly. 'Cold town'.

Ballyfoyle (*Baile an Phoill*) Kilkenny. 'Townland of the pool'.

Ballygalley (*Baile Geithligh*) Antrim. *Ballegelly* 1635. Possibly 'Geithleach's townland'.

Ballygar (*Béal Átha Ghártha*) Galway. 'Ford-mouth of the garden'.

Ballygarrett (*Baile Ghearóid*) Wexford. 'Gearóid's townland'.

Ballygarvan (*Baile Garbháin*) Cork. 'Garbhán's townland'.

Ballygawley (*Baile Uí Dhálaigh*) Sligo, Tyrone. 'Ó Dálaigh's townland'.

Ballyginniff (*Baile Gainimh*) Antrim. 'Sandy townland'.

Ballyglass (*Béal Átha Glas*) Westmeath. 'Ford-mouth of the streams'.

Ballyglass (*An Baile Glas*) Mayo. 'The grey townland'.

Ballyglunin (*Béal Átha Glúinín*) Galway. 'Ford-mouth of the little bend'.

Ballygomartin (*Baile Gharraí Mhairtín*) Antrim. 'Homestead of Martin's garden'.

Ballygorey (*Baile Guaire*) Kilkenny. 'Guaire's townland'.

Ballygormani (*Baile Uí Ghormáin*) Donegal. *Balligorman c*.1660. 'Ó Gormáin's townland'.

Ballygowan (*Baile Mhic Gabhann*) Down. *Balle-McGowen* 1623. 'Mac Gabhann's townland'.

Ballygrainey (*Baile na Gréine*) Down. *Ballinegrene* 1630. 'Townland of the sun'.

Ballygub (*Baile Gob*) Kilkenny. 'Townland of the snout'.

Ballyhack (*Baile Chac*) Wexford. 'Townland of excrement'.

Ballyhackamore (*Baile Hacamar*) Down. *Ballcakamer* 1620. 'Townland of the slob land'.

Ballyhaght (*Baile an Chéachta*) Limerick. 'Townland of the plough'.

Ballyhahill (*Baile dhá Thuile*) Limerick. 'Townland of the two floods'.

Ballyhalbert (*Baile Thalbóid*) Down. *(Ecclesia de) Talbetona* 1306. 'Talbot's townland'.

Ballyhale (*Baile Héil*) Kilkenny. 'Howel's townland'.

Ballyhar (*Baile Uí Aichir*) Kerry. 'Ó hAichir's townland'.

Ballyhean (*Béal Átha hÉin*) Mayo. 'Ford-mouth of the bird'.

Ballyheelan (*Bealach an Chaoláin*) Cavan. *Ballagheelan* 1734. 'Pass of the marshy stream'.

Ballyheerin (*Baile Uí Shírín*) Donegal. *Ballyherrinmore, Ballyherrinbegg c*.1660. 'Ó Sírín's townland'.

Ballyheige (*Baile Uí Thaidhg*) Kerry. 'Ó Taidhg's townland'.

Ballyhisky (*Bealach Uisce*) Tyrone. 'Pass of water'.

Ballyholme (*Baile Hóm*) Down. *Ballehum* 1603. 'Hóm's townland'.

Ballyhooly (*Baile Átha hÚlla*) Cork. 'Pass of the ford of the apple-trees'.

Ballyhornan (*Baile Uí Chornáin*) Down. *Ballyhornan* 1636. 'Ó Cornáin's townland'.

Ballyhoura (*Bealach Abhradh*) Cork, Limerick. 'Pass of Feabhra'.

Ballyhugh (*Bealach Aodha*) Cavan. *Bellaghea* 1610. 'Aodh's pass'.

Ballyjamesduff (*Baile Shéamais Dhuibh*) Cavan. *Bally James Doough or Black James' town c*.1744. 'Townland of James Duff'. *James Duff*, Earl of Fife, was granted land here in the early 17th cent.

Ballykean (*Baile Uí Chéin*) Offaly. 'Ó Céin's townland'.

Ballykeel (*An Baile Caol*) Down. *Ballikeeleloghaghery* 1632. 'The narrow townland'.

Ballykeeran (*Bealach Caorthainn*) Westmeath. 'Pass of the rowan-tree'.

Ballykelly (*Baile Uí Cheallaigh*) Derry. *Ballykellye* 1613. 'Ó Ceallaigh's townland'.

Ballykilbeg (*Baile na gCeall Beag*) Down. *Ballenagallbee* 1512. 'Townland of the little churches'.

Ballykillare (*Baile Cille Láir*) Down. 'Townland of the central church'.

Ballykinler (*Baile Coinnleora*) Down. *Ballicanlor* 1542. 'Townland of the candlestick'. Lands here were granted in *c.*1200 to Christchurch Cathedral, Dublin, for the upkeep of a perpetual light before the crucifix there.

Ballykinsella (*Baile an Chinsealaigh*) Waterford. 'Ó Cinsealaigh's townland'.

Ballyknockan (*Buaile an Chnocáin*) Wicklow. 'Milking place of the hillock'.

Ballylanders (*Baile an Londraigh*) Limerick. 'de Londra's townland'.

Ballylaneen (*Baile Uí Laithnín*) Waterford. 'Ó Laithnín's townland'.

Ballylar (*Baile Láir*) Donegal. 'Townland of the threshing floor'.

Ballyleny (*Baile Léana*) Armagh. *Ballylaney* 1661. 'Townland of the wet meadow'.

Ballylesson (*Baile na Leasán*) Down. *Ballenelassan*. 'Townland of the little forts'.

Ballylickey (*Béal Átha Leice*) Cork. 'Ford-mouth of the flagstone'.

Ballyliffin (*Baile Lifín*) Donegal. *Ballylaffin* 1608. 'Townland of the halfpenny'. The reference is perhaps to a land division.

Ballylinan (*Baile Uí Laigheanáin*) Laois. 'Ó Laigheanán's townland'.

Ballylintagh (*An Baile Linnteach*) Derry. *Bellylintach* 1663. 'The townland of pools'.

Ballylongford (*Béal Átha Longfoirt*) Kerry. 'Ford-mouth of the fortress'.

Ballylooby (*Béal Átha Lúbaigh*) Tipperary. 'Ford-mouth of the winding (river)'.

Ballyloughbeg (*Baile an Locha Beag*) Antrim. 'Townland of the small lake'.

Ballylumford (*Baile an Longfoirt*) Antrim. *Ballylemford* 1669. 'Townland of the fortress'.

Ballymacarbry (*Baile Mhac Cairbre*) Waterford. 'Mac Carbre's townland'.

Ballymacarret (*Baile Mhic Gearóid*) Down. *Bally McCarritt* 1623. 'Mac Gearóid's townland'.

Ballymacart (*Baile Mhac Airt*) Waterford. 'Mac Art's townland'.

Ballymacaw (*Baile Mhac Dháith*) Waterford. 'Mac Dáith's townland'.

Ballymacelligott (*Baile Mhic Eileagóid*) Kerry. 'Mac Eileagód's townland'.

Ballymackey (*Baile Uí Mhacaí*) Tipperary. 'Ó Macaí's townland'.

Ballymaconnelly (*Baile Mhic Conaíle*) Antrim. *Ballymaconnally* 1635. 'Mac Conaíle's townland'.

Ballymacurly (*Baile Mhic Thorlaigh*) Roscommon. 'Mac Thorlaigh's townland'.

Ballymacward (*Baile Mhic an Bhaird*) Galway. 'Townland of Mac an Bhard'.

Ballymadog (*Baile Mhadóg*) Cork. 'Madóg's townland'.

Ballymagan (*Baile MhicCionaoith*) Donegal. *Ballimcganny c.*1635. 'Mac Cionaoith's townland'.

Ballymagarry (*Baile mo Gharraí*) Antrim. 'Townland of my garden'.

Ballymagorry (*Baile Mhic Gofraidh*) Tyrone. *Ballemagorie* 1616. 'Mac Gofraidh's townland'.

Ballymagrorty (*Baile Mhic Robhartaigh*) Derry. *Ballym'roartie* 1604. 'Mac Robhartaigh's townland'.

Ballymaguigan (*Baile Mhic Guaigín*) Derry. *Ballymccuggin c.*1659. 'Mac Guaigín's townland'.

Ballymahon (*Baile Uí Mhatháin*) Longford. 'Ó Matháin's townland'.

Ballymakeery (*Baile Mhic Íre*) Cork. 'Mac Íre's townland'.

Ballymartin (*Baile Mhic Giolla Mhártain*)
Down. *Ballymicgyll Mertyn* 1552. 'Townland of
Mac Giolla Mhártain'.

Ballymartle (*Baile Mhairtéal*) Cork.
'Mairtéal's townland'.

Ballymascanlan (*Baile Mhic Scanláin*)
Louth. 'Mac Scanlán's townland'.

Ballymena (*An Baile Meánach*) Antrim.
Ballymeanagh 1626. 'The middle townland'.

Ballymoe (*Béal Átha Mó*) Galway,
Roscommon. 'Ford-mouth of Mogh'.

Ballymoney (*Baile Monaidh*) Antrim. *Bali
Monaid* 1412. 'Townland of the bog'.

Ballymoon (*Baile Móin*) Carlow. 'Townland
of the bog'.

Ballymore (*Baile Mór*) Cork, Donegal,
Kildare, Westmeath. 'Big townland'.

Ballymorris (*Baile Mhuiris*) Waterford.
'Muiris's townland'.

Ballymote (*Baile an Mhóta*) Sligo.
'Townland of the castle mound'.

Ballymoyle (*Baile Maol*) Wicklow. 'Bald
townland'.

Ballymullen (*Baile an Mhuilinn*) Kerry.
'Townland of the mill'.

Ballymurn (*Baile Uí Mhurúin*) Wexford.
'Ó Murúin's townland'.

Ballymurphy (*Baile Uí Mhurchú*) Carlow.
'Ó Murchú's townland'.

Ballymurragh (*Baile Mhurchadha*)
Limerick. 'Murchadh's townland'.

Ballymurray (*Baile Uí Mhuirigh*)
Roscommon. 'Ó Muireadhaigh's townland'.

Ballynabola (*Baile na Buaile*) Wexford.
'Townland of the milking place'.

Ballynabrackey (*Buaile na Bréamhaí*)
Meath. 'Milking place of the wolf plain'.

Ballynabragget (*Baile na Brád*) Down.
Ballynebrade 1612. 'Townland of the valley'.

Ballynacally (*Baile na Caillí*) Clare.
'Townland of the hag'.

Ballynacargy (*Baile na Carraige*)
Westmeath. 'Townland of the rock'.

Ballynacarriga (*Béal na Carraige*) Cork.
'Mouth of the rock'.

Ballynacarrigy (*Baile na Carraige*)
Westmeath. 'Townland of the rock'.

Ballynacole (*Baile Niocóil*) Cork. 'Niocól's
townland'.

Ballynacorr (*Baile na Cora*) Armagh.
Ballinecorrowe 1610. 'Townland of the weir'.

Ballynacorra (*Baile na Cora*) Cork.
'Townland of the weir'.

Ballynacourty (*Baile na Cúirte*) Waterford.
'Townland of the mansion'.

Ballynadrumny (*Baile na Droimní*) Kildare.
'Townland of the ridge'.

Ballynafa (*Baile na Faiche*) Kildare.
'Townland of the green'.

Ballynafeigh (*Baile na Faiche*) Down.
Ballinefeigh 1605. 'Townland of the green'.

Ballynafey (*Baile na Faiche*) Antrim.
'Townland of the green'.

Ballynafid (*Baile na Feide*) Westmeath.
'Townland of the runnel'.

Ballynafie (*Baile na Faiche*) Antrim.
'Townland of the green'.

Ballynagarrick (*Baile na gCarraig*) Down.
Ballynegaricke 1611. 'Townland of the rocks'.

Ballynagaul (*Baile na nGall*) Waterford.
'Townland of the stones'.

Ballynageeragh (*Baile na gCaorach*) Down.
'Townland of the sheep'.

Ballynagore (*Béal Átha na nGabhar*)
Westmeath. 'Ford-mouth of the goats'.

Ballynagree (*Baile na Graí*) Cork.
'Townland of the stud'.

Ballynaguilkee (*Baile na Giolcaí*)
Waterford. 'Townland of the reeds'.

Ballynahatinna (*Baile na hAitinn*) Galway.
'Townland of the gorse'.

Ballynahatten (*Baile na hAitinn*) Down,
Louth. 'Townland of the gorse'.

Ballynahinch (*Baile na hInse*) Down,
Galway, Tipperary. 'Townland of the river-
meadow'.

Ballynahow (*Baile na hAbha*) Kerry. 'Townland of the river'.

Ballynahowen (*Buaile na hAbhann*) Westmeath. 'Milking place of the river'.

Ballynahown (*Buaile na hAbhann*) Galway, Westmeath. 'Milking place of the river'.

Ballynakill (*Baile na Cille*) Carlow. 'Townland of the church'.

Ballynamallaght (*Béal Átha na Mallacht*) Tyrone. 'Ford-mouth of the curses'.

Ballynameen (*Baile na Míne*) Derry. 'Townland of the smooth place'.

Ballynamona (*Baile na Móna*) Cork. 'Townland of the bog'.

Ballynamult (*Béal na Molt*) Waterford. 'Estuary of the wether'.

Ballynanty (*Baile Uí Neachtain*) Limerick. 'Ó Neachtain's townland'.

Ballynascreen (*Baile na Scrín*) Derry. 'Townland of the shrine'.

Ballynashannagh (*Baile na Seanach*) Donegal. *Ballysheany* 1612. Possibly 'townland of the fox'.

Ballynaskreena (*Bbaile na Scríne*) Kerry. 'Townland of the shrine'.

Ballyneaner (*Baile an Aonfhir*) Tyrone. *Baile an Aoin-fhir c.*1675. 'Townland of the lone man'.

Ballynease (*Baile Naosa*) Derry. *Bellinees* 1654. 'Naois's townland'.

Ballyneety (*Baile an Fhaoitigh*) Limerick. 'de Faoite's townland'.

Ballyneill (*Baile Uí Néill*) Tipperary. 'Ó Néill's townland'.

Ballyness (*Baile an Easa*) Derry. 'Townland of the waterfall'.

Ballynoe (*An Baile Nua*) Cork. 'The new townland'.

Ballynure (*Baile an Iúir*) Antrim. *Ballinower* 1605. 'Townland of the yew-tree'.

Ballyoran (*Baile Uaráin*) Armagh. *B:uoran* 1609. 'Townland of the spring'.

Ballyorgan (*Baile Uí Argáin*) Limerick. 'Ó hArgáin's townland'.

Ballyote (*Baile Fhóid*) Westmeath. 'Townland of the sod'.

Ballypatrick (*Baile Phádraig*) Tipperary. 'Pádraig's townland'.

Ballyphehane (*Baile Féitheán*) Cork. 'Townland of the osiers'.

Ballyphilip (*Baile Philib*) Waterford. 'Philip's townland'.

Ballyporeen (*Béal Átha Póirín*) Tipperary. 'Ford-mouth of the round stones'.

Ballyquin (*Baile Uí Choinn*) Waterford. 'Ó Coinn's townland'.

Ballyquirk (*Baile Uí Chuirc*) Tipperary. 'Ó Cuirc's townland'.

Ballyragget (*Béal Átha Ragad*) Kerry. 'Ford-mouth of the churl'.

Ballyrashane (*Baile Ráth Singean*) Derry. *Singayton al. Rathsyne* 1542. 'Townland of St John's fort'.

Ballyrawer (*Baile Ramhar*) Down. 'Fertile townland'.

Ballyreagh (*An Baile Riabhach*) Tyrone. 'The grey townland'.

Ballyree (*Baile an Fhraoigh*) Down. 'Townland of the heather'.

Ballyrobert (*Baile Roibeaird*) Antrim. *Ballyrobert c.*1659. 'Robert's townland'.

Ballyronan (*Baile Uí Rónáin*) Derry. *Two Ballioronans* 1654. 'Ó Rónáin's townland'.

Ballyroney (*Baile Uí Ruanaí*) Down. *Ballyronowe* 1611. 'Ó Ruanaí's townland'.

Ballyroosky (*Baile Rusgaidh*) Donegal. 'Townland of the marsh'.

Ballysadare (*Baile Easa Dara*) Sligo. 'Townland of the waterfall of the oak'.

Ballysakeery (*Baile Easa Caoire*) Mayo. 'Townland of the waterfall of the berry'.

Ballysallagh (*Baile Sealbhach*) Down. *Ballyshallagh* 1614. 'Townland of herds'.

Ballysally (*Baile Uí Shalaigh*) Derry. *Ballyosallye* 1543. 'Ó Salaigh's townland'.

Ballyscullion (*Baile Uí Scoillín*) Derry. *Balle Oskullyn* 1397. 'Ó Scoillín's townland'.

Ballyshannon (*Béal Átha Seanaidh*) Donegal. *(for) Bhel Atha Senaigh* 1398. 'Ford-mouth of the slope'.

Ballyshrule (*Baile Sruthail*) Galway. 'Townland of the stream'.

Ballysillan (*Baile na Saileán*) Antrim. *Ballynysillan* 1604. 'Townland of the willow grove'.

Ballysimon (*Béal Átha Síomoin*) Limerick. 'Mouth of Síomon's ford'.

Ballyskeagh (*Baile na Sceiche*) Down. 'Townland of the thorn'.

Ballysteen (*Baile Stiabhna*) Limerick. 'Stiabhna's townland'.

Ballystrew (*Baile Sruth*) Down. 'Townland of the stream'.

Ballystrudder (*Baile Strudair*) Antrim. *Ballytredder* 1669. 'Strudar's townland'.

Ballytarsna (*Baile Trasna*) Roscommon, Tipperary. 'Townland across'.

Ballyvade (*Baile Bháid*) Westmeath. 'Townland of the boat'.

Ballyvaghan (*Baile Uí Bheacháin*) Clare. 'Ó Beacháin's townland'.

Ballyvaldon (*Baile Bhalduin*) Wexford. 'Baldwin's townland'.

Ballyvaltron (*Baile Bhaltairín*) Wicklow. 'Little Walter's townland'.

Ballyvangour (*Baile Bheanna Gabhar*) Carlow. 'Townland of the peaks of the goats'.

Ballyvary (*Béal Átha Bhearaigh*) Mayo. 'Ford-mouth of the heifer'.

Ballyvoge (*Baile Uí Bhuaigh*) Cork. 'Ó Buaigh's townland'.

Ballyvourney (*Baile Bhuirne*) Cork. 'Townland of the stony place'.

Ballyvoy (*Baile Bhuí*) Antrim. *Ballyvoy* c.1657. 'Yellow townland'.

Ballyvoyle (*Baile Uí Bhaoil*) Waterford. 'Ó Baoill's townland'.

Ballywalter (*Baile Bháltair*) Down. *Ballywalter* 1661. 'Walter's townland'.

Ballyward (*Baile Mhic an Bhaird*) Down. *Ballymc-Ewarde* 1611. 'Mac an Bhaird's townland'.

Ballywater (*Baile Uachtar*) Wexford. 'Upper townland'.

Ballywee (*Baile Uaimh*) Antrim. 'Townland of the cave'.

Ballywildrick (*Baile Ualraic*) Derry. *Ba:Wolrick* 1613. 'Ualrac's townland'.

Ballywilliam (*Baile Liam*) Wexford. 'Liam's townland'.

Balmoral Aber. *Bouchmorale* 1451. Gaelic *both* 'hut'; second element uncertain.

Balmoral (*Baile Mhoireil*) Antrim. The name was adopted from BALMORAL, Scotland, *see* previous name.

Balnakeil Highland. Gaelic *Baile na Cill* 'village of the church'.

Balnamore (*Béal an Átha Móir*) Antrim. 'Mouth of the big ford'.

Balne N. Yorks. *Balne* 12th cent. Probably from Latin *balneum* 'a bathing place'.

Balrath (*Baile na Rátha*) Meath. 'Townland of the fort'.

Balroe (*Baile Rua*) Westmeath. 'Red townland'.

Balrothery (*Baile an Ridire*) Dublin. 'Townland of the knight'.

Balsall Common Solhll. *Beleshale* 1185. 'Nook of land of a man called *Bæll(i)'. OE pers. name + *halh*. **Temple Balsall** is so called because it belonged to the Knights Templar from 1185.

Balscote Oxon. *Berescote* [*sic*] 1086 (DB), *Belescot* c.1190. 'Cottage(s) of a man called Bæll(i)'. OE pers. name + *cot*.

Balsham Cambs. *Bellesham* 974, *Belesham* 1086 (DB). 'Homestead or village of a man called Bæll(i)'. OE pers. name + *hām*.

Balterley Staffs. *Baltrytheleag* 1002, *Baltredelege* 1086 (DB). 'Woodland clearing of a woman called *Baldthrýth'. OE pers. name + *lēah*.

Baltimore (*Dún na Séad*) Cork. 'Townland of the big house'. The English name represents

Irish *Baile na Tighe Mór*, while the present Irish name means 'fort of the jewels'.

Baltinglass (*Bealach Conglais*) Dublin. 'Pass of Conglas'.

Baltonsborough Somerset. *Balteresberghe* 744, *Baltunesberge* 1086 (DB). 'Hill or mound of a man called Bealdhūn'. OE pers. name + *beorg*.

Bamber Bridge Lancs. *Bymbrig* in an undated medieval document. Probably 'tree-trunk bridge'. OE *bēam* + *brycg*.

Bamburgh Northum. *Bebbanburge* c.710–20. 'Stronghold of a queen called Bebbe'. OE pers. name + *burh*.

Bamford Derbys. *Banford* 1086 (DB). 'Tree-trunk ford'. OE *bēam* + *ford*.

Bampton, usually 'farmstead made of beams or by a tree', OE *bēam* + *tūn*: **Bampton** Cumbria. *Bampton* c.1160. **Bampton** Oxon. *Bemtun* 1069, *Bentone* 1086 (DB). **Bampton, Little** Cumbria. *Parua Bampton* 1227.
  However the following has a different origin: **Bampton** Devon. *Badentone* 1086 (DB). 'Farmstead of the dwellers by the pool'. OE *bæth* + *hǣme* + *tūn*.

Banada (*Muine na Fede*) Sligo. 'Thicket of the whistle'.

Banagher (*Beannchar*) Derry, Offaly. 'Place of peaks'.

Banbridge Down. *Bann Br.* 1743. 'Bridge over the (river) Bann'. The equivalent Irish name is *Droichead na Banna*.

Banbury Oxon. *Banesberie* 1086 (DB). 'Stronghold of a man called *Ban(n)a*. OE pers. name + *burh* (dative *byrig*).

Banchory Aber. 'Horn-cast'. Gaelic *beannchar*.

Bandon (*Droichead na Bandan*) Cork. The Irish name of the town means 'bridge on the Bandon', while the river-name itself may mean 'goddess'.

Banemore (*An Bán Mór*) Kerry. 'The large pasture-land'.

Banff Aber. *Banb* c.1150. '(Place on the) River Banff'. The river-name means 'piglet' (Gaelic *banbh*), as a nickname for the present River Deveron, seen as 'rooting' its way to the coast.

Bangor (*Beannchar*) Down. *Bennchuir* 555. 'Place of points'. The name probably refers to a pointed wattle enclosure around the original monastic settlement.

Bangor Gwyd. *Benchoer* 634. 'Wattled fence'. Welsh *bangor*. The reference is probably to the wattled fence that enclosed the monastery founded in 525.

Bangor Erris (*Beannchar Iorrais*) Mayo. 'Peaked hill of Iorrais'.

Bangor Is-coed. *See* BANGOR ON DEE.

Bangor on Dee (*Bangor Is-coed*) Wrex. *Bancor* 8th cent., *Bangor* 1277, *Bangor monachorum* 1607. '(Place of the people of) Bangor on the River Dee'. The monastery is said to have been founded by St Deiniol, founder of BANGOR, with the river-name ('goddess') added for distinction. The Welsh name means 'Bangor below the wood'. The third form of the name is Latin for 'Bangor of the monks'.

Banham Norfolk. *Benham* 1086 (DB). 'Homestead or enclosure where beans are grown'. OE *bēan* + *hām* or *hamm*.

Bann (*An Bhanna*) (river) Antrim, Armagh, Derry, Down. *Banda* c.800. 'The goddess'.

Banna (*Beanna*) Kerry. 'Peaks'.

Bannau Brycheiniog. *See* BRECON BEACONS.

Banningham Norfolk. *Banincham* 1086 (DB). 'Homestead or village of the family or followers of a man called *Ban(n)a*. OE pers. name + *-inga-* + *hām*.

Bannockburn Stir. *Bannockburn* 1314, *Bannokburne* 1654. '(Place on the) Bannock Burn'. OE *burna*. The Celtic river-name means 'peaked, horned', referring to the hill from which the stream flows.

Bannow Bay (*Cuan Bhanú*) Wexford. 'Bay of the sucking pig'.

Banogue (*An Bhánog*) Limerick. 'The small green plot'.

Bansha (*An Bháinseach*) Tipperary. 'The green'.

Banstead Surrey. *Benestede* 1086 (DB). 'Place where beans are grown'. OE *bēan* + *stede*.

Banteer (*Bántír*) Cork. 'White land'.

Bantry (*Beanntraí*) Cork. '(District of) Beanntraí'. The district name probably means 'Beann's people', from a tribal name.

Banwell N. Som. *Bananwylle* 904, *Banwelle* 1086 (DB). 'Spring or stream of the murderer, or containing water thought to be poisonous'. OE *bana* 'killer' + *wella*.

Bapchild Kent. *Baccancelde* 696–716. 'Spring of a man called Bacca'. OE pers. name + *celde*.

Barbon Cumbria. *Berebrune* 1086 (DB). 'Stream frequented by bears or beavers'. OE *bere* + *burna* or OScand. *bjórr* + *brunnr*.

Barby Northants. *Berchebi* 1086 (DB). 'Farmstead or village on the hill(s)'. OScand. *berg* + *bý*.

Barcheston Warwicks. *Berricestone* 1086 (DB), 'Farmstead of a man called Beaduríc or *Bedríc*'. OE pers. name + *tūn*.

Barcombe E. Sussex. *Bercham* [sic] 1086 (DB), *Berecampe* 12th cent. 'Enclosed land used for barley'. OE *bere* + *camp*.

Bard Head Shet. (Bressay). 'Headland of the extremity'. OScand. *barth*. Bard Head is the southernmost point of Bressay.

Barden N. Yorks. near Leyburn. *Bernedan* 1086 (DB). Probably 'valley where barley is grown'. OE *bere* 'barley', *beren* 'growing with barley' + *denu*.

Bardfield, Great, Bardfield, Little & Bardfield Saling Essex. *Byrdefelda* 1086 (DB), *Berdeford Saling* 13th cent. 'Open land by a bank or border'. OE *byrde* + *feld*. Affix from the neighbouring parish of GREAT SALING.

Bardney Lincs. *Beardaneu* 731, *Bardenai* 1086 (DB). 'Island, or dry ground in marsh, of a man called *Bearda*'. OE pers. name (genitive -*n*) + *ēg*.

Bardsea Cumbria. *Berretseige* 1086 (DB). 'Island of a man called Beornrǣd'. OE pers. name + *ēg* or OScand. *ey*.

Bardsey Leeds. *Berdesei* 1086 (DB). Probably 'island (of higher land) of a man called Beornrǣd'. OE pers. name + *ēg*.

Bardsey Island (*Ynys Enlli*) Gwyd. 'Bardr's island'. OScand. pers. name + *ey*. The Welsh name has been popularly associated with a legendary giant, *Benlli*, but is more likely to be 'island of currents' (Welsh *ynys* + *an* + *lli*).

Bardsley Oldham. *Berdesley* 1422. 'Woodland clearing of a man called Beornrǣd'. OE pers. name + *lēah*.

Bardwell Suffolk. *Berdeuuella* 1086 (DB). Probably 'spring or stream of a man called *Bearda*'. OE pers. name + *wella*.

Barford, usually 'barley ford', i.e. 'ford used at harvest time', OE *bere* + *ford*; examples include: **Barford** Norfolk. *Bereforda* 1086 (DB). **Barford** Warwicks. *Bereforde* 1086 (DB). **Barford, Great** Beds. *Bereforde* 1086 (DB). **Barford St Martin** Wilts. *Bereford* 1086 (DB), *Berevord St Martin* 1304. Affix from the dedication of the church. **Barford St Michael** Oxon. *Bereford* 1086 (DB), *Bereford Sancti Michaelis* c.1250. Affix from the dedication of the church.

However the following has a different origin: **Barford, Little** Beds. *Bereforde* [sic] 1086 (DB), *Berkeford* 1202. 'Ford where birch-trees grow'. OE *beorc* + *ford*.

Barfrestone Kent. *Berfrestone* 1086 (DB). Probably 'farmstead of a man called Beorhtfrith or Beornfrith'. OE pers. name + *tūn*.

Bargoed Cphy. '(Place on the) River Bargoed'. The river-name means 'boundary' (Welsh *bargod*).

Barham Cambs. *Bercheham* 1086 (DB). 'Homestead or enclosure on a hill'. OE *beorg* + *hām* or *hamm*.

Barham Kent. *Bioraham* 799, *Berham* 1086 (DB). 'Homestead or village of a man called *Be(o)ra*'. OE pers. name + *hām*.

Barham Suffolk. *Bercham* 1086 (DB). Identical in origin with BARHAM (Cambs.).

Barholm Lincs. *Berc(a)ham* 1086 (DB). 'Homestead or enclosure on a hill'. OE *beorg* + *hām* or *hamm*.

Barkby Leics. *Barchebi* 1086 (DB). 'Farmstead or village of a man called Bọrkr or Barki'. OScand. pers. name + *bý*.

Barkestone Leics. *Barchestone* 1086 (DB). 'Farmstead of a man called Barkr or Bọrkr'. OScand. pers. name + OE *tūn*.

Barkham Wokhm. *Beorchamme* 952, *Bercheham* 1086 (DB). 'Enclosure or river-meadow where birch-trees grow'. OE *beorc* + *hamm*.

Barking, '(settlement of) the family or followers of a man called *Berica', OE pers. name + *-ingas*: **Barking** Gtr. London. *Berecingum* 731, *Berchinges* 1086 (DB). **Barking** Suffolk. *Berchinges* c.1050, *Berchingas* 1086 (DB). **Barking Tye** contains dialect *tye* 'a large common pasture'.

Barkisland Calder. *Barkesland* 1246. 'Cultivated land of a man called Barkr'. OScand. pers. name + *land*.

Barkston, 'farmstead of a man called Barkr or Bǫrkr', OScand. pers. name + OE *tūn*: **Barkston** Lincs. *Barchestune* 1086 (DB). **Barkston** N. Yorks. *Barcestune* c.1030, *Barchestun* 1086 (DB).

Barkway Herts. *Bercheuuei* 1086 (DB). 'Birch-tree way'. OE *beorc* + *weg*.

Barkwith, East & Barkwith, West Lincs. *Barcuurde* 1086 (DB). Possibly 'enclosure of a man called Barki'. OScand. pers. name + OE *worth*.

Barlaston Staffs. *Beorelfestun* 1002, *Bernulvestone* 1086 (DB). 'Farmstead of a man called Beornwulf'. OE pers. name + *tūn*.

Barlavington W. Sussex. *Berleventone* 1086 (DB). Probably 'estate associated with a man called Beornlāf'. OE pers. name + *-ing-* + *tūn*.

Barlborough Derbys. *Barleburh* c.1002, *Barleburg* 1086 (DB). Probably 'stronghold near the wood frequented by boars'. OE *bār* + *lēah* + *burh*.

Barlby N. Yorks. *Bardulbi* 1086 (DB). 'Farmstead or village of a man called *Beardwulf or Bardulf'. OE or OGerman pers. name + OScand. *bý*.

Barlestone Leics. *Beorelfestune* 1002–4, *Berulvestone* 1086 (DB). 'Farmstead or estate of a man called Beorwulf'. OE pers. name + *tūn*.

Barley Herts. *Beranlei* c.1050, *Berlai* 1086 (DB). Probably 'woodland clearing of a man called *Be(o)ra'. OE pers. name + *lēah*.

Barley Lancs. *Bayrlegh* 1324. 'Woodland clearing frequented by boars, or where barley is grown'. OE *bār* or *bere* + *lēah*.

Barling Essex. *Bærlingum* 998, *Berlinga* 1086 (DB). '(Settlement of) the family or followers of a man called *Bǣrla'. OE pers. name + *-ingas*.

Barlings Lincs. *Berlinge, Berlinga* 1086 (DB), *Berlinges* 1123. Identical in origin with the previous name.

Barlow Derbys. *Barleie* 1086 (DB). 'Woodland clearing frequented by boars, or where barley is grown'. OE *bār* or *bere* + *lēah*.

Barlow N. Yorks. *Bernlege* c.1030, *Berlai* 1086 (DB). 'Woodland clearing with a barn, or where barley is grown'. OE *bereærn* or *beren* + *lēah*.

Barmby, probably 'farmstead of the children, i.e. one held jointly by a number of heirs' from OScand. *barn* + *bý*; alternatively 'farmstead of a man called Barni or Bjarni' from OScand. pers. name + *bý*: **Barmby Moor** E. R. Yorks. *Barnebi* 1086 (DB), *Barneby in the More* 1371. Affix is OE *mōr* 'moor'. **Barmby on the Marsh** E. R. Yorks *Bærnabi* c.1050, *Barnebi* 1086 (DB). Affix is OE *mersc* 'marsh'.

Barmer Norfolk. *Benemara* [sic] 1086 (DB), *Beremere* 1202. 'Pool frequented by bears', or 'pool of a man called *Bera'. OE *bera* or pers. name + *mere*.

Barming, East & Barming, West Kent. *Bermelinge, Bermelie* 1086 (DB), *Estbarmeling* 1240, *West Barmlynge* 1308. Possibly 'dwellers at the outlying woodland clearing'. OE *bearm* 'edge' + *lēah* + *-ingas*.

Barmouth (*Y Bermo*) Gwyd. *Abermowth* 1410. 'Mouth of the River Mawddach'. Welsh *aber*. The river-name, originally *Mawdd*, probably derives from a pers. name. Its estuary was *Abermawdd*, giving the Welsh name (with *y*, 'the'). The English name has been influenced by *mouth*.

Barmston E. R. Yorks. *Benestone* 1086 (DB). 'Farmstead of a man called Beorn'. OE pers. name + *tūn*.

Barna (*Bearna*) Galway, Limerick, Offaly. 'Gap'.

Barnack Peterb. *Beornican* c.980, *Bernac* 1086 (DB). Probably '(place at) the oak-tree(s) of the warriors'. OE *beorn* + *āc*.

Barnacle Warwicks. *Bernhangre* 1086 (DB). 'Wooded slope by a barn'. OE *bereærn* + *hangra*.

Barnacullia (*Barr na Coille*) Dublin. 'Top of the wood'.

Barnaderg (*Bearna Dhearg*) Galway. 'Red gap'.

Barnageeha (*Bearna Gaoithe*) Limerick. 'Windy gap'.

Barnakillew (*Barr na Coille*) Mayo. 'Top of the wood'.

Barnakilly (*Barr na Coille*) Derry. 'Top of the wood'.

Barnard Castle Durham. *Castellum Bernardi* 1200. 'Castle of a baron called Bernard'. He was here in the 12th cent.

Barnardiston Suffolk. *Bernardeston* 1194. 'Farmstead of a man called Beornheard'. OE pers. name + *tūn*.

Barnatra (*Barr na Trá*) Mayo. 'Top of the strand'.

Barnby, probably 'farmstead of the children, i.e. one held jointly by a number of heirs', OScand. *barn* + *bý*; alternatively the first element of some names may be the OScand. pers. name *Barni* or *Bjarni*; examples include: **Barnby** Suffolk. *Barnebei* 1086 (DB). **Barnby Dun** Donc. *Barnebi* 1086 (DB), *Barneby super Don* 1285. Affix means 'on the River Don'. **Barnby, East & Barnby, West** N. Yorks. *Barnebi* 1086 (DB). **Barnby in the Willows** Notts. *Barnebi* 1086 (DB). Affix means 'among the willow-trees'. **Barnby Moor** Notts. *Barnebi* 1086 (DB). Affix means 'on the moor'.

Barnes Gtr. London. *Berne* 1086 (DB). '(Place by) the barn or barns'. OE *bere-ærn*.

Barnes (*Bearnas*) Tyrone. 'Gap'.

Barnesmore (*An Bearnas Mór*) Donegal. 'The great gap'.

Barnet, Chipping, Barnet, East & Barnet, Friern Gtr. London. *Barneto* c.1070, *Chepyng Barnet* 1321, *Est Barnet* c.1275, *Frerenbarnet* 1274. 'Land cleared by burning'. OE *bærnet*. Distinguishing affixes are OE *cīeping* 'market', *ēast* 'east' and ME *freren* 'of the brothers' (referring to early possession by the Knights of St John of Jerusalem). Chipping Barnet has also been known as *High Barnet* since the 17th cent.

Barnetby le Wold N. Lincs. *Bernedebi* 1086 (DB). 'Farmstead or village of a man called Beornnōth or *Beornede*'. OE pers. name + OScand. *bý*. Affix means 'on the wold(s)', referring to its situation at the northern edge of the Lincolnshire WOLDS.

Barney Norfolk. *Berlei* [*sic*] 1086 (DB), *Berneie* 1198. Possibly 'island, or dry ground in marsh, of a man called **Bera*'. OE pers. name (genitive -*n*) + *ēg*. Alternatively the first element may be OE *bere-ærn* 'barn' or *beren* 'growing with barley'.

Barnham Suffolk. *Byornham* c.1000, *Bernham* 1086 (DB). 'Warrior homestead', or 'homestead of a man called Beorn'. OE *beorn* or OE pers. name + *hām*.

Barnham W. Sussex. *Berneham* 1086 (DB). 'Homestead or enclosure of the warriors, or of a man called Beorna'. OE *beorn* or OE pers. name + *hām* or *hamm*.

Barnham Broom Norfolk. *Bernham* 1086 (DB). Identical in origin with BARNHAM (Suffolk). Affix is OE *brōm* 'broom'.

Barningham, 'homestead or village of the family or followers of a man called Beorn', OE pers. name + -*inga*- + *hām*: **Barningham** Durham. *Berningham* 1086 (DB). **Barningham** Suffolk. *Bernincham* 1086 (DB). **Barningham, Little** Norfolk. *Berningeham* 1086 (DB).

Barnmeen (*Bearn Mhín*) Down. *Ballybarnemyne* 1612. 'Smooth gap'.

Barnoldby le Beck NE. Lincs. *Bernulfbi* 1086 (DB). 'Farmstead or village of a man called Bjǫrnulfr'. OScand. pers. name + OScand. *bý*. Affix means 'on the stream' from OScand. *bekkr*.

Barnoldswick Lancs. *Bernulfesuuic* 1086 (DB). 'Dwelling or (dairy) farm of a man called Beornwulf or Bjǫrnulfr'. OE or OScand. pers. name + OE *wīc*.

Barnsbury Gtr. London. *Bernersbury* 1406. 'Manor of the *de Berners* family', from ME *bury*. This family held land in Islington from the 13th cent.

Barnsley Barns. *Berneslai* 1086 (DB). 'Woodland clearing of a man called Beorn'. OE pers. name + *lēah*.

Barnsley Glos. *Bearmodeslea* c.802, *Bernesleis* 1086 (DB). 'Woodland clearing of a man called Beornmōd'. OE pers. name + *lēah*.

Barnstaple Devon. *Beardastapol* late 10th cent., *Barnestaple* 1086 (DB). 'Post or pillar of the battle-axe', probably signifying the site of a meeting-place. OE **bearde* (genitive -*an*) + *stapol*.

Barnston Essex. *Bernestuna* 1086 (DB). 'Farmstead of a man called Beorn'. OE pers. name + *tūn*.

Barnston Wirral. *Bernestone* 1086 (DB). 'Farmstead of a man called Beornwulf'. OE pers. name + *tūn*.

Barnstone Notts. *Bernestune* 1086 (DB). 'Farmstead of a man called Beorn'. OE pers. name + *tūn*.

Barnton Ches. *Bertintune* 1086 (DB), *Bertherton* 1313, *Berneton* 1319. 'Farmstead of a woman called Beornthrȳth'. OE pers. name + *tūn*.

Barnwell All Saints Northants. *Byrnewilla* c.980, *Bernewelle* 1086 (DB). Possibly 'spring or stream of the warriors, or of a man called Beorna'. OE *beorn* or OE pers. name + *wella*. Alternatively the first element may be OE *byrgen* 'burial place'. Affix from the dedication of the church.

Barnwood Glos. *Berneuude* 1086 (DB). 'Wood of the warriors, or of a man called Beorna'. OE *beorn* or OE pers. name + *wudu*.

Barnycarroll (*Bearna Chearúill*) Mayo. 'Cearúll's gap'.

Barr, Great Sandw. *Bearre* 957, *Barre* 1086 (DB). Celtic **barr* 'hill-top'.

Barra (island) W. Isles. *Barru* c.1090, *Barey* c.1200. Probably pre-Celtic name, perhaps with OScand. *ey* 'island' added.

Barraduff (*Barra Dubh*) Kerry. 'Black ridge'.

Barrasford Northum. *Barwisford* 1242. 'Ford by a grove'. OE *bearu* (genitive *bearwes*) + *ford*.

Barrhead E. Renf. 'Hill-top headland'. Gaelic *barr* + ModE *head*. The English word would have been added when the Gaelic original was no longer understood.

Barri, Y. *See* BARRY.

Barrington Cambs. *Barentone* 1086 (DB). Probably 'farmstead of a man called *Bāra'. OE pers. name (genitive *-n*) + *tūn*.

Barrington Somerset. *Barintone* 1086 (DB). Probably 'estate associated with a man called *Bāra'. OE pers. name + *-ing-* + *tūn*.

Barrington, Great & Barrington, Little Glos. *Berni(n)tone* 1086 (DB). 'Estate associated with a man called Beorn(a)'. OE pers. name + *-ing-* + *tūn*.

Barrow, usually '(place at) the wood or grove', OE *bearu* (in a dative form *bearwe*); examples include: **Barrow** Suffolk. *Baro* 1086 (DB). **Barrow, Great & Barrow, Little** Ches. *Barue* 958, *Bero* 1086 (DB). **Barrow Gurney** N. Som. *Berue* 1086 (DB), *Barwe Gurnay* 1283. Affix from possession by *Nigel de Gurnai* in 1086. **Barrow, North & Barrow, South** Somerset. *Berue, Berrowene* 1086 (DB). **Barrow upon Humber** N. Lincs. *Ad Baruae* 731, *Barewe* 1086 (DB). For the river-name, see HUMBER. *Ad* in the early form is Latin 'at'. **Barrow upon Soar** Leics. *Barhou* 1086 (DB). Soar is a Celtic or pre-Celtic river-name probably meaning 'flowing one'. **Barrow upon Trent** Derbys. *Barewe* 1086 (DB). For the river-name, see TRENTHAM.

However some Barrows have a different origin: **Barrow** Rutland, near Oakham. *Berc* 1197. '(Place at) the hill or burial mound'. OE *beorg*. **Barrow in Furness** Cumbria. *Barrai* 1190. 'Promontory island'. Celtic **barr* + OScand. *ey*. The old district name Furness (*Fuththernessa* c.1150) means 'headland by the rump-shaped island', OScand. *futh* (genitive *-ar*) + *nes*.

Barrowby Lincs. *Bergebi* 1086 (DB). 'Farmstead or village on the hill(s)'. OScand. *berg* + *bý*.

Barrowden Rutland. *Berchedone* 1086 (DB). 'Hill with barrows or tumuli'. OE *beorg* + *dūn*.

Barrowford Lancs. *Barouforde* 1296. 'Ford by the grove'. OE *bearu* + *ford*.

Barry (*Y Barri*) Vale Glam. *Barri* c.1190. 'Hill'. Welsh *barr*. The name properly refers to Barry Island. The Welsh name has Welsh *y*, 'the'.

Barryroe (*Barraigh Rua*) Cork. '(District of the) Red Barraigh'.

Barsby Leics. *Barnesbi* 1086 (DB). 'Farmstead or village of the child or young heir, or of a man called Barn'. OScand. *barn* or pers. name + *bý*.

Barsham, 'homestead or village of a man called Bār', OE pers. name + *hām*: **Barsham** Norfolk. *Barseham* 1086 (DB). **Barsham** Suffolk. *Barsham* 1086 (DB).

Barston Solhll. *Bertanestone* 1086 (DB), *Berestanestona* 1185. Probably 'farmstead or estate of a man called Beorhtstān'. OE pers. name + *tūn*.

Bartestree Herefs. *Bertoldestreu* 1086 (DB). 'Tree of a man called Beorhtwald'. OE pers. name + *trēow*.

Barthomley Ches. *Bertemeleu* [*sic*] 1086 (DB), *Bertamelegh* 13th cent. Possibly 'woodland clearing of the dwellers at a place called *Brightmead* or *Brightwell* or the like'. The first element of an older place name (possibly OE *beorht* 'bright') + *hǣme* + *lēah*.

Bartley, 'birch-tree wood or clearing', OE *beorc* + *lēah*: **Bartley** Hants. *Berchelai* 1107. **Bartley Green** Birm. *Berchelai* 1086 (DB).

Bartlow Cambs. *Berkelawe* 1232. 'Mounds or tumuli where birch-trees grow'. OE *beorc* + *hlāw*.

Barton, a common name, usually OE *beretūn*, *bær-tūn* 'barley farm, outlying grange where corn is stored'; examples include: **Barton** Cambs. *Barton* 1060, *Bertone* 1086 (DB). **Barton** Glos., near Guiting. *Berton* 1158. **Barton** Lancs., near Preston. *Bartun* 1086 (DB). **Barton** Torbay. *Bertone* 1333. **Barton Bendish** Norfolk. *Bertuna* 1086 (DB), *Berton Binnedich* 1249. Affix means 'inside the ditch' (OE *binnan* + *dīc*) referring to Devil's Dyke. **Barton, Earls** Northants. *Bartone* 1086 (DB), *Erlesbarton* 1261. Affix from the Earl of Huntingdon who held the manor in the 12th cent. **Barton, Great** Suffolk. *Bertuna* 945, 1086 (DB), *Magna Bertone* 1254. Early affix is Latin *magna* 'great'. **Barton in Fabis** Notts. *Bartone* 1086 (DB), *Barton in le Benes* 1388. Latin affix means 'where beans are grown'. **Barton le Clay** Beds. *Bertone* 1086 (DB), *Barton-in-the-Clay* 1535. Affix means 'on clay soil'. **Barton Mills** Suffolk. *Bertona* 1086 (DB), *Parva Bertone* 1254. Early affix is Latin *parva* 'little'. **Barton Seagrave** Northants. *Bertone* 1086 (DB), *Barton Segrave* 1321. Manorial affix from the *de Segrave* family, here in the 13th cent. **Barton Stacey** Hants. *Bertune* c.1000, *Bertune* 1086 (DB), *Berton Sacy* 1302. Manorial affix from the *de Saci* family, here in the 12th cent. **Barton Turf** Norfolk. *Bertona* 1086 (DB), *Berton Turfe* 1394. Affix is ME *turf*, presumably because good turf was cut here. **Barton under Needwood** Staffs. *Barton* 942, *Bertone* 1086 (DB). Affix means 'near or within the forest of NEEDWOOD'. **Barton upon Humber** N. Lincs. *Bertune* 1086 (DB). For the river-name, *see* HUMBER.
   However the following has a different origin: **Barton on Sea** Hants. *Bermintune* 1086 (DB). 'Estate associated with a man called *Beorma'. OE pers. name + *-ing-* + *tūn*.

Barugh, Great & Barugh, Little N. Yorks. *Berg, Berch* 1086 (DB). '(Place at) the hill'. OE *beorg*.

Barway Cambs. *Bergeia* 1155. 'Island, or dry ground in marsh, with barrows or tumuli on it'. OE *beorg* + *ēg*.

Barwell Leics. *Barewelle* 1086 (DB). 'Spring or stream frequented by wild boars'. OE *bār* + *wella*.

Barwick, 'barley farm, outlying part of an estate', OE *bere-wīc*: **Barwick** Somerset. *Berewyk* 1219. **Barwick in Elmet** Leeds. *Bereuuith* 1086 (DB). *Elmet* is an ancient district name, obscure in origin and meaning, first recorded in the 7th cent. as *Elmed*.

Baschurch Shrops. *Bascherche* 1086 (DB). 'Church of a man called Bas(s)a'. OE pers. name + *cirice*.

Bascote Warwicks. *Bachecota* 1174. Probably 'cottage(s) of a man called *Basuca'. OE pers. name + *cot*.

Basford Staffs. *Bechesword* [*sic*] 1086 (DB), *Barkeford* 1199. Probably 'ford of a man called Beorcol'. OE pers. name + *ford*.

Bashall Eaves Lancs. *Bacschelf* 1086 (DB). 'Ridge shelf'. OE *bæc* + *scelf*. Later addition *Eaves* (from 16th cent.) is from OE *efes* 'edge of a wood'.

Bashley Hants. *Bageslucesleia* 1053, *Bailocheslei* 1086 (DB). 'Woodland clearing of a man called Bægloc'. OE pers. name + *lēah*.

Basildon Essex. *Berlesduna* [*sic*] 1086 (DB), *Bertlesdon* 1194. 'Hill of a man called Beorhtel'. OE pers. name + *dūn*.

Basing Hants. *Basengum* 871, *Basinges* 1086 (DB). '(Settlement of) the family or followers of a man called *Basa'. OE pers. name + *-ingas*. The same tribal group is referred to in the next name.

Basingstoke Hants. *Basingastoc* 990, *Basingestoches* 1086 (DB). 'Secondary settlement or outlying farmstead of the family or followers of a man called Basa'. OE pers. name + *-inga-* + *stoc*.

Baslick (*Baisleac*) Monaghan. 'Church'.

Baslickane (*Baisleacán*) Kerry. 'Small church'.

Baslow Derbys. *Basselau* 1086 (DB). 'Burial
mound of a man called Bassa'. OE pers. name +
hlāw.

Bassenthwaite Cumbria. *Bastunthuait*
*c.*1175. 'Clearing or meadow of a family called
Bastun'. ME surname + OScand. *thveit*.

Bassetlaw (district) Notts. *Bernesedelaue*
1086 (DB). Possibly 'mound or hill of the
dwellers on land cleared by burning'. OE *bærnet*
+ *sǣte* + *hlāw*.

Bassingbourn Cambs. *Basingborne* 1086
(DB). 'Stream of the family or followers of a man
called Bas(s)a'. OE pers. name + *-inga-* + *burna*.

Bassingfield Notts. *Basingfelt* 1086 (DB).
'Open land of the family or followers of a man
called Bas(s)a'. OE pers. name + *-inga-* + *feld*.

Bassingham Lincs. *Basingeham* 1086 (DB).
'Homestead or village of the family or followers
of a man called Bas(s)a'. OE pers. name + *-inga-*
+ *hām*.

Bassingthorpe Lincs. *Torp* 1086 (DB),
Basewinttorp 1202. 'Outlying farmstead or
hamlet'. OScand. *thorp* + later manorial
addition from possession by the *Basewin* family.

Baston Lincs. *Bacstune* 1086 (DB). 'Farmstead
of a man called Bak'. OScand. pers. name + OE
tūn.

Bastwick Norfolk. *Bastwic* 1044–7, *Bastuuic*
1086 (DB). 'Farm or building where bast (the
bark of the lime-tree used for rope-making) is
stored'. OE *bæst* + *wīc*.

Batcombe, 'valley of a man called Bata', OE
pers. name + *cumb*: **Batcombe** Dorset.
Batecumbe 1201. **Batcombe** Somerset, near
Bruton. *Batancumbæ* 10th cent., *Batecumbe*
1086 (DB).

Bath B. & NE. Som. *Bathum* 796, *Bade* 1086
(DB). '(Place at) the (Roman) baths'. OE *bæth* in
a dative plural form. *See also* AKEMAN STREET.

Bathampton B. & NE. Som. *Hamtun* 956,
Hantone 1086 (DB). OE *hām-tūn* 'home farm,
homestead' with later addition from its
proximity to BATH.

Bathealton Somerset. *Badeheltone* 1086
(DB). Possibly 'farmstead of a man called
Beaduhelm'. OE pers. name + *tūn*.

Batheaston B. & NE. Som. *Estone* 1086 (DB),
Batheneston 1258. 'East farmstead or village'.

OE *ēast* + *tūn* with later addition from its
proximity to BATH.

Bathford B. & NE. Som. *Forda* 957, *Forde*
1086 (DB), *Bathford* 1575. '(Place at) the ford'.
OE *ford* with later addition from its proximity to
BATH.

Bathgate W. Loth. *Batket c.*1160. 'Boar
wood'. Cumbric *badd + **ceto-*.

Bathley Notts. *Badeleie* 1086 (DB). 'Woodland
clearing with springs used for bathing'. OE *bæth*
+ *lēah*.

Batley Kirkl. *Bathelie* 1086 (DB) 'Woodland
clearing of a man called Bata'. OE pers. name +
lēah.

Batsford Glos. *Bæccesore* 727–36, *Beceshore*
1086 (DB). 'Hill-slope of a man called *Bæcci'.
OE pers. name + *ōra*.

Battersby N. Yorks. *Badresbi* 1086 (DB).
'Farmstead or village of a man called Bǫthvarr'.
OScand. pers. name + *bý*.

Battersea Gtr. London. *Badrices ege* 693
(11th cent.), *Patricesy* 1086 (DB). 'Island, or dry
ground in marsh, of a man called Beadurīc'. OE
pers. name + *ēg*.

Batterstown (*Baile an Bhóthair*) Meath.
'Homestead of the road'.

Battisford Suffolk. *Betesfort* 1086 (DB). 'Ford
of a man called *Bætti'. OE pers. name + *ford*.
Battisford Tye contains dialect *tye* 'a large
common pasture'.

Battle E. Sussex. *La Batailge* 1086 (DB).
'(Place of) the battle'. OFrench *bataille*. The
abbey here was founded to commemorate the
battle of Hastings in 1066.

Battlefield Shrops. *Batelfeld* 1415. 'Field of
battle'. OFrench *bataille*. A college of secular
canons was founded here to commemorate the
battle of Shrewsbury in 1403.

Battlesden Beds. *Badelesdone* 1086 (DB).
'Hill of a man called *Bæddel'. OE pers. name +
dūn.

Baughurst Hants. *Beaggan hyrste* 909.
'Wooded hill of a man called *Beagga'. OE
pers. name + *hyrst*. Alternatively the first
element may be OE *bagga 'a badger'.

Baulking Oxon. *Bathalacing* 963.
'The stream called *Lācing* ('the playful
one') with pools'. OE *bæth* (genitive

plural *batha*) 'bath, pool used for bathing' + *lāc* 'play' + *-ing*.

Baumber Lincs. *Badeburg* 1086 (DB). Possibly 'stronghold of a man called Badda'. OE pers. name + *burh*.

Baunton Glos. *Baudintone* 1086 (DB). 'Estate associated with a man called Balda'. OE pers. name + *-ing-* + *tūn*.

Bauteogue (*Báiteog*) Laois. 'Morass'.

Bautregaum (*Barr Trí gCom*) Kerry. 'Top of three hollows'.

Bavan (*Bádhún*) Donegal, Down. 'Cow fortress'.

Baverstock Wilts. *Babbanstoc* 968, *Babestoche* 1086 (DB). 'Outlying farmstead or hamlet of a man called Babba'. OE pers. name + *stoc*.

Bavington, Great & Bavington, Little Northum. *Babington* 1242. 'Estate associated with a man called Babba'. OE pers. name + *-ing-* + *tūn*.

Bawburgh Norfolk. *Bauenburc* 1086 (DB). 'Stronghold of a man called *Bēawa*'. OE pers. name + *burh*.

Bawdeswell Norfolk. *Baldereswella* 1086 (DB). 'Spring or stream of a man called Baldhere'. OE pers. name + *wella*.

Bawdrip Somerset. *Bagetrepe* 1086 (DB). 'Place where badgers are trapped or snared'. OE *bagga* + *træppe*.

Bawdsey Suffolk. *Baldereseia* 1086 (DB). 'Island, or dry ground in marsh, of a man called Baldhere'. OE pers. name + *ēg*.

Bawnboy (*An Bábhún Buí*) Cavan. *Bawnboy* 1664. 'The yellow fortified enclosure'.

Bawtry Donc. *Baltry* 1199. Probably 'tree rounded like a ball'. OE *ball* + *trēow*.

Baxenden Lancs. *Bastanedenecloch* 1194. 'Valley where flat stones for baking are found'. OE *bæc-stān* + *denu*, with OE *clōh* 'ravine' in the early form.

Baxterley Warwicks. *Basterleia* c.1170. 'Woodland clearing belonging to the baker'. OE *bæcestre* + *lēah*.

Baycliff Cumbria. *Belleclive* 1212. Possibly 'cliff where a signal-fire is lit'. OE *bēl* + *clif*.

Alternatively the first element could be OE *belle* 'a bell' used as a hill-name or *bel* 'a glade'.

Baydon Wilts. *Beidona* 1146. 'Hill where berries grow'. OE *beg* + *dūn*.

Bayford Herts. *Begesford* [sic] 1086 (DB), *Begeford* c.1090. 'Ford of a man called Bæga'. OE pers. name + *ford*.

Baylham Suffolk. *Beleham* 1086 (DB). Probably 'homestead or enclosure at a river-bend'. OE *bēgel* + *hām* or *hamm*.

Baylin (*Béal Linne*) Westmeath. 'Mouth of the pool'.

Bayston Shrops. *Begestan* 1086 (DB). 'Stone of a woman called Bēage or of a man called Bæga'. OE pers. name + *stān*. **Bayston Hill** is *Beystaneshull* 1301, from OE *hyll*.

Bayswater Gtr. London. *Bayards Watering Place* 1380. 'Watering place for horses, or belonging to a family called Bayard'. ME *bayard* (or ME surname from this word) + *water(ing)* 'stream, pond'.

Bayton Worcs. *Betune* 1086 (DB). 'Farmstead of a woman called Bēage or of a man called Bæga'. OE pers. name + *tūn*.

Beachampton Bucks. *Bechentone* 1086 (DB). 'Home farm by a stream'. OE *bece* + *hām-tūn*.

Beachamwell Norfolk. *Bicham* 1086 (DB), *Bichham Welles* 1212. 'Homestead or village of a man called Bicca'. OE pers. name + *hām*, with the later addition of ME *welle(s)* 'spring(s)'.

Beachley Glos. *Beteslega* 12th cent. 'Woodland clearing of a man called Betti'. OE pers. name + *lēah*.

Beachy Head E. Sussex. *Beuchef* 1279. 'Beautiful headland'. OFrench *beau* + *chef*, with the (tautological) addition of *head* in recent times.

Beacon Devon. *Bekyn* 1469. '(Place by) the beacon or signal-fire', from OE *bēacen*, no doubt with reference to the high situation of this hamlet.

Beaconsfield Bucks. *Bekenesfelde* 1184. 'Open land near a beacon or signal-fire'. OE *bēacen* + *feld*.

Beadlam N. Yorks. *Bodlum* 1086 (DB). '(Place at) the buildings'. OE *bōthl* in a dative plural form *bōthlum*.

Beadnell Northum. *Bedehal* 1161. 'Nook of land of a man called Bēda'. OE pers. name (genitive -*n*) + *halh*.

Beaford Devon. *Baverdone* [*sic*] 1086 (DB), *Beuford* 1242. 'Ford infested with gadflies'. OE *bēaw* + *ford*.

Beagh (*Beitheach*) Leitrim. 'Birch land'.

Beaghmore (*Beatheach Mór*) Tyrone. 'Large birch land'.

Beal Northum. *Behil* 1208–10. 'Hill frequented by bees'. OE *bēo* + *hyll*.

Beal N. Yorks. *Begale* 1086 (DB). 'Nook of land in a river-bend'. OE *bēag* + *halh*.

Bealadangan (*Béal an Daingin*) Galway. 'Opening of the stronghold'.

Bealaha (*Béal Átha*) Clare. 'Ford-mouth'.

Bealings, Great & Bealings, Little Suffolk. *Belinges* 1086 (DB). Possibly '(settlement of) the dwellers in the glade, or by the funeral pyre'. OE **bel-* or *bēl* + -*ingas*.

Beaminster Dorset. *Bebingmynster* 862, *Beiminstre* 1086 (DB). 'Large church of a woman called Bebbe'. OE pers. name + *mynster*.

Beamish Durham. *Bewmys* 1288. 'Beautiful mansion'. OFrench *beau* + *mes*.

Beamsley N. Yorks. *Bedmesleia* 1086 (DB), *Bethmesleia* 1185. 'Pasture or meadow at or by the valley bottom'. OE *lēah* with an OE **bethme* (a side-form of **bothm*).

Beane (river) Herts. *See* BENINGTON.

Beanley Northum. *Benelega* c.1150. 'Clearing where beans are grown'. OE *bēan* + *lēah*.

Bearley Warwicks. *Burlei* 1086 (DB). 'Woodland clearing near a fortified place'. OE *burh* + *lēah*.

Bearpark Durham. *Beaurepayre* 1267. 'Beautiful retreat'. OFrench *beau* + *repaire*.

Bearsden E. Dunb. Said to be 'valley inhabited by wild boars', OE *bār* + *denu*, but this is doubtful.

Bearsted Kent. *Berghamstyde* 695. 'Homestead on a hill'. OE *beorg* + *hām-stede*.

Beauchief Sheff. *Beuchef* 12th cent. 'Beautiful headland or hill-spur'. OFrench *beau* + *chef*.

Beaufort (*Cendl*) Blae. The Duke of *Beaufort* owned lands here in the 18th cent. The Welsh name is from Edward *Kendall*, the ironmaster who was granted a lease of the site by the Duke in 1780.

Beaulieu Hants. *Bellus Locus Regis* 1205, *Beulu* c.1300. 'Beautiful place (of the king)'. OFrench *beau* + *lieu* (often rendered in Latin, as in the first spelling).

Beauly Highland. *Prioratus de bello loco* 1230. 'Beautiful place'. OFrench *beau* + *lieu*.

Beaumaris (*Biwmares*) Angl. *Bello Marisco* 1284. 'Beautiful marsh'. OFrench *beau* + *marais*.

Beaumont, 'beautiful hill', OFrench *beau* or *bel* + *mont*: **Beaumont** Cumbria. *Beumund* c.1240. **Beaumont** Essex. *Fulepet* 1086 (DB), *Bealmont* 12th cent. The earlier name means 'foul pit' from OE *fūl* + *pytt*.

Beausale Warwicks. *Beoshelle* [*sic*] 1086 (DB), *Beausala* 12th cent. 'Nook of land of a man called Bēaw'. OE pers. name + *halh*.

Beaworthy Devon. *Begeurde* 1086 (DB). 'Enclosure of a woman called Bēage or of a man called Bǣga'. OE pers. name + *worth*.

Bebington Wirral. *Bebinton* c.1100. 'Estate associated with a woman called Bebbe or a man called **Bebba*'. OE pers. name + -*ing*- + *tūn*.

Bebside Northum. *Bibeshet* 1198. Probably 'projecting piece of land of a man called **Bibba*'. OE pers. name + *scēat*.

Beccles Suffolk. *Becles* 1086 (DB). Probably 'pasture by a stream', OE *bece* + *lǣs*. Alternatively perhaps 'little court', Celtic **bacc* + **līss*.

Becconsall Lancs. *Bekaneshou* 1208. 'Burial mound of a man called Bekan'. OIrish pers. name + OScand. *haugr*.

Beckbury Shrops. *Becheberie* 1086 (DB). 'Stronghold or manor of a man called Becca'. OE pers. name + *burh* (dative *byrig*).

Beckenham Gtr. London. *Beohha hammes gemǣru* 973, *Bacheham* 1086 (DB). 'Homestead or enclosure of a man called **Beohha*'. OE pers. name (genitive -*n*) + *hām* or *hamm*. The early form contains OE (*ge*)*mǣre* 'boundary'.

Beckford Worcs. *Beccanford* 803, *Beceford* 1086 (DB). 'Ford of a man called Becca'. OE pers. name + *ford*.

Beckham Norfolk. *Beccheham* 1086 (DB).
'Homestead or village of a man called Becca'.
OE pers. name + *hām*.

Beckhampton Wilts. *Bachentune* 1086 (DB).
'Home farm near the ridge'. OE *bæc* + *hām-tūn*.

Beckingham, 'homestead or enclosure of
the family or followers of a man called Becca or
Beohha', OE pers. name + *hām* or *hamm*:
Beckingham Lincs. *Bekingeham* 1177.
Beckingham Notts. *Bechingeham* 1086 (DB).

Beckington Somerset. *Bechintone* 1086 (DB).
'Estate associated with a man called Becca'. OE
pers. name + *-ing-* + *tūn*.

Beckley, 'woodland clearing of a man called
Becca', OE pers. name + *lēah*: **Beckley** E.
Sussex. *Beccanlea c.*880. **Beckley** Oxon.
Beccalege 1005–12, *Bechelie* 1086 (DB).

Beckton Gtr. London. Modern name for a
district developed in the late 19th cent., so
called from Simon Adams *Beck*, governor of the
Gas, Light and Coke Co., which opened a large
works here in 1870.

Becontree Gtr. London. *Beuentreu* [*sic*] 1086
(DB), *Beghentro* 12th cent. 'Tree of a man called
Beohha'. OE pers. name (genitive *-n*) + *trēow*.
The tree marked the Hundred meeting-place.

Bedale N. Yorks. *Bedale* 1086 (DB). 'Nook of
land of a man called Bēda'. OE pers. name +
halh.

Bedburn Durham. *Bedeburne* 1243. 'Stream
of a man called Bēda'. OE pers. name + *burna*.

Beddgelert Gwyd. *Bedkelert* 1281. 'Celert's
grave'. Irish pers. name + Welsh *bedd*. Local
legend derives the pers. name from *Gelert*, a
hound slain by its master, Prince Llewellyn,
when he thought it had killed his baby son,
although it had actually killed a wolf that
threatened the child.

Beddingham E. Sussex. *Beadyngham c.*800,
Bedingeham 1086 (DB). 'Promontory of the
family or followers of a man called Bēada'. OE
pers. name + *-inga-* + *hamm*.

Beddington Gtr. London. *Beaddinctun*
901–8, *Beddintone* 1086 (DB). 'Estate
associated with a man called *Beadda*'. OE pers.
name + *-ing-* + *tūn*.

Bedfield Suffolk. *Berdefelda* [*sic*] 1086 (DB).
Bedefeld 12th cent. 'Open land of a man called
Bēda'. OE pers. name + *feld*.

Bedfont, East (Gtr. London) **& Bedfont,
West** (Surrey) *Bedefunt* 1086 (DB). Probably
'spring provided with a drinking-vessel'. OE
byden + **funta*.

Bedford Beds. *Bedanford* 880, *Bedeford* 1086
(DB). Possibly 'ford of a man called Bīeda'. OE
pers. name + *ford*. Alternatively the first element
may be OE *byden* 'vessel, tub', also 'hollow,
deep valley'. **Bedfordshire** (OE *scīr* 'district')
is first referred to in the 11th cent.

Bedhampton Hants. *Betametone* 1086 (DB).
Possibly 'farmstead of the dwellers where beet is
grown'. OE *bēte* + *hǣme* + *tūn*.

Bedingfield Suffolk. *Bedingefelda* 1086 (DB).
'Open land of the family or followers of a man
called Bēda'. OE pers. name + *-inga-* + *feld*.

Bedlington Northum. *Bedlingtun c.*1050.
Probably 'estate associated with a man called
*Bēdla or **Bētla*'. OE pers. name + *-ing-* + *tūn*.

Bedmond Herts. *Bedesunta* [*sic* for *-funta*]
1331, *Bedmont c.*1550. Probably 'spring
provided with a drinking-vessel'. OE *byden* +
*funta (later changed as if from OFrench *mont*
'hill').

Bednall Staffs. *Bedehala* 1086 (DB). 'Nook or
hollow of a man called Bēda'. OE pers. name
(genitive *-n*) + *halh*.

Bedstone Shrops. *Betietetune* 1086 (DB),
Bedeston 1176. Probably 'farmstead of a man
called **Bedgēat*'. OE pers. name + *tūn*.

Bedwas Cphy. *Bedewas c.*1102. 'Grove of
birch-trees'. Welsh *bedwos*.

Bedworth Warwicks. *Bedeword* 1086 (DB).
'Enclosure of a man called Bē(a)da'. OE pers.
name + *worth*.

Bedwyn, Great & Bedwyn, Little Wilts.
Bedewinde 778, *Bedvinde* 1086 (DB). Probably
'place where bindweed or convolvulus grows'.
OE **bedwinde*.

Beeby Leics. *Bebi* 1086 (DB). 'Farmstead
or village where bees are kept'. OE *bēo*
+ OScand. *bý*.

Beech Staffs. *Le Bech* 1285. '(Place at) the
beech-tree'. OE *bēce*.

Beech Hill W. Berks. *Le Bechehulle* 1384.
'Hill where beech-trees grow'. OE *bēce* + *hyll*.

Beechingstoke Wilts. *Stoke* 941,
Bichenestoch 1086 (DB). 'Outlying farmstead

where bitches or hounds are kept'. OE *bicce* (genitive plural *ena*) + *stoc*.

Beeding, Lower & Beeding, Upper W. Sussex. *Beadingum* c.880, *Bedinges* 1086 (DB). '(Settlement of) the family or followers of a man called Bēada'. OE pers. name + *-ingas*.

Beedon W. Berks. *Bydene* 965, *Bedene* 1086 (DB). '(Place at) the tub-shaped valley'. OE *byden*.

Beeford E. R. Yorks. *Biuuorde* 1086 (DB). '(Place) by the ford', or 'ford where bees are found'. OE *bī* or *bēo* + *ford*.

Beeley Derbys. *Begelie* 1086 (DB). 'Woodland clearing of a woman called Bēage or a man called *Bēga*'. OE pers. name + *lēah*.

Beelsby NE. Lincs. *Belesbi* 1086 (DB). 'Farmstead or village of a man called Beli'. OScand. pers. name + *bý*.

Beenham W. Berks. *Benham* 12th cent. 'Homestead or enclosure where beans are grown'. OE *bēan* + *hām* or *hamm*.

Beer Devon. *Bera* 1086 (DB). '(Place by) the grove'. OE *bearu*.

Beer Crocombe Somerset. *Bere* 1086 (DB). 'The grove', or 'the woodland pasture'. OE *bearu* or *bǣr* + manorial affix from the *Craucombe* family, here in the 13th cent.

Beer Hackett Dorset. *Bera* 1176, *Berehaket* 1362. 'The grove', or 'the woodland pasture'. OE *bearu* or *bǣr* + manorial affix from a 12th-cent. owner called *Haket*.

Beesby Lincs. *Besbi* 1086 (DB). Probably 'farmstead or village of a man called Besi'. OScand. pers. name + *bý*.

Beeston, usually 'farmstead where bent-grass grows', OE **bēos* + *tūn*: **Beeston** Beds. *Bistone* 1086 (DB). **Beeston** Leeds. *Bestune* 1086 (DB). **Beeston** Norfolk. *Bestone* 1254. **Beeston** Notts. *Bestune* 1086 (DB). **Beeston Regis** Norfolk. *Besetune* 1086 (DB). Affix is Latin *regis* 'of the king'.
    However the following has a different origin: **Beeston** Ches. *Buistane* 1086 (DB). Probably 'stone or rock where commerce takes place'. OE *byge* + *stān*.

Beetham Cumbria. *Biedun* 1086 (DB). Probably '(place by) the embankments'. OScand. **beth* in a dative plural form **bjǫthum*.

Beetley Norfolk. *Betellea* 1086 (DB). Possibly 'clearing where beet is grown'. OE *bēte* + *lēah*.

Begbroke Oxon. *Bechebroc* 1086 (DB). 'Brook of a man called Becca'. OE pers. name + *brōc*.

Beginish (*Beag Inis*) Kerry. 'Little island'.

Beglieve (*Beagshliabh*) Cavan. *Begleive* 1586. 'Little mountain'.

Behaghane (*Beitheachán*) Kerry. 'Little place of birches'.

Behy (*Beitheach*) Donegal. 'Birch land'.

Beighton Norfolk. *Begetuna* 1086 (DB). 'Farmstead of a woman called Bēage or of a man called Bǣga'. OE pers. name + *tūn*.

Beighton Sheff. *Bectune* c.1002, 1086 (DB). 'Farmstead by the stream'. OE *bece* + *tūn*.

Beith N. Ayr. Probably '(place of) birches'. Gaelic *beith*.

Bekesbourne Kent. *Burnes* 1086 (DB), *Bekesborne* 1280. 'Estate on the river called *Burna*' (from OE *burna* 'stream' referring to the Little Stour), with later manorial affix from the *de Beche* family, here in the late 12th cent.

Belaugh Norfolk. *Belahe* 1044–7, *Belaga* 1086 (DB). Possibly 'enclosure where the dead are cremated'. OE *bēl* + *haga*. Alternatively the first element may be OE **bel* 'a glade or clearing'.

Belbroughton Worcs. *Beolne*, *Broctun* 817, *Bellem*, *Brocton* 1086 (DB), *Bellebrocton* 1292. Originally two distinct names. *Bell* is an old river-name, probably from OE *beolone* 'henbane'; *Broughton* is 'farmstead on the brook', OE *brōc* + *tūn*.

Belchamp Otten, Belchamp St Paul & Belchamp Walter Essex. *Bylcham* c.940, *Belcham*, *Belcamp* 1086 (DB), *Belcham Otes* 1256, *Belchampe of St Paul* 1451, *Waterbelcham* 1297. Probably 'homestead with a beamed or vaulted roof'. OE **belc* + *hām*. Distinguishing affixes from early possession by a man called *Otto*, by St Paul's Cathedral, and by a man called *Walter*.

Belchford Lincs. *Beltesford* 1086 (DB). Possibly 'ford of a man called **Belt*'. OE pers. name + *ford*. Alternatively the first element may be OE *belt* 'belt' used in a transferred topographical sense such as 'long narrow strip'.

Belclare (*Béal Chláir*) Galway. 'Mouth of the plain'.

Belcoo (*Béal Cú*) Fermanagh. *Beallacoungamore, Beallacoungabegg* 1607. 'Mouth of the narrow'.

Belderg (*Béal Deirg*) Mayo. 'Mouth of the (river) Derg'.

Belfast (*Béal Feirste*) Antrim. (*bellum*) *Fertsi* 668. 'Ford-mouth of the sandbank'.

Belford Northum. *Beleford* 1242. Possibly 'ford by the bell-shaped hill, or near a funeral pyre'. OE *belle* or *bēl* + *ford*.

Belgooly (*Béal Guala*) Cork. 'Mouth of the ridge'.

Bellacorick (*Béal Átha Chomraic*) Mayo. 'Ford-mouth of the confluence'.

Bellaghy (*Baile Eachaidh*) Derry. *Baile Eachaidh c.*1645. 'Eochadh's homestead'.

Bellahy (*Béal Lathaí*) Sligo. 'Mouth of the miry place'.

Bellanagare (*Béal Átha na gCarr*) Roscommon. 'Ford-mouth of the carts'.

Bellanaleck (*Bealach na Leice*) Fermanagh. *Bellanaleck* 1837. 'Pass of the flagstone'.

Bellanamallard (*Béal Átha na Mallacht*) Fermanagh. *Béal Átha na Mallacht* 1645. 'Ford-mouth of the curses'. St Columba is said to have cursed some cocks here in the 6th cent.

Bellanamore (*Béal an Átha Móir*) Donegal. *Beallanaymore* 1621. 'Mouth of the big ford'.

Bellananagh (*Béal Átha na nEach*) Cavan. *Ballinenagh c.*1657. 'Ford-mouth of the horses'.

Bellaneeny (*Béal Átha an Aonaigh*) Roscommon. 'Ford-mouth of the fair'.

Bellanode (*Béal Átha an Fhóid*) Monaghan. *Ballynode* 1835. 'Ford-mouth of the sod'.

Bellarena (*Baile an Mhargaidh*) Derry. *Bellarena* 1835. Possibly 'beautiful strand'. French *belle* + Latin *arena*. The Irish name means 'townland of the market'.

Bellavary (*Béal Átha Bhearaigh*) Mayo. 'Ford-mouth of Bearach'.

Belleau Lincs. *Elgelo* 1086 (DB). 'Meadow of a man called Helgi'. OScand. pers. name + *ló*. The modern form, as if from a French name meaning 'beautiful water', is unhistorical.

Belleek (*Béal Leice*) Fermanagh. *Bel-leice* 1409. 'Mouth of the flagstone'.

Belleeks (*Béal Leice*) Armagh. *Bellick* 1657. 'Mouth of the flagstone'.

Bellerby N. Yorks. *Belgebi* 1086 (DB). 'Farmstead or village of a man called Belgr'. OScand. pers. name + *bý*.

Bellew (*Bile*) Meath. 'Sacred tree'.

Bellia (*Bile*) Clare. 'Sacred tree'.

Bellingham Gtr. London. *Beringaham* 998, *Belingeham* 1198. 'Homestead or enclosure of the family or followers of a man called *Bera'. OE pers. name + *-inga-* + *hām* or *hamm*.

Bellingham Northum. *Bellingham* 1254. 'Homestead of the dwellers at the bell-shaped hill', or simply 'homestead at the bell-shaped hill'. OE *belle* + *-inga-* + *hām*, or OE **belling* + *hām*.

Bellshill N. Lan. Apparently named from a farm called *Belziehill* or *Balziehill*. Origin obscure.

Belmesthorpe Rutland. *Beolmesthorp c.*1050, *Belmestorp* 1086 (DB). 'Outlying farmstead or hamlet of a man called Beornhelm'. OE pers. name + OScand. *thorp*.

Belmont Down. *Bellmount* 1834. 'Beautiful hill'. OFrench *bel* + *mont*. The Irish name of Belmont is *An Cnoc Álainn*, 'the beautiful hill'.

Belmont Gtr. London. near Sutton, not on record before the early 19th cent., 'beautiful hill' from OFrench *bel* + *mont*. The same name occurs in other counties, for example **Belmont** Lancs. which is *Belmunt* 1212.

Belmullet (*Béal an Mhuirthead*) Mayo. 'Sea loop'.

Belper Derbys. *Beurepeir* 1231. 'Beautiful retreat'. OFrench *beau* + *repaire*.

Belsay Northum. *Bilesho* 1163, *Belesho* 1171. Probably 'hill-spur used for a beacon or funeral pyre'. OE *bēl* + *hōh*.

Belses Sc. Bord. 'Beautiful seat'. OFrench *bel* + *assis*.

Belsize Herts. not on record before the mid-19th cent., but a common name-type found as **Bellasis, Bellasize**, etc. in other counties and meaning 'beautiful seat or residence', from OFrench *bel* + *assis*.

Belstead Suffolk. *Belesteda* 1086 (DB). 'Place in a glade' or 'place of a funeral pyre'. OE **bel* or *bēl* + *stede*.

Belstone Devon. *Bellestam* [sic] 1086 (DB), *Belestan* 1167. '(Place at) the bell-shaped stone'. OE *belle* + *stān*.

Beltany (*Bealtaine*) Donegal. 'Summer festival'.

Beltoft N. Lincs. *Beltot* 1086 (DB). Possibly 'homestead near a funeral pyre, or on dry ground in marsh'. OE *bēl* or **bel* + OScand. *toft*.

Belton, meaning uncertain, 'farmstead in a glade or on dry ground in marsh', or 'farmstead near a beacon or funeral pyre', OE **bel* or *bēl* + *tūn*; **Belton** Leics., near Shepshed. *Beltona* c.1125. **Belton** Lincs. *Beltone* 1086 (DB). **Belton** Norfolk. *Beletuna* 1086 (DB). **Belton** N. Lincs. *Beltone* 1086 (DB). **Belton** Rutland. *Belton* late 11th cent., *Bealton* 1167.

Beltra (*Béal Trá*) Mayo, Sligo. 'Mouth of the strand'.

Belturbet (*Béal Tairbirt*) Cavan. *Bél Tarbert* 1621. 'Mouth of the isthmus'.

Belvelly (*Béal an Bhealaigh*) Cork. 'Mouth of the pass'.

Belvoir Down. *Belvoir* 1744. 'Beautiful view'. OFrench *bel* + *voir*.

Belvoir Leics. *Belveder* 1130. 'Beautiful view'. OFrench *bel* + *vedeir*.

Bembridge I. of Wight. *Bynnebrygg* 1316. '(Place lying) inside (i.e. this side of) the bridge'. OE *binnan* + *brycg*.

Bemerton Wilts. *Bimertone* 1086 (DB). 'Farmstead of the trumpeters' (or possibly 'of the bitterns'). OE *bēmere* (West Saxon *bȳmere*) + *tūn*.

Bempton E. R. Yorks. *Bentone* 1086 (DB). 'Farmstead made of beams, or by a tree'. OE *bēam* + *tūn*.

Ben Cruachan (mountain) Arg. *Crechanben* c.1375. 'Mountain of the stacks'. Gaelic *beinn* + *cruach*.

Ben Glas (mountain) Stir. 'Grey mountain'. Gaelic *beinn* + *glas*.

Ben Gorm (*An Bhinn Ghorm*) (mountain) Mayo. 'The blue peak'.

Ben Lomond. Gaelic *beinn*, see LOMOND.

Ben More (mountain) Stir. 'Big mountain'. Gaelic *beinn* + *mór*.

Ben Nevis (mountain) Highland. *Gleann Nibheis* 16th cent. 'Mountain by the River Nevis'. Gaelic *beinn*. The river-name derives from an early Celtic root *nebh* 'moist, water'. The river also gives its name to **Glen Nevis** and **Loch Nevis**.

Ben Rhydding Brad. not recorded until 1858, from the pers. name *Ben* and **rydding* 'a clearing'.

Ben Vorlich (mountain) Arg. Gaelic *beinn* 'mountain' + obscure element.

Benacre Suffolk. *Benagra* 1086 (DB). 'Cultivated plot where beans are grown'. OE *bēan* + *æcer*.

Benagh (*Beitheanach*) Down. 'Place of birch-trees'.

Benbane Head (*An Bhinn Bhán*) Antrim. 'The white headland'.

Benbecula W. Isles. *Beanbeacla* 1449. Possibly 'hill of the fords'. Gaelic *beinn-na-fhaodla*.

Benbeg (*Beann Beag*) Galway. 'Little peak'.

Benbo (*Beann Bó*) Leitrim. 'Peak of the cow'.

Benbrack (*Beann Breac*) Cavan. 'Speckled peak'.

Benbrack (mountain) Dumf. 'Speckled mountain'. Gaelic *beinn* + *breac*.

Benbulbin (*Beann Ghulbain*) Sligo. 'Gulban's peak'.

Benburb (*An Bhinn Bhorb*) Tyrone. *Beunn Bhoruib* 1621. 'The bold peak'.

Benderloch Arg. *Bintaloch* 1654. 'Hill between two lochs'. Gaelic *beann* + *eader* + *da* + *loch*.

Bendooragh (*Bun Dúraí*) Antrim. *Bun Dubhroighe* c.1645. 'Bottom of black soil'.

Benefield, Upper & Benefield, Lower Northants. *Beringafeld* 10th cent., *Benefeld* 1086 (DB). 'Open land of the family or followers of a man called Bera'. OE pers. name + *-inga-* + *feld*.

Benenden Kent. *Bingdene* 993, *Benindene* 1086 (DB). 'Woodland pasture associated with a man called Bionna'. OE pers. name + *-ing-* + *denn*.

Benfleet Essex. *Beamfleote* 10th cent., *Benflet* 1086 (DB). 'Tree-trunk creek', perhaps referring to a bridge. OE *bēam* + *flēot*.

Bengore (*Beann Gabhar*) Antrim. 'Peak of the goats'.

Benhall Green Suffolk. *Benenhala* 1086 (DB). 'Nook of land where beans are grown'. OE *bēanen* + *halh*.

Beningbrough N. Yorks. *Benniburg* 1086 (DB). 'Stronghold associated with a man called Beonna'. OE pers. name + *-ing-* + *burh*.

Benington Herts. *Benington* 1086 (DB). 'Farmstead by the River Beane'. Pre-English river-name (of uncertain origin and meaning) + OE *-ing-* + *tūn*.

Benington Lincs. *Benington* 1166. 'Farmstead associated with a man called Beonna'. OE pers. name + *-ing-* + *tūn*.

Benllech Angl. 'Head of the stone'. Welsh *pen* + *llech*. The name refers to the capstone of a cromlech.

Benmore (*Beann Mór*) Antrim. 'Large peak'.

Bennington, Long Lincs. *Beningtun* 1086 (DB). Identical in origin with BENINGTON (Lincs.).

Benniworth Lincs. *Beningurde* 1086 (DB). 'Enclosure of the family or followers of a man called Beonna'. OE pers. name + *-inga-* + *worth*.

Benone (*Bun Abhann*) Derry. *Bunowne* 1654. 'Foot of the river'.

Benson Oxon. *Bænesingtun* c.900, *Besintone* 1086 (DB). 'Estate associated with a man called *Benesa*. OE pers. name + *-ing-* + *tūn*.

Benthall Shrops. near Broseley. *Benethala* 12th cent. 'Nook of land where bent-grass grows'. OE *beonet* + *halh*.

Bentham, 'homestead or enclosure where bent-grass grows', OE *beonet* + *hām* or *hamm*: **Bentham** Glos. *Benetham* 1220. **Bentham, High & Bentham, Lower** N. Yorks. *Benetain* [*sic*] 1086 (DB), *Benetham* 1214.

Bentley, a common name, 'woodland clearing where bent-grass grows', OE *beonet* + *lēah*; examples include: **Bentley** Donc.

Benedleia 1086 (DB). **Bentley** Hants., near Alton. *Beonetleh* c.965, *Benedlei* 1086 (DB). **Bentley** E. R. Yorks. *Benedlage* 1086 (DB). **Bentley, Fenny** Derbys. *Benedlege* 1086 (DB), *Fennibenetlegh* 1272. Affix is OE *fennig* 'marshy'. **Bentley, Great & Bentley, Little** Essex. *Benetleye* c.1040, *Benetlea* 1086 (DB).

Bentworth Hants. *Binteworda* 1130. Probably 'enclosure of a man called *Binta*. OE pers. name + *worth*.

Benwee Head (*An Bhinn Bhuí*) Mayo. 'The yellow peak'.

Benwick Cambs. *Beymwich* 1221. 'Farm where beans are grown', or 'farm by a tree-trunk'. OE *bēan* or *bēam* + *wīc*.

Beoley Worcs. *Beoleah* 972, *Beolege* 1086 (DB). 'Wood or clearing frequented by bees'. OE *bēo* + *lēah*.

Bepton W. Sussex. *Babintone* 1086 (DB). 'Estate associated with a woman called Bebbe or a man called *Bebba*. OE pers. name + *-ing-* + *tūn*.

Beragh (*Bearach*) Tyrone. *Berhagh* 1631. 'Place of peaks'.

Berden Essex. *Berdane* 1086 (DB). Probably 'valley with a woodland pasture'. OE *bẽr* + *denu*.

Bere Alston Devon. *Alphameston* 1339, *Berealmiston* c.1450. 'Farmstead of a man called Ælfhelm'. OE pers. name + *tūn*, with *Bere* from BERE FERRERS.

Bere Ferrers Devon. *Birlanda* [*sic*] 1086 (DB), *Ber* 1242, *Byr Ferrers* 1306. 'Woodland pasture', or 'wood, grove', OE *bẽr* or *bearu* (in the first spelling with *land* 'estate'). Manorial affix from the *de Ferers* family, here in the 13th cent.

Bere Regis Dorset. *Bere* 1086 (DB), *Kyngesbyre* 1264. 'Woodland pasture', or 'wood, grove', OE *bẽr* or *bearu*. Affix is Latin *regis* 'of the king'.

Bergholt, 'Wood on or by a hill', OE *beorg* + *holt*: **Bergholt, East** Suffolk. *Bercolt* 1086 (DB). **Bergholt, West** Essex. *Bercolt* 1086 (DB).

Berkeley Glos. *Berclea* 824, *Berchelai* 1086 (DB). 'Birch-tree wood or clearing'. OE *beorc* + *lēah*.

Berkhamsted, Great Herts. *Beorhthanstædæ* 10th cent., *Berchehamstede*

1086 (DB). Probably 'homestead on or near a hill'. OE *beorg* + *hām-stede*.

Berkhamsted, Little Herts. *Berchehamstede* 1086 (DB). Probably 'homestead where birch-trees grow'. OE *beorc* + *hām-stede*.

Berkley Somerset. *Berchelei* 1086 (DB). 'Birch-tree wood or clearing'. OE *beorc* + *lēah*.

Berkshire (the county). *Berrocscire* 893. An ancient Celtic name meaning 'hilly place' + OE *scīr* 'shire, district'.

Berkswell Solhll. *Berchewelle* 1086 (DB). 'Spring or stream of a man called Beorcol'. OE pers. name + *wella*.

Bermo, Y. See BARMOUTH.

Bermondsey Gtr. London. *Vermundesei* [sic] c.712, *Bermundesye* 1086 (DB). 'Island, or dry ground in marsh, of a man called Beornmund'. OE pers. name + *ēg*.

Bernera, Great (island) W. Isles. *Bjarnarey* c.1250. 'Bjarni's island'. OScand. pers. name + *ey*.

Berrick Salome Oxon. *Berewiche* 1086 (DB), *Berwick Sullame* 1571. 'Barley farm', or 'outlying part of an estate'. OE *bere-wīc* + manorial affix from early possession by the *de Suleham* family.

Berrier Cumbria. *Berghgerge* 1166. 'Shieling or pasture on a hill'. OScand. *berg* + *erg*.

Berrington Northum. *Berigdon* 1208–10. 'Hill with a fortification'. OE *burh* (genitive or dative *byrig*) + *dūn*.

Berrington Shrops. *Beritune* 1086 (DB). 'Farmstead associated with a fortification'. OE *burh* (genitive or dative *byrig*) + *tūn*.

Berrow Somerset. *Burgh* 973, *Berges* 1196. '(Place at) the hill(s) or mound(s)'. OE *beorg*, with reference to the sand-dunes here.

Berrow Green Worcs. *Berga* 1275. '(Place at) the hill or mound'. OE *beorg*.

Berry Pomeroy Devon. *Beri* 1086 (DB), *Bury Pomery* 1281. Identical in origin with the following name. Manorial affix from the *de Pomerei* family, here from the 11th cent.

Berrynarbor Devon. *Biria* c.1150, *Bery Narberd* 1244. '(Place at) the fortification'. OE *burh* (dative *byrig*) + manorial affix from the *Nerebert* family, here in the 13th cent.

Bersted W. Sussex. *Beorganstede* 680. Probably 'homestead by a tumulus'. OE *beorg* + *hām-stede*.

Berwick, a common name, from OE *bere-wīc* 'barley farm, outlying part of an estate'; examples include: **Berwick Bassett** Wilts. *Berwicha* 1168, *Berewykbasset* 1321. Manorial affix from the *Basset* family, here in the 13th cent. **Berwick, North** E. Loth., see NORTH BERWICK. **Berwick St James** Wilts. *Berewyk Sancti Jacobi* c.1190. Affix *St James* (Latin *Jacobus*) from the dedication of the church. **Berwick St John** Wilts. *Berwicha* 1167, *Berewyke S. Johannis* 1265. Affix from the dedication of the church. **Berwick St Leonard** Wilts. *Berewica* 12th cent., *Berewyk Sancti Leonardi* 1291. Affix from the dedication of the church. **Berwick upon Tweed** Northum. *Berewich* 1167, *Berewicum super Twedam* 1229. For the river-name, see TWEEDMOUTH.

Besford Worcs. *Bettesford* 972, *Beford* [sic] 1086 (DB). 'Ford of a man called Betti'. OE pers. name + *ford*.

Bessacarr Donc. *Beseacra* 1182. 'Cultivated plot where bent-grass grows'. OE **bēos* + *æcer*.

Bessbrook Armagh. *Bessbrook* 1888. 'Bess's brook'. The name is that of Elizabeth (*Bess*) Nicholson, wife of Joseph Nicholson, whose family carried on a linen business here in the early 19th cent. The Irish name of Bessbrook is *An Sruthán*, 'the stream'.

Bessels Leigh Oxon. See LEIGH.

Bessingham Norfolk. *Basingeham* 1086 (DB). 'Homestead of the family or followers of a man called **Basa*'. OE pers. name + *-inga-* + *hām*.

Besthorpe, 'outlying farmstead or hamlet of a man called Bōsi, or where bent-grass grows', OScand. pers. name or OE **bēos* + OScand. *thorp*: **Besthorpe** Norfolk. *Besethorp* 1086 (DB). **Besthorpe** Notts. *Bestorp* 1147.

Beswick E. R. Yorks. *Basewic* 1086 (DB). 'Dwelling or (dairy) farm of a man called Bōsi or Bessi'. OScand. pers. name + OE *wīc*.

Betchworth Surrey. *Becesworde* 1086 (DB). 'Enclosure of a man called **Becci*'. OE pers. name + *worth*.

Bethersden Kent. *Baedericesdaenne* c.1100. 'Woodland pasture of a man called **Beadurīc*'. OE pers. name + *denn*.

Bethesda Gwyd. The town arose around a Welsh Nonconformist chapel built in 1820 that was named after the biblical pool of *Bethesda*, where Jesus healed the sick (John 5:1–10).

Bethnal Green Gtr. London. *Blithehale* 13th cent., *Blethenalegrene* 1443. 'Nook of land of a man called *Blītha*'. OE pers. name (genitive -*n*) + *halh*, with the later addition of ME *grene* 'village green'. Alternatively the first element may be OE *blīthe* (dative -*an*) 'pleasant' or an OE stream-name *Blīthe* 'pleasant one'.

Betley Staffs. *Betelege* 1086 (DB). 'Woodland clearing of a woman called *Bette*'. OE pers. name + *lēah*.

Betteshanger Kent. *Betleshangre* 1176. Probably 'wooded slope by a house or building'. OE (*ge*)*bytle* + *hangra*. Alternatively the first element may be an OE pers. name *Byttel*.

Bettiscombe Dorset. *Bethescomme* 1129. 'Valley of a man called Betti'. OE pers. name + *cumb*.

Betton, probably 'farmstead or estate where beech-trees grow', OE *bēce* + *tūn*: **Betton** Shrops, near Binweston. *Betune* 1086 (DB). **Betton** Shrops., near Market Drayton. *Baitune* 1086 (DB), *Bectona* 1121.

Bettyhill Highland. 'Betty's hill'. The settlement arose in *c.*1820 and was named from Elizabeth (*Betty*), Countess of Sutherland and Marchioness of Stafford (1765–1839).

Betws-y-coed Conwy. *Betus* 1254, *Bettws y Coed* 1727. 'Chapel in the wood'. Welsh *betws* + *y* + *coed*. Welsh *betws* is borrowed from OE *bed-hūs*, 'oratory' (literally 'bead house').

Bevercotes Notts. *Beurecote* 1165. 'Place where beavers have built their nests'. OE *beofor* + *cot*.

Beverley E. R. Yorks. *Beferlic c.*1025, *Bevreli* 1086 (DB). Probably a Celtic name meaning 'beaver lodge' (from a British *bebros* + the ancestor of Welsh *llech* 'covert, hiding-place').

Beverstone Glos. *Beurestane* 1086 (DB). Probably '(boundary) stone of a man called *Beofor*'. OE pers. name + *stān*.

Bewaldeth Cumbria. *Bualdith* 1255. 'Homestead or estate of a woman called Aldgȳth'. OScand. *bú* + OE pers. name. The order of elements is Celtic.

Bewcastle Cumbria. *Bothecastre* 12th cent. 'Roman fort within which shelters or huts were situated'. OScand. *búth* + OE *ceaster*.

Bewdley Worcs. *Beuleu* 1275. 'Beautiful place'. OFrench *beau* + *lieu*.

Bewerley N. Yorks. *Beurelie* 1086 (DB). 'Woodland clearing frequented by beavers'. OE *beofor* + *lēah*.

Bewholme E. R. Yorks. *Begun* 1086 (DB). '(Place at) the river-bends'. OE *bēag* or OScand. *bjúgr* in a dative plural form *bēagum* or *bjúgum*.

Bexhill E. Sussex. *Bixlea* 772, *Bexelei* 1086 (DB). Probably 'wood or clearing where box-trees grow'. OE *byxe* + *lēah*.

Bexington Dorset. *Bessintone* 1086 (DB), *Buxinton* 1212. Probably 'farmstead or village where box-trees grow'. OE *byxen* + *tūn*.

Bexley Gtr. London. *Byxlea* 814. Probably 'wood or clearing where box-trees grow'. OE *byxe* + *lēah*. **Bexleyheath** is a 19th-cent. name with the addition *heath*.

Bexwell Norfolk. *Bekeswella* 1086 (DB). 'Spring or stream of a man called *Bēac*'. OE pers. name + *wella*.

Beyton Suffolk. *Begatona* 1086 (DB). Probably 'farmstead of a woman called Bēage or of a man called Bǣga'. OE pers. name + *tūn*.

Bibury Glos. *Beaganbyrig* 8th cent., *Begeberie* 1086 (DB). 'Stronghold or manor house of a woman called Bēage'. OE pers. name + *burh* (dative *byrig*). Bēage is named as leasing the estate in a document dated 718–45.

Bicester Oxon. *Bernecestre* 1086 (DB). 'Fort of the warriors, or of a man called Beorna'. OE *beorn* or OE pers. name + *ceaster*.

Bickenhall Somerset. *Bichalle* 1086 (DB). Probably 'hall of a man called Bica'. OE pers. name (genitive -*n*) + *heall*.

Bickenhill Solhll. *Bichehelle* 1086 (DB), *Bikenhulle* 1202. Probably 'hill with a point, projecting hill'. OE *bica* (genitive -*n*) + *hyll*.

Bicker Lincs. *Bichere* 1086 (DB). Possibly 'village marsh', from OScand. *bý* + *kjarr*. Alternatively '(place) by the marsh', with OE *bī* as first element.

Bickerstaffe Lancs. *Bikerstad* late 12th cent. Probably 'place or site of the bee-keepers'. OE *bīcere (genitive plural *bīcera) + OScand. stathr.

Bickerton, 'farmstead of the bee-keepers', OE *bīcere (genitive plural *bīcera) + tūn: **Bickerton** Ches. Bicretone 1086 (DB). **Bickerton** N. Yorks. Bicretone 1086 (DB).

Bickington, 'estate associated with a man called Beocca', OE pers. name + -ing- + tūn: **Bickington** Devon, near Ashburton. Bechintona 1107. **Bickington, Abbots** Devon. Bicatona 1086 (DB), Abbots Bekenton 1580. Affix from early possession by Hartland Abbey. **Bickington, High** Devon. Bichentona 1086 (DB), Heghebuginton 1423. Affix is OE hēah 'high'.

Bickleigh, Bickley, probably 'woodland clearing on or near a pointed ridge', OE *bica + lēah: **Bickleigh** Devon, near Plymouth. Bicheleia 1086 (DB). **Bickleigh** Devon, near Tiverton. Bicanleag 904, Bichelia 1086 (DB). **Bickley** Gtr. London. Byckeleye 1279. **Bickley Moss** Ches. Bichelei 1086 (DB). Later addition is OE mos 'peat-bog'.

Bicknacre Essex. Bikenacher 1186. Possibly 'cultivated plot of a man called Bica'. OE pers. name (genitive -n) + æcer. Alternatively the first element may be OE *bica (genitive -n) 'point'.

Bicknoller Somerset. Bykenalre 1291. 'Alder-tree of a man called Bica'. OE pers. name (genitive -n) + alor.

Bicknor Kent. Bikenora 1186. Probably 'slope below the pointed hill'. OE *bica (genitive -n) + ōra.

Bicknor, English Glos. Bicanofre 1086 (DB), Englise Bykenore 1248. Probably 'ridge with a point'. OE *bica (genitive -n) + *ofer. Affix from its situation on the English side of the River Wye.

Bicknor, Welsh Herefs. Bykenore Walens 1291. Identical in origin with the previous name, with distinguishing affix from its situation on the Welsh side of the River Wye.

Bickton Hants. Bichetone 1086 (DB). Probably 'farmstead of a man called Bica'. OE pers. name + tūn.

Bicton Shrops. near Shrewsbury. Bichetone [sic] 1086 (DB), Bikedon 1204. Probably 'hill with a pointed ridge'. OE *bica + dūn.

Bidborough Kent. Bitteberga c.1100. 'Hill or mound of a man called *Bitta'. OE pers. name + beorg.

Biddenden Kent. Bidingden 993. 'Woodland pasture associated with a man called *Bida'. OE pers. name + -ing- + denn.

Biddenham Beds. Bidenham 1086 (DB). 'Homestead, or land in a river-bend, of a man called Bīeda'. OE pers. name (genitive -n) + hām or hamm.

Biddestone Wilts. Bedestone 1086 (DB), Bedeneston 1187. Probably 'farmstead of a man called *Bīedin or *Bīede'. OE pers. name + tūn.

Biddisham Somerset. Biddesham 1065. 'Homestead or enclosure of a man called *Biddi'. OE pers. name + hām or hamm.

Biddlesden Bucks. Betesdene [sic] 1086 (DB), Bethlesdena 12th cent. Probably 'valley with a house or building'. OE *bythle + denu. Alternatively the first element may be an OE pers. name *Byttel.

Biddlestone Northum. Bidlisden 1242. Probably identical in origin with the previous name.

Biddulph Staffs. Bidolf 1086 (DB). '(Place) by the pit or quarry'. OE bī + *dylf.

Bideford Devon. Bedeford 1086 (DB). Possibly 'ford at the stream called Bȳd'. Celtic river-name (of uncertain origin and meaning) + OE ēa + ford. Alternatively the first element may be OE byden 'vessel, tub', also 'hollow, deep valley'.

Bidford on Avon Warwicks. Budiford 710, Bedeford 1086 (DB). Probably identical in origin with the previous name. For the river-name, see AVON.

Bidston Wirral. Budeston, Bediston 1260, Budestan 1286, Bidelston 1298. Possibly 'rocky hill with a house or building'. OE *bythle + stān.

Bielby E. R. Yorks. Belebi 1086 (DB). 'Farmstead or village of a man called Beli'. OScand. pers. name + bý.

Bierley, East Kirkl. Birle 1086 (DB), Birel 1303. Possibly 'woodland clearing by the stronghold'. OE burh (genitive byrh) + lēah. Alternatively from an OE *bȳrel 'small dwelling'.

Bierton Bucks. Bortone 1086 (DB). 'Farmstead near the stronghold'. OE byrh-tūn.

Big Mancot Flin. Manecote 1284, Great Mancott 1547. 'Mana's cottage'. OE pers. name + cot. Big for distinction from adjoining Little Mancot.

Bigbury Devon. *Bicheberie* 1086 (DB). 'Stronghold of a man called Bica'. OE pers. name + *burh* (dative *byrig*).

Bigby Lincs. *Bechebi* 1086 (DB). Probably 'farmstead or village of a man called *Bekki*. OScand. pers. name + *bý*.

Biggar S. Lan. *Bigir* 1170. Meaning uncertain.

Biggin Hill Gtr. London. *Byggunhull* 1499. 'Hill with or near a building'. ME *bigging* + *hull* (OE *hyll*).

Biggleswade Beds. *Pichelesuuade* [*sic*] 1086 (DB), *Bicheleswada* 1132. 'Ford of a man called *Biccel*. OE pers. name + *wæd*.

Bighton Hants. *Bicincgtun* 959, *Bighetone* 1086 (DB). 'Estate associated with a man called Bica'. OE pers. name + *-ing-* + *tūn*.

Bignor W. Sussex. *Bigenevre* 1086 (DB). 'Hill brow of a man called *Bicga*. OE pers. name (genitive *-n*) + *yfer*.

Bilborough Nott. *Bileburch* 1086 (DB). 'Stronghold of a man called *Bila* or *Billa*. OE pers. name + *burh*.

Bilbrook Staffs. *Bilrebroch* 1086 (DB). 'Brook where water-cress grows'. OE *billere* + *brōc*.

Bilbrough N. Yorks. *Mileburg* [*sic*] 1086 (DB), *Billeburc* 1167. Identical in origin with BILBOROUGH.

Bildeston Suffolk. *Bilestuna* 1086 (DB). Identical in origin with BILSTONE.

Billericay Essex. *Byllyrica* 1291. Probably from a medieval Latin word **bellerīca* meaning 'dyehouse or tanhouse'.

Billesdon Leics. *Billesdone* 1086 (DB). 'Hill of a man called Bil'. OE pers. name + *dūn*. Alternatively the first element may be OE *bill* 'sword' used of a pointed hill.

Billesley Warwicks. *Billeslæh* 704–9, *Billeslei* 1086 (DB). 'Woodland clearing of a man called Bil'. OE pers. name + *lēah*. Alternatively the first element may be OE *bill* 'pointed hill'.

Billing, Great & Billing, Little Northants. *Bel(l)inge* 1086 (DB). Possibly '(settlement of) the family or followers of a man called Bil or *Billa*. OE pers. name + *-ingas*. Alternatively 'hill place', from an OE **belling* or **billing* 'hill, ridge'.

Billingborough Lincs. *Billingeburg* 1086 (DB). Possibly 'stronghold of the family or followers of a man called Bil or *Billa*. OE pers. name + *-inga-* + *burh*. Alternatively the first element may be OE **billing* 'hill, ridge'.

Billinge St Hel. *Billing* 1202. Probably OE **billing* 'hill, ridge'.

Billingham Stock. on T. *Billingham* c.1040. 'Homestead or village on the hill called **Billing*'. OE **billing* 'hill, ridge' + *hām*.

Billinghay Lincs. *Belingei* 1086 (DB). Possibly 'island, or dry ground in marsh, of the family or followers of a man called Bil or *Billa*. OE pers. name + *-inga-* + *ēg*. Alternatively the first element may be OE **billing* 'hill, ridge'.

Billingley Barns. *Bilingeleia* 1086 (DB). Probably 'woodland clearing of the family or followers of a man called Bil or *Billa*. OE pers. name + *-inga-* + *lēah*.

Billingshurst W. Sussex. *Bellingesherst* 1202. Probably 'wooded hill of a man called Billing'. OE pers. name + *hyrst*. Alternatively the first element may be OE **billing* 'hill, ridge'.

Billington Beds. *Billendon* 1196. Probably 'hill of a man called **Billa*. OE pers. name (genitive *-n*) + *dūn*. Alternatively the first element may be OE **billing* 'hill, ridge'.

Billington Lancs. *Billingduna* 1196. Probably 'hill called **Billing*'. OE **billing* 'hill, ridge' + *dūn*.

Billockby Norfolk. *Bithlakebei* 1086 (DB). Possibly 'farmstead or village of a man called **Bithil-Áki*. OScand. pers. name + *bý*.

Bilney, probably 'island site of a man called Bil(l)a', OE pers. name (genitive *-n*) + *ēg*: **Bilney, East** Norfolk. *Billneye* 1254. **Bilney, West** Norfolk. *Bilenei* 1086 (DB).

Bilsborrow Lancs. *Billesbure* 1187. Probably 'stronghold of the promontory'. OE *bill*.

Bilsby Lincs. *Billesbi* 1086 (DB). 'Farmstead or village of a man called Billi'. OScand. pers. name + *bý*.

Bilsington Kent. *Bilsvitone* 1086 (DB). 'Farmstead of a woman called Bilswīth'. OE pers. name + *tūn*.

Bilsthorpe Notts. *Bildestorp* 1086 (DB). 'Outlying farmstead or hamlet of a man called Bildr'. OScand. pers. name + *thorp*. Alternatively

the first element may be OScand. *bildr* 'angle'
used figuratively for 'hill, promontory'.

Bilston Wolverh. *Bilsetnatun* 996, *Billestune*
1086 (DB). 'Farmstead of the dwellers at the
sharp ridge'. OE *bill* + *sǣte* (genitive plural
sǣtna) + *tūn*.

Bilstone Leics. *Bildestone* 1086 (DB).
Probably 'farmstead of a man called Bildr'.
OScand. pers. name + OE *tūn*. Alternatively the
first element may be OScand. *bildr* 'angle' used
figuratively for 'hill, promontory'.

Bilton, 'farmstead of a man called *Billa',
OE pers. name + *tūn*: **Bilton** E. R. Yorks. *Bil(l)
etone* 1086 (DB). **Bilton** N. Yorks. *Biletone* 1086
(DB).
    However with a different origin are: **Bilton**
Northum. *Bylton* 1242. 'Farmstead on a ridge'.
OE *bill* + *tūn*. **Bilton** Warwicks. *Beltone,
Bentone* 1086 (DB). Possibly 'farmstead where
henbane grows'. OE *beolone* + *tūn*.

Binbrook Lincs. *Binnibroc* 1086 (DB). '(Place)
enclosed by the brook', or 'brook of a man
called Bynna'. OE *binnan* or OE pers. name +
brōc. OE *binn(e)* 'manger, stall', perhaps used in
a figurative sense 'valley', would also be a
possible first element.

Bincombe Dorset. *Beuncumbe* 987, *Beincome*
1086 (DB). Probably 'valley where beans are
grown'. OE *bēan* + *cumb*.

Binegar Somerset. *Begenhangra* 1065.
Probably 'wooded slope of a woman called
Bēage'. OE pers. name (genitive -*n*) + *hangra*.
Alternatively the first element may be OE **begen*
'growing with berries'.

Binevenagh (*Binn Fhoibhne*) Derry.
'Foibhne's peak'.

Binfield Brack. For. *Benetfeld* c.1160. 'Open
land where bent-grass grows'. OE *beonet* + *feld*.

Binfield Heath Oxon. *Benifeld* 1177.
Probably identical in origin with the previous
name, but possibly 'open land of a man called
*Beona'. OE pers. name + *feld*. The addition
heath is found from the 16th cent.

Bingfield Northum. *Bingefeld* 1181. Probably
'open land of the family or followers of a man
called Bynna'. OE pers. name + -*inga*- + *feld*.
Alternatively in this and the following two
names the first element may be an OE **bing*
'a hollow'.

Bingham Notts. *Bingheham* 1086 (DB).
Probably 'homestead of the family or followers
of a man called Bynna'. OE pers. name + -*inga*-
+ *hām*. But *see* BINGFIELD.

Bingley Brad. *Bingelei* 1086 (DB). Probably
'woodland clearing of the family or followers of
a man called Bynna'. OE pers. name + -*inga*- +
lēah. But *see* BINGFIELD.

Binham Norfolk. *Binneham* 1086 (DB).
'Homestead or enclosure of a man called
Bynna'. OE pers. name + *hām* or *hamm*.

Binley Covtry. *Bilnei* 1086 (DB). Probably
'island site of a man called Bil(l)a'. OE pers.
name (genitive -*n*) + *ēg*.

Binstead I. of Wight. *Benestede* 1086 (DB).
'Place where beans are grown'. OE *bēan* + *stede*.

Binsted Hants. *Benestede* 1086 (DB). Identical
in origin with the previous name.

Binton Warwicks. *Bynningtun* c.1005,
Beninton 1086 (DB). 'Estate associated with a
man called Bynna'. OE pers. name + -*ing*- + *tūn*.

Bintree Norfolk. *Binnetre* 1086 (DB). 'Tree of a
man called Bynna'. OE pers. name + *trēow*.

Binweston Shrops. *Binneweston* 1292.
Probably 'west farmstead of a man called
Bynna'. OE pers. name + *west* + *tūn*.

Birch Essex. *Bric(ce)iam* 1086 (DB), *Brich* 1194.
Probably identical in origin and meaning with
the next name.

Birch, Much & Birch, Little Herefs. *Birches*
1252. '(Place at) the birch-tree(s).' OE *birce*. The
affix *Much* is from OE *micel* 'great'.

**Bircham, Great, Bircham Newton,
Bircham Tofts** Norfolk. *Brecham* 1086 (DB).
'Homestead by newly cultivated ground'. OE
brēc + *hām*. Bircham Newton is *Niwetuna* 1086
(DB), 'new farmstead', OE *nīwe* + *tūn*. Bircham
Tofts is *Toftes* 1205, OScand. *toft* 'homestead'.

Birchanger Essex. *Bilichangra* [*sic*] 1086
(DB), *Birichangre* 12th cent. 'Wooded slope
growing with birch-trees'. OE *birce* + *hangra*.

Bircher Herefs. *Burchoure* 1212. 'Ridge where
birch-trees grow'. OE *birce* + **ofer*.

Birchington Kent. *Birchenton* 1240.
'Farmstead where birch-trees grow'. OE **bircen*
+ *tūn*.

Birchover Derbys. *Barcovere* [*sic*] 1086 (DB), *Birchoure* 1226. 'Ridge where birch-trees grow'. OE *birce* + **ofer*.

Birdbrook Essex. *Bridebroc* 1086 (DB). 'Brook frequented by young birds'. OE *bridd* + *brōc*.

Birdham W. Sussex. *Bridham* 683, *Brideham* 1086 (DB). 'Homestead or enclosure frequented by young birds'. OE *bridd* + *hām* or *hamm*.

Birdlip Glos. *Bridelepe* 1221. Probably 'steep place frequented by young birds'. OE *bridd* + **hlēp*.

Birdsall N. Yorks. *Brideshala* 1086 (DB). Probably 'nook of land of a man called Bridd'. OE pers. name + *halh*. Alternatively the first element may be the OE noun *bridd* 'young bird' used in a collective sense.

Birkby N. Yorks. *Bretebi* 1086 (DB). 'Farmstead or village of the Britons'. OScand. *Bretar* + *bý*. The same name occurs in W. Yorks. and Cumbria.

Birkdale Sefton. *Birkedale* c.1200. 'Valley where birch-trees grow'. OScand. *birki* + *dalr*.

Birkenhead Wirral. *Bircheveth* c.1200, *Birkheued* 1260. 'Headland where birch-trees grow'. OE *birce*, **bircen* (with Scand. *-k-*) + *hēafod*.

Birkenshaw Kirkl. *Birkenschawe* 1274. 'Small wood or copse where birch-trees grow'. OE **bircen* (with Scand. *-k-*) + *sceaga*.

Birkin N. Yorks. *Byrcene* c.1030, *Berchine* 1086 (DB). 'Place growing with birch-trees'. OE **bircen* (with Scand. *-k-*).

Birley Herefs. *Burlei* 1086 (DB). 'Woodland clearing near a stronghold'. OE *burh* + *lēah*.

Birling, probably '(settlement of) the family or followers of a man called **Bærla*', OE pers. name + *-ingas*: **Birling** Kent. *Boerlingas* 788, *Berlinge* 1086 (DB). **Birling** Northum. *Berlinga* 1187.

Birlingham Worcs. *Byrlingahamm* 972, *Berlingeham* 1086 (DB). 'Land in a river-bend of the family or followers of a man called **Byrla*'. OE pers. name + *-inga-* + *hamm*.

Birmingham Birm. *Bermingeham* 1086 (DB). 'Homestead of the family or followers of a man called **Beorma*', or 'homestead at the place associated with **Beorma*'. OE pers. name + *-inga-* or *-ing* + *hām*.

Birnie Moray. *Brennach* c.1190. 'Marshy place'. Gaelic *braonach*.

Birr (*Biorra*) Offaly. 'Stream'.

Birra (*Biorra*) Donegal. 'Stream'.

Birstall Leics. *Burstele* 1086 (DB). OE *burhstall* 'the site of a stronghold'.

Birstall Kirkl. *Birstale* 12th cent. Identical in meaning with the previous name, but from OE *byrh-stall*.

Birstwith N. Yorks. *Beristade* 1086 (DB). Possibly 'farm built on the site of a lost farm'. OScand. *býjar-stathr*. Alternatively 'landing-place of the fort', from OE *burh* (genitive *byrg*) + *stæth*.

Birtley, 'bright clearing'. OE *beorht* + *lēah*: **Birtley** Gatesd. *Britleia* 1183. **Birtley** Northum. *Birtleye* 1229.

Bisbrooke Rutland. *Bitlesbroch* 1086 (DB). Possibly 'brook of a man called **Bitel* or **Byttel*', OE pers. name + *brōc*. Alternatively 'brook infested by water-beetles', from an OE **bitel* 'beetle'.

Bisham Winds. & Maid. *Bistesham* 1086 (DB), *Bistlesham* 1199. 'Homestead or enclosure of a man called **Byssel*'. OE pers. name + *hām* or *hamm*.

Bishampton Worcs. *Bisantune* 1086 (DB). Possibly 'homestead of a man called **Bisa*'. OE pers. name + *hām-tūn*.

Bishop, Bishops as affix. *See* main name, *e.g.* for **Bishop Auckland** (Durham) *see* AUCKLAND.

Bishop's Castle Shrops. *Castrum Episcopi* 1255, *Bisshopescastel* 1282. Named from the castle (Latin *castrum*, ME *castel*) erected c.1127 by the Bishop of Hereford.

Bishopsbourne Kent. *Burnan* 799, *Burnes* 1086 (DB), *Biscopesburne* 11th cent. 'Estate on the river called *Burna*' (from OE *burna* 'stream' referring to the Little Stour), with later affix (OE *biscop*) from its possession by the Archbishop of Canterbury.

Bishopsteignton Devon. *Taintona* 1086 (DB), *Teynton Bishops* 1341. 'Farmstead on the River Teign'. Celtic river-name (*see* TEIGNMOUTH) + OE *tūn*, with manorial affix from its possession by the Bishop of Exeter in 1086.

Bishopstoke Hants. *Stoches* 1086 (DB), *Stoke Episcopi c.*1270. 'Outlying farmstead or hamlet (of the bishop)'. OE *stoc* with later addition from its possession by the Bishop of Winchester.

Bishopstone, 'the bishop's estate', OE *biscop* + *tūn*; examples include: **Bishopstone** Bucks. *Bissopeston* 1227. **Bishopstone** E. Sussex. *Biscopestone* 1086 (DB). **Bishopstone** Herefs. *Biscopestone* 1166. **Bishopstone** Swindn. *Bissopeston* 1186. **Bishopstone** Wilts., near Wilton. *Bissopeston* 1166.

Bishopsworth Brist. *Biscopewrde* 1086 (DB). 'The bishop's enclosure'. OE *biscop* + *worth*.

Bishopthorpe York. *Torp* 1086 (DB), *Biscupthorp* 1275. 'Outlying farmstead or hamlet held by the bishop'. OE *biscop* + OScand. *thorp*.

Bishopton Darltn. *Biscoptun* 1104–8. 'The bishop's estate'. OE *biscop* + *tūn*.

Bisley Glos. *Bislege* 986, *Biselege* 1086 (DB). 'Woodland clearing of a man called *Bisa'. OE pers. name + *lēah*.

Bisley Surrey. *Busseleghe* 933. 'Woodland clearing with bushes, or of a man called *Byssa'. OE **bysce* or pers. name + *lēah*.

Bispham, 'the bishop's estate', OE *biscop* + *hām*: **Bispham** Bpool. *Biscopham* 1086 (DB). **Bispham Green** Lancs. *Biscopehaim c.*1200.

Bisterne Hants. *Betestre* [sic] 1086 (DB), *Budestorn* 1187. Probably 'thorn-tree of a man called *Bytti'. OE pers. name + *thorn*.

Bitchfield Lincs. *Billesfelt* 1086 (DB). 'Open land of a man called Bill'. OE pers. name + *feld*. Alternatively the first element may be OE *bill* 'sword' used figuratively for 'hill, promontory'.

Bittadon Devon. *Bedendone* [sic] 1086 (DB), *Bettenden* 1205. 'Valley of a man called *Beotta'. OE pers. name + *denu*.

Bittering Norfolk. *Britringa* 1086 (DB). '(Settlement of) the family or followers of a man called Beorhthere'. OE pers. name + *-ingas*.

Bitterley Shrops. *Buterlie* 1086 (DB). 'Pasture which produces good butter'. OE *butere* + *lēah*.

Bitterne Sotn. *Byterne c.*1090. Possibly 'building near a river-bend'. OE *byht* + *ærn*. Alternatively the first element may be OE *bit* 'bit' (perhaps referring to horse-tackle and the like).

Bitteswell Leics. *Betmeswelle* 1086 (DB). Possibly 'spring or stream in a broad valley'. OE **bytm* + *wella*.

Bitton S. Glos. *Betune* 1086 (DB). 'Farmstead on the River Boyd'. Pre-English river-name (of uncertain origin and meaning) + OE *tūn*.

Biwmares. *See* BEAUMARIS.

Bix Oxon. *Bixa* 1086 (DB). '(Place at) the box-tree wood'. OE **byxe*.

Blaby Leics. *Bladi* [sic] 1086 (DB), *Blabi* 1175. Probably 'farmstead or village of a man called Blár'. OScand. pers. name + *bý*.

Black as affix. *See* main name, e.g. for **Black Bourton** (Oxon.) *see* BOURTON.

Black Mountains (*Y Mynydd Du*) Powys. 'Dark mountains'. The Welsh name corresponds to the English but is singular.

Blackawton Devon. *Auetone* 1086 (DB), *Blakeauetone* 1281. 'Farmstead of a man called Afa'. OE pers. name + *tūn*. Affix is OE *blæc* 'dark-coloured' (referring to soil or vegetation).

Blackborough, 'dark-coloured hill', OE *blæc* + *beorg*: **Blackborough** Devon. *Blacaberga* 1086 (DB). **Blackborough End** Norfolk. *Blakeberge c.*1150.

Blackburn Black. w. Darw. *Blacheburne* 1086 (DB). 'Dark-coloured stream'. OE *blæc* + *burna*.

Blackden Heath Ches. *Blakedene* 1287. 'Dark valley'. OE *blæc* + *denu*.

Blackfield Hants. a recent name, self-explanatory.

Blackford Somerset. near Wedmore. *Blacford* 1227. 'Dark ford'. OE *blæc* + *ford*.

Blackfordby Leics. *Blakefordebi c.*1125. 'Farmstead at the dark ford'. OE *blæc* + *ford* + OScand. *bý*.

Blackgang I. of Wight. first recorded in 1781, 'the dark path or track', from dialect *gang*.

Blackheath Gtr. London. *Blachehedfeld* 1166. 'Dark-coloured heathland'. OE *blæc* + *hǣth* (with *feld* 'open land' in the early form).

Blackland Wilts. *Blakeland* 1194. 'Dark-coloured cultivated land'. OE *blæc* + *land*.

Blackley Manch. *Blakeley* 1282. 'Dark wood or clearing'. OE *blæc* + *lēah*.

Blacklion Cavan. *Black Lion Inn* 1778. The former Irish name was *An Leargaidh*, 'the slope'.

Blackmoor Hants. *Blachemere* 1168. 'Dark-coloured pool'. OE *blæc* + *mere*.

Blackmoor Forest & Blackmoor Vale Dorset. *Boscum de Blakemor(e)* 1212. Named from former manor of *Blakemor(e)* 1258, 'dark-coloured moor or marshy ground'. OE *blæc* + *mōr*.

Blackmore Essex. *Blakemore* 1213. 'Dark-coloured marshland'. OE *blæc* + *mōr*.

Blackpool Bpool. *Pul* c.1260, *Blackpoole* 1602. 'Dark-coloured pool'. OE *blæc* + **pull*.

Blackrod Bolton. *Blacherode* c.1189. 'Dark clearing'. OE *blæc* + *rodu*.

Blackskull Antrim. *Blackskull* 1898. The name originated from an inn with a Negro-head sign.

Blackstaff (*Abhainn Bheara*) (river) Antrim. 'River of the staff'. English *black* presumably describes the *staff* or beam that formed a primitive bridge over the river.

Blackthorn Oxon. *Blaketorn* 1190. '(Place at) the blackthorn or sloe-tree'. OE **blæc-thorn*.

Blackthorpe Suffolk. Apparently a modern name, not recorded until 1837.

Blacktoft E. R. Yorks. *Blaketofte* c.1160. 'Dark-coloured homestead'. OE *blæc* + OScand. *toft*.

Blackwater (*An Abhainn Mór*) (river) Armagh, Cork, Tyrone. 'The big river'.

Blackwatertown Armagh. The village lies on the River BLACKWATER. Its Irish name is *An Port Mór*, 'the big landing-place'.

Blackwell, 'dark-coloured spring or stream', OE *blæc* + *wella*: **Blackwell** Darltn. *Blakewell* 1183. **Blackwell** Derbys., near Buxton. *Blachewelle* 1086 (DB). **Blackwell** Warwicks. *Blacwælle* 964, *Blachewelle* 1086 (DB).

Blackwood (*Coed-duon*) Cphy. *Coed-dduon* 1833, *Blackwood* 1856. 'Dark wood'. The Welsh name corresponds to the English.

Blacon Ches. *Blachehol* [*sic*] 1086 (DB), *Blachenol* 1093. 'Dark-coloured hill'. OE *blæc* + *cnoll*.

Bladon Oxon. *Blade* 1086 (DB). A pre-English river-name of uncertain origin and meaning, an old name of the River EVENLODE.

Blaenafon. *See* BLAENAVON.

Blaenau. *See* BLAINA.

Blaenau Ffestiniog Gwyd. *Festynyok* c.1420. 'Uplands of Ffestiniog'. Welsh *blaen* (plural *blaenau*). *Ffestiniog* means 'territory of Ffestin' or possibly 'defensive (place)' (Welsh *ffestiniog*). The town is a 19th-cent. industrial development.

Blaenavon (*Blaenafon*) Torf. *Blaen Avon* 1532, *Blaen-Avon, or Avon* 1868. 'Headstream of the river'. Welsh *blaen* + *afon*. The river is the Sychan.

Blagdon, 'dark-coloured hill', OE *blæc* + *dūn*: **Blagdon** N. Som. *Blachedone* 1086 (DB). **Blagdon** Somerset. *Blakedona* 12th cent. **Blagdon** Torbay. *Blakedone* 1242.

Blaina (*Blaenau*) Blae. 'Uplands'. Welsh *blaen* (plural *blaenau*).

Blair Atholl Perth. *Athochlach* c.970, *Athfoithle* c.1050. 'Plain in Atholl'. Gaelic *blàr*. *Atholl* means 'New Ireland' (Gaelic *ath* + *Fótla*, a poetic name for Ireland). The name was given by the Gaels when they came from Ireland to settle in this part of Scotland.

Blairgowrie Perth. *Blare* 13th cent., *Blair in Gowrie* 1604. 'Plain in Gowrie'. Gaelic *blàr*. *Gowrie* is '(territory of) Gabran', after a 6th-cent. Gaelic king, for distinction from BLAIR ATHOLL.

Blaisdon Glos. *Blechedon* 1186. 'Hill of a man called **Blæcci*'. OE pers. name + *dūn*.

Blakemere Herefs. *Blakemere* 1249. 'Dark-coloured pool'. OE *blæc* + *mere*.

Blakeney, 'dark-coloured island or dry ground in marsh', OE *blæc* (dative *blacan*) + *ēg*: **Blakeney** Glos. *Blakeneia* 1196. **Blakeney** Norfolk. *Blakenye* 1242.

Blakenhall Ches. *Blachenhale* 1086 (DB). 'Dark nook of land'. OE *blæc* (dative *blacan*) + *halh*.

Blakenham, Great & Blakenham, Little Suffolk. *Blac(he)ham* 1086 (DB). Probably 'homestead or enclosure of a man called Blaca'. OE pers. name (genitive *-n*) + *hām* or *hamm*.

Blakesley Northants. *Blaculveslei* 1086 (DB). 'Woodland clearing of a man called *Blæcwulf*'. OE pers. name + *lēah*.

Blanchardstown (*Baile Bhlainséir*) Dublin. 'Blanchard's town'.

Blanchland Northum. *Blanchelande* 1165. 'White woodland glade'. OFrench *blanche* + *launde*.

Blandford Forum, Blandford St Mary Dorset. *Blaneford* 1086 (DB), *Blaneford Forum* 1297, *Blaneford St Mary* 1254. Probably 'ford where blay or gudgeon are found'. OE *blǣge* (genitive plural *blǣgna*) + *ford*. Distinguishing affixes from Latin *forum* 'market' and from the dedication of the church.

Blaney (*Bléinigh*) Fermanagh. *Bleny* 1659. 'Inlet'.

Blankney Lincs. *Blachene* [*sic*] 1086 (DB), *Blancaneia* 1157. 'Island, or dry ground in marsh, of a man called *Blanca*'. OE pers. name (genitive -*n*) + *ēg*. Alternatively the first element may be OE *blanca* 'white horse'.

Blantyre S. Lan. *Blantir* 1289. Gaelic *tir* 'land' with uncertain first element.

Blaris (*Blárás*) Down. 'Place of open fields'.

Blarney (*An Bhlarna*) Cork. 'The small field'.

Blaston Leics. *Bladestone* 1086 (DB). 'Farmstead of a man called *Blath* or *Blēath*'. Oseand. or OE pers. name + *tūn*.

Blatchington, 'estate associated with a man called Blæcca', OE pers. name + -*ing*- + *tūn*: **Blatchington, East** E. Sussex. *Blechinton* 1169. **Blatchington, West** Bright. & Hove. *Blacinctona* 1121.

Blatherwycke Northants. *Blarewiche* 1086 (DB). Possibly 'farm where bladder-plants grow'. OE *blǣdre* + *wīc*.

Blawith Cumbria. *Blawit* 1276. 'Dark wood'. OScand. *blár* + *vithr*.

Blaxhall Suffolk. *Blaccheshala* 1086 (DB). 'Nook of land belonging to a man called *Blæc*'. OE pers. name + *halh*.

Blaxton Donc. *Blacston* 1213. 'Black boundary stone'. OE *blæc* + *stān*.

Blaydon Gatesd. *Bladon* 1340. Probably 'cold or cheerless hill'. OScand. *blár* + OE *dūn*.

Bleadon N. Som. *Bleodun* 956, *Bledone* 1086 (DB). 'Variegated hill'. OE **blēo* + *dūn*.

Blean Kent. *Blean* 724, *Blehem* [*sic*] 1086 (DB). Probably '(place in) the rough ground'. OE **blēa* (dative -*n*).

Bleasby Notts. *Blisetune* 956, *Blesby* 1268. Probably 'farmstead or village of a man called Blesi'. OScand. pers. name + *bý* (OE *tūn* in the earliest form).

Bledington Glos. *Bladintun* 1086 (DB). 'Farmstead on the River *Bladon*'. Pre-English river-name of uncertain origin and meaning (an old name of RIVER EVENLODE) + OE *tūn*.

Bledlow Bucks. *Bleddanhlæw* 10th cent., *Bledelai* 1086 (DB). 'Burial-mound of a man called **Bledda*'. OE pers. name + *hlāw*.

Blencarn Cumbria. *Blencarn* 1159. 'Cairn or rock summit'. Celtic **blain* + **carn*.

Blencogo Cumbria. *Blencoggou c.*1190. 'Cuckoos' summit'. Celtic **blain* + *cog* (plural *cogow*).

Blencow, Great & Blencow, Little Cumbria. *Blenco* 1231, *Blenkhaw* 1255. Celtic **blain* 'summit' with an obscure second element, to which OScand. *haugr* 'hill' has been added.

Blendworth Hants. *Blednewrthie c.*1170. Probably 'enclosure of a man called **Blǣdna*'. OE pers. name + *worth*.

Blenheim Palace Oxon. So named to commemorate the first Duke of Marlborough's victory over the French and Bavarians at *Blenheim* (i.e. Blindheim in Bavaria) in 1704.

Blennerhasset Cumbria. *Blennerheiseta* 1188. Celtic **blain* 'summit' with an obscure second element, to which OScand. *hey* 'hay' and *sǽtr* 'shieling' have been added.

Blessington (*Baile Coimhín*) Wicklow. The Irish name, meaning 'Ó Comáin's homestead', was originally anglicized as *Ballycomin*. This was taken to represent *baile comaoine*, 'homestead of favour', and was accordingly misrendered in English as *Blessington*, 'town of blessing'.

Bletchingdon Oxon. *Blecesdone* 1086 (DB). 'Hill of a man called **Blecci*'. OE pers. name + *dūn*.

Bletchingley Surrey. *Blachingelei* 1086 (DB). 'Woodland clearing of the family or followers of a man called Blæcca'. OE pers. name + -inga- + lēah.

Bletchley Milt. K. Blechelai 12th cent. 'Woodland clearing of a man called Blæcca'. OE pers. name + lēah.

Bletchley Shrops. Blecheslee 1222, Blecheleya 13th cent. 'Woodland clearing of a man called *Blecca or *Blecci'. OE pers. name + lēah.

Bletsoe Beds. Blechesho 1086 (DB). 'Hillspur of a man called *Blecci'. OE pers. name + hōh.

Blewbury Oxon. Bleobyrig 944, Blidberia [sic] 1086 (DB). 'Hill-fort with variegated soil'. OE *blēo + burh (dative byrig).

Blickling Norfolk. Blikelinges 1086 (DB). '(Settlement of) the family or followers of a man called *Blīcla'. OE pers. name + -ingas.

Blidworth Notts. Blideworde 1086 (DB). 'Enclosure of a man called *Blītha'. OE pers. name + worth.

Blindcrake Cumbria. Blenecreyc 12th cent. 'Rock summit'. Celtic *blain + *creig.

Blisland Cornwall. Bleselonde 1284. OE land 'estate' with an obscure first element.

Blisworth Northants. Blidesworde 1086 (DB). 'Enclosure of a man called *Blīth'. OE pers. name + worth.

Blithbury Staffs. Blidebire 1200. 'Stronghold on the River Blythe'. OE river-name (from blīthe 'gentle, pleasant') + burh (dative byrig). The same river gives name to **Blithfield, Blythe Bridge**, and **Blythe Marsh**.

Blo Norton Norfolk. See NORTON.

Blockley Glos. Bloccanleah 855, Blochelei 1086 (DB). 'Woodland clearing of a man called *Blocca'. OE pers. name + lēah.

Blofield Norfolk. Blafelda 1086 (DB). Possibly 'exposed open country'. OE *blāw + feld.

Bloody Foreland (Cnoc Fola) Donegal. 'Hill of blood'. The name is said to describe the red sunsets seen here.

Bloomsbury Gtr. London. Blemondesberi 1291. 'Manor held by the de Blemund family'. ME bury (from OE byrig).

Blore Staffs. near Ilam. Blora 1086 (DB). Possibly 'exposed bank or ridge'. OE *blāw + ōra or *ofer.

Bloxham Oxon. Blochesham 1086 (DB). 'Homestead of a man called *Blocc'. OE pers. name + hām.

Bloxholm Lincs. Blochesham 1086 (DB). Identical in origin with the previous name.

Bloxwich Wsall. Blocheswic 1086 (DB). 'Dwelling or (dairy) farm of a man called *Blocc'. OE pers. name + wīc.

Bloxworth Dorset. Blacewyrthe 987, Blocheshorde 1086 (DB). 'Enclosure of a man called *Blocc'. OE pers. name + worth.

Blubberhouses N. Yorks. Bluberhusum 1172. '(Place at) the houses by the bubbling spring'. ME bluber + OE hūs (dative plural -um).

Blue Stack Mountains (Cruacha Gorma) Donegal. Blue Stack 1837, 'Blue stacks'. The original (singular) name, Irish An Chruach Ghorm, 'the blue stack', was that of the highest peak.

Blundeston Suffolk. Blundeston 1203. 'Farmstead of a man called *Blunt'. OE pers. name + tūn.

Blunham Beds. Blunham 1086 (DB). Possibly 'homestead, or land in a river-bend, of a man called *Blūwa'. OE pers. name (genitive -n) + hām or hamm.

Blunsdon St Andrew, Broad Blunsdon Swindn. Bluntesdone 1086 (DB), Bluntesdon Seynt Andreu 1281, Bradebluntesdon 1234. 'Hill of a man called *Blunt'. OE pers. name + dūn. Affixes from the dedication of the church and from OE brād 'broad, great'.

Bluntisham Cambs. Bluntesham c.1050, 1086 (DB). 'Homestead or enclosure of a man called *Blunt'. OE pers. name + hām or hamm.

Blyborough Lincs. Bliburg 1086 (DB). OE burh 'stronghold' with OE pers. name *Blītha or stream-name Blīthe 'gentle or pleasant one'.

Blyford Suffolk. Blitleford [sic] c.1060, Blideforda 1086 (DB). 'Ford over the River Blyth'. OE river-name ('the gentle or pleasant one' from OE blīthe) + ford.

Blymhill Staffs. Brumhelle [sic] 1086 (DB), Blumehil 1167. Possibly 'hill where plum-trees grow'. OE plȳme + hyll.

Blyth Northum. Blida 1130. Named from the River Blyth (OE blīthe 'the gentle or pleasant one').

Blyth Notts. *Blide* 1086 (DB). Identical in origin with the previous name, *Blyth* being the old name of the River Ryton.

Blythburgh Suffolk. *Blideburh* 1086 (DB). 'Stronghold on the River Blyth'. OE river-name ('the gentle or pleasant one' from OE *blīthe*) + *burh*.

Blythe Bridge & Blythe Marsh Staffs. *See* BLITHBURY.

Blyton Lincs. *Blitone* 1086 (DB). Possibly 'farmstead of a man called *Blītha'. OE pers. name + *tūn*.

Boa Island (*Inis Badhbha*) Fermanagh. *Badhba* 1369. 'Badhbh's island'.

Boardmills Antrim. *Boardmills* 1904. The reference is to former wooden corn mills here. The equivalent Irish name is *An Muileann Adhmaid*, 'the timber mill'.

Boarhunt Hants. *Byrhfunt* 10th cent., *Borehunte* 1086 (DB). 'Spring of the stronghold or manor'. OE *burh* (genitive *byrh*) + *funta*.

Boarstall Bucks. *Burchestala* 1158. OE *burh-stall* 'the site of a stronghold'.

Boasley Cross Devon. *Borslea* c.970, *Bosleia* 1086 (DB). 'Woodland clearing where spiky plants grow'. OE *bors + *lēah*.

Boat of Garten Highland. 'Ferry by the corn field'. English *boat* + Gaelic *gairtean*.

Bobbing Kent. *Bobinge* c.1100. '(Settlement of) the family or followers of a man called Bobba'. OE pers. name + *-ingas*.

Bobbington Staffs. *Bubintone* 1086 (DB). 'Estate associated with a man called Bubba'. OE pers. name + *-ing- + tūn*.

Bocking Essex. *Boccinge(s)* c.995, *Bochinges* 1086 (DB). '(Settlement of) the family or followers of a man called *Bocca', or '*Bocca's place'. OE pers. name + *-ingas* or *-ing*.

Boddington Glos. *Botingtune* 1086 (DB). 'Estate associated with a man called Bōta'. OE pers. name + *-ing- + tūn*.

Boddington Northants. *Botendon* 1086 (DB). 'Hill of a man called Bōta'. OE pers. name (genitive *-n*) + *dūn*.

Bodenham Herefs. *Bodeham* 1086 (DB). 'Homestead or river-bend land of a man called Boda'. OE pers. name (genitive *-n*) + *hām* or *hamm*.

Bodenham Wilts. *Boteham* 1249. 'Homestead or enclosure of a man called Bōta'. OE pers. name (genitive *-n*) + *hām* or *hamm*.

Boderg (*Both Derg*) Leitrim, Roscommon. 'Red hut'.

Bodham Norfolk. *Bod(en)ham* 1086 (DB). 'Homestead or enclosure of a man called Boda'. OE pers. name (genitive *-n*) + *hām* or *hamm*.

Bodiam E. Sussex. *Bodeham* 1086 (DB). Identical in origin with the previous name.

Bodicote Oxon. *Bodicote* 1086 (DB). 'Cottage(s) associated with a man called Boda'. OE pers. name + *-ing- + cot*.

Bodmin Cornwall. *Bodmine* c.975, 1086 (DB). Probably 'dwelling by church-land'. OCornish *bod + *meneghi*.

Bodney Norfolk. *Budeneia* 1086 (DB). 'Island, or dry ground in marsh, of a man called *Beoda'. OE pers. name (genitive *-n*) + *ēg*.

Bodoney (*Both Domhnaigh*) Tyrone. *Bodony* 1613. 'Hut of the church'.

Bofeenaun (*Both Faonáin*) Mayo. 'Faonán's hut'.

Bogare (*Both Chearr*) Kerry. 'Crooked hut'.

Bogay (*Both Ghé*) Donegal. 'Goose hut'.

Bognor Regis W. Sussex. *Bucganora* c.975. 'Shore of a woman called Bucge'. OE pers. name + *ōra*. Latin affix *regis* 'of the king' is only recent, alluding to the stay of George V here in 1929.

Bohacogram (*Both an Chograim*) Kerry. 'Hut of the whispering'.

Bohacullia (*Botha Coille*) Kerry. 'Huts of the wood'.

Bohaun (*Bothán*) Mayo. 'Little hut'.

Boheeshil (*Both Íseal*) Kerry. 'Low hut'.

Boher (*Bóthar*) Limerick. 'Road'.

Boheraphuca (*Bóthar an Phúca*) Offaly. 'Road of the sprite'.

Boherard (*Bóthar Ard*) Cork, Waterford. 'High road'.

Boherboy (*Bóthar Buí*) Cork. 'Yellow road'.

Boherbue (*Bóthar Buí*) Cork. 'Yellow road'.

Bohereen (*Bóithrín*) Limerick. 'Little road'.

Boherlahan (*Bóthar Leathan*) Tipperary. 'Broad road'.

Bohermeen (*An Bóthar Mín*) Meath. 'The smooth road'.

Bohermore (*An Bóthar Mór*) Galway. 'The big road'.

Boho (*Botha*) Fermanagh. *Botha* 1432. 'Huts'.

Bohoge (*Bothóg*) Mayo. 'Little hut'.

Bohola (*Both Chomhla*) Mayo. 'Comla's hut'.

Bolam, '(place at) the tree-trunks', OE *bol* or OScand. *bolr* in a dative plural form *bolum*: **Bolam** Durham. *Bolum* 1235. **Bolam** Northum. *Bolum* 1155.

Bolas, Great Tel. & Wrek. *Belewas* 1198, *Boulewas* 1199 (1265). OE **wæsse* 'riverside land liable to flood' with an uncertain first element, possibly an OE **bogel* 'small river-bend or meander' in a genitive plural form.

Bold Heath St Hel. *Bolde* 1204. OE *bold* 'a special house or building'.

Boldon S. Tyne. *Boldune* c.1140. Probably 'rounded hill'. OE **bol* + *dūn*. Alternatively the first element could be OE *bol* 'tree-stump' or **bole* 'smelting-place'.

Boldre Hants. *Bovre* [*sic*] 1086 (DB), *Bolre* 1152. Origin and meaning uncertain, possibly an old name of the Lymington river.

Boldron Durham. *Bolrum* c.1180. 'Clearing used for bulls'. OScand. *boli* + *rúm*.

Bole Notts. *Bolun* 1086 (DB). '(Place at) the tree-trunks'. OE **bola* or OScand. *bolr* in a dative plural form *bolum*.

Bolea (*Both Liath*) Derry. *Boleah* 1613. 'Grey hut'.

Boleran (*Baile Uí Shírín*) Derry. *Ballyirin* 1613. 'Ó Sírín's townland'.

Bolingbroke, Old & Bolingbroke, New Lincs. *Bolinbroc* 1086 (DB), *Bulingbroc* 1202. Probably 'brook associated with a man called **Bula*'. OE pers. name + connective -*ing*- + *brōc*.

Bollington, 'farmstead on the River Bollin', old river-name of uncertain origin and meaning

+ OE *tūn*: **Bollington** Ches., near Altrincham. *Bolinton* c.1222. **Bollington** Ches., near Macclesfield. *Bolynton* 1270.

Bolney W. Sussex. *Bolneye* 1263. 'Island, or dry ground in marsh, of a man called Bola'. OE pers. name (genitive -*n*) + *ēg*.

Bolnhurst Beds. *Bulehestre* [*sic*] 1086 (DB), *Bollenhirst* 11th cent. 'Wooded hill where bulls are kept'. OE *bula* (genitive plural *bulena*) + *hyrst*.

Bolsover Derbys. *Belesovre* [*sic*] 1086 (DB), *Bolesoura* 12th cent. Probably 'ridge of a man called Boll or **Bull*'. OE pers. name + **ofer*.

Bolsterstone Sheff. *Bolstyrston* 1398. 'Stone on which criminals are beheaded'. OE *bolster* + *stān*.

Bolstone Herefs. *Boleston* 1193. 'Stone of a man called Bola'. OE pers. name + *stān*.

Boltby N. Yorks. *Boltebi* 1086 (DB). 'Farmstead or village of a man called Boltr or **Bolti*'. OScand. pers. name + *bý*.

Bolton, a common name in the North of England, from OE **bōthl-tūn* 'settlement with a special building'; examples include: **Bolton** Bolton. *Boelton* 1185. **Bolton by Bowland** Lancs. *Bodeltone* 1086 (DB). The district-name Bowland (*Boelanda* 1102) probably means 'district within the curved valley (of the River Hodder)', OE *boga* 'bow, bend' + *land*. **Bolton, Castle** N. Yorks. *Bodelton* 1086 (DB). Affix from the castle built here in 1379. **Bolton le Sands** Lancs. *Bodeltone* 1086 (DB). Affix means 'on the sands'. **Bolton Percy** N. Yorks. *Bodeltune* 1086 (DB), *Bolton Percy* 1305. Manorial affix from its possession by the *de Percy* family (from 1086). **Bolton upon Dearne** Barns. *Bodeltone* 1086 (DB). The river-name Dearne is possibly from OE *derne* 'hidden', but may be of Celtic origin.

Bonby N. Lincs. *Bundebi* 1086 (DB). 'Farmstead or village of the peasant farmers'. OScand. *bóndi* + *bý*.

Bonchurch I. of Wight. *Bonecerce* 1086 (DB). 'Church of **Bona*'. OE pers. name + *cirice*. The pers. name may be a short-form of (St) Boniface, to whom the church here is dedicated.

Bondleigh Devon. *Bolenei* [*sic*] 1086 (DB), *Bonlege* 1205. Probably 'woodland clearing of a man called Bola'. OE pers. name (genitive -*n*) + *lēah*.

Bonehill Staffs. *Bolenhull* 1230. Probably 'hill where bulls graze'. OE **bula* (genitive plural **bulena*) + *hyll*.

Bo'ness Falk. *Berwardeston c.*1335, *Nes* 1494, *Burnstounnes* 1532, *Borrowstownness,* or *Bo'ness* 1868. 'Promontory of *Borrowstoun*'. OE *næss. Borrowstoun* is 'Beornweard's farm' (OE masculine pers. name + *tūn*).

Bonhill W. Dunb. *Buchlul* 1225, *Buthelulle c.*1270, *Buchnwl c.*1320. 'House by the stream'. Gaelic *bot* + *an* + *allt* (genitive *uillt*).

Boningale Shrops. *Bolynghale* 12th cent. Probably 'nook of land associated with a man called Bola'. OE pers. name + *-ing-* + *halh*.

Bonnington Kent. *Bonintone* 1086 (DB). 'Estate associated with a man called Buna'. OE pers. name + *-ing-* + *tūn*.

Bonnybridge Falk. 'Bridge over Bonny Water'. The river-name, recorded in 1682 as *aquae de Boine,* is said to derive from Scottish English *bonny,* 'beautiful'.

Bonnyrigg Midloth. *Bannockrig* 1773. 'Bannock-shaped ridge'. ModE *bannock* + Scottish *rig*.

Bonsall Derbys. *Bunteshale* 1086 (DB). 'Nook of land of a man called **Bunt'. OE pers. name + *halh*.

Bont-faen, Y. *See* COWBRIDGE.

Bonvilston (*Tresimwn*) Vale Glam. *Boleuilston c.*1160, *Bonevillestun c.*1206. 'de Bonville's farm'. OE *tūn* (Welsh *tref*). The Welsh name comes from Simon de *Bonville*.

Bookham, Great & Bookham, Little Surrey. *Bocheham* 1086 (DB). 'Homestead where beech-trees grow'. OE *bōc* + *hām*.

Boola (*Buaile*) Waterford. 'Milking place'.

Boolakennedy (*Buaile Uí Chinnéide*) Tipperary. 'Ó Cinnéide's milking place'.

Boolananave (*Buaile na nDamh*) Kerry. 'Milking place of the oxen'.

Boolavogue (*Baile Mhaodhóg*) Wexford. 'Maodóg's townland'.

Boot Cumbria. *Bout, the Bought* 1587. Probably '(place at) the river-bend'. ME *bought* (OE **buht*).

Booterstown (*Baile an Bhóthair*) Dublin. 'Town of the road'.

Boothby, 'farmstead or village with booths or shelters', OScand. *bōth* + *bý*: **Boothby Graffoe** Lincs. *Bodebi* 1086 (DB). Affix is an old wapentake-name, 'spur of land with a grove', OE *grāf* + *hōh*. **Boothby Pagnell** Lincs. *Bodebi* 1086 (DB). Manorial affix from the *Paynel* family, here in the 14th cent.

Boothferry E. R. Yorks. *Booth's Ferry* 1651. 'Ferry at Booth (near Howden)', from OScand. *ferja*. The place name **Booth** (originally *Botheby* 1550) was named from a family who came from one of the places called BOOTHBY (*see* previous names).

Bootle, 'the special building', OE *bōtl*: **Bootle** Cumbria. *Bodele* 1086 (DB). **Bootle** Sefton. *Boltelai* [*sic*] 1086 (DB), *Botle* 1212.

Boraston Shrops. *Bureston* 1188, *Buraston* 1256. Possibly 'eastern farmstead or estate', OE *ēast* + *tūn*, with the addition of OE *burh* 'fortification'.

Borden Kent. *Bordena* 1177. 'Valley or woodland pasture by a hill'. OE **bor* + *denu* or *denn*.

Bordley N. Yorks. *Borelaie* [*sic*] 1086 (DB), *Bordeleia c.*1140. Probably 'woodland clearing where boards are got'. OE *bord* + *lēah*.

Boreham, probably 'homestead or enclosure on or by a hill', OE **bor* + *hām* or *hamm*: **Boreham** Essex. *Borham c.*1045, 1086 (DB). **Boreham Street** E. Sussex. *Borham* 12th cent.

Borehamwood Herts. *Borham* 1188, *Burhamwode* 13th cent. Identical in origin with the previous names + OE *wudu* 'wood'.

Borley Essex. *Barlea* 1086 (DB). 'Woodland clearing frequented by boars'. OE *bār* + *lēah*.

Bornacoola (*Barr na Cúile*) Leitrim. 'Top of the hollow'.

Borough Green Kent. *Borrowe Grene* 1575. From OE *burh* 'manor, borough' or *beorg* 'hill, mound'.

Boroughbridge N. Yorks. *pontem de Burgo* (Latin) 1155, *Burbrigg* 1220. 'Bridge near *Burgh* (= ALDBOROUGH)'. OE *burh* 'fort, stronghold' + *brycg*.

Borris (*An Bhuiríos*) Carlow. 'The borough'.

Borris-in-Ossory (*Buiríos Mór Osraí*) Laois. 'Big borough of Osraí'. Osraí (Ossory) is an ancient territory here.

Borrisbeg (*Buiríos Beag*) Kilkenny. 'Small borough'.

Borrisnoe (*Buiríos Nua*) Tipperary. 'New borough'.

Borrisokane (*Buiríos Uí Chéin*) Tipperary. 'Borough of Ó Céin'.

Borrisoleigh (*Buiríos Ó Luigheach*) Tipperary. 'Borough of Uí Luigheach'.

Borrowby N. Yorks. near Leake. *Bergebi* 1086 (DB). 'Farmstead or village on the hill(s)'. OScand. *berg* + *bý*.

Borrowdale Cumbria. near Keswick. *Borgordale* c.1170. 'Valley of the fort river'. OScand. *borg* (genitive -*ar*) + *á* + *dalr*.

Borth Cergn. '(Place of the) ferry'. Welsh *porth*.

Borth-y-gest Gwyd. *Gest harbour* 1748. 'Harbour of the paunch'. Welsh *porth* + *y* + *cest*. The village takes its name from nearby *Moel y Gest*, 'bare hill of the paunch', a mountain so named for its shape.

Borwick Lancs. *Bereuuic* 1086 (DB). 'Barley farm' or 'outlying part of an estate'. OE *bere-wīc*.

Bosbury Herefs. *Bosanbirig* early 12th cent., *Boseberge* 1086 (DB). 'Stronghold of a man called Bōsa'. OE pers. name + *burh* (dative *byrig*).

Boscastle Cornwall. *Boterelescastel* 1302. 'Castle of a family called Boterel'. OFrench surname + *castel*.

Boscombe, probably 'valley overgrown with spiky plants', OE **bors* + *cumb*: **Boscombe** Bmouth. *Boscumbe* 1273. **Boscombe** Wilts. *Boscumbe* 1086 (DB).

Bosham W. Sussex. *Bosanham*(*m*) 731, *Boseham* 1086 (DB). 'Homestead or promontory of a man called Bōsa'. OE pers. name + *hām* or *hamm*.

Bosherston Pemb. *Stakep' bosser* 1291, *Bosherston* (*alias Stacpoll Bosher*) 1594. 'Bosher's farm'. OFrench pers. name + ME *toun*. The village was originally *Stackpole Bosher*, the manorial name distinguishing it from *Stackpole Elidor*, an alternative name for nearby CHERITON.

Bosley Ches. *Boselega* 1086 (DB). 'Woodland clearing of a man called Bōsa or Bōt'. OE pers. name + *lēah*.

Bossall N. Yorks. *Bosciale* 1086 (DB). Probably 'nook of land of a man called Bōt or **Bōtsige*'. OE pers. name + *halh*.

Bossiney Cornwall. *Botcinnii* 1086 (DB). 'Dwelling of a man called Kyni'. OCornish **bod* + pers. name.

Bostock Green Ches. *Botestoch* 1086 (DB). 'Outlying farmstead or hamlet of a man called Bōta'. OE pers. name + *stoc*.

Boston Lincs. *Botuluestan* 1130. 'Stone (marking a boundary or meeting-place) of a man called Bōtwulf'. OE pers. name + *stān*. Identification of Bōtwulf with the 7th-cent. missionary St Botulf is quite probable.

Boston Spa Leeds. *Bostongate* 1799. A recent name, perhaps so called from a family called *Boston* from BOSTON (Lincs.). The affix *Spa* refers to the mineral spring discovered here in 1744.

Bosworth, Husbands Leics. *Baresworde* 1086 (DB). Probably 'enclosure of a man called Bār'. OE pers. name + *worth*. Affix probably means 'of the farmers or husbandmen' (from late OE *hūsbonda*).

Bosworth, Market Leics. *Boseworde* 1086 (DB). 'Enclosure of a man called Bōsa'. OE pers. name + *worth*. Affix (found from the 16th cent.) alludes to the important market here.

Botany Bay Gtr. London. Recorded thus in 1819, a transferred name from *Botany Bay* in Australia (the site of an early convict settlement in the late 18th cent. near to what is now Sydney), here bestowed on a spot perhaps considered rather remote and inaccessible in the middle of Enfield Chase. There are examples of the same name in Ches., Essex, Wilts., and several other counties.

Botesdale Suffolk. *Botholuesdal* 1275. 'Valley of a man called Bōtwulf'. OE pers. name + *dæl*.

Bothal Northum. *Bothala* 12th cent. 'Nook of land of a man called Bōta'. OE pers. name + *halh*.

Bothamsall Notts. *Bodmescel* 1086 (DB), *Bodmeshil* c.1150. Etymology obscure, but possibly 'shelf by a broad river-valley'. OE **bothm* + *scelf* (alternating with OE *hyll* 'hill').

Bothel Cumbria. *Bothle* c.1125. OE *bōthl* 'a special house or building'.

Bothenhampton Dorset. *Bothehamton* 1268. 'Home farm in a valley'. OE **bothm* + *hām-tūn*.

Bothwell S. Lan. *Botheuill* c.1242, *Bothvile* c.1300. Etymology obscure.

Botley Bucks. *Bottlea* 1167. 'Woodland clearing of a man called Botta'. OE pers. name + *lēah*.

Botley Hants. *Botelie* 1086 (DB). 'Woodland clearing of a man called Bōta, or where timber is obtained'. OE pers. name or OE *bōt* + *lēah*.

Botley Oxon. *Boteleam* 12th cent. Identical in origin with the previous name.

Botolph Claydon Bucks. *see* CLAYDON.

Botolphs W. Sussex. *Sanctus Botulphus* 1288. From the dedication of the parish church to St Botolph.

Bottesford, 'ford by the house or building', OE *bōtl* + *ford*: **Bottesford** N. Lincs. *Budlesforde* 1086 (DB). **Bottesford** Leics. *Botesford* 1086 (DB), *Botlesford* c.1130.

Bottisham Cambs. *Bodekesham* 1060, *Bodichessham* 1086 (DB). 'Homestead or enclosure of a man called **Boduc*'. OE pers. name + *hām* or *hamm*.

Bottlehill Antrim. *Botle hill* c.1657. 'Bottle hill'. *Bottle* ('bundle of hay') implies a good hay harvest.

Botusfleming Cornwall. *Bothflumet* 1259. OCornish **bod* 'dwelling' with an obscure second element.

Boughadoon (*Both an Dúin*) Mayo. 'Hut of the fort'.

Boughton, found in various counties, has two distinct origins. Some are 'farmstead of a man called Bucca, or where bucks (male deer or he-goats) are kept'. OE pers. name or OE *bucca* + *tūn*; examples include: **Boughton** Northants., near Moulton. *Buchetone* 1086 (DB). **Boughton** Notts. *Buchetone* 1086 (DB).
    Other Boughtons are either 'farmstead where beech-trees grow' or 'farmstead held by charter', OE *bōc* + *tūn*; examples include: **Boughton Aluph** Kent. *Boltune* [*sic*] 1086 (DB), *Boctun* c.1020, *Botun Alou* 1237. Manorial affix from a 13th-cent. owner called *Alulf*. **Boughton Malherbe** Kent. *Boltune* [*sic*]

1086 (DB), *Boctun Malerbe* 1275. Manorial affix from early possession by the *Malherbe* family.
Boughton Monchelsea Kent. *Boltone* [*sic*] 1086 (DB), *Bocton Monchansy* 1278. Manorial affix from the *de Montchensie* family, here in the 13th cent.

Bouladuff (*An Bhuaile Dhubh*) Tipperary. 'The black milking place'.

Bouldon Shrops. *Bolledone* 1086 (DB), *Bullardone* 1166. OE *dūn* 'hill' with an uncertain first element.

Boulge Suffolk. *Bulges* 1086 (DB), *Bulge* 1254. From OFrench *bouge* 'uncultivated land covered with heather'.

Boulmer Northum. *Bulemer* 1161. 'Pond used by bulls'. OE *bula* + *mere*.

Boultham Lincs. *Buletham* 1086 (DB). 'Homestead or enclosure where ragged robin or the cuckoo flower grows'. OE *bulut* + *hām* or *hamm*.

Bourn, Bourne, '(place at) the spring(s) or stream(s)', OScand *brunnr*: **Bourn** Cambs. *Brune* 1086 (DB). **Bourne** Lincs. *Brune* 1086 (DB).
    The following has the same meaning but is from the cognate OE *burna*: **Bourne, St Mary** Hants. *Borne* 1185, *Maryborne* 1476. Affix from the original dedication of the chapel here.

Bourne End Bucks. *Burnend* 1236. 'End of the stream' (here where the River Wye meets the Thames). OE *burna* + *ende*.

Bournemouth Bmouth. *La Bournemowthe* 1407. 'The mouth of the stream'. OE *burna* + *mūtha*.

Bournville Birm. 'Town on a river called Bourne' (from OE *burna* 'stream'). A recent name incorporating French *ville* for the estate built 1879 by George Cadbury.

Bourton, usually OE *burh-tūn* 'fortified farmstead' or 'farmstead near or belonging to a stronghold or manor'; examples include: **Bourton** Dorset. *Bureton* 1212. **Bourton, Black** Oxon. *Burtone* 1086 (DB). Affix possibly refers to the black habit of the canons of Osney Abbey who had lands here. **Bourton, Flax** N. Som. *Buryton* 1260. Probably from the variant *byrh-tūn* with similar meaning. Affix from the growing of flax here. **Bourton-on-the-Water** Glos. *Burchtun* 714, *Bortune* 1086 (DB). In this name reference is to the nearby hill-fort. Affix refers to the River Windrush which flows through the village.

Bovedy (*Both Mhíde*) Derry. *Bovidie* 1654. 'Míde's hut'.

Bovevagh (*Boith Mhéabha*) Derry. (*derteach*) *Bothe Medba* 1100. 'Maeve's hut'.

Bovey Tracey, North Bovey Devon. *Bovi* 1086 (DB), *Bovy Tracy* 1276, *Northebovy* 1199. Named from the River Bovey, a pre-English river-name of uncertain origin and meaning. Manorial affix from the *de Tracy* family, here in the 13th cent.

Bovingdon Herts. *Bovyndon c.*1200. Probably 'hill associated with a man called Bofa'. OE pers. name + *-ing-* + *dūn*.

Bovington Dorset. *Bovintone* 1086 (DB). 'Estate associated with a man called Bofa'. OE pers. name + *-ing-* + *tūn*.

Bow, '(place at) the arched bridge', OE *boga*: **Bow** Devon. *Limet* [*sic*] 1086 (DB), *Nymetboghe* 1270, *la Bogh* 1281. *Nymet* is the old name of the River Yeo, *see* NYMET. **Bow** Gtr. London. *Stratford* 1177, *Stratford atte Bowe* 1279. Its earlier name means 'ford on a Roman road', OE *strǣt* + *ford*. ME *atte* means 'at the'.

Bow Brickhill Milt. K. *see* BRICKHILL.

Bowburn Durham. Named from Bowburn Beck, 'winding stream', OE *boga* + *burna*.

Bowcombe I. of Wight. *Bovecome* 1086 (DB). 'Valley of a man called Bofa', or '(place) above the valley'. OE pers. name or OE *bufan* + *cumb*.

Bowden, Great & Bowden, Little Leics. *Bugedone* 1086 (DB). 'Hill of a woman called Bucge or of a man called Būga'. OE pers. name + *dūn*.

Bowdon Traffd. *Bogedone* 1086 (DB). 'Curved hill'. OE *boga* + *dūn*.

Bowerchalke Wilts. *See* CHALKE.

Bowers Gifford Essex. *Bure* 1065, *Bura* 1086 (DB), *Buresgiffard* 1315. 'The dwellings or cottages'. OE *būr* + manorial affix from the *Giffard* family, here in the 13th cent.

Bowes Durham. *Bogas* 1148. 'The river-bends'. OE *boga* or OScand. *bogi*.

Bowland (old district and forest) Lancs./N. Yorks., *see* BOLTON BY BOWLAND.

Bowley Herefs. *Bolelei* 1086 (DB). 'Woodland clearing of a man called Bola, or where there are tree-trunks'. OE pers. name or **bola* + *lēah*.

Bowling Brad. *Bollinc* 1086 (DB). 'Place at a hollow'. OE *bolla* + *-ing*.

Bowness-on-Solway Cumbria. *Bounes c.*1225. 'Rounded headland'. OE *boga* + *næss*, or OScand. *bogi* + *nes*. Solway (*Sulewad* 1218) probably means 'estuary of the pillar ford', OScand. *súl* + *vath*, *see* SOLWAY FIRTH.

Bowness-on-Windermere Cumbria. *Bulnes* 1282. 'Bull headland'. OE *bula* + *næss*. *See* WINDERMERE.

Bowsden Northum. *Bolesdon* 1195. Probably 'hill of a man called **Boll*'. OE pers. name + *dūn*.

Bowthorpe Norfolk. *Boethorp* 1086 (DB). 'Outlying farmstead or hamlet of a man called Búi'. OScand. pers. name + *thorp*.

Box, '(place at) the box-tree', OE *box*: **Box** Glos. *la Boxe* 1260. **Box** Wilts. *Bocza* 1144.

Boxford W. Berks. *Boxora* 821, *Bousore* 1086 (DB). 'Slope where box-trees grow'. OE *box* + *ōra*.

Boxford Suffolk. *Boxford* 12th cent. 'Ford where box-trees grow'. OE *box* + *ford*. The river-name **Box** is a back-formation from the place name.

Boxgrove W. Sussex. *Bosgrave* 1086 (DB). 'Box-tree grove'. OE *box* + *grāf*.

Boxley Kent. *Boseleu* [*sic*] 1086 (DB), *Boxlea c.*1100. 'Wood or clearing where box-trees grow'. OE *box* + *lēah*.

Boxted Essex. *Bocstede* 1086 (DB). 'Place where beech-trees grow'. OE *bōc* + *stede*.

Boxted Suffolk. *Boesteda* [*sic*] 1086 (DB), *Bocstede* 1154. 'Place where beech-trees or box-trees grow'. OE *bōc* or *box* + *stede*.

Boxworth Cambs. *Bochesuuorde* 1086 (DB). 'Enclosure of a man called **Bucc*, or where bucks (male deer or he-goats) are kept'. OE pers. name or OE *bucc* + *worth*.

Boyd (river) S. Glos. *See* BITTON.

Boyle (*Mainistir na Búille*) Roscommon. 'Monastery of the River Búill'.

Boylestone Derbys. *Boilestun* 1086 (DB). Probably 'farmstead at the rounded hill'. OE *boga* + *hyll* + *tūn*.

Boynton E. R. Yorks. *Bouintone* 1086 (DB).
'Estate associated with a man called Bōfa'. OE
pers. name + *-ing-* + *tūn*.

Boyounagh (*Buíbheanach*) Galway. 'Yellow
marsh'.

Boyton, 'farmstead of a man called Boia' or
'farmstead of the boys or servants', OE pers.
name or OE **boia* + *tūn*: **Boyton** Cornwall.
Boietone 1086 (DB). **Boyton** Suffolk. *Boituna*
1086 (DB). **Boyton** Wilts. *Boientone* 1086 (DB).

Bozeat Northants. *Bosiete* 1086 (DB). 'Gate or
gap of a man called Bōsa'. OE pers. name + *geat*.

Brabling Green Suffolk. *Babylon Green*
1837. The biblical name *Babylon*, often used
rhetorically to mean 'a great and luxurious city',
here applied ironically and humorously to a
small place. There are other instances of the
name **Babylon** in Clwyd, Dorset (with
spellings from 1531), and Kent.

Brabourne Kent. *Bradanburna c.*860,
Bradeburne 1086 (DB). '(Place at) the broad
stream'. OE *brād* + *burna*.

Brabstermire Highland. *Brabustare* 1492,
Brabastermyre 1538. 'Marsh at Brabster'.
OScand. *mýrr*. Brabster is 'broad dwelling'
(OScand. *breithr* + *bólstathr*).

Braceborough Lincs. *Braseborg* 1086 (DB).
Possibly 'strong fortress', or 'fortress of a man
called **Bræsna*'. OE *bræsen* or pers. name +
burh.

Bracebridge Lincs. *Brachebrige, Bragebruge*
1086 (DB), *Bracebrige* 12th cent. OE *brycg*
'bridge' with an uncertain first element, possibly
an early instance of OFrench *brace* 'arch,
support' referring to the structure of the bridge.

Braceby Lincs. *Breizbi* 1086 (DB). 'Farmstead
or village of a man called Breithr'. OScand. pers.
name + *bý*.

Bracewell Lancs. *Braisuelle* 1086 (DB).
'Spring or stream of a man called Breithr'.
OScand. pers. name + OE *wella*.

Brackaharagh (*Brá Chatrach*) Kerry. 'Neck
of the stone fort'.

Brackenfield Derbys. *Brachentheyt* 1269.
'Bracken clearing'. OScand. **brækni* + *thveit*.

Brackley Northants. *Brachelai* 1086 (DB).
Probably 'woodland clearing of a man called
*Bracca'. OE pers. name + *lēah*.

Brackloon (*Breac Chluain*) Kerry, Mayo.
'Speckled pasture'.

Bracklyn (*Breaclainn*) Westmeath. 'Speckled
place'.

Bracknagh (*Breacánach*) Offaly. 'Speckled
place'.

Bracknahevla (*Breacach na hAibhle*)
Westmeath. 'Speckled land of the orchard'.

Bracknell Brack. For. *Braccan heal* 942.
'Nook of land of a man called **Bracca*'. OE pers.
name (genitive *-n*) + *halh*.

Bracon Ash Norfolk. *Brachene* 1175. '(Place
amid) the bracken'. ON **brækni* or OE **bræcen*
with the later addition of *ash* 'ash-tree'.

Bradbourne Derbys. *Bradeburne* 1086 (DB).
'(Place at) the broad stream'. OE *brād* + *burna*.

Bradden Northants. *Bradene* 1086 (DB).
'Broad valley'. OE *brād* + *denu*.

Braddock Cornwall. *Brodehoc* 1086 (DB).
'Broad oak', or 'broad hook of land'. OE *brād* +
āc or *hōc*.

Bradenham, 'broad homestead or
enclosure', OE *brād* (dative *-an*) + *hām* +
hamm: **Bradenham** Bucks. *Bradeham* 1086
(DB). **Bradenham** Norfolk. *Bradenham* 1086
(DB).

Bradenstoke Wilts. *Bradenestoche* 1086
(DB). 'Settlement dependent on Braydon forest'.
Pre-English forest-name of obscure origin and
meaning + OE *stoc*.

Bradfield, 'broad stretch of open land', OE
brād + *feld*; examples include: **Bradfield**
Essex. *Bradefelda* 1086 (DB). **Bradfield**
Norfolk. *Bradefeld* 1177. **Bradfield** Sheff.
Bradesfeld 1188. **Bradfield** W. Berks.
Bradanfelda 990–2, *Bradefelt* 1086 (DB).
**Bradfield Combust, Bradfield St Clare,
& Bradfield St George** Suffolk. *Bradefelda*
1086 (DB). Distinguishing affixes from ME
combust 'burnt', from early possession by the
Seyncler family, and from dedication of the
church to St George.

Bradford, a fairly common name, '(place at)
the broad ford', OE *brād* + *ford*; examples
include: **Bradford** Brad. *Bradeford* 1086 (DB).
Bradford Abbas Dorset. *Bradanforda* 933,
Bradeford 1086 (DB), *Braddeford Abbatis* 1386.
Affix is Latin *abbas* 'abbot', alluding to early
possession by Sherborne Abbey. **Bradford on**

Avon Wilts. *Bradanforda be Afne* c.900, *Bradeford* 1086 (DB). Avon is a Celtic river-name meaning simply 'river'. **Bradford Peverell** Dorset. *Bradeford* 1086 (DB), *Bradeford Peuerel* 1244. Manorial affix from the *Peverel* family, here in the 13th cent.

Brading I. of Wight. *Brerdinges* 683, *Berardinz* 1086 (DB). '(Settlement of) the dwellers on the hill-side'. OE *brerd* + *-ingas*.

Bradley, a common name, usually 'broad wood or clearing', OE *brād* + *lēah*; examples include: **Bradley** Derbys. *Braidelei* 1086 (DB). **Bradley, Maiden** Wilts. *Bradelie* 1086 (DB), *Maydene Bradelega* early 13th cent. Affix means 'of the maidens' and refers to the nuns of Amesbury who had a cell here. **Bradley, North** Wilts. *Bradlega* 1174.
   However the following has a different origin: **Bradley in the Moors** Staffs. *Bretlei* 1086 (DB), 'Wood where boards or planks are got'. OE *bred* + *lēah*.

Bradmore Notts. *Brademere* 1086 (DB). 'Broad pool'. OE *brād* + *mere*.

Bradninch Devon. *Bradenese* 1086 (DB). '(Place at) the broad ash-tree or oak-tree'. OE *brād* (dative *-an*) + *æsc* or *āc* (dative *ǣc*).

Bradnop Staffs. *Bradenhop* 1219. 'Broad valley'. OE *brād* (dative *-an*) + *hop*.

Bradoge (*Bráideog*) Donegal. 'Little throat'.

Bradox (*Na Bráideoga*) Monaghan. 'The little throats'.

Bradpole Dorset. *Bratepolle* 1086 (DB). 'Broad pool'. OE *brād* + *pōl*.

Bradshaw Bolton. *Bradeshaghe* 1246. 'Broad wood or copse'. OE *brād* + *sceaga*.

Bradstone Devon. *Bradan stane* c.970, *Bradestana* 1086 (DB). '(Place at) the broad stone'. OE *brād* + *stān*.

Bradwell, '(place at) the broad spring or stream', OE *brād* + *wella*; examples include: **Bradwell** Derbys. *Bradewelle* 1086 (DB). **Bradwell** Essex. *Bradewell* 1238. **Bradwell** Milt. K. *Bradewelle* 1086 (DB). **Bradwell** Norfolk. *Bradewell* 1211. **Bradwell** Staffs. *Bradewull* 1227. **Bradwell-on-Sea** Essex. *Bradewella* 1194.

Bradworthy Devon. *Brawardine* [sic] 1086 (DB), *Bradewurtha* 1175. 'Broad enclosure'. OE *brād* + *worthign* or *worthig*.

Braemar Aber. *the Bray of Marre* 1560. 'Upper part of Mar'. Gaelic *braigh*. See MAR(R).

Brafferton, 'farmstead by the broad ford', OE *brād* + *ford* + *tūn*: **Brafferton** Darltn. *Bradfortuna* 1091. **Brafferton** N. Yorks. *Bradfortune* 1086 (DB).

Brafield-on-the-Green Northants. *Bragefelde* 1086 (DB). 'Open country by higher ground'. OE **bragen* + *feld*. The affix dates from the 16th cent.

Braid (*Braghad*) (river) Antrim. 'Throat'.

Brailes Warwicks. *Brailes* 1086 (DB). Possibly from an OE **brægels* 'burial place, tumulus'. Alternatively a Celtic name 'hill court' from **breȝ* + **līss*.

Brailsford Derbys. *Brailesford* 1086 (DB). Possibly 'ford by a burial place'. OE **brægels* + *ford*. Alternatively the first part of this name may be Celtic, *see* the previous name.

Braintree Essex. *Branchetreu* 1086 (DB). 'Tree of a man called **Branca*'. OE pers. name + *trēow*. The river-name **Brain** is a back-formation from the place name.

Braiseworth Suffolk. *Briseworde* 1086 (DB). 'Enclosure infested with gadflies or belonging to a man called **Brīosa*'. OE *brīosa* or pers. name + *worth*.

Braishfield Hants. *Braisfelde* c.1235. OE *feld* 'open land' with an uncertain first element, possibly an OE **bræsc* 'small branches or brushwood'.

Braithwaite, 'broad clearing', OScand. *breithr* + *thveit*: **Braithwaite** Brad. *Braytweyt* 1276. **Braithwaite** Cumbria, near Keswick. *Braithait* c.1160. **Braithwaite, Low** Cumbria. *Braythweyt* 1285.

Braithwell Donc. *Bradewelle* 1086 (DB). '(Place at) the broad spring or stream'. OE *brād* (replaced by OScand. *breithr*) + *wella*.

Bramber W. Sussex. *Bremre* 956, *Brembre* 1086 (DB). OE *brēmer* 'bramble thicket'.

Bramcote Notts. *Brunecote* [sic] 1086 (DB), *Bramcote* c.1156. 'Cottage(s) where broom grows'. OE *brōm* + *cot*.

Bramdean Hants. *Bromdene* 824, *Brondene* 1086 (DB). 'Valley where broom grows'. OE *brōm* + *denu*.

Bramerton Norfolk. *Brambretuna* 1086 (DB). Possibly 'farmstead by the bramble thicket'. OE *brēmer*, **brǣmer* + *tūn*.

Bramfield Herts. *Brandefelle* [*sic*] 1086 (DB), *Brantefeld* 12th cent. Possibly 'steep open land'. OE *brant* + *feld*. Alternatively the first element may be OE *bærned* (ME *brand*) 'burnt'.

Bramfield Suffolk. *Brunfelda* [*sic*] 1086 (DB), *Bramfeld* 1166. 'Open land where broom grows'. OE *brōm* + *feld*.

Bramford Suffolk. *Bromford* 1040, *Branfort* 1086 (DB). 'Ford where broom grows'. OE *brōm* + *ford*.

Bramhall Stockp. *Bramale* 1086 (DB). 'Nook of land where broom grows'. OE *brōm* + *halh*.

Bramham Leeds. *Bram(e)ham* 1086 (DB). 'Homestead or enclosure where broom grows'. OE *brōm* + *hām* or *hamm*.

Bramhope Leeds. *Bramhop* 1086 (DB). 'Valley where broom grows'. OE *brōm* + *hop*.

Bramley, 'woodland clearing where broom grows', OE *brōm* + *lēah*: **Bramley** Hants. *Brumlei* 1086 (DB). **Bramley** Rothm. *Bramelei* 1086 (DB). **Bramley** Surrey. *Bronlei* 1086 (DB).

Brampford Speke Devon. *Branfort* 1086 (DB), *Bramford Spec* 1275. Apparently 'ford where broom grows'. OE *brōm* + *ford*. However if the form *Brenteforlond* 944 belongs here, the first element may be a Celtic or OE hill-name (*see* BRENT). Manorial affix from the *Espec* family, here in the 12th cent.

Brampton, a fairly common name, 'farmstead where broom grows', OE *brōm* + *tūn*; examples include: **Brampton** Cambs. *Brantune* 1086 (DB). **Brampton** Cumbria, near Irthington. *Brampton* 1169. **Brampton** Suffolk. *Bramtuna*, *Brantuna* 1086 (DB).
Brampton Bryan Herefs. *Brantune* 1086 (DB), *Bramptone Brian* 1275. Manorial affix from a 12th-cent. owner called *Brian*.
Brampton, Chapel & **Brampton, Church** Northants. *Brantone* 1086 (DB). The distinguishing affixes occur from the 13th cent.

Bramshall Staffs. *Branselle* [*sic*] 1086 (DB), *Bromschulf* 1327. 'Shelf of land where broom grows'. OE *brōm* + *scelf*.

Bramshaw Hants. *Bramessage* 1086 (DB), *Brumesaghe* 1186, *Brambelshagh* 1272. OE *sceaga* 'small wood, copse' with a first element

varying between OE *brǣmel*, *brēmel* 'bramble' and OE *brōm* 'broom'.

Bramshill Hants. *Bromeselle* 1086 (DB). 'Hill of the broom', i.e. 'hill where broom grows'. OE *brōm* (genitive -*es*) + *hyll*.

Bramshott Hants. *Brenbresete* 1086 (DB). 'Projecting piece of land where brambles grow'. OE *brǣmel* + *scēat*.

Bramwith, Kirk Donc. *Branuuet* [*sic*] 1086 (DB), *Branwyth* 1200, *Kyrkbramwith* 1341. 'Wood overgrown with broom'. OE *brōm* + OScand. *vithr*. Affix is OScand. *kirkja* 'church'.

Brancaster Norfolk. *Bramcestria* c.960, *Broncestra* 1086 (DB). 'Roman station at *Branodunum*'. Reduced form of ancient Celtic name (probably 'crow fort') + OE *ceaster*.

Brancepeth Durham. *Brantespethe* c.1170. 'Path or road of a man called Brandr'. OScand. pers. name + OE *pæth*.

Brandesburton E. R. Yorks. *Brantisburtone* 1086 (DB). 'Fortified farmstead of a man called Brandr'. OScand. pers. name + OE *burh-tūn*.

Brandeston Suffolk. *Brantestona* 1086 (DB). 'Farmstead of a man called **Brant*'. OE pers. name + *tūn*.

Brandiston Norfolk. *Brantestuna* 1086 (DB). Identical in origin with the previous name.

Brandon, usually 'hill where broom grows', OE *brōm* + *dūn*: **Brandon** Durham. *Bromdune* c.1190. **Brandon** Northum. *Bremdona* c.1150, *Bromdun* 1236. Here the first element alternates with OE **brēmen* 'broomy'. **Brandon** Suffolk. *Bromdun* 11th cent., *Brandona* 1086 (DB). **Brandon** Warwicks. *Brandune* 1086 (DB). **Brandon Parva** Norfolk. *Brandun* 1086 (DB).
However the following may have a different origin: **Brandon** Lincs. *Branthon* 1060–66, *Brandune* 1086 (DB). Probably 'hill by the River Brant'. OE river-name (from OE *brant* 'steep, deep') + *dūn*.

Brandon (*Cé Bhréanainn*) Kerry. 'Quay of Bréanann'.

Brandsby N. Yorks. *Branzbi* 1086 (DB). 'Farmstead or village of a man called Brandr'. OScand. pers. name + *bý*.

Branksome Poole. a 19th-cent. name taken from a house called *Branksome Tower* (built 1855) which in turn was probably named from

the setting of Sir Walter Scott's Lay of the Last
Minstrel (1805).

Branscombe Devon. *Branecescumbe* 9th
cent., *Branchescome* 1086 (DB). 'Valley of a man
called Branoc'. Celtic pers. name + OE *cumb*.

Bransford Worcs. *Branesford* 716,
Bradnesford 1086 (DB). Probably 'ford by the
hill'. OE *brægen* + *ford*.

Branston, 'farmstead of a man called *Brant',
OE pers. name + *tūn*: **Branston** Leics.
Brantestone 1086 (DB). **Branston** Staffs.
Brontiston 942, *Brantestone* 1086 (DB).
   However the following may have a different
origin: **Branston** Lincs. *Branztune* 1086 (DB).
Probably 'farmstead of a man called Brandr'.
OScand. pers. name + OE *tūn*.

Brant Broughton Lincs. *See* BROUGHTON.

Brantham Suffolk. *Brantham* 1086 (DB).
Possibly 'steep homestead or enclosure'. OE
brant (adjective) + *hām* or *hamm*.

Branthwaite Cumbria. near Workington.
Bromthweit 1210. 'Clearing where broom
grows'. OE *brōm* + OScand. *thveit*.

Brantingham E. R. Yorks. *Brentingeham*
1086 (DB). 'Homestead of the family or
followers of a man called *Brant', from OE pers.
name + *-inga-* + *hām*. Alternatively
'homestead of those dwelling on the steep
slopes', from OE *brant* + *-inga-* + *hām*.

Branton Northum. *Bremetona* c.1150.
'Farmstead overgrown with broom'. OE
brēmen + *tūn*.

Branton Donc. *Brantune* 1086 (DB).
'Farmstead where broom grows'. OE *brōm* +
tūn.

Branxton Northum. *Brankeston* 1195.
'Farmstead of a man called Branoc'. Celtic pers.
name + OE *tūn*.

Brassington Derbys. *Branzinctun* 1086 (DB).
'Estate associated with a man called
*Brandsige'. OE pers. name + *-ing-* + *tūn*.

Brasted Kent. *Briestede* [*sic*] 1086 (DB),
Bradestede c.1100. 'Broad place'. OE *brād* +
stede.

Bratoft Lincs. *Breietoft* 1086 (DB). 'Broad
homestead'. OScand. *breithr* + *toft*.

Brattleby Lincs. *Brotulbi* 1086 (DB).
'Farmstead or village of a man called *Brot-Ulfr'.
OScand. pers. name + *bý*.

Bratton, usually 'farmstead by newly
cultivated ground', OE *brǣc* + *tūn*: **Bratton**
Wilts. *Bratton* 1177. **Bratton Clovelly** Devon.
Bratona 1086 (DB), *Bratton Clavyle* 1279.
Manorial affix from the *de Clavill* family, here in
the 13th cent. **Bratton Fleming** Devon.
Bratona 1086 (DB). Manorial affix from the
Flemeng family, here in the 13th cent.
   However the following have a different origin,
'farmstead by a brook', OE *brōc* + *tūn*: **Bratton**
Tel. & Wrek. *Brochetone* 1086 (DB). **Bratton
Seymour** Somerset. *Broctune* 1086 (DB).
Manorial affix from the *Saint Maur* family, here
c.1400.

Braughing Herts. *Breahingas* 825–8,
Brachinges 1086 (DB). '(Settlement of) the family
or followers of a man called *Breahha'. OE pers.
name + *-ingas*.

Braunston, Braunstone, 'farmstead of a
man called *Brant', OE pers. name + *tūn*:
Braunston Northants. *Brantestun* 956,
Brandestone 1086 (DB). **Braunston** Rutland.
Branteston 1167. **Braunstone** Leics.
Brantestone 1086 (DB).

Braunton Devon. *Brantona* 1086 (DB).
'Farmstead where broom grows'. OE *brōm* +
tūn.

Brawby N. Yorks. *Bragebi* 1086 (DB).
'Farmstead or village of a man called Bragi'.
OScand. pers. name + *bý*.

Braxted, Great Essex. *Brachestedam* 1086
(DB). Probably 'place where fern or bracken
grows'. OE *bracu* + *stede*.

Bray Winds. & Maid. *Brai* 1086 (DB). Probably
OFrench *brai* 'mud'.

Bray (*Bré*) Wicklow. *Bree* (no date). 'Hill'.

Braybrooke Northants. *Bradebroc* 1086
(DB). '(Place at) the broad brook'. OE *brād* +
brōc.

Brayfield, Cold Milt. K. *Bragenfelda* 967.
'Open land by higher ground'. OE *bragen* +
feld. Affix means 'bleak, exposed'.

Braystones Cumbria. *Bradestanes* 1247.
'Broad stones'. OE *brād* (replaced by OScand.
breithr) + *stān*.

Brayton N. Yorks. *Breithe-tun* c.1030, *Bretone* 1086 (DB). 'Broad farmstead' or 'farmstead of a man called Breithi'. OScand. *breithr* or pers. name + OE *tūn*.

Breadalbane (district) Perth, Stir. *Bredalban* c.1600. 'Upper part of Alba'. Gaelic *bràghad*. *Alba* is the old name for Scotland.

Breadsall Derbys. *Brægdeshale* 1002, *Braideshale* 1086 (DB). 'Nook of land of a man called *Brægd'. OE pers. name + *halh*.

Breadstone Glos. *Bradelestan* [*sic*] 1236, *Bradeneston* 1273. '(Place at) the broad stone'. OE *brād* (dative *-an*) + *stān*.

Bready (*An Bhréadaigh*) Tyrone. 'The (place of) broken ground'.

Breage Cornwall. *Egglosbrec* c.1170. 'Church of St Breage'. From the female patron saint of the church (with Cornish *eglos* 'church' in the early form).

Breaghwy (*Bréachmhaigh*) Mayo, Sligo. 'Wolf plain'.

Breamore Hants. *Brumore* 1086 (DB). 'Moor or marshy ground where broom grows'. OE *brōm* + *mōr*.

Brean Somerset. *Brien* 1086 (DB). Possibly a Celtic name containing a derivative of *brez* 'hill'.

Brearton N. Yorks. *Braretone* 1086 (DB). 'Farmstead amongst the briars'. OE *brēr* + *tūn*.

Breaston Derbys. *Braidestune* 1086 (DB). 'Farmstead of a man called *Brægd'. OE pers. name + *tūn*.

Brechin Ang. *Brechin* c.1145. '(Place of) Brychan'. Brychan also gave the name of BRECON.

Breckland (district) Norfolk. 'Area in which ground has been broken up for cultivation', from dialect *breck*, a name first used in the 19th cent.

Breckles Norfolk. *Brecchles* 1086 (DB). 'Meadow by newly cultivated land'. OE *brēc* + *lǣs*.

Brecknock. *See* BRECON.

Brecon (*Aberhonddu*) Powys. *Brecheniauc* 1100. '(Place of) Brychan'. *Brychan* was a 5th-cent. prince. The Welsh name means 'mouth of the River Honddu' (OWelsh *aber*),

the river-name meaning 'pleasant' (Welsh *hawdd*). In the variant form *Brecknock*, the second element represents the Welsh 'territorial' suffix *-iog*.

Brecon Beacons (*Bannau Brycheiniog*) (mountains) Powys. 'Beacons of Brecon'. The mountains were used for signal-fires in medieval times. The Welsh name means 'peaks of Brycheiniog' i.e. 'territory of Brychan' (*see* BRECON).

Brecqhou (island) Channel Islands. *Brekehoc* c.1540. 'Steep island'. OScand. *brekka* + *holmr*. Its ancient name, recorded as *Besargia* in the 6th cent., is probably of Celtic origin with a meaning 'by or under *Sargia* = SARK'.

Bredagh (*Brédach*) Westmeath. 'Broken ground'.

Bredbury Stockp. *Bretberie* 1086 (DB). 'Stronghold or manor house built of planks'. OE *bred* + *burh* (dative *byrig*).

Brede E. Sussex. *Brade* 1161, *Brede* 1202. OE *brǣdu* 'breadth', here probably referring to the broad valley called Brede Level through which flows the River **Brede** (a back-formation from the place name).

Bredfield Suffolk. *Bredefelda* 1086 (DB). 'Broad stretch of open country'. OE *brǣdu* + *feld*.

Bredgar Kent. *Bradegare* c.1100. 'Broad triangular plot'. OE *brād* + *gāra*.

Bredhurst Kent. *Bredehurst* 1240. 'Wooded hill where boards are obtained'. OE *bred* + *hyrst*.

Bredon Worcs. *Breodun* 772, 1086 (DB). 'Hill called *Bre'. Celtic *brez* 'hill' + explanatory OE *dūn*.

Bredon's Norton Worcs. *See* NORTON.

Bredwardine Herefs. *Brocheurdie* [*sic*] 1086 (DB), *Bredewerthin* late 12th cent. OE *worthign* 'enclosure', probably with *bred* 'board, plank' or *brǣdu* 'broad stretch of land'.

Bredy, Long & Littlebredy Dorset. *Bridian* 987, *Langebride*, *Litelbride* 1086 (DB). Named from the River Bride, a Celtic river-name meaning 'gushing or surging stream'. Distinguishing affixes are OE *lang* 'long' and *lȳtel* 'little'.

Bree (*Brí*) Wexford. 'Hilly place'.

Breedoge (*Bráideog*) Roscommon. 'Little throat'.

Breedon on the Hill Leics. *Briudun* 731. 'Hill called *Bre*'. Celtic **breʒ* 'hill' + explanatory OE *dūn* 'hill'. With the more recent affix the name thus contains three different words for 'hill'.

Breighton E. R. Yorks. *Bricstune* 1086 (DB). Possibly 'bright farmstead' or 'farmstead of a man called **Beorhta*'. OE *beorht* or pers. name + *tūn*.

Bremhill Wilts. *Bre(o)mel* 937, *Breme* [sic] 1086 (DB). 'Bramble thicket'. OE *brēmel*.

Brenchley Kent. *Braencesle* c.1100. Probably 'woodland clearing of a man called Brenci'. Cornish pers. name + *lēah*.

Brendon Devon. *Brandone* 1086 (DB), *Bremdon* 12th cent. 'Hill where broom grows'. OE *brōm* (possibly influenced by OE **brēme* 'broom place') + *dūn*.

Brendon Hills Somerset. *Brunedun* 1204. An OE hill-name **Brūne* or **Brūna* (a derivative of OE *brūn* 'dark, brown'), here used of the chain of hills, with the later addition of OE *dūn* 'hill'.

Brenkley Newc. upon T. *Brinchelawa* 1178. Possibly 'hill or mound of a man called Brynca'. OE pers. name + *hlāw*. Alternatively the first element may be OE **brince* 'brink or edge'.

Brent, either a Celtic hill-name meaning 'high one, height', or an OE **brente* 'steep or high place' (an *i*-mutated derivative of OE *brant* 'steep'): **Brent, East** Somerset. *Brente* 663, *Brentemerse* 1086 (DB). With OE *mersc* 'marsh' in the Domesday form. **Brent Knoll** Somerset. *Brenteknol* 1289, sometimes known as South Brent, *Sudbrente* 1196. With OE *cnoll* 'hill-top', *sūth* 'south'. **Brent, South** Devon. *Brenta* 1086 (DB).

Brent Eleigh Suffolk. *See* ELEIGH.

Brent Pelham Herts. *See* PELHAM.

Brentford Gtr. London. *Breguntford* 705. 'Ford over the River Brent'. Celtic river-name (meaning 'holy one') + OE *ford*. The London borough of **Brent** takes its name from the river.

Brentor Devon. *Brentam* c.1170, *Brentetor* 1232. Celtic or OE hill-name (*see* BRENT) + OE *torr*.

Brentwood Essex. *Boscus arsus* 1176, *Brendewode* 1274. 'The burnt wood'. OE *berned*

(ME *brend*) + *wudu*. In the earliest form the name has been translated into Latin.

Brenzett Kent. *Brensete* 1086 (DB). 'The burnt fold or stable'. OE *berned* + (*ge*)*set*.

Brereton Ches. *Bretone* [sic] 1086 (DB), *Brereton* c.1100. 'Farmstead amongst the briars'. OE *brēr* + *tūn*.

Brereton Staffs. *Breredon* 1279. 'Hill where briars grow'. OE *brēr* + *dūn*.

Bressay (island) Shet. *Bressa* 1654. Originally 'broad island'. OScand. *breithr* + *ey*.

Bressingham Norfolk. *Bresingaham* 1086 (DB). 'Homestead of the family or followers of a man called **Brīosa*'. OE pers. name + *-inga-* + *hām*.

Bretford Warwicks. *Bretford* early 11th cent. Probably 'ford provided with planks'. OE *bred* + *ford*.

Bretforton Worcs. *Bretfertona* 709, *Bratfortune* 1086 (DB). 'Farmstead near the plank ford'. OE *bred* + *ford* + *tūn*.

Bretherdale Head Cumbria. *Britherdal* 12th cent. 'Valley of the brother(s)'. OScand. *bróthir* + *dalr*.

Bretherton Lancs. *Bretherton* 1190. 'Farmstead of the brother(s)'. OE *brōthor* or OScand. *bróthir* + OE *tūn*.

Brettenham, 'homestead of a man called **Bretta* or **Beorhta*', OE pers. name (genitive *-n*) + *hām*: **Brettenham** Norfolk. *Bretham* 1086 (DB). **Brettenham** Suffolk. *Bretenhama* 1086 (DB). The name of the River **Brett**, which rises near here, is a 'back-formation' from the place name.

Bretton, 'farmstead of the Britons'. OE *Brettas* (genitive *Bretta*) + *tūn*: **Bretton** Flin. *Bretton* c.1310. **Bretton, Monk** Barns. *Brettone* 1086 (DB), *Munkebretton* 1225. Affix from OE *munuc* 'monk' referring to the monks of Bretton Priory. **Bretton, West** Wakefd. *Bretone* 1086 (DB), *West Bretton* c.1200.

Brewham Somerset. *Briweham* 1086 (DB). 'Homestead or enclosure on the River Brue'. Celtic river-name (*see* BRUTON) + OE *hām* or *hamm*.

Brewood Staffs. *Breude* 1086 (DB). 'Wood by the hill called *Bre*'. Celtic **breʒ* 'hill' + OE *wudu*.

Briantspuddle Dorset. *Pidele* 1086 (DB), *Brianis Pedille* 1465. 'Estate on the River Piddle held by a man called Brian'. OE river-name (*see* PIDDLEHINTON) with manorial affix from 14th-cent. lord of the manor.

Bricett, Great Suffolk. *Brieseta* 1086 (DB). Possibly 'fold or stable infested with gadflies'. OE *brīosa* + (*ge*)*set*.

Bricket Wood Herts. *Bruteyt* 1228. 'Bright-coloured small island or piece of marshland'. OE *beorht* + *ēgeth*.

Brickhill, Bow (Milt. K.), **Brickhill, Great** (Bucks.) & **Brickhill, Little** (Milt. K.) *Brichelle* 1086 (DB), *Bolle Brichulle, Magna Brikehille, Parua Brichull* 1198. 'Hill called *Brig*'. Celtic **brig* 'hill-top' + explanatory OE *hyll*. Distinguishing affixes from OE pers. name *Bolla* (no doubt an early tenant), Latin *magna* 'great' and *parva* 'little'.

Bricklehampton Worcs. *Bricstelmestune* 1086 (DB). 'Estate associated with a man called Beorhthelm'. OE pers. name + -*ing*- + *tūn*.

Bricklieve (*Bricshliabh*) Sligo. 'Speckled mountain'.

Bride (river) Dorset. *See* BREDY.

Bridekirk Cumbria. *Bridekirke* c.1210. 'Church of St Bride or Brigid'. Irish saint's name + OScand. *kirkja*.

Bridestowe Devon. *Bridestou* 1086 (DB). 'Holy place of St Bride or Brigid'. Irish saint's name + OE *stōw*.

Bridford Devon. *Brideforda* 1086 (DB). Possibly 'ford suitable for brides', i.e. a shallow ford easy to cross. OE *brȳd* + *ford*. Alternatively the first element may be OE **brȳde* 'surging stream'.

Bridge Kent. *Brige* 1086 (DB). '(Place at) the bridge'. OE *brycg*.

Bridge Hewick N. Yorks. *See* HEWICK.

Bridge of Allan Stir. 'Bridge over the River Allan'. The Celtic river name means 'flowing water'.

Bridge of Alvah. *See* ALVAH.

Bridge of Weir Renf. 'Bridge by the weir'. The weir is on the River Gryfe.

Bridge Sollers Herefs. *Bricge* 1086 (DB), *Bruges Solers* 1291. '(Place at) the bridge'. OE *brycg* + manorial affix from the *de Solers* family, here in the 12th cent.

Bridgend (*Ceann an Droichid*) Donegal. 'End of the bridge'.

Bridgend (*Pen-y-bont ar Ogwr*) Bri. *Byrge End* 1535. 'End of the bridge'. OE *brycg* + *ende*. A Norman castle is said to have protected the crossing here over the River Ogmore. The Welsh name means 'head of the bridge on the Ogmore' (Welsh *pen* + *y* + *pont*). For the river-name, *see* OGMORE.

Bridgerule Devon. *Brige* 1086 (DB), *Briggeroald* 1238. '(Place at) the bridge held by a man called Ruald'. OE *brycg* + manorial affix from OScand. *Róaldr* (tenant in 1086).

Bridgford, 'ford by the bridge', OE *brycg* + *ford*: **Bridgford, East** Notts. *Brugeford* 1086 (DB). **Bridgford, West** Notts. *Brigeforde* 1086 (DB).

Bridgham Norfolk. *Brugeham* c.1050. 'Homestead or enclosure by a bridge'. OE *brycg* + *hām* or *hamm*.

Bridgnorth Shrops. *Brug* 1156, *Brugg North* 1282. '(Place at) the bridge'. OE *brycg* + later affix *north*.

Bridgwater Somerset. *Brugie* 1086 (DB), *Brigewaltier* 1194. '(Place at) the bridge held by a man called Walter'. OE *brycg* + manorial affix from an early owner.

Bridlington E. R. Yorks. *Bretlinton* 1086 (DB). 'Estate associated with a man called Berhtel'. OE pers. name + -*ing*- + *tūn*.

Bridport Dorset. *Brideport* 1086 (DB). 'Harbour or market town belonging to (Long) BREDY'. OE *port*. The river-name **Brit** is a back-formation from the place name.

Bridstow Herefs. *Bridestowe* 1277. 'Holy place of St Bride or Brigid'. Irish saint's name + OE *stōw*.

Brierfield Lancs. a self-explanatory name of 19th-cent. origin, no doubt influenced by the nearby **Briercliffe** (*Brerecleve* 1193) which is 'bank where briars grow', OE *brēr* + *clif*.

Brierley, 'woodland clearing where briars grow', OE *brēr* + *lēah*: **Brierley** Barns. *Breselai* [*sic*] 1086 (DB), *Brerelay* 1194. **Brierley Hill** Dudley. *Brereley* 14th cent.

Brigg N. Lincs. earlier *Glanford Brigg* 1235. 'Bridge at the ford where people assemble for revelry or games'. OE *glēam* + *ford* + *brycg*.

Brigh (*Bríoch*) Tyrone. *Breigh* 1633. 'Hilly place'.

Brigham, 'homestead or enclosure by a bridge', OE *brycg* + *hām* or *hamm*: **Brigham** Cumbria, near Cockermouth. *Briggham c.*1175. **Brigham** E. R. Yorks. *Bringeham* [*sic*] 1086 (DB), *Brigham* 12th cent.

Brighouse Calder. *Brighuses* 1240. 'Houses by the bridge'. OE *brycg* + *hūs*.

Brighstone I. of Wight. *Brihtwiston* 1212. 'Farmstead of a man called Beorhtwīg'. OE pers. name + *tūn*.

Brighthampton Oxon. *Byrhtelmingtun* 984, *Bristelmestone* 1086 (DB). 'Farmstead of a man called Beorhthelm'. OE pers. name + *tūn*.

Brightling E. Sussex. *Byrhtlingan* 1016–20, *Brislinga* 1086 (DB). '(Settlement of) the family or followers of a man called Beorhtel'. OE pers. name + *-ingas*.

Brightlingsea Essex. *Brictriceseia* 1086 (DB). 'Island of a man called Beorhtrīc or *Beorhtling'. OE pers. name + *ēg*.

Brighton Bright. & Hove. *Bristelmestune* 1086 (DB). 'Farmstead of a man called Beorhthelm'. OE pers. name + *tūn*.

Brighton, New Wirral. 19th-cent. resort named after BRIGHTON.

Brightwalton W. Berks. *Beorhtwaldingtune* 939, *Bristoldestone* 1086 (DB). 'Estate associated with a man called Beorhtwald'. OE pers. name (+ *-ing-*) + *tūn*.

Brightwell, 'bright or clear spring', OE *beorht* + *wella*: **Brightwell** Oxon. *Beorhtawille* 854, *Bricsteuuelle* 1086 (DB). **Brightwell** Suffolk. *Brithwelle c.*1050, *Brithtewella* 1086 (DB). **Brightwell Baldwin** Oxon. *Berhtanwellan* 887, *Britewelle* 1086 (DB). Manorial affix from possession by Sir *Baldwin* de Bereford in the late 14th cent.

Brignall Durham. *Bring(en)hale* 1086 (DB). Possibly 'nook of the family or followers of a man called Brȳni'. OE pers. name + *-inga-* + *halh*.

Brigsley NE. Lincs. *Brigeslai* 1086 (DB). 'Woodland clearing by a bridge'. OE *brycg* + *lēah*.

Brigsteer Cumbria. *Brigstere* early 13th cent. 'Bridge of a family called Stere, or one used for bullocks'. OE *brycg* (influenced by OScand. *bryggja*) + ME surname or OE *stēor*. The order of elements is Celtic.

Brigstock Northants. *Bricstoc* 1086 (DB). Probably 'outlying farm or hamlet by a bridge'. OE *brycg* + *stoc*.

Brill Bucks. *Bruhella* 1072, *Brunhelle* [*sic*] 1086 (DB). Probably 'hill called *Bre*'. Celtic *brez 'hill' + explanatory OE *hyll*.

Brilley Herefs. *Brunlege* 1219. Probably 'woodland clearing where broom grows'. OE *brōm* + *lēah*.

Brimfield Herefs. *Bromefeld* 1086 (DB). 'Open land where broom grows'. OE *brōm* + *feld*.

Brimington Derbys. *Brimintune* 1086 (DB). 'Estate associated with a man called Brēme'. OE pers. name + *-ing-* + *tūn*.

Brimpsfield Glos. *Brimesfelde* 1086 (DB). 'Open land of a man called Brēme'. OE pers. name + *feld*.

Brimpton W. Berks. *Bryningtune* 944, *Brintone* 1086 (DB). 'Estate associated with a man called Brȳni'. OE pers. name + *-ing-* + *tūn*.

Brims Ness Highland. *Brymmis* 1559. Probably 'headland of the surf'. OScand. *brim* + *nes*.

Brind E. R. Yorks. *Brende* 1188. OE *brende 'place destroyed or cleared by burning'.

Brindle Lancs. *Burnhull* 1206. Probably 'hill by a stream'. OE *burna* + *hyll*.

Brineton Staffs. *Brunitone* 1086 (DB). Probably 'estate associated with a man called Brȳni'. OE pers. name + *-ing-* + *tūn*.

Bringhurst Leics. *Bruninghyrst* 1188. Probably 'wooded hill of the family or followers of a man called Brȳni'. OE pers. name + *-inga-* + *hyrst*.

Brington, 'estate associated with a man called Brȳni', OE pers. name + *-ing-* + *tūn*: **Brington** Cambs. *Brynintune* 974, *Breninctune* 1086 (DB). **Brington, Great & Brington, Little** Northants. *Brinintone* 1086 (DB).

Briningham Norfolk. *Bruningaham* 1086 (DB). 'Homestead of the family or followers of a man called Brȳni'. OE pers. name + -inga- + hām.

Brinkhill Lincs. *Brincle* 1086 (DB). 'Woodland clearing of a man called Brynca, or on the brink of a hill'. OE pers. name or OE **brince* + lēah.

Brinkley Cambs. *Brinkelai* late 12th cent. 'Woodland clearing of a man called Brynca'. OE pers. name + lēah.

Brinklow Warwicks. *Brinckelawe* c.1155. 'Burial mound of a man called Brynca, or on the brink of a hill'. OE pers. name or OE **brince* + hlāw.

Brinkworth Wilts. *Brinkewrtha* 1065, *Brenchewrde* 1086 (DB). 'Enclosure of a man called Brynca'. OE pers. name + worth.

Brinlack (*Bun na Leaca*) Donegal. 'Foot of the flagstones'.

Brinscall Lancs. *Brendescoles* c.1200. 'Burnt huts'. ME *brend* + OScand. *skáli*.

Brinsley Notts. *Brunesleia* 1086 (DB). 'Woodland clearing of a man called Brūn'. OE pers. name + lēah.

Brinsop Herefs. *Hope* 1086 (DB), *Bruneshopa* c.1130. 'Enclosed valley of a man called Brūn or Brȳni'. OE pers. name + hop.

Brinsworth Rothm. *Brinesford* 1086 (DB). 'Ford of a man called Brȳni'. OE pers. name + ford.

Brinton Norfolk. *Bruntuna* 1086 (DB). 'Estate associated with a man called Brȳni'. OE pers. name + -ing- + tūn.

Briska (*Brioscach*) Waterford. 'Brittle land'.

Brisley Norfolk. *Bruselea* c.1105. 'Woodland clearing infested with gadflies'. OE *briosa* + lēah.

Brislington Brist. *Brihthelmeston* 1199. 'Farmstead of a man called Beorhthelm'. OE pers. name + tūn.

Bristol Brist. *Brycg stowe* 11th cent., *Bristou* 1086 (DB). 'Assembly-place by the bridge'. OE *brycg* + stōw.

Briston Norfolk. *Burstuna* 1086 (DB). 'Farmstead by a landslip or broken ground'. OE *byrst* + tūn.

Britannia Bridge (railway bridge) Gwyd. The bridge opened in 1850 and took its name from the *Britannia* Rock here.

Britford Wilts. *Brutford* 826, *Bredford* 1086 (DB). Possibly 'ford of the Britons'. OE *Bryt* + ford. Alternatively the first element may be a lost stream-name related to OE *bryttian* 'to dispense', *brytta* 'giver, lord'.

British (*Briotás*) Antrim. 'Wooden palisade'.

Briton Ferry (*Llansawel*) Neat. *Brigeton* 1201, *Brytton* 1315, *Britan Ferry caullid in Walsche Llanisauel* 1536. 'Ferry at the farm by the bridge'. OE *brycg* + tūn + ModE *ferry*. The Welsh name means 'church of Sawel' (Welsh *llan*). Compare LLANSAWEL.

Brittas (*An Briotás*) Dublin, Wicklow. 'The wooden palisade'.

Britway (*Breachmhaí*) Cork. 'Wolf plain'.

Britwell Salome Oxon. *Brutwelle* 1086 (DB), *Brutewell Solham* 1320. Possibly 'spring or stream of the Britons'. OE *Bryt* + wella. Alternatively the first element may be a lost stream-name related to OE *bryttian* 'to dispense', *brytta* 'giver, lord'. Manorial affix from the *de Suleham* family, here in the 13th cent.

Brixham Torbay. *Briseham* [sic] 1086 (DB), *Brikesham* 1205. 'Homestead or enclosure of a man called Brioc'. Celtic pers. name + OE *hām* or *hamm*.

Brixton Devon. *Brisetona* [sic] 1086 (DB), *Brikeston* 1200. Probably 'farmstead of a man called Brioc'. Celtic pers. name + OE *tūn*.

Brixton Gtr. London. *Brixiges stan* 1062, *Brixiestan* 1086 (DB). 'Stone (probably marking a Hundred meeting-place) of a man called Beorhtsige'. OE pers. name + stān.

Brixton Deverill Wilts. *Devrel* 1086 (DB), *Britricheston* 1229. 'Estate on the River *Deverill* held by a man called Beorhtrīc'. OE pers. name + tūn with Celtic river-name (meaning 'watery'), an old name for the River Wylye.

Brixworth Northants. *Briclesworde* 1086 (DB). 'Enclosure of a man called Beorhtel or **Bricel*'. OE pers. name + worth.

Brize Norton Oxon. *See* NORTON.

Broad as affix. *See* main name, e.g. for **Broad Blunsdon** (Wilts.), *see* BLUNSDON.

Broad Haven Pemb. *Brode Hauen* 1578. 'Wide harbour'. *Broad* distinguishes the village from nearby LITTLE HAVEN.

Broad Town Wilts. *Bradetun* 12th cent. 'Broad or large farmstead'. OE *brād* + *tūn*.

Broadbottom Tamesd. *Brodebothem* 1286. 'Broad valley-bottom'. OE *brād* + **bothm*.

Broadford (*Áth Leathan*) Clare. 'Broad ford'.

Broadford Highland. (Skye). 'Broad ford'. The name translates Gaelic *an t-àth leathan*.

Broadheath, Upper & Broadheath, Lower Worcs. *Hethe* 1240, *Broad Heath* 1646. '(Place at) the broad heath'. OE *hǣth* with the later addition of *broad*.

Broadhembury Devon. *Hanberia* 1086 (DB), *Brodehembyri* 1273. 'High or chief fortified place'. OE *hēah* (dative *hēan*) + *burh* (dative *byrig*). Later affix is OE *brād* 'broad, great'.

Broadhempston Devon. *Hamistone* 1086 (DB), *Brodehempstone* 1362. 'Farmstead of a man called **Hǣme* or Hemme'. OE pers. name + *tūn*. Affix is OE *brād* 'large' to distinguish this place from LITTLEHEMPSTON.

Broadland (district) Norfolk. *Broadland* 1889. 'District of the Broads'. See BROADS.

Broadmayne Dorset. *Maine* 1086 (DB), *Brademaene* 1202. Celtic **main* 'a rock, a stone', with later affix from OE *brād* 'broad, great'.

Broads, The (district) Norfolk, named from over thirty 'broads', i.e. extensive pieces of fresh water formed by the broadening out of rivers.

Broadstairs Kent. *Brodsteyr* 1435. 'Broad stairway or ascent'. OE *brād* + *stǣger*.

Broadstone Poole. a recent name for a parish formed in 1906, self-explanatory.

Broadwas Worcs. *Bradeuuesse* 779, *Bradewesham* 1086 (DB). 'Broad tract of alluvial land'. OE *brād* + **wæsse* (with *hām* 'homestead' in the 1086 form).

Broadwater W. Sussex. *Bradewatre* 1086 (DB). '(Place at) the broad stream'. OE *brād* + *wæter*.

Broadway, '(place at) the broad way or road', OE *brād* + *weg*: **Broadway** Somerset. *Bradewei* 1086 (DB). **Broadway** Worcs. *Bradanuuege* 972, *Bradeweia* 1086 (DB).

Broadwell, '(place at) the broad spring or stream', OE *brād* + *wella*: **Broadwell** Glos. *Bradewelle* 1086 (DB). **Broadwell** Oxon.

Bradewelle 1086 (DB). **Broadwell** Warwicks. *Bradewella* 1130.

Broadwey Dorset. *Wai(a)* 1086 (DB), *Brode Way* 1243. Named from the River Wey, *see* WEYMOUTH. Affix is OE *brād* 'broad, great', referring either to the width of the river here or to the size of the manor.

Broadwindsor Dorset. *Windesore* 1086 (DB), *Brodewyndesore* 1324. 'River-bank with a windlass'. OE **windels* + *ōra*, with *brād* 'great'.

Broadwoodkelly Devon. *Bradehoda* [sic] 1086 (DB), *Brawode Kelly* 1261. 'Broad wood'. OE *brād* + *wudu* + manorial affix from the *de Kelly* family, here in the 13th cent.

Broadwoodwidger Devon. *Bradewode* 1086 (DB), *Brodwode Wyger* 1310. Identical in origin with the previous name. Manorial affix from the *Wyger* family, here in the 13th cent.

Brobury Herefs. *Brocheberie* 1086 (DB). 'Stronghold or manor near a brook'. OE *brōc* + *burh* (dative *byrig*).

Brockagh (*Brocach*) Westmeath. 'Badger den'.

Brockdish Norfolk. *Brodise* [sic] 1086 (DB), *Brochedisc* c.1095. 'Pasture by the brook'. OE *brōc* + *edisc*.

Brockenhurst Hants. *Broceste* [sic] 1086 (DB), *Brokenhurst* early 12th cent. 'Broken wooded hill'. OE *brocen* 'broken up' (with reference to ploughed or uneven land or to the broken and undulating outline of terrain dissected by several small streams) + *hyrst*.

Brockford Street Suffolk. *Brocfort* 1086 (DB). 'Ford over the brook'. OE *brōc* + *ford*.

Brockhall Northants. *Brocole* 1086 (DB). OE *brocc-hol* 'a badger hole, a sett'.

Brockham Surrey. *Brocham* 1241. 'River-meadow by the brook, or frequented by badgers'. OE *brōc* or *brocc* + *hamm*.

Brockhampton, 'homestead by the brook', OE *brōc* + *hām-tūn*: **Brockhampton**, Glos., near Sevenhampton. *Brochamtone* 1166. **Brockhampton** Herefs., near Bromyard. *Brockampton* 1251.

Brocklesby Lincs. *Brochelesbi* 1086 (DB). 'Farmstead or village of a man called **Bróklauss*'. OScand. pers. name + *bý*.

Brockley Gtr. London. *Brocele* 1182, *Brocleg* 1226. Identical in origin with one or other of the following names.

Brockley N. Som. *Brochelie* 1086 (DB). Probably 'woodland clearing frequented by badgers'. OE *brocc* (genitive plural *brocca*) + *lēah*.

Brockley Suffolk. *Broclega* 1086 (DB). Probably 'woodland clearing by a brook'. OE *brōc* + *lēah*.

Brockton, Brocton, 'farmstead by a brook', OE *brōc* + *tūn*: **Brockton** Shrops., near Lilleshall. *Brochetone* 1086 (DB). **Brockton** Shrops., near Madeley. *Broctone* 1086 (DB). **Brockton** Shrops., near Worthen. *Brockton* 1272. **Brocton** Staffs., near Stafford. *Broctone* 1086 (DB).

Brockweir Glos. *Brocwere* c.1145. 'Weir by the brook'. OE *brōc* + *wer*.

Brockworth Glos. *Brocowardinge* 1086 (DB). 'Enclosure by the brook'. OE *brōc* + *worthign*.

Brocton Staffs. *See* BROCKTON.

Brodick N. Ayr. (Arran). *Brathwik* 1306, *Bradewik* 1450. 'Broad bay'. OScand. *breithr* + *vík*. The town is on the bay of the same name.

Brodie Moray. *Brothie* 1380. 'Muddy place'. Gaelic *brothaith*.

Brodsworth Donc. *Brodesworde* 1086 (DB). 'Enclosure of a man called Broddr or Brord'. OScand. or OE pers. name + *worth*.

Brokenborough Wilts. *Brokene beregge* 956, *Brocheneberge* 1086 (DB). 'Broken barrow' (probably referring to a tumulus that had been broken into). OE *brocen* + *beorg*.

Bromborough Wirral. *Brunburg* early 12th cent. 'Stronghold of a man called Brūna'. OE pers. name + *burh*.

Brome Suffolk. *Brom* 1086 (DB). 'Place where broom grows'. OE *brōm*.

Bromeswell Suffolk. *Bromeswella* 1086 (DB). 'Rising ground where broom grows'. OE *brōm* + **swelle*.

Bromfield Cumbria. *Brounefeld* c.1125. 'Brown open land, or open land where broom grows'. OE *brūn* or *brōm* + *feld*.

Bromfield Shrops. *Bromfelde* 1061, *Brunfelde* 1086 (DB). 'Open land where broom grows'. OE *brōm* + *feld*.

Bromham, 'homestead or enclosure where broom grows', OE *brōm* + *hām* or *hamm*: **Bromham** Beds. *Bruneham* 1086 (DB). Alternatively the first element in this name may be the OE pers. name *Brūna*. **Bromham** Wilts. *Bromham* 1086 (DB).

Bromley, usually 'woodland clearing where broom grows', OE *brōm* + *lēah*: **Bromley** Herts. *Bromlegh* 1248. **Bromley** Gtr. London, near Beckenham. *Bromleag* 862, *Bronlei* 1086 (DB). **Bromley, Abbots** Staffs. *Bromleage* 1002. Affix from its early possession by Burton Abbey. **Bromley, Great** & **Bromley, Little** Essex. *Brumleiam* 1086 (DB). **Bromley, Kings** Staffs. *Bromelei* 1086 (DB), *Bramlea Regis* 1167. Affix is Latin *regis* 'of the king', alluding to a royal manor.

However the following has a different origin: **Bromley** Gtr. London, near Bow. *Bræmbelege* c.1000. 'Woodland clearing where brambles grow'. OE *bræmbel* + *lēah*.

Brompton, usually 'farmstead where broom grows', OE *brōm* + *tūn*: **Brompton** N. Yorks., near Northallerton. *Bromtun* c.1050, *Bruntone* 1086 (DB). **Brompton** N. Yorks., near Snainton. *Bruntun* 1086 (DB). **Brompton on Swale** N. Yorks. *Brunton* 1086 (DB). On the River Swale (probably OE **swalwe* 'rushing water'). **Brompton, Patrick** N. Yorks. *Brunton* 1086 (DB), *Patricbrunton* 1157. Manorial affix from its early possession by a man called *Patric* (an OIrish pers. name). **Brompton, Potter** N. Yorks. *Brunetona* 1086 (DB). Affix probably alludes to early potmaking here.

However the following have a different origin: **Brompton Ralph** & **Brompton Regis** Somerset. *Burnetone, Brunetone* 1086 (DB), *Brompton Radulphi* 1274, *Brompton Regis* 1291. 'Farmstead by the chain of hills called **Brūne* or **Brūna*'. OE hill-name (here applied to the BRENDON HILLS) + *tūn*. Manorial affixes from early possession by a man called *Ralph* (Latin *Radulphus*) and by the king (Latin *regis* 'of the king').

Bromsgrove Worcs. *Bremesgrefan* 804, *Bremesgrave* 1086 (DB). 'Grove or coppiced wood of a man called Brēme'. OE pers. name + *græfe, grāf*.

Bromwich, 'dwelling or farm where broom grows', OE *brōm* + *wīc*: **Bromwich, Castle**

Solhll. *Bramewice* 1168, *Castelbromwic* 13th cent. Affix refers to a 12th-cent. earthwork. **Bromwich, West** Sandw. *Bromwic* 1086, *Westbromwich* 1322.

Bromyard Herefs. *Bromgeard c.*840, *Bromgerde* 1086 (DB). 'Enclosure where broom grows'. OE *brōm* + *geard*.

Brondesbury Gtr. London. *Bronnesburie* 1254. 'Manor of a man called Brand'. ME pers. name or surname + *bury* (from OE *byrig*, dative of *burh*).

Brook, Brooke, '(place at) the brook', OE *brōc*: **Brook** I. of Wight. *Broc* 1086 (DB). **Brook** Kent. *Broca* 11th cent. **Brooke** Norfolk. *Broc* 1086 (DB). **Brooke** Rutland. *Broc* 1176.

Brookeborough Fermanagh. *Brookeborough* 1835. The village arose in the 19th cent. on land granted to Sir Henry *Brooke* in 1666.

Brookland Kent. *Broke* 1254, *Broklande* 1262. Possibly 'cultivated land by a brook'. OE *brōc* + *land*. However OE *brōc* may here have the sense 'marsh' also found in the ME compound *brok-land* 'marsh-land'.

Brookmans Park Herts. *Brokemanes* 1468. Named from a local family called *Brokeman*.

Brookthorpe Glos. *Brostorp* [*sic*] 1086 (DB), *Brocthrop* 12th cent. 'Outlying farmstead or hamlet by a brook'. OE *brōc* + *throp*.

Brookwood Surrey. *Brocwude* 1225. 'Wood by a brook'. OE *brōc* + *wudu*.

Broom, Broome, 'place where broom grows', OE *brōm*; examples include: **Broom** Beds. *Brume* 1086 (DB). **Broom** Durham. *Brom c.*1170. **Broom** Warwicks. *Brome* 710, 1086 (DB). **Broome** Norfolk. *Brom* 1086 (DB). **Broome** Worcs. *Brom* 1169.

Broom, Loch Highland. *Braon* 1227. 'Loch of (the river) Broom'. Gaelic *loch* + *braon* 'drop, shower'.

Broomfield, 'open land where broom grows', OE *brōm* + *feld*: **Broomfield** Essex. *Brumfeldam* 1086 (DB). **Broomfield** Kent, near Maidstone. *Brunfelle* [*sic*] 1086 (DB), *Brumfeld c.*1100. **Broomfield** Somerset. *Brunfelle* 1086 (DB).

Broomfleet E. R. Yorks. *Brungareflet* 1150–4. 'Stretch of river belonging to a man called Brūngār'. OE pers. name + *flēot*.

Brooms, High Kent. earlier *Bromgebrug* 1270, *Bromelaregg* 1318, from OE *brōm* 'broom' with either *brycg* 'bridge' or *hrycg* 'ridge'.

Brora Highland. *Strabroray* 1499. 'River of the bridge'. OScand. *brú* (genitive *brúar*) + *á*. The bridge, at the mouth of the river of the same name, was long the only one in Sutherland. The form of the name above has Gaelic *srath*, 'valley'.

Broseley Shrops. *Burewardeslega* 1177. 'Woodland clearing of the fort-keeper, or of a man called Burgweard'. OE *burh-weard* or pers. name + *lēah*.

Brotherton N. Yorks. *Brothertun c.*1030. 'Farmstead of the brother, or of a man called Bróthir'. OE *bróthor* or OScand. pers. name + OE *tūn*.

Brotton Red. & Cleve. *Broctune* 1086 (DB). 'Farmstead by a brook'. OE *brōc* + *tūn*.

Brough, 'stronghold or fortification', OE *burh*; examples include: **Brough** Cumbria. *Burc* 1174. **Brough** Derbys. *Burc* 1195. **Brough** E. R. Yorks. *Burg c.*1200. **Brough** Notts. *Burgh* 1525.

Broughderg (*Bruach Dearg*) Tyrone. *Brugh Derge* 1666. 'Red bank'.

Broughnamaddy (*Bruach na Madadh*) Down. 'Bank of the dogs'.

Broughshane (*Bruach Sheáin*) Antrim. *Bruaghshane c.*1655. 'Séan's bank'.

Broughton, a common name, usually 'farmstead by a brook', OE *brōc* + *tūn*; examples include: **Broughton Astley** Leics. *Broctone* 1086 (DB), *Broghton Astele* 1322. Manorial affix from the *de Estle* family, here in the 13th cent. **Broughton Gifford** Wilts. *Broctun* 1001, *Broctone* 1086 (DB), *Brocton Giffard* 1288. Manorial affix from the *Giffard* family, here in the 13th cent. **Broughton, Great & Broughton, Little** Cumbria. *Broctuna* 12th cent. **Broughton Hackett** Worcs. *Broctun* 972, *Broctune* 1086 (DB), *Broctone Haket* 1275. Manorial affix from the *Hackett* family, here in the 12th cent. **Broughton in Furness** Cumbria. *Brocton* 1196. For the district name Furness, *see* BARROW. **Broughton Poggs** Oxon. *Brotone* 1086 (DB), *Broughton Pouges* 1526. Manorial affix from early possession of lands here by the *Pugeys* family.

However other Broughtons have a different origin: **Broughton** Hants. *Brestone* 1086 (DB),

Burchton 1173. 'Farmstead by a hill or mound'. OE *beorg* + *tūn*. **Broughton** N. Lincs. *Bertone* 1086 (DB). Identical in origin with previous name. **Broughton** Northants. *Burtone* 1086 (DB). 'Fortified farmstead' or 'farmstead near a fortification'. OE *burh-tūn*. **Broughton, Brant** Lincs. *Burtune* 1086 (DB), *Brendebrocton* 1250. Identical in origin with the previous name. Affix is ME *brend* 'burnt, destroyed by fire'.

Broughton (*Brychdyn*) Flin. *Brochetune* 1086 (DB). 'Farmstead by the brook'. OE *brōc* + *tūn*.

Brown Candover Hants. *See* CANDOVER.

Brown Edge Lancs. *Browneegge* 1551. 'Brown edge or ridge'. OE *brūn* + *ecg*.

Brownhills Wsall. *Brown Hill* 1749. A recent self-explanatory name.

Brownsea Island Dorset. *Brunkeseye* 1241. 'Island of a man called *Brūnoc'. OE pers. name + *ēg*.

Brownston Devon. *Brunardeston* 1219. 'Farmstead of a man called *Brūnweard'. OE pers. name + *tūn*.

Broxbourne Herts. *Brochesborne* 1086 (DB). 'Stream frequented by badgers'. OE *brocc* + *burna*.

Broxburn W. Loth. *Broxburne* 1638. 'Stream frequented by badgers'. OE *brocc* + *burna*.

Broxted Essex. *Brocheseued* c.1050, *Brocchesheuot* 1086 (DB). Probably 'badger's head', i.e. 'hill frequented by badgers', or 'hill resembling a badger's head'. OE *brocc* + *hēafod*. Alternatively perhaps 'the head of a stream', with OE *brōc* 'brook' as first element.

Broxton Ches. *Brosse* 1086 (DB), *Brexin, Broxun* 13th cent. An obscure name, possibly from an OE *burgæsn* 'burial place'. In any case the later *-ton* (found from 1260) is unhistorical.

Broxtowe (district) Notts. *Brocolvestou, Brochelestou* 1086 (DB). 'Assembly-place of a man called *Brocwulf'. A revival of an old wapentake-name. OE pers. name + *stōw*.

Bruera Ches. *Bruera* c.1150, *Heeth* c.1175. Latin *bruer(i)a* 'heath' (alternating in medieval and later spellings with ME *hethe*, OE *hǣth*).

Bruff (*An Brú*) Limerick. 'The palace'.

Bruisyard Suffolk. *Buresiart* 1086 (DB). 'Peasant's enclosure'. OE (*ge*)*būr* + *geard*.

Brumby N. Lincs. *Brunebi* 1086 (DB). 'Farmstead or village of a man called Brúni'. OScand. pers. name + *bý*. Alternatively the first element may be OScand. *brunnr* 'spring'.

Brundall Norfolk. *Brundala* 1086 (DB). Possibly 'broomy nook of land'. OE **brōmede* + *halh*.

Brundish Suffolk. *Burnedich* 1177. 'Pasture on a stream'. OE *burna* + *edisc*.

Brunton Northum. *Burneton* 1242. 'Farmstead by a stream'. OE *burna* + *tūn*.

Bruree (*Brú Rí*) Limerick. 'Palace of the king'.

Brushford, 'ford by the bridge', OE *brycg* + *ford*: **Brushford** Devon. *Brigeford* 1086 (DB). **Brushford** Somerset. *Brigeford* 1086 (DB).

Bruton Somerset. *Briwetone* 1086 (DB). 'Farmstead on the River Brue'. Celtic river-name (meaning 'brisk') + OE *tūn*.

Bryansford (*Áth Bhriain*) Down. *Bryansford* 1743. 'Brian's ford'.

Bryanston Dorset. *Blaneford Brian, Brianeston* 1268. 'Brian's estate', from OE *tūn*. *Brian* de Insula held this manor (originally called *Blaneford* = BLANDFORD) in the early 13th cent.

Brychdyn. *See* BROUGHTON.

Bryher I. of Scilly. *Braer* 1319. Probably 'the hills'. Cornish **bre* in a plural form.

Brymbo Wrex. *Brynbawe* 1391, *Brinbawe* 1412, *Brymbo* 1480. 'Hill of dirt'. Welsh *bryn* + *baw*.

Brympton Somerset. *Brunetone* 1086 (DB), *Brimpton* 1264, *Bromton* 1331. Probably 'farmstead where broom grows'. OE *brōm* + *tūn*.

Bryn Shrops. *Bren* 1272. Celtic **brïnn* 'hill'.

Bryn-mawr Blae. *Bryn-mawr* 1832. 'Big hill'. Welsh *bryn* + *mawr*. Until the early 19th cent. Bryn-Mawr was known as *Gwaun-helygen*, 'moorland of the willow-tree' (Welsh *gwaun* + *helygen*).

Brynamman Carm. *Bryn Amman* 1844. 'Hill of the River Amman'. Welsh *bryn*. For the river-name, *see* AMMANFORD.

Brynbuga. *See* USK.

Bubbenhall Warwicks. *Bubenhalle* [*sic*] 1086 (DB), *Bubenhull* 1211. 'Hill of a man called Bubba'. OE pers. name (genitive *-n*) + *hyll*.

Bubwith E. R. Yorks. *Bobewyth* 1066–9, *Bubvid* 1086 (DB). 'Wood (or dwelling) of a man called Bubba'. OE pers. name + OScand. *vithr* (perhaps replacing OE *wīc*).

Buccleuch Sc. Bord. *Bockcleugh* c.1590. 'Valley frequented by bucks'. OE *bucc* + **clōh*.

Buchan (district) Aber. *Buchan* c.1150, *Bouwan* c.1295. Said to be 'place of cows', OWelsh *buwch*, but this is doubtful.

Buchanhaven Aber. Recorded thus in 1775. 'Harbour in Buchan'. *See* BUCHAN.

Buck's Cross Devon. *Bochewis* 1086 (DB). 'Measure of land granted by charter'. OE *bōc* + *hīwisc*.

Buckby, Long Northants. *Buchebi* 1086 (DB), *Longe Bugby* 1565. 'Farmstead or village of a man called Bukki or Bucca'. OScand. or OE pers. name + OScand. *bý*. Alternatively the first element may be OScand. *bukkr* 'buck (male deer or he-goat)'. Affix refers to the length of the village.

Buckden Cambs. *Bugedene* 1086 (DB). 'Valley of a woman called Bucge'. OE pers. name + *denu*.

Buckden N. Yorks. *Buckeden* 12th cent. 'Valley frequented by bucks (male deer or he-goats)'. OE *bucca* + *denu*.

Buckenham Norfolk. near Hassingham. *Buc(h)anaham* 1086 (DB), *Bokenham* 1451. Probably 'homestead where bucks (male deer or he-goats) are kept'. OE *bucca* (genitive plural *buccena*) + *hām*.

Buckenham, New & Buckenham, Old Norfolk. *Buc(he)ham* 1086 (DB). 'Homestead of a man called Bucca, or where bucks (male deer or he-goats) are kept'. OE pers. name or OE *bucca* + *hām*.

Buckerell Devon. *Bucherel* 1165. Obscure in origin and meaning.

Buckfast Devon. *Bucfæsten* 1046, *Bucfestre* 1086 (DB). 'Place of shelter for bucks (male deer or he-goats)'. OE *bucca* + *fæsten*.

Buckfastleigh Devon. *Leghe Bucfestre* 13th cent. 'Wood or woodland clearing near BUCKFAST'. OE *lēah*.

Buckhaven Fife. *Bukhavin* 1549. 'Harbour where bucks are found'. OE *bucc* + *hæfen*.

Buckhorn Weston Dorset. *See* WESTON.

Buckhurst Hill Essex. *Bocherst* 1135. 'Wooded hill growing with beeches'. OE *bōc* + *hyrst*.

Buckie Moray. *Buky* 1362. The name was originally that of the stream here, the *Burn of Buckie*. 'Buck river', from *Bocaidh*, a derivative of Gaelic *boc* 'buck'.

Buckingham Bucks. *Buccingahamme* early 10th cent., *Bochingeham* 1086 (DB). 'River-bend land of the family or followers of a man called Bucca'. OE pers. name + *inga-* + *hamm*.

Buckinghamshire (OE *scīr* 'district') is first referred to in the 11th cent.

Buckland, a common name, from OE *bōc-land* 'charter land', i.e. 'estate with certain rights and privileges created by an Anglo-Saxon royal diploma'; examples include: **Buckland** Surrey. *Bochelant* 1086 (DB). **Buckland Brewer** Devon. *Bochelanda* 1086 (DB), *Boclande Bruere* 1290. Manorial affix from the *Briwerre* family, here in the 13th cent. **Buckland Dinham** Somerset. *Boclande* 951, *Bochelande* 1086 (DB), *Bokelonddynham* 1329. Manorial affix from the *de Dinan* family, here in the 13th cent. **Buckland, Egg** Plym. *Bochelanda* 1086 (DB), *Eckebokelond* 1221. Manorial affix from its possession by a man called *Heca* in 1086. **Buckland Filleigh** Devon. *Bochelan* 1086 (DB), *Bokelondefilleghe* 1333. Manorial affix from the *de Fyleleye* family, here in the 13th cent. **Buckland in the Moor** Devon. *Bochelanda* 1086 (DB), *Bokelaund in the More* 1318. Affix from its situation on the edge of DARTMOOR. **Buckland Monachorum** Devon. *Boclande* c.970, *Bochelanda* 1086 (DB), *Boclonde Monachorum* 1291. Latin affix 'of the monks', referring to an abbey founded here in 1278. **Buckland Newton** Dorset. *Boclonde* 941, *Bochelande* 1086 (DB), *Newton Buckland* 1576. Relatively late addition Newton is from STURMINSTER NEWTON.

Bucklebury W. Berks. *Borgeldeberie* 1086 (DB). 'Stronghold of a woman called Burghild'. OE pers. name + *burh* (dative *byrig*).

Bucklers Hard Hants. first so recorded in 1789, named from the *Buckler* family here in 1664, with dialect *hard* 'firm landing-place'.

Bucklesham Suffolk. *Bukelesham* 1086 (DB). 'Homestead of a man called *Buccel'. OE pers. name + *hām*.

Buckley (*Bwcle*) Flin. *Bokkeley* 1294, *Bukkelee* 1301. 'Woodland clearing where bucks graze'. OE *bucc* + *lēah*.

Buckminster Leics. *Bucheminstre* 1086 (DB). 'Large church of a man called Bucca'. OE pers. name + *mynster*.

Buckna (*Bocshnámh*) Antrim. *Boughna* 1669. 'Ford used by stags'.

Bucknall, 'nook of land of a man called Bucca, or where bucks (male deer or he-goats) graze', OE pers. name or *bucca* (genitive -*n*) + *halh*: **Bucknall** Lincs. *Bokenhale* 806, *Buchehale* 1086 (DB). **Bucknall** Stoke. *Bucenhole* 1086 (DB).

Bucknell, 'hill of a man called Bucca, or where bucks (male deer or he-goats) graze', OE pers. name or *bucca* (genitive -*n*) + *hyll*: **Bucknell** Oxon. *Buchehelle* 1086 (DB). **Bucknell** Shrops. *Buchehalle* [*sic*] 1086 (DB), *Bukenhull* 1209.

Buckode (*Bocóid*) Leitrim. 'Spot'.

Buck's Cross Devon. *Bochewis* 1086 (DB). 'Measure of land granted by charter'. OE *bōc* + *hīwisc*.

Buckton, 'farmstead of a man called Bucca, or where bucks (male deer or he-goats) are kept', OE pers. name or OE *bucca* + *tūn*: **Buckton** E. R. Yorks. *Bochetone* 1086 (DB). **Buckton** Herefs. *Buctone* 1086 (DB). **Buckton** Northum. *Buketun* 1208–10.

Buckworth Cambs. *Buchesworde* 1086 (DB). 'Enclosure of a man called *Bucc, or where bucks (male deer or he-goats) are kept'. OE pers. name or OE *bucc* + *worth*.

Budbrooke Warwicks. *Budebroc* 1086 (DB). 'Brook of a man called Budda, or one infested with beetles'. OE pers. name or OE *budda* + *brōc*.

Budby Notts. *Butebi* 1086 (DB). 'Farmstead or village of a man called Butti'. OScand. pers. name + *bý*.

Bude Cornwall. *Bude* 1400. Perhaps originally a river-name, of uncertain origin and meaning.

Budle Northum. *Bolda* 1166. 'The special house or building'. OE *bōthl*.

Budleigh, East Devon. *Bodelie* 1086 (DB). 'Woodland clearing of a man called Budda, or one infested with beetles'. OE pers. name or OE *budda* + *lēah*.

Budleigh Salterton Devon. *See* SALTERTON.

Budock Water Cornwall. 'Church of *Sanctus Budocus*' 1208. From the patron saint of the church, St Budock. Later affix is *water* in the sense 'stream'.

Budworth, 'enclosure of a man called Budda', OE pers. name + *worth*: **Budworth, Great** Ches. *Budewrde* 1086 (DB). **Budworth, Little** Ches. *Bodeurde* 1086 (DB).

Buerton Ches. near Audlem. *Burtune* 1086 (DB). 'Enclosure belonging to a fortified place'. OE *byrh-tūn*.

Bugbrooke Northants. *Buchebroc* 1086 (DB). 'Brook of a man called Bucca, or where bucks (male deer or he-goats) graze'. OE pers. name or OE *bucca* + *brōc*.

Buggan (*Bogán*) Fermanagh. 'Soft place'.

Bugthorpe E. R. Yorks. *Bugetorp* 1086 (DB). 'Outlying farmstead or hamlet of a man called Buggi'. OScand. pers. name + *thorp*.

Buildwas Shrops. *Beldewas* 1086 (DB). OE *wæsse* 'alluvial land liable to flood' with an uncertain first element.

Builth Wells (*Llanfair-ym-Muallt*) Powys. *Buelt* 10th cent. 'Cow pasture with springs'. Welsh *bu* + *gellt* (later *gwellt*). *Wells* was added in the 19th cent. when chalybeate springs were discovered. The Welsh name means 'St Mary's church in *Buallt*' (Welsh *llan* + *Mair* + *yn*), *Buallt* giving *Builth*.

Bulby Lincs. *Bolebi* 1086 (DB). Probably 'farmstead or village of a man called Boli or Bolli'. OScand. pers. name + *bý*.

Bulford Wilts. *Bulte*(*s*)*ford* 12th cent. Possibly 'ford by the island where ragged robin or the cuckoo flower grows'. OE *bulut* + *īeg* + *ford*.

Bulkeley Ches. *Bulceleia* 1170. 'Clearing or pasture where bullocks graze'. OE *bulluc* + *lēah*.

Bulkington, 'estate associated with a man called *Bulca*', OE pers. name + -*ing*- + *tūn*: **Bulkington** Warwicks. *Bochintone* 1086 (DB). **Bulkington** Wilts. *Boltintone* 1086 (DB).

Bulkworthy Devon. *Buchesworde* 1086 (DB), *Bulkewurthi* 1228. Possibly 'enclosure of a man

85   **Bunwell**

called *Bulca'. OE pers. name + *worth, worthig.*
Alternatively the first element may be OE *bulluc*
'bullock'.

Bullaun (*Ballán*) Galway. 'Round hillock'.

Bullers of Buchan (rocky coastal recess)
Aber. 'Roaring place of Buchan'. Scottish
English *buller. See* BUCHAN.

Bulley Glos. *Bulelege* 1086 (DB). 'Woodland
clearing where bulls graze'. OE *bula* + *lēah.*

Bullingham, Lower Herefs. *Boninhope* [*sic*]
1086 (DB), *Bullingehope* 1242. 'Marsh enclosure
associated with a man called *Bulla', or 'marsh
enclosure at *Bulla's place'. OE pers. name +
-ing- or *-ing* + *hop* (later replaced by *hamm*).

Bulmer, 'pool where bulls drink', OE *bula*
(genitive plural *bulena*) + *mere*: **Bulmer** Essex.
Bulenemera 1086 (DB). **Bulmer** N. Yorks.
Bolemere 1086 (DB).

Bulphan Thurr. *Bulgeuen* 1086 (DB). 'Fen
near a fortified place'. OE *burh* + *fenn.* The
spelling with *-l-* is due to Norman influence.

Bulverhythe E. Sussex. *Bulwareheda* 12th
cent. 'Landing-place of the town-dwellers (of
Hastings)'. OE *burh-ware* + *hȳth.*

Bulwell Nott. *Buleuuelle* 1086 (DB). 'Spring or
stream of a man called *Bula, or where bulls
drink'. OE pers. name or OE *bula* + *wella.*

Bulwick Northants. *Bulewic* 1162. 'Farm
where bulls are reared'. OE *bula* + *wīc.*

Bumble's Green Essex. Earlier *Brummers
Green* 1777, from Modern English *green* 'village
green' with a surname.

**Bumpstead, Helions & Bumpstead,
Steeple** Essex. *Bumesteda* 1086 (DB),
Bumpsted Helyun 1238, *Stepilbumstede* 1261.
'Place where reeds grow'. OE *bune* + *stede.*
Distinguishing affixes from *Tihel de Helion* who
held one manor in 1086, and from OE *stēpel*
'steeple, tower'.

Bun (*Bun*) Offaly. 'Bottom'.

Bunaw (*Bun Abha*) Kerry. 'River-mouth'.

Bunbeg (*An Bun Beag*) Donegal. 'The little
river-mouth'.

Bunbrusna (*Bun Brosnaí*) Westmeath.
'Mouth of the (river) Brosna'.

Bunbury Ches. *Boleberie* [*sic*] 1086 (DB),
Bonebury 12th cent. 'Stronghold of a man called
Būna'. OE pers. name + *burh* (dative *byrig*).

Bunclody (*Bun Clóidí*) Carlow, Wexford.
'Mouth of the (river) Clóideach'.

Buncrana (*Bun Cranncha*) Donegal.
Boncranagh 1601. 'Mouth of the (river)
Crannach'.

Bundoran (*Bun Dobhráin*) Donegal.
Bundorin 1802. 'Mouth of the little water'. The
Dobhrán was probably an earlier name for the
present river *Bradoge.*

Bundorragha (*Bun Dorcha*) Mayo. 'Mouth
of the dark river'.

Bunduff (*Bun Dubh*) Leitrim. 'Mouth of the
black (river)'.

Bungay Suffolk. *Bunghea* 1086 (DB). Probably
'island of the family or followers of a man called
Būna'. OE pers. name + *-inga-* + *ēg.*

Bunmahon (*Bun Machan*) Waterford.
'Mouth of the (river) Machain'.

Bunnacurry (*Bun an Churraigh*) Mayo. 'Foot
of the swamp'.

Bunnahowen (*Bun na hAbhna*) Mayo.
'Mouth of the river'.

Bunnanaddan (*Bun an Fheadáin*) Sligo.
'Mouth of the stream'.

Bunny Notts. *Bonei* 1086 (DB). Probably
'island, or dry ground in marsh, where reeds
grow'. OE *bune* + *ēg.*

Bunnyconnellan (*Muine Chonalláin*)
Mayo. 'Conallán's thicket'.

Bunowen (*Bun Abhann*) Mayo. 'Mouth of
the river'.

Bunratty (*Bun Raite*) Clare. 'Mouth of the
(river) Raite'.

Bunree (*Bun Rí*) Mayo. 'Mouth of the (river)
Rí'.

Buntingford Herts. *Buntingeford* 1185. 'Ford
frequented by buntings or yellow-hammers'.
ME *bunting* + *ford.*

Bunwell Norfolk. *Bunewell* 1198. Probably
'spring or stream where reeds grow'. OE *bune* +
wella. Alternatively the first element may be OE
bune 'drinking-vessel, pitcher'.

Burbage Derbys. *Burbache* 1417. 'Stream or ridge by a fortified place'. OE *burh* + *bece* or *bæc* (dative **bece*).

Burbage Leics. *Burhbeca* 1043, *Burbece* 1086 (DB). Probably 'ridge by a fortified place'. OE *burh* + *bæc* (dative **bece*).

Burbage Wilts. *Burhbece* 961, *Burbetce* 1086 (DB). Probably 'stream by a fortified place'. OE *burh* + *bece*.

Burcombe Wilts. *Brydancumb* 937, *Bredecumbe* 1086 (DB). Probably 'valley of a man called **Brȳda*'. OE pers. name + *cumb*.

Burcot Oxon. *Bridicote* 1198. Probably 'cottage(s) associated with a man called **Brȳda*'. OE pers. name (+ *-ing-*) + *cot*.

Burdale N. Yorks. *Bredhalle* 1086 (DB). 'Hall or house made of planks'. OE *bred* + *hall*.

Bures St Mary Suffolk **& Mount Bures** Essex. *Bure, Bura* 1086 (DB), *Buras* c.1180, *Bures atte Munte* 1328. 'The dwellings or cottages'. OE *būr* (plural *būras*). Distinguishing affixes from the dedication of the church and from ME *munt* ('at the mount or hill').

Burford, 'ford by the fortified place', OE *burh* + *ford*: **Burford** Oxon. *Bureford* 1086 (DB). **Burford** Shrops. *Bureford* 1086 (DB).

Burgess Hill W. Sussex. *Burges Hill* 1597. Named from a family called *Burgeys*, here in the 13th cent.

Burgh, a common name, usually OE *burh* 'fortification, stronghold, fortified manor'; examples include: **Burgh** Suffolk. *Burc* 1086 (DB). **Burgh-by-Sands** Cumbria. *Burch* c.1180, *Burg en le Sandes* 1292. This is an old Roman fort on the coast. **Burgh le Marsh** Lincs. *Burg* 1086 (DB). Affix means 'in the marshland'.
  However the following has a different origin: **Burgh, Great & Burgh Heath** Surrey. *Berge* 1086 (DB), *Borow heth* 1545. '(Place at) the barrow(s)', from OE *beorg*, with the later addition of *hǣth* 'heath'.

Burghclere Hants. *Cleran* 749, *Burclere* 1171. OE *burh* in one of its meanings 'fortification, manor, borough, market town' added to the original name found also in HIGHCLERE.

Burghfield W. Berks. *Borgefel* 1086 (DB). 'Open land by the hill'. OE *beorg* + *feld*.

Burghill Herefs. *Burgelle* 1086 (DB). 'Hill with a fort'. OE *burh* + *hyll*.

Burghwallis Donc. *Burg* 1086 (DB), *Burghwaleys* 1283. 'The stronghold or fortified manor', OE *burh*, with later manorial affix from the *Waleys* family, here from the 12th cent.

Burham Kent. *Burhham* 10th cent., *Borham* 1086 (DB). 'Homestead near the fortified place'. OE *burh* + *hām*.

Buriton Hants. *Buriton* 1227. 'Enclosure near or belonging to a fortified place, or dependent on a manor'. OE *byrh-tūn*.

Burland Ches. *Burlond* 1260. 'Cultivated land of the peasants'. OE *(ge)būr* + *land*.

Burlescombe Devon. *Berlescoma* [sic] 1086 (DB), *Burewoldescumbe* 12th cent. 'Valley of a man called Burgweald'. OE pers. name + *cumb*.

Burleston Dorset. *Bordelestone* 934. 'Farmstead of a man called Burdel'. OE pers. name + *tūn*.

Burley, 'woodland clearing by or belonging to a fortified place', OE *burh* + *lēah*: **Burley** Hants. *Burgelea* 1178. **Burley** Leeds. *Burcheleia* c.1200. **Burley** Rutland. *Burgelai* 1086 (DB). **Burley in Wharfedale** Brad. *Burhleg* c.972, *Burghelai* 1086 (DB). Affix is 'valley of the River Wharfe', Celtic river-name (meaning 'winding one') + OScand. *dalr*.

Burleydam Ches. *Burley* c.1130, *Burleydam* 1643. 'Woodland clearing of the peasants'. OE *(ge)būr* + *lēah* with the later addition of ME *damme* 'a milldam'.

Burlingham Norfolk. *Berlingeham* 1086 (DB). 'Homestead of the family or followers of a man called **Bǣrla* or **Byrla*'. OE pers. name + *-inga-* + *hām*.

Burlton Shrops. *Burghelton* 1241. 'Farmstead by a hill with a fort'. OE *burh* + *hyll* + *tūn*.

Burmarsh Kent. *Burwaramers* c.848, *Burwarmaresc* 1086 (DB). 'Marsh of the town-dwellers (of Canterbury)'. OE *burh-ware* + *mersc*.

Burmington Warwicks. *Burdintone* [sic] 1086 (DB), *Burminton* late 12th cent. 'Estate associated with a man called Beornmund or **Beorma*'. OE pers. name + *-ing-* + *tūn*.

Burn N. Yorks. *Byrne* c.1030. Probably 'place cleared by burning'. OE *bryne*.

Burnaston Derbys. *Burnulfestune* 1086 (DB). 'Farmstead of a man called *Brūnwulf or Brynjólfr'. OE or OScand. pers. name + OE *tūn*.

Burnby E. R. Yorks. *Brunebi* 1086 (DB). 'Farmstead or village by a spring or stream'. OScand. *brunnr* + *bý*.

Burncourt (*An Chúirt Dóite*) Tipperary. 'The burnt court'. The house *Clogheen* ('small stones'), built here in 1641, was burnt down by Cromwell in 1650.

Burneside Cumbria. *Brunoluesheued* c.1180. 'Headland or hill of a man called *Brūnwulf or Brunulf'. OE or OGerman pers. name + OE *hēafod*.

Burneston N. Yorks. *Brennigston* 1086 (DB). Probably 'farmstead of a man called Brýningr'. OScand. pers. name + *tūn*.

Burnett B. & NE. Som. *Bernet* 1086 (DB). 'Land cleared by burning'. OE *bærnet*.

Burnfoot (*Bun n hAbhainn*) Donegal. *Burnfoot Bridge* 1762. 'Foot of the stream'.

Burnham, usually 'homestead or village on a stream', OE *burna* + *hām*: **Burnham** Bucks. *Burneham* 1086 (DB). **Burnham Deepdale & Burnham Market**, **Burnham Norton & Burnham Overy**, **Burnham Thorpe** Norfolk. *Brun(e)ham*, *Depedala* 1086 (DB), *Brunham Norton* 1457, *Brunham Overhe* 1457, *Brunhamtorp* 1199. Distinguishing affixes are 'deep valley' (OE *dēop* + *dæl*), 'market', 'north farm' (OE *north* + *tūn*), 'over the river' (OE *ofer* + *ēa*) and 'outlying farmstead' (OScand. *thorp*). **Burnham on Crouch** Essex. *Burneham* 1086 (DB). Affix is from the River Crouch (not recorded before 16th cent., probably from OE *crūc* 'a cross').
    However two Burnhams have a different origin: **Burnham** N. Lincs. *Brune* [*sic*] 1086 (DB), *Brunum* c.1115. '(Place at) the springs or streams'. OScand. *brunnr* in a dative plural form *brunnum*. **Burnham on Sea** Somerset. *Burnhamm* c.880, *Burneham* 1086 (DB). 'Enclosure by a stream'. OE *burna* + *hamm*.

Burniston N. Yorks. *Brinnistun* 1086 (DB). 'Farmstead of a man called Brýningr'. OScand. pers. name + OE *tūn*.

Burnley Lancs. *Brunlaia* 1124. Probably 'woodland clearing by the River Brun'. OE rivername *Brūn(e)* (a derivative of OE *brūn* 'brown, shining') + *lēah*: the modern form of the river-

name with a short vowel is probably a back-formation from the place name.

Burnsall N. Yorks. *Brineshale* 1086 (DB). 'Nook of land of a man called Brȳni'. OE pers. name + *halh*.

Burnswark Dumf. *Burnyswarke* 1542. ME *wark* (OE *weorc*) 'fortification' (here referring to a Roman camp), first element uncertain, possibly ME *burn* (OE *burna*) 'spring'.

Burntisland Fife. *Brynt Iland* 1540. 'Burnt island'. OE *brende* + ModE *island*. A 'burnt island' is one where buildings were destroyed by fire or land cleared for cultivation by burning.

Burpham, 'homestead near the stronghold or fortified place', OE *burh* + *hām*: **Burpham** Surrey. *Borham* 1086 (DB). **Burpham** W. Sussex. *Burhham* c.920, *Bercheham* 1086 (DB).

Burradon, 'hill with a fort', OE *burh* + *dūn*: **Burradon** Northum. *Burhedon* early 13th cent. **Burradon** N. Tyne. *Burgdon* 12th cent.

Burren (*Boirinn*) Clare, Down. 'Stony district'.

Burrenbane (*Boireann Bán*) Down. 'White stony district'.

Burrenreagh (*Boireann Riabhach*) Down. 'Grey stony district'.

Burrill N. Yorks. *Borel* 1086 (DB). Probably 'hill with a fort'. OE *burh* + *hyll*.

Burringham N. Lincs. *Burringham* 1199. Possibly 'homestead of the family or followers of a man called Burgrēd or Burgrīc'. OE pers. name + -*inga*- + *hām*.

Burrington Devon. *Bernintone* 1086 (DB). 'Estate associated with a man called Beorn'. OE pers. name + -*ing*- + *tūn*.

Burrington Herefs. *Boritune* 1086 (DB). 'Farmstead by a fortified place'. OE *burh* (genitive or dative *byrig*) + *tūn*.

Burrington N. Som. *Buringtune* 12th cent. 'Farmstead by a fortified place'. OE *burh* (genitive or dative *byrig*) + *tūn*.

Burrough Green Cambs. *Burg* c.1045, *Burch* 1086 (DB), *Boroughegrene* 1571. 'The fortified place'. OE *burh* with the later addition of ME *grene* 'village green'.

Burrough on the Hill Leics. *Burg* 1086 (DB). 'The fortified place'. OE *burh*, referring to an Iron Age hill-fort.

Burrow Bridge Somerset. *Æt tham Beorge* 1065. '(Place) at the hill'. OE *beorg*.

Burrow, Nether & Burrow, Over Lancs. *Borch* 1086 (DB). 'The fortified place', here referring to a Roman fort. OE *burh*.

Burry Port Carm. *Pembrey . . . also called Burry Port* 1897. 'Port of the sand dune'. ModE dialect *burry* + ModE *port*.

Burscough Lancs. *Burscogh* c.1190. 'Wood by the fort'. OE *burh* + OScand. *skógr*.

Burshill E. R. Yorks. *Bristehil* 12th cent. 'Hill with a landslip or rough ground'. OE *byrst* + *hyll*.

Bursledon Hants. *Brixendona* c.1170. Probably 'hill associated with a man called Beorhtsige', OE pers. name + -*ing*- + *dūn*, alternating with 'hill of a man called *Beorsa (a pet-form of Beorhtsige)', OE pers. name (genitive -*n*) + *dūn*.

Burslem Stoke. *Barcardeslim* [sic] 1086 (DB), *Borewardeslyme* 1242. 'Estate in *Lyme* belonging to the fort-keeper, or to a man called Burgweard'. OE *burh-weard* or pers. name with old district name *Lyme*, perhaps from Latin *līmen* 'threshold, lintel' (figuratively 'escarpment').

Burstall Suffolk. *Burgestala* 1086 (DB). 'Site of a fort or stronghold'. OE *burh-stall*.

Burstead, Great & Bursted, Little Essex. *Burgestede* c.1000, *Burghesteda* 1086 (DB). 'Site of a fort or stronghold'. OE *burh-stede*.

Burstock Dorset. *Burewinestoch* 1086 (DB). 'Outlying farmstead or hamlet of a woman called Burgwynn or of a man called Burgwine'. OE pers. name + *stoc*.

Burston Norfolk. *Borstuna* 1086 (DB). Possibly 'farmstead by a landslip or rough ground'. OE *byrst* + *tūn*.

Burston Staffs. *Burouestone* 1086 (DB). Possibly 'farmstead of a man called Burgwine or Burgwulf'. OE pers. name + *tūn*.

Burstow Surrey. *Burestou* 12th cent. 'Place by a fort or stronghold'. OE *burh* (genitive *byrh*) + *stōw*.

Burstwick E. R. Yorks. *Brostewic* 1086 (DB). Probably 'dwelling or farm of a man called Bursti'. OScand. pers. name + OE *wīc*.

Burt (*An Beart*) Donegal. *Droma Bertach* c.830. Possibly 'the heaped ridge'.

Burton, a common name, usually OE *burh-tūn* 'fortified farmstead', or 'farmstead near or belonging to a stronghold or manor'; examples include: **Burton Agnes** E. R. Yorks. *Bortona* 1086 (DB), *Burton Agneys* 1231. Manorial affix from its possession by *Agnes* de Percy in the late 12th cent. **Burton, Bishop** E. R. Yorks. *Burton* 1086 (DB), *Bisshopburton* 1376. Manorial affix from its early possession by the Archbishops of York. **Burton, Cherry** E. R. Yorks. *Burtone* 1086 (DB), *Cheriburton* 1444. Affix is ME *chiri* 'cherry', no doubt referring to cherry-trees growing here. **Burton, Constable** N. Yorks. *Bortone* 1086 (DB), *Burton Constable* 1301. Manorial affix from its possession by the Constables of Richmond Castle in the 12th cent. **Burton Dassett** Warwicks., *see* DASSETT. **Burton Fleming** E. R. Yorks. *Burtone* 1086 (DB), *Burton Flemeng* 1234. Manorial affix from the *Fleming* family, here in the 12th cent. **Burton Hastings** Warwicks. *Burhtun* 1002, *Bortone* 1086 (DB), *Burugton de Hastings* 1313. Manorial affix from the *de Hasteng* family, here in the 13th cent. **Burton Latimer** Northants. *Burtone* 1086 (DB), *Burton Latymer* 1482. Manorial affix from the *le Latimer* family, here in the 13th cent. **Burton upon Stather** N. Lincs. *Burtone* 1086 (DB), *Burtonstather* 1275. Affix means 'by the landing-places' from OScand. *stoth* in a plural form *stothvar*.

However the following have a different origin: **Burton Bradstock** Dorset. *Bridetone* 1086 (DB). 'Farmstead on the River Bride'. Celtic river-name (*see* BREDY) + OE *tūn*. The later addition Bradstock is from the Abbey of BRADENSTOKE which held the manor from the 13th cent. **Burton Joyce** Notts. *Bertune* 1086 (DB), *Birton Jorce* 1327. OE *byrh-tūn* 'farmstead of the fortified place or stronghold'. Manorial affix from the *de Jorz* family, here in the 13th cent. **Burton Salmon** N. Yorks. *Brettona* c.1160, *Burton Salamon* 1516. 'Farmstead of the Britons'. OE *Brettas* (genitive *Bretta*) + *tūn*. Manorial affix from a man called *Salamone* who had lands here in the 13th cent. **Burton upon Trent** Staffs. *Byrtun* 1002, *Bertone* 1086 (DB). Identical in origin with Burton Joyce. For the river-name, *see* TRENTHAM.

Burtonport Donegal. 'Burton's port'. *Burton Port* 1835. The village arose by a fishing port founded in 1785 by William *Burton*. The Irish name is *Ailt an Chorráin*, 'ravine of the curve'.

Burtonwood Warrtn. *Burtoneswod* 1228. 'Wood by the fortified farmstead'. OE *burh-tūn* + *wudu*.

Burwardsley Ches. *Burwardeslei* 1086 (DB). 'Woodland clearing of the fort-keeper, or of a man called Burgweard'. OE *burh-weard* or pers. name + *lēah*.

Burwarton Shrops. *Burertone* [sic] 1086 (DB), *Burwardton* 1194. 'Farmstead of the fort-keeper, or of a man called Burgweard'. OE *burh-weard* or pers. name + *tūn*.

Burwash E. Sussex. *Burhercse* 12th cent. 'Ploughed field by the fort'. OE *burh* + *ersc*.

Burwell, 'spring or stream by the fort', OE *burh* + *wella*: **Burwell** Cambs. *Burcwell* 1060, *Buruuella* 1086 (DB). **Burwell** Lincs. *Buruelle* 1086 (DB).

Burwick Orkn. (South Ronaldsay). *Bardvik* c.1225. 'Bay of the extremity'. OScand. *barth* + *vík*. Burwick is at the southernmost point of the island.

Bury, '(place by) the fort or stronghold', OE *burh* (dative *byrig*): **Bury** Bury. *Biri* 1194. **Bury** Cambs. *Byrig* 974. **Bury** W. Sussex. *Berie* 1086 (DB).

Bury St Edmunds Suffolk. *Sancte Eadmundes Byrig* 1038. 'Town associated with St Ēadmund'. OE saint's name (a 9th-cent. king of East Anglia) + OE *burh* (dative *byrig*). The original name of Bury was *Bæderices wirde* 945, *Beadriceswyrth* c.1030, 'enclosed farmstead of a man called Beadurīc', OE pers. name + *wyrth* (a variant of *worth*).

Burythorpe N. Yorks. *Bergetorp* 1086 (DB). Probably 'outlying farmstead or hamlet of a woman called Bjǫrg'. OScand. pers. name + *thorp*.

Busby, Great N. Yorks. *Buschebi* 1086 (DB). 'Farmstead or village of a man called *Buski, or among the bushes or shrubs'. OScand. pers. name or *buskr, *buski + *bý*.

Buscot Oxon. *Boroardescote* 1086 (DB). 'Cottage(s) of the fort-keeper, or of a man called Burgweard'. OE *burh-weard* or pers. name + *cot*.

Bushbury Wolvh. *Byscopesbyri* 996, *Biscopesberie* 1086 (DB). 'The bishop's fortified manor'. OE *biscop* + *burh* (dative *byrig*).

Bushey Herts. *Bissei* 1086 (DB). 'Enclosure near a thicket, or hedged with box-trees'. OE *bysce* or *byxe* + *hæg*.

Bushley Worcs. *Biselege* 1086 (DB). 'Woodland clearing with bushes, or of a man called *Byssa'. OE *bysce* or pers. name + *lēah*.

Bushmills Antrim. *Bushmills* 1636. 'Mills on the (river) Bush'. The equivalent Irish name is *Muileann na Buaise*.

Bushton Wilts. *Bissopeston* 1242. 'The bishop's farmstead'. OE *biscop* + *tūn*.

Buston, High & Buston, Low Northum. *Buttesdune* 1166, *Butlesdon* 1249. Possibly 'hill of a man called *Buttel', OE pers. name + *dūn*. Alternatively the first element may be OE *butt 'stumpy hill'.

Butcombe N. Som. *Budancumb* c.1000, *Budicome* 1086 (DB). 'Valley of a man called Bud(d)a, or one infested with beetles'. OE pers. name or OE *budda* + *cumb*.

Bute (island) Arg. *Bot* 1093, *Boot* 1292. '(Island of) fire'. Gaelic *bód*. The name may refer to signal-fires.

Butleigh Somerset. *Budecalech* 725, *Boduchelei* 1086 (DB). 'Woodland clearing of a man called *Budeca'. OE pers. name + *lēah*.

Butlers Bridge Cavan. *Butlers Br.* 1728. 'Butler's bridge'. The bridge over the Annalea River is named from the *Butler* family, whose ancestor Sir Stephen Butler was granted land here in 1610. The equivalent Irish name is *Droichead an Bhuitléaraigh*.

Butley Suffolk. *Butelea* 1086 (DB). 'Woodland clearing of a man called *Butta'. OE pers. name + *lēah*. Alternatively the first element may be an OE *butte 'mound, hill'.

Butterleigh Devon. *Buterlei* 1086 (DB). 'Clearing with good pasture'. OE *butere* + *lēah*.

Buttermere, 'lake or pool with good pasture', OE *butere* + *mere*: **Buttermere** Cumbria. *Butermere* 1230. **Buttermere** Wilts. *Butermere* 931–9, *Butremere* 1086 (DB).

Butterton Staffs. near Leek. *Buterdon* 1200. 'Hill with good pasture'. OE *butere* + *dūn*.

Butterwick, 'dairy farm where butter is made', OE *butere* + *wīc*: **Butterwick** Durham. *Boterwyk* 1131. **Butterwick** Lincs. *Butruic* 1086 (DB). **Butterwick** N. Yorks., near Foxholes. *Butruid* [sic] 1086 (DB), *Butterwic* *c.*1130. **Butterwick** N. Yorks., near Hovingham. *Butruic* 1086 (DB). **Butterwick, East & Butterwick, West** N. Lincs. *Butreuuic* 1086 (DB).

Buttevant (*Cill na Mallach*) Cork. 'Defensive outwork'. OFrench *botavant*. The Irish name means 'church of the summits'.

Buxhall Suffolk. *Bucyshealæ c.*995, *Buckeshala* 1086 (DB). 'Nook of land of a man called *Bucc, or where bucks (male deer or he-goats) graze'. OE pers. name or OE *bucc* + *halh*.

Buxted E. Sussex. *Boxted* 1199. 'Place where beech-trees or box-trees grow'. OE *bōc* or *box* + *stede*.

Buxton Derbys. *Buchestanes c.*1100. Probably 'the rocking stones or loganstones'. OE *būg-stān*. Alternatively perhaps simply 'buck stones, stones where bucks are seen', from OE *bucc(a)* 'buck (male deer or he-goat)' + *stān*.

Buxton Norfolk. *Bukestuna* 1086 (DB). 'Farmstead of a man called *Bucc, or where bucks (male deer or he-goats) are kept'. OE pers. name or OE *bucc* + *tūn*.

Buxworth Derbys. *Buggisworth* 1275. 'Enclosure of a man called *Bucg'. OE pers. name + *worth*.

Bwcle. *See* BUCKLEY.

Bweeng (*Na Boinn*) Cork. 'The swelling'.

Byers Green Durham. *Byres* 1183, *Byres Greine* 1562. 'The byres or cowsheds'. OE *bȳre* with the later addition of ME *grene* 'village green, hamlet'.

Byfield Northants. *Bifelde* 1086 (DB). Possibly '(place) by the open country'. OE *bī* + *feld*.

Alternatively the first element may be OE *byge* 'river-bend'.

Byfleet Surrey. *Biflete* 933, *Biflet* 1086 (DB). Probably '(place) by the stream', OE *bī* + *flēot*. Alternatively from an OE compound *bī-flēot* 'small area of land cut off by the changing course of a stream'.

Byford Herefs. *Buiford* 1086 (DB). 'Ford near the river-bend'. OE *byge* + *ford*.

Bygrave Herts. *Bigravan* 973, *Bigrave* 1086 (DB). '(Place) by the grove or by the trench'. OE *bī* + *grāfa* or *grafa*.

Byker Newc. upon T. *Bikere* 1196. Identical in origin with BICKER.

Byley Ches. *Bevelei* 1086 (DB). 'Woodland clearing of a man called Bēofa'. OE pers. name + *tūn*.

Bytham, Castle & Bytham, Little Lincs. *Bytham c.*1067, *Bitham* 1086 (DB). OE *bythme* 'valley bottom, broad valley', either as a simplex name or with OE *hām* 'homestead'.

Bythorn Cambs. *Bitherna c.*960, *Bierne* [*sic*] 1086 (DB). Probably '(place) by the thorn-bush'. OE *bī* + *thyrne*. Alternatively perhaps OE *byht* 'river-bend' + *hyrne* 'angle or corner of land'.

Byton Herefs. *Boitune* 1086 (DB). 'Farmstead by the river-bend'. OE *byge* + *tūn*.

Bywell Northum. *Biguell* 1104-8, *Biewell* 1195. 'Spring by the river-bend'. OE *byge* + *wella*.

Byworth W. Sussex. *Begworth* 1279. 'Enclosure of a woman called Bēage or a man called Bǣga'. OE pers. name + *worth*.

Cabinteely (*Cábán tSíle*) Dublin. 'Síle's cabin'.

Cabourne Lincs. *Caburne* 1086 (DB). 'Stream frequented by jackdaws'. OE *cā + burna*.

Cabra (*An Chabrach*) Down. *Ballinecabre* 1605. 'The bad land'.

Cabragh (*An Chabrach*) Dublin, Tyrone. 'The bad land'.

Cadamstown (*Baile Mhic Ádaim*) Kildare, Offaly. 'Mac Ádaim's homestead'.

Cadboll Highland. *Kattepol* 1281. 'Farm frequented by wild-cats'. OScand. *köttr + ból* (*stathr*).

Cadbury, 'fortified place or stronghold of a man called Cada', OE pers. name + *burh* (dative *byrig*): **Cadbury** Devon. *Cadebirie* 1086 (DB). **Cadbury, North & Cadbury, South** Somerset. *Cadanbyrig* c.1000, *Cadeberie* 1086 (DB).

Caddington Beds. *Caddandun* c.1000, *Cadendone* 1086 (DB). 'Hill of a man called Cada'. OE pers. name (genitive -*n*) + *dūn*.

Caddy (*Cadaigh*) Antrim. *Cady* c.1657. Possibly 'land held by treaty'.

Cadeby, 'farmstead or village of a man called Káti', OScand. pers. name + *bý*: **Cadeby** Donc. *Catebi* 1086 (DB). **Cadeby** Leics. *Catebi* 1086 (DB).

Cadeleigh Devon. *Cadelie* 1086 (DB). 'Woodland clearing of a man called Cada'. OE pers. name + *lēah*.

Cader Idris (mountain) Gwyd. 'Seat of Idris'. Celtic pers. name + OWelsh *cadeir*. *Idris* is a legendary giant and magician.

Cadian (*Céidín*) Tyrone. 'Little hill'.

Cadishead Salford. *Cadewalesate* 1212. 'Dwelling or fold by the stream of a man called Cada'. OE pers. name + *wælla + set*.

Cadmore End Bucks. *Cademere* 1236. 'Estate boundary or pool of a man called Cada'. OE pers. name + *mēre* or *mere*.

Cadnam Hants. *Cadenham* 1272. 'Homestead or enclosure of a man called Cada'. OE pers. name (genitive -*n*) + *hām* or *hamm*.

Cadney N. Lincs. *Catenai* 1086 (DB). 'Island, or dry ground in marsh, of a man called Cada'. OE pers. name (genitive -*n*) + *ēg*.

Cadoxton (*Tregatwg*) Vale Glam. *Caddokeston* 1254, *Cadoxston* 1535. 'Cadog's farmstead'. OE *tūn* (Welsh *tref*). From the dedication of the church to St Cadog. The Welsh name is the equivalent.

Caenby Lincs. *Couenebi* 1086 (DB). Probably 'farmstead or village of a man called *Kafni*'. OScand. pers. name + *bý*.

Caerdydd. *See* CARDIFF.

Caerffili. *See* CAERPHILLY.

Caerfyrddin. *See* CARMARTHEN.

Caergwrle Flin. *Caergorlei* 1327. 'Fort by the clearing where cranes are seen'. Welsh *caer* + OE *corn + lēah*. The 'fort' was a Roman station.

Caergybi. *See* HOLYHEAD.

Caerleon (*Caerllion-ar-Wysg*) Newpt. *castra Legionis* c.150, *Caerleion* 1086 (DB). 'Fort of the legion'. Welsh *caer* + Latin *legio* (genitive *legionis*). The reference is to the Second Legion, stationed here after moving from *Glevum* (Gloucester). The Roman fort was *Isca Legionis*, 'Isca of the legion', *Isca* being the River USK. The Welsh name means 'Caerleon on the Usk'.

Caerllion-ar-Wysg. *See* CAERLEON.

Caernarfon Gwyd. *Kairarvon* 1191, *Kaer yn Arvon* 1258. 'Fort in Arfon'. Welsh *caer + yn*. The district name *Arfon* is Welsh *ar Fôn*, 'opposite Môn', i.e. ANGLESEY.

Caerphilly (*Caerffili*) Cphy. *Kaerfili* 1271, *Kaerphilly* 1314. 'Ffili's fort'. Welsh *caer* + Celtic pers. name.

Caersŵs Powys. *Caerswys* 14th cent. 'Swys's fort'. Welsh *caer* + Celtic pers. name.

Caer-went Mon. *Cair Guent c.*800. 'Fort of *Gwent*'. Welsh *caer*. See GWENT. The Roman station here was *Venta Silurum*, 'Venta of the Silures'.

Caher (*An Chathair*) Clare, Tipperary. 'The stone fort'.

Caheragh (*Cathrach*) Cork. 'Abounding in stone forts'.

Caherbarnagh (*An Chathair Bhearnach*) Cork. 'The gapped stone fort'.

Caherconlish (*Cathair Chinn Lis*) Limerick. 'Stone fort of the head of the ford'.

Caherconree (*Cathair Conraoi*) Kerry. 'Stone fort of Cúrí'.

Caherdaniel (*Cathair Dónall*) Kerry. 'Dónall's stone fort'.

Caherelly (*Cathair Ailí*) Limerick. 'Stone fort of the boulder'.

Cahergal (*Cathair Geal*) Kerry. 'White stone fort'.

Caherloughlin (*Cathair Lochlainn*) Clare. 'Stone fort of Lochlann'.

Cahermore (*Cathair Mhór*) Cork, Galway. 'Big stone fort'.

Cahernageeha Mountain (*Cathair na Gaoithe*) Kerry. 'Stone fort of the wind'.

Cahersiveen (*Cathair Saidhbhín*) Kerry. 'Stone fort of Saidhbhín'.

Caime (*Céim*) Wexford. 'Gap'.

Cainscross Glos. not on record until 1776, probably from *cross* 'cross-roads' with the surname *Cain*.

Cairngorm (mountains) Moray. 'Blue rocky hill'. Gaelic *carn + gorm*. The name is properly that of *Cairn Gorm*, the highest peak in the group.

Caister, Caistor, 'Roman camp or town', OE *cæster*. **Caister-on-Sea** Norfolk. *Castra* 1044–7, 1086 (DB). **Caistor** Lincs. *Castre* 1086 (DB). **Caistor St Edmund** Norfolk. *Castre c.*1025, *Castrum* 1086 (DB), *Castre Sancti Eadmundi* 1254. Affix from its early possession by the Abbey of Bury St Edmunds.

Caistron Northum. *Cers c.*1160, *Kerstirn* 1202. 'Thornbush by the fen'. ME *kers* + OScand. *thyrnir*.

Caithness (district) Highland. *Kathenessia c.*970. 'Promontory of the Cats'. OScand. *nes*. It is not known why the early Celtic tribe here were called 'cats'; the cat may have been their token animal.

Calbourne I. of Wight. *Cawelburne* 826, *Cavborne* 1086 (DB). 'Stream called *Cawel*', or 'stream where cole or cabbage grows'. Celtic river-name (of uncertain meaning) or OE *cāwel + burna*.

Calceby Lincs. *Calesbi* 1086 (DB). 'Farmstead or village of a man called Kalfr'. OScand. pers. name + *bý*.

Calcethorpe Lincs. *Cheilestorp c.*1115. 'Outlying farmstead or hamlet of a man called *Cægel*'. OE pers. name + OScand. *thorp*.

Calcot Row W. Berks. not on record before the 18th century, but probably identical in origin with CALDECOTE.

Calcutt Wilts. *Colecote* 1086 (DB). Probably 'cottage of a man called Cola'. OE pers. name + *cot*.

Caldbeck Cumbria. *Caldebek* 11th cent. 'Cold stream'. OScand. *kaldr + bekkr*.

Caldbergh N. Yorks. *Caldeber* 1086 (DB). 'Cold hill'. OScand. *kaldr + berg*.

Caldecote, Caldecott, a place name found in various counties, meaning 'cold cottage(s)', with reference to poor construction, exposed situation, or clay soil, from OE *cald + cot*; examples include: **Caldecote** Cambs. *Caldecote* 1086 (DB). **Caldecote** Herts. *Caldecota* 1086 (DB). **Caldecote** Northants. *Caldecot* 1202. **Caldecote Hill** Warwicks. *Caldecote* 1086 (DB). **Caldecott** Northants. *Caldecote* 1086 (DB). **Caldecott** Rutland. *Caldecote* 1086 (DB).

Calder, an old Celtic river-name meaning 'rapid stream'. Of the four examples in the North West of England, that in Cumbria gives name to **Calder Bridge** & **Calder Hall**, *Calder* 1178. **Calderdale** (unitary authority) is named from the West Yorkshire river so called, and **Calder Vale** Lancs. from one of the two Lancs. rivers with this name. *See also* KIELDER Northum., named from a river of identical origin and meaning.

Calder Water S. Lan. *Caldouer* 1265. 'Hard water'. British **caleto-* + **dubro-*.

Caldicot Mon. *Caldecote* 1086 (DB), *Caldicote* 1268. 'Cold shelter'. OE *cald* + *cot*.

Caldragh (*Cealtragh*) Fermanagh. 'Graveyard'.

Caldwell N. Yorks. *Caldeuuella* 1086 (DB). 'Cold spring or stream'. OE *cald* + *wella*.

Caldy Wirral. *Calders* 1086 (DB), *Caldei* 1182. 'Cold island', earlier 'cold rounded hill'. OE *cald* + *ēg* (replacing *ears* 'arse, buttock').

Caldy Island (*Ynys Byr*) Pemb. *Caldea c.*1120. 'Cold island'. OScand. *kald* + *ey*. The Welsh name means 'Pyr's island' (Welsh *ynys* + Celtic pers. name).

Cale (river) Dorset-Somerset. *See* WINCANTON.

Caledon Tyrone. *Chind Aird* 1500. The original name of the village was *Kinaird* (*Cionn Aird*), 'head of the hill'. The present name is from James Alexander, who bought land here in 1778 and was created Earl of *Caledon* in 1800.

Caledonian Canal Highland. 'Canal in *Caledonia*'. The canal and name date from 1803.

Calf of Flotta (island) Orkn. 'Small island next to Flotta'. OScand. *kalfr*. The allusion is to a calf's dependence on its mother cow. *See* FLOTTA.

California Norfolk. A transferred name from the American state so called, here bestowed on a place perhaps considered rather remote or inaccessible. There are examples of the same name in Falk. and Suffolk.

Calke Derbys. *Calc* 1132. '(Place on) the limestone'. OE *calc*.

Callaly Northum. *Calualea* 1161. 'Clearing where calves graze'. OE *calf* (genitive plural *calfra*) + *lēah*.

Callander Stir. *Kalentare* 1504. '(Place by) hard water'. British **caleto-* + **dubro-*. The river-

name may have been that of the present River Teith.

Callater Burn Aber. *Callendar* 1652. 'Hard water'. British **caleto-* + **dubro-*.

Callender Highland. *Kalenter c.*1150. '(Place by) hard water'. British **caleto-* + **dubro*.

Callerton, 'hill where calves graze', OE *calf* (genitive plural *calfra*) + *dūn*: **Callerton, Black** Newc. upon T. *Calverdona* 1212. Affix is OE *blæc* 'dark-coloured'. **Callerton, High** Northum. *Calverdon* 1242.

Callington Cornwall. *Calwetone* 1086 (DB). Probably 'farmstead by the bare hill'. OE *calu* (in a dative form) + *tūn*.

Callow Herefs. *Calua* 1180. 'The bare hill'. OE *calu* in a dative form.

Callow (*An Caladh*) Mayo, Roscommon. 'The riverside land'.

Calmsden Glos. *Kalemundesdene* 852. 'Valley of a man called *Calumund'. OE pers. name + *denu*.

Calne Wilts. *Calne* 955, 1086 (DB). A pre-English river-name of uncertain meaning.

Calow Derbys. *Calehale* 1086 (DB). 'Bare nook of land', or 'nook of land at a bare hill'. OE *calu* + *halh*.

Calshot Hants. *Celcesoran* 980, *Celceshord* 1011. OE *ord* 'point or spit of land' (replacing *ōra* 'shore' in the earliest spelling) with an uncertain first element, possibly OE *cælic* 'cup, chalice' used in some topographical sense.

Calstock Cornwall. *Kalestoc* 1086 (DB). The second element is OE *stoc* 'outlying farm, secondary settlement', the first is uncertain.

Calstone Wellington Wilts. *Calestone* 1086 (DB), *Caulston Wellington* 1568. Probably 'farmstead or village by CALNE', from OE *tūn*. The manorial addition is from a family called *de Wilinton*, here from the 13th century.

Calthorpe Norfolk. *Calethorp* 1044-7, *Caletorp* 1086 (DB). 'Outlying farmstead or hamlet of a man called Kali'. OScand. pers. name + *thorp*.

Calthwaite Cumbria. *Caluethweyt* 1272. 'Clearing where calves are kept'. OE *calf* or OScand. *kalfr* + *thveit*.

Calton, 'farm where calves are reared', OE *calf* + *tūn*: **Calton** N. Yorks. *Caltun* 1086 (DB). **Calton** Staffs. *Calton* 1238.

Caltra (*An Chealtrach*) Galway. 'The graveyard'.

Caltraghlea (*Cealtrach Lia*) Galway. 'Grey graveyard'.

Calveley Ches. *Calueleg c.*1235. 'Clearing where calves are pastured'. OE *calf* + *lēah*.

Calver Derbys. *Calvoure* 1086 (DB). 'Slope or ridge where calves graze'. OE *calf* + *ofer*.

Calverhall Shrops. *Cavrahalle* 1086 (DB), *Caluerhale* 1256. 'Nook of land where calves graze'. OE *calf* (genitive plural *calfra*) + *halh*.

Calverleigh Devon. *Calodelie* [*sic*] 1086 (DB), *Calewudelega* 1194. 'Clearing in the bare wood'. OE *calu* + *wudu* + *lēah*.

Calverley Leeds. *Caverleia* 1086 (DB). 'Clearing where calves are pastured'. OE *calf* (genitive plural *calfra*) + *lēah*.

Calverton, 'farm where calves are reared', OE *calf* (genitive plural *calfra*) + *tūn*: **Calverton** Milt. K. *Calvretone* 1086 (DB). **Calverton** Notts. *Caluretone* 1086 (DB).

Cam Glos. *Camma* 1086 (DB). Named from the River Cam, an old Celtic river-name meaning 'crooked'.

Cam Beck (river) Cumbria. *See* KIRKCAMBECK.

Camaross (*Camros*) Wexford. 'Crooked grove'.

Camber E. Sussex. *Camere* 1375, *Portus Camera* 1397, *Caumbre* 1442. From OFrench *cambre* (Latin *camera*) 'a room, an enclosed space', perhaps originally with reference to a small harbour here before the silting up of the Rother estuary.

Camberley Surrey. an arbitrary name of recent origin, altered from *Cambridge Town* which was named in 1862 from the Duke of Cambridge.

Camberwell Gtr. London. *Cambrewelle* 1086 (DB), *Camerewelle* 1199. OE *wella* 'spring or stream' with an obscure first element, possibly an early borrowed form of Latin *camera* 'vault, room' in allusion to a building or other structure at the spring.

Camblesforth N. Yorks. *Camelesforde* 1086 (DB). Possibly 'ford associated with a man called *Camel(e)*'. OE pers. name + *ford*.

Cambo Northum. *Camho* 1230. 'Hill-spur with a crest or ridge'. OE *camb* + *hōh*.

Cambois Northum. *Cammes c.*1050. A Celtic name, a derivative of Celtic *camm* 'crooked' and originally referring to the bay here.

Camborne Cornwall. *Camberon* 1182. 'Crooked hill'. Cornish *camm* + *bronn*.

Cambrian Mountains Wales. 'Mountains of Cambria'. *Cambria*, the Roman name for Wales, came from the people's name for themselves, Modern Welsh *Cymry*.

Cambridge Cambs. *Grontabricc c.*745, *Cantebrigie* 1086 (DB). 'Bridge on the River Granta'. Celtic river-name (*see* GRANTCHESTER) + OE *brycg*. The change from *Grant-* to *Cam-* is due to Norman influence. **Cambridgeshire** (OE *scīr* 'district') is first referred to in the 11th cent. The later river-name **Cam** is a back-formation from the place name.

Cambridge Glos. *Cambrigga* 1200–10. 'Bridge over the River Cam'. Celtic river-name (*see* CAM) + OE *brycg*.

Cambus O'May Aber. *Cames i maye* 1600. '(River)-bend of the plain'. Gaelic *camas* + *magh*.

Cambuskenneth Stir. *Cambuskynneth* 1147. 'Cinaed's (river)-bend'. Gaelic *camas* + Celtic pers. name.

Cambuslang S. Lan. *Camboslanc* 1296. '(River)-bend of the ships'. Gaelic *camas* + *long*. The river is the Clyde.

Camden Town Gtr. London. so named in 1795 from Earl *Camden* (died 1794) who came into possession of the manor of KENTISH TOWN of which this formed part.

Camel, Queen & Camel, West Somerset. *Cantmæl* 995, *Camelle* 1086 (DB). Possibly a Celtic name from *canto-* 'border or district' and *mēl* 'bare hill'. Affix *Queen* from its possession by Queen Eleanor in the 13th cent.

Camelford Cornwall. *Camelford* 13th cent. 'Ford over the River Camel'. Celtic river-name (possibly 'crooked one' from a derivative of Celtic *camm*) + OE *ford*.

Camerton B. & NE. Som. *Camelartone* 954, *Camelertone* 1086 (DB). 'Farmstead or estate on

Cam Brook'. Celtic river-name (earlier *Cameler*, probably a derivative of **camm* 'crooked') + OE *tūn*.

Camerton Cumbria. *Camerton c.*1150. OE *tūn* 'farmstead, estate' with an obscure first element.

Camlough (*Camloch*) Armagh. *Loch Chamloch c.*1840, 'Crooked lake'.

Cammeringham Lincs. *Camelingeham* [*sic*] 1086 (DB), *Cameryngham c.*1115. Possibly 'homestead of the family or followers of a man called **Cāfmǣr* or **Cantmǣr*'. OE pers. name + *-inga-* + *hām*.

Camowen (*Camabhainn*) Tyrone. 'Crooked river'.

Camp (*An Com*) Kerry. 'The hollow'.

Campbeltown Arg. 'Campbell's town'. Archibald *Campbell*, Earl of Argyle, was granted the site here in 1667 for the erection of a burgh of barony.

Campden, Broad & Campden, Chipping Glos. *Campedene* 1086 (DB), *Bradecampedene* 1224, *Chepyng Campedene* 1287. 'Valley with enclosures'. OE *camp* + *denu*. Affixes are OE *brād* 'broad' and OE *cēping* 'market'.

Campile (*Ceann Poill*) Wexford. 'Head of the creek'.

Camport (*Camport*) Mayo. 'Crooked shore'.

Camps, Castle & Camps, Shudy Cambs. *Canpas* 1086 (DB), *Campecastel* 13th cent., *Sudekampes* 1219. 'The fields or enclosures', from OE *camp*, plural *campas*. The affix *Castle* is from a medieval castle, *Shudy* is probably an OE **scydd* 'shed, hovel'.

Campsall Donc. *Cansale* [*sic*] 1086 (DB), *Camshale* 12th cent. Possibly 'nook of land of a man called Cam'. OE pers. name + *halh*. Alternatively the first element may be a derivative of Celtic **camm* 'crooked' used in some topographical sense.

Campsea Ash Suffolk. *Campeseia* 1086 (DB). 'Island, or dry ground in marsh, with a field or enclosure'. OE *camp* + *ēg*. *Ash* was originally a separate place, mentioned as *Esce* in 1086 (DB), from OE *æsc* 'ash-tree'.

Campsey (*Camsan*) Derry. *Camsan* 1613. 'River-bends'.

Campsie Fells Stir. 'Hills of *Campsie*'. Northern English *fell* from OScand. *fjall*. The range is named from the single hill *Campsie*, which is etymologically obscure.

Campton Beds. *Chambeltone* 1086 (DB). Probably 'farmstead by a river called *Camel*'. Lost Celtic river-name (possibly 'crooked one' from a derivative of Celtic **camm*) + OE *tūn*.

Camross (*Camros*) Laois. 'Crooked copse'.

Candlesby Lincs. *Calnodesbi* 1086 (DB). Probably 'farmstead or village of a man called **Cal(u)nōth*'. OE pers. name + OScand. *bý*.

Candover, Brown & Candover, Preston Hants. *Cendefer c.*880, *Candovre* 1086 (DB), *Brunkardoure* 1296, *Prestecandevere c.*1270. Named from the stream here, a Celtic river-name meaning 'pleasant waters'. Distinguishing affixes are manorial, from early possession by a family called *Brun* and by priests (ME genitive plural *prestene*).

Canewdon Essex. *Carenduna* [*sic*] 1086 (DB), *Canuedon* 1181. Possibly 'hill of the family or followers of a man called Cana'. OE pers. name + *-inga-* + *dūn*.

Canfield, Great Essex. *Canefelda* 1086 (DB). 'Open land of a man called Cana'. OE pers. name + *feld*.

Canford Magna, Little Canford Poole. *Cheneford* [*sic*] 1086 (DB), *Kaneford* 1195, *Lytel Canefford* 1381, *Greate Canford* 1612. 'Ford of a man called Cana'. OE pers. name + *ford*.

Canisbay Highland. *Cananesbi c.*1240. 'Conan's farm'. Gaelic pers. name + OScand. *bý*.

Cann Dorset. *Canna* 12th cent. OE *canne* 'a can, a cup', used topographically for 'a hollow, a deep valley'.

Canna (island) Highland. *Kannay* 1549. Doubtful first element + OScand. *ey*.

Cannings, All & Cannings, Bishops Wilts. *Caninge* 1086 (DB), *Aldekanning* 1205, *Bisshopescanyngges* 1314. '(Settlement of) the family or followers of a man called Cana'. OE pers. name + *-ingas*. Affixes are OE *eald* 'old' and *biscop* 'bishop' (referring to early possession by the Bishop of Salisbury).

Cannington Somerset. *Cantuctun c.*880, *Cantoctona* 1086 (DB). 'Estate or village by the QUANTOCK HILLS'. Celtic hill-name + OE *tūn*.

Cannock Staffs. *Chenet* [sic] 1086 (DB), *Canoc* 12th cent. 'The small hill, the hillock'. OE **cnocc*. It gives name to **Cannock Chase**, from ME *chace* 'tract of land for breeding and hunting wild animals'.

Canon Frome Herefs. *See* FROME.

Canonbury Gtr. London. *Canonesbury* 1373. 'Manor of the canons', from ME *canoun* and *bury*, referring to the canons of St Bartholomew's Smithfield who were granted land in ISLINGTON before 1253 (probably in the 12th cent.).

Canons Ashby Northants. *See* ASHBY.

Canterbury Kent. *Cantuarabyrg* 805–10, *Cantwaraburg* c.900, *Canterburie* 1086 (DB). 'Stronghold or fortified town of the people of KENT'. Ancient Celtic name + OE *-ware* + *burh* (dative *byrig*).

Cantley, probably 'woodland clearing of a man called *Canta', OE pers. name + *lēah*: **Cantley** Donc. *Canteleia* 1086 (DB). **Cantley** Norfolk. *Cantelai* 1086 (DB).

Cantlop Shrops. *Cantelop* 1086 (DB). OE *hop* 'enclosed place' with an obscure first element.

Cantsfield Lancs. *Cantesfelt* 1086 (DB). Probably 'open land by the River Cant'. Celtic river-name (of uncertain meaning) + OE *feld*.

Canvey Essex. *Caneveye* 1255. Possibly 'island of the family or followers of a man called Cana'. OE pers. name + *-inga-* + *ēg*.

Canwick Lincs. *Canewic* 1086 (DB). 'Dwelling or dairy-farm of a man called Cana'. OE pers. name + *wīc*.

Cape Wrath (headland) Highland. *Wraith* 1583. 'Turning-point'. OScand. *hvarf*. The cape marks the point where ships altered course to follow the coast.

Capel, a place name found in several counties, from ME *capel* 'a chapel': **Capel** Surrey. *Capella* 1190. **Capel le Ferne** Kent. *Capel ate Verne* 1377. The addition means 'at the ferny place' from OE **(ge)ferne*. **Capel St Andrew** Suffolk. *Capeles* 1086 (DB). Addition from the dedication of the chapel. **Capel St Mary** Suffolk. *Capeles* 1254. Addition from the dedication of the chapel.

Capel Curig Conwy. *Capel Kiryg* 1536, *Capel Kerig* 1578. 'Curig's chapel'. Welsh *capel* + Celtic

pers. name. From the dedication of the church to St Curig.

Capenhurst Ches. *Capeles* [sic] 1086 (DB), *Capenhurst* 13th cent. Probably 'wooded hill at a look-out place'. OE **cape* (genitive *-an*) + *hyrst*.

Capernwray Lancs. *Coupmanwra* c.1200. 'The merchant's nook or corner of land'. OScand. *kaup-mathr* + *vrá*.

Capheaton Northum. *Magna Heton* 1242, *Cappitheton* 1454. 'High farmstead, farmstead situated on high land'. OE *hēah* + *tūn*. The affix *Cap-* is from Latin *caput* 'head, chief'.

Cappagh (*An Cheapaigh*) Waterford. 'The plot of land'.

Cappamore (*An Cheapach Mhór*) Limerick. 'The big plot of land'.

Cappanacush (*Ceapach na Coise*) Kerry. 'Plot of land at the foot'.

Capparoe (*An Cheapach Rua*) Tipperary. 'The red plot of land'.

Cappataggle (*Ceapaigh an tSeagail*) Galway. 'Plot of rye'.

Cappeen (*Caipín*) Cork. 'Cap'.

Cappoquin (*Ceapach Choinn*) Waterford. 'Conn's plot of land'.

Capton Devon. *Capieton* 1278. Probably 'estate of a family called Capia'. ME surname + OE *tūn*.

Car Colston Notts. *See* COLSTON.

Caradon (district) Cornwall. *Carnetone* 1086 (DB), *Carnedune* c.1160. 'Hill at *Carn* (the tor or rock)'. Cornish *carn* + OE *dūn*. The DB spelling may be an error, or contain OE *tūn* 'farmstead'.

Carbis Bay Cornwall. a 19th-cent. village named from a farm called Carbis, recorded as *Carbons* 1391, from OCornish **car-bons* 'causeway' (literally 'cart-bridge').

Carbrook Sheff. *Kerebroc* 1200–18. Probably 'stream in the marsh'. OScand. *kjarr* + OE *brōc*. Alternatively the first element may be an old Celtic river-name.

Carbrooke Norfolk. *Cherebroc* 1086 (DB). Identical in origin with the previous name.

Carburton Notts. *Carbertone* 1086 (DB). Probably a Celtic name, 'village of the Britons', Brittonic **cair* + **Brïtton*.

Carcroft Donc. *Kercroft* 12th cent. 'Enclosure near the marsh'. OScand. *kjarr* + OE *croft*.

Cardenden Fife. *Cardenane* 14th cent., *Cardwane* 1516. 'Hollow by *Carden*'. OE *denu*. *Carden* is a name of Celtic origin meaning 'thicket' (Welsh *cardden*).

Cardeston Shrops. *Cartistune* 1086 (DB). Probably 'farm or estate of a man called *Card'. OE pers. name + *tūn*.

Cardiff (*Caerdydd*) Card. *Kairdif* 1106, *o gaer dydd* 1566, *Caer Didd* 1698. 'Fort on the River Taf'. Welsh *caer*. For the river-name (here in the genitive case), *see* LLANDAFF.

Cardigan (*Aberteifi*) Cergn. *Kerdigan* 1194. 'Ceredig's land'. *See also* CEREDIGION. The Welsh name means 'mouth of the River Teifi' (OWelsh *aber*), the river-name being of uncertain origin.

Cardington Beds. *Chernetone* [*sic*] 1086 (DB), *Kerdinton c.*1190. Probably 'estate associated with a man called *Cærda'. OE pers. name + *-ing-* + *tūn*.

Cardington Shrops. *Cardintune* 1086 (DB). Probably 'estate associated with a man called *Card(a)'. OE pers. name + *-ing-* + *tūn*.

Cardinham Cornwall. *Cardinan c.*1180. Both parts of the name mean 'fort', Cornish **ker* + **dinan*.

Cardurnock Cumbria. *Cardrunnock* 13th cent. A Celtic place name, from *cair* 'fortified town' + **durnōg* 'pebbly'.

Careby Lincs. *Careby* 1199. 'Farmstead or village of a man called Kári'. OScand. pers. name + *bý*.

Cargan (*An Carraigín*) Antrim. *Carrigan c.*1655. 'The little rock'.

Cargo Cumbria. *Cargaou c.*1178. Celtic **carreg* 'rock' + OScand. *haugr* 'hill'.

Carham Northum. *Carrum c.*1040. '(Place) by the rocks'. OE *carr* in a dative plural form *carrum*.

Carhampton Somerset. *Carrum* 10th cent., *Carentone* 1086 (DB). 'Farm at the place by the rocks'. OE *carr* (dative plural *carrum*) + *tūn*.

Carisbrooke I. of Wight. *Caresbroc* 12th cent. Possibly 'the brook called *Cary*'. Lost Celtic river-name + OE *brōc*.

Cark Cumbria. *Karke* 1491. Celtic **carreg* 'a stone, a rock'.

Cark (*Cearc*) Donegal. 'Hen'.

Carlby Lincs. *Carlebi* 1086 (DB). Probably 'homestead or village of the peasants or freemen'. OScand. *karl* + *bý*.

Carlecotes Barns. *Carlecotes* 13th cent. 'Cottages of the freemen'. OScand. *karl* + OE *cot*.

Carlesmoor N. Yorks. *Carlesmore* 1086 (DB). 'Moorland of the freeman or of a man called Karl'. OScand. *karl* or pers. name + *mór*.

Carleton, Carlton, a common place name in the old Danelaw areas of the Midlands and the North, usually 'farmstead or estate of the freemen or peasants', from OScand. *karl* (often no doubt replacing OE *ceorl*) + OE *tūn*; examples include: **Carleton** Cumbria. *Karleton* 1250. **Carleton** N. Yorks. *Carlentone* 1086 (DB). **Carleton Forehoe** Norfolk. *Carletuna* 1086 (DB), *Karleton Fourhowe* 1268. Affix from the nearby Forehoe Hills, from OE *fēower* 'four' and OScand. *haugr* 'hill'. **Carleton Rode** Norfolk. *Carletuna* 1086 (DB), *Carleton Rode* 1201. Manorial addition from the *de Rode* family, here in the 14th cent. **Carlton** Beds. *Carlentone* 1086 (DB). **Carlton** N. Yorks., near Snaith. *Carletun* 1086 (DB). **Carlton** Notts. *Karleton* 1182. **Carlton Colville** Suffolk. *Carletuna* 1086 (DB), *Carleton Colvile* 1346. Manorial addition from the *de Colevill* family, here in the 13th cent. **Carlton Curlieu** Leics. *Carletone* 1086 (DB), *Carleton Curly* 1273. Manorial addition from the *de Curly* family, here in the 13th cent. **Carlton Husthwaite** N. Yorks. *Carleton* 1086 (DB), *Carlton Husthwat* 1516. Affix from its proximity to HUSTHWAITE. **Carlton in Lindrick** Notts. *Carletone* 1086 (DB), *Carleton in Lindric* 1212. Affix from the district called Lindrick, which means 'strip of land where lime-trees grow', from OE *lind* + **ric*. **Carlton le Moorland** Lincs. *Carletune* 1086 (DB). Affix from its situation 'in the moorland'. **Carlton Miniott** N. Yorks. *Carletun* 1086 (DB), *Carleton Mynyott* 1579. Manorial addition from the *Miniott* family, here in the 14th cent. **Carlton on Trent** Notts. *Carletune* 1086 (DB). For the river-name, *see* TRENTHAM. **Carlton Scroop** Lincs. *Carletune* 1086 (DB). Manorial addition from the *Scrope* family, here in the 14th cent.

Carlingcott B. & NE. Som. *Credelincote* 1086 (DB). 'Cottage associated with a man called *Cridela'. OE pers. name + *-ing-* + *cot*.

Carlingford (*Cairlinn*) Louth. 'Bay of the hag'. *an Carrlongphort* 1213. OScand. *kerling* + *fjǫrthr*. The 'hag' is probably (one of) the three mountain-tops here known as *The Three Nuns*.

Carlisle Cumbria. *Luguvalio* 4th cent., *Carleol* c.1106. An old Celtic name meaning '(place) belonging to a man called *Luguvalos', to which Celtic *cair* 'fortified town' was added after the Roman period.

Carlow (*Ceatharlach*) Carlow. 'Quadruple lake'.

Carlton, See CARLETON.

Carluke S. Lan. *Carlug* 1304, *Cerneluke* c.1320. British *cair* 'stockaded house' + obscure element.

Carmarthen (*Caerfyrddin*) Carm. *Cair Mirdin* 1130. 'Fort at Maridunum'. Welsh *caer*. The Roman town of *Maridunum* has a Celtic name meaning 'fort by the sea' (British *mari-* + *duno-*).

Carmavy (*Carn Méibhe*) Antrim. 'Maeve's cairn'.

Carmel Gwyd. The Welsh Nonconformist chapel here was named after the biblical Mount *Carmel* (1 Kings 18:19, etc).

Carn (*Carn*) Westmeath. 'Cairn'.

Carna (*Carna*) Galway, Wexford. 'Cairns'.

Carnaby E. R. Yorks. *Cherendebi* 1086 (DB). Possibly 'homestead or village of a man called *Kærandi* or *Keyrandi'. OScand. pers. name + *bý*.

Carnagh (*Carranach*) Armagh. 'Place of cairns'.

Carnalbanagh (*Carn Albanach*) Antrim. *Carnalbanagh* 1780. 'Cairn of the Scotsmen'.

Carnanmore (*Carnán Mór*) Antrim. 'Big little cairn'.

Carncastle (*Carn an Chaistéil*) Antrim. *Carlcastel* 1279. 'Cairn of the castle'.

Carndonagh (*Carn Domhnach*) Donegal. *Carnedony* 1620. 'Cairn of the church'.

Carnew (*Carn an Bhua*) Wicklow. 'Cairn of the victory'.

Carney (*Fearann Uí Chearnaigh*) Sligo. 'Territory of Ó Ciarnaigh'.

Carnforth Lancs. *Chreneforde* 1086 (DB). Probably 'ford frequented by cranes or herons'. OE *cran* + *ford*.

Carnkenny (*Carn Cainnech*) Tyrone. 'Cainnech's cairn'.

Carnlough (*Carnlach*) Antrim. *Carnalloch* 1780. 'Place of cairns'.

Carnmoney (*Carn Monaidh*) Antrim. *(Ecclesia de) Coole of Carnmonie* 1615. 'Cairn of the bog'.

Carnoneen (*Carn Eoghainín*) Galway. 'Eoghainín's cairn'.

Carnoustie Ang. *Donaldus Carmusy* 1493. Meaning uncertain. The first element may be Gaelic *cathair*, 'fort', *càrr*, 'rock', or *carn*, 'cairn'. The second has not been satisfactorily explained.

Carnteel (*Carn tSiail*) Tyrone. (*cath*) *Cairn tSiadhail* 1239. 'Sial's cairn'.

Carntierna (*Carn Tighernaigh*) Cork. 'Tighernach's cairn'.

Carntogher (*Carn Tóchair*) Derry. (*mullach*) *an Cháirn* c.1740. 'Cairn of the causeway'.

Carperby N. Yorks. *Chirprebi*. Possibly 'homestead or village of a man called Cairpre'. OIrish pers. name + OScand. *bý*.

Carra (*Cairthe*) Mayo. 'Standing stone'.

Carracastle (*Ceathrú an Chaisil*) Roscommon. 'Quarter of the stone fort'.

Carraroe (*Ceathrú Rua*) Galway. 'Red quarter'.

Carrauntoohil (*Corrán Tuathail*) Kerry. 'Inverted crescent'.

Carreg yr Esgob (island) Pemb. *Bishops rock* 1602. 'Bishop's rock'. Welsh *carreg* + *yr* + *esgob*. The name was translated from the English.

Carrichue (*Carraig Aodha*) Derry. 'Aodh's rock'.

Carrick (district) Cornwall. *Caryk c.*1540. 'The rock', from Cornish *carrek*. The old name of Black Rock in Carrick Roads, in the Fal estuary.

Carrick (*An Charraig*) Donegal, Wexford. 'The rock'.

Carrick (district) S. Ayr. *karrio c.*1140. '(Region of) rocks'. Gaelic *carraig*, from British *carrecc* 'rock'.

Carrick-on-Suir (*Carraig na Siúre*) Tipperary. 'Rock on the (river) Suir'.

Carrickaboy (*Carraighigh Buí*) Cavan. *Careghaboy* 1629. 'Yellow rocky place'.

Carrickahorig (*Carraig an Chomhraic*) Tipperary. 'Rock of the confluence'.

Carrickanoran (*Carraig an Uardáin*) Kilkenny, Monaghan. 'Rock of the spring'.

Carrickart (*Carraig Airt*) Donegal. *Carrowfiggart* 1609. 'Art's rock'. The name probably developed from an original Irish form *Ceathrú Fhiodhghoirt*, 'quarter of the field of the wood'.

Carrickbeg (*Carraig Bheag*) Waterford. 'Little rock'.

Carrickfergus (*Carraig Fhearghais*) Antrim. (*go*) *Carraic Ferghusa* 1204. 'Rock of Fergus'.

Carrickmacross (*Carraig Mhachaire Rois*) Monaghan. *Rosse als. Machair Roysse* 1541. 'Rock of the plain of the grove'.

Carrickmore (*An Charraig Mhór*) Tyrone. 'The big rock'.

Carrickroe (*An Charraig Rua*) Monaghan. 'The red rock'.

Carrig (*Carraig*) Tipperary. 'Rock'.

Carrigafoyle (*Carraig an Phoill*) Tipperary. 'Rock of the pool'.

Carrigaholt (*Carraig an Chabhaltaigh*) Clare. 'Rock of the fleet'.

Carrigahorig (*Carraig an Chomhraic*) Tipperary. 'Rock of the confluence'.

Carrigaline (*Carraig Uí Leighin*) Cork. 'Ó Leighin's rock'.

Carrigallen (*Carraig Álainn*) Leitrim. 'Beautiful rock'.

Carrigan (*An Carraigín*) Cavan. *Carrigin* 1699. 'The little rock'.

Carriganimmy (*Carraig an Ime*) Cork. 'Rock of the butter'.

Carrigans (*An Carraigín*) Donegal. *Cairrccín* 1490. 'The little rock'.

Carriganurra (*Carraig an Fhoraidh*) Kilkenny. 'Rock of the mound'.

Carrigatogher (*Carraig an Tóchair*) Tipperary. 'Rock of the causeway'.

Carrigcannon (*Carraig Cheannann*) Kerry. 'White-headed rock'.

Carrigfada (*Carraig Fhada*) Cork. 'Long rock'.

Carrigkerry (*Carraig Chiarraí*) Limerick. 'Rock of Ciarraí'.

Carrignavar (*Carraig na bhFear*) Cork. 'Rock of the men'.

Carrigtogill (*Carraig Thuathail*) Cork. 'Tuathal's rock'.

Carrington Lincs. First recorded in 1812, and named after Robert Smith, Lord *Carrington* (1752–1838), who had lands here.

Carrington Traffd. *Carrintona* 12th cent. Possibly 'estate associated with a man called *Cāra'. OE pers. name + *-ing-* + *tūn*. Alternatively the first element may be an OE *caring* 'tending, herding' or an OE *cǣring* 'river-bend'.

Carrock, Castle Cumbria. *Castelcairoc c.*1165. 'Fortified castle'. Celtic *castel* + an adjectival derivative of Celtic *cair* 'fort'.

Carron (*Carn*) Clare. 'Cairn'.

Carrowbeg (*An Ceathrú Bheag*) Mayo. 'The small quarter'.

Carrowbehy (*Ceathrú Bheithí*) Roscommon. 'Quarter of the birches'.

Carrowdoan (*Ceathrú Domhain*) Donegal. 'Deep quarter'.

Carrowdore (*Ceathrú Dobhair*) Down. *Kerrowe Dorne* 1627. 'Quarter of the water'.

Carrowholly (*Ceathrú Chalaidh*) Mayo. 'Quarter of the landing-place'.

Carrowkeel (*An Cheathrú Chaol*) Donegal. *Carrowkeele* 1639. 'The narrow quarter'.

Carrowmena (*Ceathrú Meánach*) Donegal. 'Middle quarter'.

Carrowmore (*Ceathrú Mhór*) Galway, Mayo, Sligo. 'Big quarter'.

Carrowmoreknock (*Ceathrú Mhór an Chnoic*) Galway. 'Big quarter of the hill'.

Carrownacon (*Ceathrú na Con*) Mayo. 'Quarter of the hound'.

Carrownedan (*Ceathrú an Éadain*) Sligo. 'Quarter of the brow'.

Carrownisky (*Ceathrú an Uisce*) Mayo. 'Quarter of the water'.

Carrowreagh (*Ceathrú Riabhach*) Roscommon. 'Striped quarter'.

Carrowtawy (*Ceathrú an tSamhaidh*) Sligo. 'Quarter of the sorrel'.

Carrowteige (*Ceathrú Thaidhg*) Mayo. 'Taidg's quarter'.

Carryduff (*Ceathrú Aodha Dhuibh*) Down. *Carow-Eduffe al. Carow-Hugh-Duffe al. Tyduffe* 1623. 'Black Aodh's quarter'.

Carse of Gowrie (district) Perth. *Carse de Gowrie c.*1200. 'Alluvial plain of the goat place'. Scots *carse* + Gaelic *gabhar* (genitive *gaibhre*).

Carshalton Gtr. London. *Aultone* 1086 (DB), *Cresaulton* 1235. 'Farm by the river-spring where water-cress grows'. OE *æwell* + *tūn* with the later addition of OE *cærse*.

Carsington Derbys. *Ghersintune* 1086 (DB). Probably 'farmstead where cress grows'. OE **cærsen* + *tūn*.

Carstairs S. Lan. *Casteltarres* 1170. 'Tarres's castle'. ME *castel* + Celtic pers. name.

Carswell Marsh Oxon. *Chersvelle* 1086 (DB), *Carsewell Merssh* 1467. '(Marsh at) the spring or stream where water-cress grows'. OE *cærse* + *wella* with the later addition of ME *mershe*.

Carterton Oxon. a recent name for the village founded by one William *Carter* in 1901.

Carthorpe N. Yorks. *Caretorp* 1086 (DB). 'Outlying farmstead or hamlet of a man called Kári'. OScand. pers. name + *thorp*.

Cartington Northum. *Cretenden* 1220. Probably 'hill associated with a man called **Certa*'. OE pers. name + *-ing-* + *dūn*.

Cartmel Cumbria. *C(e)artmel* 12th cent. 'Sandbank by rough stony ground'. OScand. **kartr* + *melr*.

Cartronlahan (*Cartúr Leathan*) Galway. 'Broad quarter'.

Cary Fitzpaine & Castle Cary Somerset. *Cari* 1086 (DB), *Castelkary* 1237. Named from the River Cary, an ancient Celtic or pre-Celtic river-name. Distinguishing affixes from the *Fitz Payn* family, here in the 13th cent., and from ME *castel* with reference to the Norman castle.

Cas-gwent. *See* CHEPSTOW.

Cas-mael. *See* PUNCHESTON.

Cas-wis. *See* WISTON.

Cashel (*Caiseal*) Galway, Tipperary. 'Stone fort'.

Cashelbane (*Caiseal Bán*) Tyrone. 'White stone fort'.

Cashelgarran (*Caiseal an Ghearráin*) Sligo. 'Stone fort of the horse'.

Cashelmore (*An Caiseal Mór*) Donegal. 'The large stone fort'.

Cashla (*Caisle*) Galway. 'Stream'.

Caslai. *See* HAYSCASTLE.

Casllwchwr. *See* LOUGHOR.

Casnewydd-ar-Wysg. *See* NEWPORT (Newpt).

Cassagh (*Ceasach*) Wexford. 'Wicker causeway'.

Cassington Oxon. *Cersetone* 1086 (DB), *Kersinton* 12th cent. 'Farmstead where cress grows'. OE **cærsen* + *tūn*.

Cassop Durham. *Cazehoppe* 1183. Possibly 'valley of the wild-cat's stream'. OE *catt* + *ēa* + *hop*.

Castell-nedd. *See* NEATH.

Castellhaidd. *See* HAYSCASTLE.

Castellnewydd Emlyn. *See* NEWCASTLE EMLYN.

Casterton, 'farmstead near the (Roman) fort', OE *cæster* + *tūn*: **Casterton** Cumbria. *Castretune* 1086 (DB). **Casterton, Great & Casterton, Little** Rutland. *Castretone* 1086 (DB).

Castle as affix. *See* main name, e.g. for **Castle Bolton** (N. Yorks) *see* BOLTON.

Castle Douglas Dumf. 'Douglas's castle'. The castle is Threave Castle. In 1789 Sir William *Douglas* bought the village of Carlingwerk and developed it into a burgh of barony.

Castle Morpeth (district) Northum. A modern name combining *Castle*, the name of two wards which included NEWCASTLE UPON TYNE, and MORPETH.

Castle Point (district) Essex. A modern name, combining the two local features HADLEIGH *Castle* and CANVEY *Point*.

Castlebar (*Caisleán an Bharraigh*) Mayo. 'de Barra's castle'.

Castlebay W. Isles. (Barra). '(Town by) Castle Bay'. The bay takes its name from Kiessimul Castle. The town arose in the 19th cent. as a fishing port. The name is a translation of Gaelic *Bàgh a' Chaisteil.*

Castleblayney (*Baile na Lorgan*) Monaghan. *Castleblayney* 1663. 'Blayney's castle'. Sir Edward *Blayney* built a castle here on land granted him by James I in the early 17th cent. The Irish name means 'town of the strip of land'.

Castlecat (*Caiseal Cait*) Antrim. *Castlecat* 1780. 'Stone fort of the cat'.

Castlecaulfield (*Baile Uí Dhonnaíle*) Tyrone. *Castle-Caufield* 1618. 'Caulfield's castle'. Sir Toby *Caulfield* built a castle here in *c.*1612. The Irish name means 'townland of Ó Donnaíle'.

Castlecomer (*Caisléan an Chomair*) Kilkenny. 'Castle of the confluence'.

Castleconnell (*Caisleán Uí Chonaill*) Limerick. 'Ó Conaill's castle'.

Castleconnor (*Caisleán MhicChonchobhair*) Sligo. 'Mac Conchuir's castle'.

Castlecor (*Caisleán na Cora*) Cork. 'Castle of the weir'.

Castledawson Derry. *Castledawson al. Dawson's Bridge* 1677. 'Dawson's castle'.

Sir Joshua *Dawson* founded the village in 1710 on land owned by his family since 1622. Its Irish name is *An Seanmhullach*, 'the old hilltop'.

Castlederg (*Caisleán na Deirge*) Tyrone. *Caislén na Deirce* 1497. 'Castle of the (river) Derg'.

Castledermot (*Caisleán Díseart Diarmada*) Kildare. 'Castle of the hermitage of Diarmaid'.

Castlefinn (*Caisleán na Finne*) Donegal. *Castleffynne* 1617. 'Castle of the (river) Finn'.

Castleford Wakefd. *Ceaster forda* late 11th cent. 'Ford by the Roman fort'. OE *cæster* + *ford*.

Castlegal (*Caisle Geala*) Sligo. 'White inlet'.

Castlegar (*Caisleán Gearr*) Galway. 'Short castle'.

Castleknock (*Caisleán Cnucha*) Dublin. 'Castle of the hill'.

Castlelyons (*Caisleán Ó Liatháin*) Cork. 'Uí Liatháin's castle'.

Castlemaine (*Caisleán na Mainge*) Kerry. 'Castle of the (river) Maine'.

Castlemartin Pemb. *Castro Sancti Martini* 1290, *Castlemartin* 1341. 'Fort by St Martin's church'. ME *castel*.

Castlemartyr (*Caisleán na Martra*) Cork. 'Castle of the relics'.

Castlemorton Worcs. *Mortun* 1235, *Castell Morton* 1346. 'Farmstead in the marshy ground'. OE *mōr* + *tūn*, with the later addition of ME *castel* 'castle'.

Castlepollard (*Baile na gCros*) Westmeath. Walter *Pollard* was granted a licence to hold a weekly market here in 1674. The Irish name means 'town of the crosses'.

Castleraghan (*Caisleán Rathain*) Cavan. 'Castle of ferns'.

Castlereagh (*Caisleán Riabhach*) Down, Roscommon. 'Striped castle'.

Castlerock (*Carraig Ceasail*) Derry. 'Castles' rock'. Robert *Castles* and his crew perished on a nearby rock in 1826. The Irish name gives the meaning as 'castle rock'.

Castleshane (*Caisleán an tSiáin*) Monaghan. 'Castle of the fairy hill'.

Castleside Durham. A recent name, recorded thus from 1864, from ModE *castle* (reference uncertain) + *side* 'hill-side'.

Castleton, 'farmstead or village by a castle', from OE *castel* + *tūn*: **Castleton** Derbys. *Castelton* 13th cent. **Castleton** Rochdl. *Castelton* 1246. **Castleton** N. Yorks. *Castelton* 1577.

Castletown Isle of Man. *villa castelli* c.1370. 'Village of the castle'. ME *castel* + *toun*, with reference to the 14th-cent. castle built on the site of an earlier castle dating from the 11th cent.

Castletown Sundld. Modern estate named from the 14th-cent. Hylton Castle (now in ruins), *see* HYLTON, SOUTH.

Castletown Kinneigh (*Baile Chaisleáin Chinn Eich*) Cork. 'Town of the castle of the horse's head'.

Castlewellan (*Caisleán Uidhilín*) Down. *Ballycaslanwilliam* 1605. 'Uidhilín's castle'.

Castley N. Yorks. *Castelai* 1086 (DB). 'Wood or clearing by a heap of stones'. OE *ceastel* + *lēah*.

Caston Norfolk. *Catestuna* 1086 (DB). 'Farmstead or estate of a man called *Catt* or *Káti*'. OE or OScand. pers. name + OE *tūn*.

Castor Peterb. *Cæstre* 948, *Castre* 1086 (DB). 'The Roman fort'. OE *cæster*.

Catcleugh Northum. *Cattechlow* 1279. 'Deep valley or ravine frequented by wild-cats'. OE *catt* + *clōh*.

Catcott Somerset. *Cadicote* 1086 (DB). 'Cottage of a man called Cada'. OE pers. name + *cot*.

Caterham Surrey. *Catheham* 1179, *Katerham* 1200. Probably 'homestead or enclosure at the hill called *Cadeir*', from Celtic *cadeir* (literally 'chair' but used of lofty places) + OE *hām* or *hamm*. However just possibly 'homestead or enclosure of a man called *Catta*', with an OE pers. name as first element.

Catesby Northants. *Catesbi* 1086 (DB). 'Farmstead or village of a man called Kátr or Káti'. OScand. pers. name + *bý*.

Catfield Norfolk. *Catefelda* 1086 (DB). Possibly 'open land frequented by wild-cats', OE *catt* + *feld*. Alternatively 'open land of a man called Káti', from an OScand. pers. name.

Catford Gtr. London. *Catteford* 1240. 'Ford frequented by wild-cats'. OE *catt* + *ford*.

Catforth Lancs. *Catford* 1332. Identical in origin with the previous name.

Cathays Card. *Catt Hays* 1699. 'Enclosure inhabited by wild-cats'. OE *catt* + *haga*.

Catherington Hants. *Cateringatune* c.1015. Possibly 'farmstead of the people living by the hill called *Cadeir*', from Celtic *cadeir* 'chair' + *-inga-* + *tūn*. Alternatively 'farmstead of the family or followers of a man called *Cat(t)or*', from OE pers. name + *-inga-* + *tūn*.

Catherston Leweston Dorset. *Chartreston* 1268, *Lesterton* 1316. Originally names of adjacent estates belonging to families called Charteray and Lester respectively, from ME *toun* 'village, manor'.

Catherton Shrops. *Carderton* 1316. OE *tūn* 'farmstead, village' with an obscure first element.

Catmore W. Berks. *Catmere* 10th cent., 1086 (DB). 'Pool frequented by wild-cats'. OE *catt* + *mere*.

Catmose, Vale of (district) Rutland. *The Val of Catmouse* 1576. 'Bog or marsh frequented by wild-cats'. OE *catt* + *mos*, with ME *vale* 'valley'.

Caton Lancs. *Catun* 1086 (DB). 'Farmstead or village of a man called Káti'. OScand. pers. name + OE *tūn*.

Catsfield E. Sussex. *Cedesfeld* [sic] 1086 (DB), *Cattesfeld* 12th cent. 'Open land of a man called *Catt* or frequented by wild-cats'. OE pers. name or OE *catt* + *feld*.

Catsgore Somerset. Not on early record, but perhaps 'point of land frequented by wild-cats', OE *catt* + *gāra*.

Catshill Worcs. *Catteshull* 1199. 'Hill of a man called *Catt* or frequented by wild-cats'. OE pers. name or OE *catt* + *hyll*.

Cattal N. Yorks. *Catale* 1086 (DB). Probably 'nook of land frequented by wild-cats'. OE *catt* + *halh*.

Cattawade Suffolk. *Cattiwad* 1247. 'Ford or crossing-place frequented by wild-cats'. OE *catt* + *(ge)wæd*.

Catterall Lancs. *Catrehala* 1086 (DB). Possibly OScand. *kattar-hali* 'cat's tail' with reference to the shape of some lost feature.

Alternatively OE *halh* 'nook of land' with an obscure first element.

Catterick N. Yorks. *Katouraktónion c.*150, *Catrice* 1086 (DB). From Latin *cataracta* 'waterfall', though apparently through a misunderstanding of the original Celtic place name meaning '(place of) battle ramparts'.

Catterlen Cumbria. *Kaderleng* 1158. Celtic **cadeir* 'chair' (here 'hill') with an obscure second element, possibly a pers. name.

Catterton N. Yorks. *Cadretune* 1086 (DB). Probably 'farmstead at the hill called *Cadeir*'. Celtic **cadeir* 'chair' + OE *tūn*.

Catthorpe Leics. *Torp* 1086 (DB), *Torpkat* 1276. OScand. *thorp* 'outlying farmstead or hamlet', with a manorial addition from a family called *le Cat*(*t*).

Cattishall Suffolk. *Catteshale* 1187. Probably 'nook of land of a man called *Catt or frequented by wild-cats'. OE pers. name or OE *catt* + *halh*.

Cattistock Dorset. *Stoche* 1086 (DB), *Cattestok* 1288. OE *stoc* 'outlying farm buildings, secondary settlement', with a manorial addition from a person or family called *Cat*(*t*).

Catton, usually 'farmstead of a man called Catta or Káti', OE or OScand. pers. name + OE *tūn*: **Catton** Norfolk. *Catetuna* 1086 (DB). **Catton** N. Yorks. *Catune* 1086 (DB). **Catton, High & Catton, Low** E. R. Yorks. *Caton, Cattune* 1086 (DB). However the following has a different origin: **Catton** Northum. *Catteden* 1229. 'Valley frequented by wild-cats'. OE *catt* + *denu*.

Catwick E. R. Yorks. *Catingeuuic* 1086 (DB). 'Dwelling or (dairy) farm associated with a man called Catta'. OE pers. name (+ -*ing*-) + *wīc*.

Catworth Cambs. *Catteswyrth* 10th cent., *Cateuuorde* 1086 (DB). 'Enclosure of a man called *Catt or Catta'. OE pers. name + *worth*.

Caulcott Oxon. *Caldecot* 1199. Identical in origin with CALDECOTE.

Cauldon Staffs. *Celfdun* 1002, *Caldone* 1086 (DB). 'Hill where calves graze'. OE *celf, cælf* + *dūn*.

Cauldwell Derbys. *Caldewællen* 942, *Caldewelle* 1086 (DB). 'Cold spring or stream'. OE *cald* + *wella*.

Caundle, Bishop's, Purse Caundle & Stourton Caundle Dorset. *Candel* 1086 (DB), *Purscaundel* 1241, *Caundel Bishops* 1294, *Sturton Candell* 1569-74. The meaning of Caundle is obscure, though it may have been a name for the range of hills here. Affix *Bishop's* is from the possession of this manor by the Bishop of Salisbury. *Purse* is probably a manorial affix from a family of this name. *Stourton* refers to possession by the Lords *Stourton* from the 15th cent. until 1727.

Caunton Notts. *Calnestune* 1086 (DB). Probably 'farmstead of a man called *Cal(u)nōth'. OE pers. name + *tūn*.

Causey Park Northum. *La Chauce* 1242. ME *cauce* 'embankment, raised way'.

Cavan (*An Cabhán*) Cavan. (*i mainistir in*) *Cabhain* 1330. 'The hollow'.

Cavanagarden (*Cabhán an Gharraí*) Donegal. 'Hollow of the garden'.

Cave, North & Cave, South E. R. Yorks. *Cave* 1086 (DB). Probably from the stream here, OE **Cāfe* 'the fast-flowing one', from OE *cāf* 'quick, swift'.

Cavendish Suffolk. *Kauanadisc* 1086 (DB). 'Enclosure or enclosed park of a man called **Cāfna*'. OE pers. name + *edisc*.

Cavenham Suffolk. *Kanauaham* [sic] 1086 (DB), *Cauenham* 1198. 'Homestead or enclosure of a man called **Cāfna*'. OE pers. name + *hām* or *hamm*.

Caversfield Oxon. *Cavrefelle* 1086 (DB). 'Open land of a man called *Cāfhere'. OE pers. name + *feld*.

Caversham Readg. *Caueresham* 1086 (DB). 'Homestead or enclosure of a man called *Cāfhere'. OE pers. name + *hām* or *hamm*.

Caverswall Staffs. *Cavreswelle* 1086 (DB). 'Spring or stream of a man called *Cāfhere'. OE pers. name + *wella*.

Cavil E. R. Yorks. *Cafeld* 959, *Cheuede* 1086 (DB). 'Open land frequented by jackdaws'. OE **cā* + *feld*.

Cawdor Highland. *Kaledor* 1280. '(Place by) hard water'. British **caleto-* + **dubro-*.

Cawkwell Lincs. *Calchewelle* 1086 (DB). 'Chalk spring or stream'. OE *calc* + *wella*.

Cawood N. Yorks. *Kawuda* 963. 'Wood frequented by jackdaws'. OE **cā + wudu*.

Cawston, 'farmstead or village of a man called Kalfr', OScand. pers. name + OE *tūn*: **Cawston** Norfolk. *Caustuna, Caluestune* 1086 (DB). **Cawston** Warwicks. *Calvestone* 1086 (DB).

Cawthorne Barns. *Caltorne* 1086 (DB). 'Cold (i.e. exposed) thorn-tree'. OE *cald + thorn*.

Cawthorpe, 'outlying farmstead or hamlet of a man called Kali', OScand. pers. name + *thorp*: **Cawthorpe** Lincs. *Caletorp c.*1086 (DB). **Cawthorpe, Little** Lincs. *Calethorp c.*1150.

Cawton N. Yorks. *Caluetun* 1086 (DB). 'Farm where calves are reared'. OE *calf + tūn*.

Caxton Cambs. *Caustone [sic]* 1086 (DB), *Kakestune c.*1150. Probably 'farmstead of a man called *Kakkr'. OScand. pers. name + OE *tūn*.

Caynham Shrops. *Caiham* 1006 (DB), *Cainham* 1255. Probably 'homestead or enclosure of a man called *Cǣga'. OE pers. name (genitive -*n*) + *hām* or *hamm*.

Caythorpe, 'outlying farmstead or hamlet of a man called Káti', OScand. pers. name + *thorp*: **Caythorpe** Lincs. *Catorp* 1086 (DB). **Caythorpe** Notts. *Cathorp* 1177.

Cayton N. Yorks. *Caitun(e)* 1086 (DB). 'Farmstead of a man called *Cǣga'. OE pers. name + *tūn*.

Cefn-mawr Wrex. 'Big ridge'. Welsh *cefn + mawr*.

Ceinewydd. See NEW QUAY.

Cemaes Angl. *Kemmeys* 1291. '(Place of) bends' (Welsh *camas*, plural *cemais*). The village is frequently referred to as Cemaes Bay.

Cendl. See BEAUFORT.

Ceredigion (the unitary authority). *Cereticiaun* 12th cent. 'Ceredig's land'. Ceredig was one of the sons of Cunedda, who gave the name of GWYNEDD. His name is followed by the Welsh territorial suffix -*ion*. Ceredigion itself gave the name of CARDIGAN.

Cerne Abbas Dorset. *Cernel* 1086 (DB), *Cerne Abbatis* 1288. Named from the River Cerne, an old Celtic river-name from Celtic **carn* 'cairn, heap of stones'. Affix is Latin *abbas* 'an abbot', with reference to the abbey here.

Cerne, Nether & Cerne, Up Dorset. *Nudernecerna* 1206, *Obcerne* 1086 (DB). Named from the same river as Cerne Abbas. OE *neotherra* 'lower down' and OE *upp* 'higher up' with reference to their situation on the river.

Cerney, North & Cerney, South, Cerney Wick Glos. *Cyrnea* 852, *Cernei* 1086 (DB), *Northcerneye* 1291, *Suthcerney* 1285, *Cernewike* 1220. Named from the River Churn (an old Celtic river-name derived from the same root as the first element of CIRENCESTER) + OE *ēa* 'stream'. Wick is from OE *wīc* 'farm, dairy farm'.

Cerrigydrudion Conwy. *Kericdrudion* 1254. 'Stones of the heroes'. Welsh *carreg* (plural *cerrig*) + *y* + *drud* (plural *drudion*).

Chaceley Glos. *Ceatewesleah* 972. Possibly a derivative of Celtic **cę̄d* 'wood' + OE *lēah* 'wood, clearing'.

Chackmore Bucks. *Chakemore* 1241. Possibly 'marshy ground of a man called *Ceacca'. OE pers. name + *mōr*. Alternatively the first element may be OE **ceacce* 'hill'.

Chacombe Northants. *Cewecumbe* 1086 (DB). 'Valley of a man called *Ceawa'. OE pers. name + *cumb*.

Chadderton Oldham. *Chaderton c.*1200. Possibly 'farmstead at the hill called *Cadeir*'. Celtic **cadeir* 'chair' (here 'hill') + OE *tūn*.

Chaddesden Derby. *Cedesdene* 1086 (DB). 'Valley of a man called *Ceadd'. OE pers. name + *denu*.

Chaddesley Corbett Worcs. *Ceadresleahge* 816, *Cedeslai* 1086 (DB). Possibly 'wood or clearing at the hill called *Cadeir*'. Celtic **cadeir* 'chair' + OE *lēah*, with manorial addition from the *Corbet* family here in the 12th cent.

Chaddleworth W. Berks. *Ceadelanwyrth* 960, *Cedeledorde [sic]* 1086 (DB). 'Enclosure of a man called *Ceadela'. OE pers. name + *worth*.

Chadlington Oxon. *Cedelintone* 1086 (DB). 'Estate associated with a man called *Ceadela'. OE pers. name + -*ing*- + *tūn*.

Chadshunt Warwicks. *Ceadeles funtan* 949, *Cedeleshunte* 1086 (DB). 'Spring of a man called *Ceadel'. OE pers. name + **funta*.

Chadwell, 'cold spring or stream', OE *c(e)ald + wella*: **Chadwell** Leics. *Caldeuuelle* 1086 (DB). **Chadwell St Mary** Thurr. *Celdeuuella*

1086 (DB). Affix from the dedication of the church.

Chadwick Green St Hel. *Chaddewyk c.*1180. '(Dairy) farm of a man called Ceadda'. OE pers. name + *wīc*.

Chaffcombe Somerset. *Caffecome* 1086 (DB). Probably 'valley of a man called *Ceaffa'. OE pers. name + *cumb*.

Chaffpool (*Lochán na Cáithe*) Sligo. 'Little lake of the chaff'.

Chagford Devon. *Chageford* 1086 (DB). 'Ford where broom or gorse grows'. OE *ceagge* + *ford*.

Chailey E. Sussex. *Cheagele* 11th cent. 'Clearing where broom or gorse grows'. OE *ceagge* + *lēah*.

Chalbury Dorset. *Cheoles burge* 946. 'Fortified place associated with a man called Cēol'. OE pers. name + *burh*.

Chaldon, 'hill where calves graze', OE *cealf* + *dūn*: **Chaldon** Surrey. *Calvedone* 1086 (DB). **Chaldon Herring or East Chaldon** Dorset. *Celvedune* 1086 (DB), *Chaluedon Hareng* 1243. Manorial affix from the *Harang* family, here from the 12th cent.

Chale I. of Wight. *Cela* 1086 (DB). OE *ceole* 'throat' used in a topographical sense 'gorge, ravine'.

Chalfont St Giles & Chalfont St Peter Bucks. *Celfunte* 1086 (DB), *Chalfund Sancti Egidii* 1237, *Chalfhunte Sancti Petri* 1237–40. 'Spring frequented by calves'. OE *cealf* + **funta*, with distinguishing affixes from the dedications of the churches at the two places.

Chalford, 'chalk or limestone ford', OE *cealc* + *ford*: **Chalford** Glos. *Chalforde c.*1250. **Chalford** Oxon. *Chalcford* 1185–6.

Chalgrave Beds. *Cealhgrǽfan* 926, *Celgraue* 1086 (DB). 'Chalk pit'. OE *cealc* + *grǽf*.

Chalgrove Oxon. *Celgrave* 1086 (DB). Identical in origin with the previous name.

Chalk Kent. *Cealca c.*975, *Celca* 1086 (DB). '(Place on) the chalk'. OE *cealc*.

Chalk, Broad & Bowerchalke Wilts. *Cealcan* 826, *Ceolcum* 955, *Chelche* 1086 (DB), *Brode Chalk* 1380, *Burchelke* 1225. 'Chalk place'. OE **cealce*. Distinguishing affixes are OE *brād* 'great' and either *būra* 'of the peasants', *būr* 'dwelling' or *burh* 'stronghold'.

Chalk Farm Gtr. London. *Chaldecote* 1253. 'Cold cottage(s)'. OE *ceald* + *cot*. The 'worn-down' form *Chalk* first appears in 1746.

Challacombe Devon. near Lynton. *Celdecomba* 1086 (DB). 'Cold valley'. OE *ceald* + *cumb*.

Challock Lees Kent. *Cealfalocum* 824. 'Enclosure(s) for calves'. OE *cealf* + *loca*, with the later addition of *lǽs* 'pasture'.

Challow, East & Challow, West Oxon. *Ceawanhlǽwe* 947, *Ceveslane* [*sic*] 1086 (DB). 'Tumulus of a man called *Ceawa'. OE pers. name + *hlǽw, hlāw*.

Chalton Beds. near Toddington. *Chaltun* 1131. 'Farm where calves are reared'. OE *cealf* + *tūn*.

Chalton Hants. *Cealctun* 1015, *Ceptune* 1086 (DB). 'Farmstead on chalk'. OE *cealc* + *tūn*.

Chalvington E. Sussex. *Calvintone* 1086 (DB). 'Estate associated with a man called *Cealf(a)'. OE pers. name + *-ing-* + *tūn*.

Chandler's Ford Hants. on record from 1759, named from the *Chaundler* family, in the area from the 14th cent.

Chanonrock (*Carraig na gCanónach*) Louth. 'Rock of the canons'.

Chapel as affix. *See* main name, e.g. for **Chapel Allerton** (Leeds) *see* ALLERTON.

Chapel en le Frith Derbys. *Capella de le Frith* 1272. 'Chapel in the sparse woodland'. ME *chapele* + OE *fyrhth*, with retention of OFrench preposition and definite article.

Chapelizod (*Séipéal Iosóid*) Dublin. 'Iseult's chapel'.

Chapeltown Down. *Chapeltown* 1886. The village is named from the Roman Catholic chapel erected here in 1791. The equivalent Irish name is *Baile an tSéipéil*.

Chapeltown Sheff. *Le Chapel* 13th cent., *Chappeltown* 1707. '(Hamlet by) the chapel'. ME *chapel*.

Chapmanslade Wilts. *Chepmanesled* 1245. 'Valley of the merchants'. OE *ceap-mann* + *slæd*.

Chard, South Chard Somerset. *Cerdren* 1065, *Cerdre* 1086 (DB). Possibly 'house in rough ground'. OE *ceart* + *ærn*. Or perhaps from Brittonic **cerdīnen* 'rowan'.

Chardstock Devon. *Cerdestoche* 1086 (DB). 'Secondary settlement belonging to CHARD'. OE *stoc*.

Charfield S. Glos. *Cirvelde* 1086 (DB). 'Open land by a bending road, or with a rough surface'. OE *cearr(e)* or *ceart* + *feld*.

Charford, North Hants. *Cerdicesford* late 9th cent., *Cerdeford* 1086 (DB). 'Ford associated with the chieftain called Cerdic'. OE pers. name + *ford*.

Charing Kent. *Ciorrincg* 799, *Cheringes* 1086 (DB). Probably OE **cerring* 'a bend in a road'. Alternatively 'place associated with a man called Ceorra', OE pers. name + *-ing*.

Charing Cross Gtr. London. *Cyrringe* c.1000, *La Charryngcros* 1360. From OE **cerring* (*see* previous name), with reference either to the bend in the River Thames here or (more probably) to the well-marked bend in the old main road from the City of London to the West. *Cross* refers to the 'Eleanor Cross' set up here in 1290 by Edward I in memory of his queen, *see* WALTHAM CROSS.

Charingworth Glos. *Chevringavrde* 1086 (DB). Probably 'enclosure of the family or followers of a man called *Ceafor'. OE pers. name + *-inga-* + *worth*.

Charlbury Oxon. *Ceorlingburh* c.1000. 'Fortified place associated with a man called Ceorl'. OE pers. name + *-ing-* + *burh* (dative *byrig*).

Charlcombe B. & NE. Som. *Cerlecume* 1086 (DB). 'Valley of the freemen or peasants'. OE *ceorl* + *cumb*.

Charlecote Warwicks. *Cerlecote* 1086 (DB). 'Cottage(s) of the freemen or peasants'. OE *ceorl* + *cot*.

Charlemont Armagh. *Achadh in dá Charad* c.1645. The village is named from *Charles* Blount, 8th baron Mountjoy, for whom a fort was erected here in 1602. The original Irish name was *Achadh an dá Chora*, 'field of the two weirs'.

Charles Devon. *Carmes* [sic] 1086 (DB), *Charles* 1244. Possibly 'rock-court'. Cornish *carn* + **lys*.

Charlestown Armagh. The village is named after *Charles* Brownlow, who built houses here in c.1830.

Charlestown Cornwall. first so recorded from 1800, a small china-clay port named after its sponsor *Charles* Rashleigh, a local industrialist.

Charlesworth Derbys. *Cheuenwrde* [sic] 1086 (DB), *Chauelisworth* 1286. Probably 'enclosure of a man called *Ceafl', OE pers. name + *worth*. Alternatively the first element may be OE *ceafl* 'jaw' here used in the sense 'ravine'.

Charleton Devon. *Cherletone* 1086 (DB). 'Farmstead of the freemen or peasants'. OE *ceorl* + *tūn*.

Charlinch Somerset. *Cerdeslinc* 1086 (DB). Possibly 'ridge of a man called Cēolrēd'. OE pers. name + *hlinc*.

Charlton, a common place name, usually 'farmstead of the freemen or peasants', from OE *ceorl* (genitive plural *-a*) + *tūn*; examples include: **Charlton** Gtr. London. *Cerletone* 1086 (DB). **Charlton** Hants. *Cherleton* 1192. **Charlton** Wilts., near Malmesbury. *Ceorlatunæ* 10th cent., *Cerletone* 1086 (DB). **Charlton Abbots** Glos. *Cerletone* 1086 (DB), *Charleton Abbatis* 1535. Affix from its possession by Winchcomb Abbey. **Charlton Adam** Somerset. *Cerletune* 1086 (DB), *Cherleton Adam* 13th cent. Manorial addition from the *fitz Adam* family, here in the 13th cent. **Charlton Horethorne** Somerset. *Ceorlatun* c.950. Affix from the old Hundred of Horethorne, meaning 'grey thorn-bush' from OE *hār* + *thyrne*. **Charlton Kings** Glos. *Cherletone* 1160, *Kynges Cherleton* 1245. Affix 'Kings' because it was ancient demesne of the Crown. **Charlton Mackrell** Somerset. *Cerletune* 1086 (DB), *Cherletun Makerel* 1243. Manorial affix from the *Makerel* family. **Charlton Marshall** Dorset. *Cerletone* 1086 (DB), *Cherleton Marescal* 1314. Manorial addition from the *Marshall* family, here in the 13th cent. **Charlton Musgrove** Somerset. *Cerletone* 1086 (DB), *Cherleton Mucegros* 1225. Manorial addition from the *Mucegros* family, here in the 13th cent. **Charlton, North & Charlton, South** Northum. *Charleton del North, Charleton del Suth* 1242. **Charlton on Otmoor** Oxon. *Cerlentone* 1086 (DB), *Cherleton upon Ottemour* 1314. Affix from nearby Ot Moor, 'marshy ground of a man called *Otta', from OE pers. name + *mōr*. **Charlton, Queen** B. & NE. Som. *Cherleton* 1291. Affix because given to Queen Catherine Parr by Henry VIII.

However the following has a different origin:
Charlton Surrey. *Cerdentone* 1086 (DB).
Probably 'estate associated with a man called
Cēolrēd'. OE pers. name + *-ing-* + *tūn*.

Charlwood Surrey. *Cherlewde* 12th cent.
'Wood of the freemen or peasants'. OE *ceorl* +
wudu.

Charminster Dorset. *Cerminstre* 1086 (DB).
'Church on the River Cerne'. Celtic river-name
(*see* CERNE) + OE *mynster*.

Charmouth Dorset. *Cernemude* 1086 (DB).
'Mouth of the River Char'. Celtic river-name
(identical in origin with CERNE) + OE *mūtha*.

Charndon Bucks. *Credendone* [sic] 1086 (DB),
Charendone 1227. Probably 'hill called *Carn*'.
Celtic hill-name (from **carn* 'cairn, heap of
stones') + OE *dūn*.

Charney Bassett Oxon. *Ceornei* 821, *Cernei*
1086 (DB). 'Island on a river called *Cern*'. Lost
Celtic river-name (identical in origin with
CERNE) + OE *ēg*, with manorial addition from a
family called *Bass(es)*.

Charnock Richard Lancs. *Chernoch* 1194,
Chernok Richard 1288. Possibly a derivative of
Celtic **carn* 'cairn, heap of stones' + manorial
addition from a certain *Richard* here in the 13th
cent.

Charnwood Forest Leics. *Cernewoda* 1129.
'Wood called *Charn* ('rocky area'). Celtic **carn*
'heap of stones' + OE *wudu*.

Charsfield Suffolk. *Ceresfelda* 1086 (DB). OE
feld 'tract of open country', possibly with a
Celtic river-name *Char* as first element.

Chart, Great & Chart, Little Kent. *Cert*
762, *Certh*, *Litelcert* 1086 (DB), *Magna Chert*
13th cent. OE *cert* 'rough ground'.

Chart Sutton Kent. *Cært* 814, *Certh* 1086
(DB), *Chert juxta Suthon* 1280. OE *cert* 'rough
ground'. Affix from nearby SUTTON VALENCE.

Charterhouse Somerset. *Chartuse* 1243.
OFrench *chartrouse* 'a house of Carthusian
monks'.

Chartham Kent. *Certham* c.871, *Certeham*
1086 (DB). 'Homestead in rough ground'. OE
cert + *hām*.

Chartridge Bucks. *Charderuge* 1191–4.
Possibly 'ridge of a man called *Cearda'. OE
pers. name + *hrycg*.

Charwelton Northants. *Cerweltone* 1086
(DB). 'Farmstead on the River Cherwell'.
River-name (probably 'winding stream' from
OE **cearr(e)* + *wella*) + OE *tūn*.

Chastleton Oxon. *Ceastelton* 777, *Cestitone*
1086 (DB). 'Farmstead by the ruined prehistoric
camp'. OE *ceastel* + *tūn*.

Chatburn Lancs. *Chatteburn* 1242. 'Stream of
a man called *Ceatta'. OE pers. name + *burna*.

Chatcull Staffs. *Ceteruille* [sic] 1086 (DB),
Chatculne 1199. 'Kiln of a man called *Ceatta'.
OE pers. name + *cyln*.

Chatham, 'homestead or village in or by the
wood', Celtic **cęd* + OE *hām*: **Chatham**
Medway. *Cetham* 880, *Ceteham* 1086 (DB).
Chatham Green Essex. *Cetham* 1086 (DB).

Chatsworth Derbys. *Chetesuorde* 1086 (DB),
Chattesworth 1276. 'Enclosed settlement of a
man called *Ceatt'. OE pers. name + *worth*.

Chattenden Medway. *Chatendune* c.1100.
Possibly 'hill of a man called *Ceatta'. OE pers.
name (genitive *-n*) + *dūn*.

Chatteris Cambs. *Cæateric* 974, *Cietriz* 1086
(DB). Probably 'raised strip or ridge of a man
called *Ceatta'. OE pers. name + **ric*. However
the first element may be Celtic **cęd* 'wood'.

Chattisham Suffolk. *Cetessam* 1086 (DB).
Probably 'homestead or enclosure of a man
called *Ceatt'. OE pers. name + *hām* or *hamm*.

Chatton Northum. *Chetton* 1178. 'Farmstead
of a man called *Ceatta'. OE pers. name + *tūn*.

Chatwell, Great Staffs. *Chattewell* 1203.
'Well or spring of a man called *Ceatta'. OE pers.
name + *wella*.

Chawleigh Devon. *Calvelie* 1086 (DB).
'Clearing where calves are pastured'. OE *cealf* +
lēah.

Chawston Beds. *Calnestorne* [sic] 1086 (DB),
Caluesterne 1167. 'Thorn-tree where calves
graze, or thorn-tree of a man called *Cealf'. OE
cealf or OE pers. name (genitive *-es*) + *thyrne*.

Chawton Hants. *Celtone* 1086 (DB).
'Farmstead where calves are reared', or
'farmstead on chalk'. OE *cealf* or *cealc* + *tūn*.

Cheadle, from Celtic **cęd* 'wood' to which an
explanatory OE *lēah* 'wood' has been added:
Cheadle Staffs. *Celle* 1086 (DB). **Cheadle**
Stockp. *Cedde* 1086 (DB), *Chedle* c.1165.

Cheadle Hulme Stockp. *Hulm* 12th cent. 'Water-meadow belonging to CHEADLE'. OScand. *holmr*.

Cheam Gtr. London. *Cegham* 967, *Ceiham* 1086 (DB). Probably 'homestead or village by the tree-stumps'. OE **ceg* + *hām*.

Chearsley Bucks. *Cerdeslai* 1086 (DB). 'Wood or clearing of a man called Cēolrēd'. OE pers. name + *lēah*.

Chebsey Staffs. *Cebbesio* 1086 (DB). 'Island, or dry ground in marsh, of a man called **Cebbi*'. OE pers. name + *ēg*.

Checkendon Oxon. *Cecadene* 1086 (DB). 'Valley of a man called **Cæcca* or **Ceacca*', or 'valley by the hill'. OE pers. name or OE **cæcce* or **ceacce* (genitive *-an*) + *denu*.

Checkley Ches. *Chackileg* 1252. 'Wood or clearing of a man called **Ceaddica*'. OE pers. name + *lēah*.

Checkley Herefs. *Chakkeleya* 1195. 'Wood or clearing of a man called **Ceacca*', or 'wood or clearing on or near a hill'. OE pers. name or OE **ceacce* + *lēah*.

Checkley Staffs. *Cedla* [*sic*] 1086 (DB), *Checkeleg* 1196. Identical in origin with the previous name.

Chedburgh Suffolk. *Cedeberia* 1086 (DB). 'Hill of a man called Cedda'. OE pers. name + *beorg*.

Cheddar Somerset. *Ceodre c.*880, *Cedre* 1086 (DB). Probably OE **cēodor* 'ravine', with reference to Cheddar Gorge.

Cheddington Bucks. *Cete(n)done* 1086 (DB). 'Hill of a man called **Cetta*'. OE pers. name (genitive *-n*) + *dūn*.

Cheddleton Staffs. *Celtetone* 1086 (DB), *Chetilton* 1201. 'Farmstead in a valley'. OE *cetel* + *tūn*.

Cheddon Fitzpaine Somerset. *Succedene* [*sic*] 1086 (DB), *Chedene* 1182. OE *denu* 'valley', possibly with Celtic **cēd* 'wood'. Manorial addition from the *Fitzpaine* family, here in the 13th cent.

Chedgrave Norfolk. *Scatagraua* 1086 (DB), *Chategrave* 1165–70. Probably 'pit of a man called Ceatta'. OE pers. name + *græf*. The river-name **Chet** is a back-formation from the place name.

Chedington Dorset. *Chedinton* 1194, *Cedindun* 1226. 'Estate associated with a man called Cedd or Cedda'. OE pers. name + *-ing-* + *tūn* (often alternating with *dūn* 'hill' in early spellings).

Chediston Suffolk. *Cedestan* 1086 (DB). 'Stone of a man called Cedd'. OE pers. name + *stān*.

Chedworth Glos. *Ceddanwryde* [*sic*] 862, *Cedeorde* 1086 (DB). 'Enclosure of a man called Cedda'. OE pers. name + *worth*.

Chedzoy Somerset. *Chedesie* 729. 'Island, or dry ground in marsh, of a man called Cedd'. OE pers. name + *ēg*.

Cheekpoint (*Pointe na Síge*) Waterford. 'Streak point'.

Cheetham Hill Manch. *Cheteham* late 12th cent. 'Homestead or village by the wood called *Chet*'. Celtic **cēd* 'forest' + OE *hām*.

Cheetwood Salfd. *Chetewode* 1489. Near CHEETHAM; an explanatory *wudu* 'wood' has been added to the same first element.

Chelborough Dorset. *Celberge* 1086 (DB). Possibly 'hill of a man called Cēola'. OE pers. name + *beorg*. Alternatively the first element may be OE *ceole* 'throat, gorge' or *cealc* 'chalk' (referring to chalk cap of nearby Castle Hill).

Cheldon Barton Devon. *Chadeledona* 1086 (DB). 'Hill of a man called **Ceadela*'. OE pers. name + *dūn*. Barton is from OE *bere-tūn* 'barley-farm, demesne farm'.

Chelford Ches. *Celeford* 1086 (DB). Probably 'ford of a man called Cēola'. OE pers. name + *ford*. Alternatively the first element could be OE *ceole* 'throat', used in a topographical sense 'channel, gorge'.

Chell Heath Staffs. *Chelle* 1227. Probably 'wood of a man called Cēola'. OE pers. name + *lēah*.

Chellaston Derby. *Celerdestune* 1086 (DB). 'Farmstead of a man called Cēolheard'. OE pers. name + *tūn*.

Chellington Beds. *Chelewentone* 1219. 'Farmstead of a woman called Cēolwynn'. OE pers. name + *tūn*.

Chelmarsh Shrops. *Celmeres* 1086 (DB). 'Marsh marked out with posts or poles'. OE **cegel* + *mersc*.

Chelmondiston Suffolk. *Chelmundeston* 1174. 'Farmstead of a man called Cēolmund'. OE pers. name + *tūn*.

Chelmorton Derbys. *Chelmerdon(e)* 12th cent. Probably 'hill of a man called Cēolmǣr'. OE pers. name + *dūn*.

Chelmsford Essex. *Celmeresfort* 1086 (DB). 'Ford of a man called Cēolmǣr'. OE pers. name + *ford*. The river-name **Chelmer** is a back-formation from the place name.

Chelsea Gtr. London. *Caelichyth* 767, *Celchyth* 789, *Chelched* 1086 (DB). Probably 'landing-place for chalk or limestone'. OE *cealc*, *c(i)elce* + *hȳth*. But earliest spelling suggests OE *cælic* 'cup, chalice' (perhaps used in some transferred topographical sense).

Chelsfield Gtr. London. *Cillesfelle* 1086 (DB), *Chilesfeld* 1087. 'Open land of a man called Cēol'. OE pers. name + *feld*.

Chelsworth Suffolk. *Ceorleswyrthe* 962, *Cerleswrda* 1086 (DB). 'Enclosure of the freeman or of a man called Ceorl'. OE *ceorl* or OE pers. name + *worth*.

Cheltenham Glos. *Celtanhomme* 803, *Chintenehā* [sic] 1086 (DB). Probably 'enclosure or river-meadow by a hill-slope called *Celte*'. OE or pre-English hill-name + *hamm*. Alternatively the first element may be an OE pers. name *Celta*. The river-name **Chelt** is a back-formation from the place name.

Chelveston Northants. *Celuestone* 1086 (DB). 'Farmstead of a man called Cēolwulf'. OE pers. name + *tūn*.

Chelvey N. Som. *Calviche* 1086 (DB). 'Farm where calves are reared'. OE *cealf* + *wīc*.

Chelwood B. & NE. Som. *Celeworde* 1086 (DB). 'Enclosure of a man called Cēola'. OE pers. name + *worth*.

Chenies Bucks. *Isenhamstede* 12th cent., *Ysenamstud Cheyne* 13th cent. Probably 'homestead of a man called *Īsa*'. OE pers. name + *hām-stede* + manorial addition (which is now used alone) from the *Cheyne* family, here in the 13th cent. Alternatively the first element may be an old river-name.

Chepstow (*Cas-gwent*) Mon. *Strigull* [sic] 1224, *Chepstowe* 1308, *Chapestowe* 1338. 'Market-place'. OE *cēap* + *stōw*. The Welsh name means 'castle in Gwent' (Welsh *cas*). See GWENT.

Cherhill Wilts. *Ciriel* 1155. Possibly a Celtic name, from Brittonic *cĭrch* 'haunt, resort' + adjectival suffix *-(j)ol*.

Cherington, probably 'village with a church', OE *cirice* + *tūn*: **Cherington** Glos. *Cerintone* 1086 (DB). **Cherington** Warwicks. *Chiriton* 1199.

Cheriton, 'village with a church', OE *cirice* + *tūn*: **Cheriton** Devon. *Ciretone* 1086 (DB). **Cheriton** Hants. *Cherinton* 1167. **Cheriton** Kent. *Ciricetun c*.1090. **Cheriton, North & Cheriton, South** Somerset. *Ciretona*, *Cherintone* 1086 (DB). **Cheriton Bishop** Devon. *Ceritone* 1086 (DB), *Bishops Churyton* 1370. Affix from the Bishop of Exeter, granted land here in the 13th cent. **Cheriton Fitzpaine** Devon. *Cerintone* 1086 (DB), *Cheriton Fitz Payn* 1335. Manorial addition from the *Fitzpayn* family, here in the 13th cent.

Cheriton Pemb. *Cheriton* 1813. 'Village with a church'. OE *cirice* + *tūn*. Cheriton is also known as *Stackpole Elidor*, 'Elidir's (estate by the) pool near the rock' (OScand. *stakkr* + *pollr*).

Cherrington Tel. & Wrek. *Cerlintone* 1086 (DB), *Cherington* 1230. Possibly 'estate associated with a man called Ceorra', OE pers. name + *-ing-* + *tūn*. Alternatively 'settlement by a river-bend', from OE *cerring* + *tūn*.

Cherry Burton E. R. Yorks. *See* BURTON.

Cherry Hinton Cambs. *See* HINTON.

Cherry Willingham Lincs. *See* WILLINGHAM.

Chertsey Surrey. *Cerotaesei* 731, *Certesy* 1086 (DB). 'Island of a man called *Cerot*'. Celtic pers. name + OE *ēg*.

Cherwell (river) Northants.–Oxon. *See* CHARWELTON.

Cheselbourne Dorset. *Chiselburne* 869, *Ceseburne* 1086 (DB). 'Gravel stream'. OE *cisel* + *burna*.

Chesham, Chesham Bois Bucks. *Cæstæleshamme* 1012, *Cestreham* 1086 (DB), *Chesham Boys* 1339. 'River-meadow by a heap of stones'. OE *ceastel* + *hamm*, with manorial affix from the *de Bois* family, here in the 13th cent. The river-name **Chess** is a back-formation from the place name.

Cheshire (the county). *Cestre Scire* 1086 (DB). 'Province of the city of CHESTER'. OE *scīr* 'district'.

Cheshunt Herts. *Cestrehunt* 1086 (DB). Probably 'spring by the old (Roman) fort'. OE *ceaster* + **funta*.

Chesil Beach Dorset. *Chisille bank c.*1540. From OE *cisel* 'shingle'. It gives name to the village of **Chesil**, *Chesill* 1608.

Cheslyn Hay Staffs. *Haya de Chistlin* 1236. Probably 'coffin ridge', i.e. 'ridge where a coffin was found'. OE *cest* + *hlinc*, with *hæg* 'enclosure'.

Chessington Gtr. London. *Cisendone* 1086 (DB). 'Hill of a man called Cissa'. OE pers. name (genitive *-n*) + *dūn*.

Chester Ches. *Deoua c.*150, *Legacæstir* 735, *Cestre* 1086 (DB). OE *ceaster* 'Roman town or city'. Originally called *Deoua* from its situation on the River DEE, later *Legacæstir* meaning 'city of the legions'.

Chester le Street Durham. *Ceastre c.*1104, *Cestria in Strata* 1406. 'Roman fort on the Roman road'. OE *ceaster* + *strǣt* (Latin *strata*). The French definite article *le* remains after loss of the preposition.

Chesterblade Somerset. *Cesterbled* 1065. A difficult name, but perhaps from Brittonic **Kastr Bleith* 'stronghold of the wolf (possibly used as a pers. name)'.

Chesterfield Derbys. *Cesterfelda* 955, *Cestrefeld* 1086 (DB). 'Open land near a Roman fort or settlement'. OE *ceaster* + *feld*.

Chesterford, Great & Chesterford, Little Essex. *Ceasterford* 1004, *Cestreforda* 1086 (DB). 'Ford by a Roman fort'. OE *ceaster* + *ford*.

Chesterton, 'farmstead or village by a Roman fort or town', OE *ceaster* + *tūn*: **Chesterton** Cambs. *Cestretone* 1086 (DB). **Chesterton** Oxon. *Cestertune* 1005, *Cestretone* 1086 (DB). **Chesterton** Staffs. *Cestreton* 1214. **Chesterton** Warwicks. *Cestretune* 1043, *Cestretone* 1086 (DB).

Cheswardine Shrops. *Ciseworde* 1086 (DB), *Chesewordin* 1160. Probably 'enclosed settlement where cheese is made'. OE *cēse* + *worthign*.

Cheswick Northum. *Chesewic* 1208–10. 'Farm where cheese is made'. OE *cēse* + *wīc*.

Chetnole Dorset. *Chetenoll* 1242. 'Hill-top or hillock of a man called **Ceatta*'. OE pers. name + *cnoll*.

Chettiscombe Devon. *Chetelescome* 1086 (DB). Probably OE *cetel* 'deep valley' + explanatory OE *cumb* 'valley'.

Chettisham Cambs. *Chetesham c.*1170. Probably 'homestead or enclosure of the wood called *Chet*'. Celtic **cēd* + OE *ham* or *hamm*. Or the first element may be an OE masculine pers. name **Ceatt*.

Chettle Dorset. *Ceotel* 1086 (DB). Probably OE **ceotol* 'deep valley'.

Chetton Shrops. *Catinton* 1086 (DB), *Chetintone* 1210–12. Probably 'estate associated with a man called **Ceatta*'. OE pers. name + *-ing-* + *tūn*.

Chetwode Bucks. *Cetwuda* 949, *Ceteode* 1086 (DB). Celtic **cēd* 'wood' to which has been added an explanatory OE *wudu* 'wood'.

Chetwynd Tel. & Wrek. *Catewinde* 1086 (DB). Probably 'winding ascent of a man called Ceatta'. OE pers. name + *(ge)wind*.

Chetwynd Aston Tel. & Wrek. *See* ASTON.

Cheveley Cambs. *Cæafle c.*1000, *Chauelai* 1086 (DB). OE *lēah* 'wood' with either OE *ceaf* 'chaff' (perhaps also 'fallen twigs') or an OE **ceaf* 'chaffinch'.

Chevening Kent. *Chivening* 1199. Possibly '(settlement of) the dwellers at the ridge'. Celtic **ceˇvn* + OE *-ingas*.

Cheverell, Great & Cheverell, Little Wilts. *Chevrel* 1086 (DB). Probably a Celtic name meaning 'small piece of joint tillage', from the ancestor of Middle Welsh *kyfa(i)r* + diminutive suffix *-ell*.

Chevington, West Northum. *Chiuingtona* 1236. 'Estate associated with a man called **Cifa*'. OE pers. name + *-ing-* + *tūn*.

Cheviot (Hills) Northum. –Sc. Bord. *Chiuet* 1181. Meaning uncertain. The pre-English name is that of the single mountain here called *The Cheviot*.

Chevithorne Devon. *Cheuetorna* 1086 (DB). 'Thorn-tree of a man called Ceofa'. OE pers. name + *thorn*.

Chew Magna B. & NE. Som. *Ciw* 1065, *Chiwe* 1086 (DB). Named from the River Chew, which is a Celtic river-name, with affix from Latin *magna* 'great'.

Chew Stoke B. & NE. Som. *Stoche* 1086 (DB). 'Secondary settlement belonging to CHEW', from OE *stoc.*

Chewton Mendip Somerset. *Ciwtun c.*880, *Ciwetune* 1086 (DB), *Cheuton by Menedep* 1313. 'Estate on the River Chew'. Celtic river-name (*see* CHEW) + OE *tūn* + affix from the MENDIP HILLS.

Chicheley Milt. K. *Cicelai* 1086 (DB). 'Wood or clearing of a man called *Cicca'. OE pers. name + *lēah.*

Chichester W. Sussex. *Cisseceastre* 895, *Cicestre* 1086 (DB). Probably 'Roman town of a chieftain called Cissa', OE pers. name + *ceaster.* Alternatively the first element could be an OE word *cisse* (genitive -*an*) 'a gravelly feature'.

Chickerell Dorset. *Cicherelle* 1086 (DB). This name remains obscure in origin and meaning.

Chicklade Wilts. *Cytlid c.*912. Possibly 'gate or slope by the wood'. Celtic *cẹd* + OE *hlid* or *hlid.*

Chidden Hants. *Cittandene* 956. Possibly 'valley of a man called *Citta'. OE pers. name + *denu.* However the forms *cittanware* 956, *citwara* 958–75 denoting 'dwellers at Chidden' may suggest that the first element is rather Celtic *cẹd* 'wood'.

Chiddingfold Surrey. *Chedelingefelt* 1130, *Chidingefaud* 12th cent. Possibly 'fold of the family or followers of a man called *Cēodel or *Cid(d)el'. OE pers. name + -*inga*- + *fald.*

Chiddingly E. Sussex. *Cetelingei* [*sic*] 1086 (DB), *Chitingeleghe c.*1230. Probably 'wood or clearing of the family or followers of a man called *Citta'. OE pers. name + -*inga*- + *lēah.*

Chiddingstone Kent. *Cidingstane c.*1110. Possibly 'stone associated with a man called *Cidd or Cidda'. OE pers. name + -*ing*- + *stān.*

Chideock Dorset. *Cidihoc* 1086 (DB). 'Wooded place', from a derivative *cẹdiǫg 'wooded' of Celtic *cẹd* 'wood'. The river-name **Chid** is a back-formation from the place name.

Chidham W. Sussex. *Chedeham* 1193. Probably 'homestead or peninsula near the bay'. OE *cēod*(*e*) + *hām* or *hamm.*

Chieveley W. Berks. *Cifanlea* 951, *Civelei* 1086 (DB). 'Wood or clearing of a man called *Cifa'. OE pers. name + *lēah.*

Chignall St James & Chignall Smealy Essex. *Cingehala* [*sic*] 1086 (DB), *Chikenhale Iacob* 1254, *Chigehale Smetheleye* 1279. Possibly 'nook of land of a man called *Cicca'. OE pers. name (genitive -*n*) + *halh.* Alternatively the first element may be OE *cīcen* 'chicken'. Affixes from the dedication of the church to St James (Latin *Iacobus*), and from a nearby place meaning 'smooth clearing' from OE *smēthe* + *lēah.*

Chigwell Essex. *Cingheuuella* 1086 (DB), *Chiggewell* 1187. Possibly 'spring or stream of a man called *Cicca'. OE pers. name + *wella.*

Chilbolton Hants. *Ceolboldingtun* 909, *Cilbode*(*n*)*tune* 1086 (DB). 'Estate associated with a man called Cēolbeald'. OE pers. name + -*ing*- + *tūn.*

Chilcombe Dorset. *Ciltecombe* 1086 (DB). Possibly 'valley at a hill-slope called *Cilte'. OE or pre-English hill-name + OE *cumb.*

Chilcompton Somerset. *Comtuna* 1086 (DB), *Childecumpton* 1227. 'Valley farmstead (or village) of the young (noble)men'. OE *cild* + *cumb* + *tūn.*

Chilcote Leics. *Cildecote* 1086 (DB). 'Cottage(s) of the young (noble)men'. OE *cild* + *cot.*

Child Okeford Dorset. *See* OKEFORD.

Childer Thornton Ches. *See* THORNTON.

Childrey Oxon. *Cillarthe* 950, *Celrea* 1086 (DB). Named from Childrey Brook which probably means 'stream of a man called *Cilla or of a woman called Cille'. OE pers. name + *rīth.* Alternatively the first element may be an OE *cille* 'a spring'.

Child's Ercall Shrops. *See* ERCALL.

Childswickham Worcs. *Childeswicwon* 706, *Wicvene* 1086 (DB). Possibly Celtic *wīg 'wood' + *waun* 'marsh, moor, upland pasture', with OE *cild* 'young nobleman'.

Childwall Lpool. *Cildeuuelle* 1086 (DB). 'Spring or stream where young people assemble'. OE *cild* + *wella.*

Chilfrome Dorset. *Frome* 1086 (DB), *Childefrome* 1206. 'Estate on the River Frome belonging to the young (noble)men'. Celtic river-name (*see* FROME) with OE *cild*.

Chilgrove W. Sussex. *Chelegrave* 1200. 'Grove in a gorge or gulley', or 'grove of a man called Cēola'. OE *ceole* or OE pers. name + *grāf*.

Chilham Kent. *Cilleham* 1032, 1086 (DB). 'Homestead or village of a man called *Cilla or of a woman called Cille'. OE pers. name + *hām*. Alternatively the first element may be an OE *cille 'a spring'.

Chillenden Kent. *Ciollandene* c.833, *Cilledene* 1086 (DB). 'Valley of a man called Ciolla'. OE pers. name (genitive -*n*) + *denu*.

Chillerton I. of Wight. *Celertune* 1086 (DB). 'Farmstead of a man called Cēolheard', or 'enclosed farmstead in a valley'. OE pers. name + *tūn*, or OE *ceole* + *geard* + *tūn*.

Chillesford Suffolk. *Cesefortda* [*sic*] 1086 (DB), *Chiselford* 1211. 'Gravel ford'. OE *ceosol* + *ford*.

Chillingham Northum. *Cheulingeham* 1187. 'Homestead or village of the family or followers of a man called *Ceofel'. OE pers. name + -*inga-* + *hām*.

Chillington Devon. *Cedelintone* 1086 (DB). 'Estate associated with a man called *Ceadela'. OE pers. name + -*ing-* + *tūn*.

Chillington Somerset. *Cheleton* 1261. 'Farmstead of a man called Cēola'. OE pers. name + *tūn*.

Chilmark Wilts. *Cigelmerc* 984, *Chilmerc* 1086 (DB). 'Boundary made with poles or posts'. OE *cigel* + *mearc*.

Chilson Oxon. *Cildestuna* c.1200. 'Estate of the young (noble)man, or of a man called Cild'. OE *cild* or pers. name + *tūn*.

Chilsworthy Devon. *Chelesworde* 1086 (DB). 'Enclosure of a man called Cēol'. OE pers. name + *worth*.

Chiltern Hills Bucks. *Ciltern* 1009. Probably identical in origin with the following name, here used of a district.

Chilthorne Domer Somerset. *Cilterne* 1086 (DB), *Chilterne Dunmere* 1280. Possibly a derivative of an OE or pre-English word *celte or *cilte which may have meant 'hill-slope'.

Manorial addition from the *Dummere* family, here in the 13th cent.

Chiltington, East E. Sussex. *Childetune* 1086 (DB). Probably identical with the following name.

Chiltington, West W. Sussex. *Cillingtun* 969, *Cilletone* 1086 (DB). Possibly 'farmstead at a hill-slope called *Cilte'. OE or pre-English hill-name + -*ing* or -*ing-* + *tūn*. Alternatively the first element may be an ancient district-name *Ciltine* derived from the same word.

Chilton, a place name found in various counties, usually 'farm of the young (noble)men', from OE *cild* + *tūn*; for example: **Chilton** Bucks. *Ciltone* 1086 (DB). **Chilton** Durham. *Ciltona* 1091. **Chilton** Oxon. *Cylda tun* c.895, *Cilletone* 1086 (DB). **Chilton Cantelo** Somerset. *Childeton* 1201, *Chiltone Cauntilo* 1361. Manorial addition from the Cantelu family, here in the 13th cent. **Chilton Foliat** Wilts. *Cilletone* 1086 (DB), *Chilton Foliot* 1221. Manorial addition from the *Foliot* family, here in the 13th cent. **Chilton Street** Suffolk. *Chilton* 1254. Affix *street* has the sense 'hamlet, straggling village'. **Chilton Trinity** Somerset. *Cildetone* 1086 (DB), *Chilton Sancte Trinitatis* 1431. Affix from the dedication of the church.

However the following Chilton has a different origin: **Chilton Polden** Somerset. *Ceptone* 1086 (DB), *Chauton* 1303. Possibly 'farmstead on chalk or limestone', from OE *cealc* + *tūn*, with affix from the nearby Polden Hills (*Poeldune* 725, from OE *dūn* 'hill' added to a Celtic name *Bouelt* 705 which probably means 'cow pasture').

Chilvers Coton Warwicks. *See* COTON.

Chilwell Notts. *Chideuuelle* 1086 (DB). 'Spring or stream where young people assemble'. OE *cild* + *wella*.

Chilworth, 'enclosure of a man called Cēola'. OE pers. name + *worth*: **Chilworth** Hants. *Celeorde* 1086 (DB). **Chilworth** Surrey. *Celeorde* 1086 (DB).

Chimney Oxon. *Ceommanyg* 1069. 'Island, or dry ground in marsh, of a man called *Ceomma'. OE pers. name (genitive -*n*) + *ēg*.

Chineham Hants. *Chineham* 1086 (DB). 'Homestead or enclosure in a deep valley'. OE *cinu* + *hām* or *hamm*.

Chingford Gtr. London. *Cingefort* [*sic*] 1086 (DB), *Chingelford* 1242. 'Shingle ford'. OE **cingel* + *ford*.

Chinley Derbys. *Chynleye* 1285. 'Wood or clearing in a deep valley'. OE *cinu* + *lēah*.

Chinnock, East & Chinnock, West Somerset. *Cinnuc c.*950, *Cinioch* 1086 (DB). Probably named from Chinnock Brook, a Celtic river-name meaning 'hound stream'.

Chinnor Oxon. *Chennore* 1086 (DB). 'Slope of a man called **Ceonna*'. OE pers. name + *ōra*.

Chipnall Shrops. *Ceppacanole* 1086 (DB), *Chippeknol c.*1250. 'Knoll of a man called **Cippa*', or 'knoll where logs are got'. OE pers. name or OE *cipp* (genitive plural *cippa*) + *cnoll*.

Chippenham, probably 'river-meadow of a man called **Cippa*', OE pers. name (genitive -*n*) + *hamm*: **Chippenham** Cambs. *Chipeham* 1086 (DB). **Chippenham** Wilts. *Cippanhamme c.*900, *Chipeham* 1086 (DB).

Chipperfield Herts. *Chiperfeld* 1375. Probably 'open land where traders or merchants meet'. OE *cēapere* + *feld*.

Chipping Lancs. *Chippin* 1203. OE *cēping* 'a market, a market-place'.

Chipping as affix. *See* main name, e.g. for **Chipping Barnet** (Gtr. London) *see* BARNET.

Chipshop Devon. Recorded thus from 1765, possibly 'shed or workshop where rods, beams or pegs are made'. ME *chippe* + *shoppe*.

Chipstead, 'market-place', OE *cēap-stede*: **Chipstead** Kent. *Chepsteda* 1191. **Chipstead** Surrey. *Tepestede* [*sic*] 1086 (DB), *Chepstede* 1100–29.

Chirbury Shrops. *Cyricbyrig* mid-11th cent., *Cireberie* 1086 (DB). 'Fortified place or manor with or near a church'. OE *cirice* + *burh* (dative *byrig*).

Chirk (*Y Waun*) Wrex. *Chirk* 1295, *Cheyrk* 1309. '(Place on the) River Ceiriog'. The Celtic river-name may mean 'favoured one'. The Welsh name means 'the moorland' (Welsh *y* + *gwaun*).

Chirton Wilts. *Ceritone* 1086 (DB). 'Village with a church'. OE *cirice* + *tūn*.

Chisbury Wilts. *Cissanbyrig* early 10th cent., *Cheseberie* 1086 (DB). Possibly 'pre-English earthwork associated with a man called Cissa'.

OE pers. name + *burh* (dative *byrig*). Alternatively the first element could be an OE word **cisse* (genitive -*an*) 'a gravelly feature'.

Chiselborough Somerset. *Ceoselbergon* 1086 (DB). 'Gravel mound or hill'. OE *cisel* + *beorg*.

Chiseldon Swindn. *Cyseldene c.*880, *Chiseldene* 1086 (DB). 'Gravel valley'. OE *cisel* + *denu*.

Chisenbury Wilts. *Chesigeberie* 1086 (DB), *Chisingburi* 1202. Possibly 'stronghold of the dwellers on the gravel'. OE **cis* + -*inga*- + *burh* (dative *byrig*). Alternatively the first element could be an OE word **cising* (genitive plural -*a*) 'gravelly place'.

Chishill, Great & Chishill, Little Cambs. *Cishella* 1086 (DB). 'Gravel hill'. OE **cis* + *hyll*.

Chisholme Sc. Bord. *Chesehome* 1254. 'Cheese barn'. OE *cēse* + OScand. *helm*. The name implies a rich pasture.

Chislehampton Oxon. *Hentone* 1086 (DB), *Chiselentona* 1147. Originally 'high farm, farm situated on high land'. OE *hēah* (dative *hēan*) + *tūn*, with the later addition of OE *cisel* 'gravel'.

Chislehurst Gtr. London. *Cyselhyrst* 973. 'Gravelly wooded hill'. OE *cisel* + *hyrst*.

Chislet Kent. *Cistelet* 605 (13th-cent. copy), 1086 (DB). Possibly OE **cest, cist* 'a chest' (perhaps here 'a cistern') + (*ge*)*lǣt* 'a water conduit'. A hypocausted Roman building has been found here.

Chiswellgreen Herts. first recorded in 1782, but possibly 'the gravelly spring or stream', from OE **cis* + *wella*, with the later addition of *green*.

Chiswick Gtr. London. *Ceswican c.*1000. 'Specialized farm where cheese is made'. OE **cīese* + *wīc*.

Chisworth Derbys. *Chisewrde* 1086 (DB). Possibly 'enclosure of a man called Cissa'. OE pers. name + *worth*. Alternatively the first element could be an OE word **cisse* (genitive -*an*) 'a gravelly feature'.

Chithurst W. Sussex. *Titesherste* [*sic*] 1086 (DB), *Chyteherst* 1279. Possibly 'wooded hill of a man called **Citta*'. OE pers. name + *hyrst*. Alternatively the first element may be Celtic **cę̄d* 'wood'.

Chitterne Wilts. *Chetre* 1086 (DB), *Chytterne* 1268. Probably Celtic **cēd* 'wood' + **tre* 'farm' (later associated with OE *ærn* 'house').

Chittlehamholt Devon. *Chitelhamholt* 1288. 'Wood of the dwellers in the valley'. OE *cietel* + *hǣme* + *holt*.

Chittlehampton Devon. *Citremetona* 1086 (DB), *Chitelhamtone* 1176. 'Farmstead of the dwellers in the valley'. OE *cietel* + *hǣme* + *tūn*.

Chittoe Wilts. *Chetewe* 1167. Celtic **cēd* 'wood' + the ancestor of Welsh *tew* 'thick'. The order of elements is Celtic.

Chivelstone Devon. *Cheueletona* 1086 (DB). 'Farmstead of a man called *Ceofel'. OE pers. name + *tūn*.

Chobham Surrey. *Cebeham* 1086 (DB). 'Homestead or enclosure of a man called *Ceabba'. OE pers. name + *hām* or *hamm*.

Cholderton Wilts. *Celdretone* 1086 (DB). 'Estate associated with a man called Cēolhere or Cēolrēd'. OE pers. name + *-ing-* + *tūn*.

Chollerton Northum. *Choluerton c.*1175. Probably 'farmstead of a man called Cēolferth'. OE pers. name + *tūn*.

Cholmondeley Ches. *Calmundelei* 1086 (DB), *Chelmundeleia c.*1200. 'Woodland clearing of a man called Cēolmund'. OE pers. name + *lēah*.

Cholsey Oxon. *Ceolesig c.*895, *Celsei* 1086 (DB). 'Island, or dry ground in marsh, of a man called Cēol'. OE pers. name + *īeg*.

Cholstrey Herefs. *Cerlestreu* 1086 (DB). 'Tree of the freeman or of a man called Ceorl'. OE *ceorl* or OE pers. name + *trēow*.

Choppington Northum. *Cebbington c.*1050. 'Estate associated with a man called *Ceabba'. OE pers. name + *-ing-* + *tūn*.

Chopwell Gatesd. *Cheppwell c.*1155. Probably 'spring where trading takes place'. OE *cēap* + *wella*.

Chorley, 'clearing of the freemen or peasants', OE *ceorl* (genitive plural *-a*) + *lēah*: **Chorley** Ches., near Nantwich. *Cerlere* 1086 (DB). **Chorley** Lancs. *Cherleg* 1246. **Chorley** Staffs. *Cherlec* 1231.

Chorleywood Herts. *Cherle* 1278, *Charlewoode* 1524. 'Clearing of the freemen or peasants'. OE *ceorl* (genitive plural *-a*) + *lēah*, with the later addition of *wood*.

Chorlton, usually 'farmstead of the freemen or peasants', OE *ceorl* (genitive plural *-a*) + *tūn*: **Chorlton** Ches., near Nantwich. *Cerletune* 1086 (DB). **Chorlton Lane** Ches. *Cherlton* 1283. **Chorlton, Chapel** Staffs. *Cerletone* 1086 (DB).

However the following Chorlton has a different origin: **Chorlton cum Hardy** Manch. *Cholreton* 1243. 'Farmstead of a man called Cēolfrith'. OE pers. name + *tūn*. Now united with Hardy (*Hardey* 1555, probably 'island of a man called *Hearda', from an OE pers. name + *ēg*) as indicated by the Latin preposition *cum* 'with'.

Chowley Ches. *Celelea* 1086 (DB). 'Wood or clearing of a man called Cēola'. OE pers. name + *lēah*.

Chrishall Essex. *Cristeshala* 1086 (DB). 'Nook of land dedicated to Christ'. OE *Crist* + *halh*.

Christchurch Dorset. *Christecerce c.*1125. 'Church of Christ'. OE *Crist* + *cirice*. Earlier called *Twynham*, 'place between the rivers'. OE *betwēonan* + *ēa* (dative plural *ēam*).

Christchurch (*Eglwys y Drindod*) Newpt. *Christi Ecclesia* 1290, *Christeschurche* 1291. 'Church of Christ'. The Welsh name means 'church of the Trinity', a more recent church dedication.

Christian Malford Wilts. *Cristemaleford* 937, *Cristemeleford* 1086 (DB). 'Ford by a cross'. OE *cristel-mǣl* + *ford*.

Christleton Ches. *Cristetone* 1086 (DB), *Cristentune* 12th cent. 'Christian farmstead', or 'farmstead of the Christians'. OE *Cristen* (as adjective or noun) + *tūn*.

Christmas Common Oxon. hamlet recorded as *Christmas (Green)* in the early 18th cent., so called from *Christmas Coppice* 1617 which may have been a coppice where holly-trees (traditionally associated with Christmas) were abundant. The similar **Christmaspie** Surrey is so called from *Christmas Pie Farm* 1823, a whimsical name probably to be associated with the *Christmas* family mentioned in 16th-cent. records.

Christon N. Som. *Crucheston* 1197. 'Farmstead at the hill'. Celtic **crūg* + OE *tūn*.

Christow Devon. *Cristinestowe* 1244. 'Christian place'. OE *Cristen* + *stōw*.

Chudleigh Devon. *Ceddelegam* c.1150. 'Clearing of a man called Ciedda', or 'clearing in a hollow'. OE pers. name or OE *cēod(e)* + *lēah*.

Chulmleigh Devon. *C(h)almonleuga* 1086 (DB). 'Clearing of a man called Cēolmund'. OE pers. name + *lēah*.

Chunal Derbys. *Ceolhal* 1086 (DB). 'Nook of land of a man called Cēola', or 'nook of land near the ravine'. OE pers. name or OE *ceole* + *halh*.

Church Lancs. *Chirche* 1202. '(Place at) the church'. OE *cirice*.

Church as affix. *See* main name, e.g. for **Church Fenton** (N. Yorks.) *see* FENTON.

Church Hill Fermanagh. *Church-Hill* 1837. The village was founded in the early 17th cent. and named after a Church of Ireland church.

Church Knowle Dorset. *Cnolle* 1086 (DB), *Churchecnolle* 1346. OE *cnoll* 'hill-top' with the later addition of *cirice* 'church'.

Churcham Glos. *Hamme* 1086 (DB), *Churchehamme* 1200. OE *hamm* 'river-meadow' with the later addition of *cirice* 'church'.

Churchdown Glos. *Circesdune* 1086 (DB). Probably Celtic **crŭg* 'hill' with explanatory OE *dūn* 'hill', but the first element could perhaps be OE *cirice* 'church' with an analogical *-s* genitive.

Churchill, sometimes from Celtic **crŭg* 'hill' with explanatory OE *hyll* 'hill'; examples include: **Churchill** Devon, near Barnstaple. *Cercelle* 1086 (DB). **Churchill** Worcs., near Kidderminster. *Cercehalle* [sic] 1086 (DB), *Circhul* 11th cent.
    However other examples are probably simply 'hill with or near a church', OE *cirice* + *hyll*: **Churchill** Oxon. *Cercelle* 1086 (DB). **Churchill** N. Som. *Cherchille* 1201. **Churchill** Worcs., near Worcester. *Circehille* 1086 (DB).

Churchover Warwicks. *Wavre* 1086 (DB), *Chirchewavre* 12th cent. Originally a river-name (an earlier name for the River Swift) meaning 'winding stream' from OE *wæfre* 'wandering', with the later addition of *cirice* 'church'.

Churchstanton Somerset. *Stantone* 1086 (DB), *Cheristontone* 13th cent. 'Farmstead on stony ground'. OE *stān* + *tūn*, with later addition of OE *cirice* 'church'.

Churchstow Devon. *Churechestowe* 1242. 'Place with a church'. OE *cirice* + *stōw*.

Churn (river) Glos. *See* CERNEY.

Churt Surrey. *Cert* 685–7. OE *cert* 'rough ground'.

Churton Ches. *Churton* 12th cent. Possibly 'hill village'. Celtic **crŭg* + OE *tūn*.

Churwell Leeds. *Cherlewell* 1226. 'Spring or stream of the freemen or peasants'. OE *ceorl* + *wella*.

Cilgerran Pemb. *Kilgerran* 1165, *Chilgerran* 1166. 'Cerran's corner of land'. Welsh *cil* + Celtic pers. name.

Cilgeti. *See* KILGETTY.

Cinderford Glos. *Sinderford* 1258. 'Ford built up with cinders or slag (from iron-smelting)'. OE *sinder* + *ford*.

Cirencester Glos. *Korinion* c.150, *Cirenceaster* c.900, *Cirecestre* 1086 (DB). 'Roman camp or town called *Corinion*'. OE *ceaster* added to the reduced form of a Celtic name of uncertain origin and meaning.

Clabby (*Clabaigh*) Fermanagh. *Clabby* 1611. Possibly 'place of pock-marked land'.

Clackmannan Clac. *Clacmanan* 1147. 'Stone of Manau'. Welsh *clog*. Manau is the name of the district. The 'stone' is a glacial rock preserved in the middle of the town.

Clacton on Sea, Great Clacton & Little Clacton Essex. *Claccingtune* c.1000, *Clachintune* 1086 (DB). Probably 'estate associated with a man called *Clacc'. OE pers. name + *-ing-* + *tūn*.

Claddach Kirkibost W. Isles. (North Uist). *Kirkabol* 1654. 'Shore by the homestead of the church'. Gaelic *claddach* + OScand. *kirkja* + *bólstathr*.

Claddagh (*Cladach*) Galway, Kerry. 'Shore'.

Claddaghduff (*An Cladach Dubh*) Galway. 'The black shore'.

Clady (*Cladach*) Donegal, Tyrone. 'Shore'.

Claggan (*Cloigeann*) Donegal. 'Head'.

Claines Worcs. *Cleinesse* 11th cent. 'Clay headland'. OE *clǣg* + *næss*.

Clanabogan (*Cluain Uí Bhogáin*) Tyrone. *Clonyboggan* c.1655. 'Ó Bogáin's pasture'.

Clandeboy (*Clann Aodha Buí*) Down. *Cloinn Aedha buidhe* 1319. 'Clan of yellow Aodh'.

Clandon, East & Clandon, West Surrey. *Clanedun* 1086 (DB). 'Clean hill', i.e. 'hill free from weeds or other unwanted growth'. OE *clǣne* + *dūn*.

Clane (*Claonadh*) Kildare. 'Slanted ford'.

Clanfield, 'clean open land', i.e. 'open land free from weeds or other unwanted growth', OE *clǣne* + *feld*: **Clanfield** Hants. *Clanefeld* 1207. **Clanfield** Oxon. *Chenefelde* [*sic*] 1086 (DB), *Clenefeld* 1196.

Clanville Hants. *Clavesfelle* [*sic*] 1086 (DB), *Clanefeud* 1259. Identical in origin with the previous names.

Clapham, usually 'homestead or enclosure near a hill or hills', OE **clopp(a)* + *hām* or *hamm*: **Clapham** Beds. *Cloppham* 1060, *Clopeham* 1086 (DB). **Clapham** Gtr. London. *Cloppaham* c.880, *Clopeham* 1086 (DB). **Clapham** W. Sussex. *Clopeham* 1086 (DB).
However the following Clapham has a different origin: **Clapham** N. Yorks. *Clapeham* 1086 (DB). 'Homestead or enclosure by the noisy stream'. OE **clæpe* + *hām* or *hamm*.

Clappersgate Cumbria. *Clappergate* 1588. 'Road or gate by the rough bridge'. ME *clapper* + OScand. *gata* or OE *geat*.

Clapton, 'farmstead or village near a hill or hills', OE **clopp(a)* + *tūn*; examples include: **Clapton** Glos. *Clopton* 1171–83. **Clapton** Northants. *Cloptun* c.960, *Clotone* 1086 (DB). **Clapton** Somerset, near Crewkerne. *Clopton* 1243. **Clapton in Gordano** N. Som. *Clotune* [*sic*] 1086 (DB), *Clopton* 1225. Gordano is an old district name (*Gordeyne* 1270), probably 'dirty or muddy valley', OE *gor* + *denu*.

Clara (*Clóirtheach*) Offaly. 'Level place'.

Clarborough Notts. *Claureburgh* 1086 (DB). 'Old fortification overgrown with clover'. OE *clǣfre* + *burh*.

Clare (*Clár*) Armagh, Down, Tyrone. 'Plain'. **Clare** (the county) has the same origin and meaning.

Clare Suffolk. *Clara* 1086 (DB). Probably from a Latin stream-name *Clāra* 'clear stream', identical with that found in HIGHCLERE.

Clare Island (*Cliara*) Mayo. 'Clergy island'.

Clareen (*An Cláirín*) Offaly. 'The little plain'.

Claregalway (*Baile Chláir*) Galway. 'Plain of Galway'.

Claremorris (*Clár Chlainne Muiris*) Mayo. 'Plain of the children of Muiris'.

Clarina (*Clár Aidhne*) Limerick. 'Plank bridge of Aidhne'.

Clarinbridge (*Droichead an Chláirín*) Galway. 'Bridge of the little plank'.

Clas-ar-Wy. *See* GLASBURY.

Clash (*Clais*) Tipperary, Wicklow. 'Ravine'.

Clashmore (*Clais Mhór*) Waterford. 'Big ravine'.

Clatford, Goodworth & Clatford, Upper Hants. *Cladford, Godorde* 1086 (DB). 'Ford where burdock grows'. OE *clāte* + *ford*. Goodworth (originally a separate name) means 'enclosure of a man called Gōda', OE pers. name + *worth*.

Clatworthy Somerset. *Clateurde* 1086 (DB), *Clatewurthy* 1243. 'Enclosure where burdock grows'. OE *clāte* + *worth* (replaced by *worthig*).

Claudy (*Clóidigh*) Derry. *Cládech* c.1680. 'Strong-flowing one'. The name probably refers to a section of the River Faughan here.

Claughton, 'settlement on or by a hill', OE **clæcc* + *tūn*: **Claughton** Lancs., near Garstang. *Clactune* 1086 (DB). **Claughton** Lancs., near Lancaster. *Clactun* 1086 (DB). **Claughton** Wirral. *Clahton* 1260.

Claverdon Warwicks. *Clavendone* [*sic*] 1086 (DB), *Claverdona* 1123. 'Hill where clover grows'. OE *clǣfre* + *dūn*.

Claverham N. Som. *Claveham* 1086 (DB). 'Homestead or enclosure where clover grows'. OE *clǣfre* + *hām* or *hamm*.

Clavering Essex. *Clæfring* c.1000, *Clauelinga* 1086 (DB). 'Place where clover grows'. OE *clǣfre* + *-ing*.

Claverley Shrops. *Claverlege* 1086 (DB). 'Clearing where clover grows'. OE *clǣfre* + *lēah*.

Claverton B. & NE. Som. *Clatfordtun* c.1000, *Claftertone* 1086 (DB). 'Farmstead by the ford where burdock grows'. OE *clāte* + *ford* + *tūn*.

Clawdd Offa. *See* OFFA'S DYKE.

Clawson, Long Leics. *Clachestone* 1086 (DB). 'Farmstead of a man called Klakkr'. OScand. pers. name + OE *tūn*.

Clawton Devon. *Clavetone* 1086 (DB). 'Farmstead at the tongue of land'. OE *clawu* + *tūn*.

Claxby, 'farmstead of a man called Klakkr', OScand. pers. name + *bý*: **Claxby** Lincs., near Alford. *Clachesbi* 1086 (DB). **Claxby** Lincs., near Market Rasen. *Cleaxbyg* 1066–8, *Clachesbi* 1086 (DB).

Claxton, 'farmstead on a hill, or of a man called Klakkr', OE **clacc* or OScand. pers. name + OE *tūn*: **Claxton** Norfolk. *Clakestona* 1086 (DB). **Claxton** N. Yorks. *Claxtorp* 1086 (DB), *Clakeston* 1176. Second element originally OScand. *thorp* 'outlying farmstead'.

Clay Coton Northants. *Cotes* 12th cent., *Cleycotes* 1284. 'Cottages in the clayey district'. OE *cot* (dative plural *cotum* or a ME plural *coten*) with the later addition of *clǣg*.

Clay Cross Derbys. a late name, first recorded in 1734, probably named from a local family called *Clay*.

Claybrooke Magna Leics. *Clǣg broc* *Claibroc* 1086 (DB). 'Clayey brook'. OE *clǣg* + *brōc*, with Latin affix *magna* 'great'.

Claydon, 'clayey hill', OE *clǣgig* (dative *-an*) + *dūn*: **Claydon** Oxon. *Cleindona* 1109. **Claydon** Suffolk. *Clainduna* 1086 (DB). **Claydon, Botolph, Claydon, East, & Claydon, Middle** Bucks. *Claindone* 1086 (DB), *Botle Cleidun* 1224, *Est Cleydon* 1247, *Middelcleydon* 1242. Affixes from OE *bōtl* 'house, building', *ēast* 'east', and *middel* 'middle'. **Claydon, Steeple** Bucks. *Claindone* 1086 (DB), *Stepel Cleydon* 13th cent. Affix from OE *stēpel* 'steeple, tower'.

Claygate Surrey. *Claigate* 1086 (DB). 'Gate or gap in the clayey district'. OE *clǣg* + *geat*.

Clayhanger Wsall. *Cleyhungre* 13th cent. 'Clayey wooded slope'. OE *clǣg* + *hangra*. The same name occurs in Ches. and Devon.

Clayhidon Devon. *Hidone* 1086 (DB), *Cleyhidon* 1485. 'Hill where hay is made'. OE *hī(e)g* + *dūn*, with the later addition of OE *clǣg* 'clay'.

Claypole Lincs. *Claipol* 1086 (DB). 'Clayey pool'. OE *clǣg* + *pōl*.

Clayton, 'farmstead on clayey soil', OE *clǣg* + *tūn*; examples include: **Clayton** Brad. *Claitone* 1086 (DB). **Clayton** Donc. *Claitone* 1086 (DB). **Clayton** Staffs. *Claitone* 1086 (DB). **Clayton**

W. Sussex. *Claitune* 1086 (DB). **Clayton-le-Moors** Lancs. *Cleyton* 1243. Affix 'on the moorland' from OE *mōr*, the French definite article *le* remaining after loss of the preposition.

Clayton-le-Woods Lancs. *Cleitonam* 1160. Affix 'in the woodland' from OE *wudu*, with French *le* as in previous name. **Clayton West** Kirkl. *Claitone* 1086 (DB).

Clayworth Notts. *Clauorde* 1086 (DB). 'Enclosure on the low curving hill'. OE *clawu* + *worth*.

Cleadon S. Tyne. *Clyuedon* 1183. 'Hill of the cliffs'. OE *clif* (genitive plural *clifa*) + *dūn*.

Clearwell Glos. *Clouerwalle* c.1282. 'Spring or stream where clover grows'. OE *clāfre* + *wella*.

Cleasby N. Yorks. *Clesbi* 1086 (DB). 'Farmstead or village of a man called Kleppr or *Kleiss'. OScand. pers. name + *bý*.

Cleatlam Durham. *Cletlum* c.1200. '(Place at) the clearings where burdock grows'. OE **clēte* + *lēah* (dative plural *lēaum*). An earlier form *Cletlinga* c.1040 may mean 'burdock place', from OE **clēte* + suffix *-ling*.

Cleator (Moor) Cumbria. *Cletergh* c.1200. Probably 'hill pasture where burdock grows'. OE **clēte* + OScand. *erg*.

Cleckheaton Kirkl. *Hetun* 1086 (DB), *Claketon* 1285. 'High farmstead, farmstead situated on high land'. OE *hēah* + *tūn*. The affix *Cleck-* is from OE **clǣcc* or OScand. *klakkr* 'hill'.

Clee St Margaret Shrops. *Cleie* 1086 (DB), *Clye Sancte Margarete* 1285. Named from the Clee Hills, probably an OE **clēo* 'ball-shaped, rounded hill'. Affix from the dedication of the church.

Cleethorpes NE. Lincs. *Thorpe* 1406, *Clethorpe* 1552. 'Outlying settlement (OScand. *thorp*) near Clee'. The latter, now **Old Clee**, is *Cleia* in 1086 (DB), from OE *clǣg* 'clay' with reference to the soil.

Cleeton St Mary Shrops. *Cleotun* 1241. 'Farmstead by the hills called Clee', *see* CLEE ST MARGARET. OE *tūn*.

Cleeve, Cleve, '(place) at the cliff or bank', OE *clif* (dative *clife*); examples include: **Cleeve** N. Som. *Clive* 1243. **Cleeve, Bishop's** Glos. *Clife* 8th cent., *Clive* 1086 (DB), *Bissopes Clive* 1284. Affix from its early possession by the Bishops of Worcester. **Cleeve, Old** Somerset. *Clive* 1086 (DB). **Cleeve Prior** Worcs. *Clive*

1086 (DB), *Clyve Prior* 1291. Affix from its early possession by the Prior of Worcester.

Cleggan (*An Cloigeann*) Galway. 'The head'.

Clehonger Herefs. *Cleunge* 1086 (DB). 'Clayey wooded slope'. OE *clǣg* + *hangra*.

Clench Wilts. *Clenche* 1289. OE **clenc* 'hill'.

Clenchwarton Norfolk. *Ecleuuartuna* [sic] 1086 (DB), *Clenchewarton* 1196. Probably 'farmstead or village of the dwellers at the hill'. OE **clenc* + *-ware* + *tūn*.

Clent Worcs. *Clent* 1086 (DB). OE **clent* 'rock, rocky hill'.

Cleobury Mortimer & Cleobury North Shrops. *Cleberie, Claiberie* 1086 (DB), *Clebury Mortimer* 1272, *Northclaibiry* 1222. 'Fortified place or manor near the hills called Clee', *see* CLEE ST MARGARET. OE *burh* (dative *byrig*). Manorial addition from the *Mortemer* family, here in the 11th cent., the other distinguishing affix being ME *north*.

Clerkenwell Gtr. London. *Clerkenwell* c.1150. 'Well or spring frequented by scholars or students'. ME *clerc* (plural *-en*) + OE *wella*.

Clevancy Wilts. *Clive* 1086 (DB), *Clif Wauncy* 1231. '(Place) at the cliff', from OE *clif* (dative *clife*), with manorial addition from the *de Wancy* family, here in the 13th cent.

Clevedon N. Som. *Clivedon* 1086 (DB). 'Hill of the cliffs'. OE *clif* (genitive plural *clifa*) + *dūn*.

Cleveland (district) Red. & Cleve. *Clivelanda* c.1110. 'District of cliffs, hilly district'. OE *clif* (genitive plural *clifa*) + *land*.

Cleveleys Lancs. not on record before early this cent., probably a manorial name from a family called *Cleveley* who may have come from Cleveley near Garstang (*Cliueleye* c.1180, 'woodland clearing near a cliff or bank', OE *clif* + *lēah*).

Clewer Somerset. *Cliveware* 1086 (DB). 'The dwellers on a river-bank'. OE *clif* + *-ware*. The same name occurs in Berks.

Cley next the Sea Norfolk. *Claia* 1086 (DB). OE *clǣg* 'clay, place with clayey soil'.

Cliburn Cumbria. *Clibbrun* c.1140. 'Stream by the cliff or bank'. OE *clif* + *burna*.

Cliddesden Hants. *Cleresden* [sic] 1086 (DB), *Cledesdene* 1194. Possibly 'valley of the rock or rocky hill'. OE **clyde* + *denu*.

Cliff, Cliffe, Clyffe, '(place at) the cliff or bank', OE *clif*; examples include: **Cliff** Warwicks. *Cliva* 1166. **Cliffe** Medway. *Cliua* 10th cent., *Clive* 1086 (DB). **Cliffe** N. Yorks. *Clive* 1086 (DB). **Cliffe, King's** Northants. *Clive* 1086 (DB). A royal manor in 1086. **Cliffe, North & Cliffe, South** E. R. Yorks. *Cliue* 1086 (DB). **Cliffe, West** Kent. *Wesclive* 1086 (DB). With OE *west* 'west'. **Clyffe Pypard** Wilts. *æt Clife* 983, *Clive* 1086 (DB), *Clive Pipard* 1291. Manorial affix from the *Pipard* family, here in the 13th cent.

Cliffony (*Cliafuine*) Sligo. 'Hurdle thicket'.

Clifford, 'ford at a cliff or bank', OE *clif* + *ford*: **Clifford** Herefs. *Cliford* 1086 (DB). **Clifford** Leeds. *Cliford* 1086 (DB). **Clifford Chambers** Warwicks. *Clifforda* 922, *Clifort* 1086 (DB), *Chaumberesclifford* 1388. Affix from the chamberlain (ME *chamberere*) of St Peter's Gloucester, given the manor in 1099.

Clifton, a common place name, 'farmstead on or near a cliff or bank', OE *clif* + *tūn*; examples include: **Clifton** Beds. *Cliftune* 10th cent., *Cliftone* 1086 (DB). **Clifton** Brist. *Clistone* 1086 (DB). **Clifton** Cumbria. *Clifton* 1204. **Clifton** Derbys. *Cliftune* 1086 (DB). **Clifton** Nott. *Cliftone* 1086 (DB). **Clifton Campville** Staffs. *Clyfton* 942, *Cliftune* 1086 (DB). Manorial addition from the *de Camvill* family, here in the 13th cent. **Clifton Hampden** Oxon. *Cliftona* 1146. The affix first occurs in 1836, and may be from a family called *Hampden*. **Clifton, North & Clifton, South** Notts. *Cliftone* 1086 (DB), *Nort Clifton* c.1160, *Suth Clifton* 1280. **Clifton Reynes** Milt. K. *Cliftone* 1086 (DB), *Clyfton Reynes* 1383. Manorial addition from the *de Reynes* family, here in the early 14th cent. **Clifton upon Dunsmore** Warwicks. *Cliptone* 1086 (DB), *Clifton super Donesmore* 1306. For the affix, *see* RYTON-ON-DUNSMORE. **Clifton upon Teme** Worcs. *Cliftun ultra Tamedam* 934, *Clistune* [sic] 1086 (DB). Teme is an old Celtic or pre-Celtic river-name, *see* TENBURY WELLS.

Cliftonville Kent. a modern name invented for this 19th-cent. resort.

Climping W. Sussex. *Clepinges* [sic] 1086 (DB), *Clympinges* 1228. '(Settlement of) the family or followers of a man called **Climp*'. OE pers. name + *-ingas*.

Clint N. Yorks. *Clint* 1208. OScand. *klint* 'a rocky cliff, a steep bank'.

Clippesby Norfolk. *Clepesbei* 1086 (DB). 'Farmstead or village of a man called Klyppr or *Klippr'. OScand. pers. name + *bý*.

Clipsham Rutland. *Kilpesham* 1203. Probably 'homestead or enclosure of a man called *Cylp'. OE pers. name + *hām* or *hamm*.

Clipston(e), 'farmstead of a man called Klyppr or *Klippr'. OScand. pers. name + OE *tūn*: **Clipston** Northants. *Clipestune* 1086 (DB). **Clipston** Notts. *Cliston* 1198. **Clipstone** Notts. *Clipestune* 1086 (DB).

Clitheroe Lancs. *Cliderhou* 1102. 'Hill with loose stones'. OE *clȳder, *clider* + OE *hōh* or OScand. *haugr*.

Clive Shrops. *Clive* 1255. '(Place) at the cliff or bank'. OE *clif* (dative *clife*).

Clodock Herefs. *Ecclesia Sancti Clitauci* c.1150. '(The church of) St. Clydog'. From the patron saint of the church.

Clogh (*An Chloch*) Antrim, Kilkenny. 'The stone castle'.

Clogh Mills (*Muileann na Cloiche*) Antrim. 'Mill of the stone'.

Cloghan (*An Clochán*) Donegal, Offaly, Westmeath. 'The stony place'.

Cloghane (*An Clochán*) Kerry. 'The stony place'.

Cloghaneely (*Cloich Chionnaola*) (district) Donegal. *Cloiche Chinnfhaoladh* 1284. 'Cionnaola's stone'.

Cloghboley (*Clochbhuaile*) Sligo. 'Stone of the milking place'.

Cloghbrack (*An Chloch Bhreac*) Mayo. 'The speckled stone'.

Clogheen (*An Chloichín*) Tipperary, Waterford. 'The little stone'.

Clogher (*An Chlochar*) Mayo, Tyrone. 'The stony place'.

Cloghernach (*Clocharnach*) Waterford. 'Stony place'.

Cloghjordan (*Cloch Shiurdáin*) Tipperary. 'Jordan's stone castle'.

Cloghmore (*An Chloch Mhór*) Mayo, Monaghan. 'The big stone'.

Cloghore (*Cloich Óir*) Donegal. 'Gold stone'.

Cloghran (*Clochrán*) Dublin. 'Stepping-stones'.

Cloghroe (*Cloch Rua*) Cork. 'Red stone'.

Cloghy (*Clochaigh*) Down. *Cloghie* 1623. 'Stony place'.

Clohamon (*Cloch Amainn*) Wexford. 'Hamon's stone castle'.

Clomantagh (*Cloch Mhantach*) Kilkenny. 'Gapped stone'.

Clonadacasey (*Cluain Fhada Uí Chathasaigh*) Laois. 'Ó Cathasaigh's long pasture'.

Clonakenny (*Cluain Uí Chionaoith*) Tipperary. 'Ó Cionaoith's pasture'.

Clonakilty (*Cloich na Coillte*) Cork. 'Stone of the woods'.

Clonalvy (*Cluain Ailbhe*) Meath. 'Ailbhe's pasture'.

Clonard (*Cluain Ioraird*) Meath. 'Iorard's pasture'.

Clonard (*Cluain Ard*) Wexford. 'High pasture'.

Clonaslee (*Cluain na Slí*) Laois. 'Pasture of the path'.

Clonbanin (*Cluain Báinín*) Cork. 'Pasture of the white tweed'.

Clonbern (*Cluain Bheirn*) Galway. 'Bearn's pasture'.

Clonbulloge (*Cluain Bolg*) Offaly. 'Pasture of the bulges'.

Cloncagh (*Cluain Cath*) Limerick. 'Pasture of battles'.

Cloncreen (*Cluain Crainn*) Offaly. 'Pasture of the tree'.

Cloncurry (*Cluain Curraigh*) Kildare. 'Pasture of the marsh'.

Clondalkin (*Cluain Dolcáin*) Dublin. 'Dolcán's pasture'.

Clondarragh (*Cluain Darach*) Wexford. 'Pasture of the oak-tree'.

Clondaw (*Cluain Dáith*) Wexford. 'Dáith's pasture'.

Clondrohid (*Cluain Droichead*) Cork. 'Pasture of the bridges'.

Clondulane (*Cluain Dalláin*) Cork. 'Dallán's pasture'.

Clonea (*Cluain Fhia*) Waterford. 'Pasture of the deer'.

Clonee (*Cluain Aodha*) Meath. 'Aodh's pasture'.

Cloneen (*An Cluainín*) Tipperary. 'The little pasture'.

Clonegall (*Cluain na nGall*) Carlow. 'Pasture of the stones'.

Clonenagh (*Cluain Eidhneach*) Laois. 'Ivied pasture'.

Clones (*Cluain Eois*) Monaghan. *(abb) Cluana hEois* 701. 'Pasture of Eos'.

Cloney (*Cluanaidh*) Kildare. 'Pasture'.

Cloneygowan (*Cluain na nGamhan*) Offaly. 'Pasture of the calves'.

Clonfert (*Cluain Fearta*) Galway. 'Pasture of the grave'.

Clonfinlough (*Cluain Fionnlocha*) Offaly. 'Pasture of the white lake'.

Clongorey (*Cluain Guaire*) Kildare. 'Guaire's pasture'.

Clonkeen (*Cluain Chaoin*) Kerry. 'Pleasant pasture'.

Clonleigh (*Cluain Lao*) Donegal. 'Pasture of the calf'.

Clonlost (*Cluain Loiste*) Westmeath. 'Burnt meadow'.

Clonmacnoise (*Cluain Mic Nois*) Offaly. 'Pasture of the descendants of Noas'.

Clonmany (*Cluain Maine*) Donegal. *Clonmane* 1397. 'Maine's meadow'.

Clonmel (*Cluain Meala*) Tipperary. 'Pasture of honey'.

Clonmellon (*Cluain Mileáin*) Westmeath. 'Mileán's meadow'.

Clonmore (*Cluain Mhór*) Carlow, Tipperary. 'Large pasture'.

Clonmult (*Cluain Molt*) Cork. 'Pasture of wethers'.

Clonoe (*Cluain Eo*) Tyrone. *Clondeo c.*1306. 'Meadow of the yew-tree'.

Clonoulty (*Cluain Ultaigh*) Tipperary. 'Pasture of the Ulstermen'. The official Irish name is *Cluain Abhla*, 'meadow of the apple-tree'.

Clonroche (*Cluain an Róistigh*) Wexford. 'Roches' pasture'.

Clonroosk (*Cluain Rúisc*) Offaly. 'Pasture of the bark'.

Clonsilla (*Cluain Saileach*) Dublin. 'Pasture of the willow'.

Clontallagh (*Cluain tSalach*) Donegal. 'Dirty pasture'.

Clontarf (*Cluain Tarbh*) Dublin. 'Pasture of bulls'.

Clontibret (*Cluain Tiobrad*) Monaghan. (*Mocumae cruimthir*) *Clúana Tiprat c.*830. 'Pasture of the well'.

Clontoe (*Cluain Tó*) Tyrone. 'Tó's pasture'.

Clontuskert (*Cluain Tuaiscirt*) Galway. 'North pasture'.

Clontygora (*Cluainte Ó gCorra*) Armagh. *Clontegora* 1609. 'Pastures of the descendants of Corra'.

Clonygowan (*Cluain na nGamhan*) Offaly. 'Pasture of the calves'.

Cloon (*Cluain*) Clare, Mayo. 'Pasture'.

Cloonacool (*Cluain na Cúile*) Sligo. 'Pasture of the nook'.

Cloonagh (*Cluanach*) Westmeath. 'Meadow-land'.

Cloonaghmore (*Cluaineach Mór*) Mayo. 'Large horse meadow'.

Cloondaff (*Cluain Damh*) Mayo. 'Ox meadow'.

Cloondara (*Cluain dá Ráth*) Longford. 'Pasture of the two forts'.

Cloone (*An Chluain*) Leitrim. 'The pasture'.

Clooney (*An Chluanaidh*) Donegal. 'The pasture-land'.

Cloonfad (*Cluain Fhada*) Roscommon. 'Long pasture'.

Cloonkeavy (*Cluain Chiabhaigh*) Sligo. 'Hazy pasture'.

Cloonlara (*Cluain Lára*) Clare. 'Pasture of the mare'.

Cloonlough (*Cluain Lua*) Sligo. 'Lua's pasture'.

Cloontia (*Na Cluainte*) Mayo. 'The meadows'.

Cloonymorris (*Cluain Uí Mhuiris*) Galway. 'Ó Muiris's pasture'.

Clophill Beds. *Clopelle* 1086 (DB). 'Lumpy hill'. OE **clopp(a) + hyll*.

Clopton Suffolk. *Clop(e)tuna* 1086 (DB). 'Farmstead or village near a hill or hills'. OE **clopp(a) + tūn*.

Clorhane (*Clochrán*) Offaly. 'Stepping-stones'.

Closkelt (*Cloch Scoilte*) Down. 'Cleft rock'.

Clothall Herts. *Clatheala c.*1060, *Cladhele* 1086 (DB). 'Nook of land where burdock grows'. OE *clāte + healh*.

Clotton Ches. *Clotone* 1086 (DB). 'Farmstead at a dell or deep valley'. OE **clōh + tūn*.

Cloud, Temple B. & NE. Som. *La Clude* 1199. 'The (rocky) hill'. OE *clūd*. Affix probably refers to lands here held by the Knights Templars.

Clough (*Cloch*) Down. *Cloghmagherechat* 1634. 'Stone castle'. An earlier Irish name was *Cloch Mhachaire Cat*, 'stone castle of the plain of the cats'.

Cloughmore (*Cloch Mhór*) Down. 'Big stone'.

Cloughoge (*Clochóg*) Armagh. *Claghoge* 1661. 'Stony place'.

Cloughton N. Yorks. *Cloctune* 1086 (DB). 'Farmstead at a dell or deep valley'. OE **clōh + tūn*.

Clovelly Devon. *Cloveleia* 1086 (DB), *Clofely* 1296. Probably 'earthworks associated with a man called Fele(c)'. Cornish *cleath* 'dyke, bank' (referring to the ancient earthworks of Clovelly Dykes) + pers. name (which may also survive in the name of nearby **Velly**, *Felye* 1287).

Clowne Derbys. *Clune c.*1002, 1086 (DB). Named from the river which rises nearby; it is an old river-name identical with that in CLUN (Shrops.).

Cloyfin (*An Chloich Fhionn*) Derry. *Illannecloghfynny* 1637. 'The white stone'.

Cloyne (*Cluain*) Cork. 'Pasture'.

Clun Shrops. *Clune* 1086 (DB). Named from the River Clun, which is an ancient pre-English river-name of uncertain meaning.

Clunbury Shrops. *Cluneberie* 1086 (DB). 'Fortified place on the River Clun'. Pre-English river-name + OE *burh* (dative *byrig*).

Clungunford Shrops. *Clone* 1086 (DB), *Cloune Goneford* 1242. 'Estate on the River Clun held by a man called Gunward'. Pre-English river-name + manorial addition from the name of the lord of the manor in the time of Edward the Confessor.

Clunton Shrops. *Clutone* [sic] 1086 (DB), *Cluntune* 12th cent. 'Farmstead or village on the River Clun'. Pre-English river-name + OE *tūn*.

Clutton, 'farmstead or village at a hill', from OE *clūd + tūn*: **Clutton** B. & NE. Som. *Cluttone* 851, *Clutone* 1086 (DB). **Clutton** Ches. *Clutone* 1086 (DB).

Clwyd (the historic county). *Cloid fluvium* 1191. '(District of the) River Clwyd'. The river-name means 'hurdle' (Welsh *clwyd*), perhaps referring to a ford or causeway made of hurdles.

Clydach Swan. *Cleudach* 1208. '(Place on the) River Clydach'. The Celtic river-name means 'cleansing one' (British **clouta*).

Clyde. *See* CLYDEBANK.

Clydebank W. Dunb. '(Place on the) bank of the River Clyde'. The Celtic river-name means 'cleansing one'. Burgh in 1886.

Clyffe Pypard Wilts. *See* CLIFF.

Clynnog-fawr Gwyd. *Kelynnauk c.*1291, *Kellynawc* 1346. 'Greater (place) abounding in holly'. Welsh *celyn + -og + mawr*. The name contrasts with *Clynnog Fechan* (Welsh *fechan*, 'small'), Angl.

Clyst, Broad Devon. *Clistone* 1086 (DB), *Clistun c.*1100, *Brodeclyste* 1372. Like the following Devon places, named from the River Clyst, a Celtic river-name probably meaning 'sea inlet'. Originally 'farmstead on the River

Clyst' from OE *tūn*, later 'large estate on the River Clyst' from OE *brād*.

Clyst Honiton Devon. *See* HONITON.

Clyst Hydon Devon. *Clist* 1086 (DB), *Clist Hydone* 1268. 'Estate on the River Clyst held by the *de Hidune* family' (here in the 13th cent.). Celtic river-name + manorial addition.

Clyst St George Devon. *Clisewic* 1086 (DB), *Clystwik Sancti Georgii* 1327. Originally 'dairy farm on the River Clyst'. Celtic river-name + OE *wīc*, with addition from the dedication of the church.

Clyst St Lawrence Devon. *Clist* 1086 (DB), *Clist Sancti Laurencii* 1203. 'Estate on the River Clyst', with addition from the dedication of the church.

Clyst St Mary Devon. *Clist Sancte Marie* 1242. Identical in origin with the previous name.

Coa (*Cuach*) Fermanagh. *Coagh* 1609. 'Hollow'.

Coachford (*Áth an Chóiste*) Cork. 'Ford of the coach'. The Irish name appears to have been devised to match the English, which itself may be a corruption of some other Irish name.

Coagh (*An Cuach*) Tyrone. *Coagh* 1639. 'The hollow'.

Coalbrookdale Tel. & Wrek. *Caldebrok* 1250. '(Valley of) the cold brook'. OE *cald* + *brōc* with the later addition of *dale*.

Coalbrookvale. *See* NANT-Y-GLO.

Coaley Glos. *Couelege* 1086 (DB). 'Clearing with a hut or shelter'. OE *cofa* + *lēah*.

Coalisland Tyrone. *Coal Island* 1837. The town was the inland port for the local coalfield. The equivalent Irish name is *Oileán an Ghuail*.

Coalville Leics. a name of 19th-cent. origin, so called because it is in a coal-mining district.

Coan (*An Cuan*) Kilkenny. 'The recess'.

Coatbridge N. Lan. *Coittis* 1584. 'Bridge at Coats'. *Coats* is '(place with) cottages' (OE *cot*).

Coate Wilts. near Devizes. *Cotes* 1255. 'The cottage(s) or hut(s)'. OE *cot*.

Coates, 'the cottages or huts', from OE *cot*; examples include: **Coates** Cambs. *Cotes* c.1280. **Coates** Glos. *Cota* 1175. **Coates** Notts. *Cotes* 1200. **Coates, Great** NE. Lincs. *Cotes* 1086 (DB). **Coates, North** Lincs. *Nordcotis* c.1115. 'Cottages to the north (of FULSTOW)', OE *north*.

Coatham Red. & Cleve. *Cotum* 1123–8. '(Place) at the cottages or huts'. OE *cot* in a dative plural form *cotum*.

Coatham Mundeville Darltn. *Cotum* 13th cent., *Cotum Maundevill* 1344. Identical in origin with the previous name. Manorial affix from the *de Amundevilla* family, here in the 13th cent.

Coberley Glos. *Culberlege* 1086 (DB). 'Wood or clearing of a man called Cūthbeorht'. OE pers. name + *lēah*.

Cobh (*Cóbh*) Cork. *Cove* 1659. 'Cove'. The Irish name is an adoption of English *cove*.

Cobham Kent. *Cobba hammes mearce* 939, *Cobbeham* 1197. 'Enclosure or homestead of a man called *Cobba*. OE pers. name + *hamm* or *hām*. The early form contains OE *mearc* 'boundary'.

Cobham Surrey. *Covenham* 1086 (DB). 'Homestead or enclosure of a man called *Cofa*, or with a hut or shelter'. OE pers. name or OE *cofa* + *hām* or *hamm*.

Cockayne Hatley Beds. *See* HATLEY.

Cockenzie N. Ayr. *Cowkany* 1590. Meaning unknown. A pers. name may be involved.

Cockerham Lancs. *Cocreham* 1086 (DB). 'Homestead or enclosure on the River Cocker'. Celtic river-name (with a meaning 'crooked') + OE *hām* or *hamm*.

Cockerington, North & Cockerington, South Lincs. *Cocrintone* 1086 (DB). Possibly 'farmstead by a stream called *Cocker*'. Celtic river-name (perhaps an earlier name for the River Lud) + connective *-ing-* + *tūn*.

Cockermouth Cumbria. *Cokyrmoth* c.1150. 'Mouth of the River Cocker'. Celtic river-name (with a meaning 'crooked') + *mūtha*.

Cockfield Durham. *Kokefeld* 1223. 'Open land frequented by cocks (of wild birds), or belonging to a man called *Cocca*'. OE *cocc* or OE pers. name + *feld*.

Cockfield Suffolk. *Cochanfelde* 10th cent. 'Open land of a man called *Cohha*'. OE pers. name + *feld*.

Cockfosters Gtr. London. *Cokfosters* 1524. 'The chief forester's (place or estate)', from eModE *cock* + *for(e)ster*.

Cocking W. Sussex. *Cochinges* 1086 (DB). 'Dwellers at the hillock', or '(settlement of) the family or followers of a man called *Cocc(a)*'. OE *cocc* or OE pers. name + *-ingas*.

Cockington Torbay. *Cochintone* 1086 (DB). Probably 'estate associated with a man called *Cocc(a)*'. OE pers. name + *-ing-* + *tūn*.

Cockley Cley Norfolk. *Cleia* 1086 (DB), *Coclikleye* 1324. OE *clǣg* 'clay, place with clayey soil' + affix which may be manorial from a family called *Cockley* or a local name meaning 'wood frequented by woodcocks' from OE *cocc* + *lēah*.

Cockshutt Shrops. *La Cockesete* 1270. From OE **cocc-scyte* 'a woodland glade where nets were stretched to catch woodcock'.

Cockthorpe Norfolk. *Torp* 1086 (DB), *Coketorp* 1254. Originally OScand. *thorp* 'secondary settlement, outlying farmstead' + affix which may be manorial from a family called *Co(c)ke*, or may indicate 'where cocks are reared' from OE *cocc*.

Coddenham Suffolk. *Codenham* 1086 (DB). 'Homestead or enclosure of a man called **Cod(d)a*'. OE pers. name (genitive *-n*) + *ham* or *hamm*.

Coddington, 'estate associated with a man called *Cot(t)a*', from OE pers. name + *-ing-* + *tūn*: **Coddington** Ches. *Cotintone* 1086 (DB). **Coddington** Herefs. *Cotingtune* 1086 (DB). **Coddington** Notts. *Cotintone* 1086 (DB).

Codford St Mary & Codford St Peter Wilts. *Codan ford* 901, *Coteford* 1086 (DB), *Codeford Sancte Marie, Sancti Petri* 1291. 'Ford of a man called **Cod(d)a*'. OE pers. name + *ford*. Affixes from the dedications of the churches.

Codicote Herts. *Cutheringcoton* 1002, *Codicote* 1086 (DB). 'Cottage(s) associated with a man called Cūthhere'. OE pers. name + *-ing-* + *cot*.

Codnor Derbys. *Cotenoure* 1086 (DB). 'Ridge of a man called **Cod(d)a*'. OE pers. name (genitive *-n*) + **ofer*.

Codrington S. Glos. *Cuderintuna* 12th cent. 'Estate associated with a man called Cūthhere'. OE pers. name + *-ing-* + *tūn*.

Codsall Staffs. *Codeshale* 1086 (DB). 'Nook of land of a man called **Cōd*'. OE pers. name + *halh*.

Coed-duon. *See* BLACKWOOD.

Coed-poeth Wrex. *Coid poch* 1391, *Coed Poeth* 1412. 'Burnt wood'. Welsh *coed* + *poeth*.

Coffinswell Devon. *Willa* 1086 (DB), *Coffineswell* 1249. '(Place at) the spring or stream'. OE *w(i)ella* + manorial addition from the *Coffin* family, here in the 12th cent.

Cofton Hackett Worcs. *Coftune* 8th cent., *Costune* [sic] 1086 (DB), *Corfton Hakett* 1431. 'Farmstead with a hut or shelter'. OE *cofa* + *tūn* + manorial addition from the *Haket* family, here in the 12th cent.

Cogan Vale Glam. *Cogan* c.1139. Meaning uncertain. The origin may lie in a stream-name.

Cogenhoe Northants. *Cugenho* 1086 (DB). 'Hill-spur of a man called **Cugga*'. OE pers. name (genitive *-n*) + *hōh*.

Cogges, High Cogges Oxon. *Coges* 1086 (DB). 'The cog-shaped hills', from OE **cogg* 'cog wheel', here probably used figuratively in a topographical sense.

Coggeshall Essex. *Kockeshale* c.1060, *Cogheshala* 1086 (DB). Possibly 'nook of a man called **Cocc* or **Cogg*'. OE pers. name + *halh*. Alternatively the first element may be OE *cocc* 'cock, woodcock', **cocc* 'heap, hillock', or **cogg* 'cog wheel', also perhaps 'hill'.

Coker, East, Coker, North, & Coker, West Somerset. *Cocre* 1086 (DB). Originally the name of the stream here, a Celtic river-name with a meaning 'crooked, winding'.

Colaboll Highland. 'Kol's homestead'. OScand. pers. name + *bólstathr*.

Colan Cornwall. *Sanctus Colanus* 1205. 'Church of St Colan'. From the patron saint of the church.

Colaton Raleigh Devon. *Coletone* 1086 (DB), *Coleton Ralegh* 1316. 'Farmstead of a man called Cola'. OE pers. name + *tūn* + manorial addition from the *de Ralegh* family, here in the 13th cent.

Colbost Highland. (Skye). 'Kol's homestead'. OScand. pers. name + *bólstathr*.

Colburn N. Yorks. *Corburne* [sic] 1086 (DB), *Coleburn* 1198. 'Cool stream'. OE *cōl* + *burna*.

Colby Cumbria. *Collebi* 12th cent. 'Farmstead of a man called Kolli', or 'hill farmstead'. OScand. pers. name or *kollr* + *bý*.

Colby Norfolk. *Colebei* 1086 (DB). 'Farmstead or village of a man called Koli'. OScand. pers. name + *bý*.

Colchester Essex. *Colneceastre* early 10th cent., *Colecestra* 1086 (DB). 'Roman town on the River Colne'. Ancient pre-English river-name (*see* COLNE ENGAINE) + OE *ceaster*. Alternatively the first element may be a reduced form of Latin *colonia* 'Roman colony for retired legionaries' (the Romano-British name of Colchester being *Colonia Camulodunum*, this last being a British name meaning 'fort of the Celtic war-god Camulos').

Cold as affix. *See* main name, e.g. for **Cold Brayfield** (Milt. K.) *see* BRAYFIELD.

Coldham Cambs. *Coldham* 1251. 'Cold enclosure'. OE *cald* + *hamm*.

Coldharbour Surrey. First recorded in the late 17th cent., 'cold or cheerless dwelling', from ModE *coldharbour* (OE *cald* + *here-beorg*). The same name occurs in several other counties and usually has a derogatory sense for a spot considered inhospitable.

Coldred Kent. *Colret* 1086 (DB). 'Clearing where coal is found, or where charcoal is made'. OE *col* + **ryde*.

Coldridge Devon. *Colrige* 1086 (DB). 'Ridge where charcoal is made'. OE *col* + *hrycg*.

Coldstream Sc. Bord. *Kaldestrem* c.1178, *Caldestream* c.1207, *Coldstreme* 1290. '(Place on the) cold stream'. Coldstream is on the Tweed, but the smaller Leet Water, which joins it here, may have been the original 'cold stream'.

Coldwaltham W. Sussex. *Waltham* 10th cent., *Cold Waltham* 1340. 'Homestead or village in a forest'. OE *w(e)ald* + *hām*, with later affix meaning 'bleak, exposed'.

Cole Somerset. *Colna* 1212. Originally the name of the stream here, a pre-English river-name of uncertain meaning.

Colebatch Shrops. *Colebech* 1176. Probably 'stream valley of a man called Cola'. OE pers. name + *bece*. Alternatively the first element may be OE *col* 'charcoal'.

Colebrook Devon. near Cullompton. *Colebroca* 1086 (DB). 'Cool brook'. OE *cōl* + *brōc*.

Colebrooke Devon. *Colebroc* 12th cent. Identical in origin with the previous name.

Coleby N. Lincs. *Colebi* 1086 (DB). 'Farmstead or village of a man called Koli'. OScand. pers. name + *bý*.

Coleford Devon. *Colbrukeforde* 1330. 'Ford at COLEBROOKE'. OE *ford*.

Coleford Glos. *Coleforde* 1282. 'Ford across which coal is carried'. OE *col* + *ford*.

Coleford Somerset. near Frome. *Culeford* 1234. 'Ford across which coal or charcoal is carried'. OE *col* + *ford*.

Colehill Dorset. *Colhulle* 1431. OE *hyll* 'hill' with either *col* 'charcoal' or **coll* 'hill'.

Coleman's Hatch E. Sussex. *Colmanhacche* 1495. 'Forest gate (OE *hæcc*) associated with a family called *Coleman*' (recorded in this area from the 13th cent.).

Colemere Shrops. *Colesmere [sic]* 1086 (DB), *Culemere* 1203. Probably 'pool of a man called **Cūla*'. OE pers. name + *mere*.

Coleorton Leics. *Ovretone* 1086 (DB), *Cole Orton* 1571. 'Farmstead on the Flat-topped ridge'. OE **ofer* + *tūn*. Affix is ME *col* 'coal' with reference to mining there.

Coleraine (*Cúil Raithin*) Derry. *Cuile Raithin* c.630. 'Nook of the ferns'.

Colerne Wilts. *Colerne* 1086 (DB). 'Building where charcoal is made or stored'. OE *col* + *ærn*.

Colesbourne Glos. *Col(l)esburnan* 9th cent., *Colesborne* 1086 (DB). 'Stream of a man called **Col* or *Coll*'. OE pers. name + *burna*.

Coleshill Bucks. *Coleshyll* 1507. Possibly identical in origin with the following name, but perhaps rather 'hill of a man called **Col*' with a different pers. name. Its earlier name in medieval times was *Stoke* 1175, *Stokke* 1224, from OE *stoc* 'secondary settlement'.

Coleshill Oxon. *Colleshylle* 10th cent., *Coleselle* 1086 (DB). Probably OE **coll* 'hill' with the addition of explanatory OE *hyll*, alternatively 'hill of a man called *Coll*', from OE pers. name + *hyll*.

Coleshill Warwicks. *Colleshyl* 799, *Coleshelle* 1086 (DB). Possibly identical in origin with the previous names, but more likely 'hill on the River Cole', Celtic river-name (of uncertain meaning) + OE *hyll*.

Colgate W. Sussex. *la Collegate* 1279. 'The charcoal gate'. ME *col* + *gate* (referring to a gate into St Leonard's forest where charcoal was once made).

Colkirk Norfolk. *Colechirca* 1086 (DB). 'Church of a man called Cola or Koli'. OE or OScand. pers. name + *kirkja*.

Coll (island) Arg. Probably a pre-Celtic name.

Collaton St Mary Torbay. *Coletone* 1261. 'Farmstead of a man called Cola'. OE pers. name + *tūn* + affix from the dedication of the church.

Collier Row Gtr. London. *Colyers rewe* 1440. 'Row of houses occupied by charcoal burners'. ME *colier* + *rewe* or *rowe*.

Collinamuck (*Caladh na Muc*) Galway. 'Riverside meadow of the pigs'.

Collingbourne Ducis & Collingbourne Kingston Wilts. *Colengaburnam* 903, *Colingeburne* 1086 (DB). Probably 'stream of the family or followers of a man called *Col or Cola'. OE pers. name + *-inga-* + *burna*. Manorial additions from possession by the Dukes of Lancaster (Latin *ducis* 'of the duke') and the king (OE *cyning* + *tūn*).

Collingham, 'homestead or village of the family or followers of a man called *Col or Cola', OE pers. name + *-inga-* + *hām*: **Collingham** Leeds. *Col(l)ingeham* 1167. **Collingham** Notts. *Colingeham* 1086 (DB).

Collington Herefs. *Col(l)intune* 1086 (DB). 'Estate associated with a man called *Col or Cola'. OE pers. name + *-ing-* + *tūn*.

Collingtree Northants. *Colentreu* 1086 (DB). 'Tree of a man called Cola'. OE pers. name (genitive *-n*) + *trēow*.

Collon (*Collon*) Louth. 'Place of hazels'.

Collooney (*Cúil Mhuine*) Sligo. 'Nook of the thicket'.

Collyweston Northants. *Westone* 1086 (DB), *Colynweston* 1309. 'West farmstead'. OE *west* + *tūn* + manorial affix from *Nicholas* (of which *Colin* is a pet-form) de Segrave, here in the 13th cent.

Colmworth Beds. *Colmeworde* 1086 (DB). Possibly 'enclosure of a man called *Culma'. OE pers. name + *worth*.

Coln Rogers, Coln St Aldwyns, & Coln St Dennis Glos. *Cungle* 962, *Colne* 1086 (DB), *Culna Rogeri* 13th cent., *Culna Sancti Aylwini* 12th cent., *Colne Seint Denys* 1287. '(Places by) the River Coln'. Pre-English river-name of uncertain origin. Additions from possession by *Roger* de Gloucester (died 1106), from church dedication to *St Athelwine*, and from possession by the church of *St Denis* of Paris (in the 11th cent.).

Colnbrook Slough. *Colebroc* 1107. 'Cool brook', or 'brook of a man called Cola'. OE *cōl* or OE pers. name + *brōc*. The spelling has been later influenced by the name of the River Colne which flows nearby (for the river-name *see* COLNEY, LONDON).

Colne Cambs. *Colne* 1086 (DB). Originally the name (now lost) of the stream here, a pre-English river-name of uncertain meaning.

Colne Lancs. *Calna* 1124. '(Place by) the River Colne'. A pre-English river-name of uncertain meaning, identical with River CALNE in Wilts.

Colne Engaine, Earls Colne, Wakes Colne, & White Colne Essex. *Colne* c.950, *Colun* 1086 (DB), *Colum Engayne* 1254, *Erlescolne* 1358, *Colne Wake* 1375, *Whyte Colne* 1285. '(Places by) the River Colne', an ancient pre-English river-name of uncertain meaning. Manorial additions from possession in medieval times by the *Engayne* family, the *Earls* of Oxford, and the *Wake* family. White Colne was held by Dimidius *Blancus* in 1086 (DB).

Colney Norfolk. *Coleneia* 1086 (DB). 'Island, or dry ground in marsh, of a man called Cola'. OE pers. name (genitive *-n*) + *ēg*.

Colney, London Herts. *Colnea* 1209–35, *London Colney* 1555. 'Island by the River Colne'. Pre-English river-name + OE *ēg*, with affix from its situation on the main road to London.

Colonsay (island) Arg. *Colyunsay* 14th cent. 'Kolbein's island'. OScand. pers. name + *ey*.

Colsterdale N. Yorks. *Colserdale* 1301. 'Valley of the charcoal burners'. OE **colestre* + OE *dæl* or ON *dalr*.

Colsterworth Lincs. *Colsteuorde* 1086 (DB). 'Enclosure of the charcoal burners'. OE **colestre* + *worth*.

Colston, 'farmstead of a man called Kolr', OScand. pers. name + OE *tūn*: **Colston, Car** Notts. *Colestone* 1086 (DB), *Kyrcoluiston* 1242. Affix from OScand. *kirkja* 'church'. **Colston Bassett** Notts. *Coletone* 1086 (DB), *Coleston Bassett* 1228. Manorial addition from the *Basset* family, here in the 12th cent.

Coltishall Norfolk. *Coketeshala* 1086 (DB). Possibly 'nook of land of a man called *Cohhede or *Coccede'. OE pers. name + *halh*.

Colton Cumbria. *Coleton* 1202. 'Farmstead on a river called *Cole*'. Celtic river-name (of uncertain meaning) + OE *tūn*.

Colton Norfolk. *Coletuna* 1086 (DB). 'Farmstead of a man called Cola or Koli'. OE or OScand. pers. name + OE *tūn*.

Colton N. Yorks. *Coletune* 1086 (DB). Probably identical in origin with the previous name.

Colton Staffs. *Coltone* 1086 (DB). Probably 'farmstead of a man called Cola'. OE pers. name + *tūn*.

Colwall Herefs. *Colewelle* 1086 (DB). 'Cool spring or stream'. OE *cōl* + *wælla*.

Colwell Northum. *Colewel* 1236. 'Cool spring or stream'. OE *cōl* + *wella*.

Colwich Staffs. *Colewich* 1240. Probably 'building where charcoal is made or stored'. OE *col* + *wīc*.

Colworth W. Sussex. *Coleworth* 10th cent. 'Enclosure of a man called Cola'. OE pers. name + *worth*.

Colwyn Bay (*Bae Colwyn*) Conwy. *Coloyne* 1334. 'Bay of the River Colwyn'. The river-name means 'puppy' (Welsh *colwyn*), alluding to its small size. The Welsh name corresponds to the English (Welsh *bae*).

Colyford Devon. *Culyford* 1244. 'Ford over the River Coly'. Celtic river-name (possibly 'narrow') + OE *ford*.

Colyton Devon. *Culintona* 946, *Colitone* 1086 (DB). 'Farmstead by the River Coly'. Celtic river-name + OE *tūn*.

Combe, Coombe, 'the valley', from OE *cumb*, a common place name, especially in the South West; examples include: **Combe** Oxon. *Cumbe* 1086 (DB). **Combe, Abbas** Somerset. *Cumbe* 1086 (DB), *Coumbe Abbatisse* 1327. Manorial addition from Latin *abbatisse* 'of the abbess', alluding to early possession by

Shaftesbury Abbey. **Combe, Castle** Wilts. *Come* 1086 (DB), *Castelcumbe* 1270. Affix *castel* referring to the Norman castle here. **Combe Florey** Somerset. *Cumba* 12th cent., *Cumbeflori* 1291. Manorial addition from the *de Flury* family, here in the 12th cent. **Combe Hay** B. & NE. Som. *Come* 1086 (DB), *Cumbehawya* 1249. Manorial addition from the family of *de Haweie*, here in the 13th cent. **Combeinteignhead** Devon. *Comba* 1086 (DB), *Cumbe in Tenhide* 1227. Affix from a district called *Tenhide* because it contained 'ten hides of land' (OE *tēn* + *hīd*). **Combe Martin** Devon. *Comba* 1086 (DB), *Cumbe Martini* 1265. Manorial addition from one *Martin* whose son held the manor in 1133. **Combe, Monkton** B. & NE. Som. *Cume* 1086 (DB). Affix means 'estate of the monks' (OE *munuc* + *tūn*). **Combe Raleigh** Devon. *Cumba* 1237, *Comberalegh* 1383. Manorial addition from the *de Ralegh* family, here in the 13th cent. **Combe St Nicholas** Somerset. *Cumbe* 1086 (DB). Affix from its possession by the Priory of St Nicholas in Exeter. **Combe, Temple** Somerset. *Come* 1086 (DB), *Cumbe Templer* 1291. Affix from its possession by the Knights Templars at an early date. **Coombe** Hants. *Cumbe* 1086 (DB). **Coombe Bissett** Wilts. *Come* 1086 (DB), *Coumbe Byset* 1288. Manorial addition from the *Biset* family, here in the 12th cent. **Coombe Keynes** Dorset. *Cume* 1086 (DB), *Combe Kaynes* 1299. Manorial addition from the *de Cahaignes* family, here in the 12th cent.

Comber (*An Comar*) Down. (*ab*) *Comair* 1221. 'The confluence'.

Comberbach Ches. *Combrebeche* 12th cent. 'Valley or stream of the Britons or of a man called Cumbra'. OE *Cumbre* or OE pers. name + *bece*.

Comberford Staffs. *Cumbreford* 1187. 'Ford of the Britons'. OE *Cumbre + ford*.

Comberton Cambs. *Cumbertone* 1086 (DB). 'Farmstead of a man called Cumbra'. OE pers. name + *tūn*.

Comberton, Great & Comberton, Little Worcs. *Cumbrincgtun* 972, *Cumbrintune* 1086 (DB). 'Estate associated with a man called Cumbra'. OE pers. name + *-ing-* + *tūn*.

Combrook Warwicks. *Cumbroc* 1217. 'Brook in a valley'. OE *cumb + brōc*.

Combs Derbys. *Cumbes* 1251. 'The valleys'. OE *cumb*.

Combs Suffolk. *Cambas* 1086 (DB). 'The hill-crests or ridges'. OE *camb*.

Commeen (*An Coimín*) Donegal. *Cuimín* 1835. 'The common land'.

Commondale N. Yorks. *Colemandale* 1272. 'Valley of a man called Colman'. OIrish pers. name + OScand. *dalr*.

Compton, a common place name, 'farmstead or village in a valley', from OE *cumb* + *tūn*; examples include: **Compton** Hants. *Cuntone* 1086 (DB). **Compton** Surrey, near Guildford. *Contone* 1086 (DB). **Compton** W. Berks. *Contone* 1086 (DB). **Compton Abbas** Dorset. *Cumtune* 956, *Cuntone* 1086 (DB), *Cumpton Abbatisse* 1293. Manorial addition from Latin *abbatisse* 'of the abbess', alluding to early possession by Shaftesbury Abbey. **Compton Abdale** Glos. *Contone* 1086 (DB), *Apdale Compton* 1504. Affix probably manorial from a family called *Apdale*. **Compton Bassett** Wilts. *Contone* 1086 (DB), *Cumptone Basset* 1228. Manorial addition from the *Basset* family, here in the 13th cent. **Compton Beauchamp** Oxon. *Cumtune* 955, *Contone* 1086 (DB), *Cumton Beucamp* 1236. Manorial addition from the *de Beauchamp* family, here in the 13th cent. **Compton Bishop** Somerset. *Cumbtune* 1067, *Compton Episcopi* 1332. Affix *Bishop* (Latin *episcopus*) from its early possession by the Bishop of Wells. **Compton Chamberlayne** Wilts. *Contone* 1086 (DB), *Compton Chamberleyne* 1316. Manorial affix from a family called *Chamberlain*, here from the 13th cent. **Compton Dando** B. & NE. Som. *Contone* 1086 (DB), *Cumton Daunon* 1256. Manorial addition from the *de Auno* or *Dauno* family, here in the 12th cent. **Compton, Fenny** Warwicks. *Contone* 1086 (DB), *Fennicumpton* 1221. The affix is OE *fennig* 'muddy, marshy'. **Compton, Little** Warwicks. *Contone parva* 1086 (DB). The affix *Little* (Latin *parva*) distinguishes it from Long Compton. **Compton, Long** Warwicks. *Cuntone* 1086 (DB), *Long Compton* 1299. The affix *Long* refers to the length of the village. **Compton Martin** B. & NE. Som. *Comtone* 1086 (DB), *Cumpton Martin* 1228. Manorial addition from one *Martin* de Tours, whose son held the manor in the early 12th cent. **Compton, Nether & Compton, Over** Dorset. *Cumbtun* 998, *Contone* 1086 (DB), *Nethercumpton* 1288, *Ouerecumton* 1268. The additions *Nether* and *Over* are from OE *neotherra* 'lower' and *uferra* 'higher'. **Compton Pauncefoot** Somerset. *Cuntone*

1086 (DB), *Cumpton Paunceuot* 1291. Manorial addition from a family called *Pauncefote*. **Compton Valence** Dorset. *Contone* 1086 (DB), *Compton Valance* 1280. Manorial addition from William *de Valencia*, Earl of Pembroke, here in the 13th cent. **Compton, West** Dorset. *Comptone* 934, *Contone* 1086 (DB). Earlier called Compton Abbas but now called West Compton to distinguish it from the other Dorset place of this name.

Comrie Perth. *Comry* 1268. 'Confluence'. Gaelic *comar*. The village stands at the junction of three rivers.

Conair (*Conair*) Kerry. 'Path'.

Conderton Worcs. *Cantuaretun* 875, *Canterton* 1201. 'Farmstead of the Kent dwellers or Kentishmen'. OE *Cantware* + *tūn*.

Condicote Glos. *Cundicotan* c.1052, *Condicote* 1086 (DB). 'Cottage associated with a man called Cunda'. OE pers. name + *-ing-* + *cot*.

Condover Shrops. *Conedoure* 1086 (DB). 'Flat-topped ridge by Cound Brook'. Celtic river-name (*see* COUND) + OE **ofer*.

Coney Hall Gtr. London. 1930s estate named from a farm called *Coney Hall* 1769. ME *coni* 'rabbit' + *hall* 'large dwelling'.

Coney Island (*Coiní*) Clare, Down. 'Rabbit island'.

Coney Weston Suffolk. *Cunegestuna* 1086 (DB). 'The king's manor, the royal estate'. OScand. *konungr* + OE *tūn*.

Coneyhurst W. Sussex. *Coneyhurst* 1574. 'Wooded hill frequented by rabbits'. ME *coni* + *hurst* (OE *hyrst*).

Coneysthorpe N. Yorks. *Coningestorp* 1086 (DB). 'The king's farmstead or hamlet'. OScand. *konungr* + *thorp*.

Cong (*Conga*) Mayo. 'Narrow stretch of water between two larger stretches'.

Congerstone Leics. *Cuningestone* 1086 (DB). Identical in origin with CONEY WESTON (Suffolk).

Congham Norfolk. *Congheham* 1086 (DB). Possibly 'homestead or village at the hill'. OE **cung* + *hām*.

Congleton Ches. *Cogeltone* [sic] 1086 (DB), *Congulton* 13th cent. Probably 'farmstead at the round-topped hill'. OE **cung* + *hyll* + *tūn*.

Congresbury N. Som. *Cungresbyri* 9th cent., *Cungresberie* 1086 (DB). 'Fortified place or manor associated with a saint called Congar'. Celtic pers. name + OE *burh* (dative *byrig*).

Coningsby Lincs. *Cuningesbi* 1086 (DB). 'The king's manor or village'. OScand. *konungr* + *bý*.

Conington, 'the king's manor, the royal estate', OScand. *konungr* + OE *tūn*: **Conington** Cambs., near St Ives. *Cunictune* 10th cent., *Coninctune* 1086 (DB). **Conington** Cambs., near Sawtry. *Cunningtune* 10th cent., *Cunitone* 1086 (DB).

Conisbrough Donc. *Cunugesburh* c.1003, *Coningesburg* 1086 (DB). 'The king's fortification'. OScand. *konungr* + OE *burh*.

Coniscliffe, High & Coniscliffe, Low Darltn. *Cingcesclife* 1040. 'The king's cliff or bank'. OE *cyning* (with later influence from OScand. *konungr*) + *clif*.

Conisholme Lincs. *Cunyngesholme* c.1155. 'Island, or dry ground in marsh, belonging to the king'. OScand. *konungr* + *holmr*.

Coniston, Conistone, 'the king's manor, the royal estate', OScand. *konungr* + OE *tūn*: **Coniston** Cumbria. *Coningeston* 12th cent. **Coniston** E. R. Yorks. *Coningesbi* [*sic*] 1086 (DB), *Cuningeston* 1190. OE *tūn* replaced OScand. *bý* 'village' as second element. **Coniston Cold** N. Yorks. *Cuningestone* 1086 (DB), *Calde Cuningeston* 1202. Affix *Cold* (from OE *cald*) from its exposed situation. **Conistone** N. Yorks. *Cunestune* 1086 (DB).

Conlig (*An Choinleac*) Down. (*feall*) *na Coinnleice* 1744. 'The hound stone'. The sense of 'hound' is uncertain.

Connacht (*Cúige Chonnacht*) (the province). 'Province of the Connachta people'.

Connah's Quay Flin. *Connas Quay* 1791. 'Connah's quay'. *Connah* may have been the local innkeeper.

Connaught. See CONNACHT.

Connemara (*Conamara*) (district) Galway. *Conmaicne-mara* n.d. 'Conmaicne of the sea'. Irish tribal name + *muir*.

Connor (*Coinnire*) Antrim. (*espoc*) *Condere* 514. 'Dog oak wood'.

Conock Wilts. *Kunek* 1242, *Cunnuc* c.1250. Possibly Celtic or pre-Celtic **cunuc* of uncertain meaning (*see* CONSETT).

Cononley N. Yorks. *Cutnelai* 1086 (DB), *Conanlia* 12th cent. OE *lēah* 'wood, clearing' with an uncertain first element, possibly an old Celtic river-name or an OIrish pers. name.

Consett Durham. *Conekesheued* 1183. Probably 'head or end of **Cunuc*'. Uncertain Celtic or pre-Celtic element **cunuc* (possibly a hill-name) + OE *hēafod*.

Constable Burton N. Yorks. *See* BURTON.

Constantine Cornwall. *Sanctus Constantinus* 1086 (DB). 'Church of St Constantine'. From the patron saint of the church.

Contin Highland. *Conten* 1226. 'Confluence'. Precise origin unknown.

Convoy (*Conmhaigh*) Donegal. *Convoigh* 1610. 'Hound plain'.

Conwal (*Congbháil*) Donegal, Leitrim. 'Foundation'.

Conwy Conwy. *Conguoy* 12th cent., *Aberconuy* 12th cent., *Conway als. Aberconway* 1698, *Conway*, or *Conwy* 1868. '(Place on the) River Conwy'. The Celtic river-name, meaning, 'reedy one', gave the name of *Canovium*, the Roman town at nearby Caerhun. Conwy was formerly known as *Aberconwy*, 'mouth of the Conwy' (OWelsh *aber*).

Cookbury Devon. *Cukebyr* 1242. 'Fortification of a man called **Cuca*'. OE pers. name + *burh* (dative *byrig*).

Cookham Winds. & Maid. *Coccham* 798, *Cocheham* 1086 (DB). Possibly 'cook village', i.e. 'village noted for its cooks'. OE *cōc* + *hām*. However the first element may be an OE *cōc(e)* 'hill'.

Cookham Dean Winds. & Maid. *la Dene* 1220. 'The valley (at COOKHAM)'. OE *denu*.

Cookhill Worcs. *Cochilla* 1156. OE **cōc(e)* 'hill' with explanatory OE *hyll*.

Cookley Worcs. *Culnan clif* 964, *Culleclive* 11th cent. Possibly 'cliff of a man called **Cūlna*'. OE pers. name + *clif*.

Cookley Suffolk. *Cokelei* 1086 (DB), *Kukeleia* late 12th cent. Possibly 'wood or clearing of a man called **Cuca*', OE pers. name + *lēah*, but the first element may rather be an OE **cucu* 'cuckoo'.

Cookstown Tyrone. *a' Corr Críochach* c.1645. 'Cook's town'. Alan *Cook* founded a settlement here in 1609. The town's Irish name is *An Chorr Chríochach*, 'the boundary hill'.

Coolaney (*Cúil Mhaine*) Sligo. 'Nook of the thicket'.

Coolattin (*Cúl Aitinn*) Limerick, Wexford, Wicklow. 'Hill of gorse'.

Coolbaun (*An Cúl Bán*) Kilkenny, Tipperary. 'The white hill'.

Coolboy (*An Cúl Buí*) Donegal, Wicklow. 'The yellow hill'.

Coolcarrigan (*Cúil Charraigín*) Kildare. 'Nook of the little rock'.

Coolcor (*Cúil Chorr*) Offaly. 'Nook of the hills'.

Coolcullen (*Cúl an Chuillinn*) Kilkenny. 'Back of the steep slope'.

Coolderry (*Cúl Doire*) Monaghan, Offaly. 'Back of the oak wood'.

Coole (*An Chúil*) Westmeath. 'The nook'.

Coolea (*Cúil Aodha*) Cork. 'Aodh's nook'.

Cooley Point (*Cuailnge*) Louth. Old district name of unknown meaning (now Cuaille), with English *point*.

Coolgrange (*Cúil Ghráinseach*) Kilkenny. 'Nook of the monastic farm'.

Coolgreany (*Cúil Ghréine*) Wexford. 'Nook of the sun'.

Cooligrain (*Cúl le Gréin*) Leitrim. 'Nook of the sun'.

Cooling Medway. *Culingas* 808, *Colinges* 1086 (DB). '(Settlement of) the family or followers of a man called *Cūl* or *Cūla*'. OE pers. name + *-ingas*.

Coolkeeragh (*Cúil Chaorach*) Derry. 'Nook of the sheep'.

Coolkenna (*Cúil Uí Chionaoith*) Wicklow. 'Nook of the Ó Cionaoith'.

Coolmeen (*Cúil Mhín*) Clare. 'Smooth nook'.

Coolmore (*Cúl Mór*) Donegal. 'Big hill'.

Coolock (*An Chúlóg*) Dublin. 'The little corner'.

Coolrain (*Cúil Ruáin*) Laois. 'Nook of the red ground'.

Coolsallagh (*Cúil Saileach*) Down. 'Nook of the willows'.

Coomatloukane (*Com an tSleabhcáin*) Kerry. 'Hollow of the edible seaweed'.

Coombe, *See* COMBE.

Coomnagoppul (*Com na gCapall*) Kerry. 'Hollow of the horses'.

Cooneen (*An Cúinnín*) Fermanagh. *Cunen* 1659. 'The little corner'.

Cooraclare (*Cuar an Chláir*) Clare. 'Curve of the plain'.

Cootehill Cavan. *Coote Hill* 1725. The town is named from Thomas *Coote*, who acquired lands here in the early 17th cent., and his future wife, Frances *Hill*. The Irish name is *Muinchille*, 'sleeve'.

Copdock Suffolk. *Coppedoc* 1195. 'Pollarded oak-tree, i.e. oak-tree with its top removed'. OE **coppod* + *āc*.

Copeland (district) Cumbria. *Coupland* c.1282. 'Purchased land'. OScand. *kaupa-land*. An old name also preserved in Copeland Forest.

Copeland Island Down. *Kaupmanneyjar* 1230. 'Merchants' island'. OScand. *kaup-mathr* (genitive plural *kaup-manna*) + *ey*. Alternatively 'Kaupman's island', with an OScand. surname, later influenced by the Anglo-Norman surname *Copeland*. The equivalent Irish name is *Oileán Chóplainn*.

Copford Green Essex. *Coppanforda* 995, *Copeforda* 1086 (DB). 'Ford of a man called **Coppa*'. OE pers. name + *ford* with the later addition of *grene* 'a village green'.

Cople Beds. *Cochepol* 1086 (DB). 'Pool of a man called **Cocca*', or 'pool frequented by cocks (of wild birds)'. OE pers. name or *cocc* + *pōl*.

Copmanthorpe York. *Copeman Torp* 1086 (DB). 'Outlying farmstead or hamlet belonging to the merchants'. OScand. *kaup-mathr* (genitive plural *-manna*) + *thorp*.

Coppeen (*An Caipín*) Cork. 'Cap'.

Coppenhall Staffs. *Copehale* 1086 (DB). 'Nook of land of a man called **Coppa*'. OE pers. name (genitive *-n*) + *halh*.

Coppingford Cambs. *Copemaneforde* 1086
(DB). 'Ford used by merchants'. OScand. *kaup-mathr* (genitive plural *-manna*) + OE *ford*.

Copplestone Devon. *Copelan stan* 974.
'Peaked or towering stone'. OE **copel* + *stān*.

Coppull Lancs. *Cophill* 1218. 'Hill with a
peak'. OE *copp* + *hyll*.

Copt Hewick N. Yorks. *See* HEWICK.

Copthorne W. Sussex. *Coppethorne* 1437.
'Pollarded thorn-tree, i.e. thorn-tree with its top
removed'. OE **coppod* + *thorn*.

Corballa (*An Corrbhaile*) Sligo. 'The odd
homestead'.

Corbally (*Corrbhaile*) Clare. 'Odd
homestead'.

Corbet (*Carbad*) Down. *Corbudd* 1609. 'Jaw',
referring to some unidentified feature.

Corbo (*Corr Bhó*) Tyrone. 'Round hill of the
cow'.

Corbridge Northum. *Corebricg c.*1050.
'Bridge near Corchester'. OE *brycg* 'bridge' with
a shortened form of the old Celtic name of
Corchester (*Corstopitum*) which is near here.

Corby Northants. *Corbei* 1086 (DB).
'Farmstead or village of a man called Kori'.
OScand. pers. name + *bý*.

Corby Glen Lincs. *Corbi* 1086 (DB). Identical
in origin with previous name. The affix is from
the River Glen, a Celtic river-name meaning 'the
clean one'.

Corby, Great Cumbria. *Chorkeby c.*1115.
'Farmstead or village of a man called Corc'.
OIrish pers. name + OScand. *bý*.

Corclogh (*Corr Cloch*) Mayo. 'Snout of the
stone'.

Corcreaghy (*An Chorr Chríochach*) Louth.
'The snout of the boundary'.

Coreley Shrops. *Cornelie* 1086 (DB).
'Woodland clearing frequented by cranes or
herons'. OE *corn* + *lēah*.

Corfe, 'the cutting, the gap or pass', from OE
*corf: **Corfe** Somerset. *Corf* 1243. **Corfe
Castle** Dorset. *Corf* 955, *Corffe Castell* 1302.
The affix refers to the Norman castle here.
Corfe Mullen Dorset. *Corf* 1086 (DB), *Corf le*

Mulin 1176. The affix is from OFrench *molin* 'a
mill'.

Corfton Shrops. *Cortune* 1086 (DB), *Corfton*
1222. 'Settlement by the River Corve'. OE river-
name (from **corf* 'a pass') + *tūn*.

Corglass (*Corr Ghlas*) Leitrim. 'Grey-green
hill'.

Corhampton Hants. *Cornhamton* 1201.
'Home farm or settlement where grain is
produced'. OE *corn* + *hām-tūn*.

Cork (*Corcaigh*) Cork. 'Swamp'.

Corkey (*Corcaigh*) Antrim. 'Swamp'.

Corlead (*Corr Liath*) Longford. 'Grey
rounded hill'.

Corlesmore (*Corrlios Mór*) Cavan. 'Big
rounded hill of the fort'.

Corley Warwicks. *Cornelie* 1086 (DB).
'Clearing frequented by cranes or herons'. OE
corn + *lēah*.

Cornafanog (*Corr na bhFeannóg*)
Fermanagh. *Cornafenoge* 1751. 'Round hill of
the crows'.

Cornafean (*Corr na Féinne*) Cavan. 'Hill of
the Fianna'.

Cornafulla (*Corr na Fola*) Roscommon. 'Hill
of the blood'.

Cornamona (*Corr na Móna*) Galway. 'Hill of
the bog'.

Cornard, Great & Cornard, Little Suffolk.
Cornerda 1086 (DB). 'Cultivated land used for
corn'. OE *corn* + *erth*.

Corndarragh (*Carn Darach*) Offaly. 'Heap of
the oak-tree'.

Corney Cumbria. *Corneia* 12th cent. 'Island
frequented by cranes or herons'. OE *corn* + *ēg*.

Cornforth Durham. *Corneford c.*1116. 'Ford
frequented by cranes or herons'. OE *corn* + *ford*.

Cornhill-on-Tweed Northum. *Cornehale*
12th cent. 'Nook of land frequented by cranes or
herons'. OE *corn* + *halh*. For the river-name, *see*
TWEEDMOUTH.

Cornsay Durham. *Corneshowe* 1183. 'Hill-
spur frequented by cranes or herons'. OE *corn* +
hōh.

Cornwall (the county). *Cornubia* c.705, *Cornwalas* 891, *Cornualia* 1086 (DB). '(Territory of) the Britons or Welsh of the *Cornovii* tribe'. Celtic tribal name (meaning 'peninsula people') + OE *walh* (plural *walas*).

Cornwell Oxon. *Cornewelle* 1086 (DB). 'Stream frequented by cranes or herons'. OE *corn* + *wella*.

Cornwood Devon. *Cornehuda* [*sic*] 1086 (DB), *Curnwod* 1242. Probably 'wood frequented by cranes'. OE *corn* + *wudu*.

Cornworthy Devon. *Corneorda* 1086 (DB). 'Enclosure frequented by cranes, or where corn is grown'. OE *corn* + *worth, worthig*.

Corofin (*Cora Finne*) Clare. 'Weir of the white (water)'.

Corpusty Norfolk. *Corpestih* 1086 (DB). 'Raven path', or 'path of a man called Korpr'. OScand. *korpr* or pers. name + *stígr*.

Corracullin (*Corr an Chuillinn*) Offaly, Westmeath. 'Hill of the steep slope'.

Corrandulla (*Cor an Dola*) Galway. 'Hill of the loop'.

Corranny (*Corr Eanaigh*) Fermanagh. *Corroney* 1659. 'Hill of the bog'.

Corraun (*Corrán*) Mayo. 'Crescent-shaped place'.

Corrib, Lough (*Loch Coirib*) Galway. 'Lake of Oirbse'.

Corriga (*Carraigigh*) Leitrim. 'Abounding in rocks'.

Corrigeenroe (*Carraigín Rua*) Roscommon. 'Little red rock'.

Corringham Lincs. *Coringeham* 1086 (DB). 'Homestead of the family or followers of a man called *Cora*'. OE pers. name + *-inga-* + *hām*.

Corringham Thurr. *Currincham* 1086 (DB). 'Homestead of the family or followers of a man called *Curra*'. OE pers. name + *-inga-* + *hām*.

Corrofin (*Cora Finne*) Clare, Galway. 'Weir of the white (water)'.

Corry (*An Choraigh*) Leitrim. 'The pleasant place'.

Corscombe Dorset. *Corigescumb* 1014, *Coriescumbe* 1086 (DB). 'Valley of a stream called *Cori*'. Old pre-English river-name of

uncertain origin and meaning (for the stream rising south of the village) + OE *cumb*.

Corsham Wilts. *Coseham* 1001, *Cosseham* 1086 (DB). 'Homestead or village of a man called *Cosa* or *Cossa*'. OE pers. name + *hām*.

Corsley Wilts. *Corselie* 1086 (DB). 'Wood or clearing by the marsh'. Celtic *cors* + OE *lēah*.

Corston B. & NE. Som. *Corsantune* 941, *Corstune* 1086 (DB). 'Farmstead or estate on the River Corse'. Old river-name (from Celtic *cors* 'marsh') + OE *tūn*.

Corston Wilts. *Corstuna* 1065, *Corstone* 1086 (DB). 'Farmstead or estate on Gauze Brook'. Old river-name (from Celtic *cors* 'marsh') + OE *tūn*.

Corstorphine Edin. *Crostorfin* c.1140. 'Thorfinn's crossing'. OScand. *kros* + pers. name.

Corton Suffolk. *Karetuna* 1086 (DB). 'Farmstead of a man called Kári'. OScand. pers. name + OE *tūn*.

Corton Wilts. *Cortitone* 1086 (DB). 'Estate associated with a man called *Cort(a)*'. OE pers. name + *-ing-* + *tūn*.

Corton Denham Somerset. *Corfetone* 1086 (DB). 'Farmstead or village at a pass'. OE *corf* + *tūn* + manorial affix from the *de Dinan* family, here in the early 13th cent.

Cortown (*An Baile Corr*) Meath. 'The homestead of the twist'.

Cortubber (*Corr Tobair*) Monaghan. 'Hill of the well'.

Corvally (*Cor an Bhealaigh*) Monaghan. 'Twist of the pass'.

Corwen Denb. *Corvaen* 1254, *Korvaen* 14th cent., *Corwen* 1443. 'Sanctuary stone'. Welsh *côr* + *maen*. The church has a reputedly ancient stone called *Carreg-y-Big* ('pointed stone') built into the porch, and this may be the 'sanctuary stone'.

Coryton Devon. *Cur(r)itun* c.970, *Coriton* 1086 (DB). Probably 'farmstead on the River *Curi*'. Pre-English river-name (an old name for the River Lyd) + OE *tūn*.

Coryton Thurr. a modern name from the oil refinery established here by *Cory* & Co. in 1922.

Cosby Leics. *Cossebi* 1086 (DB). 'Farmstead or village of a man called *Cossa'. OE pers. name + OScand. *bý*.

Coseley Dudley. *Colseley* 1357. Possibly 'clearing of the charcoal burners'. OE *colestre* + *lēah*.

Cosgrove Northants. *Covesgrave* 1086 (DB). 'Grove of a man called *Cōf'. OE pers name + *grāf*.

Cosham Portsm. *Cos(s)eham* 1086 (DB). 'Homestead or enclosure of a man called *Cossa'. OE pers. name + *hām* or *hamm*.

Cossall Notts. *Coteshale* 1086 (DB). 'Nook of land of a man called *Cott'. OE pers. name + *halh*.

Cossington, 'estate associated with a man called *Cosa or Cusa', OE pers. name + *-ing-* + *tūn*: **Cossington** Leics. *Cosintone* 1086 (DB). **Cossington** Somerset. *Cosingtone* 729, *Cosintone* 1086 (DB).

Costessey Norfolk. *Costeseia* 1086 (DB). 'Island, or dry ground in marsh, of a man called *Cost'. OE (or OScand.) pers. name + OE *ēg*.

Costock Notts. *Cortingestoche* 1086 (DB). 'Outlying farmstead of the family or followers of a man called *Cort'. OE pers. name + *-inga-* + *stoc*.

Coston Leics. *Castone* 1086 (DB). Probably 'farmstead of a man called Kátr'. OScand. pers. name + OE *tūn*.

Cotes, 'the cottages or huts', from OE *cot* (ME plural *cotes*): **Cotes** Leics. *Cotes* 12th cent. **Cotes** Staffs. *Cota* 1086 (DB).

Cotesbach Leics. *Cotesbece* 1086 (DB). 'Stream or valley of a man called *Cott'. OE pers. name + *bece*, *bæce*.

Cotgrave Notts. *Godegrave* [sic] 1086 (DB), *Cotegrava* 1094. 'Grove or copse of a man called Cotta'. OE pers. name + *grāf*.

Cotham Notts. *Cotune* 1086 (DB). '(Place at) the cottages or huts'. OE *cot* in a dative plural form *cotum*.

Cothelstone Somerset. *Cothelestone* 1327. 'Farmstead of a man called Cūthwulf'. OE pers. name + *tūn*.

Cotherstone Durham. *Codrestune* 1086 (DB). 'Farmstead of a man called Cūthhere'. OE pers. name + *tūn*.

Cotleigh Devon. *Coteleia* 1086 (DB). 'Clearing of a man called Cotta'. OE pers. name + *lēah*.

Coton, '(place at) the cottages or huts', from OE *cot* in the dative plural form *cotum* (though some may be from a ME plural *coten*): **Coton** Cambs. *Cotis* 1086. **Coton** Northants. *Cote* 1086 (DB). **Coton** Staffs., near Milwich. *Cote* 1086 (DB). **Coton Clanford** Staffs. *Cote* 1086 (DB). Affix from a local place name, probably 'clean ford' from OE *clǣne* + *ford*. **Coton in the Elms** Derbys. *Cotune* 1086 (DB). Affix from the abundance of elm-trees here.

Coton, Chilvers Warwicks. *Celverdestoche* 1086 (DB), *Chelverdescote* 1185. 'Cottage(s) of a man called Cēolfrith'. OE pers. name + *cot* (ME plural *coten*). The DB form contains an alternative element OE *stoc* 'hamlet'.

Cotswolds Glos. /Worcs. *Codesuualt* 12th cent. OE *wald* 'high forest-land' with an uncertain first element, possibly an OE masculine pers. name *Cōd*.

Cottam, '(place at) the cottages or huts', from OE *cot* in a dative plural form *cotum*: **Cottam** E. R. Yorks. *Cottun* 1086 (DB). **Cottam** Lancs. *Cotun* 1227. **Cottam** Notts. *Cotum* 1274.

Cottenham Cambs. *Cotenham* 948, *Coteham* 1086 (DB). 'Homestead or village of a man called Cot(t)a'. OE pers. name (genitive *-n*) + *hām*.

Cottered Herts. *Chodrei* [sic] 1086 (DB), *Codreth* 1220. OE *rīth* 'stream' with an uncertain first element, possibly a pers. name or an OE *cōd* 'spawn of fish'.

Cotterstock Northants. *Codestoche* 1086 (DB), *Cotherstoke* 12th cent. Probably from an OE *corther-stoc* 'dairy farm'.

Cottesbrooke Northants. *Cotesbroc* 1086 (DB). 'Brook of a man called *Cott.' OE pers. name + *brōc*.

Cottesmore Rutland. *Cottesmore* c.976, *Cotesmore* 1086 (DB). 'Moor of a man called *Cott'. OE pers. name + *mōr*.

Cottingham, 'homestead of the family or followers of a man called *Cott or Cotta', OE pers. name + *-inga-* + *hām*: **Cottingham** E. R. Yorks. *Cotingeham* 1086 (DB). **Cottingham** Northants. *Cotingeham* 1086 (DB).

Cottingwith E. R. Yorks. *Coteuuid* [sic] 1086 (DB), *Cotingwic* 1195. 'Dairy-farm associated with a man called *Cott or Cotta'. OE pers. name + *-ing-* + *wīc* (replaced by OScand. *vithr* 'wood').

Cottisford Oxon. *Cotesforde* 1086 (DB). 'Ford of a man called *Cott'. OE pers. name + *ford*.

Cotton, Far Northants. *Cotes* 1196. 'The cottages or huts'. OE *cot* in a plural form, *see* COTON. Affix *Far* to distinguish it from other places with this name.

Cotwalton Staffs. *Cotewaltun* 1002, *Cotewoldestune* 1086 (DB). Possibly 'farmstead at the wood or stream of a man called Cotta'. OE pers. name + *wald* or *wælla* + *tūn*.

Coughton Warwicks. *Coctune* 1086 (DB). Probably 'farmstead near the hillock'. OE *cocc* + *tūn*.

Coulsdon Gtr. London. *Cudredesdune* 967, *Colesdone* 1086 (DB), *Cul(l)esdon* 12th cent. Pre-Conquest spellings indicate 'hill of a man called Cūthrǣd', from OE pers. name + *dūn*. DB and other early medieval spellings suggest a different (or alternative) first element, perhaps a hill-name *Cull from Celtic *cull 'bosom, belly'.

Coulston, East & Coulston, West Wilts. *Covelestone* 1086 (DB). 'Farmstead of a man called *Cufel'. OE pers. name + *tūn*.

Coulton N. Yorks. *Coltune* 1086 (DB). Probably 'farmstead where charcoal is made'. OE *col* + *tūn*.

Cound Shrops. *Cuneet* 1086 (DB). Named from Cound Brook, a Celtic river-name of uncertain meaning.

Coundon Durham. *Cundon* 1197. Probably 'hill where cows are pastured'. OE *cū* (genitive plural *cūna*) + *dūn*.

Countesthorpe Leics. *Torp* 1209–35, *Cuntastorp* 1242. 'Outlying farmstead or hamlet belonging to a countess'. OScand. *thorp* with manorial affix from ME *contesse*.

Countisbury Devon. *arx Cynuit* c.894, *Contesberie* 1086 (DB). 'Fort or stronghold of a man called Cynuit'. Celtic pers. name + OE *burh* (dative *byrig*).

Coupar Angus Perth. Meaning unknown. The name locates the place in ANGUS for distinction from CUPAR in Fife.

Coupland Northum. *Coupland* 1242. 'Purchased land'. OScand. *kaupa-land*.

Courteenhall Northants. *Cortenhale* 1086 (DB). Possibly 'nook of land of a man called *Corta or *Curta'. OE pers. name (genitive *-n*) + *halh*. Alternatively the first element may be OE *corte (genitive *-n*) 'enclosed farmstead'.

Courtmacsherry (*Cúirt Mhic Shéafraidh*) Cork. 'Mansion of Mac Shéafraidh'. The personal name is the Irish form of Geoffrey.

Cove, from OE *cofa* which meant 'hut or shelter', also 'recess or cove': **Cove** Devon. *La Kove* 1242. **Cove** Hants. *Coue* 1086 (DB). **Cove, North** Suffolk. *Cove* 1204. **Cove, South** Suffolk. *Coua* 1086 (DB).

Covehithe Suffolk. *Coveheith* 1523. 'Harbour near (South) COVE', from OE *hȳth*.

Coven Staffs. *Cove* 1086 (DB). '(Place at) the huts or shelters'. OE *cofa* in the dative plural form *cofum*.

Coveney Cambs. *Coueneia* c.1060. 'Island of a man called *Cofa', or 'cove island'. OE pers. name or OE *cofa* (genitive *-n*) + *ēg*.

Covenham St Bartholomew & Covenham St Mary Lincs. *Covenham* 1086 (DB). 'Homestead of a man called *Cofa', or 'homestead with a hut or shelter'. OE pers. name or OE *cofa* (genitive *-n*) + *hām*. Affixes from the dedications of the churches.

Coventry Covtry. *Couentre* 1043, *Couentreu* 1086 (DB). 'Tree of a man called *Cofa'. OE pers. name (genitive *-n*) + *trēow*.

Coverack Cornwall. *Covrack* 1588. Probably originally the name of the stream here, meaning unknown.

Coverham N. Yorks. *Covreham* 1086 (DB). 'Homestead or village on the River Cover'. Celtic river-name ('welling stream') + OE *hām*.

Covington Cambs. *Covintune* 1086 (DB). 'Estate associated with a man called *Cofa'. OE pers. name + *-ing-* + *tūn*.

Cow Honeybourne Worcs. *See* HONEYBOURNE.

Cowarne, Much & Cowarne, Little Herefs. *Cuure* [sic] 1086 (DB), *Couern* 1255. 'Cow house, dairy farm'. OE *cū* + *ærn*. Affix *Much* is from OE *mycel* 'great'.

Cowbit Lincs. *Coubiht* 1267. 'River-bend where cows are pastured'. OE *cū* + *byht*.

Cowbridge (*Y Bont-faen*) Vale Glam. *Coubrugge* 1263, *Bontvaen* c.1500. 'Cattle bridge'. OE *cu* + *brycg*. The Welsh name means 'the stone bridge', referring to another bridge nearby.

Cowden Kent. *Cudena* c.1100, 'Pasture for cows'. OE *cū* + *denn*.

Cowden, Great E. R. Yorks. *Coledun* 1086 (DB). 'Hill where charcoal is made'. OE *col* + *dūn*.

Cowdenbeath Fife. *terris de Baithe-Moubray alias Cowdounes-baithe* 1626. 'Cowden's part of Beath ("birch-trees")'. Gaelic *beith*. *Moubray* and *Cowdoun/-den* are probably surnames.

Cowes, East Cowes I. of Wight. settlements on either side of the River Medina estuary named from two former sandbanks referred to as *Estcowe* and *Westcowe* in 1413, so-called like other coastal features from their fancied resemblance to animals. OE *cū* 'a cow'.

Cowesby N. Yorks. *Cahosbi* 1086 (DB). 'Farmstead or village of a man called Kausi'. OScand. pers. name + *bý*.

Cowfold W. Sussex. *Coufaud* 1232. 'Small enclosure for cows'. OE *cū* + *fald*.

Cowick, East & Cowick, West E. R. Yorks. *Cuwich* 1197, 'Cow farm, dairy farm'. OE *cū* + *wīc*.

Cowley Devon. near Exeter. *Couelegh* 1237. 'Clearing of a man called *Cofa* or *Cufa*'. OE pers. name + *lēah*.

Cowley Glos. *Kulege* 1086 (DB). 'Clearing where cows are pastured'. OE *cū* + *lēah*.

Cowley Gtr. London. *Cofenlea* 959, *Covelie* 1086 (DB). 'Clearing of a man called *Cofa*'. OE pers. name + *lēah*.

Cowley Oxon. *Couelea* 1004, *Covelie* 1086 (DB). 'Clearing of a man called *Cofa* or *Cufa*'. OE pers. name + *lēah*.

Cowling N. Yorks. near Bedale. *Torneton* 1086 (DB), *Thornton Colling* 1202. Originally 'farmstead where thorn-trees grow' from OE *thorn* + *tūn*, with manorial affix from a person or family called *Colling*. The first part of the name went out of use in the 15th cent.

Cowling N. Yorks. near Glusburn. *Collinghe* 1086 (DB). 'Place characterized by a hill'. OE **coll* + *-ing*.

Cowlinge Suffolk. *Culinge* 1086 (DB). Probably 'place associated with a man called **Cūl* or **Cūla*'. OE pers. name + *-ing*.

Cowpen Bewley Stock. on T. *Cupum* c.1160. '(Place by) the coops or baskets (for catching fish)'. OE **cūpe* in a dative plural form **cūpum*. Affix from its possession by the manor of Bewley ('beautiful place', OFrench *beau* + *lieu*).

Cowplain Hants. recorded from 1859, self-explanatory.

Cowton, East & Cowton, North N. Yorks. *Cudtone, Cotun* 1086 (DB). Probably 'cow farm, dairy farm'. OE *cū* + *tūn*.

Coxhoe Durham. *Cokishow* c.1240. Probably 'hill-spur of a man called **Cocc*'. OE pers. name + *hōh*.

Coxley Somerset. *Cokesleg* 1207. 'Wood or clearing belonging to the cook'. OE *cōc* + *lēah*. The wife of a cook in the royal household is recorded as holding lands near Wells in 1086 (DB).

Coxwell, Great & Coxwell, Little Oxon. *Cocheswelle* 1086 (DB), *Cokeswell* 1227. Probably 'spring or stream of a man called **Cocc*'. OE pers. name + *wella*.

Coxwold N. Yorks. *Cuhawalda* [sic] 758, *Cucualt* 1086 (DB), *Cukewald* 1154–89. Probably 'woodland haunted by the cuckoo'. OE **cucu* + *wald*.

Craanford (*Áth an Chorráin*) Wexford. 'Ford of the crescent'.

Crackenthorpe Cumbria. *Cracantorp* late 12th cent. 'Outlying farmstead or hamlet of a man called **Krakandi*, or one frequented by crows or ravens'. OScand. pers. name or OE **crāca* (genitive plural *-ena*) + OScand. *thorp*.

Crackpot N. Yorks. *Crakepote* 1298. 'Limestone cleft frequented by crows or ravens'. OScand. *kráka* + ME *potte*.

Cracoe N. Yorks. *Crakehou* 12th cent. 'Spur of land or hill frequented by crows or ravens'. OScand. *kráka* + OE *hōh* or OScand. *haugr*.

Cradley Herefs. *Credelaie* 1086 (DB). 'Woodland clearing of a man called Creoda'. OE pers. name + *lēah*.

Craggagh (*An Chreagach*) Clare. 'The rocky place'.

Craig (*An Chreig*) Tyrone. 'The rock'.

Craigatuke (*Creig an tSeabhaic*) Tyrone. 'Rock of the hawk'.

Craigavad (*Creig an Bháda*) Down. 'Rock of the boat'.

Craigavon Armagh. Craigavon was designated a New Town in 1965 and named after James *Craig*, first prime minister of Northern Ireland, whose title was Viscount *Craigavon*.

Craigban (*Creig Bán*) Antrim. 'White rock'.

Craigbane (*Creig Bán*) Derry. 'White rock'.

Craigboy (*Creig Buí*) Galway. 'Yellow rock'.

Craigbrack (*Creig Breac*) Derry. 'Speckled rock'.

Craigdoo (*Creig Dubh*) Donegal. 'Black rock'.

Craigie Perth. *Cragyn* 1266. '(Place) by the rock'. Gaelic *creag*.

Craiglea (*Creig Liath*) Derry. 'Grey rock'.

Craigmore (*Creig Mór*) Antrim, Derry. 'Big rock'.

Craigmore Stir. (Bute). 'Big rock'. Gaelic *creag + mór*.

Craignagat (*Creig na gCat*) Antrim. 'Rock of the cats'.

Craigs (*Na Creaga*) Antrim. 'The rocks'.

Crail Fife. *Cherel c.*1150, *Caraile* 1153. May contain British *cair* 'fortified place'.

Crakehall, Great & Crakehall, Little N. Yorks. *Crachele* 1086 (DB). 'Nook of land frequented by crows or ravens'. OScand. *kráka* or *krákr* + OE *halh*.

Crambe N. Yorks. *Crambom* 1086 (DB). '(Place at) the river-bends'. OE **cramb* in a dative plural form **crambum*.

Cramlington Northum. *Cramlingtuna* c.1130. Possibly 'farmstead of the people living at the cranes' stream'. OE *cran + wella + -inga- + tūn*.

Cran (*Crann*) Cavan, Fermanagh. 'Tree'.

Crana (*Crannach*) (river) Donegal. 'Abounding in trees'.

Cranage Ches. *Croeneche* [sic] 1086 (DB), *Cranlach* 12th cent. 'Boggy place or stream frequented by crows'. OE *crāwe* (genitive plural *crāwena*) + **læc(c)*.

Cranagh (*An Chrannóg*) Tyrone. 'The place abounding in wood'.

Cranborne Dorset. *Creneburne* 1086 (DB). 'Stream frequented by cranes or herons'. OE *cran + burna*. It gives name to **Cranborne Chase**, first recorded in the 13th cent., from ME *chace* 'tract of land for breeding and hunting wild animals'. The river-name **Crane** is a back-formation from the place name.

Cranbrook Kent. *Cranebroca* 11th cent. 'Brook frequented by cranes or herons'. OE *cran + brōc*. The river-name **Crane** is a back-formation from the place name.

Crancam (*Crann Cam*) Roscommon. 'Crooked tree'.

Cranfield Beds. *Crangfeldæ* 1060, *Cranfelle* 1086 (DB). 'Open land frequented by cranes or herons'. OE *cran, cranuc + feld*.

Cranford, 'ford frequented by cranes or herons', OE *cran + ford*: **Cranford** Gtr. London. *Cranforde* 1086 (DB). **Cranford St Andrew & Cranford St John** Northants. *Craneford* 1086 (DB), *Craneford Sancti Andree*, *Craneford Sancti Iohannis* 1254. Affixes from the dedications of the churches.

Cranford (*Creamhgort*) Donegal. *Crawarte* 1603. 'Garlic field'.

Cranham Glos. *Craneham* 12th cent. 'Enclosure or river-meadow frequented by cranes or herons'. OE *cran + hamm*.

Cranham Gtr. London. *Craohv* [sic] 1086 (DB), *Crawenho* 1201. 'Spur of land frequented by crows'. OE *crāwe* (genitive plural *crāwena*) + *hōh* (later spelling *-ham* perhaps reflecting a dative plural *hōum*).

Cranleigh Surrey. *Cranlea* 1166. 'Woodland clearing frequented by cranes or herons'. OE *cran + lēah*.

Cranmore, East & Cranmore, West Somerset. *Cranemere* 10th cent., *Crenemelle* [sic] 1086 (DB). 'Pool frequented by cranes or herons'. OE *cran + mere*.

Crann (*Crann*) Armagh. 'Tree'.

Crannagh (*Crannach*) Galway, Laois, Mayo, Roscommon, Tyrone. 'Abounding in trees'.

Crannogue (*Crannóg*) Tyrone. 'Artificial island'.

Crannogueboy (*Crannóg Buí*) Donegal. 'Yellow artificial island'.

Cranny (*Crannach*) Clare, Derry, Donegal, Tyrone. 'Abounding in trees'.

Cranoe Leics. *Craweho* 1086 (DB). 'Spur of land frequented by crows'. OE *crāwe* (genitive plural *crāwena*) + *hōh*.

Cransford Suffolk. *Craneforda* 1086 (DB). 'Ford frequented by cranes or herons'. OE *cran* + *ford*.

Cransley, Great Northants. *Cranslea* 956, *Craneslea* 1086 (DB). 'Woodland clearing frequented by cranes or herons'. OE *cran* + *lēah*.

Cranswick E. R. Yorks. *See* HUTTON CRANSWICK.

Crantock Cornwall. *Sanctus Carentoch* 1086 (DB). 'Church of St Carantoc'. From the patron saint of the church.

Cranwell Lincs. *Craneuuelle* 1086 (DB). 'Spring or stream frequented by cranes or herons'. OE *cran* + *wella*.

Cranwich Norfolk. *Cranewisse* 1086 (DB). 'Marshy meadow frequented by cranes or herons'. OE *cran* + *wisc*.

Cranworth Norfolk. *Cranaworda* 1086 (DB). 'Enclosure frequented by cranes or herons'. OE *cran* + *worth*.

Craster Northum. *Craucestre* 1242. 'Old fortification or earthwork haunted by crows'. OE *crāwe* + *ceaster*.

Craswall Herefs. *Cressewell* 1231. 'Spring or stream where water-cress grows'. OE *cærse* + *wella*.

Cratfield Suffolk. *Cratafelda* 1086 (DB). 'Open land of a man called *Cræta*'. OE pers. name + *feld*.

Crathie Aber. Origin unknown.

Crathorne N. Yorks. *Cratorne* 1086 (DB). Possibly 'thorn-tree in a nook of land'. OScand. *krá* + *thorn*.

Cratloe (*An Chreatalach*) Clare. 'The place of frames'.

Craughwell (*Creachmhaoil*) Galway. 'Garlic wood'.

Craven (district) N. Yorks. *Crave* 1086 (DB), *Cravena* 12th cent. Possibly a Celtic name meaning 'garlic place' or from pre-Celtic **cravona* 'stony region'.

Craven Arms Shrops. a 19th-cent. name for the town that grew up at this important railway junction, taken from the *Craven Arms* inn built *c*.1800 and itself named from the Earls of Craven.

Crawcrook Gatesd. *Crawecroca* 1130. 'Bend (in the River Tyne), or nook of land, frequented by crows'. OE *crāwe* + OScand. *krókr* or OE **crōc*.

Crawford S. Lan. *Crauford c*.1150. 'Ford frequented by crows'. OE *crāwe* + *ford*.

Crawfordsburn Down. *Crawford's Burn* 1744. 'Crawford's stream'. The *Crawford* family settled here in the early 17th cent. The equivalent Irish name is *Sruth Chráfard*.

Crawley, usually 'wood or clearing frequented by crows', OE *crāwe* + *lēah*; examples include: **Crawley** Hants. *Crawanlea* 909, *Crawelie* 1086 (DB). **Crawley** Oxon. *Croule* 1214. **Crawley** W. Sussex. *Crauleia* 1203. **Crawley Down**, *Crauledun* 1272, has the addition of OE *dūn* 'hill, down'. **Crawley, North** Milt. K. *Crauelai* 1086 (DB).

Crawshaw Booth Lancs. *Croweshagh* 1324. 'Small wood or copse frequented by crows'. OE *crāwe* + *sc(e)aga*, with the later addition of OScand. *bōth* 'cowhouse, herdsman's hut'.

Cray, Foots, Cray, North, Cray, St Mary, & Cray, St Pauls Gtr. London. *Cræga(n)* 10th cent., *Crai(e)* 1086 (DB), *Fotescraei, Northcraei c*.1100, *Creye sancte Marie* 1257, *Craye Paulin* 1258. Named from the River Cray, a Celtic river-name meaning 'rough, turbulent'. Foots Cray was held by a man called *Fot* at the time of Domesday. The affixes in St Mary & St Pauls Cray are from the dedications of the churches to St Mary and St Paulinus.

Crayford Gtr. London. *Crecganford* [sic] 457, *Creiford* 1199. 'Ford over the River Cray'. Celtic river-name (*see* CRAY) + OE *ford*. The earliest spelling (an error for *Cræganford*) is from the Anglo-Saxon Chronicle (late 9th cent.), according to which in 457 Crayford was the site of a decisive battle between the invading Jutes and the native Britons.

Crayke N. Yorks. *Creic* 10th cent., 1086 (DB). Celtic **creig* 'a rock, a cliff'.

Creacombe Devon. *Crawecome* 1086 (DB). 'Valley frequented by crows'. OE *crāwe* + *cumb*.

Creagh (*Créach*) Fermanagh. *Creagh* 1752. 'Coarse pasture'.

Creake, North & Creake, South Norfolk. *Creic, Suthcreich* 1086 (DB), *Northcrec* 1211. Celtic **creig* 'a rock, a cliff'.

Creaton Northants. *Cretone* 1086 (DB). Probably 'farmstead at the rock or cliff'. Celtic **creig* + OE *tūn*.

Crecora (*Craobh Chomhartha*) Limerick. 'Palace of the sign'.

Credenhill Herefs. *Cradenhille* 1086 (DB). 'Hill of a man called Creoda'. OE pers. name (genitive -*n*) + *hyll*.

Crediton Devon. *Cridiantune* 930, *Chritetona* 1086 (DB). 'Farmstead or estate on the River Creedy'. Celtic river-name (meaning 'dwindler', i.e 'weakly flowing river') + OE *tūn*.

Creech, from Celtic **crūg* 'a mound or hill': **Creech, East** Dorset. *Criz* 1086 (DB). Referring originally to Creech Barrow. **Creech St Michael** Somerset. *Crice* 1086 (DB). Affix from the dedication of the church.

Creed Cornwall. *Sancta Crida c.*1250. 'Church of St Cride'. From the patron saint of the church.

Creedy (river) Devon. *See* CREDITON.

Creegh (*Críoch*) Clare. 'Boundary'.

Creenagh (*Críonach*) Antrim. 'Things dry and rotten with age'.

Creeslough (*An Craoslach*) Donegal. *Creslagh* 1610. 'The gullet lake'.

Creeting St Mary Suffolk. *Cratingas* 1086 (DB), *Creting Sancte Marie* 1254. '(Settlement of) the family or followers of a man called **Crǣta*'. OE pers. name + -*ingas*, with affix from the dedication of the church.

Creeton Lincs. *Cretone* 1086 (DB). Probably 'farmstead of a man called **Crǣta*'. OE pers. name + *tūn*.

Creevagh (*Craobhach*) Clare, Donegal, Galway, Mayo, Offaly, etc. 'Place of sacred trees'.

Creeve (*Craobh*) Antrim, Armagh, Donegal, Down, Longford, Mayo, Monaghan, etc. 'Sacred tree'.

Creevelea (*Craobh Liath*) Leitrim. 'Grey sacred tree'.

Creeveroe (*Craobh Ruadh*) Armagh. 'Red branch'.

Creeves (*Craobha*) Limerick. 'Sacred trees'.

Cregagh (*An Chreagaigh*) Down. *Craigogh c.*1659. 'The rocky place'.

Cregboy (*Creig Buí*) Galway. 'Yellow rock'.

Cregduff (*Creig Dubh*) Mayo. 'Black rock'.

Cregg (*Creig*) Sligo. 'Rock'.

Creggan (*Creagán*) Armagh, Derry, Westmeath. 'Little rocky place'.

Cregganbaun (*An Creagán Bán*) Mayo. 'The white rocky place'.

Creggs (*Na Creaga*) Galway, Roscommon. 'The rocks'.

Cremorne (*Críoch Mughdorn*) (district) Monaghan. *Crioch-Mughdhorn* n.d. 'Territory of the Mughdhorna tribe'. The tribal name means 'people of Mughdhorn', meaning the descendants of this man.

Crendon, Long, Bucks. *Credendone* 1086 (DB). 'Hill of a man called Creoda'. OE pers. name (genitive -*n*) + *dūn*. Affix (in use since 17th cent.) refers to length of village.

Cressage Shrops. *Cristesache* 1086 (DB). 'Christ's oak-tree'. OE *Crist* + *āc* (dative *ǣc*), probably with reference to a spot where preaching took place.

Cressing Essex. *Cressyng* 1136. 'Place where water-cress grows'. OE **cærsing*.

Cressingham, Great & Cressingham, Little Norfolk. *Cressingaham* 1086 (DB). Possibly 'homestead of the family or followers of a man called **Cressa*'. OE pers. name + -*inga*- + *hām*. Alternatively 'homestead with cress-beds', with a first element OE **cærsing*.

Cresswell, Creswell, 'spring or stream where water-cress grows', from OE *cærse* + *wella*: **Cresswell** Northum. *Kereswell* 1234. **Cresswell** Staffs. *Cressvale* 1086 (DB). **Creswell** Derbys. *Cressewella* 1176.

Cretingham Suffolk. *Gretingaham* 1086 (DB). 'Homestead of the people from a gravelly district'. OE *grēot* + -*inga*- + *hām*.

Crew (*Craobh*) Antrim, Derry, Tyrone. 'Sacred tree'.

Crew Hill (*Craobh*) Tyrone. 'Hill of the sacred tree'.

Crewe Ches. *Creu* 1086 (DB). Celtic **crïu* 'a fish-trap, a weir'.

Crewe Ches. near Farndon. *Creuhalle* 1086 (DB). Identical in origin with the previous name + OE *hall* 'a hall, a manor house'.

Crewkerne Somerset. *Crucern* 9th cent., *Cruche* 1086 (DB). Probably 'house or building at the hill', Celtic **crũg* + OE *ærn*, although -*ern* may rather represent a Celtic suffix.

Criccieth. *See* CRICIETH.

Crich Derbys. *Cryc* 1009, *Crich* 1086 (DB). Celtic *crũg* 'a mound or hill'.

Crichel, Long & Crichel, Moor Dorset. *Circel* 1086 (DB), *Langecrechel* 1208, *Mor Kerchel* 1212. 'Hill called *Crich*'. Celtic *crũg* 'hill' + explanatory OE *hyll*. Affixes are OE *lang* 'long' and OE *mōr* 'marshy ground'.

Cricieth Gwyd. *Crukeith* 1273. 'Mound of the captives'. Welsh *crug* + *caith*. The name refers to the Norman castle, built in 1230 on a headland ('mound').

Crick Northants. *Crec* 1086 (DB). Celtic **creig* 'a rock, a cliff'.

Crickadarn (*Crucadarn*) Powys. *Kruc kadarn* 1550, *Crucadarne* 1566, *Crickadarne* 1578, *Crickadarn* 1727, *Crickadarn*, or *Cerrigcadarn* 1868. 'Strong mound'. Welsh *crug* + *cadarn*.

Cricket Malherbie & Cricket St Thomas Somerset. *Cruchet, Cruche* 1086 (DB), *Cryket Malherbe* 1320, *Cruk Thomas* 1291. Celtic **crũg* 'a mound or hill', with OFrench suffix -*et* 'little' originally applied to the smaller of the two settlements, later transferred to Cricket St Thomas. Affixes from the *Malherbe* family here in the 13th cent., and from the dedication of the church.

Crickheath Shrops. *Gruchet* 1272. Celtic **crũg* 'a mound or hill' with an uncertain second element, but this name may have the same origin as CRICIETH.

Crickhowell (*Crucywel*) Powys. *Crichoel* 1263, *Crukhowell* 1281. 'Hywel's mound'. Welsh *crug* + pers. name.

Cricklade Wilts. *Crecgelade, Cracgelade* 10th cent., *Crichelade* 1086 (DB). OE *gelād* 'difficult river-crossing' with an obscure first element, possibly Celtic **creig* 'rock'.

Cricklewood Gtr. London. *Le Crikeldwode* 1294. 'The wood with indented outline'. ME **crikeled* + *wode*.

Cridling Stubbs N. Yorks. *Credeling* 12th cent., *Credlingstubbes* 1480. 'Place of a man called **Cridela*'. OE pers. name + -*ing*, with the later addition of Stubbs, 'tree-stumps', from OE *stubb*.

Crieff Perth. '(Place at) the tree'. Gaelic *Craoibh*, from *craobh* 'tree'.

Criffell (mountain) Dumf. *Crefel* 1330. Possibly 'split mountain'. OScand. *kryfja* + *fjall*; first element doubtful.

Crigglestone Wakefd. *Crigestone* 1086 (DB). 'Farmstead at the hill called *Crik*'. Celtic **crũg* 'hill' + explanatory OE *hyll* + OE *tūn*.

Crilly (*Crithligh*) Tyrone. *Crelly* 1608. 'Quaking bog'.

Crimplesham Norfolk. *Crepelesham* [*sic*] 1086 (DB), *Crimplesham* 1200. 'Homestead of a man called **Crympel*'. OE pers. name + *hām*.

Cringleford Norfolk. *Kringelforda* 1086 (DB). 'Ford by the round hill'. OScand. *kringla* + OE *ford*.

Crinkle (*Crionchoill*) Offaly. 'Withered wood'.

Croagh (*Cruach*) Donegal, Limerick. 'Rick'.

Croaghan (*Cruachán*) Donegal. 'Little rick'.

Croaghanmoira (*Cruach na Machaire*) Wicklow. 'Rick of the plain'.

Croaghbeg (*Cruach Beag*) Donegal. 'Little rick'.

Croaghpatrick (*Cruach Phádraig*) Mayo. 'Pádraig's rick'.

Crockerton Wilts. *Crokerton* 1249. 'Estate of the potters or of a family called Crocker'. OE **croccere* or ME surname + OE *tūn*.

Crocknagapple (*Cnoc na gCapall*) Donegal. 'Hill of the horses'.

Croft, usually from OE *croft* 'a small enclosed field'; examples include: **Croft** Lincs. *Croft* 1086 (DB). **Croft** N. Yorks. *Croft* 1086 (DB).

However the following has a different origin:
Croft Leics. *Craeft* 836, *Crec* [*sic*] *Crebre* [*sic*] 1086 (DB). OE *creft* 'a machine, an engine', perhaps referring to some kind of mill or windlass.

Crofton Gtr. London. *Crop tun* 973. 'Farmstead on a rounded hill'. OE *cropp* + *tūn*.

Crofton Wakefd. *Scroftune* [*sic*] 1086 (DB), *Croftona* 12th cent. 'Farmstead with a croft or enclosure'. OE *croft* + *tūn*.

Croghan (*Cruachán*) Offaly, Roscommon, Wicklow. 'Little rick'.

Croglin Cumbria. *Crokelyn* c.1140. 'Torrent with a bend in it'. OE **crōc* + *hlynn*.

Crohane (*Cruachán*) Kerry. 'Little rick'.

Crom (*Crom*) Fermanagh. 'Bend'.

Cromane (*An Cromán*) Kerry. 'The hip'.

Cromarty Highland. *Crumbathyn* 1264. 'Crooked (place)'. OGaelic *crumb* + doubtful element.

Crombie Aber. '(Place) at the crooked place'. Gaelic *crombach* (locative *crombaich*).

Cromer Norfolk. *Crowemere* 13th cent. 'Lake frequented by crows'. OE *crāwe* + *mere*.

Cromford Derbys. *Crunforde* 1086 (DB). 'Ford by the river-bend'. OE **crumbe* + *ford*.

Cromhall S. Glos. *Cromhal* 1086 (DB). 'Nook of land in the river-bend'. OE **crumbe* + *halh*.

Cromkill (*Cromchoill*) Antrim. 'Bent wood'.

Cromwell Notts. *Crunwelle* 1086 (DB). 'Crooked stream'. OE *crumb* + *wella*.

Crondall Hants. *Crundellan* c.880, *Crundele* 1086 (DB). '(Place at) the chalk-pits or quarries'. OE *crundel* in a dative plural form.

Cronton Knows. *Crohinton* 1242. Possibly 'farmstead at the place with a nook'. OE **crōh* + *-ing* + *tūn*.

Crook Cumbria. *Croke* 12th cent. OScand. *krókr* or OE **crōc* 'land in a bend, secluded corner of land'.

Crook Durham. *Crok* 1309. Identical in origin with the previous name.

Crookham Northum. *Crucum* 1244. '(Place) at the river-bends'. OScand. *krókr* in a dative plural form.

Crookham W. Berks. *Crocheham* 1086 (DB). Probably 'homestead or village by the river-bends'. OE **crōc* + *hām*.

Crookham, Church Hants. *Crocham* 1248. Identical in origin with CROOKHAM (W. Berks.).

Crookhaven (*An Cruachán*) Cork. 'The haven of the little rick'.

Croom (*Cromadh*) Limerick. 'Crooked ford'.

Croome, Earls Worcs. *Cromman* 10th cent., *Crumbe* 1086 (DB), *Erlescrombe* 1495. Originally the name of the stream here, a Celtic river-name meaning 'the crooked or winding one'. Manorial affix from its early possession by the Earls of Warwick.

Cropredy Oxon. *Cropelie* [*sic*] 1086 (DB), *Croprithi* c.1275. 'Small stream of a man called **Croppa*, or near a hill'. OE pers. name or *crop(p)* + *rīthig*.

Cropston Leics. *Cropeston* 12th cent. Probably 'Farmstead of a man called **Cropp* or *Kroppr*'. OE or OScand. pers. name + OE *tūn*. Alternatively the first element may be OE *crop(p)* 'hill'.

Cropthorne Worcs. *Croppethorne* 8th cent., *Cropetorn* 1086 (DB). 'Thorn-tree near a hill'. OE *crop(p)* + *thorn*.

Cropton N. Yorks. *Croptune* 1086 (DB). 'Farmstead near a hill'. OE *crop(p)* + *tūn*.

Cropwell Bishop & Cropwell Butler Notts. *Crophille* 1086 (DB), *Bischopcroppehill* 1280, *Croppill Boteiller* 1265. 'Rounded hill'. OE *crop(p)* + *hyll*. Manorial additions from possession by the Archbishop of York and by the *Butler* family, here from the 12th cent.

Crory (*Cruaire*) Wexford. 'Hard land'.

Crosby, usually 'village where there are crosses'. from OScand. *krossa-bý*; examples include: **Crosby** Cumbria, near Maryport. *Crossby* 12th cent. **Crosby, Great & Crosby, Little** Sefton. *Crosebi* 1086 (DB), *Magna Crossby* c.1190, *Parva Crosseby* 1242. Affixes are Latin *magna* 'great', and *parva* 'little'. **Crosby Garrett** Cumbria. *Crosseby* 1200, *Crossebi Gerard* 1206. Manorial affix from a person or family called *Gerard*. **Crosby, Low** Cumbria. *Crossebi* c.1200. **Crosby Ravensworth** Cumbria. *Crosseby Raveneswart* 12th cent. Manorial affix from the OScand. pers. name *Rafnsvartr*.

The following name probably has the same origin: **Crosby** N. Lincs. *Cropesbi* [*sic*] 1086 (DB), *Crosseby* 1206. The DB spelling suggests 'village of a man called Kroppr', from OScand. pers. name + *bý*, but may be an error. Later spellings support 'village with crosses'.

Croscombe Somerset. *Correges cumb* 705, *Coriscoma* 1086 (DB). Identical in origin with CORSCOMBE Dorset.

Cross (*Crois*) Clare, Mayo. 'Cross'.

Cross Keys Cavan. *Crosskeys* 1728. The name was that of an inn here. The Irish name of the village is *Carraig an Tobair*, 'rock of the well'.

Crossabeg (*Na Crosa Beaga*) Wexford. 'The little crosses'.

Crossakeel (*Crosa Caoil*) Meath. 'Crosses of Caol'.

Crossbarry (*Crois an Bharraigh*) Cork. 'Cross of the Barries'.

Crosscanonby Cumbria. *Crosseby Canoun* 1285. 'Village where there are crosses'. OScand. *krossa-bý*. *Canon* refers to lands here given to the canons of Carlisle.

Crossdoney (*Cros Domhnaigh*) Cavan. *Crossdony* c.1660. 'Cross of the church'.

Crossens Sefton. *Crossenes* c.1250. 'Headland or promontory with crosses'. OScand. *kross* + *nes*.

Crosserlough (*Crois ar Loch*) Cavan. *Croserlagh* 1601. 'Cross on the lake'.

Crossgar (*An Chrois Ghearr*) Down. 'The short cross'.

Crossmaglen (*Crois Mhic Lionnáin*) Armagh. *Crosmoyglan* 1609. 'Mac Lionnáin's cross'.

Crossmolina (*Crois Mhaoilíona*) Mayo. 'Maoilíona's cross'.

Crossooha (*Crois Uathaidh*) Galway. 'Single cross'.

Crosspatrick (*Crois Phádraig*) Kilkenny, Wicklow. 'Pádraig's cross'.

Crossreagh (*An Chros Riabhach*) Cavan. 'The grey cross' or 'the brindled cross'.

Crosthwaite Cumbria. *Crosthwait* 12th cent. 'Clearing with a cross'. OScand. *kross* + *thveit*.

Croston Lancs. *Croston* 1094. 'Farmstead or village with a cross'. OScand. *kross* or OE *cros* + *tūn*.

Crostwick Norfolk. *Crostueit* 1086 (DB). 'Clearing with a cross'. OScand. *kross* + *thveit*.

Crostwight Norfolk. *Crostwit* 1086 (DB). Identical in origin with the previous name.

Crouch (river) Essex. *See* BURNHAM ON CROUCH.

Croughton Northants. *Creveltone* 1086 (DB). 'Farmstead or village on the fork of land'. OE **creowel* + *tūn*.

Crow Hants. *Croue* 1086 (DB). Celtic **crüw* 'weir, fish-trap' or **crou* 'sty'.

Crowan Cornwall. *Eggloscrauuen* c.1170. 'Church of St Cravenna'. From the patron saint of the church (in the early spelling with Cornish *eglos* 'church').

Crowbane (*Cruach Bán*) Donegal. 'White rick'.

Crowborough E. Sussex. *Cranbergh* (*sic*, for *Craubergh*) 1292. 'Hill or mound frequented by crows'. OE *crāwe* + *beorg*.

Crowcombe Somerset. *Crawancumb* 10th cent., *Crawecumbe* 1086 (DB). 'Valley frequented by crows'. OE *crāwe* + *cumb*.

Crowdecote Derbys. *Crudecote* 13th cent. 'Cottage of a man called **Crūda*'. OE pers. name + *cot*.

Crowfield Suffolk. *Crofelda* 1086 (DB). Probably 'open land near the nook or corner'. OE **crōh* + *feld*.

Crowhurst E. Sussex. *Croghyrste* 772, *Croherst* 1086 (DB). Probably 'wooded hill near the nook or corner'. OE **crōh* + *hyrst*.

Crowhurst Surrey. *Crouhurst* 12th cent. 'Wooded hill frequented by crows'. OE *crāwe* + *hyrst*.

Crowland Lincs. *Cruwland* 8th cent., *Croiland* 1086 (DB). Probably 'estate or tract of land at the river-bend'. OE **crūw* + *land*.

Crowle N. Lincs. *Crule* 1086 (DB), *Crull* c.1070. Originally the name of a river here (now gone through draining), from OE **crull* meaning 'winding'.

Crowle Worcs. *Crohlea* 9th cent., *Croelai* 1086 (DB). 'Woodland clearing by the nook or corner'. OE **crōh + lēah*.

Crowmarsh Gifford Oxon. *Cravmares* 1086 (DB), *Cromershe Giffard* 1316. 'Marsh frequented by crows'. OE *crāwe + mersc*, with manorial affix from a man called *Gifard* who held the manor in 1086.

Crownthorpe Norfolk. *Congrethorp*, *Cronkethor* 1086 (DB). OScand. *thorp* 'outlying farmstead or hamlet' with an uncertain first element, possibly a pers. name.

Crowthorne Wokhm. first recorded in 1607, 'thorn-tree frequented by crows'.

Crowton Ches. *Crouton* 1260. 'Farmstead frequented by crows'. OE *crāwe + tūn*.

Croxall Staffs. *Crokeshalle* 942, *Crocheshalle* 1086 (DB). 'Nook of land of a man called Krókr, or near a bend'. OScand. pers. name or OE **crōc + OE *halh*.

Croxdale Durham. *Crocestail* c.1190. 'Projecting piece of land of a man called Krókr'. OScand. pers. name + OE *tægl*.

Croxden Staffs. *Crochesdene* 1086 (DB). Possibly 'valley of a man called Krókr'. OScand. pers. name + OE *denu*. Alternatively the first element may be OE **crōc* 'nook, bend'.

Croxley Green Herts. *Crokesleya* 1166. 'Woodland clearing of a man called Krókr'. OScand. pers. name + OE *lēah*.

Croxton, 'farmstead in a nook, or of a man called Krókr', from OE **crōc* or OScand. pers. name + OE *tūn*: **Croxton** Cambs. *Crochestone* 1086 (DB). **Croxton** N. Lincs. *Crochestune* 1086 (DB). **Croxton** Norfolk, near Thetford. *Crokestuna* 1086 (DB). **Croxton** Staffs. *Crochestone* 1086 (DB). **Croxton Kerrial** Leics. *Crohtone* 1086 (DB), *Croxton Kyriel* 1247. Manorial addition from the *de Cryoll* family, here in the 13th cent. **Croxton, South** Leics. *Crochestone* 1086 (DB), *Sudcroxton* 1212. OE *sūth* 'south'.

Croyde Devon. *Cridehold a* [sic] 1086 (DB), *Crideho* 1242. Possibly OE **crȳde* 'headland' referring to the promontory called Croyde Hoe (with explanatory OE *hōh* 'hill-spur'), or Celtic **crūd* 'cradle' with reference to the trough-like valley in which the village lies.

Croydon Cambs. *Crauuedene* 1086 (DB). 'Valley frequented by crows'. OE *crāwe + denu*.

Croydon Gtr. London. *Crogedene* 809, *Croindene* 1086 (DB). 'Valley where wild saffron grows'. OE *croh, *crogen + denu*.

Crucadarn. *See* CRICKADARN.

Cruckmeole Shrops. *Mele* 1255, *Crokemele* 1292. Named from Meole Brook (*see* MEOLE BRACE), first element as in following name.

Cruckton Shrops. *Crokton, Crukton* 1272. Probably 'farmstead or estate by the river-bend'. OE **crōc + tūn*.

Crucywel. *See* CRICKHOWELL.

Crudgington Tel. & Wrek. *Crugetone* [sic] 1086 (DB), *Crugelton* 12th cent. Probably 'farmstead by a hill called *Cruc*'. Celtic **crūg* + explanatory OE *hyll* + OE *tūn*.

Crudwell Wilts. *Croddewelle* 9th cent., *Credvelle* 1086 (DB). 'Spring or stream of a man called Creoda'. OE pers. name + *wella*.

Cruit (*An Chruit*) (island) Donegal. 'The hump'.

Crumlin (*Cromghlinn*) Antrim, Dublin. 'Crooked valley'.

Crundale Kent. *Crundala* c.1100. OE *crundel* 'a chalk-pit, a quarry'.

Crusheen (*Croisín*) Clare. 'Little cross'.

Cruwys Morchard Devon. *See* MORCHARD.

Crux Easton Hants. *See* EASTON.

Crymych Pemb. *Crymmych* 1584, *Nant Crymich* 1652. '(Place by the) crooked stream'. Welsh *crwm*. The village probably took its name directly from the *Crymych Arms* inn here.

Crystal Palace Gtr. London. district named from the spectacular building, originally erected in Hyde Park for the Great Exhibition of 1851 but moved here later that year and destroyed by fire in 1936. The structure was first dubbed *Crystal Palace* by the magazine Punch on account of its large areas of glass.

Cubbington Warwicks. *Cobintone* 1086 (DB). 'Estate associated with a man called *Cubba'. OE pers. name + *-ing- + tūn*.

Cubert Cornwall. *Sanctus Cubertus* 1269. 'Church of St Cuthbert'. From the patron saint of the church.

Cublington Bucks. *Coblincote* [sic] 1086 (DB), *Cubelintone* 12th cent. 'Estate associated with a man called *Cubbel'. OE pers. name + -ing- + tūn (with alternative cot 'cottage' in the Domesday spelling).

Cuckfield W. Sussex. *Kukefeld, Kukufeld* c.1095. Probably 'open land haunted by the cuckoo'. OE *cucu + feld.

Cucklington Somerset. *Cocintone [sic]* 1086 (DB), *Cukelingeton* 1212. 'Estate associated with a man called *Cucol or *Cucola'. OE pers. name + -ing- + tūn.

Cuddesdon Oxon. *Cuthenesdune* 956, *Codesdone* 1086 (DB). 'Hill of a man called *Cūthen'. OE pers. name + dūn.

Cuddington, 'estate associated with a man called Cud(d)a', OE pers. name + -ing- + tūn: **Cuddington** Bucks. *Cudintuna* 12th cent. **Cuddington** Ches. *Codynton* c.1235. **Cuddington Heath** Ches. *Cuntitone [sic]* 1086 (DB), *Cudyngton Hethe* 1532.

Cudham Gtr. London. *Codeham* 1086 (DB). 'Homestead or enclosure of a man called Cuda'. OE pers. name + hām or hamm.

Cudliptown Devon. *Cudelipe* early 12th cent. 'Steep place of a man called Cuda'. OE pers. name + hlīep, with the later addition of *town* in the sense 'hamlet' from the 17th cent.

Cudworth Barns. *Cutheworthe* 12th cent. 'Enclosure of a man called Cūtha'. OE pers. name + worth.

Cudworth Somerset. *Cudeworde* 1086 (DB). 'Enclosure of a man called Cuda'. OE pers. name + worth.

Cuffley Herts. *Kuffele* 1255. Probably 'woodland clearing of a man called *Cuffa'. OE pers. name + lēah.

Cuilmore (*An Choill Mhór*) Mayo. 'The big wood'.

Culcavy (*Cúil Chéibhe*) Down. *Cowlecavy* 1632. 'Corner of the long grass'.

Culcheth Warrtn. *Culchet* 1201. Probably 'narrow wood'. Celtic *cūl + *cēd.

Culcrum (*An Choill Chrom*) Antrim. 'The crooked wood'.

Culdaff (*Cúil Dabhcha*) Donegal. *Culldavcha* 1429. 'Corner of the vat'.

Culfadda (*Coill Fhada*) Sligo. 'Long wood'.

Culford Suffolk. *Culeforda* 1086 (DB). 'Ford of a man called *Cūla'. OE pers. name + ford.

Culgaith Cumbria. *Culgait* 12th cent. Identical in origin with CULCHETH.

Culham Oxon. *Culanhom* 821. 'River-meadow of a man called *Cūla'. OE pers. name + hamm.

Culkeeny (*Cúil Chaonaigh*) Donegal. 'Wood of the moss'.

Culkerton Glos. *Cvlcortone* 1086 (DB). 'Farmstead with a water-hole, or belonging to a man called *Culcere'. OE *culcor or OE pers. name + tūn.

Culky (*Cuilcigh*) Fermanagh. *Colkie* 1609. 'Abounding in reeds'.

Cullahill (*An Chúlchoill*) Laois. 'The back of the wood'.

Culleens (*Na Coillíní*) Mayo. 'The small hazel woods'.

Cullen (*Cuilleann*) Cork, Tipperary. 'Steep slope'.

Cullentra (*An chuil eanntrach*) Tyrone. 'The holly place'.

Cullercoats N. Tyne. *Culvercoats* c.1600. 'Dove-cots'. OE *culfre + cot.

Cullingworth Brad. *Colingauuorde* 1086 (DB). 'Enclosure of the family or followers of a man called *Cūla'. OE pers. name + -inga- + worth.

Cullion (*Cuilleann*) Tyrone. 'Steep slope'.

Cullompton Devon. *Columtune* c.880, *Colump [sic]* 1086 (DB). 'Farmstead on the Culm river'. Celtic river-name (meaning 'dove', i.e. 'gentle stream') + OE tūn.

Culloville (*Baile Mhic Cullach*) Armagh. *Cullovill* 1766. 'Mac Cullach's townland'.

Cullybackey (*Cúil na Baice*) Antrim. *Ballycholnabacky* 1607. 'Corner of the river-bend'.

Cullyhanna (*Coilleach Eanach*) Armagh. *Collyhanagh* 1640. 'Wood of the swamp'.

Culmington Shrops. *Comintone [sic]* 1086 (DB), *Culminton* 1197. Possibly 'farmstead on a stream called Culm ('the dove', i.e. 'gentle stream')'. Celtic river-name + -ing- + tūn.

Culmore (*An Chúil Mhór*) Derry. (*isin*) *Chúil móir* 1600. 'The big corner'.

Culmore (*Coill Mhór*) Mayo. 'Big wood'.

Culmstock Devon. *Culumstocc* 938, *Culmestoche* 1086 (DB). 'Outlying farmstead on the Culm river'. Celtic river-name (*see* CULLOMPTON) + OE *stoc*.

Culnady (*Cúil Chnáidí*) Derry. (*go*) *Cúil Cnáimhdhidhe* c.1740. 'Nook of the place of burrs'.

Culross Fife. *Culenross* 12th cent. Probably 'holly point'. Gaelic *cuileann* + *ros*.

Cults Abdn. *Qhylt* 1450, *Cuyltis* 1456. '(Place in the) nook'. Gaelic *cùillte* + English plural *-s*.

Culverthorpe Lincs. *Torp* 1086 (DB), *Calewarthorp* 1275. Originally simply 'outlying farmstead or hamlet' from OScand. *thorp*, later with the name of an owner added, form uncertain.

Culworth Northants. *Culeorde* 1086 (DB). 'Enclosure of a man called *Cūla*'. OE pers. name + *worth*.

Cumberland (historical county). *Cumbra land* 945. 'Region of the *Cymry* or Cumbrian Britons'. OE **Cumbre* + *land*.

Cumbernauld N. Lan. *Cumbrenald* c.1295, *Cumyrnald* 1417. 'Meeting of the streams'. Gaelic *comar-nan-allt*.

Cumberworth, 'enclosure of a man called Cumbra, or of the Britons', OE pers. name or OE **Cumbre* + *worth*: **Cumberworth** Lincs. *Combreuorde* 1086 (DB). **Cumberworth, Lower & Cumberworth, Upper** Kirkl. *Cumbreuurde* 1086 (DB).

Cumbrae (island) N. Ayr. *Cumberays* 1264. 'Island of the Cymry'. OScand. *ey*. Great *Cumbrae* and *Little Cumbrae* are probably named after the Welsh (Cumbrian) inhabitants of southern Scotland, *Cymry* being the name of the Welsh for themselves (*see* WALES).

Cumbria (new county). *Cumbria* 8th cent. 'Territory of the *Cymry* or Cumbrian Britons'. A Latinization of the OE tribal name **Cumbre*, see CUMBERLAND.

Cummer (*Comar*) Galway. 'Confluence'.

Cummersdale Cumbria. *Cumbredal* 1227. 'Valley of the Cumbrian Britons'. OE **Cumbre* (genitive plural *-a*) + OScand. *dalr*.

Cumnock E. Ayr. *Comnocke* 1297, *Comenok* 1298, *Cumnock* 1300. Meaning uncertain.

Cumnor Oxon. *Cumanoran* 931, *Comenore* 1086 (DB). 'Hill-slope of a man called **Cuma*'. OE pers. name (genitive *-n*) + *ōra*.

Cumrew Cumbria. *Cumreu* c.1200. 'Valley by the hill-slope'. Celtic **cumm* + **riu*.

Cumwhinton Cumbria. *Cumquintina* c.1155. 'Valley of a man called Quintin'. Celtic **cumm* + OFrench pers. name.

Cumwhitton Cumbria. *Cumwyditon* 1278. 'Valley at a place called *Whytington* (estate associated with a man called Hwīta)'. Celtic **cumm* added to an English place name (from OE pers. name + *-ing-* + *tūn*).

Cunard (*Cionn Ard*) Dublin. 'High head'.

Cundall N. Yorks. *Cundel* 1086 (DB). Possibly 'hollow of the cows'. OE *cū* (genitive plural *cūna*) + *dæl*.

Cunnister Shet. (Yell). *Cynnyngsitter* 1639. 'King's summer pasture'. OScand. *konungr* + *sǣtr*.

Cupar Fife. *Cupre* 1183, *Coper* 1294. Meaning uncertain. The name is British. *Compare* COUPAR ANGUS.

Curbridge Oxon. *Crydan brigce* 956. 'Bridge of a man called Creoda'. OE pers. name + *brycg*.

Curdridge Hants. *Cuthredes hricgæ* 901. 'Ridge of a man called Cūthrǣd'. OE pers. name + *hrycg*.

Curland Somerset. *Curiland* 1252. Probably 'cultivated land belonging to CURRY'. OE *land*.

Curlew Mountains (*An Corrshliabh*) Roscommon, Sligo. 'The pointed mountains'.

Curr (*Corr*) Tyrone. *Corr* 1666. 'Round hill'.

Currabeha (*An Chorr Bheithe*) Cork. 'The round hill of the birch'.

Curracloe (*Currach Cló*) Wexford. 'Marsh of the impression'.

Curragh (*Currach*) Kildare. 'Marsh'.

Curragh (*Currach*) Waterford. 'Marsh'.

Curraghboy (*An Currach Buí*) Roscommon. 'The yellow marsh'.

Curraghlawn (*Currach Leathan*) Wexford. 'Broad marsh'.

Curraghmore (*Currach Mór*) Waterford. 'Big marsh'.

Curraghroe (*An Currach Rua*) Roscommon. 'The red marsh'.

Curraglass (*Cora Ghlas*) Cork. 'Green weir'.

Curraglass (*Currach Glas*) Cork. 'Grey marsh'.

Curran (*Corrán*) Antrim, Derry. 'Crescent-shaped place'.

Currandrum (*Corr an Droma*) Galway. 'Pointed hill of the ridge'.

Currane (*An Corrán*) Mayo. 'The crescent-shaped place'.

Currans (*Corráin*) Kerry. 'Crescent-shaped places'.

Curreeny (*Na Coirríní*) Tipperary. 'The little pointed hills'.

Currie Edin. *Curey* 1210, *Curry* 1213, *Curri* 1246. '(Place) in the marshland'. Gaelic locative of *currach*.

Currow (*Corra*) Kerry. 'Round hills'.

Curry (*An Choraidh*) Sligo. 'The place of the weir'.

Curry Mallet & Curry Rivel, North Curry Somerset. *Curig* 9th cent., *Curi, Nortcuri* 1086 (DB), *Curi Malet, Curry Revel* 1225. 'Estates on the river called *Curi*'. Pre-English river-name (of uncertain origin and meaning), with distinguishing affixes from families called *Malet* and *Revel*, here from the late 12th cent., and from OE *north*.

Curryglass (*Currach Glas*) Cork. 'Grey marsh'.

Curthwaite, East & Curthwaite, West Cumbria. *Kyrkthwate* 1272. 'Clearing near or belonging to a church'. OScand. *kirkja* + *thveit*.

Cury Cornwall. *Egloscuri* 1219. 'Church of St Cury' (a pet-form of St Corentin). From the patron saint of the church, with Cornish *eglos* 'church' in the early form.

Cush (*Cois*) Limerick. 'Foot'.

Cushendall (*Cois Abhann Dalla*) Antrim. *Bun an Dalla c.*1854. 'Foot of the River Dall'. The village was earlier *Bun Abhann Dalla*, 'foot of the River Dall', the present name being an anglicized form of the alternative Irish name.

Cushendun (*Cois Abhann Duinne*) Antrim. *Bun-abhann Duine* 1567. 'Foot of the River Dun'. The village was earlier *Bun Abhann Duinne*, 'foot of the River Dun', the present name being an anglicized form of the alternative Irish name.

Cushina (*Cois Eidhní*) Offaly. 'Beside the ivy (river)'.

Cusop Herefs. *Cheweshope* 1086 (DB). OE *hop* 'small enclosed valley', first element possibly a lost Celtic river-name *Cyw*.

Cutsdean Glos. *Codestune* 10th cent., 1086 (DB), *Cottesdena* 12th cent. OE *tūn* 'farmstead' and/or OE *denu* 'valley' with an uncertain first element, possibly an OE masculine pers. name *Cōd*.

Cuxham Oxon. *Cuces hamm* 995, *Cuchesham* 1086 (DB). 'River-meadow of a man called *Cuc*'. OE pers. name + *hamm*.

Cuxton Medway. *Cucolanstan* 880, *Coclestane* 1086 (DB). '(Boundary) stone of a man called *Cucola*'. OE pers. name + *stān*.

Cuxwold Lincs. *Cucuwalt* 1086 (DB), *Cucuwald c.*1115. Probably 'woodland haunted by the cuckoo'. OE *cucu* + *wald*.

Cwmafan Neat. 'Valley of the River Afan'. Welsh *cwm*. The Afan flows through the town to enter the sea at ABERAVON.

Cwmbrân Newpt. *Cwmbran* 1707. 'Valley of the River Brân'. Welsh *cwm*. The Celtic river name means 'raven' (Welsh *brân*), referring to its dark waters. Cwmbran was developed as a New Town in 1949.

Cydweli. *See* KIDWELLY.

Cymru. *See* WALES.

Cynon (river) Rhon., giving name to **Abercynon**, 'mouth of the River Cynon', Welsh *aber*. The river-name derives from a pers. name, also found in **Tregynon**, 'farm of Cynon', Welsh *tre*.

Daar (*Abhainn na Darach*) (river) Limerick. 'River of the oak-tree'.

Daars (*Dairghe*) Kildare. 'Oaks'.

Dacorum (district) Herts. *Danais* 1086 (DB), *hundredo Dacorum* 1196. '(Hundred) belonging to the Danes', from the genitive plural of Latin *Daci* 'the Dacians' (erroneously used of the Danes in medieval times).

Dacre Cumbria. *Dacor* c.1125. Named from the stream called Dacre Beck, a Celtic river-name meaning 'the trickling one'.

Dacre N. Yorks. *Dacre* 1086 (DB). Originally a Celtic river-name for the stream here, identical in origin with the previous name.

Dadford Bucks. *Dodeforde* 1086 (DB). 'Ford of a man called Dodda'. OE pers. name + *ford*.

Dadlington Leics. *Dadelintona* c.1190. Probably 'estate associated with a man called *Dæd(d)el*'. OE pers. name + *-ing-* + *tūn*.

Dagenham Gtr. London. *Deccanhaam* c.690, *Dakenham* 1261. 'Homestead or village of a man called *Dæcca*'. OE pers. name (genitive *-n*) + *hām*.

Daggons Dorset. *Daggans* 1553. Named from the family of Richard *Dagon*, here in the 14th cent.

Daglingworth Glos. *Daglingworth* c.1150. 'Enclosure of the family or followers of a man called *Dæggel* or *Dæccel*', or 'enclosure associated with the same man'. OE pers. name + *-inga-* or *-ing-* + *worth*.

Dagnall Bucks. *Dagenhale* 1196. 'Nook of land of a man called *Dægga*'. OE pers. name (genitive *-n*) + *healh*.

Daingean (*Daingean*) Offaly. 'Fortress'.

Dalbeattie Dumf. *Dalbaty* 1469. 'Haugh by the birch wood'. Gaelic *dail* + *beithigh* (genitive of *beitheach*).

Dalby, 'farmstead or village in a valley', OScand. *dalr* + *bý*; examples include: **Dalby, Great & Dalby, Little** Leics. *Dalbi* 1086 (DB). **Dalby, Low** N. Yorks. *Dalbi* 1086 (DB).

Dalderby Lincs. *Dalderby* c.1150. Possibly 'farmstead or village in a small valley', from OScand. *dæld* (genitive *-ar*) + *bý*. Alternatively 'valley farmstead where deer are kept', from OScand. *dalr* + *djúr* + *bý*.

Dale Derbys. *La Dale* late 12th cent. '(Place in) the valley'. OE *dæl*.

Dalgleish Sc. Bord. 'Green alluvial valley'. Gaelic *dail* + *ghlais*.

Dalham Suffolk. *Dalham* 1086 (DB). 'Homestead or village in a valley'. OE *dæl* + *hām*.

Dalkeith Midloth. *Dolchet* 1144. 'Meadow by a wood'. Pictish **dol* + **cēd*.

Dalkey (*Deilginis*) Dublin. 'Thorn island'. OScand. *dalkr* + *ey*.

Dallas Moray. *Dolays* 1232. 'Meadow station'. British **dol* + **foss*.

Dalling, Field & Dalling, Wood Norfolk. *Dallinga* 1086 (DB), *Fildedalling* 1272, *Wode Dallinges* 1198. '(Settlement of) the family or followers of a man called Dalla'. OE pers. name + *ingas*. The distinguishing affixes are *feld* or **filden* (denoting 'in the open country') and *wudu* (denoting 'in the woodland').

Dallinghoo Suffolk. *Dallingahou* 1086 (DB). 'Hill-spur of the family or followers of a man called Dalla'. OE pers. name + *-inga-* + *hōh*.

Dallington E. Sussex. *Dalintone* 1086 (DB). 'Estate associated with a man called Dalla'. OE pers. name + *-ing-* + *tūn*.

Dalry N. Ayr. *Dalry* 1315. 'Heather meadow'. Gaelic *dail fhraoich* (genitive of *fraoch*).

Dalston Cumbria. *Daleston* 1187. Probably 'farmstead of a man called *Dall'. OE pers. name + *tūn*.

Dalston Gtr. London. *Derleston* 1294. 'Farmstead of a man called Dēorlāf'. OE pers. name + *tūn*.

Dalton, a common place name especially in the northern counties, 'farmstead or village in a valley', from OE *dæl* + *tūn*; examples include: **Dalton** Dumf. *Delton* c.1280. **Dalton** Lancs. *Daltone* 1086 (DB). **Dalton** Northum., near Hexham. *Dalton* 1256. **Dalton** Northum., near Stamfordham. *Dalton* 1201. **Dalton** N. Yorks., near Richmond. *Daltun* 1086 (DB). **Dalton** N. Yorks., near Thirsk. *Deltune* 1086 (DB). **Dalton** Rothm. *Daltone* 1086 (DB). **Dalton in Furness** Cumbria. *Daltune* 1086 (DB), *Dalton in Fournais* 1332. For the district name Furness, *see* BARROW IN FURNESS. **Dalton-le-Dale** Durham. *Daltun* 8th cent. The addition means 'in the valley', from OE *dæl*. **Dalton, North** E. R. Yorks. *Dalton* 1086 (DB), *Northdaltona* c.1155. **Dalton-on-Tees** N. Yorks. *Dalton* 1204. On the River Tees. **Dalton Piercy** Hartlepl. *Daltun* c.1150, *Dalton Percy* 1370. Manorial affix from the *Percy* family, here up to the 14th cent. **Dalton, South** E. R. Yorks. *Delton* 1086 (DB), *Suthdalton* 1260.

Dalua (*Abhainn dá Lua*) (river) Cork. 'River of two waters'.

Dalwhinnie Highland. 'Valley of the champions'. Gaelic *dail* + *chuinnidh* (genitive).

Dalwood Devon. *Dalewude* 1195. 'Wood in a valley'. OE *dæl* + *wudu*.

Dalziel N. Lan. 'White meadow'. Gaelic *dail* + *ghil* (dative of *geal*).

Damerham Hants. *Domra hamme* c.880, *Dobreham* [*sic*] 1086 (DB). 'Enclosure or river-meadow of the judges'. OE *dōmere* + *hamm*.

Damma (*Dá Mhagh*) Kerry. 'Two plains'.

Danbury Essex. *Danengeberiam* 1086 (DB). Probably 'stronghold of the family or followers of a man called Dene'. OE pers. name + *-inga-* + *burh* (dative *byrig*).

Danby N. Yorks. *Danebi* 1086 (DB). 'Farmstead or village of the Danes'. OScand. *Danir* (genitive plural *Dana*) + *bý*.

Danby Wiske N. Yorks. *Danebi* 1086 (DB), *Daneby super Wiske* 13th cent. Identical in

origin with the previous name. Situated on the River Wiske (*see* APPLETON WISKE).

Dane (river) Ches. *See* DANEBRIDGE, DAVENHAM.

Danebridge Ches. *Dauenbrugge* 1357. 'Bridge over the River Dane'. Celtic river-name (meaning 'trickling stream') + OE *brycg*.

Danehill E. Sussex. *Denne* 1279, *Denhill* 1437. 'Hill by the woodland pasture'. OE *denn* with the later addition of *hyll*.

Darcy Lever Bolton. *See* LEVER.

Darenth Kent. *Daerintan* 10th cent., *Tarent* 1086 (DB). '(Estate on) the River Darent'. Darent is a Celtic river-name meaning 'river where oak-trees grow'.

Daresbury Halton. *Deresbiria* 12th cent. 'Stronghold of a man called Dēor'. OE pers. name + *burh* (dative *byrig*).

Darfield Barns. *Dereuueld* 1086 (DB). 'Open land frequented by deer or other wild animals'. OE *dēor* + *feld*.

Dargate Kent. *Deregate* 1275. 'Deer gate'. ME *dere* (OE *dēor*) + *gate*.

Darkley (*Dearclaigh*) Armagh. *Darkeley* 1657. 'Place of caves'.

Darlaston Wsall. *Derlaveston* 1262. 'Farmstead or village of a man called Dēorlāf'. OE pers. name + *tūn*.

Darley, 'woodland clearing frequented by deer or wild animals', OE *dēor* + *lēah*: **Darley Abbey** Derby. *Derlega* 12th cent. **Darley Dale** Derbys. *Dereleie* 1086 (DB).

Darlingscott Warwicks. *Derlingescot* 1210. 'Cottage(s) of a man called *Dēorling'. OE pers. name + *cot*.

Darlington Darltn. *Dearthingtun* c.1040, *Dearningtun* c.1104. 'Estate associated with a man called Dēornōth'. OE pers. name + *-ing-* + *tūn*.

Darliston Shrops. *Derloueston* 1199. 'Farmstead of a man called Dēorlāf'. OE pers. name + *tūn*.

Darlton Notts. *Derluuetun* 1086 (DB). 'Farmstead of a woman called *Dēorlufu'. OE pers. name + *tūn*.

Darragh (*An Darach*) Clare. 'The oak-tree'.

Darraragh (*Dairbhre*) Mayo. 'Oak wood'.

Darrery (*Dairbhre*) Cork, Galway, Limerick. 'Oak wood'.

Darrington Wakefd. *Darni*(*n*)*tone* 1086 (DB). Probably 'estate associated with a man called Dēornōth'. OE pers. name + *-ing-* + *tūn*.

Darrynane (*Doire Fhíonáin*) Kerry. 'Fíonán's oak wood'.

Darsham Suffolk. *Dersham* 1086 (DB). 'Homestead or village of a man called Dēor'. OE pers. name + *hām*.

Dartford Kent. *Tarentefort* 1086 (DB). 'Ford over the River Darent'. Celtic river-name (*see* DARENTH) + OE *ford*.

Dartington Devon. *Dertrintona* 1086 (DB). 'Farmstead on the River Dart'. Celtic river-name (meaning 'river where oak-trees grow') + *-ing-* + *tūn*.

Dartmoor Devon. *Dertemora* 1182. 'Moor in the Dart valley'. Celtic river-name (*see* DARTINGTON) + OE *mōr*.

Dartmouth Devon. *Dertamuthan* 11th cent. 'Mouth of the River Dart'. Celtic river-name (*see* DARTINGTON) + OE *mūtha*.

Darton Barns. *Dertun* 1086 (DB). 'Enclosure for deer, deer park'. OE *dēor-tūn*.

Darwen Black. w. Darw. *Derewent* 1208. '(Estate on) the River Darwen'. Darwen is a Celtic river-name meaning 'river where oak-trees grow'.

Dassett, Avon & Dassett, Burton Warwicks. *Derceto*(*ne*) 1086 (DB), *Auene Dercete* 1202, *Magna Dercet* 1242, *Chepynge Dorset* 1397, *Dassett Magna alias Burton Dassett* 1604. Probably 'oak forest', from Celtic **derw* + **cēd*, referring to western edge of once heavily wooded district. The distinguishing affixes are from the old Celtic name of the stream here (*Avon* meaning 'river, water') and (for Burton Dassett) from Latin *magna* 'great', OE *cēping* 'market' and from a nearby place Burton (*Buriton* 1327, from OE *burh-tūn* 'fortified farmstead').

Datchet Winds. & Maid. *Deccet* 10th cent., *Daceta* 1086 (DB). Probably an old Celtic name from **cēd* 'wood' with an obscure first element.

Datchworth Herts. *Dacewrthe* 969, *Daceuuorde* 1086 (DB). 'Enclosure of a man called **Dæcca*'. OE pers. name + *worth*.

Dauntsey Wilts. *Dometesig* 850, *Dantesie* 1086 (DB). Probably 'island or well-watered land of a man called **Dōmgeat*'. OE pers. name + *īeg*.

Davenham Ches. *Deveneham* 1086 (DB). 'Homestead or village on the River Dane'. Celtic river-name (meaning 'trickling stream') + OE *hām*.

Daventry Northants. *Daventrei* 1086 (DB). 'Tree of a man called **Dafa*'. OE pers. name (genitive *-n*) + *trēow*.

Davidson's Mains Edin. 'Davidson's home farm'. William *Davidson* acquired the home farm ('mains') of the Muirhouse estate here in 1776.

Davidstow Cornwall. 'Church of *Sanctus David* (*alias Dewstow*)' 1269. 'Holy place of St David'. Saint's name (Cornish *Dewy*) + OE *stōw*.

Dawley Tel. & Wrek. *Dalelie* 1086 (DB), *Dalilea* c.1200. Probably 'woodland clearing associated with a man called Dealla'. OE pers. name + *-ing-* + *lēah*.

Dawlish Devon. *Douelis* 1086 (DB). Originally the name of the stream here, a Celtic river-name meaning 'dark stream'. **Dawlish Warren** is recorded from the 13th cent., from ME *wareine* 'a game preserve'.

Dawros Head (*Ceann Dhamrois*) Donegal. 'Headland of the oxen'.

Daylesford Glos. *Dæglesford* 718, *Eilesford* [*sic*] 1086 (DB). 'Ford of a man called **Dægel*'. OE pers. name + *ford*.

Deal Kent. *Addelam* [*sic*] 1086 (DB), *Dela* 1158. '(Place at) the hollow or valley', from OE *dæl* (with Latin preposition *ad* 'at' prefixed in the earliest spelling).

Dean, Deane, a common place name, '(place in) the valley', from OE *denu*: **Deane** Bolton. *Dene* 1292. **Dean** Cumbria. *Dene* c.1170. **Deane** Hants. *Dene* 1086 (DB). **Dean, East & Dean, West** Hants. *Dene* 1086 (DB). **Dean, East & Dean, West** W. Sussex. *Dene* 8th cent., *Estdena*, *Westdena* 1150. **Dean, Forest of** Glos. *foresta de Dene* 12th cent. **Dean, Lower & Dean, Upper** Beds. *Dene* 1086 (DB), *Netherdeane*, *Overdeane* 1539. OE *neotherra* 'lower' and *uferra* 'upper'. **Dean Prior** Devon. *Denu* 1086 (DB), *Dene Pryour* 1415. Affix from its possession by Plympton Priory from the 11th cent.

Deanshanger Northants. *Dinneshangra* 937. 'Sloping wood of a man called Dynni'. OE pers. name + *hangra*.

Dearham Cumbria. *Derham* c.1160. 'Homestead or enclosure where deer are kept'. OE *dēor* + *hām* or *hamm*.

Dearne (river). *See* BOLTON UPON DEARNE.

Debach Suffolk. *Depebecs* 1086 (DB). 'Valley or ridge near the river called *Dēope* (the deep one)'. OE river-name (*see* DEBENHAM) + *bece* or *bæc* (dative **bece*).

Debden Essex. *Deppedana* 1086 (DB). 'Deep valley'. OE *dēop* + *denu*.

Debden (Green) Essex. *Tippedene* 1062, *Tippedana* 1086 (DB). 'Valley of a man called **Tippa*'. OE pers. name + *denu*, with the addition of *green* from the 18th cent.

Debenham Suffolk. *Depbenham* 1086 (DB). 'Homestead or village by the river called *Dēope* (the deep one, or stream in a deep valley)'. OE river-name + *hām*. The river referred to is now called the **Deben**, this being a relatively late back-formation from the place name.

Deddington Oxon. *Dædintun* 1050–2, *Dadintone* 1086 (DB). 'Estate associated with a man called Dǣda'. OE pers. name + *-ing-* + *tūn*.

Dedham Essex. *Delham* [*sic*] 1086 (DB), *Dedham* 1166. 'Homestead or village of a man called **Dydda*'. OE pers. name + *hām*.

Dee (river) Abdn. 'Goddess'. British *deva*, identical with next name.

Dee (river) Ches. *Deoua* c.150. An ancient Celtic river-name meaning 'the goddess, the holy one'.

Deehommed (*Deachoimheád*) Down. *Dycovead* 1583. 'Good view'.

Deel (*Daoil*) (river) Limerick, Mayo, Westmeath. 'Black one'.

Deele (*Daoil*) (river) Donegal. 'Black one'.

Deene Northants. *Den* 1065, *Dene* 1086 (DB). '(Place in) the valley'. OE *denu*.

Deenethorpe Northants. *Denetorp* 1169. 'Outlying farmstead or secondary settlement dependent on DEENE'. OScand. *thorp*.

Deenish Island (*Duibhinis*) Kerry. 'Black island'.

Deepdale Cumbria. *Depedale* 1433. 'Deep valley'. OE *dēop* + *dæl*.

Deeping Gate Peterb. *Depynggate* 1390. 'Road to DEEPING (*see* next name). OScand. *gata*.

Deeping St James & Deeping St Nicholas, Market Deeping & West Deeping Lincs. *Estdepinge, West Depinge* 1086 (DB). 'Deep or low place'. OE **dēoping*. Distinguishing affixes from the dedications of the churches, and from *market, east,* and *west*.

Deerhurst Glos. *Deorhyrst* 804, *Derheste* [*sic*] 1086 (DB). 'Wooded hill frequented by deer'. OE *dēor* + *hyrst*.

Defford Worcs. *Deopanforda* 972, *Depeford* 1086 (DB). 'Deep ford'. OE *dēop* + *ford*.

Deganwy Conwy. *Arx Decantorum* 812, *Dugannu* 1191, *Diganwy* 1254. '(Place of the) Decantae'. The *Decantae* (perhaps 'noble ones') were a British tribe who occupied this region of north Wales.

Deighton, 'farmstead surrounded by a ditch', from OE *dīc* + *tūn*: **Deighton** N. Yorks., near Northallerton. *Dictune* 1086 (DB). **Deighton** York. *Distone* [*sic*] 1086 (DB), *Dicton* 1176. **Deighton, Kirk & Deighton, North** N. Yorks. *Distone* [*sic*] 1086 (DB), *Kirke Dighton* 14th cent., *Northdictun* 12th cent. Distinguishing affixes are OScand. *kirkja* 'church' and OE *north*.

Deiniolen Gwyd. '(Church of) St Deiniolen'. The village was originally *Ebeneser*, after the Welsh Nonconformist chapel here, from biblical *Eben-ezer*, where the Israelites were defeated by the Philistines (1 Samuel 4:1). The present name was adopted in the early 20th cent. from the parish name *Llanddeiniolen* (Welsh *llan*, 'church').

Delabole Cornwall. *Deliou* 1086 (DB), *Delyou Bol* 1284. The name of the Domesday manor *Deliou* (possibly from a Cornish word meaning 'leaves' and surviving as nearby **Deli**) with the later addition of Cornish *pol* 'pit' referring to the great slate quarry here.

Delamere Ches. named from the ancient Forest of Delamere, *foresta de la Mare* 13th cent., 'forest of the lake', from OE *mere* with the OFrench preposition and article *de la*.

Delgany (*Deilgne*) Wicklow. 'Thorny place'.

Delph, New Delph Oldham. *Delfe* 1544. From OE (*ge*)*delf* 'a quarry'.

Delvin (*Dealbhna*) Westmeath. '(Territory of the) descendants of Dealbaeth'.

Dembleby Lincs. *Dembelbi* 1086 (DB). 'Farmstead or village with a pool'. OScand. *dembil + *bý*.

Denaby Donc. *Denegebi, Degenebi* 1086 (DB), *Deneby* 1219. 'Farmstead or village of the Danes'. OE *Dene* (genitive plural *Deniga, Dena*) + OScand. *bý*.

Denbigh (*Dinbych*) Denb. *Dinbych* 1269. 'Little fortress'. Welsh *din* + *bach*. The 'little fortress' would have stood where the ruins of the 12th-cent. castle now lie. Compare TENBY.

Denbury Devon. *Deveneberie* 1086 (DB). 'Earthwork or fortress of the Devon people'. OE *Defnas* (from Celtic *Dumnonii*) + *burh* (dative *byrig*).

Denby Derbys. *Denebi* 1086 (DB). 'Farmstead or village of the Danes'. OE *Dene* (genitive plural *Dena*) + OScand. *bý*.

Denby Dale Kirkl. *Denebi* 1086 (DB). Identical in origin with the previous name, with the later addition of *dale* 'valley'.

Denchworth Oxon. *Deniceswurthe* 947, *Denchesworde* 1086 (DB). 'Enclosure of a man called *Denic'. OE pers. name + *worth*.

Denford Northants. *Deneforde* 1086 (DB). 'Ford in a valley'. OE *denu* + *ford*.

Denge Marsh Kent. *See* DUNGENESS.

Dengie Essex. *Deningei* c.707, *Daneseia* 1086 (DB). Probably 'island or well-watered land associated with a man called Dene, or of Dene's people'. OE pers. name + *-ing-* or *-inga-* + *ēg*.

Denham, 'homestead or village in a valley', OE *denu* + *hām*: **Denham** Bucks. *Deneham* 1066, *Daneham* 1086 (DB). **Denham** Suffolk, near Bury St Edmunds. *Denham* 1086 (DB). **Denham** Suffolk, near Eye. *Denham* 1086 (DB).

Denholme Brad. *Denholme* 1252. 'Water-meadow in the valley'. OE *denu* + OScand. *holmr*.

Denmead Hants. *Denemede* 1205. 'Meadow in the valley'. OE *denu* + *mæd*.

Dennington Suffolk. *Dingifetuna* 1086 (DB). 'Farmstead of a woman called *Denegifu'. OE pers. name + *tūn*.

Denny Falk. *Litill Dany* 1510. Meaning unknown.

Denshaw Oldham. Recorded thus from 1635, 'the small wood or copse in the valley'. ME *dene* + *shaw* (OE *sceaga*).

Denston Suffolk. *Danerdestuna* 1086 (DB). 'Farmstead of a man called Deneheard'. OE pers. name + *tūn*.

Denstone Staffs. *Denestone* 1086 (DB). 'Farmstead of a man called Dene'. OE pers. name + *tūn*.

Dent Cumbria. near Sedbergh. *Denet* 1202–8. Possibly an old river-name, origin and meaning uncertain. **Dentdale** (recorded from 1577) contains OScand. *dalr* 'valley'.

Denton, a common place name, usually 'farmstead or village in a valley', from OE *denu* + *tūn*; examples include: **Denton** Cambs. *Dentun* 10th cent., *Dentone* 1086 (DB). **Denton** Darltn. *Denton* 1200. **Denton** E. Sussex. *Denton* 9th cent. **Denton** Kent, near Dover. *Denetun* 799, *Danetone* 1086 (DB). **Denton** Lincs. *Dentune* 1086 (DB). **Denton** Norfolk. *Dentuna* 1086 (DB). **Denton** N. Yorks. *Dentun* c.972, 1086 (DB). **Denton** Oxon. *Denton* 1122. **Denton** Tamesd. *Denton* c.1220. **Denton, Upper & Denton, Nether** Cumbria. *Denton* 1169.

However the following Denton has a different origin: **Denton** Northants. *Dodintone* 1086 (DB). 'Estate associated with a man called Dodda or Dudda'. OE pers. name + *-ing-* + *tūn*.

Denver Norfolk. *Danefella* [*sic*] 1086 (DB), *Denever* 1200. 'Ford or passage used by the Danes'. OE *Dene* (genitive plural *Dena*) + *fær*.

Denwick Northum. *Den*(*e*)*wyc* 1242. 'Dwelling or dairy-farm in a valley'. OE *denu* + *wīc*.

Deopham Norfolk. *Depham* 1086 (DB). 'Homestead or village near the deep place or lake'. OE *dēope* + *hām*.

Depden Suffolk. *Depdana* 1086 (DB). 'Deep valley'. OE *dēop* + *denu*.

Deptford, 'deep ford', OE *dēop* + *ford*: **Deptford** Gtr. London. *Depeford* 1293. **Deptford** Wilts. *Depeford* 1086 (DB).

Derby Derby. *Deoraby* 10th cent., *Derby* 1086 (DB). 'Farmstead or village where deer are kept'. OScand. *djúr* + *bý*. **Derbyshire** (OE *scīr* 'district') is first referred to in the 11th cent.

Derby, West Lpool. *Derbei* 1086 (DB), *Westderbi* 1177. Identical in origin with the previous name.

Dereham, East & Dereham, West Norfolk. *Der(e)ham* 1086 (DB), *Estderham* 1428, *Westderham* 1203. 'Homestead or enclosure where deer are kept'. OE *dēor* + *hām* or *hamm*.

Derg (*Dearg*) (river) Tyrone. *The Red river al. Dearg*. 'Red one'.

Derg, Lough (*Loch Dearg*) Clare, Donegal, Galway, Tipperary. 'Red lake'.

Dernagree (*Doire na Graí*) Clare. 'Oak grove of the stud of horses'.

Dernish (*Dair Inis*) Clare, Fermanagh, Sligo. 'Oak island'.

Derrad (*Doire Fhada*) Westmeath. 'Long oak grove'.

Derradda (*Doire Fhada*) Leitrim. 'Long oak wood'.

Derragh (*Darach*) Clare, Longford, Mayo. 'Oak'.

Derrane (*Doireán*) Roscommon. 'Little grove'.

Derravaragh, Lough (*Loch Dairbhreach*) Westmeath. *Logh Dervaragh* 1412. 'Lake of the oak wood'.

Derreen (*Doirín*) (river) Carlow. '(River of the) little oak grove'.

Derreen (*Doirín*) Galway. 'Oak grove'.

Derreenacarrin (*Doire an Chairn*) Clare. 'Oak grove of the cairn'.

Derreenaclaurig (*Doirín an Chláraigh*) Kerry. 'Oak grove of the planking'.

Derreenafoyle (*Doirín an Phoill*) Kerry. 'Oak grove of the pool'.

Derreenavurrig (*Doirín an Mhuirigh*) Kerry. 'Oak grove of the mariner'.

Derreendrislagh (*Doirín Drisleach*) Kerry. 'Brambly oak grove'.

Derreennamucklagh (*Doirín na Muclach*) Kerry. 'Oak grove of the swineherds'.

Derreensillagh (*Doirín Saileach*) Kerry. 'Oak grove of willows'.

Derries (*Doirí*) Offaly. 'Oak woods'.

Derringstone Kent. *Dieringestune* early 13th cent. 'Farmstead or estate of a man called Dēoring'. OE pers. name + *tūn*.

Derrington Staffs. *Dodintone* 1086 (DB). 'Estate associated with a man called Dod(d)a or Dud(d)a'. OE pers. name + *-ing-* + *tūn*.

Derrinlaur (*Doire an Láir*) Waterford. 'Middle oak grove'.

Derry (*Doire*) Derry. *Doire Calgaigh* 535. 'Oak grove'. The original Irish name was *Doire Chalgaigh*, 'Calgach's oak grove'. A later name was *Doire Cholm Cille*, 'Columcille's oak grove', after a monastery founded by St Columba in 546. In 1613 Derry was renamed *Londonderry* by a group of merchants from *London*.

Derry (*Doire an Bhile*) Down. 'Oak grove of the sacred tree'.

Derryadd (*Doire Fhada*) Armagh. *Derriada* 1609. 'Long oak grove'.

Derryaghy (*Doire Achaidh*) Antrim. *Ardrachi* 1306. 'Oak grove of the field'. An earlier Irish name was *Ard Achaidh*, 'height of the field'.

Derryanvil (*Doire Chonamhail*) Armagh. *Derrehanavile*. 'Oak grove of Conamhail'.

Derrybawn (*Doire Bán*) Wicklow. 'White oak grove'.

Derrybeg (*Doirí Beaga*) Donegal. 'Small oak groves'.

Derryboy (*Doire Buí*) Down. *Dirryboy* 1634. 'Yellow oak grove'.

Derrycaw (*Doire an Chatha*) Armagh. 'Oak wood of the battle'.

Derrychrier (*Doire an Chriathair*) Derry. *Derkreayr* 1397. 'Oak wood of the quagmire'.

Derrycoffey (*Doire Uí Chofaigh*) Offaly. 'Oak wood of Ó Cofaigh'.

Derrycoogh (*Doire Cuach*) Tipperary. 'Oak grove of the cuckoos'.

Derrycooly (*Doire Cúile*) Offaly. 'Oak wood of the nook'.

Derrydamph (*Doire Damh*) Cavan. 'Oak grove of the oxen'.

Derryfubble (*Doire an Phobail*) Tyrone. *Dirripuble* 1609. 'Oak wood of the people'.

Derrygarrane (*Doire an Ghearráin*) Kerry. 'Oak wood of the gelding'.

Derrygarrane North (*Doire an Ghearráin Thuaidh*) Kerry. 'Northern oak wood of the gelding'.

Derrygarrane South (*Doire an Ghearráin Theas*) Kerry. 'Southern oak wood of the gelding'.

Derrygolan (*Doire an Ghabláin*) Westmeath. 'Oak grove of the fork'.

Derrygrath (*Deargráth*) Tipperary. 'Red fort'.

Derrygrogan (*Doire Grógáin*) Offaly. 'Oak wood of the little heap'.

Derryharney (*Doire Charna*) Fermanagh. *Derriharne* 1659. 'Oak wood of the cairn'.

Derryhaw (*Doire an Chatha*) Armagh. *Derricah* 1661. 'Oak wood of the battle'.

Derryhowlaght (*Doire Thaimleacht*) Fermanagh. 'Oak grove of the grave'.

Derrykeevan (*Doire Uí Chaomháin*) Armagh. *Derrykivyn* 1657. 'Ó Caomhán's oak grove'.

Derrykeighan (*Doire Chaocháin*) Antrim. *Daire Chaechain c.*800. 'Caochán's oak grove'.

Derrylahard (*Doire Leath Ard*) Cork. 'Oak grove of the half height'.

Derrylane (*Doire Leathan*) Cavan. 'Broad oak wood'.

Derrylaney (*Doire Léana*) Fermanagh. 'Oak grove of the meadow'.

Derrylard (*Doire Leath Ard*) Armagh. 'Oak grove of the half height'.

Derryleagh (*Doire Liath*) Kerry. 'Grey oak wood'.

Derrylester (*Doire an Leastair*) Fermanagh. 'Oak wood of the small boat'.

Derrylicka (*Doire Lice*) Kerry. 'Oak wood of the flagstone'.

Derrylin (*Doire Fhlainn*) Fermanagh. *Derrelin* 1609. 'Flann's oak grove'.

Derrymihin (*Doire Mheithean*) Cork. 'Oak wood of saplings'.

Derrymore (*Doire Mór*) Armagh. 'Large oak wood'.

Derrynablaha (*Doire na Blátha*) Kerry. 'Oak wood of flowers'.

Derrynacaheragh (*Doirín na Cathrach*) Cork. 'Oak grove of the stone fort'.

Derrynacleigh (*Doire na Cloiche*) Galway. 'Oak grove of the stone'.

Derrynafeana (*Doire na bhFiann*) Kerry. 'Oak grove of the Fianna'.

Derrynafunsha (*Doire na Fuinse*) Kerry. 'Oak grove of the ash'.

Derrynagree (*Doire na Graí*) Kerry. 'Oak grove of the stud'.

Derrynanaff (*Doire na nDamh*) Mayo. 'Oak grove of the oxen'.

Derrynasaggart Mountains (*Cnoic Dhoire na Sagart*) Kerry. 'Mountains of the oak grove of the priests'.

Derrynashaloge (*Doire na Sealg*) Tyrone. 'Oak grove of the hunt'.

Derryness (*Doirinis*) Donegal. 'Oak island'.

Derryrush (*Doire Iorrais*) Galway. 'Oak grove of the headland'.

Derrythorpe N. Lincs. *Dudingthorp c.*1184. Probably 'outlying farmstead or hamlet of a man called Dudding'. OE pers. name + OScand. *thorp.*

Derrytrasna (*Doire Trasna*) Armagh. *Derrytrasney* 1661. 'Transverse oak grove'.

Derryveagh Mountains (*Sléibhte Dhoire Bheitheach*) Donegal. 'Mountains of the oak grove of the birches'.

Derryvohy (*Doire Bhoithe*) Mayo. 'Oak grove of the hut'.

Derrywillow (*Doire Bhaile*) Leitrim. 'Oak grove of the homestead'.

Derrywinny (*Doire Bhainne*) Cavan. 'Oak grove of the milk'.

Dersingham Norfolk. *Dersincham* 1086 (DB). 'Homestead of the family or followers of a man called Dēorsige'. OE pers. name + *-inga-* + *hām.*

Dervock (*Dearbhóg*) Antrim. *Deroogg c.*1659. 'Oak plantation'.

Derwent (river), four examples, in Cumbria, Derbys., Durham, and Yorks.; a Celtic river-name recorded from the 8th cent. in the form *Deruuentionis fluvii* 'river where oak-trees grow abundantly'.

Derwentside (district) Durham. A modern name, from the River DERWENT (*see* previous entry).

Desborough Northants. *Dereburg* 1086 (DB), *Deresburc* 1166. 'Stronghold or fortress of a man called Dēor'. OE pers. name + *burh*.

Desert (*Díseart*) Monaghan. 'Hermitage'.

Desertcreat (*Díseart dá Chríoch*) Tyrone. 'Hermitage of two territories'.

Desertegny (*Díseart Éignigh*) Donegal. 'Éigneach's hermitage'.

Desertmartin (*Díseart Mhártain*) Derry. *Dísiort Mhártain c.*1645. 'Mártan's hermitage'.

Desertoghill (*Díseart Uí Thuathail*) Derry. 'Ó Tuathal's hermitage'.

Desford Leics. *Deresford* 1086 (DB). 'Ford of a man called Dēor'. OE pers. name + *ford*.

Detchant Northum. *Dichende* 1166. 'End of the ditch or dike'. OE *dīc* + *ende*.

Dethick Derbys. *Dethec* 1154–9. 'Death oak-tree', i.e. 'oak-tree on which criminals were hanged'. OE *dēath* + *āc*.

Detling Kent. *Detlinges* 11th cent. '(Settlement of) the family or followers of a man called *Dyttel'. OE pers. name + *-ingas*.

Devenish (*Daimhinis*) (island) Fermanagh. (*Molassí*) *Daminis c.*830. 'Island of oxen'.

Deveron (river) Banff. *Douern* 1273. Gaelic *dubh* 'black' + pre-Celtic 'river'.

Devizes Wilts. *Divises* 11th cent. '(Place on) the boundaries'. OFrench *devise*.

Devlin (*Duibh Linn*) Donegal, Mayo, Monaghan. 'Black pool'.

Devon (the county). *Defena, Defenascir* late 9th cent. '(Territory of) the Devonians, earlier called the *Dumnonii'. OE tribal name *Defnas* (from Celtic *Dumnonii*) + *scīr* 'district'.

Devonport Plym. a modern self-explanatory name only in use since 1824.

Dewchurch, Much & Dewchurch, Little Herefs. *Lann Deui* 7th cent., *Dewischirche c.*1150. 'Church of St Dewi'. Saint's name (the Welsh form of David) + OE *cirice* (with OWelsh *lann* in the early form). Distinguishing affixes are OE *micel* 'great' and *lȳtel* 'little'.

Dewlish Dorset. *Devenis* [sic] 1086 (DB), *Deueliz* 1194. Originally the name of the stream here, a Celtic river-name meaning 'dark stream'.

Dewsbury Kirkl. *Deusberia* 1086 (DB). 'Stronghold of a man called Dewi'. OWelsh pers. name + OE *burh* (dative *byrig*).

Dian (*Daingean*) Monaghan, Tyrone. 'Fortress'.

Dibden (Purlieu) Hants. *Depedene* 1086 (DB). 'Deep valley'. OE *dēop* + *denu*. The addition is from ME *purlewe* 'outskirts of a forest'.

Dicker, Lower & Dicker, Upper E. Sussex. *Diker* 1229. From ME *dyker* 'ten', probably alluding to a plot of land for which this number of iron rods was paid in rent.

Dickleburgh Norfolk. *Dicclesburc* 1086 (DB). Possibly 'stronghold of a man called *Dicel or *Dicla'. OE pers. name + *burh*.

Didbrook Glos. *Duddebrok* 1248. 'Brook of a man called *Dyd(d)a'. OE pers. name + *brōc*.

Didcot Oxon. *Dudecota* 1206. 'Cottage(s) of a man called Dud(d)a'. OE pers. name + *cot*.

Diddington Cambs. *Dodinctun* 1086 (DB). 'Estate associated with a man called Dod(d)a or Dud(d)a'. OE pers. name + *-ing-* + *tūn*.

Diddlebury Shrops. *Dodeleberia c.*1090. 'Stronghold or manor of a man called *Dud(d)ela'. OE pers. name + *burh* (dative *byrig*).

Didley Herefs. *Dodelegie* 1086 (DB). 'Woodland clearing of a man called Dod(d)a or Dud(d)a'. OE pers. name + *lēah*.

Didmarton Glos. *Dydimeretune* 972, *Dedmertone* 1086 (DB). Probably 'farmstead by Dyd(d)a's pool'. OE pers. name + *mere* + *tūn*.

Didsbury Manch. *Dedesbiry* 1246. 'Stronghold of a man called *Dyd(d)i'. OE pers. name + *burh* (dative *byrig*).

Diffreen (*Dubh Thrian*) Leitrim. 'Black third'.

Diffwys (mountain) Gwyd. 'Steep slope'. Welsh *diffwys*.

Digby Lincs. *Dicbi* 1086 (DB). 'Farmstead or village at the ditch'. OE *dīc* or OScand. *díki* + *bý*.

Dilham Norfolk. *Dilham* 1086 (DB). 'Homestead or enclosure where dill grows'. OE *dile* + *hām* or *hamm*.

Dilhorne Staffs. *Dulverne* 1086 (DB). 'House or building by a pit or quarry'. OE **dylf* + *ærn*.

Dilston Northum. *Deuelestune* 1172. 'Farmstead by the dark stream'. Celtic river-name + OE *tūn*.

Dilton Marsh Wilts. *Dulinton* 1190. Probably 'farmstead of a man called *Dulla'. OE pers. name (genitive -*n*) + *tūn*.

Dilwyn Herefs. *Dilven* 1086 (DB). '(Settlement at) the shady or secret places'. OE *dīgle* in a dative plural form *dīglum*.

Dinas Powys Vale Glam. *Dinaspowis* 1187, *Dinas Powis* c.1262. 'Fort of Powis'. Welsh *dinas*. *Powys* apparently has the same origin as POWYS.

Dinbych. *See* DENBIGH.

Dinbych-y-pysgod. *See* TENBY.

Dinchope Shrops. *Doddinghop* 12th cent. 'Enclosed valley associated with a man called Dudda', or 'valley at Dudda's place'. OE pers. name + -*ing*- or -*ing* + *hop*.

Dinder Somerset. *Dinre* 1174. Probably 'hill with a fort'. Celtic **din* + **breʒ*.

Dinedor Herefs. *Dunre* 1086 (DB). Possibly identical in origin with the previous name.

Dingin (*Daingean*) Cavan. 'Fortress'.

Dingle (*An Daingean*) Kerry. 'The fortress'.

Dingley Northants. *Dinglei* 1086 (DB). Possibly 'woodland clearing with hollows'. OE **dyngel* + *lēah*.

Dingwall Highland. *Dingwell* 1227. 'Field of the assembly'. OScand. *thing-vǫllr*. The assembly was the Scandinavian 'parliament'. *See also* THINGWALL.

Dinis (*Duibhinis*) Cork. 'Black island'.

Dinmore Herefs. *See* HOPE.

Dinnington Newc. upon T. *Donigton* 1242. Probably identical in origin with the next name.

Dinnington Rothm. *Dunintone* 1086 (DB). 'Estate associated with a man called Dunn(a)'. OE pers. name + -*ing*- + *tūn*.

Dinnington Somerset. *Dinnitone* 1086 (DB). 'Estate associated with a man called Dynne'. OE pers. name + -*ing*- + *tūn*.

Dinorwic. *See* PORT DINORWIC.

Dinsdale, Low Darltn. *Ditneshal* c.1190. 'Nook of land belonging to a man called *Dyttīn'. OE pers. name + *halh*.

Dinton, 'estate associated with a man called Dunn(a)', OE pers. name + -*ing*- + *tūn*: **Dinton** Bucks. *Danitone* [sic] 1086 (DB), *Duninton* 1208. **Dinton** Wilts. *Domnitone* [sic] 1086 (DB), *Dunyngtun* 12th cent.

Diptford Devon. *Depeforde* 1086 (DB). 'Deep ford'. OE *dēop* + *ford*.

Dipton Durham. *Depedene* c.1190. 'Deep valley'. OE *dēop* + *denu*.

Diptonmill Northum. *Depeden* 1269. Identical in origin with the previous name, with the later addition of *mill*.

Disert (*Díseart*) Donegal. 'Hermitage'.

Diserth. *See* DYSERTH.

Diseworth Leics. *Digtheswyrthe* c.972, *Diwort* 1086 (DB). 'Enclosure of a man called *Digoth'. OE pers. name + *worth*.

Dishforth N. Yorks. *Disforde* 1086 (DB). 'Ford across a ditch'. OE *dīc* + *ford*.

Disley Ches. *Destesleg* c.1251. OE *lēah* 'wood, woodland clearing' with an obscure first element, possibly an OE **dystels* 'mound or heap'.

Diss Norfolk. *Dice* 1086 (DB). '(Place at) the ditch or dike'. OE *dīc*.

Distington Cumbria. *Dustinton* c.1230. OE *tūn* 'farmstead, estate' with an obscure first element, possibly a pers. name.

Ditchburn Northum. *Dicheburn* 1236. 'Ditch stream'. OE *dīc* + *burna*.

Ditcheat Somerset. *Dichesgate* 842, *Dicesget* 1086 (DB). 'Gap in the dyke or earthwork'. OE *dīc* + *geat*.

Ditchingham Norfolk. *Dicingaham* 1086 (DB). 'Homestead of the dwellers at a ditch or dyke', or 'homestead of the family or followers of a man called *Dic(c)a or *Dīca'. OE *dīc* or OE pers. name + *-inga-* + *hām*.

Ditchling E. Sussex. *Dicelinga* 765, *Dicelinges* 1086 (DB). '(Settlement of) the family or followers of a man called *Diccel or *Dīcel'. OE pers. name + *-ingas*.

Dittisham Devon. *Didasham* 1086 (DB). 'Enclosure or promontory of a man called *Dyddi'. OE pers. name + *hamm*.

Ditton, usually 'farmstead by a ditch or dike', OE *dīc* + *tūn*: **Ditton** Halton. *Ditton* 1194. **Ditton** Kent. *Dictun* 10th cent., *Dictune* 1086 (DB). **Ditton, Fen** Cambs. *Dictunæ* c.975, *Fen Dytton* 1286. Affix is OE *fenn* 'fen, marshland'. **Ditton, Long & Ditton, Thames** Surrey. *Dictun* 1005, *Ditune* 1086 (DB), *Longa Dittone* 1242, *Temes Ditton* 1235. Distinguishing affixes are OE *lang* 'long' (alluding to the length of the village) and THAMES from the river.
    However the following has a different origin: **Ditton Priors** Shrops. *Dodintone* 1086 (DB). 'Estate associated with a man called Dod(d)a or Dud(d)a'. OE pers. name + *-ing-* + *tūn*. Manorial affix from its possession by Wenlock Priory.

Dixton Glos. *Dricledone* 1086 (DB), *Diclisdon* 1169. Probably 'down of the hill with dikes or earthworks'. OE *dīc* + *hyll* + *dūn*. Alternatively the first element may be an OE pers. name *Dīcel or *Diccel.

Doagh Beg (*Dumhaigh Bhig*) Donegal. *Dowaghbegg* c.1655. 'Little sandbank'.

Doagh Isle (*Dumhachoileán*) Donegal. 'Sandhill island'.

Dobcross Oldham. a late name, first recorded as *Dobcrosse* 1662. 'Cross (or cross-roads) associated with a man called Dobbe'. ME pers. name (a pet-form of *Robert*) or surname + *cross*.

Docking Norfolk. *Doccynge* c.1035, *Dochinga* 1086 (DB). 'Place where docks or water-lilies grow'. OE *docce* + *-ing*.

Docklow Herefs. *Dockelawe* 1291. 'Mound or hill where docks grow'. OE *docce* + *hlāw*.

Dockray Cumbria. *Docwra* 1278. 'Nook of land where docks or water-lilies grow'. OE *docce* + OScand. *vrá*.

Doddinghurst Essex. *Doddenhenc* [*sic*] 1086 (DB), *Duddingeherst* 1218. 'Wooded hill of the family or followers of a man called Dudda or Dodda'. OE pers. name + *-inga-* + *hyrst*.

Doddington, Dodington, 'estate associated with a man called Dud(d)a or Dod(d)a', OE pers. name + *-ing-* + *tūn*: **Doddington** Cambs. *Dundingtune* [*sic*] c.975, *Dodinton* 1086 (DB). **Doddington** Kent. *Duddingtun* c.1100. **Doddington** Lincs. *Dodin(c)tune* 1086 (DB). **Doddington** Northum. *Dodinton* 1207. **Doddington** Shrops. *Dodington* 1285. **Doddington, Dry** Lincs. *Dodintune* 1086 (DB). Affix from OE *drȳge* 'dry'. **Doddington, Great** Northants. *Dodintone* 1086 (DB), *Great Dodington* 1290. **Dodington** S. Glos. *Dodintone* 1086 (DB).

Doddiscombsleigh Devon. *Leuga* 1086 (DB), *Doddescumbeleghe* 1309. Originally 'the woodland clearing', OE *lēah*. Later addition from the *Doddescumb* family, here in the 13th cent.

Dodford, 'ford of a man called Dodda', OE pers. name + *ford*: **Dodford** Northants. *Doddanford* 944, *Dodeforde* 1086 (DB). **Dodford** Worcs. *Doddeford* 1232.

Dodington S. Glos. *See* DODDINGTON.

Dodleston Ches. *Dodestune* [*sic*] 1086 (DB), *Dodleston* 1153. 'Farmstead of a man called *Dod(d)el'. OE pers. name + *tūn*.

Dodworth Barns. *Dodesuu(o)rde* 1086 (DB). 'Enclosure of a man called Dod(d) or Dod(d)a'. OE pers. name + *worth*.

Doe Castle (*Caisleán na dTuath*) Donegal. 'Castle of the territories'.

Dogdyke Lincs. *Dockedic* 12th cent. 'Ditch where docks or water-lilies grow'. OE *docce* + *dīc*.

Dolgarrog Conwy. *Dolgarrog* 1534. 'Water-meadows of the fast-flowing stream'. Welsh *dôl* + *carrog*.

Dolgellau Gwyd. *Dolkelew* 1254, *Dolgethly* 1338. 'Water-meadow of cells'. Welsh *dôl* + *cell* (plural *cellau*). The 'cells' were possibly monastic cells or perhaps they were merchants' booths.

Dolla (*Dolla*) Tipperary. 'Loops'.

Dollar Clac. *Doler* 1461. 'Meadow'. Celtic *dol* + *ar*.

Dollingstown Down. 'Dolling's town'.
Apparently from the Revd Boghey *Dolling*, a
local rector early in the 19th cent.

Dollymount (*Cnocán Doirinne*) Dublin.
'Doireanne's mount'.

Dolphinholme Lancs. *Dolphineholme* 1591.
'Island or water-meadow of a man called
Dolgfinnr'. OScand. pers. name + *holmr*.

Dolton Devon. *Duueltone* 1086 (DB). Possibly
'farmstead in the open country frequented by
doves'. OE *dūfe* + *feld* + *tūn*.

Donabate (*Domhnach Bat*) Dublin. 'Church
of the boat'.

Donacarney (*Domhnach Cearnaigh*) Meath.
'Cearnach's church'.

Donadea (*Domhnach Dheá*) Kildare.
'Deadha's church'.

Donagh (*Domhnach*) Fermanagh. *Donagh*
1659. 'Church'.

Donaghadee (*Domhnach Daoi*) Down.
Donanachti 1204. Possibly 'Daoi's church'.

Donaghanie (*Domhnach an Eich*) Tyrone.
Domhach an Eich 1518. 'Church of the horse'.

Donaghcloney (*Domhnach Cluana*) Down.
*Domhnach Cluana c.*1645. 'Church of the
meadow'.

Donaghedy (*Domhnach Caoide*) Tyrone.
'Caoide's church'.

Donaghey (*Dún Eachaidh*) Tyrone.
'Eachaidh's fort'.

Donaghmore (*Domhnach Mór*) Laois,
Meath, Tipperary. 'Large church'.

Donaghmoyne (*Domhnach Maighean*)
Monaghan. 'Church of the little plain'.

Donaghpatrick (*Domhnach Phádraig*)
Meath. 'Pádraig's church'.

Donaghrisk (*Domhnach Riascadh*) Tyrone.
'Church of the marsh'.

Donamon (*Dún Iomain*) Roscommon.
'Ioman's fort'.

Donard (*Dún Ard*) Wexford, Wicklow. 'High
fort'.

Donaskeagh (*Dún na Sciath*) Tipperary.
'Fort of the shields'.

Doncaster Donc. *Doneceastre* 1002,
Donecastre 1086 (DB). 'Roman fort on the River
Don'. Celtic river-name (meaning simply
'water, river') + OE *ceaster*.

Donegal (*Dún na nGall*) Donegal. (*mainistir*)
Dúin na nGall 1474. 'Fort of the foreigners',
referring to the Danes active here in the
9th cent. The county of Donegal is sometimes
known by the Irish name *Tír Chonaill*,
'Conaill's territory'.

Doneraile (*Dún ar Aill*) Cork. 'Fort on the
cliff'.

**Donhead St Andrew & Donhead
St Mary** Wilts. *Dunheved* 871, *Duneheve* 1086
(DB), *Dounheved Sanct Andree* 1302, *Donheved
Sancte Marie* 1298. 'Head or end of the down'.
OE *dūn* + *hēafod*. Affixes from the dedications
of the churches.

Donibristle Fife. *Donibrysell c.*1165.
'Breasal's fort'. Gaelic *dúnadh* + pers. name.

Donington, Donnington, usually 'estate
associated with a man called Dun(n) or
Dun(n)a', OE pers. name + *-ing-* + *tūn* (possibly
in some cases 'estate at the hill-place', OE
dūning* + *tūn*); examples include: **Donington
Lincs. *Duninctune* 1086 (DB). **Donington,
Castle** Leics. *Duni(n)tone* 1086 (DB), *Castel
Donyngton* 1302. Affix refers to a former castle
here. **Donington le Heath** Leics. *Duntone*
1086 (DB), *Donygton super le heth* 1462. Affix
means 'on the heath', from OE *hēth*.
Donington on Bain Lincs. *Duninctune* 1086
(DB), *Donyngton super Beyne* 13th cent. Affix from
its situation on the River Bain ('the short one',
from OScand. *beinn*). **Donnington** Glos.
*Doninton c.*1195. **Donnington** Herefs.
Dunninctune 1086 (DB). **Donnington**
Shrops., near Eaton Constantine. *Dunniton*
1180. **Donnington** Shrops., near Oakengates.
Donnyton 1272. **Donnington** W. Berks.
Deritone [*sic*] 1086 (DB), *Dunintona* 1167.
   However the following has a different origin:
Donnington W. Sussex. *Dunketone* 966,
Cloninctune [*sic*] 1086 (DB). 'Farmstead of a man
called *Dunnuca'. OE pers. name + *tūn*.

Donisthorpe Leics. *Durandestorp* 1086 (DB).
'Outlying farmstead or hamlet of a man called
Durand'. OFrench pers. name + OScand. *thorp*.

Donnington. *See* DONINGTON.

Donnybrook (*Domhnach Broc*) Dublin.
'Broc's church'.

Donnycarney (*Domhnach Cearna*) Dublin. 'Cearnach's church'.

Donohill (*Dún Eochaille*) Tipperary. 'Fort of the yew wood'.

Donore (*Dún Uabhair*) Meath. 'Fort of pride'.

Donoughmore (*Domhnach Mór*) Cork. 'Big church'.

Donyatt Somerset. *Duunegete* 8th cent., *Doniet* 1086 (DB). Probably 'gate or gap of a man called Dun(n)a'. OE pers. name + *geat*.

Doo, Lough (*Loch Dúlocha*) Clare, Mayo. 'Black lake'.

Dooagh (*Dumha Acha*) Mayo. 'Mound of the field'.

Doocastle (*Caisleán an Dumha*) Mayo. 'Mound castle'.

Doochary (*An Dúchoraidh*) Donegal. 'The black weir'.

Doocreggaun (*Dubh Creagán*) Mayo. 'Little black crag'.

Doogary (*An Dúgharraí*) Cavan. 'The black garden'.

Dooghbeg (*Dumha Beag*) Mayo. 'Small mound'.

Doogort (*Dumha Gort*) Mayo. 'Salty mound'.

Doohamlet (*Dúthamhlacht*) Monaghan. 'Black burial mound'.

Doohat (*Dútháite*) Monaghan. 'Black site'.

Doohoma (*Dumha Thuama*) Mayo. 'Mound of the tomb'.

Dooish (*Dubhais*) Donegal, Tyrone. 'Black ridge'.

Dooleeg (*Dumha Liag*) Mayo. 'Black flagstone'.

Doolin (*Dúlainn*) Clare. 'Black pool'.

Doon (*Dún*) Galway, Limerick. 'Fort'.

Doon (river) E. Ayr. *Don* 1197. '(River of the) river goddess'. British *Deuona*.

Doonaha (*Dún Átha*) Clare. 'Fort of the ford'.

Doonally (*Dún Aille*) Sligo. 'Fort of the cliff'.

Doonbeg (*An Dún Beag*) Clare. 'The small fort'.

Doonfeeny (*Dún Fine*) Mayo. 'Fine's fort'.

Dooraheen (*Dubhrathain*) Westmeath. 'Black ferny ground'.

Dorchester Dorset. *Durnovaria* 4th cent., *Dornwaraceaster* 864, *Dorecestre* 1086 (DB). 'Roman town called *Durnovaria*'. Reduced form of Celtic name (perhaps meaning 'place with fist-sized pebbles') + OE *ceaster*.

Dorchester Oxon. *Dorciccaestræ* 731, *Dorchecestre* 1086 (DB). 'Roman town called *Dorcic*'. Celtic name (obscure in origin and meaning) + OE *ceaster*.

Dordon Warwicks. *Derdon* 13th cent. 'Hill frequented by deer or other wild animals'. OE *dēor* + *dūn*.

Dore Sheff. *Dore* late 9th cent., 1086 (DB). '(Place at) the gate or narrow pass'. OE *dor*.

Dore, Abbey Herefs. *Dore* 1147. Named from the River Dore, a Celtic river-name meaning simply 'the waters'. Affix from the former Cistercian Abbey here.

Dorking Surrey. *Dorchinges* 1086 (DB). Possibly '(settlement of) the family or followers of a man called *Deorc*', OE pers. name + *-ingas*. Alternatively, perhaps '(settlement of) the people living by the River *Dorce*', this last being a postulated old name of Celtic origin (meaning 'bright stream') for the upper Mole (for which *see* MOLESEY).

Dormansland Surrey. *Deremanneslond* 1263. 'Cultivated land or estate of the Dereman family'. OE *land*, with the surname of a local family.

Dormanstown Red. & Cleve. a modern name for a new town planned in 1918 for the employees of the manufacturing firm Dorman Long.

Dormington Herefs. *Dorminton* 1206. 'Estate associated with a man called Dēormōd or Dēormund'. OE pers. name + *-ing-* + *tūn*.

Dorney Bucks. *Dornei* 1086 (DB). 'Island, or dry ground in marsh, frequented by bumble-bees'. OE *dora* (genitive plural *dorena*) + *ēg*.

Dornoch Highland. *Durnach* 1145. 'Place of fist-stones'. Gaelic *dornach*.

Dornock Dumf. *Durnack* 1182. 'Place of fist-stones'. OWelsh *durnauc*.

Dorridge Solhll. *Derrech* 1400. 'Ridge frequented by deer or wild animals'. OE *dēor* + *hrycg*.

Dorrington Lincs. *Derintone* 1086 (DB). 'Estate associated with a man called Dēor(a)'. OE pers. name + *-ing-* + *tūn*.

Dorrington Shrops. near Condover. *Dodinton* 1198. 'Estate associated with a man called Dod(d)a'. OE pers. name + *-ing-* + *tūn*.

Dorset (the county). *Dornsætum* late 9th cent. '(Territory of) the people around *Dorn*'. Reduced form of *Dornwaraceaster* (*see* DORCHESTER) + OE *sǣte*.

Dorsey (*Na Doirse*) Armagh. 'The gateways'. The present name is a short form of Irish *Doirse Eamhna*, 'gateways of Navan', after *Navan*, the ancient capital of Ulster.

Dorsington Warwicks. *Dorsintune* 1086 (DB). 'Estate associated with a man called Dēorsige'. OE pers. name + *-ing-* + *tūn*.

Dorstone Herefs. *Dodintune* [*sic*] 1086 (DB), *Dorsington* c.1138. Possibly identical in origin with the previous name.

Dorton Bucks. *Dortone* 1086 (DB). 'Farmstead or village at the narrow pass'. OE *dor* + *tūn*.

Dosthill Staffs. *Dercelai* [*sic*] 1086 (DB), *Dercetehull* 1195. OE *hyll* 'hill', probably added to a Celtic name *Dercet* meaning 'oak wood' identical with DASSETT.

Doughton Glos. *Ductune* 775–8. 'Duck farmstead'. OE *dūce* + *tūn*.

Douglas (*Dúglas*) Cork, Laois. 'Black stream'.

Douglas S. Lan. *Duuelglas* c.1150. '(Place on the) Douglas Water'. The Celtic river-name means 'black stream' (OGaelic *dub* + *glais*).

Douglas Isle of Man. *Dufglas* c.1257. Identical in origin with previous two names.

Doulting Somerset. *Dulting* 725, *Doltin* 1086 (DB). Originally the old name of the River Sheppey on which Doulting stands, probably Celtic **dōlad* 'flood' + OE suffix *-ing*.

Dove (river) Derbys., an old Celtic river-name first recorded as *Dufan* in the 10th cent. and meaning 'black, dark'. It gives name to **Dovedale** (*Duvesdale* 1269), from OScand. *dalr* 'valley'. The two rivers called **Dove** in Yorkshire are identical in origin.

Dovea (*An Dubhfhéith*) Tipperary. 'The black bog'.

Dovenby Cumbria. *Duuaneby* 1230. 'Farmstead or village of a man called Dufan'. OIrish pers. name + OScand. *bý*.

Dover Kent. *Dubris* 4th cent., *Dofras* c.700, *Dovere* 1086 (DB). Named from the stream here, now called the Dour, a Celtic river-name **dubrās* meaning simply 'the waters'.

Dovercourt Essex. *Douorcortae* c.1000, *Druurecurt* [*sic*] 1086 (DB). Possibly 'enclosed farmyard by the river called *Dover*'. Celtic river-name (meaning 'the waters') + OE **cort(e)* (perhaps from Latin *cohors, cohortem*).

Doverdale Worcs. *Douerdale* 706, *Lunvredele* 1086 [*sic*] (DB). 'Valley of the river called *Dover*'. Celtic river-name (meaning 'the waters') + OE *dæl*.

Doveridge Derbys. *Dubrige* 1086 (DB), *Duvebruge* 1252. 'Bridge over the River Dove'. Celtic river-name (meaning 'the dark one') + OE *brycg*.

Dowdeswell Glos. *Dogodeswellan* 8th cent., *Dodesuuelle* 1086 (DB). 'Spring or stream of a man called **Dogod*'. OE pers. name + *wella*.

Dowlais Mer. T. '(Place by the) black stream'. Welsh *du* + *glais*.

Dowland Devon. *Duuelande* 1086 (DB). Possibly 'estate in the open country frequented by doves'. OE *dūfe* + *feld* + *land*.

Dowlish Wake Somerset. *Duuelis* 1086 (DB), *Duueliz Wak* 1243. Named from the stream here, a Celtic river-name meaning 'dark stream'. Manorial addition from the *Wake* family, here in the 12th cent.

Down (*An Dún*) (the county). 'The fort'. The fort in question is that of DOWNPATRICK.

Down, Downe, '(place at) the hill', OE *dūn*: **Down, East & Down, West** Devon. *Duna* 1086 (DB), *Estdoune* 1260, *Westdone* 1273. **Down St Mary** Devon. *Done* 1086 (DB), *Dune St Mary* 1297. Affix from the dedication of the church. **Downe** Gtr. London. *Dona* 1283, *Doune* 1316.

Down Ampney Glos. *See* AMPNEY.

Down Hatherley Glos. *See* HATHERLEY.

Downend S. Glos. *Downe ende* 1573. Self-explanatory, 'end of the down', OE *dūn*.

Downham, usually 'homestead on or near a hill', OE *dūn* + *hām*: **Downham** Cambs. *Duneham* 1086 (DB). **Downham** Essex. *Dunham* 1168. **Downham Market** Norfolk. *Dunham c.*1050, 1086 (DB). Affix from the market here, referred to as early as the 11th cent.
  However the following have a different origin, '(place at) the hills', from OE *dūn* in the dative plural form *dūnum*: **Downham** Lancs. *Dunum* 1194. **Downham** Northum. *Dunum* 1186.

Downhead Somerset. *Duneafd* 851, *Dunehefde* 1086 (DB). 'Head or end of the down'. OE *dūn* + *hēafod*.

Downhill (*Dún*) Derry. (*go*) *Dún Bó* 1182. 'Fort of the hill'. The original name was *Dún Bó*, 'fort of the cows', but English *hill* then replaced the second word of this.

Downholme N. Yorks. *Dune* 1086 (DB), *Dunum* 12th cent. '(Place at) the hills'. OE *dūn* in a dative plural form *dūnum*.

Downings (*Dúnaibh*) Donegal. *Na nDuini* 1603. 'Forts'.

Downpatrick (*Dún Pádraig*) Down. (*expugnatio*) Dúin Lethghlaisse *496. 'Pádraig's fort'. The town's original Irish name was Dún Lethglaise, possibly 'fort of the side of the stream'. The present name is not recorded before the 17th cent.*

Downton, 'farmstead on or by the hill or down', OE *dūn* + *tūn*: **Downton** Wilts. *Duntun* 672, *Duntone* 1086 (DB). **Downton on the Rock** Herefs. *Duntune* 1086 (DB). Affix from ME *rokke* 'a rock or peak'.

Dowra (*Damhshraith*) Cavan, Leitrim. 'River-meadow of the ox'.

Dowsby Lincs. *Dusebi* 1086 (DB). 'Farmstead or village of a man called Dúsi'. OScand. pers. name + *bý*.

Doxey Staffs. *Dochesig* 1086 (DB). 'Island, or dry ground in marsh, of a man called *Docc'. OE pers. name + *ēg*.

Doynton S. Glos. *Didintone* 1086 (DB). 'Estate associated with a man called *Dydda'. OE pers. name + *-ing-* + *tūn*.

Draperstown Derry. *Cross* 1813. 'Drapers' town'. The name was originally applied to the town of *Moneymore*, founded by the Drapers' Company of London in the early 17th cent.

Draperstown was founded in 1798 and took its current name in 1818. Its Irish name is *Baile na Croise*, 'town of the cross', referring to the crossroads here.

Draughton, 'farmstead on a slope used for dragging down timber and the like', OScand. *drag* + OE *tūn*: **Draughton** Northants. *Dractone* 1086 (DB). **Draughton** N. Yorks. *Dractone* 1086 (DB).

Drax N. Yorks. *Drac* 1086 (DB), *Drachs* 11th cent. 'The portages, or places where boats are dragged overland or pulled up from the water'. OScand. *drag*.

Drax, Long N. Yorks. *Langrak* 1208. 'Long stretch of river'. OE *lang* + **racu*.

Draycote, Draycott, a common place name, probably 'shed where drays or sledges are kept', OE *dræg* + *cot*; examples include: **Draycote** Warwicks. *Draicote* 1203. **Draycott** Derbys. *Draicot* 1086 (DB). **Draycott** Somerset, near Cheddar. *Draicote* 1086 (DB). **Draycott in the Clay** Staffs. *Draicote* 1086 (DB). Affix 'in the clayey district', from OE *clǣg*. **Draycott in the Moors** Staffs. *Draicot* 1251. Affix 'in the moorland', from OE *mōr*.

Drayton, a common place name, 'farmstead at or near a portage or slope used for dragging down loads', or 'farmstead where drays or sledges are used', OE *dræg* + *tūn*; examples include: **Drayton** Norfolk. *Draituna* 1086 (DB). **Drayton** Oxon., near Banbury. *Draitone* 1086 (DB). **Drayton** Oxon., near Didcot. *Draitune* 958, *Draitone* 1086 (DB). **Drayton** Somerset, near Curry Rivel. *Drayton* 1243. **Drayton Bassett** Staffs. *Draitone* 1086 (DB), *Drayton Basset* 1301. Manorial affix from the *Basset* family, here in the 12th cent. **Drayton Beauchamp** Bucks. *Draitone* 1086 (DB), *Drayton Belcamp* 1239. Manorial affix from the *Beauchamp* family, here in the 13th cent. **Drayton, Dry** Cambs. *Draitone* 1086 (DB), *Driedraiton* 1218. Affix from OE *drȳge* to distinguish it from Fen Drayton. **Drayton, East & Drayton, West** Notts. *Draitone*, *Draitun* 1086 (DB), *Est Draiton* 1276, *West Draytone* 1269. **Drayton, Fen** Cambs. *Drægtun* 1012, *Draitone* 1086 (DB), *Fendreiton* 1188. Affix from OE *fenn* 'marshland' in contrast to Dry Drayton. **Drayton, Fenny** Leics. *Draitone* 1086 (DB), *Fenedrayton* 1465. Affix is OE *fennig* 'muddy, marshy'. **Drayton, Market** Shrops. *Draitune* 1086 (DB). Affix from the important market here. **Drayton**

Parslow Bucks. *Draitone* 1086 (DB), *Drayton Passelewe* 1254. Manorial affix from the *Passelewe* family, here in the 11th cent.

Drayton, West Gtr. London. *Drægtun* 939, *Draitone* 1086 (DB), *Westdrayton* 1465.

Dreen (*Draighean*) Derry. *Drien* 1613. 'Place of blackthorn'.

Drehid (*Droichead*) Kildare. 'Bridge'.

Drenewydd, Y. *See* NEWTOWN.

Drewsteignton Devon. *Taintone* 1086 (DB), *Teyngton Drue* 1275. 'Farmstead or village on the River Teign'. Celtic river-name (*see* TEIGNMOUTH) + OE *tūn* + manorial affix from a man called *Drew*, here in the 13th cent.

Driby Lincs. *Dribi* 1086 (DB). 'Dry farmstead or village'. OE *drȳge* + OScand. *bý*.

Driffield, 'open land characterized by dirt, or by stubble', OE *drit* or *drīf* + *feld*: **Driffield** Glos. *Drifelle* [*sic*] 1086 (DB), *Driffeld* 1190. **Driffield, Great & Driffield, Little** E. R. Yorks. *Drifeld* 1086 (DB).

Drigg Cumbria. *Dreg* 12th cent. 'The portage, or place where boats are dragged overland or pulled up from the water'. OScand. **dræg*.

Drighlington Leeds. *Dreslin(g)tone* [*sic*] 1086 (DB), *Drichtlington* 1202. 'Estate associated with a man called **Dryhtel* or **Dyrhtla*'. OE pers. name + -ing- + *tūn*.

Drimfries (*Droim Fraoigh*) Donegal. 'Heather ridge'.

Driminidy (*Droim Inide*) Cork. 'Shrovetide ridge'.

Drimnagh (*Droimeanach*) Dublin. 'Ridged land'.

Drimoleague (*Drom dhá Liag*) Cork. 'Ridge of two flagstones'.

Drimpton Dorset. *Dremeton* 1244. 'Farmstead of a man called **Drēama*'. OE pers. name + *tūn*.

Drinagh (*Draighneach*) Cork, Wexford. 'Place of blackthorn'.

Drinan (*Draighneán*) Dublin. 'Place of blackthorn'.

Drinkstone Suffolk. *Drincestune c.*1050, *Drencestuna* 1086 (DB). 'Farmstead of a man called Drengr'. OScand. pers. name + OE *tūn*.

Droghed (*Droichead*) Derry. 'Bridge'.

Drogheda (*Droichead Átha*) Louth. 'Bridge of the ford'.

Droichead Nua (*Droichead Nua*) Kildare. 'New bridge'.

Droimlamph (*Droim Leamh*) Derry. 'Elm ridge'.

Drointon Staffs. *Dregetone* [*sic*] 1086 (DB), *Drengeton* 1199. Probably 'farmstead of the free tenants'. OScand. drengr + OE *tūn*.

Droitwich Worcs. *Wich* 1086 (DB), *Drihtwych* 1347. 'Dirty or muddy salt-works'. OE wīc with the later addition of OE drit 'dirt'.

Drom (*Drom*) Kerry, Tipperary. 'Ridge'.

Dromacomer (*Drom an Chomair*) Limerick. 'Ridge of the confluence'.

Dromada (*Droim Fhada*) Limerick, Mayo. 'Long ridge'.

Dromahair (*Drom dhá Thiar*) Leitrim. 'Ridge of two demons'.

Dromalivaun (*Drom Leamháin*) Kerry. 'Elm ridge'.

Dromara (*Droim Bearach*) Down. *Drumberra* 1306. 'Ridge of heifers'.

Dromard (*An Droim Ard*) Sligo. 'The high ridge'.

Drombane (*Drom Bán*) Tipperary. 'White ridge'.

Drombofinny (*Droim Bó Fnne*) Cork. 'Ridge of the white cow'.

Dromcolliher (*Drom Collachair*) Limerick. 'Ridge of the hazel wood'.

Dromin (*Droim Ing*) Limerick, Louth. 'Ridge of the difficulty'.

Dromina (*Drom Aithne*) Cork. 'Ridge of the commandment'.

Dromindoora (*Drom an Dúdhoire*) Clare. 'Ridge of the dark oak grove'.

Dromineer (*Drom Inbhir*) Tipperary. 'Ridge of the estuary'.

Dromiskin (*Droim Ineasclainn*) Louth. 'Ridge of the torrent'.

Dromkeen (*Drom Caoin*) Limerick. 'Beautiful ridge'.

Dromlusk (*Droim Loiscthe*) Kerry. 'Burnt ridge'.

Drommahane (*Drom Átháin*) Cork. 'Ridge of the little ford'.

Dromod (*Dromad*) Leitrim. 'Ridge'.

Dromore (*Droim Mór*) Cork, Down, Tyrone, etc. 'Big ridge'.

Dronfield Derbys. *Dranefeld* 1086 (DB). 'Open land infested with drones'. OE *drān* + *feld*.

Droxford Hants. *Drocenesforda* 826, *Drocheneford* 1086 (DB). OE *ford* 'ford', with an obscure first element, possibly an OE *drocen 'dry place'.

Droylsden Tamesd. *Drilisden c.*1250. Possibly 'valley of the dry spring or stream'. OE *drȳge* + *well* + *denu*.

Druidston Pemb. *Drewyston* 1393. 'Drew's farm'. Anglo-Norman pers. name + OE *tūn*.

Druim Fada (mountain ridge) Highland. 'Long ridge'. Gaelic *druim* + *fada*.

Drum (*Droim*) Monaghan. *Driyme* 1591. 'Ridge'.

Drumacrib (*Droim Mhic Roib*) Monaghan. 'Mac Rob's ridge'.

Drumadrihid (*Droim an Droichid*) Clare. 'Ridge of the bridge'.

Drumahoe (*Droim na hUamha*) Derry. *Dromhoghs* 1622. 'Ridge of the cave'.

Drumakill (*Droim na Coille*) Monaghan. 'Ridge of the wood'.

Drumanee (*Droim an Fhiaidh*) Derry. 'Ridge of the deer'.

Drumanespic (*Droim an Easpaig*) Cavan. 'Ridge of the bishop'.

Drumaness (*Droim an Easa*) Down. *Drumenessy c.*1710. 'Ridge of the waterfall'.

Drumaney (*Droim Eanaigh*) Tyrone. 'Ridge of the bog'.

Drumannon (*Droim Meannáin*) Armagh. 'Ridge of the kid'.

Drumantee (*Dromainn Tí*) Armagh. 'Ridge of the house'.

Drumaroad (*Droim an Róid*) Down. *Ballydromerode* 1635. 'Ridge of the route'.

Drumarraght (*Droim Arracht*) Fermanagh. 'Ridge of the apparition'.

Drumatober (*Droim an Tobair*) Galway. 'Ridge of the well'.

Drumbane (*An Drom Bán*) Tipperary. 'The white ridge'.

Drumbeg (*Drom Beag*) Down. 'Little ridge'.

Drumbo (*Droim Bó*) Down. (*ab*) *Dromma Bó c.*830. 'Ridge of cows'.

Drumboy (*Drom Buí*) Donegal. 'Yellow ridge'.

Drumbrughas (*Droim Brughas*) Cavan, Fermanagh. 'Ridge of the mansion'.

Drumcar (*Droim Chora*) Louth. 'Ridge of the weir'.

Drumcliff (*Droim Cliabh*) Sligo. 'Ridge of the baskets'.

Drumcollogher (*Drom Collachair*) Limerick. 'Ridge of the hazel wood'.

Drumcondra (*Droim Conrach*) Dublin, Meath. 'Ridge of the path'.

Drumconrath (*Droim Conrach*) Meath. 'Ridge of the path'.

Drumcose (*Droim Cuas*) Fermanagh. *Drumcoose* 1609. 'Ridge of hollows'.

Drumcree (*Droim Cria*) Westmeath. 'Ridge of the clay'.

Drumcroone (*Droim Cruithean*) Derry. *Drim Crum* 1813. 'Ridge of the Cruithin'. The tribal name is related to *Briton*.

Drumcullen (*Droim Cuilinn*) Offaly. 'Holly ridge'.

Drumdeevin (*Droim Diomhaoin*) Donegal. 'Idle ridge'.

Drumderaown (*Droim 'ir dhá Abhainn*) Cork. 'Ridge between two rivers'.

Drumenny (*Droim Eanaigh*) Tyrone. *Dromany* 1615. 'Ridge of the marsh'.

Drumfad (*Droim Fhada*) Donegal, Down, Sligo, Tyrone. 'Long ridge'.

Drumfadda (*Droim Fhada*) Cork, Kerry. 'Long ridge'.

Drumfea (*Droim Féich*) Carlow. 'Ridge of the raven'.

Drumfin (*Droim Fionn*) Sligo. 'White ridge'.

Drumfree (*Droim Fraoigh*) Donegal. *Druim freigh* 1620. 'Ridge of the heather'.

Drumgath (*Droim gCath*) Down. *Drumgaa* 1435. 'Ridge of battles'.

Drumgoose (*Droim gCaas*) Armagh, Monaghan, Tyrone. 'Ridge of the cave'.

Drumhawan (*Droim Shamhuin*) Monaghan. 'Ridge of the winter festival'.

Drumhuskert (*Droim Thuaisceart*) Mayo. 'Northern ridge'.

Drumkeen (*Droim Caoin*) Donegal. 'Smooth ridge'.

Drumkeeran (*Droim Caorthainn*) Leitrim. 'Ridge of the rowans'.

Drumlane (*Droim Leathan*) Cavan. 'Wide ridge'.

Drumlea (*Droim Léith*) Tyrone. *Dromma Leithi* 1391. 'Grey ridge'.

Drumlee (*Droim Lao*) Down. *Dromlee* 1659. 'Ridge of the calf'.

Drumleevan (*Droim Leamháin*) Leitrim. 'Elm ridge'.

Drumlegagh (*Droim Liagach*) Tyrone. *Dromlegah* 1609. 'Ridge of stones'.

Drumlish (*Droim Lis*) Longford. 'Ridge of the fort'.

Drumloose (*Droim Lus*) Westmeath. 'Ridge of herbs'.

Drummin (*An Dromainn*) Mayo. 'The little ridge'.

Drummond (*Dromainn*) Laois. 'Ridge'.

Drummond Perth. *Droman* 1296. 'Ridge'. Gaelic *drumainn*.

Drummullin (*Droim an Mhuilinn*) Roscommon. 'Ridge of the mill'.

Drumnacarra (*Droim na Chairthe*) Louth. 'Ridge of the pillar stone'.

Drumnadrochit Highland. 'Ridge of the bridge'. Gaelic *druim* + *na* + *drochaid*.

Drumnafinnagle (*Droim na Fionghal*) Donegal. 'Ridge of the fratricide'.

Drumnafinnila (*Droim na Fionghal*) Leitrim. 'Ridge of the fratricide'.

Drumnaheark (*Droim na hAdhairce*) Donegal. 'Ridge of the horn'.

Drumnahoe (*Droim na hUamha*) Antrim, Tyrone. 'Ridge of the cave'.

Drumnakilly (*Droim na Coille*) Tyrone. *Dromnekillye* 1613. 'Ridge of the wood'.

Drumnanaliv (*Droim na nDealbh*) Monaghan. 'Ridge of the phantoms'.

Drumnashaloge (*Droim na Sealg*) Tyrone. 'Ridge of the hunt'.

Drumquin (*Droim Caoin*) Tyrone. *Druimchaoin* 1212. 'Smooth ridge'.

Drumragh (*Droim Rátha*) Tyrone. 'Ridge of the ring fort'.

Drumralla (*Droim Rálach*) Fermanagh. 'Ridge of oak'.

Drumraney (*Droim Raithne*) Westmeath. 'Ridge of ferns'.

Drumree (*Droim Rí*) Meath. 'Ridge of the king'.

Drumshanbo (*Droim Seanbhó*) Leitrim. 'Ridge of the old cow'.

Drumskinny (*Droim Scine*) Fermanagh. *Dromskynny* 1639. 'Ridge of the knife'.

Drumsna (*Drom ar Snámh*) Leitrim. 'Ridge of the swimming place'.

Drumsru (*Droim Sruth*) Kildare. 'Ridge of the streams'.

Drumsurn (*Droim Sorn*) Derry. *Drumsoren* 1613. 'Ridge of the furnace'.

Drung (*Drong*) Cavan. 'Folk'.

Dry Doddington Lincs. *See* DODDINGTON.

Dry Drayton Cambs. *See* DRAYTON.

Drybeck Cumbria. *Dribek* 1256. 'Stream which sometimes dries up'. OE *drȳge* + OScand. *bekkr*.

Drybrook Glos. *Druybrok* 1282. 'Brook which sometimes dries up'. OE *drȳge* + *brōc*.

Dryburgh Sc. Bord. *Drieburh* c.1160. 'Dry fortress'. OE *drȳge* + *burh*. The reference is presumably to a former dried-up stream, or a position unreachable by floods.

Duagh (*Dubháth*) Kerry. 'Black ford'.

Duarrigle (*Dubh Aireagal*) Cork. 'Black oratory'.

Dublin (*Dubh Linn*) Dublin. *Eblana* c.150, *Dyflin* c.1000, *Duibh-linn* 12th cent. 'Black pool'. Irish *dubh* + *linn*. The official Irish name of Dublin is *Baile Átha Cliath*, 'town of the hurdle ford'. Both names refer to the River Liffey.

Duckington Ches. *Dochintone* 1086 (DB). 'Estate associated with a man called *Ducc(a)*'. OE pers. name + *-ing-* + *tūn*.

Ducklington Oxon. *Duclingtun* 958, *Dochelintone* 1086 (DB). Probably 'estate associated with a man called *Ducel*'. OE pers. name + *-ing-* + *tūn*.

Duckmanton, Long Derbys. *Ducemannestune* c.1002, *Dochemanestun* 1086 (DB). 'Farmstead of a man called *Ducemann*'. OE pers. name + *tūn*.

Duddenhoe End Essex. *Dudenho* 12th cent. 'Spur of land or ridge of a man called Dudda'. OE pers. name (genitive *-n*) + *hōh*, with the addition of *end* 'district' from the 18th cent.

Duddington Northants. *Dodintone* 1086 (DB). 'Estate associated with a man called Dud(d)a or Dod(d)a'. OE pers. name + *-ing-* + *tūn*.

Duddo Northum. *Dudehou* 1208–10. 'Spur of land or ridge of a man called Dud(d)a'. OE pers. name + *hōh*.

Duddon Ches. *Dudedun* 1185. 'Hill of a man called Dud(d)a'. OE pers. name + *dūn*.

Duddon (river) Lancs., Cumbria. *See* DUNNERDALE.

Dudleston Shrops. *Dodeleston* 1267. 'Farmstead of a man called Dud(d)el or *Dod(d)el*'. OE pers. name + *tūn*.

Dudley Dudley. *Dudelei* 1086 (DB). 'Woodland clearing of a man called Dud(d)a'. OE pers. name + *lēah*.

Duffield, 'open land frequented by doves', OE *dūfe* + *feld*: **Duffield** Derbys. *Duvelle* [*sic*] 1086 (DB), *Duffeld* 12th cent. **Duffield, North & Duffield, South** N. Yorks. *Dufeld*, *Nortdufelt, Suddufeld* 1086 (DB).

Dufftown Moray. 'Duff's town'. The town was founded in 1817 by James *Duff*, 4th Earl of Fife (1776–1857).

Dufton Cumbria. *Dufton* 1256. 'Farmstead where doves are kept'. OE *dūfe* + *tūn*.

Duggleby N. Yorks. *Difgelibi* 1086 (DB). 'Farmstead or village of a man called Dubgall or Dubgilla'. OIrish pers. name + OScand. *bý*.

Dugort (*Dumha Goirt*) Mayo. 'Black field'.

Duhallow (*Duthaigh Ealla*) Cork. 'District of the (river) Ealla'.

Dukestown Blae. 'Duke's town'. The township arose in the 19th cent. around the Duke's Pit coalmine, opened on land owned by the *Duke* of Beaufort.

Dukinfield Tamesd. *Dokenfeld* 12th cent. 'Open land where ducks are found'. OE *dūce* (genitive plural *dūcena*) + *feld*.

Dulane (*Tulán*) Meath. 'Little hill'.

Duleek (*Damhliag*) Meath. 'Stone church'.

Dullingham Cambs. *Dullingham* c.1045, *Dullingeham* 1086 (DB). 'Homestead of the family or followers of a man called *Dull(a)*'. OE pers. name + *-inga-* + *hām*.

Duloe Cornwall. *Dulo* 1283. 'Place between the two rivers called Looe'. Cornish *dew* 'two' + river-names from Cornish *logh* 'pool'.

Dulverton Somerset. *Dolvertune* 1086 (DB). Possibly 'farmstead near the hidden ford'. OE *dīegel* + *ford* + *tūn*.

Dulwich Gtr. London. *Dilwihs* 967. 'Marshy meadow where dill grows'. OE *dile* + *wisc*.

Dumbarton W. Dunb. *Dumbrethan* c.1290. 'Fort of the Britons'. Gaelic *dùn*. The name was applied by the neighbouring Gaels to the stronghold occupied from the 5th cent. by the Britons, who called their fortress *Alclut*, 'rock of the Clyde'.

Dumbleton Glos. *Dumbeltun* 995, *Dubentune* [*sic*] 1086 (DB). Probably 'farmstead near a shady glen or hollow'. OE *dumbel + *tūn*.

Dumfries Dumf. *Dunfres* c.1183. 'Woodland stronghold'. Gaelic *dùn* + *preas* 'copse, thicket'.

Dummer Hants. *Dunmere* 1086 (DB). 'Pond on a hill'. OE *dūn* + *mere*.

Dummoher (*Droim Mothar*) Limerick. 'Ridge of the ruined fort'.

Dun Laoghaire (*Dún Laoghaire*) Dublin. 'Laoghaire's fort'.

Dunadry (*Dún Eadradh*) Antrim. 'Middle fort'.

Dunaff (*Dún Damh*) Donegal. *Duneffe* c.1659. 'Fort of oxen'.

Dunany (*Dún Áine*) Louth. 'Áine's fort'.

Dunbar E. Loth. *Dynbaer* 709. 'Summit fort'. Gaelic *dùn* + *barr*, based on British *din-bar.

Dunbarton Down. 'Dunbar's town'. Linen mills were founded here in c.1838 by Hugh *Dunbar*.

Dunbeg (*Dún Beag*) Galway. 'Small fort'.

Dunbell (*Dún Bile*) Kilkenny. 'Fort of the sacred tree'.

Dunblane Stir. *Dumblann* c.1200. 'Blaan's hill'. Gaelic *dùn*. *Blaan* was a 6th-cent. bishop who had a monastery here.

Dunboyne (*Dún Búinne*) Meath. 'Báethán's fort'.

Duncannon (*Dún Canann*) Wexford. 'Canainn's fort'.

Duncansby Head Highland. *Dungalsbaer* c.1225. 'Headland by Dungal's farm'. OScand. *bý*.

Dunchideock Devon. *Dvnsedoc* 1086 (DB). 'Wooded fort'. Celtic *din (influenced by OE *dūn* 'hill') + **cẹ̄diǭg*.

Dunchurch Warwicks. *Donecerce* [*sic*] 1086 (DB), *Duneschirche* c.1150. Probably 'church of a man called Dun(n)'. OE pers. name + *cirice*.

Duncormick (*Dún Chormaic*) Wexford. 'Cormac's fort'.

Duncote Northants. *Dunecote* 1227. 'Cottage(s) of a man called Dun(n)a'. OE pers. name + *cot*.

Duncrun (*Dún Cruithen*) Derry. 'Fort of the Cruithin'.

Duncton W. Sussex. *Donechitone* 1086 (DB). 'Farmstead of a man called *Dunnuca'. OE pers. name + *tūn*.

Dundalk (*Dún Dealgan*) Louth. 'Dealga's fort'.

Dundareirke (*Dún dá Radharc*) Cork. 'Fort of two prospects'.

Dundaryark (*Dún dá Rhadharc*) Kilkenny. 'Fort of two prospects'.

Dundee Dund. *Dunde* c.1180. 'Fort of *Daigh*'. Gaelic *dùn*. *Daigh* may be a pers. name.

Dunderry (*Dún Doire*) Meath. 'Fort of the oak grove'.

Dundon Somerset. *Dondeme* [*sic*] 1086 (DB), *Dunden* 1236. 'Valley by the hill'. OE *dūn* + *denu*.

Dundonald (*Dún Dónaill*) Down. *Dondouenald* c.1183. 'Dónall's fort'.

Dundraw Cumbria. *Drumdrahrigg* 1194. 'Slope of the ridge'. Celtic *drum* + OScand. *drag* (with OScand. *hryggr* 'ridge' in the 12th-cent. form).

Dundrod (*Dún dTreodáin*) Antrim. *Doonkilltroddon* c.1672. 'Fort of Treodán'. The present name is a shortened form of *Dún Cille Treodáin*, 'fort of the church of St Treodán'.

Dundrum (*Dún Droma*) Down, Dublin, Laois, Tipperary. 'Fort of the ridge'.

Dundry N. Som. *Dundreg* 1065. Possibly 'slope of the hill (used for dragging down loads)', OE *dūn* + *dræg*. Alternatively perhaps a Celtic name for Dundry Hill from *din 'fort' with another element.

Duneane (*Dún dá Éan*) Antrim. 'Fort of two birds'.

Duneel (*Dún Aoil*) Westmeath. 'Lime fort'.

Duneight (*Dún Echdach*) Down. 'Eochaid's fort'.

Dunfanaghy (*Dún Fionnachaidh*) Donegal. *Dunfenoghy* 1679. 'Fort of the white field'.

Dunfermline Fife. *Dumfermelyn* 11th cent., *Dumferlin* 1124. Meaning uncertain. The first element is Gaelic *dùn*, 'fort, hill'. The rest is unexplained.

Dunfore (*Dún Fuar*) Sligo. 'Cold fort'.

Dungall (*Dún Gall*) Antrim. 'Fort of the strangers'.

Dungannon (*Dún Geanainn*) Tyrone. (*caislén*) Dúingenainn *1505*. '*Geanann's fort*'.

Dunganstown (*Baile Uí Dhonnagáin*) Wexford. 'Donnagán's homestead'.

Dungarvan (*Dún Garbháin*) Kilkenny, Waterford. 'Garbhán's fort'.

Dungeness Kent. *Dengenesse* 1335. 'Headland (OE *næss*) near Denge Marsh', the latter being *Dengemersc* 774, possibly 'marsh of the pasture district' from OE *denn* + **gē* + *mersc*, alternatively 'marsh with manured land' from OE *dyncge* (Kentish *dencge*).

Dungiven (*Dún Geimhin*) Derry. (*o*) *Dún Geimin c.830*. Possibly 'Geimhean's fort'.

Dunglow (*An Clochán Liath*) Donegal. *Doungloo* 1600. 'The grey stepping stones'. The present name represents Irish *Dún gCloiche*, 'fort of stone'.

Dungourney (*Dún Guairne*) Cork. 'Guairne's fort'.

Dunham, 'homestead or village at a hill', OE *dūn* + *hām*: **Dunham** Notts. *Duneham* 1086 (DB). **Dunham, Great & Dunham, Little** Norfolk. *Dunham* 1086 (DB). **Dunham on the Hill** Ches. *Doneham* 1086 (DB), *Dunham on the Hill* 1534. Affix from OE *hyll*. **Dunham Town** Traffd. *Doneham* 1086 (DB).

Dunhampton Worcs. *Dunhampton* 1222. 'Homestead or home farm by the hill'. OE *dūn* + *hām-tūn*.

Dunhill (*Dún Aill*) Waterford. 'Fort cliff'.

Dunholme Lincs. *Duneham* 1086 (DB). Probably 'homestead or village of a man called Dunna'. OE pers. name + *hām*.

Dunineny (*Dún an Aonaigh*) Antrim. 'Fort of the assembly'.

Duniry (*Dún Doighre*) Galway. 'Fort of the blast'.

Dunkeld Perth. *Duncalden* 10th cent. 'Fort of the Caledonians'. Gaelic *dùn*. The 'Caledonians' are the Picts who occupied this region.

Dunkerrin (*Dún Cairin*) Offaly. 'Caire's fort'.

Dunkerron (*Dún Ciaráin*) Kerry. 'Ciarán's fort'.

Dunkeswell Devon. *Doducheswelle* [*sic*] 1086 (DB), *Dunekeswell* 1219. 'Hedgesparrow's, or **Dunnuc's*, spring'. OE **dunnuc* (possibly used as pers. name) + *wella*.

Dunkineely (*Dún Cionnaola*) Donegal. *Duncanally c.1659*. 'Cionnaola's fort'.

Dunkirk Kent. first recorded in 1790, a transferred name from Dunkerque in France. There are examples of the same name, originally given to places considered lawless or remote, in Ches., Notts., S. Glos., Staffs., and Wilts.

Dunlavin (*Dún Luáin*) Wicklow. 'Luán's fort'.

Dunleer (*Dún Léire*) Louth. 'Léire's fort'.

Dunlewy (*Dún Lúiche*) Donegal. *Dunluy* 1654. 'Lughaidh's fort'.

Dunloy (*Dún Lathaí*) Antrim. *Dunlogh c.1672*. 'Fort of the mire'.

Dunluce (*Dúnlios*) Antrim. *Dhúinlis* 1513. 'Fortified residence'.

Dunmanus (*Dún Mánais*) Cork. 'Mánas's fort'.

Dunmoon (*Dún Móin*) Waterford. 'Món's fort'.

Dunmore (*Dún Mór*) Donegal, Galway, Waterford. 'Big fort'.

Dunmow, Great & Dunmow, Little Essex. *Dunemowe* 951, *Dommawa* 1086 (DB). 'Meadow on the hill'. OE *dūn* + **māwe*.

Dunmoyle (*An Dún Maol*) Tyrone. *Dunmoyle c.1655*. 'The bald fort'.

Dunmurry (*Dún Muírígh*) Antrim. *Ballydownmory* 1604. 'Muiríoch's fort'.

Dunnalong (*Dún na Long*) Derry. 'Fortress of the ships'.

Dunnamaggan (*Dún na mBogán*) Kildare, Kilkenny. 'Fort of the soft ground'.

Dunnamanagh (*Dún na Manach*) Tyrone. *Downeamannogh c.*1655. 'Fort of the monks'.

Dunnamark (*Dún na mBarc*) Kerry. 'Fort of the boats'.

Dunnamore (*Domhnach Mór*) Tyrone. *Donach Mór c.*930. 'Big church'.

Dunnerdale, Hall Cumbria. *Dunerdale* 1293. 'Valley of the River Duddon'. River-name of uncertain origin and meaning + OScand. *dalr*. Affix probably *hall* 'manor house'.

Dunnet Head. *See* THURSO.

Dunnington E. R. Yorks. *Dodintone* 1086 (DB). 'Estate associated with a man called Dud(d)a'. OE pers. name + *-ing-* + *tūn*.

Dunnington York. *Donniton* 1086 (DB). 'Estate associated with a man called Dun(n)a'. OE pers. name + *-ing-* + *tūn*.

Dunnockshaw Lancs. *Dunnockschae* 1296. 'Small wood or copse frequented by hedge-sparrows'. OE *dunnoc* + *sceaga*.

Dunoon Arg. *Dunnon c.*1240, *Dunhoven* 1270. 'Fort by the river'. Gaelic *Dùn Obhainn*. The river is the Clyde.

Dunquin (*Dún Chaoin*) Kerry. 'Pleasant fort'.

Dunraymond (*Dún Réamainn*) Monaghan. 'Réamann's fort'.

Dunree (*Dún Riabhach*) Donegal. 'Striped fort'.

Duns Sc. Bord. *Duns* 1150. From either OE *dūn* 'hill' or Gaelic *dùn* 'fort, mound' + English *-s*.

Duns Tew Oxon. *See* TEW.

Dunsallagh (*Dún Salach*) Clare. 'Dirty fort'.

Dunsby Lincs. near Rippingale. *Dunesbi* 1086 (DB). 'Farmstead or village of a man called Dun(n)'. OE pers. name + OScand. *bý*.

Dunsden Green Oxon. *Dunesdene* 1086 (DB), *Donsden grene* 1589. 'Valley of a man called Dyn(n)e'. OE pers. name + *denu*, with *grene* 'village green' added from the 16th cent.

Dunseverick (*Dún Sobhairce*) Antrim. *Dun Sobhairce* 351. 'Sobhairce's fort'.

Dunsfold Surrey. *Duntesfaude* 1259. 'Fold or small enclosure of a man called *Dunt*. OE pers. name + *fald*.

Dunsford Devon. *Dun(n)esforda* 1086 (DB). 'Ford of a man called Dun(n)'. OE pers. name + *ford*.

Dunsforth, Lower & Dunsforth, Upper N. Yorks. *Doneford(e)*, *Dunesford* 1086 (DB). Identical in origin with the previous name.

Dunshaughlin (*Dún Seachlainn*) Meath. 'Seachlann's fort'.

Dunsley N. Yorks. *Dunesle* 1086 (DB). 'Woodland clearing of a man called Dun(n)'. OE pers. name + *lēah*.

Dunsmore (old district) Warwicks. *See* RYTON-ON-DUNSMORE.

Dunstable Beds. *Dunestaple* 1123. 'Boundary post of a man called Dun(n)a'. OE pers. name + *stapol*.

Dunstall Staffs. *Tunstall* 13th cent. 'Site of a farm, a farmstead'. OE **tūn-stall*.

Dunstan Northum. *Dunstan* 1242. 'Stone or rock on a hill'. OE *dūn* + *stān*.

Dunster Somerset. *Torre* 1086 (DB), *Dunestore* 1138. 'Craggy hill-top of a man called Dun(n)'. OE pers. name + *torr*.

Dunston, 'farmstead of a man called Dun(n)', OE pers. name + *tūn*: **Dunston** Lincs. *Dunestune* 1086 (DB). **Dunston** Norfolk. *Dunestun* 1086 (DB). **Dunston** Staffs. *Dunestone* 1086 (DB).

Dunterton Devon. *Dondritone* 1086 (DB). OE *tūn* 'farmstead', possibly added to an old Celtic name meaning 'fort village' (**din* + *tre*).

Duntinny (*Dún Teine*) Donegal. 'Fort of the fire'.

Duntisbourne Abbots, Duntisbourne Leer, & Duntisbourne Rouse Glos. *Duntesburne* 1055, *Dantesborne*, *Tantesborne*, *Duntesborne* 1086 (DB), *Duntesbourn Abbatis* 1291, *Duntesbourn Lyre* 1307, *Duntesbourn Rus* 1287. 'Stream of a man called **Dunt*'. OE pers. name + *burna*. Manorial affixes from possession in medieval times by the Abbot of St Peter's Abbey at Gloucester, by the Abbey of Lire in Normandy, and by the family of *le Rous*.

Duntish Dorset. *Dunhethis* 1249. 'Pasture on a hill'. OE *dūn* + **etisc*.

Dunton Beds. *Donitone* 1086 (DB). Probably 'farmstead on a hill'. OE *dūn* + *tūn*.

Dunton Bucks. *Dodintone* 1086 (DB). 'Estate associated with a man called Dud(d)a or Dod(d)a'. OE pers. name + *-ing-* + *tūn*.

Dunton Norfolk. *Dontuna* 1086 (DB). 'Farmstead on a hill'. OE *dūn* + *tūn*.

Dunton Bassett Leics. *Donitone* 1086 (DB), *Dunton Basset* 1418. Probably identical with the previous name. Manorial affix from the *Basset* family, here in the 12th cent.

Dunton Green Kent. *Dunington* 1244. 'Estate associated with a man called Dun(n) or Dun(n)a'. OE pers. name + *-ing-* + *tūn*.

Dunwich Suffolk. *Duneuuic* 1086 (DB). Probably not to be identified with the Celtic name *Domnoc* in Bede (731), perhaps rather 'trading centre at the dunes' from OE *dūn* (genitive plural *dūna*) + *wīc*.

Durdle Door Dorset. *Dirdale Door* 1811. In spite of the lack of early spellings the first element may be OE *thyrelod* 'pierced' with *duru* 'door, opening'.

Durham Durham. *Dunholm* 1056. 'Hill island or promontory'. OE *dūn* + OScand. *holmr*.

Durleigh Somerset. *Derlege* 1086 (DB). 'Wood or clearing frequented by deer'. OE *dēor* + *lēah*.

Durley Hants. *Deorleage* 901, *Derleie* 1086 (DB). Identical in origin with the previous name.

Durness Highland. *Dyrnes* 1230. 'Promontory frequented by deer'. OScand *dýr* + *nes*.

Durnford, Great Wilts. *Diarneford* 1086 (DB). 'Hidden ford'. OE *dierne* + *ford*.

Durrington, 'estate associated with a man called Dēor(a)'. OE pers. name + *-ing-* + *tūn*: **Durrington** W. Sussex. *Derentune* 1086 (DB). **Durrington** Wilts. *Derintone* 1086 (DB).

Durrow (*Darú*) Laois. 'Oak plain'.

Durrus (*Dúras*) Cork. 'Black grove'.

Dursley Glos. *Dersilege* 1086 (DB). 'Woodland clearing of a man called Dēorsige'. OE pers. name + *lēah*.

Durston Somerset. *Derstona* 1086 (DB). 'Farmstead of a man called Dēor'. OE pers. name + *tūn*.

Durweston Dorset. *Derwinestone* 1086 (DB). 'Farmstead of a man called Dēorwine'. OE pers. name + *tūn*.

Duston Northants. *Dustone* 1086 (DB). 'Farmstead on a mound', or 'farmstead with dusty soil'. OE **dus* or *dūst* + *tūn*.

Dutton Ches. *Duntune* 1086 (DB). 'Farmstead at a hill'. OE *dūn* + *tūn*.

Duxford Cambs. *Dukeswrthe* c.950, *Dochesuuorde* 1086 (DB). 'Enclosure of a man called **Duc(c)'. OE pers. name + *worth*.

Dwygyfylchi Conwy. 'Two circular forts'. Welsh *dwy* + *cyfylchi*.

Dyan (*Daingean*) Tyrone. *Dingan* 1613. 'Stronghold'.

Dyce Abdn. *Dys* 1329. Meaning unknown.

Dyfed (the historic county). '(District of the) Demetae'. The *Demetae*, with name of unknown meaning, were a pre-Roman people who inhabited the part of Wales corresponding to modern Pembrokeshire.

Dyffryn Vale Glam. *The Differin* 1596. '(Place in the) valley'. Welsh *dyffryn*.

Dyke Lincs. *Dic* 1086 (DB). '(Place at) the ditch or dike'. OE *dīc* or OScand. *dík*.

Dymchurch Kent. *Deman circe* c.1100. Probably 'church of the judge'. OE *dēma* + *cirice*.

Dymock Glos. *Dimoch* 1086 (DB). Probably 'fort or stronghold of pigs', from Celtic **dīn* 'fort' + **mocc* 'pigs'.

Dyrham S. Glos. *Deorhamme* 950, *Dirham* 1086 (DB). 'Enclosed valley frequented by deer'. OE *dēor* + *hamm*.

Dysart Fife. *Disard* c.1210. 'Hermitage'. Gaelic *diseart*.

Dysart (*An Díseart*) Roscommon, Westmeath. 'The hermitage'.

Dysert (*Díseart*) Clare. 'Hermitage'.

Dyserth (*Diserth*) Denb. *Dissard* 1086 (DB), *Dyssart* 1320. 'Hermitage'. Welsh *diserth*.

E

Eagle Lincs. *Aclei, Aycle* 1086 (DB). 'Wood or woodland clearing where oak-trees grow'. OE *āc* (replaced by OScand. *eik*) + OE *lēah*.

Eaglesfield Cumbria. *Eglesfeld c.*1170. 'Open land near a Romano-British Christian church'. Celtic **eglẹ̄s* + OE *feld*.

Eakring Notts. *Ecringhe* 1086 (DB). 'Ring or circle of oak-trees'. OScand. *eik* + *hringr*.

Ealand N. Lincs. *Aland* 1316. 'Cultivated land by water or by a river'. OE *ēa-land*.

Ealing Gtr. London. *Gillingas c.*698. '(Settlement of) the family or followers of a man called **Gilla'. OE pers. name + *-ingas*.

Eamont Bridge Cumbria. *Eamotum* 11th cent., *Amotbrig* 1362. 'Bridge at the river confluences'. OE *ēa-mōt* (replaced by OScand. *á-mót*) + *brycg*.

Earby Lancs. *Eurebi* 1086 (DB). 'Upper farmstead', or 'farmstead of a man called Jǫfurr'. OScand. *efri* or pers. name + *bý*.

Eardington Shrops. *Eardigtun c.*1030, *Ardintone* 1086 (DB). 'Estate associated with a man called **Earda'. OE pers. name + *-ing-* + *tūn*.

Eardisland Herefs. *Lene* 1086 (DB), *Erleslen* 1230. 'Nobleman's estate in *Leon'. OE *eorl* added to old Celtic name for the district (*see* LEOMINSTER).

Eardisley Herefs. *Herdeslege* 1086 (DB). 'Woodland clearing of a man called Ægheard'. OE pers. name + *lēah*.

Eardiston Shrops. *Erdeston* 1203. Possibly 'farmstead or estate of a man called Ēorēd'. OE pers. name + *tūn*.

Eardiston Worcs. *Eardulfestun c.*957, *Ardolvestone* 1086 (DB). 'Farmstead or estate of a man called Eardwulf'. OE pers. name + *tūn*.

Earith Cambs. *Herheth* 1244. 'Muddy or gravelly landing-place'. OE *ēar* + *hȳth*.

Earl or Earl's as affix. *See* main name, e.g. for **Earl's Barton** (Northants.) *see* BARTON.

Earle Northum. *Yherdhill* 1242. 'Hill with a fence or enclosure'. OE *geard* + *hyll*.

Earlham Norfolk. *Erlham* 1086 (DB). Possibly 'homestead of a nobleman'. OE *eorl* + *hām*.

Earlston Sc. Bord. *Erchildun c.*1144, *Ercildune c.*1180, *Earlston, or Ercildon* 1868. 'Earcil's hill'. OE *dūn*.

Earn (river) Perth. *Erne* 1190. 'Flowing one'. Pre-Celtic.

Earnley W. Sussex. *Earneleagh* 8th cent. 'Wood or woodland clearing where eagles are seen'. OE *earn* + *lēah*.

Earsdon N. Tyne. *Erdesdon* 1233. Probably 'hill of a man called Ēanrǣd or Ēorǣd'. OE pers. name + *dūn*.

Earsham Norfolk. *Ersam* 1086 (DB). Possibly 'homestead or village of a man called Ēanhere'. OE pers. name + *hām*.

Earswick York. *Edresuuic* [*sic*] 1086 (DB), *Ethericewyk* 13th cent. 'Dwelling or farm of a man called Æthelrīc'. OE pers. name + *wīc*.

Eartham W. Sussex. *Ercheham* [*sic*] 12th cent., *Ertham* 1279. 'Homestead or enclosure with ploughed land'. OE *erth* + *hām* or *hamm*.

Easby N. Yorks. near Stokesley. *Esebi* 1086 (DB). 'Farmstead or village of a man called Ēsi'. OScand. pers. name + *bý*.

Easebourne W. Sussex. *Eseburne* 1086 (DB). 'Stream of a man called Ēsa'. OE pers. name + *burna*.

Easenhall Warwicks. *Esenhull* 1221. 'Hill of a man called Ēsa'. OE pers. name (genitive -*n*) + *hyll*.

Eashing Surrey. *Æscengum* late 9th cent., *Essinge* 1272. '(Settlement of) the ash-tree folk', or '(settlement of) the family or followers of a man called Æsc'. OE *æsc* or pers. name + *-ingas*.

Easington, usually 'estate associated with a man called Ēsa', OE pers. name + *-ing-* + *tūn*: **Easington** Bucks. *Hesintone* 1086 (DB). **Easington** Durham. *Esingtun c.*1040. **Easington** E. R. Yorks. *Esintone* 1086 (DB). **Easington** Red. & Cleve. *Esingetun* 1086 (DB).
However the following have different origins: **Easington** Northum. *Yesington* 1242. 'Farmstead or estate on the stream called *Yesing*'. OE **gēosing* 'gushing stream' + *tūn*. **Easington** Oxon., near Cuxham. *Esidone* 1086 (DB). 'Hill of a man called Ēsa'. OE pers. name (genitive *-n*) + *dūn*.

Easingwold N Yorks. *Eisincewald* 1086 (DB). 'High forest-land of the family or followers of a man called Ēsa'. OE pers. name + *-inga-* + *wald*.

Easky (*Iascach*) Sligo. 'Abounding in fish'.

Easole Street Kent. *Oesewalum* 824, *Eswalt* 1086 (DB). Possibly 'ridges or banks associated with a god or gods'. OE *ēs*, *ōs* + *walu*.

East as affix. *See* main name, e.g. for **East Allington** (Devon) *see* ALLINGTON.

East Anglia, a Latinized term for the territory of 'the East Angles' (OE *Ēast Engle*). The Anglian tribe so called settled in NORFOLK and SUFFOLK and established a kingdom here during the 7th, 8th, and 9th cents.

East Kilbride S. Lan. *Kellebride* 1180. 'Eastern (place by) St Brigid's church'. Gaelic *cill*. *East* distinguishes the town from WEST KILBRIDE, N. Ayr.

East Linton E. Loth. *Lintun* 1127. 'Eastern flax farm'. OE *līn* + *tūn*. *East* distinguishes the place from *West Linton*, Peebles.

East Lothian. *See* MIDLOTHIAN.

East Wemyss Fife. *Wemys* 1239, *Easter Weimes* 1639. 'Eastern (place by) caves'. ME *east* + Gaelic *uamh* + English plural *-s*.

East Williamston Pemb. *Williamston* 1541. 'Eastern (place called) William's farm'. OE *tūn*. *East* distinguishes the village from *West Williamston*.

Eastbourne E. Sussex. *Burne* 1086 (DB), *Estbourne* 1310. '(Place at) the stream'. OE *burna*, with the later addition of *ēast* 'east' to distinguish it from WESTBOURNE.

Eastbridge Suffolk. Recorded thus from 1837, earlier *Briges* 1086 (DB), *Brigge* 1275, '(place at) the bridge'. OE *brycg*, with reference to a crossing of Minsmere River.

Eastburn Brad. *Estbrune* 1086 (DB). 'East stream', or '(land lying) east of the stream'. OE *ēast* or *ēastan* + *burna*.

Eastbury W. Berks. *Eastbury c.*1090. 'East manor'. OE *ēast* + *burh* (dative *byrig*).

Eastchurch Kent. *Eastcyrce c.*1100. 'East church'. OE *ēast* + *cirice*.

Eastcote, Eastcott, 'eastern cottage(s)', OE *ēast* + *cot*; examples include: **Eastcote** Gtr. London. *Estcote* 1248. **Eastcott** Wilts., near Potterne. *Estcota* 1167.

Eastcourt Wilts. near Crudwell. *Escote* 901. Identical in origin with the previous two names.

Eastdean E. Sussex. *Esdene* 1086 (DB). 'East valley'. OE *ēast* + *denu*. 'East' in relation to WESTDEAN.

Easter Ardross Highland. 'Eastern promontory of the height'. English *easter* + Gaelic *àird* + *ros*.

Easter Quarff Shet. *Quharf* 1569. 'Eastern nook'. English *easter* + OScand. *hvarf*. The name contrasts with nearby *Wester Quarff*.

Easter Ross (district) Highland. 'Eastern Ross'. English *easter*. *See* ROSS.

Easter, Good & Easter, High Essex. *Estre* 11th cent., *Estra* 1086 (DB), *Godithestre* 1200, *Heyestre* 1254. '(Place at) the sheep-fold'. OE *eowestre*. Distinguishing affixes from possession in Anglo-Saxon times by a woman called *Gōdgȳth* or *Gōdgifu*, and from OE *hēah* 'high'.

Eastergate W. Sussex. *Gate* 1086 (DB), *Estergat* 1263. '(Place at) the gate or gap'. OE *geat*, with the later addition of *ēasterra* 'more easterly' to distinguish it from WESTERGATE.

Easterton Wilts. *Esterton* 1348. 'More easterly farmstead'. OE *ēasterra* + *tūn*.

Eastham Wirral. *Estham* 1086 (DB). 'East homestead or enclosure'. OE *ēast* + *hām* or *hamm*.

Easthampstead Brack. For. *Lachenestede* 1086 (DB), *Yethamstede* 1176. 'Homestead by the gate or gap'. OE *geat* + *hām-stede*.

Easthope Shrops. *Easthope* 901, *Stope* [*sic*] 1086 (DB). 'Eastern enclosed valley'. OE *ēast* + *hop*.

Easthorpe Essex. *Estorp* 1086 (DB). 'Eastern outlying farmstead or hamlet'. OE *ēast* + OScand. *thorp*.

Easthouses Midloth. *Esthus* 1241, *Esthouse* 1345, *Eisthousis* 1590. 'Eastern house'. ME *east* + *hous*. The plural *-s* is a relatively recent development.

Eastington Glos. near Stonehouse. *Esteueneston* 1220. 'Farmstead of a man called Ēadstān'. OE pers. name + *tūn*.

Eastleach Martin & Eastleach Turville Glos. *Lecche* 862, *Lec(c)e* 1086 (DB), *Estleche Sancti Martini* 1291, *Estleche Roberti de Tureuill* 1221. 'Eastern estate on the River Leach'. OE *ēast* + river-name from OE **lœc(c)*, **lece* 'stream flowing through boggy land'. Affixes from the dedication of the church and from the *de Turville* family, here from the 13th cent.

Eastleigh Hants. *East lea* 932, *Estleie* 1086 (DB). 'East wood or clearing'. OE *ēast* + *lēah*.

Eastling Kent. *Eslinges* 1086 (DB). '(Settlement of) the family or followers of a man called Ēsla'. OE pers. name + *-ingas*.

Eastney Portsm. *Esteney* 1242. '(Place in) the east of the island'. OE *ēastan* + *ēg*.

Eastnor Herefs. *Astenofre* 1086 (DB). '(Place to) the east of the ridge'. OE *ēastan* + **ofer*.

Eastoft N. Lincs. *Eschetoft* c.1170. 'Homestead or curtilage where ash-trees grow'. OScand. *eski* + *toft*.

Easton, a very common place name, usually 'east farmstead or village', i.e. one to the east of another settlement, OE *ēast* + *tūn*; examples include: **Easton** Cambs. *Estone* 1086 (DB). **Easton** Cumbria, near Netherby. *Estuna* 12th cent. **Easton** Hants., near Winchester. *Eastun* 825, *Estune* 1086 (DB). **Easton** I. of Wight. *Estetune* 1244. **Easton** Lincs. *Estone* 1086 (DB). **Easton** Norfolk. *Estuna* 1086 (DB). **Easton** Suffolk, near Framlingham. *Estuna* 1086 (DB). **Easton, Crux** Hants. *Eastune* 801, *Estune* 1086 (DB), *Eston Croc* 1242. Manorial addition from a family called *Croc(h)*, here in the 11th cent. **Easton, Great** Leics. *Estone* 1086 (DB). **Easton Grey** Wilts. *Estone* 1086 (DB), *Eston Grey* 1281. Manorial addition from the *de Grey* family, here in the 13th cent. **Easton in**

Gordano N. Som. *Estone* 1086 (DB), *Eston in Gordon* 1293. Affix is an old district name, see CLAPTON. **Easton Maudit** Northants. *Estone* 1086 (DB), *Estonemaudeut* 1298. Manorial affix from the *Mauduit* family, here in the 12th cent. **Easton on the Hill** Northants. *Estone* 1086 (DB). Affix from its situation on the brow of a hill. **Easton Royal** Wilts. *Estone* 1086 (DB). Affix from OFrench *roial* 'royal' referring to its situation on the edge of an old royal forest. **Easton, Ston** Somerset. *Estone* 1086 (DB), *Stonieston* 1230. Affix from OE *stānig* 'stony' referring to stony ground.

However some Eastons have a different origin, among them: **Easton, Great & Easton, Little** Essex. *E(i)stanes* 1086 (DB). Probably 'stones by the island or well-watered land'. OE *ēg* + *stān*.

Eastrea Cambs. *Estereie* c.1020. 'Eastern part of the island (of WHITTLESEY)'. OE **ēastor* + *ēg*.

Eastriggs Dumf. 'Eastern arable land'. OE *ēast* + OScand. *hryggr* or OE *hrycg* 'ridge'.

Eastrington E. R. Yorks. *Eastringatun* 959, *Estrincton* 1086 (DB). Probably 'farmstead of those living to the east (of HOWDEN)'. OE **ēastor* + *-inga-* + *tūn*.

Eastry Kent. *Eastorege* 9th cent., *Estrei* 1086 (DB). 'Eastern district or region'. OE **ēastor* + **gē*.

Eastwell Leics. *Estwelle* 1086 (DB). 'Eastern spring or stream'. OE *ēast* + *wella*.

Eastwick Herts. *Esteuuiche* 1086 (DB). 'East dwelling or (dairy) farm'. OE *ēast* + *wīc*.

Eastwood Notts. *Estewic* [*sic*] 1086 (DB), *Estweit* 1165. 'East clearing'. OE *ēast* + OScand. *thveit*.

Eastwood Sthend. *Estuuda* 1086 (DB). 'Eastern wood'. OE *ēast* + *wudu*.

Eathorpe Warwicks. *Ethorpe* 1232. 'Outlying farmstead or hamlet on the river'. OE *ēa* + OScand. *thorp*.

Eaton, a common place name with two different origins. Most are 'farmstead or estate on a river' (probably denoting a settlement which performed a special local function in relation to the river), from OE *ēa* + *tūn*, among them: **Eaton** Ches., near Congleton. *Yeiton* c.1262. **Eaton** Norfolk. *Ettune* 1086 (DB). **Eaton** Notts. *Etune* 1086 (DB). **Eaton** Oxon. *Eatun* 9th cent., *Eltune* [*sic*] 1086 (DB). **Eaton** Shrops., near Bishop's Castle. *Eton* 1252. **Eaton**

Shrops., near Ticklerton. *Eton* 1227. **Eaton
Bishop** Herefs. *Etune* 1086 (DB), *Eton Episcopi*
1316. Affix from its possession by the Bishop
(Latin *episcopus*) of Hereford. **Eaton, Castle**
Swindn. *Ettone* 1086 (DB). The affix is found
from the 15th cent. **Eaton Constantine**
Shrops. *Etune* 1086 (DB), *Eton Costentyn* 1285.
Manorial affix from the *de Costentin* family, here
in the 13th cent. **Eaton Hastings** Oxon. *Etone*
1086 (DB), *Eton Hastinges* 1298. Manorial affix
from the *de Hastinges* family, here in the 12th
cent. **Eaton Socon** Cambs. *Etone* 1086 (DB).
Affix from OE *sōcn* 'district with a right of
jurisdiction'. **Eaton upon Tern** Shrops. *Eton*
*c.*1223. On the River Tern, a Celtic river-name
meaning 'the strong one'.

However other Eatons have a different origin,
'farmstead on a spur of land, or on dry ground
in marsh, or on well-watered land', from OE *ēg* +
tūn, among them: **Eaton** Ches., near
Tarporley. *Eyton* 1240. **Eaton** Leics. *Aitona*
*c.*1130. **Eaton Bray** Beds. *Eitone* 1086 (DB).
Manorial affix from the *Bray* family, here in the
15th cent. **Eaton, Little** Derbys. *Detton* [*sic*]
1086 (DB), *Little Eton* 1392. **Eaton, Long**
Derbys. *Aitune* 1086 (DB), *Long Eyton* 1288. Affix
refers to length of village.

Ebberston N. Yorks. *Edbriztune* 1086 (DB).
'Farmstead of a man called Ēadbeorht'. OE pers.
name + *tūn*.

Ebbesborne Wake Wilts. *Eblesburna* 826,
Eblesborne 1086 (DB), *Ebbeleburn Wak* 1249.
Probably 'stream of a man called *Ebbel'. OE
pers. name + *burna*, with manorial affix from
the *Wake* family, here in the 12th cent.

Ebbw Vale (*Glynebwy*) Blae. 'Valley of the
River Ebwy'. The river-name means 'colt'
(Welsh *ebol*), perhaps because horses regularly
drank or forded the river here or because the
water was 'frisky'. The Welsh name is the
equivalent of the English (Welsh *glyn*, 'valley').

Ebchester Durham. *Ebbescestr* 1230. 'Roman
fort of a man called Ebba or a woman called
Æbbe'. OE pers. name + *ceaster*.

Ebrington Glos. *Bristentune* [*sic*] 1086 (DB),
Edbrihttona 1155. 'Farmstead of a man called
Ēadbeorht'. OE pers. name + *tūn*.

Ecchinswell Hants. *Eccleswelle* 1086 (DB).
Probably 'spring or stream of a man called
*Eccel'. Celtic pers. name + OE *wella*.

Ecclefechan Dumf. *Eglesfeghan* 1303.
'St Fechin's church'. British **egles*.

Eccles, from Celtic **eglēs* 'Romano-British
Christian church'; examples include: **Eccles**
Kent. *Æcclesse c.*975, *Aiglessa* 1086 (DB). **Eccles**
Salford. *Eccles c.*1200.

Ecclesfield Sheff. *Eclesfeld* 1086 (DB). 'Open
land near a Romano-British Christian church'.
Celtic **eglēs* + OE *feld*.

Eccleshall Staffs. *Ecleshelle* [*sic*] 1086 (DB),
Eccleshale 1227. 'Nook of land near a Romano-
British Christian church'. Celtic **eglēs* + OE
halh.

Eccleston, 'farmstead by a Romano-British
Christian church', Celtic **eglēs* + OE *tūn*:
Eccleston Ches. *Eclestone* 1086 (DB).
Eccleston Lancs. *Aycleton* 1094. **Eccleston**
St Hel. *Ecclistona* 1190. **Eccleston, Great &
Eccleston, Little** Lancs. *Eglestun* 1086 (DB),
Great Ecleston 1285, *Parua Eccliston* 1261.

Eccup Leeds. *Echope* 1086 (DB). 'Enclosed
valley of a man called Ecca'. OE pers. name +
hop.

Eckington, 'estate associated with a man
called Ecca or Ecci', OE pers. name + *-ing-* + *tūn*:
Eckington Derbys. *Eccingtune c.*1002,
Eckintune 1086 (DB). **Eckington** Worcs.
Eccyncgtun 972, *Aichintune* 1086 (DB).

Ecton Northants. *Echentone* 1086 (DB).
'Farmstead of a man called Ecca'. OE pers.
name + *tūn*.

Edale Derbys. *Aidele* 1086 (DB). 'Valley with
an island or well-watered land'. OE *ēg* + *dæl*.

Edburton W. Sussex. *Eadburgeton* 12th cent.
'Farmstead of a woman called Ēadburh'. OE
pers. name + *tūn*.

Eden (*Éadan*) Antrim. *Eden or Edengrenny*
1837. 'Brow'. The earlier full Irish name was
Éadan Gréine, 'sunny brow'.

Eden (district) Cumbria. A modern adoption
of the river-name (*see* EDENHALL).

Eden, Castle Durham. *Geodene, Iodene*
*c.*1040, *Casteleden* 1248. Named from Eden
Burn, a Celtic river-name meaning simply
'water' (with OE *burna* 'stream'). The 13th-cent.
affix is ME *castel* 'castle'.

Edenbridge Kent. *Eadelmesbregge c.*1100.
'Bridge of a man called Ēadhelm'. OE pers.
name + *brycg*. The river-name Eden is a
back-formation from the place name.

Edenderry (*Éadan Doire*) Antrim, Offaly. 'Brow of the oak wood'.

Edendork (*Éadan na dTorc*) Tyrone. *Adanadorg* 1609. 'Hill brow of the pigs'.

Edenfield Lancs. *Aytounfeld* 1324. 'Open land by the island farmstead, or by the farmstead on well-watered land'. OE *ēg* + *tūn* + *feld*.

Edenhall Cumbria. *Edenhal* 1159. 'Nook of land by the River Eden'. Celtic river-name (meaning simply 'water') + OE *halh*.

Edenham Lincs. *Edeneham* 1086 (DB). 'Homestead or enclosure of a man called Ēada'. OE pers. name (genitive -*n*) + *hām* or *hamm*.

Edensor Derbys. *Edensoure* 1086 (DB). 'Sloping bank or ridge of a man called *Ēadin*'. OE pers. name + *ofer*.

Edentrillick (*Éadan Trilic*) Down. *Balliedentrillicke* 1585. 'Hill brow of the megalithic tomb'.

Ederney (*Eadarnaidh*) Fermanagh. *Edernagh* 1610 'Middle place'.

Edgbaston Birm. *Celboldeston* [sic] 1086 (DB), *Egbaldestone* 1184. 'Farmstead of a man called Ecgbald'. OE pers. name + *tūn*.

Edgcott Bucks. *Achecote* 1086 (DB). Probably 'cottage(s) made of oak'. OE *æcen* + *cot*.

Edge Shrops. *Egge* 1255. '(Place at) the edge or escarpment'. OE *ecg*.

Edgefield Norfolk. *Edisfelda* 1086 (DB). 'Open land by an enclosure or enclosed park'. OE *edisc* + *feld*.

Edgeworthstown (*Meathas Troim*) Longford. 'Edgeworth's town'. The *Edgeworth* family were here from the 16th cent., Richard Edgeworth building Edgeworthstown House in the late 18th cent. The Irish name means 'frontier of the elder-tree'.

Edgmond Tel. & Wrek. *Edmendune* [sic] 1086 (DB), *Egmundun* 1155. 'Hill of a man called Ecgmund'. OE pers. name + *dūn*.

Edgton Shrops. *Egedune* 1086 (DB). Probably 'hill of a man called Ecga'. OE pers. name + *dūn*.

Edgware Gtr. London. *Ægces wer c.*975. 'Weir or fishing-enclosure of a man called Ecgi'. OE pers. name + *wer*.

Edgworth Black. w. Darw. *Eggewrthe* 1212. Probably 'enclosure on an edge or hillside'. OE *ecg* + *worth*.

Edinburgh Edin. *Eidyn c.*600, *Edenburge* 1126. 'Fortification at Eidyn'. OE *burh*. The meaning of *Eidyn* is unknown.

Edingale Staffs. *Ednunghale* 1086 (DB). 'Nook of land of the family or followers of a man called *Ēadin*'. OE pers. name + -*inga*- + *halh*.

Edingley Notts. *Eddyngleia c.*1180. 'Woodland clearing associated with a man called Eddi'. OE pers. name + -*ing*- + *lēah*.

Edingthorpe Norfolk. *Ædidestorp* 1177, *Edinestorp* (probably for *Ediuestorp*) 1198. 'Outlying farmstead or hamlet of a woman called Ēadgȳth'. OE pers. name + OScand. *thorp*.

Edington Somerset. *Eduuintone* 1086 (DB). 'Farmstead of a man called Ēadwine or of a woman called Ēadwynn'. OE pers. name + *tūn*.

Edington Wilts. *Ethandune* late 9th cent., *Edendone* 1086 (DB). 'Uncultivated hill', or 'hill of a man called Ētha'. OE *ēthe* (genitive -*an*) or OE pers. name (genitive -*n*) + *dūn*.

Edith Weston Rutland. *See* WESTON.

Edlesborough Bucks. *Eddinberge* [sic] 1086 (DB), *Eduluesberga* 1163. 'Hill or barrow of a man called Ēadwulf'. OE pers. name + *beorg*.

Edlingham Northum. *Eadwulfincham c.*1050. 'Homestead of the family or followers of a man called Ēadwulf', or 'homestead at Ēadwulf's place'. OE pers. name + -*inga*- or -*ing* + *hām*.

Edlington, Lincs. *Ellingetone* [sic] 1086 (DB), *Edlingtuna c.*1115. 'Estate associated with a man called *Ēdla*'. OE pers. name + -*ing*- + *tūn*.

Edmondsham Dorset. *Amedesham* 1086 (DB). 'Homestead or enclosure of a man called *Ēadmōd* or Ēadmund'. OE pers. name + *hām* or *hamm*.

Edmondthorpe Leics. *Edmerestorp* 1086 (DB). 'Outlying farmstead or hamlet of a man called Ēadmǣr'. OE pers. name + OScand. *thorp*.

Edmonton Gtr. London. *Adelmetone* 1086 (DB). 'Farmstead of a man called Ēadhelm'. OE pers. name + *tūn*.

estaston**Edstaston** Shrops. *Stanestune* [*sic*] 1086 (DB), *Edestaneston* 1256. 'Farmstead of a man called Ēadstān'. OE pers. name + *tūn*.

Edstone, Great N. Yorks. *Micheledestun* 1086 (DB). 'Farmstead of a man called *Ēadin'. OE pers. name + *tūn*. Affix in the early form is OE *micel* 'great'.

Edvin Loach Herefs. *Gedeuen* 1086 (DB), *Yedefen Loges* 1242. 'Fen or marshland of a man called *Gedda'. OE pers. name + *fenn*. Manorial affix from the *de Loges* family, here in the 13th cent.

Edwalton Notts. *Edvvoltone* 1086 (DB). 'Farmstead of a man called Ēadweald'. OE pers. name + *tūn*.

Edwardstone Suffolk. *Eduardestuna* 1086 (DB). 'Farmstead of a man called Ēadweard'. OE pers. name + *tūn*.

Edwinstowe Notts. *Edenestou* 1086 (DB). 'Holy place of St Ēadwine'. OE pers. name + *stōw*.

Effingham Surrey. *Epingeham* [*sic*] 1086 (DB), *Effingeham* 1180. 'Homestead of the family or followers of a man called *Effa'. OE pers. name + *-inga-* + *hām*.

Egerton Kent. *Eardingtun* [*sic*] c.1100, *Egarditon* 1203. 'Estate associated with a man called Ecgheard'. OE pers. name + *-ing-* + *tūn*.

Egerton Manch. A recent name, recorded thus from 1843. From the family surname *Egerton* of the Earls of Bridgewater who held land here.

Egerton Green Ches. *Eggerton* 1259. Probably 'farmstead of a man called Ecghere'. OE pers. name + *tūn*. *Green* is added from the 18th cent.

Egg Buckland Plym. *See* BUCKLAND.

Eggborough, High & Eggborough, Low N. Yorks. *Egeburg* 1086 (DB). 'Stronghold of a man called Ecga'. OE pers. name + *burh*.

Eggington Beds. *Ekendon* 1195. Probably 'hill of a man called Ecca'. OE pers. name (genitive *-n*) + *dūn*.

Egginton Derbys. *Ecgintune* 1012, *Eghintune* 1086 (DB). 'Estate associated with a man called Ecga'. OE pers. name + *-ing-* + *tūn*.

Egglescliffe Stock. on T. *Eggesclive* c.1185, *Egglescliue* 1197. Probably 'cliff or bank of a man

called Ecgwulf'. OE pers. name (alternating with the pet-form *Ecgi*) + *clif*.

Eggleston Durham. *Egleston* 1197. Probably 'farmstead of a man called Ecgwulf'. OE pers. name + *tūn*.

Egham Surrey. *Egeham* 933, 1086 (DB). 'Homestead or village of a man called Ecga'. OE pers. name + *hām*.

Egleton Rutland. *Egoluestun* 1218. 'Farmstead of a man called Ecgwulf'. OE pers. name + *tūn*.

Eglingham Northum. *Ecgwulfincham* c.1050. 'Homestead of the family or followers of a man called Ecgwulf', or 'homestead at Ecgwulf's place'. OE pers. name + *inga-* or *-ing* + *hām*.

Eglinton Derry. The original name of the village was *Muff*, from Irish *An Mhagh*, 'the plain'. The present name was adopted in 1858 when the Earl of *Eglinton* was Lord Lieutenant of Ireland.

Eglish (*Eaglais*) Tyrone. 'Church'.

Egloshayle Cornwall. *Egloshail* 1166. 'Church on an estuary'. Cornish *eglos* + *heyl*.

Egloskerry Cornwall. *Egloskery* c.1145. 'Church of St Keri'. Cornish *eglos* + saint's name.

Eglwys Lwyd, Yr. *See* LUDCHURCH.

Eglwys y Drindod. *See* CHRISTCHURCH.

Egmanton Notts. *Agemuntone* 1086 (DB). 'Farmstead of a man called Ecgmund'. OE pers. name + *tūn*.

Egremont Cumbria. *Egremont* c.1125. 'Sharp-pointed hill'. OFrench *aigre* + *mont*.

Egton N. Yorks. *Egetune* 1086 (DB). 'Farmstead of a man called Ecga'. OE pers. name + *tūn*. **Egton Bridge** is named from the 19th-cent. railway bridge here.

Eigg (island) Highland. *Egg* 1654. '(Island with an) indentation'. Gaelic *Eilean Eige*.

Eighter (*Íochtar*) Cavan. 'Lower portion'.

Eightercua (*Íochtar Cua*) Kerry. 'Lower hollow'.

Eilean Donnan (island) Highland. *Elandonan* c.1425. 'St Donan's island'. Gaelic *eilean*.

Eilean Siar. *See* WESTERN ISLES.

Éire, the Irish name for Ireland. *See* IRELAND.

Eirk (*Adharc*) Kerry. 'Peak'.

Elberton S. Glos. *Eldbertone* [*sic*] 1086 (DB), *Albricton* 1186. 'Farmstead of a man called Æthelbeorht'. OE pers. name + *tūn*.

Elburton Plym. *Aliberton* 1254. Identical in origin with the previous name.

Elcombe Swindn. *Elecome* 1086 (DB). 'Valley where elder-trees grow', or 'valley of a man called Ella'. OE *elle(n)* or OE pers. name + *cumb*.

Eldersfield Worcs. *Yldresfeld* 972, *Edresfelle* [*sic*] 1086 (DB). Probably 'open land of the elder-tree'. OE **hyldre* + *feld*.

Eldwick Brad. *Helguic* 1086 (DB). 'Dwelling or (dairy) farm of a man called Helgi'. OScand. pers. name + OE *wīc*.

Eleigh, Brent & Eleigh, Monks Suffolk. *Illeyge* 946–c.951, *Illanlege* 1000–1002, *Illeleia* 1086 (DB), *Brendeylleye* 1312, *Monekesillegh* 1304. 'Woodland clearing of a man called **Illa*'. OE pers. name + *lēah*. Affixes are from ME *brend* 'burnt, destroyed by fire' and OE *munuc* 'a monk' (alluding to possession by Christ Church, Canterbury).

Elford, 'ford where elder-trees grow', or 'ford of a man called Ella', OE *elle(n)* or OE pers. name + *ford*: **Elford** Northum. *Eleford* 1256. **Elford** Staffs. *Elleford* 1002, *Eleford* 1086 (DB).

Elgin Moray. *Elgin* 1136. Meaning unknown, compare GLENELG.

Elham Kent. *Alham* 1086 (DB). 'Homestead or enclosure where eels are found'. OE *ǣl* + *hām* or *hamm*.

Elie Fife. *Elye* 1491, *The Alie c.*1600. '(Place of the) tomb'. Gaelic *ealaidh*, dative of *ealadh*.

Eling Hants. *Edlinges* 1086 (DB). '(Settlement of) the family or followers of a man called **Ēadla* or Æthel'. OE pers. name + *ingas*.

Elkesley Notts. *Elchesleie* 1086 (DB). 'Woodland clearing of a man called Ēalāc'. OE pers. name + *lēah*.

Elkington, North & Elkington, South Lincs. *Alchinton* 1086 (DB). *Northalkinton* 12th cent., *Sudhelkinton* early 13th cent. 'Estate associated with a man called Ēadlāc'. OE pers. name + *-ing-* + *tūn*.

Elkstone Glos. *Elchestane* 1086 (DB). 'Boundary stone of a man called Ēalāc'. OE pers. name + *stān*.

Elkstone, Lower & Elkstone, Upper Staffs. *Elkesdon* 1227. 'Hill of a man called Ēalāc'. OE pers. name + *dūn*.

Ella, Kirk & Ella, West E. R. Yorks. *Aluengi* [*sic*] 1086 (DB), *Kirk Elley* 15th cent., *Westeluelle* 1305. 'Woodland clearing of a man called Ælf(a)'. OE pers. name + *lēah*. Distinguishing affixes are from OScand. *kirkja* 'church' and *west*.

Elland Calder. *Elant* 1086 (DB). 'Cultivated land, or estate, by the river'. OE *ēa-land*.

Ellastone Staffs. *Edelachestone* 1086 (DB). 'Farmstead of a man called Ēadlāc'. OE pers. name + *tūn*.

Ellen (river) Cumbria. *See* ALLERDALE.

Ellenhall Staffs. *Linehalle* [*sic*] 1086 (DB), *Ælinhale c.*1200. Possibly 'nook of land associated with a man called Ælle or Ella', from OE pers. name + *-ing-* + *halh*. Alternatively 'nook (by the river) where flax is grown', from OE *līn* + *halh* with the later addition of *ēa*.

Ellerbeck N. Yorks. *Elrebec* 1086 (DB). 'Stream where alders grow'. OScand. *elri* + *bekkr*.

Ellerby N. Yorks. *Elwordebi* 1086 (DB). 'Farmstead or village of a man called Ælfweard'. OE pers. name + OScand. *bý*.

Ellerdine Heath Tel. & Wrek. *Elleurdine* 1086 (DB). 'Enclosure of a man called Ella'. OE pers. name + *worthign*, with the later addition of *heath*.

Ellerker E. R. Yorks. *Alrecher* 1086 (DB). 'Marsh where alders grow'. OScand. *elri* + *kjarr*.

Ellerton E. R. Yorks. *Elreton* 1086 (DB). 'Farmstead by the alders'. OScand. *elri* + OE *tūn*.

Ellerton Shrops. *Athelarton* 13th cent. 'Farmstead of a man called Æthelheard'. OE pers. name + *tūn*.

Ellesborough Bucks. *Esenberge* [*sic*] 1086 (DB), *Eselbergh* 1195. Probably 'hill where asses are pastured'. OE *esol* + *beorg*.

Ellesmere Shrops. *Ellesmeles* [*sic*] 1086 (DB), *Ellismera* 1177. 'Lake or pool of a man called Elli'. OE pers. name + *mere*.

Ellesmere Port Ches. a modern name, only in use since the early 19th cent., so called because the Ellesmere Canal (from the previous name) joins the Mersey here.

Ellingham, 'homestead of the family or followers of a man called Ella', or 'homestead at Ella's place', OE pers. name + *-inga-* or *-ing* + *hām*: **Ellingham** Norfolk. *Elincham* 1086 (DB). **Ellingham, Great & Ellingham, Little** Norfolk. *Elin(c)gham* 1086 (DB), *Magna Elingham, Parva Elingham* 1242. Distinguishing affixes are Latin *magna* 'great' and *parva* 'little'. **Ellingham** Northum. *Ellingeham* c.1130. However the following contains a different pers. name: **Ellingham** Hants. *Adelingeham* 1086 (DB). 'Homestead of the family or followers of a man called Æthel'. OE pers. name + *-inga-* + *hām*.

Ellingstring N. Yorks. *Elingestrengge* 1198. 'Water-course at the place where eels are caught', or 'water-course at the place associated with a man called Ella or Eli'. OScand. *strengr* with either OE *ǣl, ēl* + *-ing* or OE pers. name + *-ing*.

Ellington, 'farmstead at the place where eels are caught', or 'farmstead associated with a man called Ella or Eli', OE *tūn* with either OE *ǣl, ēl* + *-ing* or OE pers. name + *-ing-*: **Ellington** Cambs. *Elintune* 1086 (DB). **Ellington** Northum. *Elingtona* 1166. **Ellington, High & Ellington, Low** N. Yorks. *Ellintone* 1086 (DB).

Ellisfield Hants. *Esewelle* [sic] 1086 (DB), *Elsefeld* 1167. Probably 'open land of a man called *Ielfsa'. OE pers. name + *feld*.

Ellon Aber. *Eilan* c.1150. Meaning unknown.

Ellough Suffolk. *Elga* 1086 (DB), *Elgh* 1286. Etymology uncertain, but possibly '(place at) the heathen temple'. OE *ealh* (dative *ealge*).

Elloughton E. R. Yorks. *Elgendon* 1086 (DB), *Helgedon* 1196. Possibly 'hill of a man called Helgi'. OScand. pers. name + OE *dūn*, but first element uncertain.

Elm, '(place at) the elm-tree(s)', OE *elm* (dative plural *elmum*): **Elm** Cambs. *Elm, Eolum* 10th cent. **Elm, Great** Somerset. *Telma* 1086 (DB). *T-* in the early form is from the OE preposition *æt* 'at'.

Elmbridge (district) Surrey. *Amelebrige* 1086 (DB). 'Bridge over the River *Emel'*. A revival of an old hundred-name. Pre-Celtic river-name (of uncertain meaning, an old name for the River Mole for which *see* MOLESEY) + OE *brycg*.

Elmbridge Worcs. *Elmerige* 1086 (DB). 'Ridge where elm-trees grow'. OE *elm, *elmen* + *hrycg*.

Elmdon Essex. *Elmenduna* 1086 (DB). 'Hill where elm-trees grow'. OE **elmen* + *dūn*.

Elmdon Solhll. *Elmedone* 1086 (DB). 'Hill of the elm-trees'. OE *elm* + *dūn*.

Elmesthorpe Leics. *Ailmerestorp* 1199. *Torp* 1086 (DB). 'Outlying farmstead or hamlet of a man called Æthelmǣr'. OE pers. name + OScand. *thorp*.

Elmet (old district), *See* BARWICK IN ELMET.

Elmham, 'homestead or village where elm-trees grow', OE *elm, *elmen* + *hām*: **Elmham, North** Norfolk. *Ælmham* c.1035, *Elmenham* 1086 (DB). **Elmham, South** Norfolk. *Almeham* 1086 (DB). The parishes of All Saints, St James, St Margaret, and St Michael South Elmham are named from the dedications of the churches. However the affix in St Cross South Elmham is from *Sancroft* 1254, 'sandy enclosure', OE *sand* + *croft*.

Elmley, 'elm-tree wood or clearing', OE *elm* + *lēah*: **Elmley Castle** Worcs. *Elmlege* 780, *Castel Elmeleye* 1327. Affix from the former castle here. **Elmley Lovett** Worcs. *Ælmeleia* 1086 (DB), *Almeleye Lovet* 1275. Manorial addition from the *Lovett* family, here in the 13th cent.

Elmore Glos. *Elmour* 1176. 'River-bank or ridge where elm-trees grow'. OE *elm* + *ōfer* or **ofer*.

Elmsall, North & Elmsall, South Wakefd. *Ermeshale* [sic] 1086 (DB), *North Elmesale* 1320, *Suthelmeshal* 1230. 'Nook of land by the elm-tree'. OE *elm* + *halh*.

Elmsett Suffolk. *Ylmesæton* c.995, *Elmeseta* 1086 (DB). '(Settlement of) the dwellers among the elm-trees'. OE **elme* + *sǣte*.

Elmstead Gtr. London. *Elmsted* 1320. 'Place by the elm-trees'. OE *elm* + *stede*.

Elmstead Market Essex. *Elmesteda* 1086 (DB), *Elmested Market* 1475. 'Place where elm-trees grow'. OE **elme* or **elmen* + *stede*. Affix *market* from the important early market here.

Elmsted Kent. *Elmanstede* 811. 'Homestead by the elm-trees'. OE *elm* + *hām-stede*.

Elmstone Kent. *Ailmereston* 1203.
'Farmstead of a man called Æthelmǣr'. OE pers.
name + *tūn*.

Elmstone Hardwicke Glos. *Almundestan*
1086 (DB). 'Boundary stone of a man called
Alhmund'. OE pers. name + *stān*. Hardwicke
has been added from a nearby place, *see*
HARDWICKE.

Elmswell Suffolk. *Elmeswella* 1086 (DB).
'Spring or stream where elm-trees grow'.
OE *elm* + *wella*.

Elmton Derbys. *Helmetune* 1086 (DB).
'Farmstead where elm-trees grow'. OE *elm*,
*elmen + *tūn*.

Elphin (*Ail Finn*) Roscommon. 'Fionn's stone'.

Elsdon Northum. *Eledene* 1226. Probably
'valley of a man called El(l)i'. OE pers. name +
denu.

Elsenham Essex. *Elsenham* 1086 (DB).
'Homestead or village of a man called Elesa'.
OE pers. name (genitive -*n*) + *hām*.

Elsfield Oxon. *Esefelde* 1086 (DB), *Elsefeld*
*c.*1130. 'Open land of a man called Elesa'.
OE pers. name + *feld*.

Elsham N. Lincs. *Elesham* 1086 (DB).
'Homestead or village of a man called El(l)i'.
OE pers. name + *hām*.

Elsing Norfolk. *Helsinga* 1086 (DB).
'(Settlement of) the family or followers of a man
called Elesa'. OE pers. name + -*ingas*.

Elslack N. Yorks. *Eleslac* 1086 (DB). 'Stream or
valley of a man called El(l)i'. OE pers. name +
lacu or OScand. *slakki*.

Elstead Surrey. *Helestede* 1128. 'Place where
elder-trees grow'. OE *elle*(*n*) + *stede*.

Elsted W. Sussex. *Halestede* [*sic*] 1086 (DB),
Ellesteda 1180. Identical in origin with the
previous name.

Elston Notts. *Elvestune* 1086 (DB). Probably
'farmstead of a man called Eiláfr'. OScand. pers.
name + *tūn*.

Elstow Beds. *Elnestou* 1086 (DB). 'Assembly-
place of a man called **Ellen'. OE pers. name +
stōw.

Elstree Herts. *Tithulfes treow* 11th cent.
'Boundary tree of a man called Tīdwulf'. OE
pers. name + *trēow*. Initial *T*- disappeared in the

13th cent. due to confusion with the preposition
at.

Elstronwick E. R. Yorks. *Asteneuuic* [*sic*] 1086
(DB), *Elstanwik c.*1265. 'Dwelling or (dairy) farm
of a man called Ælfstān'. OE pers. name + *wīc*.

Elswick Lancs. *Edelesuuic* 1086 (DB).
'Dwelling or (dairy) farm of a man called
Æthelsige'. OE pers. name + *wīc*.

Elsworth Cambs. *Eleswurth* 974, *Elesuuorde*
1086 (DB). 'Enclosure of a man called El(l)i'.
OE pers. name + *worth*.

Elterwater Cumbria. *Heltewatra c.*1160.
'Lake frequented by swans'. OScand. *elptr* +
OE *wæter*.

Eltham Gtr. London. *Elteham* 1086 (DB).
'Homestead or river-meadow frequented by
swans', or (perhaps rather) 'of a man called
**Elta'. OE *elfitu* or OE pers. name + *hām* or
hamm.

Eltisley Cambs. *Hecteslei* [*sic*] 1086 (DB),
Eltesle 1228. 'Woodland clearing of a man called
Elti'. OE pers. name + *lēah*.

Elton, sometimes probably 'farmstead where
eels are got', OE *ǣl* + *tūn*: **Elton** Ches., near
Ellesmere Port. *Eltone* 1086 (DB). **Elton** Derbys.
Eltune 1086 (DB). **Elton** Stock. on T. *Eltun*
*c.*1040.
     However other Eltons have a different origin:
Elton Cambs. *Æthelingtun* 10th cent.,
Adelintune 1086 (DB). 'Farmstead of the
princes', or 'farmstead associated with a man
called Æthel'. OE *ætheling* + *tūn*, or OE pers.
name + -*ing*- + *tūn*. **Elton** Herefs. *Elintune* 1086
(DB). Probably 'farmstead of a man called Ella'.
OE pers. name + *tūn*. **Elton** Notts. *Ailetone* [*sic*]
1086 (DB), *Elleton* 1088. Probably identical with
the previous name.

Elvaston Derbys. *Ælwoldestune* 1086 (DB).
'Farmstead of a man called Æthelweald'. OE
pers. name + *tūn*.

Elveden Suffolk. *Eluedena* 1086 (DB). 'Swan
valley' or 'valley haunted by elves or fairies'. OE
elfitu or *elf* (genitive plural -*a*) + *denu*.

Elvetham Hants. *Elfteham* 727, *Elveteham*
1086 (DB). 'River-meadow frequented by
swans'. OE *elfitu* + *hamm*.

Elvington York. *Aluuintone* 1086 (DB).
'Farmstead of a man called Ælfwine or a woman
called Ælfwynn'. OE pers. name + *tūn*.

Elwick Hartlepl. *Ellewic c.*1150. 'Dwelling or (dairy) farm of a man called Ella'. OE pers. name + *wīc.*

Elwick Northum. *Ellewich* 12th cent. 'Dwelling or (dairy) farm of a man called Ella'. OE pers. name + *wīc.*

Elworth Ches. *Ellewrdth* 1282. 'Enclosure of a man called Ella'. OE pers. name + *worth.*

Elworthy Somerset. *Elwrde* 1086 (DB). 'Enclosure of a man called Ella'. OE pers. name + *worth* (later replaced by *worthig*).

Ely Cambs. *Elge* 731, *Elyg* 1086 (DB). 'District where eels are to be found'. OE *ǣl, ēl* + **gē.*

Emberton Milt. K. *Ambretone* 1086 (DB), *Emberdestone* 1227. 'Farmstead of a man called Ēanbeorht'. OE pers. name + *tūn.*

Embleton Cumbria. *Emelton* 1195, *Embelton* 1233. 'Farmstead of a man called Ēanbald'. OE pers. name + *tūn.*

Embleton Northum. *Emlesdone* 1212. 'Hill infested by caterpillars', or 'hill of a man called Æmele'. OE *emel* or OE pers. name + *dūn.*

Emborough Somerset. *Amelberge* [*sic*] 1086 (DB), *Emeneberge* 1200. 'Flat-topped mound or hill'. OE *emn* + *beorg.*

Embsay N. Yorks. *Embesie* 1086 (DB). Probably 'enclosure of a man called Embe'. OE pers. name + *hæg.*

Emlagh (*Imleach*) Mayo. 'Borderland'.

Emlaghfad (*Imleach Fada*) Sligo. 'Long borderland'.

Emlaghmore (*An tImleach Mór*) Kerry. 'The large borderland'.

Emley Kirkl. *Ameleie* 1086 (DB). 'Woodland clearing of a man called *Em(m)a'. OE pers. name + *lēah.*

Emly (*Imleach*) Tipperary. 'Borderland'.

Emmington Oxon. *Amintone* 1086 (DB). 'Estate associated with a man called Eama'. OE pers. name + *-ing-* + *tūn.*

Emneth Norfolk. *Anemetha* 1170. Possibly 'river-confluence of a man called Ēana'. OE pers. name + (*ge*)*mȳthe*. Alternatively the second element may be OE *mǣth* 'mowing grass, meadow'.

Emo (*Ioma*) Laois. 'Image'.

Empingham Rutland. *Epingeham* [*sic*] 1086 (DB), *Empingeham* 12th cent. 'Homestead of the family or followers of a man called *Empa'. OE pers. name + *-inga-* + *hām.*

Empshott Hants. *Hibesete* [*sic*] 1086 (DB), *Himbeset c.*1170. 'Corner of land frequented by swarms of bees'. OE *imbe* + *scēat* or **scīete.*

Emsworth Hants. *Emeleswurth* 1224. 'Enclosure of a man called Æmele'. OE pers. name + *worth.*

Emyvale (*Ioma*) Monaghan. 'Bed of the saint'. *Emyvale or Skernagerragh* 1779. Irish *ioma* + English *vale*. A saint of unknown name is said to have lived and slept here. An alternative name was *Scarnageeragh*, from Irish *Scairbh na gCaorach*, 'shallow ford of the sheep'.

Enborne W. Berks. *Aneborne* [*sic*] 1086 (DB), *Enedburn* 1220. 'Duck stream'. OE *ened* + *burna.*

Enderby Leics. *Andretesbie, Endrebi* 1086 (DB). 'Farmstead or village of a man called Eindrithi'. OScand. pers. name + *bý.*

Enderby, Bag, Enderby, Mavis, & Enderby, Wood Lincs. *Andrebi, Endrebi* 1086 (DB), *Bagenderby* 1291, *Enderby Malbys* 1302, *Wodenderby* 1198. Probably identical in origin with the previous name. The affix *Mavis* is manorial, from the *Malebisse* family here in the 13th cent. The affix *Bag* is perhaps from ME *bagge* 'a bag' used in a transferred topographical sense with reference to the shape of the village. *Wood* (from ME *wode* 'wood') indicates a situation in a once wooded area.

Endon Staffs. *Enedun* 1086 (DB). 'Hill of a man called Ēana, or where lambs are reared'. OE pers. name or OE **ēan* + *dūn.*

Enfield Gtr. London. *Enefelde* 1086 (DB). 'Open land of a man called Ēana, or where lambs are reared'. OE pers. name or OE **ēan* + *feld.*

Enford Wilts. *Enedford* 934, *Enedforde* 1086 (DB). 'Duck ford'. OE *ened* + *ford.*

England *Englaland c.*890. 'Land of the Angles'. OE *Engle* (genitive plural *Engla*) 'the Angles' (i.e. the people from the continental homeland of *Angel* in Schleswig) + *land.*

Englefield W. Berks. *Englafelda c.*900, *Englefel* 1086 (DB). 'Open land of the Angles'. OE *Engle* + *feld.*

Englefield Green Surrey. *Ingelfeld* 1282. Possibly 'open land of a man called *Ingel or Ingweald'. OE pers. name + *feld*. *Green* is added from the 17th cent.

English Bicknor Glos. *See* BICKNOR.

English Frankton Shrops. *See* FRANKTON.

Englishcombe B. & NE. Som. *Ingeliscuma* 1086 (DB). Probably 'valley of a man called *Ingel or Ingweald'. OE pers. name + *cumb*.

Enham Alamein Hants. *Eanham* early 11th cent., *Etham* [sic] 1086 (DB). 'Homestead or enclosure where lambs are reared'. OE *ēan + *hām* or *hamm*. Affix from 1945, commemorating the battle of El Alamein and referring to the centre for disabled ex-servicemen here. Formerly called **Knight's Enham**, *Knyghtesenham* 1389, from the knight's fee held here by Matthew de Columbers in the mid-13th cent.

Enmore Somerset. *Animere* 1086 (DB). 'Duck pool'. OE *ened + *mere*.

Ennell, Lough (*Loch Ainnínne*) Westmeath. 'Ainneann's lake'.

Ennerdale Bridge Cumbria. *Anenderdale* c.1135, *Eghnerdale* 1321. 'Valley of a man called Anundr'. OScand. pers. name (genitive *-ar*) + *dalr*. Later the first element was replaced by the river-name Ehen (of obscure origin).

Ennis (*Inis*) Clare. 'Island'.

Enniskean (*Inis Céin*) Cork. 'Cian's island'.

Enniskerry (*Áth na Sceire*) Wicklow. 'Ford of the rocky place'.

Enniskillen (*Inis Ceithleann*) Fermanagh. (*Caislén*) *Insi Ceithlenn* 1439. 'Ceithleann's island'.

Ennistimmon (*Inis Díomáin*) Clare. 'Díomán's island'.

Enoch Dumf. '(Place by the) marsh'. Gaelic *eanach*.

Enstone Oxon. *Henestan* 1086 (DB). 'Boundary stone of a man called Enna'. OE pers. name + *stān*.

Enville Staffs. *Efnefeld* 1086 (DB). 'Smooth or level open land'. OE *efn + *feld*.

Epperstone Notts. *Eprestone* 1086 (DB). Probably 'farmstead of a man called Eorphere'. OE pers. name + *tūn*.

Epping Essex. *Eppinges* 1086 (DB). Probably '(settlement of) the people of the upland or higher ground'. OE *yppe* + *-ingas*.

Eppleby N. Yorks. *Aplebi* 1086 (DB). 'Farmstead where apple-trees grow'. OE *æppel* or OScand. *epli* + *bý*.

Epsom Surrey. *Ebbesham* c.973, *Evesham* [sic] 1086 (DB). 'Homestead or village of a man called Ebbe or Ebbi'. OE pers. name + *hām*.

Epwell Oxon. *Eoppan wyllan* 956. 'Spring or stream of a man called Eoppa'. OE pers. name + *wella*.

Epworth N. Lincs. *Epeurde* 1086 (DB). 'Enclosure of a man called Eoppa'. OE pers. name + *worth*.

Ercall, Child's & Ercall, High Shrops. *Arcalun* [sic], *Archelov* 1086 (DB), *Childes Ercalewe, Magna Ercalewe* 1327. Possibly an old hill-name *Earcaluw* from OE *ēar* 'gravel, mud' and *calu(w)* 'bare hill', first applied to a settlement and then to a district (the two Ercalls are some six miles apart). Affixes are OE *cild* 'son of a noble family' and Latin *magna* 'great'.

Erdington Birm. *Hardintone* 1086 (DB). 'Estate associated with a man called *Earda'. OE pers. name + *-ing-* + *tūn*.

Erewash (district) Derbys. *Irewys* c.1145. A modern adoption of the OE river-name, which means 'winding (marshy) stream'. OE *irre* + *wisc* (influenced by modern *wash*).

Eriboll Highland. *Eribull* 1499. 'Farm on a gravel bank'. OScand. *eyri* + *ból*(*stathr*).

Ericht, Loch Highland. 'Loch of assemblies'. Gaelic *loch* + *eireachd*.

Eridge Green E. Sussex. *Ernerigg* 1202. 'Ridge frequented by eagles'. OE *earn* + *hrycg*.

Eriskay (island) W. Isles. *Eriskeray* 1549. 'Erikr's island'. OScand. pers. name + *ey*, 'island'.

Eriswell Suffolk. *Hereswella* 1086 (DB). Possibly 'spring or stream of a man called *Here'. OE pers. name + *wella*.

Erith Gtr. London. *Earhyth* 677, *Earhith* c.960, *Erhede* 1086 (DB). 'Muddy or gravelly landing-place'. OE *ēar* + *hýth*.

Erlestoke Wilts. *Erlestoke* 12th cent. 'Outlying farmstead belonging to the nobleman'. OE *eorl* + *stoc*.

Ermine Street (Roman road from London to the Humber). *Earninga strǣt* 955. 'Roman road of the family or followers of a man called *Earn(a)'. OE pers. name + *-inga-* + *strǣt*. No doubt originally applied to a stretch of the road near ARRINGTON (Cambs.) before the name was extended to the whole length. The name Ermine Street was later transferred to another Roman road, that from Silchester to Gloucester.

Ermington Devon. *Ermentona* 1086 (DB). Probably 'estate associated with a man called *Earma'. OE pers. name + *-ing-* + *tūn*.

Erne, Lough (*Loch Éirne*) Fermanagh. 'Érann's lake'. The name is that of a goddess.

Erpingham Norfolk. *Erpingham* 1044–7, (*H*)*erpincham* 1086 (DB). 'Homestead of the family or followers of a man called *Eorp'. OE pers. name + *-inga-* + *hām*.

Errigal (*Earagail*) (mountain) Donegal. 'Oratory'.

Errigal Keerogue (*Earagail Do Chiaróg*) Tyrone. 'Do Chiaróg's oratory'.

Erris (*Iorras*) Mayo. 'Promontory'.

Errislannan (*Iorras Fhlannáin*) Galway. 'Flannan's head'.

Erskine Renf. *Erskin* 1225, *Yrskin* 1227, *Ireskin* 1262, *Harskin* c.1300. Meaning uncertain. The name may be Celtic, but early forms are inconsistent.

Erwarton Suffolk. *Eurewardestuna* 1086 (DB). 'Farmstead of a man called *Eoforweard'. OE pers. name + *tūn*.

Eryholme N. Yorks. *Argun* 1086 (DB). '(Place at) the shielings or summer pastures'. OScand. *erg* in a dative plural form *ergum*.

Eryri. *See* SNOWDONIA.

Escomb Durham. *Ediscum* 10th cent. '(Place at) the enclosed parks or pastures'. OE *edisc* in a dative plural form.

Escrick N. Yorks. *Ascri* [*sic*] 1086 (DB), *Eskrik* 1169. 'Strip of land or narrow ridge where ash-trees grow'. OScand. *eski* + OE *ric*.

Esh Durham. *Es, Esse* 12th cent. '(Place at) the ash-tree'. OE *æsc*. The affix in nearby **Esh Winning** is *winning* 'a coal-mine'.

Esher Surrey. *Æscæron* 1005, *Aissele* [*sic*] 1086 (DB). 'District where ash-trees grow'. OE *æsc* + *scearu*.

Eshnadarragh (*Ais na Darrach*) Fermanagh. 'Ridge of the oak'.

Eshnadeelada (*Ais na Diallaite*) Fermanagh. 'Ridge of the saddle'.

Eshott Northum. *Esseta* 1187. 'Clump of ash-trees' from OE *æscet*, or 'corner of land growing with ash-trees' from OE *æsc* + *scēat*.

Eshton N. Yorks. *Estune* 1086 (DB). 'Farmstead by the ash-tree(s)'. OE *æsc* + *tūn*.

Esk (river) Dumf. *Ask* c.1200. 'Water'. British *isca*.

Eskdale Green Cumbria. *Eskedal* 1294. 'Valley of the River Esk'. Celtic river-name (meaning simply 'the water') + OScand. *dalr*.

Eske, Lough (*Loch Iasc*) Donegal. 'Lake abounding in fish'.

Esker (*Eiscir*) Dublin, Longford. 'Gravel ridge'.

Eskerridge (*Eiscir*) Tyrone. 'Gravel ridge'.

Eskine (*Eisc Dhoimnin*) Kerry. 'Deep fissure'.

Eskra (*Eiscrach*) Tyrone. 'Place of gravel ridges'.

Esnadarra (*Ais na Darach*) Fermanagh. 'Side of the place abounding in oaks'.

Esprick Lancs. *Eskebrec* c.1210. 'Hill-slope where ash-trees grow'. OScand. *eski* + *brekka*.

Essendine Rutland. *Esindone* [*sic*] 1086 (DB), *Esenden* 1230. 'Valley of a man called Ēsa'. OE pers. name (genitive *-n*) + *denu*.

Essendon Herts. *Eslingadene* 11th cent. 'Valley of the family or followers of a man called *Ēsla'. OE pers. name + *-inga-* + *denu*.

Essex (the county). *East Seaxe* late 9th cent., *Exsessa* 1086 (DB). '(Territory of) the East Saxons'. OE *ēast* + *Seaxe*.

Essington Staffs. *Esingetun* 996, *Eseningetone* 1086 (DB). 'Farmstead of the family or followers of a man called Esne'. OE pers. name + *-inga-* + *tūn*.

Eston Red. & Cleve. *Astun* 1086 (DB). 'East farmstead or village'. OE *ēast* + *tūn*.

Etal Northum. *Ethale* 1232. Probably 'nook of land used for grazing'. OE *ete* + *halh*. Alternatively the first element could be the OE masculine pers. name *Ēata*.

Etchilhampton Wilts. *Echesatingetone* 1086 (DB), *Ehelhamton* 1196. Possibly 'farmstead of the dwellers at the oak-tree hill'. OE *āc* (genitive *ǣc*) + *hyll* + *hǣme* (earlier *sǣte*) + *tūn*.

Etchingham E. Sussex. *Hechingeham* 1158. 'Homestead or enclosure of the family or followers of a man called Ecci'. OE pers. name + *-inga-* + *hām* or *hamm*.

Etive, Loch Arg. 'Eite's loch'. Gaelic *loch*. *Eiteag* 'the little horrid one' is the goddess of the loch.

Eton Winds. & Maid. *Ettone* 1086 (DB). 'Farmstead or estate on the river' (probably denoting a settlement which performed a special local function in relation to the river). OE *ēa* + *tūn*.

Etruria Stoke. named from a house called *Etruria Hall* built by Josiah Wedgwood, who founded his famous pottery here in 1769, the allusion being to pottery from ancient Etruria (modern Tuscany).

Ettington Warwicks. *Etendone* 1086 (DB). Probably 'hill of a man called Ēata', OE pers. name (genitive *-n*) + *dūn*. Alternatively the first element may be OE *eten* 'grazing, pasture'.

Etton, probably 'farmstead of a man called Ēata', OE pers. name + *tūn*: **Etton** E. R. Yorks. *Ettone* 1086 (DB). **Etton** Peterb. *Ettona* 1125–8.

Ettrick Forest Sc. Bord. *Ethric c.*1235. 'Forest of Ettrick Water'. The river-name is of uncertain meaning.

Etwall Derbys. *Etewelle* 1086 (DB). Probably 'spring or stream of a man called Ēata'. OE pers. name + *wella*.

Euston Suffolk. *Euestuna* 1086 (DB). 'Farmstead of a man called Efe'. OE pers. name + *tūn*.

Euxton Lancs. *Eueceston* 1187. 'Farmstead of a man called *Eofoc'. OE pers. name + *tūn*.

Evanton Highland. 'Evan's village'. The village was founded in the early 19th cent. by *Evan* Fraser of Balconie.

Evedon Lincs. *Evedune* 1086 (DB). 'Hill of a man called Eafa'. OE pers. name + *dūn*.

Evenley Northants. *Evelaia* [*sic*] 1086 (DB), *Euenlai* 1147. 'Level woodland clearing'. OE *efen* + *lēah*.

Evenlode Glos. *Euulangelade* 772, *Eunilade* 1086 (DB). 'Difficult river-crossing of a man called *Eowla'. OE pers. name (genitive *-n*) + *gelād*. The river-name **Evenlode** is a 'back-formation' from the place name; the old name of the river was *Bladon, see* BLADON and BLEDINGTON.

Evenwood Durham. *Efenwuda c.*1040. 'Level woodland'. OE *efen* + *wudu*.

Evercreech Somerset. *Evorcric* 1065, *Evrecriz* 1086 (DB). Celtic *crūg* 'hill' with an uncertain first element, possibly OE *eofor* 'wild boar' or a Celtic word for some kind of plant or tree.

Everdon, Great Northants. *Eferdun* 944, *Everdone* 1086 (DB). 'Hill frequented by wild boars'. OE *eofor* + *dūn*.

Everingham E. R. Yorks. *Yferingaham c.*972, *Evringham* 1086 (DB). 'Homestead of the family or followers of a man called Eofor'. OE pers. name + *-inga-* + *hām*.

Everleigh Wilts. *Eburleagh* 704. 'Wood or clearing frequented by wild boars'. OE *eofor* + *lēah*.

Everley N. Yorks. *Eurelai* 1086 (DB). Identical in origin with the previous name.

Eversden, Great & Eversden, Little Cambs. *Euresdone* 1086 (DB), *Everesdon Magna, Parva* 1240. 'Hill of the wild boar, or of a man called Eofor'. OE *eofor* or OE pers. name + *dūn*. Affixes are Latin *magna* 'great', *parva* 'little'.

Eversholt Beds. *Eureshot* [*sic*] 1086 (DB), *Euresholt* 1185. 'Wood of the wild boar'. OE *eofor* + *holt*.

Evershot Dorset. *Teversict* [*sic*] 1202, *Evershet* 1286. Probably 'corner of land frequented by wild boars'. OE *eofor* + *scēat* or **scīete*. Initial *T-* in the first form may be from the preposition *at*.

Eversley Hants. *Euereslea c.*1050, *Evreslei* 1086 (DB). 'Wood or clearing of the wild boar, or of a man called Eofor'. OE *eofor* or OE pers. name + *lēah*.

Everton, 'farmstead where wild boars are seen', OE *eofor* + *tūn*: **Everton** Beds. *Euretone*

1086 (DB). **Everton** Lpool. *Evretona* 1094.
Everton Notts. *Evretone* 1086 (DB).

Evesbatch Herefs. *Sbech* [sic] 1086 (DB),
Esbeche 12th cent. 'Stream valley of a man called
Ēsa'. OE pers. name + *bece, bæce*.

Evesham Worcs. *Eveshomme* 709, *Evesham*
1086 (DB). 'Land in a river-bend belonging to a
man called Ēof'. OE pers. name + *hamm*.

Evington Leic. *Avintone* 1086 (DB). 'Estate
associated with a man called Eafa'. OE pers.
name + *-ing-* + *tūn*.

Ewell Surrey. *Euuelle* 933, *Etwelle* [sic] 1086
(DB). '(Place at) the river-source'. OE *æwell*.

Ewell Minnis & Temple Ewell Kent.
Æwille c.772, *Ewelle* 1086 (DB). Identical in
origin with the previous name. The affix *Minnis*
is from OE *mænnes* 'common land', *Temple*
alludes to possession by the Knights Templar
from the 12th cent.

Ewelme Oxon. *Auuilme* 1086 (DB). '(Place at)
the river-source'. OE *æwelm*.

Ewen Glos. *Awilme* 931. Identical in origin
with the previous name, here with reference to
the source of River THAMES.

Ewerby Lincs. *Ieresbi* 1086 (DB), *Iwarebi*
1185. 'Farmstead or village of a man called
Ívarr'. OScand. pers. name + *bý*.

Ewhurst Surrey. *Iuherst* 1179. 'Yew-tree
wooded hill'. OE *īw* + *hyrst*.

Ewloe Flin. *Ewlawe* 1281. 'Hill with a river-
source'. OE *æwell* + *hlāw*.

Eworthy Devon. *Yworthy* 1468. 'Enclosure
where yew-trees grow'. OE *īw* + *worthig*.

Ewshot Hants. *Hyweshate* 1236. 'Corner or
angle of land where yew-trees grow'. OE *īw* +
scēat.

Ewyas Harold Herefs. *Euuias* c.1150,
Euuiasharold 1176. A Welsh name meaning
'sheep district'. Affix from a nobleman called
Harold who held the manor in the 11th cent.

Exbourne Devon. *Hechesburne* [sic] 1086
(DB), *Yekesburne* 1242. 'Stream of the cuckoo, or
of a man called *Gēac'. OE *gēac* or OE pers.
name + *burna*.

Exbury Hants. *Teocreberie* [sic] 1086 (DB),
Ykeresbirie 1196. 'Fortified place of a man called
*Eohhere'. OE pers. name + *burh* (dative *byrig*).

Exe, Nether & Exe, Up Devon. *Niresse* [sic],
Ulpesse [sic] 1086 (DB), *Nitherexe* 1196, *Uphexe*
1238. Named from the River Exe, a Celtic river-
name meaning simply 'the water'. Affixes are
OE *neotherra* 'lower (down river)' and *upp*
'higher (up river)'.

Exebridge Somerset. *Exebrigge* 1255. 'Bridge
over the River Exe'. Celtic river-name + OE
brycg.

Exelby N. Yorks. *Aschilebi* 1086 (DB).
'Farmstead or village of a man called Eskil'.
OScand. pers. name + *bý*.

Exeter Devon. *Iska* c.150, *Exanceaster* c.900,
Execestre 1086 (DB). 'Roman town on the River
Exe'. Celtic river-name (*see* EXE) + OE *ceaster*.

Exford Somerset. *Aisseford* [sic] 1086 (DB),
Exeford 1243. 'Ford over the River Exe'. Celtic
river-name + OE *ford*.

Exhall, 'nook of land near a Romano-British
Christian church'. Celtic **eglēs* + OE
halh. **Exhall** Warwicks., near Coventry.
Eccleshale 1144. **Exhall** Warwicks.,
near Alcester. *Eccleshale* 710, *Ecleshelle*
1086 (DB).

Exminster Devon. *Exanmynster* c.880,
Esseminstre 1086 (DB). 'Monastery by the River
Exe'. Celtic river-name + OE *mynster*.

Exmoor Devon/Somerset. *Exemora* 1204.
'Moorland on the River Exe'. Celtic river-name +
OE *mōr*.

Exmouth Devon. *Exanmutha* c.1025. 'Mouth
of the River Exe'. Celtic river-name + OE *mūtha*.

Exning Suffolk. *Essellinge* [sic] 1086 (DB),
Exningis 1158. '(Settlement of) the family or
followers of a man called *Gyxen'. OE pers.
name + *-ingas*.

Exton Devon. *Exton* 1242. 'Farmstead on the
River Exe'. Celtic river-name + OE *tūn*.

Exton Hants. *East Seaxnatune* 940,
Essessentune 1086 (DB). 'Farmstead of the East
Saxons'. OE *Ēastseaxe* + *tūn*.

Exton Rutland. *Exentune* 1086 (DB). Probably
'farmstead where oxen are kept'. OE *oxa*
(genitive plural **exna*) + *tūn*.

Exton Somerset. *Exton* 1216. 'Farmstead on
the River Exe'. Celtic river-name + OE *tūn*.

Eyam Derbys. *Aiune* 1086 (DB). '(Place at) the islands, or the pieces of land between streams'. OE *ēg* in a dative plural form *ēgum*.

Eydon Northants. *Egedone* 1086 (DB). 'Hill of a man called *Æga'. OE pers. name + *dūn*.

Eye, '(place at) the island, or well-watered land, or dry ground in marsh', OE *ēg*: **Eye** Herefs. *Eia c.*1175. **Eye** Peterb. *Ege* 10th cent. **Eye** Suffolk. *Eia* 1086 (DB).

Eyemouth Sc. Bord. *Aymouthe* 1250. 'Mouth of the Eye Water'. OE *mūtha*. The river-name means simply 'river' (OE *ēa*).

Eyke Suffolk. *Eik* 1185. '(Place at) the oak-tree'. OScand. *eik*.

Eynesbury Cambs. *Eanulfesbyrig c.*1000, *Einuluesberie* 1086 (DB). 'Stronghold of a man called Ēanwulf'. OE pers. name + OE *burh* (dative *byrig*).

Eynsford Kent. *Æinesford c.*960. 'Ford of a man called *Ægen'. OE pers. name + *ford*.

Eynsham Oxon. *Egenes homme* 864, *Eglesham* 1086 (DB). Possibly 'enclosure or river-meadow of a man called *Ægen'. OE pers. name + *hamm*.

Eype Dorset. *Yepe* 1365. 'Steep place'. OE *gēap.

Eythorne Kent. *Heagythethorne* 9th cent. 'Thorn-tree of a woman called *Hēahgȳth'. OE pers. name + *thorn*.

Eyton upon the Weald Moors Tel. & Wrek. *Etone* 1086 (DB), *Eyton super le Wildmore* 1344. 'Farmstead on dry ground in marsh, or on well-watered land'. OE *ēg* + *tūn*. The affix means 'in the wild moorland', from OE *wilde* + *mōr*.

Faccombe Hants. *Faccancumb* 863, *Facumbe* 1086 (DB). 'Valley of a man called *Facca'. OE pers. name + *cumb*.

Faceby N. Yorks. *Feizbi* 1086 (DB). 'Farmstead or village of a man called Feitr'. OScand. pers. name + *bý*.

Fad, Lough (*Loch Fada*) Donegal. 'Long lake'.

Fada, Loch Highland. 'Long loch'. Gaelic *loch* + *fada*.

Faddiley Ches. *Fadilee* c.1220. 'Woodland clearing of a man called *Fad(d)a'. OE pers. name + *lēah*.

Fadmoor N. Yorks. *Fademora* 1086 (DB). 'Moor of a man called *Fad(d)a'. OE pers. name + *mōr*.

Faes (*Féá*) Limerick. 'Wood'.

Faha (*Faiche*) Kerry, Waterford. 'Green'.

Fahamore (*Faiche Mhór*) Kerry. 'Big green'.

Fahan (*Fathain*) Donegal. *Fathunmurra* 1311. 'Burial place'.

Fahanasoodry (*Faiche na Súdaire*) Limerick. 'Green of the tanners'.

Faheeran (*Faiche Chairáin*) Offaly. 'Ciarán's green'.

Fahy (*Faiche*) Offaly. 'Green'.

Faiaflannan (*Faiche Flannáin*) Donegal. 'Flannan's green'.

Failsworth Oldham. *Fayleswrthe* 1212. Possibly 'enclosure with a special kind of fence'. OE **fēgels* + *worth*.

Fair Isle Shet. *Fridarey* (no date), *Fároy* 1350. The earlier name was reinterpreted as 'island of sheep'. OScand. *fár* + *ey*.

Fairburn N. Yorks. *Farenburne* c.1030, *Fareburne* 1086 (DB). 'Stream where ferns grow'. OE *fearn* + *burna*.

Fairfield Worcs. *Forfeld* 817. 'Open land where hogs are pastured'. OE *fōr* + *feld*.

Fairford Glos. *Fagranforda* 862, *Fareforde* 1086 (DB). 'Fair or clear ford'. OE *fæger* + *ford*.

Fairlie N. Ayr. Unexplained.

Fairlight E. Sussex. *Farleghe* c.1175. 'Woodland clearing where ferns grow'. OE *fearn* + *lēah*.

Fairstead Essex. *Fairstedam* 1086 (DB). 'Fair or pleasant place'. OE *fæger* + *stede*.

Fakenham, 'homestead of a man called *Facca', OE pers. name (genitive -*n*) + *hām*: **Fakenham** Norfolk. *Fachenham* 1086 (DB). **Fakenham Magna & Little Fakenham** Suffolk. *Fakenham* c.1060, *Fachenham*, *Litla Fachenham* 1086 (DB), *Fakenham Magna*, -*Parva* 1254. Affixes are OE *lytel*, Latin *magna* 'great' and *parva* 'small'.

Fala Midloth. *Faulawe* 1250. 'Variegated hill'. OE *fāg* + *hlāw*.

Falcarragh (*Fál Carrach*) Donegal. *Fál Carrach* 1835. 'Rough enclosure'.

Faldingworth Lincs. *Falding(e)urde* 1086 (DB). 'Enclosure for folding livestock'. OE **falding* + *worth*.

Falfield S. Glos. *Falefeld* 1227. 'Pale brown or fallow-coloured open land'. OE *fealu* + *feld*.

Fali, Y. See VALLEY.

Falkenham Suffolk. *Faltenham* 1086 (DB). Probably 'homestead of a man called *Falta'. OE pers. name (genitive -*n*) + *hām*.

Falkirk Falk. *Egglesbreth* 1065, *varia capella* 1166, *Varie Capelle* 1253, *Faukirke* 1298. '(Place with a) speckled church'. OE *fāg* + *cirice*. A 'speckled church' is one with mottled stone. The first three forms above, respectively Gaelic, Latin, and OFrench, have the same meaning.

Falkland Fife. *Falleland* 1128, *Falecklen* 1160. The first element is obscure.

Fallagloon (*Fáladh Lúan*) Derry. *Fallow Lowne* 1613. 'Enclosure of the lambs'.

Falleenadatha (*Faillín an Deatha*) Limerick. 'Little cliff of the smoke'.

Fallowfield Manch. *Fallufeld* 1317. Probably identical in origin with FALFIELD.

Falmer E. Sussex. *Falemere* 1086 (DB). Probably 'fallow-coloured pool'. OE *fealu* + *mere*.

Falmouth Cornwall. *Falemuth* 1235. 'Mouth of the River Fal'. River-name (of uncertain origin and meaning) + OE *mūtha*.

Fambridge, North & Fambridge, South Essex. *Fanbruge* 1086 (DB), *North Fambregg, Suthfambregg* 1291. 'Bridge by a fen or marsh', or perhaps rather 'causeway across a fen or marsh'. OE *fenn* + *brycg*.

Fangdale Beck N. Yorks. *Fangedala* 12th cent. 'Stream in the valley good for fishing'. OScand. *fang* + *dalr* + *bekkr*.

Fangfoss E. R. Yorks. *Frangefos* [*sic*] 1086 (DB), *Fangefosse* 12th cent. Possibly 'ditch used for fishing'. OScand. *fang* + OE **foss*.

Far Cotton Northants. *See* COTTON.

Far Sawrey Cumbria. *See* SAWREY.

Fara (island) Orkn. 'Island of sheep'. OScand. *fár* + *ey*.

Faraid Head Highland. Gaelic *An Fharaird* 'projecting headland'.

Farcet Cambs. *Faresheued* 10th cent. 'Bull's headland or hill'. OE *fearr* + *hēafod*.

Fardrum (*Fordroim*) Westmeath. 'Ridge top'.

Fareham Hants. *Fearnham* c.970, *Fernham* 1086 (DB). 'Homestead where ferns grow'. OE *fearn* + *hām*.

Farewell Staffs. *Fagerwell* 1200. 'Pleasant spring or stream'. OE *fæger* + *wella*.

Fargrim (*Fordroim*) Fermanagh, Leitrim. 'Ridge top'.

Faringdon, Farringdon, 'fern-covered hill', OE *fearn* + *dūn*: **Faringdon** Oxon. *Færndunæ* c.971, *Ferendone* 1086 (DB). **Farringdon** Devon. *Ferhendone* 1086 (DB). **Farringdon** Hants. *Ferendone* 1086 (DB).

Farington Lancs. *Farinton* 1149. 'Farmstead where ferns grow'. OE *fearn* + *tūn*.

Farleigh, 'woodland clearing growing with ferns', OE *fearn* + *lēah*: **Farleigh** Gtr. London. *Ferlega* 1086 (DB). **Farleigh, East & Farleigh, West** Kent. *Fearnlege* 9th cent., *Ferlaga* 1086 (DB). **Farleigh Hungerford** Somerset. *Fearnlæh* 987, *Ferlege* 1086 (DB), *Farlegh Hungerford* 1404. Affix from the *Hungerford* family, here in the 14th cent. **Farleigh, Monkton** Wilts. *Farnleghe* 1001, *Farlege* 1086 (DB), *Monekenefarlegh* 1321. Affix means 'of the monks' from OE *munuc*, alluding to a priory founded here in 1125. **Farleigh Wallop** Hants. *Ferlege* 1086 (DB). Affix from the *Wallop* family, here in the 14th cent.

Farlesthorpe Lincs. *Farlestorp* 1190. 'Outlying farmstead or hamlet of a man called Faraldr'. OScand. pers. name + *thorp*.

Farleton Cumbria. *Farelton* 1086 (DB). 'Farmstead of a man called **Færela* or **Faraldr*'. OE or OScand. pers. name + OE *tūn*.

Farley, 'woodland clearing growing with ferns', OE *fearn* + *lēah*; examples include: **Farley** Staffs. *Fernelege* 1086 (DB). **Farley** Wilts. *Farlege* 1086 (DB). **Farley Hill** Wokhm. *Ferlega* 1167.

Farlington N. Yorks. *Ferlintun* 1086 (DB). Probably 'estate associated with a man called **Færela*'. OE pers. name + *-ing-* + *tūn*.

Farlow Shrops. *Ferlau* 1086 (DB), *Farnlawe* 1222. 'Fern-covered mound or tumulus'. OE *fearn* + *hlāw*.

Farmborough B. & NE. Som. *Fearnberngas* [*sic*] 901, *Ferenberge* 1086 (DB). 'Hill(s) or mound(s) growing with ferns'. OE *fearn* + *beorg*.

Farmcote Glos. *Fernecote* 1086 (DB). 'Cottage(s) among the ferns'. OE *fearn* + *cot*.

Farmington Glos. *Tormentone* [*sic*] 1086 (DB), *Tormerton* 1182. Probably 'farmstead near the pool where thorn-trees grow'. OE *thorn* + *mere* + *tūn*.

Farnagh (*Farnocht*) Westmeath. 'Bare hill'.

Farnaght (*Farnocht*) Leitrim. 'Bare hill'.

Farnalore (*Fearann an Lobhair*) Westmeath. 'Land of the leper'.

Farnanes (*Na Fearnáin*) Cork. 'The place producing alders'.

Farnaught (*Farnocht*) Sligo. 'Bare hill'.

Farnborough, 'hill(s) or mound(s) growing with ferns', OE *fearn* + *beorg*: **Farnborough** Gtr. London. *Ferenberga* 1180. **Farnborough** Hants. *Ferneberga* 1086 (DB). **Farnborough** Warwicks. *Feornebeorh* c.1015, *Ferneberge* 1086 (DB). **Farnborough** W. Berks. *Fearnbeorgan* c.935, *Fermeberge* 1086 (DB).

Farncombe Surrey. *Fernecome* 1086 (DB). 'Valley where ferns grow'. OE *fearn* + *cumb*.

Farndish Beds. *Fernadis, Fernedis* 1086 (DB). 'Enclosure or enclosed park where ferns grow'. OE *fearn* + *edisc*.

Farndon, 'hill growing with ferns', OE *fearn* + *dūn*: **Farndon** Ches. *Fearndune* 924, *Ferentone* [*sic*] 1086 (DB). **Farndon** Notts. *Farendune* 1086 (DB). **Farndon, East** Northants. *Ferendone* 1086 (DB).

Farne Islands Northum. *Farne* c.700, *Farneheland* 1257. From OIrish *ferann* 'domain' (probably with reference to LINDISFARNE), later with OE *ēaland* 'island'.

Farney (*Fearnmagh*) Monaghan. 'Alder plain'.

Farnham, 'homestead or enclosure where ferns grow', OE *fearn* + *hām* or *hamm*: **Farnham** Dorset. *Fernham* 1086 (DB). **Farnham** Essex. *Phernham* 1086 (DB). **Farnham** N. Yorks. *Farneham* 1086 (DB). **Farnham** Suffolk. *Farnham* 1086 (DB). **Farnham** Surrey. *Fernham* c.686, *Ferneham* 1086 (DB). The second element of this name is probably OE *hamm* in the sense 'river-meadow'. **Farnham Royal** Bucks. *Ferneham* 1086 (DB), *Fernham Riall* 1477. Affix (from OFrench *roial*) because it was held by the grand serjeanty of supporting the king's right arm at his coronation.

Farningham Kent. *Ferningeham* 1086 (DB). Possibly 'homestead of the dwellers among the ferns'. OE *fearn* + *-inga-* + *hām*.

Farnley, 'woodland clearing growing with ferns', OE *fearn* + *lēah*: **Farnley** Leeds. *Ferneleia* 1086 (DB). **Farnley** N. Yorks. *Fernleage* c.1030, *Fernelai* 1086 (DB). **Farnley Tyas** Kirkl. *Fereleia* [*sic*] 1086 (DB), *Farnley Tyas*

1322. Manorial affix from the family of *le Tyeis*, here in the 13th cent.

Farnsfield Notts. *Fearnesfeld* 958, *Farnesfeld* 1086 (DB). 'Open land where ferns grow'. OE *fearn* + *feld*.

Farnworth, 'enclosure where ferns grow', OE *fearn* + *worth*: **Farnworth** Bolton. *Farnewurd* 1185. **Farnworth** Halton. *Farneword* 1324.

Farra (*Farrach*) Armagh. 'Meeting-place'.

Farragh (*Farrach*) Cavan. 'Meeting-place'.

Farraghroe (*Farracha Ruadha*) Westmeath. 'Red meeting-places'.

Farran (*Fearann*) Cork, Kerry, Westmeath. 'Territory'.

Farranboley (*Fearann Buaile*) Dublin. 'Territory of the milking place'.

Farranfadda (*Fearann Fada*) Cork. 'Long territory'.

Farranfore (*Fearann Fuar*) Kerry. 'Cold land'.

Farranmacbride (*Fearann Mhic Bhride*) Donegal. 'Mac Bride's land'.

Farringdon. *See* FARINGDON.

Farrington Gurney B. & NE. Som. *Ferentone* 1086 (DB). 'Farmstead where ferns grow'. OE *fearn* + *tūn*. Manorial affix from the *de Gurnay* family, here in the 13th cent.

Farsetmore (*Fearsad Mór*) Donegal. 'Large sandbank'.

Farsid (*Fearsad*) Cork. 'Sandbank'.

Farsley Leeds. *Ferselleia* 1086 (DB). Possibly 'clearing used for heifers'. OE **fers* + *lēah*.

Farta (*Fearta*) Galway. 'Grave mounds'.

Fartagh (*Feartach*) Cavan, Fermanagh. 'Place of grave mounds'.

Farthinghoe Northants. *Ferningeho* 1086 (DB). Probably 'hill-spur of the dwellers among the ferns'. OE *fearn* +*-inga-* + *hōh*.

Farthingstone Northants. *Fordinestone* 1086 (DB). Probably 'farmstead of a man called Farthegn'. OScand. pers. name + OE *tūn*.

Farway Devon. *Farewei* 1086 (DB). Possibly 'ford or stream crossing way'. OE *fær* + *weg*.

Alternatively the first element may be OE *fǽr* 'danger'.

Fasnakyle Highland. Gaelic *Fas na Coille*. 'Level place by a wood'. Gaelic *fas* + *coille*.

Fauldhouse W. Loth. *Fawlhous* 1523, *Falhous c.*1540, *Faldhous* 1559. 'House on fallow land'. OE *falh* + *hūs*.

Faulkbourne Essex. *Falcheburna* 1086 (DB). 'Stream frequented by falcons'. OE **falca* + *burna*.

Faulkland Somerset. *Fouklande* 1243. 'Folk-land', i.e. 'land held according to folk-right'. OE *folc-land*.

Fauls Shrops. *Le Faall* 1301, *le Falles* 1363. 'The place(s) where trees were felled, the forest clearing(s)'. OE *(ge)fall*.

Faversham Kent. *Fefresham* 811, *Faversham* 1086 (DB). 'Homestead or village of the smith'. OE **fæfer* + *hām*.

Fawkham Green Kent. *Fealcnaham* 10th cent., *Fachesham* [*sic*] 1086 (DB). 'Homestead or village of a man called **Fealcna*'. OE pers. name + *hām*.

Fawler Oxon. *Fauflor* 1205. 'Variegated floor', i.e. 'tessellated pavement'. OE *fāg* + *flōr*.

Fawley Bucks. *Falelie* 1086 (DB). 'Fallow-coloured woodland clearing', or 'clearing with ploughed land'. OE *fealu* or *fealg* + *lēah*.

Fawley Hants. *Falegia, Falelei* 1086 (DB). Probably identical in origin with the previous name.

Fawley W. Berks. *Faleslei* 1086 (DB). Probably 'wood frequented by the fallow deer'. OE *fealu* (used as noun) + *lēah*.

Fawley Chapel Herefs. *Falileiam* 1142. 'Woodland clearing where hay is made'. OE *fælethe* + *lēah*.

Faxfleet E. R. Yorks. *Faxflete* 1190. 'Stream of a man called Faxi', or 'stream near which coarse grass grows'. OScand. pers. name or OE *feax* + *flēot*.

Fazakerley Lpool. *Fasacre, Fasacrelegh* 1325. 'Woodland clearing by the border acre'. OE *fæs* 'fringe, border' + *æcer* 'cultivated land' + *lēah*.

Fazeley Staffs. *Faresleia c.*1142. 'Clearing used for bulls'. OE *fearr* + *lēah*.

Feagarrid (*Féith Ghairid*) Waterford. 'Short stream'.

Feakle (*Fiacall*) Clare. 'Tooth'.

Feale (*Feil*) (river) Kerry, Limerick. 'Fial's river'.

Fearby N. Yorks. *Federbi* 1086 (DB). OScand. *bý* 'farmstead, village' with a doubtful first element, possibly OE *fether*, OScand. *fjǫthr* 'feather' (perhaps referring to a place frequented by flocks of birds).

Fearn Highland. *Ferne* 1529. '(Place of) alders'. Gaelic *fearna*. The Gaelic name is *Manachainn Rois* 'the monastery of Ross'.

Fearn Hill (*Fearn*) Donegal. 'Alder hill'.

Fearnhead Warrtn. *Ferneheued* 1292. 'Fern-covered hill'. OE *fearn* + *hēafod*.

Featherstone, '(place at) the four stones, i.e. a tetralith', OE *feother-* + *stān*: **Featherstone** Staffs. *Feother(e)stan* 10th cent., *Ferdestan* 1086 (DB). **Featherstone** Wakefd. *Fredestan* 1086 (DB).

Feckenham Worcs. *Feccanhom* 804, *Fecheham* 1086 (DB). 'Enclosure or water-meadow of a man called **Fecca*'. OE pers. name + *hamm*.

Fedamore (*Feadamair*) Limerick. 'Place of streams'.

Fedany (*Feadanach*) Down. 'Place of streams'.

Feeagh, Lough (*Loch Fíoch*) Mayo. 'Wooded lake'.

Feebane (*Fioda Bán*) Monaghan. 'White wood'.

Feenagh (*Fíonach*) Limerick. 'Wooded place'.

Feenish (*Fiodh-Inis*) Clare. 'Woody island'.

Feeny (*Na Fíneadha*) Derry. *Nefenne* 1613. 'The woods'.

Feering Essex. *Feringas* 1086 (DB). '(Settlement of) the family or followers of a man called **Fēra*'. OE pers. name + *-ingas*.

Feetham N. Yorks. *Fytun* 1242. '(Place at) the riverside meadows'. OScand. *fit* in a dative plural form *fitjum*.

Felbridge Surrey. *Feltbruge* 12th cent. 'Bridge by the open land'. OE *feld* + *brycg*.

Felbrigg Norfolk. *Felebruge* 1086 (DB). 'Bridge made of planks'. OScand. *fjǫl* + OE *brycg*.

Felcourt Surrey. *Feldecote* 1403. 'Cottage(s) by the open land'. OE *feld* + *cot*.

Felinheli, Y. See PORT DINORWIC.

Felixkirk N. Yorks. *Felicekirke* 13th cent. 'Church dedicated to St Felix'. Saint's name + OScand. *kirkja*.

Felixstowe Suffolk. *Filchestou* 1254. Probably 'holy place or meeting-place of a man called *Filica'. OE pers. name + *stōw*. The pers. name is probably an anglicization of (St) Felix, first Bishop of East Anglia, with whom of course the place is associated.

Felling Gatesd. *Fellyng* c.1220. OE *felling* 'woodland clearing' or *felging* 'fallow land'.

Felmersham Beds. *Falmeresham* 1086 (DB). 'Homestead or enclosure by a fallow-coloured pool, or of a man called *Feolomǽr'. OE *fealu* + *mere* or OE pers. name + *hām* or *hamm*.

Felmingham Norfolk. *Felmincham* 1086 (DB). 'Homestead of the family or followers of a man called *Feolma'. OE pers. name + *-inga-* + *hām*.

Felpham W. Sussex. *Felhhamm* c.880, *Falcheham* 1086 (DB), 'Enclosure with fallow land'. OE *felh* + *hamm*.

Felsham Suffolk. *Fealsham* 1086 (DB). 'Homestead or village of a man called *Fǽle'. OE pers. name + *hām*.

Felsted Essex. *Felstede* 1086 (DB). 'Open-land place'. OE *feld* + *stede*.

Feltham Gtr. London. *Feltham* 969, *Felteham* 1086 (DB). Probably 'homestead or enclosure where mullein or a similar plant grows'. OE *felte* + *hām* or *hamm*. Alternatively the first element may be OE *feld* 'open land'.

Felthorpe Norfolk. *Felethorp* 1086 (DB). Probably 'outlying farmstead or hamlet of a man called *Fǽla'. OE pers. name + OScand. *thorp*.

Felton, usually 'farmstead or village in open country', OE *feld* + *tūn*; examples include: **Felton** Herefs. *Felton* 1086 (DB). **Felton** Northum. *Feltona* 1167. **Felton Butler** Shrops. *Feltone* 1086 (DB), *Felton Butiler* 13th cent. Manorial affix from the *Buteler* family, here in the 12th cent. **Felton, West** Shrops. *Feltone* 1086 (DB).

Feltrim (*Fealdruim*) Down. 'Wolf ridge'.

Feltwell Norfolk. *Feltuuella* 1086 (DB). Probably 'spring or stream where mullein or a similar plant grows'. OE *felte* + *wella*.

Fen Ditton Cambs. See DITTON.

Fen Drayton Cambs. See DRAYTON.

Fenagh (*Fiodhnach*) Leitrim. 'Wooded place'.

Fenby, Ashby cum NE. Lincs. See ASHBY.

Fenham Northum. near Fenwick. *Fennum* c.1085. '(Place in) the fens'. OE *fenn* in a dative plural form *fennum*.

Fenit (*An Fhianait*) Kerry. 'The wild place'.

Feniton Devon. *Finetone* 1086 (DB). 'Farmstead by Vine Water'. Celtic river-name (meaning 'boundary stream') + OE *tūn*.

Fenland (district) Cambs. A modern name, referring to the local topography.

Fennagh (*Fionnmhach*) Carlow. 'White plain'.

Fenni, Y. See ABERGAVENNY.

Fennor (*Fionnúir*) Waterford. 'Place by white water'.

Fenny as affix. See main name, e.g. for **Fenny Bentley** (Derbys.) see BENTLEY.

Fenrother Northum. *Finrode* 1189. 'Clearing by a mound or heap'. OE *fin* + *rother*.

Fenstanton Cambs. *Stantun* 1012, *Stantone* 1086 (DB), *Fenstanton* 1260. 'Farmstead on stony ground in a marshy district'. OE *fenn* + *stān* + *tūn*.

Fenton, 'farmstead or village in a fen or marshland', OE *fenn* + *tūn*; examples include: **Fenton** Cambs. *Fentun* 1236. **Fenton** Lincs., near Claypole. *Fentun* 1212. **Fenton** Lincs., near Kettlethorpe. *Fentuna* c.1115. **Fenton** Stoke. *Fentone* 1086 (DB). **Fenton, Church & Fenton, Little** N. Yorks. *Fentune* 963, *Fentun* 1086 (DB), *Kirkfenton* 1338. Affix from OE *cirice*, OScand. *kirkja*. **Fenton Town** Northum. *Fenton* 1242.

Fenwick, 'dwelling or (dairy) farm in a fen or marsh', OE *fenn* + *wīc*: **Fenwick** Donc. *Fenwic*

1166. Fenwick Northum., near Kyloe. *Fenwic* 1208. **Fenwick** Northum., near Stamfordham. *Fenwic* 1242.

Fenwick E. Ayr. 'Dairy farm by a bog'. Probably OE *fenn* + *wīc*.

Feock Cornwall. *Lanfioc* 12th cent. 'Church of St Fioc'. From the patron saint of the church, with Cornish **lann* in the early form.

Feohanagh (*An Fheothanach*) Limerick. 'The windy place'.

Ferbane (*An Féar Bán*) Offaly. 'The white grass'.

Fermanagh (*Fear Manach*) (the county). (*tigherna*) *Fermanach* 1010. '(Place of the) men of the Manaigh tribe'. The tribe took their name from their chief.

Fermoy (*Mainistir Fhear Maighe*) Cork. 'Monastery of the district of Fir Mhaí'. The district name means 'men of the plain'.

Fermoyle (*Formael*) Kerry. 'Round hill'.

Fern Ang. '(Place of) alders'. Gaelic *fearn*.

Fern, Lough (*Loch Fearna*) Donegal. 'Lake of alders'.

Ferndown Dorset. *Fyrne* 1321. OE *fergen* 'wooded hill' or **fierne* 'ferny place', with the later addition of *dūn* 'down, hill'.

Fernham Oxon. *Fernham* 9th cent. 'River-meadow where ferns grow'. OE *fearn* + *hamm*.

Fernhurst W. Sussex. *Fernherst* c.1200. 'Fern-covered wooded hill'. OE *fearn* + *hyrst*.

Fernilee Derbys. *Ferneley* 12th cent. 'Woodland clearing where ferns grow'. OE *fearn* + *lēah*.

Ferns (*Fearna*) Wexford. 'Place of alders'.

Ferrensby N. Yorks. *Feresbi* [*sic*] 1086 (DB), *Feringesby* 13th cent. Probably 'farmstead or village of the man from the Faroe Islands'. OScand. *færeyingr* + *bý*.

Ferriby, North (E. R. Yorks.) **& Ferriby, South** (N. Lincs.) *Ferebi* 1086 (DB), *North Feribi* 1284, *Suthferebi* c.1130. 'Farmstead or village near the ferry'. OScand. *ferja* + *bý*. *North* and *South* with reference to their situation on opposite banks of the Humber.

Ferring W. Sussex. *Ferring* 765, *Feringes* 1086 (DB). Probably '(settlement of) the family or

followers of a man called **Fēra*'. OE pers. name + *-ingas*.

Ferrybridge Wakefd. *Ferie* 1086 (DB), *Ferybrig* 1198. 'Bridge by the ferry'. OScand. *ferja* + OE *brycg*. The first bridge here across the River Aire was built in the late 12th cent.

Ferryhill Durham. *Feregenne* 10th cent., *Ferye on the Hill* 1316. OE *fergen* 'wooded hill', with the later addition of *hyll* 'hill'.

Fersfield Norfolk. *Fersafeld* c.1035, *Ferseuella* [*sic*] 1086 (DB). Possibly 'open land where heifers graze'. OE **fers* + *feld*.

Ferta (*Fearta*) Kerry. 'Grave mounds'.

Fertagh (*Feartach*) Leitrim, Meath. 'Place of graves'.

Fetcham Surrey. *Fecham* 10th cent., *Feceham* 1086 (DB). Probably 'homestead or village of a man called **Fecca*'. OE pers. name + *hām*.

Fethard (*Fiodh Ard*) Tipperary, Wexford. 'High wood'.

Fetlar (island) Shet. *Fetilár* c.1250. Pre-Norse island name of unknown meaning.

Fettercairn Aber. *Fotherkern* c.970, *Fettercairn* c.1350. 'Slope by a thicket'. Gaelic *foithir* + Pictish *carden*.

Fews (*Feá*) Armagh, Waterford. 'Woods'.

Fewston N. Yorks. *Fostune* 1086 (DB). 'Farmstead of a man called Fótr'. OScand. pers. name + OE *tūn*.

Ffestiniog. See BLAENAU FFESTINIOG.

Fflint, Y. See FLINT.

Ffontygari. See FONT-Y-GARY.

Fforest Fawr (district) Powys. 'Great forest'. Welsh *fforest* + *mawr*.

Fiddington Glos. *Fittingtun* 1004, *Fitentone* 1086 (DB). 'Estate associated with a man called **Fita*'. OE pers. name + *-ing-* + *tūn*.

Fiddington Somerset. *Fitintone* 1086 (DB). Probably identical in origin with the previous name.

Fiddleford Dorset. *Fitelford* 1244. 'Ford of a man called Fitela'. OE pers. name + *ford*.

Fiddown (*Fiodh Dúin*) Kilkenny. 'Wood of the fort'.

Field Staffs. *Felda* 1130. '(Place at) the open land'. OE *feld*.

Field Dalling Norfolk. *See* DALLING.

Fieries (*Foidhrí*) Kerry. 'Slopes'.

Fife (the unitary authority). *Fib* c.1150, *Fif* 1165. '(Territory of) Fib'. *Fib* was one of the seven sons of Cruithe, legendary father of the Picts. But the pers. name dates later than the territory associated with it, so some earlier name must be involved.

Fifehead Magdalen & Fifehead Neville Dorset. *Fifhide* 1086 (DB), *Fifyde Maudaleyne* 1388, *Fyfhud Neuyle* 1287. '(Estate of) five hides of land'. OE *fíf* + *híd*. Affix *Magdalen* from the dedication of the church, *Neville* from the *de Nevill* family, here in the 13th cent.

Fifield, '(estate of) five hides of land', OE *fíf* + *híd*: **Fifield** Oxon. *Fifhide* 1086 (DB). **Fifield** Winds. & Maid. *Fifhide* 1316. **Fifield Bavant** Wilts. *Fifhide* 1086 (DB), *Fiffehyde Beaufaunt* 1436. Manorial affix from the *de Bavent* family, here in the 14th cent.

Figheldean Wilts. *Fisgledene* [sic] 1086 (DB), *Figelden* 1227. 'Valley of a man called *Fygla*'. OE pers. name + *denu*.

Figile (*Abhainn Fhiodh Gaibhle*) (river) Offaly. 'River of the wood of the fork'.

Filby Norfolk. *Filebey* 1086 (DB). 'Farmstead or village of a man called *Fili* or Fila'. OScand. or OE pers. name + *bý*. Alternatively the first element may be OScand. *fíli* 'planks' (perhaps referring to a bridge or other structure).

Filey N. Yorks. *Fiuelac* [sic] 1086 (DB), *Fivelai* 12th cent. Possibly 'promontory shaped like a sea monster'. OE *fífel* + *ēg*. The allusion would be to Filey Brigg (OScand. *bryggja* 'jetty'), a ridge of rock, half a mile long, projecting into the sea. Alternatively 'the five clearings', from OE *fíf* + *lēah*.

Filgrave Milt. K. *Filegrave* 1241. 'Pit or grove of a man called *Fygla*'. OE pers. name + *grǣf* or *gráf*.

Filkins Oxon. *Filching* 12th cent. Probably '(settlement of) the family of followers of a man called *Filica*'. OE pers. name + *-ingas*.

Filleigh Devon. near Barnstaple. *Filelei* 1086 (DB). Probably 'woodland clearing where hay is made'. OE *filethe* + *lēah*.

Fillingham Lincs. *Figelingeham* 1086 (DB). 'Homestead of the family or followers of a man called *Fygla*'. OE pers. name + *-inga-* + *hām*.

Fillongley Warwicks. *Filingelei* 1086 (DB). 'Woodland clearing of the family or followers of a man called *Fygla*'. OE pers. name + *-inga-* + *lēah*.

Filton S. Glos. *Filton* 1187. 'Farm or estate where hay is made'. OE *filethe* + *tūn*.

Fimber E. R. Yorks. *Fym(m)ara* 12th cent. 'Pool by a wood pile, or amidst the coarse grass'. OE *fín* or *finn* + *mere*.

Fin, Lough (*Loch Fionn*) Clare. 'White lake'.

Finaghy (*An Fionnachadh*) Antrim. *Ballyfinnaghey* 1780. 'The white field'.

Finborough Suffolk. *Fineberga* 1086 (DB). 'Hill or mound frequented by woodpeckers'. OE *fína* + *beorg*.

Fincham Norfolk. *P(h)incham* 1086 (DB). 'Homestead or enclosure frequented by finches'. OE *finc* + *hām* or *hamm*.

Finchampstead Wokhm. *Finchamestede* 1086 (DB). 'Homestead frequented by finches'. OE *finc* + *hām-stede*.

Finchdean Hants. *Finchesdene* 1167. 'Valley of the finch, or of a man called Finc'. OE *finc* or OE pers. name (from the same word) + *denu*.

Finchingfield Essex. *Fincingefelda* 1086 (DB). 'Open land of the family or followers of a man called Finc'. OE pers. name + *-inga-* + *feld*.

Finchley Gtr. London. *Fincheleе* c.1208. 'Woodland clearing frequented by finches'. OE *finc* + *lēah*.

Findern Derbys. *Findre* 1086 (DB). An obscure name, still not satisfactorily explained.

Findhorn Moray. '(Place on the river) Findhorn'. Gaelic *fionn* 'white' + a pre-Celtic river-name; *compare* DEVERON.

Findlater Aber. *Finletter* 1266. 'White slope'. Gaelic *fionn* + *leitir*.

Findon W. Sussex. *Findune* 1086 (DB). 'Hill with a heap or mound', or 'heap-shaped hill'. OE *fín* + *dūn*. Alternatively the first element may be OE *finn* 'coarse grass'.

Findrum (*Fionn Droim*) Donegal, Tyrone. 'White ridge'.

Finea (*Fiodh an Átha*) Westmeath. 'Wood of the ford'.

Finedon Northants. *Tingdene* 1086 (DB). 'Valley where assemblies meet'. OE *thing* + *denu*.

Fingal's Cave Arg. (Staffa). *Fingal* is the legendary Irish giant Fionn Mac Cumhail. The Gaelic name of the cave is *An Uamh Binn*, 'the melodious cave', referring to the eerie sound made by the sea among the basalt pillars.

Fingest Bucks. *Tingeherst* 12th cent. 'Wooded hill where assemblies are held'. OE *thing* + *hyrst*.

Finghall N. Yorks. *Finegala* [*sic*] 1086 (DB), *Finyngale* 1157. Probably 'nook of land of the family or followers of a man called *Fin(a)*'. OE pers. name + *-inga-* + *halh*.

Finglas (*Fionnghlas*) Dublin. 'Bright stream'.

Finglesham Kent. *Thenglesham* 832, *Flengvessam* [*sic*] 1086 (DB). 'Homestead or village of the prince, or of a man called *Thengel*'. OE *thengel* or OE pers. name + *hām*.

Fingringhoe Essex. *Fingringaho* 10th cent. Possibly 'hill-spur of the dwellers on the finger of land'. OE *finger* +*-inga-* + *hōh*.

Finisk (*Fionnuisce*) (river) Waterford. 'White water'.

Finmere Oxon. *Finemere* 1086 (DB). 'Pool frequented by woodpeckers'. OE *fīna* + *mere*.

Finn, Lough (*Loch Finne*) Donegal. 'White lake'.

Finnart Perth. *Fynnard* 1350. 'White or holy headland'. Gaelic *fionn* + *àrd*.

Finnea (*Fiodh an Átha*) Cavan, Westmeath. 'Wood of the ford'.

Finningham Suffolk. *Finingaham* 1086 (DB). 'Homestead of the family or followers of a man called *Fīn(a)*'. OE pers. name + *-inga-* + *hām*.

Finningley Donc. *Feniglei* [*sic*] 1086 (DB), *Feningelay* 1175. 'Woodland clearing of the dwellers in the fen'. OE *fenn* + *-inga-* + *lēah*.

Finny (*Fionnaithe*) Mayo. 'White kiln'.

Finsbury Gtr. London. *Finesbire* 1235, *Finesbury* 1254. 'Manor of a man called Finn'. OScand. pers. name + ME *bury* (from OE *byrig*, dative of *burh*).

Finsthwaite Cumbria. *Fynnesthwayt* 1336. 'Clearing of a man called Finn'. OScand. pers. name + *thveit*.

Finstock Oxon. *Finestochia* 12th cent. 'Outlying farmstead frequented by woodpeckers'. OE *fīna* + *stoc*.

Fintona (*Fionntamhnach*) Tyrone. (*hi f*) *fionntamhnach* 1488. 'White clearing'.

Fintown (*Baile na Finne*) Donegal. *Fintown* 1835. 'Homestead of the (river) Finn'.

Fintra (*Fionn traigh*) Clare. 'White strand'.

Fintragh (*Fionn traigh*) Donegal. 'White strand'.

Finvoy (*An Fhionnbhoith*) Antrim. *Le Fynmaugh* 1333. 'The white hut'.

Firbeck Rothm. *Fritebec* 12th cent. 'Woodland stream'. OE *fyrhth* + OScand. *bekkr*.

Firle, West E. Sussex. *Ferle* 1086 (DB). OE *fierel* 'oak place' or Latin *ferālia* 'wilds'.

Firsby Lincs. near Spilsby. *Frisebi* 1202. 'Farmstead or village of the Frisians'. OScand. *Frísir* (genitive *Frísa*) + *bý*.

Firth of Forth (sea inlet) Fife, E. Loth. *Forthin* c.970. 'Estuary of the River Forth'. ME *firth*. The Celtic river-name probably means 'the slow-flowing one'.

Fishbourne, Fishburn, 'fish stream, stream where fish are caught', OE *fisc* + *burna*: **Fishbourne** I. of Wight. *Fisseburne* 1267. **Fishbourne** W. Sussex. *Fiseborne* 1086 (DB). **Fishburn** Durham. *Fisseburn* c.1183.

Fishguard (*Abergwaun*) Pemb. *Fissigart* 1200, *Fissegard, id est, Aber gweun* 1210. 'Fish yard'. OScand. *fiskr* + *garthr*. A 'fish yard' is an enclosure for catching fish or for keeping them in when caught. The Welsh name means 'mouth of the River Gwaun' (OWelsh *aber*), the river-name meaning 'marsh' (Welsh *gwaun*).

Fishlake Donc. *Fiscelac* 1086 (DB). 'Fish stream'. OE *fisc* + *lacu*.

Fishtoft Lincs. *Toft* 1086 (DB), *Fishtoft* 1416. OScand. *toft* 'building site, curtilage'. The addition *Fish-* (from 15th cent.) may be a surname or indicate a connection with fishing.

Fiskerton, 'farmstead or village of the fishermen', OE *fiscere* (replaced by OScand. *fiskari*) + *tūn*: **Fiskerton** Lincs. *Fiskertuna*

1060, *Fiscartone* 1086 (DB). **Fiskerton** Notts. *Fiscertune* 956, *Fiscartune* 1086 (DB).

Fittleton Wilts. *Viteletone* 1086 (DB). 'Farmstead of a man called Fitela'. OE pers. name + *tūn*.

Fittleworth W. Sussex. *Fitelwurtha* 1168. 'Enclosure of a man called Fitela'. OE pers. name + *worth*.

Fitz Shrops. *Witesot* [sic] 1086 (DB), *Fittesho* 1194. 'Hill-spur of a man called *Fitt*'. OE pers. name + *hōh*.

Fitzhead Somerset. *Fifhida* 1065. '(Estate of) five hides'. OE *fīf* + *hīd*.

Five Ashes E. Sussex. Recorded thus *c.*1512, '(place at) the five ash-trees'. OE *fīf* + *æsc*.

Fivehead Somerset. *Fifhide* 1086 (DB). Identical in origin with FITZHEAD.

Fivemiletown Tyrone. *Fivemiletown* 1831. The town is said to be five miles from Clogher, Brookeborough, and Tempo respectively. Its original Irish name was *Baile na Lorgan*, 'homestead of the long low ridge'.

Flackwell Heath Bucks. *Flakewelle* 1227. Possibly 'spring or stream of a man called *Flæcca*'. OE pers. name + *wella*.

Fladbury Worcs. *Fledanburg* late 7th cent., *Fledebirie* 1086 (DB). 'Stronghold or manor house of a woman called *Flǣde*'. OE pers. name + *burh* (dative *byrig*).

Flagg Derbys. *Flagun* 1086 (DB). Probably 'place where turfs are cut'. OScand. *flag* in a dative plural form *flagum*.

Flagstaff Armagh. The name is said to refer to a flag raised on a hill nearby as a guide to shipping. The Irish name of the hamlet was *Barr an Fheadáin*, 'top of the watercourse'.

Flamborough E. R. Yorks. *Flaneburg* 1086 (DB). 'Stronghold of a man called Fleinn'. OScand. pers. name + OE *burh*. Flamborough Head (from OE *hēafod* 'headland') is first recorded in the 14th cent.

Flamstead Herts. *Fleamstede* 990, *Flamestede* 1086 (DB). 'Place of refuge'. OE *flēam* + *stede*.

Flannan Isles W. Isles. 'Islands of St Flannan'.

Flansham W. Sussex. *Flennesham* 1220. OE *hām* 'homestead' or *hamm* 'enclosure' with an obscure first element, probably a pers. name.

Flasby N. Yorks. *Flatebi* 1086 (DB). 'Farmstead or village of a man called Flatr'. OScand. pers. name + *bý*.

Flash Staffs. *The Flasshe* 1586. 'The swampy pool'. ME *flasshe* (OScand. *flask*).

Flat Holm (island) Vale Glam. *Les Holmes* 1358, *Flotholm* 1375. 'Fleet island'. OScand. *floti* + *holmr*. The island was used as a Viking naval base.

Flaunden Herts. *Flawenden* 13th cent. Probably 'flagstone valley'. OE **flage* + *denu*.

Flawborough Notts. *Flodberge* [sic] 1086 (DB), *Flouberge* 12th cent. 'Hill with stones'. OE *flōh* + *beorg*.

Flawith N. Yorks. *Flathwayth c.*1190. Possibly 'ford of the female troll or witch'. OScand. *flagth* + *vath*. Alternatively the first element may be OScand. **flatha* 'flat meadow' or OE **fleathe* 'water-lily'.

Flaxby N. Yorks. *Flatesbi* 1086 (DB). 'Farmstead or village of a man called Flatr'. OScand. pers. name + *bý*.

Flaxley Glos. *Flaxlea* 1163. 'Clearing where flax is grown'. OE *fleax* + *lēah*.

Flaxton N. Yorks. *Flaxtune* 1086 (DB). 'Farmstead where flax is grown'. OE *fleax* + *tūn*.

Fleckney Leics. *Flechenie* 1086 (DB). Probably 'dry ground in marsh of a man called **Flecca*'. OE pers. name (genitive *-n*) + *ēg*.

Flecknoe Warwicks. *Flechenho* 1086 (DB). Possibly 'hill-spur of a man called **Flecca*'. OE pers. name (genitive *-n*) + *hōh*.

Fleet, '(place at) the stream, pool, or creek', OE *flēot*; examples include: **Fleet** Hants. *Flete* 1313. **Fleet** Lincs. *Fleot* 1086 (DB).

Fleetwood Lancs. a modern name, from Sir Peter *Fleetwood* who laid out the town in 1836.

Flempton Suffolk. *Flemingtuna* 1086 (DB). Possibly 'farmstead of the Flemings (people from Flanders)'. OE *Fleming* + *tūn*.

Fletching E. Sussex. *Flescinge(s)* 1086 (DB). '(Settlement of) the family or followers of a man called **Flecci*'. OE pers. name + *-ingas*.

Fletton, Old Peterb. *Fletun* 1086 (DB). 'Farmstead on a stream'. OE *flēot* + *tūn*.

Flexford Surrey. *Flexwere* 1317. 'The weir where flax grows'. OE *fleax* + *wer* (later confused with *ford*).

Flimby Cumbria. *Flemyngeby* 12th cent. 'Farmstead or village of the Flemings (people from Flanders)'. OScand. *Flǽmingr* + *bý*.

Flimwell E. Sussex. *Flimenwelle* 1210. 'The spring of the fugitives'. OE *flíema* (genitive plural *-ena*) + *wella*. On the county boundary, so the fugitives would have been from Kent.

Flint (*Y Fflint*) Flin. *Le Chaylou* 1277, *le Fflynt* 1300. '(Place of) hard rock'. ME *flint*. The name refers to the stone platform on which Flint Castle was built in 1277. The first form above is the French equivalent.

Flintham Notts. *Flintham* 1086 (DB). 'Homestead or enclosure of a man called *Flinta*. OE pers. name + *hām* or *hamm*.

Flinton E. R. Yorks. *Flintone* 1086 (DB). 'Farmstead where flints are found'. OE *flint* + *tūn*.

Flitcham Norfolk. *Flicham* 1086 (DB). 'Homestead or village where flitches of bacon are produced'. OE *flicce* + *hām*.

Flitton Beds. *Flittan* c.985, *Flictham* [*sic*] 1086 (DB). Obscure in origin and meaning.

Flitwick Beds. *Flicteuuiche* 1086 (DB). OE *wīc* 'dwelling, (dairy) farm' with an uncertain first element.

Flixborough N. Lincs. *Flichesburg* 1086 (DB). 'Stronghold of a man called Flík'. OScand. pers. name + OE *burh*.

Flixton, 'farmstead or village of a man called Flík', OScand. pers. name + OE *tūn*: **Flixton** N. Yorks. *Fleustone* [*sic*] 1086 (DB), *Flixtona* 12th cent. **Flixton** Suffolk, near Bungay. *Flixtuna* 1086 (DB). **Flixton** Traffd. *Flixton* 1177.

Flockton Kirkl. *Flochetone* 1086 (DB). 'Farmstead of a man called Flóki'. OScand. pers. name + OE *tūn*.

Flodden Northum. *Floddoun* 1517. Possibly 'hill with stones'. OE *flōh* + *dūn*.

Flookburgh Cumbria. *Flokeburg* 1246. Probably 'stronghold of a man called Flóki'. OScand. pers. name + OE *burh*.

Flordon Norfolk. *Florenduna* 1086 (DB). Probably 'hill with a floor or pavement'. OE *flōre* + *dūn*.

Flore Northants. *Flore* 1086 (DB). '(Place at) the floor', probably with reference to a lost tessellated pavement. OE *flōr(e)*.

Florence Court (*Mullach na Seangán*) Fermanagh. The mansion was originally owned by Lord Mount *Florence*, created Earl of Enniskillen in 1784. The Irish name means 'height of the ants'.

Flotta (island) Orkn. *Flotey* c.1250. 'Flat island'. OScand. *flatr* + *ey*.

Flotterton Northum. *Flotweyton* 12th cent. Possibly 'farmstead by the road liable to flood'. OE *flot* + *weg* + *tūn*.

Flowton Suffolk. *Flochetuna* 1086 (DB). 'Farmstead of a man called Flóki'. OScand. pers. name + OE *tūn*.

Flushing Cornwall. Recorded from 1698, so named after the port of Flushing (Vlissingen) in Holland, probably because it was founded by Dutch settlers.

Flyford Flavell Worcs. *Fleferth* 10th cent. Possibly OE *fyrhth* 'sparse woodland' with an uncertain first element. The affix *Flavell* is simply a Normanized form of Flyford, added to distinguish this place from GRAFTON FLYFORD.

Fobbing Thurr. *Phobinge* 1086 (DB). '(Settlement of) the family or followers of a man called *Fobba*, or '*Fobba's* place'. OE pers. name + *-ingas* or *-ing*.

Fochabers Moray. *Fochoper* 1124, *Fochabyr* 1238. First element obscure, but second may be Pictish *aber* 'confluence', with English *-s*.

Fockerby N. Lincs. *Fulcwardby* 12th cent. 'Farmstead or village of a man called Folcward'. OGerman pers. name + OScand. *bý*.

Fofannybane (*Fofannaigh Bhán*) Down. 'White thistle land'.

Fofannyreagh (*Fofannaigh Riabhach*) Down. 'Striped thistle land'.

Foggathorpe E. R. Yorks. *Fulcartorp* 1086 (DB). 'Outlying farmstead or hamlet of a man called Folcward'. OGerman pers. name + OScand. *thorp*.

Foilnaman (*Faill na mBan*) Tipperary. 'Cliff of the women'.

Foleshill Covtry. *Focheshelle* [*sic*] 1086 (DB), *Folkeshulla* 12th cent. 'Hill of the people', or 'hill of a man called *Folc'. OE *folc* or OE pers. name + *hyll*.

Folke Dorset. *Folk* 1244. '(Land held by) the people'. OE *folc*.

Folkestone Kent. *Folcanstan c.*697, *Fulchestan* 1086 (DB). Probably 'stone (marking a Hundred meeting-place) of a man called *Folca'. OE pers. name + *stān*.

Folkingham Lincs. *Folchingeham* 1086 (DB). 'Homestead of the family or followers of a man called *Folc(a)'. OE pers. name + *-inga-* + *hām*.

Folkington E. Sussex. *Fochintone* [*sic*] 1086 (DB), *Folkintone c.*1150. 'Estate associated with a man called *Folc(a)'. OE pers. name + *-ing-* + *tūn*.

Folksworth Cambs. *Folchesworde* 1086 (DB). 'Enclosure of a man called *Folc'. OE pers. name + *worth*.

Folkton N. Yorks. *Fulcheton* 1086 (DB). 'Farmstead of a man called Folki or *Folca'. OScand. or OE pers. name + *tūn*.

Follifoot N. Yorks. *Pholifet* 12th cent. '(Place of) the horse-fighting' (alluding to a Viking sport). OE *fola* + *feoht*.

Font-y-gary (*Ffontygari*) Vale Glam. *Fundygary* 1587. Origin uncertain. The first element may be English *font*, referring to Font-y-gary Well here.

Fonthill Bishop & Fonthill Gifford Wilts. *Funtial* 901, *Fontel* 1086 (DB), *Fontel Episcopi*, *Fontel Giffard* 1291. Probably Celtic, 'spring place', Brittonic *funt + adjectival suffix *-(*j*)*ol*. Manorial affixes from possession by the Bishop (Latin *episcopus*) of Winchester and the *Gifard* family at the time of Domesday.

Fontmell Magna & Fontmell Parva Dorset. *Funtemel* 877, *Fontemale* 1086 (DB), *Parva Funtemel* 1360, *Magnam Funtemell* 1391. Originally a Celtic river-name (meaning 'spring by the bare hill'). Distinguishing affixes are Latin *magna* 'great' and *parva* 'small'.

Foolow Derbys. *Foulowe* 1269. Probably 'hill frequented by birds'. OE *fugol* + *hlāw*.

Foots Cray Gtr. London. *See* CRAY.

Forcett N. Yorks. *Forset* 1086 (DB). Probably 'fold by a ford'. OE *ford* + *set*.

Ford, a common name, '(place by) the ford', from OE *ford*; examples include: **Ford** Northum. *Forda* 1224. **Ford** Shrops. *Forde* 1086 (DB). **Ford** W. Sussex. *Fordes c.*1194. In this name the form was originally plural.

Fordham, 'homestead or enclosure by a ford', OE *ford* + *hām* or *hamm*: **Fordham** Cambs. *Fordham* 10th cent., *Fordeham* 1086 (DB). **Fordham** Essex. *Fordeham* 1086 (DB). **Fordham** Norfolk. *Fordham* 1086 (DB).

Fordingbridge Hants. *Fordingebrige* 1086 (DB). 'Bridge of the people living by the ford'. OE *ford* + *-inga-* + *brycg*.

Fordon E. R. Yorks. *Fordun* 1086 (DB). '(Place) in front of the hill'. OE *fore* + *dūn*.

Fordwich Kent. *Fordeuuicum* 675, *Forewic* [*sic*] 1086 (DB). 'Trading settlement at the ford'. OE *ford* + *wīc*.

Fore ([*Baile*] *Fhobhair*) Westmeath. '(Townland of the) spring'.

Foreland, North Kent. *Forland* 1326. 'Promontory', OE *fore* + *land*. **South Foreland** near Dover has the same origin.

Foremark Derbys. *Fornewerche* 1086 (DB). 'Old fortification'. OScand. *forn* + *verk*.

Forest Heath (district) Suffolk. A modern name, referring to the local topography.

Forest Hill Oxon. *Fostel* [*sic*] 1086 (DB), *Forsthulle* 1122. 'Hill with a ridge'. OE *forst + *hyll*.

Forest Row E. Sussex. *Forstrowe* 1467. 'Row (of trees or houses) in ASHDOWN FOREST'. ME *forest* + *row*.

Forfar Ang. *Forfare c.*1200. Unexplained.

Forkill (*Foirceal*) Armagh. (*i*)*nOircel c.*1350. 'Hollow'.

Formal (*Formael*) Meath. 'Bare round hill'.

Formby Sefton. *Fornebei* 1086 (DB). 'Old farmstead', or 'farmstead of a man called Forni'. OScand. *forn* or OScand. pers. name + *bý*.

Formil (*Formael*) Tyrone. 'Round hill'.

Formweel (*Formael*) Galway. 'Round hill'.

Forncett St Mary & Forncett St Peter Norfolk. *Fornesseta* 1086 (DB). 'Dwelling or fold of a man called Forni'.

OScand. pers. name + OE (ge)set. Affixes from the dedications of the churches.

Fornham All Saints & Fornham St Martin Suffolk. *Fornham* 1086 (DB), *Fornham Omnium Sanctorum, Fornham Sancti Martini* 1254. 'Homestead or village where trout are caught'. OE *forne* + *hām*. Affixes from the church dedications.

Forres Moray. *Forais* c.1195. Gaelic *Farais* 'little shrubbery', from Gaelic *fo* 'under' + *ras* 'shrubbery'.

Forsbrook Staffs. *Fotesbroc* 1086 (DB). 'Brook of a man called Fótr'. OScand. pers. name + OE *brōc*.

Forston Dorset. *Fosardeston* 1236. 'Estate of the Forsard family'. ME *toun*. This family was here from the early 13th cent.

Fort Augustus Highland. The village grew up around the garrison enlarged by General Wade in the 1730s and named after William *Augustus*, Duke of Cumberland (1721–65). The Gaelic name is *Cill Chuimein* 'Cuimein's Church'.

Fort George Highland. The fortress was established after the 1745 Jacobite rising and named after *George* II (1683–1760). The Gaelic name is *An Gearasdan* 'the garrison'.

Fort William Highland. The fortress was originally built in 1655 then rebuilt in 1690 as a garrison and named after the reigning monarch, *William* III (1650–1702). The Gaelic name is *An Gearasdan, compare* FORT GEORGE.

Forteviot Perth. *Fothuirtabaicht* c.970. 'Slope of Tobacht'. Gaelic *fothair* 'slope'.

Forth. *See* FIRTH OF FORTH.

Forthampton Glos. *Forhelmentone* 1086 (DB). 'Estate associated with a man called Forthhelm'. OE pers. name + -*ing*- + *tūn*.

Forton, 'farmstead or village by a ford', OE *ford* + *tūn*: **Forton** Hants. *Forton* 1312. **Forton** Lancs. *Fortune* 1086 (DB). **Forton** Shrops. *Fordune* [*sic*] 1086 (DB), *Forton* 1240. **Forton** Staffs. *Forton* 1198.

Fortrose Highland. *Forterose* 1455. '(Place) beneath the headland'. Gaelic *foter* + *ros*. The Gaelic name is *A' Chananaich* 'the chanonry'.

Fortwilliam (*Dún Liam*) Antrim. 'William's fort'. A former fort here was perhaps named

after *William* de Burgh, Earl of Ulster, murdered nearby in 1333.

Fosbury Wilts. near Tidcombe. *Fostesberge* 1086 (DB). 'Stronghold of the ridge'. OE **forst* + *burh* (dative *byrig*).

Fosdyke Lincs. *Fotesdic* 1183. 'Ditch or water channel of a man called Fótr'. OScand. pers. name + *dík*.

Fosse Way (Roman road from Lincoln to Bath). *Foss* 8th cent. From OE **foss* 'ditch', so called from its having had a prominent ditch on either side.

Foston, 'farmstead or village of a man called Fótr', OScand. pers. name + OE *tūn*: **Foston** Derbys. *Fostun* 12th cent. Recorded as *Farulveston* 1086 (DB), 'farmstead of a man called Farulfr', OScand. pers. name + OE *tūn*. **Foston** Lincs. *Foztun* 1086 (DB). **Foston** N. Yorks. *Fostun* 1086 (DB). **Foston on the Wolds** E. R. Yorks. *Fodstone* 1086 (DB), *Foston on le Wolde* 1609. For the affix, *see* WOLDS.

Fotherby Lincs. *Fodrebi* 1086 (DB). 'Farmstead of a man called Fótr'. OScand. pers. name + *bý*.

Fotheringhay Northants. *Fodringeia* 1086 (DB). Probably 'island or well-watered land used for grazing'. OE **fōdring* + *ēg*.

Foula Shet. 'Bird island'. OScand. *fugl* + *ey*.

Foulden Norfolk. *Fugalduna* 1086 (DB). 'Hill frequented by birds'. OE *fugol* + *dūn*.

Foulness Essex. *Fughelnesse* 1215. 'Promontory frequented by birds'. OE *fugol* + *næss*.

Foulridge Lancs. *Folric* 1219. 'Ridge where foals graze'. OE *fola* + *hrycg*.

Foulsham Norfolk. *Folsam* 1086 (DB). 'Homestead of a man called Fugol'. OE pers. name + *hām*.

Fountains Abbey N. Yorks. *Fonteyns* 1275. 'The fountains or springs', from OFrench *fontein*, referring to the six springs within the 12th-cent. abbey site.

Four Marks Hants. *Fowremerkes* 1548. From OE *fēower* + *mearc* 'boundary', so called because the boundaries of four ancient parishes meet here.

Four Throws Kent. *Fourtrowes* 1790. 'The four trees'. OE *fēower* + *trēow*.

Fourstones Northum. *Fourstanys* 1236.
'Four stones', here describing a tetralith. OE
fēower + *stān*.

Fovant Wilts. *Fobbefunte* 901, *Febefonte* [*sic*]
1086 (DB). 'Spring of a man called *Fobba'. OE
pers. name + *funta*.

Foveran Aber. *Furene* c.1150. '(Place with a)
little spring'. Gaelic *fuaran*.

Fowey Cornwall. *Fawi* c.1223. Named from
the River Fowey, a Cornish name probably
meaning 'beech-tree river'.

Fowlmere Cambs. *Fuglemære* 1086 (DB).
'Mere or lake frequented by birds'. OE *fugol* +
mere.

Fownhope Herefs. *Hope* 1086 (DB),
Faghehope 1242. OE *hop* 'small enclosed valley'
with the later addition of *fāg* (dative *fāgan*)
'variegated, multi-coloured'.

Foxearth Essex. *Focsearde* 1086 (DB). 'The
fox's earth, the fox-hole'. OE *fox* + *eorthe*.

Foxham Wilts. *Foxham* 1065. 'Homestead or
enclosure where foxes are seen'. OE *fox* + *hām*
or *hamm*.

Foxholes N. Yorks. *Fox(o)hole* 1086 (DB).
'The fox-hole(s), the fox's earth(s)'. OE *fox-hol*.

Foxley, 'woodland clearing frequented by
foxes', OE *fox* + *lēah*: **Foxley** Norfolk. *Foxle*
1086 (DB). **Foxley** Wilts. *Foxelege* 1086 (DB).

Foxt Staffs. *Foxwiss* 1176. 'The fox's den'. OE
fox + *wist*.

Foxton, usually 'farmstead where foxes are
often seen', OE *fox* + *tūn*: **Foxton** Cambs.
Foxetune 1086 (DB). **Foxton** Leics. *Foxtone*
1086 (DB).
  However the following has a different second
element: **Foxton** Durham. *Foxedene* c.1170.
'Valley frequented by foxes'. OE *fox* + *denu*.

Foxwist Green Ches. *Foxwyste* 1475. 'The
fox's den'. OE *fox* + *wist*, with *Green* added from
the 19th cent.

Foy Herefs. *Lann Timoi* c.1150. '(Church of)
St Moi'. From the patron saint of the church,
originally with Welsh *llan* 'church'.

Foyers Highland. *Fyers* 1769. 'Terraced slope'.
Gaelic *foithir* + English plural -*s*.

Foyfin (*Faiche Fionn*) Donegal. 'White lawn'.

Foyle, Lough (*Loch Feabhail*) Donegal.
Possibly 'lake of the lip'.

Foynes (*Faing*) Limerick. Possibly 'raven'.

Fradswell Staffs. *Frodeswelle* 1086 (DB).
'Spring or stream of a man called Frōd'. OE
pers. name + *wella*.

Fraisthorpe E. R. Yorks. *Frestintorp* 1086
(DB). 'Outlying farmstead or hamlet of a man
called *Freistingr or *Freysteinn'. OScand. pers.
name + *thorp*.

Framfield E. Sussex. *Framelle* [*sic*] 1086 (DB),
Fremefeld 1257. 'Open land of a man called
*Frem(m)a or *Fremi'. OE pers. name + *feld*.

Framilode Glos. *Framilade* 1086 (DB).
'Difficult crossing over the River Frome', Celtic
river-name (meaning 'fair, fine') + OE *gelād*.

**Framingham Earl & Framingham
Pigot** Norfolk. *Framingaham* 1086 (DB),
Framelingham Comitis, *Framelingham Picot*
1254. 'Homestead of the family or followers of a
man called Fram'. OE pers. name + -*inga*- +
hām. Manorial affixes from early possession by
the Earl of Norfolk (Latin *comitis* 'of the earl')
and by the *Picot* family.

Framlingham Suffolk. *Fram(e)lingaham*
1086 'Homestead of the family or followers of a
man called *Framela'. OE pers. name + -*inga*- +
hām.

Frampton, usually 'farmstead or village on
the River Frome' (of which there are several),
Celtic river-name (meaning 'fair, fine') + OE
tūn: **Frampton** Dorset. *Frantone* 1086 (DB).
Frampton Cotterell S. Glos. *Frantone* 1086
(DB), *Frampton Cotell* 1257. Manorial affix from
the *Cotel* family, here in the 12th cent.
Frampton Mansell Glos. *Frantone* 1086
(DB), *Frompton Maunsel* 1368. Manorial affix
from the *Maunsel* family, here in the 13th cent.
Frampton on Severn Glos. *Frantone* 1086
(DB), *Fromton upon Severne* 1311.
  However the following has a different origin:
Frampton Lincs. *Franetone* 1086 (DB).
'Farmstead of a man called *Frameca or *Fráni'.
OE or OScand. pers. name + OE *tūn*.

Framsden Suffolk. *Framesdena* 1086 (DB).
'Valley of a man called Fram'. OE pers. name +
denu.

Framwellgate Moor Durham. *Framwelgat*
1352. 'Street by the strongly gushing spring'.
OE *fram* + *wella* + OScand. *gata*.

Franche Worcs. *Frenesse* 1086 (DB). 'Ash-tree of a man called *Frēa'. OE pers. name (genitive -*n*) + *æsc*.

Frankby Wirral. *Frankeby* 13th cent. 'Farm belonging to a Frenchman'. ME *Franke* + OScand. *bý*.

Frankley Worcs. *Franchelie* 1086 (DB). 'Woodland clearing of a man called Franca'. OE pers. name + *lēah*.

Frankton, 'farmstead or village of a man called Franca', OE pers. name + *tūn*: **Frankton** Warwicks. *Franchetone* 1086 (DB). **Frankton, English & Frankton, Welsh** Shrops. *Franchetone* 1086 (DB), *Englyshe Frankton, Welsch Francton* 1577. The distinguishing affixes are self-explanatory, Welsh Frankton being some five miles nearer to the Welsh border.

Fransham Norfolk. *Frandesham* 1086 (DB). OE *hām* 'homestead' or *hamm* 'enclosure' with a pers. name of uncertain form.

Frant E. Sussex. *Fyrnthan* 956. 'Place overgrown with fern or bracken'. OE **fiernthe*.

Fraserburgh Aber. *The toun and burghe of Faythlie, now callit Fraserburghe* 1597. 'Fraser's chartered town'. ME *burh*. Alexander *Fraser*, 7th Laird of Philorth, was granted a charter to raise the town of Faithlie into a burgh of barony here in 1546.

Frating Green Essex. *Fretinge* c.1060, *Fratinga* 1086 (DB). '(Settlement of) the family or followers of a man called *Frǣt(a)', or '*Frǣt(a)'s place'. OE pers. name + -*ingas* or -*ing*.

Fratton Portsm. *Frodin(c)gtune* 982, *Frodinton* 1086 (DB). 'Estate associated with a man called Frōda'. OE pers. name + -*ing*- + *tūn*.

Freaghduff (*Fraech Dubh*) Tipperary. 'Black heathery place'.

Freaghmore (*Fraech Mór*) Westmeath. 'Large heathery place'.

Freckenham Suffolk. *Frekeham* 895, *Frakenaham* 1086 (DB). 'Homestead or village of a man called *Freca'. OE pers. name (genitive -*n*) + *hām*.

Freckleton Lancs. *Frecheltun* 1086 (DB). Probably 'farmstead of a man called *Frecla'. OE pers. name + *tūn*.

Freeby Leics. *Fredebi* 1086 (DB). 'Farmstead or village of a man called Fræthi'. OScand. pers. name + *bý*.

Freeduff (*Fraech Dubh*) Armagh, Cavan. 'Black heathery place'.

Freethorpe Norfolk. *Frietorp* 1086 (DB). 'Outlying farmstead or hamlet of a man called Fræthi'. OScand. pers. name + *thorp*.

Freiston Lincs. near Boston. *Fristune* 1086 (DB). 'Farmstead or village of the Frisians'. OE *Frīsa* + *tūn*.

Fremington, 'estate associated with a man called *Fremi or *Frem(m)a', OE pers. name + -*ing*- + *tūn*: **Fremington** Devon. *Framintone* 1086 (DB). **Fremington** N. Yorks. *Fremington* 1086 (DB).

Frensham Surrey. *Fermesham* 10th cent. 'Homestead or village of a man called *Fremi'. OE pers. name + *hām*.

Freshfield Sefton. a recent name, apparently called after a man named *Fresh* who reclaimed the land after encroachment by sand.

Freshford B. & NE. Som. *Ferscesford* c.1000. 'Ford over the freshwater stream'. OE *fersc* 'fresh' (here used as a noun) + *ford*.

Freshford (*Achadh Úr*) Kilkenny. 'Fresh field'. The English name mistranslates the Irish, confusing *achadh*, 'field', with *áth*, 'ford'.

Freshwater I. of Wight. *Frescewatre* 1086 (DB). 'River with fresh water'. OE *fersc* + *wæter*.

Fressingfield Suffolk. *Fessefelda* [*sic*] 1086 (DB), *Frisingefeld* 1185. Possibly 'open land of the family or followers of a man called Frīsa ("the Frisian")'. OE pers. name + -*inga*- + *feld*.

Freston Suffolk. *Fresantun* 1000–2, *Fresetuna* 1086 (DB). 'Farmstead or village of the Frisian'. OE *Frēsa* (possibly used as a pers. name) + *tūn*.

Frettenham Norfolk. *Fretham* 1086 (DB). 'Homestead or village of a man called *Frǣta'. OE pers. name (genitive -*n*) + *hām*.

Fridaythorpe E. R. Yorks. *Fridagstorp* 1086 (DB). Probably 'outlying farmstead or hamlet of a man called *Frīgedæg'. OE pers. name + OScand. *thorp*.

Friern Barnet Gtr. London. *See* BARNET.

Friesthorpe Lincs. *Frisetorp* 1086 (DB). 'Outlying farmstead of the Frisians'. OScand. *Frísir* (genitive *Frísa*) + *thorp*.

Frilford Oxon. *Frieliford* 1086 (DB). 'Ford of a man called *Frithela*'. OE pers. name + *ford*.

Frilsham W. Berks. *Frilesham* 1086 (DB). 'Homestead or village of a man called *Frithel*'. OE pers. name + *hām*.

Frimley Surrey. *Fremle* 1203. 'Woodland clearing of a man called *Frem(m)a*'. OE pers. name + *lēah*.

Frindsbury Medway. *Freondesberiam* 764, *Frandesberie* 1086 (DB). 'Stronghold of a man called *Frēond*'. OE pers. name + *burh* (dative *byrig*).

Fring Norfolk. *Frainghes* 1086 (DB). Probably '(settlement of) the family or followers of a man called *Frēa*'. OE pers. name + *-ingas*.

Fringford Oxon. *Feringeford* 1086 (DB). Probably 'ford of the family or followers of a man called *Fēra*'. OE pers. name + *-inga-* + *ford*.

Frinsted Kent. *Fredenestede* 1086 (DB). 'Place of protection'. OE **frithen* + *stede*.

Frinton on Sea Essex. *Frientuna* 1086 (DB). 'Farmstead of a man called *Fritha*', or 'protected farmstead'. OE pers. name (genitive *-n*) or OE **frithen* + *tūn*.

Frisby on the Wreake Leics. *Frisebie* 1086 (DB). 'Farmstead or village of the Frisians'. OScand. *Frísir* (genitive *Frísa*) + *bý*. Wreake is an OScand. river-name meaning 'twisted, winding'.

Friskney Lincs. *Frischenei* 1086 (DB). 'River with fresh water'. OE *fersc* (dative *-an*, with Scand. *-sk-*) + *ēa*.

Friston E. Sussex. *Friston* 1200. Possibly 'farmstead of a man called *Frēo*'. OE pers. name + *tūn*.

Friston Suffolk. *Frisetuna* 1086 (DB). 'Farmstead or village of the Frisians'. OE *Frīsa* + *tūn*.

Fritham Hants. *Friham* 1212. Probably 'enclosure in sparse woodland'. OE *fyrhth* + *hamm*.

Frithelstock Devon. *Fredeletestoc* 1086 (DB). 'Outlying farmstead of a man called *Frithulāc*'. OE pers. name + *stoc*.

Frithville Lincs. *Le Frith* 1331. OE *fyrhth* 'sparse woodland' with the later addition of *-ville* (from French *ville* 'village, town').

Frittenden Kent. *Friththingden* 9th cent. 'Woodland pasture associated with a man called Frith'. OE pers. name + *-ing-* + *denn*.

Fritton, 'farmstead offering safety or protection', or 'farmstead of a man called Frithi', OE *frith* or OScand. pers. name + OE *tūn*: **Fritton** Norfolk, near Gorleston. *Fridetuna* 1086 (DB). **Fritton** Norfolk, near Morningthorpe. *Fridetuna* 1086 (DB).

Fritwell Oxon. *Fertwelle* 1086 (DB). Possibly 'spring used for divination'. OE *freht* + *wella*.

Frizington Cumbria. *Frisingaton* c.1160. Probably 'estate of the family or followers of a man called Frīsa ("the Frisian")'. OE pers. name + *-inga-* + *tūn*.

Frobost W. Isles. (South Uist). 'Fróthi's homestead'. OScand. pers. n. Fróthi + *bólstathr*.

Frocester Glos. *Frowecestre* 1086 (DB). 'Roman town on the River Frome'. Celtic river-name (meaning 'fair, fine') + OE *ceaster*.

Frodesley Shrops. *Frodeslege* 1086 (DB). 'Woodland clearing of a man called Frōd'. OE pers. name + *lēah*.

Frodingham, 'homestead of the family or followers of a man called Frōd(a)', OE pers. name + *-inga-* + *hām*: **Frodingham** N. Lincs., near Scunthorpe. *Frodingham* 12th cent. **Frodingham, North** E. R. Yorks. *Frotingham* 1086 (DB), *North Frothyngham* 1297.

Frodsham Ches. *Frotesham* 1086 (DB). 'Homestead or promontory of a man called Frōd'. OE pers. name + *hām* or *hamm*.

Froghanstown (*Baile Fraocháin*) Westmeath. 'Homestead of bilberries'.

Frogmore, usually 'pool frequented by frogs', OE *frogga* + *mere*; for example **Frogmore** Hants. *Frogmore* 1294.

Frome, named from the River Frome (of which there are several), a Celtic river-name meaning 'fair, fine, brisk': **Frome** Somerset. *Froom* 8th cent. **Frome, Bishop's, Frome, Canon, & Frome, Castle** Herefs. *Frome* 1086 (DB), *Frume al Evesk* 1252, *Froma Canonicorum, Froma Castri* 1242. Affixes 'Bishop's' (OFrench *eveske*) and 'Canon' (Latin

Canonicorum 'of the canons') refer to possession by the Bishop of Hereford and the canons of Lanthony in medieval times, 'Castle' refers to a Norman castle. **Frome St Quintin** Dorset. *Litelfrome* 1086 (DB), *Fromequintin* 1288. At first called 'little' (OE *lytel*) to distinguish it from other manors on the same river, later given a manorial affix from the *St Quintin* family, here in the 13th cent.

Frosses (*Na Frosa*) Antrim. *Frasses c.*1657. 'Marshy place' (literally 'the showers').

Frostenden Suffolk. *Froxedena* 1086 (DB). Probably 'valley frequented by frogs'. OE *frosc* + *denu*.

Frosterley Durham. *Forsterlegh* 1239. 'Woodland clearing or pasture of the forester'. ME *forester* + OE *lēah*.

Froxfield, 'open land frequented by frogs', OE *frosc* + *feld*: **Froxfield** Wilts. *Forscanfeld* 9th cent. **Froxfield Green** Hants. *Froxafelda* 10th cent.

Fryerning Essex. *Inga* 1086 (DB), *Friering* 1469. Originally '(settlement of) the people of the district'. OE **gē* + *-ingas*. Later distinguished from other manors so called by the affix *Freren-* 'of the brethren' (from ME *frere*) referring to possession by the Knights Hospitallers in the 12th cent.

Fryston, Monk N. Yorks. *Fristun c.*1030, *Munechesfryston* 1166. 'Farmstead of the Frisians'. OE *Frīsa* + *tūn*. Affix from OE *munuc* 'monk' referring to possession by Selby Abbey in the 11th cent.

Fryton N. Yorks. *Frideton* 1086 (DB). 'Farmstead offering safety or protection', or 'farmstead of a man called Frithi'. OE *frith* or OScand. pers. name + OE *tūn*.

Fryup N. Yorks. *Frehope* 12th cent., *Frihop* 1223. Possibly 'small enclosed valley of a man called **Frīga*'. OE pers. name + *hop*.

Fuerty (*Fiodharta*) Roscommon. 'Wooded place'.

Fugglestone St Peter Wilts. *Fughelistone* 1236. 'Farmstead of a man called Fugol'. OE pers. name + *tūn*. Affix from the dedication of the church.

Fulbeck Lincs. *Fulebec* 1086 (DB). 'Foul or dirty stream'. OE *fūl* + OScand. *bekkr*.

Fulbourn Cambs. *Fuulburne c.*1050, *Fuleberne* 1086 (DB). 'Stream frequented by birds'. OE *fugol* + *burna*.

Fulbrook Oxon. *Fulebroc* 1086 (DB). 'Foul or dirty brook'. OE *fūl* + *brōc*.

Fulford, 'foul or dirty ford', OE *fūl* + *ford*: **Fulford** Somerset. *Fuleforde* 1327. **Fulford** Staffs. *Fuleford* 1086 (DB). **Fulford** York. *Fuleford* 1086 (DB).

Fulham Gtr. London. *Fulanham c.*705, *Fuleham* 1086 (DB). 'River-bend land of a man called **Fulla*'. OE pers. name + *hamm*.

Fulking W. Sussex. *Fochinges* [*sic*] 1086 (DB), *Folkinges c.*1100. '(Settlement of) the family or followers of a man called **Folca*'. OE pers. name + *-ingas*.

Full Sutton E. R. Yorks. *See* SUTTON.

Fullerton Hants. *Fugelerestune* 1086 (DB). 'Farmstead or village of the fowlers or bird-catchers'. OE *fuglere* + *tūn*.

Fulletby Lincs. *Fullobi* 1086 (DB), *Fuletebi c.*1115. OScand. *bý* 'farmstead, village' with an obscure first element, possibly a pers. name.

Fulmer Bucks. *Fugelmere* 1198. 'Mere or lake frequented by birds'. OE *fugol* + *mere*.

Fulmodeston Norfolk. *Fulmotestuna* 1086 (DB). 'Farmstead of a man called Fulcmod'. OGerman pers. name + *tūn*.

Fulnetby Lincs. *Fulnedebi* 1086 (DB). OScand. *bý* 'farmstead, village' with an uncertain first element, possibly OScand. **full-nautr* 'one who has a full share'.

Fulstow Lincs. *Fugelestou* 1086 (DB). 'Holy place or meeting-place of a man called Fugol'. OE pers. name + *stōw*. Alternatively the first element may be OE *fugol* 'bird', hence 'place where birds abound'.

Fulwell Sundld. *Fulewella* 12th cent. 'Foul or dirty spring or stream'. OE *fūl* + *wella*.

Fulwood, 'foul or dirty wood', OE *fūl* + *wudu*: **Fulwood** Lancs. *Fulewde* 1199. **Fulwood** Notts. *Folewode* 13th cent.

Funshion (*Fuinseann*) (river) Cork, Limerick. 'Ash-producing one'.

Funtington W. Sussex. *Fundintune* 12th cent. Possibly 'farmstead at the place characterized by springs'. OE **funting* + *tūn*.

Furness (old district) Cumbria. *See* BARROW IN FURNESS.

Furneux Pelham Herts. *See* PELHAM.

Furraleigh (*Foradh Liath*) Waterford. 'Grey mound'.

Fybagh (*An Fhadhbach*) Kerry. 'The lumpy place'.

Fyfield, '(estate of) five hides of land', OE *fíf* + *híd*: **Fyfield** Essex. *Fifhidam* 1086 (DB). **Fyfield** Glos. *Fishide* 12th cent.

Fyfield Hants. *Fifhidon* 975, *Fifhide* 1086 (DB). **Fyfield** Oxon. *Fif Hidum* 956, *Fivehide* 1086 (DB). **Fyfield** Wilts. *Fifhide* 1086 (DB).

Fyfin (*Faiche Fionn*) Tyrone. *Foifin* 1609. 'White green'.

Fylde, The (district) Lancs.
See POULTON-LE-FYLDE.

Fyne, Loch. Arg. 'Loch of the River Fyne'. The river-name means 'wine'. Gaelic *fìon*.

G

Gaddesby Leics. *Gadesbi* 1086 (DB). 'Farmstead or village of a man called Gaddr'. OScand. pers. name + *bý*. Alternatively the first element may be OScand. *gaddr* 'spur of land'.

Gaddesden, Great & Gaddesden, Little Herts. *Gætesdene* 10th cent., *Gatesdene* 1086 (DB). 'Valley of a man called *Gǣte(n)*'. OE pers. name + *denu*. The river-name **Gade** is a back-formation from the place name.

Gaggin (*Géagánach*) Cork. 'Arm'.

Gagingwell Oxon. *Gadelingwelle* c.1173. 'Spring or stream of the kinsmen or companions'. OE *gædeling* + *wella*.

Gaick Forest Highland. 'Forest in a cleft'. Gaelic *gàg* (dative *gàig*).

Gailey Staffs. *Gageleage* c.1002, *Gragelie* [*sic*] 1086 (DB). 'Woodland clearing where bog-myrtle grows'. OE *gagel* + *lēah*.

Gaineamhach, Loch Highland. 'Sandy lake'. Gaelic *loch* + *gaineamh* + -*ach*.

Gainford Durham. *Geg(e)nforda* c.1040. 'Direct ford', i.e. 'ford on a direct route'. OE *gegn* + *ford*, with initial *G-* due to Scand. influence.

Gainsborough Lincs. *Gainesburg* 1086 (DB). 'Stronghold of a man called *Gegn*'. OE pers. name + *burh*.

Gairloch Highland. *Gerloth* 1275, *Gerloch* 1366. '(Place on) Gair Loch'. The loch name means 'short inlet of the sea' (Gaelic *geàrr* + *loch*).

Gairsay (island) Orkn. 'Garek's island'. OScand. *ey*.

Gaisgill Cumbria. *Gasegille* 1310. 'Ravine frequented by wild geese'. OScand. *gás* + *gil*.

Gaitsgill Cumbria. *Geytescall* 1278, *Gaytescales* 1279. 'Shieling(s) where goats are kept'. OScand. *geit* + *skáli*.

Gala Water. See GALASHIELS.

Galashiels Sc. Bord. *Galuschel* 1237. 'Huts by Gala Water'. ME *schele*. The river-name is of uncertain origin.

Galbally (*Gallbhaile*) Limerick, Tyrone, Wexford, etc. 'Foreigner's homestead'.

Galbooly (*Gallbhuaile*) Tipperary. 'Stone milking place'.

Galby (or **Gaulby**) Leics. *Galbi* 1086 (DB). Probably 'farmstead or village on poor soil'. OScand. *gall* + *bý*.

Galgate Lancs. *Galwaithegate* c.1190. Possibly '(place by) the Galloway road', i.e. 'the road used by cattle drovers from Galloway'. Scottish place name (meaning 'territory of the stranger-Gaels') + OScand. *gata*.

Galhampton, Galmpton, 'farmstead of rent-paying peasants', OE *gafol-mann* + *tūn*: **Galhampton** Somerset. *Galmeton* 1199. **Galmpton** Devon, near Salcombe. *Walementone* [*sic*] 1086 (DB). **Galmpton** Torbay. *Galmentone* 1086 (DB).

Gallan (*Gallán*) Tyrone. 'Standing stone'.

Gallan Head W. Isles. (Lewis). Perhaps 'pillar headland', from Gaelic *gallan*. Gaelic *An Gallan Uigeach*.

Gallane (*Gallán*) Cork. 'Standing stone'.

Gallen (*Gailenga*) Mayo. '(Territory of the) descendants of Gaileng'.

Galleywood Essex. *Gauelwode* 1250. 'Wood which pays a rent or tax'. OE *gafol* + *wudu*.

Galliagh (*Baile na gCailleach*) Derry. *Ballinecalleagh* 1604. 'Settlement of the nuns'.

Galloway (district) Dumf. *Galweya* c.970. '(Territory among the) stranger Gaels'. Gaelic *gall* + *Ghaidel*. The 'stranger Gaels' were people of mixed Irish and Scandinavian descent who settled here in the 9th cent.

Galmoy (*Gabhalmhaigh*) Tipperary. 'Fork of the plain'.

Galphay N. Yorks. *Galghagh* 12th cent. 'Enclosure where a gallows stands'. OE *galga* + *haga*.

Galston E. Ayr. *Gauston* 1260. OE *tūn* 'farmstead, village'. First element obscure.

Galtrim (*Cala Trium*) Meath. 'River-meadow of the elder'.

Galty Mountains (*Na Gaibhlte*) Limerick, Tipperary. Perhaps 'forted valleys'.

Galway (*Gaillimh*) Galway. 'Stony (river)'.

Galwolie (*Gallbhuaile*) Donegal. 'Milking place of the foreigner'.

Gamallt (mountain) Cergn. 'Crooked hillside'. Welsh *cam* + *allt*.

Gamblesby Cumbria. *Gamelesbi* 1177. 'Farmstead or village of a man called Gamall'. OScand. pers. name + *bý*.

Gamlingay Cambs. *Gamelingei* 1086 (DB). 'Enclosure or well-watered land associated with a man called *Gamela, or of *Gamela's people'. OE pers. name + *-ing-* or *-inga-* + *hæg* or *ēg*.

Gamston, 'farmstead of a man called Gamall', OScand. pers. name + OE *tūn*: **Gamston** Notts., near East Retford. *Gamelestune* 1086 (DB). **Gamston** Notts., near Nottingham. *Gamelestune* 1086 (DB).

Ganarew Herefs. *Genoreu* c.1150, *Guenerui* 1186. From the Welsh pers. name *Gwynwarwy* (St Gunguarui).

Ganstead E. R. Yorks. *Gagenestad* 1086 (DB). Probably 'homestead of a man called *Gagni or *Gagne'. OScand. pers. name + *stathr*.

Ganthorpe N. Yorks. *Gameltorp* [*sic*] 1086 (DB), *Galmestorp* 1169. 'Outlying farmstead or hamlet of a man called Galmr'. OScand. pers. name + *thorp*.

Ganton N. Yorks. *Galmeton* 1086 (DB). Probably 'farmstead of a man called Galmr'. OScand. pers. name + OE *tūn*.

Gara, Lough (*Loch Uí Gadhra*) Sligo. 'Ó Gadhra's lake'.

Garadice (*Garbhros*) Leitrim. 'Rough copse'.

Garbally (*Gearrbhaile*) Galway. 'Short homestead'.

Garboldisham Norfolk. *Gerboldesham* 1086 (DB). 'Homestead or village of a man called *Gǣrbald'. OE pers. name + *hām*.

Gardrum (*Gearrdroim*) Fermanagh, Tyrone. 'Short ridge'.

Garelochhead Arg. *Gerloch* 1272, *Keangerloch* c.1345. 'Head of the short lake'. Gaelic *gearr* + *loch* + English *head* (translating Gaelic *cearn* 'head').

Garford Oxon. *Garanforda* 940, *Wareford* [*sic*] 1086 (DB). 'Ford of a man called *Gāra', or 'ford at the triangular plot of ground'. OE pers. name or *gāra* + *ford*.

Garforth Leeds. *Gereford* 1086 (DB). 'Ford of a man called *Gǣra', or 'ford at the triangular plot of ground'. OE pers. name or *gāra* (influenced by OScand. *geiri*) + *ford*.

Gargrave N. Yorks. *Geregraue* 1086 (DB). 'Grove in a triangular plot of ground'. OScand. *geiri* (replacing OE *gāra*) + OE *grāf*.

Garnavilla (*Garrán an Bhile*) Tipperary. 'Grove of the sacred tree'.

Garnish (*Gar-inis*) Cork. 'Rough island'.

Garrabost W. Isles. (Lewis). 'Farmstead in an enclosure'. OScand. *garthr* + *bólstathr*.

Garranamanagh (*Garrán na Manach*) Kerry. 'Grove of the monks'.

Garranard (*Garrán Ard*) Mayo. 'High grove'.

Garrane (*Garrán*) Tipperary. 'Grove'.

Garranlahan (*An Garrán Leathan*) Roscommon. 'The broad grove'.

Garraun (*Garrán*) Clare, Galway. 'Grove'.

Garrigill Cumbria. *Gerardgile* 1232. 'Deep valley of a man called Gerard'. OGerman pers. name + OScand. *gil*.

Garriskil (*Garascal*) Westmeath. 'Rough nook'.

Garrison (*An Garastún*) Fermanagh. The village is named from a barrack erected here by William III after the battle of Aughrim in 1691.

Garroch Head Arg. *Garrach* 1449. 'Rough place (headland)'. Gaelic *garbh-ach* + English *head*.

Garron (river) Herefs. *See* LLANGARRON.

Garron Point (*Gearrán*) Antrim. 'Horse point'.

Garry (*Garraí*) Antrim. 'Garden'.

Garryduff (*Garraí Dubh*) Cork. 'Black garden'.

Garryhill (*An Gharbhchoill*) Carlow. 'The rough wood'.

Garrynafela (*Garraí na Féile*) Westmeath. 'Garden of the hospitality'.

Garryowen (*Garraí Eoghain*) Limerick. 'Eoghan's garden'.

Garryspillane (*Garraí Uí Spealáin*) Limerick. 'Ó Spealáin's garden'.

Garsdale Cumbria. *Garcedale c.*1240. 'Valley of a man called Garthr', or 'grass valley'. OScand. pers. name + *dalr*, or OE *gærs* + *dæl*.

Garsdon Wilts. *Gersdune* 701, *Gardone* 1086 (DB). 'Grass hill'. OE *gærs* + *dūn*.

Garshall Green Staffs. *Garnonshale* 1310. Possibly 'nook of land of a family called Garnon'. ME surname + OE *halh*.

Garsington Oxon. *Gersedun* 1086 (DB). 'Grassy hill'. OE **gærsen* + *dūn*.

Garstang Lancs. *Cherestanc* [*sic*] 1086 (DB), *Gairstang c.*1195. 'Spear post', probably signifying the site of a meeting-place. OScand. *geirr* + *stǫng*.

Garston Lpool. *Gerstan* 1094. Possibly 'the great stone'. OE *grēat* + *stān*.

Garston, East W. Berks. *Esgareston* 1180. 'Estate of a man called Esgar'. OScand. pers. name + OE *tūn*. The modern form is a 'folk etymology' that completely disguises the original meaning of the name.

Garswood St Hel. *Grateswode* 1367. OE *wudu* 'wood' with an obscure first element, possibly a pers. name.

Gartan (*Gártán*) Donegal. *nGartan c.*1630. 'Little field'.

Garthorpe Leics. *Garthorp c.*1130. Possibly 'outlying farmstead or hamlet within an enclosure'. OScand. *garthr* + *thorp*. Alternatively the first element may be OE *gāra* 'triangular plot of ground'.

Garthorpe N. Lincs. *Gerulftorp* 1086 (DB). 'Outlying farmstead or hamlet of a man called Geirulfr or Gairulf'. OScand. or OGerman pers. name + OScand. *thorp*.

Garton, 'farmstead in or near the triangular plot of ground', OE *gāra* + *tūn*: **Garton** E. R. Yorks. *Gartun* 1086 (DB). **Garton on the Wolds** E. R. Yorks. *Gartune* 1086 (DB), *Garton in Wald* 1347. For the affix, *see* WOLDS.

Garvagh (*Garbhachadh*) Derry. *Garvaghy* 1634. 'Rough field'.

Garvaghey (*Garbhachadh*) Tyrone. *Garaghye* 1629. 'Rough field'.

Garvaghy (*Garbhachadh*) Down. *Garwaghadh* 1428. 'Rough field'.

Garvellachs (islands) Arg. *Garbeallach* 1390. 'Rough rocks'. Gaelic *garbh* + *eileach*, + English plural *-s*.

Garvellan Rocks Dumf. *Garvellan* 1580. 'Rough island'. Gaelic *garbh* + *eilean*, + English *rocks*.

Garvery (*Garbhaire*) Fermanagh. *Garvorey* 1659. 'Rough land'.

Garvestone Norfolk. *Gerolfestuna* 1086 (DB). 'Farmstead or village of a man called Geirulfr or Gairulf'. OScand. or OGerman pers. name + OE *tūn*.

Garway Herefs. *Garou* 1137, *Langarewi* 1189. Probably 'church of a man called Guoruoe'. Welsh *llan* + pers. name.

Gasper Wilts. *Gayespore* 1280. Possibly 'hillspur of a man called **Gǣga*'. OE pers. name + *spora*.

Gastard Wilts. *Gatesterta* 1154. 'Tail of land where goats are kept'. OE *gāt* + *steort*.

Gasthorpe Norfolk. *Gadesthorp* 1086 (DB). 'Outlying farmstead or hamlet of a man called Gaddr'. OScand. pers. name + *thorp*.

Gatcombe I. of Wight. *Gatecome* 1086 (DB). 'Valley where goats are kept'. OE *gāt* + *cumb*.

Gate Helmsley N. Yorks. *See* HELMSLEY.

Gateforth N. Yorks. *Gæiteford c.*1030. 'Goats' ford', i.e. 'ford used when moving goats'. OScand. *geit* + OE *ford*.

Gateholm (island) Pemb. *Goteholme* 1480. 'Island where goats graze'. OScand. *geit* + *holmr*.

Gatehouse of Fleet Dumf. 'Gatehouse by the River Fleet'. A 'gatehouse' is a roadside house. The river-name means simply 'stream' (OScand. *fljót* or OE *flēot*). Burgh of barony in 1795.

Gateley Norfolk. *Gatelea* 1086 (DB). 'Clearing or pasture where goats are kept'. OE *gāt* + *lēah*.

Gatenby N. Yorks. *Ghetenesbi* 1086 (DB). Possibly 'farmstead or village of a man called Gaithan'. OIrish pers. name + OScand. *bý*.

Gateshead Gatesd. *Gateshevet* c.1150, *Gatesheued* 1183. 'Goat's headland or hill'. OE *gāt* + *hēafod*.

Gathabawn (*An Geata Bán*) Kilkenny. 'The white gate'.

Gathurst Wigan. *Gatehurst* 1547. Probably 'wooded hill where goats are kept'. OE *gāt* + *hyrst*.

Gatley Stockp. *Gateclyve* 1290. 'Cliff or bank where goats are kept'. OE *gāt* + *clif*.

Gatwick W. Sussex. *Gatwik* 1241. 'Farm where goats are kept'. OE *gāt* + *wīc*.

Gaulby Leics. *See* GALBY.

Gaunless (river) Durham. *See* AUCKLAND.

Gautby Lincs. *Goutebi* 1196. 'Farmstead or village of a man called Gauti'. OScand. pers. name + *bý*.

Gawber Barns. *Galgbergh* 1304. 'Gallows hill'. OE *galga* + *beorg*.

Gawcott Bucks. *Chauescote* [*sic*] 1086 (DB), *Gauecota* 1090. 'Cottage(s) for which rent is payable'. OE *gafol* + *cot*.

Gawley's Gate Antrim. The hamlet is named from a 17th-cent. tollgate, manned by one *Gawley*. The equivalent Irish name is *Geata Mhic Amhlaí*.

Gawsworth Ches. *Govesurde* 1086 (DB). Probably 'enclosure of the smith'. Welsh *gof* + OE *worth*.

Gay Island (*Inis na nGédh*) Fermanagh. 'Goose island'.

Gaydon Warwicks. *Gaidone* 1194. 'Hill of a man called *Gǣga*. OE pers. name + *dūn*.

Gayhurst Milt. K. *Gateherst* 1086 (DB). 'Wooded hill where goats are kept'. OE *gāt* + *hyrst*.

Gayles N. Yorks. *Gales* 1534. '(Place at) the ravines'. OScand. *geil*.

Gayton, usually 'farmstead where goats are kept', from OScand. *geit* + *tūn*: **Gayton** Norfolk. *Gaituna* 1086 (DB). **Gayton** Wirral. *Gaitone* 1086 (DB). **Gayton le Marsh** Lincs. *Geiton* 1206. Affix means 'in the marsh' from OE *mersc* with loss of preposition. **Gayton le Wold** Lincs. *Gettone* 1086 (DB). Affix means 'on the wold(s)' with loss of preposition, *see* WOLDS.

   However the following have a different origin, 'farmstead of a man called *Gǣga*, from OE pers. name + *tūn*: **Gayton** Northants. *Gaiton* 1162. **Gayton** Staffs. *Gaitone* 1086 (DB).

Gayton Thorpe Norfolk. *Torp* 1086 (DB), *Geytonthorp* 1402. 'Outlying farmstead or hamlet dependent on GAYTON'. OScand. *thorp*.

Gaywood Norfolk. *Gaiuude* 1086 (DB). 'Wood of a man called *Gǣga*. OE pers. name + *wudu*.

Gazeley Suffolk. *Gaysle* 1219. 'Woodland clearing of a man called *Gǣgi*. OE pers. name + *lēah*.

Gearha (*Gaortha*) Kerry. 'Wooded valley'.

Gearhameen (*Gaortha Mín*) Kerry. 'Smooth wooded valley'.

Gearhasallagh (*Gaortha Sailech*) Kerry. 'Wooded valley of willows'.

Gedding Suffolk. *Gedinga* 1086 (DB). '(Settlement of) the family or followers of a man called *Gydda*. OE pers. name + *-ingas*.

Geddington Northants. *Geitentone* 1086 (DB). Possibly 'estate associated with a man called *Gǣte or Geiti*. OE or OScand. pers. name + OE *-ing-* + *tūn*.

Gedling Notts. *Ghellinge* [*sic*] 1086 (DB), *Gedlinges* 1187. '(Settlement of) the family or followers of a man called *Gēdel*. OE pers. name + *-ingas*.

Gedney Lincs. *Gadenai* 1086 (DB). Probably 'island or well-watered land of a man called *Gǣda*. OE pers. name (genitive -*n*) + *ēg*.

Geldeston Norfolk. *Geldestun* 1242. 'Farmstead or village of a man called *Gyldi*. OE pers. name + *tūn*.

Gelli Gandryll, Y. *See* HAY-ON-WYE.

Gelston Lincs. *Cheuelestune* 1086 (DB). Probably 'farmstead or village of a man called *Gjǫfull'. OScand. pers. name + OE *tūn*.

George Nympton Devon. *See* NYMPTON.

Georgeham Devon. *Hama* 1086 (DB), *Hamme Sancti Georgii* 1356. Originally 'the well-watered valley' from OE *hamm*. Later affix from the dedication of the church to St George.

Georgemas Highland. 'St George's mass'. OE *mǣsse*. The allusion is to an annual fair on St George's Day (23 April).

Germansweek Devon. *Wica* 1086 (DB), *Wyke Germyn* 1458. Originally 'the dwelling or (dairy) farm' from OE *wīc*. Later affix from the dedication of the church to St Germanus.

Germoe Cornwall. 'Chapel of *Sanctus Germoch*' 12th cent. From the patron saint of the chapel.

Gerrans Cornwall. 'Church of *Sanctus Gerentus*' 1202. 'Church of St Gerent'. From the patron saint of the church.

Gerrards Cross Bucks. *Gerards Cross* 1692. Named from a local family called *Jarrard* or *Gerrard*.

Gestingthorpe Essex. *Gyrstlingathorpe* late 10th cent., *Ghestingetorp* 1086 (DB). 'Outlying farmstead of the family or followers of a man called *Gyrstel*'. OE pers. name + *-inga-* + *throp*.

Giant's Causeway (*Clochán na bhFomhórach*) (columnar rock formation) Antrim. *Clochán na bhfogmharach c.*1675. 'Causeway of the Fomorians'. The English name appears to be a loose translation of the Irish, while the modern Irish name, *Clochán an Aifir*, is apparently a corruption of the earlier name. The Fomorians were legendary giant sea rovers.

Gidding, Great, Gidding, Little, & Gidding, Steeple Cambs. *Geddinge* 1086 (DB), *Magna Giddinge* 1220, *Gydding Parva* 13th cent., *Stepelgedding* 1260. Probably '(settlement of) the family or followers of a man called *Gydda*'. OE pers. name + *-ingas*. Distinguishing affixes are Latin *magna* 'great', *parva* 'little' and OE *stēpel* 'steeple, tower'.

Gidea Park Gtr. London. *La Gidiehall* 1258, *Guydie hall parke* 1668. Literally 'the foolish or crazy hall', from ME *gidi* + *hall*, perhaps alluding to a building of unusual design or construction.

Gidleigh Devon. *Gideleia* 1156. 'Woodland clearing of a man called *Gydda*'. OE pers. name + *lēah*.

Giggleswick N. Yorks. *Ghigeleswic* 1086 (DB). 'Dwelling or (dairy) farm of a man called Gikel or Gichel'. OE or ME pers. name (probably a short form of the biblical name *Judichael*) + *wīc*.

Gigha (island) Arg. *Guthey, Gudey* 13th cent. 'God's isle or good isle'. OScand. *guth* or *góthr* + *ey*.

Gilberdyke E. R. Yorks. *Dyc* 1234, *Gilbertdike* 1376. '(Place at) the ditch or dike'. OE *dīc* with manorial addition from a person or family called *Gilbert*.

Gilcrux Cumbria. *Killecruce c.*1175. Probably 'retreat by a hill'. Celtic *cil* + *crŭg*.

Gildersome Leeds. *Gildehusum* 1181. '(Place at) the guild-houses'. OScand. *gildi-hús* in a dative plural form.

Gildingwells Rothm. *Gildanwell* 13th cent. Probably 'gushing spring'. OE *gyldande* + *wella*.

Gilford (*Átha Mhic Giolla*) Down. 'Magill's ford'. *Gilford* 1678. The village is named from Captain John *Magill*, who acquired lands here in the 17th cent.

Gill, Lough (*Loch Gile*) Kerry, Leitrim, Sligo. 'Lake of brightness'.

Gillamoor N. Yorks. *Gedlingesmore* [*sic*] 1086 (DB), *Gillingamor* late 12th cent. 'Moorland belonging to the family or followers of a man called *Gȳthla* or *Gētla*'. OE pers. name + *-inga-* + *mōr*.

Gilling '(settlement of) the family or followers of a man called *Gȳthla* or *Gētla*', OE pers. name + *-ingas*: **Gilling East** N. Yorks. *Ghellinge* 1086 (DB). **Gilling West** N. Yorks. *Ingetlingum* 731, *Ghellinges* 1086 (DB).

Gillingham, 'homestead of the family or followers of a man called *Gylla*', OE pers. name + *-inga-* + *hām*: **Gillingham** Dorset. *Gelingeham* 1086 (DB). **Gillingham** Medway. *Gyllingeham* 10th cent., *Gelingeham* 1086 (DB). **Gillingham** Norfolk. *Kildincham* [*sic*] 1086 (DB), *Gelingeham* 12th cent.

Gilmorton Leics. *Mortone* 1086 (DB), *Gilden Morton* 1327. 'Farmstead in marshy ground'.

OE *mōr* + *tūn*, with later affix from ME *gilden* 'wealthy, splendid'.

Gilsland (Northum.), **Gilsland Spa** (Cumbria). *Gillesland* 12th cent. 'Estate of a man called Gille or Gilli'. OIrish or OScand. pers. name + *land*.

Gilston Herts. *Gedeleston* 1197. 'Farmstead or village of a man called *Gēdel or *Gydel'. OE pers. name + *tūn*.

Gimingham Norfolk. *Gimingeham* 1086 (DB). 'Homestead of the family or followers of a man called *Gymi or *Gymma'. OE pers. name + *-inga-* + *hām*.

Ginge, East & Ginge, West Oxon. *Gæging* 10th cent., *Gainz* 1086 (DB), *Estgeyng*, *Westgenge* 13th cent. Originally an OE river-name meaning 'one that turns aside', from the stem of OE *gǣgan* + *-ing*.

Gipping Suffolk. *Gippinges* 1154–89. '(Settlement of) the family or followers of a man called Gyppi or *Gyppa'. OE pers. name + *-ingas*. The river-name **Gipping** is a back-formation from the place name.

Girsby N. Yorks. *Grisebi* 1086 (DB). 'Farmstead of a man called Gríss', or 'farmstead where young pigs are reared'. OScand. pers. name or OScand. *gríss* + *bý*.

Girton, 'farmstead or village on gravelly ground', OE *grēot* + *tūn*: **Girton** Cambs. *Grittune c.*1060, *Gretone* 1086 (DB). **Girton** Notts. *Gretone* 1086 (DB).

Girvan S. Ayr. *Girven* 1275. Derived from the Water of *Girvan*, of unknown origin.

Gisburn Lancs. *Ghiseburne* [sic] 1086 (DB), *Giselburn* 12th cent. Probably 'gushing stream'. OE **gysel* + *burna*. Alternatively the first element may be an OE pers. name **Gysla*.

Gisleham Suffolk. *Gisleham* 1086 (DB). 'Homestead or village of a man called *Gysla'. OE pers. name + *hām*.

Gislingham Suffolk. *Gyselingham c.*1060, *Gislingaham* 1086 (DB). 'Homestead of the family or followers of a man called *Gysla'. OE pers. name + *-inga-* + *hām*.

Gissing Norfolk. *Gessinga* 1086 (DB). '(Settlement of) the family or followers of a man called *Gyssa or *Gyssi'. OE pers. name + *-ingas*.

Gittisham Devon. *Gidesham* 1086 (DB). 'Homestead or enclosure of a man called Gyddi'. OE pers. name + *hām* or *hamm*.

Givendale, Great E. R. Yorks. *Ghiuedale* 1086 (DB). Probably 'valley of a river called **Gævul'. OScand. river-name (meaning 'good for fishing') + *dalr*.

Glaisdale N. Yorks. *Glasedale* 12th cent. 'Valley of a river called **Glas'. Celtic river-name (meaning 'grey-green') + OScand. *dalr*.

Glamis Ang. *Glamenes* 1178, *Glames* 1187. Meaning unknown.

Glamorgan (*Morgannwg*) (the historic county). 'Morgan's shore'. Welsh *glan*. Morgan was a 7th-cent. prince of Gwent. The Welsh name means 'Morgan's territory'.

Glanaman Carm. 'Bank of the River Aman'. Welsh *glan*. For the river-name, *see* AMMANFORD.

Glanaruddery Mountains (*Sléibhte Ghleann an Ridire*) Kerry. 'Mountains of the valley of the knight'.

Glandford Norfolk. *Glamforda* 1086 (DB). Probably 'ford where people assemble for revelry or games'. OE *glēam* + *ford*.

Glandore (*Gleann Dor*) Cork. 'Valley of the doors'.

Glanmire (*Gleann Maghair*) Cork. 'Valley of the plain'.

Glanmore (*Gleann Mór*) Kerry. 'Big valley'.

Glannavaddoge (*Gleann na bhFeadóg*) Galway. 'Valley of the plovers'.

Glantane (*An Gleanntán*) Cork. 'The little valley'.

Glanton Northum. *Glentendon* 1186. 'Hill frequented by birds of prey or used as a look-out place'. OE **glente* + *dūn*.

Glanvilles Wootton Dorset. *See* WOOTTON.

Glanworth (*Gleannúir*) Cork. 'Valley of the yew'.

Glapthorn Northants. *Glapethorn* 12th cent. 'Thorn-tree of a man called Glappa'. OE pers. name + *thorn*.

Glapwell Derbys. *Glappewelle* 1086 (DB). 'Stream of a man called Glappa', or 'stream where the buckbean plant grows'. OE pers. name or OE *glæppe* + *wella*.

Glasbury (*Clas-ar-Wy*) Powys. *Clastbyrig* 1056, *Glesburia* 1191, *Classebury* 1322. 'Town of the monastic community'. Welsh *clas* + OE *burh*. The Welsh name means 'monastic community on the River Wye' (a name of unknown meaning).

Glascarn (*Glascharn*) Westmeath. 'Green cairn'.

Glascoed Mon. 'Green wood'. Welsh *glas* + *coed*.

Glascote Staffs. *Glascote* 12th cent. 'Hut where glass is made'. OE *glæs* + *cot*.

Glasdrumman (*An Ghlasdromainn*) Down. *Glasedrommyn* 1540. 'The grey-green ridge'.

Glasgow Glas. *Glasgu* 1136. 'Green hollow'. British **glas-* + **cau*.

Glashaboy (*Glaise Buí*) Cork. 'Yellow stream'.

Glaslough (*Glasloch*) Monaghan. *Glaslagh* 1591. 'Grey-green lake'.

Glasnevin (*Glas Naíon*) Dublin. 'Stream of the child'.

Glassavullaun (*Glas an Mhulláin*) Dublin. 'Streamlet of the small summit'.

Glasshouses N. Yorks *Glassehouse* 1387. 'House(s) where glass was made'. ME *glasse* + *hous*. There is another **Glasshouse** in Glos. (recorded from 1755).

Glassleck (*Glasleic*) Cavan. 'Grey flagstone'.

Glasson Cumbria. *Glassan* 1259. Probably a Celtic river-name containing a derivative of **glas* 'grey-green'.

Glasson Lancs. *Glassene* c.1265. Perhaps originally a river-name meaning 'clear or bright one' from OE **glǣsne* or **glǣsen*.

Glassonby Cumbria. *Glassanebi* 1177. 'Farmstead or village of a man called Glassán'. OIrish pers. name + OScand. *bý*.

Glasthule (*Glas Tuathail*) Dublin. 'Tuathal's streamlet'.

Glaston Rutland. *Gladestone* 1086 (DB). Probably 'farmstead of a man called **Glathr*'. OScand. pers. name + OE *tūn*.

Glastonbury Somerset. *Glastingburi* 725, *Glæstingeberia* 1086 (DB). 'Stronghold of the people living at *Glaston*'. Celtic name (possibly meaning 'woad place') + OE *-inga-* + *burh* (dative *byrig*).

Glastry (*Glasrach*) Down. *Balliglassarie* 1604. 'Green grassy place'.

Glatton Cambs. *Glatune* 1086 (DB). 'Pleasant farmstead'. OE *glæd* + *tūn*.

Glazebrook Warrtn. *Glasbroc* 1227. Named from Glaze Brook, a Celtic river-name (meaning 'grey-green') + OE *brōc*.

Glazebury Warrtn. a late name of recent origin formed from GLAZEBROOK.

Glazeley Shrops. *Gleslei* 1086 (DB), *Glasele* 1255. Possibly 'bright clearing', OE **glǣs* + *lēah*. Alternatively the first element may be a lost stream-name from OE *glæs* 'glass'.

Gleadless Sheff. *Gledeleys* 13th cent. Probably 'woodland clearings frequented by kites'. OE *gleoda* + *lēah*.

Gleaston Cumbria. *Glassertun* [*sic*] 1086 (DB), *Gleseton* 1269. Probably 'bright farmstead or village'. OE **glǣs* + *tūn*. The first element may be used as a stream-name 'the bright one'.

Gledhow Leeds. *Gledhou* 1334–7. 'Hill frequented by kites'. OE *gleoda* + OScand. *haugr*.

Glemham, Great & Glemham, Little Suffolk. *Glaimham* 1086 (DB). Probably 'homestead or village noted for its revelry or games'. OE *glēam* + *hām*. The river-name **Glem** is a back-formation from the place name.

Glemsford Suffolk. *Glemesford* c.1050, *Clamesforda* [*sic*] 1086 (DB). Probably 'ford where people assemble for revelry or games'. OE *glēam* + *ford*. The river-name **Glem** is a back-formation from the place name.

Glen (*Gleann*) Cavan, Kilkenny, Waterford. 'Valley'.

Glen Affric Highland. 'Valley of the River Affric'. Gaelic *gleann*. The river-name means 'very dappled' (Gaelic *ath* + *breac*).

Glen Parva & Great Glen Leics. *Glenne* 849, *Glen* 1086 (DB), *Parva Glen* 1242, *Magna Glen* 1247. Probably '(place at) the valley'. OE *glenn* (from Celtic **glïnn*). Distinguishing affixes are Latin *parva* 'small' and *magna* 'great'.

Glen of Imail (*Gleann Ó Maoil*) Wicklow. 'Valley of Uí Maoil'.

Glenade (*Gleann Éada*) Sligo. 'Éada's valley'.

Glenageary (*Gleann na gCaorach*) Dublin. 'Valley of the sheep'.

Glenagowr (*Gleann na nGabhar*) Limerick. 'Valley of the goats'.

Glenamaddy (*Gleann na Madadh*) Roscommon. 'Valley of the dogs'.

Glenamoy (*Gleann na Muaidhe*) Mayo. 'Valley of the (river) Moy'.

Glenanne (*Gleann Anna*) Armagh. *Glen Anne* 1828. 'Anne's valley'. The name was originally that of the house owned by George Gray, who named it after his wife, Eliza Anne, née Henry.

Glenard (*Gleann Aird*) Antrim. 'High valley'.

Glenariff (*Gleann Aireamh*) Antrim. *Glenarthac* 1279. 'Valley of arable land'.

Glenarm (*Gleann Arma*) Antrim. 'Valley of arms'.

Glenasmole (*Gleann na Smól*) Dublin. 'Valley of the thrushes'.

Glenavy (*Lann Abhaigh*) Antrim. (*o*) *Lainn abhaich c.*1450. 'Church of the dwarf'. According to legend, St Patrick built a church here and left it in charge of his disciple Daniel, nicknamed 'dwarf' for his small size.

Glenbeg (*Gleann Beag*) Cork. 'Small valley'.

Glenbeigh (*Gleann Beithe*) Kerry. 'Valley of the birch'.

Glenboy (*Gleann Buí*) Sligo. 'Yellow valley'.

Glenbryan (*Gleann Bhriain*) Wexford. 'Brian's valley'.

Glenbush (*Gleann na Buaise*) Antrim. 'Valley of the (river) Bush'.

Glencairn (*Gleann an Chairn*) Waterford. 'Valley of the cairn'. The current Irish name is *Baile an Gharráin*, as for BALLINGARRANE.

Glencar (*Gleann an Chairthe*) Kerry, Sligo. 'Valley of the standing stone'.

Glencoe Highland. *Glenchomure* 1343, *Glencole* 1491. 'Valley of the River Coe'. Gaelic *gleann*. The meaning of the river-name is unknown.

Glencolumbkille (*Gleann Cholm Cille*) Donegal. *Glend Colaim Cilli* 1532. 'Valley of Colm Cille'.

Glencovet (*Gleann Coimheada*) Donegal. 'Valley of the lookout post'.

Glencullen (*Gleann Cuilinn*) Dublin. 'Valley of holly'.

Glendalough (*Gleann dá Loch*) Wicklow. 'Valley of two lakes'.

Glendavagh (*Gleann dá Mhaighe*) Tyrone. 'Valley of two plains'.

Glenderry (*Gleann Doire*) Kerry. 'Valley of the oak grove'.

Glendowan (*Gleann Domhain*) (district) Donegal. 'Deep valley'.

Glenduff (*Gleann Dubh*) Cork. 'Black valley'.

Glendun (*Gleann Doinne*) (valley) Antrim. *Glendun* 1832. 'Valley of the (river) Dun'.

Gleneagles Perth. *Gleninglese c.*1165, *Glenegas* 1508. Gaelic *gleann* 'valley' + unknown second element.

Gleneany (*Gleann Eidneach*) Donegal. 'Valley of the river abounding in ivy'.

Gleneask (*Gleann Iasc*) Sligo. 'Fish valley'.

Gleneely (*Gleann Daoile*) Donegal. (*a*) *nGleann Daoile c.*1630. 'Valley of the (river) Daoil'.

Glenelg Highland. *Glenelg* 1292. Gaelic *Gleann Eilge*. Gaelic *gleann* 'valley', otherwise obscure. Compare ELGIN.

Glenfarne (*Gleann Fearna*) Leitrim, Sligo. 'Valley of alders'.

Glenfield Leics. *Clanefelde* 1086 (DB). 'Clean open land', i.e. 'open land free from weeds or other unwanted growth'. OE *clǣne* + *feld*.

Glenflesk (*Gleann Fleisce*) Kerry. 'Valley of the hoop'.

Glengarriff (*Gleann Garbh*) Cork. 'Rough valley'.

Glengesh (*Gleann Gheise*) Donegal. *Glengeish c.*1655. 'Valley of the taboo'.

Glengevlin (*Gleann Gaibhle*) Cavan. 'Gaibhle's valley'.

Glenhull (*Gleann Choll*) Tyrone. 'Valley of hazels'.

Glenmalure (*Gleann Molúra*) Wicklow. 'Molúra's valley'.

Glenmore (*Gleann Mór*) Kerry, Kilkenny. 'Big valley'.

Glennageragh (*Gleann na gCaorach*) Tyrone. 'Valley of the sheep'.

Glennamaddy (*Gleann na Madadh*) Galway. 'Valley of the dogs'.

Glennascaul (*Gleann an Scáil*) Galway. 'Valley of the phantom'.

Glennawoo (*Gleann na bhFuath*) Sligo. 'Valley of the spectres'.

Glenoe (*Gleann Ó*) Antrim. *Glenoe* 1833. 'Valley of the lump'.

Glenone (*Cluain Eoghain*) Derry. *Cloynon* 1654. 'Eoghan's pasture'.

Glenquin (*Gleann an Chuin*) Limerick. 'Valley of the hollow'.

Glenroe (*An Gleann Rua*) Limerick. 'The red valley'.

Glenrothes Fife. 'Valley of Rothes'. Glenrothes was designated a New Town in 1948 and named after the Earl of *Rothes*, the local laird. *See* ROTHES.

Glensmoil (*Gleann an Smóil*) Donegal. 'Valley of the thrush'.

Glensooska (*Gleann Samhaisce*) Kerry. 'Valley of the heifer'.

Glentane (*Gleanntán*) Galway. 'Little valley'.

Glentham Lincs. *Glentham* 1086 (DB). Possibly 'homestead frequented by birds of prey or at a look-out place'. OE *glente* + *hām*.

Glenties (*Na Gleannta*) Donegal. *Na Gleanntaidh* 1835. 'The valleys'.

Glentogher (*Gleann Tóchair*) Donegal. *ye causeway of Clantogher c.*1655. 'Valley of the causeway'.

Glentworth Lincs. *Glentewrde* 1086 (DB). Possibly 'enclosure frequented by birds of prey or at a look-out place'. OE *glente* + *worth*.

Glenvar (*Gleann Bhairr*) Donegal. *Glinvar* 1796. 'Valley of the top'.

Glenwherry (*Gleann an Choire*) Antrim. 'Valley of the cauldron'.

Glewstone Herefs. *Gleanston* (*sic* for *Gleauston*) 1212. 'Farmstead or estate of a man called Glēaw'. OE pers. name + *tūn*.

Glin (*An Gleann*) Limerick. 'The valley'.

Glinton Peterb. *Clinton* 1060, *Glintone* 1086 (DB). Possibly 'fenced farmstead'. OE *glind* + *tūn*.

Glooston Leics. *Glorstone* 1086 (DB). 'Farmstead of a man called Glōr'. OE pers. name + *tūn*.

Glossop Derbys. *Glosop* 1086 (DB), *Glotsop* 1219. 'Valley of a man called *Glott*. OE pers. name + *hop*.

Gloucester Glos. *Coloniae Glev'* 2nd cent., *Glowecestre* 1086 (DB). 'Roman town called *Glevum*'. Celtic name (meaning 'bright place') + OE *ceaster*. The early form contains Latin *colonia* 'Roman colony for retired legionaries'. **Gloucestershire** (OE *scīr* 'district') is first referred to in the 11th cent.

Glounthaune (*An Gleanntán*) Cork. 'The small valley'.

Glusburn N. Yorks. *Glusebrun* 1086 (DB). '(Place at) the bright or shining stream'. OScand. *glus(s)* + *brunnr*.

Glympton Oxon. *Glimtuna c.*1050, *Glintone* 1086 (DB). 'Farmstead on the River Glyme'. Celtic river-name (meaning 'bright stream') + OE *tūn*.

Glyn Ceiriog Wrex. *Lansanfreit* 1291, *llansanffraid y glyn* 1590, *ll.sain ffred glyn Kerioc* 1566, *Lansantffraid Glynn Ceiriog* 1795, *Llansaintffraid-Glyn-Ceiriog* 1868. 'Valley of the River Ceiriog'. For the river-name, *see* CHIRK. The formal name of the village is *Llansanffraid Glynceiriog*. Compare LLANSANFFRAID GLAN CONWY.

Glyn-neath (*Glyn-nedd*) Neat. *Glynneth* 1281, *Glyn Nedd* 15th cent. 'Valley of the River Neath'. Welsh *glyn*. For the river-name, *see* NEATH.

Glyn-nedd. *See* GLYN-NEATH.

Glynde E. Sussex. *Glinda* 1165. '(Place at) the fence or enclosure'. OE **glind*.

Glyndebourne E. Sussex. *Burne juxta Glynde* 1288. 'Stream near GLYNDE'. OE *burna*.

Glynebwy. *See* EBBW VALE.

Glynn (*Gleann*) Antrim, Carlow, Wexford. 'Valley'.

Gneevgullia (*Gníomh go Leith*) Kerry. 'A gneeve and a half'. A gneeve is a land measure equal to one-twelfth of a ploughland.

Gnosall Staffs. *Geneshale* 1086 (DB), *Gnowesala* 1140. Probably 'nook of land at Genow'. Celtic **genow* 'mouth (of a valley)' + OE *halh*.

Goadby, 'farmstead or village of a man called Gauti', OScand. pers. name + *bý*: **Goadby** Leics. *Goutebi* 1086 (DB). **Goadby Marwood** Leics. *Goutebi* 1086 (DB). Manorial affix from the *Maureward* family, here in the 13th cent.

Goatacre Wilts. *Godacre* 1242, *Gotacre* 1268. 'Plot of cultivated land where goats are kept'. OE *gāt* + *æcer*.

Goathill Dorset. *Gatelme* [*sic*] 1086 (DB), *Gathulla* 1176. 'Hill where goats are pastured'. OE *gāt* + *hyll*.

Goathland N. Yorks. *Godelandia* c.1110. 'Cultivated land of a man called Gōda', or 'good cultivated land'. OE pers. name or OE *gōd* + *land*.

Goathurst Somerset. *Gahers* [*sic*] 1086 (DB), *Gothurste* 1292. 'Wooded hill where goats are kept'. OE *gāt* + *hyrst*.

Gobbins, The (*Na Gobáin*) (cliffs) Antrim. *the Gabbon* 1683. 'The peaks'.

Gobowen Shrops. *Goebowens* 1699. Origin uncertain, but possibly 'Owen's embankment', from Welsh *cob* and the pers. name or surname *Owen*.

Godalming Surrey. *Godelmingum* c.880, *Godelminge* 1086 (DB). '(Settlement of) the family or followers of a man called *Godhelm*'. OE pers. name + *-ingas*.

Goddington Gtr. London. *Godinton* 1240. A manorial name from the *de Godinton* family (from Godinton near Ashford in Kent, which

is 'estate associated with a man called Gōda', OE pers. name + *-ing-* + *tūn*).

Godmanchester Cambs. *Godmundcestre* 1086 (DB). 'Roman station associated with a man called Godmund'. OE pers. name + *ceaster*.

Godmanstone Dorset. *Godemanestone* 1166. 'Farmstead of a man called Godmann'. OE pers. name + *tūn*.

Godmersham Kent. *Godmeresham* 822, *Gomersham* 1086 (DB). 'Homestead or village of a man called Godmær'. OE pers. name + *hām*.

Godney Somerset. *Godeneia* 10th cent. 'Island or well-watered land of a man called Gōda'. OE pers. name (genitive *-n*) + *ēg*.

Godolphin Cross Cornwall. *Wulgholgan* 1194. An obscure name, origin and meaning uncertain.

Godshill, 'hill associated with a heathen god or with the Christian God', OE *god* + *hyll*: **Godshill** Hants. *Godeshull* 1230. **Godshill** I. of Wight. *Godeshella* 12th cent.

Godstone Surrey. *Godeston* 1248, *Codeston* 1279. Probably 'farmstead of a man called **Cōd*'. OE pers. name + *tūn*.

Gola (*Gabhla*) (island) Donegal. *Goolagh* 1614. 'Place of the fork'.

Golborne Wigan. *Goldeburn* 1187. 'Stream where marsh marigolds grow'. OE *golde* + *burna*.

Golcar Kirkl. *Gudlagesarc* 1086 (DB). 'Shieling or hill-pasture of a man called Guthleikr or Guthlaugr'. OScand. pers. name + *erg*.

Golden (*An Gabhailín*) Tipperary. 'The little fork'.

Goldhanger Essex. *Goldhangra* 1086 (DB). Probably 'wooded slope where marigolds or other yellow flowers grow'. OE *golde* + *hangra*.

Golding Shrops. *Goldene* 1086 (DB). 'Gold valley', referring to yellow flowers or to soil colour. OE *gold* + *denu*.

Goldsborough N. Yorks. near Knaresborough. *Godenesburg* [*sic*] 1086 (DB), *Godelesburc* 1170. 'Stronghold of a man called Godel'. OE (or OGerman) pers. name + *burh*.

Goldthorpe Barns. *Goldetorp* 1086 (DB). 'Outlying farmstead or hamlet of a man called Golda'. OE pers. name + OScand. *thorp*.

Goleen (*An Góilín*) Cork. 'The inlet'.

Golspie Highland. *Goldespy* 1330. OScand. *bý* 'farmstead, village', probably with an OE or OScand. pers. name.

Gomeldon Wilts. *Gomeledona* 1189. Probably 'hill of a man called *Gumela'. OE pers. name + *dūn*.

Gomersal Kirkl. *Gomershale* 1086 (DB). 'Nook of land of a man called *Gūthmǣr'. OE pers. name + *halh*.

Gomshall Surrey. *Gomeselle* [sic] 1086 (DB), *Gumeselva* 1168. 'Shelf or terrace of land of a man called *Guma'. OE pers. name + *scelf*.

Gonalston Notts. *Gunnulvestune* 1086 (DB). 'Farmstead of a man called Gunnulf'. OScand. pers. name + OE *tūn*.

Gonerby, Great Lincs. *Gunfordebi* 1086 (DB). 'Farmstead or village of a man called Gunnfrøthr'. OScand. pers. name + *bý*.

Good Easter Essex. *See* EASTER.

Gooderstone Norfolk. *Godestuna* [sic] 1086 (DB), *Gutherestone* 1254. 'Farmstead of a man called Gūthhere'. OE pers. name + *tūn*.

Goodleigh Devon. *Godelege* 1086 (DB). 'Woodland clearing of a man called Gōda'. OE pers. name + *lēah*.

Goodmanham E. R. Yorks. *Godmunddingaham* 731, *Gudmundham* 1086 (DB). 'Homestead of the family or followers of a man called Gōdmund'. OE pers. name + *-inga-* + *hām*.

Goodmayes Gtr. London. *Goodmayes* 1456. Named from the *Godemay* family who had lands here in the 14th cent.

Goodnestone, 'farmstead of a man called Gōdwine', OE pers. name + *tūn*: **Goodnestone** Kent, near Aylesham. *Godwineston* 1196. **Goodnestone** Kent, near Faversham. *Godwineston* 1208.

Goodrich Herefs. *Castellum Godric* 1102. Originally 'castle of a man called Gōdrīc'. OE pers. name (that of a land-holder in 1086 (DB)) with Latin *castellum*.

Goodrington Torbay. *Godrintone* 1086 (DB). 'Estate associated with a man called Gōdhere'. OE pers. name + *-ing-* + *tūn*.

Goodworth Clatford Hants. *See* CLATFORD.

Goole E. R. Yorks. *Gulle* 1362. 'The stream or channel'. ME *goule*.

Goosey Oxon. *Goseie* 9th cent., *Gosei* 1086 (DB). 'Goose island'. OE *gōs* + *ēg*.

Goosnargh Lancs. *Gusansarghe* 1086 (DB). 'Shieling or hill-pasture of a man called Gussān'. OIrish pers. name + OScand. *erg*.

Goostrey Ches. *Gostrel* [sic] 1086 (DB), *Gorestre c.*1200. 'The gorse bush'. OE *gorst-trēow*.

Gordano (old district) N. Som. *See* CLAPTON IN GORDANO.

Gordonstoun Moray. 'Gordon's estate'. ME *toun*. The estate was acquired in 1638 by Sir Robert *Gordon*.

Gorey (*Guaire*) Wexford. 'Sandbank'.

Goring, '(settlement of) the family or followers of a man called *Gāra', OE pers. name + *-ingas*: **Goring** Oxon. *Garinges* 1086 (DB). **Goring by Sea** W. Sussex. *Garinges* 1086 (DB).

Gorleston on Sea Norfolk. *Gorlestuna* 1086 (DB). Probably 'farmstead of a man called *Gurl'. OE pers. name + *tūn*.

Gorran Cornwall. *Sanctus Goranus* 1086 (DB). '(Church of) St Goran'. From the patron saint of the church.

Gorseinon Swan. 'Eynon's marsh'. Welsh *cors*. The identity of the named man is unknown.

Gorsley Glos. *Gorstley* 1228. 'Woodland clearing where gorse grows'. OE *gorst* + *lēah*.

Gort (*Gort*) Galway. 'Tilled field'.

Gortaclare (*Gort an Chláir*) Tyrone. *Gortclare* 1620. 'Tilled field of the plain'.

Gortahill (*Gort an Choill*) Cavan. 'Tilled field of the wood'.

Gortamullin (*Gort an Mhuilinn*) Kerry. 'Tilled field of the mill'.

Gortanear (*Gort an Fhéir*) Westmeath. 'Garden of the grass'.

Gortarevan (*Gort an Riabháin*) Offaly. 'Tilled field of the little stripe'.

Gortatlea (*Gort an tSléibhe*) Kerry. 'Tilled field of the mountain'.

Gortavalla (*Gort an Bhaile*) Limerick. 'Tilled field of the homestead'.

Gortavilly (*Gort an Bhile*) Tyrone. 'Tilled field of the sacred tree'.

Gortavoy (*Gort an Bheathaigh*) Tyrone. *Gortavaghy* 1666. 'Tilled field of the cow'.

Gorteen (*Goirtín*) Galway, Sligo, Waterford, etc. 'Little tilled field'.

Gorteeny (*Goirtíní*) Galway. 'Little tilled fields'.

Gorticastle (*Gort an Chaisil*) Tyrone. *Gorte-Castell* 1666. 'Tilled field of the stone fort'.

Gortin (*Goirtín*) Tyrone. *Goirtín c.*1613. 'Little tilled field'.

Gortinagin (*Gortín na gCeann*) Tyrone. 'Tilled field of the heads'.

Gortinure (*Gort an Iuir*) Derry. 'Field of the yew'.

Gortmore (*Gort Mhór*) Tipperary. 'Large tilled field'.

Gortnagappul (*Gort na gCapall*) Clare, Kerry. 'Field of the horses'.

Gortnagoyne (*Gort na gCadhan*) Galway, Roscommon. 'Field of the ducks'.

Gortnagrour (*Gort na gCreabhar*) Limerick. 'Field of the woodcock'.

Gortnahaha (*Gort na hÁithe*) Clare, Tipperary. 'Field of the kiln'.

Gortnahoe (*Gort na hUamha*) Tipperary. 'Tilled field of the cave'.

Gortnahoimna (*Gort na hOmna*) Cavan. 'Field of the oak'.

Gortnahoo (*Gort na hUamha*) Tipperary. 'Tilled field of the cave'.

Gortnananny (*Gort an Eanaigh*) Galway. 'Field of the marsh'.

Gortnavea (*Gort na bhFiadh*) Galway. 'Field of the deer'.

Gortnaveigh (*Gort na bhFiadh*) Tipperary. 'Field of the deer'.

Gorton Manch. *Gorton* 1282. 'Dirty farmstead'. OE *gor* + *tūn*.

Gortreagh (*Gort Riabhach*) Tyrone. *Gortreagh c.*1655. 'Striped field'.

Gortymadden (*Gort Uí Mhadaín*) Galway. 'Ó Madaín's tilled field'.

Gorvagh (*Garbhach*) Sligo. 'Rough place'.

Gosbeck Suffolk. *Gosebech* 1179. 'Stream frequented by geese'. OE *gōs* + *bece* (replaced by OScand. *bekkr*).

Gosberton Lincs. *Gosebertechirche* 1086 (DB), *Gosburton* 1487. 'Church of a man called Gosbert'. OGerman pers. name + OE *cirice* (replaced by ME *toun* 'village' from 15th cent.).

Gosfield Essex. *Gosfeld* 1198. 'Open land frequented by (wild) geese'. OE *gōs* + *feld*.

Gosford, 'ford frequented by geese', OE *gōs* + *ford*; e.g. **Gosford** Devon. *Goseford* 1249.

Gosforth, 'ford frequented by geese', OE *gōs* + *ford*; **Gosforth** Cumbria. *Goseford c.*1150. **Gosforth** Newc. upon T. *Goseford* 1166.

Gosport Hants. *Goseport* 1250. 'Market town where geese are sold'. OE *gōs* + *port*.

Goswick Northum. *Gossewic* 1202. 'Farm where geese are kept'. OE *gōs* + *wīc*.

Gotham Notts. *Gatham* 1086 (DB). 'Homestead or enclosure where goats are kept'. OE *gāt* + *hām* or *hamm*.

Gotherington Glos. *Godrinton* 1086 (DB). 'Estate associated with a man called Gūthhere'. OE pers. name + *-ing-* + *tūn*.

Goudhurst Kent. *Guithyrste* 11th cent. Probably 'wooded hill of a man called Gūtha'. OE pers. name + *hyrst*.

Gougane Barra (*Guagán Barra*) Cork. 'Mountain recess of Barra'.

Goulceby Lincs. *Colchesbi* 1086 (DB). 'Farmstead or village of a man called *Kolkr'. OScand. pers. name + *bý*.

Gourock Invclyd. *Ouir et Nether Gowrockis* 1661. '(Place by the) rounded hillock'. Gaelic *guireag* 'pimple'.

Govan Glas. *Gvuan c.*1150. Meaning not known.

Gowdall E. R. Yorks. *Goldale* 12th cent. 'Nook of land where marigolds grow'. OE *golde* + *halh*.

Gower (*Gŵyr*) (peninsula) Swan. 'Curved (promontory)'. Welsh *gŵyr*.

Gowerton (*Tre-gŵyr*) Swan. 'Town of the Gower (peninsula)'. *See* GOWER. OE *tūn* (Welsh *tref*).

Gowlane (*Gabhlán*) Kerry. 'Fork'.

Gowlaun (*Gabhlán*) Galway. 'Fork'.

Gowna, Lough (*Loch Gamhna*) Cavan. 'Lake of the calf'.

Gowran (*Gabhrán*) Kerry. '(Pass of) Gobhrán'.

Goxhill, probably from OScand. **gausli* '(place at) the gushing spring': **Goxhill** E. R. Yorks. *Golse* [*sic*] 1086 (DB), *Gosla, Gousele* 12th cent. **Goxhill** N. Lincs. *Golse* [*sic*] 1086 (DB), *Gousle, Gousel* 12th cent.

Gracehill Antrim. 'Hill of grace'. *Gracehill* 1835. The village was founded by the Moravians in 1756. Its Irish name is *Baile Uí Chinnéide*, 'homestead of Ó Cinnéide'.

Graffham W. Sussex. *Grafham* 1086 (DB). 'Homestead or enclosure in or by a grove'. OE *grāf* + *hām* or *hamm*.

Grafham Cambs. *Grafham* 1086 (DB). Identical in origin with the previous name.

Grafton, a common name, usually 'farmstead in or by a grove', OE *grāf* + *tūn*; examples include: **Grafton** Herefs., near Hereford. *Crafton* 1303. **Grafton** N. Yorks. *Graftune* 1086 (DB). **Grafton** Oxon. *Graptone* [*sic*] 1086 (DB), *Graftona* 1130. **Grafton, East & Grafton, West** Wilts. *Graftone* 1086 (DB). **Grafton Flyford** Worcs. *Graftun* 9th cent., *Garstune* [*sic*] 1086 (DB). Affix from the nearby FLYFORD FLAVELL. **Grafton Regis** Northants. *Grastone* [*sic*] 1086 (DB), *Graftone* 12th cent. Latin affix *regis* 'of the king' because it was a royal manor. **Grafton Underwood** Northants. *Grastone* [*sic*] 1086 (DB), *Grafton Underwode* 1367. Affix 'under or near the wood' (OE *under* + *wudu*) refers to Rockingham Forest.

However the following has a different origin: **Grafton, Temple** Warwicks. *Greftone* 10th cent., *Grastone* [*sic*] 1086 (DB), *Temple Grafton* 1363. 'Farmstead by the pit or trench'. OE *græf* + *tūn*. Affix refers to early possession by the Knights Templars or Hospitallers.

Grageen (*Gráigín*) Wexford. 'Little hamlet'.

Graig (*Gráig*) Cork, Galway, Limerick, Tyrone. 'Hamlet'.

Graigeen (*Gráigín*) Limerick. 'Little hamlet'.

Graignagower (*Gráig na nGabhar*) Kerry. 'Hamlet of the goats'.

Graigue (*Gráig*) Cork, Galway, Sligo, Tipperary. 'Hamlet'.

Graiguecullen (*Gráig Cguillin*) Carlow. 'Hamlet of the steep slope'.

Graiguenamanagh (*Gráig na Manach*) Kerry. 'Hamlet of the monks'.

Grain Medway. *Grean* c.1100. 'Gravelly, sandy ground'. OE **grēon*. **Isle of Grain** (formerly an actual island) is *Ile of Greane* 1610.

Grainsby Lincs. *Grenesbi* 1086 (DB). Probably 'farmstead or village of a man called Grein'. OScand. pers. name + *bý*.

Grainthorpe Lincs. *Germund(s)torp* 1086 (DB). 'Outlying farmstead of a man called Geirmundr or Germund'. OScand. or OGerman pers. name + OScand. *thorp*.

Grampians (mountains) Aber, Highland, Perth. Meaning uncertain. Claimed by some to be the *Mons Grampius* of the Romans.

Grampound Cornwall. *Grauntpount* 1302. '(Place at) the great bridge'. OFrench *grant* + *pont*.

Granagh (*Greanach*) Limerick. 'Gravelly place'.

Granborough Bucks. *Grenebeorge* c.1060, *Grenesberga* [*sic*] 1086 (DB). 'Green hill'. OE *grēne* + *beorg*.

Granby Notts. *Granebi* 1086 (DB). 'Farmstead or village of a man called Grani'. OScand. pers. name + *bý*.

Graney (*Greanach*) Kildare. 'Gravelly place'.

Grange, usually 'outlying farm belonging to a religious house', ME *grange*; examples include: **Grange** Cumbria, near Keswick. *The Grange* 1576. Belonged to the monks of the Furness Abbey. **Grange-over-Sands** Cumbria. *Grange* 1491. Belonged to Cartmel Priory, affix meaning 'across the sands of Morecambe Bay'.

However the following has a different origin: **Grange** Medway. *Grenic* c.1100. Probably

identical in origin and meaning with
GREENWICH.

Grange (*Gráinseach*) Cavan, Dublin, Galway,
etc. 'Monastic grange'.

Grangecon (*Gráinseach Coinn*) Wicklow.
'Conn's grange'.

Grangemouth Falk. 'Mouth of the Grange
Burn'. The river is named from the *grange* of
Newbattle Abbey.

Grangetown Red. & Cleve. 19th-cent. iron
and steel town, named from nearby ESTON
Grange which belonged to Fountains Abbey (for
the meaning of *grange see* GRANGE).

Grangicam (*Gráinseach Cam*) Down.
'Crooked grange'.

**Gransden, Great & Gransden,
Little** Cambs. *Grantandene* 973,
Grante(s)dene 1086 (DB). 'Valley of a man called
*Granta or *Grante'. OE pers. name + *denu*.

Gransha (*Gráinseach*) Donegal, Down, etc.
'Monastic grange'.

Gransmoor E. R. Yorks. *Grentesmor* 1086
(DB). 'Marshland of a man called *Grante or
*Grentir'. OE or OScand. pers. name + OE *mōr*
or OScand. *mór*.

Granston (*Treopert*) Pemb. *Villa Grandi*
1291, *Grandiston* 1535. 'Grand's farm'. French
pers. name + OE *tūn*. The Welsh name means
'Robert's farm' (French pers. name + Welsh
tref).

Grantchester Cambs. *Granteseta* 1086 (DB).
'Settlers on the River Granta'. Celtic river-name
(etymology obscure) + OE *sǣte*.

Grantham Lincs. *Grantham* 1086 (DB).
Possibly 'homestead or village of a man called
*Granta'. OE pers. name + *hām*. Alternatively
the first element may be OE *grand* 'gravel'.

Grantley, High N. Yorks. *Grantelege c.*1030,
Grentelai 1086 (DB). 'Woodland clearing of a
man called *Granta or *Grante'. OE pers. name
+ *lēah*.

Granton Edin. *Grendun c.*1200. 'Farm built
on gravel'. OE *grand* + *tūn*.

Grantown-on-Spey Highland. 'Grant's
town on the River Spey'. The town arose as a
model village planned by James *Grant* in 1765.
The river has a pre-Celtic name of unknown
meaning.

Grappenhall Warrtn. *Gropenhale* 1086 (DB).
'Nook of land at a ditch or drain'. OE *grōpe*
(genitive -*an*) + *halh*.

Grasby Lincs. *Gros(e)bi* 1086 (DB). OScand. *bý*
'farmstead, village', first element possibly
OScand. *gróthsamr* 'fertile'.

Grasmere Cumbria. *Gressemere* 1245.
Probably 'mere called grass lake'. OE *gres* + *sǣ*
'lake' with explanatory *mere*.

Grassendale Lpool. *Gresyndale* 13th cent.
'Grassy valley', or 'valley used for grazing'. OE
gærsen or *gærsing* + *dæl* or OScand. *dalr*.

Grassholm (island) Pemb. *Insula Grasholm*
15th cent. 'Grassy island'. OScand. *gras* + *holmr*.

Grassington N. Yorks. *Ghersintone* 1086
(DB). 'Grazing or pasture farm'. OE *gærsing* +
tūn.

Grassthorpe Notts. *Grestorp* 1086 (DB).
'Grass farmstead or hamlet'. OScand. *gres* +
thorp.

Grately Hants. *Greatteleiam* 929. 'Great wood
or clearing'. OE *grēat* + *lēah*.

Gratwich Staffs. *Crotewiche* [*sic*] 1086 (DB),
Grotewic 1176. '(Dairy) farm by the gravelly
place'. OE *grēote* + *wīc*.

Graveley Cambs. *Greflea* 10th cent., *Gravelei*
1086 (DB). Possibly 'woodland clearing by the
pit or trench'. OE *græf* + *lēah*. Alternatively
identical in origin with the following name.

Graveley Herts. *Gravelai* 1086 (DB). 'Clearing
by a grove or copse'. OE *grǣfe* or *grǣf(a)* + *lēah*.

Graveney Kent. *Grafonaea* 9th cent. '(Place
at) the ditch stream'. OE *grafa* (genitive -*n*) +
ēa.

Gravenhurst Beds. *Grauenhurst* 1086 (DB).
'Wooded hill with a coppice or a ditch'. OE *grāfa*
or *grafa* (genitive -*n*) + *hyrst*.

Gravesend Kent. *Gravesham* [*sic*] 1086 (DB),
Grauessend 1157. '(Place at) the end of the grove
or copse'. OE *grāf* + *ende*. The DB form seems to
contain OE *hām* 'homestead'.

Gravesham (district) Kent. A modern
adoption of the Domesday Book name of
GRAVESEND.

Grayingham Lincs. *Graingeham* 1086 (DB).
'Homestead of the family or followers of a man
called *Grǣg(a)'. OE pers. name + -*inga*- + *hām*.

Grayrigg Cumbria. *Grarigg* 12th cent. 'Grey ridge'. OScand. *grár* + *hryggr*.

Grays Thurr. *Turruc* 1086 (DB), *Turrokgreys* 1248. Originally Grays THURROCK, with manorial affix from the *de Grai* family, here in the 12th cent.

Grayshott Hants. *Grauesseta* 1185. 'Corner of land near a grove'. OE *grāf* + *scēat*.

Grazeley Wokhm. *Grǣgsole c.*950. Probably 'wolves' wallowing-place'. OE **grǣg* + *sol*.

Greasbrough Rothm. *Gersebroc* 1086 (DB). Probably 'grassy brook'. OE *gærs*, **gærsen* + *brōc*.

Greasby Wirral. *Gravesberie* 1086 (DB), *Grauisby c.*1100. 'Stronghold at a grove or copse'. OE *grǣfe* + *burh* (dative *byrig*) (replaced by OScand. *bý* 'farmstead').

Greasley Notts. *Griseleia* 1086 (DB). OE *lēah* 'wood or woodland clearing' with an uncertain first element, possibly OE *grēosn* 'gravel'.

Great as affix. *See* main name, e.g. for **Great Abington** (Cambs.) *see* ABINGTON.

Great Cumbrae. *See* CUMBRAE.

Greatford Lincs. *Greteford* 1086 (DB). 'Gravelly ford'. OE *grēot* + *ford*.

Greatham, 'gravelly homestead or enclosure', OE *grēot* + *hām* or *hamm*: **Greatham** Hants. *Greteham* 1086 (DB). **Greatham** Hartlepl. *Gretham* 1196.

Greatstone-on-Sea Kent. a recent name, taken from a rocky headland called *Great Stone* 1801, a shoreline feature before coastal changes took place.

Greatworth Northants. *Grentevorde* [sic] 1086 (DB), *Gretteworth* 12th cent. 'Gravelly enclosure'. OE *grēot* + *worth*.

Green Hammerton N. Yorks. *See* HAMMERTON.

Greenan (*Grianán*) Antrim, Donegal, Fermanagh, etc. 'Sunny place'.

Greenan Arg. (Bute). *Grenan* 1400. 'Sunny spot'. Gaelic *grianan*.

Greenane (*Grianán*) Kerry, Wicklow. 'Sunny place'.

Greenans (*Grianáin*) Donegal. 'Sunny places'.

Greenanstown (*Baile Uí Ghrianáin*) Meath. 'Homestead of Ó Grianáin'.

Greenaun (*Grianán*) Clare, Leitrim. 'Sunny place'.

Greencastle Antrim. Donegal, Down, Tyrone. 'Green castle'. The Irish name for these places varies. A 'green castle' is usually one in verdant countryside.

Greenford Gtr. London. *Grenan forda* 845, *Greneforde* 1086 (DB). '(Place at) the green ford'. OE *grēne* + *ford*.

Greenham W. Berks. *Greneham* 1086 (DB). 'Green enclosure or river-meadow'. OE *grēne* + *hamm*.

Greenhaugh Northum. *Le Grenehalgh* 1326. 'The green nook of land'. OE *grēne* + *halh*.

Greenhead Northum. *Le Greneheued* 1290. 'The green hill'. OE *grēne* + *hēafod*.

Greenhithe Kent. *Grenethe* 1264. 'Green landing-place'. OE *grēne* + *hȳth*.

Greenhow Hill N. Yorks. *Grenehoo* 1540. 'Green mound or hill'. OE *grēne* + OScand. *haugr*, with the later addition of *hill*.

Greenlaw Sc. Bord. *Grenlawe* 1250. 'Green hill'. OE *grēne* + *hlāw*.

Greenock Invclyd. *Grenok c.*1395. 'Sunny hillock'. Gaelic *grianag*.

Greenodd Cumbria. *Green Odd* 1774. 'Green point or tongue of land'. From OScand. *oddi*.

Greenore (*An Grianfort*) Louth. 'The sunny port'.

Greens Norton Northants. *See* NORTON.

Greenstead Essex. *Grenstede* 10th cent., *Grenesteda* 1086 (DB). 'Green place', i.e. 'pasture used for grazing'. OE *grēne* + *stede*.

Greensted Essex. *Gernesteda* 1086 (DB). Identical in origin with the previous name.

Greenwich Gtr. London. *Grenewic* 964, *Grenviz* 1086 (DB). 'Green trading settlement or harbour'. OE *grēne* + *wīc*.

Greet Glos. *Grete* 12th cent. 'Gravelly place'. OE **grēote*.

Greete Shrops. *Grete* 1183. Identical in origin with the previous name.

Greetham, 'gravelly homestead or enclosure', OE *grēot* + *hām* or *hamm*: **Greetham** Lincs. *Gretham* 1086 (DB). **Greetham** Rutland. *Gretham* 1086 (DB).

Greetland Calder. *Greland* [*sic*] 1086 (DB), *Greteland* 13th cent. 'Rocky cultivated land'. OScand. *grjót* + *land*.

Greetwell, North Lincs. *Grentewelle* [*sic*] 1086 (DB), *Gretwella* c.1115. 'The gravelly spring'. OE *grēot* + *wella*.

Greinton Somerset. *Graintone* 1086 (DB). 'Farmstead of a man called *Grǣga*. OE pers. name (genitive -*n*) + *tūn*.

Grenagh (*Greanach*) Cork. 'Gravelly place'.

Grendon, usually 'green hill', OE *grēne* + *dūn*: **Grendon** Northants. *Grendone* 1086 (DB). **Grendon** Warwicks. *Grendone* 1086 (DB). **Grendon Underwood** Bucks. *Grennedone* 1086 (DB). Affix means 'in or near the wood'.
   However the following has a different second element: **Grendon** Herefs. *Grenedene* 1086 (DB). 'Green valley'. OE *grēne* + *denu*.

Gresffordd. *See* GRESFORD.

Gresford (*Gresfford*) Wrex. *Gretford* [*sic*] 1086 (DB), *Gresford* 1273. 'Grassy ford'. OE *græs* + *ford*.

Gresham Norfolk. *Gressam* 1086 (DB). 'Grass homestead or enclosure'. OE *græs* + *hām* or *hamm*.

Gresley, Castle & Gresley, Church Derbys. *Gresele* c.1125, *Castelgresele* 1252, *Churchegreseleye* 1363. OE *lēah* 'woodland clearing' with an uncertain first element, possibly OE *grēosn* 'gravel'. Distinguishing affixes from the former castle and the church here.

Gressenhall Norfolk. *Gressenhala* 1086 (DB). 'Grassy or gravelly nook of land'. OE *gærsen* or *grēosn* + *halh*.

Gressingham Lancs. *Ghersinctune* 1086 (DB), *Gersingeham* 1183. 'Homestead or enclosure with grazing or pasture'. OE *gærsing* + *hām* or *hamm* (replaced by *tūn* 'farmstead' in the Domesday form).

Greta (river), three examples, in Cumbria, Lancs., and N. Yorks.; an OScand. river-name recorded from the 13th cent., 'stony stream', from OScand. *grjót* + *á*.

Gretna Green Dumf. *Gretenho* 1223, *Gretenhou* c.1240, *Gratnay* 1576. 'Green by Gretna'. *Gretna* is 'gravelly hill' (OE **grēoten* + *hōh*).

Gretton Glos. *Gretona* 1175. 'Farmstead near GREET'. OE *tūn*.

Gretton Northants. *Gretone* 1086 (DB). 'Gravel farmstead'. OE *grēot* + *tūn*.

Gretton Shrops. *Grotintune* 1086 (DB). 'Farmstead on gravelly ground'. OE **grēoten* + *tūn*.

Grewelthorpe N. Yorks. *Torp* 1086 (DB), *Gruelthorp* 1281. OScand. *thorp* 'outlying farmstead' with later manorial affix from a family called *Gruel*.

Greyabbey (*Mainistir Liath*) Down. *Monasterlech* 1193. 'Grey abbey'. The reference is to a Cistercian abbey founded here in 1193.

Greysouthen Cumbria. *Craykesuthen* c.1187. 'Rock or cliff of a man called Suthán'. Celtic **creig* + OIrish pers. name.

Greystoke Cumbria. *Creistoc* 1167. Probably 'secondary settlement by a river once called *Cray*'. Lost Celtic river-name (meaning 'rough, turbulent') + OE *stoc*.

Greywell Hants. *Graiwella* 1167. Probably 'spring or stream frequented by wolves', OE **grǣg* + *wella*.

Grianan (*Grianán*) Donegal. 'Sunny place'.

Griff Warwicks. *Griva* 12th cent. '(Place at) the deep valley or hollow'. OScand. *gryfja*.

Griffithstown Torf. Henry *Griffiths*, the first stationmaster of Pontypool Road, founded a new settlement here c.1856.

Grimley Worcs. *Grimanleage* 9th cent., *Grimanleh* 1086 (DB). 'Wood or glade haunted by a spectre or goblin'. OE *grīma* + *lēah*.

Grimoldby Lincs. *Grimoldbi* 1086 (DB). 'Farmstead or village of a man called Grimaldi'. OScand. pers. name + *bý*.

Grimsargh Lancs. *Grimesarge* 1086 (DB). 'Hill-pasture of a man called Grímr'. OScand. pers. name + *erg*.

Grimsby, 'farmstead or village of a man called Grímr', OScand. pers. name + *bý*: **Grimsby** NE. Lincs. *Grimesbi* 1086 (DB). **Grimsby, Little** Lincs. *Grimesbi* 1086 (DB).

Grimscote Northants. *Grimescote* 12th cent. 'Cottage(s) of a man called Grimr'. OScand. pers. name + OE *cot*.

Grimsetter Orkn. 'Grímr's homestead'. OScand. pers. name + *setr*.

Grimshader W. Isles. (Lewis). 'Grímr's homestead'. OScand. pers. name + *setr*.

Grimstead, East & Grimstead, West Wilts. *Gremestede* 1086 (DB). Probably 'green homestead'. OE *grēne* + *hām-stede*.

Grimsthorpe Lincs. *Grimestorp* 1212. 'Outlying farmstead or hamlet of a man called Grímr'. OScand. pers. name + *thorp*.

Grimston, 'farmstead or estate of a man called Grímr', OScand. pers. name + OE *tūn*; examples include: **Grimston** Leics. *Grimestone* 1086 (DB). **Grimston** Norfolk. *Grimastun* c.1035, *Grimestuna* 1086 (DB). **Grimston, North** N. Yorks. *Grimeston* 1086 (DB).

Grindale E. R. Yorks. *Grendele* 1086 (DB). 'Green valley'. OE *grēne* + *dæl*.

Grindleton Lancs. *Gretlintone* 1086 (DB), *Grenlington* 1251. Probably 'farmstead near the gravelly stream'. OE **grendel* + *-ing* + *tūn*.

Grindley Staffs. *Grenleg* 1251. 'Green woodland clearing'. OE *grēne* + *lēah*.

Grindlow Derbys. *Grenlawe* 1199. 'Green hill or mound'. OE *grēne* + *hlāw*.

Grindon, 'green hill', OE *grēne* + *dūn*: **Grindon** Northum. *Grandon* 1210. **Grindon** Staffs. *Grendone* 1086 (DB).

Gringley on the Hill Notts. *Gringeleia* 1086 (DB). Possibly 'woodland clearing of the people living at the green place'. OE *grēne* + *-inga-* + *lēah*.

Grinsdale Cumbria. *Grennesdale* c.1180. Probably 'valley by the green promontory'. OE *grēne* + *næss* + OScand. *dalr*.

Grinshill Shrops. *Grivelesul* [*sic*] 1086 (DB), *Grineleshul* 1242. OE *hyll* 'hill' with an uncertain first element, possibly a word **grynel* (genitive *-es*), a derivative of OE *grin, gryn* 'a snare for animals'.

Grinstead, 'green place', i.e. 'pasture used for grazing', OE *grēne* + *stede*: **Grinstead, East** W. Sussex. *Grenesteda* 1121, *Estgrenested* 1271.

Grinstead, West W. Sussex. *Grenestede* 1086 (DB), *Westgrenested* 1280.

Grinton N. Yorks. *Grinton* 1086 (DB). 'Green farmstead'. OE *grēne* + *tūn*.

Gristhorpe N. Yorks. *Grisetorp* 1086 (DB). 'Outlying farmstead or hamlet of a man called Gríss, or where young pigs are reared'. OScand. pers. name or *gríss* + *thorp*.

Griston Norfolk. *Gristuna* 1086 (DB). Possibly 'farmstead of a man called Gríss, or where young pigs are reared'. OScand. pers. name or *gríss* + OE *tūn*. Alternatively the first element may be OE *gres* 'grass'.

Grittenham Wilts. *Gruteham* 850. 'Gravelly homestead or enclosure'. OE **grīeten* + *hām* or *hamm*.

Grittleton Wilts. *Grutelington* 940, *Gretelintone* 1086 (DB). Possibly 'estate associated with a man called **Grytel*'. OE pers. name + *-ing-* + *tūn*.

Grizebeck Cumbria. *Grisebek* 13th cent. 'Brook by which young pigs are kept'. OScand. *gríss* + *bekkr*.

Grizedale Cumbria. *Grysdale* 1336. 'Valley where young pigs are kept'. OScand. *gríss* + *dalr*.

Groby Leics. *Grobi* 1086 (DB). Probably 'farmstead near a hollow or pit'. OScand. *gróf* + *bý*.

Groeslon Gwyd. *Croes-lôn* 1838. 'Crossroads'. Welsh *croes* + *lôn*.

Grogan (*Grógán*) Laois. 'Little heap'.

Gronant Flint. *Gronant* 1086 (DB). 'Gravel brook'. Welsh *gro* + *nant*.

Groombridge E. Sussex. *Gromenebregge* 1239. 'Bridge where young men congregate'. ME *grome* (genitive plural *-ene*) + OE *brycg*.

Groomsport (*Port an Ghiolla Ghruama*) Down. 'Port of the gloomy fellow'. *Mollerytoun* 1333. The name appears to refer to a person surnamed *Mallory*, meaning 'unhappy' (OFrench *malheure*).

Grosmont N. Yorks. *Grosmunt* 1228. Originally the name of the priory here, in turn named from the mother Priory of Grosmont in France ('big hill' from OFrench *gros* + *mont*).

Grosmont (*Y Grysmwnt*) Mon. *Grosse Monte* 1187, *Grosmont* 1232. 'Big hill'. OFrench *gros* + *mont*.

Groton Suffolk. *Grotena* 1086 (DB). Probably 'sandy or gravelly stream'. OE **groten* + *ēa*.

Grove, a common name, '(place at) the grove or coppiced wood', OE *grāf(a)*; examples include: **Grove** Notts. *Grava* 1086 (DB). **Grove** Oxon. *la Graue* 1188.

Grundisburgh Suffolk. *Grundesburch* 1086 (DB). Possibly 'stronghold or manor of a man called *Grund'. OE pers. name + *burh*. Alternatively the first element may be OE *grund* 'ground, stretch of land', also 'foundation of a building'.

Grutness Shet. 'Gravel promontory'. OScand. *grjót* + *nes*.

Grysmwnt, Y. See GROSMONT (Mon.).

Guay Perth. *Guay* 1457. '(Place by the) marsh'. Gaelic *gaoth*.

Gubaveeny (*Gob an Mhianaigh*) Louth. 'Mouth of the mine'.

Gubbacrock (*Gob dhá Chnoc*) Fermanagh. 'Beak of two hills'.

Guernsey (island) Channel Islands. *Greneroy* early 11th cent., *Ghernesei* 1168. OScand. *ey* 'island' with an uncertain first element, possibly an OScand. masculine pers. name *Grani* or *Warinn* (genitive *-ar*), or OScand. *grǫn* (genitive *granar*) 'pine-tree'. The spelling with medial *-s-* probably originated in early English official usage, influenced by the name JERSEY. In the 4th-cent. *Maritime Itinerary*, Guernsey is recorded as *Lisia*, possibly a Celtic name meaning 'sedge place' or 'place of a man called *Liscos'.

Guestling E. Sussex. *Gestelinges* 1086 (DB). '(Settlement of) the family or followers of a man called *Gyrstel'. OE pers. name + *-ingas*.

Guestwick Norfolk. *Geghestueit* 1086 (DB). 'Clearing belonging to GUIST'. OScand. *thveit*.

Guilden Morden Cambs. See MORDEN.

Guilden Sutton Ches. See SUTTON.

Guildford Surrey. *Gyldeforda* c.880, *Gildeford* 1086 (DB). Probably 'ford by the gold-coloured (i.e. sandy) hill', from OE **gylde* + *ford*, see HOG'S BACK.

Guileen (*Gaibhlín*) Cork. 'Little fork'.

Guilsborough Northants. *Gildesburh* c.1070, *Gisleburg* [sic] 1086 (DB). 'Stronghold of a man called *Gyldi'. OE pers. name + *burh*.

Guisachan Highland. *Gulsackyn* 1221, *Guisachane* 1578. 'Pine forests'. Gaelic *guithsachan*.

Guisborough Red. & Cleve. *Ghigesburg* 1086 (DB). Probably 'stronghold of a man called Gígr'. OScand. pers. name + OE *burh*.

Guiseley Leeds. *Gislicleh* c.972, *Gisele* 1086 (DB). 'Woodland clearing of a man called *Gīslic'. OE pers. name + *lēah*.

Guist Norfolk. *Gæssæte* c.1035, *Gegeseta* 1086 (DB). Possibly 'dwelling of a man called *Gæga or *Gægi'. OE pers. name + *sǣte*.

Guiting Power & Temple Guiting Glos. *Gythinge* 814, *Getinge* 1086 (DB), *Gettinges Poer* 1220, *Guttinges Templar* 1221. Originally a river-name, 'running stream, stream with a good current', OE *gyte* + *-ing*. Manorial affixes from early possession by the *le Poer* family and by the Knights Templars.

Guldeford, East E. Sussex. *Est Guldeford* 1517. Named from the *Guldeford* family (so called because they came from GUILDFORD).

Gulladoo (*Guala Dhubh*) Leitrim. 'Black shoulder'.

Gulladuff (*Guala Dubh*) Derry. *Galladow* 1609. 'Black shoulder'.

Gulval Cornwall. 'Church of *Sancta Welvela*' 1328. '(Church of) St Gwelvel'. From the patron-saint of the church.

Gumfreston Pemb. *Villa Gunfrid* 1291, *Gonnfreiston* 1364. 'Gunfrid's farm'. OGerman pers. name + OE *tūn*.

Gumley Leics. *Godmundesleah* 8th cent., *Godmundelai* 1086 (DB). 'Woodland clearing of a man called Godmund'. OE pers. name + *lēah*.

Gunby E. R. Yorks. *Gunelby* 1066–9. 'Farmstead or village of a woman called Gunnhildr'. OScand. pers. name + *bý*.

Gunby Lincs. *Gunnebi* 1086 (DB). 'Farmstead or village of a man called Gunni'. OScand. pers. name + *bý*.

Gunnersbury Gtr. London. *Gounyldebury* 1334. 'Manor house of a woman called Gunnhildr'. OScand. pers. name + ME *bury* (from OE *byrig*, dative of *burh*).

Gunnerton Northum. *Gunwarton* 1170.
Probably 'farmstead of a woman called
Gunnvǫr'. OScand. pers. name + OE *tūn*.

Gunness N. Lincs. *Gunnesse* 1199.
'Headland of a man called Gunni'. OScand.
pers. name + *nes*.

Gunnislake Cornwall. *Gonellake* 1485.
Probably 'stream of a man called Gunni'.
OScand. pers. name + OE *lacu*.

Gunthorpe, usually 'outlying farmstead of a
man called Gunni', OScand. pers. name + *thorp*:
Gunthorpe Norfolk. *Gunestorp* 1086 (DB).
Gunthorpe Peterb. *Gunetorp* 1130.
    However the following has a different origin:
Gunthorpe Notts. *Gulnetorp* 1086 (DB).
'Outlying farmstead of a woman called
Gunnhildr'. OScand. pers. name + *thorp*.

Gunwalloe Cornwall. 'Chapel of *Sanctus
Wynwolaus*' 1332. From the patron saint of the
church or chapel, St Winwaloe.

Gurranebraher (*Garrán na mBráthar*)
Cork. 'Grove of the brothers'.

**Gussage All Saints & Gussage St
Michael** Dorset. *Gyssic* 10th cent., *Gessic* 1086
(DB), *Gussich All Saints* 1245, *Gussich St Michael*
1297. Probably 'gushing stream', originally the
name of the river here. OE **gyse* + *sīc*. Affixes
from the dedications of the churches.

Gusserane (*Ráth na gCosarán*) Wexford.
'Fort of the trampling'.

Guston Kent. *Gocistone* 1086 (DB).
'Farmstead of a man called **Gūthsige*'. OE pers.
name + *tūn*.

Guyhirn Cambs. *La Gyerne* 1275. OE
hyrne 'angle or corner of land' with OFrench
guie 'a guide' (with reference to controlling
tidal flow) or 'a salt-water ditch'.

Guyzance Northum. *Gynis* 1242. A manorial
name, 'estate of the family called *Guines*'.

Gwalchmai Angl. *Trefwalkemyth* 1291,
Trefwalghmey 1350, *Gwalghmey* 1352.
'(Place of) Gwalchmai'. The name is that
of the Welsh court poet *Gwalchmai* (*fl.*1130–80),
whose name is said to mean 'May hawk'
(Welsh *gwalch* + *Mai*). The first two forms
above have Welsh *tref* 'township'.

Gwaunysgor Flin. *Wenescol* 1086
(DB). 'Meadow by the fort'. Welsh *gwaun*
+ *ysgor*.

Gweebarra (*Gaoth Beara*) (district) Donegal.
'Inlet of the water'.

Gweedore (*Gaoth Dobhair*) Donegal.
Gydower al. Lower Dower 1657. 'Inlet of the
water'.

Gweek Cornwall. *Wika* 1201. Cornish **gwig*
'village' or OE *wīc* 'hamlet'.

Gweesalia (*Gaoth Sáile*) Mayo. 'Inlet of the
sea'.

Gwenfô. *See* WENVOE.

Gwennap Cornwall. 'Church of *Sancta
Wenappa*' 1269. '(Church of) St Wynup'. From
the patron saint of the church.

Gwent (the historic county). 'Trading place'.
British **venta*.

Gwithian Cornwall. 'Parish of *Sanctus
Goythianus*' 1334. From the patron saint of the
church, St Gothian.

Gwynedd (the unitary authority). 'Territory
of the Venedoti'. The ancient kingdom derives
its name from the Venedoti tribe.

Gwyr. *See* GOWER.

Gyleen (*Gaibhlín*) Cork. 'Little fork'.

Haa Shet. (Whalsay). 'Hall'. OScand. *hallr*.

Habblesthorpe Notts. *Happelesthorp* 1154. Probably 'outlying farmstead of a man called *Hæppel*'. OE pers. name + OScand. *thorp*.

Habost W. Isles. (Lewis). 'High farm'. OScand. *hár* + *bólstathr*.

Habrough NE. Lincs. *Haburne* [*sic*] 1086 (DB), *Haburg* 1202. 'High or chief stronghold'. OE *hēah* (replaced by OScand. *hár*) + *burh*.

Habton, Great N. Yorks. *Habetun* 1086 (DB). 'Farmstead of a man called *Hab(b)a*'. OE pers. name + *tūn*.

Haccombe Devon. *Hacome* 1086 (DB), *Hakcumbe* c.1200. Probably 'valley with a hatch or fence'. OE *hæcc* + *cumb*.

Haceby Lincs. *Hazebi* 1086 (DB), *Hathsebi* 1115. 'Farmstead or village of a man called Haddr'. OScand. pers. name + *bý*.

Hacheston Suffolk. *Hacestuna* 1086 (DB). 'Farmstead of a man called *Hæcci*'. OE pers. name + *tūn*.

Hackford Norfolk. near Wymondham. *Hakeforda* 1086 (DB). 'Ford with a hatch or by a bend'. OE *hæcc* or *haca* + *ford*.

Hackforth N. Yorks. *Acheford* 1086 (DB). Identical in origin with the previous name.

Hackleton Northants. *Hachelintone* 1086 (DB). 'Estate associated with a man called *Hæccel*'. OE pers. name + *-ing-* + *tūn*.

Hackness N. Yorks. *Hacanos* 731, *Hagenesse* 1086 (DB). 'Hook-shaped or projecting headland'. OE *haca* + *nose* (replaced by *næss*).

Hackney Gtr. London. *Hakeneia* 1198. Possibly 'island, or dry ground in marsh, of a man called *Haca*'. OE pers. name (genitive *-n*) + *ēg*. Alternatively the first element may be OE *haca* (genitive *-n*) 'hook-shaped ridge or tongue of land'.

Hackthorn Lincs. *Haggethorn* 968, *Hagetorne* 1086 (DB). '(Place at) the hawthorn or prickly thorn-tree'. OE *hagu-thorn* or rather *haca-thorn*.

Hackthorpe Cumbria. *Hakatorp* c.1150. Possibly 'outlying farmstead or hamlet of a man called Haki'. OScand. pers. name + *thorp*. Alternatively the first element may be OScand. *haki* or OE *haca* 'hook-shaped promontory'.

Haconby Lincs. *Hacunesbi* 1086 (DB). 'Farmstead or village of a man called Hákon'. OScand. pers. name + *bý*.

Haddenham, 'homestead or village of a man called Hǣda', OE pers. name + *hām*: **Haddenham** Bucks. *Hedreham* [*sic*] 1086 (DB), *Hedenham* 1142. **Haddenham** Cambs. *Hǣdan ham* 970, *Hadreham* 1086 (DB).

Haddington E. Loth. *Hadynton* 1098. 'Farmstead associated with a man called Hada'. OE pers. name + *-ing-* + *tūn*.

Haddington Lincs. *Hadinctune* 1086 (DB). 'Estate associated with a man called Headda or Hada'. OE pers. name + *-ing-* + *tūn*.

Haddiscoe Norfolk. *Hadescou* 1086 (DB). 'Wood of a man called Haddr or Haddi'. OScand. pers. name + *skógr*.

Haddlesey, Chapel & Haddlesey, West N. Yorks. *Hathel-sǣ* c.1030, *Chappel Haddlesey* 1605, *Westhathelsay* 1280. Probably 'marshy pool in a hollow'. OE *hathel* + *sǣ*. Distinguishing affixes from ME *chapel* and *west*.

Haddon, usually 'heath hill, hill where heather grows', OE *hǣth* + *dūn*: **Haddon, East & Haddon, West** Northants. *Eddone*, *Hadone* 1086 (DB), *Esthaddon* 1220, *Westhaddon* 12th cent. **Haddon, Nether & Haddon, Over** Derbys. *Hadun(e)* 1086 (DB), *Nethir Haddon* 1248, *Uverehaddon* 1206. Distinguishing affixes are OE *neotherra* 'lower' and *uferra* 'higher'.

However the following may have a different origin: **Haddon** Cambs. *Haddedun* 951, *Adone* 1086 (DB). Probably 'hill of a man called Headda'. OE pers. name + *dūn*.

Hadfield Derbys. *Hetfelt* 1086 (DB). 'Heathy open land, or open land where heather grows'. OE *hǣth* + *feld*.

Hadham, Much & Hadham, Little Herts. *Hǣdham* 957, *Hadam, Parva Hadam* 1086 (DB), *Muchel Hadham* 1373. Probably 'heath homestead'. OE *hǣth* + *hām*. Distinguishing affixes from OE *mycel* 'great' and *lȳtel* 'little' (Latin *parva*).

Hadleigh, Hadley, 'heath clearing, clearing where heather grows', OE *hǣth* + *lēah*: **Hadleigh** Essex. *Hǣthlege* c.1000, *Leam* [*sic*] 1086 (DB). **Hadleigh** Suffolk. *Hǣdleage* c.995, *Hetlega* 1086 (DB). **Hadley** Tel. & Wrek. *Hatlege* 1086 (DB). **Hadley, Monken** Gtr. London. *Hadlegh* 1248, *Monken Hadley* 1489. Affix means 'of the monks' (from ME *monk*, plural *monken*) referring to early possession by the Benedictine monks of Walden Abbey.

Hadlow Kent. *Haslow* [*sic*] 1086 (DB), *Hadlou* 1235. Probably 'mound or hill where heather grows'. OE *hǣth* + *hlāw*.

Hadlow Down E. Sussex. *Hadleg* 1254. Probably 'woodland clearing where heather grows'. OE *hǣth* + *lēah*.

Hadnall Shrops. *Hadehelle* [*sic*] 1086 (DB), *Hadenhale* 1242. 'Nook of land of a man called Headda'. OE pers. name (genitive *-n*) + *halh*.

Hadrian's Wall Cumbria.-Northum.-Tyneside, Roman fortification built AD 122–6, called after Roman Emperor *Hadrian* and giving name to WALL (Northum.), HEDDON ON THE WALL, and WALLSEND.

Hadstock Essex. *Hadestoc* 11th cent. 'Outlying farmstead of a man called Hada'. OE pers. name + *stoc*.

Hadzor Worcs. *Hadesore* 1086 (DB). 'Ridge of a man called *Headd'. OE pers. name + *ofer*.

Hafren Forest Powys. 'Forest of the River Severn'. *See* SEVERN.

Hagbourne, East & Hagbourne, West Oxon. *Haccaburna* c.895, *Hacheborne* 1086 (DB). Probably '(place by) the stream of a man called *Hacca'. OE pers. name + *burna*.

Haggerston Gtr. London. *Hergotestane* 1086 (DB). 'Boundary stone of a man called Hærgod'. OE pers. name + *stān*.

Haggerston Northum. *Agardeston* 1196. Probably 'estate of a family called Hagard'. ME (from OFrench) surname + OE *tūn*.

Hagley Worcs. near Kidderminster. *Hageleia* 1086 (DB). 'Woodland clearing where haws grow'. OE **hagga* + *lēah*.

Hagworthingham Lincs. *Hacberdingeham* [*sic*] 1086 (DB), *Hagwrthingham* 1198. Possibly 'homestead of the family or followers of a man called *Haguweard'. OE pers. name + *-inga-* + *hām*.

Haigh, 'the enclosure', OScand. *hagi* or OE *haga*: **Haigh** Barns. *Hagh* 1379. **Haigh** Wigan. *Hage* 1194.

Haighton Green Lancs. *Halctun* 1086 (DB). 'Farmstead in a nook of land'. OE *halh* + *tūn*.

Hail Weston Cambs. *See* WESTON.

Hailes Glos. *Heile* 1086 (DB), *Heilis* 1114. Perhaps a British folk-name derived from an old name **Heil* of the stream here.

Hailey, 'clearing where hay is made', OE *hēg* + *lēah*: **Hailey** Herts. *Hailet* [*sic*] 1086 (DB), *Heile* 1235. **Hailey** Oxon. *Haylegh* 1241.

Hailsham E. Sussex. *Hamelesham* [*sic*] 1086 (DB), *Helesham* 1189. 'Homestead or enclosure of a man called *Hægel'. OE pers. name + *hām* or *hamm*.

Hainault Gtr. London. *Henehout* 1221, *Hyneholt* 1239. 'Wood belonging to a religious community' (here the Abbey of BARKING). OE *hīwan* (genitive plural *hīgna*) + *holt*.

Hainford Norfolk. *Hemfordham* (for *Hein-*) c.1060, *Han-, Hamforda* 1086 (DB), *Heinford* 12th cent. 'Ford near an enclosure'. OE **hægen* + *ford*.

Hainton Lincs. *Haintone* 1086 (DB). 'Farmstead in an enclosure'. OE **hægen* + *tūn*.

Haisthorpe E. R. Yorks. *Ascheltorp* 1086 (DB). 'Outlying farmstead of a man called Hǫskuldr'. OScand. pers. name + *thorp*.

Halam Notts. *Healum* 958. '(Place at) the nooks or corners of land'. OE *halh* in a dative plural form *halum*.

Halberton Devon. *Halsbretone* 1086 (DB). Possibly 'farmstead by a hazel wood'. OE *hæsel* + *bearu* + *tūn*.

Halden, High Kent. *Hadinwoldungdenne* c.1100. 'Woodland pasture associated with a man called Heathuwald'. OE pers. name + *-ing-* + *denn*.

Hale, a common name, '(place at) the nook or corner of land', OE *halh* (dative *hale*); examples include: **Hale** Halton. *Halas* 1094. Originally in a plural form. **Hale** Hants. *Hala* 1161. **Hale** Traffd. *Hale* 1086 (DB). **Hale, Great & Hale, Little** Lincs. *Hale* 1086 (DB).

Hales, 'the nooks or corners of land', OE *halh* in a plural form: **Hales** Norfolk. *Hals* 1086 (DB). **Hales** Staffs. *Hales* 1291.

Halesowen Dudley. *Hala* 1086 (DB), *Hales Ouweyn* 1276. 'Nooks or corners of land'. OE *halh* in a plural form, with manorial affix from the Welsh prince called *Owen* who held the manor in the early 13th cent.

Halesworth Suffolk. *Healesuurda* 1086 (DB). Probably 'enclosure of a man called *Hæle*'. OE pers. name + *worth*.

Halewood Knows. *Halewode* c.1200. 'Wood near HALE'. OE *wudu*.

Halford Shrops. *Hauerford* 1155, *Haleford* 1221–2. OE *ford* 'a ford' with an uncertain first element, possibly OE *halh* 'nook of land, land in a river-bend'.

Halford Warwicks. *Halchford* 12th cent. 'Ford by a nook or corner of land'. OE *halh* + *ford*.

Halifax Calder. *Halyfax* c.1095. Possibly 'area of coarse grass in a nook of land'. OE *halh* + *gefeaxe*.

Halkyn (*Helygain*) Flin. *Helchene* 1086 (DB), *Helygen* 1315. 'Place of willow-trees'. Welsh *helygen*.

Hallam Sheff. *Hallun* 1086 (DB). Possibly identical in origin with the next name, but alternatively '(place at) the rocks', from OScand. *hallr* or OE *hall* in a dative plural form *hallum*. **Hallamshire**, *Halumsira* 1161, contains OE *scīr* 'district'.

Hallam, Kirk & Hallam, West Derbys. *Halun* 1086 (DB), *Kyrkehallam* 12th cent., *Westhalum* 1230. '(Place at) the nooks of land'. OE *halh* in a dative plural form *halum*.

Distinguishing affixes from OScand. *kirkja* 'church' and OE *west*.

Hallaton Leics. *Alctone* 1086 (DB). 'Farmstead in a nook of land or narrow valley'. OE *halh* + *tūn*.

Hallatrow B. & NE. Som. *Helgetrev* 1086 (DB). '(Place by) the holy tree'. OE *hālig* + *trēow*.

Halling Medway. *Hallingas* 8th cent., *Hallinges* 1086 (DB). '(Settlement of) the family or followers of a man called *Heall*'. OE pers. name + *-ingas*.

Hallingbury, Great & Hallingbury, Little Essex. *Hallingeberiam* 1086 (DB). 'Stronghold of the family or followers of a man called *Heall*'. OE pers. name + *-inga-* + *burh* (dative *byrig*).

Hallington Northum. *Halidene* 1247. 'Holy valley'. OE *hālig* + *denu*.

Halloughton Notts. *Healhtune* 958. 'Farmstead in a nook of land or narrow valley'. OE *halh* + *tūn*.

Hallow Worcs. *Halhagan* 9th cent., *Halhegan* 1086 (DB). 'Enclosures in a nook or corner of land'. OE *halh* + *haga*.

Halnaker W. Sussex. *Helnache* 1086 (DB). Possibly 'half an acre'. OE *healf* (dative *-an*) + *æcer* 'plot of cultivated or arable land'.

Halsall Lancs. *Heleshale* 1086 (DB). Probably 'nook of land of a man called *Hæle*'. OE pers. name + *halh*.

Halse Northants. *Hasou* [*sic*] 1086 (DB), *Halsou* c.1160. 'Neck of land forming a ridge'. OE *hals* + *hōh*.

Halse Somerset. *Halse* 1086 (DB). '(Place at) the neck of land'. OE *hals*.

Halsham E. R. Yorks. *Halsaham* 1033, *Halsam* 1086 (DB). 'Homestead on the neck of land'. OE *hals* + *hām*.

Halstead, 'place of refuge or shelter (for livestock)', OE *h(e)ald* + *stede*: **Halstead** Essex. *Haltesteda* 1086 (DB). **Halstead** Kent. *Haltesteda* c.1100. **Halstead** Leics. *Elstede* 1086 (DB), *Haldsted* 1230.

Halstock Dorset. *Halganstoke* 998. 'Outlying farmstead belonging to a religious foundation'. OE *hālig* + *stoc*.

Halstow, 'holy place', OE *hālig* + *stōw*:
Halstow, High Medway. *Halgesto* c.1100.
Halstow, Lower Kent. *Halgastaw* c.1100.

Haltemprice E. R. Yorks. *Hautenprise* 1324.
'High or noble enterprise'. OFrench *haut* +
emprise, originally with reference to an
Augustinian Priory founded here in 1322.

Haltham Lincs. *Holtham* 1086 (DB).
'Homestead by or in a wood'. OE *holt* + *hām*.

Halton, a common name, usually 'farmstead
in a nook or corner of land', OE *halh* + *tūn*:
Halton Bucks. *Healtun* c.1033, *Haltone* 1086
(DB). **Halton** Lancs. *Haltune* 1086 (DB).
Halton Leeds. *Halletune* 1086 (DB). **Halton,
East** N. Lincs. *Haltune* 1086 (DB). **Halton
East** N. Yorks. *Haltone* 1086 (DB). **Halton
Holegate** Lincs. *Haltun* 1086 (DB). Affix
means 'road in a hollow', OScand. *holr* + *gata*.
Halton, West N. Lincs. *Haltone* 1086 (DB).
Halton West N. Yorks. *Halctun* 12th cent.
    However the following have a different origin:
Halton Halton. *Heletune* 1086 (DB), *Hethelton*
1174. Possibly 'farmstead at a heathery place'.
OE **hāthel* + *tūn*. **Halton** Northum. *Haulton*
1161. Possibly 'farmstead at the look-out hill'.
OE **hāw* + *hyll* + *tūn*.

Haltwhistle Northum. *Hautwisel* 1240. OE
twisla 'junction of two streams', possibly with
OFrench *haut* 'high'.

Halvergate Norfolk. *Halfriate* 1086 (DB).
Possibly 'land for which a half heriot (a feudal
service or payment) is due'. OE *half* +
here-geatu.

Halwell Devon. *Halganwylle* 10th cent. 'Holy
spring'. OE *hālig* + *wella*.

Halwill Devon. *Halgewilla* 1086 (DB).
Identical in origin with the previous name.

Ham, a common name, from OE *hamm* which
had various meanings, including 'enclosure,
land hemmed in by water or higher ground,
land in a river-bend'; examples include: **Ham**
Glos. *Hamma* 1194. **Ham** Gtr. London. *Hama*
c.1150. **Ham** Kent. *Hama* 1086 (DB). **Ham**
Wilts. *Hamme* 931, *Hame* 1086 (DB). **Ham,
East & Ham, West** Gtr. London. *Hamme* 958,
Hame 1086 (DB), *Estham* 1206, *Westhamma*
1186. **Ham, High** Somerset. *Hamme* 973,
Hame 1086 (DB).

Hamble-le-Rice Hants. *Hamele* 1165,
Hamele in the Rice 1391. Named from the River
Hamble on which it stands, 'crooked river', i.e.
'river with bends in it', from OE **hamel* + *ēa*.
The affix means 'in the brushwood', from OE
hrīs.

Hambleden Bucks. *Hamelan dene* 1015,
Hanbledene 1086 (DB). 'Crooked or undulating
valley'. OE **hamel* + *denu*.

Hambledon, 'crooked or flat-topped hill', OE
*hamel + *dūn*: **Hambledon** Hants.
Hamelandunæ 956, *Hamledune* 1086 (DB).
Hambledon Surrey. *Hameledune* 1086 (DB).

Hambleton, usually 'farmstead at the
crooked or flat-topped hill', OE **hamel* (used as
a noun) + *tūn*: **Hambleton** Lancs. *Hameltune*
1086 (DB). **Hambleton** N. Yorks., near Selby.
Hameltun 1086 (DB).
    However the following has the same origin as
HAMBLEDON: **Hambleton, Upper** Rutland.
Hameldun 1086 (DB).

Hambrook S. Glos. *Hanbroc* 1086 (DB).
Probably 'brook by the stone'. OE *hān* + *brōc*.

Hamdon Hill Somerset. *See* NORTON SUB
HAMDON, STOKE SUB HAMDON.

Hameringham Lincs. *Hameringam* 1086
(DB). Possibly 'homestead of the family or
followers of a man called Hathumær'. OE pers.
name + *-inga-* + *hām*.

Hamerton Cambs. *Hambertune* 1086 (DB).
'Farmstead with a smithy', or 'farmstead where
a plant such as hammer-sedge grows'. OE
hamor + *tūn*.

Hamilton Leics. *Hamelton* c.1130, *Hameldon*
1220. Perhaps 'farmstead of a man called
*Hamela', OE pers. name + *tūn*. Or '(place at)
the flat-topped hill, OE **hamel* + *dūn*.

Hamilton S. Lan. *Hamelton* 1291. The name
may have been 'imported' here in the 13th cent.
by Sir Walter *de Hameldone*, from HAMILTON,
Leics. Its earlier name was Cadzow (*Cadihow*
1150), of uncertain origin.

Hamiltonsbawn Armagh. 'Hamilton's
bawn'. *Hamilton's-Bawn* 1681. John *Hamilton*
built a bawn or fortified mansion here in 1619.
The equivalent Irish name is *Bábhún Hamaltún*.

Hammersmith Gtr. London. *Hamersmyth*
1294. '(Place with) a hammer smithy or forge'.
OE *hamor* + *smiththe*.

**Hammerton, Green & Hammerton,
Kirk** N. Yorks. *Hambretone* 1086 (DB),
Grenhamerton 1176, *Kyrkehamerton* 1226.

Probably identical in origin with HAMERTON (Cambs.). Distinguishing affixes from OE *grēne* 'village green' and OScand. *kirkja* 'church'.

Hammerwich Staffs. *Humeruuich* [*sic*] 1086 (DB), *Hamerwich* 1191. Probably 'building with a smithy'. OE *hamor + wīc*.

Hammoon Dorset. *Hame* 1086 (DB), *Hamme Moun* 1280. 'Enclosure or land in a river-bend'. OE *hamm*, with manorial affix from the *Moion* family which held the manor in 1086.

Hamnavoe Shet. (Yell). *Hafnarvag* 12th cent. 'Inlet of the harbour'. OScand. *hafn + vágr*.

Hampden, Great & Hampden, Little Bucks. *Hamdena* 1086 (DB). Probably 'valley with an enclosure'. OE *hamm + denu*.

Hampnett Glos. *Hantone* 1086 (DB), *Hamtonett* 1213. 'High farmstead'. OE *hēah* (dative *hēan*) + *tūn*, with the addition of OFrench *-ette* 'little'.

Hampole Donc. *Hanepole* 1086 (DB). 'Pool of a man called Hana', or 'pool frequented by cocks (of wild birds)'. OE pers. name or OE *hana + pōl*.

Hampshire (the county). *Hamtunscir* late 9th cent. 'District based on *Hamtun* (i.e. SOUTHAMPTON)'. OE *scīr*.

Hampstead, Hamstead, 'the homestead', OE *hām-stede*: **Hampstead** Gtr. London. *Hemstede* 959, *Hamestede* 1086 (DB). **Hampstead Norreys** W. Berks. *Hanstede* 1086 (DB), *Hampstede Norreys* 1517. Manorial affix from the *Norreys* family who bought the manor in 1448. **Hamstead** Birm. *Hamstede* 1227. **Hamstead** I. of Wight. *Hamestede* 1086 (DB). **Hamstead Marshall** W. Berks. *Hamestede* 1086 (DB), *Hamsted Marchal* 1284. Manorial affix from the *Marshal* family, here in the 13th cent.

Hampsthwaite N. Yorks. *Hamethwayt* *c*.1180. 'Clearing of a man called Hamr or Hamall'. OScand. pers. name + *thveit*.

Hampton, a common name, has no less than three different origins. Some are from OE *hām-tūn* 'home farm, homestead': **Hampton Lovett** Worcs. *Hamtona* 716, *Hamtune* 1086 (DB). Manorial affix from the *Luvet* family, here in the 13th cent. **Hampton, Meysey** Glos. *Hantone* 1086 (DB), *Meseishampton* 1287. Manorial addition from the *de Meisi* family, here from the 12th cent. **Hampton Poyle** Oxon. *Hantone* 1086 (DB), *Hampton Poile* 1428.

Manorial affix from the *de la Puile* family, here in the 13th cent.
Other Hamptons are 'farmstead in an enclosure or river-bend', OE *hamm + tūn*: **Hampton** Gtr. London. *Hamntone* 1086 (DB). **Hampton Bishop** Herefs. *Hantune* 1086 (DB), *Homptone* 1240. Manorial affix from early possession by the Bishop of Hereford. **Hampton Lucy** Warwicks. *Homtune* 781, *Hantone* 1086 (DB). Manorial affix from its possession by the *Lucy* family in the 16th cent.
Other Hamptons are 'high farmstead', OE *hēah* (dative *hēan*) + *tūn*: **Hampton** Shrops. *Hempton* 1391. **Hampton** Worcs., near Evesham. *Heantune* 780, *Hantun* 1086 (DB). **Hampton in Arden** Solhll. *Hantone* 1086 (DB). Affix refers to the medieval Forest of Arden, *see* HENLEY-IN-ARDEN. **Hampton on the Hill** Warwicks. *Hamtone* 12th cent.

Hams, South (district) Devon. *Southammes* 1396. Possibly 'cultivated areas of land to the south (of Dartmoor)'. OE *sūth + hamm*.

Hamsey E. Sussex. *Hamme* 961, *Hame* 1086 (DB), *Hammes Say* 1306. 'The enclosure, the land in a river-bend'. OE *hamm* with manorial addition from the *de Say* family, here in the 13th cent.

Hamstall Ridware Staffs. *See* RIDWARE.

Hamstead. *See* HAMPSTEAD.

Hamsterley Durham. *Hamsteleie* *c*.1190. 'Clearing infested with corn-weevils'. OE **hamstra + lēah*.

Hamworthy Dorset. *Hamme* 1236, *Hamworthy* 1463. OE *hamm* 'enclosure', here possibly 'peninsula', with the later addition of *worthig* 'enclosure'.

Hanborough, Church & Hanborough, Long Oxon. *Haneberge* 1086 (DB). 'Hill of a man called Hagena or Hana'. OE pers. name + *beorg*.

Hanbury, 'high or chief fortified place', OE *hēah* (dative *hēan*) + *burh* (dative *byrig*): **Hanbury** Staffs. *Hambury* *c*.1185. **Hanbury** Worcs. *Heanburh* *c*.765, *Hambyrie* 1086 (DB).

Hanchurch Staffs. *Hancese* [*sic*] 1086 (DB), *Hanchurche* 1212. 'High church'. OE *hēah* (dative *hēan*) + *cirice*.

Handa Island Highland. 'Sand island'. OScand. *sandr + ey*.

Handbridge Ches. *Bruge* 1086 (DB), *Honebrugge c.*1150. 'Bridge at the rock'. OE *hān* + *brycg*.

Handcross W. Sussex. recorded as *Handcrosse* 1617, perhaps 'cross used as a signpost', or 'cross-roads where five roads meet'.

Handforth Ches. *Haneford* 12th cent. 'Ford frequented by cocks (of wild birds)', or 'ford at the stones (used as markers)'. OE *hana* or *hān* + *ford*.

Handley, 'high wood or clearing', OE *hēah* (dative *hēan*) + *lēah*: **Handley** Ches. *Hanlei* 1086 (DB). **Handley, Sixpenny** Dorset. *Hanlee* 877, *Hanlege* 1086 (DB), *Sexpennyhanley* 1575. Affix added in 16th cent. from an old Hundred name *Sexpene* 'hill of the Saxons', from OE *Seaxe* + Celtic **penn*. **Handley, West** Derbys. *Henleie* 1086 (DB).

Handsacre Staffs. *Hadesacre* [*sic*] 1086 (DB), *Handesacra* 1196. 'Arable plot of a man called **Hand*'. OE pers. name + *æcer*.

Handsworth Birm. *Honesworde* 1086 (DB). 'Enclosure of a man called Hūn'. OE pers. name + *worth*.

Handsworth Sheff. *Handesuuord* 1086 (DB). 'Enclosure of a man called **Hand*'. OE pers. name + *worth*.

Hanford Stoke. *Heneford* [*sic*] 1086 (DB), *Honeford* 1212. 'Ford frequented by cocks (of wild birds)', or 'ford at the stones'. OE *hana* or *hān* + *ford*.

Hanham S. Glos. *Hanun* 1086 (DB). '(Place at) the rocks'. OE *hān* in a dative plural form *hānum*.

Hankelow Ches. *Honcolawe* 12th cent. 'Mound or hill of a man called **Haneca*'. OE pers. name + *hlāw*.

Hankerton Wilts. *Hanekyntone* 680. 'Estate associated with a man called **Haneca*'. OE pers. name + *-ing-* + *tūn*.

Hankham E. Sussex. *Hanecan hamme* 947, *Henecham* 1086 (DB). 'Enclosure, or dry ground in marsh, of a man called **Haneca*'. OE pers. name + *hamm*.

Hanley, 'high wood or clearing', OE *hēah* (dative *hēan*) + *lēah*: **Hanley** Stoke. *Henle* 1212. **Hanley Castle** Worcs. *Hanlege* 1086 (DB). Affix from the early 13th-cent. castle here.

Hanley Child & Hanley William Worcs. *Hanlege* 1086 (DB), *Cheldreshanle* 1255, *Williames Henle* 1275. Affixes from OE *cild* (plural *cildra*) 'young monk, noble-born son', and from a *William* de la Mare who held one of the manors in 1242.

Hanlith N. Yorks. *Hangelif* [*sic*] 1086 (DB), *Hahgenlid* 12th cent. 'Slope or hill-side of a man called Hagni or Hǫgni'. OScand. pers. name + *hlíth*.

Hannahstown Antrim. 'Hannah's town'. *Hannahstown* 1780. A family with the Scottish surname *Hanna(h)* were here in the 17th cent. The equivalent Irish name is *Baile Haine*.

Hanney, East & Hanney, West Oxon. *Hannige* 956, *Hannei* 1086 (DB). 'Island, or land between streams, frequented by cocks (of wild birds)'. OE *hana* + *ēg*.

Hanningfield, East, Hanningfield, South, & Hanningfield, West Essex. *Hamningefelde c.*1036, *Haningefelda* 1086 (DB). Probably 'open land of the family or followers of a man called Hana'. OE pers. name + *-inga-* + *feld*.

Hannington Hants. *Hanningtun* 1023, *Hanitune* 1086 (DB). 'Estate associated with a man called Hana'. OE pers. name + *-ing-* + *tūn*.

Hannington Northants. *Hanintone* 1086 (DB). Identical in origin with the previous name.

Hannington Swindn. *Hanindone* 1086 (DB). 'Hill frequented by cocks (of wild birds)', or 'hill of a man called Hana'. OE *hana* (genitive plural *hanena*) or OE pers. name (genitive *-n*) + *dūn*.

Hanslope Milt. K. *Hamslape* 1086 (DB). 'Muddy place or slope of a man called *Hāma*'. OE pers. name + **slæp*.

Hanthorpe Lincs. *Hermodestorp* 1086 (DB). 'Outlying farmstead or hamlet of a man called Heremōd or Hermóthr'. OE or OScand. pers. name + OScand. *thorp*.

Hanwell Gtr. London. *Hanewelle* 959, *Hanewelle* 1086 (DB). 'Spring or stream frequented by cocks (of wild birds)'. OE *hana* + *wella*.

Hanwell Oxon. *Hanewege* 1086 (DB), *Haneuell* 1236. 'Way (and stream) of a man called Hana'. OE pers. name + *weg* (replaced by *wella* in the 13th cent.).

Hanwood, Great Shrops. *Hanewde* 1086 (DB). Possibly 'wood frequented by cocks (of wild birds)', OE *hana* + *wudu*. Alternatively the first element could be OE *hān* 'rock, stone' or the OE pers. name Hana.

Hanworth Gtr. London. *Haneworde* 1086 (DB). 'Enclosure of a man called Hana'. OE pers. name + *worth*.

Hanworth Norfolk. *Haganaworda* 1086 (DB), 'Enclosure of a man called Hagena'. OE pers. name + *worth*.

Hanworth, Cold Lincs. *Haneurde* 1086 (DB), *Calthaneworth* 1322. 'Enclosure of a man called Hana'. OE pers. name + *worth*. Affix is OE *cald* 'cold, exposed'.

Happisburgh Norfolk. *Hapesburc* 1086 (DB). 'Stronghold of a man called *Hæp'. OE pers. name + *burh*.

Hapsford Ches. *Happesford* 13th cent. 'Ford of a man called *Hæp'. OE pers. name + *ford*.

Hapton Lancs. *Apton* 1242. 'Farmstead by a hill'. OE *hēap* + *tūn*.

Hapton Norfolk. *Habetuna* 1086 (DB). 'Farmstead of a man called *Hab(b)a'. OE pers. name + *tūn*.

Harberton Devon. *Herburnaton* 1108. 'Farmstead on the River Harbourne'. OE river-name ('pleasant stream' from OE *hēore* + *burna*) + *tūn*.

Harbledown Kent. *Herebolddune* 1175. 'Hill of a man called Herebeald'. OE pers. name + *dūn*.

Harborne Birm. *Horeborne* 1086 (DB). 'Dirty or muddy stream'. OE *horu* + *burna*.

Harborough (district) Leics. A modern name, with reference to MARKET HARBOROUGH.

Harborough Magna Warwicks. *Herdeberge* 1086 (DB), *Hardeburgh Magna* 1498. 'Hill of the flocks or herds'. OE *heord* + *beorg*. Affix is Latin *magna* 'great'.

Harborough, Market Leics. *Haverbergam* 1153, *Mercat Heburgh* 1312. Probably 'hill where oats are grown'. OScand. *hafri* or OE **hæfera* + OScand. *berg* or OE *beorg*. Alternatively the first element may be OE *hæfer* or OScand. *hafr* 'a he-goat'. Affix (ME *merket*) from the important market here.

Harbottle Northum. *Hirbotle* c.1220. Probably 'dwelling of the hireling'. OE *hȳra* + *bōthl*.

Harbourne (river) Devon. *See* HARBERTON.

Harbury Warwicks. *Hereburgebyrig* 1002, *Erburgeberie* 1086 (DB). 'Stronghold or manor-house of a woman called Hereburh'. OE pers. name + *burh* (dative *byrig*).

Harby, 'farmstead with a herd or flock', or 'farmstead of the herdsmen', OScand. *hjǫrth* (genitive singular *hjarthar*) or OE *heorde* (genitive plural *heorda*) + OScand. *bý*: **Harby** Leics. *Herdebi* 1086 (DB). **Harby** Notts. *Herdrebi* 1086 (DB).

Harden Brad. *Hareden* late 12th cent. 'Rock valley', or 'valley frequented by hares'. OE **hær* or *hara* + *denu*.

Hardham W. Sussex. *Heriedeham* 1086 (DB). 'Homestead or river-meadow of a woman called Heregȳth'. OE pers. name + *hām* or *hamm*.

Hardingham Norfolk. *Hardingeham* 1161. 'Homestead of the family or followers of a man called *Hearda'. OE pers. name + *-inga-* + *hām*.

Hardingstone Northants. *Hardingestone* 1086 (DB), *Hardingesthorn* 12th cent. 'Thorn-tree of a man called Hearding'. OE pers. name + *thorn*.

Hardington, 'estate associated with a man called *Hearda', OE pers. name + *-ing-* + *tūn*: **Hardington** Somerset. *Hardintone* 1086 (DB). **Hardington Mandeville** Somerset. *Hardintone* 1086 (DB). Manorial affix from its possession by the *de Mandeville* family from the 12th cent.

Hardley Hants. *Hardelie* 1086 (DB). 'Hard clearing'. OE *heard* + *lēah*.

Hardley Street Norfolk. *Hardale* 1086 (DB). Probably identical in origin with the previous name.

Hardmead Milt. K. *Herulfmede* 1086 (DB). Probably 'meadow of a man called Heoruwulf or Herewulf'. OE pers. name + *mǣd*.

Hardres, Lower & Hardres, Upper Kent. *Haredum* 785, *Hardes* 1086 (DB). '(Place at) the woods'. OE **harad* in a plural form.

Hardstoft Derbys. *Hertestaf* [*sic*] 1086 (DB), *Hertistoft* 1257. 'Homestead of a man called *Heort or Hjǫrtr'. OE or OScand. pers. name + OScand. *toft*.

Hardwick, Hardwicke, a common name, 'herd farm, farm for livestock', OE *heorde-wīc*: **Hardwick** Bucks. *Harduich* 1086 (DB). **Hardwick** Cambs. *Hardwic c.*1050, *Harduic* 1086 (DB). **Hardwick** Norfolk, near King's Lynn. *Herdwic* 1242. **Hardwick** Northants. *Heordewican c.*1067, *Herdewiche* 1086 (DB). **Hardwick** Oxon., near Bicester. *Hardewich* 1086 (DB). **Hardwick** Oxon., near Witney. *Herdewic* 1199. **Hardwick, East & Hardwick, West** Wakefd. *Harduic* 1086 (DB). **Hardwick, Priors** Warwicks. *Herdewyk* 1043, *Herdewiche* 1086 (DB), *Herdewyk Priour* 1310. Affix from its possession by Coventry Priory in the 11th cent. **Hardwicke** Glos., near Stroud. *Herdewike* 12th cent. **Hardwicke** Glos., near Tewkesbury. *Herdeuuic* 1086 (DB). **Hardwicke** Herefs., near Clifford. *La Herdewyk* 14th cent.

Hardy Manch. *See* CHORLTON.

Hareby Lincs. *Harebi* 1086 (DB). Probably 'farmstead or village of a man called Hāri'. OScand. pers. name + *bý*.

Harefield Gtr. London. *Herefelle [sic]* 1086 (DB), *Herefeld* 1206. Probably 'open land used by an army (perhaps a Viking army)'. OE *here* + *feld*.

Haresfield Glos. *Hersefel* 1086 (DB), *Hersefeld* 1213. 'Open land of a man called *Hersa'. OE pers. name + *feld*.

Harewood Leeds. *Harawuda* 10th cent., *Hareuuode* 1086 (DB). 'Grey wood', or 'wood by the rocks', or 'wood frequented by hares'. OE *hār* or *hær* or *hara* + *wudu*.

Harford Devon. near Ivybridge. *Hereford* 1086 (DB). 'Ford suitable for the passage of an army'. OE *here-ford*.

Hargrave, probably 'grey grove, or grove on a boundary', OE *hār* + *grāf* or *græfe*: **Hargrave** Ches. *Haregrave* 1086 (DB). **Hargrave** Northants. *Haregrave* 1086 (DB). **Hargrave** Suffolk. *Haragraua* 1086 (DB).

Haringey Gtr. London. *See* HORNSEY.

Harkstead Suffolk. *Herchesteda* 1086 (DB). Probably 'pasture or homestead of a man called Hereca'. OE pers. name + *stede*.

Harlaston Staffs. *Heorelfestun* 1002, *Horulvestone* 1086 (DB). 'Farmstead of a man called Heoruwulf'. OE pers. name + *tūn*.

Harlaxton Lincs. *Herlavestune* 1086 (DB). 'Farmstead of a man called *Herelāf or *Heorulāf'. OE pers. name + *tūn*.

Harlech Gwyd. *Hardelagh c.*1290. 'Fine rock'. Welsh *hardd* + *llech*. The name refers to the prominent crag on which Harlech Castle was built in the 13th cent.

Harlesden Gtr. London. *Herulvestune* 1086 (DB). 'Farmstead of a man called Heoruwulf or Herewulf'. OE pers. name + *tūn*.

Harleston, Harlestone, 'farmstead of a man called Heoruwulf or Herewulf', OE pers. name + *tūn*: **Harleston** Devon. *Harliston* 1252. **Harleston** Norfolk. *Heroluestuna* 1086 (DB). **Harleston** Suffolk. *Heroluestuna* 1086 (DB). **Harlestone** Northants. *Herolvestune* 1086 (DB).

Harley Shrops. *Harlege* 1086 (DB), *Herleia c.*1090. Probably 'rock wood or clearing'. OE *hær* + *lēah*.

Harling, East Norfolk. *Herlinge c.*1060, *Herlinga* 1086 (DB). '(Settlement of) the family or followers of a man called *Herela', or 'place of a man called *Herela'. OE pers. name + *-ingas* or *-ing*.

Harlington Beds. *Herlingdone* 1086 (DB). 'Hill of the family or followers of a man called *Herela'. OE pers. name +*-inga-* + *dūn*.

Harlington Gtr. London. *Hygereding tun* 831, *Herdintone* 1086 (DB). 'Estate associated with a man called *Hygerēd'. OE pers. name + *-ing-* + *tūn*.

Harlow Essex. *Herlawe* 1045, *Herlaua* 1086 (DB). 'Mound or hill associated with an army (perhaps a Viking army)'. OE *here* + *hlāw*.

Harlow Hill Northum. *Hirlawe* 1242. Possibly 'jay hill'. OE *higera* + *hlāw*.

Harlsey, East N. Yorks. *Herlesege* 1086 (DB). 'Island, or dry ground in marsh, of a man called *Herel'. OE pers. name + *ēg*.

Harlthorpe E. R. Yorks. *Herlesthorpia* 1150–60. 'Outlying farmstead of a man called Herleifr or Herlaugr'. OScand. pers. name + *thorp*.

Harlton Cambs. *Herletone* 1086 (DB). 'Farmstead of a man called *Herela'. OE pers. name + *tūn*.

Harmby N. Yorks. *Hernebi* 1086 (DB). 'Farmstead or village of a man called Hjarni'. OScand. pers. name + *bý*.

Harmondsworth Gtr. London. *Hermondesyeord* [sic] 781, *Hermodesworde* 1086 (DB). 'Enclosure of a man called Heremund or Heremōd'. OE pers. name + *worth*.

Harmston Lincs. *Hermodestune* 1086 (DB). 'Farmstead of a man called Heremōd or Hermóthr'. OE or OScand. pers. name + OE *tūn*.

Harnham, East & Harnham, West Wilts. *Harnham* 1115. Probably 'enclosure frequented by hares'. OE *hara* (genitive plural *harena*) + *hamm*.

Harnhill Glos. *Harehille* 1086 (DB). 'Grey hill' or 'hill frequented by hares'. OE *hār* (dative *-an*) or *hara* (genitive plural *-ena*) + *hyll*.

Harold Hill Gtr. London. named from HAROLD WOOD.

Harold Wood Gtr. London. *Horalds Wood* c.1237. Named from Earl Harold, king of England until his defeat at Hastings in 1066; he held the nearby manor of HAVERING.

Haroldston West Pemb. *Harauldyston* 1307. 'Western (place called) Harold's farm'. OScand. pers. name + OE *tūn*. *West* distinguishes the place from nearby *Haroldston St Issells* ('St Ismael').

Harome N. Yorks. *Harum* 1086 (DB). '(Place at) the rocks or stones'. OE **hær* in a dative plural form **harum*.

Harpenden Herts. *Herpedene* c.1060. Probably 'valley of the harp'. OE *hearpe* (genitive *-an*) + *denu*. Alternatively the first element may be a reduced form of OE *here-pæth* 'highway or main road'.

Harpford Devon. *Harpeford* 1167. 'Ford on a highway or main road'. OE *here-pæth* + *ford*.

Harpham E. R. Yorks. *Arpen* [sic] 1086 (DB), *Harpam* 1100–15. 'Homestead where the harp is played'. OE *hearpe* + *hām*. Alternatively the first element may have the sense 'salt-harp, sieve for making salt'.

Harpley, 'clearing of the harp', perhaps 'harp-shaped clearing', OE *hearpe* + *lēah*: **Harpley** Norfolk. *Herpelai* 1086 (DB). **Harpley** Worcs. *Hoppeleia* [sic] 1222, *Harpele* 1275.

Harpole Northants. *Horpol* 1086 (DB). 'Dirty or muddy pool'. OE *horu* + *pōl*.

Harpsden Oxon. *Harpendene* 1086 (DB). Probably 'valley of the harp'. OE *hearpe*

(genitive *-an*) + *denu*. Alternatively the first element may be a reduced form of OE *here-pæth* 'highway or main road'.

Harpswell Lincs. *Herpeswelle* 1086 (DB). Possibly 'spring or stream of the harper'. OE *hearpere* + *wella*. Alternatively the first element may be a reduced form of OE *here-pæth* 'highway or main road'.

Harptree, East & Harptree, West B. & NE. Som. *Harpetreu* 1086 (DB). Probably 'tree by a highway or main road'. OE *here-pæth* + *trēow*.

Harpurhey Manch. *Harpourhey* 1320. 'Enclosure of a man called *Harpour*'. OE *hæg* with the surname of a 14th-cent. landowner.

Harrietsham Kent. *Herigeardes hamm* 10th cent., *Hariardesham* 1086 (DB). 'River-meadow of a man called Heregeard, or near army quarters'. OE pers. name (or OE *here* + *geard*) + *hamm*.

Harringay Gtr. London. *See* HORNSEY.

Harrington Cumbria. *Haueringtona* c.1160. 'Estate associated with a man called **Hæfer*'. OE pers. name + *-ing-* + *tūn*.

Harrington Lincs. *Haringtona* 12th cent. Possibly 'farmstead associated with a man called Hærra'. OE pers. name + *-ing-* + *tūn*. Alternatively the first element may be an OE **hæring* 'stony place' or **hāring* 'grey wood'.

Harrington Northants. *Arintone* [sic] 1086 (DB), *Hederingeton* 1184. Possibly 'estate associated with a man called **Heathuhere*'. OE pers. name + *-ing-* + *tūn*.

Harringworth Northants. *Haringwrth* c.1060, *Haringeworde* 1086 (DB). Possibly 'enclosure of the dwellers at a stony place'. OE **hær* + *-inga-* + *worth*.

Harris (island district) W. Isles. *Heradh* c.1500, *Harrige* 1542, *Harreis* 1588. Gaelic *na-h-earadh*, probably from OScand. *herath* 'district'.

Harrogate N. Yorks. *Harwegate* 1332. '(Place at) the road to the cairn or heap of stones'. OScand. *hǫrgr* + *gata*.

Harrold Beds. *Hareuuelle* [sic] 1086 (DB), *Harewolda* 1163. Probably 'high forest-land on a boundary'. OE *hār* + *weald*.

Harrow Gtr. London. *Gumeninga hergae* 767, *Herges* 1086 (DB). 'The heathen shrine(s) or temple(s) (of a tribe called the *Gumeningas*)'.

From OE *hearg*, with an obscure OE tribal name in the earliest spelling.

Harrow Weald Gtr. London. *Welde* 1294, *Harewewelde* 1388. 'Woodland near HARROW'. OE *weald*.

Harrowden, Great & Harrowden, Little Northants. *Hargedone* 1086 (DB). 'Hill of the heathen shrines or temples'. OE *hearg* + *dūn*.

Harston Cambs. *Herlestone* 1086 (DB). Probably 'farmstead of a man called *Herel'. OE pers. name + *tūn*.

Harston Leics. *Herstan* [*sic*] 1086 (DB), *Harstan* 1195. 'The grey boundary stone'. OE *hār* + *stān*.

Hart (district) Hants. A modern adoption of the river-name Hart (itself a back-formation from either HARTFORDBRIDGE or HARTLEY WINTNEY).

Hart Hartlepl. *Hert* c.1170. OE *heorot* 'stag', probably an early back-formation from HARTLEPOOL.

Hartburn, 'stream frequented by harts or stags', OE *heorot* + *burna*: **Hartburn** Northum. *Herteburne* 1198. **Hartburn, East** Stock. on T. *Herteburna* c.1190.

Hartest Suffolk. *Hertest* c.1050, *Herterst* 1086 (DB). 'Wooded hill frequented by harts or stags'. OE *heorot* + *hyrst*.

Hartfield E. Sussex. *Hertevel* [*sic*] 1086 (DB), *Hertefeld* 12th cent. 'Open land frequented by harts or stags'. OE *heorot* + *feld*.

Hartford, usually 'ford frequented by harts or stags', OE *heorot* + *ford*: **Hartford** Ches. *Herford* [*sic*] 1086 (DB), *Hartford* late 12th cent. **Hartford, East** Northum. *Hertford* 1198.
   However the following has a different origin: **Hartford** Cambs. *Hereforde* 1086 (DB). 'Ford suitable for the passage of an army'. OE *here-ford*.

Hartfordbridge Hants. *Hertfordbrigge* 1327. 'Bridge at *Hertford* (ford frequented by harts or stags)'. OE *heorot* + *ford* + *brycg*. The name of the River **Hart** is a modern back-formation from either this name or HARTLEY WINTNEY.

Harthill, 'hill frequented by harts or stags', OE *heorot* + *hyll*: **Harthill** Ches. *Herthil* 1259. **Harthill** Rothm. *Hertil* 1086 (DB).

Harting, East & Harting, South W. Sussex. *Hertingas* 970, *Hertinges* 1086 (DB). '(Settlement of) the family or followers of a man called *Heort'. OE pers. name + *-ingas*.

Hartington Derbys. *Hortedun* 1086 (DB), *Hertingdon* 1244. Possibly 'hill associated with a man called *Heort(a)', OE pers. name + *-ing-* + *dūn*. Alternatively 'hill at the place where harts are found', OE *heorting* + *dūn*.

Hartland Devon. *Heortigtun* c.880, *Hertitona* 1086 (DB), *Hertilanda* 1130. 'Farmstead (or estate) on the peninsula frequented by harts or stags'. OE *heorot* + *īeg* + *tūn* (later *land*).

Hartlebury Worcs. *Heortlabyrig* 817, *Huerteberie* 1086 (DB). 'Stronghold of a man called *Heortla'. OE pers. name + *burh* (dative *byrig*).

Hartlepool Hartlepl. *Herterpol* c.1170. 'Pool or bay near the stag peninsula'. OE *heorot* + *ēg* + *pōl*.

Hartley, usually 'wood or clearing frequented by harts or stags', OE *heorot* + *lēah*; examples include: **Hartley** Kent, near Cranbrook. *Heoratleag* 843. **Hartley** Kent, near Longfield. *Erclei* [*sic*] 1086 (DB), *Hertle* 1253. **Hartley Wespall** Hants. *Harlei* 1086 (DB), *Hertlegh Waspayl* c.1270. Manorial addition from early possession by the *Waspail* family. **Hartley Wintney** Hants. *Hurtlege* 12th cent., *Hertleye Wynteneye* 13th cent. Manorial addition from its possession by the Priory of Wintney in the 13th cent.
   However other Hartleys have different origins: **Hartley** Cumbria. *Harteclo* 1176. Probably 'hard ridge of land'. OE *heard* + *clā*. **Hartley** Northum. *Hertelawa* 1167. 'Mound or hill frequented by harts or stags'. OE *heorot* + *hlāw*.

Hartlip Kent. *Heordlyp* 11th cent. 'Gate or fence over which harts or stags can leap'. OE *heorot* + *hlīep*.

Harton N. Yorks. *Heretune* 1086 (DB). Probably 'farmstead by the rocks or stones'. OE *hær* + *tūn*.

Harton S. Tyne. *Heortedun* c.1104. 'Hill frequented by harts or stags'. OE *heorot* + *dūn*.

Hartpury Glos. *Hardepiry* 12th cent. 'Pear-tree with hard fruit'. OE *heard* + *pirige*.

Hartshill Warwicks. *Ardreshille* [*sic*] 1086 (DB), *Hardredeshella* 1151. 'Hill of a man called Heardrēd'. OE pers. name + *hyll*.

Hartshorne Derbys. *Heorteshone* 1086 (DB). 'Hart's horn', i.e. 'hill thought to resemble a hart's horn'. OE *heorot* + *horn*.

Hartwell Northants. *Hertewelle* 1086 (DB). 'Spring or stream frequented by harts or stags'. OE *heorot* + *wella*.

Harvington Worcs. *Herverton* 709, *Herferthun* 1086 (DB). 'Farmstead near the ford suitable for the passage of an army'. OE **here-ford* + *tūn*.

Harwell Oxon. *Haranwylle* 956, *Harvvelle* 1086 (DB). 'Spring or stream by the hill called *Hāra* (the grey one)'. OE hill-name (from *hār* 'grey') + *wella*.

Harwich Essex. *Herewic* 1248. 'Army camp', probably that of a Viking army. OE *here-wīc*.

Harwood, 'grey wood', 'wood by the rocks', or 'wood frequented by hares', OE *hār* or **hær* or *hara* + *wudu*: **Harwood** Bolton. *Harewode* 1212. **Harwood Dale** N. Yorks. *Harewode* 1301, *Harwoddale* 1577. With OScand. *dalr* 'valley'. **Harwood, Great** Lancs. *Majori Harewuda* early 12th cent. Affix is Latin *maior* 'greater'.

Harworth Notts. *Hareworde* 1086 (DB). Probably 'enclosure on the boundary'. OE *hār* + *worth*.

Hascombe Surrey. *Hescumb* 1232. Possibly 'the witch's valley'. OE *hægtesse* + *cumb*.

Haselbech Northants. *Esbece* [sic] 1086 (DB), *Haselbech* 12th cent. 'Valley stream where hazels grow'. OE *hæsel* + *bece*.

Haselbury Plucknett Somerset. *Halberge* [sic] 1086 (DB), *Haselbare Ploukenet* 1431. 'Hazel wood or grove'. OE *hæsel* + *bearu*. Manorial affix from its possession by the *de Plugenet* family in the 13th cent.

Haseley, Great & Haseley, Little Oxon. *Hæseleia* 1002, *Haselie* 1086 (DB). 'Hazel wood or clearing'. OE *hæsel* + *lēah*.

Haselor Warwicks. *Haseloue* 1086 (DB). 'Flat-topped ridge where hazels grow'. OE *hæsel* + **ofer*.

Hasfield Glos. *Hasfelde* 1086 (DB). 'Open land where hazels grow'. OE *hæsel* + *feld*.

Hasguard Pemb. *Hiscart* c.1200. 'House cleft'. OScand. *hús* + *skarth*. The name apparently refers to the small valley that cuts into rising ground here.

Hasketon Suffolk. *Haschetuna* 1086 (DB). 'Farmstead of a man called **Haseca*'. OE pers. name + *tūn*.

Hasland Derbys. *Haselont* 1129–38. 'Hazel grove'. OScand. *hasl* + *lundr*.

Haslemere Surrey. *Heselmere* 1221. 'Pool where hazels grow'. OE *hæsel* + *mere*.

Haslingden Lancs. *Heselingedon* 1241. 'Valley growing with hazels'. OE *hæslen* + *denu*.

Haslingfield Cambs. *Haslingefeld* 1086 (DB). Probably 'open land of the family or followers of a man called **Hæsel(a)*'. OE pers. name + *-inga-* + *feld*. Alternatively the first element may be the genitive plural of OE **hæsling* 'place growing with hazels'.

Haslington Ches. *Hasillinton* early 13th cent. 'Farmstead where hazels grow'. OE *hæslen* + *tūn*.

Hassall Ches. *Eteshale* [sic] 1086 (DB), *Hatishale* 13th cent. 'The witch's nook of land'. OE *hægtesse* + *halh*.

Hassingham Norfolk. *Hasingeham* 1086 (DB). 'Homestead of the family or followers of a man called **Hasu*'. OE pers. name + *-inga-* + *hām*.

Hassocks W. Sussex. a 19th-cent. settlement named from a field called *Hassocks*, from OE *hassuc* 'clump of coarse grass'.

Hassop Derbys. *Hetesope* 1086 (DB). Probably 'the witch's valley'. OE *hægtesse* + *hop*.

Hastingleigh Kent. *Hæstingalege* 993, *Hastingelai* 1086 (DB). 'Woodland clearing of the family or followers of a man called **Hæsta*'. OE pers. name + *-inga-* + *lēah*.

Hastings E. Sussex. *Hastinges* 1086 (DB). '(Settlement of) the family or followers of a man called **Hæsta*'. OE pers. name + *-ingas*. Earlier *Hæstingaceaster* c.915, 'Roman town of **Hæsta*'s people', from OE *ceaster*.

Haswell Durham. *Heswell* late 12th cent. Possibly 'spring or stream where hazels grow'. OE *hæsel* + *wella*. Alternatively the first element may be OE *hægtesse* 'witch'.

Hatch, a common name, from OE *hæcc* 'a hatch-gate (leading to a forest)' or 'floodgate (in a stream)'; examples include: **Hatch** Beds. *La Hache* 1232. **Hatch** Hants. *Heche* 1086 (DB). **Hatch** Wilts. *Hache* 1200. **Hatch Beauchamp** Somerset. *Hache* 1086 (DB),

Hache Beauchampe 1243. Manorial affix from its possession by the *Beauchamp* family in the 13th cent. **Hatch, West** Somerset. *Hache* 1201, *Westhache* 1243.

Hatch End Gtr. London. *Le Hacchehend* 1448. 'District by the gate' (probably a gate to Pinner Park). OE *hæcc* + *ende*.

Hatcliffe NE. Lincs. *Hadecliue* 1086 (DB). 'Cliff or bank of a man called Hadda'. OE pers. name + *clif*.

Hatfield, a common name, 'heathy open land, or open land where heather grows', OE *hǣth* + *feld*; examples include: **Hatfield** Donc. *Haethfelth* 731, *Hedfeld* 1086 (DB). **Hatfield** Herefs. *Hetfelde* 1086 (DB). **Hatfield** Herts. *Haethfelth* 731, *Hetfelle* 1086 (DB). **Hatfield** Worcs. *Hadfeld* 1182. **Hatfield Broad Oak** Essex. *Hadfelda* 1086 (DB), *Hatfeld Brodehoke* c.1130. Affix from a large oak-tree, OE *brād* + *āc*. **Hatfield, Great & Hatfield, Little** E. R. Yorks. *Haifeld, Heifeld* 1086 (DB), *Haitefeld* 12th cent. First element influenced by OScand. *heithr* 'heath'. **Hatfield Peverel** Essex. *Hadfelda* 1086 (DB), *Hadfeld Peurell* 1166. Manorial affix from its possession by Ralph *Peverel* in 1086.

Hatford Oxon. *Hevaford* [sic] 1086 (DB), *Hauetford* 1176. 'Ford near a headland or hill'. OE *hēafod* + *ford*.

Hatherden Hants. *Hetherden* 1193. Probably 'hawthorn valley'. OE *hagu-thorn* + *denu*.

Hatherleigh Devon. *Hadreleia* 1086 (DB). Probably identical in origin with the next name.

Hatherley, Down & Hatherley, Up Glos. *Athelai* 1086 (DB), *Dunheytherleye* 1273, *Hupheberleg* 1221. 'Hawthorn clearing'. OE *hagu-thorn* + *lēah*. Distinguishing affixes from OE *dūne* 'lower downstream' and *upp* 'higher upstream'.

Hathern Leics. *Avederne* [sic] 1086 (DB), *Hauthirn* c.1130. '(Place at) the hawthorn'. OE **hagu-thyrne*.

Hatherop Glos. *Etherope* 1086 (DB). Probably 'high outlying farmstead'. OE *hēah* + *throp*.

Hathersage Derbys. *Hereseige* [sic] 1086 (DB), *Hauersegg* c.1220. 'He-goat's ridge', or 'ridge of a man called *Hæfer'. OE *hæfer* or OE pers. name + *ecg*.

Hatherton Ches. *Haretone* [sic] 1086 (DB), *Hatherton* 1262. 'Farmstead where hawthorn grows'. OE *hagu-thorn* + *tūn*.

Hatherton Staffs. *Hagenthorndun* 996, *Hargedone* 1086 (DB). 'Hill where hawthorn grows'. OE *hagu-thorn* + *dūn*.

Hatley, probably 'woodland clearing on the hill', OE *hætt* + *lēah*: **Hatley, Cockayne** Beds. *Hattenleia* c.960, *Hatelai* 1086 (DB), *Cocking Hatley* 1576. Manorial affix from the *Cockayne* family, here in the 15th cent. **Hatley, East & Hatley St George** Cambs. *Hatelai* 1086 (DB), *Esthatteleia* 1199, *Hattele de Sancto Georgio* 1279. Distinguishing affixes from OE *ēast* 'east' and from possession by the family *de Sancto Georgio* here in the 13th cent.

Hattingley Hants. *Hattingele* 1204. Probably 'woodland clearing at the hill place'. OE *hætt* + *-ing* + *lēah*.

Hatton, 'farmstead on a heath', OE *hǣth* + *tūn*; examples include: **Hatton** Derbys. *Hatune* 1086 (DB). **Hatton** Gtr. London. *Hatone* 1086 (DB). **Hatton** Lincs. *Hatune* 1086 (DB). **Hatton** Shrops. *Hatton* 1212. **Hatton** Warrtn. *Hattone* c.1230. **Hatton, Cold** Tel. & Wrek. & **Hatton, High** Shrops. *Hatune, Hetune* 1086 (DB), *Colde Hatton* 1233, *Heye Hatton* 1327. Distinguishing affixes from OE *cald* 'cold, exposed' and *hēah* 'high, chief'. **Hatton Heath** Ches. *Etone* [sic] 1086 (DB), *Hettun* 1185.

Haugham Lincs. *Hecham* 1086 (DB). Probably 'high or chief homestead'. OE *hēah* + *hām*.

Haughley Suffolk. *Hagele* c.1040, *Hagala* 1086 (DB). 'Wood or clearing with a hedge, or where hawthorns grow'. OE *haga* + *lēah*.

Haughton, usually 'farmstead in or by a nook of land', OE *halh* + *tūn*: **Haughton** Shrops., near Oswestry. *Halchton* 1285. **Haughton** Shrops., near Shifnal. *Halghton* 1281. **Haughton** Shrops., near Shrewsbury. *Haustone* 1086 (DB). **Haughton** Staffs. *Haltone* 1086 (DB). **Haughton Green** Tamesd. *Halghton* 1307. **Haughton-le-Skerne** Darltn. *Halhtun* c.1040. On the River Skerne (for the origin of this river-name *see* SKERNE with which it is identical in meaning).

However the following has a different origin: **Haughton** Notts. *Hoctun* 1086 (DB). 'Farmstead on a spur of land'. OE *hōh* + *tūn*.

Haunton Staffs. *Hagnatun* 942. 'Farmstead of a man called Hagena'. OE pers. name + *tūn*.

Hautbois, Little Norfolk. *Hobbesse* 1044–7, *Hobuisse* 1086 (DB). Probably 'marshy meadow with tussocks or hummocks'. OE *hobb + wisc*, *wisse*.

Hauxley Northum. *Hauekeslaw* 1204. 'Mound of the hawk, or of a man called *Hafoc'. OE hafoc or pers. name + hlāw.

Hauxton Cambs. *Hafucestune* c.975, *Hauochestun* 1086 (DB). 'Farmstead of a man called *Hafoc'. OE pers. name + tūn.

Hauxwell, East N. Yorks. *Hauocheswelle* 1086 (DB). 'Spring of the hawk, or of a man called *Hafoc'. OE hafoc or pers. name + wella.

Havant Hants. *Hamanfuntan* 935, *Havehunte* 1086 (DB). 'Spring of a man called Hāma'. OE pers. name + *funta.

Havenstreet I. of Wight. *Hethenestrete* 1255. Probably 'street (of houses) or hamlet of a man called *le Hethene'*, from ME *strete*. A certain Richard *le Hethene* is recorded in this area c.1240. Alternatively perhaps 'heather-covered street', from OE *hǣthen*.

Haverfordwest (*Hwlffordd*) Pemb. *Haverfordia* 1191, *Hareford* 1283, *Heverford West* 1448, *Herefordwest* 1471. 'Western ford used by goats'. OE *hæfer + ford + west*. *West* was added to distinguish the town from HEREFORD, Herefs. The Welsh name has developed from the original English.

Haverhill Suffolk. *Hauerhella* 1086 (DB). Probably 'hill where oats are grown'. OE *hæfera + hyll*. Alternatively the first element may be OE *hæfer* 'a he-goat'.

Haverigg Cumbria. *Haverig* c.1180. 'Ridge where oats are grown, or where he-goats graze'. OScand. *hafri* or *hafr + hryggr*.

Havering-atte-Bower Gtr. London. *Haueringas* 1086 (DB), *Hauering atte Bower* 1272. '(Settlement of) the family or followers of a man called *Hæfer'. OE pers. name + -ingas*. Affix means 'at the bower or royal residence' from ME *atte* 'at the' and *bour* (OE *būr*).

Haversham Milt. K. *Hæfæresham* 10th cent., *Havresham* 1086 (DB). 'Homestead or river-meadow of a man called *Hæfer'. OE pers. name + hām or hamm.

Haverthwaite Cumbria. *Haverthwayt* 1336. 'Clearing where oats are grown'. OScand. *hafri + thveit*.

Hawarden (*Penarlâg*) Flin. *Haordine* 1086 (DB), *Haworthyn* 1275. 'High enclosure'. OE

hēah + worthign. The Welsh name probably means 'high ground rich in cattle' (Welsh *pennardd + alafog*).

Hawes N. Yorks. *Hawes* 1614. '(Place at) the pass between the hills'. OE *hals*.

Hawick Sc. Bord. *Hawic* c.1167. 'Enclosed farm'. OE *haga + wīc*.

Hawkchurch Devon. *Hauekechierch* 1196. 'Church of a man called *Hafoc'. OE pers. name + cirice*.

Hawkedon Suffolk. *Hauokeduna* 1086 (DB). Probably 'hill frequented by hawks'. OE *hafoc + dūn*.

Hawkesbury S. Glos. *Havochesberie* 1086 (DB). 'Stronghold of the hawk, or of a man called *Hafoc'. OE hafoc or OE pers. name + burh* (dative *byrig*).

Hawkhill Northum. *Hauechil* 1178. 'Hill frequented by hawks'. OE *hafoc + hyll*.

Hawkhurst Kent. *Hauekehurst* 1254. 'Wooded hill frequented by hawks'. OE *hafoc + hyrst*.

Hawkinge Kent. *Hauekinge* 1204. 'Place frequented by hawks', or 'place of a man called *Hafoc'. OE hafoc or pers. name + -ing*.

Hawkley Hants. *Hauecle* 1207. 'Woodland clearing frequented by hawks'. OE *hafoc + lēah*.

Hawkridge Somerset. *Hauekerega* 1194. 'Ridge frequented by hawks'. OE *hafoc + hrycg*.

Hawkshead Cumbria. *Hovkesete* c.1200. 'Mountain pasture of a man called Haukr'. OScand. pers. name + *sǣtr*.

Hawkswick N. Yorks. *Hochesuuic* [sic] 1086 (DB), *Haukeswic* 1176. 'Dwelling or (dairy) farm of a man called Haukr'. OScand. pers. name + OE *wīc*.

Hawksworth Leeds. *Hafecesweorthe* c.1030, *Hauocesorde* 1086 (DB). 'Enclosure of a man called *Hafoc'. OE pers. name + worth.

Hawksworth Notts. *Hochesuorde* 1086 (DB). 'Enclosure of a man called Hōc'. OE pers. name + *worth*.

Hawkwell Essex. *Hacuuella* 1086 (DB). 'Winding stream'. OE *haca* 'hook, bend' + *wella*.

Hawley Hants. *Hallee, Halely* 1248. Possibly 'woodland clearing near a hall or large house'. OE *heall* + *lēah*. Alternatively the first element may be *healh* 'nook or corner of land'.

Hawley Kent. *Hagelei* [*sic*] 1086 (DB), *Halgeleg* 1203. 'Holy wood or clearing'. OE *hālig* + *lēah*.

Hawling Glos. *Hallinge* 1086 (DB). '(Settlement of) the people from HALLOW', or '(settlement of) the people at the nook of land'. OE *halh* + *-ingas*. Alternatively, identical in origin with HALLING (Kent).

Hawnby N. Yorks. *Halm(e)bi* 1086 (DB). 'Farmstead or village of a man called Halmi'. OScand. pers. name + *bý*. Alternatively the first element may be OScand. *halmr* 'straw'.

Haworth Brad. *Hauewrth* 1209. 'Enclosure with a hedge'. OE *haga* + *worth*.

Hawsker, High N. Yorks. *Houkesgarth* *c.*1125. 'Enclosure of a man called Haukr'. OScand. pers. name + *garthr*.

Hawstead Suffolk. *Haldsteda* 1086 (DB). 'Place of refuge or shelter'. OE *h(e)ald* + *stede*.

Hawthorn Durham. *Hagathorne c.*1115. '(Place at) the hawthorn'. OE *hagu-thorn*.

Hawton Notts. *Holtone* 1086 (DB). Probably 'farmstead in a hollow'. OE *hol* + *tūn*.

Haxby York. *Haxebi* 1086 (DB). 'Farmstead or village of a man called Hákr'. OScand. pers. name + *bý*.

Haxey N. Lincs. *Acheseia* 1086 (DB). 'Island, or dry ground in marsh, of a man called Haki'. OScand. pers. name + OE *ēg* or OScand. *ey*.

Hay-on-Wye (*Y Gelli Gandryll*) Powys. *Hagan* 958, *Haya* 1144. 'Enclosure on the River Wye'. OE *hæg*. The Welsh name means 'the woodland of a hundred plots' (Welsh *y* + *celli* + *candryll*). The river-name is of uncertain origin and meaning.

Haydock St Hel. *Hedoc* 1169. Probably a Welsh name from **heiddiog* 'barley place, corn farm'.

Haydon, usually 'hill or Down where hay is made', OE *hēg* + *dūn*: **Haydon** Dorset. *Heydone* 1163. **Haydon Wick** Swindn. *Haydon* 1242, *Haydonwyk* 1249. The addition is OE *wīc* 'dairy farm'.
   However the following has a different second element: **Haydon Bridge** Northum. *Hayden* 1236. 'Valley where hay is made'. OE *hēg* + *denu*.

Hayes, 'land overgrown with brushwood', OE **hæs(e)*: **Hayes** Gtr. London, near Bromley. *Hesa* 1177. **Hayes** Gtr. London, near Hillingdon. *linga hæse* 793, *Hesa* 1086 (DB). Prefix *linga* in earliest spelling is obscure.

Hayfield Derbys. *Hedfelt* [*sic*] 1086 (DB), *Heyfeld* 1285. Probably 'open land where hay is obtained'. OE *hēg* + *feld*.

Hayle Cornwall. *Heyl* 1265. Named from the River Hayle, a Celtic name meaning 'estuary'.

Hayling Island Hants. *Heglingaigæ* 956, *Halingei* 1086 (DB). 'Island of the family or followers of a man called *Hægel'. OE pers. name + *-inga-* + *ēg*.

Hayling, North & Hayling, South Hants. *Hailinges c.*1140. '(Settlement of) the family or followers of a man called *Hægel'. OE pers. name + *-ingas*.

Haynes Beds. *Hagenes* 1086 (DB). 'The enclosures'. OE **hægen, *hagen* in a plural form.

Hayscastle (*Cas-lai*) Pemb. *Castrum Hey* 1293, *Heyscastel* 1326. 'Hay's castle'. ME pers. name + *castel*. The Welsh name is a development of *Castell-hay* which translates the English.

Hayton, 'farmstead where hay is made or stored', OE *hēg* + *tūn*: **Hayton** Cumbria, near Brampton. *Hayton c.*1170. **Hayton** E. R. Yorks. *Haiton* 1086 (DB). **Hayton** Notts. *Heiton* 1175.

Haywards Heath W. Sussex. *Heyworth* 1261, *Haywards Hoth* 1544. 'Heath by the enclosure with a hedge'. OE *hege* + *worth*, with the later addition of *hæth*.

Haywood, 'enclosed wood', OE *hæg* or *hege* + *wudu*: **Haywood** Herefs. *Haywode* 1276. **Haywood, Great & Haywood, Little** Staffs. *Haiwode* 1086 (DB). **Haywood Oaks** Notts. *Heywod* 1232. The affix no doubt refers to some notable oak-trees.

Hazel Grove Stockp. *Hesselgrove* 1690. 'Hazel copse'. OE *hæsel* + *grāf*.

Hazelbury Bryan Dorset. *Hasebere* 1201, *Hasilbere Bryan* 1547. 'Hazel wood or grove'. OE *hæsel* + *bearu*. Manorial affix from the *Bryene* family, here in the 14th cent.

Hazeley Hants. *Heishulla* 1167. 'Brushwood hill'. OE **hæs* + *hyll*.

Hazelwood Derbys. *Haselwode* 1306. 'Hazel wood'. OE *hæsel* + *wudu*.

Hazlemere Bucks. *Heselmere* 13th cent. 'Pool where hazels grow'. OE *hæsel* + *mere*.

Hazleton Glos. *Hasedene* [sic] 1086 (DB), *Haselton* 12th cent. 'Farmstead where hazels grow'. OE *hæsel* + *tūn*.

Heacham Norfolk. *Hecham* 1086 (DB). 'Homestead with a hedge or hatch-gate'. OE *hecg* or *hecc* + *hām*.

Headbourne Worthy Hants. *See* WORTHY.

Headcorn Kent. *Hedekaruna* c.1100. Possibly 'tree-trunk (used as a footbridge) of a man called *Hydeca'. OE pers. name + *hruna*.

Headingley Leeds. *Hedingeleia* 1086 (DB). 'Woodland clearing of the family or followers of a man called Head(d)a'. OE pers. name + *-inga-* + *lēah*.

Headington Oxon. *Hedenandun* 1004, *Hedintone* 1086 (DB). 'Hill of a man called *Hedena'. OE pers. name (genitive -*n*) + *dūn*.

Headlam Durham. *Hedlum* c.1190. '(Place at) the woodland clearings where heather grows'. OE *hǣth* + *lēah* (in a dative plural form *lēaum*).

Headley, 'woodland clearing where heather grows', OE *hǣth* + *lēah*; examples include: **Headley** Hants. *Hallege* 1086 (DB), *Hetliga* c.1190. **Headley** Surrey. *Hallega* 1086 (DB).

Headon Notts. *Hedune* 1086 (DB). Probably 'high hill'. OE *hēah* + *dūn*.

Heage Derbys. *Heyheg* 1251. 'High edge or ridge'. OE *hēah* + *ecg*.

Healaugh, Healey, 'high clearing or wood', OE *hēah* + *lēah*: **Healaugh** N. Yorks. near Reeth. *Hale* [sic] 1086 (DB), *Helagh* 1200. **Healey** Northum., near Hexham. *Heley* 1268. **Healey** N. Yorks. *Helagh* c.1280.

Healing NE. Lincs. *Hegelinge* 1086 (DB). '(Settlement of) the family or followers of a man called *Hægel'. OE pers. name + *-ingas*.

Heanish Arg. (Tiree). Unidentified first element + OScand. *nes* 'headland'. Compare HYNISH.

Heanor Derbys. *Hainoure* 1086 (DB). '(Place at) the high ridge'. OE *hēah* (dative *hēan*) + *ofer*.

Heanton Punchardon Devon. *Hantone* 1086 (DB), *Heanton Punchardun* 1297. 'High (or chief) farm'. OE *hēah* (dative *hēan*) + *tūn*.

Manorial affix from its possession by the *Punchardon* family (from the 11th cent.).

Heapham Lincs. *Iopeham* 1086 (DB). 'Homestead or enclosure where rose-hips or brambles grow'. OE *hēope* or *hēopa* + *hām* or *hamm*.

Heath, a common name, '(place at) the heath', OE *hǣth*; examples include: **Heath and Reach** Beds. *La Hethe* 1276. *See* REACH. **Heath** Derbys. *Heth* 1257. Earlier called *Lunt* 1086 (DB) from OScand. *lundr* 'small wood, grove'.

Heath Hayes Staffs. *Hethhey* 1570. 'Heathy enclosure(s)'. OE *hǣth* + *hæg*.

Heathcote Derbys. *Hedcote* late 12th cent. 'Cottage(s) on a heath'. OE *hǣth* + *cot*.

Heather Leics. *Hadre* 1086 (DB). OE *hǣthor* 'heathland' or *hǣddre* 'heather place'.

Heathfield, 'heathy open land, open land overgrown with heather', OE *hǣth* + *feld*: **Heathfield** E. Sussex. *Hadfeld* 12th cent. **Heathfield** Somerset. *Hafella* [sic] 1086 (DB), *Hathfeld* 1159.

Heathrow Gtr. London. *La Hetherewe* c.1410. 'Row of houses on or near the heath'. ME *hethe* + *rewe*.

Heatley Warrtn. *Heyteley* 1525. 'Woodland clearing where heather grows'. OE *hǣth* + *lēah*.

Heaton, a common name, 'high farmstead', OE *hēah* + *tūn*; examples include: **Heaton** Brad. *Hetun* 1160. **Heaton** Newc. upon T. *Heton* 1256. **Heaton, Castle** Northum. *Heton* 1183. Affix may refer to a hall-house here in medieval times. **Heaton, Earls** Kirkl. *Etone* 1086 (DB), *Erlesheeton* 1308. Manorial affix from possession by the Earls of Warren.

Hebburn S. Tyne. *Heabyrm* (for -*byrin*) c.1104. 'High burial place or tumulus'. OE *hēah* + *byrgen*.

Hebden, 'valley where rose-hips or brambles grow', OE *hēope* or *hēopa* + *denu*: **Hebden** N. Yorks. *Hebedene* 1086 (DB). **Hebden Bridge** Calder. *Hepdenbryge* 1399. With OE *brycg* 'bridge' (over Hebden Water).

Hebrides (islands) Arg., Highland, W. Isles. *Hæbudes* 77, *Hebudes* 300. Meaning uncertain. The Roman name was *Ebudae* or *Ebudes*, and the present name resulted from a misreading of the latter, with *ri* for *u*. The OScand. name was *Suthreyar*, 'southern islands', as being south of Orkney and Shetland. The Gaelic name is *Innse Gall*.

Hebron Northum. *Heburn* 1242. Probably identical in origin with HEBBURN.

Heck, Great N. Yorks. *Hech* 1153–5. 'The hatch or gate'. OE *hæcc*.

Heckfield Hants. *Hechfeld* 1207. Possibly 'open land by a hedge or hatch-gate'. OE *hecg* or *hecc* + *feld*.

Heckingham Norfolk. *Hechingheam* 1086 (DB), *Hekingeham* 1245. 'Homestead of the family or followers of a man called Heca'. OE pers. name + *-inga-* + *hām*.

Heckington Lincs. *Hechintune* 1086 (DB). 'Estate associated with a man called Heca'. OE pers. name + *-ing-* + *tūn*.

Heckmondwike Kirkl. *Hedmundewic* [sic] 1086 (DB), *Hecmundewik* 13th cent. 'Dwelling or (dairy) farm of a man called Hēahmund'. OE pers. name + *wīc*.

Hedderwick E. Loth. *Hatheruuich* 1094. 'Farmstead among heather'. OE *hæddre* + *wīc*.

Heddington Wilts. *Edintone* 1086 (DB). 'Estate associated with a man called Hedde'. OE pers. name + *-ing-* + *tūn*.

Heddon, 'hill where heather grows', OE *hæth* + *dūn*: **Heddon, Black** Northum. *Hedon* 1271. **Heddon on the Wall** Northum. *Hedun* 1175. Affix from its situation on HADRIAN'S WALL.

Hedenham Norfolk. *Hedenaham* 1086 (DB). 'Homestead of a man called *Hedena*'. OE pers. name + *hām*.

Hedge End Hants. *Cutt Hedge End* 1759. Self-explanatory, with *cut* 'clipped' in the 18th-cent. form.

Hedgerley Bucks. *Huggeleg* 1195. 'Woodland clearing of a man called *Hycga*'. OE pers. name + *lēah*.

Hedingham, Castle & Hedingham, Sible Essex. *Hedingham* 1086 (DB), *Heyngham Sibille* 1231, *Hengham ad castrum* 1254. Probably 'homestead of the family or followers of a man called *Hyth(a)*, or of the dwellers at the landing-place'. OE pers. name or *hȳth* + *-inga-* + *hām*. Distinguishing affixes from ME *castel* (Latin *castrum*) 'castle' and from the family of a lady called *Sibil* who held land here in the 13th cent.

Hedley on the Hill Northum. *Hedley* 1242. 'Woodland clearing where heather grows'. OE *hæth* + *lēah*.

Hednesford Staffs. *Hedenesford* 13th cent. 'Ford of a man called *Heddīn*'. OE pers. name + *ford*.

Hedon E. R. Yorks. *Hedon* 12th cent. 'Hill where heather grows'. OE *hæth* + *dūn*.

Heigham, Potter Norfolk. *Echam* 1086 (DB), *Hegham Pottere* 1182. Possibly 'homestead or enclosure with a hedge or hatch-gate'. OE *hecg* or *hecc* + *hām* or *hamm*. The affix (from OE *pottere*) must allude to pot-making here at an early date.

Heighington Darltn. *Heghyngtona* 1183. Probably 'farmstead or estate on high ground'. OE *hēahing* + *tūn*.

Heighington Lincs. *Hictinton* 1242. Probably 'estate associated with a man called *Hyht*'. OE pers. name + *-ing-* + *tūn*.

Heighton, South E. Sussex. *Hectone* 1086 (DB), *Sutheghton* 1327. 'High farmstead'. OE *hēah* + *tūn*, with OE *sūth* 'south'.

Helen's Bay Down. The village was named after *Helen* Selina Sheridan by her son, Frederick Hamilton-Temple-Blackwood, 5th Lord Dufferin (1826–1902). Its equivalent Irish name is *Cuan Héilin*.

Helensburgh Arg. 'Helen's town'. Scottish English *burgh*. A new settlement was founded here in 1776 by Sir James Colquhoun of Luss, who named it after his wife, Lady *Helen* Sutherland.

Helford Cornwall. *Helleford* 1230. 'Estuary crossing-place'. Cornish *heyl* + OE *ford*.

Helhoughton Norfolk. *Helgatuna* 1086 (DB). 'Farmstead of a man called Helgi'. OScand. pers. name + OE *tūn*.

Helions Bumpstead Essex. *See* BUMPSTEAD.

Helland Cornwall. *Hellaunde* 1284. 'Old church-site'. Cornish *hen-lann*.

Hellesdon Norfolk. *Hægelisdun* c.985, *Hailesduna* 1086 (DB). 'Hill of a man called *Hægel*'. OE pers. name + *dūn*.

Hellidon Northants. *Elliden* 12th cent. Possibly 'holy, healthy, or prosperous valley'. OE *hælig* + *denu*.

Hellifield N. Yorks. *Helgefeld* 1086 (DB). Probably 'open land of a man called Helgi'. OScand. pers. name + OE *feld*.

Hellingly E. Sussex. *Hellingeleghe* 13th cent. 'Woodland clearing of the family or followers of a man called *Hielle*, or of the hill dwellers'. OE pers. name or OE *hyll* + *-inga-* + *lēah*.

Hellington Norfolk. *Halgatune* 1086 (DB). 'Farmstead of a man called Helgi'. OScand. pers. name + OE *tūn*.

Helmdon Northants. *Elmedene* 1086 (DB). 'Valley of a man called *Helma*'. OE pers. name + *denu*.

Helmingham Suffolk. *Helmingheham* 1086 (DB). 'Homestead of the family or followers of a man called Helm'. OE pers. name + *-inga-* + *hām*.

Helmington Row Durham. *Helmeraw* 1384, *Elmedenrawe* 1522. 'Row of houses at Helmington ('valley by helmet-shaped hill')'. OE *helm* 'helmet' + *denu* + *rāw*.

Helmsdale Highland. *Hjalmunddal* 1225. 'Hjalmund's valley'. OScand. pers. name + *dalr*.

Helmshore Lancs. *Hellshour* 1510. 'Steep slope with a cattle shelter'. OE *helm* + **scora*.

Helmsley N. Yorks. *Elmeslac* [*sic*] 1086 (DB), *Helmesley* 12th cent. 'Woodland clearing of a man called Helm'. OE pers. name + *lēah*.

Helmsley, Gate & Helmsley, Upper N. Yorks. *Hamelsec(h)* 1086 (DB), *Gatehemelsay* 1438, *Over Hemelsey* 1301. 'Island, or dry ground in marsh, of a man called Hemele'. OE pers. name + *ēg*. Distinguishing affixes from OScand. *gata* 'road' (here a Roman road) and OE *uferra* 'upper'.

Helperby N. Yorks. *Helperby* 972, *Helprebi* 1086 (DB). 'Farmstead or village of a woman called Hjalp'. OScand. pers. name (genitive *-ar*) + *bý*.

Helperthorpe N. Yorks. *Elpetorp* 1086 (DB). 'Outlying farmstead or hamlet of a woman called Hjalp'. OScand. pers. name (genitive *-ar*) + *thorp*.

Helpringham Lincs. *Helperi(n)cham* 1086 (DB). 'Homestead of the family or followers of a man called Helprīc'. OE pers. name + *-inga-* + *hām*.

Helpston Peterb. *Hylpestun* 948. 'Farmstead of a man called *Help*'. OE pers. name + *tūn*.

Helsby Ches. *Helesbe* [*sic*] 1086 (DB), *Hellesbi* late 12th cent. 'Farmstead or village on a ledge'. OScand. *hjallr* + *bý*.

Helston, Helstone, 'estate at an old court', Cornish **hen-lys* + OE *tūn*: **Helston** Cornwall. *Henlistone* 1086 (DB). **Helstone** Cornwall. *Henliston* 1086 (DB).

Helton Cumbria. *Helton* c.1160. Probably 'farmstead on a slope'. OE *helde* + *tūn*.

Helvellyn Cumbria. *Helvillon* 1577. Probably Cumbric **hal* 'moor' + **velyn* 'yellow'.

Helvick Waterford. 'Rock shelf bay'. OScand. *hjalli* + *vík*.

Helygain. See HALKYN.

Hemblington Norfolk. *Hemelingetun* 1086 (DB). 'Estate associated with a man called Hemele'. OE pers. name + *-ing-* + *tūn*.

Hemel Hempstead Herts. See HEMPSTEAD.

Hemingbrough N. Yorks. *Hemingburgh* 1080–6, *Hamiburg* 1086 (DB). Probably 'stronghold of a man called Hemingr'. OScand. pers. name + OE *burh*.

Hemingby Lincs. *Hamingebi* 1086 (DB). 'Farmstead or village of a man called Hemingr'. OScand. pers. name + *bý*.

Hemingford Abbots & Hemingford Grey Cambs. *Hemmingeford* 974, *Emingeford* 1086 (DB), *Hemingford Abbatis* 1276, *Hemingford Grey* 1316. 'Ford of the family or followers of a man called Hemma or Hemmi'. OE pers. name + *-inga-* + *ford*. Distinguishing affixes from early possession by the Abbot of Ramsey and the *de Grey* family.

Hemingstone Suffolk. *Hamingestuna* 1086 (DB). 'Farmstead of a man called Hemingr'. OScand. pers. name + OE *tūn*.

Hemington, 'estate associated with a man called Hemma or Hemmi', OE pers. name + *-ing-* + *tūn*: **Hemington** Leics. *Aminton* [*sic*] c.1130. *Hemminton* c. 1200. **Hemington** Northants. *Hemmingtune* 1077, *Hemintone* 1086 (DB). **Hemington** Somerset. *Hammingtona* 1086 (DB).

Hemley Suffolk. *Helmelea* 1086 (DB). 'Woodland clearing of a man called *Helma*'. OE pers. name + *lēah*.

Hempholme E. R. Yorks. *Hempholm* 12th cent. 'Raised ground in marshland where hemp is grown'. OE *hænep* + OScand. *holmr*.

Hempnall Norfolk. *Hemenhala* 1086 (DB). 'Nook of land of a man called Hemma'. OE pers. name (genitive *-n*) + *halh*.

Hempstead, usually 'the homestead', OE *hām-stede, hæm-stede*: **Hempstead** Essex. *Hamesteda* 1086 (DB). **Hempstead** Norfolk, near Lessingham. *Hemsteda* 1086 (DB). **Hempstead, Hemel** Herts. *Hamelamestede* 1086 (DB). Hemel is an old district-name first recorded *c.*705 meaning 'broken, undulating' from OE **hamel*.
    However the following has a different origin: **Hempstead** Norfolk, near Holt. *Henepsteda* 1086 (DB). 'Place where hemp is grown'. OE *hænep* + *stede*.

Hempsted Glos. *Hechanestede* 1086 (DB). 'High homestead'. OE *hēah* + *hām-stede*.

Hempton Norfolk. *Hamatuna* 1086 (DB). 'Farmstead or village of a man called Hemma'. OE pers. name + *tūn*.

Hempton Oxon. *Henton* 1086 (DB). 'High farm'. OE *hēah* (dative *hēan*) + *tūn*.

Hemsby Norfolk. *Heimesbei* 1086 (DB). Probably 'farmstead or village of a man called **Hēmer'*. OScand. pers. name + *bý*.

Hemswell Lincs. *Helmeswelle* 1086 (DB). 'Spring or stream of a man called Helm'. OE pers. name + *wella*.

Hemsworth Wakefd. *Hamelesuurde* [*sic*] 1086 (DB), *Hymeleswrde* 12th cent. 'Enclosure of a man called **Hymel'*. OE pers. name + *worth*.

Hemyock Devon. *Hamihoc* 1086 (DB). Possibly a Celtic river-name meaning 'summer stream', otherwise 'river-bend of a man called Hemma', from OE pers. name + *hōc*.

Henbury Brist. *Heanburg* 692, *Henberie* 1086 (DB). 'High or chief fortified place'. OE *hēah* (dative *hēan*) + *burh* (dative *byrig*).

Henbury Ches. *Hameteberie* 1086 (DB). OE *burh* (dative *byrig*) 'stronghold or manor house' with an uncertain first element, possibly OE *hǣmed* 'community'.

Hendon Gtr. London. *Hendun* 959, *Handone* 1086 (DB). '(Place at) the high hill'. OE *hēah* (dative *hēan*) + *dūn*.

Hendon Sundld. *Hynden* 1382. 'Valley frequented by hinds'. OE *hind* + *denu*.

Hendred, East & Hendred, West Oxon. *Hennarith* 956, *Henret* 1086 (DB). 'Stream frequented by hens (of wild birds)'. OE *henn* + *rīth*.

Hendy-gwyn. *See* WHITLAND.

Henfield W. Sussex. *Hanefeld* 770, *Hamfelde* 1086 (DB). Probably 'open land characterized by stones or rocks'. OE *hān* + *feld*.

Hengoed Carm. 'Old wood'. Welsh *hen* + *coed*.

Hengrave Suffolk. *Hemegretham* [*sic*] 1086 (DB), *Hemegrede* *c.*1095. 'Grassy meadow of a man called Hemma'. OE pers. name + **grēd* (with hām 'homestead' in the Domesday form).

Henham Essex. *Henham* *c.*1045, 1086 (DB). 'High homestead or enclosure'. OE *hēah* (dative *hēan*) + *hām* or *hamm*.

Henley, usually 'high (or chief) wood or clearing', OE *hēah* (dative *hēan*) + *lēah*; examples include: **Henley** Somerset. *Henleighe* 973. **Henley** Suffolk. *Henleia* 1086 (DB). **Henley-in-Arden** Warwicks. *Henle* *c.*1180. Affix refers to the medieval Forest of Arden (*Eardene* 1088), possibly a Celtic name meaning 'high district'. **Henley-on-Thames** Oxon. *Henleiam* *c.*1140. *See* THAMES.
    However the following has a different origin: **Henley** Shrops. *Haneluf* [*sic*] 1086 (DB), *Hennele* 1242. 'Wood or clearing frequented by hens (of wild birds)'. OE *henn* + *lēah*.

Henlow Beds. *Haneslauue* 1086 (DB). 'Hill or mound frequented by hens (of wild birds)'. OE *henn* + *hlāw*.

Hennock Devon. *Hainoc* 1086 (DB). '(Place at) the high oak-tree'. OE *hēah* (dative *hēan*) + *āc*.

Henny, Great Essex. *Heni* 1086 (DB). 'High island or land partly surrounded by water'. OE *hēah* (dative *hēan*) + *ēg*.

Hensall N. Yorks. *Edeshale* [*sic*] 1086 (DB), *Hethensale* 12th cent. 'Nook of land of a man called **Hethīn* or Hethinn'. OE or OScand. pers. name + OE *halh*.

Henshaw Northum. *Hedeneshalch* 12th cent. Identical in origin with the previous name.

Henstead Suffolk. *Henestede* 1086 (DB). Probably 'place frequented by hens (of wild birds)'. OE *henn* + *stede*.

Henstridge Somerset. *Hengstesrig* 956, *Hengest(e)rich* 1086 (DB). 'Ridge where stallions

are kept' or 'ridge of a man called Hengest'.
OE *hengest* or OE pers. name + *hrycg*.

Henton Oxon. *Hentone* 1086 (DB). 'High (or
chief) farmstead'. OE *hēah* (dative *hēan*) + *tūn*.

Henton Somerset. *Hentun* 1065. Identical in
origin with previous name, or 'farmstead where
hens are kept', OE *henn* + *tūn*.

Hepburn Northum. *Hybberndune* c.1050.
Probably identical in origin with HEBBURN
(and with the addition of OE *dūn* 'hill' in the
early spelling).

Hepple Northum. *Hephal* 1205. 'Nook of land
where rose-hips or brambles grow'. OE *hēope* or
hēopa + *halh*.

Hepscott Northum. *Hebscot* 1242. 'Cottage(s)
of a man called *Hebbi*'. OE pers. name + *cot*.

Heptonstall Calder. *Heptonstall* 1253.
'Farmstead where rose-hips or brambles grow'.
OE *hēope* or *hēopa* + *tūn-stall*.

Hepworth, probably 'enclosure of a man
called *Heppa*', OE pers. name + *worth*:
Hepworth Kirkl. *Heppeuuord* 1086 (DB).
Hepworth Suffolk. *Hepworda* 1086 (DB).

Herbrandston Pemb. *Villa Herbrandi* 13th
cent., *Herbraundistone* 1307. 'Herbrand's farm'.
OGerman pers. name + OE *tūn*.

Hereford, 'ford suitable for the passage of an
army', OE *here-ford*: **Hereford** Herefs.
Hereford 958, 1086 (DB). **Herefordshire** (OE
scīr 'district') is first referred to in the 11th cent.
Hereford, Little Herefs. *Lutelonhereford*
1086 (DB). Affix is OE *lȳtlan*, dative of *lȳtel* 'little'.

Hergest Herefs. *Hergest*(h) 1086 (DB).
Probably a Welsh name, but obscure in origin
and meaning.

Herm (island) Channel Islands. *Erm* 1087,
Erme 1361. A Brittonic development of its
ancient Roman name *Sarmia* (recorded thus in
the 4th-cent. *Maritime Itinerary*), of unknown
origin and meaning.

Hermiston Edin. *Hirdmannistoun* 1233.
'Herdsman's farmstead'. OE *hirdman* + *tūn*.

Hermitage, '(place by) the hermitage'. ME
ermitage. **Hermitage** Berks. *Le Eremytage*
1550. **Hermitage** Dorset. *the hermitage of
Blakemor* 1309, *Ermytage* 1389, *see* BLACKMOOR
FOREST. **Hermitage** W. Sussex. *Armetage*
1635.

Herne, Herne Bay Kent. *Hyrnan* c.1100.
'(Place at) the angle or corner of land'. OE *hyrne*.

Hernhill Kent. *Haranhylle* c.1100. '(Place at)
the grey hill'. OE *hār* (dative -*an*) + *hyll*.

Herriard Hants. *Henerd* [*sic*] 1086 (DB),
Herierda c.1160. Probably 'army quarters'
(perhaps of a Viking army), OE *here* + *geard*.
Alternatively this may be a Celtic name meaning
'long ridge' from *hyr* + *garth*.

Herringfleet Suffolk. *Herlingaflet* 1086 (DB).
'Creek or stream of the family or followers of a
man called *Herela*'. OE pers. name + -*inga*- +
flēot.

Herringswell Suffolk. *Hyrningwella* 1086
(DB). Possibly 'spring or stream by the horn-
shaped or curving hill'. OE *hyrning* + *wella*.

Herrington, East Sundld. *Herintune*,
Harintune c.1116. Probably 'farmstead or estate
at the stony place'. OE *hæring* + *tūn*.

Hersham Surrey. *Hauerichesham* 1175.
'Homestead or river-meadow of a man called
Hæferic'. OE pers. name + *hām* or *hamm*.

Herstmonceux E. Sussex. *Herst* 1086 (DB),
Herstmonceus 1287. OE *hyrst* 'wooded hill' +
manorial affix from possession by the *Monceux*
family in the 12th cent.

Hertford Herts. *Herutford* 731, *Hertforde* 1086
(DB). 'Ford frequented by harts or stags'. OE
heorot + *ford*. **Hertfordshire** (OE *scīr*
'district') is first referred to in the 11th cent.

Hertingfordbury Herts. *Herefordingberie*
[*sic*] 1086 (DB), *Hertfordingeberi* 1220.
'Stronghold of the people of HERTFORD'.
OE -*inga*- + *burh* (dative *byrig*).

Hertsmere (district) Herts. A modern name,
combining *Herts*, the abbreviated form of
HERTFORDSHIRE, and *mere* (OE *gemǣre*)
'boundary'.

Hesket, Hesketh, probably 'boundary land
where horses graze', but possibly 'race course
for horses', OScand. *hestr* + *skeith*; examples
include: **Hesket, High & Hesket, Low**
Cumbria. *Hescayth* 1285. **Hesketh** Lancs.
Heschath 1288.
   However the following has a different origin:
Hesket Newmarket Cumbria. *Eskeheued*
1227. 'Hill growing with ash-trees'. OScand. *eski*
+ OE *hēafod*. Affix first found in the 18th cent.
refers to the market here.

Heskin Green Lancs. *Heskyn* 1257. A Celtic name from **hesgīn* '(marshy) place where sedge grows'.

Hesleden Durham. *Heseldene c.*1040. 'Valley where hazels grow'. OE *hæsel + denu*.

Heslerton, East & Heslerton, West N. Yorks. *Heslerton* 1086 (DB). 'Farmstead where hazels grow'. OE **hæsler + tūn*.

Heslington York. *Haslinton* 1086 (DB). 'Farmstead by the hazel wood'. OE **hæsling + tūn*.

Hessay York. *Hesdesai* [*sic*] 1086 (DB), *Heslesaia* 12th cent. 'Marshland, or island, where hazels grow'. OE *hæsel* (influenced by OScand. *hesli*) + *sǣ* or *ēg*.

Hessett Suffolk. *Heteseta* [*sic*] 1086 (DB), *Heggeset* 1225. 'Fold (for animals) with a hedge'. OE *hecg + (ge)set*.

Hessle E. R. Yorks. *Hase* [*sic*] 1086 (DB), *Hesel* 12th cent. '(Place at) the hazel-tree'. OE *hæsel* (influenced by OScand. *hesli*).

Hest Bank Lancs. *Hest* 1177. OE **hǣst* 'undergrowth, brushwood'.

Heston Gtr. London. *Hestone c.*1125. 'Farmstead among the brushwood'. OE **hǣs + tūn*.

Heswall Wirral. *Eswelle* [*sic*] 1086 (DB), *Haselwell c.*1200. 'Spring where hazels grow'. OE *hæsel + wella*.

Hethe Oxon. *Hedha* 1086 (DB). OE *hǣth* 'heath, uncultivated land'.

Hethersett Norfolk. *Hederseta* 1086 (DB). Probably '(settlement of) the dwellers among the heather'. OE **hǣddre + sǣte*.

Hett Durham. *Het c.*1168. '(Place at) the hat-shaped hill'. OE *hætt* or OScand. *hetti* (dative of *hǫttr*).

Hetton N. Yorks. *Hetune* 1086 (DB). 'Farmstead on a heath'. OE *hǣth + tūn*.

Hetton le Hole Sundld. *Heppedun* 1180. 'Hill where rose-hips or brambles grow'. OE *hēope* or *hēopa + dūn*. Affix means 'in the hollow'.

Heugh Northum. *Hou* 1279. '(Place at) the ridge or spur of land'. OE *hōh*.

Heveningham Suffolk. *Heueniggeham* 1086 (DB). 'Homestead of the family or followers of a man called **Hefin*'. OE pers. name + *-inga- + hām*.

Hever Kent. *Heanyfre* 814. '(Place at) the high edge or hill-brow'. OE *hēah* (dative *hēan*) + *yfer*.

Heversham Cumbria. *Hefresham c.*1050, *Eureshaim* 1086 (DB). Probably 'homestead of a man called Hēahfrith'. OE pers. name + *hām*.

Hevingham Norfolk. *Heuincham* 1086 (DB). 'Homestead of the family or followers of a man called Hefa'. OE pers. name + *-inga- + hām*.

Hewelsfield Glos. *Hiwoldestone* 1086 (DB), *Hualdesfeld c.*1145. 'Open land of a man called Hygewald'. OE pers. name + *feld* (replacing earlier *tūn* 'farmstead').

Hewick, Bridge & Hewick, Copt N. Yorks. *Heawic* 1086 (DB), *Hewik atte brigg* 1309, *Coppedehaiwic* 1208. 'High (or chief) dairy-farm'. OE *hēah* (dative *hēan*) + *wīc*. Distinguishing affixes 'at the bridge' (from OE *brycg*) and 'with a peak or hill-top' (OE *coppede*).

Hewish, 'measure of land that would support a family', OE *hīwisc*: **Hewish** Somerset. *Hywys* 1327. **Hewish, East & Hewish, West** N. Som. *Hiwis* 1198.

Hexham Northum. *Hagustaldes ham* 685 (*c.*1120). 'The warrior's homestead'. OE *hagustald* (perhaps as a pers. name) + *hām*.

Hexton Herts. *Hegestanestone* 1086 (DB). 'Farmstead of a man called Hēahstān'. OE pers. name + *tūn*.

Hexworthy Devon. *Hextenesworth* 1317. 'Enclosed farmstead of a man called Hēahstān'. OE pers. name + *worth, worthig*. **Hexworthy** Cornwall may have the same origin.

Heybridge Essex. *Heaghbregge c.*1200. 'High (or chief) bridge'. OE *hēah + brycg*.

Heydon Cambs. *Haidenam* 1086 (DB). 'Valley where hay is made', or 'valley with an enclosure'. OE *hēg* or *hæg + denu*.

Heydon Norfolk. *Heidon* 1196. 'Hill where hay is made'. OE *hēg + dūn*.

Heyford, 'hay ford', i.e. 'ford used chiefly at hay-making time', OE *hēg + ford*: **Heyford, Lower & Heyford, Upper** Oxon. *Hegford* 1086 (DB). **Heyford, Nether & Heyford, Upper** Northants. *Heiforde* 1086 (DB).

Heysham Lancs. *Hessam* 1086 (DB). 'Homestead or village among the brushwood'. OE *hǣs* + *hām*.

Heyshott W. Sussex. *Hethsete* c.1100. 'Corner of land where heather grows'. OE *hǣth* + *scēat*.

Heytesbury Wilts. *Hestrebe* [sic] 1086 (DB), *Hehtredeberia* c.1115. 'Stronghold of a woman called *Hēahthrȳth*'. OE pers. name + *burh* (dative *byrig*).

Heythrop Oxon. *Edrope* [sic] 1086 (DB), *Hethrop* 11th cent. 'High hamlet or outlying farmstead'. OE *hēah* + *throp*.

Heywood Rochdl. *Heghwode* 1246. 'High (or chief) wood'. OE *hēah* + *wudu*.

Heywood Wilts. *Heiwode* 1225. 'Enclosed wood'. OE *hæg* + *wudu*.

Hibaldstow N. Lincs. *Hiboldestou* 1086 (DB). 'Holy place where St Hygebald is buried'. OE pers. name + *stōw*.

Hickleton Donc. *Icheltone* 1086 (DB). Probably 'farmstead frequented by woodpeckers'. OE *hicol* + *tūn*.

Hickling, '(settlement of) the family or followers of a man called *Hicel*', OE pers. name + *-ingas*: **Hickling** Norfolk. *Hikelinga* 1086 (DB). **Hickling** Notts. *Hikelinge* c.1000, *Hechelinge* 1086 (DB).

Hidcote Boyce Glos. *Hudicota* 716, *Hedecote* 1086 (DB), *Hudicote Boys* 1327. 'Cottage of a man called *Hydeca* or *Huda*'. OE pers. name + *cot*. Manorial affix from a family called *de Bosco* or *Bois*, here in the 13th cent.

Hiendley, South Wakefd. *Hindeleia* 1086 (DB). 'Wood or clearing frequented by hinds or does'. OE *hind* + *lēah*.

High as affix. *See* main name, e.g. for **High Ackworth** (Wakefd.) *see* ACKWORTH.

High Beach Essex. *Highbeach-green* 1670. Probably named from the beech-trees here.

High Peak (district) Derbys. *Alto Pecco* 1219 (here in a Latinized form). A revival of an old hundred-name in the PEAK DISTRICT.

Higham, 'high (or chief) homestead or enclosure', OE *hēah* + *hām* or *hamm*; examples include: **Higham** Derbys. *Hehham* 1155. **Higham** Kent, near Tonbridge. *Hegham* 1327. **Higham** Lancs. *Hegham* 1296. **Higham** Suffolk, near Stratford St Mary. *Hecham* c.1050,

Heihham 1086 (DB). **Higham, Cold** Northants. *Hecham* 1086 (DB), *Colehigham* 1541. Affix is OE *cald* 'cold, exposed'. **Higham Ferrers** Northants. *Hecham* 1086 (DB), *Heccham Ferrar* 1279. Manorial affix from the *Ferrers* family, here in the 12th cent. **Higham Gobion** Beds. *Echam* 1086 (DB), *Heygham Gobyon* 1291. Manorial affix from the *Gobion* family, here from the 12th cent. **Higham on the Hill** Leics. *Hecham* 1220-35. The affix is found from the 16th cent. **Higham Upshire** Kent. *Heahhaam* c.765, *Hecham* 1086 (DB). The affix means 'higher district' from OE *upp* + *scīr*.

Highampton Devon. *Hantone* 1086 (DB), *Heghanton* 1303. 'High farmstead'. OE *hēah* (dative *hēan*) + *tūn*, with the later addition of ME *heghe* 'high' after the meaning of the original first element had been forgotten.

Highbridge Somerset. *Highbridge* 1324. 'High (or chief) bridge'. OE *hēah* + *brycg*.

Highbury Gtr. London. *Heybury* c.1375. 'High manor', from ME *heghe* + *bury*.

Highclere Hants. *Clere* 1086 (DB), *Alta Clera* c.1270. Perhaps originally a Latin stream-name *Clāra* meaning 'clear stream', with the later addition of *high* (Latin *alta*).

Highcliffe Dorset. *High Clift* [sic] 1759, earlier *Black Cliffe* 1610. Self-explanatory.

Higher as affix. *See* main name, e.g. for **Higher Ansty** (Dorset) *see* ANSTY.

Highgate Gtr. London. *Le Heighgate* 1354. 'High tollgate'. ME *heghe* + *gate*.

Highland (the unitary authority). *the heland* 1529, *the High-Lands of Scotland* c.1627. 'High land'. The mountainous region of northern Scotland is distinguished from the Lowlands to the south.

Highleadon Glos. *Hineledene* 13th cent. 'Estate on the River Leadon belonging to a religious community'. OE *hīwan* (genitive plural *hīgna*) + Celtic river-name (meaning 'broad stream').

Highley Shrops. *Hugelei* 1086 (DB). 'Woodland clearing of a man called *Hugga*'. OE pers. name + *lēah*.

Highnam Glos. *Hamme* 1086 (DB), *Hinehamme* 12th cent. OE *hamm* 'river-meadow' with the later addition of OE *hīwan* (genitive plural *hīgna*) 'a religious community'.

Hightown Ches. late name, meaning simply 'the high part of the town' (of CONGLETON).

Highway Wilts. *Hiwei* 1086 (DB). 'Road used for carrying hay'. OE (West Saxon) *hīg* + *weg*.

Highworth Swindn. *Wrde* 1086 (DB), *Hegworth* 1232. OE *worth* 'enclosure, farmstead' with the later addition of *hēah* 'high'.

Hilborough Norfolk. *Hildeburhwella* 1086 (DB), *Hildeburwrthe* 1242. 'Stream or enclosure of a woman called Hildeburh'. OE pers. name + *wella* or *worth*.

Hildenborough Kent. *Hyldenn* 1240, *Hildenborough* 1389. Probably 'woodland pasture on or by a hill'. OE *hyll* + *denn* with the later addition of *burh* 'manor, borough'.

Hildersham Cambs. *Hildricesham* 1086 (DB). 'Homestead of a man called *Hildrīc'. OE pers. name + *hām*.

Hilderstone Staffs. *Hidulvestune* 1086 (DB). 'Farmstead of a man called Hildulfr or Hildwulf'. OScand. or OE pers. name + *tūn*.

Hilderthorpe E. R. Yorks. *Hilgertorp* 1086 (DB). 'Outlying farmstead of a man called Hildiger or a woman called Hildigerthr'. OScand. pers. name + *thorp*.

Hilfield Dorset. *Hylfelde* 934. 'Open land by a hill'. OE *hyll* + *feld*.

Hilgay Norfolk. *Hillingeiæ* 974, *Hidlingheia* 1086 (DB). 'Island, or dry ground in marsh, of the family or followers of a man called *Hȳthla or *Hydla'. OE pers. name + *-inga-* + *ēg*.

Hill S. Glos. *Hilla* 1086 (DB). '(Place at) the hill'. OE *hyll*.

Hill Head Hants. a recent name, self-explanatory.

Hill, North & Hill, South Cornwall. *Henle* 1238, *Northhindle* 1260, *Suthhulle* 1270, *Suthhynle* 1306. 'High wood or clearing' or 'hinds' wood or clearing'. OE *hēah* (dative *hēan*) or *hind* + *lēah*, with later replacement by *hyll* 'hill'.

Hillam N. Yorks. *Hillum* 963. '(Place at) the hills'. OE *hyll* in a dative plural form *hyllum*.

Hillesden Bucks. *Hildesdun* 949, *Ilesdone* 1086 (DB). 'Hill of a man called Hildi'. OE pers. name + *dūn*.

Hillfarrance Somerset. *Hilla* 1086 (DB), *Hull Ferun* 1253. '(Place at) the hill'. OE *hyll*. Manorial addition from the *Furon* family, here in the 12th cent.

Hillhall Down. 'Hill's hall'. Peter *Hill*, son of Sir Moses Hill of HILLSBOROUGH, built a bawn or fortified mansion here in *c*.1637.

Hillingdon Gtr. London. *Hildendune c*.1080, *Hillendone* 1086 (DB). 'Hill of a man called Hilda'. OE pers. name (genitive *-n*) + *dūn*.

Hillington Norfolk. *Helingetuna* 1086 (DB). 'Farmstead of the family or followers of a man called *Hȳthla or *Hydla'. OE pers. name + *-inga-* + *tūn*.

Hillmorton Warwicks. *Mortone* 1086 (DB), *Hulle and Morton* 1247. Originally two distinct names, now combined. *Hill* is '(place at) the hill' from OE *hyll*. *Morton* is 'marshland farmstead' from OE *mōr* + *tūn*.

Hillsborough Down. 'Hill's fortified mansion'. Sir Moses *Hill*, Provost Marshal of Ulster, built a fortress here in *c*.1610. The Irish name of Hillsborough is *Cromghlinn*, as for CRUMLIN.

Hillswick Shet. *Hildiswik c*.1250. 'Hildir's bay'. OScand. pers. name *Hildir* + *vík*.

Hilltown Down. 'Hill's town'. Wills *Hill*, 1st Marquis of Downshire, built a church here in 1766.

Hilmarton Wilts. *Helmerdingtun* 962, *Helmerintone* 1086 (DB). 'Estate associated with a man called *Helmheard'. OE pers. name + *-ing-* + *tūn*.

Hilperton Wilts. *Help(e)rinton* 1086 (DB). 'Estate associated with a man called *Hylprīc'. OE pers. name + *-ing-* + *tūn*.

Hilton, sometimes 'farmstead or village on or near a hill', OE *hyll* + *tūn*: **Hilton** Cambs. *Hiltone* 1196. **Hilton** Derbys. *Hiltune* 1086 (DB).

However other Hiltons have a different origin: **Hilton** Cumbria. *Helton* 1289. Probably 'farmstead on a slope'. OE *helde* + *tūn*. **Hilton** Dorset. *Eltone* 1086 (DB). 'Farmstead on a slope or where tansy grows'. OE *h(i)elde* or *helde* + *tūn*. **Hilton** Durham. *Helton* 1180. 'Farmstead on a slope'. OE *helde* + *tūn*.

Himbleton Worcs. *Hymeltun* 816, *Himeltun* 1086 (DB). 'Farmstead where the hop plant (or some similar plant) grows'. OE *hymele* + *tūn*.

Himley Staffs. *Himelei* 1086 (DB). 'Woodland clearing where the hop plant (or some similar plant) grows'. OE *hymele* + *lēah*.

Hincaster Cumbria. *Hennecastre* 1086 (DB). 'Old fortification or earthwork haunted by hens (of wild birds)'. OE *henn* + *ceaster*.

Hinckley Leics. *Hinchelie* 1086 (DB). 'Woodland clearing of a man called Hynca'. OE pers. name + *lēah*.

Hinckley & Bosworth (district) Leics. A modern name, combining HINCKLEY and (Market) BOSWORTH.

Hinderclay Suffolk. *Hild(e)ric(es)lea* c.1000, *Hilderclea* 1086 (DB). 'Woodland clearing of a man called *Hildrīc*. OE pers. name + *lēah*.

Hinderwell N. Yorks. *Hildrewell* 1086 (DB). Possibly 'spring or well associated with St Hild'. OE pers. name (with Scandinavianized genitive *-ar*) + *wella*. Alternatively the first element may be OE *hyldre* or *hylder* 'elder-tree'.

Hindhead Surrey. *Hyndehed* 1571. 'Hill frequented by hinds or does'. OE *hind* + *hēafod*.

Hindley Wigan. *Hindele* 1212. 'Wood or clearing frequented by hinds or does'. OE *hind* + *lēah*.

Hindlip Worcs. *Hindehlep* 966, *Hindelep* 1086 (DB). 'Gate or fence over which hinds can leap'. OE *hind* + *hlīep*.

Hindolveston Norfolk. *Hidolfestuna* 1086 (DB). 'Farmstead of a man called Hildwulf'. OE pers. name + *tūn*.

Hindon Wilts. *Hynedon* 1268. Probably 'hill belonging to a religious community'. OE *hīwan* (genitive plural *hīgna*) + *dūn*.

Hindringham Norfolk. *Hindringaham* 1068 (DB). Possibly 'homestead of the people living behind (the hills)'. OE *hinder* + *-inga-* + *hām*.

Hingham Norfolk. *Hincham* 1086 (DB), *Heingeham* 1173. Probably 'homestead of the family or followers of a man called Hega'. OE *pers. name* + *-inga-* + *hām*.

Hinksey, North & Hinksey, South Oxon. *Hengestesige* 10th cent. 'Island or well-watered land of the stallion or of a man called Hengest'. OE *hengest* or OE pers. name + *ēg*.

Hinstock Shrops. *Stoche* 1086 (DB), *Hinestok* 1242. OE *stoc* 'outlying farmstead, dependent

settlement' with the later addition of ME *hine* 'household servants'.

Hintlesham Suffolk. *Hintlesham* 1086 (DB). 'Homestead or enclosure of a man called *Hyntel*. OE pers. name + *hām* or *hamm*.

Hinton, a very common name, has two different origins. Sometimes 'high (or chief) farmstead' from OE *hēah* (dative *hēan*) + *tūn*: **Hinton** S. Glos., near Dyrham. *Heanton* 13th cent. **Hinton Ampner** Hants. *Heantun* 1045, *Hentune* 1086 (DB), *Hinton Amner* 13th cent. Affix is OFrench *aumoner* 'almoner', from its early possession by the almoner of St Swithun's Priory at Winchester. **Hinton Blewett** B. & NE. Som. *Hantone* 1086 (DB), *Hentun Bluet* 1246. Manorial affix from its early possession by the *Bluet* family. **Hinton, Broad** Wilts. *Hentone* 1086 (DB), *Brodehenton* 1319. Affix is OE *brād* 'large'. **Hinton Charterhouse** B. & NE. Som. *Hantone* 1086 (DB), *Henton Charterus* 1273. Affix is OFrench *chartrouse* 'a house of Carthusian monks' referring to a priory founded in the early 13th cent. **Hinton, Great** Wilts. *Henton* 1216. **Hinton St George** Somerset. *Hantone* 1086 (DB), *Hentun Sancti Georgii* 1246. Affix from the dedication of the church. **Hinton St Mary** Dorset. *Hamtune* 944, *Haintone* 1086 (DB), *Hinton Marye* 1627. Affix from the possession of the manor by the Abbey of St Mary at Shaftesbury. **Hinton Waldrist** Oxon. *Hentune* 1086 (DB), *Hinton Walrush* 1676. Manorial affix from the family *de Sancto Walerico*, here in the 12th cent.

However many Hintons mean 'farmstead belonging to a religious community', from OE *hīwan* (genitive plural *hīgna*) + *tūn*: **Hinton** Herefs., near Peterchurch. *Hinetune* 1086 (DB). **Hinton** Northants. *Hintone* 1086 (DB). **Hinton, Cherry** Cambs. *Hintone* 1086 (DB), *Cheryhynton* 1576. Affix is ME *chiri* 'cherry' from the number of cherry-trees formerly growing here. **Hinton in the Hedges** Northants. *Hintone* 1086 (DB). Affix (meaning 'among the hedges') is from OE *hecg*. **Hinton, Little** Wilts. *Hinneton* 1205. **Hinton Martell** Dorset. *Hinetone* 1086 (DB), *Hineton Martel* 1226. Manorial affix from the *Martell* family, here in the 13th cent. **Hinton on the Green** Worcs. *Hinetune* 1086 (DB). Affix is ME *grene* 'village green'.

Hints, '(place on) the roads or paths', Welsh *hynt*: **Hints** Shrops. *Hintes* 1242. **Hints** Staffs. *Hintes* 1086 (DB).

Hinwick Beds. *Heneuuic(h)* 1086 (DB). 'Farm where hens are kept'. OE henn + wīc.

Hinxhill Kent. *Hengestesselle c.*864, *Haenostesyle c.*1100. Possibly 'hill of the stallion or of a man called Hengest'. OE *hengest* or OE pers. name + *hyll*. Alternatively 'shed or shelter for stallions', OE *hengest* + **(ge)sell*.

Hinxton Cambs. *Hestitone* [*sic*] 1086 (DB), *Hengstiton* 1202. 'Estate associated with a man called Hengest'. OE pers. name + *-ing-* + *tūn*.

Hinxworth Herts. *Haingesteuuorde* 1086 (DB). Probably 'enclosure where stallions are kept'. OE *hengest* + *worth*.

Hipperholme Calder. *Huperun* 1086 (DB). '(Place among) the osiers'. OE **hyper* in a dative plural form **hyperum*.

Hirst Courtney, Temple Hirst N. Yorks. *Hyrst c.*1030, *Hirst Courtenay* 1303, *Templehurst* 1316. '(Place at) the wooded hill'. OE *hyrst*. Distinguishing affixes from possession by the *Courtney* family (here in the 13th cent.) and by the Knights Templar (here in the 12th cent.).

Hirwaun Rhon. 'Long moor'. *Hyrweunworgan* 1203, *Hirwen Urgan* 1536, *Hirwaun Wrgan* 1638. Welsh *hir* + *gwaun*. The forms of the name above include the name of *Gwrgan*, reputedly the last king of Glamorgan.

Histon Cambs. *Histone* 1086 (DB). Possibly 'farmstead of the sons or young men'. OE *hys(s)e* + *tūn*.

Hitcham Suffolk. *Hecham* 1086 (DB). 'Homestead with a hedge or hatch-gate'. OE *hecg* or *hecc* + *hām*.

Hitchin Herts. *Hiccam c.*945, *Hiz* 1086 (DB). '(Place in the territory of) the tribe called *Hicce*'. Old tribal name (possibly derived from a Celtic river-name meaning 'dry') in a dative plural form *Hiccum*.

Hittisleigh Devon. *Hitenesleia* 1086 (DB). 'Woodland clearing of a man called **Hyttīn*'. OE pers. name + *lēah*.

Hixon Staffs. *Hustedone* [*sic*] 1086 (DB), *Huchtesdona* 1130. 'Hill of a man called **Hyht*'. OE pers. name + *dūn*.

Hoar Cross Staffs. *Horcros* 1230. 'Grey or boundary cross'. OE *hār* + *cros*.

Hoarwithy Herefs. *La Horewythy* 13th cent. '(Place at) the whitebeam'. OE *hār* + *wīthig*.

Hoath Kent. *La hathe* 13th cent. '(Place at) the heath'. OE **hāth*.

Hoathly, 'heathy woodland clearing' or 'woodland clearing where heather grows', OE **hāth* + *lēah*: **Hoathly, East** E. Sussex. *Hodlegh* 1287. **Hoathly, West** W. Sussex. *Hadlega* 1121.

Hôb, Yr. *See* HOPE.

Hoby Leics. *Hobie* 1086 (DB). 'Farmstead or village on a spur of land'. OE *hōh* + OScand. *bý*.

Hockering Norfolk. *Hokelinka* [*sic*] 1086 (DB), *Hokeringhes* 12th cent. '(Settlement of) the people at the rounded hill'. OE **hocer* + *-ingas*.

Hockerton Notts. *Hocretone* 1086 (DB). 'Farmstead at the hump or rounded hill'. OE **hocer* + *tūn*.

Hockham, Great Norfolk. *Hocham* 1086 (DB). 'Homestead of a man called Hocca, or where hocks or mallows grow'. OE pers. name or OE *hocc* + *hām*.

Hockley Essex. *Hocheleia* 1086 (DB). 'Woodland clearing of a man called Hocca, or where hocks or mallows grow'. OE pers. name or OE *hocc* + *lēah*.

Hockley Heath Solhll. *Huckeloweheth c.*1280. 'Heath near the mound or hill of a man called **Hucca*'. OE pers. name + *hlāw* + *hēth*.

Hockliffe Beds. *Hocgan clif* 1015, *Hocheleia* [*sic*] 1086 (DB). Probably 'steep hill-side of a man called **Hocga*'. OE pers. name + *clif*.

Hockwold cum Wilton Norfolk. *Hocuuella* [sic] 1086 (DB), *Hocwolde* 1242. 'Wooded area where hocks or mallows grow'. OE *hocc* + *wald*. *See* WILTON.

Hockworthy Devon. *Hocoorde* 1086 (DB). 'Enclosure of a man called Hocca'. OE pers. name + *worth*.

Hodder (river) Lancs.-Yorkshire. Recorded from the 10th cent., a Celtic river-name meaning 'pleasant water'.

Hoddesdon Herts. *Hodesdone* 1086 (DB). 'Hill of a man called **Hod*'. OE pers. name + *dūn*.

Hoddlesden Black. w. Darw. *Hoddesdene* 1296. 'Valley of a man called **Hod* or **Hodel*'. OE pers. name + *denu*.

Hodgeston Pemb. *Villa Hogges* 1291, *Hoggeston* 1376. 'Hogge's manor'. ME pers. name or surname + *toun*.

Hodnet Shrops. *Odenet* 1086 (DB), *Hodenet* 1121. Probably a British name meaning 'pleasant valley', from the words which became Welsh *hawdd* and *nant*.

Hoe Norfolk. *Hou* 1086 (DB). '(Place at) the ridge or spur of land'. OE *hōh* (dative *hōe*).

Hoff Cumbria. *Houf* 1179. 'The heathen temple or sanctuary'. OScand. *hof*.

Hog's Back Surrey. first so called (from its shape) in 1823, earlier *Geldedon* 1195, probably 'hill called *Gylde* (the gold-coloured one)', from OE **gylde* + *dūn, see* GUILDFORD.

Hoggeston Bucks. *Hochestone* [*sic*] 1086 (DB), *Hoggeston* 1200. 'Farmstead of a man called **Hogg*'. OE pers. name + *tūn*.

Hoghton Lancs. *Hoctonam* [*sic*] *c.*1160. 'Farmstead on or near a hill-spur'. OE *hōh* + *tūn*.

Hognaston Derbys. *Ochenavestun* 1086 (DB). Possibly 'grazing farm of a man called Hocca'. OE pers. name + *æfēsn* + *tūn*.

Hogsthorpe Lincs. *Hocgestorp* 12th cent. 'Outlying farmstead or hamlet of a man called **Hogg*'. OE pers. name + OScand. *thorp*.

Holbeach Lincs. *Holebech* 1086 (DB). Probably 'hollow or concave ridge'. OE *hol* + *bæc* (locative **bece*).

Holbeck Notts. *Holebek c.*1180. 'Hollow stream, stream in a hollow'. OScand. *holr* or *hol* + *bekkr*.

Holbeton Devon. *Holbouton* 1229. 'Farmstead in the hollow bend'. OE *hol* + *boga* + *tūn*.

Holborn Gtr. London. *Holeburne* 1086 (DB). 'Hollow stream, stream in a hollow'. OE *hol* + *burna*.

Holbrook, 'hollow brook, brook in a hollow', OE *hol* + *brōc*: **Holbrook** Derbys. *Holebroc* 1086 (DB). **Holbrook** Suffolk. *Holebroc* [*sic*] 1086 (DB).

Holbury Hants. near Fawley. *Holeberi* 1187. 'Hollow stronghold', or 'stronghold in a hollow'. OE *hol* + *burh* (dative *byrig*).

Holcombe, a common name, 'deep or hollow valley', OE *hol* + *cumb*; examples include: **Holcombe** Bury. *Holecumba* early 13th cent. **Holcombe** Devon, near Dawlish. *Holacumba c.*1070, *Holcomma* 1086 (DB). **Holcombe**

Holcombe Burnell Devon. *Holecumba* 1086 (DB), *Holecumbe Bernard* 1263. Manorial affix (later modified to the modern form) from a man called *Bernard* whose son held the manor in 1242. **Holcombe Rogus** Devon. *Holecoma* 1086 (DB), *Holecombe Roges* 1281. Manorial affix from one *Rogo* who held the manor in 1086.

Holcot Northants. *Holecote* 1086 (DB). 'Cottage(s) in the hollow(s)'. OE *hol* + *cot*.

Holden Lancs. *Holedene* 1086 (DB). 'Hollow valley'. OE *hol* + *denu*.

Holdenby Northants. *Aldenesbi* 1086 (DB). 'Farmstead or village of a man called Halfdan'. OScand. pers. name + *bý*.

Holderness (district) E. R. Yorks. *Heldernesse* 1086 (DB). 'Headland ruled by a high-ranking yeoman'. OScand. *holdr* + *nes*.

Holdgate Shrops. *Castrum Helgoti* 1109–18, *Hologodescastel* 1199, *Holegod* 1294. 'Castle of a man called Helgot'. OFrench pers. name with Latin *castrum*, ME *castel* in the early forms.

Holdingham Lincs. *Haldingeham* 1202. 'Homestead of the family or followers of a man called **Hald*'. OE pers. name + *-inga-* + *hām*.

Holford Somerset. *Holeforde* 1086 (DB). 'Hollow ford, ford in a hollow'. OE *hol* + *ford*.

Holker Cumbria. *Holecher* 1086 (DB). 'Hollow marsh, marsh in a hollow'. OScand. *holr* or *hol* + *kjarr*.

Holkham Norfolk. *Holcham* 1086 (DB). 'Homestead in or near a hollow'. OE *holc* + *hām*.

Hollacombe Devon. *Holecome* 1086 (DB). 'Deep or hollow valley'. OE *hol* + *cumb*.

Holland, 'cultivated land by a hill-spur', OE *hōh* + *land*: **Holland on Sea** Essex. *Holande c.*1000, *Holanda* 1086 (DB). **Holland, Up** Lancs. *Hoiland* 1086 (DB). Affix is OE *upp* 'higher up'.

However the following, although an identical compound, may have a somewhat different meaning: **Holland** Lincs. *Hoiland* 1086 (DB). Probably 'district characterized by hill-spurs'.

Hollesley Suffolk. *Holeslea* 1086 (DB). 'Woodland clearing in a hollow, or of a man called **Hōl*'. OE *hol* or OE pers. name + *lēah*.

Hollingbourne Kent. *Holingeburna* 10th cent., *Holingeborne* 1086 (DB). 'Stream of the family or followers of a man called **Hōla*, or of the people dwelling in the hollow'. OE pers. name or *hol* + *-inga-* + *burna*.

Hollington, 'farmstead where holly grows', OE *holegn* + *tūn*: **Hollington** Derbys. *Holintune* 1086 (DB). **Hollington** E. Sussex. *Holintune* 1086 (DB). **Hollington** Staffs. *Holyngton* 13th cent.

Hollingworth Tamesd. *Holisurde* [*sic*] 1086 (DB), *Holinewurth* 13th cent. 'Holly enclosure'. OE *holegn* + *worth*.

Holloway Gtr. London. *Le Holeweye* 1307. 'The road in a hollow'. OE *hol* + *weg*.

Hollowell Northants. *Holewelle* 1086 (DB). 'Spring or stream in a hollow'. OE *hol* + *wella*.

Hollym E. R. Yorks. *Holam* 1086 (DB). Probably 'homestead or enclosure in or near the hollow'. OE *hol* + *hām* or *hamm*. Alternatively '(place at) the hollows', from OE *hol* or OScand. *holr* in a dative plural form *holum*.

Holme, a common place name, usually from OScand. *holmr* 'island, dry ground in marsh, water-meadow': **Holme** Cambs. *Hulmo* 1167. **Holme** Cumbria. *Holme* 1086 (DB). **Holme** N. Yorks. *Hulme* 1086 (DB). **Holme** Notts. *Holme* 1203. **Holme Chapel** Lancs. *Holme* 1305. **Holme Hale** Norfolk. *Holm* 1086 (DB), *Holmhel* 1267. Second element is OE *halh* (dative *hale*) 'nook of land'. **Holme next the Sea** Norfolk. *Holm* c.1035, 1086 (DB). **Holme Pierrepont** Notts. *Holmo* 1086 (DB), *Holme Peyrpointe* 1571. Manorial affix from the *de Perpount* family, here in the 14th cent. **Holme St Cuthbert** Cumbria. *Sanct Cuthbert Chappell* 1538. Named from the dedication of the chapel. **Holme, South** N. Yorks. *Holme* 1086 (DB), *Southolme* 1301. **Holme upon Spalding Moor** E. R. Yorks. *Holme* 1086 (DB), *Holm in Spaldingmor* 1293. Spalding Moor is an old district-name meaning 'moor of the people called Spaldingas', from the tribe who gave their name to SPALDING (Lincs.) + OE *mōr*.
    However some examples of Holme have other origins: **Holme** Kirkl., near Holmfirth. *Holne* 1086 (DB). '(Place at) the holly-tree'. OE *holegn*. **Holme, East & Holme, West** Dorset. *Holne* 1086 (DB), *Estholn, Westholn* 1288. Identical in origin with the previous name. **Holme Lacy** Herefs. *Hamme* 1086 (DB), *Homme Lacy* 1221. OE *hamm* 'enclosure, land in a river-bend'. Manorial affix from its possession by the *de Laci* family from 1086. **Holme on the Wolds** E. R. Yorks. *Hougon* 1086 (DB), *Holme super Wolde* 1578. '(Place at)

the mounds or hills'. OScand. *haugr* in a dative plural form *haugum*. The affix is from OE *wald* 'high forest-land'.

Holmer, 'pool in a hollow', OE *hol* + *mere*: **Holmer** Herefs. *Holemere* 1086 (DB). **Holmer Green** Bucks. *Holemere* 1208.

Holmes Chapel Ches. *Hulm* 12th cent., *Holme chapell* 1400–5. 'Chapel at the place called *Hulme* or *Holme*'. OScand. *holmr* 'water-meadow' with the later addition of ME *chapel*.

Holmesfield Derbys. *Holmesfelt* 1086 (DB). Probably 'open land near a place called *Holm*'. OE *feld* with a lost place name *Holm* from OScand. *holmr* 'island, dry ground in marsh'.

Holmfirth Kirkl. *Holnefrith* 1274. 'Sparse woodland belonging to HOLME'. OE *fyrhth*.

Holmpton E. R. Yorks. *Holmetone* 1086 (DB). 'Farmstead near the shore-meadows'. OScand. *holmr* + OE *tūn*.

Holmwood, North & Holmwood, South Surrey. *Homwude* 1241. 'Wood in or near an enclosure or river-meadow'. OE *hamm* + *wudu*.

Holne Devon. *Holle* [*sic*] 1086 (DB), *Holna* 1178. '(Place at) the holly tree'. OE *holegn*.

Holnest Dorset. *Holeherst* 1185. 'Wooded hill where holly grows'. OE *holegn* + *hyrst*.

Holsworthy Devon. *Haldeurdi* 1086 (DB). 'Enclosure of a man called *Heald'. OE pers. name + *worthig*.

Holt, a common name, '(place at) the wood or thicket', OE *holt*; examples include: **Holt** Dorset. *Winburneholt* 1185, *Holte* 1372. Originally 'wood near WIMBORNE [MINSTER]'. **Holt** Norfolk. *Holt* 1086 (DB). **Holt** Wilts. *Holt* 1242. **Holt** Worcs. *Holte* 1086 (DB). **Holt** Wrex. *Holte* 1326, *the holt* 1535. **Holt End** Hants. *Holt* 1167. **Holt, Nevill** Leics. *Holt* early 12th cent. Manorial affix from the *Nevill* family, here in the late 15th cent.

Holtby York. *Holtebi* 1086 (DB). Probably 'farmstead or village of a man called Holti'. OScand. pers. name + *bý*.

Holton, a common name, has a number of different origins: **Holton** Lincs., near Beckering. *Houtune* 1086 (DB). 'Farmstead on a spur of land'. OE *hōh* ı *tūn*. **Holton** Oxon. *Healhtun* 956, *Eltone* 1086 (DB). 'Farmstead in a nook of land'. OE *healh* + *tūn*. **Holton**

Somerset. *Healhtun c.*1000. Identical in origin with the previous name. **Holton** Suffolk. *Holetuna* 1086 (DB). 'Farmstead near a hollow, or of a man called *Hōla*'. OE *hol* or OE pers. name + *tūn*. **Holton Heath** Dorset. *Holtone* 1086 (DB). 'Farmstead near a hollow, or near a wood'. OE *hol* or *holt* + *tūn*. **Holton le Clay** Lincs. *Holtun* [*sic*] 1086 (DB), *Houtona c.*1115. 'Farmstead on a spur of land'. OE *hōh* + *tūn*. Affix means 'in the clayey district' from OE *clǣg*. **Holton le Moor** Lincs. *Hoctune* 1086 (DB). Identical in origin with the previous name. Affix means 'in the moorland' from OE *mōr*. **Holton St Mary** Suffolk. *Holetuna* 1086 (DB). 'Farmstead near a hollow, or of a man called *Hōla*'. OE *hol* or OE pers. name + *tūn*. Affix from the dedication of the church.

Holwell, a common name, has several different origins: **Holwell** Dorset, near Sherborne. *Holewala* 1188. 'Ridge or bank in a hollow'. OE *hol* + *walu*. **Holwell** Herts. *Holewelle* 969, 1086 (DB). 'Spring or stream in a hollow'. OE *hol* + *wella*. **Holwell** Leics. *Holewelle* 1086 (DB). Identical in origin with the previous name. **Holwell** Oxon. *Haliwell* 1222. 'Holy spring or stream'. OE *hālig* + *wella*.

Holwick Durham. *Holewyk* 1235. Probably 'dwelling or farm in a hollow'. OE *hol* + *wīc*.

Holworth Dorset. *Holewertthe* 934, *Holverde* 1086 (DB). 'Enclosure in a hollow'. OE *hol* + *worth*.

Holy Island (*Ynys Gybi*) Angl. 'Holy island'. OE *hālig* + *ēg-land*. The Welsh name means 'Cybi's island' (Welsh *ynys*) after the Celtic saint to whom the church at HOLYHEAD is dedicated.

Holy Island Northum. *Halieland* 1195. 'Holy island' (from its association with early Christian missionaries). OE *hālig* + *ēg-land*. *See* LINDISFARNE.

Holybourne Hants. *Haliburne* 1086 (DB). 'Holy stream'. OE *hālig* + *burna*.

Holyhead (*Caergybi*) Angl. *Haliheved* 1315, *Holyhede* 1395. 'Holy headland'. OE *hālig* + *hēafod*. The Welsh name means 'Cybi's fort' (Welsh *caer*), from the Celtic saint to whom the church is dedicated.

Holyport Winds. & Maid. *Horipord* 1220. 'Muddy or dirty market town'. OE *horig* + *port*. The first element was intentionally changed to *holy* by the late 14th cent.

Holystone Northum. *Halistan* 1242. 'Holy stone' (perhaps one at which the gospel was preached). OE *hālig* + *stān*.

Holywell Cambs. *Haliewelle* 1086 (DB). 'Holy spring or stream'. OE *hālig* + *wella*.

Holywell Fermanagh. *Holywell* 1835. 'Holy well'. A well-preserved holy well here has the Irish name *Dabhach Phádraig*, 'St Patrick's well'.

Holywell (*Treffynnon*) Flin. *Haliwel* 1093. 'Holy well'. OE *hālig* + *wella*. The name refers to the sacred well of St Winifred, who founded a nunnery here in the 7th cent. The Welsh name means 'village of the well' (Welsh *tref* + *ffynnon*).

Holywood Down. 'Holy wood'. The name originated with the Normans, who gave it to woodland adjoining the ancient church here. The Irish name of Holywood is *Ard Mhic Nasca*, 'Mac Nasca's height'.

Holywood Dumf. *de Sacro Nemore* 1252. 'Holy wood'. OE *hālig* + *wudu*. The earlier name was *Dercongal*, 'Congal's oak wood'.

Homer Shrops. *Holmere* 1550. 'Pond or pool in a hollow'. ME *hole* + *mere*.

Homersfield Suffolk. *Humbresfelda* 1086 (DB). 'Open land of a man called Hūnbeorht'. OE pers. name + *feld*.

Homerton Gtr. London. *Humburton* 1343. 'Farmstead of a woman called Hūnburh'. OE pers. name + *tūn*.

Homington Wilts. *Hummingtun* 956, *Humitone* 1086 (DB). 'Estate associated with a man called *Humma*'. OE pers. name + *-ing-* + *tūn*.

Honeyborough Pemb. *Honyburgh* 1325. 'Manor where honey is produced'. OE *hunig* + *burh*.

Honeybourne, Church & Honeybourne, Cow Worcs. *Huniburna* 709, *Huniburne, Heniberge* [*sic*] 1086 (DB), *Churchoniborne* 1535, *Calewe Honiburn* 1374. '(Places on) the stream by which honey is found'. OE *hunig* + *burna*. Affixes from OE *cirice* 'church' and *calu* 'bare, lacking vegetation'.

Honeychurch Devon. *Honechercha* 1086 (DB). Probably 'church of a man called Hūna'. OE pers. name + *cirice*.

Honiley Warwicks. *Hunilege* 1208. 'Woodland clearing where honey is found'. OE *hunig* + *lēah*.

245

Hopton

Honing Norfolk. *Hanninge* 1044–7, *Haninga* 1086 (DB). Probably '(settlement of) the people at the rock or hill'. OE *hān* + *-ingas*.

Honingham Norfolk. *Hunincham* 1086 (DB). 'Homestead of the family or followers of a man called Hūn(a)'. OE pers. name + *-inga-* + *hām*.

Honington Lincs. *Hundintone* 1086 (DB). Probably 'estate associated with a man called *Hund'. OE pers. name +*-ing-* + *tūn*.

Honington Suffolk. *Hunegetuna* 1086 (DB). Probably 'farmstead of the family or followers of a man called Hūn(a)'. OE pers. name + *-inga-* + *tūn*.

Honington Warwicks. *Hunitona* 1043, *Hunitone* 1086 (DB). 'Farmstead where honey is produced'. OE *hunig* + *tūn*.

Honiton Devon. *Honetone* 1086 (DB). 'Farmstead of a man called Hūna'. OE pers. name + *tūn*.

Honiton, Clyst Devon. *Hinatune* c.1100, *Clysthynetone* 1281. 'Farmstead (on the River Clyst) belonging to a religious community' (in this case Exeter Cathedral). OE *hīwan* (genitive plural *hīgna*) + *tūn*, to which the Celtic river-name (*see* CLYST) was later added.

Honley Kirkl. *Haneleia* 1086 (DB). 'Woodland clearing where woodcock abound, or where there are stones and rocks'. OE *hana* or *hān* + *lēah*.

Hoo, Hooe, '(place at) the spur of land', OE *hōh* (dative *hōe*): **Hoo** Kent. *Hoge* c.687, *How* 1086 (DB). **Hoo Green** Suffolk. *Ho* c.1050, *Hou* 1086 (DB). **Hoo, St Mary's** Medway. *Ho St Mary* 1272. Affix from the dedication of the church. **Hoo, Sutton** Suffolk. *Hou, Hoi* 1086 (DB). Affix from its proximity to SUTTON Suffolk. **Hooe** Plym. *Ho* 1086 (DB). **Hooe** E. Sussex. *Hou* 1086 (DB).

Hook, Hooke, usually '(place at) the hook of land, or bend in a river or hill', OE *hōc*: **Hook** Gtr. London. *Hoke* 1227. **Hook** Hants., near Basingstoke. *Hoc* 1223. **Hook** Wilts. *La Hok* 1238. **Hooke** Dorset. *Lahoc* 1086 (DB). With the French definite article *la* in the early form.

However the following have a different origin: **Hook** E. R. Yorks. *Huck* 12th cent. OE *hūc* 'river-bend'. **Hook Norton** Oxon. *Hocneratune* early 10th cent. Possibly 'farmstead of a tribe called *Hoccanēre*'. OE tribal name (meaning 'people at Hocca's hill-slope', from OE pers. name and *ōra*) + *tūn*.

Hoole Ches. *Hole* 1119. '(Place at) the hollow'. OE *hol*.

Hoole, Much Lancs. *Hulle* 1204, *Magna Hole* 1235. OE *hulu* 'a shed, a hovel'. Affix is OE *mycel* 'great' (earlier Latin *magna*).

Hooton, 'farmstead on a spur of land'. OE *hōh* + *tūn*: **Hooton** Ches. *Hotone* 1086 (DB). **Hooton Levitt** Rothm. *Hotone* 1086 (DB). Manorial affix from the *Livet* family, here in the 13th cent. **Hooton Pagnell** Donc. *Hotun* 1086 (DB), *Hotton Painel* 1192. Manorial affix from its possession by the *Painel* family from the 11th cent. **Hooton Roberts** Rothm. *Hotun* 1086 (DB), *Hoton Robert* 1285. Manorial affix from its possession by a man called *Robert* in the 13th cent.

Hope, a common name, from OE *hop* 'small enclosed valley, enclosed plot of land'; examples include: **Hope** Derbys. *Hope* 926, 1086 (DB). **Hope** Devon. *La Hope* 1281. **Hope** (Yr Hôb) Flin. *Hope* 1086 (DB). **Hope** Shrops. *Hope* 1242. **Hope Bowdler** Shrops. *Fordritishope* 1086 (DB), *Hop* 1201, *Hopebulers* 1273. The DB prefix is probably the OE pers. name *Forthrǣd*; the later manorial affix is from the *de Bulers* family, here from 1201. **Hope Green** Ches. *Hope* 1282. With the later addition of *grene* 'village green'. **Hope Mansell** Herefs. *Hope* 1086 (DB). *Hoppe Maloisel* 12th cent. Manorial affix from early possession by the *Maloisel* family, here in the 13th cent. **Hope, Sollers** Herefs. *Hope* 1086 (DB), *Hope Solers* 1242. Manorial affix from the *de Solariis* family, here in the 13th cent. **Hope under Dinmore** Herefs. *Hope* 1086 (DB), *Hope sub' Dinnemor* 1291. *Dinmore* may be a Welsh name *din mawr* meaning 'great fort', or alternatively 'marsh of a man called *Dynna*, from OE pers. name + *mōr*.

Hopeman Moray. Meaning unknown. The village was founded in 1805 by William Young of Inverugie and took its name from a local estate.

Hopton, 'farmstead in a small enclosed valley or enclosed plot of land', OE *hop* + *tūn*: **Hopton** Shrops., near Hodnet. *Hotune* [sic] 1086 (DB), *Hopton* 1242. **Hopton** Staffs. *Hoptuna* 1167. **Hopton** Suffolk, near Thetford. *Hopetuna* 1086 (DB). **Hopton Cangeford** Shrops. *Hopton* 1242, *Hopton Cangefot* 1272. Manorial affix from its early possession by the *Cangefot* family. **Hopton Castle** Shrops. *Opetune* 1086 (DB). Affix from the Norman castle here. **Hopton Wafers** Shrops. *Hopton* 1086 (DB), *Hopton Wafre* 1236. Manorial affix from its early possession by the *Wafre* family.

Hopwas Staffs. *Opewas* 1086 (DB). 'Marshy or alluvial land near an enclosure'. OE *hop* + **wæsse*.

Hopwood, 'wood near an enclosure or in a small enclosed valley', OE *hop* + *wudu*: **Hopwood** Rochdl. *Hopwode* 1278. **Hopwood** Worcs. *Hopwuda* 849.

Horam E. Sussex. *Horeham* 1813. Perhaps manorial in origin, or identical with HORHAM Suffolk.

Horbling Lincs. *Horbelinge* 1086 (DB). Probably 'muddy settlement of the family or followers of a man called Bill or *Billa'. OE *horu* + OE pers. name + *-ingas*.

Horbury Wakefd. *Horberie* 1086 (DB). 'Stronghold on muddy land'. OE *horu* + *burh* (dative *byrig*).

Horden Durham. *Horeden* c.1040. 'Dirty or muddy valley'. OE *horu* + *denu*.

Hordle Hants. *Herdel* [*sic*] 1086 (DB), *Hordhull* 1242. 'Hill where treasure was found'. OE *hord* + *hyll*.

Hordley Shrops. *Hordelei* 1086 (DB). 'Woodland clearing where treasure was found'. OE *hord* + *lēah*.

Horham Suffolk. *Horham* c.950. 'Muddy homestead'. OE *horu* + *hām*.

Horkesley, Great & Horkesley, Little Essex. *Horchesleia* c.1130. 'Woodland clearing with a shelter', or 'dirty, muddy clearing'. OE *horc* or *horsc* + *lēah*.

Horkstow N. Lincs. *Horchetou* [*sic*] 1086 (DB), *Horkestowe* 12th cent. Probably 'shelter for animals or people'. OE *horc* + *stōw*.

Horley, probably 'woodland clearing in a horn-shaped piece of land'. OE *horn*, **horna* + *lēah*: **Horley** Oxon. *Hornelie* 1086 (DB). **Horley** Surrey. *Horle* 12th cent.

Hormead, Great Herts. *Horemede* 1086 (DB). 'Muddy meadow'. OE *horu* + *mæd*.

Hornblotton Green Somerset. *Hornblawertone* 851, *Horblawetone* 1086 (DB). 'Farmstead of the hornblowers or trumpeters'. OE *horn-blāwere* + *tūn*.

Hornby, 'farmstead or village on a horn-shaped piece of land, or of a man called Horn or *Horni'. OScand. *horn* or pers. name + *bý*: **Hornby** Lancs. *Hornebi* 1086 (DB). **Hornby**

N. Yorks., near Great Smeaton. *Hornebia* 1086 (DB). **Hornby** N. Yorks., near Hackforth. *Hornebi* 1086 (DB).

Horncastle Lincs. *Hornecastre* 1086 (DB). 'Roman station or fortification on a horn-shaped piece of land'. OE **horna* + *cæster*.

Hornchurch Gtr. London. *Hornechurch* 1233. 'Church with horn-like gables'. OE *horn* (noun), *hornede* (adjective) + *cirice*.

Horncliffe Northum. *Hornecliff* 1210. 'Slope of the horn-shaped hill or piece of land'. OE **horna* + *clif*.

Horndean Hants. *Harmedene* 1199. 'Valley frequented by shrews or weasels', or 'valley of a man called *Hearma'. OE *hearma* or identical pers. name + *denu*.

Horndon on the Hill Thurr. *Horninduna* 1086 (DB). 'Horn-shaped hill'. OE **horning* + *dūn*.

Horne Surrey. *Horne* 1208. '(Place at) the horn-shaped hill or piece of land'. OE *horn* or **horna*.

Horning Norfolk. *Horningga* 1020–2, *Horningam* 1086 (DB). '(Settlement of) the people living at the horn-shaped piece of land, here in a river-bend'. OE *horn*, **horna* + *-ingas*. Alternatively 'horn-shaped pieces of land', from the plural of an OE **horning*, is formally possible.

Horninghold Leics. *Horniwale* [*sic*] 1086 (DB), *Horningewald* 1163. 'Woodland of the people living at the horn-shaped piece of land', OE *horn*, +-*inga*- +*wald*. Or 'woodland at the horn-shaped piece of land', OE **horning* +*wald*.

Horninglow Staffs. *Horninglow* 12th cent. 'Mound at the horn-shaped hill or piece of land'. OE **horning* + *hlāw*.

Horningsea Cambs. *Horninges ige* c.975, *Horningesie* 1086 (DB). 'Island, or dry ground in marsh, of a man called *Horning or by the horn-shaped hill'. OE pers. name or **horning* + *ēg*.

Horningsham Wilts. *Horningesham* 1086 (DB). Probably 'homestead of a man called *Horning'. OE pers. name + *hām*. Or the first element may be OE **horning* 'horn-shaped hill or piece of land'.

Horningtoft Norfolk. *Horninghetoft* 1086 (DB). 'Homestead of the people living at the horn-shaped hill or piece of land'. OE *horn*, **horna* + *-inga*- + OScand. *toft*.

Hornsby Cumbria. *Ormesby c.*1210. 'Farmstead or village of a man called Ormr'. OScand. pers. name + *bý*.

Hornsea E. R. Yorks. *Hornessei* 1086 (DB). 'Lake with a horn-shaped peninsula'. OScand. *horn* + *nes* + *sǽr*.

Hornsey Gtr. London. *Haringeie* 1201, *Haringesheye* 1243, *Haringay alias Hornesey c.*1580. 'Enclosure in the grey wood' or 'of a man called *Hæring*'. OE *hāring* or OE pers. name + *hæg*. Nearby HARINGEY (borough) and Harringay are different developments of the same name.

Hornton Oxon. *Hornigeton* 1194. 'Farmstead near the horn-shaped piece of land'. OE *horning* + *tūn*, or *horn(a)* + *-ing-* + *tūn*.

Horrabridge Devon. *Horebrigge* 1345. 'Grey or boundary bridge'. OE *hār* + *brycg*.

Horringer Suffolk. *Horningeserda* 1086 (DB). 'Ploughed land at the bend or headland' or 'of a man called *Horning*'. OE *horning* or OE pers. name + *erth*.

Horrington, East & Horrington, West Somerset. *Hornningdun* 1065. Possibly 'horn-shaped hill', OE *horning* + *dūn*. Alternatively 'hill of the people living at the horn-shaped piece of land', OE *horn* or *horna* + *-inga-* + *dūn*.

Horseheath Cambs. *Horseda c.*1080, *Horsei* [*sic*] 1086 (DB). Probably 'heath where horses are kept'. OE *hors* + *hǣth*.

Horsell Surrey. *Horisell* 13th cent. 'Shelter for animals in a muddy place'. OE *horu* + *(ge)sell*.

Horsey Norfolk. *Horseia* 1086 (DB). 'Island, or dry ground in marsh, where horses are kept'. OE *hors* + *ēg*.

Horsford Norfolk. *Hosforda* 1086 (DB). 'Ford which horses can cross'. OE *hors* + *ford*.

Horsforth Leeds. *Horseford* 1086 (DB). Identical in origin with the previous name.

Horsham, 'homestead or village where horses are kept', OE *hors* + *hām*: **Horsham** W. Sussex. *Horsham* 947. **Horsham St Faith** Norfolk. *Horsham* 1086 (DB). Affix from the dedication of the church.

Horsington Lincs. *Horsintone* 1086 (DB). 'Estate associated with a man called Horsa'. OE pers. name + *-ing-* + *tūn*.

Horsington Somerset. *Horstenetone* 1086 (DB). 'Farmstead of the horsekeepers or grooms'. OE *hors-thegn* + *tūn*.

Horsley, 'clearing or pasture where horses are kept', OE *hors* + *lēah*: **Horsley** Derbys. *Horselei* 1086 (DB). **Horsley** Glos. *Horselei* 1086 (DB). **Horsley** Northum. *Horseley* 1242. **Horsley, East & Horsley, West** Surrey. *Horsalæge* 9th cent., *Horslei, Orselei* 1086 (DB).

Horsmonden Kent. *Horsbundenne c.*1100. Probably 'woodland pasture near the stream where horses drink'. OE *hors* + *burna* + *denn*.

Horstead, Horsted, 'place where horses are kept', OE *hors* + *stede*: **Horstead** Norfolk. *Horsteda* 1086 (DB). **Horsted Keynes** W. Sussex. *Horstede* 1086 (DB), *Horsted Kaynes* 1307. Manorial affix from its possession by William *de Cahainges* in 1086. **Horsted, Little** E. Sussex. *Horstede* 1086 (DB), *Little Horstede* 1307.

Horton, a common name, usually 'dirty or muddy farmstead', OE *horu* + *tūn*; examples include: **Horton** Bucks., near Ivinghoe. *Hortone* 1086 (DB). **Horton** Dorset. *Hortun* 1033, *Hortune* 1086 (DB). **Horton** Northants. *Hortone* 1086 (DB). **Horton** Oxon. *Hortun* 1005–12, *Hortone* 1086 (DB). **Horton** Shrops., near Wem. *Hortune* 1086 (DB). **Horton** Staffs. *Horton* 1239. **Horton** Surrey. *Horton* 1178. **Horton** Wilts. *Horton* 1158. **Horton Green** Ches. *Horton c.*1240. **Horton-in-Ribblesdale** N. Yorks. *Hortune* 1086 (DB), *Horton in Ribbelesdale* 13th cent. Affix means 'valley of the River Ribble', OE river-name (*see* RIBBLETON) + OE *dæl* or OScand. *dalr*. **Horton Kirby** Kent. *Hortune* 1086 (DB), *Horton Kyrkeby* 1346. Manorial affix from its possession by the *de Kirkeby* family in the 13th cent.
    However the following has a different origin: **Horton** S. Glos. *Horedone* 1086 (DB). 'Hill frequented by harts or stags'. OE *heorot* + *dūn*.

Horwich Bolton. *Horewic* 1221. '(Place at) the grey wych-elm(s)'. OE *hār* + *wice*.

Horwood Devon. *Horewode, Hareoda* 1086 (DB). 'Grey wood or muddy wood'. OE *hār* or *horu* + *wudu*.

Horwood, Great & Horwood, Little Bucks. *Horwudu* 792, *Hereworde* [*sic*] 1086 (DB). 'Dirty or muddy wood'. OE *horu* + *wudu*.

Hose

Hose Leics. *Hoches, Howes* 1086 (DB). 'The spurs of land'. OE *hōh* in a plural form *hō(h)as*.

Hotham E. R. Yorks. *Hode* 963, *Hodhum* 1086 (DB). '(Place at) the shelters'. OE **hōd* in a dative plural form **hōdum*.

Hothfield Kent. *Hathfelde* c.1100. 'Heathy open land'. OE **hāth + feld*.

Hoton Leics. *Hohtone* 1086 (DB). 'Farmstead on a spur of land'. OE *hōh + tūn*.

Hough, 'the ridge or spur of land', OE *hōh*:
Hough Ches., near Willaston. *Hohc* 1241.
Hough Ches., near Wilmslow. *Le Hogh* 1289.

Hough-on-the-Hill Lincs. *Hach, Hag* 1086 (DB). OE *haga* 'enclosure'.

Hougham Lincs. *Hacham* 1086 (DB). Probably 'river-meadow belonging to HOUGH-ON-THE-HILL'. OE *hamm*.

Houghton, a common name, usually 'farmstead on or near a ridge or hill-spur', OE *hōh + tūn*: **Houghton** Cambs. *Hoctune* 1086 (DB). **Houghton** Cumbria. *Hotton* 1246.
Houghton Hants. *Hohtun* 982, *Houstun* [*sic*] 1086 (DB). **Houghton** W. Sussex. *Hohtun* 683.
Houghton Conquest Beds. *Houstone* 1086 (DB), *Houghton Conquest* 1316. Manorial affix from the *Conquest* family, here in the 13th cent.
Houghton, Great & Houghton, Little Northants. *Hohtone* 1086 (DB), *Magna Houtona* 1199, *Parva Houtone* 1233. Distinguishing affixes are Latin *magna* 'great' and *parva* 'little'.
Houghton le Side Darltn. *Hoctona* 1200, *Houghton-in-the-Syde* c.1583. Affix means 'by the hill-slope', from ME *side*. **Houghton le Spring** Sundld. *Hoctun* c.1170, *Houghton in le Spryng* 1410. Affix is originally manorial from the *Spring* family, here in the early 14th cent.
Houghton on the Hill Leics. *Hohtone* 1086 (DB). The affix is recorded from the 17th cent.
Houghton Regis Beds. *Houstone* 1086 (DB), *Kyngeshouton* 1287. Latin affix *regis* 'of the king' because it was a royal manor at an early date.
Houghton St Giles Norfolk. *Hohttune* 1086 (DB). Affix from the dedication of the church.
    However the following have a different origin, 'farmstead in a nook of land', OE *halh + tūn*:
Houghton, Great Barns. *Haltun* 1086 (DB), *Magna Halghton* 1303. Affix is Latin *magna* 'great'. **Houghton, Little** Barns. *Haltone* 1086 (DB), *Parva Halghton* 1303. Affix is Latin *parva* 'little'.

Hound Green Hants. *Hune* 1086 (DB). 'Place where the plant hoarhound grows'. OE *hūne*.

Hounslow Gtr. London. *Honeslaw* [*sic*] 1086 (DB), *Hundeslawe* 1217. 'Mound or tumulus of the hound, or of a man called **Hund*'. OE *hund* or OE pers. name (from the same word) + *hlāw*.

Houston Renf. *Villa Hugonis* c.1200, *Huston* c.1230. 'Hugo's village'. Anglo-Norman pers. name + OE *tūn*.

Hove Bright. & Hove. *La Houue* 1288. OE *hūfe* 'hood-shaped hill' or 'shelter'.

Hoveringham Notts. *Horingeham* [*sic*] 1086 (DB), *Houeringeham* 1167. 'Homestead of the people living at the hump-shaped hill'. OE *hofer + -inga- + hām*.

Hoveton Norfolk. *Houetonne* 1044–7, *Houetuna* 1086 (DB). 'Farmstead of a man called Hofa, or where the plant ale-hoof grows'. OE pers. name or OE *hōfe + tūn*.

Hovingham N. Yorks. *Hovingham* 1086 (DB). 'Homestead of the family or followers of a man called Hofa'. OE pers. name + *-inga- + hām*.

Howden E. R. Yorks. *Heafuddene* 959, *Hovedene* 1086 (DB). 'Valley by the headland or spit of land'. OE *hēafod* (replaced by OScand. *hǫfuth*) + *denu*. Alternatively the first element *hēafod* may have had the sense 'head, most important'.

Howe Highland. '(Place by the) mound'. OScand. *haugr*.

Howe Norfolk. *Hou* 1086 (DB). '(Place at) the hill or barrow'. OScand. *haugr*.

Howell Lincs. *Huuelle* 1086 (DB). Second element OE *wella* 'spring, stream', first element doubtful, possibly OE *hūfe* 'hood, covering' or *hūf* 'owl'.

Howick Northum. *Hewic* c.1100, *Hawic* 1230. Probably 'high (or chief) dairy-farm'. OE *hēah + wīc*.

Howle Tel. & Wrek. *Hugle* 1086 (DB). '(Place at) the mound or hillock'. OE **hugol*.

Howsham, '(place at) the houses or buildings', OE *hūs* or OScand. *hús* in a dative plural form: **Howsham** N. Lincs. *Usun* 1086 (DB). **Howsham** N. Yorks. *Huson* 1086 (DB).

Howth Dublin. 'Headland'. OScand. *hǫfuth*. The Irish name of Howth is *Binn Éadair*, 'Éadar's peak'.

Howton Herefs. *Huetune c.*1184. Probably 'estate of a man called Hugh'. OFrench pers. name + OE *tūn*.

Hoxa (island) Orkn. *Haugaheith c.*1390. 'Island with a mound'. OScand. *haugr* + *ey*.

Hoxne Suffolk. *Hoxne c.*950, *Hoxana* 1086 (DB). Probably OE *hōhsinu* 'a hock', here used figuratively with reference to a spur of land or other feature thought to resemble a horse's hock.

Hoxton Gtr. London. *Hochestone* 1086 (DB). 'Farmstead of a man called *Hōc'. OE pers. name + *tūn*.

Hoy (island) Orkn. *Hoye* 1492. 'High island'. OScand. *há* + *ey*.

Hoylake Wirral. *Hoylklake* [*sic*] *c.*1280, *Hyle Lake* 1687. 'Tidal lake by the Hile (the hillock or sandbank)'. OE *hygel* + ME *lake*.

Hoyland, 'cultivated land on or near a hill-spur', OE *hōh* + *land*: **Hoyland, High** Barns. *Holand* 1086 (DB), *Heyholand* 1283. Affix is OE *hēah* 'high'. **Hoyland Nether** Barns. *Ho(i)land* 1086 (DB), *Nether Holand* 1390. Affix is OE *neotherra* 'lower'. **Hoyland Swaine** Barns. *Holande* 1086 (DB), *Holandeswayn* 1266. Manorial affix from possession in the 12th cent. by a man called *Swein* (OScand. *Sveinn*).

Hubberholme N. Yorks. *Huburgheham* 1086 (DB). 'Homestead of a woman called Hūnburh'. OE pers. name + *hām*.

Hubberston Pemb. *Villa Huberti* 1291. 'Hubert's manor'. OGerman pers. name + ME *toun*.

Huby N. Yorks. near Easingwold. *Hobi* 1086 (DB). 'Farmstead on the spur of land'. OE *hōh* + OScand. *bý*.

Huby N. Yorks. near Stainburn. *Huby* 1198. 'Farmstead of a man called Hugh'. OFrench pers. name + OScand. *bý*.

Hucclecote Glos. *Hochilicote* 1086 (DB). 'Cottage(s) associated with a man called *Hucel(a)'. OE pers. name +*-ing-* + *cot*.

Hucking Kent. *Hugginges* 1195. '(Settlement of) the family or followers of a man called *Hucca'. OE pers. name + *-ingas*.

Hucklow, Great & Hucklow, Little Derbys. *Hochelai* 1086 (DB), *Magna Hockelawe* 1251, *Parva Hokelawe* 13th cent. 'Mound or hill of a man called *Hucca'. OE pers.

name + *hlāw*. Affixes are Latin *magna* 'great' and *parva* 'little'.

Hucknall, 'nook of land of a man called *Hucca', OE pers. name (genitive *-n*) + *halh*: **Hucknall** Notts. *Hochenale* 1086 (DB). **Hucknall, Ault** Derbys. *Hokenhale* 1291, *Haulte Huknall* 1535. Affix is OFrench *haut* 'high'.

Huddersfield Kirkl. *Odresfeld* 1086 (DB). Possibly 'open land of a man called *Hudræd'. OE pers. name + *feld*. Alternatively the first element may be an OE *hūder* 'a shelter'.

Huddington Worcs. *Hudintune* 1086 (DB). 'Estate associated with a man called Hūda'. OE pers. name + *-ing-* + *tūn*.

Hudswell N. Yorks. *Hudreswelle* [*sic*] 1086 (DB), *Hudeleswell* 12th cent. Probably 'spring or stream of a man called *Hūdel'. OE pers. name + *wella*.

Huggate E. R. Yorks. *Hughete* 1086 (DB). Possibly 'road to or near the mounds'. OScand. *hugr* + *gata*.

Hugh Town I. of Scilly. *Hugh Town* 17th cent., named from *the Hew Hill* 1593. Probably OE *hōh* 'spur of land'.

Hughenden Valley Bucks. *Huchedene* 1086 (DB). 'Valley of a man called *Huhha'. OE pers. name (genitive *-n*) + *denu*. The recent addition *Valley* is strictly tautological.

Hughley Shrops. *Leg'* 1203–4, *Huleye c.*1291. Originally 'the wood or clearing', from OE *lēah*. Later addition from possession by a man called *Hugh* mentioned in 1203–4.

Huish, a common name, 'measure of land that would support a family', OE *hīwisc*; examples include: **Huish** Wilts. *Iwis* 1086 (DB). **Huish Champflower** Somerset. *Hiwis* 1086 (DB), *Hywys Champflur* 1274. Manorial affix from the *Champflur* family, here in the 13th cent. **Huish Episcopi** Somerset. *Hiwissh* 973. Affix is Latin *episcopi* 'bishop's', referring to early possession by the Bishop of Wells. **Huish, North & Huish, South** Devon. *Hewis, Heuis* 1086 (DB).

Hulcott Bucks. *Hoccote* 1200, *Hulecote* 1228. Probably 'hovel-like cottages'. OE *hulu* + *cot*.

Hull. *See* KINGSTON UPON HULL.

Hull, Bishop's Somerset. *Hylle* 1033, *Hilla* 1086 (DB), *Hulle Episcopi* 1327. '(Place at) the hill'. OE *hyll*. Affix from its early possession by the Bishops of Winchester.

Hulland Derbys. *Hoilant* 1086 (DB). 'Cultivated land by a hill-spur'. OE *hōh* + *land*.

Hullavington Wilts. *Hunlavintone* 1086 (DB). 'Estate associated with a man called Hūnlāf'. OE pers. name + *-ing-* + *tūn*.

Hullbridge Essex. *Whouluebregg* 1375. Probably 'bridge with arches'. OE *hwalf* + *brycg*.

Hulme, 'the island, the dry ground in marsh, the water-meadow', OScand. *holmr*: **Hulme** Manch. *Hulm* 1246. **Hulme, Cheadle** Stockp. *Hulm* late 12th cent., *Chedle Hulm* 1345. Within the parish of CHEADLE. **Hulme End** Staffs. *Hulme* 1227. **Hulme Walfield** Ches. *Wallefeld et Hulm c.*1262. Walfield was originally a separate manor, 'open land at a spring', OE *wælla* or *wælm* + *feld*.

Hulton, Abbey Stoke. *Hulton* 1235. 'Farmstead on a hill'. OE *hyll* + *tūn*. Affix from the Cistercian abbey founded here in 1223.

Humber (river) E. R. Yorks. *Humbri fluminis c.*720, *Humbre* 9th cent. An ancient pre-English river-name of uncertain origin and meaning which also occurs elsewhere in England (as in next name).

Humber Court Herefs. *Humbre* 1086 (DB). Originally the name of Humber Brook, *see* previous name.

Humberston NE. Lincs. *Humbrestone* 1086 (DB). '(Place by) the boundary stone in the River Humber'. Ancient river-name (*see* HUMBER) + OE *stān*.

Humbleton E. R. Yorks. *Humeltone* 1086 (DB). 'Farmstead by a rounded hillock, or where hops grow', or 'farmstead of a man called Humli'. OE **humol* or *humele* or OScand. pers. name + OE *tūn*.

Humbleton Northum. *Hameldun* 1170. 'Crooked or scarred hill'. OE **hamel* + *dūn*.

Hume Sc. Bord. *Houm* 1127. '(Place at the) hill-spurs'. OE *hōh* (dative plural *hōum*).

Humshaugh Northum. *Hounshale* 1279. Probably 'nook of land of a man called Hūn'. OE pers. name + *halh*.

Huna Highland. *Hofn c.*1250. 'Harbour river'. OScand. *hafn* + *á*.

Huncoat Lancs. *Hunnicot* 1086 (DB). Possibly 'cottage(s) of a man called Hūna'. OE pers. name + *cot*. Alternatively the first element may be OE *hunig* 'honey'.

Huncote Leics. *Hunecote* 1086 (DB). Probably 'cottage(s) of a man called Hūna'. OE pers. name + *cot*.

Hunderthwaite Durham. *Hundredestoit* 1086 (DB). 'Clearing or meadow of a man called Húnrøthr'. OScand. pers. name + *thveit*. Alternatively the first element may be OScand. *hundrath* or OE *hundred* in the sense 'administrative district'.

Hundleby Lincs. *Hundelbi* 1086 (DB). 'Farmstead or village of a man called Hundulfr'. OScand. pers. name + *bý*.

Hundleton Pemb. *Hundenton* 1475. 'Farm where dogs are kept'. OE *hund* + *tūn*.

Hundon Suffolk. *Hunedana* 1086 (DB). 'Valley of a man called Hūna'. OE pers. name + *denu*.

Hungerford W. Berks. *Hungreford* 1101–18. 'Hunger ford', i.e. possibly 'ford leading to poor or unproductive land'. OE *hungor* + *ford*.

Hunmanby N. Yorks. *Hundemanebi* 1086 (DB). 'Farmstead or village of the houndsmen or dog-keepers'. OScand. **hunda-mann* (genitive plural *-a*) + *bý*.

Hunningham Warwicks. *Huningeham* 1086 (DB). 'Homestead of the family or followers of a man called Hūna'. OE pers. name + *-inga-* + *hām*.

Hunsdon Herts. *Honesdone* 1086 (DB). 'Hill of a man called Hūn'. OE pers. name + *dūn*.

Hunsingore N. Yorks. *Hulsingoure* [sic] 1086 (DB), *Hunsinghouere* 1194. 'Promontory associated with a man called Hūnsige'. OE pers. name + *-ing-* + **ofer*.

Hunslet Leeds. *Hunslet* [sic] 1086 (DB), *Hunesflete* 12th cent. 'Inlet or stream of a man called Hūn'. OE pers. name + *flēot*.

Hunsley, High E. R. Yorks. *Hund(r)eslege* 1086 (DB). 'Woodland clearing of a man called **Hund*'. OE pers. name + *lēah*.

Hunsonby Cumbria. *Hunswanby* 1292. 'Farmstead or village of the houndsmen or dog-keepers'. OScand. *hunda-sveinn* + *bý*.

Hunstanton Norfolk. *Hunstanestun* c.1035, *Hunestanestuna* 1086 (DB). 'Farmstead of a man called Hūnstān'. OE pers. name + *tūn*.

Hunstanworth Durham. *Hunstanwortha* 1183. 'Enclosure of a man called Hūnstān'. OE pers. name + *worth*.

Hunston Suffolk. *Hunterstuna* 1086 (DB). 'Farmstead of the hunter'. OE **huntere* + *tūn*.

Hunston W. Sussex. *Hunestan* 1086 (DB). 'Boundary stone of a man called Hūn(a)'. OE pers. name + *stān*.

Hunter's Quay Arg. 'Hunter's quay'. The *Hunter* family of nearby Hafton House gave the clubhouse here to the Royal Clyde Yacht Club in 1872.

Huntingdon Cambs. *Huntandun* 973, *Huntedun* 1086 (DB). 'Hill of the huntsman, or of a man called Hunta'. OE *hunta* or pers. name (genitive -*n*) + *dūn*. **Huntingdonshire** (OE *scīr* 'district') is first referred to in the 11th cent.

Huntingfield Suffolk. *Huntingafelde* 1086 (DB). 'Open land of the family or followers of a man called Hunta'. OE pers. name + -*inga*- + *feld*.

Huntington Staffs. *Huntendon* 1167. 'Hill of the huntsmen'. OE *hunta* (genitive plural *huntena*) + *dūn*.

Huntington York. *Huntindune* 1086 (DB). Probably 'hunting hill', i.e. 'hill where hunting takes place'. OE *hunting* + *dūn*.

Huntley Glos. *Huntelei* 1086 (DB). 'Huntsman's wood or clearing'. OE *hunta* + *lēah*.

Huntly Aber. Transferred from HUNTLY, Sc. Bord., in feudal times by Earls of Huntly. Applied originally to barony and castle only.

Huntly Sc. Bord. *Huntleie* 1180. 'Wood or clearing of the huntsman'. OE *hunta* + *leah*.

Hunton Kent. *Huntindone* 11th cent. 'Hill of the huntsmen'. OE *hunta* (genitive plural *huntena*) + *dūn*.

Hunton N. Yorks. *Huntone* 1086 (DB). 'Farmstead of a man called Hūna'. OE pers. name + *tūn*.

Huntsham Devon. *Honesham* 1086 (DB). 'Homestead or enclosure of a man called Hūn'. OE pers. name + *hām* or *hamm*.

Huntspill Somerset. *Honspil* 1086 (DB). 'Tidal creek of a man called Hūn'. OE pers. name + *pyll*.

Huntworth Somerset. *Hunteworde* 1086 (DB). 'Enclosure of the huntsman, or of a man called Hūnta'. OE *hunta* or pers. name + *worth*.

Hunwick Durham. *Hunewic* c.1050. 'Dwelling or (dairy) farm of a man called Hūna'. OE pers. name + *wīc*.

Hunworth Norfolk. *Hunaworda* 1086 (DB). 'Enclosure of a man called Hūna'. OE pers. name + *worth*.

Hurdsfield Ches. *Hirdelesfeld* 13th cent. Probably 'open land at a hurdle (fence)'. OE *hyrdel*, **hyrdels* + *feld*.

Hurley, 'woodland clearing in a recess in the hills', OE *hyrne* + *lēah*: **Hurley** Warwicks. *Hurle* early 11th cent., *Hurnlee* c.1180. **Hurley** Winds. & Maid. *Herlei* 1086 (DB), *Hurnleia* 1106–21.

Hurn Dorset. *Herne* 1086 (DB). '(Place at) the angle or corner of land'. OE *hyrne*.

Hursley Hants. *Hurselye* 1171. Probably 'mare's woodland clearing'. OE **hyrse* + *lēah*.

Hurst Wokhm. *Herst* 1220. '(Place at) the wooded hill'. OE *hyrst*. The same name occurs in other counties.

Hurst, Old Cambs. *See* OLD HURST.

Hurstbourne Priors & Hurstbourne Tarrant Hants. *Hysseburnan* c.880, *Esseborne* 1086 (DB). 'Stream with winding water-plants', probably referring to green canary-grass. OE *hysse* + *burna*. Distinguishing affixes from early possession of the two manors by Winchester Priory and Tarrant Abbey.

Hurstpierpoint W. Sussex. *Herst* 1086 (DB), *Herst Perepunt* 1279. OE *hyrst* 'wooded hill' + manorial addition from possession by Robert *de Pierpoint* in 1086.

Hurstwood Lancs. *Hurstwode* 1285. 'Wood on a hill'. OE *hyrst* + *wudu*.

Hurworth Darltn. *Hurdewurda* 1158. 'Enclosure made with hurdles'. OE **hurth* + *worth*.

Husabost Highland. (Skye). 'Dwelling place at the house'. OScand. *hús* + *bólstathr*.

Husbands Bosworth Leics. *See* BOSWORTH.

Husborne Crawley Beds. *Hysseburnan* 969, *Crawelai* 1086 (DB), *Husseburn Crouleye* 1276. Originally two separate places. 'Stream with winding water-plants', from OE *hysse* + *burna*, and 'wood or clearing frequented by crows', from OE *crāwe* + *lēah*.

Husthwaite N. Yorks. *Hustwait* 1167. 'Clearing with a house or houses built on it'. OScand. *hús* + *thveit*.

Huthwaite Notts. *Hodweit* 1199. 'Clearing on a spur of land'. OE *hōh* + OScand. *thveit*.

Huttoft Lincs. *Hotoft* 1086 (DB). 'Homestead on a spur of land'. OE *hōh* + OScand. *toft*.

Hutton, a common name, 'farmstead on or near a ridge or hill-spur', OE *hōh* + *tūn*: **Hutton** Cumbria. *Hoton* 1291. **Hutton** Essex. *Atahov* 1086 (DB), *Houton* 1200. In DB it is simply '(place) at the ridge or hillspur'. **Hutton** Lancs., near Penwortham. *Hoton* 12th cent. **Hutton** N. Som. *Hotune* 1086 (DB). **Hutton** Sc. Bord. *Hotun* c.1098. **Hutton Buscel** N. Yorks. *Hotun* 1086 (DB), *Hoton Buscel* 1253. Manorial affix from the *Buscel* family, here in the 12th cent. **Hutton Conyers** N. Yorks. *Hotone* 1086 (DB), *Hotonconyers* 1198. Manorial affix from the *Conyers* family, here in the early 12th cent. **Hutton Cranswick** E. R. Yorks. *Hottune* 1086 (DB). Distinguishing affix from its proximity to Cranswick (*Cranzuic* 1086 (DB)) which is possibly '(dairy) farm of a man called *Cranuc*', OE pers. name + *wīc*. **Hutton Henry** Durham. *Hotun* c.1040, *Huton Henrie* 1611. Affix from its possession by *Henry* de Essh in the 14th cent. **Hutton-le-Hole** N. Yorks. *Hotun* 1086 (DB). Affix means 'in the hollow' from OE *hol*. **Hutton Lowcross** Red. & Cleve. *Hotun* 1086 (DB). Affix from a nearby place (*Loucros* 12th cent.) which may be 'cross of a man called Logi', OScand. pers. name + *cros*. **Hutton Magna** Durham. *Hotton* 1086 (DB), *Magna Hoton* 1157. Latin *magna* 'great'. **Hutton Roof** Cumbria. *Hotunerof* 1278. Affix is probably manorial from some early owner called *Rolf* (OScand. *Hrólfr*). **Hutton Rudby** N. Yorks. *Hotun* 1086 (DB), *Hoton by Ruddeby* 1310. Affix from a nearby place (*Rodebi* 1086

(DB)) which is 'farmstead of a man called Ruthi', OScand. pers. name + *bý*. **Hutton, Sand** N. Yorks. *Hotone* 1086 (DB), *Sandhouton* 1219. Affix is OE *sand* 'sand'. **Hutton Sessay** N. Yorks. *Hottune* 1086 (DB). Affix from its proximity to SESSAY. **Hutton, Sheriff** N. Yorks. *Hotone* 1086 (DB), *Shirefhoton* c.1200. Affix from the *Sheriff* of York who held this manor from the mid-12th cent., OE *scīr-gerēfa*. **Hutton Wandesley** N. Yorks. *Hoton* c.1200, *Hotun Wandelay* 1253. Affix from a nearby place (*Wandeslage* 1086 (DB)), 'woodland clearing of a man called *Wand* or *Wandel*', OE pers. name + *lēah*.

Huxley Ches. *Huxelehe* early 13th cent. Probably 'woodland clearing of a man called *Hucc*'. OE pers. name + *lēah*.

Huyton Knows. *Hitune* 1086 (DB). 'Estate with a landing-place'. OE *hȳth* + *tūn*.

Hwlffordd. *See* HAVERFORDWEST.

Hyde, 'estate assessed at one hide, an amount of land for the support of one free family and its dependants', OE *hīd*; for example: **HydePark** Gtr. London. *Hida* 1204, *Hide Park* 1543. **Hyde** Tamesd. *Hyde* early 13th cent.

Hykeham, North & Hykeham, South Lincs. *Hicham* 1086 (DB). OE *hām* 'homestead' with a doubtful first element, possibly an OE *hīce* 'some kind of small bird' or an OE pers. name *Hīca*.

Hylton, South Sundld. *Helton* c.1170. 'Farmstead or estate on a slope'. OE *helde* + *tūn*.

Hyndburn (district) Lancs. A modern adoption of the OE river-name Hyndburn Brook, which means 'stream frequented by hinds or does'. OE *hind* + *burna*.

Hynish Arg. (Tiree). Unidentified first element + OScand. *nes* 'headland'. *Compare* HEANISH.

Hythe, 'landing-place or harbour', OE *hȳth*: **Hythe** Kent. *Hede* 1086 (DB). **Hythe** Hants. *La Huthe* 1248.

Ibberton Dorset. *Abristetone* 1086 (DB). 'Estate associated with a man called Ēadbeorht'. OE pers. name + *-ing-* + *tūn*.

Ible Derbys. *Ibeholon* 1086 (DB). 'Hollow(s) of a man called Ibba'. OE pers. name + *hol*.

Ibsley Hants. *Tibeslei* 1086 (DB), *Ibeslehe* 13th cent. 'Woodland clearing of a man called *Tibbi* or *Ibbi*'. OE pers. name + *lēah*.

Ibstock Leics. *Ibestoche* 1086 (DB). 'Outlying farmstead or hamlet of a man called Ibba'. OE pers. name + *stoc*.

Ibstone Bucks. *Ebestan* 1086 (DB). 'Boundary stone of a man called Ibba'. OE pers. name + *stān*.

Ibthorpe Hants. *Ebbedrope* 1236, *Ybethrop* 1269. 'Outlying farmstead of a man called Ibba'. OE pers. name + *throp*.

Ibworth Hants. *Hibbewrth* 1233, *Ibbewrth* 1245. 'Enclosure of a man called Ibba'. OE pers. name + *worth*.

Ickburgh Norfolk. *Iccheburc* 1086 (DB). 'Stronghold of a man called *Ic(c)a*'. OE pers. name + *burh*.

Ickenham Gtr. London. *Ticheham* 1086 (DB). 'Homestead or village of a man called Tic(c)a'. OE pers. name (genitive *-n*) + *hām*. Initial *T-* was lost in the 13th cent. due to confusion with the preposition *at*.

Ickford Bucks. *Iforde* [sic] 1086 (DB), *Ycford* 1175. 'Ford of a man called *Ic(c)a*'. OE pers. name + *ford*.

Ickham Kent. *Ioccham* 785, *Gecham* 1086 (DB). 'Homestead or village comprising a yoke (some fifty acres) of land'. OE *geoc* + *hām*.

Ickleford Herts. *Ikelineford* 12th cent. 'Ford associated with a man called Icel'. OE pers. name + *-ing-* + *ford*.

Icklesham E. Sussex. *Icoleshamme* 770. 'Promontory or river-meadow of a man called Icel'. OE pers. name + *hamm*.

Ickleton Cambs. *Icelingtune* c.975, *Hichelintone* 1086 (DB). 'Estate associated with a man called Icel'. OE pers. name + *-ing-* + *tūn*.

Icklingham Suffolk. *Ecclingaham* 1086 (DB). 'Homestead of the family or followers of a man called *Ycel*'. OE pers. name + *-inga-* + *hām*.

Icknield Way (prehistoric trackway from Norfolk to Dorset). *Icenhylte* 903. Obscure in origin and meaning, although there may be some connection with the *Iceni*, the Iron Age tribe who once inhabited Norfolk and Suffolk. The name was transferred in the 12th cent. to the Roman road from Bourton on the Water to Templeborough near Rotherham, now called **Icknield** or **Ryknild Street**.

Ickwell Green Beds. *Ikewelle* c.1170. 'Beneficial spring or stream', or 'spring or stream of a man called *Gic(c)a*'. OE *gēoc* 'help' or OE pers. name + *wella*.

Icomb Glos. *Iccacumb* 781, *Iccumbe* 1086 (DB). 'Valley of a man called *Ic(c)a*'. OE pers. name + *cumb*.

Ida (*Uí Deaghaigh*) Kilkenny. 'Descendants of Deaghadh'.

Iddesleigh Devon. *Edeslege* [sic] 1086 (DB), *Edwislega* 1107. 'Woodland clearing of a man called Ēadwīg'. OE pers. name + *lēah*.

Ide Devon. *Ide* 1086 (DB). Possibly an old river-name of pre-English origin.

Ide Hill Kent. *Edythehelle* c.1250. 'Hill of a woman called Edith (OE Ēadgȳth)'. OE pers. name + *hyll*.

Ideford Devon. *Yudaforda* 1086 (DB). 'Ford of a man called *Giedda*', or 'ford where people assemble for speeches or songs'. OE pers. name or OE *giedd* + *ford*.

Iden E. Sussex. *Idene* 1086 (DB). 'Woodland pasture where yew-trees grow'. OE **īg* + *denn*.

Idle Brad. *Idla* c.1190, *Idel* 13th cent. Probably OE *īdel* 'idel, empty' used as a noun in the sense 'uncultivated land'.

Idlicote Warwicks. *Etelincote* 1086 (DB). 'Cottage(s) associated with a man called **Yttel(a)*'. OE pers. name + *-ing-* + *cot*.

Idmiston Wilts. *Idemestone* 947. 'Farmstead of a man called **Idmǣr* or **Idhelm*'. OE pers. name + *tūn*.

Idridgehay Derbys. *Edrichesei* 1230. 'Enclosure of a man called Ēadrīc'. OE pers. name + *hæg*.

Idrigill Highland. (Skye). 'Outer gully'. OScand. *ytri* + *gil*.

Idstone Oxon. *Edwineston* 1199. 'Farmstead of a man called Ēadwine'. OE pers. name + *tūn*.

Iffley Oxon. *Gifetelea* 1004, *Givetelei* 1086 (DB). Probably 'woodland clearing frequented by the plover or similar bird'. OE **gīfete* + *lēah*.

Ifield W. Sussex. *Ifelt* 1086 (DB). 'Open land where yew-trees grow'. OE **īg* + *feld*.

Ifold W. Sussex. *Ifold* 1296. Probably 'fold or enclosure in well-watered land'. OE *īeg* + *fald*.

Iford E. Sussex. *Niworde* [*sic*] 1086 (DB), *Yford* late 11th cent. 'Ford in well-watered land', or 'ford where yew-trees grow'. OE *īeg* or **īg* + *ford*.

Ifton Heath Shrops. *Iftone* 1272. Possibly 'farmstead of a man called Ifa'. OE pers. name + *tūn*.

Ightfield Shrops. *Istefelt* 1086 (DB), *Ychtefeld* 1210–12. 'Open land by the River *Ight*'. Pre-English river-name (of uncertain meaning) + OE *feld*.

Ightham Kent. *Ehteham* c.1100. 'Homestead or village of a man called **Ehta*'. OE pers. name + *hām*.

Iken Suffolk. *Icanho* late 9th cent. (Anglo-Saxon Chronicle s.a. 654), *Ykene* 1212. 'Heel or spur of land of a man called **Ica*'. OE pers. name (genitive -*n*) + *hōh*.

Ilam Staffs. *Hilum* 1002. Possibly a Celtic river-name (an early name for River Manifold) meaning 'trickling stream'. Alternatively the name may be '(place at) the pools', from OScand. *hylr* in a dative plural form *hylum*.

Ilchester Somerset. *Givelcestre* 1086 (DB). 'Roman town on the River *Gifl*'. Celtic river-name (meaning 'forked river') + OE *ceaster*. *Gifl* was an earlier name for the River Yeo.

Ilderton Northum. *Ildretona* c.1125. 'Farmstead where elder-trees grow'. OE **hyldre* + *tūn*.

Ilen (*Abhainn Eibhlinn*) (river) Cork. 'Sparkling river'.

Ilford Gtr. London. *Ilefort* 1086 (DB). 'Ford over the river called *Hyle*'. Celtic river-name (meaning 'trickling stream') + OE *ford*. *Hyle* was an early name for the River Roding below ONGAR.

Ilfracombe Devon. *Alfreincome* 1086 (DB). 'Valley associated with a man called Ælfrēd'. OE pers. name + *-ing-* + *cumb*.

Ilkeston Derbys. *Tilchestune* [*sic*] 1086 (DB), *Elkesdone* early 11th cent. 'Hill of a man called Ēalāc'. OE pers. name + *dūn*.

Ilketshall St Andrew & Ilketshall St Margaret Suffolk. *Ilcheteleshala* 1086 (DB). 'Nook of land of a man called **Ylfketill*'. OScand. pers. name + OE *halh*. Distinguishing affixes from the dedications of the churches.

Ilkley Brad. *Hillicleg* c.972, *Illiclei* 1086 (DB). Possibly 'woodland clearing of a man called **Yllica* or **Illica*'. OE pers. name + *lēah*.

Illanfad (*Oileán Fada*) Donegal. 'Long island'.

Illaunfadda (*Oileán Fada*) Galway. 'Long island'.

Illaunmore (*Oileán Mór*) Tipperary. 'Large island'.

Illaunslea (*Oileán Shléibhe*) Kerry. 'Island of the mountain'.

Illauntannig (*Oileán tSeanaigh*) Kerry. 'Seanach's island'.

Illey Dudley. *Hillely* early 13th cent. 'Woodland clearing of a man called Hilla'. OE pers. name + *lēah*.

Illingworth Calder. *Illingworthe* 1276. 'Enclosure associated with a man called **Illa or Ylla*'. OE pers. name + *-ing-* + *worth*.

Illogan Cornwall. 'Church of *Sanctus Illoganus*' 1291. 'Church of St Illogan'. From the patron saint of the church.

Illston on the Hill Leics. *Elvestone* 1086 (DB). Possibly 'farmstead of a man called Ælfhere'. OE pers. name (perhaps replaced by OScand. *Iólfr*) + *tūn*.

Ilmer Bucks. *Imere* [*sic*] 1086 (DB), *Ilmere* 1161–3. Probably 'pool where leeches are found'. OE *īl* + *mere*.

Ilmington Warwicks. *Ylmandun* 978, *Ilmedone* 1086 (DB). 'Hill growing with elm-trees'. OE **ylme* (genitive *-an*) + *dūn*.

Ilminster Somerset. *Illemynister* 995, *Ileminstre* 1086 (DB). 'Large church on the River Isle'. Celtic river-name (of uncertain meaning) + OE *mynster*.

Ilsington Devon. *Ilestintona* 1086 (DB). 'Estate associated with a man called **Ielfstān*'. OE pers. name + *-ing-* + *tūn*.

Ilsley, East & Ilsley, West W. Berks. *Hildeslei* 1086 (DB). 'Woodland clearing of a man called Hildi'. OE pers. name + *lēah*.

Ilton N. Yorks. *Ilcheton* 1086 (DB). 'Farmstead of a man called **Yllica* or **Illica*'. OE pers. name + *tūn*.

Ilton Somerset. *Atiltone* [*sic*] 1086 (DB), *Ilton* 1243. 'Farmstead on the River Isle'. Celtic river-name (of uncertain meaning) + OE *tūn*.

Imaile (*Uí Mail*) Wicklow. 'Descendants of Mal'.

Immingham NE. Lincs. *Imungeham* [*sic*] 1086 (DB), *Immingeham* c.1115. 'Homestead of the family or followers of a man called Imma'. OE pers. name + *-inga-* + *hām*.

Impington Cambs. *Impintune* c.1050, *Epintone* 1086 (DB). 'Estate associated with a man called **Empa* or **Impa*'. OE pers. name + *-ing-* + *tūn*.

Inagh (*Eidhneach*) Clare. 'Abounding in ivy'.

Inan (*Eidhneán*) Meath. 'Ivy place'.

Inane (*Eidhneán*) Cork, Tipperary. 'Ivy place'.

Ince, 'the island', OWelsh **inis*: **Ince** Ches. *Inise* 1086 (DB). **Ince** Wigan. *Ines* 1202. **Ince Blundell** Sefton. *Hinne* [*sic*] 1086 (DB), *Ins Blundell* 1332. Manorial affix from the *Blundell* family, here in the early 13th cent.

Inch (*Inis*) Cork, Down, Kerry, Wexford. 'Island, water-meadow'.

Inchagoill (*Inis an Ghaill*) Galway. 'Island of the foreigners'.

Inchannon (*Inis Eonáin*) Cork. 'Eonán's island'.

Inchcape Rock Ang. Gaelic *inis* 'rock' + unknown element + English *rock*.

Inchcolm (island) Fife. *Insula Sancti Columbae* c.1123, *St Colmes Ynch* 1605. 'St Columba's island'. Gaelic *inis*.

Inchee (*Insí*) Kerry. 'Water-meadows'.

Inchenny (*Inis Eanaigh*) Tyrone. 'Island of the marsh'.

Inchfarrannaglerach (*Inse Fhearann na gCléireach*) Kerry. 'Water-meadows of the land of the clerics'.

Inchicore (*Inse Chór*) Dublin. 'Islands of the snout'.

Inchicronan (*Inse Cronáin*) Clare. 'Cronán's island'.

Inchideraille (*Inis Idir dá Fhaill*) Cork. 'Island between two cliffs'.

Inchigeelagh (*Inse Geimhleach*) Cork. 'Island of the prisoner'.

Inchinaleega (*Inse na Léige*) Kerry. 'Water-meadow of the stone'.

Inchinglanna (*Inse an Ghleanna*) Kerry. 'Water-meadow of the valley'.

Inchiquin, Lough (*Loch Inse Uí Chuinn*) Clare, Cork. 'Lake of Ó Cuinn's island'.

Inchkeith (island) Fife. *Insula Ked* c.1200. 'Wooded island'. Gaelic *inis* + British **ceto-*.

Inchnamuck (*Inse na Muc*) Tipperary. 'Island of the pigs'.

Indian Queens Cornwall. Named from an inn so called, recorded thus in 1802. The inn-name probably refers to the celebrated American Indian princess Pocahontas.

Ingatestone Essex. *Gynges Atteston* 1283. 'Manor called *Ing* (for which *see* FRYERNING) at the stone'. One of a group of places so called, this one distinguished by reference to a Roman milestone. OE *stān*.

Ingbirchworth Barns. *Berceuuorde* 1086 (DB), *Yngebyrcheworth* 1424. 'Enclosure where birch-trees grow'. OE *birce* + *worth* with the later addition of OScand. *eng* 'meadow'.

Ingham, possibly 'homestead or village of a man called Inga', OE pers. name + *hām*, although it has recently been suggested that the first element of these names may be a word meaning 'the Inguione', a member of the ancient Germanic tribe called the Inguiones: **Ingham** Lincs. *Ingeham* 1086 (DB). **Ingham** Norfolk. *Hincham* 1086 (DB). **Ingham** Suffolk. *Ingham* 1086 (DB).

Ingleby, 'farmstead or village of the Englishmen', OScand. *Englar* + *bý*: **Ingleby** Derbys. *Englaby* 1009, *Englebi* 1086 (DB). **Ingleby** Lincs. *Englebi* 1086 (DB). **Ingleby Arncliffe** N. Yorks. *Englebi* 1086 (DB). Affix is from a nearby place (*Erneclive* 1086), 'eagles' cliff', OE *earn* + *clif*. **Ingleby Greenhow** N. Yorks. *Englebi* 1086 (DB). Affix is from a nearby place (*Grenehou* 12th cent.), 'green mound', OE *grēne* + OScand. *haugr*.

Inglesham Swindn. *Inggeneshamme* c.950. 'Enclosure or river-meadow of a man called *Ingen* or *Ingīn*'. OE pers. name + *hamm*.

Ingleton Durham. *Ingeltun* c.1040. Probably 'farmstead of a man called Ingeld'. OScand. pers. name + OE *tūn*.

Ingleton N. Yorks. *Inglestune* 1086 (DB). 'Farmstead near the hill or peak'. OE *ingel* + *tūn*.

Inglewhite Lancs. *Inglewhite* 1662. Probably from OE *wiht* 'river-bend' with an uncertain first element.

Ingoe Northum. *Hinghou* 1229. Probably OE *ing* 'hill, peak' with the addition of *hōh* 'spur of land'.

Ingoldisthorpe Norfolk. *Torp* 1086 (DB), *Ingaldestorp* 1203. 'Outlying farmstead or hamlet of a man called Ingjaldr'. OScand. pers. name + *thorp*.

Ingoldmells Lincs. *in Guldelsmere* [sic] 1086 (DB), *Ingoldesmeles* 1180. 'Sandbanks of a man called Ingjaldr'. OScand. pers. name + *melr*.

Ingoldsby Lincs. *Ingoldesbi* 1086 (DB). 'Farmstead or village of a man called Ingjaldr'. OScand. pers. name + *bý*.

Ingram Northum. *Angerham* 1242. 'Homestead or enclosure with grassland'. OE *anger* + *hām* or *hamm*.

Ingrave Essex. *Ingam* 1086 (DB), *Gingeraufe* 1276. 'Manor called *Ing* (for which see FRYERNING) held by a man called Ralf (OGerman Radulf)'. He was a tenant in 1086.

Ingworth Norfolk. *Inghewurda* 1086 (DB). Probably 'enclosure of a man called Inga'. OE pers. name + *worth*.

Inishargy (*Inis Cairrge*) Down. *Iniscargi* 1306. 'Island of the rock'.

Inishbarra (*Inis Bearachain*) Galway. 'Island of the heifers'.

Inishbiggle (*Inis Bigil*) Mayo. 'Island of the fasting'.

Inishbofin (*Inis Bó Finne*) Donegal, Galway. 'Island of the white cow'.

Inishcarra (*Inis Cara*) Cork. 'Island of the weir'.

Inishcrone (*Inis Crabhann*) Sligo. 'Island of the gravel ridge of the river'.

Inishdadroum (*Inis dá Drom*) Clare. 'Island of two backs'.

Inishdooey (*Inis Dubhthaigh*) (island) Donegal. 'Dubhthach's island'. The current Irish name is *Oileán Dúiche*, 'Sandbank island'.

Inisheer (*Inis Oírr*) Galway. 'Eastern island'.

Inishfree (*Inis Fraoigh*) Donegal. 'Island of heather'.

Inishirrer (*Inis Oirthir*) Donegal. 'Island of the eastern area'.

Inishkeen (*Inis Caoin*) Limerick, Monaghan. 'Beautiful island'.

Inishkeeragh (*Inis Caorach*) Donegal. 'Island of sheep'.

Inishlackan (*Inis Leacan*) Galway. 'Island of the hillside'.

Inishmaan (*Inis Meáin*) Galway. 'Middle island'.

Inishmacowney (*Inish Mhic Uaithne*) Clare. 'Mac Uaithne's island'.

Inishmeane (*Inis Meáin*) Donegal. *Inismahan* 1830. 'Middle island'.

Inishmore (*Inis Mór*) Galway. 'Big island'.

Inishmurray (*Inis Muirí*) Sligo. 'Muireadhach's island'.

Inishnagor (*Inis na gCorr*) Donegal, Sligo. 'Island of the cranes'.

Inishowen (*Inis Eoghain*) Donegal. *(i)nInis Eoghain* 465. 'Eoghan's island'.

Inishrush (*Inis Rois*) Derry. *Inishroisse* 1615. 'Island of the wood'.

Inishshark (*Inis Airc*) Galway. 'Island of hardship'.

Inishsirrer (*Inis Oirthir*) Donegal. 'Eastern island'.

Inishtooskert (*Inis Tuaisceart*) Kerry. 'Northern island'.

Inishturk (*Inis Toirc*) Galway. 'Island of the boar'.

Inistioge (*Inis Tíog*) Kilkenny. 'Tíog's island'.

Inkberrow Worcs. *Intanbeorgas* 789, *Inteberge* 1086 (DB). 'Hills or mounds of a man called Inta'. OE pers. name + *beorg* (plural -*as*).

Inkpen W. Berks. *Ingepenne* c.935, *Hingepene* 1086 (DB). OE *ing* 'hill, peak' with either Celtic *penn* 'hill' or OE *penn* 'enclosure, fold'.

Innerleithen Sc. Bord. *Innerlethan* c.1160. 'Mouth of the Leithen Water'. Gaelic *inbhir* + British *Lektona* 'the dripping river'.

Innisfallen (*Inis Faithlenn*) Kerry. 'Fathlenn's island'.

Inskip Lancs. *Inscip* 1086 (DB). Probably a Celtic name, 'island of the bowl or hollow', from *ïnïs* + *cïb*.

Instow Devon. *Johannesto* 1086 (DB). 'Holy place of St John'. Saint's name + OE *stōw*. The church here is dedicated to St John the Baptist.

Inver (*Inbhear Náile*) Donegal. 'Estuary of Náile'.

Inveran (*Indreabhán*) Galway. 'Little estuary'.

Inveraray Arg. 'Mouth of the River Aray'. Gaelic *inbhir*.

Inverclyde (the unitary authority). A modern administrative formation. 'Mouth of the River Clyde'. Gaelic *inbhir*. For the river-name, *see* CLYDEBANK.

Invereela (*Inbhear Daeile*) Wicklow. 'Estuary of the (river) Deel'.

Invergordon Highland. 'Gordon's (place at the) river-mouth'. Gaelic *inbhir*. Alexander *Gordon* was landowner here c.1760. Earlier, the place was known as *Inverbreckie*, 'mouth of the Breckie', from Gaelic *inbhir* + a river name meaning 'speckled' (Gaelic *breac*).

Inverkeithing Fife. *Hinhirkethy* c.1057, *Innerkethyin* 1114. 'Mouth of the Keithing Burn'. Gaelic *inbhir*. The river-name probably means 'wooded stream' (British *ceto-*).

Inverleith Edin. *Inerlet* c.1130. 'Mouth of the Water of Leith'. Gaelic *inbhir*. For the river-name, *see* LEITH.

Inverness Highland. *Invernis* 1300. 'Mouth of the River Ness'. Gaelic *inbhir*. The river has a Celtic or pre-Celtic name from a root *ned-* 'moist'.

Inverurie Aber. *Inverurie* 1199, *Innervwry* c.1300. 'Confluence of the River Urie (and Don)'. Gaelic *inbhir* + unexplained river name.

Inwardleigh Devon. *Lega* 1086 (DB), *Inwardlegh* 1235. 'Woodland clearing'. OE *lēah*, with manorial addition from the man called *Inwar* (OScand. *Ingvar*) who held the manor in 1086.

Inworth Essex. *Inewrth* 1206. 'Enclosure of a man called Ina'. OE pers. name + *worth*.

Iona (island) Arg. *Hiiensis* 634, *Ioua insula* c.700, *Hiona-Columcille* c.1100. '(Place of) yew-trees'. OIrish *eo*. The island is associated with St Columba, Gaelic *I Chaluim Chille* or simply *I*.

Iping W. Sussex. *Epinges* 1086 (DB). '(Settlement of) the family or followers of a man called *Ipa*'. OE pers. name + -*ingas*.

Ipplepen Devon. *Iplanpenne* 956, *Iplepene* 1086 (DB). 'Fold or enclosure of a man called *Ipela*'. OE pers. name + *penn*.

Ipsden Oxon. *Yppesdene* 1086 (DB). 'Valley near the upland'. OE *yppe* + *denu*.

Ipstones Staffs. *Yppestan* 1175. Probably 'stone in the high place or upland'. OE *yppe* + *stān*.

Ipswich Suffolk. *Gipeswic* c.975, 1086 (DB). Probably 'harbour or trading centre of a man called *Gip*'. OE pers. name + *wīc*. Alternatively the first element may be an OE *gip(s)* 'opening, gap' (with reference to the wide estuary of the River ORWELL).

Irby, Ireby, 'farmstead or village of the Irishmen', OScand. *Írar* + *bý*: **Irby** Wirral. *Irreby* c.1100. **Irby in the Marsh** Lincs. *Irebi* c.1115. **Irby upon Humber** NE. Lincs. *Iribi* 1086 (DB). For the river-name, *see* HUMBER. **Ireby** Cumbria. *Irebi* c.1160. **Ireby** Lancs. *Irebi* 1086 (DB).

Irchester Northants. *Yranceaster* 973, *Irencestre* 1086 (DB). 'Roman station associated with a man called Ira or *Yra'. OE pers. name + *ceaster*.

Ireby, *See* IRBY.

Ireland (*Éire*) 'Land of the Irish'. The name amounts to OIrish *Eriu*, 'Eria' + OE *land*. The country name itself may relate to a fertility goddess.

Ireland's Eye (*Inis Mac Neasáin*) (island) Dublin. 'Island of Mic Neasáin'. The English name is a convoluted version of the earlier Irish name *Inis Ereann*, 'Eria's island'. The female personal name was taken to be the genitive of *Éire*, 'Ireland', while *Eye* represents OScand. *ey*, 'island'.

Ireleth Cumbria. *Irlid* 1190. 'Hill-slope of the Irishmen'. OScand. *Írar* + *hlíth*.

Ireton, Kirk Derbys. *Hiretune* 1086 (DB), *Kirkirton* 1370. 'Farmstead of the Irishmen'. OScand. *Írar* + OE *tūn*. Affix is OScand. *kirkja* 'church'.

Irlam Salford. *Urwelham* c.1190. 'Homestead or enclosure on the River Irwell'. OE river-name ('winding stream' from OE *irre* + *wella*) + OE *hām* or *hamm*.

Irnham Lincs. *Gerneham* 1086 (DB). 'Homestead or village of a man called *Georna'. OE pers. name + *hām*.

Iron Acton S. Glos. *See* ACTON.

Ironbridge Tel. & Wrek., a late name, so called from the famous iron bridge built here across the River Severn in 1779.

Irthington Cumbria. *Irthinton* 1169. 'Farmstead on the River Irthing'. Celtic river-name + OE *tūn*.

Irthlingborough Northants. *Yrtlingaburg* 780, *Erdi(n)burne* [*sic*] 1086 (DB). Probably 'fortified manor belonging to the ploughmen'. OE *erthling, yrthling* + *burh*.

Irton Cumbria. *Yrton* c.1225. 'Farmstead on the River Irt'. OE *tūn* with an old pre-English river-name of uncertain meaning.

Irton N. Yorks. *Iretune* 1086 (DB). 'Farmstead of the Irishmen'. OScand. *Írar* + OE *tūn*.

Irvine N. Ayr. *Hirun* c.1190. '(Place on the) River Irvine'. The Celtic river name is probably the same as the River Irfon in Ceredigion. Etymology obscure.

Irvinestown Fermanagh. *Irvinestown* or *Lowthertown* 1837. The original village here was named *Lowthertown*, after Sir Gerard *Lowther*, who founded it in 1618. In 1667 the estate was sold to Sir Christopher *Irvine*, and his name was adopted for the village in the early 19th cent. The equivalent Irish name is *Baile an Irbhinigh*.

Irwell (river) Lancs. *See* IRLAM.

Isbister Shet. (Whalsay). 'East farm'. OScand. *austr* + *bólstathr*.

Isertkelly (*Díseart Cheallaigh*) Kilkenny. 'Ceallach's hermitage'.

Isfield E. Sussex. *Isefeld* 1214. 'Open land of a man called *Isa'. OE pers. name + *feld*.

Isham Northants. *Ysham* 974, *Isham* 1086 (DB). 'Homestead or promontory by the River Ise'. Celtic river-name (meaning 'water') + OE *hām* or *hamm*.

Isis (river) Oxon., recorded as *Isa* c.1350, *Isis* 1577, an artificial formation from *Tamesis*, the early form for THAMES, which was falsely interpreted as THAME + Isis.

Island (*Oiléan*) Cork, Mayo. 'Island'.

Island Magee (*Oileán Mhic Aodha*) Antrim. *Rensevin* 1521. 'Mac Aodha's island'. An earlier name for the peninsula here was *Rinn Seimhne*, 'peninsula of (the district of) Seimhne', from a tribal name.

Islandmoyle (*Oileán Maol*) Down. 'Bare island'.

Islandreagh (*Oileán Riabhach*) Antrim. 'Striped island'.

Islay (island) Arg. *Ilea* c.690, *Ile* 800. Possibly 'swelling island'. Celtic *ili*, but probably pre-Celtic.

Isle Abbotts & Isle Brewers Somerset. *Yli* 966, *I(s)le* 1086 (DB), *Ile Abbatis* 1291, *Ile Brywer* 1275. Named from the River Isle, which is a Celtic river-name of uncertain meaning. Distinguishing affixes from early possession by Muchelney Abbey (Latin *abbatis* 'of the abbot') and by the *Briwer* family.

Isle of Dogs Gtr. London. first recorded as *the Isle of Dogs* in 1520, probably simply descriptive (but with derogatory overtones) of a marshy peninsula (ME *ile*) frequented by (wild or stray) dogs.

Isleham Cambs. *Yselham* 895, *Gisleham* 1086 (DB). 'Homestead of a man called *Gīsla'. OE pers. name + *hām*.

Isleworth Gtr. London. *Gisheresuuyrth* 677 (16th cent. copy), *Gistelesworde* 1086 (DB). 'Enclosure of a man called Gīslhere'. OE pers. name + *worth*.

Islington Gtr. London. *Gislandune* c.1000, *Iseldone* 1086 (DB). 'Hill of a man called *Gīsla'. OE pers. name (genitive -*n*) + *dūn*.

Islip Northants. *Isslepe* 10th cent., *Islep* 1086 (DB). 'Slippery place by the River Ise'. Pre-English river-name (meaning 'water') + OE *slæp*.

Islip Oxon. *Githslepe* c.1050, *Letelape* [sic] 1086 (DB). 'Slippery place by the River *Ight* (an old name for the River Ray)'. Pre-English river-name (of uncertain meaning) + OE *slæp*.

Itchen Abbas Hants. *Icene* 1086 (DB), *Ichene Monialium* 1167. Named from the River Itchen, an ancient pre-Celtic river-name of unknown origin and meaning. The Latin affix in the 12th cent. form means 'of the nuns', the later affix *Abbas* is a reduced form of Latin *abbatissa* 'abbess', both referring to possession by the Abbey of St Mary, Winchester.

Itchen Stoke Hants. *Stoche* 1086 (DB), *Ichenestoke* 1185. 'Secondary settlement from ITCHEN [ABBAS]'. OE *stoc*.

Itchenor, East & Itchenor, West W. Sussex. *Iccannore* 683, *Icenore* 1086 (DB). 'Shore of a man called *Icca'. OE pers. name (genitive -*n*) + *ōra*.

Itchingfield W. Sussex. *Ecchingefeld* 1222. 'Open land of the family or followers of a man called Ecci'. OE pers. name + *-inga-* + *feld*.

Itchington S. Glos. *Icenantune* 967, *Icetune* 1086 (DB). 'Farmstead by a stream called

Itchen'. Lost pre-Celtic river-name (of unknown meaning) + OE *tūn*.

Itchington, Bishops & Itchington, Long Warwicks. *Yceantune* 1001, *Ice(n)tone* 1086 (DB), *Bisshopesychengton* 1384, *Longa Hichenton* c.1185. 'Farmstead on the River Itchen'. Pre-Celtic river-name (of unknown meaning) + OE *tūn*. Distinguishing affixes from possession by the Bishop of Coventry and Lichfield and from OE *lang* 'long'.

Itteringham Norfolk. *Utrincham* 1086 (DB). Probably 'homestead of the family or followers of a man called *Ytra or *Ytri'. OE pers. name + *-inga-* + *hām*.

Ivegill Cumbria. *Yuegill* 1361. 'Deep narrow valley of the River Ive'. OScand. river-name (meaning 'yew stream') + *gil*.

Ivel (river) Beds.-Herts. *See* NORTHILL.

Iver Bucks. *Evreham* 1086 (DB), *Eura* c.1130. '(Place by) the brow of a hill or the tip of a promontory'. OE *yfer* (-*am* in the Domesday spelling is a Latin ending).

Iveston Durham. *Yuestan, Iuestan* 12th cent. 'Boundary stone of a man called Ifa or *Ifi'. OE pers. name + *stān*.

Ivinghoe Bucks. *Evinghehou* 1086 (DB). 'Hill-spur of the family or followers of a man called Ifa'. OE pers. name +*-inga-* + *hōh*.

Ivington Herefs. *Ivintune* 1086 (DB). 'Estate associated with a man called Ifa'. OE pers. name + *-ing-* + *tūn*.

Ivy Hatch Kent. *Heuyhatche* 1325. '(Place at) the heavy hatch or gate'. ME *hevi* + *hatche*.

Ivybridge Devon. *Ivebrugge* 1292. 'Ivy-covered bridge'. OE *īfig* + *brycg*.

Ivychurch Kent. *Iuecirce* c.1090. 'Ivy-covered church'. OE *īfig* + *cirice*.

Iwade Kent. *Ywada* 1179. 'Ford or crossing-place where yew-trees grow'. OE *īw* + *(ge)wæd*. The crossing-place is to the Isle of Sheppey.

Iwerne Courtney or Shroton, Iwerne Minster Dorset. *Ywern* 877, *Werne, Evneministre* 1086 (DB), *Iwerne Courteney alias Shyrevton* 1403. Named from the River Iwerne, a Celtic river-name possibly meaning 'yew river' or referring to a goddess. Distinguishing affixes from the *Courtenay* family, here in the 13th cent., and from OE *mynster* 'church of a monastery' in allusion to early possession by Shaftesbury Abbey. Shroton means 'sheriff's estate', OE *scīr-rēfa* + *tūn*.

Ixworth Suffolk. *Gyxeweorde* c.1025, *Giswortha* 1086 (DB). 'Enclosure of a man called *Gicsa or *Gycsa'. OE pers. name + *worth*.

Ixworth Thorpe Suffolk. *Torp* 1086 (DB), *Ixeworth thorp* 1305. 'Secondary settlement dependent on IXWORTH'. OScand. *thorp*.

Jacobstow, Jacobstowe, 'holy place of St James', saint's name (Latin *Jacobus*) + OE *stōw*. **Jacobstow** Cornwall. *Jacobestowe* 1270. **Jacobstowe** Devon. *Jacopstoue* 1331.

Jameston Pemb. *Seint Jameston* 1331. 'St James's farm'. OE *tūn*. A fair was held here annually on St James's day.

Jamestown W. Dunb. 'James's town'. Original name *Dumhead*, called *Jamestown* when accommodation was needed for those employed at the Levenbank Works.

Jarlshof Shet. 'Jarl's temple'. OScand. *jarl* + *hof*. The site was so called in 1816 by Sir Walter Scott, who used the name for the medieval farmhouse in his novel *The Pirate* (1821). A jarl is a Scandinavian chief.

Jarrow S. Tyne. *Gyruum c.*716. '(Settlement of) the fen people'. OE tribal name *Gyrwe* (dative plural *Gyrwum*) derived from OE *gyr* 'mud, marsh'.

Jaywick Sands Essex. *Clakyngeywyk* 1438, *Gey wyck* 1584. From ME *wick* (OE *wīc*) '(dairy farm', first element probably ME *Jay* (the bird), originally *clacking* ('chattering') *jay* (from ME *clacken* 'to chatter'), no doubt with reference to a place frequented by jays.

Jedburgh Sc. Bord. *Gedwearde* 800, *Geddewrde c.*1100, *Jeddeburgh c.*1160. 'Enclosure by the River Jed'. Ancient river-name (of obscure origin) + OE *worth* (replaced by ME *burgh* 'town').

Jeffreyston Pemb. *Villa Galfrid c.*1214, *Geffreiston* 1362. 'Geoffrey's farm'. OGerman pers. name + OE *tūn*. The church is dedicated to St Jeffry and St Oswald, but the former saint may have been suggested by the place name.

Jemimaville Highland. The village was named by Sir George Munro, 4th Laird of Poyntzfield, after his wife, *Jemima* Charlotte Graham, whom he married in 1822.

Jerrettspass Armagh. *Gerrard* (*Pass*) 1835. 'Jerrett's pass'. The surname is a form of *Garrett*.

Jersey (island) Channel Islands. *Gersoi, Jersoi c.*1025, *Gerseie* 1213. 'Island of a man called Geirr or *Gērr*'. OScand. pers. name + *ey*. In the 4th-cent. *Maritime Itinerary*, Jersey is recorded as *Andium*, a name of unknown origin and meaning.

Jervaulx Abbey N. Yorks. *Jorvalle c.*1145. 'Valley of the River Ure'. Ancient pre-English river-name of uncertain origin + OFrench *val*(*s*).

Jesmond Newc. upon T. *Gesemue* 1204, *Jesemuth* 1242. '(Place at) the mouth of river *Yese* (now Ouse Burn)'. OE river-name ('gushing stream') + *mūtha*.

Jethou (island) Channel Islands. *Ketehou c.*11th cent., *Gethehou* 1150. 'Goats' island'. OScand. *geit* (genitive plural *-a*) + *holmr*. In the 4th-cent. *Maritime Itinerary*, Jethou is recorded as *Barsa*, probably a Celtic name meaning 'summit place'.

Jevington E. Sussex. *Lovingetone* [*sic*] 1086 (DB), *Govingetona* 1189. 'Farmstead of the family or followers of a man called *Geofa'. OE pers. name + *-inga-* + *tūn*.

Jodrell Bank Ches. recorded from 1831, 'bank or hill-side belonging to the Jodrell family', land-owners here in the 19th cent.

John o'Groats Highland. *John o'Groat* was appointed bailee to the earls of Caithness in the 15th cent.

Johnby Cumbria. *Johannebi* 1200. 'Farmstead or village of a man called Johan'. OFrench pers. name + OScand. *bý*.

Johnston Pemb. *Villa Johannis* 1296, *Johanneston* 1393. 'John's farm'. OE *tūn*.

Johnstone Renf. *Jonestone* 1292. 'John's farm'. OE *tūn*. The town was founded in 1781 on the site of Brig ò Johnstone.

Jonesborough Armagh. 'Jones's town'.
Jones Borough 1714. The village was founded in
1706 by Roth *Jones*, a local landlord.

Joppa Edin. The name was first mentioned in
1793, itself named after the biblical town of
Joppa (Acts 9:36, etc).

Jordanston (*Trefwrdan*) Pemb. *Villa Jordahi*
[*sic*] 1291, *Jordanyston* 1326. 'Jordan's farm'.
OE *tūn* (Welsh *tref*). The Welsh name translates
the English.

Jordanstown Antrim. 'Jordan's
town'. *Bally Jurdon* 1604. An Anglo-Norman
family named *Jordan* was here in the
12th cent.

Juniper Green Edin. 'Field of
junipers'. Originally a park, first mentioned
in 1753.

Jura (island) Arg. *Doirad Eilinn* 678, *Dure*
1336. 'Deer island'. OScand. *dýr* + *ey*.

Kaber Cumbria. *Kaberge* late 12th cent. 'Hill frequented by jackdaws'. OE *cā + beorg, or OScand. *ká + berg.

Kale Water Sc. Bord. *Kalne* c.1200. 'Calling one'. British *calauna.

Kanturk (*Ceann Toirc*) Cork. 'Head of the boar'.

Katesbridge Down. 'Kate's bridge'. *McCay's Bridge* 1744. The name is that of *Kate* McKay, said to have owned a house here where workmen lodged when building a bridge in the 18th cent. The equivalent Irish name is *Droichead Cháit*.

Katrine, Loch Stir. *Loch Ketyerne* 1463. Gaelic *loch* + obscure element.

Kea, Old Kea Cornwall. *Sanctus Che* 1086 (DB). '(Church of) St Kea'. From the dedication of the church.

Keadeen (*Céidín*) (mountain) Wicklow. *Kedi* c.1238, *Kedyn* c.1310. 'Little flat-topped hill'.

Keadew (*Céideadh*) Roscommon. 'Flat-topped hill'.

Keady (*An Céide*) Armagh. *An Chéideadh* c.1854. 'The flat-topped hill'.

Keadydrinagh (*Céide Draigneach*) Sligo. 'Flat-topped hill of the blackthorn'.

Keal, East & Keal, West Lincs. *Estrecale, Westrecale* 1086 (DB). 'The ridge of hills'. OScand. *kjǫlr*. Distinguishing affixes are OE *ēasterra* 'more easterly' and *westerra* 'more westerly'.

Kealariddig (*Caol an Roidigh*) Kerry. 'Marshy stream of the red mire'.

Kealkill (*An Caolchoill*) Cork. 'The narrow wood'.

Kearsley Bolton. *Cherselawe* 1187, *Kersleie* c.1220. 'Clearing where cress grows'. OE *cærse + lēah*.

Kearstwick Cumbria. *Kesthwaite* 1547. 'Valley clearing'. OScand. *kjóss + thveit*.

Kearton N. Yorks. *Karretan* [sic] 13th cent., *Kirton* 1298. Possibly 'farmstead of a man called Kærir'. OScand. pers. name + OE *tūn*.

Keava (*Céibh*) Roscommon. 'Long grass'.

Keave (*Céibh*) Galway. 'Long grass'.

Keddington Lincs. *Cadinton* [sic] 1086 (DB), *Kedingtuna* 12th cent. Probably 'farmstead associated with a man called Cedd(a)'. OE pers. name + -ing- + tūn.

Kedington Suffolk. *Kydington* 1043–5, *Kidituna* 1086 (DB). Identical in origin with the previous name.

Kedleston Derbys. *Chetestune* 1086 (DB). 'Farmstead of a man called Ketill'. OScand. pers. name + OE *tūn*.

Keel (*An Caol*) Mayo. 'The narrow stream'.

Keelby Lincs. *Chelebi* 1086 (DB). 'Farmstead or village on or near a ridge'. OScand. *kjǫlr + bý*.

Keele Staffs. *Kiel* 1169. 'Hill where cows graze'. OE *cȳ + hyll*.

Keeloges (*Caológa*) Wicklow. *Byllock* [sic] 1619, *Culloges* 1690. 'Narrow ridges'.

Keenagh (*Caonach*) Longford. 'Moss'.

Keenaghbeg (*Caonach Beag*) Mayo. 'Little moss'.

Keeragh Islands (*Oileáin na gCaorach*) Wexford. 'Islands of the sheep'.

Keeraunnagark (*Caorán na gCearc*) Galway. 'Moor of the hens'.

Keeston (*Tregetin*) Pemb. *Villa Ketyng* 1289, *Ketingeston* 1295. 'Keting's farm'. OE *tūn* (Welsh *tref*). The English and Welsh names translate each other.

Keevil Wilts. *Kefle* 964, *Chivele* 1086 (DB). Probably 'woodland clearing in a hollow'. OE *cȳf* + *lēah*.

Kegworth Leics. *Cachewarde* [*sic*] 1086 (DB), *Caggworth* c.1125. Possibly 'enclosure of a man called *Cægga'. OE pers. name + *worth*.

Keighley Brad. *Chichelai* 1086 (DB). 'Woodland clearing of a man called *Cyhha'. OE pers. name + *lēah*.

Keimaneigh (*Céim an Fhia*) Cork. 'Gap of the deer'.

Keinton Mandeville Somerset. *Chintune* 1086 (DB), *Kyngton Maundevill* 1280. 'Royal manor'. OE *cyne-* + *tūn*. Manorial affix from the *Maundevill* family, here in the 13th cent.

Keisby Lincs. *Chisebi* 1086 (DB). 'Farmstead or village of a man called Kisi'. OScand. pers. name + *bý*. Alternatively the first element may be OScand. *kís* 'gravel, coarse sand'.

Keith Moray. *Geth* 1187, *Ket* c.1220. '(Place by the) wood'. British *ceto-*.

Kelbrook Lancs. *Chelbroc* 1086 (DB). Probably 'brook flowing in a ravine'. OE *ceole* (with Scand. *k-*) + *brōc*.

Kelby Lincs. *Chelebi* 1086 (DB). Probably 'farmstead or village on or near a ridge or wedge-shaped piece of land'. OScand. *kjǫlr* or *kæl* + *bý*.

Keld N. Yorks. *Appeltrekelde* 1301, *Kelde* 1538. 'The spring', from OScand. *kelda*, 'by the apple-tree' in the early form.

Keldholme N. Yorks. *Keldeholm* 12th cent. 'Island or river-meadow near the spring'. OScand. *kelda* + *holmr*.

Kelfield, 'chalky open land, or open land where chalk was spread', OE *calc* 'chalk' (alternating with *i*-mutated derivative *celce* 'chalky place', and with Scand. *k-*) + *feld*: **Kelfield** N. Lincs. *Chalchefeld, Calkefeld, Kelkefeld* 12th cent. **Kelfield** N. Yorks. *Chelchefeld* 1086 (DB).

Kelham Notts. *Calun* [*sic*] 1086 (DB), *Kelum* 1156. '(Place at) the ridges'. OScand. *kjǫlr* in a dative plural form.

Kelk, Great E. R. Yorks. *Chelche* 1086 (DB). 'Chalky ground'. OE *celce* (with Scand. *k-*).

Kellet, Nether & Kellet, Over Lancs. *Chellet* 1086 (DB). 'Slope with a spring'. OScand. *kelda* + *hlíth*.

Kelleth Cumbria. *Keldelith* 12th cent. Identical in origin with the previous name.

Kelling Norfolk. *Chillinge* c.970, *Kellinga* 1086 (DB). '(Settlement of) the family or followers of a man called *Cylla or Ceolla'. OE pers. name + *-ingas*.

Kellington N. Yorks. *Chellinctone* 1086 (DB). 'Estate associated with a man called Ceolla'. OE pers. name (with Scand. *k-*) + *-ing-* + *tūn*.

Kelloe Durham. *Kelflaw* c.1140. 'Hill where calves graze'. OE *celf* + *hlāw*.

Kells (*Cealla*) Antrim, Kerry, Kilkenny. 'Churches'.

Kells (*Ceanannas*) Meath. 'Great residence'.

Kelly Devon. *Chenleie* [*sic*] 1086 (DB), *Chelli* 1166. Probably Cornish *kelli* 'grove, small wood'.

Kelmarsh Northants. *Keilmerse* 1086 (DB). Probably 'marsh marked out by poles or posts'. OE *cegel* (with Scand. *k-*) + *mersc*.

Kelmscot Oxon. *Kelmescote* 1234. 'Cottage(s) of a man called Cēnhelm'. OE pers. name + *cot*.

Kelsale Suffolk. *Keleshala* 1086 (DB). 'Nook of land of a man called *Cēl(i) or Cēol'. OE pers. name + *halh*.

Kelsall Ches. *Kelsale* 1257. 'Nook of land of a man called Kell'. ME pers. name + *halh*.

Kelsey, North & Kelsey, South Lincs. *Chelsi, Nortchelesei* 1086 (DB), *Sudkeleseia* 1177. Possibly 'island, or dry ground in marsh, of a man called Cēol'. OE pers. name (with Scand. *k-*) + *ēg*.

Kelshall Herts. *Keleshelle* c.1050, *Cheleselle* 1086 (DB). 'Hill of a man called *Cylli'. OE pers. name + *hyll*.

Kelso Sc. Bord. *Calkou* 1126, *Calcehou* c.1128. 'Chalk hill-spur'. OE *calc* + *hōh*.

Kelston B. & NE. Som. *Calvestona* 1100–35. 'Farm where calves are reared, or farm of a man called *C(e)alf'. OE *c(e)alf* or OE pers. name (genitive *-es*) + *tūn*.

Kelty Fife. *Quilte* 1250, *Quilt* 1329. Possibly '(place by the) hard water'. Gaelic *Cailtidh*, from Celtic **caleto-* 'hard' + **dubro-* 'water'.

Kelvedon Essex. *Cynlauedyne* 998, *Chelleuedana* 1086 (DB). 'Valley of a man called Cynelāf'. OE pers. name + *denu*.

Kelvedon Hatch Essex. *Kylewendune* 1066, *Keluenduna* 1086 (DB), *Kelwedon Hacche* 1276. 'Speckled hill'. OE *cylu* (dative *cylwan*) + *dūn*, with the later addition of *hæcc* 'gate'.

Kelvinside Glas. '(Place) beside the River Kelvin'. The river name is said to derive from Gaelic *caol abhainn*, 'narrow water', but this is uncertain.

Kemberton Shrops. *Chenbritone* 1086 (DB). 'Farmstead of a man called Cēnbeorht'. OE pers. name + *tūn*.

Kemble Glos. *Kemele* 682, *Chemele* 1086 (DB). Probably a Celtic place name meaning 'neighbourhood, district'.

Kemerton Worcs. *Cyneburgingctun* 840, *Chenemertune* 1086 (DB). 'Farmstead associated with a woman called Cyneburg'. OE pers. name + *-ing-* + *tūn*.

Kempley Glos. *Chenepelei* 1086 (DB). Probably 'woodland clearing of a man called *Cenep(a)'. OE pers. name + *lēah*.

Kempsey Worcs. *Kemesei* 799, *Chemesege* 1086 (DB). 'Island, or dry ground in marsh, of a man called *Cymi'. OE pers. name + *ēg*.

Kempsford Glos. *Cynemæres forda* 9th cent., *Chenemeresforde* 1086 (DB). 'Ford of a man called Cynemǣr'. OE pers. name + *ford*.

Kempston Beds. *Kemestan* 1060, *Camestone* 1086 (DB). Possibly 'farmstead at the bend'. A derivative of Celtic **camm* 'crooked' + OE *tūn*.

Kempton Shrops. *Chenpitune* 1086 (DB). 'Farmstead of the warrior or of a man called *Cempa'. OE *cempa* or OE pers. name + *tūn*.

Kemsing Kent. *Cymesing* 822. 'Place of a man called *Cymesa'. OE pers. name + *-ing*.

Kenardington Kent. *Kynardingtune* 11th cent. 'Estate associated with a man called Cyneheard'. OE pers. name + *-ing-* + *tūn*.

Kenchester Herefs. *Chenecestre* 1086 (DB). 'Roman fort or town associated with a man called Cēna'. OE pers. name + *ceaster*.

Kencot Oxon. *Chenetone* [sic] 1086 (DB), *Chenicota* c.1130. 'Cottage(s) of a man called Cēna'. OE pers. name + *cot* (in the first form *tūn* 'farmstead').

Kendal Cumbria. *Cherchebi* 1086 (DB), *Kircabikendala* c.1095, *Kendale* 1452. 'Village with a church in the valley of the River Kent'. Originally OScand. *kirkju-bý* with the addition of a district name (Celtic river-name of uncertain meaning + OScand. *dalr*) which now alone survives.

Kenilworth Warwicks. *Chinewrde* [sic] 1086 (DB), *Chenildeworda* early 12th cent. 'Enclosure of a woman called Cynehild'. OE pers. name + *worth*.

Kenley, 'woodland clearing of a man called Cēna', OE pers. name + *lēah*: **Kenley** Gtr. London. *Kenele* 1255. **Kenley** Shrops. *Chenelie* 1086 (DB).

Kenmare (*Ceann Mara*) Kerry. 'Head of the sea'. The official Irish name of Kenmare is *Neidín*, 'little nest'.

Kenn, originally the name of the streams on which the places stand, a pre-English river-name of uncertain origin and meaning: **Kenn** Devon. *Chent* 1086 (DB). **Kenn** N. Som. *Chen(t)* 1086 (DB).

Kennerleigh Devon. *Kenewarlegh* 1219. 'Woodland clearing of a man called Cyneweard'. OE pers. name + *lēah*.

Kennet (district) Wilts. A modern adoption of the Celtic river-name Kennet, for which *see* KENNETT, EAST & WEST.

Kennett Cambs. *Chenet* 1086 (DB). Named from the River Kennett, a Celtic river-name of doubtful meaning.

Kennett, East & Kennett, West Wilts. *Cynetan* 939, *Chenete* 1086 (DB). Named from the River Kennet, a river-name identical in origin with that in the previous name.

Kennford Devon. *Keneford* 1300. 'Ford over the River Kenn'. Pre-English river-name (of uncertain origin and meaning) + OE *ford*.

Kenninghall Norfolk. *Keninchala* 1086 (DB). Probably 'nook of land of the family or followers of a man called Cēna'. OE pers. name + *-inga-* + *halh*.

Kennington Gtr. London. *Chenintune* 1086 (DB), *Kenintone* 1229. 'Farmstead associated with a man called Cēna'. OE pers. name + *-ing-* + *tūn*.

Kennington Kent. *Chenetone* 1086 (DB). 'Royal manor'. OE *cyne-* + *tūn*.

Kennington Oxon. *Chenitun* 821, 1086 (DB). 'Farmstead associated with a man called Cēna'. OE pers. name + *-ing-* + *tūn*.

Kennoway Fife. *Kennachin* 1183, *Kennachyn* 1250, *Kennoquhy c.*1510. Meaning uncertain.

Kennythorpe N. Yorks. *Cheretorp* [*sic*] 1086 (DB), *Kinnerthorp c.*1180. Probably 'outlying farmstead or hamlet of a man called Cēnhere'. OE pers. name + OScand. *thorp*.

Kensal Green Gtr. London. *Kingisholte* 1253, *Kynsale Green* 1550. 'The king's wood'. OE *cyning* + *holt*, with the later addition of ME *grene* 'village green'.

Kensington Gtr. London. *Chenesitun* 1086 (DB), *Kensintone* 1221. 'Estate associated with a man called Cynesige'. OE pers. name + *-ing-* + *tūn*.

Kensworth Beds. *Ceagnesworthe* 975, *Canesworde* 1086 (DB). 'Enclosure of a man called *Cǣgīn*. OE pers. name + *worth*.

Kent (the county). *Cantium* 51 BC. An ancient Celtic name, often explained as 'coastal district' but possibly 'land of the hosts or armies'.

Kent (river) Lancs., Cumbria., *See* KENDAL.

Kentchurch Herefs. *Lan Cein c.*1130, *Keynchirche* 1341. 'Church of St Ceina'. Female saint's name + OE *cirice* (with OWelsh **lann* in the early form).

Kentford Suffolk. *Cheneteforde* 11th cent. 'Ford over the River Kennett'. Celtic river-name (of uncertain origin and meaning) + OE *ford*.

Kentisbeare Devon. *Chentesbere* [*sic*] 1086 (DB), *Kentelesbere* 1242. 'Wood or grove of a man called **Centel*. OE pers. name + *bearu*.

Kentisbury Devon. *Chentesberie* [*sic*] 1086 (DB), *Kentelesberi* 1260. 'Stronghold of a man called **Centel*. OE pers. name + *burh* (dative *byrig*).

Kentish Town Gtr. London. *Kentisston* 1208. Probably 'estate held by a family called Kentish'. ME surname (meaning 'man from Kent') + OE *tūn*.

Kentmere Cumbria. *Kentemere* 13th cent. 'Pool by the River Kent'. Celtic river-name (of uncertain meaning) + OE *mere*.

Kenton Devon. *Chentone* 1086 (DB). 'Farmstead on the River Kenn'. Pre-English river-name (of uncertain origin and meaning) + OE *tūn*.

Kenton Gtr. London. *Keninton* 1232. 'Estate associated with a man called Cēna'. OE pers. name + *-ing-* + *tūn*.

Kenton Suffolk. *Chenetuna* 1086 (DB). Probably 'royal manor'. OE *cyne-* + *tūn*. Alternatively the first element may be the OE masculine pers. name *Cēna*.

Kenwick Shrops. *Kenewic* 1203. 'Dwelling or (dairy) farm of a man called Cēna'. OE pers. name + *wīc*.

Kenwyn Cornwall. *Keynwen* 1259. Probably 'white ridge'. OCornish *keyn* + *gwynn*.

Kenyon Warrtn. *Kenien* 1212. Possibly a shortened form of an OWelsh name *Cruc Einion* 'mound of a man called Einion'.

Kepwick N. Yorks. *Chipuic* 1086 (DB). 'Hamlet with a market'. OE *cēap* (with Scand. *k-*) + *wīc*.

Keresley Covtry. *Keresleia c.*1144. Possibly 'woodland clearing of a man called Cēnhere'. OE pers. name + *lēah*.

Kerne Bridge Herefs. *Kernebrigges* 1272. Probably 'bridge(s) by a mill'. OE *cweorn* + *brycg*.

Kerrier (district) Cornwall. *Kerier* 1201. A revival of an old hundred-name, perhaps meaning 'place of rounds', referring to a type of hill-fort. Cornish *ker* 'a round, a fort' with a plural suffix.

Kerry (*Ciarraí*) (the county). '(Place of) Ciar's people'.

Kerrykeel (*Ceathrú Chaol*) Donegal. 'Narrow quarter'.

Kersall Notts. *Cherueshale* 1086 (DB), *Kyrneshale* 1196. Possibly 'nook of land of a man called Cynehere'. OE pers. name + *halh*.

Kersey Suffolk. *Cæresige c.*995, *Careseia* 1086 (DB). 'Island (of higher ground) where cress grows'. OE *cærse* + *ēg*.

Kersoe Worcs. *Criddesho* 780. 'Hill-spur of a man called *Criddi'. OE pers. name + *hōh*.

Kesgrave Suffolk. *Gressegraua* [*sic*] 1086 (DB), *Kersigrave* 1231. OE *grǣf(a)* 'grove, coppiced wood' or *grǣf* 'pit, trench', possibly with OE *cærse* 'cress'.

Kesh (*Ceis*) Fermanagh, Sligo. 'Wicker causeway'.

Keshcarrigan (*Ceis Charraigín*) Leitrim. 'Wicker causeway of the little rock'.

Kessingland Suffolk. *Kessingalanda* 1086 (DB). 'Cultivated land of the family or followers of a man called *Cyssi'. OE pers. name + *-inga-* + *land*.

Kesteven Lincs. *Ceostefne* c.1000, *Chetsteven* 1086 (DB). OScand. *stefna* 'meeting-place', here probably 'administrative district', added to an ancient district name from Celtic *cēd* 'wood'.

Keston Gtr. London. *Cysse stan* 973, *Chestan* 1086 (DB). 'Boundary stone of a man called *Cyssi'. OE pers. name + *stān*.

Keswick, 'farm where cheese is made', OE *cēse* (with Scand. *k-*) + *wīc*: **Keswick** Cumbria. *Kesewic* c.1240. **Keswick** Norfolk, near Bacton. *Casewic* c.1150. **Keswick** Norfolk, near Norwich. *Chesewic* 1086 (DB). **Keswick, East** Leeds. *Chesuic* 1086 (DB).

Kettering Northants. *Cytringan* 956, *Cateringe* 1086 (DB). Probably '(settlement of) the family or followers of a man called *Cytra'. OE pers. name + *-ingas*.

Ketteringham Norfolk. *Keteringham* c.1060, *Keterincham* 1086 (DB). 'Homestead of the family or followers of a man called *Cytra'. OE pers. name + *-inga-* + *hām*.

Kettlebaston Suffolk. *Kitelbeornastuna* 1086 (DB). 'Farmstead of a man called Ketilbjǫrn'. OScand. pers. name + OE *tūn*.

Kettleburgh Suffolk. *Chetelberia*, *Ketelbiria* 1086 (DB). Probably 'hill by the deep valley'. OE *cetel* (with Scand. *k-*) + OE *beorg*.

Kettleby, Ab Leics. *Chetelbi* 1086 (DB), *Abeketleby* 1236. 'Farmstead or village of a man called Ketil'. OScand. pers. name + *bý*, with manorial affix from the name of an early owner of the estate, either the OE pers. name *Abba* or the OScand. pers. name *Abbi* (both masculine).

Kettleshulme Ches. *Ketelisholm* 1285. 'Island or water-meadow of a man called Ketil'. OScand. pers. name + *holmr*.

Kettlesing N. Yorks. *Ketylsyng* 1379. 'Meadow of a man called Ketil'. OScand. pers. name + *eng*.

Kettlestone Norfolk. *Ketlestuna* 1086 (DB). 'Farmstead of a man called Ketil'. OScand. pers. name + OE *tūn*.

Kettlethorpe Lincs. *Ketelstorp* 1220. 'Outlying farmstead or hamlet of a man called Ketil'. OScand. pers. name + *thorp*.

Kettlewell N. Yorks. *Cheteleuuelle* 1086 (DB). 'Spring or stream in a deep valley'. OE *cetel* (with Scand. *k-*) + *wella*.

Ketton Rutland. *Chetene* 1086 (DB). An old river-name, possibly a derivative of Celtic *cēd* 'wood' + OE *ēa* 'river'.

Kew Gtr. London. *Cayho* 1327. Probably 'key-shaped spur of land', OE *cǣg* + *hōh*. Alternatively the first element may be ME *key* 'quay or landing place'.

Kewstoke N. Som. *Chiwestoch* 1086 (DB). OE *stoc* 'secondary settlement' with the addition of the Celtic saint's name *Kew*.

Kexbrough Barns. *Cezeburg* [*sic*] 1086 (DB), *Kesceburg* c.1170. Probably 'stronghold of a man called Keptr'. OScand. pers. name + OE *burh*.

Kexby Lincs. *Cheftesbi* 1086 (DB). Possibly 'farmstead or village of a man called Keptr'. OScand. pers. name + *bý*.

Kexby York. *Kexebi* 12th cent. Probably 'farmstead or village of a man called Keikr'. OScand. pers. name + *bý*.

Key, Lough (*Loch Cé*) Roscomon. 'Lake of the quay'.

Keyham Leics. *Caiham* 1086 (DB). Possibly 'homestead or village of a man called *Cǣga', OE pers. name + *hām*. Alternatively the first element may be OE *cǣg* 'key' (used of a 'key-shaped ridge') or *cǣg* 'stone, boulder'.

Keyhaven Hants. *Kihavene* c.1170. 'Harbour where cows are shipped'. OE *cū* (genitive *cȳ*) + *hæfen*.

Keyingham E. R. Yorks. *Caingeham* 1086 (DB). 'Homestead of the family or followers of a man called *Cǣga'. OE pers. name + *-inga-* + *hām*.

Keymer W. Sussex. *Chemere* 1086 (DB). 'Cow's pond'. OE *cū* (genitive *cȳ*) + *mere*.

Keynsham B. & NE. Som. *Cægineshamme* c.1000, *Cainesham* 1086 (DB). Possibly 'land in a river-bend belonging to a man called *Cægin'. OE pers. name + *hamm*.

Keysoe Beds. *Caissot* 1086 (DB). Probably 'key-shaped spur of land'. OE *cæg* + *hōh*. Alternatively the first element may be OE *cæg 'a stone, a boulder'.

Keyston Cambs. *Chetelestan* 1086 (DB). 'Boundary stone of a man called Ketil'. OScand. pers. name + OE *stān*.

Keyworth Notts. *Caworde* 1086 (DB), *Kewurda* 1154–89. OE *worth* 'enclosure, enclosed settlement' with an uncertain first element, possibly OE *cæg* 'key' (used of a place that could be locked) or *cæg 'stone, boulder'.

Kibblesworth Gatesd. *Kibleswrthe* 1185. 'Enclosure of a man called *Cybbel'. OE pers. name + *worth*.

Kibworth Beauchamp & Kibworth Harcourt Leics. *Chiburde* 1086 (DB), *Kybeworth Beauchamp* 1315, *Kibbeworth Harecourt* 13th cent. 'Enclosure of a man called *Cybba'. OE pers. name + *worth*. Affixes from early possession by the *de Beauchamp* and *de Harewecurt* families.

Kidbrooke Gtr. London. *Chitebroc* c.1100, *Ketebroc* 1202. 'Brook where kites are seen'. OE *cȳta* + *brōc*.

Kidderminster Worcs. *Chideminstre* [*sic*] 1086 (DB), *Kedeleministre* 1154. 'Monastery of a man called *Cydela'. OE pers. name + *mynster*.

Kiddington Oxon. *Chidintone* 1086 (DB). 'Estate associated with a man called Cydda'. OE pers. name + *-ing-* + *tūn*.

Kidlington Oxon. *Chedelintone* 1086 (DB). 'Estate associated with a man called *Cydela'. OE pers. name + *-ing-* + *tūn*.

Kidsgrove Staffs. *Kydcrowe* 1596. Probably 'pen for young goats'. ModE *kid* 'young goat' + ModE dialect *crow, crew* (from Welsh *crau*) 'hut, pen'.

Kidwelly (*Cydweli*) Carm. *Cetgueli* 10th cent., *Cedgueli* c.1150, *Kedwely* 1191, *Kydwelly* 1458. 'Cadwal's territory'. The identity of *Cadwal* is unknown.

Kielder Northum. *Keilder* 1326. Named from the river here, a Celtic river-name meaning 'rapid stream'.

Kielduff (*An Chill Dubh*) Kerry. 'The black church'.

Kiilinaspick (*Cill an Easpaig*) Kilkenny. 'Church of the bishop'.

Kilanerin (*Coill an Iarainn*) Wexford. 'Wood of the iron'.

Kilavullen (*Cill an Mhuilinn*) Cork. 'Church of the mill'.

Kilbaha (*Cill Bheathach*) Clare. 'Church of the birches'.

Kilballyporter (*Coill an Bhealaigh*) Meath. 'Porter's wood of the pass'.

Kilbane (*An Choill Bhán*) Clare. 'The white wood'.

Kilbarrack (*Cill Bharróg*) Dublin. 'Church of little Barra'.

Kilbarron (*Cill Bharfionn*) Donegal. 'Barfionn's church'.

Kilbarry (*Cill Barra*) Cork. 'Barra's church'.

Kilbeg (*Coill Bheag*) Wicklow. *Kilbege* 1526. 'Little wood'.

Kilbeggan (*Cill Bheagáin*) Westmeath. 'Beagán's church'.

Kilbeheny (*Coill Bheithne*) Limerick. 'Wood of birches'.

Kilbennan (*Cill Bheanáin*) Galway. 'Beanán's church'.

Kilberry (*Cill Bhearaigh*) Kildare, Meath. 'Church of Bearach'.

Kilbirnie N. Ayr. *Kilbyrny* 1413. 'Church of St Brendan'. Gaelic *cill*.

Kilbrickan (*Cill Bhreacáin*) Galway. 'Church of Breacán'.

Kilbricken (*Cill Bhriocáin*) Laois. 'Church of Briocán'.

Kilbride (*Cill Bhríde*) Carlow, Down, Wicklow. 'Church of Bríd'.

Kilbride Arg. 'Church of St Bridget'. Gaelic *cill*.

Kilbride, East & Kilbride, West. *See* EAST KILBRIDE and WEST KILBRIDE.

Kilbrien (*Cill Bhriain*) Waterford. 'Church of Brian'.

Kilbrin (*Cill Bhrain*) Cork. 'Church of Bran'.

Kilbrittain (*Cill Briotáin*) Cork. 'Church of Briotán'.

Kilbrogan (*Cill Brógáin*) Cork. 'Brogan's church'.

Kilburn Derbys. *Kileburn* 1179. Possibly 'stream of a man called *Cylla*'. OE pers. name + *burna*.

Kilburn Gtr. London. *Cuneburna c.*1130, *Keleburne* 1181. Possibly identical in origin with previous name, but some early spellings show characteristic interchange of *n* and *l*, or suggest 'stream of the cows', from OE *cū* (genitive plural *cūna, cȳna*) + *burna*.

Kilburn N. Yorks. *Chileburne* 1086 (DB). Perhaps identical in origin with KILBURN (Derbys).

Kilby Leics. *Cilebi* 1086 (DB). Probably 'farmstead or village of the young (noble)men'. OE *cild* (with Scand. *k-*) + OScand. *bý*.

Kilcaimin (*Cill Chaimín*) Galway. 'Church of Caimín'.

Kilcar (*Cill Charthaigh*) Donegal. *Cill Carthaigh c.*1630. 'Carthach's church'.

Kilcarn (*Cill an Chairn*) Meath. 'Church of the cairn'.

Kilcash (*Cill Chais*) Tipperary. 'Church of Cas'.

Kilcavan (*Cill Chaomháin*) Laois. 'Church of Caomhán'.

Kilchattan Arg. (Bute). *Killecatan* 1449. 'Church of St Catan'. Gaelic *cill*.

Kilchenzie Arg. 'Church of St Kenneth'. Gaelic *cill*.

Kilchreest (*Cill Chríost*) Galway. 'Church of Christ'.

Kilclare (*Coill Chláir*) Leitrim. 'Wood of the plain'.

Kilclief (*Cill Chléithe*) Down. (*i*)*Cill Chlethi c.*1125. 'Church of wattle'.

Kilcloher (*Cill Chloichir*) Clare. 'Church of the stony place'.

Kilclone (*Cill Chluana*) Meath. 'Church of the meadow'.

Kilclonfert (*Cill Chluana Fearta*) Offaly. 'Church of the meadow of the grave'.

Kilclooney (*Cill Chluana*) Donegal. 'Church of the meadow'.

Kilcock (*Cill Choca*) Kildare. 'Church of Coca'.

Kilcogy (*Cill Chóige*) Cavan. *Kilcoaga* 1611. 'Church of the fifth'. The reference is to a land division.

Kilcolgan (*Cill Cholgáin*) Galway. 'Church of Colgán'.

Kilcolman (*Cill Cholmáin*) Cork, Limerick. 'Church of Colmán'.

Kilcomin (*Cill Chuimín*) Offaly. 'Church of Cuimín'.

Kilcommon (*Cill Chuimín*) Tipperary. 'Church of Cuimín'.

Kilconly (*Cil Chonla*) Galway. 'Conla's church'.

Kilconnell (*Cill Chonaill*) Galway. 'Church of Conall'.

Kilcoo (*Cill Chua*) Down. *Killcow* 1609. 'Cua's church'.

Kilcoole (*Cill Chomhghaill*) Wicklow. 'Church of Comhghall'.

Kilcooley (*Cill Chúile*) Tipperary. 'Church of the nook'.

Kilcormac (*Cill Chormaic*) Offaly. 'Church of Cormac'.

Kilcornan (*Cill Chornáin*) Limerick. 'Church of Cornán'.

Kilcorney (*Cill Coirne*) Cork. 'Church of Corne'.

Kilcot Glos. *Chilecot* 1086 (DB). 'Cottage(s) of a man called *Cylla*'. OE pers. name + *cot*.

Kilcotty (*Cill Chota*) Wexford. 'Church of Cota'.

Kilcrohane (*Cill Crócháin*) Cork. 'Church of Cróchán'.

Kilcullen (*Cill Chuillinn*) Kildare. 'Church of the steep slope'.

Kilcummin (*Cill Chuimín*) Kerry. 'Church of Cuimín'.

Kilcurry (*Cill an Churraigh*) Louth. 'Church of the marsh'.

Kildale N. Yorks. *Childale* 1086 (DB). Probably 'narrow valley'. OScand. *kíll* + *dalr*.

Kildalkey (*Cill Dealga*) Meath. 'Church of the thorn'.

Kildangan (*Cill Daingin*) Kildare. 'Church of the fortress'.

Kildangan (*Coill an Daingin*) Offaly. 'Wood of the fortress'.

Kildare (*Cill Dara*) Kildare. 'Church of the oak-tree'.

Kildavin (*Cill Damháin*) Carlow. 'Church of Damhán'.

Kildermody (*Cill Dhiarmada*) Waterford. 'Church of Diatmaid'.

Kildimo (*Cill Díoma*) Limerick. 'Díoma's church'.

Kildonan Highland. *Kelduninach c.*1230. 'Church of St Donan'. Gaelic *cill*.

Kildorough (*Coill Dorcha*) Cavan. 'Dusky wood'.

Kildorrery (*Cill Dairbhre*) Cork. 'Church of the place of oaks'.

Kildreenagh (*Cill Draigneach*) Carlow. 'Church of blackthorns'.

Kildrum (*Coill Droma*) Donegal. 'Wood of the ridge'.

Kildwick W. Yorks. *Childeuuic* 1086 (DB). 'Dairy-farm of the young men or attendants'. OE *cild* (with Scand. *k-*) + *wīc*.

Kildysart (*Cill an Dísirt*) Clare. 'Church of the hermitage'.

Kilfeaghan (*Cill Fhéichín*) Down. 'Féichín's church'.

Kilfeakle (*Cill Fiacal*) Tipperary. 'Church of the teeth'.

Kilfearagh (*Cill Fhiachrach*) Clare. 'Church of Fiachra'.

Kilfenora (*Cill Fionnúrach*) Clare. 'Church of Fionnúir'.

Kilfinane (*Cill Fhionáin*) Limerick. 'Church of Fionán'.

Kilfinny (*Cill na Fiodhnaí*) Limerick. 'Church of the wooded place'.

Kilflynn (*Cill Flainn*) Kerry. 'Church of Flan'.

Kilfree (*Cill Fraoigh*) Sligo. 'Church of the heather'.

Kilfullert (*Coill Fulachta*) Down. *Ballykillyfollyat* 1609. 'Wood of the cooking place'.

Kilgarvan (*Cill Gharbháin*) Kerry. 'Church of Garbhán'.

Kilgetty (*Cilgeti*) Pemb. *Kylketty* 1330. 'Ceti's corner of land'. Welsh *cil*. The identity of Ceti is unknown.

Kilglass (*Cil Ghlas*) Roscommon, Sligo. 'Green church'.

Kilglass (*Coill Ghlas*) Galway, Sligo. 'Green wood'.

Kilgobnet (*Cill Ghobnait*) Kerry, Waterford. 'Church of Gobnait'.

Kilgowan (*Coill Ghabann*) Kildare. 'Wood of the smith'.

Kilham, '(place at) the kilns', OE *cyln* in a dative plural form *cylnum*: **Kilham** E. R. Yorks. *Chillun* [*sic*] 1086 (DB). **Kilham** Northum. *Killum* 1177.

Kilkea (*Cill Chathaig*) Kildare. 'Church of Cathach'.

Kilkeasy (*Cill Chéise*) Kildare. 'Church of Céis'.

Kilkee (*Cill Chaoi*) Clare. 'Church of Caoi'.

Kilkeel (*Cill Chaoil*) Down. *Kylkeyl* 1369. 'Church of the narrow place'.

Kilkelly (*Cill Cheallaigh*) Mayo. 'Church of Ceallach'.

Kilkenneth Arg. 'Church of St Kenneth'. Gaelic *cill*.

Kilkenny (*Cill Chainnigh*) Kilkenny. 'Church of Cainneach'.

Kilkerin (*Cill Chéirín*) Clare. 'Church of Céirín'.

Kilkerrin (*Cill Chiaráin*) Galway. 'Ciarán's church'.

Kilkerrin (*Cill Choirín*) Galway. 'Church of the little turn'.

Kilkhampton Cornwall. *Chilchetone* 1086 (DB), *Kilkamton* 1194. Primitive Cornish **kelk* 'circle, ring' (here probably 'border, boundary') + OE *tūn* (alternating with *hām-tūn*) 'farmstead, estate'.

Kilkieran (*Cill Chiaráin*) Galway, Kilkenny. 'Church of Ciarán'.

Kilkinlea (*Cill Chinn Sléibhe*) Limerick. 'Church of the head of the mountain'.

Kilkishen (*Cill Chisín*) Clare. 'Church of the wicker causeway'.

Kill (*Cill*) Cavan, Kildare, Waterford. 'Church'.

Killachonna (*Cill Dachonna*) Westmeath. 'Church of Dachonna'.

Killacolla (*Cill an Coille*) Limerick. 'Church of the wood'.

Killadangan (*Coill an Daingin*) Mayo. 'Wood of the fortress'.

Killadeas (*Cill Chéile Dé*) Fermanagh. 'Church of the Céle Dé'. A *Céle Dé* (Culdee) ('servant of God') was a member of a group of monastic reformers originating in the 8th cent.

Killadoon (*Coill an Dúin*) Mayo. 'Wood of the fort'.

Killadreenan (*Cill Achaidh Draighnigh*) Wicklow. 'Church of the field of the blackthorns'.

Killadysert (*Cill an Dísirt*) Clare. 'Church of the hermitage'.

Killag (*Cill Laig*) Wexford. 'Church of the hollow'.

Killagan (*Cill Lagáin*) Antrim. *(ecclesia de) Killagan* 1622. 'Church of the small hollow'.

Killala (*Cill Ala*) Mayo. 'Church of Ala'.

Killaloe (*Cill Dalua*) Clare. 'Church of Dalua'.

Killaloo (*Coill an Lao*) Derry. *Kille Loy* 1613. 'Wood of the calf'.

Killamarsh Derbys. *Chinewoldemaresc* 1086 (DB). 'Marsh of a man called Cynewald'. OE pers. name + *mersc*.

Killamery (*Cill Lamhraí*) Kilkenny. 'Church of Lamhrach'.

Killane (*Cill Anna*) Offaly. 'Church of Anna'.

Killann (*Cill Ana*) Wexford. 'Church of Anna'.

Killannummery (*Cill an Iomaire*) Leitrim. 'Church of the ridge'.

Killans (*Coillíní*) Antrim. 'Little hills'.

Killaraght (*Cill Adrochtae*) Sligo. 'Adrochta's church'.

Killard Point (*Aird na Cille*) Down. 'Point of the church'.

Killare (*Cill Air*) Westmeath. 'Church of slaughter'.

Killarga (*Cill Fhearga*) Leitrim. 'Church of Fearga'.

Killarney (*Cill Airne*) Kerry. 'Church of the sloes'.

Killary (*Caolaire*) Galway, Mayo. 'Narrow sea'.

Killashandra (*Cill na Seanrátha*) Cavan. *Kylshanra* 1432. 'Church of the old fort'.

Killashee (*Cill Úsaille*) Longford. 'Church of Úsaile'. The official form of the Irish name is *Cill na Sí*, 'church of the fairy hill'.

Killasser (*Cill Lasrach*) Mayo. 'Lasrach's church'.

Killateeaun (*Coill an tSiáin*) Mayo. 'Wood of the fairy hill'.

Killavally (*Coill an Bhaile*) Mayo. 'Wood of the homestead'.

Killavally (*Coill an Bhealaigh*) Westmeath. 'Wood of the pass'.

Killavil (*Cill Fhábhail*) Sligo. 'Church of the journey'.

Killea (*Cill Aodha*) Tipperary, Waterford. 'Church of Aodh'.

Killeagh (*Cill Ia*) Cork. 'Church of Ia'.

Killeany (*Cill Éinne*) Galway. 'Church of Éanna'.

Killeavan (*Cill Laobháin*) Monaghan. 'Church of Laobhán'.

Killeavy (*Cill Shléibhe*) Armagh. *Cille sleibe Cuilinn* 517. 'Church of the mountain'. The church stood at the foot of Slieve Gullion.

Killedmond (*Cill Éamainn*) Carlow. 'Church of Éamonn'.

Killeedy (*Cill Íde*) Limerick. 'Church of Íde'.

Killeen (*Cillín*) Tipperary. 'Little church'.

Killeenadeema (*Cillín a Díoma*) Galway. 'Little church of Díoma'.

Killeenduff (*Cillín Dubh*) Offaly. 'Little black church'.

Killeeshill (*Cill Íseal*) Tyrone. *Killissyll* 1455. 'Low church'.

Killegar (*Coill an Ghairr*) Limerick. 'Wood of the offal'.

Killeigh (*Cill Aichidh*) Offaly. 'Church of the field'.

Killen (*Cillín*) Tyrone. *Killen c.*1655. 'Little church'.

Killenagh (*Cill Éanach*) Waterford. 'Church of the marsh'.

Killenaule (*Cill Náile*) Tipperary. 'Church of Náile'.

Killerby Darltn. *Culuerdebi* 1091. Probably 'farmstead or village of a man called *Ketilfrithr'. OScand. pers. name + *bý*.

Killerig (*Cill Dheirge*) Carlow. 'Church of the red ground'.

Killeshil (*Cill Íseal*) Offaly. 'Low church'.

Killeter (*Coill Íochtair*) Tyrone. *Keliter* 1613. 'Lower wood'.

Killiecrankie Perth. Perhaps 'wood of aspens'. Gaelic *coille* + *critheann*.

Killimer (*Cill Íomar*) Clare. 'Church of Íomar'.

Killimor (*Cill Íomair*) Galway. 'Church of Íomar'.

Killinaboy (*Cill Iníne Baoith*) Clare. 'Church of the daughter of Baoth'.

Killinardrish (*Cill an Ard-doras*) Cork. 'Church of the high door'.

Killinchy (*Cill Dhuinsí*) Down. (*o*) *Cill Dunsighe c.*1400. 'Church of Duinseach'. Duinseach was a female saint.

Killincooly (*Cillín Cúile*) Wexford. 'Little church of the nook'.

Killiney (*Cill Éinne*) Kerry. 'Church of Éanna'.

Killiney (*Cill Iníon Léinín*) Dublin. 'Church of the daughters of Léinín'.

Killinghall N. Yorks. *Chilingale* 1086 (DB). 'Nook of land of the family or followers of a man called *Cylla'. OE pers. name + *-inga-* + *halh*.

Killingholme Lincs. *Chelvingeholm* 1086 (DB). Probably 'homestead of the family or followers of a man called Cēolwulf'. OE pers. name + *-inga-* + *hām* (replaced by OScand. *holmr* 'island, dry ground in marsh').

Killington Cumbria. *Killintona* 1175. 'Estate associated with a man called *Cylla'. OE pers. name + *-ing-* + *tūn*.

Killingworth N. Tyne. *Killingwrth* 1242. 'Enclosure associated with a man called *Cylla'. OE pers. name + *-ing-* + *worth*.

Killinick (*Cill Fhionnóg*) Wexford. 'Church of Fionnóg'.

Killinierin (*Coill an Iarainn*) Wexford. 'Wood of the iron'.

Killinkere (*Cillín Chéir*) Cavan. 'Little church of the wax candles'.

Killisky (*Cill Uisce*) Wicklow. 'Church of the water'.

Killoe (*Cill Eo*) Longford. 'Church of the yew'.

Killogeenaghan (*Coill Ó gCuinneagáin*) Westmeath. 'Wood of Uí Cuinneagáin'.

Killone (*Cill Eoghain*) Clare. 'Eoghan's church'.

Killoneen (*Cill Eoghanín*) Offaly. 'Little Eoghan's church'.

Killoran (*Cill Odhráin*) Galway. 'Oran's church'.

Killorglin (*Cill Orglan*) Kerry. 'Church of Forgla'.

Killoscully (*Cill Ó Scolaí*) Tipperary. 'Ó Scolaí's church'.

Killough (*Cill Locha*) Down. *Killogh* 1710. 'Church of the lake'.

Killour (*Cill Iúir*) Mayo. 'Church of the tower'.

Killowen (*Cill Eoin*) Down. *Kelcone* 1595. 'Eoin's church'.

Killowny (*Cill Uaithne*) Offaly. 'Church of Uaithne'.

Killucan (*Cill Liúcainne*) Westmeath. 'Liúcian's church'.

Killurin (*Cill Liúráin*) Wexford. 'Church of Liúrán'.

Killusty (*Cill Loiscthe*) Tipperary. 'Burnt church'.

Killybegs (*Na Cealla Beaga*) Donegal. (*cuan*) *na cCeall mBicc* 1513. 'The little churches'. The reference is to an early monastic settlement.

Killybrone (*Coillidh Brón*) Monaghan. 'Sad woodland'.

Killycolpy (*Coill an Cholpa*) Tyrone. *Kilcolpy* 1639. 'Wood of the heifer'.

Killycomain (*Coill Mhic Giolla Mhíchíl*) Armagh. *Killykillvehall* 1657. 'Mac Giolla Mhíchíl's wood'.

Killygarn (*Coill na gCarn*) Antrim. *Ballichilnegarne* 1606. 'Wood of the cairn'.

Killygarvan (*Cill Gharbháin*) Clare. 'Church of Garbhán'.

Killyglen (*Cill Ghlinne*) Antrim. 'Church of the glen'.

Killykergan (*Coill Uí Chiaragáin*) Derry. *Killakerrigan* 1654. 'Ó Ciaragáin's wood'.

Killylea (*Coillidh Léith*) Armagh. *Killeleagh* 1657. 'Grey woodland'.

Killyleagh (*Cill Ó Laoch*) Down. *Cill Ó Laoch* c.1645. 'Church of Uí Laoch'.

Killyman (*Cill na mBan*) Tyrone. *Cill na mBan* c.1645. 'Church of the women'.

Killyon (*Cil Liain*) Offaly. 'Church of Lian'.

Kilmacanoge (*Cill Mocheanóg*) Wicklow. 'Mocheanóg's church'.

Kilmacduagh (*Cill Mhic Duach*) Clare. 'Mac Duach's church'.

Kilmacolm Invclyd. *Kilmacolme c.*1205, *Kilmalcolm* [*sic*] 1868. 'Church of my Colm'. Gaelic *cill* + *mo. Colm* is St Columba, 'my' showing a personal dedication.

Kilmacow (*Cill Mhic Bhúith*) Kilkenny. 'Church of Mac Búith'.

Kilmacrenan (*Cill Mhic Réanáin*) Donegal. (*o*) *cCill mic Némain* 1129. 'Church of Mac Réanán'. The present name evolved from original *Cill Mhac nÉanáin*, 'church of Mac Éanán'.

Kilmacthomas (*Coill Mhic Thomáisín*) Waterford. 'Wood of Mac Thomáisín'.

Kilmactranny (*Cill Mhic Treana*) Sligo. 'Church of Mac Treana'.

Kilmaine (*Cill Mheáin*) Mayo. 'Middle church'.

Kilmainham (*Cill Mhaighneann*) Dublin, Meath. 'Church of Maighne'.

Kilmaley (*Cill Mháille*) Clare. 'Máille's church'.

Kilmalin (*Cill Moling*) Wicklow. 'Church of Moling'.

Kilmallock (*Cill Mocheallóg*) Limerick. 'Church of Mocheallóg'.

Kilmanagh (*Cill Mhanach*) Kilkenny. 'Church of the monks'.

Kilmanahan (*Cill Mainchín*) Waterford. 'Manchan's church'.

Kilmarnock E. Ayr. *Kelmernoke* 1299. 'Church of my little Ernan'. Gaelic *cill* + *mo. Ernan* is St Ernan, here with the diminutive suffix -*oc* 'my' showing a personal dedication.

Kilmead (*Cill Míde*) Kildare. 'Middle church'.

Kilmeaden (*Cill Mhiodáin*) Waterford. 'Church of Miodán'.

Kilmeague (*Cill Maodhóg*) Kildare. 'Church of Maodhóg'.

Kilmeedy (*Cill Míde*) Limerick. 'Church of Íde'.

Kilmersdon Somerset. *Kunemersdon* 951, *Chenemeresdone* 1086 (DB). 'Hill of a man called Cynemǣr'. OE pers. name + *dūn*.

Kilmessan (*Cill Mheasáin*) Meath. 'Church of Measán'.

Kilmeston Hants. *Cenelemestune* 961, *Chelmestune* 1086 (DB). 'Farmstead of a man called Cēnhelm'. OE pers. name + *tūn*.

Kilmichael (*Cill Mhichíl*) Cork, Wexford. 'Church of Micheál'.

Kilmihil (*Cill Mhichíl*) Clare. 'Church of Micheál'.

Kilmington, 'estate associated with a man called Cynehelm', OE pers. name + -*ing*- + *tūn*:
Kilmington Devon. *Chenemetone* 1086 (DB).
Kilmington Wilts. *Cilemetone* 1086 (DB).

Kilmoganny (*Cill Mogeanna*) Kilkenny. 'Mogeanna's church'.

Kilmore (*Cill Mhór*) Clare, Roscommon, Wexford. 'Big church'.

Kilmoremoy (*Cellola Maghna Muaide*) Mayo. 'Big church of the River Moy'.

Kilmorony (*Cill Maolrunaidh*) Laois. 'Church of Maolrunaidh'.

Kilmovee (*Cill Mobhí*) Mayo. 'Church of Mobhí'.

Kilmoyle (*Cill Maol*) Antrim. 'Dilapidated church'.

Kilmoyly (*Cill Mhaoile*) Kerry. 'Church of the hornless cow'.

Kilmurry (*Cill Mhuire*) Clare, Cork. 'Mary's church'.

Kilmurvy (*Cill Mhuirbhigh*) Galway. 'Church of the beach'.

Kilmyshall (*Cill Mísil*) Wexford. 'Church of the low central place'.

Kilnaboy (*Cill Iníne Baoith*) Clare. 'Church of the daughter of Baoth'.

Kilnacloghy (*Coill na Cloiche*) Roscommon. 'Wood of the stone'.

Kilnagross (*Coill na gCros*) Leitrim. 'Wood of the crosses'.

Kilnalag (*Coill na Lag*) Galway. 'Wood of the hollows'.

Kilnaleck (*Cill na Leice*) Cavan. *Killnalecky* 1609. 'Church of the flagstone'.

Kilnamanagh (*Coill na Manach*) Tipperary. 'Wood of the monks'.

Kilnamona (*Cill na Móna*) Clare. 'Church of the bog'.

Kilnhurst Rothm. *Kilnhirst* 12th cent. 'Wooded hill with a kiln'. OE *cyln* + *hyrst*.

Kilnsea E. R. Yorks. *Chilnesse* 1086 (DB). 'Pool near the kiln'. OE *cyln* + *sǣ*.

Kilnsey N. Yorks. *Chileseie* [sic] 1086 (DB), *Kilnesey* 1162. Probably 'marsh near the kiln'. OE *cyln* + **sǣge*.

Kilnwick E. R. Yorks. *Chileuuit* [sic] 1086 (DB), *Killingwic* late 12th cent. 'Farm associated with a man called *Cylla'. OE pers. name + -*ing*- + *wīc*. Alternatively the first element may be OE *cyln* 'a kiln'.

Kilnwick Percy E. R. Yorks. *Chelingewic* 1086 (DB), *Killingwik Perci* 1303. 'Farm of the family or followers of a man called *Cylla'. OE pers. name + -*inga*- + *wīc*, with manorial affix from the *de Percy* family, here from the 12th cent.

Kilpeck Herefs. *Chipeete* [sic] 1086 (DB), *Cilpedec c.*1150. 'Corner or nook where snares for animals are placed'. OWelsh *cil* + **pedec*.

Kilpedder (*Cill Pheadair*) Wicklow. 'Peadar's church'.

Kilpin E. R. Yorks. *Celpene* 959, *Chelpin* 1086 (DB). Probably 'enclosure for calves'. OE *celf* (with Scand. *k-*) + *penn*.

Kilquiggan (*Cill Chomhgáin*) Wicklow. 'Comhgán's church'.

Kilraine (*Cill Riáin*) Donegal. 'Rián's church'.

Kilranelagh (*Cill Rannaileach*) Wicklow. 'Church of the verses'.

Kilraughts (*Cill Reachtais*) Antrim. (*ecclesia de*) *Kellrethi* 1306. 'Church of the legislation'.

Kilrea (*Cill Ria*) Derry. (*Coeman*) *Chilli Riada c.*900. Perhaps 'church of the journey'.

Kilreekill (*Cill Rícill*) Galway. 'Rícill's church'.

Kilronan (*Cill Rónáin*) Galway. 'Rónán's church'.

Kilrooskey (*Coill na Rúscaí*) Roscommon. 'Wood of the bark strips'.

Kilroot (*Cill Ruaidh*) Antrim. *Cilli Ruaid c.*830. 'Church of the red land'.

Kilross (*Cill Ros*) Donegal. 'Church of the wood'.

Kilross (*Cill Ros*) Tipperary. 'Church of the promontory'.

Kilrush (*Cill Rois*) Clare. 'Church of the grove'.

Kilsallagh (*Coill Salach*) Galway, Mayo. 'Willow wood'.

Kilsallaghan (*Cill Shalcháin*) Dublin. 'Church of the willows'.

Kilsaran (*Cill Saráin*) Louth. 'Sarán's church'.

Kilsby Northants. *Kildesbig* 1043, *Chidesbi* 1086 (DB). 'Farmstead or village of the young (noble)man, or of a man called Cild'. OE *cild* or pers. name (with Scand. *k-*) + OScand. *bý*.

Kilshanchoe (*Cill Seanchua*) Kildare. 'Church of the old hollow'.

Kilshanny (*Cill Seanaigh*) Clare. 'Seanach's church'.

Kilsheelan (*Cill Síoláin*) Tipperary. 'Síolán's church'.

Kilskeer (*Cill Scíre*) Meath. 'Scíre's church'.

Kilskeery (*Cill Scíre*) Tyrone. *(orcain) Cille Scire* 951. 'Scíre's church'.

Kilsyth N. Lan. *Kelvesyth* 1210, *Kelnasydhe* 1217, *Kilsyth* 1239. Meaning uncertain.

Kiltamagh (*Coillte Mach*) Mayo. 'Woods of the plain'.

Kiltealy (*Cill Tíle*) Wexford. 'Tíl's church'.

Kilteel (*Cill tSíle*) Kildare. 'Síle's church'.

Kilteely (*Cill Tíle*) Limerick. 'Tíl's church'.

Kiltegan (*Cill Téagáin*) Wicklow. 'Téagán's church'.

Kilternan (*Cill Tiarnáin*) Dublin. 'Tiarnán's church'.

Kiltober (*Coill Tobair*) Westmeath. 'Wood of the well'.

Kilton Somerset. *Cylfantun c.*880, *Chilvetune* 1086 (DB). 'Farmstead near the club-shaped hill'. OE *cylfe* + *tūn*.

Kiltoom (*Cill Tuama*) Roscommon. 'Church of the burial mound'.

Kiltormer (*Cill Tormóir*) Galway. 'Tormór's church'.

Kiltullagh (*Cill Tulach*) Galway. 'Church of the hills'.

Kilturk (*Coill Torc*) Fermanagh. *Kilturke* 1609. 'Wood of boars'.

Kiltyclogher (*Coillte Clochair*) Leitrim. 'Woods of the stony place'.

Kilve Somerset. *Clive* 1086 (DB), *Kylve* 1243. 'Club-shaped hill'. OE **cylfe*.

Kilvergan (*Cill Uí Mhuireagáin*) Armagh. 'Church of Ó Muireagáin'.

Kilvington, 'estate associated with a man called **Cylfa* or Cynelāf', OE pers. name + -*ing*- + *tūn*: **Kilvington** Notts. *Chelvinctune* 1086 (DB). **Kilvington, North** N. Yorks. *Chelvintun* 1086 (DB). **Kilvington, South** N. Yorks. *Chelvinctune* 1086 (DB).

Kilwaughter (*Cill Uachtair*) Antrim. *(ecclesia de) Killochre* 1306. 'Upper church'.

Kilwinning N. Ayr. *Killvinin c.*1160, *Ecclesia Sancti Vinini* 1184, *Kynwenyn c.*1300, *Kylvynnyne* 1357. 'Church of St Finnian'. Gaelic *cill*.

Kilworth (*Cill Uird*) Cork. 'Church of the ritual'.

Kilworth, North & Kilworth, South Leics. *Chivelesworde* 1086 (DB), *Kivelingewurthe* 1195. 'Enclosure of a man called **Cyfel*', varying with 'enclosure of **Cyfel*'s family or followers'. OE pers. name (+ -*inga*-) + *worth*.

Kimberley Norfolk. *Chineburlai* 1086 (DB). 'Woodland clearing of a woman called Cyneburg'. OE pers. name + *lēah*.

Kimberley Notts. *Chinemarleie* 1086 (DB). 'Woodland clearing of a man called Cynemǣr'. OE pers. name + *lēah*.

Kimble, Great & Kimble, Little Bucks. *Chenebelle* 1086 (DB). 'Royal bell-shaped hill'. OE *cyne-* + *belle*.

Kimblesworth Durham. *Kymblesworth c.*1225. 'Enclosure of a man called **Cymel*'. OE pers. name + *worth*.

Kimbolton, 'farmstead of a man called Cynebald', OE pers. name + *tūn*: **Kimbolton** Cambs. *Chenebaltone* 1086 (DB). **Kimbolton** Herefs. *Kimbalton* 13th cent.

Kimcote Leics. *Chenemundescote* 1086 (DB). 'Cottage(s) of a man called Cynemund'. OE pers. name + *cot*.

Kimmeridge Dorset. *Cameric* [*sic*] 1086 (DB), *Kimerich* 1212. 'Convenient track or strip of land, or one belonging to a man called Cȳma'. OE *cȳme* or OE pers. name + **ric*.

Kimmerston Northum. *Kynemereston* 1244. 'Farmstead of a man called Cynemǣr'. OE pers. name + *tūn*.

Kimpton, 'farmstead associated with a man called Cȳma', OE pers. name + *-ing-* + *tūn*: **Kimpton** Hants. *Chementune* 1086 (DB). **Kimpton** Herts. *Kamintone* 1086 (DB).

Kinawley (*Cill Náile*) Fermanagh. *Cell Náile* c.1170. 'Náile's church'.

Kinbane (*An Cionn Bán*) (headland) Antrim. *Keanbaan* 1551. 'The white head'. The headland is also known by the English name *White Head*.

Kincardine (the historic county). *Cincardin* 1195. 'Head of the copse'. Gaelic *cinn* + Pictish *carden*.

Kincardine O'Neil Aber. *Kyncardyn O Nelee* 1337. 'Uí Néill's (estate at the) head of the copse'. Gaelic *cinn* + Pictish *carden*, or *Kincardine of Neil*.

Kincasslagh (*Cionn Caslach*) Donegal. *Cancaslough* 1835. 'Head of the inlet'.

Kincora (*Cionn Coradh*) Clare. 'Head of the weir'.

Kincraig Highland. 'Head of the rock'. Gaelic *ceann* + *creag*.

Kincun (*Cionn Con*) Mayo. 'Head of the dog'.

Kindrum (*Cionn Droma*) Donegal. *Kindrom* 1608. 'Hill of the ridge'.

Kineigh (*Cionn Eich*) Kerry. 'Head of the horse'.

Kineton Glos. *Kinton* 1191. 'Royal manor'. OE *cyne-* + *tūn*.

Kineton Warwicks. *Cyngtun* 969, *Quintone* 1086 (DB). 'The king's manor'. OE *cyning* + *tūn*.

King's as affix. *See* main name, e.g. for **King's Bromley** (Staffs.) *see* BROMLEY.

King's Heath Birm. *Kyngesheth* 1511. Self-explanatory, 'the king's heath', with reference to the royal ownership of the manor of KING's NORTON in which the heathland lay.

King's Lynn Norfolk. *See* LYNN.

Kingarrow (*Ceann Garbh*) Donegal. 'Rough head'.

Kingham Oxon. *Caningeham* [*sic*] 1086 (DB), *Keingaham* 11th cent. 'Homestead of the family or followers of a man called *Cǣga'. OE pers. name + *-inga-* + *hām*.

Kinghorn Fife. *Chingor* c.1136, *Kingornum* c.1140. 'Head of the marsh'. Gaelic *cinn* + *gorn*.

Kinglassie Fife. *Kinglassin* 1224. Possibly 'head of the stream'. Gaelic *ceann* + *glais*.

Kingsbridge Devon. *Cinges bricge* 962. 'The king's bridge'. OE *cyning* + *brycg*.

Kingsbury, 'the king's manor', OE *cyning* + *burh* (dative *byrig*): **Kingsbury** Gtr. London. *Cyngesbyrig* 1003–4, *Chingesberie* 1086 (DB). **Kingsbury Episcopi** Somerset. *Cyncgesbyrig* 1065, *Chingesberie* 1086 (DB). Latin affix *episcopi* 'of the bishop' from its early possession by the Bishop of Bath.
    However the following has a different origin: **Kingsbury** Warwicks. *Chinesberie* 1086 (DB), *Kinesburi* 12th cent. 'Stronghold of a man called Cyne'. OE pers. name + *burh* (dative *byrig*).

Kingsclere Hants. *Kyngeclera* early 12th cent. OE *cyning* 'king' (denoting a royal manor) added to the original name found also in HIGHCLERE.

Kingscourt (*Dún an Rí*) Cavan. *King's Court orw. Donnaree* 1767. 'Fort of the king'.

Kingsdon, Kingsdown, 'the king's hill', OE *cyning* + *dūn*: **Kingsdon** Somerset. *Kingesdon* 1194. **Kingsdown** Kent, near Deal. *Kyngesdoune* 1318. **Kingsdown, West** Kent. *Kingesdon* 1199.

Kingsey Bucks. *Eya* 1174, *Kingesie* 1197. 'The king's island'. OE *cyning* + *ēg*.

Kingsford Worcs. *Cenungaford* 964. 'Ford of the family or followers of a man called Cēna'. OE pers. name + *-inga-* + *ford*.

Kingsgate Kent. so called because it was at this *gate* (in the sense of 'gap in the cliffs') that Charles II landed in 1683.

Kingskerswell Devon. *Carsewelle* 1086 (DB), *Kyngescharsewell* 1270. 'Spring or stream where water-cress grows'. OE *cærse* + *wella*. The manor belonged to the king in 1086.

Kingsland Herefs. *Lene* 1086 (DB), *Kingeslan* 1213. 'The king's estate in *Leon*'. OE *cyning* with old Celtic name for the district (*see* LEOMINSTER).

Kingsley, 'the king's wood or clearing', OE *cyning* + *lēah*: **Kingsley** Ches. *Chingeslie* 1086 (DB). **Kingsley** Hants. *Kyngesly c.*1210. **Kingsley** Staffs. *Chingeslei* 1086 (DB).

Kingsmill Tyrone. *Kings mills* 1833. 'King's mill'. A man named *King* owned a corn mill here.

Kingsnorth Kent. *Kingesnade* 1226. 'Detached piece of land or woodland belonging to the king'. OE *cyning* + *snād*.

Kingstanding Birm. from ME *standing* 'a hunter's station from which to shoot game', although according to tradition the mound here was where Charles I reviewed his troops in 1642. The same name occurs in other counties.

Kingsteignton Devon. *Teintona* 1086 (DB), *Kingestentone* 1274. 'Farmstead on the River Teign'. Celtic river-name (*see* TEIGNMOUTH) + OE *tūn*, with affix from its possession by the king in 1086.

Kingsthorpe Northants. *Torp* 1086 (DB), *Kingestorp* 1190. 'Outlying farmstead or hamlet belonging to the king'. OE *cyning* + OScand. *thorp*.

Kingston, a common name, usually 'the king's manor or estate', OE *cyning* + *tūn*; examples include: **Kingston** Hants. *Kingeston* 1194. **Kingston Bagpuize** Oxon. *Cyngestun c.*976, *Chingestune* 1086 (DB), *Kingeston Bagepuz* 1284. Manorial affix from the *de Bagpuize* family who held the manor from 1086. **Kingston Blount** Oxon. *Chingestone* 1086 (DB), *Kyngestone Blont* 1379. Manorial affix from the *le Blund* family, here in the 13th cent. **Kingston Deverill** Wilts. *Devrel* 1086 (DB), *Deverel Kyngeston* 1249. For the original river-name *Deverill see* BRIXTON DEVERILL. **Kingston Lacy** Dorset. *Kingestune* 1170, *Kynggestone Lacy* 1319. Manorial affix from the *de Lacy* family who held the manor in the 13th cent. **Kingston Lisle** Oxon. *Kingeston* 1220, *Kyngeston Lisle* 1322. Manorial affix from the *del Isle* family, here from the 13th cent. **Kingston near Lewes** E. Sussex. *Kyngestona c.*1100. *See* LEWES. **Kingston St Mary** Somerset. *Kyngestona* 12th cent. Affix from the dedication

of the church. **Kingston upon Hull** K. upon Hull. *Kyngeston* 1256, alternatively called simply **Hull** from the River Hull from early times (*Hul* 1228), the river-name being either OScand. (meaning 'deep one') or Celtic (meaning 'muddy one'). **Kingston upon Thames** Gtr. London. *Cyninges tun* 838, *Chingestune* 1086 (DB). *See* THAMES.
 However the following has a different origin: **Kingston on Soar** Notts. *Chinestan* 1086 (DB). 'The royal stone'. OE *cyne-* + *stān*. For the river-name, *see* BARROW UPON SOAR.

Kingstone Herefs. near Hereford. *Chingestone* 1086 (DB). 'The king's manor or estate'. OE *cyning* + *tūn*.

Kingstone Somerset. *Chingestana* 1086 (DB). 'The king's boundary stone'. OE *cyning* + *stān*.

Kingswear Devon. *Kingeswere* 12th cent. 'The king's weir'. OE *cyning* + *wer*.

Kingswinford Dudley. *Svinesford* 1086 (DB), *Kyngesswynford* 1322. 'Pig ford'. OE *swīn* + *ford*. Affix from its being a royal manor.

Kingswood, 'the king's wood', OE *cyning* + *wudu*: **Kingswood** Glos. *Kingeswoda* 1166. **Kingswood** S. Glos. *Kingeswode* 1231. **Kingswood** Surrey. *Kingeswode c.*1180.

Kington, 'royal manor or estate', OE *cyne-* (replaced by *cyning*) + *tūn*: **Kington** Herefs. *Chingtune* 1086 (DB). **Kington** Worcs. *Cyngtun* 972, *Chintune* 1086 (DB). **Kington Magna** Dorset. *Chintone* 1086 (DB), *Magna Kington* 1243. Affix is Latin *magna* 'great'. **Kington St Michael** Wilts. *Kingtone* 934, *Chintone* 1086 (DB), *Kyngton Michel* 1279. Affix from the dedication of the church. **Kington, West** Wilts. *Westkinton* 1195.

Kingussie Highland. *Kinguscy c.*1210. 'Head of the pinewood'. Gaelic *ceann* + *giuthsach*.

Kinlet Shrops. *Chinlete* 1086 (DB). 'Royal share or lot'. OE *cyne-* + *hlēt*. The manor was held in 1066 by Queen Edith, widow of Edward the Confessor.

Kinloch Highland. 'Head of the loch'. Gaelic *ceann* + *loch*.

Kinlochleven Highland. 'Head of Loch Leven'. Gaelic *ceann* + *loch*. *Leven* is 'elm river' (Gaelic *leamhain*).

Kinlough (*Cionn Locha*) Leitrim. 'Head of the lake'.

Kinmel Bay (*Bae Cinmel*) Conwy. *Kilmeyl* 1331. 'Mael's nook'. Welsh *cil*. Bay is a recent addition.

Kinnea (*Cionn Ech*) Cavan, Donegal. 'Head of the horse'.

Kinnegad (*Ceann Átha Gad*) Westmeath. 'Head of the ford of the withies'.

Kinnerley, Kinnersley, 'woodland clearing of a man called Cyneheard', OE pers. name + *lēah*: **Kinnerley** Shrops. *Chenardelei* 1086 (DB). **Kinnersley** Herefs. *Cyrdes leah c.*1030, *Curdeslege* 1086 (DB). **Kinnersley** Worcs. *Chinardeslege* 1123.

Kinnerton, 'farmstead or estate of a man called Cyneheard', OE pers. name + *tūn*: **Kinnerton** Powys. *Kynardton* 1303. **Kinnerton, Higher** (Flin.) **& Kinnerton, Lower** (Ches.). *Kynarton* 1240.

Kinnity (*Cionn Eitigh*) Offaly. 'Eiteach's headland'.

Kinoulton Notts. *Kinildetune c.*1000, *Chineltune* 1086 (DB). 'Farmstead of a woman called Cynehild'. OE pers. name + *tūn*.

Kinross Perth. *Kynros c.*1144. 'End of the promontory'. Gaelic *ceann* + *ros*.

Kinsale (*Cionn tSáile*) Cork. 'Head of the salt water'.

Kinsalebeg (*Cionn tSáile Beag*) Waterford. 'Little head of the salt water'.

Kinsaley (*Cionn tSáile*) Dublin. 'Head of the salt water'.

Kinsham Herefs. *Kingeshemede* 1216. 'Border meadow belonging to the king'. OE *cyning* + **hemm-mǣd*.

Kinsley Wakefd. *Chineslai* 1086 (DB). 'Woodland clearing of a man called Cyne'. OE pers. name + *lēah*.

Kinson Dorset. *Chinestanestone* 1086 (DB), *Kinestaneston* 1238. 'Farmstead of a man called Cynestān'. OE pers. name + *tūn*.

Kintail Forest Highland. *Keantalle* 1509. Perhaps 'forest at the head of the salt water'. Gaelic *ceann t-saile*.

Kintbury W. Berks. *Cynetanbyrig c.*935, *Cheneteberie* 1086 (DB). 'Fortified place on the River Kennet'. Celtic river-name (*see* KENNETT) + OE *burh* (dative *byrig*).

Kintyre (peninsula) Arg. *Ciunntire* 807. 'End of the land'. Gaelic *ceann* + *tire*.

Kinvarra (*Cinn Mhara*) Galway. 'Head of the sea'.

Kinver Staffs. *Cynibre* 736, *Chenevare* 1086 (DB). Celtic **breʒ* 'hill' with an obscure first element (probably taken as OE *cyne-* 'royal' at an early date).

Kippax Leeds. *Chipesch* 1086 (DB). 'Ash-tree of a man called **Cippa* or **Cyppa*'. OE pers. name (with Scand. *K-*) + OE *æsc* (replaced by OScand. *askr*).

Kippure (*Cipiúr*) (mountain) Wicklow. *Kippoore* 1604. Perhaps 'big place of rough grass'.

Kirby, a common name in the Midlands and North, usually 'village with a church', OScand. *kirkju-bý*; examples include: **Kirby Bedon** Norfolk. *Kerkebei* 1086 (DB), *Kirkeby Bydon* 1291. Manorial affix from the *de Bidun* family, here in the 12th cent. **Kirby Bellars** Leics. *Chirchebi* 1086 (DB), *Kyrkeby Beler* 1332. Manorial affix from the *Beler* family, here from the 12th cent. **Kirby Grindalythe** N. Yorks. *Chirchebi* 1086 (DB), *Kirkby in Crendalith* 1367. Affix is an old district name referring to 'slope of the valley frequented by cranes', from OE *cran* + *dæl* + OScand. *hlíth*. **Kirby le Soken** Essex. *Kyrkebi* 1181, *Kirkeby in the Sokne* 1385. Affix is from OE *sōcn* 'district with special jurisdiction'. **Kirby, West** Wirral. *Cherchebia* 1081, *Kirkeby c.*1138, *Westkyrby* 1287. Affix 'west' to distinguish this place from another *Kirkby* in WALLASEY.

However the following have a different origin, 'farmstead or village of a man called Kærir', OScand. pers. name + *bý*: **Kirby, Cold** N. Yorks. *Carebi* 1086 (DB). The affix means 'bleak, exposed'. **Kirby Muxloe** Leics. *Carbi* 1086 (DB). The affix (found from the 17th cent.) is manorial, from lands here held by a family of that name.

Kircubbin (*Cill Ghobáin*) Down. *Ballicarcubbin* 1605. 'Gobán's church'.

Kirdford W. Sussex. *Kinredeford* 1228. 'Ford of a woman called Cynethrȳth or a man called Cynerēd'. OE pers. name + *ford*.

Kirk as affix. *See* main name, e.g. for **Kirk Bramwith** (Donc.) *see* BRAMWITH.

279 **Kirkleatham**

Kirkabister Shet. (Bressay). *Kirkabuster* 1654. 'Homestead of the church'. OScand. *kirkju-bólstathr*.

Kirkandrews upon Eden Cumbria. *Kirkandres c.*1200. 'Church of St Andrew'. OScand. *kirkja*, named from the dedication of the church. Eden is a Celtic river-name.

Kirkbampton Cumbria. *Banton c.*1185, *Kyrkebampton* 1292. 'Farmstead made of beams or by a tree'. OE *bēam* + *tūn*. Later affix is OScand. *kirkja* 'church'.

Kirkbride Cumbria. *Chirchebrid* 1163. 'Church of St Bride or Brigid'. OScand. *kirkja* + Irish saint's name.

Kirkburn E. R. Yorks. *Westburne* 1086 (DB), *Kirkebrun* 1272. '(Place on) the spring or stream'. OE *burna* with OScand. *kirkja* 'church'.

Kirkburton Kirkl. *Bertone* 1086 (DB). 'Farmstead near or belonging to a fortification'. OE *byrh-tūn*, with affix OScand. *kirkja* 'church' from the 16th cent.

Kirkby, a common name in the Midlands and North, 'village with a church', OScand. *kirkju-bý*; examples include: **Kirkby** Knows. *Cherchebi* 1086 (DB). **Kirkby in Ashfield** Notts. *Chirchebi* 1086 (DB), *Kirkeby in Esfeld* 1216. The affix is an old district name, *see* ASHFIELD. **Kirkby Lonsdale** Cumbria. *Cherchebi* 1086 (DB), *Kircabi Lauenesdale* 1090–7. The affix means 'in the valley of the River Lune', Celtic river-name (*see* LANCASTER) + OScand. *dalr*. **Kirkby Malzeard** N. Yorks. *Chirchebi* 1086 (DB), *Kirkebi Malesard c.*1105. The affix means 'poor clearing', from OFrench *mal* + *assart*. **Kirkby Overblow** N. Yorks. *Cherchebi* 1086 (DB), *Kirkeby Oreblowere* 1211. The affix is from OE **or-blāwere* 'ore-blower' alluding to the early smelting of iron ore here. **Kirkby, South** Wakefd. *Cherchebi* 1086 (DB), *Sudkirkebi c.*1124. Affix *south* to distinguish this place from another Kirkby (now lost) in Pontefract. **Kirkby Stephen** Cumbria. *Cherkaby Stephan c.*1094. Affix from the dedication of the church or from the name of an early owner. **Kirkby Thore** Cumbria. *Kirkebythore* 1179. Manorial affix probably from the name (OScand. *Thórir*) of some early owner.

Kirkbymoorside N. Yorks. *Chirchebi* 1086 (DB), *Kirkeby Moresheved c.*1170. Identical in origin with the previous names, with affix meaning 'head or top of the moor', OE *mōr* + *hēafod*.

Kirkcaldy Fife. *Kircalethyn* 12th cent. 'Fort at Caledin'. Welsh *caer*. *Caledin* means 'hard hill' (Welsh *caled* + *din*). *Kirk-* was substituted for the 'fort' element when this was no longer understood.

Kirkcambeck Cumbria. *Camboc c.*1177, *Kirkecamboc c.*1280. 'Place on Cam Beck with a church'. Celtic river-name (meaning 'crooked stream') with the addition of OScand. *kirkja*.

Kirkconnel Dumf. *Kyrkconwelle* 1347, *Kirkconevel* 1354. 'Church of St Conall'. OScand. *kirkja*.

Kirkcudbright Dumf. *Kircuthbright* 1296. 'Church of St Cuthbert'. OScand. *kirkja*. *Kirk-* may have originally been *Kil-* (Gaelic *cill*), since the pers. name follows the generic word, as is usual in Celtic names, rather than preceding it.

Kirkham, 'homestead or village with a church', OE *cirice* (replaced by OScand. *kirkja*) + OE *hām*: **Kirkham** Lancs. *Chicheham* [*sic*] 1086 (DB), *Kyrkham* 1094. **Kirkham** N. Yorks. *Cherchan* 1086 (DB).

Kirkharle Northum. *Herle* 1177, *Kyrkeherl* 1242. Possibly 'woodland clearing of a man called **Herela*'. OE pers. name + *lēah*.

Kirkheaton, 'high farmstead', OE *hēah* + *tūn* with the later addition of OScand. *kirkja* 'church': **Kirkheaton** Kirkl. *Heptone* [*sic*] 1086 (DB), *Kirkheton* 13th cent. **Kirkheaton** Northum. *Heton* 1242.

Kirkibost Highland. (Skye). 'Homestead with a church'. OScand. *kirkju-bólstathr*.

Kirkintilloch E. Dunb. *Kirkintulach c.*1200. 'Fort at the head of the hillock'. Welsh *caer* + Gaelic *ceann* + *tulach* (genitive *tulaich*). It was earlier *Caerpentaloch* 10th cent., from *caer* + Welsh *pen* 'head, end' + *tulach*.

Kirkistown Down. *Kirkston* 1631. 'Kirk's town'. A man surnamed *Kirk* owned land here.

Kirkland, 'estate belonging to a church', OScand. *kirkja* + *land*: **Kirkland** Cumbria, near Blencarn. *Kyrkeland c.*1140. **Kirkland** Cumbria, near Lamplugh. *Kirkland* 1586.

Kirkleatham Red. & Cleve. *Westlidum* 1086 (DB), *Kyrkelidun* 1181. '(Place at) the slopes'. OE *hlith* or OScand. *hlíth* in a dative plural form. Affix is OE *west* replaced by OScand. *kirkja* 'church'.

Kirklees Kirkl. *Kirkelei, Kirkeleis c.*1205. 'Clearings or pastures belonging to a church'. OScand. *kirkja* + OE *lēah* (plural *lēas*). A small Cistercian nunnery was established here in the 12th cent.

Kirklevington Stock. on T. *Levetona* 1086 (DB). 'Farmstead on the River Leven'. Celtic river-name (possibly meaning 'smooth') + OE *tūn*, with later addition of OScand. *kirkja* 'church'.

Kirkley Suffolk. *Kirkelea* 1086 (DB). 'Woodland clearing near or belonging to a church'. OScand. *kirkja* + OE *lēah*.

Kirklington, 'estate associated with a man called *Cyrtla', OE pers. name + *-ing-* + *tūn*: **Kirklington** N. Yorks. *Cherdinton* [*sic*] 1086 (DB), *Chirtlintuna c.*1150. **Kirklington** Notts. *Cyrlingtune* 958, *Cherlinton* 1086 (DB).

Kirklinton Cumbria. *Leuenton c.*1170, *Kirkeleuinton* 1278. 'Farmstead by the River Lyne'. Celtic river-name (possibly 'the smooth one') + OE *tūn*, with later affix from OScand. *kirkja* 'church'.

Kirkness Orkn. 'Promontory with a church'. OScand. *kirkja* + *nes*.

Kirknewton Northum. *Niwetona* 12th cent. 'New farmstead'. OE *nīwe* + *tūn*, with later affix from OScand. *kirkja* 'church'.

Kirkoswald Cumbria. *Karcoswald* 1167. 'Church of St Oswald'. From the dedication of the church (OScand. *kirkja*) to this saint, a 7th-cent. king of Northumbria.

Kirkoswald S. Ayr. 'St Oswald's church'. OScand. *kirkja*.

Kirksanton Cumbria. *Santacherche* 1086 (DB), *Kirkesantan c.*1150. 'Church of St Sanctán'. OScand. *kirkja* with Irish saint's name.

Kirkton of Logie Buchan Aber. 'Village with a church of the hollow (place) in Buchan'. OScand. *kirkja* + OE *tūn* + Gaelic *logach* (dative *logaigh*). *See* BUCHAN.

Kirktown of Alvah. *See* ALVAH.

Kirkwall Orkn. *Kirkiuvagr c.*1225. 'Church bay'. OScand. *kirkja* + *vágr*. The bay is the Bay of Kirkwall, and the church the Cathedral of St Magnus.

Kirkwhelpington Northum. *Welpinton* 1176. 'Estate associated with a man called

*Hwelp'. OE pers. name + *-ing-* + *tūn*, with later affix from OScand. *kirkja* 'church'.

Kirmington N. Lincs. *Chernitone* 1086 (DB), *Kirningtun c.*1150. Possibly 'farmstead where butter is churned'. OE **cirning* 'churning' (with Scand. *k-*) + *tūn*.

Kirmond le Mire Lincs. *Chevremont* 1086 (DB). 'Goat hill'. OFrench *chèvre* + *mont*. Affix means 'in the marshy ground', from OScand. *mýrr*.

Kirriemuir Ang. *Kerimor* 1250. 'Big quarter'. Gaelic *ceathramh* + *mór*. A *ceathramh* was the fourth part of a *dabhach*, a variable area of land considered large enough to support a single household.

Kirstead Green Norfolk. *Kerkestede c.*1095. 'Site of a church'. OE *cirice* (replaced by OScand. *kirkja*) + OE *stede*.

Kirtling Cambs. *Chertelinge* 1086 (DB). 'Place associated with a man called *Cyrtla'. OE pers. name + *-ing*.

Kirtlington Oxon. *Kyrtlingtune c.*1000, *Certelintone* 1086 (DB). 'Estate associated with a man called *Cyrtla'. OE pers. name + *-ing-* + *tūn*.

Kirton, 'village with a church', OScand. *kirkja* (probably replacing OE *cirice*) + OE *tūn*: **Kirton** Lincs. *Chirchetune* 1086 (DB). **Kirton** Notts. *Circeton* 1086 (DB). **Kirton** Suffolk. *Kirketuna* 1086 (DB). **Kirton in Lindsey** N. Lincs. *Chirchetone* 1086 (DB). *See* LINDSEY.

Kishkeam (*Coiscéim na Caillí*) Cork. 'Footstep of the hag'.

Kislingbury Northants. *Ceselingeberie* 1086 (DB). 'Stronghold of the dwellers on the gravel, or of the family or followers of a man called *Cysel'. OE *cisel* or pers. name + *-inga-* + *burh* (dative *byrig*).

Kiveton Park Rothm. *Ciuetone* 1086 (DB). Probably 'farmstead by a tub-shaped feature'. OE *cȳf* + *tūn*.

Knaith Lincs. *Cheneide* 1086 (DB). 'Landing-place by a river-bend'. OE *cnēo* + *hȳth*.

Knaphill Surrey. *La Cnappe* 1225, *Knephull* 1440. 'The hill'. OE *cnæpp*, with later explanatory *hyll*.

Knappagh (*Cnapach*) Mayo. 'Bumpy area'.

Knapton, 'the young man's farmstead', or 'farmstead of a man called Cnapa', OE *cnapa* or pers. name + *tūn*: **Knapton** Norfolk. *Kanapatone* 1086 (DB). **Knapton** York. *Cnapetone* 1086 (DB). **Knapton, East & Knapton, West** N. Yorks. *Cnapetone* 1086 (DB).

Knapwell Cambs. *Cnapwelle* c.1045, *Chenepewelle* 1086 (DB). 'Spring or stream of the young man, or of a man called Cnapa'. OE *cnapa* or pers. name + *wella*.

Knaresborough N. Yorks. *Chenaresburg* 1086 (DB). Probably 'stronghold of a man called Cēnheard'. OE pers. name + *burh*.

Knarsdale Northum. *Knaresdal* 1254. 'Valley of Knar, the rugged rock'. OE **cnearr* + OScand. *dalr*.

Knayton N. Yorks. *Cheniueton* 1086 (DB). 'Farmstead of a woman called Cēngifu'. OE pers. name + *tūn*.

Knebworth Herts. *Chenepeworde* 1086 (DB). 'Enclosure of a man called Cnebba'. OE pers. name + *worth*.

Kneesall Notts. *Cheneshale* [*sic*] 1086 (DB), *Cneshala* 1175. Possibly 'nook of land of a man called Cynehēah'. OE pers. name + *halh*.

Kneesworth Cambs. *Cnesworth* c.1218. Possibly 'enclosure of a man called Cynehēah'. OE pers. name + *worth*.

Kneeton Notts. *Cheniueton* 1086 (DB). 'Farmstead of a woman called Cēngifu'. OE pers. name + *tūn*.

Knightcote Warwicks. *Knittecote* 1242. 'Cottage(s) of the young men or retainers'. OE *cniht* + *cot*.

Knighton, a fairly common name, 'farmstead or village of the young men or retainers', OE *cniht* + *tūn*; examples include: **Knighton** Leic. *Cnihtetone* 1086 (DB). **Knighton** Staffs., near Eccleshall. *Chnitestone* 1086 (DB). **Knighton, West** Dorset. *Chenistetone* 1086 (DB).

Knighton (*Trefyclo*) Powys. *Chenistetone* 1086 (DB), *Cnicheton* 1193. 'Farmstead of the young men or retainers'. OE *cniht* + *tūn*. The Welsh name means 'farm by the dyke' (Welsh *tref* + *y* + *clawdd*), referring to OFFA'S DYKE nearby.

Knightsbridge Gtr. London. *Cnihtebricge* c.1050. 'Bridge of the young men or retainers', i.e. where they congregate. OE *cniht* + *brycg*.

Knill Herefs. *Chenille* 1086 (DB). '(Place at) the hillock'. OE **cnylle*.

Knipton Leics. *Gniptone* 1086 (DB). 'Farmstead below the steep hillside'. OScand. *gnípa* + OE *tūn*.

Knitsley Durham. *Knicheley* 1279. 'Woodland clearing where knitches or faggots are obtained'. OE *cnycc* + *lēah*.

Kniveton Derbys. *Cheniuetun* 1086 (DB). 'Farmstead of a woman called Cēngifu'. OE pers. name + *tūn*.

Knock Cumbria. *Chonoc* 12th cent. OIrish *cnocc* 'a hillock'.

Knock (*Cnoc*) Clare, Down, Mayo. 'Hill'.

Knockaderry (*Cnoc an Doire*) Limerick. 'Hill of the oak wood'.

Knockadoon (*Cnoc an Dúin*) Cork. 'Hill of the fort'.

Knockainy (*Cnoc Áine*) Limerick. 'Áine's hill'.

Knockalough (*Cnoc an Locha*) Clare, Tipperary. 'Hill of the lake'.

Knockananna (*Cnoc an Eanaigh*) Wicklow. 'Hill of the marsh'.

Knockanarrigan (*Cnoc an Aragain*) Wicklow. 'Hill of the conflicts'.

Knockanevin (*Cnocán Aoibhinn*) Cork. 'Pleasant little hill'.

Knockanillaun (*Cnoc an Oileáin*) Mayo. 'Hill of the island'.

Knockanoran (*Cnoc an Uaráin*) Cork, Laois. 'Hill of the spring'.

Knockanore (*Cnoc an Óir*) Waterford. 'Hill of gold'.

Knockatallon (*Cnoc an tSalainn*) Monaghan. 'Hill of the salt'.

Knockaunnaglashy (*Cnocán na Glaise*) Kerry. 'Little hill of the stream'.

Knockboy (*Cnoc Buí*) Cork, Kerry, Waterford. 'Yellow hill'.

Knockbrack (*An Cnoc Breac*) Donegal. 'The speckled hill'.

Knockbreda Down. The present parish was formed in 1658 from the two separate parishes of *Knock* (Irish *An Cnoc*, 'the hill') and *Breda* (Irish *Bréadach*, 'broken land').

Knockcloghrim (*Cnoc Clochdhroma*) Derry. *Knock-loghran* 1654. 'Hill of the stony ridge'.

Knockcroghery (*Cnoc an Chrochaire*) Roscommon. 'Hill of the hangman'.

Knockdrin (*Cnoc Droinne*) Westmeath. 'Humpy hill'.

Knockeen (*Cnoicín*) Kerry. 'Little hill'.

Knockerry (*Cnoc Doire*) Clare. 'Hill of the oak wood'.

Knockgraffon (*Cnoc Rafann*) Tipperary. 'Rafainn's hill'.

Knockholt Kent. *Ocholt* 1197, *Nocholt* 1353. '(Place at) the oak wood'. OE *āc* + *holt*, with initial *N-* from the OE definite article.

Knockin Shrops. *Cnochin* 1165. Celtic *cnöccïn* 'a little hillock'.

Knocklayd (*Cnoc Leithid*) Antrim. *Cnoc Leaaid* 1542. 'Hill of the slope'.

Knocklong (*Cnoc Loinge*) Limerick. 'Hill of the ship'.

Knockmealdown Mountains (*Sléibhte Chnoc Mhaoldhomhnaigh*) Tipperary, Waterford. 'Mountains of the hill of Maoldomhnach'.

Knockmore (*An Cnoc Mór*) Mayo. 'The big hill'.

Knockmoyle (*Cnoc Maol*) Galway, Tyrone. 'Bald hill'.

Knocknacarry (*Cnoc na Cora*) Antrim. *Knockenecarry* c.1657. 'Hill of the weir'.

Knocknagashel (*Cnoc na gCaiseal*) Kerry. 'Hill of the stone forts'.

Knocknagoney (*Cnoc na gCoiníní*) Down. *Ballacknocknegonie* 1604. 'Hill of the rabbits'.

Knocknagree (*Cnoc na Graí*) Cork. 'Hill of the horse stud'.

Knocknahorn (*Cnoc na hEorna*) Tyrone. 'Hill of the barley'.

Knocknamuckly (*Cnoc na Muclaí*) Armagh. *Knocknamokally* 1609. 'Hill of the piggery'.

Knockraha (*Cnoc Rátha*) Cork. 'Hill of the fort'.

Knockrath (*Cnoc Rátha*) Wicklow. 'Hill of the fort'.

Knocktopher (*Cnoc an Tóchair*) Kilkenny. 'Hill of the causeway'.

Knockvicar (*Cnoc an Bhiocáire*) Roscommon. 'Hill of the vicar'.

Knodishall Suffolk. *Cnotesheala* 1086 (DB). 'Nook of land of a man called *Cnott'. OE pers. name + *halh*.

Knook Wilts. *Cunuche* 1086 (DB). Probably OWelsh *cnucc* 'a hillock'.

Knossington Leics. *Nossitone* [sic] 1086 (DB), *Cnossintona* c.1150. 'Estate at the hill, or of a man called *Cnossa'. OE *cnoss + -ing*, or pers. name + *-ing* + *tūn*.

Knott End on Sea Lancs. *Cnote* 13th cent. ME *knot* 'a hillock'.

Knotting Beds. *Chenotinga* 1086 (DB). Possibly '(settlement of) the family or followers of a man called *Cnotta'. OE pers. name + *-ingas*. Alternatively perhaps '(settlement of) the hill people', from OE *cnotta* 'knot, lump' + *-ingas*.

Knottingley Wakefd. *Notingeleia* 1086 (DB). 'Woodland clearing of the family or followers of a man called *Cnotta'. OE pers. name + *-inga-* + *lēah*.

Knotty Ash Lpool. recorded as simply *Ash* c.1700, 'the ash-tree', later with the adjective *knotty* 'gnarled'.

Knowle, '(place at) the hill-top or hillock', OE *cnoll*; examples include: **Knowle** Brist. *Canole* 1086 (DB). **Knowle** Solhll. *La Cnolle* 1221. *Knowle, Church* Dorset. *C(he)nolle* 1086 (DB), *Churchecnolle* 1346. Later addition is OE *cirice* 'church'.

Knowlton Kent. *Chenoltone* 1086 (DB). 'Farmstead by a hillock'. OE *cnoll* + *tūn*.

Knowsley Knows. *Chenuļueslei* 1086 (DB). 'Woodland clearing of a man called Cēnwulf or Cynewulf'. OE pers. name + *lēah*.

Knowstone Devon. *Chenutdestana* 1086 (DB). 'Boundary stone of a man called Knútr'. OScand. pers. name + OE *stān*.

Knoydart (district) Highland. *Knodworath* 1309. 'Knut's inlet'. OScand. pers. name + *fjǫrthr*. The pers. name is identical with that of

King *Knut* (Canute), who invaded Scotland in 1031.

Knoyle, East & Knoyle, West Wilts. *Cnugel* 948, *Chenvel* 1086 (DB). '(Place at) the knuckle-shaped hill'. OE **cnuwel*.

Knutsford Ches. *Cunetesford* 1086 (DB). Probably 'ford of a man called Knútr'. OScand. pers. name + OE *ford*.

Kyle of Lochalsh Highland. '(Place by the) Kyle of Lochalsh'. *Kyle* is Gaelic *caol*, 'strait'. The name of the loch may mean 'foaming one' (Gaelic *aillseach*).

Kyleakin Highland. (Skye). '(Place on) Kyle Akin'. The name of the strait means 'Haakon's strait' (Gaelic *caol*), apparently referring to *Haakon* IV of Norway, who is said to have sailed through it in 1263 after his defeat at Largs.

Kylebrack (*An Choill Bhreac*) Galway. 'The speckled wood'.

Kylemore (*An Choill Mhóir*) Galway. 'The big wood'.

Kyles of Bute (sea channel) Arg. 'Straits of Bute'. Gaelic *caolas*. *See* BUTE.

Kyloe Northum. *Culeia* 1195. 'Clearing or pasture for cows'. OE *cū* + *lēah*.

Kyme, North & Kyme, South Lincs. *Chime* 1086 (DB). '(Place) at the hollow, or at the edge or brink'. OE **cymbe* or **cimbe*.

Kyre Magna Worcs. *Cuer* 1086 (DB). Named from the stream here, a pre-English river-name of uncertain origin and meaning.

Labasheeda (*Leaba Shíoda*) Clare. 'Grave of Síoda'.

Labbacallee (*Leaba Caillighe*) Cork. 'Hag's bed'.

Labbamolaga (*Leaba Molaga*) Cork. 'Molaga's bed'.

Labby (*Leaba*) Derry, Donegal, Sligo. 'Grave'.

Labbyfirmore (*Leaba an Fhir Mhóir*) Monaghan. 'Grave of the big man'.

Laceby NE. Lincs. *Levesbi* 1086 (DB). 'Farmstead or village of a man called Leifr'. OScand. pers. name + *bý*.

Lach Dennis Ches. *Lece* 1086 (DB), *Lache Deneys* 1260. 'The boggy stream'. OE **lece*, with manorial affix from ME *danais* 'Danish' referring to a man called *Colben* who held the manor in 1086.

Lachtcarn (*Leacht Chairn*) Tipperary. 'Grave of the cairn'.

Lack (*An Leac*) Fermanagh. *Lack* 1834. 'The flagstone'.

Lackagh (*Leacach*) Kildare. 'Abounding in flagstones'.

Lackamore (*An Leaca Mhór*) Tipperary. 'The big hillside'.

Lackan (*An Leacain*) Wicklow. 'The hillside'.

Lackaroe (*Leaca Rua*) Cork, Offaly. 'Red hillside'.

Lackford Suffolk. *Lecforda* 1086 (DB). 'Ford where leeks grow'. OE *lēac* + *ford*. The river-name **Lark** is a back-formation from the place name.

Lackington, White Dorset. *Wyghtlakynton* 1354. 'Farmstead or estate associated with a man called Wihtlāc'. OE pers. name + *-ing-* + *tūn*.

Lacock Wilts. *Lacok* 854, *Lacoc* 1086 (DB). 'The small stream', OE **lacuc*.

Ladbroke Warwicks. *Hlodbroc* 998, *Lodbroc* 1086 (DB). Possibly 'brook used for divination'. OE *hlod* + *brōc*.

Ladock Cornwall. 'Church of *Sancta Ladoca*' 1268. From the patron saint of the church.

Ladybank Fife. Late 19th-cent. development, perhaps so named through connection with the monks of Limdores.

Lagan (*Abhainn an Lagáin*) (river) Antrim, Down. *Logia c.*150. 'River of the low-lying district'.

Lagan (*Lagán*) Donegal. 'Little hollow'.

Lagg N. Ayr. '(Place by a) hollow'. Gaelic *lag*.

Laggan N. Ayr. '(Place by a) little hollow'. Gaelic *lagan*.

Laghil (*Leamhchoill*) Donegal. 'Elm wood'.

Laghile (*Leamhchoill*) Clare, Tipperary. 'Elm wood'.

Laght (*Leacht*) Cork. 'Grave'.

Laghtane (*Leachtán*) Limerick. 'Little grave'.

Laghtgeorge (*Leacht Sheoirse*) Limerick. 'Seoirse's grave'.

Laghtneill (*Leacht Néill*) Cork. 'Néill's grave'.

Laghy (*An Lathaigh*) Donegal. 'The mire'.

Lagnamuck (*Lag na Muc*) Mayo. 'Hollow of the pigs'.

Lagore (*Loch Gabhra*) Meath. 'Lake of the goat'.

Laharan (*Leath Fhearann*) Cork, Kerry. 'Half a measure of land'.

Lahardaun (*Leathardán*) Mayo. 'Side of a plateau'.

Lahesseragh (*Leath Sheisreach*) Tipperary. 'Half ploughland'.

Laindon Essex. *Ligeandune c.*1000, *Leienduna* 1086 (DB). Probably 'hill by a stream called *Lea*'. Lost Celtic river-name (possibly meaning 'light river') + OE *dūn*.

Lairg Highland. *Larg c.*1230. '(Place of) the shank'. Gaelic *lorg*.

Lake Wilts. *Lake* 1289. OE *lacu* 'small stream'.

Lakeland, South (district) Cumbria. A modern name, with *Lakeland* as an alternative name for the Lake District.

Lakenham, Old & Lakenham, New Norfolk. *Lakemham* 1086 (DB). Probably 'homestead or village of a man called **Lāca*'. OE pers. name (genitive *-n*) + *hām*.

Lakenheath Suffolk. *Lacingahith* 945, *Lakingahethe* 1086 (DB). 'Landing-place of the people living by streams, or of the family or followers of a man called **Lāca*'. OE *lacu* or pers. name + *-inga-* + *hȳth*.

Laleham Surrey. *Laeleham* 1042–66, *Leleham* 1086 (DB). 'Homestead or enclosure where twigs or brushwood are found'. OE *lǣl* + *hām* or *hamm*.

Laleston (*Trelales*) Bri. *Lagelestun c.*1165, *Laweleston* 1268. 'Legeles' farm'. OE *tūn* (Welsh *tref*). Thomas and Walter *Legeles* ('lawless') are recorded in the 13th cent. as members of an Anglo-Scand. family.

Lamarsh Essex. *Lamers* 1086 (DB). 'Loam marsh'. OE *lām* + *mersc*.

Lamas Norfolk. *Lammesse* 1044–7, *Lamers* 1086 (DB). Identical in origin with the previous name.

Lambay (*Reachrainn*) (island) Dublin. 'Lamb island'. OScand. *lamb* + *ey*. The meaning of the Irish name is obscure.

Lambeg (*Lann Bheag*) Antrim. *Landebeg c.*1450. 'Little church'.

Lamberhurst Kent. *Lamburherste c.*1100. 'Wooded hill where lambs graze'. OE *lamb* (genitive plural *lambra*) + *hyrst*.

Lambeth Gtr. London. *Lambehitha* 1062, *Lamhytha* 1089. 'Landing-place for lambs'. OE *lamb* + *hȳth*.

Lambley, 'clearing or pasture for lambs', OE *lamb* + *lēah*: **Lambley** Northum. *Lambeleya* 1201. **Lambley** Notts. *Lambeleia* 1086 (DB).

Lambourn, Lambourne, probably 'stream where lambs are washed', OE *lamb* + *burna*: **Lambourn, Upper Lambourn** W. Berks. *Lambburnan c.*880, *Lamborne* 1086 (DB), *Uplamburn* 1182. Distinguishing affix is OE *upp* 'higher upstream'. **Lambourne** Essex. *Lamburna* 1086 (DB).

Lambrook, East & Lambrook, West Somerset. *Landbroc* 1065. OE *brōc* 'brook' with OE *land* 'cultivated ground, estate'.

Lambston Pemb. *Villa Lamberti* 1291, *Lamberteston* 1321. 'Lambert's farm'. Flemish pers. name + OE *tūn*.

Lamerton Devon. *Lamburnan c.*970, *Lambretona* 1086 (DB). 'Farmstead on the lamb stream or loam stream'. OE *lamb* or *lām* + *burna* + *tūn*.

Lamesley Gatesd. *Lamelay* 1297. 'Clearing or pasture for lambs'. OE *lamb* + *lēah*.

Lammermuir (district) Sc. Bord. 'Moor where lambs graze'. OE *lamb* (genitive plural *lambra*) + *mōr*.

Lammermuir Hills E. Loth. *Lombormore* 800. 'Hills by the moor where lambs graze'. OE *lamb* (genitive plural *lambra*) + *mōr*.

Lammy (*Leamhach*) Fermanagh, Tyrone. 'Abounding in elms'.

Lamonby Cumbria. *Lambeneby* 1257. 'Farmstead or village of a man called Lambin'. ME pers. name + OScand. *bý*.

Lamorran Cornwall. *Lannmoren* 969. 'Church-site of St Moren'. From Cornish **lann* with the patron saint of the church.

Lampeter (*Llanbedr Pont Steffan*) Cergn. *Lanpeder* 1284, *Lampeter Pount Steune* 1301. 'Church of St Peter'. Welsh *llan* + *Pedr*. The Welsh name means 'church of St Peter by Stephen's bridge' (Welsh *llan* + *Pedr* + *pont* + *Steffan*).

Lamplugh Cumbria. *Lamplou c.*1150. 'Bare valley'. Celtic *nant* + **bluch*.

Lamport Northants. *Langeport* 1086 (DB). 'Long village or market-place'. OE *lang* + *port*.

Lamyatt Somerset. *Lambageate* late 10th cent., *Lamieta* 1086 (DB). 'Gate for lambs'. OE *lamb* (genitive plural *-a*) + *geat*.

Lanaglug (*Lann na gClog*) Tyrone. 'Church of the bells'.

Lanark S. Lan. *Lannarc* 1188. '(Place in) the glade'. OWelsh *lannerch*.

Lancaster Lancs. *Loncastre* 1086 (DB). 'Roman fort on the River Lune'. Celtic river-name (probably meaning 'healthy, pure') + OE *cæster*. **Lancashire** (the county) is a reduced form of *Lancastreshire* 14th cent., from Lancaster + OE *scīr* 'district'.

Lanchester Durham. *Langecestr* 1196. Apparently 'long Roman fort or stronghold'. OE *lang* + *ceaster*. The first element may however preserve a reduced form of its old Romano-British name *Longovicium* (probably 'place of the ship-fighters' from Celtic **longo* 'ship').

Lancing W. Sussex. *Lancinges* 1086 (DB). '(Settlement of) the family or followers of a man called **Wlanc*'. OE pers. name + *-ingas*.

Land's End Cornwall. *Londeseynde* 1337. 'End of the mainland'. OE *land* + *ende*.

Landbeach Cambs. *Bece* 1086 (DB), *Landebeche* 1218. OE *bece* 'stream, valley' or *bæc* (locative **bece*) 'low ridge', with the later addition of *land* 'dry land' to distinguish it from WATERBEACH.

Landcross Devon. *Lanchers* 1086 (DB). Possibly OE *ears* 'buttock-shaped hill' with OE *hlanc* 'long, narrow'. Alternatively a Celtic name meaning 'church by the fen', from **lann* + **cors*.

Landford Wilts. *Langeford* [*sic*] 1086 (DB), *Laneford* 1242. Probably 'ford crossed by a lane'. OE *lanu* + *ford*.

Landkey Devon. *Landechei* 1166. 'Church-site of St Ke'. Cornish **lann* + saint's name.

Landore Swan. '(Place by the) river-bank'. Welsh *glan* + *dŵr*.

Landrake Cornwall. *Landerhtun* late 11th cent. Cornish *lannergh* 'a clearing' (with OE *tūn* 'farmstead' in the early spelling).

Landulph Cornwall. *Landelech* 1086 (DB). 'Church-site of Dilic'. Cornish **lann* + saint's name.

Landwade Cambs. *Landuuade* 11th cent. Probably 'estate or district ford'. OE *land* + *wæd*.

Laneast Cornwall. *Lanast* 1076. Cornish **lann* 'church-site' with an obscure second element, probably a pers. name.

Laneham, Church Laneham Notts. *Lanun* 1086 (DB). '(Place at) the lanes'. OE *lanu* in a dative plural form *lanum*.

Lanercost Cumbria. *Lanrecost* 1169. Celtic **lannerch* 'glade, clearing', perhaps with the pers. name *Aust* (from Latin *Augustus*).

Lanesborough (*Béal Átha Liag*) Longford. 'Lane's borough'. The name is apparently that of the English landowner. The original Irish name means 'ford-mouth of the standing stone'.

Langar Notts. *Langare* 1086 (DB). 'Long gore or point of land'. OE *lang* + *gāra*.

Langbar N. Yorks. *Langeberhe* 1199. 'The long hill'. OE *lang* + *beorg*.

Langbaurgh N. Yorks. *Langebergh(e) 1231. Identical in origin with previous name.*

Langcliffe N. Yorks. *Lanclif* 1086 (DB). 'Long cliff or bank'. OE *lang* + *clif*.

Langdale, Little Cumbria. *Langedenelittle* c.1160. 'Long valley'. OE *lang* + *denu* (replaced by OScand. *dalr*), with *lȳtel* 'little'.

Langdon, 'long hill or down', OE *lang* + *dūn*: **Langdon, East & Langdon, West** Kent. *Langandune* 861, *Estlangedoun*, *Westlangedone* 1291. **Langdon Hills** Essex. *Langenduna* 1086 (DB), *Langdon Hilles* 1485.

Langenhoe Essex. *Langhou* 1086 (DB). 'Long hill-spur'. OE *lang* (dative *-an*) + *hōh*.

Langford, usually 'long ford', OE *lang* + *ford*; examples include: **Langford** Beds. *Longaford* 944–6, *Langeford* 1086 (DB). **Langford** Essex. *Langeford* 1086 (DB). **Langford** Oxon. *Langeford* 1086 (DB). **Langford Budville** Somerset. *Langeford* 1212, *Langeford Budevill* 1305. Manorial affix from the *de Buddevill* family, here in the 13th cent.
     However the following has a different origin: **Langford** Notts. *Landeforde* 1086 (DB). Possibly 'ford of a man called **Landa*', OE pers. name + *ford*; alternatively the first element may be OE *land* in a sense such as 'boundary or district'.

Langham, usually 'long homestead or enclosure', OE *lang* + *hām* or *hamm*: **Langham** Norfolk. *Langham* 1047–70,

1086 (DB). **Langham** Rutland. *Langham* 1202.
Langham Suffolk. *Langham* 1086 (DB).
   However the following has a different origin:
Langham Essex. *Laingaham* 1086 (DB).
Possibly 'homestead of the family or followers of
a man called *Lahha'. OE pers. name + -*inga*- +
hām.

Langho Lancs. *Langale* 13th cent. 'Long nook
of land'. OE *lang* + *halh*.

Langholm Dumf. *Langholm* 1376. 'Long
island'. OScand. *langr* + *holmr*. The 'island' is
the strip of land by the River Esk.

Langley, a fairly common name, 'long wood
or clearing', OE *lang* + *lēah*; examples include:
Langley Derbys., near Heanor. *Langeleie* 1086
(DB). **Langley** Kent. *Longanleag* 814, *Langvelei*
1086 (DB). **Langley** Slough. *Langeley* 1208,
Langele Marais 1316. Manorial affix (later
Marish) from the *Mareis* family, here in the 13th
cent. **Langley, Abbots & Langley, Kings**
Herts. *Langalege* c.1060, *Langelai* 1086 (DB),
Abbotes Langele 1263, *Kyngeslangeley* 1436.
Distinguishing affixes refer to early possession
by the Abbot of St Albans and the King.
Langley Burrell Wilts. *Langelegh* 940,
Langhelei 1086 (DB), *Langele Burel* 1309.
Manorial affix from the *Burel* family, here from
the 13th cent. **Langley, Kington** Wilts.
Langhelei 1086 (DB). Affix *Kington*, first added in
the 17th cent., is from KINGTON ST MICHAEL
nearby. **Langley, Kirk** Derbys. *Langelei* 1086
(DB), *Kyrkelongeleye* 1269. Affix is OScand.
kirkja 'church'. **Langley Park** Durham.
Langeleye 1232.

Langney E. Sussex. *Langelie* [sic] 1086 (DB),
Langania 1121. 'Long island, or long piece of
dry ground in marsh'. OE *lang* (dative -*an*) + *ēg*.

Langold Notts. *Langalde* 1246. 'Long shelter
or place of refuge'. OE *lang* + *hald*.

Langport Somerset. *Longport* 10th cent.,
Lanport 1086 (DB). 'Long market-place'. OE
lang + *port*.

Langrick Lincs. *Langrak* 1243. 'Long stretch
of river'. OE *lang* + *racu.

Langridge B. & NE. Som. *Lancheris* [sic] 1086
(DB), *Langerig* 1225. 'Long ridge'. OE *lang* +
hrycg.

Langrigg Cumbria. *Langrug* 1189. 'Long
ridge'. OScand. *langr* + *hryggr*.

Langrish Hants. *Langerishe* 1236. 'Long rush-
bed'. OE *lang* + *rysce*.

Langsett Barns. *Langeside* 12th cent. 'Long
hill-slope'. OE *lang* + *sīde*.

Langstone Hants. *Langeston* 1289. 'Long (or
tall) stone'. OE *lang* + *stān*.

Langstone Newpt. *Langestone* c.1280,
Llangstone 1868. '(Place by the) long stone'.
OE *lang* + *stān*.

Langthorne N. Yorks. *Langetorp* [sic] 1086
(DB), *Langethorn* c.1100. 'Tall thorn-tree'. OE
lang + *thorn*.

Langthorpe N. Yorks. *Torp* 1086 (DB),
Langliuetorp 12th cent. 'Outlying farmstead or
hamlet of a woman called Langlif'. OScand.
pers. name + *thorp*.

Langthwaite N. Yorks. *Langethwait* 1167.
'Long clearing'. OScand. *langr* + *thveit*.

Langtoft, 'long homestead or curtilage',
OScand. *langr* + *toft*: **Langtoft** E. R. Yorks.
Langetou [sic] 1086 (DB), *Langetoft* c.1165.
Langtoft Lincs. *Langetof* 1086 (DB).

Langton, a fairly common name, usually 'long
farmstead or estate', OE *lang* + *tūn*; examples
include: **Langton, Church & Langton,
East** Leics. *Lang(e)tone* 1086 (DB), *Chirch
Langeton* 1316, *Estlangeton* 1327. **Langton,
Great** N. Yorks. *Langeton* 1086 (DB), *Great
Langeton* 1223. **Langton Herring** Dorset.
Langetone 1086 (DB), *Langeton Heryng* 1336.
Manorial affix from the *Harang* family, here in
the 13th cent. **Langton Matravers** Dorset.
Langeton 1165, *Langeton Mawtravers* 1428.
Manorial addition from the *Mautravers* family,
here from the 13th cent.
   However the following have a different origin:
Langton Durham. *Langadun* c.1040. 'Long hill
or down'. OE *lang* + *dūn*. **Langton, Tur** Leics.
Terlintone 1086 (DB). Probably 'estate
associated with a man called Tyrhtel or *Tyrli'.
OE pers. name + -*ing*- + *tūn*. The name was later
remodelled under the influence of nearby
Church & East Langton.

Langtree Devon. *Langtrewa* 1086 (DB). 'Tall
tree'. OE *lang* + *trēow*.

Langwathby Cumbria. *Langwadebi* 1159.
'Farmstead or village by the long ford'. OScand.
langr + *vath* + *bý*.

Langwith, 'long ford', OScand. *langr* + *vath*:
Langwith, Nether Notts. *Languath* c.1179,
Netherlangwat 1252. Affix is OE *neotherra* 'lower
(downstream)' to distinguish it from the next

name. **Langwith, Upper** Derbys. *Langwath* 1208.

Langworth Lincs. near Wragby. *Longwathe* c.1055. Identical in origin with the previous names.

Lanivet Cornwall. *Lannived* 1268. 'Church-site at the pagan sacred place'. Cornish **lann* + *neved*.

Lanlivery Cornwall. *Lanliveri* 12th cent. 'Church-site of **Livri*'. Cornish **lann* + pers. name.

Lanreath Cornwall. *Lanredoch* 1086 (DB). 'Church-site of Reydhogh'. Cornish **lann* + pers. name.

Lansallos Cornwall. *Lansaluus* 1086 (DB). 'Church-site of **Salwys*'. Cornish **lann* + pers. name.

Lanteglos Cornwall. near Fowey. *Lanteglos* 1249. 'Valley of the church'. Cornish *nans* + *eglos*.

Lanton Northum. *Langeton* 1242. 'Long farmstead or estate'. OE *lang* + *tūn*.

Laois (*Laois*) (the county). '(Place of the) people of Laeighseach'. Lughaidh *Laeighseach* was granted lands here after he had driven invading forces from Munster. The name is pronounced 'Leesh', giving the former anglicized form *Leix*.

Lapford Devon. *Slapeforda* [*sic*] 1086 (DB), *Lapeford* 1107. Possibly 'ford of a man called **Hlappa*', OE pers. name + *ford*. Alternatively the first element may be an OE word **(h)lēape* or **(h)lǣpe* 'lapwing'.

Lapley Staffs. *Lappeley* 1061, *Lepelie* 1086 (DB). Possibly 'woodland clearing at the end of the estate or parish'. OE *læppa* + *lēah*.

Lapworth Warwicks. *Hlappawurthin* 816, *Lapeforde* [*sic*] 1086 (DB). Probably 'enclosure of a man called **Hlappa*'. OE pers. name + *worthign* (replaced by *worth*).

Laracor (*Láithreach Cora*) Meath. 'Site of the weir'.

Laragh (*Láithreach*) Laois, Monaghan, Wicklow. 'Site'.

Laragh (*Leath Ráth*) Longford. 'Half a ring fort'.

Larbert Falk. *Lethbert* 1195. 'Half-wood'. Pictish *lled-bert*. Compare PERTH.

Largan (*Leargain*) Mayo. 'Slope'.

Larganreagh (*Leargan Riabhach*) Donegal. 'Striped slope'.

Largo Fife. *Largauch* 1250. 'Steep (place)'. Gaelic *leargach*.

Largs N. Ayr. *Larghes* 1140. '(Place by) slopes'. Gaelic *learg*, with English *-s* plural.

Largy (*Leargaidh*) Antrim, Derry, Fermanagh, etc. 'Slope'.

Largybrack (*Leargaidh Breac*) Donegal. 'Speckled slope'.

Largydonnell (*Learga Uí Dhónaill*) Sligo. 'Ó Dónaill's slope'.

Largyreagh (*Leargaidh Riabhach*) Derry. 'Striped slope'.

Larkfield Kent. *Lavrochesfel* 1086 (DB). 'Open land frequented by larks'. OE *lāwerce* + *feld*.

Larkhall S. Lan. *Laverockhall* 1620. Meaning uncertain. *Laverock* in the early form of the name represents a Scottish equivalent of English *lark*, but there is no evidence that the bird gave the name. It is possible that *Laverock* is a surname here.

Larling Norfolk. *Lurlinga* 1086 (DB). '(Settlement of) the family or followers of a man called **Lyrel*'. OE pers. name + *-ingas*.

Larne (*Latharna*) Antrim. 'Descendants of Lathar'. According to legend, Lathar was one of the 25 sons of Úgaine Mór (Hugony), a pre-Christian king of Ireland.

Larnog. *See* LAVERNOCK.

Lartington Durham. *Lyrtingtun* c.1050, *Lertinton* 1086 (DB). 'Estate associated with a man called **Lyrti*'. OE pers. name + *-ing-* + *tūn*.

Lasham Hants. *Esseham* [*sic*] 1086 (DB), *Lasham* 1175. 'Smaller homestead, or homestead of a man called **Leassa*'. OE *lǣssa* or pers. name + *hām*.

Lasswade Midloth. *Laswade* 1148, *Lesswade* c.1150. 'Pasture ford'. OE *lǣs* (genitive *lǣswe*) + *wæd*.

Lastingham N. Yorks. *Lestingeham* 1086 (DB). 'Homestead of the family or followers of a man called *Lǣsta'. OE pers. name + -inga- + hām.

Latchingdon Essex. Læcendune c.1050, Lacenduna 1086 (DB). Probably 'hill by the well-watered ground'. OE *læcen + dūn.

Latchley Cornwall. Lacchislegh 1318. 'Wood or woodland clearing in or near marshy ground'. OE *læc(c) + lēah.

Lathbury Milt. K. Lateberie 1086 (DB). 'Fortification built with laths or beams'. OE lætt + burh (dative byrig).

Latheron Highland. Lagheryn 1287. Probably 'miry place'. Gaelic láthach. The extended form of the name is Latheronwheel, adding Gaelic a' Phuill, 'of the pool'.

Latimer Bucks. Yselhamstede 1220, Isenhampstede Latymer 1389. For its original name, see CHENIES. The manorial addition (now used alone) is from the Latymer family, here in the 14th cent.

Latnamard (Leacht na mBard) Monaghan. 'Grave of the bards'.

Latteragh (Leatracha) Tipperary. 'Hill-slopes'.

Latteridge S. Glos. Laderugga 1176. Possibly 'ridge by a water course'. OE lād + hrycg.

Lattiford Somerset. Lodereforda 1086 (DB). 'Ford frequented by beggars'. OE loddere + ford.

Lattin (Laitean) Tipperary. 'Valley bottom'.

Latton Wilts. Latone 1086 (DB). 'Leek or garlic enclosure, herb garden'. OE lēac-tūn.

Latton (Leatón) Monaghan. 'Valley bottom'.

Lauder Sc. Bord. Louueder 1208. '(Place on the) Leader Water'. The Celtic river name may mean 'cleansing river' (British *lou- + *dubro-).

Laugharne (Talacharn) Carm. Talacharn 1191, Laugharne, or Tal-Llacharn 1868. 'End (of the place)'. Welsh tâl. The English form of the name has dropped the Ta- of the Welsh equivalent. The second element is of obscure origin.

Laughil (Leamhchoill) Galway, Longford, Mayo, etc. 'Elm wood'.

Laughill (Leamhchoill) Fermanagh, Laois. 'Elm wood'.

Laughterton Lincs. ?Leuggtricdun c.680, Lactertun 1227. Probably 'farmstead where lettuce grows'. OE leahtric + tūn. Identification of the first spelling with this place is uncertain.

Laughton, usually 'leek or garlic enclosure, herb garden', OE lēac-tūn: **Laughton** E. Sussex. Lestone 1086 (DB). **Laughton** Leics. Lachestone 1086 (DB). **Laughton en le Morthen** Rothm. Lastone 1086 (DB), Latton in Morthing 1230. Affix means 'in the (district called) Morthen', a name first recorded in the 12th cent. from OE mōr or OScand. mór 'moorland' + thing 'assembly'.
However the following has a different origin: **Laughton** Lincs., near Folkingham. Loctone 1086 (DB). 'Enclosure that can be locked'. OE loc + tūn.

Launcells Cornwall. Landseu [sic] 1086 (DB), Lanceles 1204. Cornish *lann 'church-site' with an uncertain second element.

Launceston Cornwall. Lanscavetone 1086 (DB). 'Estate near the church-site of St Stephen'. Cornish *lann + saint's name + OE tūn.

Laune River (Leamháin) Kerry. 'Elm river'.

Launton Oxon. Langtune c.1050, Lantone 1086 (DB). 'Long farmstead or estate'. OE lang + tūn.

Lauragh (An Láithreach) Kerry. 'The site'.

Laurelvale Armagh. Laurel Vale 1835. The name was invented by Thomas Sinton in c.1830 to apply to his house here. The Irish name is Tamhnaigh Bhealtaine, 'cultivated spot of the May festival'.

Laurencekirk Aber. 'St Laurence's church'. The town, founded c.1770, was originally known as Kirkton of St Laurence, Kirkton being 'village with a church' (OScand. kirkja + OE tūn). It later adopted its present name, after St Laurence of Canterbury.

Lavagh (Leamhach) Donegal, Sligo. 'Abounding in elms'.

Lavally (Leth Baile) Clare, Cork, Galway, etc. 'Half townland'.

Lavant, East Lavant W. Sussex. Loventone 1086 (DB), Lavent 1227. Originally 'farmstead on the River Lavant', with OE tūn, later taking name from the river alone, a Celtic river-name meaning 'gliding one'.

Lavendon Milt.K. *Lavendene* 1086 (DB). 'Valley of a man called Lāfa'. OE pers. name (genitive *-n*) + *denu*.

Lavenham Suffolk. *Lauanham c.*995, *Lauenham* 1086 (DB). 'Homestead of a man called Lāfa'. OE pers. name (genitive *-n*) + *hām*.

Laver, High, Laver, Little, & Laver, Magdalen Essex. *Lagefare c.*1010, *Lagafara* 1086 (DB), *Laufar la Magdelene* 1263. 'Water passage or crossing'. OE *lagu* + *fær*. Affix *Magdalen* from the dedication of the church.

Lavernock (*Larnog*) Vale Glam. *Lawernach* 13th cent. 'Lark hill'. OE *lāwerce* + *cnocc*. The Welsh name is a phonetic form of the English.

Laverstock Wilts. *Lavvrecestoches* 1086 (DB). 'Outlying farmstead or hamlet frequented by larks'. OE *lāwerce* + *stoc*.

Laverstoke Hants. *Lavrochestoche* 1086 (DB). Identical in origin with the previous name.

Laverton, usually probably 'farmstead frequented by larks', OE *lāwerce* + *tūn*: **Laverton** Glos. *Lawertune c.*1160. **Laverton** Somerset. *Lavretone* 1086 (DB).
    However the following has a different origin: **Laverton** N. Yorks. *Lavretone* 1086 (DB). 'Farmstead on the River Laver'. Celtic river-name (meaning 'babbling brook') + OE *tūn*.

Lavey (*Leamhach*) Cavan, Derry. 'Abounding in elms'.

Lavington, probably 'estate associated with a man called Lāfa', OE pers. name + *-ing-* + *tūn*: **Lavington, East & Lavington, West** W. Sussex. *Levitone* 1086 (DB). **Lavington, Market & Lavington, West** Wilts. *Laventone* 1086 (DB). Distinguishing affixes *market* and *west* first found in the 17th cent.

Lawford, probably 'ford of a man called *Lealla*, OE pers. name + *ford*: **Lawford** Essex. *Lalleford* 1045, *Laleforda* 1086 (DB). **Lawford, Church & Lawford, Long** Warwicks. *Lelleford* 1086 (DB), *Chirche Lalleford, Long Lalleford* 1235. Distinguishing affixes are OE *cirice* 'church' and *lang* 'long'.

Lawhitton Cornwall. *Landuuithan* 10th cent., *Languitetone* 1086 (DB). Probably 'valley or church-site of a man called Gwethen', Cornish *nans* or **lann* + pers. name, with the later addition of OE *tūn* 'farmstead'.

Lawkland N. Yorks. *Laukeland* 12th cent. 'Arable land where leeks are grown'. OScand. *laukr* + *land*.

Lawley Tel. & Wrek. *Lauelei* 1086 (DB). 'Woodland clearing of a man called Lāfa'. OE pers. name + *lēah*.

Lawrencetown Down. *Laurencetown* formerly Hall's Mill *c.*1834. 'Lawrence's town'. The *Lawrence* family were here before the 18th cent., when the village was known as *Hall's Mill*, after Francis *Hall*, who leased land from the Lawrences in 1674. The corresponding Irish name is *Baile Labhráis*.

Lawrenny Pemb. *Leurenni c.*1200. '(Place on) low ground'. Welsh *llawr*. The second element is of uncertain origin.

Lawshall Suffolk. *Lawessela* 1086 (DB). 'Dwelling or shelter by a hill or mound'. OE *hlāw* + *sele* or **sell*.

Lawton, 'farmstead by a hill or mound', OE *hlāw* + *tūn*: **Lawton** Herefs. *Lavtune* 1086 (DB). **Lawton, Church** Ches. *Lautune* 1086 (DB), *Church Laughton* 1331.

Laxay W. Isles. (Lewis). '(Place by) the salmon river'. OScand. *lax* + *á*.

Laxey Isle of Man. Named from the River Laxey on which it is situated, 'salmon river', from OScand. *lax* + *á*.

Laxfield Suffolk. *Laxefelda* 1086 (DB). 'Open land of a man called **Leaxa*'. OE pers. name + *feld*.

Laxton, 'estate associated with a man called **Leaxa*', OE pers. name + *-ing-* + *tūn*: **Laxton** E. R. Yorks. *Laxinton* 1086 (DB). **Laxton** Notts. *Laxintune* 1086 (DB).
    However the following may have a slightly different meaning: **Laxton** Northants. *Lastone* 1086 (DB). '**Leaxa*'s estate'. OE pers. name + *tūn*.

Laycock Brad. *Lacoc* 1086 (DB). 'The small stream'. OE **lacuc*.

Layer Breton, Layer de la Haye & Layer Marney Essex. *Legra* 1086 (DB), *Leyre Bretoun* 1254, *Legra de Haya* 1236, *Leyre Marini* 1254. Probably originally a river-name *Leire* of Celtic origin. Distinguishing affixes from early possession by families called *Breton, de Haia* and *de Marinni*.

Layham Suffolk. *Hligham c.*995, *Leiham* 1086 (DB). 'Homestead with a shelter'. OE **hlīg* + *ham*.

Laytham E. R. Yorks. *Ladon* 1086 (DB). '(Place at) the barns'. OScand. *hlatha* in a dative plural form *hlathum*.

Layton, East & Layton, West N. Yorks. *Latton, Lastun* 1086 (DB). 'Leek enclosure, herb garden'. OE *lēac-tūn*.

Lazenby Red. & Cleve. *Lesingebi* 1086 (DB). 'Farmstead of the freedmen, or of a man called **Leysingr*'. OScand. *leysingi* or pers. name + *bý*.

Lazonby Cumbria. *Leisingebi* 1165. Identical in origin with the previous name.

Lea (or Lee) (river) Beds.–Herts.–Essex. *See* LEYTON.

Lea, '(place at) the wood or woodland clearing', OE *lēah*; examples include: **Lea** Derbys. *Lede* [*sic*] 1086 (DB), *Lea c.*1155. **Lea** Herefs. *Lecce* [*sic*] 1086 (DB), *La Lee* 1219. **Lea** Lincs. *Lea* 1086 (DB). **Lea** Wilts. *Lia* 1190. **Lea Town** Lancs. *Lea* 1086 (DB).

Leabeg (*Liath Beag*) Offaly. 'Small grey place'.

Leach (river) Glos. *See* EASTLEACH.

Leaden Roding Essex. *See* RODING.

Leadenham Lincs. *Ledeneham* 1086 (DB). Probably 'homestead or village of a man called **Lēoda*'. OE pers. name (genitive *-n*) + *hām*.

Leadgate Durham. *Lydyate* 1404. OE *hlid-geat* 'a swing-gate'.

Leadhills S. Lan. '(Place by) the lead hills'. The reference is to lead mines.

Leadon (river) Glos.–Herefs.–Worcs., *See* HIGHLEADON.

Leafield Oxon. *La Felde* 1213. 'The open land'. OE *feld* with the OFrench definite article.

Leafin (*Liathmhuine*) Meath. 'Grey thicket'.

Leafonny (*Liathmhuine*) Sligo. 'Grey thicket'.

Leagrave Luton. *Littegraue* 1224, *Lihte-, Littlegraue* 1227. Possibly 'grove or coppiced wood of a man called **Lihtla*', OE pers. name + *grāf(a)*. Alternatively the first element may be OE *lēoht* 'light-coloured'.

Leahan (*Leathan*) Donegal. 'Broad space'.

Leahanmore (*Leathan Mór*) Donegal. 'Big broad space'.

Leake, '(place at) the brook', OScand. *lœkr*: **Leake** N. Yorks *Lec(h)e* 1086 (DB). **Leake, East & Leake, West** Notts. *Lec(c)he* 1086 (DB). **Leake, Old** Lincs. *Leche* 1086 (DB).

Lealholm N. Yorks. *Lelun* 1086 (DB). '(Place among) the twigs or brushwood'. OE *læl* in a dative plural form *lǣlum*.

Leamington, 'farmstead on the River Leam', Celtic river-name (meaning 'elm river', or 'marshy river') + OE *tūn*: **Leamington Hastings** Warwicks. *Lunnitone* 1086 (DB), *Lemyngton Hasting* 1285. Manorial affix from the *Hastings* family, here in the 13th cent. **Royal Leamington Spa** Warwicks. *Lamintone* 1086 (DB). The affix was granted by Queen Victoria in 1838, *Spa* referring to the medicinal springs here.

Leamlara (*Léim Lára*) Cork. 'Leap of the mare'.

Leap (*Léim*) Cork. 'Leap'.

Learmore (*Ladhar Mór*) Tyrone. 'Large fork'.

Learmount (*Ladhar*) Tyrone. 'Mount of the fork'.

Learmouth, East & Learmouth, West Northum. *Leuremue* 1177. 'Mouth of the stream where rushes grow'. OE *lǣfer* + *mūtha*.

Leasgill Cumbria. *Lesegill* 1458. 'Bright or bare ravine', or 'ravine of a woman called Ljósa'. OScand. *ljóss* (adjective) or pers. name + *gil*.

Leasingham Lincs. *Leuesingham* 1086 (DB). 'Homestead of the family or followers of a man called Lēofsige'. OE pers. name + *-inga-* + *hām*.

Leatherhead Surrey. *Leodridan c.*880, *Leret* [*sic*] 1086 (DB). 'Grey ford'. Celtic **lēd* + **rïd*.

Leathley N. Yorks. *Ledelai* 1086 (DB). 'Woodland clearing on the slopes'. OE *hlith* (genitive plural *hleotha*) + *lēah*.

Leaton Shrops. *Letone* 1086 (DB). Probably 'settlement by a wood or woodland clearing'. OE *lēah* + *tūn*.

Leaveland Kent. *Levelant* 1086 (DB). 'Cultivated land of a man called Lēofa'. OE pers. name + *land*.

Leavenheath Suffolk. *heath of Levynhey* 1292, *Levenesheth* 1351. 'Heath at Levin's enclosure'. ME pers. name or surname (from OE *Lēofwine*) + *hey* 'enclosure' with *hethe* 'heath'.

Leavening N. Yorks. *Ledlinghe* [*sic*] 1086 (DB), *Leyingges* 1242. Possibly '(settlement of) the family or followers of a man called Lēofhēah'. OE pers. name + -*ingas*.

Leaves Green Gtr. London. *Lese Green* 1500. Named from a local family called *Leigh* recorded from 1447, with ME *grene* 'village green'.

Lebberston N. Yorks. *Ledbeztun* 1086 (DB). 'Farmstead of a man called Lēodbriht'. OE pers. name + *tūn*.

Lecale (*Leath Cathail*) Down. 'Cathal's portion'.

Lecarrow (*An Leithcheathrú*) Galway. 'The half quarter'.

Lechlade Glos. *Lecelade* 1086 (DB). Probably 'river-crossing near the River Leach'. River-name (from OE **læc(c)*, **lece* 'boggy stream') + *gelād*.

Leckanvy (*Leacán Mhaigh*) Mayo. 'Hillside of the plain'.

Leckaun (*An Leacán*) Sligo. 'The hillside'.

Leckford Hants. *Leahtforda* 947, *Lechtford* 1086 (DB). 'Ford at or over a channel'. OE **leaht* + *ford*.

Leckhampstead, 'homestead where leeks are grown', OE *lēac* + *hām-stede*: **Leckhampstead** Bucks. *Lechamstede* 1086 (DB). **Leckhampstead** W. Berks. *Lechamstede* 956–9, *Lecanestede* 1086 (DB).

Leckhampton Glos. *Lechametone* 1086 (DB). 'Home farm where leeks are grown'. OE *lēac* + *hām-tūn*.

Leckwith (*Lecwydd*) Card. *Leocwtha* c.1170. '(Place of) Helygwydd'. The pers. name is apparently that of a local saint.

Leconfield E. R. Yorks. *Lachinfeld* 1086 (DB). OE *feld* 'open land' with an uncertain first element, possibly a derivative of OE **læc(c)*, **lece* 'boggy stream'.

Lecwydd. *See* LECKWITH.

Ledburn Bucks. *Leteburn* 1212. 'Stream with a conduit, or by a cross-roads'. OE (*ge*)*lǣt(e)* + *burna*.

Ledbury Herefs. *Liedeberge* 1086 (DB). Probably 'fortified place on the River Leadon'. Celtic river-name (*see* HIGHLEADON) + OE *burh* (dative *byrig*).

Ledsham Ches. *Levetesham* 1086 (DB). 'Homestead of a man called Lēofede'. OE pers. name + *hām*.

Ledsham Leeds. *Ledesham* c.1030, 1086 (DB). 'Homestead within the district of LEEDS'. OE *hām*.

Ledston Leeds. *Ledestune* 1086 (DB). 'Farmstead within the district of LEEDS'. OE *tūn*.

Ledwell Oxon. *Ledewelle* 1086 (DB). 'Spring or stream called the loud one'. OE **hlȳde* + *wella*.

Lee, '(place at) the wood or woodland clearing', OE *lēah*; examples include: **Lee** Gtr. London. *Lee* 1086 (DB). **Lee** Hants. *Ly* 1236. **Lee** Shrops. *Lee* 1327. **Lee Brockhurst** Shrops. *Lege* 1086 (DB), *Leye under Brochurst* 1285. Affix from a nearby place, 'wooded hill frequented by badgers', OE *brocc* + *hyrst*. **Lee on the Solent** Hants. *Lie* 1212. *See* SOLENT.

Lee (*Laoi*) (river) Cork. Perhaps 'water'.

Leebotwood Shrops. *Botewde* 1086 (DB), *Lega in Bottewode* c.1170–76. 'Wood of a man called Botta', OE pers. name + *wudu*, with the later addition of a place name Lee from OE *lēah* 'clearing'.

Leece Cumbria. *Lies* 1086 (DB). Probably from Celtic **līss* 'a hall, a court, the chief house in a district'.

Leeds Kent. *Esledes* 1086 (DB), *Hledes* c.1100. Possibly from a stream-name, OE **hlȳde* 'the loud one'.

Leeds Leeds. *Loidis* 731, *Ledes* 1086 (DB). A Celtic name, originally *Lādenses*, meaning 'people living by the strongly flowing river'.

Leeg (*Liag*) Monaghan. 'Pillar stone'.

Leek Staffs. *Lec* 1086 (DB), '(Place at) the brook'. OScand. *lœkr*.

Leek (*Liag*) Monaghan. 'Pillar stone'.

Leek Wootton Warwicks. *See* WOOTTON.

Leeke (*Liag*) Antrim, Derry. 'Pillar stone'.

Leeming N. Yorks. *Leming* 12th cent. Named from Leeming Beck, a river-name possibly of OE origin and meaning 'bright stream'.

Leenaun (*An Líonán*) Galway. 'The shallow sea bed'.

Lees Oldham. *The Leese* 1604. 'The woods or woodland clearings'. OE *lēah* (plural *lēas*).

Lefinn (*Liathmhuine*) Donegal. 'Grey thicket'.

Legananny (*Liagán Áine*) Down. *Ballyleganananagh* 1609. 'Pillar stone of Áine'. The stone in question is the nearby Legananny Dolmen.

Legbourne Lincs. *Lecheburne* 1086 (DB). Probably 'boggy stream'. OE **læc(c)*, **lece* + *burna*.

Leggamaddy (*Lag an Mhadaidh*) Down. *Liggamaddy* 1755. 'Hollow of the dog'.

Leggs (*Na Laig*) Fermanagh. 'The hollows'.

Legh, High Ches. *Lege* 1086 (DB). '(Place at) the wood or woodland clearing'. OE *lēah*, with the addition of *High* from the 15th cent.

Legland (*Leithghleann*) Tyrone. *Legglan* 1661. 'Side of the valley'.

Legoniel (*Lag an Aoil*) Antrim. *Ballylagaile* 1604. 'Hollow of the lime'. There are several disused lime quarries locally.

Legsby Lincs. *Lagesbi* 1086 (DB). 'Farmstead or village of a man called Leggr'. OScand. pers. name + *bý*.

Legvoy (*Lec Mhagh*) Roscommon. 'Plain full of flagstones'.

Lehinch (*Leath Inse*) Clare. 'Peninsula'. The current Irish name is *An Leacht*, 'the grave'.

Leicester Leic. *Ligera ceastre* early 10th cent., *Ledecestre* 1086 (DB). 'Roman town of the people called Ligore'. Tribal name (derived from the ancient river-name preserved in nearby LEIRE) + OE *ceaster*. **Leicestershire** (OE *scīr* 'district') is first referred to in the 11th cent.

Leigh, a common name, '(place at) the wood or woodland clearing', OE *lēah*; examples include: **Leigh** Kent. *Lega* c.1100. **Leigh** Surrey. *Leghe* late 12th cent. **Leigh** Wigan. *Legh* 1276. **Leigh** Worcs. *Beornothesleah* 972, *Lege* 1086 (DB). 'Belonging to a man called Beornnōth' in the 10th cent. spelling. **Leigh, Abbots** N. Som. *Lege* 1086 (DB). Affix from its early possession by the Abbot of St Augustine's Bristol. **Leigh, Bessels** Oxon. *Leie* 1086 (DB), *Bessilles Lee* 1538. Manorial affix from the *Besyles* family, here in the 15th cent. **Leigh, North & Leigh, South** Oxon. *Lege* 1086 (DB), *Northleg* 1225, *Suthleye* early 13th cent. **Leigh-on-Sea** Sthend. *Legra* [sic] 1086 (DB), *Legha* 1226. **Leigh Sinton** Worcs., see SINTON. **Leigh-upon-Mendip** Somerset. *Leage* c.1100. See MENDIP HILLS.

Leighlin (*Leithghleann*) Carlow. 'Side of a valley'.

Leighmoney (*Liathmhuine*) Cork. 'Grey thicket'.

Leighs, Great & Leighs, Little Essex. *Lega* 1086 (DB), *Leyes* 1271. OE *lēah* 'wood, woodland clearing'.

Leighterton Glos. *Lettrintone* 12th cent. 'Farmstead where lettuce grows'. OE *leahtric* + *tūn*.

Leighton, 'leek or garlic enclosure, herb garden', OE *lēac-tūn*: **Leighton** Shrops. *Lestone* 1086 (DB). **Leighton Bromswold** Cambs. *Lectone* 1086 (DB), *Letton super Bruneswald* 1254. Affix is from a nearby place, 'high forest-land of a man called Brūn', OE pers. name + *weald*. **Leighton Buzzard** Beds. *Lestone* 1086 (DB), *Letton Busard* 1254. Manorial affix is from a family called *Busard*, no doubt landowners here in the 13th cent.

Leinster (*Cúige Laighean*) (the province). *Laynster* 1515. 'District of the Lagin people'. OScand. possessive *-s* + Irish *tír*. The tribal name may mean 'spear'.

Leinthall Earls & Leinthall Starkes Herefs. *Lentehale* 1086 (DB), *Leintall Comites* 1275, *Leinthale Starkare* 13th cent. 'Nook of land by the River *Lent*'. Lost Celtic river-name (meaning 'torrent, stream') + OE *halh*. Manorial affixes from early possession by an *earl* (Latin *comitis* 'of the earl') and a man called *Starker*.

Leintwardine Herefs. *Lenteurde* 1086 (DB). 'Enclosure on the River *Lent*'. Lost Celtic river-name (see previous name) + *worth* (replaced by *worthign*).

Leire Leics. *Legre* 1086 (DB). Probably an ancient pre-English river-name for the small tributary of the River Soar on which the place stands.

Leiston Suffolk. *Leistuna* 1086 (DB).
Possibly 'farmstead near a beacon-fire'. OE *lēg*
+ *tūn*.

Leith Edin. *Inverlet c.*1130, *Inverlethe c.*1315.
Originally Inverleith, 'mouth of the River Leith'.
Gaelic *inbhir*, 'river-mouth' + **Lekta* 'dripping
(water)'.

Leitrim (*Liatroim*) Down, Leitrim. 'Grey
ridge'.

Leix. *See* LAOIS.

Leixlip Kildare. 'Salmon leap'. OScand. *leax* +
hlaup. The equivalent Irish name is *Léim an
Bhradáin*.

Lelant Cornwall. *Lananta c.*1170.
'Church-site of Anta'. Cornish **lann* + female
saint's name.

Lelley E. R. Yorks. *Lelle* 1246. 'Wood or
clearing where twigs or brushwood are found'.
OE *lǣl* + *lēah*.

Lemanaghan (*Liath Mancháin*) Offaly.
'Grey land of Manchán'.

Lemington, Lower Glos. *Limentone* 1086
(DB). Probably 'farmstead near a stream called
Limen'. Lost Celtic river-name (meaning 'elm
river') + OE *tūn*.

Lemlara (*Léim Lára*) Cork. 'Leap of the
mare'.

Lemybrien (*Léim Uí Bhriain*) Waterford.
'Leap of Ó Briain'.

Lenaderg (*Láithreach Dhearg*) Down.
Laireachtdyke 1427. 'Red site'.

Lenadoon Point (*Léana Dúin*) Sligo.
'Meadow of the fort'.

Lench, from OE **hlenc* 'an extensive hill-
slope', originally the name of a district and
giving name to: **Lench, Abbots** Herefs.
Abeleng 1086 (DB). Manorial affix originally
from possession by a man with the OE pers.
name *Abba*. **Lench, Atch** Herefs. *Achelenz*
1086 (DB). Manorial affix from possession by a
man with the OE pers. name * Æcci*. **Lench,
Church** Herefs. *Lench* 9th cent., *Chirichlench*
1054. Affix is OE *cirice* 'church'. **Lench, Rous**
Herefs. *Lenc* 983, *Biscopesleng* 1086 (DB), *Rous
Lench* 1445. Manorial affix from the *Rous* family,
here in the 14th cent. (held by the Bishop of
Worcester in 1086).

Lendrick Perth. *Lennerick* 1669. '(Place by)
the glade'. OWelsh *lannerch*.

Lene, Lough (*Loch Léibhinn*) Westmeath.
'Léibhinn's lake'.

Lenham Kent. *Leanaham* 858, *Lerham* [*sic*]
1086 (DB). 'Homestead or village of a man called
*Lēana'. OE pers. name + *hām*. The river-name
Len is a back-formation from the place name.

Lennoxtown E. Dunb. 'Lennox's town'. The
town arose in the 1780s and took its name from
the family of the earls and dukes of *Lennox*,
whose title comes from the ancient territory of
Lennox (Gaelic *leamhanach*, 'abounding in
elms').

Lenton Lincs. *Lofintun c.*1067, *Lavintone* 1086
(DB). Probably 'estate associated with a man
called *Lāfa*'. OE pers. name + *-ing-* + *tūn*.

Lenwade Norfolk. *Langewade* 1199. 'Ford
crossed by a lane'. OE *lanu* + *gewæd*.

Leny (*Léana*) Westmeath. 'Meadow'.

Leominster Herefs. *Leomynster* 10th cent.,
Leominstre 1086 (DB). 'Church in *Leon*'. Old
Celtic name for the district (meaning 'at the
streams') + OE *mynster*.

Leonard Stanley Glos. *See* STANLEY.

Leppington N. Yorks. *Lepinton* 1086 (DB).
'Estate associated with a man called Leppa'. OE
pers. name + *-ing-* + *tūn*.

Lepton Kirkl. *Leptone* 1086 (DB). 'Farmstead
on a hill-slope'. OE **hlēp* + *tūn*.

Lerrig (*Leirg*) Kerry. 'Slope'.

Lerwick Shet. *Lerwick* 1625. 'Mud bay'.
OScand. *leirr* + *vík*.

Lesbury Northum. *Lechesbiri c.*1190.
'Fortified house of the leech or physician'. OE
lǣce + *burh* (dative *byrig*).

Leslie Aber. *Lesslyn c.*1180, *Lescelin c.*1232.
A difficult name. **Leslie** in Fife is
transferred from this name.

Lesmahagow S. Lan. *Lesmahagu c.*1130.
'Enclosure of beloved Fechin'. Gaelic *leas*. The
saint's name is in the affectionate diminutive
form *Mo-Fhegu* (Gaelic *mo*, 'my').

Lesnewth Cornwall. *Lisniwen* [*sic*] 1086 (DB),
Lisneweth 1201. 'New court'. Cornish **lys* +
nowydh.

Lessingham Norfolk. *Losincham* 1086 (DB).
Possibly 'homestead of the family or
followers of a man called Lēofsige'. OE pers.
name + *-inga-* + *hām*.

Letchworth Herts. *Leceworde* 1086.
Probably 'enclosure that can be locked'.
OE **lycce* + *worth*.

Letcombe Bassett & Letcombe Regis
Oxon. *Ledecumbe* 1086 (DB). 'Valley of a man
called Lēoda'. OE pers. name + *cumb*.
Distinguishing affixes from early possession by
the *Bassett* family and by the Crown.

Letheringham Suffolk. *Letheringaham* 1086
(DB). Probably 'homestead of the family or
followers of a man called Lēodhere'. OE pers.
name + *-inga-* + *hām*.

Letheringsett Norfolk. *Leringaseta* 1086
(DB). Probably 'dwelling or fold of the family or
followers of a man called Lēodhere'. OE pers.
name + *-inga-* + (*ge*)*set*.

Letter (*Leitir*) Fermanagh, Kerry. 'Hillside'.

Letterbarrow (*Leitir Barra*) Donegal. 'Top
hillside'.

Letterbreen (*Leitir Bhrúin*) Fermanagh.
Letterbreen 1751. Possibly 'hillside of the fairy
fort'.

Letterbrick (*Leitir Bruic*) Mayo. 'Hillside of
the badger'.

Letterfearn Highland. *Letterfearn* 1509.
'Hill-slope growing with alders'. Gaelic *leitir* +
fearn.

Letterfinnish (*Leitir Fionnuisce*) Kerry.
'Hillside of bright water'.

Lettergesh (*Leitir Geis*) Galway. 'Hillside of
the taboo'.

Letterkenny (*Leitir Ceanainn*) Donegal.
Latterkeny 1685. 'Hillside of the white top'.

Lettermacaward (*Leitir Mhic an Bhaird*)
Donegal. *Lettermacaward* 1608. 'Mac an Bard's
hillside'.

Lettermore (*Leitir Móir*) Galway. 'Big
hillside'.

Lettermullan (*Leitir Mealláin*) Galway.
'Hill-slope of Meallán'.

Letternadarriv (*Leitir na dTarbh*) Kerry.
'Hillside of the bulls'.

Letterston (*Treletert*) Pemb. *Villa Letardi*
1230, *Lettardston* 1332. 'Letard's farm'.
OGerman pers. name + OE *tūn* (Welsh *tref*). The
Welsh name translates the English.

Letton, probably 'leek enclosure, herb
garden', OE *lēac-tūn*: **Letton** Herefs., near
Eardisley. *Letune* 1086 (DB). **Letton** Herefs.,
near Walford. *Lectune* 1086 (DB).

Letwell Rothm. *Lettewelle* c.1150. Possibly
'spring or stream with an obstructed flow'.
ME *lette* + OE *wella*.

Leuchars Fife. *Locres* 1300. '(Place of) rushes'.
Gaelic *luachar* (genitive *luachair*) + English
plural *-s*.

Levally (*Leathbhaile*) Galway, Mayo. 'Half
townland'.

Leven (river), Stock. on T. *See*
KIRKLEVINGTON.

Leven E. R. Yorks. *Leuene* 1086 (DB).
Probably originally the name of the stream
here, a lost Celtic river-name possibly
meaning 'smooth one'.

Leven Fife. *Levin* c.1535. '(Place on) the River
Leven'. The river-name means 'elm river'
(Gaelic *leamhain*).

Levens Cumbria. *Lefuenes* 1086 (DB).
Probably 'promontory of a man called Lēofa'.
OE pers. name (genitive *-n*) + *næss*.

Levenshulme Manch. *Lewyneshulm* 1246.
'Island of a man called Lēofwine'. OE pers.
name + OScand. *holmr*.

Lever, Darcy & Lever, Little Bolton.
Parua Lefre 1212, *Darcye Lever* 1590. 'Place
where rushes grow'. OE *læfer*. Distinguishing
affixes are from possession by the *D'Arcy* family,
here c.1500, and from Latin *parva* 'little'.

Leverington Cambs. *Leverington* c.1130.
'Estate associated with a man called Lēofhere'.
OE pers. name + *-ing-* + *tūn*.

Leverton Lincs. *Leuretune* 1086 (DB).
Probably 'farmstead where rushes grow'.
OE *læfer* + *tūn*.

Leverton, North & Leverton, South
Notts. *Legretone* 1086 (DB). Probably 'farmstead
on a stream called *Legre*'. Celtic river-name
(of uncertain meaning) + OE *tūn*.

Levington Suffolk. *Leuentona* 1086 (DB). Probably 'farmstead of a man called Lēofa'. OE pers. name (genitive -*n*) + *tūn*.

Levisham N. Yorks. *Leuecen* [*sic*] 1086 (DB), *Leuezham* c.1230. Probably 'homestead or village of a man called Lēofgēat'. OE pers. name + *hām*.

Lew Oxon. *Hlæwe* 984, *Lewa* 1086 (DB). '(Place at) the mound or tumulus'. OE *hlæw*.

Lewannick Cornwall. *Lanwenuc* c.1125. 'Church-site of Gwenek'. Cornish **lann* + pers. name.

Lewes E. Sussex. (*to*) *Læwe* early 10th cent., c.960, *Lewes* 1086 (DB). From the rare OE word *læw* 'wound, incision', here used in a topographical sense 'gap'. The spelling with plural -*s* in DB and other post-Conquest sources is due to Anglo-Norman influence.

Lewis (island) W. Isles. *Leodus* c.1100, *Leoghuis* 1449. Pre-Norse name of unknown linguistic origin.

Lewisham Gtr. London. *Lievesham* 918, *Levesham* 1086 (DB). 'Homestead or village of a man called **Lēofsa*'. OE pers. name + *hām*. Also referred to earlier in the phrase *Liofshema mearc* 862, 'boundary of the people of Lewisham'.

Lewknor Oxon. *Leofecanoran* 990, *Levecanole* [*sic*] 1086 (DB). 'Hill-slope of a man called Lēofeca'. OE pers. name (genitive -*n*) + *ōra*.

Lewtrenchard Devon. *Lewe* 1086 (DB), *Lyu Trencharde* 1261. Originally the name of the river here, a Celtic river-name meaning 'bright one'. Manorial affix from the *Trenchard* family, here in the 13th cent.

Lexham, East & Lexham, West Norfolk. *Lecesham* 1086 (DB). 'Homestead or village of the leech or physician'. OE *læce* + *hām*.

Leybourne Kent. *Lillanburna* 10th cent., *Leleburne* 1086 (DB). 'Stream of a man called **Lylla*'. OE pers. name + *burna*.

Leyburn N. Yorks. *Leborne* 1086 (DB). OE *burna* 'stream', possibly with **hlēg* 'shelter'.

Leyland Lancs. *Lailand* 1086 (DB). 'Estate with untilled ground'. OE **læge* + *land*.

Leysdown on Sea Kent. *Legesdun* c.1100. Probably 'hill with a beacon-fire'. OE *lēg* + *dūn*.

Leyton Gtr. London. *Lugetune* c.1050, *Leintune* 1086 (DB). 'Farmstead on the River Lea (or Lee)'. Celtic river-name (possibly meaning 'bright river' or 'river dedicated to the god Lugus') + OE *tūn*.

Lezant Cornwall. *Lansant* c.1125. 'Church-site of Sant'. Cornish **lann* + pers. name.

Liafin (*Liathmhuine*) Donegal. 'Grey thicket'.

Libberton S. Lan. *Libertun* c.1186. Probably 'barley farm on a hillside'. OE *hlith* + *bere-tūn*.

Liberton Edin. *Libertun* 1128. Probably 'barley farm on a hillside'. OE *hlith* + *bere-tūn*.

Lichfield Staffs. *Licitfelda* c.710–20. 'Open land near *Letocetum*'. OE *feld* added to a Celtic place-name meaning 'grey wood', Celtic **lēd*, **luid* + **cēd*.

Lickeen (*Licín*) Kildare. 'Small flagstone'.

Lickowen (*Leic Eoghain*) Cork. 'Eogha's flagstone'.

Liddesdale Sc. Bord. *Lidelesdale* 1179. 'Valley of the River Liddel'. The river-name means 'valley of the River Lid', itself meaning 'loud one' (OE **hlȳde* + *dæl*).

Liddington Swindn. *Lidentune* 940, *Ledentone* 1086 (DB). 'Farmstead on the noisy stream'. OE **hlȳde* + *tūn*.

Lidgate Suffolk. *Litgata* 1086 (DB). '(Place at) the swing-gate'. OE *hlid-geat*.

Lidlington Beds. *Litincletone* 1086 (DB). 'Farmstead of the family or followers of a man called **Lȳtel*'. OE pers. name + -*inga*- + *tūn*.

Lifford (*Leifear*) Donegal. (*caislén*) *Leithbhir* 1527. 'Side of the water'.

Lifton Devon. *Liwtune* c.880, *Listone* [*sic*] 1086 (DB). 'Farmstead on the River Lew'. Celtic river-name (*see* LEWTRENCHARD) + OE *tūn*.

Lighthorne Warwicks. *Listecorne* [*sic*] 1086 (DB), *Litthethurne* 1235. 'Light-coloured thorn-tree'. OE *lēoht* + *thyrne*.

Lilbourne Northants. *Lilleburne* 1086 (DB). 'Stream of a man called Lilla'. OE pers. name + *burna*.

Lilburn Northum. *Lilleburn* 1170. Identical in origin with the previous name.

Lilleshall Tel. & Wrek. *Linleshelle [sic]* 1086 (DB), *Lilleshull* 1162. 'Hill of a man called Lill'. OE pers. name + *hyll*.

Lilley Herts. *Linleia* 1086 (DB). 'Woodland clearing where flax is grown'. OE *līn* + *lēah*.

Lilling, East & Lilling, West N. Yorks. *Lilinge* 1086 (DB). '(Settlement of) the family or followers of a man called Lilla'. OE pers. name + -*ingas*.

Lillingstone Dayrell & Lillingstone Lovell Bucks. *Lillingestan* 1086 (DB), *Litlingestan Daireli* 1166. 'Boundary stone of the family or followers of a man called *Lȳtel*'. OE pers. name + -*inga*- + *stān*. Manorial affixes from early possession by the *Dayrell* and *Lovell* families.

Lillington Dorset. *Lilletone* 1166, *Lullinton* 1200. 'Estate associated with a man called *Lylla*'. OE pers. name +-*ing*- + *tūn*.

Lillington Warwicks. *Lillintone* 1086 (DB). 'Estate associated with a man called Lilla'. OE pers. name + -*ing*- + *tūn*.

Lilliput Dorset, first recorded 1783, named from the imaginary island peopled by pygmies in Jonathan Swift's *Gulliver's Travels* (1726).

Lilstock Somerset. *Lulestoch* 1086 (DB). 'Outlying farmstead or hamlet of a man called *Lylla*'. OE pers. name + *stoc*.

Lim (river) Devon–Dorset. *See* LYME REGIS.

Limavady (*Léim an Mhadaidh*) Derry. *(i) Leim an Mhadaidh* 1542. 'Leap of the dog'. The present town was founded in the early 17th cent. under the name *Newtown Limavady*.

Limber, Great Lincs. *Lindbeorhge c.*1067, *Lindberge* 1086 (DB). 'Hill where lime-trees grow'. OE *lind* + *beorg*.

Limbury Luton. *Lygeanburg* late 9th cent. 'Stronghold on the River Lea'. Celtic river-name (*see* LEYTON) + OE *burh* (dative *byrig*).

Limehouse Gtr. London. *Le Lymhostes* 1367. 'The lime oasts or kilns'. OE *līm* + *āst*.

Limerick (*Luimneach*) Limerick. *Luimneach* 11th cent. 'Bare area'.

Limpenhoe Norfolk. *Limpeho* 1086 (DB). Probably 'hill-spur of a man called *Limpa*'. OE pers. name (genitive -*n*) + *hōh*.

Limpley Stoke Wilts. *See* STOKE.

Limpsfield Surrey. *Limenesfelde* 1086 (DB). 'Open land at *Limen*'. OE *feld* added to a Celtic place name or river-name, a derivative of the British word for 'elm'.

Linby Notts. *Lidebi [sic]* 1086 (DB), *Lindebi* 1163. 'Farmstead or village where lime-trees grow'. OScand. *lind* + *bý*.

Linchmere W. Sussex. *Wlenchemere* 1186. 'Pool of a man called *Wlenca*'. OE pers. name + *mere*.

Lincoln Lincs. *Lindon c.*150, *Lindum colonia* late 7th cent., *Lincolia* 1086 (DB). 'Roman colony (for retired legionaries) by the pool', referring to the broad pool in the River Witham. Celtic **lindo*- + Latin *colonia*. **Lincolnshire** (OE *scīr* 'district') is first referred to in the 11th cent.

Lindal in Furness Cumbria. *Lindale c.*1220. 'Valley where lime-trees grow'. OScand. *lind* + *dalr*. For the old district name Furness, *see* BARROW IN FURNESS.

Lindale Cumbria. *Lindale* 1246. Identical in origin with the previous name.

Lindfield W. Sussex. *Lindefeldia c.*765. 'Open land where lime-trees grow'. OE *linden* + *feld*.

Lindisfarne Northum. *Lindisfarnae c.*700. 'Domain at (a stream called) *Lindis*', from a derivative of OIrish **lind* 'lake' + *ferann* 'land'. Also called HOLY ISLAND.

Lindrick (old district) Notts. *See* CARLTON IN LINDRICK.

Lindridge Worcs. *Lynderycge* 11th cent. 'Ridge where lime-trees grow'. OE *lind* + *hrycg*.

Lindsell Essex. *Lindesela* 1086 (DB). 'Dwelling among lime-trees'. OE *lind* + *sele*.

Lindsey Lincs. (*prouincia*) *Lindissi c.*704–14, *Lindesi* 1086 (DB). Ancient district name, probably 'island, or dry ground in marsh, of the tribe called the Lindēs', from OE *īg*, *ēg* added to *Lindēs* 'the people of Lincoln or of the pool', *see* LINCOLN.

Lindsey Suffolk. *Lealeseia c.*1095. 'Island, or dry ground in marsh, of a man called **Lelli*'. OE pers. name + *ēg*. **Lindsey Tye** contains dialect *tye* 'a large common pasture'.

Linford, Great Milt. K. *Linforde* 1086 (DB). Probably 'ford where maple-trees grow'. OE *hlyn* + *ford*.

Lingen Herefs. *Lingen* 704–9, *Lingham* 1086 (DB). Probably originally the name of the brook here, perhaps a Celtic name meaning 'clear stream'.

Lingfield Surrey. *Leangafelda* 9th cent. Probably 'open land of the dwellers in the wood or clearing'. OE *lēah* + -*inga*- + *feld*.

Lingwood Norfolk. *Lingewode* 1199. 'Wood on a bank or slope'. OE *hlinc* + *wudu*.

Linkenholt Hants. *Linchehou* [*sic*] 1086 (DB), *Lynkeholte* c.1145. 'Wood by the banks or terraces'. OE *hlinc*, **hlincen* (adjective, 'with terraces') + *holt*.

Linkinhorne Cornwall. *Lankinhorn* c.1175. 'Church-site of a man called Kenhoarn'. Cornish **lann* + pers. name.

Linley, probably 'lime-tree wood or clearing', OE *lind* + *lēah*: **Linley** Shrops. near Bridgnorth. *Linlega* c.1135, *Lindleg* 1204. **Linley** Shrops. near Norbury. *Linlega* c.1150, *Lindele* 1209.

Linlithgow W. Loth. *Linlidcu* c.1138. '(Place by) Linlithgow Loch'. The name of the loch means 'lake in the damp hollow' (OWelsh **linn* + Welsh *llaith* + *cau*).

Linns (*Lan*) Louth. 'Church'.

Linshiels Northum. *Lynsheles* 1292. Probably 'shieling(s) where lime-trees grow'. OE *lind* + **scēla*.

Linslade Beds. *Hlincgelad* 966, *Lincelada* 1086 (DB). 'Difficult river-crossing by a terrace way'. OE *hlinc* + *gelād*.

Linstead Magna & Linstead Parva Suffolk. *Linestede* 1086 (DB). 'Place where flax is grown, or where maple-trees grow'. OE *līn* or *hlyn* + *stede*. Affixes are Latin *magna* 'great' and *parva* 'little'.

Linstock Cumbria. *Linstoc* 1212. 'Outlying farmstead or hamlet where flax is grown'. OE *līn* + *stoc*.

Linthwaite Kirkl. *Lindthait* late 12th cent., *Linthwait* 1208. 'Clearing where flax is grown, or where lime-trees grow'. OScand. *lín* or *lind* + *thveit*.

Linton, usually 'farmstead where flax is grown', OE *līn* + *tūn*; examples include: **Linton** Cambs. *Lintune* 1008, *Lintone* 1086 (DB). **Linton** Derbys. *Lintone* 942, *Linctune* [*sic*] 1086 (DB). **Linton** Herefs., near Ross. *Lintune* 1086 (DB). **Linton** N. Yorks. *Lipton* [*sic*] 1086 (DB),

Linton 12th cent. Alternatively the first element may be OE *hlynn* 'rushing stream'. **Linton** Leeds. *Lintone* 1086 (DB). **Linton, East** E. Loth., *see* EAST LINTON. **Linton-on-Ouse** N. Yorks. *Luctone* [*sic*] 1086 (DB), *Linton* 1176. For the river-name, *see* OUSEBURN.

However the following has a quite different origin: **Linton** Kent. *Lilintuna* c.1100. 'Estate associated with a man called Lill or Lilla'. OE pers. name + -*ing*- + *tūn*.

Linwood, 'lime-tree wood', OE *lind* + *wudu*: **Linwood** Hants. *Lindwude* 1200. **Linwood** Lincs., near Market Rasen. *Lindude* 1086 (DB).

Liphook Hants. *Leophok* 1364. Probably 'angle of land by the deer-leap or steep slope'. OE *hlīep* + *hōc*.

Lisacul (*Lios an Choill*) Roscommon. 'Fort of the wood'.

Lisbane (*Lios Bán*) Down, Limerick. 'White fort'.

Lisbeg (*Lios Beag*) Galway. 'Little fort'.

Lisbellaw (*Lios Béal Átha*) Fermanagh. 'Fort of the mouth of the ford'.

Lisboduff (*Lios Bó Dubh*) Cavan. *Lisbodowe* 1610. 'Fort of the black cows'.

Lisboy (*Lios Buí*) Antrim, Cork, Down, etc. 'Yellow fort'.

Lisburn Antrim. *Lysnecarvagh* 1683. The first element is probably from the earlier name *Lisnagarvey*, Irish *Lios na gCearrbhach*, 'fort of the gamblers'. The second element is dubiously derived from the burning of the town and castle in the siege of 1641.

Liscannor (*Lios Ceannúir*) Clare. 'Fort of Ceannúr'.

Liscard Wirral. *Lisnekarke* c.1260. An OIrish name, from the ancestor of *lios na carraige* 'hall at the rock'.

Liscarney (*Lios Cearnaigh*) Mayo. 'Carnach's fort'.

Liscarroll (*Lios Cearúill*) Cork. 'Fort of Cearúll'.

Liscloon (*Lios Claon*) Tyrone. *Liscleene* 1661. 'Sloping fort'.

Liscolman (*Lios Cholmáin*) Antrim. *Slaightcolminan* 1669. 'Colmán's fort'.

Lisdeen (*Lios Duibhinn*) Clare. 'Fort of the chief'.

Lisdoonvarna (*Lios Dúin Bhearna*) Clare. 'Fort of the gap'.

Lisdowney (*Lios Dúnadhaigh*) Kilkenny. 'Fort of Dúnadhach'.

Lisduff (*Lios Dubh*) Cavan, Mayo. 'Black fort'.

Lisgoole (*Lios Gabhail*) Fermanagh. 'Fort of the fork'.

Liskeard Cornwall. *Lys Cerruyt c.*1010, *Liscarret* 1086 (DB). Probably 'court of a man called Kerwyd'. Cornish *lys* + pers. name.

Lislea (*Lios Liath*) Armagh, Derry. 'Grey fort'.

Lislevane (*Lios Leamháin*) Cork. 'Elm fort'.

Lismacaffry (*Lios Mhic Gofraidh*) Westmeath. 'Fort of Mac Gofraidh'.

Lismahon (*Lios Mócháin*) Down. 'Móchán's fort'.

Lismore (*Lios Mór*) Waterford. 'Big fort'.

Lismoyney (*Lios Maighne*) Westmeath. 'Fort of the precinct'.

Lisnacask (*Lios na Cásca*) Westmeath. 'Easter fort'.

Lisnacree (*Lios na Crí*) Down. *Lisnechrehy* 1613. 'Fort of the boundary'.

Lisnadill (*Lios na Daille*) Armagh. *Lisnedolle* 1609. 'Fort of the blindness'.

Lisnagade (*Lios na gCéad*) Down. 'Fort of the hundreds'.

Lisnageer (*Lios na gCaor*) Cavan. 'Fort of the berries'.

Lisnagelvin (*Lios na nGealbhán*) Derry. *Lisnegellwan* 1663. 'Fort of the sparrows'.

Lisnagry (*Lios na Graí*) Limerick. 'Fort of the stud'.

Lisnakill (*Lios na Cille*) Waterford. 'Fort of the church'.

Lisnalinchy (*Lios Uí Loingsigh*) Antrim. 'Fort of Ó Loingseach'.

Lisnalong (*Lios na Long*) Monaghan. 'Fort of the boats'.

Lisnamuck (*Lios na Muc*) Derry. *Lisnamuc c.*1834. 'Fort of the pigs'.

Lisnarrick (*Lios na nDaróg*) Fermanagh. *Lisnarrogue* 1659. 'Fort of the oaks'.

Lisnaskea (*Lios na Scéithe*) Fermanagh. (*ag*) *Scéith Ghabhra* 1589. 'Fort of the shield'. The final element of the name, the genitive form of Irish *sciath*, 'shield', is from *Sciath Ghabhra*, the name of the inauguration site of the Maguires, perhaps understood as 'fortress of the (white) horses'.

Lisnastrean (*Lios na Srian*) Down. *Ballelisneshrean* 1623. 'Fort of the bridles'.

Lispole (*Lios Póil*) Kerry. 'Fort of the pool'.

Lisrodden (*Lios Rodáin*) Antrim. *Ballylessraddan* 1605. 'Rodán's fort'.

Lisroe (*Lios Ruadh*) Clare, Cork, Kerry, etc. 'Red fort'.

Lisronagh (*Lios Ruanach*) Tyrone. 'Seal fort'.

Lisryan (*Lios Riáin*) Longford. 'Rián's fort'.

Liss Hants. *Lis* 1086 (DB). Celtic **līss* 'a court, chief house in a district'.

Lissacreasig (*Lios an Chraosaigh*) Cork. 'Fort of the glutton'.

Lissalway (*Lios Sealbhaigh*) Roscommon. 'Fort of Sealbhach'.

Lissan (*Leasán*) Tyrone. *Leassan* 1455. 'Little fort'.

Lissaniska (*Lios an Uisce*) Cork, Kerry, Limerick. 'Fort of the water'.

Lissanisky (*Lios an Uisce*) Cork, Roscommon, Tipperary. 'Fort of the water'.

Lissardagh (*Lios Ardachaidh*) Cork. 'Fort of the high field'.

Lissatava (*Lios an tSamhaidh*) Mayo. 'Fort of the sorrel'.

Lisselton (*Lios Eiltín*) Kerry. 'Fort of the little doe'.

Lissett E. R. Yorks. *Lessete* 1086 (DB). Probably 'dwelling near the pasture-land'. OE *lǣs* + (*ge*)*set*.

Lissinagroagh (*Lisín na gCruach*) Sligo. 'Little fort of the ridges'.

Lissington Lincs. *Lessintone* 1086 (DB).
Probably 'estate associated with a man called
Lēofsige'. OE pers. name +*-ing-* + *tūn*.

Lissue (*Lios Aedha*) Antrim. 'Aedh's fort'.

Lissycasey (*Lios Uí Chathasaigh*) Clare.
'Ó Cathasaigh's fort'.

Listooder (*Lios an tSudáire*) Down.
Balle-Listowdrie 1623. 'Fort of the tanner'.

Listowel (*Lios Tuathail*) Kerry. 'Fort of
Tuathal'.

Litcham Norfolk. *Licham* 1086 (DB). Possibly
'homestead or village with an enclosure'. OE
*lycce + *hām*.

Litchborough Northants. *Liceberge* 1086
(DB). Probably 'hill with an enclosure'. OE **lycce*
+ *beorg*.

Litchfield Hants. *Liveselle* [*sic*] 1086 (DB),
Lieueselva 1168. Possibly 'shelf of land with a
shelter'. OE **hlīf* + *scelf* or *scylf*.

Litherland Sefton. *Liderlant* 1086 (DB).
'Cultivated land on a slope'. OScand. *hlíth*
(genitive *-ar*) + *land*.

Litlington Cambs. *Litlingeton* 1183,
Lidlingtone 1086 (DB). 'Farmstead of the family
or followers of a man called **Lȳtel*'. OE pers.
name + *-inga-* + *tūn*.

Litlington E. Sussex. *Litlinton* 1191. 'Little
farmstead' from OE *lȳtel* (dative *-an*) + *tūn*, or
'farmstead associated with a man called
*Lȳtel(a)' from OE pers. name + *-ing-* + *tūn*.

Little as affix. *See* main name, e.g. for **Little
Abington** (Cambs.) *see* ABINGTON.

Little Bridge Antrim. *Little bridge* 1836. The
'little bridge' over the Lissan Water here is so
called by contrast with a 'big bridge' to the
south.

Little Cumbrae. *See* CUMBRAE.

Little England Beyond Wales (district)
Pemb. *Ynglond be yond Walys* 1519, *Little
England beyond Wales* 1594. The
southern part of Pembrokeshire has been
chiefly English-speaking since the Viking
invasion of the 9th cent.

Little Haven Pemb. *the Lytel hauen* 1578.
'Little harbour'. OE *hæfen*. *Little* distinguishes
the place from nearby BROAD HAVEN.

Little Mancot. *See* BIG MANCOT.

Little Minch. *See* MINCH.

Littleborough, 'little fort or stronghold', OE
lȳtel + *burh*: **Littleborough** Notts. *Litelburg*
1086 (DB). **Littleborough** Rochdl.
Littlebrough 1577.

Littlebourne Kent. *Littelburne* 696,
Liteburne 1086 (DB). 'Little estate on the river
called *Burna*' (from OE *burna* 'stream' referring
to the Little Stour), with OE *lȳtel*.

Littlebredy Dorset. *See* BREDY.

Littlebury Essex. *Lytlanbyrig c.*1000,
Litelbyria 1086 (DB). 'Little fort or stronghold'.
OE *lȳtel* + *burh* (dative *byrig*).

Littledean Glos. *Dene* 1086 (DB), *Parva Dene*
1220. OE *denu* 'valley', with later affix *little*
(Latin *parva*) to distinguish this place from
MITCHELDEAN.

Littleham Devon. near Exmouth.
Lytlanhamme 1042, *Liteham* 1086 (DB). 'Little
enclosure or river-meadow'. OE *lȳtel* + *hamm*.

Littlehampton W. Sussex. *Hantone* 1086
(DB), *Lyttelhampton* 1482. OE *hām-tūn* 'home
farm, homestead' with the later addition of *little*
perhaps to distinguish it from SOUTHAMPTON.

Littlehempston Devon. *Hamistone* 1086
(DB), *Parua Hæmeston* 1176. 'Farmstead of a
man called **Hæme* or Hemme'. OE pers. name
+ *tūn*. Affix *little* (Latin *parva*) to distinguish this
place from BROADHEMPSTON.

Littlehoughton Northum. *Houcton Parva*
1242. 'Farmstead on or near a ridge or hill-spur'.
OE *hōh* + *tūn*. Affix *little* (Latin *parva*) to
distinguish this place from LONGHOUGHTON.

Littlemore Oxon. *Luthlemoria c.*1130. 'Little
marsh'. OE *lȳtel* + *mōr*.

Littleover Derby. *Parva Ufre* 1086 (DB).
'(Place at) the ridge'. OE **ofer* with the affix *little*
(Latin *parva*) to distinguish this place from
MICKLEOVER.

Littleport Cambs. *Litelport* 1086 (DB). 'Little
(market) town'. OE *lȳtel* + *port*.

Littlestone-on-Sea Kent. a 19th-cent.
name from a rocky headland called *Little Stone*
1801, formerly a coastal feature near here.

Littleton, 'little farmstead or estate', OE *lȳtel* +
tūn; examples include: **Littleton** Ches.

Litelton 1435. Earlier *Parva Cristentona* c.1125, 'the smaller part of CHRISTLETON'. **Littleton** Hants., near Winchester. *Lithleton* 1285.

Littleton Somerset. *Liteltone* 1086 (DB).

Littleton Surrey. *Liteltone* 1086 (DB),

Littleton Drew Wilts. *Liteltone* 1086 (DB), *Littleton Dru* 1311. Manorial affix from the *Driwe* family, here in the 13th cent. **Littleton, High** B. & NE. Som. *Liteltone* 1086 (DB), *Heghelitleton* 1324. Affix is OE *hēah* 'high'. **Littleton, Middle**, **Littleton, North** & **Littleton, South** Worcs. *Litletona* 709, *Liteltune* 1086 (DB). **Littleton Pannell** Wilts. *Liteltone* 1086 (DB), *Lutleton Paynel* 1317. Manorial affix from the *Paynel* family, here in the 13th cent. **Littleton upon Severn** S. Glos. *Lytletun* 986, *Liteltone* 1086 (DB). For the river-name, *see* SEVERN.

Littlewick Green Winds. & Maid. *Lidlegewik* c.1060. 'Farm by the woodland clearing with a gate or on a slope'. OE *hlid* or **hlid* + *lēah* + *wīc*.

Littleworth Oxon. *Wyrthæ* c.971, *Ordia [sic]* 1086 (DB), *Lytleworth* 1284. OE *worth* 'enclosure', later with the affix *lȳtel* 'little' to distinguish this place from LONGWORTH.

Litton Derbys. *Litun* 1086 (DB). Possibly 'farmstead on a slope'. OE *hlith* + *tūn*.

Litton N. Yorks. *Litone* 1086 (DB). Probably identical in origin with the previous name.

Litton Somerset. *Hlytton* c.1060, *Litune* 1086 (DB). Possibly 'farmstead at a gate or by a slope'. OE *hlid* or **hlid* + *tūn*.

Litton Cheney Dorset. *Lideton* 1204. Probably 'farmstead by a noisy stream'. OE **hlȳde* + *tūn*. Manorial affix from the *Cheyne* family, here in the 14th cent.

Livermere, Great Suffolk. *Leuuremer* c.1050, *Liuermera* 1086 (DB). 'Liver-shaped pool, or pool with thick or muddy water'. OE *lifer* + *mere*.

Liverpool Lpool. *Liuerpul* c.1190. 'Pool or creek with thick or muddy water'. OE *lifer* + *pōl*.

Liversedge Kirkl. *Livresec* 1086 (DB). Probably 'edge or ridge of a man called Lēofhere'. OE pers. name + *ecg*.

Liverton Red. & Cleve. *Liuretun* 1086 (DB). 'Farmstead of a man called Lēofhere, or on a stream with thick or muddy water'. OE pers. name or *lifer* + *tūn*.

Livingston W. Loth. *Villa Leuing* c.1128, *Leuinestun* c.1150. 'Leving's farmstead'. ME pers. name + *toun*.

Lixnaw (*Leic Snámha*) Kerry. 'Flagstone of the swimming place'.

Lizard Cornwall. *Lisart* 1086 (DB). 'Court on a height'. Cornish **lys* + **ardh*.

Llan Sain Siôr. *See* ST GEORGE's.

Llanandras. *See* PRESTEIGNE.

Llanasa Flin. *Llanassa* 1254. 'Church of St Asaph'. Welsh *llan*.

Llanbadarn Fawr Cergn. *Lampadervaur* 1181, *Llanbadarn Vawr* 1557. 'Great church of St Padarn'. Welsh *llan* + *mawr*.

Llanbedr Pont Steffan. *See* LAMPETER.

Llanberis Gwyd. *Lanperis* 1283. 'Church of St Peris'. Welsh *llan*.

Llanbethery (*Llanbydderi*) Vale Glam. *Landebither* 13th cent. Possibly 'church of deaf ones', compare LLANYBYDDER.

Llancarfan Vale Glam. *Lancaruan* 1106, *Nant Carban* 1136. Possibly 'valley of the River Carfan'. Welsh *nant*. Early forms of the name confirm that original Welsh *nant* was replaced by *llan*, 'church'.

Llandaff (*Llandaf*) Card. *Lanntaf* c.1150, *Landaph* 1191. 'Church on the River Taf'. Welsh *llan*. The Celtic or pre-Celtic river-name may mean 'dark' or simply 'stream'. The church is the cathedral church of St Teilo, whose name lies behind LLANDEILO. Compare CARDIFF.

Llanddewi Brefi Cergn. *Landewi Brevi* 13th cent. 'Church of St David on the River Brefi'. Welsh *llan* + *Dewi*. The river-name means 'noisy one' (Welsh *bref*).

Llanddewi Nant Hoddni. *See* LLANTHONY.

Llanddewi Velfrey Pemb. *Landwei in Wilfrey* 1488. 'Church of St David in Efelffre'. Welsh *llan* + *Dewi*. The name of the commote *Efelffre* is of uncertain origin, although *-fre* may represent Welsh *bre*, 'hill'.

Llanddulas Conwy. *Llan Dhylas* c.1700. 'Church on the River Dulas'. Welsh *llan*. The river-name may mean 'black stream' (Welsh *du* + *glais*).

Llandegfan Angl. *Llandegvan* 1254. 'Church of St Tegfan'. Welsh *llan*.

Llandeilo Carm. *Lanteliau Maur* 1130, *Llandilo Vawr* 1656. 'Church of St Teilo'. Welsh *llan*. The second word of the early forms is Welsh *mawr*, 'great'.

Llandeloy Pemb. *Landalee* 1291. 'Tylwyf's church'. Welsh *llan*. The pers. name is that of the owner of the site.

Llandinabo Herefs. *Lann Hunapui c.*1130. 'Church-site of a man called Iunapui'. Welsh *llan* + pers. name.

Llandochau. *See* LLANDOUGH.

Llandough (*Llandochau*) Vale Glam. *Landochan* 1106, *Landough* 1427. 'Docco's church'. Welsh *llan*.

Llandovery (*Llanymddyfri*) Carm. *Llanamdewri* 12th cent., *Lanandeveri* 1194, *Lanymdevery* 1383. 'Church near the waters'. Welsh *llan* + *am* + *dwfr* (plural *dyfri*).

Llandrillo-yn-Rhos. *See* RHOS-ON-SEA.

Llandrindod Wells Powys. *Lando* 1291, *Llandrindod* 1554. 'Church of the Trinity with wells'. Welsh *llan* + *trindod*. The original name of the church was *Llanddwy*, 'church of God'. *Wells* was added from the local chalybeate springs, first exploited in the 18th cent.

Llandudno Conwy. *Llandudno* 1291, *Lantudenou* 1376. 'Church of St Tudno'. Welsh *llan*.

Llandudoch. *See* ST DOGMAELS.

Llandysul Cergn. *Landessel* 1291, *Lantissill* 1299. 'Church of St Tysul'. Welsh *llan*.

Llanelli Carm. *Lann elli c.*1173. 'Church of St Elli'. Welsh *llan*. Elli is said to have been one of the daughters of Brychan, who gave the name of BRECON.

Llanelwy. *See* ST ASAPH.

Llanerch-y-medd Angl. *llan'meth* 1445, *Lannerch y medd* 15th cent. 'Glade in which mead is made'. Welsh *llannerch* + *y* + *medd*.

Llanfair Caereinion Powys. *Llanveyr* 1279, *Llanvair in Krynion* 1579. 'Church of St Mary in Caereinion'. Welsh *llan* + *Mair*. The name of the cantref *Caereinion* means 'fort of Einion' (Welsh *caer* + pers. n).

Llanfair Pwllgwyngyll Angl. *Llan Vair y pwyll Gwinghill* 1536, *ll.fair ymhwll gwingill* 1566. 'Church of St Mary in Pwllgwyngyll'. Welsh *llan* + *Mair*. The township name *Pwllgwyngyll* means 'pool of the white hazels' (Welsh *pwll* + *gwyn* + *cyll*). A fanciful addition of *gogerychwyrndrobwllllantysiliogogogoch* was further made in the mid-19th cent., meaning 'fairly near the rapid whirlpool by the church of St Tysilio at the red place' (Welsh *go* + *ger* + *y* + *chwyrn* + *trobwll* + *llantysilio* + *coch*), based on local names or features, giving an overall *Llanfairpwllgwyngyllgogerychwyrndro bwllllantysiliogogogoch*.

Llanfair Waterdine Shrops. *Watredene* 1086 (DB), *Thlanveyr* 1284. 'St Mary's church in the watery (or wet) valley'. Welsh *llan* + saint's name, with the addition from OE *wæter* + *denu*.

Llanfair-ym-Muallt. *See* BUILTH WELLS.

Llanfairfechan Conwy. *Lanueyr* 1284, *Llanbayar* 1291, *Llanvair Vechan* 1475. 'Little church of St Mary'. Welsh *llan* + *Mair* + *bechan*. The church is 'little' compared to the 'big' church of St Mary at Conwy.

Llanfihangel-ar-arth Carm. *Llanfihangel Orarth* 1291. 'St Michael's church on the high hill'. Welsh *llan* + *Mihangel* + *ar* + *garth*. The second word of the early form represents Welsh *gor-* super- + *garth* 'hill', referring to the height of the hill.

Llanfihangel Dyffryn Arwy. *See* MICHAELCHURCH-ON-ARROW.

Llanfihangel-y-pwll. *See* MICHAELSTON-LE-PIT.

Llanfyllin Powys. *Llanvelig* 1254, *Lanvyllyn* 1291, *Lanvethlyng* 1309. 'Church of St Myllin'. Welsh *llan*.

Llanfyrnach Pemb. *Ecclesia Sancti Bernaci* 1291. 'Church of St Brynach'. Welsh *llan*.

Llangadog Carm. *Lancadauc* 1281, *Lankadoc* 1284. 'Church of St Cadog'. Welsh *llan*.

Llangarron Herefs. *Lann Garan c.*1130. 'Church-site on the River Garron'. Welsh *llan* + Celtic river-name meaning 'crane stream'.

Llangefni Angl. *Llangevni* 1254. 'Church on the River Cefni'. Welsh *llan*.

Llangeitho Cergn. *Lankethau* 1284, *Langeytho* 1290. 'Church of St Ceitho'. Welsh *llan.* Ceitho is one of the saints to whom LLANPUMSAINT is dedicated.

Llangollen Denb. *Lancollien* 1234, 'Church of St Collen'. Welsh *llan.*

Llangolman Pemb. *Llangolman* 1394. 'Church of St Colman'. Welsh *llan.*

Llangrove Herefs. *Longe grove* 1372. 'Long grove or copse'. OE *lang* + *grāf.*

Llangurig Powys. *Lankiric* 1254. 'Church of St Curig'. Welsh *llan.*

Llangwm Pemb. *Llandegumme* 1301. 'Church in the valley'. Welsh *llan* + *cwm.*

Llangybi Cergn. *Lankeby* 1284. 'Church of St Cybi'. Compare HOLYHEAD. Welsh *llan.*

Llanidloes Powys. *Lanidloes* 1254. 'Church of St Idloes'. Welsh *llan.*

Llanilltud Fawr. *See* LLANTWIT MAJOR.

Llanllyfni Gwyd. *Thlauthleueny* [*sic*] 1352, *Llanllyfne* 1432. 'Church on the River Llyfni'. Welsh *llan.* The river-name means 'smooth-flowing one' (Welsh *llyfn*).

Llanpumsaint Carm. 'Church of the five saints'. Welsh *llan* + *pump* + *saint.* The five saints are Ceitho, Celynnen, Gwyn, Gwynno, and Gwynoro.

Llanrothal Herefs. *Lann Ridol* c.1130. Welsh *llan* 'church-site', probably with a pers. name.

Llanrug Gwyd. *ll.v'el yn Ruc* 1566, *Llanvihengle in Rug* 1614. 'Church in Rug'. Welsh *llan.* The township name *Rug* means 'heather' (Welsh *grug*). The full name of the village was *Llanfihangel-yn-Rug*, 'church of St Michael in Rug'.

Llanrwst Conwy. *lhannruste* 1254, *Lannwrvst* 1291. 'Church of St Gwrwst'. Welsh *llan.*

Llansanffraid Glan Conwy Conwy. *Llansanfraid* 1535, *Lhan St Ffraid* c.1700, *Llan Sanfraed Glan Conwy* 1713. 'Church of St Ffraid on the bank of the River Conwy'. Welsh *llan* + *sant* + *glan.* Ffraid is a Welsh form of Brigid (Bridget). For the river-name, *see* CONWY.

Llansanffraid Glynceiriog. *See* GLYN CEIRIOG.

Llansawel Carm. *Lansawyl* 1265, *Lansawel* 1301. 'Church of St Sawel'. The saint's name is a form of Samuel. Compare BRITON FERRY. Welsh *llan.*

Llansawel. *See* BRITON FERRY.

Llanstadwell Pemb. *Lanstadhewal* 12th cent. 'Church of St Tudwal'. Welsh *llan.*

Llanthony Mon. *Lantony* 1160. 'Church on the River Honddi'. Welsh *llan.* The name is usually said to be an abbreviated form of *Llanddewi Nant Hoddni,* 'church of St David on the River Hoddni' (Welsh *llan* + *Dewi* + *nant*), but more likely *Llanthony* is a form of the river-name, *Nant Hoddni,* with *llan* replacing *nant.* The river-name means 'pleasant' (Welsh *hawdd*).

Llantrisant Mon. *Landtrissen* 1246. 'Church of three saints'. Welsh *llan* + *tri* + *sant.* The three saints are Dyfodwg, Gwynno, and Illtud.

Llantrithyd Vale Glam. *Landrirede* c.1126, *Nantririd* 1545. 'Rhirid's valley'. Welsh *nant.*

Llantwit Major Vale Glam. *Landiltuit* 1106, *Lannyltwyt* 1291, *Lantwyt* 1480. 'Main church of St Illtud'. Welsh *llan,* 'church'. The name is a part translation, part contraction of Welsh *Llanilltud Fawr,* with Welsh *mawr* represented by Latin *major.* The place is 'great' in relation to *Llantwit Fardre* (Welsh *Llanilltud Faerdre*), Rhon., and *Llantwit-juxta-Neath* (Welsh *Llanilltud Nedd*), Neath.

Llantydewi. *See* ST DOGWELLS.

Llanwarne Herefs. *Ladgvern* [*sic*] 1086 (DB), *Lann Guern* c.1130. 'Church by the alder grove'. Welsh *llan* + *gwern.*

Llanwnda Pemb. *Lanwndaph* 1202. 'Church of St Gwyndaf'. Welsh *llan.*

Llanwrda Carm. *Llanwrdaf* 1302. 'Church of St Gwrdaf'. Welsh *llan.*

Llanwrtyd Wells Powys. *Llanwrtid* 1553, *Ll. Wrtyd* 1566. 'Church of St Gwrtud with wells'. Welsh *llan.* The town developed as a spa in the 19th cent.

Llanyblodwel Shrops. *Llanblodwell* 1535. 'Church of *Blodwell*'. Welsh *llan* with an older English name first recorded as *Blodwelle* c.1200, meaning 'blood spring or stream' from OE *blōd* + *wella,* with reference to the colour of the water or to some other quality or superstition.

Llanybydder Carm. *Lannabedeir* 1319, *Llanybyddeyr* 1401. 'Church of the deaf ones'. Welsh *llan* + *y* 'the' + *byddar* (plural *byddair*). The name may allude to a congregation who were deaf to the call preached to them, or whose hitherto deaf ears were opened by it.

Llan-y-cefn Pemb. *Llanekevyn* 1499. 'Church on the ridge'. Welsh *llan* + *y* + *cefn*.

Llanychaer Pemb. *Launerwayth* 1291, *Llanerchaeth* 1408. Meaning uncertain. Early forms suggest that the first element may represent Welsh *llannerch*, 'glade'. The rest is obscure.

Llanymddyfri. *See* LLANDOVERY.

Llanymynech Powys./Shrops. *Llanemeneych* 1254. 'Church of the monks'. Welsh *llan* + *mynach*.

Lleyn (Peninsula) Gwyd. (*Penrhyn Llŷn*). Meaning uncertain. The name may have originally been that of a Celtic people 'the Lageni' hence *Lleyn* and later *Llŷn*, their own name perhaps from *Leinster*, Ireland, whence they may have emigrated.

Llwyneliddon. *See* ST LYTHANS.

Llyfni (river) Gwyd. *Lleueni* 1578. 'Smooth-flowing one'. Welsh *llyfn*. *Compare* LLYNFI.

Llŷn. *See* LLEYN PENINSULA.

Llyn Tegid. *See* BALA.

Llynfi (river) Bri. *Thleweny* 1314. 'Smooth-flowing one'. Welsh *llyfn*. Compare LLYFNI.

Llys-y-frân Pemb. *Lysurane* 1326, *Lysfrane* 1402. 'Brân's court', or from *brân* 'crow'. Welsh *llys*.

Load, Long Somerset. *La Lade* 1285. 'The watercourse or drainage channel'. OE *lād*.

Loanends Antrim. 'Lane ends'. Scottish *loan* + English *ends*. Three minor roads join the main road to Antrim here.

Loanhead Midloth. *Loneheid* 1618. 'End of the lane'. Scots *loan* 'lane' + *head* 'upper end'.

Lochaber (district) Highland. 'Loch at the confluence'. Gaelic *loch* + Pictish *aber*.

Lochalsh. *See* KYLE OF LOCHALSH.

Lochgelly Fife. *Lochgellie* 1606. '(Place by) Loch Gelly'. Gaelic *loch* + *Gealaidh* (from *geal* 'white').

Lochgilphead Arg. *Lochgilpshead* 1650. 'Head of Loch Gilp'. Gaelic *loch*. The name of the loch may mean 'chisel-shaped loch' (Gaelic *gilb*).

Lochinvar (loch) Dumf. *Lochinvar* 1540. Possibly 'loch of the height'. Gaelic *Loch an Bharr*.

Lochinver Highland. '(Place by) Loch Inver'. Gaelic *loch*. The loch is a sea inlet at the mouth (Gaelic *inbhir*) of the river of the same name.

Lochmaben Dumf. *Locmaban* 1166, *Lochmalban c.*1320, *Lochmabane* 1502. 'Loch by Maben'. Gaelic *loch*. The meaning of *Maben* is uncertain.

Lochnagar (mountain) Aber. *Loch Garr* 1640, *Loch na Garr* 1807. The name of the mountain has been transferred from the loch. The loch has a name of uncertain origin.

Lochy, Loch Highland. 'Loch of the River Lochy'. Gaelic *loch*. The river-name means 'dark one' (Gaelic *lochaidh*).

Lockerbie Dumf. *Lockardebie* 1306. 'Locard's farm'. OScand. pers. name + *bý*.

Lockeridge Wilts. *Locherige* 1086 (DB). Possibly 'ridge with folds or enclosures'. OE *loc*(*a*) + *hrycg*.

Lockerley Hants. *Locherlega* 1086 (DB). Probably 'woodland clearing of a keeper or shepherd'. OE **lōcere* + *lēah*.

Locking N. Som. *Lockin* 1212. 'Place associated with a man called Locc', OE pers. name + *-ing*, or from an OE **locing* 'fold or enclosure'.

Lockinge, East & Lockinge, West Oxon. *Lacinge* 868, *Lachinges* 1086 (DB), *Estloking*, *Westloking* 1327. Originally the name of the stream here (now called Lockinge Brook) meaning 'the playful one', from OE *lāc* 'play', *lācan* 'to play' + *-ing*.

Lockington E. R. Yorks. *Locheton* 1086 (DB). 'Estate associated with a man called **Loca*'. OE pers. name + *-ing-* + *tūn*. Alternatively the first element may be an OE **locing* 'fold, enclosure'.

Lockton N. Yorks. *Lochetun* 1086 (DB). Probably 'farmstead of a man called **Loca*', OE pers. name + *tūn*. Alternatively the first element may be OE *loca* 'enclosure'.

Loddington Leics. *Ludintone* 1086 (DB). Probably 'estate associated with a man called Luda'. OE pers. name +*-ing-* + *tūn*.

Loddington Northants. *Lodintone* 1086 (DB). 'Estate associated with a man called Lod(a)'. OE pers. name + *-ing-* + *tūn*.

Loddiswell Devon. *Lodeswille* 1086 (DB). 'Spring or stream of a man called *Lod', OE pers. name + *wella*.

Loddon (river) Berks.–Hants. *See* SHERFIELD.

Loddon Norfolk. *Lodne* 1043, *Lotna* 1086 (DB). Originally a Celtic river-name (possibly meaning 'muddy stream'), an old name for the River Chet.

Lode Cambs. *Lada* 12th cent. OE *lād* 'watercourse or drainage channel'.

Loders Dorset. *Lodre(s)* 1086 (DB). Probably a Celtic name, from *lo-dre* (the ancestor of Welsh *llodre*) 'habitation site, homestead'.

Lodsworth W. Sussex. *Lodesorde* 1086 (DB). 'Enclosure of a man called *Lod'. OE pers. name + *worth*.

Lofthouse Leeds. *Loftose* 1086 (DB). 'House(s) with a loft or upper floor'. OScand. *loft-hús* in a dative plural form.

Loftus Red & Cleve. *Loctehusum* 1086 (DB). Identical in origin with the previous name.

Loghill (*Leamhchoill*) Limerick. 'Elm wood'.

Logie Buchan. *See* KIRKTON OF LOGIE BUCHAN.

Lolworth Cambs. *Lollesworthe* 1034, *Lolesuuorde* 1086 (DB). 'Enclosure of a man called *Loll or Lull'. OE pers. name + *worth*.

Loman (river) Devon. *See* UPLOWMAN.

Lomond, Loch Arg. *Lochlomond* c.1340. '(Loch by) the River Leven'. Gaelic *loch*. The loch takes its name from the river, which has the same origin as the River Leven in Fife, *see* LEVEN.

Londesborough E. R. Yorks. *Lodenesburg* 1086 (DB). 'Stronghold of a man called Lothinn'. OScand. pers. name + OE *burh*.

London *Londinium* c.115. Earlier attempts to explain this ancient name have been rejected by Celtic scholars. The latest view is that it may be derived from two pre-Celtic ('Old European') roots with added Celtic suffixes describing a location on the lower course of the Thames (below its lowest fordable point at WESTMINSTER), with a meaning something like 'place at the navigable or unfordable river'.

London Apprentice Cornwall. Village named from a 19th-cent. inn called *The London Apprentice* (there are three examples of this inn-name in London itself).

London Colney Herts. *See* COLNEY.

Londonderry. *See* DERRY.

Londonthorpe Lincs. *Lundertorp* 1086 (DB). 'Outlying farmstead or hamlet by a grove'. OScand. *lundr* (genitive *lundar*) + *thorp*.

Long as affix. *See* under main name, e.g. for **Long Ashton** (N. Som.) *see* ASHTON.

Long, Loch Arg. 'Ship loch'. Gaelic *loch* + *long*.

Longbenton N. Tyne. *Bentune* c.1190, *Magna Beneton* 1256. 'Farmstead where bent-grass grows, or where beans are grown'. OE *beonet* or *bēan* + *tūn*, with later affix *long*.

Longborough Glos. *Langeberge* 1086 (DB). 'Long hill or barrow'. OE *lang* + *beorg*.

Longbridge Birm. *Long Bridge* 1831. Named from a crossing of the River Rea.

Longbridge Deverill Wilts. *Devrel* 1086 (DB), *Deverill Langebrigge* 1316. 'Estate on the River *Deverill* with the long bridge'. OE *lang* + *brycg* with Celtic river-name (for which *see* BRIXTON DEVERILL).

Longburton Dorset. *Burton* 1244, *Langebourton* 1460. 'Fortified farm', or 'farm near a fortification'. OE *burh-tūn*. Affix is OE *lang* referring to the length of the village.

Longcot Oxon. *Cotes* 1233, *Longcote* 1332. 'The cottages', from OE *cot*, with the later addition of *lang* 'long'.

Longden Shrops. *Langedune* 1086 (DB). 'Long hill or down'. OE *lang* + *dūn*.

Longdendale Derbys. Tamesd. *See* MOTTRAM.

Longdon, 'long hill or down', OE *lang* + *dūn*: **Longdon** Staffs. *Langandune* 1002. **Longdon** Worcs. *Langandune* 969, *Longedun* 1086 (DB). **Longdon upon Tern** Tel. & Wrek. *Languedune* 1086 (DB). For the river-name, *see* EATON UPON TERN.

Longfield Kent. *Langafelda* 10th cent., *Langafel* 1086 (DB). 'Long stretch of open country'. OE *lang* + *feld*.

Longford, 'long ford', OE *lang* + *ford*: **Longford** Derbys. *Langeford* 1197. **Longford** Glos. *Langeford* 1107. **Longford** Shrops., near Market Drayton. *Langeford* 13th cent. **Longford** Tel. & Wrek. *Langanford* 1002, *Langeford* 1086 (DB).

Longford (*An Longfort*) Longford. 'The fortified place'.

Longframlington Northum. *Fremelintun* 1166. 'Estate associated with a man called *Framela'. OE pers. name + -ing- + tūn. Affix *long* from the length of the village.

Longham Norfolk. *Lawingham* 1086 (DB). Probably 'homestead of the family or followers of a man called *Lāwa'. OE pers. name + -inga- + hām.

Longhirst Northum. *Langherst* 1200. 'Long wooded hill'. OE *lang* + *hyrst*.

Longhope Glos. *Hope* 1086 (DB), *Langehope* 1248. 'The valley'. OE *hop* with the later addition of *lang* 'long'.

Longhorsley Northum. *Horsleg* 1196. 'Woodland clearing where horses are kept'. OE *hors* + *lēah*, with the later affix *long*.

Longhoughton Northum. *Houcton Magna* 1242. 'Farmstead on or near a ridge or hill-spur'. OE *hōh* + *tūn*. Affix *long* (earlier Latin *magna* 'great') to distinguish this place from LITTLEHOUGHTON.

Longleat Wilts. *Langelete* 1235. 'Long stream or channel'. OE *lang* + (*ge*)*lǣt*.

Longney Glos. *Longanege* 972, *Langenei* 1086 (DB). 'Long island'. OE *lang* (dative *-an*) + *ēg*.

Longnor Shrops. *Longenalra* 1155. 'Tall alder-tree', or 'long alder-copse'. OE *lang* (dative *-an*) + *alor*.

Longnor Staffs. near Buxton. *Langenoure* 1227. 'Long ridge'. OE *lang* (dative *-an*) + *ofer*.

Longparish Hants. *Langeparisshe* 1389. Self-explanatory. Earlier called Middleton, *Middeltune* 1086 (DB), 'middle farmstead or estate', from OE *middel* + *tūn*.

Longridge Lancs. *Langrig* 1246. 'Long ridge'. OE *lang* + *hrycg*.

Longsdon Staffs. *Longesdon* 1242. Probably 'hill called *Lang* ('the long one')'. OE *lang* + *dūn*.

Longstanton Cambs. *See* STANTON.

Longstock Hants. *Stoce* 982, *Stoches* 1086 (DB), *Langestok* 1233. OE *stoc* 'outlying farm or hamlet', with the later addition of *lang* 'long'.

Longstone, Great & Longstone, Little Derbys. *Langesdune* 1086 (DB). Probably 'hill called *Lang* ('the long one')'. OE *lang* + *dūn*.

Longton, Longtown, 'long farmstead or estate', OE *lang* + *tūn*: **Longton** Lancs. *Langetuna* c.1155. **Longton** Stoke. *Longeton* 1212. **Longtown** Cumbria. *Longeton* 1267. **Longtown** Herefs. *Longa villa* 1540, earlier *Ewias* 1086 (DB), *Ewyas Lascy* 13th cent. (manorial affix from Roger *de Laci*, owner in 1086, *see* EWYAS HAROLD).

Longville in the Dale Shrops. *Longefewd* 1255. 'Long stretch of open country'. OE *lang* + *feld*, with later affix from *dale* 'valley'.

Longwitton Northum. *Wttun* 1236, *Langwotton* 1340. 'Farmstead in or by a wood'. OE *wudu* + *tūn*, with later affix *lang* 'long'.

Longworth Oxon. *Wurthe* 958, *Langewrth* 1284. OE *worth* or *wyrth* 'enclosure', with the later addition of *lang* 'long'.

Looe, East & Looe, West Cornwall. *Loo* c.1220. Cornish *logh* 'pool, inlet'.

Loop Head (*Ceann Léime*) Clare. 'Head of the leap'. OScand. *hlaup*.

Loose Kent. *Hlose* 11th cent. '(Place at) the pig-sty'. OE *hlōse*.

Loosley Row Bucks. *Losle* 1241. 'Woodland clearing with a pig-sty'. OE *hlōse* + *lēah*.

Lopen Somerset. *Lopen*(*e*) 1086 (DB). Celtic *penn* 'hill' or OE *penn* 'fold, enclosure' with an uncertain first element.

Lopham, North & Lopham, South Norfolk. *Lopham* 1086 (DB). 'Homestead or village of a man called *Loppa'. OE pers. name + *hām*.

Loppington Shrops. *Lopitone* 1086 (DB). 'Estate associated with a man called *Loppa'. OE pers. name + -ing- + tūn.

Lorn (district) Arg. *Lorne* 1304. '(Territory of) the people of *Loarn Mór*', 'the great fox', active in the 6th cent.

Lorton, High & Lorton, Low Cumbria. *Loretona c.*1150. Probably 'farmstead on a stream called **Hlóra*'. OScand. river-name (meaning 'roaring one') + OE *tūn*.

Loscoe Derbys. *Loscowe* 1277. 'Wood with a lofthouse'. OScand. *loft* + *skógr*.

Loskeran (*Loscreán*) Waterford. 'Burnt ground'.

Losset (*Losad*) Cavan. 'Productive field'.

Lossiemouth Moray. 'Mouth of the River Lossie'. The river-name means 'water of herbs' (Gaelic *Uisge Lossa*).

Lostock, 'outlying farmstead with a pig-sty', OE *hlōse* + *stoc*: **Lostock** Bolton. *Lostok* 1205. **Lostock Gralam** Ches. *Lostoch c.*1100, *le Lostoke Graliam* 1288. Manorial affix from a 12th-cent. owner called *Gralam*.

Lostwithiel Cornwall. *Lostwetiel* 1194. 'Tail-end of the woodland'. Cornish *lost* + *gwydhyel*.

Lothian. *See* MIDLOTHIAN.

Loudwater Bucks. *La Ludewatere* 1241. Originally a river-name, 'the noisy stream', OE *hlūd* + *wæter*.

Lough. *See* next word as name, e.g. for Lough Neagh, *see* NEAGH, LOUGH.

Loughan (*Lochán*) Meath. 'Little lake'.

Loughanavally (*Lochán an Bhealaigh*) Westmeath. 'Little lake of the routeway'.

Loughanure (*Loch an Iúir*) Donegal. 'Lake of the yew-tree'.

Loughborough Leics. *Lucteburne* [*sic*] 1086 (DB), *Lucteburga* 12th cent. 'Fortified house of a man called Luhhede'. OE pers. name + *burh*.

Loughbrickland (*Loch Bricleann*) Down. (*ic*) *Loch B(r)icreann c.*830. 'Lake of Bricle (earlier Bricriu)'.

Loughduff (*An Lathaigh Dhubh*) Cavan. 'The black mire'.

Lougher (*Luachair*) Kerry. 'Rushy place'.

Loughermore (*Luachair Mhór*) Derry. 'Big rushy place'.

Loughgall (*Loch gCál*) Armagh. An ancient name, meaning unknown.

Loughglinn (*Loch Glinne*) Roscommon. 'Lake of the valley'.

Loughguile (*Loch gCaol*) Antrim. *Lochkele* 1262. 'Lake of the narrows'.

Loughil (*Leamhchoill*) Limerick. 'Elm wood'.

Loughinisland (*Loch an Oileáin*) Down. *Loghnellan* 1616. 'Lake of the island'.

Loughlinstown (*Baile Uí Lachnáin*) Dublin. 'Homestead of Ó Lachnán'.

Loughmacrory (*Loch Mhic Ruairí*) Tyrone. 'Mac Ruairí's lake'.

Loughor (*Casllwchwr*) Swan. *Caslougher* 1691. '(Place by the river) Loughor'. The Celtic river name means 'shining one' (Welsh *llychwr*). The Welsh name means 'castle by the Loughor' (Welsh *cas*).

Loughrea (*Loch Riach*) Galway. 'Grey lake'.

Loughros Point (*Pointe Luachrois*) Donegal. *(hi) Luachros* 1509. 'Rushy point'.

Loughton Milt. K. *Lochintone* 1086 (DB). 'Estate associated with a man called Luhha'. OE pers. name (+ *-ing-*) + *tūn*.

Loughton Essex. *Lukintone* 1062, *Lochetuna* 1086 (DB). 'Estate associated with a man called **Luca*'. OE pers. name (+ *-ing-*) + *tūn*.

Loughton Shrops. *Lochetona* 1138. 'Farmstead by a pool'. OE *luh* + *tūn*.

Lound, 'small wood or grove', OScand. *lundr*: **Lound** Lincs. *Lund* 1086 (DB). **Lound** Notts. *Lund* 1086 (DB). **Lound** Suffolk. *Lunda* 1086 (DB). **Lound, East** N. Lincs. *Lund* 1086 (DB), *Estlound* 1370. Addition is OE *ēast* 'east'.

Loup, The (*An Lúb*) Derry. *The Loop Lodge* 1798. 'The loop'. The reference is presumably to a bend in a stream here.

Louth Lincs. *Lude* 1086 (DB). Named from the River Lud, OE **Hlūde* 'the loud one, the noisy stream'.

Louth (*Lú*) Louth. *Lughmhaigh* n.d. Perhaps 'Lú's penis', referring to a standing stone. Lú is a sun god. The name formerly had a second element, long explained as Irish *magh*, 'plain'.

Lovat Highland. *Loveth c.*1235. 'Swampy place'. Gaelic *lobh* + *-aid*.

Lover Wilts. Not on early record, origin uncertain, but thought to be a corruption (or rather a local dialect pronunciation) of *Lower* REDLYNCH (the name Lover rhymes with Dover).

Loversall Donc. *Loureshale* 1086 (DB). 'Nook of a man called Lēofhere'. OE pers. name + *halh*.

Loveston Pemb. *Lovelleston* 1362. 'Lovell's farm'. French pers. name + OE *tūn*.

Lovington Somerset. *Lovintune* 1086 (DB). 'Estate associated with a man called Lufa'. OE pers. name + *-ing-* + *tūn*.

Low as affix. *See* main name, e.g. for **Low Ackworth** (Wakefd.) *see* ACKWORTH.

Lowdham Notts. *Ludham* 1086 (DB). 'Homestead or village of a man called *Hlūda'. OE pers. name + *hām*.

Lower as affix. *See* main name, e.g. for **Lower Aisholt** (Somerset) *see* AISHOLT.

Lowery (*Leamhraidhe*) Donegal, Fermanagh. 'Place of elms'.

Lowesby Leics. *Glowesbi* [*sic*] 1086 (DB), *Lousebia c.*1125. Possibly 'farmstead or village of a man called Lauss or Lausi'. OScand. pers. name + *bý*.

Lowestoft Suffolk. *Lothu Wistoft* 1086 (DB). 'Homestead of a man called Hlothvér'. OScand. pers. name + *toft*.

Loweswater Cumbria. *Lousewater c.*1160. Named from the lake, 'leafy lake', from OScand. *lauf* + *scér* with the addition of OE *wæter* 'expanse of water'.

Lowick Cumbria. *Lofwik* 1202. Probably 'leafy creek or river-bend'. OScand. *lauf* + *vík*.

Lowick Northants. *Luhwic* 1086 (DB). 'Dwelling or (dairy) farm of a man called Luhha or *Luffa'. OE pers. name + *wīc*.

Lowick Northum. *Lowich* 1181. 'Dwelling or (dairy) farm on the River Low'. River-name (probably from OE *luh* 'pool') + OE *wīc*.

Lowther Cumbria. *Lauder c.*1175. Named from the River Lowther, possibly an OScand. river-name meaning 'foamy river', OScand. *lauthr* + *á*, but perhaps Celtic in origin.

Lowther Hills Dumf. Meaning uncertain. The name may be related to LAUDER.

Lowthorpe E. R. Yorks. *Log(h)etorp* 1086 (DB). 'Outlying farmstead or hamlet of a man called Lági or Logi'. OScand. pers. name + *thorp*.

Lowton Wigan. *Lauton* 1202. 'Farmstead by a mound or hill'. OE *hlāw* + *tūn*.

Loxbeare Devon. *Lochesbere* 1086 (DB). 'Wood or grove of a man called Locc'. OE pers. name + *bearu*.

Loxhore Devon. *Lochesore* 1086 (DB). 'Hill-slope of a man called Locc'. OE pers. name + *ōra*.

Loxley Warwicks. *Locheslei* 1086 (DB). 'Woodland clearing of a man called Locc'. OE pers. name + *lēah*.

Loxton N. Som. *Lochestone* 1086 (DB). 'Farmstead on the stream called Lox Yeo'. Celtic river-name (of uncertain meaning) + OE *ēa* + *tūn*.

Lubenham Leics. *Lubanham, Lobenho* 1086 (DB). 'Hill-spur(s) of a man called *Lubba'. OE pers. name (genitive *-n*) + *hōh* (dative plural *hōum*).

Lucan (*Leamhcán*) Dublin. 'Place of elms'.

Luccombe, probably 'valley of a man called Lufa', OE pers. name + *cumb*; alternatively the first element may be OE *lufu* 'love' with reference to courtship and lovemaking: **Luccombe** Somerset. *Locumbe* 1086 (DB). **Luccombe Village** I. of Wight. *Lovecumbe* 1086 (DB).

Lucker Northum. *Lucre* 1170. Probably 'marsh with a pool'. OE *luh* + OScand. *kjarr*.

Luckington Wilts. *Lochintone* 1086 (DB). 'Estate associated with a man called *Luca'. OE pers. name + *-ing-* + *tūn*.

Lucton Herefs. *Lugton* 1185. 'Farmstead on the River Lugg'. Celtic river-name (*see* LUGWARDINE) + OE *tūn*.

Ludborough Lincs. *Ludeburg* 1086 (DB). Probably 'fortified place belonging to or associated with Louth', from OE *burh*, *see* LOUTH. Alternatively 'fortified place of a man called Luda', OE pers. name + *burh*.

Ludchurch (*Yr Eglwys Lwyd*) Pemb. *Ecclesia de Loudes* 1324, *Loudeschirch* 1353. The modern Welsh name means 'the grey church' (Welsh *yr* + *eglwys* + *llwyd*) but is probably a mistranslation of a medieval English surname *Loud*.

Luddenden Calder. *Luddingdene* 1284. 'Valley of the stream called **Hlūding* (the loud one)'. OE name for the stream now called Luddenden Brook + *denu*. **Luddenden Foot** is *Luddingdeneforth* 1307, *Luddendenfoote* 1601, 'ford near Luddenden', from OE *ford* (changed to *foot* by folk etymology).

Luddesdown Kent. *Hludesduna* 10th cent., *Ledesdune* 1086 (DB). 'Hill of a man called **Hlūd*'. OE pers. name + *dūn*.

Luddington N. Lincs. *Ludintone* 1086 (DB). 'Estate associated with a man called Luda'. OE pers. name + *-ing-* + *tūn*.

Ludford Shrops. *Ludeford* 1086 (DB). 'Ford on the noisy stream' (referring to a crossing of the River Teme). OE **hlūde* + *ford*.

Ludgershall, 'nook with a trapping-spear', OE **lūte-gār* + *halh*: **Ludgershall** Bucks. *Lotegarser* [*sic*] 1086 (DB), *Lutegareshale* 1164. **Ludgershall** Wilts. *Lutegaresheale* 1015, *Litlegarsele* [*sic*] 1086 (DB).

Ludgvan Cornwall. *Luduhan* 1086 (DB). Possibly 'place of ashes, burnt place'. Cornish *lusow*, *ludw* + suffix.

Ludham Norfolk. *Ludham* 1021–4, 1086 (DB). 'Homestead or village of a man called Luda'. OE pers. name + *hām*.

Ludlow Shrops. *Ludelaue* 1138. 'Mound or tumulus by the noisy stream' (referring to the River Teme). OE **hlūde* + *hlāw*.

Ludwell Wilts. *Luddewell* early 12th cent. 'Loud spring or stream'. OE *hlūd* + *wella*.

Ludworth Durham. *Ludewrth c.*1160. Probably 'enclosure of a man called Luda'. OE pers. name + *worth*. Alternatively the first element may be OE **hlūde* 'noisy stream'.

Luffenham, North & Luffenham, South Rutland. *Luf(f)enham* 1086 (DB). 'Homestead or village of a man called **Luffa*'. OE pers. name + *hām*.

Luffincott Devon. *Lughyngecot* 1242. 'Cottage(s) of the family or followers of a man called Luhha'. OE pers. name + *-inga-* + *cot*.

Lugacaha (*Log an Chatha*) Westmeath. 'Hollow of the battle'.

Luggacurren (*Log an Churraigh*) Laois. 'Hollow of the marsh'.

Lugnaquillia (*Log na Coille*) (mountain) Wicklow. *Logney O Nill* 1617, *Lugnaquilla* 1760. 'Hollow of the wood'.

Lugwardine Herefs. *Lucvordine* 1086 (DB). 'Enclosure on the River Lugg'. Celtic river-name (meaning 'bright stream') + OE *worthign*.

Lullingstone Kent. *Lolingestone* 1086 (DB). 'Farmstead or estate of a man called Lulling'. OE pers. name + *tūn*.

Lullington, 'estate associated with a man called Lulla', OE pers. name + *-ing-* + *tūn*: **Lullington** Derbys. *Lullitune* 1086 (DB). **Lullington** Somerset. *Loligtone* 1086 (DB).

Lulsley Worcs. *Lulleseia* 12th cent. 'Island, or dry ground in marsh, of a man called Lull'. OE pers. name + *ēg*.

Lulworth, East & Lulworth, West Dorset. *Lulvorde* 1086 (DB). 'Enclosure of a man called Lulla'. OE pers. name + *worth*.

Lumb Lancs. *Le Lome* 1534. 'The pool'. OE **lumm*.

Lumby N. Yorks. *Lundby* 963. 'Farmstead in or near a grove'. OScand. *lundr* + *bý*.

Lumley, Great Durham. *Lummalea c.*1040. 'Woodland clearing by the pools.' OE **lumm* + *lēah*.

Lund, 'small wood or grove', OScand. *lundr*: **Lund** E. R. Yorks. *Lont* 1086 (DB). **Lund** N. Yorks., near Barlby. *Lund* 1066–9, *Lont* 1086 (DB).

Lundy Island Devon. *Lundeia* 1189. 'Puffin island'. OScand. *lundi* + *ey*.

Lune (river) Lancs., Cumbria. *See* LANCASTER.

Lunt Sefton. *Lund* 1251. Identical in origin with LUND.

Luppitt Devon. *Lovapit* 1086 (DB). 'Pit or hollow of a man called **Lufa*'. OE pers. name + *pytt*.

Lupton Cumbria. *Lupetun* 1086 (DB). 'Farmstead of a man called **Hluppa*'. OE pers. name + *tūn*.

Lurga (*An Lorgain*) Mayo. 'The ridge'.

Lurgan (*Lorgain*) Armagh, Donegal, Westmeath. 'Ridge'.

Lurgangreen (*Lorgain*) Louth. 'Green of the ridge'.

Lurganreagh (*Lorgan Riabhach*) Down. 'Striped ridge'.

Lurgashall W. Sussex. *Lutegareshale* 12th cent. 'Nook with a trapping-spear'. OE *lūte-gār + halh*.

Lusby Lincs. *Luzebi* 1086 (DB). 'Farmstead or village of a man called Lútr'. OScand. pers. name + *bý*.

Lusk (*Lusca*) Dublin. Meaning uncertain.

Luss Arg. *Lus* 1225. '(Place of) plants'. Gaelic *lus*.

Lustleigh Devon. *Leuestelegh* 1242. Probably 'woodland clearing of a man called *Lēofgiest*'. OE pers. name + *lēah*.

Luston Herefs. *Lustone* 1086 (DB). Probably 'insignificant, or louse-infested, farmstead'. OE *lūs + tūn*.

Luton Devon. near Ideford. *Leueton* 1238. Possibly 'farmstead of a woman called Lēofgifu'. OE pers. name + *tūn*.

Luton Luton. *Lygetun* 792, *Loitone* 1086 (DB). 'Farmstead on the River Lea'. Celtic river-name (*see* LEYTON) + OE *tūn*.

Luton Medway. *Leueton* 1240. 'Farmstead of a man called Lēofa'. OE pers. name + *tūn*.

Lutterworth Leics. *Lutresurde* 1086 (DB). Probably 'enclosure on a stream called *Hlūtre* (the clear or pure one)'. OE river-name (perhaps an earlier name for the River Swift) + *worth*.

Lutton Lincs. *Luctone* 1086 (DB). 'Farmstead by a pool'. OE *luh + tūn*.

Lutton Northants. *Lundingtun* [sic] *c*.970, *Luditone* 1086 (DB). 'Estate associated with a man called Luda'. OE pers. name + *-ing- + tūn*.

Lutton, East & Lutton, West N. Yorks. *Ludton* 1086 (DB). Probably 'estate of a man called Luda'. OE pers. name + *tūn*.

Luxborough Somerset. *Lolochesberie* 1086 (DB). 'Stronghold, or hill, of a man called Lulluc'. OE pers. name + *burh* or *beorg*.

Luxulyan Cornwall. *Luxulian* 1282. 'Chapel of Sulyen'. Cornish *log* + pers. name.

Lybster Highland. *Libister* 1538. 'Dwelling place by a slope'. OScand. *hlíth + bólstathr*.

Lydbrook, Lower & Lydbrook, Upper Glos. *Luidebroc* 1224. 'Brook called *Hlȳde* or the loud one'. OE river-name + *brōc*.

Lydbury North Shrops. *Lideberie* 1086 (DB). Probably 'manor house by the stream called *Hlȳde* (the noisy one)'. OE river-name + *burh* (dative *byrig*). The affix *north* distinguishes it from LEDBURY which however has a quite different origin.

Lydd Kent. *Hlidum* 774. Possibly '(place at) the gates'. OE *hlid* in a dative plural form.

Lydden (river) Dorset. *See* LYDLINCH.

Lydden Kent. near Dover. *Hleodaena c*.1100. Probably 'valley with a shelter, or sheltered valley'. OE *hlēo + denu*.

Lyddington Rutland. *Lidentone* 1086 (DB). Possibly 'farmstead or estate associated with a man called Hlȳda', from OE pers. name + *-ing- + tūn*. Alternatively 'farmstead on the noisy stream', OE *hlȳde, *hlȳding +tūn*.

Lyde Herefs. *Lude* 1086 (DB). Named from the stream here. OE *Hlȳde* 'the loud one'.

Lydeard St Lawrence & Bishop's Lydeard Somerset. *Lidegeard* 854, *Lidiard, Lediart* 1086 (DB). Celtic *garth* 'ridge' with an uncertain first element, possibly *lẹ̄d* 'grey'. Distinguishing affixes from the dedication of the church and from early possession by the Bishop of Wells.

Lydford Devon. *Hlydanforda c*.1000, *Lideforda* 1086 (DB). 'Ford over the River Lyd'. OE river-name (from *hlȳde* 'noisy stream') + *ford*.

Lydford Somerset. *Lideford* 1086 (DB). 'Ford over the noisy stream'. OE *hlȳde + ford*.

Lydham Shrops. *Lidum* 1086 (DB). Possibly '(place at) the gates or the slopes'. OE *hlid* or *hlid* in a dative plural form.

Lydiard Millicent Wilts. **& Lydiard Tregoze** Swindn. *Lidgeard* 901, *Lidiarde* 1086 (DB), *Lidiard Milisent* 1275, *Lidiard Tregoz* 1196. Identical in origin with LYDEARD. Manorial affixes from a woman called *Millisent* and from the *Tresgoz* family, here in the late 12th cent.

Lydiate Sefton. *Leiate* [sic] 1086 (DB), *Liddigate* 1202. '(Place at) the swing-gate'. OE *hlid-geat*.

Lydlinch Dorset. *Lidelinz* 1182. 'Ridge by, or bank of, the River Lydden'. Celtic river-name (probably meaning 'the broad one') + OE *hlinc*.

Lydney Glos. *Lideneg c.*853, *Ledenei* 1086 (DB). 'Island or river-meadow of the sailor, or of a man called *Lida'. OE *lida* or pers. name (genitive -*n*) + *ēg*.

Lye, Lower & Lye, Upper Herefs. *Lege* 1086 (DB). '(Place at) the wood or woodland clearing'. OE *lēah*.

Lyford Oxon. *Linforda* 944, *Linford* 1086 (DB). 'Ford where flax grows'. OE *līn* + *ford*.

Lyme (old district) Lancs., Ches., Staffs. *See* ASHTON-UNDER-LYNE, AUDLEM, BURSLEM, NEWCASTLE UNDER LYME

Lyme Regis Dorset. *Lim* 774, *Lime* 1086 (DB), *Lyme Regis* 1285. Named from the River Lim, a Celtic river-name meaning simply 'stream'. Affix is Latin *regis* 'of the king'.

Lyminge Kent. *Liminge* 689, *Leminges* 1086 (DB). 'District around the River *Limen*'. Celtic river-name (*see* LYMPNE) + OE *gē*.

Lymington Hants. *Lentune* [sic] 1086 (DB), *Limington* 1186. Probably 'farmstead on a river called *Limen*'. Lost Celtic river-name (meaning 'elm river' or 'marshy river') + OE *tūn*.

Lyminster W. Sussex. *Lullyngmynster c.*880, *Lolinminstre* 1086 (DB). 'Monastery or large church associated with a man called Lulla'. OE pers. name + -*ing*- + *mynster*.

Lymm Warrtn. *Lime* 1086 (DB). 'The noisy stream or torrent'. OE *hlimme*.

Lympne Kent. *Lemanis* 4th cent. 'Elm-wood place', or 'marshy place'. Celtic place name related to the River *Limen* (the old name for the East Rother), a Celtic river-name meaning 'elm-wood river' or 'marshy river'.

Lympsham Somerset. *Limpelesham* 1189. OE *hām* 'homestead' or *hamm* 'enclosure' with an uncertain first element, possibly an OE pers. name *Limpel*.

Lympstone Devon. *Levestone* [sic] 1086 (DB), *Leveneston* 1238. 'Farmstead of a man called Lēofwine'. OE pers. name + *tūn*.

Lynally (*Lann Eala*) Offaly. 'Church of (Colmán) Eala'.

Lyndhurst Hants. *Linhest* 1086 (DB). 'Wooded hill growing with lime-trees'. OE *lind* + *hyrst*.

Lyndon Rutland. *Lindon* 1167. 'Hill where flax is grown, or where lime-trees grow'. OE *līn* or *lind* + *dūn*.

Lyne (river) Cumbria. *See* KIRKLINTON.

Lyne Surrey. near Chertsey. *La Linde* 1208. 'The lime-tree'. OE *lind*.

Lyneham, 'homestead or enclosure where flax is grown', OE *līn* + *hām* or *hamm*:
Lyneham Oxon. *Lineham* 1086 (DB).
Lyneham Wilts. *Linham* 1224.

Lynemouth Northum. *Lynmuth* 1342. 'Mouth of the River Lyne'. Celtic river-name (meaning 'stream') + OE *mūtha*.

Lyng Norfolk. *Ling* 1086 (DB). Probably OE *hlinc* 'bank or terrace'.

Lyng Somerset. *Lengen* early 10th cent., *Lege* [sic] 1086 (DB). OE *lengen* 'length', that is 'long place', referring to a long narrow settlement.

Lynmouth Devon. *Lymmouth* 1330. 'Mouth of the River Lyn'. OE river-name (from *hlynn* 'torrent') + *mūtha*.

Lynn (*Lann*) Westmeath. 'Church'.

Lynn, King's & Lynn, West Norfolk. *Lena, Lun* 1086 (DB). 'The pool'. Celtic *līnn*. Affix *King's* dates from the 16th cent.

Lynsted Kent. *Lindestede* 1212–20. 'Place where lime-trees grow'. OE *lind* + *stede*.

Lynton Devon. *Lintone* 1086 (DB). 'Farmstead on the River Lyn'. OE river-name (*see* LYNMOUTH) + *tūn*.

Lyonshall Herefs. *Lenehalle* 1086 (DB). 'Nook of land in *Leon*'. Celtic district name (meaning 'at the streams') + OE *halh*.

Lyracrumpane (*Ladhar an Chrompáin*) Kerry. 'Fork of the river valley'.

Lyre (*An Ladhar*) Cork. 'The fork'.

Lyrenageeha (*Ladhar na Gaoithe*) Cork. 'Fort of the fork'.

Lyrenagreena (*Ladhar na Gréine*) Limerick. 'Fort of the fork'.

Lytchett Matravers & Lytchett Minster Dorset. *Lichet* 1086 (DB), *Lichet Mautrauers* 1280, *Licheminster* 1244. 'Grey wood'. Celtic **lẹ̄d* + **cẹ̄d*. Distinguishing affixes from the *Maltrauers* family, here in 1086, and from OE *mynster* 'large church'.

Lytham St Anne's Lancs. *Lidun* 1086 (DB). '(Place at) the slopes or dunes'. OE *hlith* in a dative plural form *hlithum*. Affix from the dedication of the church.

Lythe N. Yorks. *Lid* 1086 (DB). 'The slope'. OScand. *hlíth*.

Maam (*Mám*) Galway. 'Mountain pass'.

Maas (*Más*) Donegal. 'Buttock'.

Mabe Burnthouse Cornwall. *Lavabe* 1524, *Mape* 1549. 'Church-site of Mab'. Cornish **lann* + pers. name. Addition *Burnthouse*, no doubt alluding to a fire here, is first found in early 19th cent.

Mablethorpe Lincs. *Malbertorp* 1086 (DB). 'Outlying farmstead of a man called Malbert'. OGerman pers. name + OScand. *thorp*.

Mabrista (*Má Briste*) Westmeath. 'Broken plain'.

Macclesfield Ches. *Maclesfeld* 1086 (DB). Probably 'open country of a man called **Maccel*'. OE pers. name + *feld*.

Macduff Aber. The settlement was rebuilt in 1783 by James *Duff*, 2nd Earl of Fife, who claimed to be descended from the semi-mythical Macduff in Shakespeare's *Macbeth*, and he named it after his father, William *Duff*, 1st Earl of Fife, who had acquired land here.

Mace (*Más*) Galway, Mayo. 'Buttock'.

Macgillycuddy's Reeks (*Na Cruacha Dubha*) (mountains) Kerry. 'Mac Giolla Chuda's ricks'. The Irish name means 'the black peaks'.

Machars, The (district) Dumf. 'The plains'. Gaelic *machair* + English plural *-s*.

Machen Cphy. *Mahhayn* 1102. 'Cein's plain'. Welsh *ma*.

Machynlleth Powys. *Machenleyd* 1254. 'Cynllaith's plain'. Welsh *ma*.

Mackworth Derbys. *Macheuorde* 1086 (DB). 'Enclosure of a man called **Macca*'. OE pers. name + *worth*.

Macnean, Lough (*Loch Mac nÉan*) Cavan, Fermanagh, Leitrim. 'Mac Nean's lake'.

Macosquin (*Maigh Choscáin*) Derry (*abb mainistre*) *Mhaighe Coscrain* 1505. 'Coscán's plain'.

Macroom (*Maigh Chromtha*) Cork. 'Plane of the crooked ford'.

Madabawn (*An Maide Bán*) Cavan. 'The white plant'.

Maddadoo (*Maide Dubh*) Westmeath. 'Black plant'.

Madehurst W. Sussex. *Medliers* 1188, *Medhurst* 1255. Possibly 'wooded hill near meadow-land', OE *mǣd* + *hyrst*. Alternatively the first element may be OE *mæthel* 'assembly, meeting'.

Madeley, 'woodland clearing of a man called **Māda*', OE pers. name + *lēah*: **Madeley** Staffs., near Newcastle. *Matdanlieg* 975, *Madelie* 1086 (DB). **Madeley** Tel. & Wrek. *Madelie* 1086 (DB).

Madingley Cambs. *Madingelei* 1086 (DB). 'Woodland clearing of the family or followers of a man called **Māda*'. OE pers. name + *-inga-* + *lēah*.

Madley Herefs. *Medelagie* 1086 (DB), *Madele*, *Maddeleia* c.1200. OE *lēah* 'woodland clearing' with an uncertain first element, possibly an OE pers. name **Māda* as in MADELEY.

Madresfield Worcs. *Madresfeld* c.1086, *Metheresfeld* 1192. 'Open land of the mower or of a man called Mæthhere'. OE *mǣthere* or pers. name + *feld*.

Madron Cornwall. '(Church of) *Sanctus Madernus*' 1203. From the patron saint of the church.

Maenclochog Pemb. *Maenclochawc* 1257. 'Stone sounding like a bell'. Welsh *maen* + *clochog*.

Maenorbyr. *See* MANORBIER.

Maentwrog Gwyd. *Mayntwroc* 1292. 'Rock of St Twrog'. Welsh *maen*.

Maer Staffs. *Mere* 1086 (DB). '(Place at) the pool'. OE *mere*.

Maesbrook Shrops. *Meresbroc* 1086 (DB), *Maysbrok* 1272. Possibly 'brook of the boundary', OE *mǣre* (genitive -*s*) + *brōc*, but the first element may rather be Welsh *maes* 'open field'.

Maesbury Shrops. *Meresberie* 1086 (DB). 'Fort or manor house near the boundary'. OE *mǣre* (genitive -*s*) + *burh* (dative *byrig*).

Maesteg Bri. *Maes tege issa* 1543. 'Fair field'. Welsh *maes* + *teg*.

Maesyfed. *See* NEW RADNOR.

Magdalen Laver Essex. *See* LAVER.

Magharee Islands (*Machairí*) Kerry. 'Flat places'.

Maghera (*Machaire Rátha*) Derry. *Rátha Lúiricch* 814. 'Plain of the fort'. An earlier name was *Ráth Lúraigh*, 'Lúrach's fort', after a 6th-cent. saint.

Maghera (*Machaire*) Down. 'Plain'.

Magherabeg (*Machaire Beag*) Donegal. 'Small plain'.

Magheraboy (*Machaire Buí*) Mayo. 'Yellow plain'.

Magheracloone (*Machaire Cluana*) Monaghan. 'Plain of the pasture'.

Magheradernan (*Machaire Ua dTighearnáin*) Westmeath. 'Plain of Uí Tighearnáin'.

Magherafelt (*Machaire Fíolta*) Derry. *Teeoffigalta* 1426. 'Plain of Fíolta'. An earlier name was *Teach Fíolta*, 'Fíolta's house'.

Magheragall (*Machaire na gCeall*) Antrim. *Magheregall* 1661. 'Plain of the churches'.

Magheralin (*Machaire Lainne*) Down. (*ó*) *Lainn Ronain c.*830. 'Plain of the church'. An earlier name was *Lann Rónáin Fhinn*, 'church of fair Rónan'.

Magheramorne (*Machaire Morna*) Antrim. (*Domnach mor*) *Maigi Domóerna c.*900. 'Plain of the Morna'. The tribal name is of obscure origin.

Magheranageeragh (*Machaire na bCaorach*) Fermanagh, Tyrone. 'Plain of the sheep'.

Magheranerla (*Machaire an Iarla*) Westmeath. 'Field of the earl'.

Magheree Islands (*Machairí*) Kerry. 'Flat places'.

Maghery (*An Machaire*) Donegal. 'The plain'.

Maghull Sefton. *Magele* [sic] 1086 (DB), *Maghal* 1219. Probably 'nook of land where mayweed grows'. OE *mægthe* + *halh*.

Magilligan (*Aird Mhic Ghiollagáin*) Derry, 'Point of Mac Giollagán'.

Maguiresbridge Fermanagh. 'Maguire's bridge'. *Magwyre's bridge* 1639. The village grew up after 1760 when Brian *Maguire* of the local family of this name was given a grant to hold a market here. The corresponding Irish name is *Droichead Mhig Uidhir*.

Mahoonagh (*Caislean Maí Tamhnach*) Limerick. 'Stone fort of the plain of the fields'.

Maida Hill & Maida Vale Gtr. London. Recorded thus in the early 19th cent. So named to commemorate the battle of *Maida* in southern Italy, the site of a British victory against the French in 1806.

Maiden as affix. *See* main name, e.g. for **Maiden Bradley** (Wilts.) *see* BRADLEY.

Maiden Castle Dorset. *Mayden Castell* 1607. Iron Age hill-fort, perhaps so named to mean 'fortification that has never been taken', or 'fortification so strong it could be defended by young women'. There are examples of the same name for prehistoric fortifications in Cumbria and N. Yorks.

Maiden Wells Pemb. *Mayden Welle* 1336. 'Spring of the maidens, i.e. where they gathered'. OE *mægden* + *wella*.

Maidencombe Torbay. *Medenecoma* 1086 (DB). 'Valley of the maidens, i.e. where they gathered'. OE *mægden* + *cumb*.

Maidenhead Winds. & Maid. *Maidenhee* 1202. 'Landing-place of the maidens'. OE *mægden* + *hÿth*.

Maidenwell Lincs. *Welle* 1086 (DB), *Maidenwell* 1212. Originally '(place at) the spring or stream', from OE *wella*, with the later addition of *mægden* 'maiden'.

Maidford Northants. *Merdeford* [*sic*] 1086 (DB), *Maideneford* 1166. 'Ford of the maidens, i.e. where they gathered'. OE *mægden* + *ford*.

Maids Moreton Bucks. *See* MORETON.

Maidstone Kent. *Mægthan stan* late 10th cent., *Meddestane* 1086 (DB). Probably 'stone of the maidens, i.e. where they gathered'. OE *mægth, mægden* + *stān*.

Maidwell Northants. *Medewelle* 1086 (DB). 'Spring or stream of the maidens, i.e. where they gathered'. OE *mægden* + *wella*.

Maigue (*Maigh*) (river) Limerick. 'River of the plain'.

Main (*An Mhin*) (river) Antrim. 'The water'.

Maine (*Maighin*) (river) Kerry. Meaning uncertain.

Mainland (island) Orkn. *Meginland* c.1150. 'Chief district'. OScand. *megin* + *land*.

Mainstone Shrops. *Meyneston* 1284. 'The strength stone, i.e. one that requires great strength to lift'. OE *mægen* + *stān*.

Maisemore Glos. *Mayesmora* 1138. Probably 'the great field'. Welsh *maes* + *mawr*.

Makerfield (old district) Lancs. *See* ASHTON-IN-MAKERFIELD.

Malahide (*Mullach Íde*) Dublin. 'Íde's summit'.

Malborough Devon. *Malleberge* 1249. 'Hill or mound of a man called *Mǣrla, or where gentian grows'. OE pers. name or *meargealla* + *beorg*.

Malden, New Malden Gtr. London. *Meldone* 1086 (DB). 'Hill with a cross or crucifix'. OE *mǣl* + *dūn*.

Maldon Essex. *Mældune* early 10th cent., *Malduna* 1086 (DB). Identical in origin with the previous name.

Malham N. Yorks. *Malgun* 1086 (DB). '(Settlement) by the gravelly places'. OScand. *malgi* in a dative plural form.

Malin (*Málainn*) Donegal. 'Brow'.

Malin Beg (*Málainn Bhig*) Donegal. 'Small brow'.

Malin Head (*Cionn Mhálanna*) Donegal. 'Headland of the brow'.

Malin More (*Málainn Mhóir*) Donegal. 'Large brow'.

Mallaig Highland. Perhaps 'bay of gulls'. OScand. *már* + *vágr*.

Mallaranny (*An Mhala Raithní*) Mayo. 'The brow of the ferns'.

Malling, '(settlement of) the family or followers of a man called *Mealla', OE pers. name + *-ingas*: **Malling, East & Malling, West** Kent. *Meallingas* 942–6, *Mellingetes* [*sic*] 1086 (DB). **Malling, South** E. Sussex. *Mallingum* 838, *Mellinges* 1086 (DB).

Mallow (*Mala*) Cork. 'Plain of the (river) Allow'. *See* ALLOW.

Mallusk (*Maigh Bhloisce*) Antrim. *Manyblos* 1231. 'Blosce's plain'.

Malmesbury Wilts. *Maldumesburg* 685, *Malmesberie* 1086 (DB). 'Stronghold of a man called Maeldub'. OIrish pers. name + OE *burh* (dative *byrig*).

Malone (*Maigh Luain*) Antrim. *Mylon* 1604. 'Luan's plain'.

Malpas, 'difficult passage', OFrench *mal* + *pas*; examples include: **Malpas** Ches. *Malpas* c.1125. **Malpas** Newpt. *Malo Passu* 1291.

Maltby, 'farmstead or village of a man called Malti, or where malt is made', OScand. pers. name or *malt* + *bý*: **Maltby** Rothm. *Maltebi* 1086 (DB). **Maltby** Stock. on T. *Maltebi* 1086 (DB). **Maltby le Marsh** Lincs. *Maltebi* 1086 (DB). Affix 'in the marshland' from OE *mersc*.

Malton N. Yorks. *Maltune* 1086 (DB). Possibly 'farmstead where an assembly was held'. OE *mæthel* + *tūn*. Alternatively the first element may be OScand. *methal* 'middle'.

Malvern, Great & Malvern, Little Worcs. *Mælfern* c.1030, *Malferna* 1086 (DB). 'Bare hill'. Celtic *mēl* + *brïnn*. **Malvern Link**, *Link* 1215, is from OE *hlinc* 'ledge, terrace'.

Mamble Worcs. *Mamele* 1086 (DB). Probably a derivative of Celtic *mamm* 'breast-like hill'.

Man, Isle of. An ancient name, linked in legend with that of an Irish god, Manannan mac Lir, but its origin is obscure. In Roman times it was called *Monapia*, possibly to be associated with the Roman name *Mona* (of unknown origin and meaning) for ANGLESEY. The Manx name for the island is *Ellan Vannan* meaning 'island of Man'.

Manaccan Cornwall. '(Church of) *Sancta Manaca*' 1309. Probably from the patron saint of the church.

Manaton Devon. *Manitone* 1086 (DB). 'Farmstead held communally, or by a man called Manna'. OE (*ge*)*mǣne* or pers. name + *tūn*.

Manby Lincs. near Louth. *Mannebi* 1086 (DB). 'Farmstead or village of a man called Manni, or of the men'. OScand. pers. name or *mathr* (genitive plural *manna*) + *bý*.

Mancetter Warwicks. *Manduessedum* 4th cent., *Manacestre* 1195. OE *ceaster* 'Roman fort or town' added to a reduced form of the original Celtic name which probably means 'horse chariot' (perhaps alluding to a topographical feature).

Manchester Manch. *Mamucio* 4th cent., *Mamecestre* 1086 (DB). OE *ceaster* 'Roman fort or town' added to a reduced form of the original Celtic name (meaning obscure, but probably containing Celtic **mamm* 'breast-like hill').

Mancot Royal Flin. The hamlet arose as a housing estate built near BIG MANCOT in 1917 for workers in the nearby Royal Ordnance Factory.

Manea Cambs. *Moneia* 1177. OE *ēg* 'island, well-watered land, dry ground in marsh' with a doubtful first element, possibly OE (*ge*)*mǣne* 'held in common'.

Maney Birm. *Maney(e)* 1285. Probably identical in origin with previous name.

Manfield N. Yorks. *Mannefelt* 1086 (DB). 'Open land of a man called Manna, or held communally'. OE pers. name or OE (*ge*)*mǣne* + *feld*. **Mangotsfield** S. Glos. *Manegodesfelle* 1086 (DB). 'Open land of a man called Mangod'. OGerman pers. name + OE *feld*.

Manley Ches. *Menlie* 1086 (DB). 'Common wood or clearing'. OE (*ge*)*mǣne* + *lēah*.

Manningford Bohune & Manningford Bruce Wilts. *Maningaford* 987, *Maniford* 1086 (DB), *Manyngeford Bon*, *Manyngeford Breuse* 1279. 'Ford of the family or followers of a man called Manna'. OE pers. name + *-inga-* + *ford*.

Manorial additions from the *Boun* and *Breuse* families, here in the 13th cent.

Mannington Dorset. *Manitone* 1086 (DB). 'Estate associated with a man called Manna'. OE pers. name + *-ing-* + *tūn*.

Manningtree Essex. *Manitre* 1248. 'Many trees', or 'tree of a man called Manna'. OE *manig* or OE pers. name + *trēow*.

Manorbier (*Maenorbyr*) Pemb. *Mainour pir* 1136, *Manerbire* 1331. 'Pyr's manor'. Welsh *maenor*.

Manorcunningham Donegal. *The Manor of Fort Cownyngham* 1629. The village is named after James *Cunningham* of Ayrshire, Scotland, who was granted land here in 1610. The equivalent Irish name is *Mainéar Uí Chuinneagáin*.

Manordeifi Pemb. *Mamardeyvi* 1219, *Menordaun* 1291. 'Estate on the River Teifi'. Welsh *maenor*. The river-name means simply 'stream'.

Manorhamilton Leitrim. Lands here were granted by Charles I to Sir Frederick *Hamilton*. The Irish name is *Cluainín*, short for *Cuainín Uí Ruairc*, 'little meadow of Ó Ruairc'.

Manorowen Pemb. *Maynornawan* 1326. 'Gnawan's manor'. Welsh *maenor*.

Mansell Gamage & Mansell Lacy Herefs. *Mælueshylle c.*1045, *Malveselle* 1086 (DB), *Maumeshull Gamages*, *Maumeshull Lacy* 1242. Probably 'hill of the gravel ridge'. OE **malu* + *hyll*. Manorial affixes from the *de Gamagis* family, here in the 12th cent., and from the *de Lacy* family, here in 1086.

Mansfield Notts. *Mamesfelde* 1086 (DB). 'Open land by the River Maun'. Celtic river-name (from Celtic **mamm* 'breast-like hill') + OE *feld*.

Mansfield Woodhouse Notts. *Wodehuse* 1230, *Mamesfeud Wodehus* 1280. 'Woodland hamlet near MANSFIELD'. OE *wudu* + *hūs*.

Manston, 'farmstead of a man called Mann', OE pers. name + *tūn*: **Manston** Dorset. *Manestone* 1086 (DB). **Manston** Kent. *Manneston* 1254.

Manswood Dorset. *Mangewood* 1774. Possibly 'mange wood, i.e wood thought to be diseased', from ModE *mange*.

Manthorpe Lincs. near Thurlby. *Mannetorp* 1086 (DB). 'Outlying farmstead or village of a man called Manni, or of the men'. OScand. pers. name or *mathr* (genitive plural *manna*) + *thorp*.

Manton N. Lincs. *Malmetun* 1060–6, *Mameltune* 1086 (DB). 'Farmstead or estate on sandy ground'. OE **malm* + *tūn*.

Manton Rutland. *Mannatonam* 1120–9, *Manatona c.*1130. Probably 'farmstead or estate of a man called Manna'. OE pers. name + *tūn*.

Manton Wilts. *Manetune* 1086 (DB). 'Farmstead held communally, or by a man called Manna'. OE *(ge)mǣne* or OE pers. name + *tūn*.

Mantua (*An Móinteach*) Roscommon. 'The moorland'.

Manuden Essex. *Magghedana* [*sic*] 1086 (DB), *Manegedan c.*1130. Probably 'valley of the family or followers of a man called Manna'. OE pers. name + *-inga-* + *denu*.

Maplebeck Notts. *Mapelbec* 1086 (DB). 'Stream where maple-trees grow'. OE **mapel* + OScand. *bekkr*.

Mapledurham Oxon. *Mapeldreham* 1086 (DB). 'Homestead where maple-trees grow'. OE *mapuldor* + *hām*.

Mapledurwell Hants. *Mapledrewell* 1086 (DB). 'Spring or stream where maple-trees grow'. OE *mapuldor* + *wella*.

Maplestead, Great & Maplestead, Little Essex. *Mapulderstede* 1042, *Mapledestedam* 1086 (DB). 'Place where maple-trees grow'. OE *mapuldor* + *stede*.

Mapperley Derbys. *Maperlie* 1086 (DB). 'Maple-tree wood or clearing'. OE *mapuldor* + *lēah*.

Mapperton Dorset. near Beaminster. *Malperetone* 1086 (DB). 'Farmstead where maple-trees grow'. OE *mapuldor* + *tūn*.

Mappleborough Green Warwicks. *Mepelesbarwe* 848, *Mapelberge* 1086 (DB). 'Hill where maple-trees grow'. OE **mapel* + *beorg*.

Mappleton E. R. Yorks. *Mapletone* 1086 (DB). 'Farmstead where maple-trees grow'. OE **mapel* + *tūn*.

Mappowder Dorset. *Mapledre* 1086 (DB). '(Place at) the maple-tree'. OE *mapuldor*.

Mar(r) (district) Aber. '(Land of the) Mar people'. The tribal name probably comes from a pers. name *Marsos*.

Marazion Cornwall. *Marghasbigan c.*1265. 'Little market'. Cornish *marghas* + *byghan*.

Marbury Ches. near Whitchurch. *Merberie* 1086 (DB). 'Fortified place near a lake'. OE *mere* + *burh* (dative *byrig*).

March Cambs. *Merche* 1086 (DB). '(Place at) the boundary'. OE *mearc* in an old locative form.

Marcham Oxon. *Merchamme* 900, *Merceham* 1086 (DB). 'Enclosure or river-meadow where smallage (wild celery) grows'. OE *merece* + *hamm*.

Marchamley Shrops. *Marcemeslei* 1086 (DB), *Marchemeleg* 1227. Possibly 'woodland clearing of a man called Merchelm', OE pers. name + *lēah*. Alternatively 'woodland clearing at (the territory of) the Mercians', with a first element OE *Mierce* in a dative plural form *Miercum*.

Marchington Staffs. *æt Mærcham* 951, *æt Mærchamtune* 1002–4, *Merchametone* 1086 (DB). Probably 'farmstead at the river-meadow where smallage (wild celery) grows'. OE *merece* + *hamm* + *tūn*.

Marchwood Hants. *Merceode* 1086 (DB). 'Wood where smallage (wild celery) grows'. OE *merece* + *wudu*.

Marcle, Much & Marcle, Little Herefs. *Merchelai* 1086 (DB). 'Wood or clearing on a boundary'. OE *mearc* + *lēah*. Affix *Much* is from OE *mycel* 'great'.

Marden Herefs. *Maurdine* [*sic*] 1086 (DB), *Magewurdin* 1177. 'Enclosed settlement in the district called MAUND'. OE *worthign*.

Marden Kent. *Maeredaen c.*1100. 'Woodland pasture for mares, or near a boundary'. OE *mere* or *(ge)mǣre* + *denn*.

Marden Wilts. *Merhdæne* 963, *Meresdene* 1086 (DB). Probably 'fertile valley'. OE *mearg* 'marrow' (here used in a figurative sense) + *denu*.

Marden, East, Marden, North, & Marden, West W. Sussex. *Meredone* 1086 (DB). 'Boundary hill'. OE *(ge)mǣre* + *dūn*.

Marefield Leics. *Merdefelde* 1086 (DB). 'Open land frequented by martens or weasels'. OE *mearth* + *feld*.

Mareham le Fen Lincs. *Marun* 1086 (DB), *Marum* c.1150, *Mareham in the ffenne* 1644. '(Place at) the pools'. OE *mær(e)* in dative plural form *mærum*. Affix *le Fen* means 'in the fen or marshland'.

Mareham on the Hill Lincs. *Meringhe* 1086 (DB), *Maringes* 12th cent., *Maring of the hill* 1517. '(Settlement of) the pond dwellers'. OE *mær(e)* + *-ingas*. Development to *-ham* from 19th cent.

Maresfield E. Sussex. *Mersfeld* 1234. 'Open land by a marsh or pool'. OE *mersc* or *mere* + *feld*.

Marfleet K. upon Hull. *Mereflet* 1086 (DB). 'Pool creek or stream'. OE *mere* + *flēot*.

Margaret Roding Essex. *See* RODING.

Margaretting Essex. *Ginga* 1086 (DB), *Gynge Margarete* 1291. 'Manor called Ing (for which *see* FRYERNING) of St Margaret'. From the dedication of the church.

Margate Kent. *Meregate* 1254. Probably 'gate or gap leading to the sea, or by a pool'. OE *mere* + *geat*.

Marham Norfolk. *Merham* c.1050, *Marham* 1086 (DB). 'Homestead by or with a pond'. OE *mere* + *hām*.

Marhamchurch Cornwall. *Maronecirce* 1086 (DB). 'Church of St Marwen or Merwenn'. Female saint's name + OE *cirice*.

Marholm Peterb. *Marham* c.1060. 'Homestead by or with a pond'. OE *mere* + *hām*.

Mariansleigh Devon. *Lege* 1086 (DB), *Marinelegh* 1238. 'The wood or clearing' from OE *lēah*, with the later addition of the saint's name *Marina* or *Marion* (a diminutive of *Mary*).

Marishes, High & Marishes, Low N. Yorks. *Aschilesmares, Chiluesmares, Ouduluesmersc* 1086 (DB). 'The marshes', from OE *mersc*. Different manors originally distinguished by the names of early owners, OScand. *Ásketill, Ketilfrøthr,* and *Authulfr*.

Mark Somerset. *Mercern* 1065. 'House or building near a boundary'. OE *mearc* + *ærn*.

Mark Cross E. Sussex. *Markecross* 1509. 'Cross on the (parish) boundary'. ME *marke* (OE *mearc*) + *cross*.

Markby Lincs. *Marchebi* 1086 (DB). Possibly 'farmstead or village of a man called Marki'.

OScand. pers. name + *bý*. Alternatively the first element may be OScand. *mǫrk* 'frontier area, wilderness'.

Market as affix. *See* main name, e.g. for **Market Bosworth** (Leics.) *see* BOSWORTH.

Markethill Armagh. 'Hill with a market'. *Markethill* c.1640. The town, with its steep main street and still thriving livestock markets, grew up in the early 17th cent. The equivalent Irish name is *Cnoc an Mhargaigh*.

Markfield Leics. *Merchenefeld* 1086 (DB). 'Open land of the Mercians'. OE *Merce* (genitive plural *Mercna*) + *feld*.

Markham, East & Markham, West Notts. *Marcham* 1086 (DB), *Estmarcham* 1192. 'Homestead or village on a boundary'. OE *mearc* + *hām*.

Markinch Fife. *Marchinke* 1055, *Markynchs* c.1290. 'Horse meadow'. Gaelic *marc* + *innis*.

Markington N. Yorks. *Mercingtune* c.1030, *Merchinton* 1086 (DB). Possibly 'farmstead of the Mercians', OE *Merce* (genitive plural *Mercna*) + *tūn*. Alternatively 'farmstead of the boundary-dwellers, or by a boundary'. OE *(ge)merce* (with Scand. *-k-*) + *-inga-* or *-ing* + *tūn*.

Marks Tey Essex. *See* TEY.

Marksbury B. & NE. Som. *Merkesburi* 936, *Mercesberie* 1086 (DB). Possibly 'stronghold of a man called *Mæðrec* or *Mearc*'. OE pers. name + *burh* (dative *byrig*).

Markyate Herts. *Markȝate* 12th cent. 'Gate at the (county) boundary'. OE *mearc* + *geat*.

Marland, Peters Devon. *Mirlanda* 1086 (DB). 'Cultivated land by a pool'. OE *mere* + *land*.

Marlborough Wilts. *Merleberge* 1086 (DB). 'Hill or mound of a man called *Mæðrla*, or where gentian grows'. OE pers. name or *meargealla* + *beorg*.

Marlcliff Warwicks. *Mearnanclyfe* 872. Possibly 'cliff of a man called *Mearna*'. OE pers. name + *clif*.

Marldon Devon. *Mergheldone* 1307. 'Hill where gentian grows'. OE *meargealla* + *dūn*.

Marlesford Suffolk. *Merlesforda* 1086 (DB). Probably 'ford of a man called *Mæðrel*'. OE pers. name + *ford*.

Marlingford Norfolk. *Marthingforth c.*1000, *Merlingeforda, Marthingeforda* 1086 (DB). Possibly 'ford of the family or followers of a man called *Mearthel'. OE pers. name + *-inga-* + *ford*.

Marlow Bucks. *Merelafan* 1015, *Merlaue* 1086 (DB). 'Land remaining after the draining of a pool'. OE *mere* + *lāf*.

Marnham, High & Marnham, Low Notts. *Marneham* (DB). Possibly 'homestead or village of a man called *Mearna'. OE pers. name + *hām*.

Marnhull Dorset. *Marnhulle* 1267. Possibly 'hill of a man called *Mearna'. OE pers. name + *hyll*.

Marple Stockp. *Merpille* early 13th cent. 'Pool or stream at the boundary'. OE *(ge)mære* + *pyll*.

Marr Donc. *Marra* 1086 (DB). '(Place at) the marsh'. OScand. *marr*.

Marrick N. Yorks. *Marige* 1086 (DB). OE *hrycg* 'ridge' or **ric* 'raised strip of land' with an uncertain first element, possibly OE *(ge)mære* 'boundary' or OScand. *marr* 'marshy ground'.

Marsco (mountain) Highland (Skye). Meaning obscure.

Marsden Kirkl. *Marchesden* 12th cent. Probably 'boundary valley'. OE *mercels* + *denu*.

Marsh, '(place at) the marsh', OE *mersc*: **Marsh** Shrops. *Mersse* 1086 (DB). **Marsh Gibbon** Bucks. *Merse* 1086 (DB), *Mersh Gibwyne* 1292. Manorial affix from the *Gibwen* family, here in the 12th cent.

Marsh Baldon Oxon. *See* BALDON.

Marsham Norfolk. *Marsam* 1086 (DB). 'Homestead or village by a marsh'. OE *mersc* + *hām*.

Marshborough Kent. *Masseberge* 1086 (DB). 'Hill or barrow of a man called *Mæssa'. OE pers. name + *beorg*.

Marshchapel Lincs. *Mersch Chapel* 1347. 'Chapel in the marshland'. ME *mersh* + *chapel*.

Marshfield S. Glos. *Meresfelde* 1086 (DB). Probably 'open land on the boundary' OE *(ge)mære* + *feld*. Alternatively the first element may be OE *mersc* 'marsh'.

Marshwood Dorset. *Merswude* 1188. 'Wood by a marsh'. OE *mersc* + *wudu*.

Marske, '(place at) the marsh', OE *mersc* (with Scand. *-sk*): **Marske** N. Yorks. *Mersche* 1086 (DB). **Marske-by-the-Sea** Red. & Cleve. *Mersc* 1086 (DB).

Marston, a common name, 'farmstead in or by a marsh', OE *mersc* + *tūn*; examples include: **Marston** Oxon. *Mersttune c.*1069. **Marston, Broad** Worcs. *Merestune* 1086 (DB), *Brademerston* 1224. Affix is OE *brād* 'broad' referring to the shape of the village site. **Marston, Long** N. Yorks. *Mersetone* 1086 (DB). **Marston, Long** Warwicks. *Merstuna* 1043, *Merestone* 1086 (DB), *Longa Merston* 1285. Affix refers to the length of the village. Earlier often called *Dry Marston* or *Marston Sicca* (from Latin *sicca* 'dry'). **Marston Magna** Somerset. *Merstone* 1086 (DB), *Great Merston* 1248. Affix is Latin *magna* 'great'. **Marston Meysey** Wilts. *Merston* 1199, *Merston Meysi* 1259. Manorial affix from the *de Meysi* family, here in the 13th cent. **Marston Moretaine** Beds. *Merestone* 1086 (DB), *Merston Morteyn* 1383. Manorial affix from its early possession by the *Morteyn* family. **Marston, Priors** Warwicks. *Merston* 1236, *Prioris Merston* 1316. The manor was held by the Prior of Coventry in 1242. **Marston Trussell** Northants. *Mersitone* 1086 (DB), *Merston Trussel* 1235. Manorial affix from the *Trussel* family, here in the 13th cent.

Marstow Herefs. *Lann Martin c.*1130, *Martinestowe* 1291. 'Church or holy place of St Martin'. Saint's name + OE *stōw* (replacing Welsh *llan* in the early form).

Marsworth Bucks. *Mæssanwyrth* 10th cent., *Missevorde* [sic] 1086 (DB). 'Enclosure of a man called *Mæssa'. OE pers. name + *worth*.

Marten Wilts. *Mertone* 1086 (DB). 'Farmstead near a boundary, or by a pool'. OE *(ge)mære* or *mere* + *tūn*.

Marthall Ches. *Marthale* late 13th cent. 'Nook of land frequented by martens or weasels'. OE *mearth* + *halh*.

Martham Norfolk. *Martham* 1086 (DB). 'Homestead or enclosure frequented by martens or weasels'. OE *mearth* + *hām* or *hamm*.

Martin, 'farmstead near a boundary, or by a pool', OE *(ge)mære* or *mere* + *tūn*: examples include: **Martin** Hants. *Mertone* 946. **Martin** Lincs., near Horncastle. *Mærtune* 1060, *Martone* 1086 (DB). **Martin** Lincs., near Metheringham. *Martona* 12th cent. **Martin Hussingtree** Worcs. *Meretun, Husantreo* 972, *Husentre* 1086

(DB), *Marten Hosentre* 1535. Originally two separate manors, Hussingtree being 'tree of a man called Hūsa', from OE pers. name (genitive *-n*) + *trēow*.

Martinhoe Devon. *Matingeho* 1086 (DB). 'Hill-spur of the family or followers of a man called *Matta'. OE pers. name + *-inga-* + *hōh*.

Martinscroft Warrtn. *Martinescroft* 1332. 'Enclosure of a man called Martin'. ME pers. name + *croft*.

Martinstown Dorset. *Wintreburne* 1086 (DB), *Wynterburn Seynt Martyn* 1280, *Martyn towne* 1494. Originally 'estate on the River WINTERBORNE with a church dedicated to St Martin'. The alternative name came into use in the late 15th cent.

Martinstown Antrim. 'Martin's town'. The village takes its name from a man surnamed *Martin*. The corresponding Irish name is *Baile Uí Mháirtín*.

Martlesham Suffolk. *Merlesham* [*sic*] 1086 (DB), *Martelsham* 1254. Possibly 'homestead by a woodland clearing frequented by martens', OE *mearth* + *lēah* + *hām*. Alternatively the first element may be an OE pers. name *Mearthel*.

Martletwy Pemb. *Martletwye* c.1230. 'Grave of a saint'. Welsh *merthyr*. The second element represents the name of the saint, possibly Tywai or Tyfai.

Martley Worcs. *Mærtleag* c.1030, *Mertelai* 1086 (DB). 'Woodland clearing frequented by martens'. OE *mearth* + *lēah*.

Martock Somerset. *Mertoch* 1086 (DB). Possibly 'outlying farmstead or hamlet by a pool'. OE *mere* + *stoc*.

Marton, a common name, usually 'farmstead by a pool', OE *mere* + *tūn*, although some names may be 'farmstead near a boundary' with OE (*ge*)*mære* as first element; examples include: **Marton** Lincs. *Martone* 1086 (DB). **Marton** Warwicks. *Mortone* [*sic*] 1086 (DB), *Merton* 1151. **Marton-le-Moor** N. Yorks. *Marton* 1198, *Marton on the Moor* 1292. Affix is from OE *mōr* 'moor'. **Marton, Long** Cumbria. *Meretun* c.1170. Affix refers to the length of the parish.

Martyr Worthy Hants. *See* WORTHY.

Marwood Devon. *Mereuda* 1086 (DB). Probably 'wood near a boundary'. OE (*ge*)*mære* + *wudu*.

Mary Tavey Devon. *See* TAVY.

Maryborough. *See* PORT LAOISE.

Marylebone Gtr. London. *Maryburne* 1453. '(Place by) St Mary's stream'. OE *burna*. Named from the dedication of the church built in the 15th cent. The *-le-* is intrusive and dates from the 17th cent.

Maryport Cumbria. *Mary-port* 1762. A modern name, the harbour here built in the 18th cent. by Humphrey Senhouse being named after his wife *Mary*.

Marystow Devon. *Sancte Marie Stou* 1266. 'Holy place of St Mary'. OE *stōw*.

Masham N. Yorks. *Massan* [*sic*] 1086 (DB), *Masham* 1153. 'Homestead or village of a man called *Mæssa'. OE pers. name + *hām*.

Mashbury Essex. *Maisseberia* 1068, *Massebirig* 1086 (DB). 'Stronghold of a man called *Mæssa or *Mæcca'. OE pers. name + *burh* (dative *byrig*).

Massingham, Great & Massingham, Little Norfolk. *Masingeham* 1086 (DB). 'Homestead of the family or followers of a man called *Mæssa'. OE pers. name + *-inga-* + *hām*.

Mastergeehy (*Máistir Gaoithe*) Kerry. 'Churning wind'.

Matching Essex. *Matcinga* 1086 (DB). '(Settlement of) the family or followers of a man called *Mæcca', or '*Mæcca's place'. OE pers. name + *-ingas* or *-ing*.

Matehy (*Má Teithe*) Cork. 'Smooth plain'.

Matfen Northum. *Matefen* 1159. Probably 'fen of a man called *Matta'. OE pers. name + *fenn*.

Matfield Kent. *Mattefeld* c.1230. 'Open land of a man called *Matta'. OE pers. name + *feld*.

Mathon Herefs. *Matham* 1014, *Matma* 1086 (DB). 'The treasure or gift'. OE *māthm*.

Mathry Pemb. *Marthru* c.1150. Possibly 'grave of a saint'. Welsh *merthyr*.

Matlask Norfolk. *Matelasc* 1086 (DB). 'Ash-tree where meetings are held'. OE *mæthel* + *æsc* or OScand. *askr*.

Matlock Derbys. *Meslach* [*sic*] 1086 (DB), *Matlac* 1196. 'Oak-tree where meetings are held'. OE *mæthel* + *āc*.

Matterdale End Cumbria. *Mayerdale [sic] c.*1250, *Matherdal* 1323. 'Valley where madder grows'. OScand. *mathra* + *dalr*.

Mattersey Notts. *Madressei* 1086 (DB). 'Island, or well-watered land, of a man called Mæthhere'. OE pers. name + *ēg*.

Mattingley Hants. *Matingelege* 1086 (DB). 'Woodland clearing of the family or followers of a man called *Matta'. OE pers. name + -*inga-* + *lēah*.

Mattishall Norfolk. *Mateshala* 1086 (DB). Probably 'nook of land of a man called *Matt'. OE pers. name + *halh*.

Mauchline E. Ayr. *Machline c.*1130, *Mauhhelin c.*1177, *Mauchlyn c.*1200. Possibly 'plain by the pool'. Gaelic *magh* + *linn*.

Maugersbury Glos. *Meilgaresbyri* 714, *Malgeresberie* 1086 (DB). 'Stronghold of a man called *Mæthelgār'. OE pers. name + *burh* (dative *byrig*).

Maulagallane (*Meall an Ghalláin*) Kerry. 'Knoll of the standing stone'.

Maulden Beds. *Meldone* 1086 (DB). 'Hill with a crucifix'. OE *mǣl* + *dūn*.

Maulds Meaburn Cumbria. *See* MEABURN.

Maulnagower (*Meall na nGabhar*) Kerry. 'Knoll of the goats'.

Maulnahorna (*Meall na hOrna*) Kerry. 'Knoll of the barley'.

Maulyneill (*Meall Uí Néill*) Kerry. 'Ó Néill's knoll'.

Maumakeogh (*Mám an Cheo*) Mayo. 'Pass of the mist'.

Maumtrasna (*Mám Trasna*) Mayo. 'Mountain pass'.

Maumturk (*Mám Tuirc*) Galway. 'Pass of the boar'.

Maun (river) Notts. *See* MANSFIELD.

Maunby N. Yorks. *Mannebi* 1086 (DB). 'Farmstead or village of a man called Magni'. OScand. pers. name + *bý*.

Maund Bryan Herefs. *Magene* 1086 (DB), *Magene Brian* 1242. A difficult name, but possibly '(place at) the hollows', from OE *maga* 'stomach' (used in a topographical sense) in a dative plural form *magum*. Alternatively *Maund*

(also a district-name) may represent the survival of an ancient Celtic name *Magnis* (probably 'the rocks'). Manorial affix from a 12th-cent. owner called *Brian*.

Mausrower (*Más Ramhar*) Kerry. 'Fat haunch-shaped hill'.

Mausrevagh (*Más Riabhach*) Galway. 'Striped thigh'.

Mautby Norfolk. *Malteby* 1086 (DB). 'Farmstead or village of a man called Malti, or where malt is made'. OScand. pers. name or *malt* + *bý*.

Mavis Enderby Lincs. *See* ENDERBY.

Mawddach. *See* BARMOUTH.

Mawdesley Lancs. *Madesle* 1219. 'Woodland clearing of a woman called Maud'. OFrench pers. name (from OGerman *Mahthildis*) + OE *lēah*.

Mawgan Cornwall. *Scanctus Mawan* 1086 (DB). '(Church of) St Mawgan'. From the patron saint of the church.

Mawnan Cornwall. '(Church of) *Sanctus Maunanus*' 1281. From the patron saint of the church.

Maxey Peterb. *Macuseige c.*963. 'Island, or dry ground in marsh, of a man called Maccus'. OScand. pers. name + *ēg*.

Maxstoke Warwicks. *Makestoka* 1169. 'Outlying farmstead or hamlet of a man called *Macca'. OE pers. name + *stoc*.

Maxwelltown Dumf. *Mackeswel* 1144. 'Village at Maxwell'. OE *tūn*. *Maxwell* is 'Maccus's pool' (OE *wella*). The first element may be a surname. Now part of Dumfries.

May, Isle of Fife. *Mai* 1143, *Maeyar* 1250. Origin unknown.

Maybole S. Ayr. *Mayboill* 1275. 'Kinswomen's dwelling'. OE *mǣge* + *bōthl*.

Mayfair Gtr. London. District named from the great 'May Fair' once held on the open land here, on record from 1686: the fair began on the 1st of May and lasted for 15 days.

Mayfield E. Sussex. *Magavelda* 12th cent. 'Open land where mayweed grows'. OE *mægthe* + *feld*.

Mayfield Staffs. *Medevelde* [*sic*] 1086 (DB), *Matherfeld* c.1180. 'Open land where madder grows'. OE *mæddre* + *feld*.

Mayford Surrey. *Maiford* 1212. Possibly 'maidens' ford', or 'ford where mayweed grows'. OE *mægth* or *mægthe* + *ford*.

Mayglass (*Magh Glas*) Wexford. 'Green plain'.

Mayland Essex. *La Mailanda* 1188. Possibly 'land or estate where mayweed grows', from OE *mægthe* + *land*. Alternatively '(place at) the island', from OE *ēg-land*, with *M*- from the dative case of the OE definite article.

Maynooth (*Maigh Nuadhat*) Kildare. *Magh Nuaddat* n.d. 'Plain of Nuadha'. Nuadha was the supreme leader of the gods.

Mayo (*Maigh Eo*) (the county). 'Plain of yew-trees'.

Mayobridge Down. 'Bridge of Mayo'. *Droichead Mhuigheó* 1901. The village is named from the townland of *Mayo* (Irish *Maigh Eo*, 'plain of yew-trees').

Mayogall (*Maigh Ghuala*) Derry. 'Plain of the shoulder'.

Maze (*An Mhaigh*) Down. *the 5 towns of the Mew* 1585. 'The plain'.

Meaburn, King's & Meaburn, Maulds Cumbria. *Maiburne* 12th cent., *Meburne Regis* 1279, *Meburnemaud* 1278. 'Meadow stream'. OE *mǣd* + *burna*. Distinguishing affixes from possession by the Crown (Latin *regis* 'of the king') and by a woman called *Maud* here in the 12th cent.

Mealaghans (*Maelachán*) Offaly. 'Little bare hills'.

Meall Dearg (mountain) Highland. 'Red mountain'. Gaelic *meall* + *dearg*.

Meall Dubh (mountain) Highland. 'Black mountain'. Gaelic *meall* + *dubh*.

Meall Garbh (mountain) Highland. 'Rough mountain'. Gaelic *meall* + *garbh*.

Meall Gorm (mountain) Highland. 'Blue mountain'. Gaelic *meall* + *gorm*.

Meall Meadhonach (mountain) Highland. 'Middle mountain'. Gaelic *meall* + *meadhon*.

Meall Mór (mountain) Highland. 'Big mountain'. Gaelic *meall* + *mór*.

Meare Somerset. *Mere* 1086 (DB). '(Place at) the pool or lake'. OE *mere*.

Mearns (district) E. Renf. *Moerne* 12th cent. Gaelic *An Mhaoirne* 'the Stewartry', from *maor* 'steward', referring to an area administered by an official with delegated authority.

Mears Ashby Northants. *See* ASHBY.

Measham Leics. *Messeham* 1086 (DB). 'Homestead or village on the River Mease'. OE river-name ('mossy river' from OE *mēos*) + *hām*.

Meath (*An Mhí*) (the county). *Mide* 9th cent. 'The middle (place)'. Meath was formerly the fifth province of Ireland.

Meathop Cumbria. *Midhop* c.1185. 'Middle enclosure in marsh', or '(place) among the enclosures'. OE *midd* (adjective) or *mid* (preposition), replaced by OScand. *mithr*) + *hop*.

Meaux E. R. Yorks. *Melse* 1086 (DB). 'Sand-bank pool'. OScand. *melr* + *sǽr*.

Meavy Devon. *Mæwi* 1031, *Mewi* 1086 (DB). Named from the River Meavy, probably a Celtic river-name meaning 'lively stream'.

Medbourne Leics. *Medburne* 1086 (DB). 'Meadow stream'. OE *mǣd* + *burna*.

Meddon Devon. *Madone* 1086 (DB). 'Meadow hill'. OE *mǣd* + *dūn*.

Medina I. of Wight. district named from the River Medina, *Medine* 1196, 'the middle one', from OE *medume*.

Medmenham Bucks. *Medmeham* 1086 (DB). 'Middle or middle-sized homestead or enclosure'. OE *medume* (dative *-an*) + *hām* or *hamm*.

Medstead Hants. *Medested* 1202. 'Meadow place', or 'place of a man called *Mēde*'. OE *mǣd* or pers. name + *stede*.

Medway (river) Sussex-Kent. *Medeuuæge* 8th cent. Ancient pre-English river-name *Wey* (of obscure etymology), possibly compounded with Celtic or OE *medu* 'mead' with reference to the colour or the sweetness of the water. Medway is now also the name of a unitary authority.

Meeldrum (*Maoldruim*) Westmeath. 'Bald ridge'.

Meelin (*An Mhaolinn*) Cork. 'The hillock'.

Meenaclady (*Mín an Chladaigh*) Donegal. 'Smooth place of the shore'.

Meenacross (*Mín na Croise*) Donegal. 'Smooth place of the cross'.

Meenagorp (*Mín na gCorp*) Tyrone. 'Smooth place of the corpses'.

Meenaneary (*Mín an Aoire*) Donegal. *Mininarie* 1755. 'Smooth place of the shepherd'.

Meenavean (*Mín na bhFiann*) Donegal. 'Smooth place of the Fianna'.

Meeniska (*Meadhón Uisge*) Westmeath. 'Middle water'.

Meerbrook Staffs. *Merebroke* 1338. 'Boundary brook'. OE (*ge*)*mære* + *brōc*.

Meesden Herts. *Mesdone* 1086 (DB). 'Mossy or boggy hill'. OE *mēos* + *dūn*.

Meeth Devon. *Meda* 1086 (DB). OE (*ge*)*mȳthe* 'confluence of rivers' or *mæth* 'place where corn or grass is cut'.

Megdale Dumf. 'Marshy meadow'. Cumbric *mig-* + *dól*.

Meigh (*Maigh*) Armagh. *Moye* 1657. 'Plain'.

Meigle Perth. *Migdele* c.1200, *Mygghil* 1378. 'Marshy meadow'. Pictish *mig-* + *dól*.

Meir Stoke. *Mere* 1242. Probably '(place on) the boundary'. OE (*ge*)*mære*.

Meirionnydd. *See* MERIONETH.

Melbourn Cambs. *Meldeburna* 970, *Melleburne* 1086 (DB). Possibly 'stream where orach or a similar plant grows'. OE *melde* + *burna*.

Melbourne Derbys. *Mileburne* 1086 (DB). Probably 'mill stream'. OE *myln* + *burna*.

Melbourne E. R. Yorks. *Middelburne* 1086 (DB). 'Middle stream'. OE *middel* (replaced by OScand. *methal*) + OE *burna*.

Melbury, probably 'multi-coloured fortified place', OE *mæle* + *burh* (dative *byrig*): **Melbury Abbas** Dorset. *Meleburge* 956, *Meleberie* 1086 (DB), *Melbury Abbatisse* 1291. Affix is Latin *abbatissa* 'abbess' from its early possession by Shaftesbury Abbey. **Melbury Bubb**, **Melbury Osmond**, & **Melbury Sampford** Dorset. *Mele*(*s*)*berie* 1086 (DB),

Melebir Bubbe 1244, *Melebur Osmund* 1283, *Melebury Saunford* 1312. Manorial affixes from the *Bubbe* family, from a man called *Osmund*, and from the *Saunford* family, all here in the 13th cent.

Melchbourne Beds. *Melceburne* 1086 (DB). 'Stream by pastures yielding good milk'. OE **melce* + *burna*.

Melcombe Regis Dorset. *Melecumb* 1223. Probably 'valley where milk is produced'. OE *meoluc* + *cumb*.

Meldon Devon. *Meledon* 1175. 'Multi-coloured hill'. OE *mæle* + *dūn*.

Meldon Northum. *Meldon* 1242. 'Hill with a crucifix'. OE *mæl* + *dūn*.

Meldreth Cambs. *Melrede* 1086 (DB). 'Mill stream'. OE *myln* + *rīth*.

Melford, Long Suffolk. *Melaforda* 1086 (DB). Probably 'ford by a mill'. OE *myln* + *ford*.

Meliden (*Allt Melyd*) Denb. *Altmelyden* 1291. 'Hill of St Melydn'. The word for 'hill' is missing in the English form of the name, but present in the Welsh (*allt*).

Melkinthorpe Cumbria. *Melcanetorp* c.1150. 'Outlying farmstead or hamlet of a man called Maelchon'. OIrish pers. name + OScand. *thorp*.

Melkridge Northum. *Melkrige* 1279. 'Ridge where milk is produced'. OE *meoluc* + *hrycg*.

Melksham Wilts. *Melchesham* 1086 (DB). Possibly 'homestead or enclosure where milk is produced'. OE *meoluc* + *hām* or *hamm*. Or from an OE byname *Melc*.

Mellifont Louth. 'Fountain of honey'. Latin *mel* + *fons* (genitive *fontis*).

Melling, probably '(settlement of) the family or followers of a man called *Mealla', or '*Mealla's place', OE pers. name + -*ingas* or -*ing*: **Melling** Lancs. *Mellinge* 1086 (DB). **Melling** Sefton. *Melinge* 1086 (DB).

Mellis Suffolk. *Melles* 1086 (DB). 'The mills'. OE *myln*.

Mellor, 'bare or smooth-topped hill', Celtic **mēl* + **breȝ*: **Mellor** Lancs. *Malver* c.1130. **Mellor** Stockp. *Melver* 1283.

Mells Somerset. *Milne* 942, *Mulle* 1086 (DB). 'The mill(s)'. OE *myln*.

Melmerby, 'farmstead or village of a man called Maelmuire', OIrish pers. name + OScand. *bý*: **Melmerby** Cumbria. *Malmerbi* 1201. **Melmerby** N. Yorks., near Coverham. *Melmerbi* 1086 (DB). **Melmerby** N. Yorks., near Ripon. *Malmerbi* 1086 (DB). First element alternatively OScand. *malmr* 'sandy field'.

Melmore (*An Meall Mór*) Donegal. 'The big lump'.

Melplash Dorset. *Melpleys* 1155. Probably 'multi-coloured pool'. OE *mǣle* + **plæsc*.

Melrose Sc. Bord. *Mailros c.*700. 'Bare promontory'. British *mailo-* + *ros*.

Melsonby N. Yorks. *Malsenebi* 1086 (DB). Possibly 'farmstead or village of a man called Maelsuithan'. OIrish pers. name + OScand. *bý*.

Meltham Kirkl. *Meltham* 1086 (DB). Possibly 'homestead or village where smelting is done'. OE **melt* + *hām*.

Melton, usually 'middle farmstead', OE *middel* (replaced by OScand. *methal*) + *tūn*: **Melton, Great & Melton, Little** Norfolk. *Middilton c.*1060, *Meltuna* 1086 (DB). **Melton, High** Donc. *Middeltun* 1086 (DB). **Melton Mowbray** Leics. *Medeltone* 1086 (DB), *Melton Moubray* 1284. Manorial affix from the *de Moubray* family, here in the 12th cent. **Melton Ross** N. Lincs. *Medeltone* 1086 (DB), *Melton Roos* 1375. Manorial affix from the *de Ros* family, here in the 14th cent.
However the following may have a different origin: **Melton** Suffolk. *Meltune c.*1050, *Meltuna* 1086 (DB). Perhaps rather 'farmstead with a crucifix', OE *mǣl* + *tūn*. **Melton Constable** Norfolk. *Maeltuna* 1086 (DB), *Melton Constable* 1320. Possibly identical with the previous name. Manorial affix from possession by the *Constable* of the Bishop of Norwich in the 12th cent.

Melton (district) Leics. A modern name, with reference to MELTON MOWBRAY.

Melverley Shrops. *Meleurlei* [*sic*] 1086 (DB), *Milverlegh* 1311. Probably 'woodland clearing by the mill ford'. OE *myln* + *ford* + *lēah*.

Melvin, Lough (*Loch Meilbhe*) Fermanagh. 'Lake of Meilbhe'.

Membury Devon. *Manberia* 1086 (DB). OE *burh* (dative *byrig*) 'fortified place' with an uncertain first element, possibly Celtic **main* 'stone' or OE *mǣne* 'common'.

Menai Bridge (*Porthaethwy*) Angl. 'Bridge over the Menai Strait'. A bridge was built across the Menai Strait in 1826 and the town developed soon after. The name of the strait itself is of uncertain origin. The Welsh name means 'ferry of Daethwy' (Welsh *porth*), from a tribal name.

Mendham Suffolk. *Myndham c.*950, *Mendham* 1086 (DB). 'Homestead or village of a man called **Mynda*'. OE pers. name + *hām*.

Mendip Hills Somerset. *Menedepe* 1185. Probably Celtic **mönïth* 'mountain, hill' with an uncertain second element, perhaps OE *yppe* in the sense 'upland, plateau'.

Mendlesham Suffolk. *Mundlesham* 1086 (DB). 'Homestead or village of a man called **Myndel*'. OE pers. name + *hām*.

Menheniot Cornwall. *Mahiniet* 1260. 'Land or plain of **Hynyed*'. Cornish **ma* + pers. name.

Menlough (*Mionlach*) Galway. 'Small place'.

Menlough (*Mionloch*) Galway. 'Small lake'.

Menston Brad. *Mensinctun c.*972, *Mersintone* 1086 (DB). 'Estate associated with a man called **Mensa*'. OE pers. name + *-ing-* + *tūn*.

Menstrie Clac. *Mestryn* 1261, *Mestry* 1315. 'Farmstead in the plain'. Welsh *maes* + *tre*.

Mentmore Bucks. *Mentemore* 1086 (DB). 'Moor of a man called **Menta*'. OE pers. name + *mōr*.

Meole Brace Shrops. *Mela(m)* 1086 (DB), *Melesbracy* 1273. Named from Meole Brook (the old name of Rea Brook) which may derive from OE *melu, meolu* 'meal' perhaps used figuratively of a stream with cloudy water. Manorial affix from the *de Braci* family, here in the 13th cent.

Meols, Great & Meols, Little Wirral. *Melas* 1086 (DB). 'The sandhills'. OScand. *melr*.

Meon, East & Meon, West Hants. *Meone c.*880, *Mene*, *Estmeone* 1086 (DB). Named from the River Meon, a Celtic or pre-Celtic river-name of uncertain origin and meaning.

Meonstoke Hants. *Menestoche* 1086 (DB). 'Outlying farmstead on the River Meon or dependent on MEON'. OE *stoc*.

Meopham Kent. *Meapaham* 788, *Mepeham* 1086 (DB). 'Homestead or village of a man called **Mēapa*'. OE pers. name + *hām*.

Mepal Cambs. *Mepahala* 12th cent. 'Nook of land of a man called *Mēapa'. OE pers. name + *halh*.

Meppershall Beds. *Maperteshale* 1086 (DB). Probably 'nook of the maple-tree'. OE *mæpel-trēow* + *halh*.

Mercia, a Latinized form of the OE tribal name *Merce* 'the Mercians', this being a derivative of OE *mearc* 'march, boundary' with the sense 'people of the march or boundary'. The Anglian tribe so called settled in the area roughly corresponding with the West Midlands and established a kingdom here during the 7th, 8th, and 9th centuries (the 'march' referring to the Welsh border).

Merchiston Edin. *Merchinstoun* 1266, *Merchammeston* 1278, *Merchenstoun* 1371. 'Merchiaun's farm'. British pers. name + OE *tūn*.

Mere, '(place at) the pool or lake', OE *mere*: **Mere** Ches. *Mera* 1086 (DB). **Mere** Wilts. *Mere* 1086 (DB).

Mereworth Kent. *Meranworth* 843. *Marovrde* 1086 (DB). 'Enclosure of a man called *Mæra'. OE pers. name + *worth*.

Meriden Solhll. *Mereden* 1230. 'Pleasant valley', or 'valley where merry-making takes place'. OE *myrge* + *denu*.

Merioneth (*Meirionnydd*) (the historic county). 'Seat of Meirion'. *Meirion* (Meriaun) is said to be the grandson of Cunedda.

Merrington Shrops. *Muridon* 1254. 'Pleasant hill', or 'hill where merry-making takes place'. OE *myrge* + *dūn*.

Merrington, Kirk Durham. *Mærintun* c.1123, *Kyrke Merington* 1331. Probably 'estate associated with a man called *Mæra'. OE pers. name + *-ing-* + *tūn*. Alternatively the first element may be an OE *mæring 'conspicuous place', or 'boundary place'. Affix is OScand. *kirkja* 'church'.

Merriott Somerset. *Meriet* 1086 (DB). Possibly 'boundary gate'. OE *mære* + *geat*.

Merrow Surrey. *Marewe* 1185. Possibly OE *mearg* 'marrow' in a figurative sense such as 'fertile ground'. Alternatively a Celtic name meaning 'marl-place'.

Mersea, East & Mersea, West Essex. *Meresig* early 10th cent., *Meresai* 1086 (DB). 'Island of the pool'. OE *mere* (genitive *-s*) + *ēg*.

Mersey (river) Ches.–Lancs. *Mærse* 1002. 'Boundary river'. OE *mære* (genitive *-s*) + *ēa*. The Mersey forms the old boundary between Cheshire and Lancashire.

Mersham Kent. *Mersaham* 858, *Merseham* 1086 (DB). 'Homestead or village of a man called *Mærsa'. OE pers. name + *hām*.

Merstham Surrey. *Mearsætham* 947, *Merstan* 1086 (DB). 'Homestead near a trap for martens or weasels'. OE *mearth* + *sǣt* + *hām*.

Merston W. Sussex. *Mersitone* 1086 (DB). 'Farmstead on marshy ground'. OE *mersc* + *tūn*.

Merstone I. of Wight. *Merestone* 1086 (DB). Identical in origin with the previous name.

Merther Cornwall. *Eglosmerthe* 1201. 'Church at the saint's grave'. Cornish *eglos* + *merther*.

Merthyr Dyfan Vale Glam. *Mertherdevan*. 'Dyfan's grave'. Welsh *merthyr*. The identity of the named saint is uncertain.

Merthyr Tydfil Mer. T. *Merthir* 1254, *Merthyr Tutuil* 13th cent. 'Tudful's grave'. Welsh *merthyr*. Tudful is said to have been one of the daughters of Brychan, who gave the name of BRECON.

Merton, 'farmstead by the pool', OE *mere* + *tūn*: **Merton** Devon. *Mertone* 1086 (DB). **Merton** Gtr. London. *Mertone* 949, *Meretone* 1086 (DB). **Merton** Norfolk. *Meretuna* 1086 (DB). **Merton** Oxon. *Meretone* 1086 (DB).

Meshaw Devon. *Mauessart* 1086 (DB). Probably 'bad or infertile clearing'. OFrench *mal* + *assart*.

Messing Essex. *Metcinges* 1086 (DB). '(Settlement of) the family or followers of a man called *Mæcca'. OE pers. name + *-ingas*.

Messingham N. Lincs. *Mæssingaham* c.1067, *Messingeham* 1086 (DB). 'Homestead of the family or followers of a man called *Mæssa'. OE pers. name + *-inga-* + *hām*.

Metfield Suffolk. *Medefeld* 1214. 'Open land with meadow'. OE *mǣd* + *feld*.

Metheringham Lincs. *Medricesham* 1086 (DB), *Mederingeham* 1193. Possibly 'homestead of a man called *Mēdrīc or of his people'. OE pers. name (+ -inga-) + hām.

Methley Leeds. *Medelai* 1086 (DB). 'Clearing where grass is mown', OE *mǣth* + *lēah*, or 'middle land between rivers', OScand. *methal* + OE *ēg*.

Methven Perth. *Methfen* 1211. 'Middle stone'. Welsh *medd* + *maen*.

Methwold Norfolk. *Medelwolde* c.1050, *Methelwalde* 1086 (DB). 'Middle forest'. OScand. *methal* + OE *wald*.

Methwold Hythe Norfolk. *Methelwoldehythe* 1277. 'Landing-place near METHWOLD'. OE *hȳth*.

Mettingham Suffolk. *Metingaham* 1086 (DB). Probably 'homestead of the family or followers of a man called *Metti'. OE pers. name + -inga- + hām.

Mevagissey Cornwall. *Meffagesy* c.1400. From the patron saints of the church, Saints Meva and Issey (the medial -ag- in the place-name being from Cornish *hag* 'and').

Mexborough Donc. *Mechesburg* 1086 (DB). 'Stronghold of a man called *Mēoc or Mjúkr'. OE or OScand. pers. name + OE *burh*.

Mey Highland. '(Place in) the plain'. Gaelic *magh*.

Meysey Hampton Glos. See HAMPTON.

Michaelchurch, 'church dedicated to St Michael', saint's name + OE *cirice*: **Michaelchurch** Herefs. *Lann mihacgel* c.1150. Welsh *llan* 'church'. **Michaelchurch Escley** Herefs. *Michaeleschirche* c.1257. Affix is from Escley Brook, probably a Celtic river-name *Esk* ('the water') + OE *hlynn* 'torrent'.

Michaelchurch-on-Arrow (*Llanfihangel Dyffryn Arwy*) Powys. 'St Michael's church by the River Arrow'. OE *cirice*. The Celtic or pre-Celtic river name means simply 'stream'. The Welsh name means 'St Michael's church in the valley of the River Arrow' (Welsh *llan* + *Mihangel* + *dyffryn*).

Michaelston-le-Pit (*Llanfihangel-y-pwll*) Vale Glam. *Michelstowe* c.1291, *Mighelstowe* 1483, *Michaelstown* 1535, *Mighelston in le Pitt* 1567. 'St Michael's holy place in the pit'. OE *stōw* + OFrench *le* + OE *pytt*. The church is dedicated to St Michael, and the 'pit' is the wide depression in which the village lies. The

addition distinguishes the place from nearby *Michaelston-super-Ely*, 'Michaelston on the (river) Ely'. The Welsh name means 'St Michael's church in the pit' (Welsh *llan* + *Mihangel* + *y* + *pwll*).

Michaelstow Cornwall. *Mighelestowe* 1302. 'Holy place of St Michael'. Saint's name + OE *stōw*.

Micheldever Hants. *Mycendefr* 862, *Miceldevere* 1086 (DB). Named from the stream here, probably a Celtic name meaning 'boggy waters', the second element being the Celtic word found in DOVER. At an early date, the first element was confused with OE *micel* 'great'.

Michelmersh Hants. *Miclamersce* 985. 'Large marsh'. OE *micel* + *mersc*.

Mickfield Suffolk. *Mucelfelda* 1086 (DB). 'Large tract of open country'. OE *micel* + *feld*.

Mickle Trafford Ches. See TRAFFORD.

Mickleby N. Yorks. *Michelbi* 1086 (DB). 'Large farmstead'. OScand. *mikill* + *bý*.

Mickleham Surrey. *Micleham* 1086 (DB). 'Large homestead or river-meadow'. OE *micel* + *hām* or *hamm*.

Mickleover Derby. *Vfre* 1011, *Ufre* 1086 (DB), *Magna Oufra* c.1100. '(Place at) the ridge'. OE *ofer*, with the affix *micel* 'great' (earlier Latin *magna*) to distinguish this place from LITTLEOVER.

Mickleton, 'large farmstead', OE *micel* + *tūn*: **Mickleton** Durham. *Micleton* 1086 (DB). **Mickleton** Glos. *Micclantun* 1005, *Muceltvne* 1086 (DB).

Mickley Northum. *Michelleie* c.1190. 'Large wood or clearing'. OE *micel* + *lēah*.

Middle as affix. See main name, e.g. for **Middle Assendon** (Oxon.) see ASSENDON.

Middleham, 'middle homestead or enclosure', OE *middel* + *hām* or *hamm*: **Middleham** N. Yorks. *Medelai* [sic] 1086 (DB), *Midelham* 12th cent. **Middleham, Bishop** Durham. *Middelham* 12th cent. Affix from early possession by the Bishop of Durham.

Middlesbrough Middlesbr. *Midelesburc* c.1165. 'Middlemost stronghold'. OE *midlest* + *burh*.

Middlesex (historical county). *Middelseaxan* 704, *Midelsexe* 1086 (DB). '(Territory of) the

Middle Saxons', originally a tribal name from OE *middel* + *Seaxe.*

Middlestown Wakefd. *Midle Shitlington* 1325, *Myddleston* 1551. A contracted form of 'Middle Shitlington' (now Sitlington), which is *Schelintone* 1086 (DB), 'estate associated with a man called *Scyttel', OE pers. name + *-ing-* + *tūn.*

Middleton, a very common name, usually 'middle farmstead or estate', OE *middel* + *tūn*; examples include: **Middleton** Norfolk. *Mideltuna* 1086 (DB). **Middleton** Rochdl. *Middelton* 1194. **Middleton Cheney** Northants. *Mideltone* 1086 (DB), *Middleton Cheyndut* 1342. Manorial affix from the *de Chendut* family, here in the 12th cent. **Middleton in Teesdale** Durham. *Middeltona c.*1180, *Mideltone in Tesedale* 1236. Affix means 'in the valley of the River Tees', Celtic river-name ('surging river') + OScand. *dalr.* **Middleton on Sea** W. Sussex. *Middeltone* 1086 (DB). **Middleton on the Wolds** E. R. Yorks. *Middeltun* 1086 (DB). *See* WOLDS. **Middleton St George** Darltn. *Middilton* 1259, *Middelton Seint George* 1382. Affix from the dedication of the church. **Middleton Scriven** Shrops. *Middeltone* 1086 (DB), *Skrevensmyddleton* 1577. Manorial affix is probably the ME surname *Scriven* from a family of this name with lands here. **Middleton Tyas** N. Yorks. *Middeltun* 1086 (DB), *Midilton Tyas* 14th cent. Manorial affix from the *le Tyeis* family who may have had lands here at an early date.
  However the following have a different origin: **Middleton on the Hill** Herefs. *Miceltune* 1086 (DB). 'Large farmstead or estate'. OE *micel* + *tūn.* **Middleton Priors** Shrops. *Mittelington* 1221. Possibly 'farmstead at stream-junction'. OE **(ge)mȳthel* + connective *-ing-* + *tūn.* Affix from its early possession by Wenlock Priory.

Middletown Armagh. 'Middle settlement'. *Killaninane* 1657. The earlier Irish name was *Coillidh Chanannáin*, 'Canannán's wood'.

Middlewich Ches. *Wich, Mildestuich* 1086 (DB). 'Middlemost salt-works'. OE *midlest* + *wīc.*

Middlezoy Somerset. *Soweie* 725, *Sowi* 1086 (DB), *Middlesowy* 1227. 'Island on the River *Sow*'. Lost pre-English river-name (of doubtful etymology) with the later affix *middle.*

Midge Hall Lancs. *Miggehalgh* 1390. 'Nook of land infested with midges'. OE *mycg* + *halh.*

Midgham W. Berks. *Migeham* 1086 (DB). 'Homestead or enclosure infested with midges'. OE *mycg* + *hām* or *hamm.*

Midgley, 'wood or clearing infested with midges', OE *mycg* + *lēah*: **Midgley** Calder. *Micleie* 1086 (DB). **Midgley** Wakefd. *Migelaia* 12th cent.

Midhurst W. Sussex. *Middeherst* 1186. 'Middle wooded hill', or '(place) among the wooded hills'. OE *midd* (adjective) or *mid* (preposition) + *hyrst.*

Midlothian (the unitary authority). *Loonia c.*970, *Lodoneó* 1098, *Louthion c.*1200. 'Middle Lothian'. The name, that of the historic county between *East Lothian* and *West Lothian*, is ultimately that of the district of *Lothian*, named after one Leudonus.

Midsomer Norton B. & NE. Som. *See* NORTON.

Milborne, Milbourne, Milburn, 'mill stream', OE *myln* + *burna*: **Milborne Port** Somerset. *Mylenburnan c.*880, *Meleburne* 1086 (DB), *Milleburnport* 1249. Affix is OE *port* 'market town'. **Milborne St Andrew** Dorset. *Muleburne* 934, *Meleburne* 1086 (DB), *Muleburne St Andrew* 1294. Affix from the dedication of the church. **Milbourne** Northum. *Meleburna* 1158. **Milburn** Cumbria. *Milleburn* 1178.

Milcombe Oxon. *Midelcumbe* 1086 (DB). 'Middle valley'. OE *middel* + *cumb.*

Milden Suffolk. *Mellinga* [*sic*] 1086 (DB), *Meldinges c.*1130. '(Settlement of) the family or followers of a man called *Melda', OE pers. name + *-ingas*, or 'place where orach grows', OE *melde* + *-ing.*

Mildenhall Suffolk. *Mildenhale c.*1050, *Mitdenehalla* 1086 (DB), *Middelhala* 1130. Probably identical in origin with the next name (some spellings showing confusion with ME *middel* 'middle').

Mildenhall Wilts. *Mildanhald* 803–5, *Mildenhalle* 1086 (DB). 'Nook of land of a woman called Milde or a man called *Milda'. OE pers. name (genitive *-n*) + *halh.*

Mile End, 'district or hamlet a mile away (from a larger place)', OE *mīl* + *ende*: **Mile End** Essex. *Milende* 1156–80. **Mile End** Gtr. London. *La Mile Ende* 1288.

Milebush Antrim. 'Mile bush'. The name apparently refers to a bush one mile from Carrickfergus.

Mileham Norfolk. *Meleham* 1086 (DB). 'Homestead or village with a mill'. OE *myln* + *hām*.

Milford, 'ford by a mill', OE *myln* + *ford*: **Milford** Derbys. *Muleford* 1086 (DB). **Milford** Surrey. *Muleford* 1235. **Milford on Sea** Hants. *Melleford* 1086 (DB). **Milford, South** N. Yorks. *Myleford* c.1030.

Milford Armagh. 'Ford by a mill'. A linen mill was formerly here by a ford over the River Callan.

Milford Haven (*Aberdaugleddau*) Pemb. *de Milverdico portu* c.1191, *Mellferth* 1207, *Milford* 1219, *Aber Dav Gleddyf* 15th cent. 'Harbour at Milford'. The original site was *Milford*, 'sandy inlet' (OScand. *melr* + *fjorthr*). English *Haven* was then added when the meaning of this was forgotten. The Welsh name means 'mouth of the two Rivers Cleddau' (Welsh *aber* + *dau*). The River Cleddau, divided here into the Eastern Cleddau and Western Cleddau, is the Welsh word *cleddau* meaning 'sword'.

Mill Hill Gtr. London. *Myllehill* 1547. Self-explanatory, 'hill with a windmill'.

Millbrook, 'brook by a mill', OE *myln* + *brōc*: **Millbrook** Beds. *Melebroc* 1086 (DB). **Millbrook** Sotn. *Melebroce* 956, *Melebroc* 1086 (DB).

Millbrook Antrim. *Millbrook* 1840. 'Brook by a mill'. A cotton mill was formerly here.

Millford Donegal. 'Ford by a mill'. A corn mill formerly stood by the stream here.

Millington E. R. Yorks. *Milleton* 1086 (DB). 'Farmstead with a mill'. OE *myln* + *tūn*.

Millmeece Staffs. *Mess* 1086 (DB), *Mulnemes* 1289. 'Mossy or boggy place'. OE *mēos* with the later addition of *myln* 'mill'.

Millom Cumbria. *Millum* c.1180. '(Place at) the mills'. OE *myln* in a dative plural form *mylnum*.

Millport N. Ayr. 'Port with a mill'.

Milltown Antrim. Armagh, Cavan, Derry, etc. 'Town with a mill'.

Milngavie E. Dunb. *Mylngavie* 1669. '(Place by) the windmill', Gaelic *muileann gaoithe*, or *meall na gaoithe* 'windy hill'.

Milnrow Rochdl. *Mylnerowe* 1554. 'Row of houses by a mill'. OE *myln* + *rāw*.

Milnthorpe Cumbria. *Milntorp* 1272. 'Outlying farmstead or hamlet with a mill'. OE *myln* + OScand. *thorp*.

Milson Shrops. *Mulstone* 1086 (DB). Possibly 'farmstead of a man called *Myndel'. OE pers. name + *tūn*. Alternatively derivation may be from OE *myln-stān* 'millstone'.

Milstead Kent. *Milstede* late 11th cent. Possibly 'middle place'. OE *middel* + *stede*.

Milston Wilts. *Mildestone* 1086 (DB). Probably 'middlemost farmstead'. OE *midlest* + *tūn*.

Milton, a very common name, usually 'middle farmstead or estate', OE *middel* + *tūn*; examples include: **Milton** Cambs. *Middeltune* c.975, *Middeltone* 1086 (DB). **Milton** Oxon., near Didcot. *Middeltun* 956, *Middeltune* 1086 (DB). **Milton Abbas** Dorset. *Middeltone* 934, *Mideltune* 1086 (DB), *Middelton Abbatis* 1268. Latin affix means 'of the abbot' with reference to the abbey here. **Milton Abbot** Devon. *Middeltone* 1086 (DB), *Middelton Abbots* 1297. Affix from its early possession by Tavistock Abbey. **Milton Bryan** Beds. *Middelton* 1086 (DB), *Mideltone Brian* 1303. Manorial affix from a 12th-cent. owner called *Brian*. **Milton Clevedon** Somerset. *Mideltune* 1086 (DB), *Milton Clyvedon* 1408. Manorial affix from a family called *de Clyvedon*, here c.1200. **Milton Damerel** Devon. *Mideltone* 1086 (DB), *Middelton Aubemarle* 1301. Manorial affix from Robert *de Albemarle* who held the manor in 1086. **Milton Ernest** Beds. *Middeltone* 1086 (DB), *Middelton Orneys* 1330. Manorial affix from early possessions here of a man called *Erneis*. **Milton, Great & Milton, Little** Oxon. *Mideltone* 1086 (DB). **Milton Keynes** Milt. K. *Middeltone* 1086 (DB), *Middeltone Kaynes* 1227. Manorial affix from the *de Cahaignes* family, here from the 12th cent. **Milton Malsor** Northants. *Mideltone* 1086 (DB), *Milton alias Middleton Malsor* 1781. Manorial affix from the *Malsor* family, here in the 12th cent. **Milton Regis** Kent. *Middeltun* 893, *Middeltone* 1086 (DB). Affix is Latin *regis* 'of the king'. **Milton under Wychwood** Oxon. *Mideltone* 1086 (DB). For the affix, *see* ASCOTT.

However some Miltons have a different origin, 'farmstead or village with a mill', OE *myln + tūn*; examples include: **Milton** Cumbria. *Milneton* 1285. **Milton** Stoke. *Mulneton* 1227.

Milverton, 'farmstead or village by the mill ford', OE *myln + ford + tūn*: **Milverton** Somerset. *Milferton* 11th cent., *Milvertone* 1086 (DB). **Milverton** Warwicks. *Malvertone [sic]* 1086 (DB), *Mulverton* 1123.

Milwich Staffs. *Mulewiche* 1086 (DB). 'Farmstead with a mill'. OE *myln + wīc*.

Mimms, North & Mimms, South Herts. *Mimes* 1086 (DB). Possibly '(territory of) a tribe called the *Mimmas*', although this tribal name is obscure in origin and meaning.

Minch (sea strait) W. Isles, Highland. Gaelic *a' Mhaoil*. The strait is divided into North Minch and the more southerly Little Minch.

Minchinhampton Glos. *Hantone* 1086 (DB), *Minchenhamtone* 1221. 'High farmstead'. OE *hēah* (dative *hēan*) + *tūn*, with affix *myncena* 'of the nuns' (with reference to possession by the nunnery of Caen in the 11th cent.).

Mindrum Northum. *Minethrum* 1050. 'Mountain ridge'. Celtic **mönïth + *drum*.

Mine Head (*Mionn Ard*) Waterford. 'High crown'.

Minehead Somerset. *Mynheafdon* 1046, *Maneheve* 1086 (DB). Pre-English hill-name (possibly Celtic **mönïth* 'mountain') + OE *hēafod* 'projecting hill-spur'.

Minety Wilts. *Mintig* 844. Possibly 'island, or dry ground in marsh, where mint grows'. OE *minte + ēg*. Alternatively perhaps a Celtic name from **mīn* 'edge' or **mïn* 'kid' + **tï* 'house'.

Minginish (district) Highland (Skye). 'Great headland'. OScand. *megin + nes*.

Mingulay (island) W. Isles. *Megala* 1580. 'Great island'. OScand. *mikill + ey*.

Miningsby Lincs. *Melingesbi [sic]* 1086 (DB), *Mithingesbia* 1142. Probably 'farmstead or village of a man called Mithjungr'. OScand. pers. name + *bý*.

Minions Cornwall. Recorded thus from 1897, so named from the nearby tumulus called *Mimiens borroughe* 1613, 'Minion's barrow', derivation uncertain.

Minshull, Church Ches. *Maneshale [sic]* 1086 (DB), *Chirchemunsulf* late 13th cent. 'Shelf or ledge of a man called Monn'. OE pers. name + *scelf*. Affix is OE *cirice* 'church'.

Minsmere Haven Suffolk. *Amynnesmere Havynne* 1452, named from the lost village of Minsmere, *Mensemara* 1086 (DB), 'river-mouth lake', OScand. *(ā)mynni* + OE *mere*, with ME *haven* 'harbour'. The river-name **Minsmere** is a back-formation from the place name.

Minskip N. Yorks. *Minescip* 1086 (DB). OE *(ge)mǣnscipe* 'a community, a communal holding'.

Minstead Hants. *Mintestede* 1086 (DB). 'Place where mint grows or is grown'. OE *minte + stede*.

Minster, 'the monastery or large church', OE *mynster*: **Minster** Kent, near Ramsgate. *Menstre* 694. **Minster** Kent, near Sheerness. *Menstre* 1203. **Minster Lovell** Oxon. *Minstre* 1086 (DB), *Ministre Lovel* 1279. Manorial affix from the *Luvel* family, here in the 13th cent.

Minsterley Shrops. *Menistrelie* 1086 (DB). 'Woodland clearing near or belonging to a minster church' (probably referring to that at Westbury). OE *mynster + lēah*.

Minsterworth Glos. *Mynsterworthig* c.1030. 'Enclosure of the monastery' (here St Peter's Gloucester). OE *mynster + worth* (*worthig* in the early form).

Minterne Magna Dorset. *Minterne* 987. 'House near the place where mint grows'. OE *minte + ærn*. Affix is Latin *magna* 'great'.

Minting Lincs. *Mentinges* 1086 (DB). '(Settlement of) the family or followers of a man called *Mynta'. OE pers. name + *-ingas*.

Minto Sc. Bord. *Myntowe* 1296. '(Place by) the mountain'. The village takes its name from the nearby Minto Hills, 'mountain hill-spur' (OWelsh *minid* + OE *hōh*). The OE word was added when the original was no longer understood.

Minton Shrops. *Munetune* 1086 (DB). 'Farmstead or estate by the mountain (Long Mynd)'. Welsh *mynydd* + OE *tūn*.

Mirfield Kirkl. *Mirefeld* 1086 (DB). 'Pleasant open land', or 'open land where festivities are held'. OE *myrge* or *myrgen + feld*.

Miserden Glos. *Musardera* 1186. 'Musard's manor'. OFrench surname + *suffix -ere*. In 1086 (DB) the manor (then called *Grenhamstede* 'the

green homestead' from OE *grēne* + *hām-stede*)
was already held by the *Musard* family.

**Missenden, Great & Missenden,
Little**   Bucks. *Missedene* 1086 (DB). Probably
'valley where water-plants or marsh-plants
grow'. OE **mysse* + *denu*.

Misson   Notts. *Misne* 1086 (DB). Possibly
'mossy or marshy place', from a derivative of OE
mos 'moss, marsh'.

Misterton, 'farmstead or estate with a church
or belonging to a monastery', OE *mynster* + *tūn*:
Misterton Leics. *Minstretone* 1086 (DB).
Misterton Notts. *Ministretone* 1086 (DB).
Misterton Somerset. *Mintreston* 1199.

Mistley   Essex. *Mitteslea* [sic] 1086 (DB),
Misteleg 1225. Probably 'wood or clearing where
mistletoe grows'. OE *mistel* + *lēah*.

Mitcham   Gtr. London. *Michelham* 1086 (DB),
Michham 1178. 'Large homestead or village'. OE
micel + *hām*.

Mitcheldean   Glos. *Dena* 1220, *Muckeldine*
1224. '(Place at) the valley'. OE *denu* with the
later addition of *micel* 'great'.

Mitchell   Cornwall. *Meideshol* 1239. Probably
'maiden's hollow'. OE *mægd(en)* + *hol*.

Mitchelstown   (*Baile Mhistéala*) Cork. *Villa
Michel* 13th cent. 'Mitchel's townland'. The
name is probably that of a Welsh-Norman
landowner.

Mitford   Northum. *Midford* 1196. 'Ford where
two streams join'. OE *(ge)mȳthe* + *ford*.

Mitton, 'farmstead where two rivers join', OE
(ge)mȳthe + *tūn*; examples include: **Mitton,
Great** Lancs. *Mitune* 1086 (DB). **Mitton,
Upper** Worcs. *Myttun* 841, *Mettune* 1086 (DB).

Mixbury   Oxon. *Misseberie* 1086 (DB).
'Fortified place near a dunghill'. OE *mixen* +
burh (dative *byrig*).

Moate   (*An Móta*) Westmeath. 'The mound'.

Mobberley   Ches. *Motburlege* 1086 (DB).
'Clearing at the fortification where meetings are
held'. OE *mōt* + *burh* + *lēah*.

Moccas   Herefs. *Mochros* c.1130, *Moches* 1086
(DB). 'Moor where swine are kept'. Welsh *moch*
+ *rhos*.

Mockerkin   Cumbria. *Moldcorkyn* 1208.
Probably 'hill-top of a man called Corcán'.
OIrish pers. name + OScand. **moldi*.

Modbury   Devon. *Motberia* 1086 (DB).
'Fortification where meetings are held'. OE *mōt*
+ *burh* (dative *byrig*).

Moddershall   Staffs. *Modredeshale* 1086
(DB). 'Nook of a man called Mōdrēd'. OE pers.
name + *halh*.

Modelligo   (*Má Deilge*) Waterford. 'Plain of
the thorn'.

Modreeny   (*Má Dreimhne*) Tipperary. 'Plain
of fury'.

Moelfre   Angl. *Moylvry* 1399, *Moelvre* 1528.
'Bare hill'. Welsh *moel* + *bre*.

Moffat   Dumf. *Moffet* 1179, *Moffete* 1296. Of
unknown origin.

Mogeely   (*Maigh Dhíle*) Cork. 'Díle's plain'.

Mogerhanger   Beds. *Mogarhangre* 1216. OE
hangra 'wooded slope', perhaps added to a
Celtic name meaning 'wall, ruin'.

Moglass   (*Maigh Ghlas*) Tipperary. 'Green
plain'.

Mohill   (*Maothail*) Leitrim. 'Soft place'.

Moidart   (district) Highland. *Muddeward*
1292. 'Mundi's inlet'. OScand. pers. name +
fjǫrthr.

Moira   (*Maigh Ráth*) Down. (*cath*) *Maighe
Rath* 634. 'Plain of wheels', perhaps referring to
a meeting of routes.

Mold   (*Yr Wyddgrug*) Flin. *de Monte alto* 1267,
Montem Altum 1278, *Moald* 1284. 'High hill'.
OFrench *mont* + *hault*. The reference is to Bailey
Hill, where there was a British and Norman
fortification and possibly a Roman one. The
Welsh name means 'the burial mound' (Welsh
yr + *gŵydd* + *crug*), referring to the same hill.

Mole Valley   (district) Surrey. A modern
name, from the River Mole (for which *see*
MOLESEY).

Molendinar Burn   Glas. *Mellindonor* 1185.
Obscure name.

Molescroft   E. R. Yorks. *Molescroft* 1086 (DB).
'Enclosure of a man called Mūl'. OE pers. name
+ *croft*.

Molesey, East & Molesey, West   Surrey.
Muleseg 672–4, Molesham [*sic*] 1086 (DB).
'Island, or dry ground in marsh, of a man called
Mūl'. OE pers. name + *ēg*. Alternatively the first
element may be OE *mūl* 'mule' (from which the
pers. name is derived). The river-name **Mole** is
a back-formation from the place-name.

Molesworth   Cambs. *Molesworde* 1086 (DB).
'Enclosure of a man called Mūl'. OE pers. name
+ *worth*.

Molland   Devon. *Mollande* 1086 (DB). OE
land 'cultivated land, estate' with an uncertain
first element, possibly a pre-English hill-name.

Mollington, 'estate associated with a man
called Moll', OE pers. name + *-ing-* + *tūn*;
Mollington Ches. *Molintone* 1086 (DB).
Mollington Oxon. *Mollintune c.*1015,
Mollitone 1086 (DB).

Molton, North & Molton, South   Devon.
Nortmoltone, Sudmoltone 1086 (DB). OE *tūn*
'farmstead, estate' with the same first element
as in MOLLAND. Distinguishing affixes are OE
north and *sūth*. The river-name **Mole** is a
back-formation from the place name; the old
name of the river was *Nymet, see* NYMPTON.

Môn. *See* ANGLESEY.

Monadhliath Mountains   Highland.
'Grey-blue moors'. Gaelic *monadh* + *liath.*

Monaghan   (*Muineachán*) Monaghan.
Muineachán 1462. 'Place of thickets'.

Monamolin   (*Muine Moling*) Wexford.
'Thicket of Moling'.

Monard   (*Móin Ard*) Tipperary. 'High bog'.

Monaseed   (*Móin na Saighead*) Wexford.
'Bog of the arrows'.

Monaster   (*An Mhainistir*) Limerick. 'The
monastery'.

Monasteraden   (*Mainistir Réadáin*) Sligo.
'Réadán's monastery'.

Monasteranenagh   (*Mainistir an Aonaigh*)
Limerick. 'Monastery of the assembly'.

Monasterboice   (*Mainistir Bhuite*) Louth.
'Buite's monastery'.

Monasterevin   (*Mainistir Eimhín*) Kildare.
'Eimhín's monastery'.

Monasterleigh   (*Mainistir Liath*) Down.
'Grey abbey'.

Monavullagh   (*Móin an Mhullaigh*)
(mountains) Waterford. 'Bog of the summit'.

Monea   (*Maigh Niadh*) Fermanagh. (*Monua*)
*Maigi Niad c.*830. 'Plain of warriors'.

Moness   Perth. '(Place) at waterfalls'. Gaelic *i
mbun eas.*

Monewden   Suffolk. *Munegadena* 1086 (DB).
Probably 'valley of the family or followers
of a man called *Munda'. OE pers. name
+ *-inga-* + *denu.*

Moneyflugh   (*Muine Fliuch*) Kerry. 'West
thicket'.

Moneygall   (*Muine Gall*) Offaly. 'Thicket of
the stones'.

Moneygashel   (*Muine na gCaiseal*) Cavan.
Monygashell 1619. 'Thicket of the stone forts'.

Moneyglass   (*Muine Glas*) Antrim.
Ballymoyneglass 1605. 'Green thicket'.

Moneylea   (*Muine Liath*) Westmeath. 'Grey
thicket'.

Moneyneany   (*Móin na nIonadh*) Derry.
Monaneney 1622. 'Bog of the wonders'.

Moneynick   (*Muine Chnoic*) Antrim.
Ballivonykie 1605. 'Thicket of the hill'.

Moneyreagh   (*Mónaidh Riabhach*) Down.
Ballymoyne-Righ 1606. 'Grey bog'.

Moneystown   (*An Muine*) Wicklow. 'The
town of the thicket'.

**Mongeham, Great & Mongeham,
Little**   Kent. *Mundelingeham* 761,
Mundingeham 1086 (DB). 'Homestead of the
family or followers of a man called *Mundel', or
'homestead at *Mundel's place'. OE pers. name
+ *-inga-* or *-ing* + *hām.*

Monifieth   Ang. *Munifod* 1178. Gaelic *moine*
'moor' + unknown element.

Monivea   (*Muine Mheá*) Galway. 'Thicket of
the mead'.

Monk as affix. *See* main name, e.g. for **Monk
Bretton** (Barns.) *see* BRETTON.

Monkhopton   Shrops. *Hopton* 1255,
Munkehopton 1577. 'Farmstead in a valley . OE

hop + tūn. Later affix from its possession by Wenlock Priory.

Monkland Herefs. *Leine* 1086 (DB), *Munkelen* c.1180. 'Estate in *Leon* held by the monks'. Old Celtic name for the district (*see* LEOMINSTER) with OE *munuc*. The manor belonged before 1086 to Conches Abbey in Normandy.

Monkleigh Devon. *Lega* 1086 (DB), *Munckenelegh* 1244. 'Wood or clearing of the monks'. OE *munuc* + *lēah*, with reference to possession by Montacute Priory from the 12th cent.

Monkokehampton Devon. *Monacochamentona* 1086 (DB). 'Estate on the River Okement held by the monks'. OE *munuc* + Celtic river-name (possibly 'swift stream') + OE *tūn*. It once belonged to Glastonbury Abbey.

Monks as affix. *See* main name, e.g. for **Monks Eleigh** (Suffolk) *see* ELEIGH.

Monkseaton N. Tyne. *Seton Monachorum* 1380. 'Farmstead by the sea'. OE *sǣ + tūn*. Affix 'of the monks' (Latin *monachorum* in the early form) from its early possession by the monks of Tynemouth.

Monksilver Somerset. *Selvere* 1086 (DB), *Monkesilver* 1249. Probably an old river-name from OE *seolfor* 'silver', i.e. 'clear or bright stream'. Affix is OE *munuc* alluding to early possession by the monks of Goldcliff Priory.

Monkstown (*Baile na Manach*) Antrim. *Ballynemanagh* 1605. 'Townland of the monks'.

Monkton, 'farmstead of the monks', OE *munuc + tūn*; examples include: **Monkton** Devon. *Muneketon* 1244, **Monkton** Kent. *Munccetun* c.960, *Monocstune* 1086 (DB). **Monkton, Bishop** N. Yorks. *Munecatun* c.1030, *Monucheton* 1086 (DB). Affix from its possession by the Archbishops of York. **Monkton, Moor** & **Monkton, Nun** N. Yorks. *Monechetone* 1086 (DB), *Moremonketon* 1402, *Nunmonkton* 1397. Distinguishing affixes are OE *mōr* 'moor' and *nunne* 'nun' (from the nunnery founded here in the 12th cent.). **Monkton, West** Somerset. *Monechetone* 1086 (DB).

Monkton as affix. *See* main name, e.g. for **Monkton Combe** (B. & NE. Som.), *see* COMBE.

Monkwearmouth Sundld. *Uuiremutha* c.730, *Wermuth Monachorum* 1291. 'The

mouth of the River Wear'. Celtic or pre-Celtic river-name (probably meaning simply 'water, river' or 'the bending one') + OE *mūtha*. Affix *Monk-* (Latin *monachorum* 'of the monks') refers to early possession by the monks of the Jarrow-Monkwearmouth monastery.

Monmouth (*Trefynwy*) Mon. *Munwi Mutha* 11th cent., *Munemuda* 1190, *Monmouth* 1267. 'Mouth of the River Monnow'. OE *mūtha*. The Celtic river name may mean 'fast-flowing'. The Welsh name means 'homestead on the River Mynwy' (Welsh *tref*).

Monnington on Wye Herefs. *Manitune* 1086 (DB). 'Estate held communally, or by a man called Manna'. OE (*ge*)*mǣne* or pers. name (genitive *-n*) + *tūn*. For the river-name, *see* ROSS ON WYE.

Monroe (*Móin Ruadh*) Westmeath. 'Red bog'.

Montacute Somerset. *Montagud* 1086 (DB). 'Pointed hill'. OFrench *mont + aigu*.

Montford Shrops. *Maneford* 1086 (DB), *Moneford* 1204. Possibly 'ford where people gather', from OE *gemāna* 'fellowship, association' + *ford*.

Montgomery (*Trefaldwyn*) Powys. *Montgomeri* 1086 (DB). The Norman lord Roger de Montgomery built a castle here and named it after his other castle at Montgommery, Normandy. In 1102 this castle passed to the Norman *Baldwin* de Bollers, who gave the Welsh name, 'Baldwin's town' (Welsh *tref*).

Montrose Ang. *Munros* c.1178. 'Moor of the promontory'. Gaelic *mon* (a short form of *monadh* 'moor') + *ros*.

Monxton Hants. *Anna de Becco* c.1270, *Monkestone* 15th cent. 'Estate (on the River Ann) held by the monks (of Bec Abbey)'. OE *munuc + tūn*, *see* ANN.

Monyash Derbys. *Maneis* 1086 (DB). 'Many ash-trees'. OE *manig + æsc*.

Monzievaird Perth. *Muithauard* c.1200, *Moneward* 1203. 'Plain of the bards'. Gaelic *magh + bard*.

Mooncoin (*Móin Choinn*) Kildare. 'Conn's bog'.

Moone (*Maoin*) Kildare. 'Gift'.

Moor Crichel Dorset. *See* CRICHEL.

Moor Monkton N. Yorks. *See* MONKTON.

Moorby Lincs. *Morebi* 1086 (DB). 'Farmstead or village in the moor'. OScand. *mór* + *bý*.

Moore Halton. *Mora* 12th cent. 'The marsh'. OE *mōr*.

Moorfoot Hills Sc. Bord. *Morthwait c.*1142. 'Hills by the moorland clearing'. OScand. *mór* + *thveit* + OE *hyll*.

Moorlinch Somerset. *Mirieling* 971. 'Pleasant ledge or terrace'. OE *myrge* + *hlinc*.

Moors, West Dorset. *La More* 1310, *Moures* 1407. 'The marshy ground(s)'. OE *mōr*.

Moorsholm Red & Cleve. *Morehusum* 1086 (DB). '(Place at) the houses on the moor'. OE *mōr* + *hūs*, or OScand. *mór* + *hús*, in a dative plural form in *-um*.

Moorsley, Low Sundld. *Mor(l)eslau* 12th cent. Possibly 'hill of a man called Morulf'. Continental Germanic pers. name + OE *hlāw*.

Morar (district) Highland. *Morderer* 1292, *Mordhowar* 14th cent. 'Great water'. Gaelic *mor* + *dobhar*.

Moray (the unitary authority). *Moreb c.*970, *Morauia* 1124, *Murewe c.*1185. 'Sea settlement'. Celtic **mori-* + **treb*.

Morborne Cambs. *Morburne* 1086 (DB). 'Marsh stream'. OE *mōr* + *burna*.

Morchard, 'great wood or forest', Celtic **mǭr* + **cę̄d*: **Morchard Bishop** Devon. *Morchet* 1086 (DB), *Bisschoppesmorchard* 1311. Affix from its possession by the Bishop of Exeter in 1086. **Morchard, Cruwys** Devon. *Morchet* 1086 (DB), *Cruwys Morchard c.*1260. Manorial affix from the *de Crues* family, here in the 13th cent.

Morcombelake Dorset. *Mortecumbe* 1240, *Morecomblake* 1558. OE *cumb* 'valley' with OE pers. name **Morta* or **mort* 'young salmon or similar fish' + *lacu* 'stream'.

Morcott Rutland. *Morcote* 1086 (DB). 'Cottage(s) in the marshland'. OE *mōr* + *cot*.

Morden, 'hill in marshland', OE *mōr* + *dūn*: **Morden** Dorset. *Mordune* 1086 (DB). **Morden** Gtr. London. *Mordune* 969, *Mordone* 1086 (DB). **Morden, Guilden & Morden, Steeple** Cambs. *Mordune* 1015, *Mordune* 1086 (DB), *Gildene Mordon* 1204, *Stepelmordun* 1242.

Distinguishing affixes are OE *gylden* 'wealthy, splendid' and *stēpel* 'church steeple'.

Mordiford Herefs. *Mord(e)ford* 12th cent. OE *ford* 'a ford' with an uncertain first element.

Mordon Durham. *Mordun c.*1040. 'Hill in marshland'. OE *mōr* + *dūn*.

More Shrops. *La Mora* 1181. 'The marsh'. OE *mōr*.

Morebath Devon. *Morbatha* 1086 (DB). 'Bathing-place in marshy ground'. OE *mōr* + *bæth*.

Morecambe Lancs. a modern revival of an old Celtic name for the Lune estuary, *Morikámbē* 'the curved inlet', first recorded *c.* AD 150.

Moreleigh Devon. *Morlei* 1086 (DB). 'Woodland clearing on or near a moor'. OE *mōr* + *lēah*.

Morenane (*Boirneán*) Limerick. 'Little stony place'.

Moresby Cumbria. *Moresceby c.*1160. 'Farmstead or village of a man called Maurice'. OFrench pers. name + OScand. *bý*.

Morestead Hants. *Morstede* 1172. 'Place by a moor'. OE *mōr* + *stede*.

Moreton, Morton, a common name, 'farmstead in moorland or marshy ground', OE *mōr* + *tūn*; examples include: **Moreton** Dorset. *Mortune* 1086 (DB). **Moreton** Wirral. *Moreton* 1278. **Moreton-in-Marsh** Glos. *Mortun* 714, *Mortune* 1086 (DB), 'Morton in Hennemersh* 1253. Affix is an old district-name meaning 'marsh frequented by wild hen-birds (moorhens or the like)', OE *henn* + *mersc*. **Moreton, Maids** Bucks. *Mortone* 1086 (DB), *Maidenes Morton c.*1488. Affix from the tradition that the church here was built by two maiden ladies in the 15th cent. **Moreton Morrell** Warwicks. *Mortone* 1086 (DB), *Merehull* 1279, *Morton Merehill* 1285. Originally two distinct names first combined in the 13th cent. Morrell means 'boundary hill', OE (*ge*)*mǣre* + *hyll*. **Moreton Pinkney** Northants. *Mortone* 1086 (DB). Manorial affix (first used in the 16th cent.) from the *de Pinkeny* family who held the manor in the 13th cent. **Moreton Say** Shrops. *Mortune* 1086 (DB), *Morton de Say* 1255. Manorial affix from the *de Sai* family, here in 1199. **Morton** Derbys. *Mortun* 956, *Mortune* 1086 (DB). **Morton** Lincs., near Bourne. *Mortun* 1086 (DB), **Morton** Lincs., near

Gainsborough. *Mortune* 1086 (DB). **Morton Bagot** Warwicks. *Mortone* 1086 (DB), *Morton Bagod* 1262. Manorial affix from the *Bagod* family, here from the 12th cent.

Moretonhampstead Devon. *Mortone* 1086 (DB), *Morton Hampsted* 1493. Identical in origin with the previous names, with a later addition which may be a family name or from a nearby place (from OE *hām-stede* 'homestead').

Morfa Cergn. 'Plain by the sea'. Welsh *morfa*.

Morgannwg. *See* GLAMORGAN.

Morland Cumbria. *Morlund* c.1140. 'Grove in the moor or marsh'. OScand. *mór* + *lundr*.

Morley, 'woodland clearing in or near a moor or marsh', OE *mōr* + *lēah*: **Morley** Derbys. *Morlege* 1002, *Morleia* 1086 (DB). **Morley** Durham. *Morley* 1312. **Morley** Leeds. *Moreleia* 1086 (DB). **Morley** Norfolk. *Morlea* 1086 (DB).

Morningside Edin. 'Morning slope'. The name is first recorded for an estate formed here in 1657 on the amalgamation of two adjoining properties.

Morningthorpe Norfolk. *Torp, Maringatorp* 1086 (DB). Possibly 'outlying farmstead of the dwellers by the lake or by the boundary'. OE *mere* or *mǣre* + *-inga-* + OScand. *thorp*.

Mornington (*Baile Uí Mhornáin*) Meath. 'Ó Mornán's town'.

Moroe (*Maigh Rua*) Limerick. 'Red plain'.

Morpeth Northum. *Morthpath* c.1200. 'Path where a murder took place'. OE *morth* + *pæth*.

Morriston (*Treforys*) Swan. 'Morris's town'. The settlement was built in c.1768 by Sir John Morris to provide housing for local workers. The Welsh name has the same sense (Welsh *tref* + *Morys*).

Morston Norfolk. *Merstona* 1086 (DB). 'Farmstead by a marsh'. OE *mersc* + *tūn*.

Mortehoe Devon. *Morteho* 1086 (DB). Possibly 'hill-spur called *Mort* or the stump'. OE **mort* + *hōh*.

Mortimer W. Berks. from the manorial affix of STRATFIELD MORTIMER.

Mortlake Gtr. London. *Mortelage* 1086 (DB). Possibly 'stream of a man called **Morta*'. OE pers. name + *lacu*. Alternatively the first element

may well be OE **mort* 'young salmon or similar fish'.

Morton. *See* MORETON.

Morvah Cornwall. *Morveth* 1327. 'Grave by the sea'. Cornish *mor* + *bedh*.

Morval Cornwall. *Morval* 1238. Probably Cornish *mor* 'sea' with an obscure second element.

Morville Shrops. *Membrefelde* 1086 (DB), *Mamerfeld, Momerfeld* c.1138. OE *feld* 'open land' with an uncertain first element, possibly a derivative of Celtic **mamm* 'breast-like hill'.

Morwenstow Cornwall. *Morwestewe* 1201. 'Holy place of St Morwenna'. Cornish saint's name + OE *stōw*.

Mosborough Sheff. *Moresburh* c.1002, *Moresburg* 1086 (DB). 'Fortified place in the moor'. OE *mōr* + *burh*.

Mosedale Cumbria. *Mosedale* 1285. 'Valley with a bog'. OScand. *mosi* + *dalr*.

Moseley Birm. *Museleie* 1086 (DB). 'Woodland clearing infested with mice'. OE *mūs* + *lēah*.

Moseley Wolverh. *Moleslei* 1086 (DB). 'Woodland clearing of a man called Moll'. OE pers. name + *lēah*.

Moss Donc. *Mose* 1416. 'The swamp or bog'. OE *mos* or OScand. *mosi*.

Moss-side Antrim. *Mosside* c.1657. The name may be English, 'side of the peat bog', or represent Irish *Maigh Saighead*, 'plain of arrows', denoting a battle site.

Mossley Tamesd. *Moselegh* 1319. 'Woodland clearing by a swamp or bog'. OE *mos* or OScand. *mosi* + OE *lēah*.

Mossley Antrim. *Mossley* 1839. The name was adopted from previous name.

Mosterton Dorset. *Mortestorne* 1086 (DB). 'Thorn-tree of a man called **Mort*'. OE pers. name + *thorn*.

Moston, usually 'moss or marsh farmstead', OE *mos* + *tūn*; for example: **Moston** Shrops. *Mostune* 1086 (DB).

Motcombe Dorset. *Motcumbe* 1244. 'Valley where meetings are held'. OE *mōt* + *cumb*.

Mothel (*Maothail*) Waterford. 'Soft place'.

Motherwell N. Lan. *Matervelle c.*1250, *Moydirwal* 1265. 'Our Lady's well'. ME *moder* + *welle*.

Mottingham Gtr. London. *Modingahamm* 973, *Modingeham* 1044. 'Homestead or enclosure of the family or followers of a man called *Mōda'. OE pers. name + -inga- + hām or hamm.

Mottisfont Hants. *Mortesfunde* [*sic*] 1086 (DB), *Motesfont* 1167. 'Spring near a river-confluence', or 'spring where meetings are held'. OE *mōt* + *funta.

Mottistone I. of Wight. *Modrestan* 1086 (DB). 'Stone of the speaker(s) at a meeting'. OE *mōtere* + *stān.

Mottram, possibly 'speakers' place' or 'place where meetings are held', OE *mōtere* or *mōt* + *rūm*: **Mottram in Longdendale** Tamesd. *Mottrum c.*1220, *Mottram in Longedenedale* 1308. Affix is a district name, 'dale of the long valley', OE *lang* + *denu* + OScand. *dalr*. **Mottram St Andrew** Ches. *Motre* [*sic*] 1086 (DB), *Motromandreus* 1351. Affix from the dedication of a chapter here recorded in the 15th cent.

Mouldsworth Ches. *Moldeworthe* 12th cent. 'Enclosure at a hill'. OE *molda* + *worth.

Moulsecoomb Bright. **&** Hove. *Mulescumba c.*1100. 'Valley of a man called *Mūl'. OE pers. name + *cumb.

Moulsford Oxon. *Muleforda c.*1110. Probably 'ford of a man called *Mūl'. OE pers. name + *ford.

Moulsoe Milt. K. *Moleshou* 1086 (DB). 'Hill-spur of a man called *Mūl'. OE pers. name + *hōh.

Moulton, 'farmstead of a man called *Mūla, or where mules are kept', OE pers. name or OE *mūl* + *tūn*: **Moulton** Ches. *Moletune* 1086 (DB). **Moulton** Lincs. *Multune* 1086 (DB). **Moulton** Northants. *Multone* 1086 (DB). **Moulton** N. Yorks. *Moltun* 1086 (DB). **Moulton** Suffolk. *Muletuna* 1086 (DB). **Moulton St Michael** Norfolk. *Mulantun c.*1035, *Muletuna* 1086 (DB). Affix from the dedication of the church.

Moulton Vale Glam. *Molton* 1533. Probably identical with previous names.

Mount Bures Essex. *See* BURES.

Mount Nugent Cavan. *Mount Nugent* 1837. The *Nugent* family were granted lands here in 1597.

Mountain Ash (*Aberpennar*) Rhon. The town arose as an industrial development in the 19th cent. and took its name from an inn, itself named after a prominent mountain ash (rowan). The Welsh name means 'confluence of the River Pennar' (Welsh *aber*), with the river named after a nearby mountain, *Cefn Pennar*, 'ridge of the height' (Welsh *cefn* + *pennardd*).

Mountbolus (*Cnocán Bhólais*) Offaly. 'Hill of Bolas'.

Mountcharles Donegal. 'Charles's mount', *town* (land) *of Mount Charles al. Tawnytallan* 1676. The village is named after *Charles* Conyngham, landlord here in the 17th cent.

Mountfield E. Sussex. *Montifelle* [*sic*] 1086 (DB), *Mundifeld* 12th cent. 'Open land of a man called *Munda'. OE pers. name + *feld.

Mountjoy Tyrone. *Mountjoy forest* 1834. The village grew up in the early 19th cent. on land owned by Charles John Gardiner, 2nd Viscount *Mountjoy.

Mountmellick (*Móinteach Mílic*) Laois. 'Moor of the wet ground'.

Mountnessing Essex. *Ginga* 1086 (DB), *Gynges Munteny* 1237. 'Manor called *Ing* (for which *see* FRYERNING) held by the *de Mounteney* family'.

Mountnorris Armagh. (*fort of*) *Mountnorris* orw. *Aghnecranchie* 1606. The name is that of Sir John *Norris*, who built a fort here in the late 16th cent.

Mountrath (*Móin Rátha*) Laois. 'Bog of the fort'.

Mountsorrel Leics. *Munt Sorel* 1152. 'Sorrel-coloured hill', referring to the pinkish-brown stone here. OFrench *mont* + *sorel.

Mourne Mountains (*Múrna*) Down. 'Mountains of the Múrna'. The *Múrna* (earlier *Mughdhorna*) tribe originated in modern Co. Monaghan, where they gave the name of CREMORNE, but migrated to Co. Down in the 12th cent.

Mousa (island) Shet. *Mosey c.*1150. 'Moor island'. OScand. *mór* + *ey.

Mousehole Cornwall. *Musehole* 1284. Self-explanatory. OE *mūs* + *hol*, originally referring to a large cave here.

Moveen (*Má Mhín*) Clare. 'Smooth plain'.

Movilla (*Maigh Bhile*) Down. *Moige Bile* 850. 'Plain of the sacred tree'.

Moville (*Magh Bhile*) Donegal. (*for lār*) *Maighe Bile* 1167. 'Plain of the sacred tree'.

Mow Cop Ches. /Staffs. *Mowel* c.1270, *Mowle-coppe* 1621. 'Heap hill'. OE *mūga* + *hyll* with the later addition of *copp* 'hill-top'.

Mowhan (*Much Bhán*) Armagh. *Mowhan* 1838. 'White plain'.

Mowsley Leics. *Muselai* 1086 (DB). 'Wood or clearing infested with mice'. OE *mūs* + *lēah*.

Moy (*An Mhaigh*) Tyrone. *Madh* c.1645. 'The plain'.

Moy (*Muaidh*) (river) Mayo, Sligo. 'Stately one'.

Moy Highland. *Muy* 1235. '(Place) in the plain'. Gaelic *magh* (locative *maigh*).

Moyallan (*Maigh Alúine*) Down. *Moynalvin* c.1657. 'Plain of alum'.

Moyard (*Maigh Ard*) Galway. 'High plain'.

Moyarget (*Maigh Airgid*) Antrim. *Myerget* c.1657. 'Plain of silver'.

Moyasta (*Maigh Sheasta*) Clare. 'Upright plain'.

Moycarky (*Maigh Chairce*) Limerick. 'Plain of oats'.

Moycashel (*Magh Chaisil*) Westmeath. 'Plain of the stone fort'.

Moycullen (*Magh Cuilinn*) Galway. 'Plain of holly'.

Moydow (*Maigh Dumha*) Longford. 'Plain of the mound'.

Moydrum (*Magh Droma*) Westmeath. 'Plain of the ridge'.

Moygashel (*Maigh gCaisil*) Tyrone. *Moygashell* 1609. 'Plain of the stone fort'.

Moyle (*Mael*) Tipperary. 'Hill'.

Moylgrove (*Trewyddel*) Pemb. *Ecclesia de grava Matild'* 1291. 'Matilda's grove'. OGerman pers. name + OE *grāfa*, 'grove'. The Welsh name means 'grove farm' (Welsh *tref* + *gwyddel*).

Moylough (*Maigh Locha*) Galway. 'Plain of the lake'.

Moymore (*Má Mhór*) Clare. 'Big plain'.

Moynalty (*Maigh nEalta*) Meath. 'Plain of the flocks of birds'.

Moyne (*Maighean*) Mayo, Tipperary, Wicklow. 'Precinct'.

Moyne (*Maighín*) Mayo, Tipperary. 'Little plain'.

Moyness Highland. 'Meadow in the plain'. Gaelic *magh* + *inis*.

Moyola (*Maigh Fhoghlach*) (river) Derry. 'Plain of plundering'. The original name of the river was *Bior*, 'water'. Later, it was known as *Abhainn na Scríne*, 'river of Ballynascreen', the parish name meaning 'townland of the shrine'.

Moyvally (*Magh Bhealaigh*) Kildare. 'Plain of the pass'.

Moyvore (*Magh Mhora*) Westmeath. 'Mora's plain'.

Moyvoughly (*Má Bhachla*) Westmeath. 'Plain of the crosier'.

Much as affix. *See* main name, e.g. for **Much Birch** (Herefs.) *see* BIRCH.

Muck (island) Highland. *Helantmok* 1370. '(Island of) pigs'. Gaelic *muc*. The early form has Gaelic *eilean*, 'island'.

Muckamore (*Maigh Chomair*) Antrim. 'Plain of the confluence'.

Mucking Thurr. *Muc*(*h*)*inga* 1086 (DB). '(Settlement of) the family or followers of a man called Mucca', or 'Mucca's place'. OE pers. name + *-ingas* or *-ing*.

Muckish (*An Mhucais*) (mountain) Donegal. 'The ridge of the pig'. The ridge is shaped like a pig's back.

Muckle Flugga (island) Shet. 'Great cliff island'. OScand. *mikill* + *flugey*.

Mucklestone Staffs. *Moclestone* 1086 (DB). 'Farmstead of a man called Mucel'. OE pers. name + *tūn*.

Muckno, Lough (*Loch Mucshnámha*) Monaghan. 'Lake of the swimming place of the pigs'.

Muckros (*Mucros*) Donegal. 'Wood of the pigs'.

Muckross (*Mucros*) Kerry. 'Wood of the pigs'.

Muckton Lincs. *Machetone* [*sic*] 1086 (DB), *Muketun* 12th cent. 'Farmstead of a man called Muca'. OE pers. name + *tūn*.

Muddiford Devon. *Modeworthi* 1303. Probably 'enclosure of a man called *Mōda'. OE pers. name + *worthig* (later replaced by *ford*).

Mudeford Dorset. *Modeford* 13th cent. Probably 'mud ford', ME *mode, mudde* + *ford*. The river-name **Mude** is a back-formation from the place name.

Mudford Somerset. *Mudiford* 1086 (DB). 'Muddy ford'. OE *muddig* + *ford*.

Mudgley Somerset. *Mudesle* 1157. OE *lēah* 'wood or clearing' with an uncertain first element, possibly a pers. name.

Muff (*Magh*) Donegal. *Mough* 1621. 'Plain'.

Mugginton Derbys. *Mogintun* 1086 (DB). 'Estate associated with a man called *Mogga or *Mugga'. OE pers. name + *-ing-* + *tūn*.

Muggleswick Durham. *Muclincgwic* c.1195, *Mokeleswyk* 1304. 'Farmstead of a man called Mucel'. OE pers. name (+ *-ing-*) + *wīc*.

Muine Bheag (*Muine Bheag*) Carlow. 'Little thicket'.

Muir of Ord Highland. 'Moorland of the rounded hill'. OE *mōr* + Gaelic *òrd*.

Muirkirk E. Ayr. 'Moorland church'. OScand. *mór* + *kirkja*.

Muker N. Yorks. *Meuhaker* 1274. 'Narrow cultivated plot'. OScand. *mjór* + *akr*.

Mulbarton Norfolk. *Molkebertuna* 1086 (DB). 'Outlying farm where milk is produced'. OE *meoluc* + *bere-tūn*.

Muldonagh (*Maol Domhnaigh*) Derry. *Meldony* 1613. 'Round hill of the church'.

Mull (island) Arg. *Malaios* c.150. Pre-Celtic island name.

Mull of Oa Arg. (Islay). 'Headland of Oa'. Gaelic *maol na h-Otha*.

Mullagh (*Mullach*) Cavan, Galway, Meath. 'Summit'.

Mullaghanattin (*Mullach an Aitinn*) Kerry. 'Summit of the gorse'.

Mullaghareirk (*Mullach an Radhairc*) (mountains) Cork, Limerick. 'Summit of the prospect'.

Mullaghbawn (*An Mullach Bán*) Armagh. 'White summit'.

Mullaghcarn (*Mullach Chairn*) Tyrone. 'Summit of the cairn'.

Mullaghcleevaun (*Mullach Cliabháin*) Wicklow. 'Summit of the cradle'.

Mullaghmore (*Mullach Mór*) Derry, Sligo. 'Big summit'.

Mullan (*An Muileann*) Monaghan. 'The mill'.

Mullan (*Mullán*) Tyrone. 'Little hill'.

Mullenakill (*Muileann na Cille*) Kilkenny. 'Mill of the church'.

Mullinacuff (*Muileann Mhic Dhuibh*) Wicklow. 'Mill of Mac Duibh'.

Mullinahone (*Muileann na hUamhan*) Tipperary. 'Mill of the cave'.

Mullinavat (*Muileann an Bhata*) Kilkenny. 'Mill of the stick'.

Mullingar (*Muileann Cearr*) Westmeath. *Mullin Cirr* 1305, *Muilenn Cearr* 1450. 'Crooked mill'.

Mullinoran (*Muileann an Fhuaráin*) Westmeath. 'Mill of the spring'.

Mullion Cornwall. 'Church of *Sanctus Melanus*' 1262. From the patron saint of the church.

Mulrany (*Mala Raithne*) Mayo. 'Ferny brae'.

Mulroy Bay (*Maol Rua*) Donegal. *Moyroy* 1608. 'Red current'.

Multyfarnham (*Muilte Farannáin*) Westmeath. 'Farannán's mills'.

Mumbles Head Swan. *Mommulls* 1549. '(Place of the) headland'. OScand. *múli*. The first element is obscure.

Mumby Lincs. *Mundebi* 1086 (DB). Possibly 'farmstead or village of a man called Mundi'. OScand. pers. name + *bý*. Alternatively the first element may be OE or OScand. *mund* in the sense 'protection' or 'hedge'.

Mundesley Norfolk. *Muleslai* 1086 (DB). 'Woodland clearing of a man called Mūl or *Mundel'. OE pers. name + *lēah*.

Mundford Norfolk. *Mundeforda* 1086 (DB). 'Ford of a man called *Munda'. OE pers. name + *ford*.

Mundham, 'homestead or enclosure of a man called *Munda', OE pers. name + *hām* or *hamm*: **Mundham** Norfolk. *Mundaham* 1086 (DB). **Mundham** W. Sussex. *Mundhame* c.692, *Mundreham* 1086 (DB).

Mundon Essex. *Munduna* 1086 (DB). 'Protection hill', i.e. probably 'raised ground safe from flooding, or protected by fencing'. OE *mund* + *dūn*.

Mungrisdale Cumbria. *Grisedale* 1285, *Mounge Grieesdell* 1600. 'Valley where young pigs are kept'. OScand. *gríss* + *dalr*, with the later addition of the saint's name *Mungo* from the dedication of the church.

Munlough (*Móin Loch*) Cavan. 'Lake of the bog'.

Munsley Herefs. *Muleslage, Muneslai* 1086 (DB). 'Woodland clearing of a man called Mūl or *Mundel'. OE pers. name + *lēah*.

Munslow Shrops. *Mosselawa* 1167, *Munselowe* 1252. OE *hlāw* 'mound, tumulus', with an uncertain first element.

Munster (*Cúige Mumhan*) (the province). 'District of the Mumu tribe'. OScand. possessive -*s* + Irish *tír*.

Munterburn (*Muintir Bhirn*) Tyrone. 'Family of Birn'.

Munterloney (*Muintir Luinigh*) Tyrone. 'Family of Luinigh'.

Murcott Oxon. *Morcot* c.1191. 'Cottage(s) in marshy ground'. OE *mōr* + *cot*.

Murlough (*Murbholg*) Down. 'Sea swell'.

Murragh (*Murbhach*) Cork. 'Salt marsh'.

Murrayfield Edin. 'Murray's field'. Archibald *Murray*, later Lord Henderland, bought an estate here in 1733 and named it after himself.

Murreagh (*Muiríoch*) Kerry. 'Beach'.

Murroe (*Magh Rua*) Limerick. 'Red plain'.

Murroogh (*Murúch*) Clare. 'Beach'.

Murrow Cambs. *Morrowe* 1376, 'Row (of houses) in marshy ground'. OE *mōr* + *rāw*.

Mursley Bucks. *Muselai* [*sic*] 1086 (DB), *Murselai* 12th cent. Probably 'woodland clearing of a man called *Myrsa'. OE pers. name + *lēah*.

Murton, 'farmstead in moorland or marshy ground', OE *mōr* + *tūn*: **Murton** Cumbria. *Morton* 1235. **Murton** Durham, near Seaham. *Morton* c.1200. **Murton** Northum. *Morton* 1204. **Murton** York. *Mortun* 1086 (DB).

Musbury Devon. *Musberia* 1086 (DB). 'Old fortification infested with mice'. OE *mūs* + *burh* (dative *byrig*).

Muscoates N. Yorks. *Musecote* c.1160. 'Cottages infested with mice'. OE *mūs* + *cot*.

Musgrave, Great & Musgrave, Little Cumbria. *Musegrave* 12th cent. 'Grove or copse infested with mice'. OE *mūs* + *grāf*.

Muskham, North & Muskham, South Notts. *Muscham, Nordmuscham* 1086 (DB). Possibly 'homestead or village of a man called *Musca'. OE pers. name + *hām*.

Musselburgh E. Loth. *Muselburge* c.1100. 'Mussel town'. OE *mus(c)le* + *burh*. There is a famous mussel-bed here.

Muston Leics. *Mustun* 12th cent. 'Mouse-infested farmstead'. OE *mūs* + *tūn*.

Muston N. Yorks. *Mustone* 1086 (DB). Identical with previous name, or 'farmstead of a man called Músi', OScand. pers. name + OE *tūn*.

Muswell Hill Gtr. London. *Mosewella* c.1155. 'Mossy or boggy spring'. OE *mēos* + *wella*, with the addition of *hill* from the 17th cent.

Mweelahorna (*Maol na hEorna*) Waterford. 'Hill of the barley'.

Mweelrea (*Cnoc Maol Réidh*) (mountain) Mayo. 'Grey hill'.

Mybster Highland. 'Marshy farmstead'. OScand. *mýrr* + *bólstathr*.

Myddle Shrops. *Mulleht* [*sic*] 1086 (DB), *Muthla* 1121. Possibly '(place at) the confluence of streams', from a derivative *(ge)mýthel* of OE *(ge)mýthe*.

Mylor Bridge Cornwall. 'Church of *Sanctus Melorus*' 1258. From the patron saint of the church.

Myndtown Shrops. *Munete* 1086 (DB), *Myntowne* 1577. 'Place at the mountain (Long Mynd)'. Welsh *mynydd*, with the later addition of *town*.

Mynydd Bach (mountain) Cergn. 'Little mountain'. Welsh *mynydd* + *bach*.

Mynydd Du (mountain) Cergn. 'Black mountain'. Welsh *mynydd* + *du*.

Myroe (*Maigh na Rua*) Derry. *Moyrowe* 1613. 'Plain of the (river) Roe'.

Myshall (*Míseal*) Carlow. 'Low central place'.

Mytholmroyd Calder. *Mithomrode* late 13th cent. 'Clearing at the river-mouths'. OE (*ge*)*mȳthe* (dative plural (*ge*)*mȳthum*) + *rodu*.

Myton, Mytton, 'farmstead where two rivers join', OE (*ge*)*mȳthe* + *tūn*: **Myton on Swale** N. Yorks. *Mytun* 972, *Mitune* 1086 (DB). On the River Swale (probably OE **swalwe* 'rushing water') near to where it joins the Ure. **Mytton** Shrops. *Mutone* 1086 (DB).

Naas (*An Nás*) Kildare. 'The assembly'.

Naburn York. *Naborne* 1086 (DB). Possibly 'stream called the narrow one'. OE *naru* + *burna*.

Nackington Kent. *Natyngdun* late 10th cent. *Latintone* [*sic*] 1086 (DB). Probably 'hill at the wet place'. OE **næting* + *dūn*.

Nacton Suffolk. *Nachetuna* 1086 (DB). Possibly 'farmstead of a man called Hnaki'. OScand. pers. name + OE *tūn*.

Nacung, Loch (*Loch na Cuinge*) Donegal. 'Lake of the yoke'.

Nad (*Nead an Iolair*) Cork. 'Eagle's nest'.

Nafferton E. R. Yorks. *Nadfartone* 1086 (DB). Probably 'farmstead of a man called Náttfari'. OScand. pers. name + OE *tūn*.

Nailsea N. Som. *Nailsi* 1196. 'Island, or dry ground in marsh, of a man called **Nægl*'. OE pers. name + *ēg*.

Nailstone Leics. *Naylestone* 1225. 'Farmstead of a man called **Nægl*'. OE pers. name + *tūn*.

Nailsworth Glos. *Nailleswurd* 1196. 'Enclosure of a man called **Nægl*'. OE pers. name + *worth*.

Nairn Highland. *Inuernaren* c.1195, *Narne* 1382. 'Mouth of the River Nairn'. The river has a Celtic or pre-Celtic name possibly meaning 'penetrating one'. The first form above has Gaelic *inbhir*, 'river-mouth'.

Nantwich Ches. *Wich* 1086 (DB), *Nametwihc* 1194. 'The salt-works'. OE *wīc* with the later addition of ME *named* 'renowned, famous'.

Nant-y-glo Blae. *Nantygloe* 1752. 'Valley of the coal'. Welsh *nant* + *y* + *glo*.

Nappa N. Yorks. near Hellifield. *Napars* [*sic*] 1086 (DB), *Nappai* 1182. Possibly 'enclosure in a bowl-shaped hollow'. OE *hnæpp* + *hæg*.

Napton on the Hill Warwicks. *Neptone* 1086 (DB). Probably 'farmstead on a hill thought to resemble an inverted bowl'. OE *hnæpp* + *tūn*.

Narberth (*Arberth*) Pemb. *Nerberth* 1291, *la Nerbert* 1331. '(Place) near the hedge'. Welsh *yn* + *ar* + *perth*. The *N-* comes from Welsh *yn Arberth*, 'in Arberth'.

Narborough Leics. *Norburg* c.1220. 'North stronghold'. OE *north* + *burh*.

Narborough Norfolk. *Nereburh* 1086 (DB). Possibly 'stronghold near a narrow place or pass'. OE **neru* + *burh*. The river-name **Nar** is a back-formation from the place name.

Narraghmore (*An Fhorrach Mhór*) Kildare. 'The big area of land'.

Nart (*An Fheart*) Monaghan. 'Grave mound'.

Naseby Northants. *Navesberie* 1086 (DB). 'Stronghold of a man called Hnæf'. OE pers. name + *burh* (dative *byrig*), replaced by OScand. *bý* 'village' in the 12th cent.

Nash, 'place at the ash-tree', OE *æsc*, with initial *N-* from ME *atten* 'at the'; examples include: **Nash** Bucks. *Esse* 1231. **Nash** Herefs. *Nasche* 1239. **Nash** Shrops. *La Esse* 1242, *Nazsche* 1391.

Nassington Northants. *Nassingtona* 1017–34, *Nassintone* 1086 (DB). Probably 'farmstead on the promontory'. OE *næss* + *-ing* + *tūn*.

Nasty Herts. *Nasthey* 1294. '(Place at) the east enclosure'. OE *ēast* + *hæg*, with *N-* from ME *atten* 'at the'.

Nateby, 'farmstead or village where nettles grow, or of a man called **Nati*', OScand. **nata* or pers. name + *bý*: **Nateby** Cumbria. *Nateby* 1242. **Nateby** Lancs. *Nateby* 1204.

Nately, Up Hants. *Nataleie* 1086 (DB), *Upnateley* 1274. 'Wet wood or clearing'. OE **næt* + *lēah*. Affix is OE *upp* 'higher up'.

Natland Cumbria. *Natalund c.*1175. 'Grove where nettles grow, or of a man called *Nati'. OScand. *nata or pers. name + *lundr*.

Naughton Suffolk. *Nawelton c.*1150. Possibly 'farmstead of a man called Nagli'. OScand. pers. name + OE *tūn*.

Naul (*An Aill*) Dublin. 'The cliff'.

Naunton, 'new farmstead or estate', OE *nīwe* (dative *nīwan*) + *tūn*: **Naunton** Glos., near Winchcombe. *Niwetone* 1086 (DB). **Naunton** Worcs. *Newentone c.*1120. **Naunton Beauchamp** Worcs. *Niwantune* 972, *Newentune* 1086 (DB), *Newenton Beauchamp* 1370. Manorial affix from the *Beauchamp* family, here from the 11th cent.

Navan (*An Eamhain*) Armagh. Meaning uncertain. The name was first applied to Navan Fort, known in Irish as *Eamhain Mhacha*, said to mean 'Macha's brooch' (*eomhuin*, 'neckpin'), referring to the pin used by the goddess Macha to draw a circle around herself marking the outline of the future fort.

Navan (*An Uaimh*) Meath. 'The cave'.

Navenby Lincs. *Navenebi* 1086 (DB). 'Farmstead or village of a man called Nafni'. OScand. pers. name + *bý*.

Navestock Essex. *Nasingestok* 867, *Nassestoca* 1086 (DB). Probably 'outlying farmstead on the promontory'. OE *næss* + *-ing* + *stoc*. Alternatively 'outlying farmstead belonging to the promontory people or to NAZEING'.

Nawton N. Yorks. *Nagletune* 1086 (DB). 'Farmstead of a man called Nagli'. OScand. pers. name + OE *tūn*.

Nayland Suffolk. *Eilanda* 1086 (DB), *Neiland* 1227. '(Place at) the island'. OE *ēg-land*, with *N-* from ME *atten* 'at the'.

Nazeing Essex. *Nassingan* 1062, *Nasinga* 1086 (DB). '(Settlement of) the people of the promontory'. OE *næss* + *-ingas*.

Neagh, Lough (*Loch nEathach*) Antrim, Armagh, Derry, Down, Tyrone. *Loch nEchach c.*600. 'Eochu's lake'.

Neale (*An Éill*) Mayo. 'The flock of birds'.

Near Sawrey Cumbria. *See* SAWREY.

Neasden Gtr. London. *Neosdune c.*1000. 'Nose-shaped hill'. OE *neosu + dūn*.

Neasham Darltn. *Nesham* 1158. 'Homestead or enclosure by the projecting piece of land in a river-bend'. OE *neosu + hām* or *hamm*.

Neath (*Castell-Nedd*) Neat. *Neth* 1191, *Neeth* 1306. '(Place on the) River Neath'. The Celtic river name perhaps means 'shining one'. The Welsh name means 'castle on the Nedd' (Welsh *castell*).

Neatishead Norfolk. *Netheshird* 1020–22, *Snateshirda* [sic] 1086 (DB). Probably 'household of a retainer'. OE *genēat + hīred*.

Necton Norfolk. *Nechetuna* 1086 (DB). Probably 'farmstead by a neck of land'. OE *hnecca + tūn*.

Ned (*Nead*) Cavan, Derry, Fermanagh. 'Nest'.

Neddans (*Na Feadáin*) Tipperary. 'The streamlets'.

Nedeen (*Nadín*) Cork, Kerry. 'Little nest'.

Nedging Tye Suffolk. *Hnyddinge c.*995, *Niedinga* 1086 (DB). 'Place associated with a man called *Hnydda'. OE pers. name + *-ing*. Later affix is *tye* 'common pasture'.

Needham, 'poor or needy homestead', OE *nēd + hām*: **Needham** Norfolk. *Nedham* 1352. **Needham Market** Suffolk. *Nedham* 13th cent., *Nedeham Markett* 1511. **Needham Street** Suffolk. *Nedham c.*1185. Affix *street* here means 'hamlet'.

Needingworth Cambs. *Neddingewurda* 1161. 'Enclosure of the family or followers of a man called *Hnydda'. OE pers. name + *-inga-* + *worth*.

Needles, The I. of Wight. *Nedlen* 1333, *les Nedeles* 1409. From ME *nedl* 'needle' (plural *-en*, *-es*) with reference to the pointed shape of these chalk stacks.

Needwood Staffs. *Nedwode* 1248. 'Poor wood', or 'wood resorted to in need (as a refuge)'. OE *nēd + wudu*.

Neen Savage & Neen Sollars Shrops. *Nene* 1086 (DB), *Nenesauvage* 13th cent., *Nen Solers* 1274. Originally the Celtic or pre-Celtic name of the river here (an old name for the River Rea). Manorial affixes from the *le Savage* family here in the 13th cent., and from the *de Solers* family here in the late 12th cent.

Neenton Shrops. *Newenton* [sic] 1086 (DB), *Nenton* 1242. 'Farmstead or estate on the River Neen', OE *tūn*, *see* previous name. The

Domesday form suggests confusion of the first element with OE *nīwe* (dative *-an*) 'new'.

Nefyn Gwyd. *Newin* 1254, *Nefyn* 1291. Origin uncertain. The name may represent a pers. name.

Nelson Lancs. a 19th-cent. name for the new textile town, taken from the Lord Nelson Inn.

Nempnett Thrubwell B. & NE. Som. *Emnet c.*1200, *Trubewel* 1201. Originally two places. Nempnett is '(place at) the level ground' from OE *emnet* with N- from ME *atten* 'at the'. Thrubwell is from OE *wella* 'spring, stream' with an uncertain first element.

Nenagh (*An tAonach*) Tipperary. 'The assembly'.

Nendrum (*Naendruim*) Down. 'Nine ridges'.

Nene (river) Northants.–Lincs. *Nyn* 948. An ancient Celtic or pre-Celtic name, of obscure etymology.

Nenthead Cumbria. *Nentheade* 1631. 'Place at the source of the River Nent'. Celtic river-name (from *nant* 'a glen') + OE *hēafod*.

Nesbit Northum. near Doddington. *Nesebit* 1242. 'Promontory river-bend'. OE **neosu* + *byht*.

Ness, '(place at) the promontory or projecting ridge', OE *næss*: **Ness** Ches. *Nesse* 1086 (DB). **Ness, East & Ness, West** N. Yorks. *Ne(i)sse* 1086 (DB). **Ness, Great & Ness, Little** Shrops. *Nessham, Nesse* 1086 (DB).

Ness, Loch Highland. 'Loch of the River Ness'. *See* INVERNESS.

Neston Ches. *Nestone* 1086 (DB). 'Farmstead near the promontory'. OE *næss* + *tūn*.

Nether as affix. *See* main name, e.g. for **Nether Cerne** (Dorset) *see* CERNE.

Netheravon Wilts. *Nigravre* [*sic*] 1086 (DB), *Nederauena c.*1150. '(Settlement) lower down the River Avon'. OE *neotherra* + Celtic river-name (meaning simply 'river').

Netherbury Dorset. *Niderberie* 1086 (DB). 'Lower fortified place'. OE *neotherra* +*burh* (dative *byrig*).

Netherfield E. Sussex. *Nedrefelle* 1086 (DB). 'Open land infested with adders'. OE *nǣddre* + *feld*.

Netherhampton Wilts. *Notherhampton* 1242. 'Lower homestead'. OE *neotherra* + *hām-tūn*.

Netherseal & Overseal Derbys. *Scel(l)a* 1086 (DB), *Nether Scheyle, Overe Scheyle* 13th cent. 'Small wood or copse'. OE **scegel*. Distinguishing affixes are OE *neotherra* 'lower' and *uferra* 'upper'.

Netherthong Kirkl. *Thoying* 13th cent., *Nethyrthonge* 1448. 'Narrow strip of land'. OE *thwang*, with *neotherra* 'lower' to distinguish it from UPPERTHONG.

Netherton, a common name, 'lower farmstead or estate', OE *neotherra* + *tūn*; examples include: **Netherton** Northum. *Nedertun c.*1050. **Netherton** Worcs., near Evesham. *Neotheretun* 780, *Neotheretune* 1086 (DB).

Netley Hants. *Lætanlia* 955–8, *Latelei* 1086 (DB). Probably 'neglected clearing, or clearing left fallow'. OE **lǣte* + *lēah*, with change to initial N- only from the 14th cent. perhaps through confusion with the following name.

Netley Marsh Hants. *Natanleaga* late 9th cent., *Nutlei* [*sic*] 1086 (DB), *Nateleg* 1248. Probably 'wet wood or clearing'. OE **næt* + *lēah*, with the addition of *Marsh* from the mid-18th cent.

Nettlebed Oxon. *Nettlebed* 1246. 'Plot of ground overgrown with nettles'. OE *netele* + *bedd*.

Nettlecombe Dorset. *Netelcome* 1086 (DB). 'Valley where nettles grow'. OE *netele* + *cumb*.

Nettleden Herts. *Neteleydene c.*1200. 'Valley growing with nettles'. OE *netelig* + *denu*.

Nettleham Lincs. *Netelham* 1086 (DB). 'Homestead or enclosure where nettles grow'. OE *netele* + *hām* or *hamm*.

Nettlestead Kent. *Netelamstyde* 9th cent., *Nedestede* [*sic*] 1086 (DB). 'Homestead where nettles grow'. OE *netele* + *hāmstede*.

Nettlestone I. of Wight. *Hoteleston* [*sic*] 1086 (DB), *Nutelastone* 1248. 'Farmstead by the nut pasture or nut wood'. OE *hnutu* + *lǣs* or *lēah* (genitive *lēas*) + *tūn*.

Nettleton Lincs. *Neteltone* 1086 (DB). 'Farmstead where nettles grow'. OE *netele* + *tūn*.

Nettleton Wilts. *Netelingtone* 940, *Niteletone* 1086 (DB). 'Farmstead at the place overgrown with nettles'. OE *netele* + *-ing* + *tūn*.

Nevendon Essex. *Nezendena* [*sic*] 1086 (DB), *Neuendene* 1218. '(Place at) the level valley'. OE *efen* + *denu*, with *N-* from ME *atten* 'at the'.

Nevill Holt Leics. *See* HOLT.

Nevis, Ben. *See* BEN NEVIS.

New as affix. *See* main name, e.g. for **New Alresford** (Hants.) *see* ALRESFORD.

New Birmingham (*Gleann an Ghuail*) Tipperary. The site was exploited for coal in the 18th cent. and named after *Birmingham*, then in a coalmining region. The Irish name means 'valley of the coal'.

New Ferry Wirral. a district named from a ferry across the River MERSEY established here in the 18th cent.

New Forest Hants. *Nova Foresta* 1086 (DB). 'The new forest' created by William the Conqueror in the 11th cent. for hunting game.

New Galloway Dumf. The town dates from 1629, when a royal charter was granted by Charles I to Sir John Gordon, who chose a name to mark his family ties with GALLOWAY.

New Inn Cavan. The name was originally that of a stagecoach inn.

New Invention Wsall. first on record in 1663, of doubtful origin but possibly referring to some contrivance used in the clay-mining industry, or perhaps derived from an inn here so called. Origin as an inn-name is probably more likely for the identical **New Invention** Shrops.

New Mills Derbys. *New Miln* 1625. Self-explanatory, although the plural form is recent. Earlier called *Midelcauel* 1306, 'the middle allotment of land', from ME *cauel*.

New Quay (*Ceinewydd*) Cergn. 'New quay'. A new quay was built here in 1835 to replace an older and smaller harbour. The Welsh name means the same as the English (Welsh *cei* + *newydd*).

New Radnor (*Maesyfed*) Powys. *Raddrenoue* 1086 (DB), *Radenoura* 1191, *Radnore* 1201, *New Radenore* 1298. 'New Radnor'. The original *Radnor*, 'red bank' (OE *rēad* + *ōra*), was the present *Old Radnor*, called in Welsh *Pencraig*,

'head of the rock' (Welsh *pen* + *craig*). In 1064 the present Radnor was granted rights as the new centre of the region. Its Welsh name means 'Hyfaidd's field' (Welsh *maes*).

New Tredegar Cphy. The town arose in the 19th cent. around a colliery which was regarded as 'new' by comparison with the ironworks at TREDEGAR.

New York Lincs. Recorded thus from 1824, a transferred name from the American city (perhaps prompted by the proximity of BOSTON some eight miles to the south.). There are examples of the same name in Arg. (**Newyork**) and N. Tyne.

Newark, 'new fortification or building', OE *nīwe* + *weorc*: **Newark** Peterb. *Nieuyrk* 1189. **Newark on Trent** Notts. *Niweweorce* c.1080, *Neuuerche* 1086 (DB). For the river-name, *see* TRENTHAM.

Newbald, North E. R. Yorks. *Neoweboldan* 972, *Niuuebold* 1086 (DB). 'New building'. OE *nīwe* + *bold*.

Newbiggin, a common name in the North of England, 'new building or house', OE *nīwe* + ME *bigging*; examples include: **Newbiggin** Cumbria, near Appleby. *Neubigging* 1179. **Newbiggin** Durham, near Middleton. *Neubigging* 1316. **Newbiggin by the Sea** Northum. *Niwebiginga* 1187.

Newbold, a common name in the Midlands, 'new building', OE *nīwe* + *bold*: examples include: **Newbold** Derbys. *Newebold* 1086 (DB). **Newbold** Leics. *Neoboldia* 12th cent. **Newbold on Avon** Warwicks. *Neobaldo* 1077, *Newebold* 1086 (DB). Avon is a Celtic river-name meaning simply 'river'. **Newbold on Stour** Warwicks. *Nioweboldan* 991. Stour is a Celtic or OE river-name probably meaning 'the strong one'. **Newbold Pacey** Warwicks. *Niwebold* 1086 (DB), *Neubold Pacy* 1235. Manorial affix from the *de Pasci* family, here in the 13th cent.

Newborough Staffs. *Neuboreg* 1280. 'New market town or borough'. OE *nīwe* + *burh*. The borough was probably founded in 1263.

Newborough (*Niwbwrch*) Angl. *Novus Burgus* 1305, *Neuburgh* 1324, *Newborough* 1379. 'New borough'. OE *nīwe* + *burh*. The 'new borough' was established in the late 13th cent. by Edward I.

Newbottle, 'new building', OE *nīwe + bōthl:* **Newbottle** Northants. *Niwebotle* 1086 (DB). **Newbottle** Sundld. *Neubotl* 1196.

Newbourne Suffolk. *Neubrunna* 1086 (DB). Possibly 'new stream', i.e. 'stream which has changed course', OE *nīwe + burna.* Alternatively perhaps '(place with) seven springs', OScand. *níu + brunnr.*

Newbrough Northum. *Nieweburc* 1203. 'New fortification'. OE *nīwe + burh.*

Newburgh Lancs. *Neweburgh* 1431. 'New market town or borough'. OE *nīwe + burh.*

Newburgh Fife. *Niwanbyrig* c.1130, *Novus burgus* 1266. 'New borough'. OE *nīwe + burh.* The town dates from at least the 12th cent.

Newburn Newc. upon T. *Neuburna* 1121-9. 'New stream', i.e. 'stream which has changed course'. OE *nīwe + burna.*

Newbury W. Berks. *Neuberie* c.1080. 'New market town or borough'. OE *nīwe + burh* (dative *byrig*). The original name of the manor was *Ulvritune* 1086 (DB), 'estate associated with a man called Wulfhere', OE pers. name + *-ing- + tūn.*

Newby, a common name in the North of England, 'new farmstead or village', OE *nīwe + OScand. bý:* examples include: **Newby** Cumbria, near Appleby. *Neubi* 12th cent. **Newby East** Cumbria. *Neuby* c.1190. **Newby West** Cumbria. *Neuby* c.1200. **Newby Wiske** N. Yorks. *Neuby* 1157. Affix from the River Wiske, *see* APPLETON WISKE.

Newcastle, 'new castle', OE *nīwe + castel* (often in Latin in early spellings): **Newcastle** Shrops. *Novum castrum* 1284. **Newcastle under Lyme** Staffs. *Novum castellum subtus Lymam* 1173. *Lyme* is an old district name, possibly 'escarpment', *see* BURSLEM. **Newcastle upon Tyne** Newc. upon T. *Novum Castellum* 1130. For the river-name, *see* TYNEMOUTH.

Newcastle (*An Caisleán Nua*) Down. (*ag fersait*) an chaisléin nui *1433. 'The new castle'.*

Newcastle Emlyn (*Castelnewydd Emlyn*) Carm. *Novum castrum de Emlyn* c.1240, *Emlyn with New Castle* 1257, *Newcastle Emlyn* 1295. 'New castle in Emlyn'. OE *nīwe + castel.* The cantref name *Emlyn* means 'around the valley' (Welsh *am + glyn*).

Newchurch, 'new church', OE *nīwe + cirice:* **Newchurch** I. of Wight. *Niechirche* 12th cent. **Newchurch** Kent. *Nevcerce* 1086 (DB).

Newdigate Surrey. *Niudegate* c.1167. 'Gate by the new wood'. OE *nīwe + wudu + geat.*

Newenden Kent. *Newedene* 1086 (DB). 'New woodland pasture'. OE *nīwe* (dative *-an*) + *denn.*

Newent Glos. *Noent* 1086 (DB). Probably a Celtic name meaning 'new place'.

Newhall, 'new hall or manor house', OE *nīwe + hall:* **Newhall** Ches. *La Nouehall* 1252. **Newhall** Derbys. *Niewehal* 1197.

Newham Gtr. London. a recent name created in 1965 for the new London borough comprising EAST HAM & WEST HAM.

Newham Northum. near Bamburgh. *Neuham* 1242 'New homestead or enclosure'. OE *nīwe + hām* or *hamm.*

Newhaven E. Sussex. *Newehaven* 1587. 'Newly built harbour', from ME *haven* (OE *hæfen*). Its old name was *Mechingas* 1121, *Mecinges* 1204, '(settlement of) the family or followers of a man called **Mēce*', from OE pers. name + *-ingas.*

Newhaven Edin. 'Newly built harbour'. Newhaven dates from *c.*1505.

Newholm N. Yorks. *Neueham* 1086 (DB). 'New homestead or enclosure'. OE *nīwe + hām* or *hamm.*

Newick E. Sussex. *Niwicha* 1121. 'New dwelling or farm'. OE *nīwe + wīc.*

Newington, 'the new farmstead or estate', OE *nīwe* (dative *nīwan*) + *tūn:* **Newington** Kent, near Hythe. *Neventone* 1086 (DB). **Newington** Kent, near Sittingbourne. *Newetone* 1086 (DB). **Newington** Oxon. *Niwantun* c.1045, *Nevtone* 1086 (DB). **Newington, North** Oxon. *Newinton* 1200. **Newington, South** Oxon. *Niwetone* 1086 (DB). **Newington, Stoke** Gtr. London. *Neutone* 1086 (DB), *Neweton Stoken* 1274. Affix means 'by the tree-stumps' or 'made of logs', from OE *stoccen.*

Newland, Newlands, 'new arable land', OE *nīwe + land:* **Newland** Glos. *Nova terra* 1221, *Neweland* 1248. **Newland** N. Yorks. *Newland* 1234. **Newland** Worcs. *La Newelande* 1221. **Newlands** Northum. *Neuland* 1343.

Newlyn Cornwall. *Nulyn* 1279, *Lulyn* 1290. Probably 'pool for a fleet of boats'. Cornish *lu* + *lynn*.

Newlyn East Cornwall. 'Church of *Sancta Niwelina*' 1259. From the patron saint of the church.

Newmarket Suffolk. *Novum Forum* 1200, *la Newmarket* 1418. 'New market town'. ME *market* (rendered by Latin *forum* in the earliest form).

Newmills Tyrone. *New Mills* 1837. The reference is to a former corn mill.

Newnham, a fairly common name, 'the new homestead or enclosure', OE *nīwe* (dative *-an*) + *hām* or *hamm*; examples include: **Newnham** Glos. *Nevneham* 1086 (DB). **Newnham** Herts. *Neuham* 1086 (DB). **Newnham** Kent. *Newenham* 1177. **Newnham** Northants. *Niwanham* 1021–3.

Newnton, North Wilts. *Northniwetune* 892, *Newetone* 1086 (DB). 'New farmstead'. OE *nīwe* (dative *nīwan*) + *tūn*.

Newport, 'new market town', OE *nīwe* + *port*: **Newport** Devon. *Neuport* 1295. **Newport** Essex. *Neuport* 1086 (DB). **Newport** I. of Wight. *Neweport* 1202. **Newport** Tel. & Wrek. *Novus Burgus* 12th cent., *Newport* 1237. **Newport Pagnell** Milt. K. *Neuport* 1086 (DB), *Neuport Paynelle* 1220. Manorial affix from the *Paynel* family, here in the 12th cent.

Newport (*Casnewydd-ar-Wysg*) Newpt. *Novus Burgus* 1138, *Nova Villa* 1290, *Neuborh* 1291. 'New town'. OE *nīwe* + *port*. The 12th-cent. castle here was regarded as 'new' by comparison with the Roman fort at Caerleon. The Welsh name means 'new castle on the River Usk' (Welsh *cas* + *newydd* + *ar*). *See* USK.

Newport (*Trefdraeth*) Pemb. *Nuport* 1282, *Newburgh* 1296, *Novus Burgus* 1316. 'New town'. OE *nīwe* + *port*. The original town with market rights here may have been regarded as 'new' by comparison with Fishguard. The Welsh name means 'town by the shore' (Welsh *tref* + *traeth*).

Newport-on-Tay Fife. 'New port on the River Tay'. OE *nīwe* + *port*. For the river-name, *see* TAYPORT.

Newquay Cornwall. *Newe Kaye* 1602. Named from the new quay (ME *key*) built here in the 15th cent.

Newry (*An tIúr*) Down. *Iobhar Chind Trachta* 1089. 'The yew-tree'. An earlier name was *Iúr Cinn Trá*, 'yew-tree at the head of the strand'.

Newsham, Newsholme, a fairly common name in the North of England, '(place at) the new houses', OE *nīwe* + *hūs* in a dative plural form *nīwum hūsum*; examples include: **Newsham** Northum. *Neuhusum* 1207. **Newsham** N. Yorks., near Ravensworth. *Neuhuson* 1086 (DB). **Newsholme** E. R. Yorks. near Howden. *Neuhusam* 1086 (DB). **Newsholme** Lancs., near Barnoldswick. *Neuhuse* 1086 (DB).

Newstead Northum. *Novus Locus* 13th cent., *Newstede* 1339. 'New farmstead'. OE *nīwe* + *stede*.

Newstead Notts. *Novus Locus* 12th cent., *Newstede* 1302. 'New monastic site'. OE *nīwe* + *stede*.

Newstead Sc. Bord. *Novo Loco* 1189. 'New place'. OE *nīwe* + *stede*.

Newthorpe N. Yorks. *Niwan-thorp* c.1030. 'New outlying farm'. OE *nīwe* + OScand. *thorp*.

Newton, a very common name, 'the new farmstead, estate, or village', OE *nīwe* + *tūn*; examples include: **Newton** Suffolk. *Niwetuna*, *Neuuetona* 1086 (DB). **Newton Abbot** Devon. *Nyweton* c.1200, *Nyweton Abbatis* 1270. The Latin affix meaning 'of the abbot' refers to its possession by Torre Abbey in the 12th cent. **Newton Arlosh** Cumbria. *Arlosk* 1185, *Neutonarlosk* 1345. Arlosh is probably a Celtic name meaning 'burnt place'. **Newton Blossomville** Milt. K. *Niwetone* 1175, *Newenton Blosmevill* 1254. Manorial affix from the *de Blosseville* family, here in the 13th cent. **Newton Burgoland** Leics. *Neutone* 1086 (DB), *Neuton Burgilon* 1390. Manorial affix from early possession of lands here by the *Burgilon* family. **Newton Ferrers** Devon. *Niwetone* 1086 (DB), *Neweton Ferers* 1303. Manorial affix from the *de Ferers* family, here in the 13th cent. **Newton Flotman** Norfolk. *Niwetuna* 1086 (DB), *Neuton Floteman* 1291. Manorial affix from the ME surname *Floteman*, itself derived from OE *flotmann* 'sailor, pirate'. **Newton-le-Willows** St Hel. *Neweton* 1086 (DB). Affix means 'by the willow-trees'. **Newton Longville** Bucks. *Nevtone* 1086 (DB), *Newenton Longevile* 1254. Affix from its possession by the church of St Faith of Longueville in France from the mid-12th cent. **Newton, Maiden** Dorset. *Newetone* 1086

(DB), *Maydene Neweton* 1288. Addition means 'of the maidens', perhaps referring to early possession of the manor by nuns. **Newton, North** Somerset. *Newetune* 1086 (DB). **Newton, Old** Suffolk. *Neweton* 1196, *Eldneuton* 1418. With ME *elde* (OE *eald*) 'old'. **Newton Poppleford** Devon. *Poplesford* 1226, *Neweton Popilford* 1305. Poppleford means 'pebble ford', OE **popel* + *ford*. **Newton St Cyres** Devon. *Niwantune* c.1070, *Nywetone Sancti Ciricii* 1330. Affix from the dedication of the church to St Ciricius. **Newton under Roseberry** Red. & Cleve. *Newetun, Nietona* 1086 (DB), *Newetunie sub Ohtnebercg* mid-12th cent. Affix from its situation under the prominent hill known as ROSEBERRY TOPPING.

In contrast to the above 'new settlements' of medieval origin, the following makes use of the old name type for a modern development: **Newton Aycliffe** Durham, a recent 'new town' named from AYCLIFFE.

Newton Mearns E. Renf. *Newtoun de Mernis* 1609. 'New town in Mearns'. *See* MEARNS.

Newton Stewart Dumf. 'New village of Stewart'. ME *newe* + *toun*. The village was laid out in 1677 by William *Stewart*, son of the 2nd Earl of Galloway.

Newtongrange Midloth. 'Grange of the new estate'. ME *newe* + *toun* + *grange*. The grange was so named for distinction from the one at *Prestongrange*, near PRESTONPANS.

Newtown I. of Wight. *Niwetune* c.1200, *nouum burgum de Francheuile* 1254. 'New town or borough', OE *nīwe* + *tūn*. The alternative name *Francheville*, in use up to the 16th cent., is 'free town', OFrench *franche* + *ville*.

Newtown (*Y Drenewydd*) Powys. *Newentone* 1250, *the Newtown* 1360. 'New village'. OE *nīwe* + *tūn*. The original 'new town' was designated a New Town in 1967. The Welsh name means 'the new town' (Welsh *y* + *tref* + *newydd*).

Newtown Crommelin Antrim. *Newtown Cromlin* 1833. The village was founded in 1824 by the Huguenot Nicholas *Crommelin*.

Newtown Cunningham Donegal. *Newtowne Cunningham* c.1655. Land here was granted in 1610 to John *Cunningham*, whose family also gave the name of MANORCUNNINGHAM.

Newtown Forbes Longford. The name comes from the Scotsman Sir Arthur *Forbes*, who settled here in the early 17th cent.

Newtown Linford Leics. *Neuton* 1325, *Neuton Lynforthe* 1446. 'New village', ME *newe* + *toun*. Affix from a place name *Lyndeford* 1293, 'ford where lime-trees grow', OE *linden* + *ford*.

Newtownabbey Antrim. The town was created in 1958 out of seven existing villages, one of which, *Whiteabbey*, gave the basis of the name.

Newtownards (*Baile Nua na hAirde*) Down. *New Town of Blathewyc* 1333. 'New town of the promontory'. The current name refers to the ARDS PENINSULA. The original name was *New Town of Blathewic*, from *Uí Bhlathmhaic*, 'descendants of Blathmhac', a personal name meaning 'famous son'.

Newtownbreda (*Baile Nua na Bréadaí*) Down. *Newtownbreda* 1832. 'New town of Breda'. The townland name *Breda* represents Irish *Bréadach*, 'broken land'.

Newtownbutler Fermanagh. *Neetowne* 1622. The *Butler* family settled in Ireland in the 17th cent., and the village of *Newtown* took its present name when Theophilus *Butler* was created Baron of Newtownbutler in 1715.

Newtownhamilton Armagh. *Newtown Hamilton* 1792. The village was founded in c.1770 by Alexander *Hamilton*, a descendant of the John Hamilton who founded HAMILTONSBAWN.

Newtownmountkennedy (*Baile an Chinnéidigh*) Wicklow. 'Kennedy's town'. Sir Robert *Kennedy* was granted the manor here in the second half of the 17th cent.

Newtownstewart Tyrone. Sir William *Stewart* was granted extensive lands in Tyrone and Donegal in the early 17th cent.

Neyland Pemb. *Nailand* 1596, *New Milford, or Neyland* 1896. '(Place at the) island'. OE *ēg-land. N-* is from preceding ME *atten*, 'at the'. Neyland was formerly also known as *New Milford*, by contrast with MILFORD HAVEN.

Nibley, 'woodland clearing near the peak', OE **hnybba* + *lēah*: **Nibley** S. Glos. *Nubbelee* 1189. **Nibley, North** Glos. *Hnibbanlege* 940.

Nidd N. Yorks. *Nith* 1086 (DB). Named from the River Nidd, a Celtic or pre-Celtic river-name of uncertain etymology.

Nier (*An Uidhír*) (river) Waterford. 'The dun-coloured one'.

Nigg Highland. *Nig* 1257. '(Place by) the notch'. Gaelic *eig*. *N-* is from preceding Gaelic *an*, 'the'.

Ninch (*An Inse*) Meath. 'The island'.

Ninebanks Northum. *Ninebenkes* 1228. 'The nine banks or hills'. OE *nigon* + *benc*.

Ninfield E. Sussex. *Nerewelle* [*sic*] 1086 (DB), *Nimenefeld* 1255. 'Newly taken-in open land'. OE *nīwe* + *numen* + *feld*.

Ningwood I. of Wight. *Lenimcode* [*sic*] 1086 (DB), *Ningewode* 1189. 'The enclosed wood, or the wood taken into cultivation'. OE **niming* + *wudu*.

Nithsdale (valley) Dumf. *Stranit* *c.*1124, *Stranud* 1181, *Niddesdale* 1256. 'Valley of the River Nith'. OScand. *dalr*. The river-name means 'new one' (British **nouio-*). The first two forms above have Gaelic *srath*, 'valley'.

Niton I. of Wight. *Neeton* 1086 (DB). 'New farmstead'. OE *nīwe* + *tūn*.

Niwbwrch. See NEWBOROUGH.

No Man's Heath, 'heath no one owns', ME *noman* + *hethe*: **No Man's Heath** Ches. *Nomonheth* 1483. **No Man's Heath** Warwicks. Recorded thus 1835, referring to its situation on the county boundary.

Noah's Ark Kent. a late name appearing on the first Ordnance Survey map of 1819 and apparently transferred from a house so called built *c.*1700. Other examples of this fanciful name, perhaps alluding to an abundance or diversity of animals, occur in Cheshire and Derbyshire.

Nobber (*An Obair*) Meath. 'The construction'.

Nobottle Northants. *Neubote* [*sic*] 1086 (DB), *Neubottle* 12th cent. 'New building'. OE *nīwe* + *bōthl*.

Nocton Lincs. *Nochetune* 1086 (DB). Probably 'farmstead where wether sheep are kept'. OE *hnoc* + *tūn*.

Noke Oxon. *Ac(h)am* 1086 (DB), *Noke* 1382. '(Place at) the oak-tree'. OE *āc*, with *N-* from ME *atten* 'at the'.

Nolton Pemb. *Noldeton* 1317, *Nolton* 1403. 'Old farm'. OE *ald* + *tūn*. The *N-* comes from ME *atten*, 'at the'.

Nomansland, 'land no one owns', usually land on a boundary, ME *noman* + *land*: **Nomansland** Devon. Recorded thus in 1809. **Nomansland** Wilts. First recorded 1811.

Noneley Shrops. *Nuneleg* 1221. 'Woodland clearing of a man called Nunna'. OE pers. name + *lēah*.

Nonington Kent. *Nunningitun* *c.*1100. 'Estate associated with a man called Nunna'. OE pers. name + *-ing-* + *tūn*.

Norbiton Gtr. London. *Norberton* 1205. 'Northern grange or outlying farm'. OE *north* + *bere-tūn*. *See* SURBITON.

Norbury, 'northern stronghold or manor-house', OE *north* + *burh* (dative *byrig*): **Norbury** Ches., near Marbury. *Norberie* 1086 (DB). **Norbury** Derbys. *Nortberie* 1086 (DB). **Norbury** Gtr. London. *Northbury* 1359. **Norbury** Shrops. *Norbir* 1237. **Norbury** Staffs. *Nortberie* 1086 (DB).

Nore (*An Fheoir*) (river) Laois, Tipperary. 'The Feoir'.

Norfolk (the county). *Nordfolc* 1086 (DB). '(Territory of) the northern people (of the East Angles)'. OE *north* + *folc*, *see* SUFFOLK.

Norham Northum. *Northham* *c.*1050. 'Northern homestead or enclosure'. OE *north* + *hām* or *hamm*.

Norley Ches. *Northleg* 1239. 'Northern glade or clearing'. OE *north* + *lēah*.

Normanby, 'farmstead or village of the Northmen or Norwegian Vikings', OE *Northman* + OScand. *bý*: **Normanby** Lincs. *Normanebi* 1086 (DB). **Normanby** N. Lincs. *Normanebi* 1086 (DB). **Normanby** N. Yorks. *Normanebi* 1086 (DB). **Normanby** Red. & Cleve. *Northmannabi* *c.*1050. **Normanby le Wold** Lincs. *Normane(s)bi* 1086 (DB). Affix means 'on the wolds', from OE *wald* 'high forest-land'.

Normandy Surrey. first on record as *Normandie* 1656, apparently taking its name from an inn here called 'The Duke of Normandy'.

Normanton, a fairly common name in the North and North Midlands, 'farmstead of the

Northmen or Norwegian Vikings', OE
Northman + tūn; examples include:
Normanton Derby. *Normantun* 1086 (DB).
Normanton Wakefd. *Normantone* 1086 (DB).
Normanton, South Derbys. *Normentune*
1086 (DB). **Normanton, Temple** Derbys.
Normantune 1086 (DB), *Normanton Templer*
1330. Manorial affix from its possession by the
Knights Templars in the late 12th cent.

Norrington Common Wilts. *Northinton*
1227. '(Place lying) north in the village'. OE
north + in + tūn.

North as affix. *See* main name, e.g. for **North
Anston** (Rothm.) *see* ANSTON.

North Berwick E. Loth. *Beruvik c.*1225,
Northberwyk 1250. 'Northern barley farm'. OE
bere-wīc. The town is 'north' in relation to
BERWICK-UPON-TWEED, Northum.

North Minch. *See* MINCH.

North Ronaldsay (island) Orkn. *Rinansey*,
Rinarsey 13th cent. 'Northern Ringan's island'.
OScand. *ey. Ringan* is said to be a form of the
name of St Ninian (or Finian), while *North*
distinguishes this island from SOUTH
RONALDSAY, where the pers. name is different.

North Uist. *See* UIST.

Northallerton N. Yorks. *Aluretune* 1086
(DB), *North Alverton* 1293. 'Farmstead of a man
called Ælfhere'. OE pers. name + *tūn*, with the
affix *north* from the 13th cent.

Northam, 'northern enclosure or
promontory', OE *north + hamm*: **Northam**
Devon. *Northam* 1086 (DB). **Northam** Sotn.
Northam 1086 (DB).

Northampton Northants. *Hamtun* early
10th cent., *Northantone* 1086 (DB). 'Home farm,
homestead'. OE *hām-tūn*, with prefix *North* to
distinguish this place from SOUTHAMPTON
(which has a different origin).
Northamptonshire (OE *scīr* 'district') is first
referred to in the 11th cent.

Northaw Herts. *North Haga* 11th cent.
'Northern enclosure'. OE *north + haga*.

Northborough Peterb. *Northburh* 12th cent.
'Northern fortification'. OE *north + burh*.

Northbourne Kent. *Nortburne* 618,
Norborne 1086 (DB). 'Northern stream'. OE
north + burna.

Northchurch Herts. *Ecclesia de North
Berchamstede* 1254, *le Northcherche* 1347.
Self-explanatory, 'the north church' of
BERKHAMSTED, from OE *north + cirice*.

Northenden Manch. *Norwordine* 1086 (DB).
'Northern enclosure'. OE *north + worthign*.

Northfield Birm. *Nordfeld* 1086 (DB), 'Open
land lying to the north' (relative to King's
Norton). OE *north + feld*.

Northfleet Kent. *Flyote* 10th cent., *Norfluet*
1086 (DB). '(Place at) the stream'. OE *flēot* with
the addition of *north* to distinguish it from
SOUTHFLEET.

Northiam E. Sussex. *Hiham* 1086 (DB),
*Nordhyam c.*1200. Probably 'promontory where
hay is grown'. OE *hīg + hamm*, with the later
addition of *north*. Alternatively the original first
element may be OE *hēah* 'high'.

Northill Beds. *Nortgiuele* 1086 (DB).
'Northern settlement of the tribe called *Gifle*',
see SOUTHILL. OE *north* + tribal name derived
from the River lvel, a Celtic river-name meaning
'forked stream'.

Northington Hants. *Northametone* 903.
'Farmstead of the dwellers to the north' (of
Alresford or Winchester). OE *north + hǣme +
tūn*.

Northleach Glos. *Lecce* 1086 (DB),
Northlecche 1200. 'Northern estate on the River
Leach'. OE *north* + river-name (*see* EASTLEACH).

Northleigh Devon. *Lege* 1086 (DB),
Northleghe 1291. '(Place at) the wood or
clearing'. OE *lēah*, with *north* to distinguish it
from SOUTHLEIGH.

Northlew Devon. *Leuia* 1086 (DB), *Northlyu*
1282. Named from the River Lew, a Celtic
river-name meaning 'bright stream'.

Northmoor Oxon. *More* 1059, *Northmore*
1367. 'The marsh'. OE *mōr* with the later
addition of *north*.

Northolt Gtr. London. *Northhealum* 960,
Northala 1086 (DB). 'Northern nook(s) of land'.
OE *north + halh*, *see* SOUTHALL.

Northowram Calder. *Ufrun* 1086 (DB),
Northuuerum 1202. '(Place at) the ridges'. OE
*ofer, *ufer in a dative plural form **uferum*, with
north to distinguish it from SOUTHOWRAM.

Northrepps Norfolk. *Norrepes* 1086 (DB), *Nordrepples* 1185. Probably 'north strips of land'. OE *north* + **reopul*.

Northumberland (the county). *Norhumberland* 1130. 'Territory of the *Northhymbre* (i.e. those living north of the River Humber)'. OE tribal name + *land*. In Anglo-Saxon times the territory of the tribe and kingdom of the *Northhymbre* was much larger than the present county.

Northwich Ches. *Wich, Norwich* 1086 (DB). 'North salt-works'. OE *north* + *wīc*.

Northwick S. Glos. *Northwican* 955–9. 'Northern (dairy) farm'. OE *north* + *wīc*.

Northwold Norfolk. *Northuuold* 970, *Nortwalde* 1086 (DB). 'Northern forest'. OE *north* + *wald*.

Northwood, 'northern wood', OE *north* + *wudu*: **Northwood** Gtr. London. *Northwode* 1435. **Northwood** I. of Wight. *Northewde* early 13th cent.

Norton, a very common name, 'north farmstead or village', i.e. one to the north of another settlement, OE *north* + *tūn*; examples include: **Norton** N. Yorks. *Nortone* 1086 (DB). **Norton** Suffolk. *Nortuna* 1086 (DB). **Norton** Worcs., near Evesham. *Nortona* 709, *Nortune* 1086 (DB). **Norton Bavant** Wilts. *Nortone* 1086 (DB), *Nortonbavent* 1381. Manorial affix from the *de Bavent* family, here in the 14th cent. **Norton, Blo** Norfolk. *Nortuna* 1086 (DB), *Blonorton* 1291. Affix is probably ME *blo* 'bleak, exposed'. **Norton, Bredon's** Worcs. *Nortune* 1086 (DB). Affix from nearby Bredon Hill, *see* BREDON. **Norton, Brize** Oxon. *Nortone* 1086 (DB), *Northone Brun c.*1266. Manorial affix from a William *le Brun* who had land here in 1200. **Norton Canon** Herefs. *Nortune* 1086 (DB), *Norton Canons* 1327. Affix from its early possession by the canons of Hereford Cathedral. **Norton, Chipping & Norton, Over** Oxon. *Nortone* 1086 (DB), *Chepingnorthona* 1224, *Overenorton* 1302. Distinguishing affixes are OE *cēping* 'market' and *uferra* 'higher'. **Norton Disney** Lincs. *Nortune* 1086 (DB), Norton *Isny* 1331. Manorial affix from the *de Isney* family, here in the 12th cent. **Norton Fitzwarren** Somerset. *Nortone* 1086 (DB). Manorial affix from lands here held by a family called *Fitzwarren*. **Norton, Greens** Northants. *Nortone* 1086 (DB), *Grenesnorton* 1465. Manorial affix from the *Grene* family, here in the 14th cent. **Norton in Hales** Shrops. *Nortune*

1086 (DB), *Norton in Hales* 1291. Affix refers to the district called *Hales*, a name surviving also in HALES Stoke. **Norton-in-the-Moors** Stoke. *Nortone* 1086 (DB), *Norton super le Mores* 1285. Affix is from OE *mōr* 'moor, marshy ground'. **Norton, King's** Birm. *Nortune* 1086 (DB), *Kinges Norton* 1221. A royal manor at the time of Domesday Book. **Norton, King's** Leics. *Nortone* 1086 (DB), *Kynges Norton* 1189–99. Described in Domesday Book as parcel of the royal demesne appandant to the manor of GREAT BOWDEN. **Norton, Midsomer** B. & NE. Som. *Midsomeres Norton* 1248. Affix from the festival of St John the Baptist, patron saint of the church, on Midsummer Day. **Norton St Philip** Somerset. *Nortune* 1086 (DB), *Norton Sancti Phillipi* 1316. Affix from the dedication of the church. **Norton sub Hamdon** Somerset. *Nortone* 1086 (DB), *Norton under Hamedon* 1246. Affix refers to Hamdon Hill, *see* STOKE SUB HAMDON.

Norton, Hook Oxon. *See* HOOK.

Norwell Notts. *Nortwelle* 1086 (DB). 'North spring or stream'. OE *north* + *wella*.

Norwich Norfolk. *Northwic* 10th cent., *Noruic* 1086 (DB). 'North harbour or trading centre'. OE *north* + *wīc*.

Norwick Shet. (Unst). 'Northern inlet'. OScand. *north* + *vík*.

Norwood, 'north wood', OE *north* + *wudu*: **Norwood Green** Gtr. London. *Northuuda* 832. **Norwood Hill** Surrey. *Norwode* 1250. **Norwood, South & Norwood, West** Gtr. London. *Norwude* 1176.

Noseley Leics. *Noveslei* 1086 (DB). 'Woodland clearing of a man called Nōthwulf'. OE pers. name + *lēah*.

Noss Head Highland. 'Promontory'. OScand. *nós*.

Noss Mayo Devon. *Nesse Matheu* 1286. 'Matheu's headland'. OE *næss* with OFrench pers. name.

Nosterfield N. Yorks. *Nostrefeld* 1204. Probably 'open land with a sheep-fold or by a hillock'. OE *eowestre* or **ōster* + *feld*, with *N-* from ME *atten* 'at the'.

Notgrove Glos. *Natangrafum* 716–43, *Nategraua* 1086 (DB). 'Wet grove or copse'. OE **næt* + *grāf*.

Notley, Black & Notley, White Essex. *Hnutlea* 998, *Nutlea* 1086 (DB), *Blake Nuteleye* 1252, *White Nuteleye* 1235. 'Wood or clearing where nut-trees grow'. OE *hnutu* + *lēah*. Distinguishing affixes *blæc* and *hwīt* may refer to soil colour or vegetation.

Notting Hill Gtr. London. *Knottynghull* 1356. ME *hull* 'hill' may have been added to an earlier *Knottyng* which could be an old hill name (OE *cnotta* 'knot, lump' + *-ing*), or '*Cnotta's place' (OE masculine pers. name + *-ing*). Alternatively *Knottyng* may be a ME surname from KNOTTING (Beds.).

Nottingham Nott. *Snotengaham* late 9th cent., *Snotingeham* 1086 (DB). 'Homestead of the family or followers of a man called Snot'. OE pers. name + *-inga-* + *hām*, with loss of S- in the 12th cent. due to Norman influence.
Nottinghamshire (OE *scīr* 'district') is first referred to in the 11th cent.

Nottington Dorset. *Notinton* 1212. 'Estate associated with a man called *Hnotta'. OE pers. name + *-ing-* + *tūn*.

Notton Wakefd. *Notone* 1086 (DB). 'Farmstead where wether sheep are kept'. OE *hnoc* + *tūn*.

Notton Wilts. *Natton* 1232. 'Cattle farm'. OE *nēat* + *tūn*.

Noughaval (*Nuachabhail*) Cork. 'New foundation'.

Nuffield Oxon. *Tofeld* c.1181. Probably 'tough open land'. OE *tōh* + *feld*, with change to initial *N-* only from the 14th cent.

Nun Monkton N. Yorks. *See* MONKTON.

Nunburnholme E. R. Yorks. *Brunha'* 1086 (DB), *Nonnebrynholme* 1530. '(Place at) the springs or streams'. OScand. *brunnr* in a dative plural form *brunnum*, with later affix from the Benedictine nunnery here.

Nuneaton Warwicks. *Etone* 1086 (DB), *Nonne Eton* 1247. 'Farmstead by a river'. OE *ēa* + *tūn*, with later affix from the Benedictine nunnery founded here in the 12th cent.

Nuneham Courtenay Oxon. *Neuham* 1086 (DB), *Newenham Courteneye* 1320. 'New homestead or village'. OE *nīwe* (dative *nīwan*) + *hām*. Manorial affix from the *Curtenay* family, here in the 13th cent.

Nunney Somerset. *Nuni* 954, *Nonin* [sic] 1086 (DB). Probably 'island, or dry ground in marsh, of a man called Nunna', OE pers. name + *ēg*. Alternatively the first element may be OE *nunne* 'a nun'.

Nunnington N. Yorks. *Noningtune* 1086 (DB). 'Estate associated with a man called Nunna'. OE pers. name + *-ing-* + *tūn*.

Nunthorpe Middlesbr. *Torp* 1086 (DB), *Nunnethorp* 1240. 'Outlying farmstead or hamlet of the nuns'. OScand. *thorp*, with later affix from the medieval nunnery here.

Nunton Wilts. *Nunton(a)* 1209. Probably 'farmstead or estate of a man called Nunna'. OE pers. name + *tūn*.

Nure (*An Iubhar*) Westmeath. 'The yew-tree'.

Nurney (*An Urnaí*) Carlow, Kildare. 'The oratory'.

Nursling Hants. *Nhutscelle* c.800, *Notesselinge* 1086 (DB). 'Nutshell place', probably indicating a small abode or settlement. OE *hnutu* + *scell* + *-ing*.

Nutbourne W. Sussex. near Pulborough. *Nordborne* 1086 (DB). 'North stream'. OE *north* + *burna*.

Nutfield Surrey. *Notfelle* 1086 (DB). 'Open land where nut-trees grow'. OE *hnutu* + *feld*.

Nuthall Notts. *Nutehale* 1086 (DB). 'Nook of land where nut-trees grow'. OE *hnutu* + *halh*.

Nuthampstead Herts. *Nuthamstede* c.1150. 'Homestead where nut-trees grow'. OE *hnutu* + *hām-stede*.

Nuthurst W. Sussex. *Nothurst* 1228. 'Wooded hill where nut-trees grow'. OE *hnutu* + *hyrst*.

Nutley E. Sussex. *Nutleg* 1249. 'Wood or clearing where nut-trees grow'. OE *hnutu* + *lēah*.

Nybster Highland. 'New farm'. OScand. *nýr* + *bólstathr*.

Nyetimber W. Sussex. near Pagham. *Neuetunbra* 12th cent. 'New timbered building'. OE *nīwe* + *timbre*.

Nymet Rowland & Nymet Tracey Devon. *Nymed* 974, *Limet* [sic] 1086 (DB), *Nimet Rollandi* 1242, *Nemethe Tracy* 1270. Celtic *nimet* 'holy place' (probably also an old name of the River Yeo). Manorial affixes from possession by a man called *Roland* in the 12th cent. and by the *de Trascy* family in the 13th.

Nympsfield Glos. *Nymdesfelda* 862, *Nimdesfelde* 1086 (DB). 'Open land by the holy place'. Celtic *nimet* + OE *feld*.

Nympton, 'farmstead near the river called *Nymet*' (probably an old name for the River Mole), Celtic *nimet* 'holy place' + OE *tūn*:
Nympton, Bishop's Devon. *Nimetone* 1086 (DB), *Bysshopes Nymet* 1334. Affix from its possession by the Bishop of Exeter in 1086.
Nympton, George Devon. *Nimet* 1086 (DB), *Nymeton Sancti Georgij* 1291. Affix from the dedication of the church. **Nympton, King's** Devon. *Nimetone* 1086 (DB), *Kyngesnemeton* 1254. Affix from its status as a royal manor in 1066.

Nynehead Somerset. *Nigon Hidon* 11th cent., *Nichehede* 1086 (DB). '(Estate of) nine hides'. OE *nigon* + *hīd*.

Nyton W. Sussex. *Nyton* 1327. Probably 'new farmstead', OE *nīge* + *tūn*, or 'farmstead on an island of dry ground in marsh' if the first element is *īeg* (with *N-* from ME *atten* 'at the').

Oadby Leics. *Oldebi* [*sic*] 1086 (DB), *Outheby* 1199. Probably 'farmstead or village of a man called Authi'. OScand. pers. name + *bý*.

Oake Somerset. *Acon* 11th cent., *Acha* 1086 (DB). '(Place at) the oak-trees'. OE *āc* in a dative plural form *ācum*.

Oaken Staffs. *Ache* 1086 (DB). Identical in origin with the previous name.

Oakengates Tel. & Wrek. *Okenyate* 1414. 'The gate or gap where oak-trees grow'. OE **ācen* + *geat*.

Oakenshaw Kirkl. *Akescahe* 1246. 'Copse where oak-trees grow'. OE **ācen* + *sceaga*.

Oakford Devon. *Alforda* [*sic*] 1086 (DB), *Acford* 1166. 'Ford by the oak-tree(s)'. OE *āc* + *ford*.

Oakham Rutland. *Ocham* 1067, *Ocheham* 1086 (DB). 'Homestead or enclosure of a man called Oc(c)a'. OE pers. name + *hām* or *hamm*.

Oakhanger Hants. *Acangre* 1086 (DB). 'Wooded slope where oak-trees grow'. OE *āc* + *hangra*.

Oakington Cambs. *Hochinton* 1086 (DB). 'Estate associated with a man called Hocca'. OE pers. name + *-ing-* + *tūn*.

Oakle Street Glos. *Acle* c.1270. 'Wood or clearing where oak-trees grow', OE *āc* + *lēah*. Later affix *street* probably refers to its situation on a Roman road.

Oakley, a fairly common name, 'wood or clearing where oak-trees grow', OE *āc* + *lēah*; examples include: **Oakley** Beds. *Achelai* 1086 (DB). **Oakley** Poole. *Ocle* 1327. **Oakley, Great** Essex. *Accleia* 1086 (DB).

Oakmere Ches. *Ocmare* 1277. '(Place at) the lake where oak-trees grow', OE *āc* + *mere*.

Oaksey Wilts. *Wochesie* 1086 (DB). 'Island, or well-watered land, of a man called *Wocc'. OE pers. name + *ēg*.

Oakthorpe Leics. *Achetorp* 1086 (DB). 'Outlying farmstead or hamlet of a man called Áki'. OScand. pers. name + *thorp*.

Oakworth Brad. *Acurde* 1086 (DB). 'Oak-tree enclosure'. OE *āc* + *worth*.

Oare, '(place at) the shore or hill-slope', OE *ōra*: **Oare** Kent. *Ore* 1086 (DB). **Oare** Wilts. *Oran* 934.

Oare Somerset. *Are* 1086 (DB). Named from the stream here, now called Oare Water. This is an ancient Celtic river-name identical with the River Ayr in Scotland.

Oban Arg. *Oban* 1643. '(Place by) the little bay'. Gaelic *òban*. The modern Gaelic name *An t-Òban Latharnach* means 'the little bay of Lorn' (*see* LORN).

Oborne Dorset. *Womburnan* 975, *Wocburne* 1086 (DB). '(Place at) the crooked or winding stream'. OE *wōh* + *burna*.

Occlestone Green Ches. *Aculuestune* 1086 (DB). 'Farmstead of a man called Ācwulf'. OE pers. name + *tūn*.

Occold Suffolk. *Acholt* c.1050, *Acolt* 1086 (DB). 'Oak-tree wood'. OE *āc* + *holt*.

Ochil Hills Perth. *Oychellis* 1461, *Ocelli montes* 1580. 'High ones'. British *uxello-*.

Ochiltree Dumf. *Ouchiltre* 1232. 'High homestead'. British *uxello-* + Welsh *tref*.

Ockbrook Derbys. *Ochebroc* 1086 (DB). 'Brook of a man called Occa'. OE pers. name + *brōc*.

Ockendon, North (Gtr. London) & **Ockendon, South** (Essex). *Wokendune* c.1070, *Wochenduna* 1086 (DB). 'Hill of a man called *Wocca'. OE pers. name (genitive -*n*) + *dūn*.

Ockham Surrey. *Bocheham* [*sic*] 1086 (DB), *Hocham* 1170. Possibly 'homestead or enclosure of a man called Occa'. OE pers. name + *hām* or *hamm*.

Ockley Surrey. *Hoclei* 1086 (DB). Probably 'woodland clearing of a man called Occa'. OE pers. name + *lēah*.

Ocle Pychard Herefs. *Aclea* c.1030, *Acle* 1086 (DB), *Acle Pichard* 1242. 'Oak-tree wood or clearing'. OE *āc* + *lēah*. Manorial affix from the *Pichard* family, here in the 13th cent.

Odcombe Somerset. *Udecome* 1086 (DB). Probably 'valley of a man called Uda'. OE pers. name + *cumb*.

Oddingley Worcs. *Oddingalea* 816. *Oddunclei* 1086 (DB). 'Woodland clearing of the family or followers of a man called *Odda'. OE pers. name + -*inga*- + *lēah*.

Oddington Oxon. *Otendone* 1086 (DB). 'Hill of a man called *Otta'. OE pers. name (genitive -*n*) + *dūn*.

Odell Beds. *Wadehelle* 1086 (DB). 'Hill where woad is grown'. OE *wād* + *hyll*.

Odiham Hants. *Odiham* 1086 (DB). 'Wooded homestead or enclosure'. OE *wudig* + *hām* or *hamm*.

Odstock Wilts. *Odestoche* 1086 (DB). 'Outlying farmstead or hamlet of a man called *Od(d)a'. OE pers. name + *stoc*.

Odstone Leics. *Odestone* 1086 (DB). Probably 'farmstead of a man called Oddr'. OScand. pers. name + OE *tūn*.

Offaly (*Uíbh Fhailí*) (the county). '(Place of the) descendants of Failge'. Russ *Failge* was the eldest son of Catháir Már (Cahirmore), King of Ireland from 120 to 123.

Offa's Dyke (*Clawdd Offa*) (linear rampart forming ancient boundary between England and Wales). Denb.–Mon. *Offan dic* 854, *Offedich* 1184. 'Dyke or earthwork traditionally associated with the 8th-cent. Mercian king Offa', from OE *dīc*. The Welsh name has the same sense. *See also* KNIGHTON.

Offchurch Warwicks. *Offechirch* 1139. 'Church of a man called Offa'. OE pers. name + *cirice*.

Offenham Worcs. *Offeham* 709, *Offenham* 1086 (DB). 'Homestead of a man called Offa or Uffa'. OE pers. name (genitive -*n*) + *hām*.

Offham E. Sussex. *Wocham* c.1092. Probably 'crooked homestead or enclosure'. OE *wōh* + *hām* or *hamm*.

Offham Kent. *Offaham* 10th cent. 'Homestead of a man called Offa'. OE pers. name + *hām*.

Offley, 'woodland clearing of a man called Offa', OE pers. name + *lēah*: **Offley, Bishops & Offley, High** Staffs. *Offeleia* 1086 (DB). Affix *Bishops* is from early possession by the Bishop of Lichfield. **Offley, Great** Herts. *Offanlege* 944–6, *Offelei* 1086 (DB).

Offord Cluny & Offord Darcy Cambs. *Upeforde, Opeforde* 1086 (DB), *Offord Clunye* 1257, *Offord Willelmi Daci* 1220. 'Upper ford'. OE *upp*(*e*) + *ford*. Manorial affixes from early possession by the monks of Cluny Abbey in France and by the *Dacy* (or *le Daneys*) family.

Offton Suffolk. *Offetuna* 1086 (DB). 'Farmstead of a man called Offa'. OE pers. name + *tūn*.

Offwell Devon. *Offewille* 1086 (DB). 'Spring or stream of a man called Uffa'. OE pers. name + *wella*.

Ogbourne Maizey, Ogbourne St Andrew & Ogbourne St George Wilts. *Oceburnan* 10th cent., *Ocheburne* 1086 (DB), *Ockeburn Meysey* 1242, *Okeborne Sancti Andree* 1289, *Okeburne Sancti Georgii* 1332. 'Stream of a man called Occa'. OE pers. name + *burna*. Distinguishing affixes from the *de Meysey* family (here in the 13th cent.) and from the church dedications. The river-name **Og** is a back-formation from the place name.

Oghil (*Eochill*) Galway. 'Yew wood'.

Ogle Northum. *Hoggel* 1170. Possibly 'hill of a man called Ocga'. OE pers. name + *hyll*.

Ogmore Vale Glam. '(Place on the) River Ogmore'. The river-name (Welsh *Ogwr*) means 'harrow' (Welsh *og*). *See also* BRIDGEND.

Ogonnelloe (*Tuath Ó gConaíle*) Clare. 'Territory of Uí Chonaíle'.

Ogwell, East & Ogwell, West Devon. *Wogganwylle* 956, *Wogewille* 1086 (DB). 'Spring or stream of a man called *Wocga'. OE pers. name + *wella*.

Okeford, Child & Okeford Fitzpaine Dorset. *Acford* 1086 (DB), *Childacford* 1227, *Ocford Fitz Payn* 1321. 'Oak-tree ford'. OE *āc* + *ford*. Affixes are from OE *cild* 'noble-born son' (probably with reference to some early owner) and from the *Fitz Payn* family, here from the 13th cent.

Okehampton Devon. *Ocmundtun* c.970, *Ochenemitona* 1086 (DB). 'Farmstead on the River Okement'. Celtic river-name (possibly 'swift stream') + OE *tūn*.

Old as affix. *See* main name, e.g. for **Old Alresford** (Hants.) *see* ALRESFORD.

Old Northants. *Walda* 1086 (DB). '(Place at) the woodland or forest'. OE *wald*.

Old Hurst Cambs. *Waldhirst* 1227. 'Wooded hill by the wold or forest'. OE *wald* + *hyrst*.

Old Radnor. *See* NEW RADNOR.

Old Wives Lees Kent. *Old Woodes Lease* 1569. 'Pasture at *Oldewode* (*Ealdewod* 13th cent., 'the old wood')'. OE *eald* + *wudu* + *lǣs*.

Oldberrow Warwicks. *Ulenbeorge* 709, *Oleberge* 1086 (DB). 'Hill or mound of a man called *Ulla'. OE pers. name + *beorg*.

Oldbury, 'old fortification or stronghold', OE *(e)ald* + *burh* (dative *byrig*): **Oldbury** Kent. *Ealdebery* 1302. **Oldbury** Sandw. *Aldeberia* 1174. **Oldbury** Shrops. *Aldeberie* 1086 (DB). **Oldbury** Warwicks. *Aldburia* 12th cent. **Oldbury on the Hill** Glos. *Ealdanbyri* 972, *Aldeberie* 1086 (DB). Affix 'on the Hill' first used in the 18th cent. **Oldbury upon Severn** S. Glos. *Aldeburhe* 1185.

Oldham Oldham. *Aldholm* 1226–8. 'Island at feature called *Alt*'. Celtic *alt* 'slope, cliff' + OScand. *holmr*.

Oldland S. Glos. *Aldelande* 1086 (DB). 'Old or long-used cultivated land'. OE *ald* + *land*.

Ollerton, 'farmstead where alder-trees grow', OE *alor* + *tūn*: **Ollerton** Ches. *Alretune* 1086 (DB). **Ollerton** Notts. *Alretun* 1086 (DB).

Olney Milt. K. *Ollanege* 979, *Olnei* 1086 (DB). 'Island, or dry ground in marsh, of a man called *Olla'. OE pers. name (genitive -*n*) + *ēg*.

Olton Solhll. *Alton* 1221. Possibly 'old farmstead'. OE *ald* + *tūn*.

Olveston S. Glos. *Ævestune* 955–9, *Alvestone* 1086 (DB). 'Farmstead of a man called Ælf'. OE pers. name + *tūn*.

Omagh (*An Ómaigh*) Tyrone. (*caislen*) *na hOghmaighe* c.1450. 'The virgin plain'.

Ombersley Worcs. *Ambreslege* 706. Possibly 'woodland clearing of a man called *Ambre', OE pers. name + *lēah*. Alternatively 'woodland clearing of the bunting' if the first element is rather OE *amer*.

Omeath (*Ó Méith*) Louth. '(People) of Uí Méith'.

Omey Island (*Iomaí*) Galway. 'Bed (of Feichín)'.

Ompton Notts. *Almuntone* 1086 (DB). 'Farmstead of a man called Alhmund'. OE pers. name + *tūn*.

Onaght (*Eoghanacht*) Galway. '(Place of the) descendants of Eoghan'.

Onecote Staffs. *Anecote* 1199. 'Lonely cottage(s)'. OE *āna* + *cot*.

Ongar, Chipping & Ongar, High Essex. *Aungre* 1045, *Angra* 1086 (DB), *Chepyngaungre* 1314, *Heyghangre* 1339. 'Pasture-land'. OE *anger*, with distinguishing affixes *cēping* 'market' and *hēah* 'high'.

Onibury Shrops. *Aneberie* 1086 (DB), *Onebur* 1247. River-name (possibly 'single river', i.e. one formed by the uniting of two headwaters, from OE *āna* and *ēa*) + *burh* (dative *byrig*) 'manor'.

Onn, High Staffs. *Otne* 1086 (DB), *Onna*, *Othna* c.1130. Plural form of Celtic *odn* 'kiln'.

Onneley Staffs. *Anelege* 1086 (DB). 'Woodland clearing of a man called Onna', or 'isolated clearing'. OE pers. name or *āna* + *lēah*.

Oola (*Ulla*) Limerick. 'Apple-trees'.

Oran (*Uarán*) Roscommon. 'Spring'.

Oranmore (*Órán Mór*) Galway. 'Big spring'.

Orby Lincs. *Heresbi* [sic] 1086 (DB), *Orreby* c.1115. 'Farmstead or village of a man called Orri'. OScand. pers. name + *bý*.

Orchard Portman Somerset. *Orceard* 854.
OE *orceard* 'a garden or orchard'. Manorial affix
from the *Portman* family, here in the 15th cent.

Orchard, East & Orchard, West Dorset.
Archet 939. '(Place) beside the wood'. Celtic *ar* +
cę̄d.

Orcheston Wilts. *Orc(h)estone* 1086 (DB).
Probably 'farmstead of a man called Ordrīc'. OE
pers. name + *tūn*.

Orcop Herefs. *Orcop* 1138. 'Top of the slope or
ridge'. OE *ōra* + *copp*.

Ord, East Northum. *Horde* 1196. 'Projecting
ridge'. OE *ord*.

Ore E. Sussex. *Ora* 1121–5. '(Place at) the hill-
slope or ridge'. OE *ōra*.

Orford Suffolk. *Oreford* 1164. Possibly 'ford
near the shore or low ridge'. OE *ōra* + *ford*.
Orford Ness (recorded from 1805) is from
ME *ness* 'headland'. The river-name **Ore** is a
back-formation from the place name.

Orford Warrtn. *Orford* 1332. Possibly 'upper
ford'. OE *uferra* + *ford*.

Orgreave Staffs. *Ordgraue* 1195. Probably
'pointed grove or copse'. OE *ord* + *grǣfe*.

Orkney (the unitary authority). *Orkas* 330 BC,
Orcades 1st cent., *Orkaneya* 970. 'Islands of
(the) Orcos'. OScand. *ey*. The islands may have
originally had a Celtic tribal name meaning
'boar'. This was then apparently taken by the
Vikings to mean 'seal' (OScand. *orkn*).

Orlar (*Orlár*) Mayo. 'Valley floor'.

Orleton Herefs. *Alretune* 1086 (DB).
'Farmstead where alder-trees grow'. OE *alor* +
tūn.

Orlingbury Northants. *Ordinbaro* [sic] 1086
(DB), *Ordelinberg* 1202. Probably 'hill associated
with a man called *Ordla*'. OE pers. name + -*ing*-
+ *beorg*.

Ormes Head, Great Conwy. *Ormeshede
insula* 15th cent. 'Snake headland'. OScand.
ormr + *hofuth*. Compare WORMS HEAD.

Ormesby, Ormsby, usually 'farmstead or
village of a man called Ormr', OScand. pers.
name + *bý*: **Ormesby** Middlesbr. *Ormesbi* 1086
(DB). **Ormesby St Margaret & Ormesby
St Michael** Norfolk. *Ormesby c.*1020,
Ormesbei 1086 (DB). **Ormsby, North** Lincs.
Vrmesbyg 1066–8, *Ormesbi* 1086 (DB).

However the following may have a different
origin: **Ormsby, South** Lincs. *Ormesbi* 1086
(DB), *Ormeresbi* early 12th cent. 'Farmstead of a
man called Ormarr'. OScand. pers. name + *bý*.

**Ormside, Great & Ormside,
Little** Cumbria. *Ormesheued c.*1140.
'Headland of a man called Ormr'. OScand. pers.
name + OE *hēafod*. Alternatively the first
element may be OScand. *ormr* 'snake'.

Ormskirk Lancs. *Ormeshirche c.*1190.
'Church of a man called Ormr'. OScand. pers.
name + *kirkja*.

Oronsay, 'ebb island', OScand. *orfins-ey* >
orfiris-ey (Gaelic *Orasaigh*): **Oronsay** (island)
Arg. *Orvansay* 1549. **Oronsay** (island)
Highland. *Oransay* 1549.

Orphir Orkn. 'Tidal island'. *Orfura c.*1225,
*Orfiara c.*1500. OScand. *ór* + *fjara* + *ey*. The
name, literally 'out-of-low-water island', relates
to the coastline, regarded as an island that
appears above the waterline only at low tide.

Orpington Gtr. London. *Orpedingtun* 1032,
Orpinton 1086 (DB). 'Estate associated with a
man called *Orped*'. OE pers. name + -*ing*- +
tūn.

Orrell, possibly 'hill where ore is dug', OE *ōra*
+ *hyll*: **Orrell** Sefton. *Orhul* 1299. **Orrell**
Wigan. *Horhill* 1202.

Orsedd, Yr. See ROSSETT.

Orsett Thurr. *Orseathan* 957, *Orseda* 1086
(DB). '(Place at) the pits where (iron) ore is dug'.
OE *ōra* + *sēath* (dative plural -*um*).

Orston Notts. *Oschintone* 1086 (DB). Probably
'estate associated with a man called *Ōsica*'. OE
pers. name + -*ing*- + *tūn*.

Orton, usually 'higher farmstead', or
'farmstead by a ridge or bank', OE *uferra* or *ofer*
or *ōfer* + *tūn*: **Orton** Cumbria. *Overton* 1239.
Orton Northants. *Overtone* 1086 (DB). **Orton
Longueville & Orton Waterville** Peterb.
Ofertune 958, *Ovretune* 1086 (DB), *Ouerton
Longavill* 1247, *Ouertone Wateruile* 1248.
Manorial affixes from the *de Longauilla* and *de
Waltervilla* families, here in the 12th cent.
Orton on the Hill Leics. *Wortone* [sic] 1086
(DB), *Overton c.*1215. **Orton, Water**
Warwicks. *Overton* 1262, *Water Ouerton* 1546.
Affix is ME *water* 'stream' (referring to its
situation by River Tame).
     However the following has a different origin:
Orton, Great & Orton, Little Cumbria.

Orreton 1210. 'Farmstead of a man called Orri'. OScand. pers. name + OE *tūn*.

Orwell Cambs. *Ordeuuelle* 1086 (DB). 'Spring by a pointed hill'. OE *ord* + *wella*.

Orwell (river) Suffolk. *Arewan* 11th cent., *Orewell* 1341. An ancient Celtic or pre-Celtic river-name (probably meaning 'swift one'), to which has been added OE *wella* 'stream'.

Osbaldeston Lancs. *Ossebaldiston c.*1200. 'Farmstead of a man called Ōsbald'. OE pers. name + *tūn*.

Osbaston Leics. *Sbernestun* [sic] 1086 (DB), *Osberneston* 1200. 'Farmstead of a man called Ásbjǫrn'. OScand. pers. name + OE *tūn*.

Osborne I. of Wight. *Austeburn* 1316. Probably 'stream at the sheep-fold'. OE *eowestre* + *burna*.

Osbournby Lincs. *Osbernebi* 1086 (DB). 'Farmstead or village of a man called Ásbjǫrn'. OScand. pers. name + *bý*.

Osgathorpe Leics. *Osgodtorp* 1086 (DB). 'Outlying farmstead or hamlet of a man called Ásgautr'. OScand. pers. name + *thorp*.

Osgodby, 'farmstead or village of a man called Ásgautr', OScand. pers. name + *bý*: **Osgodby** Lincs., near Market Rasen. *Osgotebi* 1086 (DB). **Osgodby** N. Yorks. *Asgozbi* 1086 (DB).

Osmaston Derbys. near Derby. *Osmundestune* 1086 (DB). 'Farmstead of a man called Ōsmund'. OE pers. name + *tūn*.

Osmington Dorset. *Osmingtone* 934, *Osmentone* 1086 (DB). 'Estate associated with a man called Ōsmund'. OE pers. name + *-ing-* + *tūn*.

Osmotherley N. Yorks. *Asmundrelac* [sic] 1086 (DB), *Osmunderle* 1088. 'Woodland clearing of a man called Ásmundr'. OScand. pers. name (genitive *-ar*) + OE *lēah*.

Ospringe Kent. *Ospringes* 1086 (DB). Possibly '(place at) the spring'. OE **or-spring* or **of-spring*.

Ossett Wakefd. *Osleset* 1086 (DB). 'Fold of a man called **Ōsla*, or 'fold frequented by blackbirds'. OE pers. name or OE *ōsle* + *set*.

Ossington Notts. *Oschintone* 1086 (DB). Probably 'estate associated with a man called **Ōsica*'. OE pers. name + *-ing-* + *tūn*.

Osterley Gtr. London. *Osterle, Ostrele* 1274. 'Woodland clearing with a sheep-fold'. OE *eowestre* + *lēah*.

Oswaldkirk N. Yorks. *Oswaldescerca* 1086 (DB). 'Church dedicated to St Ōswald'. OE pers. name + *cirice* (replaced by OScand. *kirkja*).

Oswaldtwistle Lancs. *Oswaldestwisel* 1246. 'River-junction of a man called Ōswald'. OE pers. name + *twisla*.

Oswestry Shrops. *Oswaldestroe* 1191. 'Tree of a man called Ōswald'. OE pers. name + *trēow*. The traditional connection with St Oswald, 7th-cent. king of Northumbria, is uncertain.

Otford Kent. *Otteford* 832, *Otefort* 1086 (DB). 'Ford of a man called **Otta*'. OE pers. name + *ford*.

Otham Kent. *Oteham* 1086 (DB). 'Homestead or village of a man called **Otta*'. OE pers. name + *hām*.

Othery Somerset. *Othri* 1225. Probably 'the other or second island'. OE *ōther* + *ēg*.

Otley, 'woodland clearing of a man called **Otta*', OE pers. name + *lēah*: **Otley** Leeds. *Ottanlege c.*972, *Otelai* 1086 (DB). **Otley** Suffolk. *Otelega* 1086 (DB).

Otterbourne, Otterburn, 'stream frequented by otters', OE *oter* + *burna*: **Otterbourne** Hants. *Oterburna c.*970, *Otreburne* 1086 (DB). **Otterburn** Northum. *Oterburn* 1217. **Otterburn** N. Yorks. *Otreburne* 1086 (DB).

Otterden Kent. *Otringedene* 1086 (DB). 'Woodland pasture of the family or followers of a man called Oter'. OE pers. name + *-inga-* + *denn*.

Otterham Cornwall. *Otrham* 1086 (DB). Probably 'enclosure or river-meadow by the River Ottery'. OE river-name ('otter stream', OE *oter* + *ēa*) + *hamm*.

Otterington, North & Otterington, South N. Yorks. *Otrinctun* 1086 (DB). 'Estate associated with a man called Oter'. OE pers. name + *-ing-* + *tūn*.

Ottershaw Surrey. *Otershaghe c.*890. 'Small wood frequented by otters'. OE *oter* + *sceaga*.

Otterton Devon. *Otritone* 1086 (DB). 'Farmstead by the River Otter'. OE river-name (*see* OTTERY) + *tūn*.

Ottery St Mary Devon. *Otri* 1086 (DB), *Otery Sancte Marie* 1242. Named from the River Otter, 'stream frequented by otters', OE *oter* + *ēa*. Affix from the dedication of the church.

Ottery, Venn Devon. *Fenotri* 1156. 'Marshy land by the River Otter'. OE *fenn*, *see* previous name.

Ottringham E. R. Yorks. *Otringeham* 1086 (DB). 'Homestead of the family or followers of a man called Oter'. OE pers. name + *-inga-* + *hām*.

Oughter, Lough (*Loch Uachtar*) Cavan. 'Upper lake'.

Oughterard (*Uachtar Ard*) Galway, Kildare. 'High upper place'.

Oughtibridge Sheff. *Uhtiuabrigga* 1161. Probably 'bridge of a woman called *Ūhtgifu*'. OE pers. name + *brycg*.

Oulart (*An tAbhallort*) Wexford. 'The orchard'.

Oulston N. Yorks. *Uluestun* 1086 (DB). 'Farmstead of a man called Wulf or Ulfr'. OE or OScand. pers. name + OE *tūn*.

Oulton Cumbria. *Ulveton* c.1200. 'Farmstead of a man called *Wulfa*'. OE pers. name + *tūn*.

Oulton Leeds. *Aleton* 1180. Identical in origin with OULTON Suffolk.

Oulton Norfolk. *Oulstuna* 1086 (DB). Probably 'farmstead of a man called Authulfr'. OScand. pers. name + OE *tūn*.

Oulton Suffolk. *Aleton* 1203. 'Farmstead of a man called Áli', or 'old farmstead'. OScand. pers. name or OE *ald* + *tūn*. **Oulton Broad** contains *broad* in the sense 'extensive piece of water', *see* BROADS.

Ounageeragh (*Abha na gCaorach*) Cork. 'River of the sheep'.

Oundle Northants. *Undolum* c.710–20, *Undele* 1086 (DB). '(Settlement of) the tribe called Undalas'. OE tribal name (meaning possibly 'those without a share' or 'undivided').

Ourna, Lough (*Loch Odharna*) Tipperary. 'Dun-coloured lake'.

Ousby Cumbria. *Uluesbi* 1195. 'Farmstead or village of a man called Ulfr'. OScand. pers. name + *bý*.

Ousden Suffolk. *Uuesdana* 1086 (DB). 'Valley of the owl'. OE *ūf* + *denu*.

Ouseburn, Great & Ouseburn, Little N. Yorks. *Useburne* 1086 (DB). '(Place at) the stream flowing into the River Ouse'. Celtic or pre-Celtic river-name (meaning simply 'water') + OE *burna*.

Ousefleet E. R. Yorks. *Useflete* 1100–1108. '(Place at) the channel of the River Ouse'. OE *flēot*, *see* previous name.

Ouston Durham. *Vlkilstan* 1244–9. 'Boundary stone of a man called Ulfkell'. OScand. pers. name + OE *stān*.

Out Rawcliffe Lancs. *See* RAWCLIFFE.

Outwell Norfolk. *Wellan* 963, *Utuuella* 1086 (DB). '(Place at) the spring or stream'. OE *wella*, with affix *ūte* 'outer, lower downstream' to distinguish it from UPWELL.

Outwood, 'outlying wood', that is 'wood on the outskirts of a manor or parish', OE *ūt* + *wudu*; examples include: **Outwood** Surrey, recorded thus in 1640. **Outwood** Wakefd. *Outewode* 1436.

Ovenden Calder. *Ovenden* 1219. Probably 'valley of a man called Ōfa'. OE pers. name (genitive *-n*) + *denu*.

Ovens (*Na hUamhanna*) Cork. 'The caves'.

Over as affix. *See* main name, e.g. for **Over Compton** (Dorset) *see* COMPTON.

Over, '(place at) the ridge or slope', OE **ofer*: **Over** Cambs. *Ouer* 1060, *Ovre* 1086 (DB). **Over** Ches. *Ovre* 1086 (DB). **Over** S. Glos. *Ofre* 1005.

Overbury Worcs. *Uferebiri* 875, *Ovreberie* 1086 (DB). 'Upper fortification'. OE *uferra* + *burh* (dative byrig).

Overseal Derbys. *See* NETHERSEAL.

Overstone Northants. *Oveston* 12th cent. Probably 'farmstead of a man called Ufic'. OE pers. name + *tūn*.

Overstrand Norfolk. *Othestranda* [sic] 1086 (DB), *Overstrand* 1231. 'Shore with an edge or margin', i.e. probably 'narrow shore'. OE *ōfer* + *strand*.

Overton, a fairly common name, usually 'higher farmstead', OE *uferra* + *tūn*; examples include: **Overton** Hants. *Uferantun* 909,

Ovretune 1086 (DB). **Overton, West** Wilts. *Uferantune* 939, *Ovretone* 1086 (DB).

However some Overtons may have a different origin, 'farmstead by a ridge or bank', OE **ofer* or *ōfer* + *tūn*; examples include: **Overton** Lancs. *Ouretun* 1086 (DB). **Overton** N. Yorks. *Ovretun* 1086 (DB). **Overton** (*Owrtyn*) Wrex. *Overtone* 1201. **Overton, Market** Rutland. *Overtune* 1086 (DB), *Marketesoverton* 1200. Affix is ME *merket* referring to the early market here.

Oving, '(settlement of) the family or followers of a man called Ūfa', OE pers. name + *-ingas*: **Oving** Bucks. *Olvonge* [*sic*] 1086 (DB), *Vuinges* 12th cent. **Oving** W. Sussex. *Uuinges* 956.

Ovingdean Bright. & Hove. *Hovingedene* 1086 (DB). Probably 'valley of the family or followers of a man called Ūfa'. OE pers. name + *-inga-* + *denu*.

Ovingham Northum. *Ovingeham* 1238. 'Homestead of the family or followers of a man called Ōfa', or 'homestead at Ōfa's place'. OE pers. name + *-inga-* or *-ing* + *hām*.

Ovington, usually 'estate associated with a man called Ūfa', OE pers. name + *-ing-* + *tun*: **Ovington** Essex. *Ouituna* 1086 (DB). **Ovington** Hants. *Ufinctune c.*970. **Ovington** Norfolk. *Uvinton* 1202.

However the following have a different origin: **Ovington** Durham. *Ulfeton* 1086 (DB). 'Estate associated with a man called Wulfa'. OE pers. name + *-ing-* + *tūn*. **Ovington** Northum. *Ofingadun* 699–705. 'Hill of the family or followers of a man called Ōfa'. OE pers. name + *-inga-* + *dūn*.

Ow (*Abh*) (river) Wicklow. 'River'.

Owel, Lough (*Loch Uair*) Westmeath. 'Uair's lake'.

Owenass (*Abhainn Easa*) (river) Laois. 'River of the waterfall'.

Owenavorragh (*Abhainn an Bhorraidh*) (river) Wexford. 'River liable to flood'.

Owenbeg (*An Abhainn Bheag*) Sligo. 'The little river'.

Owenboy (*Abhainn Bhuí*) Cork. 'Yellow river'.

Owendalulleegh (*Abhainn dá Loilíoch*) (river) Galway. 'River of two milch cows'.

Owenduff (*An Abhainn Dubh*) (river) Mayo. 'The black river'.

Owenea (*Abhainn an Fia*) (river) Donegal. 'River of the deer'.

Owenglin (*Abhainn Ghlinne*) (river) Galway. 'River of the valley'.

Oweniny (*Abhainn Eidhneach*) (river) Mayo. 'Ivied river'.

Owenkillew (*Abhainn Choilleadh*) (river) Tyrone. 'River of the wood'.

Owenmore (*Abhainn Mór*) (river) Cork, Mayo, Sligo. 'Big river'.

Owennacurra (*Abhainn na Cora*) (river) Cork. 'River of the weir'.

Owenrea Burn (*Abhainn Riabhach*) Tyrone. 'Grey river'.

Owentocker (*Abhainn Tacair*) (river) Donegal. 'River of the pickings'.

Owenur (*Abhainn Fhuar*) (river) Roscommon. 'Cold river'.

Owenwee (*Abhainn Bhuí*) Donegal. 'Yellow river'.

Ower Hants. near Copythorne. *Hore* 1086 (DB). '(Place at) the bank or slope'. OE *ōra*.

Ower (*Odhar*) Galway. 'Dun-coloured place'.

Owermoigne Dorset. *Ogre* 1086 (DB), *Oure Moyngne* 1314. Probably 'the wind-gap(s)', from a Celtic (Brittonic) **oir* 'cold' + **drust* (plural **dröstow*) 'door', referring to gaps in the chalk hills which funnel winds off the sea. Manorial affix from the *Moigne* family, here in the 13th cent.

Owersby, North & Owersby, South Lincs. *Aresbi, Oresbi* 1086 (DB). Possibly 'farmstead or village of a man called Ávarr'. OScand. pers. name + *bý*.

Owey Island (*Uaigh*) Donegal. *The Island of Inishowey* 1613. 'Cave'.

Owmby, probably 'farmstead or village of a man called Authunn', OScand. pers. name + *bý*, though alternatively the first element may be OScand. *authn* 'uncultivated land, deserted farm': **Owmby** Lincs. *Odenebi* 1086 (DB). **Owmby-by-Spital** Lincs. *Ounebi* 1086 (DB). *See* SPITAL.

Owrtyn. *See* OVERTON.

Owslebury Hants. *Oselbyrig c.*970. 'Stronghold of a man called **Ōsla*', or 'stronghold frequented by blackbirds'. OE pers. name or OE ōsle + burh (dative byrig).

Owston Donc. Aust(h)un 1086 (DB). 'East farmstead'. OScand. austr + OE tūn.

Owston Leics. Osulvestone 1086 (DB). 'Farmstead of a man called Ōswulf'. OE pers. name + tūn.

Owston Ferry N. Lincs. Ostone 1086 (DB). Identical in origin with OWSTON Donc.

Owstwick E. R. Yorks. Osteuuic 1086 (DB). 'Eastern outlying farm'. OScand. austr (possibly replacing OE ēast) + OE wīc.

Owthorpe Notts. Ovetorp 1086 (DB). 'Outlying farmstead or hamlet of a man called Úfi or Ūfa'. OScand. or OE pers. name + OScand. thorp.

Owvane (Abh Bhán) (river) Cork. 'White river'.

Ox Mountains (Sliabh Gamh) Sligo. The Irish name, meaning 'storm mountain' (gamh), has been taken to mean 'ox mountain' (damh).

Oxborough Norfolk. Oxenburch 1086 (DB). 'Fortification where oxen are kept'. OE oxa (genitive plural oxna) + burh.

Oxendon, Great Northants. Oxendone 1086 (DB). 'Hill where oxen are pastured'. OE oxa (genitive plural oxna) + dūn.

Oxenholme Cumbria. Oxinholme 1274. 'River-meadow where oxen are pastured'. OE oxa (genitive plural oxna) + OScand. holmr.

Oxenhope Brad. Hoxnehop 12th cent. 'Valley where oxen are kept'. OE oxa (genitive plural oxna) + hop.

Oxenton Glos. Oxendone 1086 (DB). 'Hill where oxen are pastured'. OE oxa (genitive plural oxna) + dūn.

Oxford Oxon. Oxnaforda 10th cent., Oxeneford 1086 (DB). 'Ford used by oxen'. OE oxa (genitive plural oxna) + ford. **Oxfordshire** (OE scīr 'district') is first referred to in the 11th cent.

Oxford Island (peninsula) Armagh. Oxford's Island 1835. The name is a corruption of Hawksworth's Island. Captain Robert Hawksworth is recorded as a tenant here in 1666.

Oxhey Herts. Oxangehæge 1007. 'Enclosure for oxen'. OE oxa (genitive plural oxna) + (ge)hæg.

Oxhill Warwicks. Octeselve 1086 (DB). 'Shelf or ledge of a man called Ohta'. OE pers. name + scelf.

Oxley Wolverh. Oxelie 1086 (DB). 'Woodland clearing where oxen are pastured'. OE oxa + lēah.

Oxnam Sc. Bord. Oxanaham 1152. 'Farmstead where oxen are kept'. OE oxa (genitive plural oxna) + hām.

Oxshott Surrey. Okessela 1179. 'Projecting piece of land of a man called Ocga'. OE pers. name + scēat.

Oxspring Barns. Ospring [sic] 1086 (DB), Oxspring 1154. 'Spring frequented by oxen'. OE oxa + spring.

Oxted Surrey. Acstede 1086 (DB). 'Place where oak-trees grow'. OE āc + stede.

Oxton Notts. Oxetune 1086 (DB). 'Farmstead where oxen are kept'. OE oxa + tūn.

Oxwick Norfolk. Ossuic [sic] 1086 (DB), Oxewic 1242. 'Farm where oxen are kept'. OE oxa + wīc.

Oykel (river) Highland. Okel 1365. 'High one'. British uxello-.

Oystermouth (Ystumllwynarth) Swan. The Welsh form, involving ystum 'bend (of a river)', llwyn 'grove' and garth 'promontory' (of Mumbles Hill), has long been transformed into Oystermouth, supposedly interpreted as 'estuary of oysters'.

Ozleworth Glos. Oslan wyrth 940, Osleuuorde 1086 (DB). 'Enclosure of a man called *Ōsla', or 'enclosure frequented by blackbirds'. OE pers. name or OE ōsle + worth.

Pabay (island) Highland (Skye). *Paba* 1580. 'Hermit island'. OScand. *papi + ey*.

Packington Leics. *Pakinton* 1043, *Pachintone* 1086 (DB). 'Estate associated with a man called *Pac(c)a*'. OE pers. name + *-ing- + tūn*.

Padbury Bucks. *Pateberie* 1086 (DB). 'Fortified place of a man called Padda'. OE pers. name + *burh* (dative *byrig*).

Paddington Gtr. London. *Padintune* 959, *Padington c.*1050. 'Estate associated with a man called Padda'. OE pers. name + *-ing- + tūn*.

Paddlesworth Kent. near Folkestone. *Peadleswurthe* 11th cent. 'Enclosure of a man called **Pæddel*'. OE pers. name + *worth*.

Paddock Wood Kent. *Parrok* 1279. OE *pearroc* 'small enclosure, paddock'.

Padiham Lancs. *Padiham* 1251. Possibly 'homestead or enclosure associated with a man called Padda'. OE pers. name + *-ing- + hām* or *hamm*.

Padley, Upper & Padley, Nether Derbys. *Paddeley c.*1230. 'Woodland clearing of a man called Padda', or 'one frequented by toads'. OE pers. name or OE **padde + lēah*.

Padstow Cornwall. *Sancte Petroces stow* 11th cent. 'Holy place of St Petroc'. Cornish saint's name + OE *stōw*.

Padworth W. Berks. *Peadanwurthe* 956, *Peteorde* 1086 (DB). 'Enclosure of a man called Peada'. OE pers. name + *worth*.

Pagham W. Sussex. *Pecganham* 680, *Pageham* 1086 (DB). 'Homestead or promontory of a man called **Pæcga*'. OE pers. name + *hām* or *hamm*.

Paglesham Essex. *Paclesham* 1066, *Pachesham* [*sic*] 1086 (DB). 'Homestead or village of a man called **Pæccel*'. OE pers. name + *hām*.

Paignton Torbay. *Peintone* 1086 (DB). Probably 'estate associated with a man called Pæga'. OE pers. name + *-ing- + tūn*.

Pailton Warwicks. *Pallentuna* 1077. 'Estate associated with a man called **Pægel*'. OE pers. name + *-ing- + tūn*.

Painswick Glos. *Wiche* 1086 (DB), *Painswike* 1237. 'The dwelling or dairy farm'. OE *wīc*. Later manorial affix from *Pain* Fitzjohn who held the manor in the early 12th cent.

Paisley Renf. *Passeleth* 1161, *Passelek* 1298. '(Place with a) church'. MIrish *baslec*, from Latin *basilica*.

Pakefield Suffolk. *Paggefella* 1086 (DB). 'Open land of a man called **Pacca*'. OE pers. name + *feld*.

Pakenham Suffolk. *Pakenham c.*950, *Pachenham* 1086 (DB). 'Homestead or village of a man called **Pacca*'. OE pers. name (genitive *-n*) + *hām*.

Palace (*Pailís*) Cork, Wexford. 'Stockade'.

Palatine (*Baile na bPailitíneach*) Carlow. 'Town of the Palatines'.

Palestine Hants. Recorded thus in 1840, a transferred name from the country so called, here perhaps bestowed on a place considered rather remote or inaccessible.

Palgrave Suffolk. *Palegrave* 962, *Palegraua* 1086 (DB). Probably 'grove where poles are got'. OE *pāl + grāf*.

Palgrave, Great Norfolk. *Pag(g)raua* 1086 (DB). Possibly 'grove of a man called Paga'. OE pers. name + *grāf*.

Pall Mall Gtr. London. Street recorded as *Pall Mall Walk* 1650, so called from the ball game of *pallemaille* or *pallmall* introduced into England from Italy and France and played here in the 17th cent.

Pallas (*Pailís*) Galway, Kerry. 'Stockade'.

Pallas Green (*Pailís Ghréine*) Limerick. 'Stockade of Grian'.

Pallasbeg (*Pailís Beag*) Limerick. 'Small stockade'.

Pallasboy (*Pailís Buí*) Westmeath. 'Yellow stockade'.

Pallaskenry (*Pailís Chaonraí*) Limerick. 'Stockade of Caonraí'.

Palling, Sea Norfolk. *Pallinga* 1086 (DB). '(Settlement of) the family or followers of a man called Pælli'. OE pers. name + *-ingas*. Affix from its coastal situation.

Pallis (*Pailís*) Kerry, Wexford. 'Stockade'.

Palmers Green Gtr. London. *Palmers grene* 1608. Village green (ME *grene*) named from a family called *Palmer*, known in this area from the 14th cent.

Palterton Derbys. *Paltertune* c.1002, *Paltretune* 1086 (DB). 'Farmstead on the River Poulter'. Celtic river-name (meaning 'spear-shaft') + OE *tūn*.

Pamber Hants. *Penberga* 1165. Possibly 'hill with a fold or enclosure'. OE *penn* + *beorg*. Alternatively the first element could be Celtic *penn* 'head, hill'.

Pamphill Dorset. *Pamphilla* 1168. Possibly 'hill of a man called *Pampa* or *Pempa*', OE pers. name + *hyll*, but the first element may be an OE word *pamp* meaning 'hill'.

Pampisford Cambs. *Pampesuuorde* 1086 (DB). Probably 'enclosure of a man called *Pamp*'. OE pers. name + *worth*.

Pancrasweek Devon. *Pancradeswike* 1197. 'Hamlet with a church dedicated to St Pancras'. OE *wīc*.

Pancross Vale Glam. *Pannecrosse* 1605. Probably 'Pain's crossroads'. ME pers. name + *cros*.

Panfield Essex. *Penfelda* 1086 (DB). 'Open land by the River Pant'. Celtic river-name (meaning 'valley') + OE *feld*.

Pangbourne W. Berks. *Pegingaburnan* 844, *Pangeborne* 1086 (DB). 'Stream of the family or followers of a man called Pǣga'. OE pers. name + *-inga-* + *burna*. The river-name **Pang** is a back-formation from the place name.

Pannal N. Yorks. *Panhal* 1170. 'Pan-shaped nook of land'. OE *panne* + *halh*.

Pant Shrops. Recorded as *Pant Trystan* 1837, 'Tristan's valley', from Welsh *pant* 'valley'.

Panton Lincs. *Pantone* 1086 (DB). Probably 'farmstead in a hollow'. OE *panne* 'pan' (in transferred topographical sense) + *tūn*.

Panxworth Norfolk. *Pankesford* 1086 (DB). OE *ford* 'a ford' with an uncertain first element, possibly a pers. name.

Papcastle Cumbria. *Pabecastr* 1260. 'Roman fort occupied by a hermit'. OScand. *papi* + OE *cæster*.

Papplewick Notts. *Papleuuic* 1086 (DB). 'Dwelling or (dairy) farm in the pebbly place'. OE **papol* + *wīc*.

Paps, The (*An Dá Chích*) (mountain) Kerry. 'The two breasts'.

Papworth Everard & Papworth St Agnes Cambs. *Pappawyrthe* 1012, *Papeuuorde* 1086 (DB), *Pappewrth Everard* 1254, *Anneys Papwrth* 1241. 'Enclosure of a man called **Pappa*'. OE pers. name + *worth*. Manorial affixes from 12th-cent. owners called *Evrard* and *Agnes*.

Par Cornwall. *Le Pare* 1573. Probably 'the cove or harbour', from Cornish *porth*.

Paradise, a name found in several counties, usually bestowed on a spot considered particularly pleasant, from ME *paradis*; examples include **Paradise** Glos., near Painswick, recorded as *Paradys* 1327.

Parbold Lancs. *Perebold* 1200. 'Dwelling where pears grow'. OE *peru* + *bold*.

Pardshaw Cumbria. *Perdishaw* c.1205. 'Hill or mound of a man called **Perd(i)*'. OE pers. name + OScand. *haugr*.

Parham Suffolk. *Perreham* 1086 (DB). Probably 'homestead or enclosure where pears grow'. OE *peru* + *hām* or *hamm*.

Park (*Páirc*) Derry, Mayo. 'Field'.

Parkgate Antrim. *Parkgate* 1780. The village arose at the entrance to a great park laid out in the early 17th cent. by Sir Arthur Chichester, Lord Deputy of Ireland.

Parkgate Ches. a self-explanatory name for this former port taken from *the Parkgate* 1610, this referring to the gate of the former park at NESTON.

Parkgorm (*Páirc Gorm*) Tyrone. 'Blue field'.

Parkham Devon. *Percheham* 1086 (DB). Probably 'enclosure with paddocks'. OE *pearroc* + *hamm*.

Parkhurst, Parkhurst Forest I. of Wight. *Perkehurst c.*1200. 'The wooded hill in the hunting park', from OE *hyrst* with ME *park*. This 'park' was a royal chase or forest from the time of Domesday Book.

Parkmore (*Páirc Mhór*) Antrim, Galway, Tyrone. 'Big field'.

Parkreagh (*Páirc Riabhach*) Tyrone. 'Grey field'.

Parkstone Poole. *Parkeston* 1326. 'The park (boundary) stone', from OE *stān* with ME *park*, this probably referring to a medieval hunting park in CANFORD.

Parley, West Dorset. *Perlai* 1086 (DB). 'Wood or clearing where pears grow'. OE *peru* + *lēah*.

Parndon, Great Essex. *Perenduna* 1086 (DB). 'Hill where pears grow'. OE **peren* + *dūn*.

Parracombe Devon. *Pedrecumbe, Pedracomba* 1086 (DB), *Parrecumbe* 1238, *Pearecumbe* 1297. Possibly 'valley of the pedlars', from OE **peddere* (genitive plural *peddera*) + *cumb*. Alternatively, if the DB forms are corrupt, the first element may be OE *pearroc* 'an enclosure'.

Parson Drove Cambs. *Personesdroue* 1324. 'Cattle road used by or belonging to a parson'. ME *persone* + *drove*.

Partick Glas. *Perdeyc c.*1136. Origin obscure.

Partington Traffd. *Partinton* 1260. 'Estate associated with a man called **Pearta*'. OE pers. name + *-ing-* + *tūn*.

Partney Lincs. *Peartaneu* 731, *Partenai* 1086 (DB). 'Island of higher ground, or promontory, or dry ground in marsh, of a man called **Pearta*'. OE pers. name (genitive *-n*) + *ēg*.

Parwich Derbys. *Piowerwic* 963, *Pevrewic* 1086 (DB). Possibly 'dairy farm on the River *Pever*'. Lost Celtic river-name (meaning 'bright stream') + OE *wīc*.

Pass of Keimaneigh (*Céim an Fhia*) Cork. 'Pass of the gap of the deer'.

Passenham Northants. *Passanhamme* early 10th cent., *Passeham* 1086 (DB). 'River-meadow of a man called Passa'. OE pers. name (genitive *-n*) + *hamm*.

Paston Norfolk. *Pastuna* 1086 (DB). Possibly 'farmstead of a man called **Pæcci*', OE pers. name + *tūn*. Alternatively the first element may be an OE **pæsc(e)* 'muddy place, pool'.

Patcham Bright. **&** Hove. *Piceham* [sic] 1086 (DB), *Peccham c.*1090. Possibly 'homestead of a man called **Pæcca*'. OE pers. name + *hām*.

Patching W. Sussex. *Pæccingas* 960, *Petchinges* 1086 (DB). '(Settlement of) the family or followers of a man called **Pæcc(i)*'. OE pers. name + *-ingas*.

Patchway S. Glos. *Petshagh* 1276. 'Enclosure of a man called Pēot'. OE pers. name + *haga*.

Pateley Bridge N. Yorks. *Pathlay* 1202. 'Woodland clearing near the path(s)'. OE *pæth* + *lēah*.

Patney Wilts. *Peattanige* 963. 'Island or well-watered land of a man called **Peatta*'. OE pers. name (genitive *-n*) + *ēg*.

Patrick Brompton N. Yorks. *See* BROMPTON.

Patrington E. R. Yorks. *Patringtona* 1033, *Patrictone* 1086 (DB). OE *tūn* 'farmstead, estate' with an uncertain first element, probably a pers. name or folk-name.

Patrixbourne Kent. *Borne* 1086 (DB), *Patrikesburn c.*1214. 'Estate on the river called *Burna*' (from OE *burna* 'stream' referring to the Little Stour), with later affix from William *Patricius* who held the manor in the 12th cent.

Patterdale Cumbria. *Patrichesdale c.*1180. 'Valley of a man called Patric'. OIrish pers. name + OScand. *dalr*.

Pattingham Staffs. *Patingham* 1086 (DB). 'Homestead of the family or followers of a man called **P(e)atta*', or 'homestead at **P(e)atta's* place'. OE pers. name + *-inga-* or *-ing* + *hām*.

Pattishall Northants. *Pascelle* [sic] 1086 (DB), *Patesshille* 12th cent. 'Hill of a man called **Pætti*'. OE pers. name + *hyll*.

Paul Cornwall. 'Church of *Beatus Paulus*, or of *Sanctus Paulinus*' 1259. From the patron saint of the church, St Paul or Paulinus.

Paulerspury Northants. *Pirie* 1086 (DB), *Pirye Pavely* c.1280. Originally '(place at) the pear-tree', from OE *pirige*. Manorial affix from the *de Pavelli* family who held the manor from 1086, thus distinguishing this place from neighbouring POTTERSPURY.

Paull E. R. Yorks. *Pagele* 1086 (DB). '(Place at) the stake (marking a landing-place)'. OE **pagol*.

Paulton B. & NE. Som. *Palton* 1171. 'Farmstead on a ledge or hill-slope'. OE **peall* + *tūn*.

Pavenham Beds. *Pabeneham* 1086 (DB). 'Homestead or river-meadow of a man called **Papa*'. OE pers. name (genitive -*n*) + *hām* or *hamm*.

Paxford Glos. *Paxford* 1208. 'Ford of a man called **Pæcc*'. OE pers. name + *ford*.

Paxton, Great & Paxton, Little Cambs. *Pachstone* 1086 (DB). Probably 'farmstead of a man called **Pæcc*'. OE pers. name + *tūn*.

Payhembury Devon. *Hanberie* 1086 (DB), *Paihember* 1236. 'High or chief fortified place'. OE *hēah* (dative *hēan*) + *burh* (dative *byrig*). Later manorial affix from the ME pers. name or surname *Paie*.

Paythorne Lancs. *Pathorme* 1086 (DB). Possibly 'thorn-tree of a man called *Pái*'. OScand. pers. name + *thorn*.

Peacehaven E. Sussex. a recent name for this new resort, chosen to commemorate the end of the First World War.

Peak District Derbys. *Pecsætna lond* 7th cent., *Pec* 1086. OE **pēac* 'a peak, a pointed hill', once applied to some particular hill but used of a larger area from early times. The first form means 'land of the peak dwellers', with OE *sǣte* + *land*. The medieval hunting forest here is referred to from the 13th cent. (e.g. *foresta de Pecco* 1223), this in turn giving name to the village of **Peak Forest** Derbys. (*Peake Forest* 1577).

Peakirk Peterb. *Pegecyrcan* 1016. 'Church of St Pega'. OE female saint's name + OE *cirice* (replaced by OScand. *kirkja*).

Pease Pottage W. Sussex. recorded as *Peaspottage Gate* 1724, probably a reference to soft muddy ground.

Peasedown St John B. & NE Som. Originally a coalmining village developed in the 19th cent., name not on early record but no doubt 'hill where peas grow', ME *pese* + *doun*. Affix from the dedication of the church.

Peasemore W. Berks. *Praxemere* [*sic*] 1086 (DB), *Pesemere* 1166. 'Pond by which peas grow'. OE *pise* + *mere*.

Peasenhall Suffolk. *Pesehala* 1086 (DB). 'Nook of land where peas grow'. OE **pisen* + *halh*.

Peasmarsh E. Sussex. *Pisemerse* 12th cent. 'Marshy ground where peas grow'. OE *pise* + *mersc*.

Peatling Magna & Peatling Parva Leics. *Petlinge* 1086 (DB). '(Settlement of) the family or followers of a man called **Pēotla*'. OE pers. name + -*ingas*.

Pebmarsh Essex. *Pebeners* 1086 (DB). 'Ploughed field of a man called Pybba'. OE pers. name (genitive -*n*) + *ersc*.

Pebworth Worcs. *Pebewrthe* 848, *Pebeworde* 1086 (DB). 'Enclosure of a man called **Peobba*'. OE pers. name + *worth*.

Peckforton Ches. *Pevreton* 1086 (DB). 'Farmstead at the ford by a peak or pointed hill'. OE **pēac* + *ford* + *tūn*.

Peckham, 'homestead by a peak or pointed hill', OE **pēac* + *hām*: **Peckham** Gtr. London. *Pecheham* 1086 (DB). **Peckham Rye** (-*Rithe* 1520) is from OE *rīth* 'stream'. **Peckham, East & Peckham, West** Kent. *Peccham* 10th cent., *Pecheham* 1086 (DB).

Peckleton Leics. *Pechintone* [*sic*] 1086 (DB), *Petlington* 1180. 'Estate associated with a man called **Peohtla*'. OE pers. name + -*ing*- + *tūn*.

Pedmore Dudley. *Pevemore* [*sic*] 1086 (DB), *Pubemora* 1176. 'Marsh of a man called Pybba'. OE pers. name + *mōr*.

Pedwell Somerset. *Pedewelle* 1086 (DB). 'Spring or stream of a man called Pēoda or Peda'. OE pers. name + *wella*.

Peebles Sc. Bord. *Pebles* c.1125. '(Place with) tents'. Welsh *pabell* (plural *pebyll*), with English plural -*s*.

Peel Isle of Man. Recorded as *Pelam* in 1399, 'the palisade', from Anglo-Norman and ME *pel* 'a palisade, an enclosure formed by a palisade', with reference to the ancient castle on St Patrick's Isle. Up to the 16th cent usually known as *Holmetown*, that is 'island village' from OScand. *holmr* + ME *toun*. The Manx name of the town is *Port-na-Hinsey* meaning 'port of the island'.

Pegswood Northum. *Peggiswrth* 1242. 'Enclosure of a man called **Pecg*'. OE pers. name + *worth*.

Pelaw Gatesd. *Pelow, Pelhou* 1183. Probably 'hill-spur with a palisade'. ME *pel* + *how* (OE *hōh*). Alternatively the first element may be ME *pele* 'triangular feature'.

Peldon Essex. *Piltendone c.*950, *Peltenduna* 1086 (DB). Probably 'hill of a man called **Pylta*'. OE pers. name (genitive *-n*) + *dūn*.

Pelham, Brent, Pelham, Furneux, & Pelham, Stocking Herts. *Peleham* 1086 (DB), *Barndepelham* 1230, *Pelham Furnelle* 1240, *Stokenepelham* 1235. 'Homestead or village of a man called **Pēola*'. OE pers. name + *hām*. Distinguishing affixes from OE *bærned* 'burnt, destroyed by fire', from the *de Fornellis* family here in the 13th cent., and from OE *stoccen* 'made of logs' or 'by the tree-stumps'.

Pelsall Wsall. *Peoleshale* 996, *Peleshale* 1086 (DB). 'Nook of land of a man called **Pēol*'. OE pers. name + *halh*.

Pelton Durham. *Pelton* 1243. 'Village with a palisade', or 'village by a triangular feature'. ME *pel* or *pele* + *toun*.

Pelynt Cornwall. *Plunent* 1086 (DB). 'Parish of Nennyd'. Cornish *plu* + saint's name.

Pembridge Herefs. *Penebruge* 1086 (DB). Possibly 'bridge of a man called **Pena* or **Pægna*'. OE pers. name + *brycg*.

Pembroke (*Penfro*) Pemb. *Pennbro c.*1150, *Pembroch* 1191. 'Land at the end'. Welsh *pen-* + **brog-*.

Pembroke Dock Pemb. *Pembroke Dockyard* 1817. The port was established by the government in 1814 near the head of Milford Haven estuary and took the name of nearby PEMBROKE.

Pembury Kent. *Peppingeberia c.*1100. Possibly 'fortified place of the family or followers of a man called **Pepa* or the like'. OE (?) pers. name + *-inga-* + *burh* (dative *byrig*).

Pen-y-Bont ar Ogwr. *See* BRIDGEND.

Penally (*Penalun*) Pemb. *Pennalun* 9th cent., *Penn Alun* 1136. 'Alun's headland'. Welsh *pen*.

Penare Cornwall. *Pennarth* 1284. 'The headland'. Cornish **penn-ardh*.

Penarlâg. *See* HAWARDEN.

Penarth Vale Glam. *Penarth* 1254. 'Top of the headland'. Welsh *pen* + *garth*.

Pencombe Herefs. *Pencumbe* 12th cent. Probably 'valley with an enclosure'. OE *penn* + *cumb*.

Pencoyd Herefs. *Pencoyt* 1291. 'Wood's end'. Celtic **penn* + **coid*.

Pencraig Herefs. *Penncreic c.*1150. 'Top of the crag or rocky hill'. Celtic **penn* + **creig*.

Pendle (district) Lancs. A modern adoption of an early name of Pendle Hill (for which *see* PENDLETON).

Pendlebury Salford. *Penelbiri* 1202. 'Manor by the hill called *Penn*'. Celtic **penn* 'head, end, top' + explanatory OE *hyll* + *burh* (dative *byrig*).

Pendleton Lancs. *Peniltune* 1086 (DB). 'Farmstead by the hill called *Penn* (Pendle Hill)'. Celtic **penn* 'head, end, top' + explanatory OE *hyll* + *tūn*.

Pendock Worcs. *Penedoc* 875, 1086 (DB). Possibly 'end of the barley or corn field' or 'hill where barley or corn grows'. Celtic **penn* + a derivative of **heith* 'barley' or **īd* 'corn'.

Pendoggett Cornwall. *Pendeugod* 1289. '(Place at) the top of two woods'. Cornish *pen* 'head, end, top' + *dew* 'two' + *cos* (Old Cornish *cuit*) 'wood'.

Penfro. *See* PEMBROKE.

Penge Gtr. London. *Pænge* 957, *Penceat* 1067. 'Wood's end'. Celtic **penn* + **cēd*.

Penhurst E. Sussex. *Penehest* [*sic*] 1086 (DB), *Peneherste* 12th cent. Possibly 'wooded hill of a man called **Pena*'. OE pers. name + *hyrst*.

Penicuik Midloth. *Penikok* 1250. 'Hill of the cuckoo'. OWelsh *penn* + *y* + *cog*.

Penistone Barns. *Pengestone* 1086 (DB), *Peningeston* 1199. Probably 'farmstead by a hill called *Penning*'. Celtic **penn* 'head, end, top' + OE *-ing* + *tūn*.

Penketh Warrtn. *Penket* 1242. 'Wood's end'. Celtic **penn* + **cēd*.

Penkridge Staffs. *Pennocrucium* 4th cent., *Pancriz* 1086 (DB). 'Tumulus on a headland'. Celtic **penn* + **crūg*. The river-name **Penk** is a back-formation from the place name.

Penmaenmawr Conwy. *Penmayne mawre* 1473, *Penmaen mawr* 1795. '(Place by) Penmaen Mawr'. The prominent headland *Penmaen Mawr* has a name meaning 'great stone headland' (Welsh *penmaen* + *mawr*).

Penmark Vale Glam. *Penmarc* 1153, *Penmarch* 13th cent. 'Height of the horse'. Welsh *pen* + *march*.

Penn, '(place at) the head, end or top', Celtic **penn*: **Penn** Bucks. *Penna* 1188. **Penn, Lower** Staffs. & **Penn, Upper** Wolverh. *Penne* 1086 (DB).

Pennard, East & Pennard, West Somerset. *Pengerd* 681, *Pennarminstre* 1086 (DB). Celtic **penn* 'head, end' with either **garth* 'ridge' or **arth* 'height'. The Domesday form contains OE *mynster* 'large church'.

Pennines, The, a name first appearing in the 18th cent. and of unknown origin, perhaps based partly on the Celtic element **penn* 'head, end, top' or an invention influenced by the name Apennines for the Italian mountain range.

Pennington, probably 'farmstead paying a penny rent', OE *pening* + *tūn*: **Pennington** Cumbria. *Pennigetun* 1086 (DB). **Pennington, Lower** Hants. *Peynton* 12th cent.

Pennsylvania S. Glos. A transferred name from the American state so called. There are examples of the same name in Devon and Dorset.

Penny Bridge Cumbria. a recent name, from a local family called *Penny*.

Pennycomequick Devon. *Penicomequick* 1643. Originally perhaps a nickname for a prosperous farm or productive piece of ground.

Penrhyn Gwyr. *See* WORMS HEAD.

Penrhyn Llŷn. *See* LLEYN PENINSULA.

Penrhyndeudraeth Gwyd. *Penrindeudrait* 1292. 'Promontory between two beaches'. Welsh *penrhyn* + *dau* + *traeth*. The village stands between *Traeth Mawr* ('big beach') and *Traeth Bach* ('little beach').

Penrith Cumbria. *Penrith* c.1100. 'Headland by the ford'. Celtic **penn* + **rïd*.

Penrose Cornwall. *Penros* 1286. 'End of the moorland or hill-spur'. Cornish *penn ros*.

Penruddock Cumbria. *Penreddok* 1278. Perhaps 'headland by the little ford'. Celtic **penn* + **rïdōg*.

Penryn Cornwall. *Penryn* 1236. 'The promontory or headland'. Cornish *penn rynn*.

Pensax Worcs. *Pensaxan* 11th cent. 'Hill of the Saxons'. Celtic **penn* 'head, end, top' + **Sachson*. The order of elements is Celtic.

Pensby Wirral. *Penisby* c.1229. 'Farmstead or village at a hill called *Penn*'. Celtic **penn* 'head, end, top' + OScand. *bý*.

Penselwood Somerset. *Penne* 1086 (DB), *Penne in Selewode* 1345. Celtic **penn* 'head, end, top' with the later addition 'in SELWOOD'.

Pensford B. & NE. Som. *Pensford* 1400. OE *ford* 'a ford' with an uncertain first element, possibly OE *penn* 'enclosure' or a hill-name *Penn* from Celtic **penn* 'head, top'.

Penshaw Sundld. *Pencher* 1183. Origin uncertain, but possibly 'the head of the rocks', Celtic **penn* + **cerr*.

Penshurst Kent. *Pensherst* 1072. Possibly 'wooded hill of a man called **Pefen*'. OE pers. name + *hyrst*.

Pentire, West Pentire Cornwall. *Pentir* c.1270. 'The headland'. Cornish *penn tir*.

Pentland Firth Highland. Orkn. *Pettaland fjorthr* c.1085. 'Sea inlet in the land of the Picts'. OScand. *Pett* + *land* + *fjorthr*.

Pentlow Essex. *Pentelawe* c.1045, *Pentelauua* 1086 (DB). Probably 'hill or tumulus of a man called **Penta*'. OE pers. name + *hlāw*.

Pentney Norfolk. *Penteleiet* [*sic*] 1086 (DB), *Pentenay* 1200. Possibly 'island, or dry ground in marsh, of a man called **Penta*'. OE pers. name (genitive *-n*) + *ēg*. Alternatively the first element may be a Celtic river-name *Pante* (probably from *pant* 'a hollow').

Penton Mewsey Hants. *Penitone* 1086 (DB), *Penitune Meysi* 1264. 'Farmstead paying a penny rent'. OE *pening* + *tūn*. Manorial affix from the *de Meisy* family, here in the 13th cent.

Pentraeth Angl. *Pentrayth* 1254. 'End of the beach'. Welsh *pen* + *traeth*.

Pentrich Derbys. *Pentric* 1086 (DB). 'Hill of the boar'. Celtic **penn* 'head, end, top' + **tyrch*.

Pentridge Dorset. *Pentric* 762, 1086 (DB). Identical in origin with the previous name.

Pentyrch Card. *Penntirch* 12th cent. 'Hill of the boar'. Welsh *pen* + *twrch*.

Penwith Cornwall. *See* ST JUST.

Penwortham, Higher & Penwortham, Lower Lancs. *Peneverdant* [*sic*] 1086 (DB), *Penuertham* 1149. Probably 'enclosed homestead at a hill called *Penn*'. Celtic **penn* 'head, end, top' + OE *worth* + *hām*.

Pen-y-groes Gwyd. *Pen-y-groes* 1838. 'Top of the crossroads'.

Penzance Cornwall. *Pensans* 1284. 'Holy headland'. Cornish *penn* + *sans*.

Peopleton Worcs. *Piplincgtun* 972, *Piplintune* 1086 (DB). Probably 'estate associated with a man called **Pyppel*'. OE pers. name + *-ing* + *tūn*.

Peover, Lower & Peover, Over Ches. *Pevre* 1086 (DB). Named from the River Peover, a Celtic river-name meaning 'the bright one'.

Peper Harow Surrey. *Pipereherge* 1086 (DB). Probably 'heathen temple of the pipers'. OE *pīpere* + *hearg*.

Peplow Shrops. *Papelau* 1086 (DB). 'Pebble mound or tumulus'. OE **papol* + *hlāw*.

Perivale Gtr. London. *Pyryvale* 1508. 'Pear-tree valley'. ME *perie* + *vale*.

Perranarworthal Cornwall. *Arewethel* 1181, 'church of *Sanctus Pieranus* in *Arwothel*' 1388. 'Parish of St Piran (in the place) beside the marsh'. Cornish saint's name + **ar* + **goethel*.

Perranporth Cornwall. *St Perins creeke* 1577, *Perran Porth* 1810. 'Cove or harbour of St Piran's parish', from English dialect *porth*, with reference to PERRANZABULOE.

Perranuthnoe Cornwall. *Odenol* 1086 (DB), 'church of *Sanctus Pieranus* of *Udno*' 1373. 'Parish of St Piran in the manor of *Uthno*' (meaning obscure).

Perranzabuloe Cornwall. *Lanpiran* 1086 (DB), *Peran in Zabulo* 1535. 'Parish of St Piran in the sand'. Cornish saint's name (with **lann* 'church-site' in the first form) + Latin *in sabulo*.

Perrott, named from the River Parrett, a pre-English river-name of obscure origin: **Perrott, North** Somerset. *Peddredan* c.1050, *Peret* 1086 (DB). **Perrott, South** Dorset. *Pedret* 1086 (DB), *Suthperet* 1268.

Perry Barr Birm. *Pirio* 1086 (DB). '(Place at) the pear-tree'. OE *pirige*. It is close to GREAT BARR.

Pershore Worcs. *Perscoran* 972, *Persore* 1086 (DB). 'Slope or bank where osiers grow'. OE **persc* + *ōra*.

Pertenhall Beds. *Partenhale* 1086 (DB). 'Nook of land of a man called **Pearta*'. OE pers. name (genitive *-n*) + *halh*.

Perth Perth. *Pert* c.1128, *St Johnstoun or Perth* 1220. '(Place by a) thicket or copse'. Pictish **perta-*. Formerly also *St Johnstoun*, from the dedication of the church to St John the Baptist.

Perton Staffs. *Pertona* 1167. 'Farmstead where pears grow'. OE *peru* + *tūn*.

Peter Tavy Devon. *See* TAVY.

Peterborough Peterb. *Burg* 1086 (DB), *Petreburgh* 1333. 'St Peter's town'. OE *burh*, with saint's name from the dedication of the abbey. The original monastery here, founded in the 7th cent., was called *Medeshamstede*, 'homestead of a man called **Mēde*', OE pers. name + *hām-stede*. The *Soke* of Peterborough is from OE *sōcn* 'district under a particular jurisdiction'.

Peterchurch Herefs. *Peterescherche* 1302. 'Church dedicated to St Peter'. OE *cirice*.

Peterhead Aber. *Petyrheid* 1544. 'St Peter's headland'. OE *hēafod*. The headland is named after the former St Peter's Kirk. An earlier name was *Inverugie*, 'mouth of the River Ugie' (Gaelic *inbhir*).

Peterlee Durham. a recent name for a new town founded 1948, coined to commemorate the former miners' leader Peter Lee (died 1935).

Peters Marland Devon. *See* MARLAND.

Petersfield Hants. *Peteresfeld* 1182. Probably '(settlement at) the open land with a church dedicated to St Peter'. OE *feld*.

Petersham Gtr. London. *Patricesham* 1086 (DB). 'River-bend land of a man called *Peohtrīc'. OE pers. name + *hamm*.

Peterstow Herefs. *Peterestow* 1207. 'Holy place of St Peter'. OE *stōw*.

Petham Kent. *Piteham* 1086 (DB). 'Homestead or enclosure in a hollow'. OE *pytt* + *hām* or *hamm*.

Petherick, Little Cornwall. 'Parish of *Sanctus Petrocus Minor*' 1327. From the dedication of the church to St Petrock.

Petherton, 'farmstead on the River Parrett', pre-English river-name of obscure origin + OE *tūn*: **Petherton, North** Somerset. *Nordperet, Peretune* 1086 (DB). **Petherton, South** Somerset. *Sudperetone* 1086 (DB).

Petherwin, '(church of) Padern the blessed', from the Cornish patron saint of the church + *gwynn* 'blessed': **Petherwin, North** Cornwall. *North Piderwine* 1259. **Petherwin, South** Cornwall. *Suthpydrewyn* 1269.

Petrockstow Devon. *Petrochestou* 1086 (DB). 'Holy place of St Petrock', Cornish saint's name + OE *stōw*.

Pett E. Sussex. *Pette* 1195. '(Place at) the pit'. OE *pytt*.

Pettaugh Suffolk. *Petehaga* 1086 (DB). 'Enclosure of a man called Pēota'. OE pers. name + *haga*.

Pettigoe (*Paiteago*) Donegal, Fermanagh. 'Lump'.

Pettistree Suffolk. *Petrestre* 1253. 'Tree of a man called Peohtrēd'. OE pers. name + *trēow*.

Petton Devon. *Petetona* c.1150. 'Farmstead of a man called *Peatta'. OE pers. name + *tūn*.

Petton Shrops. *Pectone* 1086 (DB). 'Farmstead on or by a pointed hill'. OE *pēac* + *tūn*.

Petts Wood Gtr. London. named from a ship-building family called *Pett* recorded as having a lease of oak woods in Chislehurst in 1577.

Petworth W. Sussex. *Peteorde* 1086 (DB). 'Enclosure of a man called Pēota'. OE pers. name + *worth*.

Pevensey E. Sussex. *Pefenesea* 947, *Pevenesel* 1086 (DB). 'River of a man called *Pefen'. OE pers. name + *ēa*.

Pewsey Wilts. *Pefesigge* c.880, *Pevesie* 1086 (DB). 'Island or well-watered land of a man called *Pefe'. OE pers. name + *ēg*.

Pexall Ches. *Pexhull* 1274. 'Hill called *Pēac'. OE *pēac* 'peak, pointed hill' + *hyll*.

Pharis (*Fáras*) Antrim. *Faras* 1780. 'Residence'.

Phillack Cornwall. 'Church of *Sancta Felicitas'* [sic] 1259. *Felok* 1388. From the dedication of the church to a St Felek.

Philleigh Cornwall. 'Church of *Sanctus Filius*' 1312. From the dedication of the church to a St Fily.

Phoenix Park (*Páirc an Fhionnuisce*) Dublin. 'Park of the clear water'. The English name, a corruption of the Irish, has been perpetuated by the Phoenix Column (1745), surmounted by a phoenix.

Piccadilly Gtr. London. Street named from *Pickadilly Hall* 1623, a house so called because its owner, a successful tailor, made his fortune from the sale of *piccadills* or *piccadillies*, a fashionable kind of collar at the time.

Pickenham, North & Pickenham, South Norfolk. *Pichenham* 1086 (DB). Probably 'homestead or village of a man called *Pīca'. OE pers. name (genitive *-n*) + *hām*.

Pickering N. Yorks. *Picheringa* 1086 (DB). Possibly '(settlement of) the family or followers of a man called *Pīcer'. OE pers. name + *-ingas*.

Pickhill N. Yorks. *Picala* 1086 (DB). 'Nook of land by the pointed hills', or 'nook of a man called *Pīca'. OE *pīc* or pers. name + *halh*.

Picklescott Shrops. *Pikelescote* 1204. Probably 'cottage(s) of a man called Pīcel', OE pers. name + *cot*. Alternatively the first element could be an OE hill-name *Pīcel 'pointed hill', a derivative of OE *pīc* 'point'.

Pickmere Ches. *Pichemere* 12th cent. 'Lake where pike are found'. OE *pīc* + *mere*.

Pickwell Leics. *Pichewelle* 1086 (DB). 'Spring or stream by the pointed hill(s)'. OE *pīc* + *wella*.

Pickwick Wilts. *Pykewyke* 1268. Probably 'dairy farm on or by a pointed spur of land'. OE *pīc* + *wīc*.

Pickworth, 'enclosure of a man called *Pīca', OE pers. name + *worth*: **Pickworth** Lincs.

Picheuuorde 1086 (DB). **Pickworth** Rutland. *Pichewurtha* 1170.

Picton, 'farmstead of a man called *Pīca', OE pers. name + *tūn*: **Picton** Ches. *Picheton* 1086 (DB). **Picton** N. Yorks. *Piketon* 1200.

Piddinghoe E. Sussex. *Pidingeho* 12th cent. 'Hill-spur of the family or followers of a man called *Pyda'. OE pers. name + *-inga-* + *hōh*.

Piddington, 'estate associated with a man called *Pyda', OE pers. name + *-ing-* + *tūn*: **Piddington** Northants. *Pidentone* 1086 (DB). **Piddington** Oxon. *Petintone* 1086 (DB).

Piddle, North & Piddle, Wyre Worcs. *Pidele(t)* 1086 (DB), *Northpidele* 1271, *Wyre Pidele* 1208. Named from Piddle Brook (OE **pidele* 'a marsh or fen'). Affix *Wyre* is probably from WYRE FOREST.

Piddlehinton Dorset. *Pidele* 1086 (DB), *Pidel Hineton* 1244. 'Estate on the River Piddle belonging to a religious community'. OE river-name (**pidele* 'a marsh or fen') + *hīwan* + *tūn*.

Piddletrenthide Dorset. *Uppidelen* 966, *Pidrie* 1086 (DB), *Pidele Trentehydes* 1212. 'Estate on the River Piddle assessed at thirty hides'. OE river-name (*see* PIDDLEHINTON) + OFrench *trente* + OE *hīd*.

Pidley Cambs. *Pydele* 1228. 'Woodland clearing of a man called *Pyda'. OE pers. name + *lēah*.

Piercebridge Darltn. *Persebrigce c.*1040. Possibly 'bridge where osiers grow'. OE **persc* + *brycg*. Alternatively the name may mean 'osier bridge or causeway', referring to a causeway of faggots across marshy ground.

Pigdon Northum. *Pikedenn* 1205. 'Valley of a man called *Pīca, or by the pointed hills'. OE pers. name or OE *pīc* + *denu*.

Pike (*Píce*) Laois, Tipperary, Waterford. 'Pike'.

Pilgrims Hatch Essex. *Pylgremeshacch* 1483. 'Hatch-gate used by pilgrims' (to the chapel of St Thomas). ME *pilegrim* + OE *hæcc*.

Pilham Lincs. *Pileham* 1086 (DB). 'Homestead made with stakes' or 'homestead of a man called Pīla'. OE *pīl* or OE pers. name + *hām*.

Pill N. Som. Recorded as *Pill or Crockern Pill* 1830, from Somerset dialect *pill* 'small stream' (OE *pyll*). *Crockern* may represent 'pottery (ME *crokkerne*) or 'of the potters' (ME *crokkerene*).

Pillaton Cornwall. *Pilatona* 1086 (DB). 'Farmstead made with stakes'. OE *pīl* + *tūn*.

Pillerton Hersey & Pillerton Priors Warwicks. *Pilardintone* 1086 (DB), *Pilardinton Hersy, Pilardinton Prior* 1247. 'Estate associated with a man called Pīlheard'. OE pers. name + *-ing-* + *tūn*. Manorial affixes from possession by the *de Hersy* family, here in the 13th cent., and by the *Prior* of Sheen.

Pilley Barns. *Pillei* 1086 (DB). 'Wood or clearing where stakes are obtained'. OE *pīl* + *lēah*.

Pilling Lancs. *Pylin c.*1195. Named from the River Pilling, probably OE *pyll* 'creek' + *-ing*.

Pilning S. Glos. *Pyllyn* 1529. Probably 'district by the creek or stream'. OE *pyll* + *ende*.

Pilsbury Derbys. *Pilesberie* 1086 (DB). 'Fortified place of a man called *Pīl'. OE pers. name + *burh* (dative *byrig*).

Pilsdon Dorset. *Pilesdone* 1086 (DB). Probably 'hill with a peak', or 'hill marked by a stake'. OE *pīl* + *dūn*.

Pilsley Derbys., near Clay Cross. *Pilleslege c.*1002, *Pinneslei* 1086 (DB). Possibly 'woodland clearing of a man called *Pinnel'. OE pers. name + *lēah*.

Pilton Northants. *Pilchetone* 1086 (DB). 'Farmstead of a man called *Pīleca'. OE pers. name + *tūn*.

Pilton Rutland. *Pilton* 1202. Probably 'farmstead by a stream'. OE *pyll* + *tūn*.

Pilton Somerset. *Piltune* 725, *Piltone* 1086 (DB). Identical in origin with the previous name.

Piltown (*Baile an Phoill*) Kilkenny. 'Townland of the creek'.

Pimlico Gtr. London. District recorded thus in 1626 and so named from an alehouse in HOXTON referred to as *Pimlyco* 1609, this in turn called after its publican Ben *Pimlico* 1598. The unusual surname and place name may well be a transferred name from the *Pamlico* Indians of North America who lived along the banks of the Pamlico River near to the abortive settlements of Sir Walter Raleigh's Virginia founded in the 1580s.

Pimperne Dorset. *Pimpern* 935, *Pinpre* 1086 (DB). Possibly 'five trees' from Celtic **pimp* + *prenn*, or 'place among the hills' from a derivative of an OE word **pimp* 'hill'.

Pinchbeck Lincs. *Pincebec* 1086 (DB). 'Minnow stream', or 'finch ridge'. OE **pinc* or **pinca* + *bece* (influenced by OScand. *bekkr*) or *bæc* (locative **bece*).

Pinhoe Devon. *Peonho* [*sic*] *c.*1050, *Pinnoc* 1086 (DB), *Pinho c.*1220. OE *hōh* 'hill-spur', possibly with OE *pinn* 'pin, peg, pointed hill'.

Pinner Gtr. London. *Pinnora* 1232. 'Peg-shaped or pointed ridge or hill'. OE *pinn* + *ōra*. The river-name **Pinn** is a back-formation from the place name.

Pinvin Worcs. *Pendefen* 1187. 'Fen of a man called Penda'. OE pers. name + *fenn*.

Pinxton Derbys. *Penkeston* 1208. Possibly 'farmstead of a man called **Penec*'. OE pers. name + *tūn*.

Pipe Herefs. *Pipe* 1086 (DB). OE *pīpe* 'a pipe, a conduit', originally referring to the stream here.

Pipe Ridware Staffs. *See* RIDWARE.

Pipewell Northants. *Pipewelle* 1086 (DB). 'Spring or stream with a pipe or conduit'. OE *pīpe* + *wella*.

Pippacott Devon. *Pippecote* 1311. 'Cottage(s) of a man called **Pyppa*'. OE pers. name + *cot*.

Pirbright Surrey. *Perifrith* 1166. 'Sparse woodland where pear-trees grow'. OE *pirige* + *fyrhth*.

Pirton, 'pear orchard, or farmstead where pear-trees grow', OE *pirige* + *tūn*: **Pirton** Herts. *Peritone* 1086 (DB). **Pirton** Worcs. *Pyritune* 972, *Peritune* 1086 (DB).

Pishanagh (*Piseánach*) Westmeath. 'Vetches'.

Pishill Oxon. *Pesehull* 1195. 'Hill where peas grow'. OE *pise* + *hyll*.

Pitcairn Perth. *Peticarne* 1247. 'Portion of the mound'. Pictish **pett* + Gaelic *càrn* (genitive *càirn*).

Pitcaple Aber. 'Portion of the mare'. Pictish **pett* + Gaelic *capull* (genitive *capuill*).

Pitchcott Bucks. *Pichecote* 1176. 'Cottage or shed where pitch is made or stored'. OE *pic* + *cot*.

Pitchford Shrops. *Piceforde* 1086 (DB). 'Ford near a place where pitch is found'. OE *pic* + *ford*.

Pitcombe Somerset. *Pidecombe* 1086 (DB). Probably 'marsh valley'. OE **pide* + *cumb*.

Pitcorthie Fife. *Pittecorthin* 1093. 'Portion with a pillar stone'. Pictish **pett* + Gaelic *coirthe*.

Pitcruvie Fife. 'Portion of the tree'. Pictish **pett* + Gaelic *craobh* (genitive *craoibhe*).

Pitfoskie Aber. 'Portion by a shelter'. Pictish **pett* + Gaelic *fosgadh*.

Pitglassie Highland. 'Portion of the meadow'. Pictish **pett* + Gaelic *ghlasaich*.

Pitliver Fife. *Petlyver* 1450. 'Portion of the book'. Pictish **pett* + Gaelic *leabhar* (genitive *leabhair*). The place may have been church property.

Pitlochry Perth. 'Portion of the stones'. Pictish **pett* + Gaelic *cloichreach*. The reference is probably to stepping stones over the River Tummel.

Pitlurg Aber. *Petnalurge* 1232. 'Portion of the shank'. Pictish **pett* + Gaelic *lorg*.

Pitmedden Aber. 'Middle portion'. Pictish **pett* + Gaelic *medon*.

Pitmillan Aber. *Pitmulen c.*1315. 'Portion of the mill'. Pictish **pett* + Gaelic *muileann*.

Pitminster Somerset. *Pipingmynstre* 938, *Pipeminstre* 1086 (DB). 'Church associated with a man called Pippa'. OE pers. name + *-ing-* + *mynster*.

Pitney Somerset. *Petenie* 1086 (DB). 'Island, or dry ground in marsh, of a man called **Pytta* or Pēota'. OE pers. name (genitive *-n*) + *ēg*.

Pitsea Essex. *Piceseia* 1086 (DB). 'Island, or dry ground in marsh, of a man called Pīc'. OE pers. name + *ēg*.

Pitsford Northants. *Pitesford* 1086 (DB). 'Ford of a man called **Peoht*'. OE pers. name + *ford*.

Pitsligo Aber. *Petslegach* 1426. 'Portion of the shelly place'. Pictish **pett* + Gaelic *sligeach*.

Pitstone Bucks. *Pincelestorne* [*sic*] 1086 (DB), *Pichelesthorne* 1220. 'Thorn-tree of a man called Pīcel'. OE pers. name + *thorn*.

Pittendreich Aber. 'Portion on the hill face'. Pictish **pett* + Gaelic *drech*.

Pittenweem Fife. *Petnaweme c.*1150. 'Portion of the cave'. Pictish **pett* + Gaelic *na h-uamha*.

Pittington Durham. *Pitindun c.*1150. 'Hill associated with a man called *Pytta'. OE pers. name + *-ing-* + *dūn*.

Pitton Wilts. *Putenton* 1165. 'Farmstead of a man called Putta, or one frequented by hawks'. OE pers. name or OE *putta* + *tūn*.

Pity Me Durham. a whimsical name bestowed in the 19th cent. on a place considered desolate, exposed, or difficult to cultivate. The same name occurs elsewhere, e.g. as **Pityme** Cornwall.

Plaish Shrops. *Plesham* 1086 (DB), *Playsse* 1255. '(Place at) the shallow pool'. OE **plæsc*.

Plaistow W. Sussex. *La Pleyestowe* 1271. 'Place for play or sport'. OE *pleg-stōw*. The same name occurs in other counties.

Plaitford Hants. *Pleiteford* 1086 (DB). Probably 'ford where play or sport takes place'. OE **pleget* + *ford*.

Plawsworth Durham. *Plaweswortha* 1174. 'Enclosure for games or sports'. OE *plaga* (ME plural *plawes*) + *worth*.

Plaxtol Kent. *Plextole* 1386. 'Place for play or sport'. OE *pleg-stōw*.

Playden E. Sussex. *Pleidena* 1086 (DB). 'Woodland pasture where play or sport takes place'. OE *plega* + *denn*.

Playford Suffolk. *Plegeforda* 1086 (DB). 'Ford where play or sport takes place'. OE *plega* + *ford*.

Plealey Shrops. *Pleyle* 1256. 'Woodland clearing where play or sport takes place'. OE *plega* + *lēah*.

Pleasington Black. w. Darw. *Plesigtuna* 1196. 'Estate associated with a man called Plēsa'. OE pers. name + *-ing-* + *tūn*.

Pleasley Derbys. *Pleseleia* 1166. 'Woodland clearing of a man called Plēsa'. OE pers. name + *lēah*.

Plenmeller Northum. *Playnmelor* 1279. Probably 'top of the bare hill'. Celtic **blain* + **mēl* + **breȝ*.

Pleshey Essex. *Plaseiz c.*1150. OFrench *plaisseis* 'an enclosure made with interlaced fencing'.

Plowden Shrops. *Plaueden* 1252. 'Valley where play or sport takes place'. OE *plaga* + *denu*.

Pluckley Kent. *Pluchelei* 1086 (DB). 'Woodland clearing of a man called Plucca'. OE pers. name + *lēah*.

Plumbland Cumbria. *Plumbelund c.*1150. 'Plum-tree grove'. OE *plūme* + OScand. *lundr*.

Plumbridge (*Droichead an Phlum*) Tyrone. *Plumb Bridge* 1837. Possibly 'bridge by the deep pool'. Scottish *plum* + English *bridge*. The bridge is in a deep valley of the Glenelly River.

Plumley Ches. *Plumleia* 1119. 'Plum-tree wood or clearing'. OE *plūme* + *lēah*.

Plumpton, 'farmstead where plum-trees grow', OE *plūme* + *tūn*: **Plumpton** Cumbria. *Plumton* 1212. **Plumpton** E. Sussex. *Pluntune* 1086 (DB). **Plumpton, Great & Plumpton, Little** Lancs. *Pluntun* 1086 (DB).

Plumstead, 'place where plum-trees grow', OE *plūme* + *stede*: **Plumstead** Gtr. London. *Plumstede* 961–9, *Plumestede* 1086 (DB). **Plumstead** Norfolk, near Holt. *Plumestede* 1086 (DB). **Plumstead, Great & Plumstead, Little** Norfolk. *Plumesteda* 1086 (DB).

Plumtree Notts. *Pluntre* 1086 (DB). '(Place at) the plum-tree'. OE *plūm-trēow*.

Plungar Leics. *Plungar, Plumgard c.*1130. 'Triangular plot where plum-trees grow', or 'plum-tree enclosure'. OE *plūme* + OE *gāra* or OScand. *garthr*.

Plush Dorset. *Plyssch* 891. '(Place at) the shallow pool'. OE **plysc*.

Plymouth Plym. *Plymmue* 1230. 'Mouth of the River Plym'. OE *mūtha* with a river-name formed from the next name by the process of back-formation.

Plympton Plym. *Plymentun c.*905, *Plintona* 1086 (DB). 'Farmstead of the plum-tree'. OE *plȳme* + *tūn*.

Plymstock Plym. *Plemestocha* 1086. Probably 'outlying farmstead or hamlet connected with PLYMPTON'. OE *stoc*.

Plymtree Devon. *Plumtrei* 1086 (DB). '(Place at) the plum-tree'. OE **plȳmtrēow*.

Plynlimon (*Pumlumon*) (mountain) Cergn. 'Five beacons'. Welsh *pum* + *llumon*.

Pockley N. Yorks. *Pochelai* [sic] 1086 (DB), *Pokelai c.*1190. 'Woodland clearing of a man called *Poca'. OE pers. name + *lēah*.

Pocklington E. R. Yorks. *Poclinton* 1086
(DB). 'Estate associated with a man called
*Pocela'. OE pers. name + *-ing-* + *tūn.*

Podimore Somerset. *Puddimore* 1811.
Probably identical in origin with PODMORE.

Podington Beds. *Podintone* 1086 (DB).
'Estate associated with a man called Poda or
Puda'. OE pers. name + *-ing-* + *tūn.*

Podmore Staffs. *Podemore* 1086 (DB). 'Marsh
frequented by toads or frogs'. ME *pode* + OE
mōr.

Pointon Lincs. *Pochinton* 1086 (DB). 'Estate
associated with a man called Pohha'. OE pers.
name + *-ing-* + *tūn.*

Poisoned Glen (valley) Tyrone. The English
name translates Irish *Cró Nimhe*, 'glen of
poison', perhaps referring to some poisonous
plant that grew here.

Polden Hills Somerset. *See* CHILTON
POLDEN.

Polebrook Northants. *Pochebroc* 1086 (DB).
'Brook by a pouch-shaped feature', OE *pohha* +
brōc. Alternatively the first element may be
an OE **poc(c)e* 'frog'.

Polegate E. Sussex. first recorded as
Powlegate Corner 1563. 'Gate by the pool', from
ME *po(o)le.*

Poles (*Pollaí*) Cavan. 'Pools'.

Polesworth Warwicks. *Polleswyrth c.*1000.
'Enclosure of a man called **Poll'. OE pers.
name + *worth.*

Polfore (*Poll Fuar*) Tyrone. 'Cold pool'.

Poling W. Sussex. *Palinges* 1199. '(Settlement
of) the family or followers of a man called **Pāl'.
OE pers. name + *-ingas.*

Pollagh (*Pollach*) Mayo, Offaly. 'Abounding
in holes'.

Pollans (*Pollán*) Donegal. 'Little hole'.

Pollatomish (*Poll an Tómais*) Mayo.
'Tómas's hole'.

Pollboy (*Poll Buí*) Galway. 'Yellow hole'.

Pollicott, Upper Bucks. *Policote* 1086 (DB).
Possibly 'cottage(s) associated with a man
called **Poll'. OE pers. name + *-ing-* + *cot.*

Pollington E. R. Yorks. *Pouelington c.*1185.
'Farmstead associated with a piece of ground
called *Pofel'. OE **pofel* (of uncertain meaning) +
-ing- + *tūn.*

Pollokshaws Glas. *Pullock* 1158. 'The woods
of Pollock', from ME *shawe* 'small wood, copse'.
The place was originally *Pollok*, 'little pool',
from British *poll* + suffix. The second element
was then added to distinguish it from nearby
Pollokshields, 'the sheds of Pollock', from ME
schele 'shed, hut'.

Pollokshields. *See* POLLOKSHAWS.

Pollremon (*Poll Réamainn*) Galway.
'Réamann's hole'.

Pollrone (*Poll Ruadhain*) Kilkenny.
'Ruadhan's hole'.

Polmadie Glas. *Polmacde c.*1185, *Polmadie*
1617. 'Pool of the sons of Daigh'. British *poll* +
Gaelic *mac* (genitive plural *mic*).

Polperro Cornwall. *Portpira* 1303. Probably
'harbour of a man called **Pyra'. Cornish *porth* +
pers. name.

Polruan Cornwall. *Porthruan* 1284. 'Harbour
of a man called Ruveun'. Cornish *porth* + pers.
name.

Polsham Somerset. *Paulesham* 1065.
'Enclosure of a man called Paul'. ME pers. name
+ OE *hamm.*

Polstead Suffolk. *Polstede c.*975, *Polesteda*
1086 (DB). 'Place by the pools'. OE *pōl* + *stede.*

Poltimore Devon. *Pultimore* 1086 (DB).
Possibly 'marshy ground of a man called **Pulta'.
OE pers. name + *mōr.*

Polyphant Cornwall. *Polefand, Polofant* 1086
(DB). 'Toad's pool'. OCornish *pol* + **lefant.*

Pomeroy Tyrone. *Pomeroy Demesne c.*1834.
Possibly 'apple orchard'. OFrench *pommeraie.*

Pont Henry (*Pont-henri*) Carm. *Pont Henry
c.*1627, *Pont Hendry* 1760. 'Henry's bridge'.
Welsh *pont.* The identity of the named man is
unknown.

Pont-y-Pwl. *See* PONTYPOOL.

Pontardawe Neat. *Tir penybont ardawe*
1583, *Ty pen y bont ar y Tawe* 1675, *Ty pen y
bont ar tawey* 1706. 'Bridge over the River
Tawe'. Welsh *pont* + *ar.* For the river-name,

see SWANSEA. The early forms relate to a house (Welsh *tŷ*).

Pontarddulais Swan. *Ponte ar theleys* 1557. 'Bridge over the River Dulais'. Welsh *pont* + *ar*. The river-name means 'dark stream' (Welsh *du* + *glais*).

Pontefract Wakefd. *Pontefracto* 1090. 'Broken bridge'. Latin *pons* + *fractus*.

Ponteland Northum. *Punteland* 1203. 'Cultivated land by the river called Pont'. Celtic river-name (meaning 'valley') + OE *ēa-land*.

Pontesbury Shrops. *Pantesberie* 1086 (DB). Probably 'fortified place of a man called Pant'. OE pers. name + *burh* (dative *byrig*).

Pont-faen Pemb. *Pons Lapideus* 13th cent., *Pontvaen* 13th cent. '(Place by the) stone bridge'. Welsh *pont* + *maen*.

Pontllan-fraith Cphy. *tre penybont llynvraith* 1492, *Pontllynfraith* 1713. 'Bridge by the speckled pool'. Welsh *pont* + *llyn* + *braith*. The first two words of the first form above mean 'farm at the end of the bridge'.

Ponton, Great & Ponton, Little Lincs. *Pamtone, Pamptune, Magna-, Parva Pantone* 1086 (DB), *Panton(e)* 12th cent. OE *tūn* 'farmstead' with an uncertain first element, possibly an OE **pamp* 'hill', but the name may be identical with PANTON.

Pontycymer Bri. 'Bridge at the confluence'. Welsh *pont* + *y* + *cymer*.

Pontypool (*Pont-y-Pŵl*) Torf. *Pont y poole* 1614. 'Bridge by the pool'. Welsh *pont* + *y* + ModE *pool*.

Pontypridd Rhon. *Pont y Ty Pridd* c.1700. 'Bridge by the earthen house'. Welsh *pont* + *y* + *tŷ* + *pridd*.

Pool, Poole, usually '(place at) the pool or creek'. OE *pōl*: **Poole** Poole. *Pole* 1183. **Poole Keynes** Glos. *Pole* 10th cent., 1086 (DB). **Pool, South** Devon. *Pole* 1086 (DB).
    However the following has a different origin: **Pool** Leeds. *Pofle* c.1030, *Pouele* 1086 (DB). From an OE word **pofel* of uncertain meaning.

Poolbeg (*Poll Beag*) Dublin. 'Little pool'.

Pooley Bridge Cumbria. *Pulhoue* 1252. 'Hill or mound by a pool'. OE *pōl* + OScand. *haugr*.

Poorton, North & Poorton, South Dorset. *Pourtone* 1086 (DB). OE *tūn* 'farmstead' with an obscure first element which may be an old river-name.

Popham Hants. *Popham* 903, *Popeham* 1086 (DB). OE *hām* 'homestead' or *hamm* 'enclosure' with an uncertain first element, possibly an OE **pop(p)* 'pebble'.

Poplar Gtr. London. *Popler* 1327. '(Place at) the poplar-tree'. ME *popler*.

Poppleton, Nether & Poppleton, Upper York. *Popeltune* c.972, *Popletone* 1086 (DB). 'Farmstead on pebbly soil'. OE **popel* + *tūn*.

Porin Highland. 'Pasture'. Gaelic *pórainn*.

Poringland Norfolk. *Porringalanda* 1086 (DB). 'Cultivated land or estate of a tribe called the Porringas'. Obscure first element (probably an OE pers. name) + *-inga-* + *land*.

Porlock Somerset. *Portloca* 10th cent., *Portloc* 1086 (DB). 'Enclosure by the harbour'. OE *port* + *loca*.

Port Dinorwic (*Y Felinheli*) Gwyd. *the port of Dinorwic* 1851, *Port-Dinorwig* 1868. 'Port of Dinorwic'. ModE *port*. The town arose as a port for the export of slate from the quarries at *Dinorwic*, 'fort of the Ordovices'. The Welsh name means 'the sea mill' (Welsh *y* + *melin* + *heli*). Today, *Y Felinheli* is the only official name.

Port Ellen Arg. 'Ellen's port'. The name is that of Lady Ellenor (*Ellen*) Campbell of Islay, widow of the Gaelic scholar W. F. Campbell, who planned the settlement in 1821.

Port Gate Northum. *Portyate* 1269. OE *port* 'gate' with explanatory OE *geat* 'gate', referring to the gap in Hadrian's Wall where Watling Street runs through it.

Port Glasgow Invclyd. 'Port for Glasgow'. The town arose on the Firth of Clyde in the 1660s with the aim of becoming a port for GLASGOW.

Port Isaac Cornwall. *Portusek* 1337. From Cornish *porth* 'cove, harbour' with an uncertain second element, possibly an adjective **usek* meaning literally 'characterized by chaff' but perhaps referring to tidal debris.

Port Láirge. See WATERFORD.

Port Laoise (*Port Laoise*) Laois. 'Port of the tribe of Laeighis'. Compare LAOIS.

Port Sunlight Wirral. a modern name for the late 19th-cent. model village, called after the brand-name of the soap made here.

Port Talbot Neat. The town dates from 1836, when docks were built on Swansea Bay on land owned by the *Talbot* family of nearby Margam Abbey.

Port William Dumf. 'William's port'. The village was founded in 1770 by Sir *William* Maxwell of Monreith.

Portacloy (*Port na Cloche*) Mayo. 'Landing place of the stone'.

Portadown (*Port an Dúnáin*) Armagh. *Port a' Dúnáin* 1646. 'Landing-place of the little fort'.

Portaferry (*Port an Pheire*) Down. *Port na Peireadh c.*1617. 'Landing-place of the ferry'.

Portarlington (*Cúil an tSúdaire*) Laois. Lands here were owned in the 17th cent. by Sir Henry Bennet, Lord *Arlington*. The Irish name means 'nook of the tanner'.

Portavogie (*Port an Bhogaigh*) Down. *Portaboggach* 1605. 'Harbour of the bog'.

Portballintrae (*Port Bhaile an Trá*) Antrim. *Portballentra* 1605. 'Landing-place of Ballintrae'. The townland name Ballintrae represents Irish *Baile an Trá*, 'townland of the strand'.

Portbury N. Som. *Porberie* 1086 (DB). 'Fortified place near the harbour'. OE *port* + *burh* (dative *byrig*).

Portchester Hants. *Porteceaster c.*960, *Portcestre* 1086 (DB). 'Roman fort by the harbour called *Port*'. OE *port* + *ceaster, see* PORTSMOUTH.

Portesham Dorset. *Porteshamme* 1024, *Portesham* 1086 (DB). 'Enclosure belonging to the town' (probably referring to ABBOTSBURY). OE *port* + *hamm*.

Portglenone (*Port Chluain Eoghain*) Antrim. *Portecomeonone* 1593. 'Landing-place of Glenone'. The townland name Glenone represents Irish *Cluain Eoghain*, 'Eoghan's meadow'.

Porth Rhon. 'Gateway'. Welsh *porth*. Porth stands at the confluence of the Rhondda and Little Rhondda, a 'gateway' to places further south in the valley.

Porthaethwy. *See* MENAI BRIDGE.

Porthallow Cornwall. near Manaccan. *Worthalaw* [*sic*] 967, *Porthaleu* 1333. 'Cove or harbour by a stream called *Alaw*'. Cornish *porth* + river-name (probably pre-Celtic) of uncertain meaning.

Porth-cawl Bri. *Portcall* 1632, *Porth Cawl* 1825. 'Harbour where seakale grows'. Welsh *porth* + *cawl*.

Porthceri. *See* PORTHKERRY.

Porthkerry (*Porthceri*) Vale Glam. *Portciri* 1254. 'Harbour of Ceri'. Welsh *porth*.

Porthleven Cornwall. *Port-levan c.*1605. 'Smooth harbour', Cornish *porth* + *leven*. Alternatively 'harbour of the stream called *Leven* (the smooth one)'.

Porthmadog Gwyd. *Portmadoc* 1838. 'Port of Madocks'. Welsh *porth*. The name is that of William Alexander *Madocks* (1772–1828), MP for Boston, Lincs, who developed nearby TREMADOG.

Portington E. R. Yorks. *Portinton* 1086 (DB). 'Farmstead belonging to a market town'. OE *port* + *-ing-* + *tūn*.

Portinscale Cumbria. *Porqeneschal c.*1160. 'Shielings of the townswomen'. OE *port-cwēn(e)* + OScand. *skáli*.

Portishead N. Som. *Portesheve* [*sic*] 1086 (DB), *Portesheved* 1200. 'Headland by the harbour'. OE *port* + *hēafod*.

Portland, Isle of Dorset. *Port* 9th cent., *Portlande* 862, *Porland* 1086 (DB). 'Estate by the harbour'. OE *port* + *land*. **Portland Bill**, recorded from 1649, is from OE *bile* 'a bill, a beak' used topographically for 'promontory'.

Portlaw (*Port Lách*) Waterford. 'Port of the hill'.

Portlemouth, East Devon. *Porlamuta* [*sic*] 1086 (DB), *Porthelemuthe* 1308. Possibly 'harbour estuary', from Cornish *porth* + **heyl* with OE *mūtha* 'mouth'. Alternatively 'mouth of the harbour stream', from OE *port* + *wella* + *mūtha*.

Portloman (*Port Lomáin*) Westmeath. 'Lomán's port'.

Portmadoc. *See* PORTHMADOG.

Portmahomack Highland. *Portmachalmok* 1678. 'Harbour of Machalmac'. Gaelic *port*. *Machalmac* means 'my little Colman' (Gaelic *ma* + saint's name *Colman* + diminutive suffix *-oc*).

Portmarnock (*Port Mearnóg*) Dublin. 'Mearnóg's port'.

Portmeirion Gwyd. 'Port of Meirion'. The name was devised by the architect Clough Williams-Ellis in the 1920s for his Italianate coastal village in Merionethshire. *Compare* MERIONETH.

Portnablahy (*Port na Bláiche*) Donegal. 'Port of the buttermilk'.

Portnoo (*Port Nua*) Donegal. *Port Noo* 1849. 'New port'.

Portobello Dublin. The name commemorates the Battle of Portobello in 1739, *see* next name.

Portobello Edin. *Porto-Bello* 1753, *Portobello* 1779. The name was originally that of the *Portobello Hut*, a house built in the mid-18th cent., itself named from the capture of Portobello, Panama, by Admiral Vernon in 1739.

Porton Wilts. *Portone* 1086 (DB). Probably identical in origin with POORTON (Dorset).

Portpatrick Dumf. *Portpatrick* 1630. 'Harbour of St Patrick'. From a chapel dedicated to St Patrick.

Portree Highland. (Skye). *Portri* 1549, *Portree, or Port a Roi* 1868. 'Harbour of the slope'. Gaelic *port* + *righe*.

Portroe (*An Port Rua*) Tipperary. 'The red port'.

Portrush (*Port Rois*) Antrim. *Portros* 1262. 'Port of the promontory'.

Portsalon (*Port an tSalainn*) Donegal. 'Port of the salt'.

Portsdown Portsm. *Portesdone* 1086 (DB). 'Hill of the harbour called *Port*'. OE *port* + *dūn*, *see* PORTSMOUTH.

Portsea Portsm. *Portesig* 982. 'Island of the harbour called *Port*'. OE *port* + *ēg*, *see* PORTSMOUTH.

Portslade by Sea Bright. & Hove. *Porteslage* [*sic*] 1086 (DB), *Portes Ladda* c.1095. Probably 'crossing-place near the harbour'. OE *port* + *gelād*.

Portsmouth Portsm. *Portesmuthan* late 9th cent. 'Mouth of the harbour called *Port*'. OE *port*

(a loan-word from Latin *portus* 'a harbour') + *mūtha*. The harbour was no doubt known as *Portus* in the Roman period.

Portstewart Derry. 'Stewart's port'. *Portstewart* 1780. The town arose in 1794 on land purchased from the *Stewart* family, who had held it from 1734. The Irish equivalent name is *Port Stíobhaird*.

Portswood Sotn. *Porteswuda* 1045. 'Wood belonging to the market town (i.e. SOUTHAMPTON)'. OE *port* + *wudu*.

Portumna (*Port Omna*) Galway. 'Bank of the tree-trunk'.

Poslingford Suffolk. *Poslingeorda* 1086 (DB). 'Enclosure of the family or followers of a man called *Possel*'. OE pers. name + *-inga-* + *worth*.

Postling Kent. *Postinges* [*sic*] 1086 (DB), *Postlinges* 1212. Possibly '(settlement of) the family or followers of a man called *Possel*'. OE pers. name + *-ingas*.

Postwick Norfolk. *Possuic* 1086 (DB). 'Dwelling or (dairy) farm of a man called *Possa*'. OE pers. name + *wīc*.

Potsgrove Beds. *Potesgrave* 1086 (DB). OE *grāf(a)* 'grove, coppiced wood' with an uncertain first element, possibly OE *pott* in the sense 'pit, hole'.

Pott Shrigley Ches. *See* SHRIGLEY.

Potter Brompton N. Yorks. *See* BROMPTON.

Potter Heigham Norfolk. *See* HEIGHAM.

Potter Street Essex. *Potters streete* 1594. Named from a family called *le Pottere* ('the potmaker') here in the 13th cent.

Potterhanworth Lincs. *Haneworde* 1086 (DB). 'Enclosure of a man called Hana'. OE pers. name + *worth*, the later affix referring to potmaking here.

Potterne Wilts. *Poterne* 1086 (DB). 'Building where pots are made'. OE *pott* + *ærn*.

Potters Bar Herts. *Potterys Barre* 1509. 'Forest-gate of a family called Potter'. ME surname + *barre*.

Potterspury Northants. *Perie* 1086 (DB), *Potterispirye* 1287. Originally '(place at) the pear-tree', from OE *pirige*, with later affix *Potters-* 'of the pot-maker(s)' from the pottery here. *See* PAULERSPURY.

Pottiaghan (*Poiteacháin*) Westmeath. 'Bad ground'.

Potto N. Yorks. *Pothow* 1202, 'Mound or hill by a pit or where pots were found'. OE *pott* + OScand. *haugr*.

Potton Beds. *Pottun c.*960, *Potone* 1086 (DB), 'Farmstead where pots are made'. OE *pott* + *tūn*.

Poughill Cornwall. *Pochehelle* [*sic*] 1086 (DB), *Pochewell* 1227. 'Spring or stream by a feature resembling a pouch or bag', or 'spring or stream of a man called Pohha'. OE *pohha* or pers. name + *wella*.

Poughill Devon. *Pochehille* 1086 (DB). 'Bag-shaped hill', or 'hill of a man called Pohha'. OE *pohha* or pers. name + *hyll*.

Poulaphuca (*Poll an Phúca*) Dublin. 'Pool of the sprite'.

Poulargid (*Poll an Airgid*) Cork. 'Silver pool'.

Poulnagat (*Poll na gCat*). Westmeath. 'Hole of the cats'.

Poulshot Wilts. *Paveshou* [*sic*] 1086 (DB), *Paulesholt* 1187. 'Wood of a man called Paul', ME pers. name + OE *holt*.

Poulton, 'farmstead by a pool or creek', OE **pull* or *pōl* + *tūn*: **Poulton** Glos. *Pultune* 855. **Poulton** Wirral. *Pulton* 1260. **Poulton-le-Fylde** Lancs. *Poltun* 1086 (DB). Affix means 'in the district called The Fylde', from OE *filde* 'a plain' (recorded as *Filde* 1246).

Poundon Bucks. *Paundon* 1255. OE *dūn* 'hill' with an uncertain first element, possibly OE *pāwa* (genitive -*n*) 'peacock'.

Poundstock Cornwall. *Pondestoch* 1086 (DB). 'Outlying farmstead or hamlet with a pound for animals'. OE **pund* + *stoc*.

Powderham Devon. *Poldreham* 1086 (DB). 'Promontory in reclaimed marshland'. OE **polra* + *hamm*.

Powerstock Dorset. *Povrestoch* 1086 (DB). OE *stoc* 'outlying farmstead' with the same obscure first element as in POORTON.

Powick Worcs. *Poincguuic* 972, *Poiwic* 1086 (DB). 'Dwelling or farm associated with a man called Pohha'. OE pers. name + -*ing*- + *wīc*.

Powys (the unitary authority). 'Provincial (place)'. Low Latin *pagensis*. The name implies that those who lived here were 'country folk',

inhabiting an open upland tract that was not protected in the same way as the regions to the north and south, with their hills and valleys.

Poxwell Dorset. *Poceswylle* 987, *Pocheswelle* 1086 (DB). 'Steeply rising ground of a man called *Poca*', or 'spring of a man called *Poc*'. OE pers. name + **swelle* or *wella*.

Poynings W. Sussex. *Puningas* 960, *Poninges* 1086 (DB). '(Settlement of) the family or followers of a man called *Pūn*(a)'. OE pers. name + -*ingas*.

Poyntington Dorset. *Ponditone* 1086 (DB). 'Estate associated with a man called *Punt*'. OE pers. name + -*ing*- + *tūn*.

Poynton Ches. *Povinton* 1249. 'Estate associated with a man called *Pofa*'. OE pers. name + -*ing*- + *tūn*.

Poynton Green Tel. & Wrek. *Peventone* 1086 (DB). 'Estate associated with a man called *Pēofa*'. OE pers. name + -*ing*- + *tūn*.

Poyntz Pass Armagh. *Poyns pass c.*1655. 'Poyntz's pass'. Lieutenant Charles *Poyntz* had defended the pass here against Hugh O'Neill, Earl of Tyrone, in 1598. The modern village was founded in 1798.

Pratt's Bottom Gtr. London, first recorded in 1779, 'valley bottom associated with a family called Pratt' (known in this area from the 14th cent.), from OE *botm*.

Prawle, East & Prawle, West Devon. *Prenla* [*sic*] 1086 (DB). *Prahulle* 1204. Possibly 'look-out hill'. OE **prāw* + *hyll*.

Prebaun (*Preabán*) Westmeath. 'Patch'.

Preen, Church Shrops. *Prene* 1086 (DB), *Chircheprene* 1256. OE *prēon* 'a brooch, a pin', perhaps used figuratively of some landscape feature here, with later affix OE *cirice*.

Prees Shrops. *Pres* 1086 (DB). '(Place by) the brushwood or thicket'. OWelsh *pres*.

Preesall Lancs. *Pressouede* 1086 (DB). 'Brushwood headland'. OWelsh *pres* + OScand. *hofuth* or OE *hēafod*.

Prendwick Northum. *Prendewic* 1242. 'Dwelling or farm of a man called Prend'. OE pers. name + *wīc*.

Prenton Wirral. *Prestune* [*sic*] 1086 (DB), *Prenton* 1260. 'Farmstead of a man called Præn'. OE pers. name + *tūn*.

Prestcot, Prestcott, 'priests' cottage(s)', OE *prēost + cot*; e.g. **Prescot** Knows. *Prestecota* 1178. **Prescott** Devon, near Culmstock. *Prestecote* 1238.

Preseli Hills Pemb. *Precelly, or Percelly* 1868. The name renders Welsh *Mynydd Preseli* (Welsh *mynydd*), probably from Welsh *prys*, 'wood'.

Prestatyn Denb. *Prestetone* 1086 (DB), *Prestatton* 1301. 'Priests' farmstead'. OE *prēost + tūn*. The name is equivalent to English PRESTON but Welsh stress has preserved the middle syllable.

Prestbury, 'manor of the priests', OE *prēost + burh* (dative *byrig*): **Prestbury** Ches. *Presteberi* 1181. **Prestbury** Glos. *Preosdabyrig c.*900, *Presteberie* 1086 (DB).

Presteigne (*Llanandras*) Powys. *Prestehemed* 1137, *Prestene* 1548, *Presteign, or Llan-Andras of the Welsh* 1868. 'Priests' border meadow'. OE *prēost + *hemm-mǣd*. The Welsh name means 'church of St Andrew' (Welsh *llan + Andras*).

Presthope Shrops. *Presthope* 1167. 'Valley of the priests'. OE *prēost + hop*.

Preston, a common name, 'farmstead of the priests', OE *prēost + tūn*; examples include: **Preston** Dorset. *Prestun* 1228. **Preston** E. Loth, *see* PRESTONPANS. **Preston** E. R. Yorks. *Prestone* 1086 (DB). **Preston** E. Sussex. *Prestetone* 1086 (DB). **Preston** Lancs. *Prestune* 1086 (DB). **Preston** Northum. *Preston* 1242. **Preston** Suffolk. *Preston c.*1060, *Prestona* 1086 (DB). **Preston Bissett** Bucks. *Prestone* 1086 (DB), *Preston Byset* 1327. Manorial affix from the *Biset* family, holders of the manor from the 12th cent. **Preston Capes** Northants. *Prestetone* 1086 (DB), *Preston Capes* 1300. Manorial affix from the *de Capes* family, here in the 13th cent. **Preston Gubbals** Shrops. *Prestone* 1086 (DB), *Preston Gobald* 1292. Manorial affix from a priest called *Godebold* who held the manor in 1086. **Preston, Long** N. Yorks. *Prestune* 1086 (DB). Affix refers to the length of the village. **Preston-under-Scar** N. Yorks. *Prestun* 1086 (DB), *Preston undescar* 1568. Affix means 'under the rocky hill' from OScand. *sker*. **Preston Wynne** Herefs. *Prestetune* 1086 (DB). Manorial affix from a family called *la Wyne* here in the 14th cent.

Preston Candover Hants. *See* CANDOVER.

Prestonpans E. Loth. *Saltprestoun* 1587, *Prestonpans* 1654. '(Salt) pans by Preston'. OE *panne*. The monks of Newbattle Abbey laid out salt pans here near PRESTON in the early 13th cent.

Prestwich Bury. *Prestwich* 1194. 'Dwelling or farm of the priest(s)'. OE *prēost + wīc*.

Prestwick Northum. *Prestwic* 1242. Identical in origin with the previous name.

Prestwick S. Ayr. *Prestwic c.*1170. Identical in origin with the previous two names.

Prickwillow Cambs. *Prickewylev* 1251. Probably 'willow-tree from which skewers or goads were made'. OE *pricca + *wilig*.

Priddy Somerset. *Pridia* 1182. Probably 'earth house'. Celtic *prith + *tiʒ*.

Priestholm. *See* PUFFIN ISLAND.

Primethorpe Leics. *Torp* 1086 (DB), *Prymesthorp* 1316. 'Outlying farmstead or hamlet of a man called Prim'. OE pers. name + OScand. *thorp*.

Princes Risborough Bucks. *See* RISBOROUGH.

Princethorpe Warwicks. *Prandestorpe* 1221. 'Outlying farmstead or hamlet of a man called Præn or Pren'. OE pers. name + OScand. *thorp*.

Princetown Devon. a modern name, called after the Prince Regent, later George IV (1820–30).

Priors Hardwick Warwicks. *See* HARDWICK.

Priors Marston Warwicks. *See* MARSTON.

Priston B. & NE. Som. *Prisctun* 931, *Prisctone* 1086 (DB). 'Farmstead near the brushwood or copse'. OWelsh *prisc + OE *tūn*.

Prittlewell Sthend. *Pritteuuella* 1086 (DB). 'Babbling spring or stream'. OE *pritol + wella*.

Privett Hants. *Pryfetes flodan* late 9th cent. '(Place at) the privet copse'. OE *pryfet* (with *flōde* 'channel, gutter' in the early form).

Probus Cornwall. *Sanctus Probus* 1086 (DB). From the dedication of the church to St Probus.

Prudhoe Northum. *Prudho* 1173. 'Hill-spur of a man called *Prūda*'. OE pers. name + *hōh*.

Puckaun (*Pocán*) Tipperary. 'Little heap'.

Puckeridge Herts. *Pucherugge* 1294. 'Raised strip or ridge haunted by a goblin'. OE *pūca + *ric* or *hrycg*.

Puckington Somerset. *Pokintuna* 1086 (DB). 'Estate associated with a man called *Pūca'. OE pers. name + *-ing-* + *tūn*.

Pucklechurch S. Glos. *Pucelancyrcan* 950, *Pulcrecerce* 1086 (DB). 'Church of a man called *Pūcela'. OE pers. name + *cirice*.

Puddington, 'estate associated with a man called Put(t)a', OE pers. name + *-ing-* + *tūn*: **Puddington** Ches. *Potitone* 1086 (DB). **Puddington** Devon. *Potitone* 1086 (DB).

Puddletown Dorset. *Pitretone* [*sic*] 1086 (DB), *Pideleton* 1212. 'Farmstead on the River Piddle'. OE river-name (*pidele* 'a marsh or fen') + *tūn*.

Pudleston Herefs. *Pillesdune* [*sic*] 1086 (DB), *Putlesdone* 1212. 'Hill of the mouse-hawk or of a man called Pyttel'. OE *pyttel* or pers. name + *dūn*.

Pudsey Leeds. *Podechesaie* 1086 (DB). Probably 'enclosure of a man called *Pudoc, or by the hill called "the Wart"'. OE pers. name or *puduc* + *hæg*.

Puffin Island (*Ynys Seiriol*) Angl. 'Island inhabited by puffins'. The island is also known as *Priestholm*, 'priests' island' (OScand. *holmr*). The Welsh name means 'St Seiriol's island' (Welsh *ynys*).

Pulborough W. Sussex. *Poleberge* 1086 (DB). 'Hill or mound by the pools'. OE *pōl*, *pull* + *beorg*.

Pulford Ches. *Pulford* 1086 (DB). 'Ford by a pool or on a stream'. OE *pōl* or *pull* + *ford*.

Pulham, 'homestead or enclosure by the pools or streams', OE *pōl* or *pull* + *hām* or *hamm*: **Pulham** Dorset. *Poleham* 1086 (DB). **Pulham Market & Pulham St Mary** Norfolk. *Polleham* c.1050, *Pullaham* 1086 (DB). Affixes from the important market here and from the dedication of the church.

Pulloxhill Beds. *Polochessele* 1086 (DB). Possibly 'hill of a man called *Pulloc'. OE pers. name + *hyll*.

Pulverbatch, Church Shrops. *Polrebec* 1086 (DB), *Puluerbach* 1262, *Chyrche Pulrebach* 1272. OE *bæce* 'stream valley' with an uncertain first element which may be an old stream-name.

Affix is ME *churche* 'church' to distinguish this place from **Castle Pulverbatch**, *Castelpolrebache* 1301, with affix ME *castel* 'castle'.

Pumlumon. *See* PLYNLIMON.

Puncheston (*Cas-mael*) Pemb. *Pounchardon* 1291. 'Bridge where thistles grow'. OFrench *pont + chardon*. The name may have been transferred from *Pontchardon*, Normandy. Alternatively, 'Ponchard's hill' (French pers. name + OE *dūn*). The Welsh name means 'Mael's fort' (Welsh *cas*).

Puncknowle Dorset. *Pomacanole* 1086 (DB). Probably 'hillock of a man called *Puma'. OE pers. name + *cnoll*.

Purbeck, Isle of Dorset. *Purbicinga* 948, *Porbi* 1086 (DB). 'Beak-shaped ridge frequented by the bittern or snipe'. OE *pūr* + *bic*. In the earliest spelling the name is combined with a form of OE *-ingas* 'dwellers in'.

Purbrook Hants. *Pukebrok* 1248. 'Brook haunted by a goblin'. OE *pūca* + *brōc*.

Purfleet Thurr. *Purteflyete* 1285. 'Creek or stream of a man called *Purta'. OE pers. name + *flēot*.

Puriton Somerset. *Peritone* 1086 (DB). 'Pear orchard, or farmstead where pear-trees grow'. OE *pirige* + *tūn*.

Purleigh Essex. *Purlea* 998, *Purlai* 1086 (DB). 'Wood or clearing frequented by the bittern or snipe'. OE *pūr* + *lēah*.

Purley Gtr. London. *Pirlee* 1200. 'Wood or clearing where pear-trees grow'. OE *pirige* + *lēah*.

Purley W. Berks. *Porlei* 1086 (DB). Identical in origin with PURLEIGH Essex.

Purse Caundle Dorset. *See* CAUNDLE.

Purslow Shrops. *Possalau* 1086 (DB). 'Tumulus of a man called *Pussa'. OE pers. name + *hlāw*.

Purston Jaglin Wakefd. *Preston* 1086 (DB), *Preston Jakelin* 1269. 'Farmstead of the priests'. OE prēost + *tūn*. Manorial affix from some early feudal tenant called *Jakelyn*.

Purton, 'pear orchard, or farmstead where pear-trees grow', OE *pirige* + *tūn*: **Purton** Glos., near Berkeley. *Piritone* 1297. **Purton**

Glos., near Lydney. *Peritone* 1086 (DB). **Purton** Wilts. *Puritone* 796, *Piritone* 1086 (DB).

Pusey Oxon. *Pesei* 1086 (DB). 'Island, or dry ground in marsh, where peas grow'. OE *pise* + *ēg*.

Putford, West Devon. *Poteforda* 1086 (DB). 'Ford of a man called Putta, or one frequented by hawks'. OE pers. name or OE **putta* + *ford*.

Putley Herefs. *Poteslepe* [*sic*] 1086 (DB), *Putelega c.*1180. 'Woodland clearing of a man called Putta, or one frequented by hawks'. OE pers. name or OE **putta* + *lēah*.

Putney Gtr. London. *Putelei* [*sic*] 1086 (DB), *Puttenhuthe* 1279. 'Landing-place of the hawk, or of a man called Putta'. OE **putta* or OE pers. name (genitive *-n*) + *hȳth*.

Puttenham, 'homestead or enclosure of a man called Putta, or one frequented by hawks', OE pers. name or OE **putta* (genitive *-n*) + *hām* or *hamm*: **Puttenham** Herts. *Puteham* 1086 (DB). **Puttenham** Surrey. *Puteham* 1199.

Puxton N. Som. *Pukereleston* 1212. 'Estate of a family called Pukerel'. OFrench surname + OE *tūn*.

Pwllheli Gwyd. *Pwllhely* 1292. '(Place by the) brine pool'. Welsh *pwll* + *heli*.

Pyecombe W. Sussex. *Picumba* late 11th cent. 'Valley infested with gnats or other insects'. OE *pīe* + *cumb*.

Pylle Somerset. *Pil* 705, *Pille* 1086 (DB). 'The tidal creek'. OE *pyll*.

Pyon, Canon & Pyon, Kings Herefs. *Pionie* 1086 (DB), *Pyone canonicorum* 1221, *Kings Pyon* 1285. 'Island infested with gnats or other insects'. OE *pīe* (genitive plural *pēona*) + *ēg*. Distinguishing affixes from early possession by the canons of Hereford and the Crown.

Pyrford Surrey. *Pyrianforda* 956, *Peliforde* [*sic*] 1086 (DB). 'Ford by a pear-tree'. OE *pirige* + *ford*.

Pyrton Oxon. *Pirigtune* 987, *Peritone* 1086 (DB). 'Pear orchard, or farmstead where pear-trees grow'. OE *pirige* + *tūn*.

Pytchley Northants. *Pihteslea* 956, *Pihteslea* 1086 (DB). 'Woodland clearing of a man called *Peoht'. OE pers. name + *lēah*.

Pyworthy Devon. *Paorda* [*sic*] 1086 (DB), *Peworthy* 1239. 'Enclosure infested with gnats or other insects'. OE *pīe* + *worthig*.

Quabbs Shrops. First recorded 1833, from ModE dialect *quab* (OE **cwabba*) 'a marsh, a bog'.

Quadring Lincs. *Quedhaveringe* 1086 (DB). 'Muddy settlement of the family or followers of a man called **Hæfer*'. OE *cwēad* + OE pers. name + *-ingas*.

Quaich (river) Perth. '(River with) potholes'. Gaelic *cuach*.

Quainton Bucks. *Chentone* [sic] 1086 (DB), *Quentona* 1167. 'The queen's farmstead or estate'. OE *cwēn* + *tūn*.

Quantock Hills Somerset. *Cantucuudu* 682. Celtic hill name, possibly a derivative of an element **canto*- 'border or district', with OE *wudu* 'wood' in the early form.

Quantoxhead, East & Quantoxhead, West Somerset. *Cantocheve* 1086 (DB). 'Projecting ridge of the QUANTOCK HILLS'. Celtic hill name + OE *hēafod*.

Quarley Hants. *Cornelea* 1167. Probably 'woodland clearing with a mill or where mill-stones are obtained'. OE *cweorn* + *lēah*.

Quarndon Derbys. *Cornun* [sic] 1086 (DB), *Querendon* c.1200. 'Hill where mill-stones are obtained'. OE *cweorn* + *dūn*.

Quarrington Lincs. *Corninctun* 1086 (DB). Probably 'farmstead at the mill-stone place'. OE *cweorn* + *-ing* + *tūn*.

Quarrington Hill Durham. *Querendune* c.1190. 'Hill where mill-stones are obtained'. OE *cweorn* + *dūn*.

Quatford Shrops. *Quatford* 1086 (DB). 'Ford in a district called *Cwat(t)*'. OE *ford* with an OE name of obscure origin.

Quatt Shrops. *Quatone* 1086 (DB), *Quatte* 1209. From the same unexplained district-name as QUATFORD, with OE *tūn* 'farmstead' in the Domesday form.

Quebec Durham. Recorded thus from 1862, originally the name of a farmhouse here, transferred from the well-known Canadian city and province.

Quedgeley Glos. *Quedesleya* c.1140. Possibly 'woodland clearing of a man called **Cwēod*'. OE pers. name + *lēah*.

Queen Camel Somerset. *See* CAMEL.

Queen Charlton B. & NE Som. *See* CHARLTON.

Queenborough Kent. *Queneburgh* 1376. 'Borough named after Queen Philippa'. OE *cwēn* + *burh*. Philippa was the wife of Edward III.

Queensbury Brad. a modern name created in 1863; earlier the village was known as *Queen's Head*, the name of an inn here.

Queensferry Edin. *Passagium S. Marg. Regine* 1183, *Queneferie* c.1295. 'Queen's river crossing'. The name honours Queen Margaret, wife of Malcolm Canmore (d.1093).

Queensferry Flin. *King's Ferry* 1828, *Queensferry* 1837. 'Ferry of the queen'. The name honours Queen Victoria.

Quendon Essex. *Kuenadana* 1086 (DB). Probably 'valley of the women'. OE *cwene* (genitive plural *cwenena*) + *denu*.

Queniborough Leics. *Cuinburg* 1086 (DB), *Queniburg* 12th cent. 'The queen's fortified manor'. OE *cwēn* + *burh*.

Quenington Glos. *Quenintone* 1086 (DB). 'Farmstead of the women', or 'farmstead associated with a woman called **Cwēn*'. OE *cwene* (genitive plural *cwenena*) + *tūn*, or OE pers. name + *-ing-* + *tūn*.

Quernmore Lancs. *Quernemor* 1228. 'Moor where mill-stones are obtained'. OE *cweor* ι + *mōr*.

Querrin (*An Cuibhreann*) Clare. 'The tilled field'.

Quethiock Cornwall. *Quedoc* 1201. 'Wooded place'. OCornish *cuidoc*.

Quidenham Norfolk. *Cuidenham* 1086 (DB). 'Homestead or village of a man called *Cwida'. OE pers. name (genitive -*n*) + *hām*.

Quidhampton, 'muddy, or well-manured, home farm', OE *cwēad* 'muck' + *hām-tūn*: **Quidhampton** Hants. *Quedementun* 1086 (DB). **Quidhampton** Wilts. *Quedhampton* 1249.

Quilly (*Coillidh*) Derry, Down, Waterford. 'Woodland'.

Quilty (*Coillte*) Clare. 'Woods'.

Quin (*Cuinche*) Clare. 'Nook'.

Quinag (mountain) Highland. 'Churn-shaped'. Gaelic *cuinneag*.

Quinton, Lower & Quinton, Upper Warwicks. *Quentone* 848, *Quenintune* 1086 (DB). 'The queen's farmstead or estate.' OE *cwēn* + *tūn*.

Quoditch Devon. *Quidhiwis* 1249. 'Muddy, or well-manured, measure of land'. OE *cwēad* 'muck' + *hīwisc* 'measure of land that would support a family'.

Quoile (*Caol*) (river) Down. *Coyle* 1618. 'Narrow water'.

Quorndon or Quorn Leics. *Querendon* 1154–89. 'Hill where mill-stones are obtained'. OE *cweorn* + *dūn*.

Quy Cambs. *See* STOW CUM QUY.

Raasay (island) Highland. *Raasa* 1263. 'Island with a ridge frequented by roedeer'. OScand. *rár* + *áss* + *ey*.

Raby Wirral. *Rabie* 1086 (DB). 'Farmstead or village at a boundary'. OScand. *rá* + *bý*.

Rackenford Devon. *Racheneforda* 1086 (DB). Possibly 'ford suitable for riding, by a stretch of river'. OE **racu* + **ærne-ford*. Alternatively 'ford of the gullies', from OE *hraca* (genitive plural *hracena*) 'throat' + *ford*.

Rackham W. Sussex. *Recham* 1166. Possibly 'homestead or enclosure with a rick or ricks'. OE *hrēac* + *hām* or *hamm*.

Rackheath Norfolk. *Racheitha* 1086 (DB). Probably 'landing-place near a gully or water-course'. OE *hraca* or **racu* + *hȳth*.

Radcliffe, Radclive, 'the red cliff or bank', OE *rēad* + *clif*: **Radcliffe** Bury. *Radecliue* 1086 (DB). **Radcliffe on Trent** Notts. *Radeclive* 1086 (DB). For the river-name Trent, *see* NEWARK. **Radclive** Bucks. *Radeclive* 1086 (DB).

Radcot Oxon. *Rathcota* 1163. 'Red cottage' or 'reed cottage' (i.e. with a roof thatched with reeds). OE *rēad* or *hrēod* + *cot*.

Rademon (*Ráth Déamain*) Down. 'Deamán's fort'.

Radford Semele Warwicks. *Redeford* 1086 (DB), *Radeford Symely* 1314. 'Red ford', i.e. where the soil is red. OE *rēad* + *ford*. Manorial affix from the *Simely* family, here in the 12th cent.

Radipole Dorset. *Retpole* 1086 (DB). 'Reedy pool'. OE *hrēod* + *pōl*.

Radlett Herts. *Radelett* 1453. 'Junction of roads'. OE *rād* + *(ge)lǣt*.

Radley Oxon. *Radelege* c.1180. 'Red wood or clearing'. OE *rēad* + *lēah*.

Radnage Bucks. *Radenhech* 1162. '(Place at) the red oak-tree'. OE *rēad* + *āc* (dative *ǣc*).

Radnor. *See* NEW RADNOR.

Radstock B. & NE. Som. *Stoche* 1086 (DB), *Radestok* 1221. 'Outlying farmstead by the road' (here the Fosse Way). OE *rād* + *stoc*.

Radstone Northants. *Rodestone* 1086 (DB). '(Place at) the rood-stone or stone cross'. OE **rōd-stān*.

Radur. *See* RADYR.

Radway Warwicks. *Radwei, Rodewei* 1086 (DB), *Radewey* 1198. Probably 'red way' from the colour of the soil here, OE *rēad* + *weg*. Alternatively 'way fit for riding on', OE *rād-weg*.

Radwell Beds. *Radeuuelle* 1086 (DB). 'Red spring or stream'. OE *rēad* + *wella*.

Radwinter Essex. *Redeuuintra* 1086 (DB). Possibly 'red vineyard', from OE *rēad* + **winter*. Alternatively 'tree of a woman called **Rædwynn*', OE pers. name + *trēow*.

Radyr (*Radur*) Card. *Radur* 1254, *Rador* 1291. 'Oratory'. Latin *oratorium*. The reference is to a small chapel for private worship.

Ragdale Leics. *Ragendele* 1086 (DB), *Rachedal* c.1125. Probably 'valley of the gully or narrow pass'. OE *hraca* + *dæl*, OScand. *dalr*.

Raglan (*Rhaglan*) Mon. *Raghelan* 1254. '(Place) before the bank'. Welsh *rhag* + *glan*. The name refers to the former castle here.

Ragnall Notts. *Ragenehil* 1086 (DB). 'Hill of a man called Ragni'. OScand. pers. name + OE *hyll*.

Rahan (*Raithean*) Offaly. 'Ferny place'.

Rahara (*Ráth Aradh*) Roscommon. 'Noble fort'.

Raharney (*Ráth Fhearna*) Westmeath. 'Fort of the alder'.

Raheen (*Ráithín*) Cork, Westmeath, Wexford. 'Little fort'.

Raheens (*Na Ráithíní*) Mayo. 'The little forts'.

Raheny (*Ráth Eanaigh*) Dublin. 'Fort of the marsh'.

Raholp (*Ráth Cholpa*) Down. *Raith Colptha* *c.*1170. 'Fort of the heifer'.

Rahugh (*Ráth Aodha*) Westmeath. 'Aodh's fort'.

Rainford St Hel. *Raineford* 1198. 'Ford of a man called *Regna, or 'boundary ford'. OE pers. name or OScand. *rein* + OE *ford*.

Rainham Gtr. London. *Renaham* 1086 (DB). Possibly identical in origin with the next name. Alternatively 'homestead of a man called *Regna', OE pers. name + *hām*.

Rainham Medway. *Roegingaham* 811. Probably 'homestead of the Roegingas'. OE tribal name (of uncertain meaning) + *hām*.

Rainhill St Hel. *Raynhull c.*1190. 'Hill of a man called *Regna, or 'boundary hill'. OE pers. name or OScand. *rein* + OE *hyll*.

Rainow Ches. *Rauenouh* 1285. 'Hill-spur frequented by ravens'. OE *hræfn* + *hōh*.

Rainton, 'estate associated with a man called *Rægen or *Regna', OE pers. name + *-ing-* + *tūn*:
Rainton N. Yorks. *Rainincton* 1086 (DB).
Rainton, East Sundld. **& Rainton, West** Durham. *Reiningtun c.*1105.

Rainworth Notts. *Reynwath* 1268. 'Clean ford', or 'boundary ford'. OScand. *hreinn* or *rein* + *vath*.

Raithby Lincs. near Louth. *Radresbi* 1086 (DB). 'Farmstead or village of a man called *Hreithi'. OScand. pers. name + *bý*.

Raithby Lincs. near Spilsby. *Radebi* 1086 (DB). 'Farmstead or village of a man called Hrathi'. OScand. pers. name + *bý*.

Rake W. Sussex. *Rake* 1296. '(Place at) the hollow or pass'. OE *hraca*.

Ram's Island (*Inis Dairgreann*) Antrim. *(primanmeara) Innsi Daircairgrenn* 1056. The present name arose from a mistaken resemblance between Irish *reithe*, 'ram', and the final element of Irish *Inis Draicrenn*, a later version of the original name, itself of unknown meaning.

Rame Cornwall. near Plymouth. *Rame* 1086 (DB). Possibly '(place at) the barrier', perhaps alluding to a fortress. OE **hrama*.

Ramore (*Ráth Mór*) Antrim. 'Big fort'.

Ramornie Fife. *Ramorgany* 1512. 'Fort of a man of Clan Morgan'. Gaelic *ràth*.

Rampisham Dorset. *Ramesham* 1086 (DB). 'Wild-garlic enclosure', or 'enclosure of the ram or of a man called Ramm'. OE *hramsa* or *ramm* or pers. name + *hamm*.

Rampside Cumbria. *Rameshede* 1292. 'Headland of the ram', perhaps from a fancied resemblance to the animal. OE *ramm* + *hēafod*.

Rampton Cambs. *Rantone* 1086 (DB). 'Farmstead where rams are kept'. OE *ramm* + *tūn*.

Rampton Notts. *Rametone* 1086 (DB). Probably identical in origin with the previous name.

Ramsbottom Bury. *Romesbothum* 1324. 'Valley of the ram, or where wild garlic grows'. OE *ramm* or *hramsa* + **bothm*.

Ramsbury Wilts. *Rammesburi* 947, *Ramesberie* 1086 (DB). 'Fortification of the raven, or of a man called *Hræfn'. OE *hræfn* or pers. name + *burh* (dative *byrig*).

Ramsden, probably 'valley where wild garlic grows', OE *hramsa* + *denu*: **Ramsden** Oxon. *Rammesden* 1246. **Ramsden Bellhouse** Essex. *Ramesdana* 1086 (DB), *Ramesden Belhous* 1254. Manorial affix from the *de Belhus* family, here in the 13th cent.

Ramsey, probably 'island where wild garlic grows', OE *hramsa* + *ēg*: **Ramsey** Cambs. *Hramesege c.*1000. **Ramsey** Essex. *Rameseia* 1086 (DB).

Ramsey Isle of Man. *Ramsa* 1257. 'Stream where wild garlic grows'. OScand. *hramsa* + *á*.

Ramsey Island (*Ynys Dewi*) Pemb. *Ramesey* 1326. 'Island where wild garlic grows'. OScand. *hramsa* + *ey*. The Welsh name means 'St David's island' (Welsh *ynys* + *Dewi*).

Ramsgate Kent. *Remmesgate* 1275. 'Gap (in the cliffs) of the raven, or of a man called *Hræfn'. OE *hræfn* or pers. name + *geat*.

Ramsgill N. Yorks. *Ramesgile* 1198. 'Ravine of the ram, or where wild garlic grows'. OE *ramm* or *hramsa* + OScand. *gil*.

Ramshorn Staffs. *Rumesoura* 1197. 'The ram's ridge', or 'ridge where wild garlic grows'. OE *ramm* or *hramsa* + **ofer*.

Ranby Notts. *Ranebi* 1086 (DB). 'Farmstead or village of a man called Hrani'. OScand. pers. name + *bý*.

Rand Lincs. *Rande* 1086 (DB). '(Place at) the border or edge'. OE *rand*.

Randalstown (*Baile Raghnaill*) Antrim. The town is named after *Randal* MacDonnell, 2nd Earl and 1st Marquis of Antrim, who in *c*.1650 married his second wife Rose O'Neill, who lived at nearby Shane's Castle.

Randwick Glos. *Rendewiche* 1121. Probably 'dwelling or farm on the edge or border'. OE **rend* + *wīc*.

Ranelagh (*Raghnallach*) Dublin. 'Raghnal's place'.

Rangeworthy S. Glos. *Rengeswurda* 1167. 'Enclosure of a man called **Hrencga*', or 'enclosure made of stakes'. OE pers. name or OE **hrynge* + *worthig*.

Rannafast (*Rann na Feirste*) Donegal. 'Point of the sandbank'.

Rannoch Moor Highland. *Rannach* 1505. '(Place of) bracken'. Gaelic *raineach*.

Ranskill Notts. *Raveschel* [sic] 1086 (DB), *Ravenskelf* 1275. 'Shelf or ledge frequented by ravens, or of a man called Hrafn'. OScand. *hrafn* or pers. name + *skjalf*.

Ranton Staffs. *Rantone* 1086 (DB). Possibly 'farmstead where rams are kept'. OE *ramm* + *tūn*.

Ranworth Norfolk. *Randwrthe c*.1045, *Randuorda* 1086 (DB). 'Enclosure by an edge or border', or 'enclosure of a man called Randi'. OE *rand* or OScand. pers. name + OE *worth*.

Ranza, Loch N. Ayr. (Arran). *Lockransay* 1433. 'Loch of the rowan-tree river'. Gaelic *loch* + OScand. **raun* + *á*.

Raphoe (*Ráth Bhoth*) Donegal. (*epscop*) *Ratha Bhoth* 813. 'Fort of the huts'.

Rasen, Market, Rasen, Middle, & Rasen, West Lincs. *æt ræsnan* 973, *Resne* 1086 (DB), *Marketrasyn* 1358, *Middelrasen* 1201, *Westrasen* 1202. '(Place at) the planks', i.e. 'plankbridge'. OE *ræsn* in dative plural form.

The river-name **Rase** is a back-formation from the place name.

Rasharkin (*Ros Earcáin*) Antrim. *Ros Earcáin c*.1400. 'Earcán's grove'.

Raskelf N. Yorks. *Raschel* 1086 (DB). 'Shelf or ledge frequented by roe-deer'. OScand. *rá* + *skjalf*.

Rastrick Calder. *Rastric* 1086 (DB). Probably 'raised strip or ridge with a resting-place'. OScand. *rọst* + OE **ric*.

Ratby Leics. *Rotebie* 1086 (DB). 'Farmstead or village of a man called Rōta'. OE pers. name + OScand. *bý*.

Ratcliffe, 'red cliff or bank', OE *rēad* + *clif*: **Ratcliffe Culey** Leics. *Redeclive* 1086 (DB). Manorial affix from the *de Culy* family, here in the 13th cent. **Ratcliffe on Soar** Notts. *Radeclive* 1086 (DB). For the river-name, *see* BARROW UPON SOAR. **Ratcliffe on the Wreake** Leics. *Radeclive* 1086 (DB). For the river-name, *see* FRISBY ON THE WREAKE.

Rath (*Ráth*) Offaly, Waterford. 'Fort'.

Rath Luirc (*Ráth Loirc*) Cork. 'Lorc's fort'. The usual Irish name is now *An Ráth*, 'the fort'.

Rathangan (*Ráth Iomgháin*) Kildare. 'Iomghán's fort'.

Rathbeg (*Ráth Beag*) Antrim. 'Little fort'.

Rathcabban (*Ráth Cabáin*) Tipperary. 'Fort of the cabin'.

Rathcarra (*Ráth Characch*) Westmeath. 'Rocky fort'.

Rathclittagh (*Ráth Cleiteach*) Westmeath. 'Fort of the feathers'.

Rathcogue (*Ráth Cogaidh*) Westmeath. 'Fort of the war'.

Rathconrath (*Ráth Conarta*) Westmeath. 'Fort of the pack of hounds'.

Rathcool (*Ráth Cúil*) Cork. 'Fort of the hill'.

Rathcoole (*Ráth Cúil*) Dublin. 'Fort of the hill'.

Rathcore (*Ráth Cuair*) Meath. 'Fort of Cuar'.

Rathcormack (*Ráth Chormaic*) Cork. 'Cormac's fort'.

Rathdangan (*Ráth Daingin*) Wicklow. 'Fort of the fortress'.

Rathdowney (*Ráth Domhnaigh*) Laois. 'Fort of the church'.

Rathdrishoge (*Ráth Driseog*) Westmeath. 'Fort of the brambles'.

Rathdrum (*Ráth Droma*) Wicklow. 'Fort of the ridge'.

Rathen Aber. *Rathyn* 1300. Origin obscure.

Rathfarnham (*Ráth Fearnáin*) Dublin. 'Fort of the alder'.

Rathfeigh (*Ráth Faiche*) Meath. 'Fort of the green'.

Rathfield (*Páirc na Rátha*) Kerry. 'Field of the ring fort'.

Rathfran (*Ráth Bhrain*) Mayo. 'Bran's fort'.

Rathfriland (*Ráth Fraoileann*) Down. *Raphrylan* 1583. 'Fraoile's fort'.

Rathfylane (*Ráth Fialáin*) Wexford. 'Fialán's fort'.

Rathgall (*Ráth Gall*) Wicklow. 'Fort of the stone'.

Rathganny (*Ráth Ghainmhe*) Westmeath. 'Sandy fort'.

Rathgar (*Ráth Garbh*) Dublin. 'Rough fort'.

Rathgormuck (*Ráth Ó gCormaic*) Waterford. 'Fort of Uí Chormaic'.

Rathillet Fife. *Radhulit* c.1195. 'Fort of the Ulstermen'. Gaelic *ràth*.

Rathkeale (*Ráth Caola*) Limerick. 'Caola's fort'.

Rathkenny (*Ráth Cheannaigh*) Antrim, Meath. 'Ceannach's fort'.

Rathlacken (*Ráth Leacan*) Mayo. 'Fort of the hillside'.

Rathlee (*Ráth Lao*) Sligo. 'Fort of the calf'.

Rathlin Island (*Reachlainn*) Antrim. (*ecclaise*) *Rechrainne* 630. Meaning uncertain.

Rathlin O'Birne Island (*Reachlainn Uí Bhirn*) Donegal. 'Ó Birn's Reachlainn'.

Rathmarega (*Ráth Mharga*) Westmeath. 'Fort of the market'.

Rathmell N. Yorks. *Rodemele* 1086 (DB). 'Red sandbank'. OScand. *rauthr* + *melr*.

Rathmelton (*Ráth Mealtain*) Donegal. *Ramalton* 1601. 'Mealtan's fort'.

Rathmore (*Ráth Mhór*) Antrim, Kerry, Kildare. 'Big fort'.

Rathmoyle (*An Ráth Mhaol*) Kilkenny. 'The dilapidated fort'.

Rathmullan (*Ráth Maoláin*) Donegal. *Ráith Mailáin* 1516. 'Maolán's fort'.

Rathmullan (*Ráth Mhaoláin*) Down. *Rathmoyln* 1306. 'Maolán's fort'.

Rathmullen (*Ráth an Mhuilinn*) Sligo. 'Fort of the mill'.

Rathnamagh (*Ráth na Mach*) Mayo. 'Fort of the plain'.

Rathnamuddagh (*Ráth na mBodach*) Westmeath. 'Fort of the churls'.

Rathnew (*Ráth Naoi*) Wicklow. 'Noé's fort'.

Rathnure (*Ráth an Iúir*) Wexford. 'Fort of the yew-tree'.

Rathowen (*Ráth Eoghain*) Westmeath. 'Eoghan's fort'.

Rathruane (*Ráth Ruáin*) Cork. 'Ruán's fort'.

Rathtoe (*Ráth Tó*) Carlow. 'Tó's fort'.

Rathtroane (*Ráth Ruáin*) Meath. 'Ruán's fort'.

Rathvilla (*Ráth Bhile*) Offaly. 'Fort of the sacred tree'.

Rathvilly (*Ráth Bhile*) Carlow. 'Fort of the sacred tree'.

Ratkeevin (*Ráth Chaoimín*) Tipperary. 'Kevin's fort'.

Ratley Warwicks. *Rotelei* 1086 (DB). 'Woodland clearing of a man called Rōta'. OE pers. name + *lēah*.

Ratlinghope Shrops. *Rotelingehope* 1086 (DB). 'Enclosed valley associated with a man called *Rōtel*, or 'valley at *Rōtel*'s place'. OE pers. name + *-ing-* or *-ing* + *hop*.

Ratoath (*Ráth Tó*) Meath. 'Tó's fort'.

Rattery Devon. *Ratreu* 1086 (DB). '(Place at) the red tree'. OE *rēad* + *trēow*.

385 **Reacashlagh**

Rattlesden Suffolk. *Rattesdene* 1042–66, *Ratlesdena* 1086 (DB). Etymology uncertain, but possibly 'valley of a man called *Rætel', OE pers. name + *denu*. Or from an OE plant-name *hratele* (cf. ModE *rattle*). The river-name **Rat** is a back-formation from the place name.

Rattoo (*Ráth Tó*) Kerry. 'Tó's fort'.

Rattray Perth. *Rotrefe* 1291. 'Farm by the fort'. Gaelic *ràth* + British *treb*.

Raubaun (*Ráth Bán*) Westmeath. 'White fort'.

Rauceby, North & Rauceby, South Lincs. *Rosbi, Roscebi* 1086 (DB). 'Farmstead or village of a man called Rauthr'. OScand. pers. name + *bý*.

Raughton Head Cumbria. *Ragton* 1182, *Raughtonheved* 1367. Probably 'farmstead where moss or lichen grows'. OE *ragu* + OE *tūn*, with the later addition of *hēafod* 'headland, hill'.

Raunds Northants. *Randan* c.980, *Rande* 1086 (DB). '(Place at) the borders or edges'. OE *rand* in a plural form.

Raveley, Great & Raveley, Little Cambs. *Ræflea* c.1060. Possibly 'woodland clearing frequented by ravens'. OE *hræfn* + *lēah*.

Ravendale, East & Ravendale, West NE. Lincs. *Ravenedal* 1086 (DB). 'Valley frequented by ravens'. OScand. *hrafn* + *dalr*.

Ravenfield Rothm. *Rauenesfeld* 1086 (DB). 'Open land of a man called *Hræfn*, or frequented by ravens'. OE pers. name or OE *hræfn* + *feld*.

Ravenglass Cumbria. *Rengles* c.1180. 'Lot or share of a man called Glas'. OIrish *rann* + pers. name.

Raveningham Norfolk. *Rauenicham* 1086 (DB). 'Homestead of the family or followers of a man called *Hræfn*'. OE pers. name + *-inga-* + *hām*.

Ravenscar N. Yorks. *Rauenesere* 1312. 'Rock frequented by ravens'. OScand. *hrafn* + *sker*.

Ravensden Beds. *Rauenesden* c.1150. 'Valley of the raven, or of a man called *Hræfn*'. OE *hræfn* or pers. name + *denu*.

Ravensthorpe, 'outlying farmstead or hamlet of a man called Hrafn', OScand. pers. name + *thorp*: **Ravensthorpe** Kirkl. *Rauenestorp* 1086 (DB). **Ravensthorpe** Northants. *Ravenestorp* 1086 (DB).

Ravenstone, 'farmstead of a man called *Hræfn or Hrafn*', OE or OScand. pers. name + *tūn*: **Ravenstone** Leics. *Ravenestun* 1086 (DB). **Ravenstone** Milt. K. *Raveneston* 1086 (DB).

Ravenstonedale Cumbria. *Rauenstandale* 12th cent. 'Valley of the raven stone'. OE *hræfn* + *stān* + OScand. *dalr*.

Ravensworth N. Yorks. *Raveneswet* [sic] 1086 (DB), *Raveneswad* 1157. 'Ford of the raven, or of a man called Hrafn'. OScand. *hrafn* or pers. name + *vath*.

Ravernet (*Ráth Bhearnáin*) Down. *Balliravarnan* 1585. 'Fort of the little gap'.

Rawcliffe, 'red cliff or bank', OScand. *rauthr* + *klif*. **Rawcliffe** E. R. Yorks. *Routheclif* c.1080. **Rawcliffe** York. *Roudeclife* 1086 (DB). **Rawcliffe, Out** Lancs. *Rodeclif* 1086 (DB). Affix is OE *ūte* 'outer'.

Rawdon Leeds. *Roudun* 1086 (DB). 'Red hill'. OScand. *rauthr* + OE *dūn*.

Rawmarsh Rothm. *Rodemesc* [sic] 1086 (DB), *Rowmareis* c.1200. 'Red marsh'. OScand. *rauthr* + OE *mersc*.

Rawreth Essex. *Raggerea* 1177. 'Stream frequented by herons'. OE *hrāgra* + *rīth*.

Rawtenstall Lancs. *Routonstall* 1324. 'Rough farmstead or cow-pasture'. OE *rūh* + *tūn-stall*.

Ray (*An Ráith*) Donegal. *Raghe* 1622. 'The fort'.

Raydon Suffolk. *Reindune* 1086 (DB). 'Hill where rye is grown'. OE *rygen* + *dūn*.

Rayleigh Essex. *Ragheleia* 1086 (DB). 'Woodland clearing frequented by female roe-deer or by she-goats'. OE *ræge* + *lēah*.

Rayne Essex. *Hrægenan* c.1000, *Raines* 1086 (DB). Possibly '(place at) the shelter or eminence'. OE *hrægene*.

Raynham Norfolk. *Reineham* 1086 (DB). Possibly 'homestead or village of a man called *Regna*'. OE pers. name + *hām*.

Rea, Lough (*Loch Riach*) Galway. 'Grey lake'.

Reacashlagh (*Réidhchaisleach*) Kerry. 'Clearing of the stone forts'.

Reach, '(place at) the raised strip of land or other linear feature', OE *rǣc*: **Reach** Beds. *Reche c.*1220. **Reach** Cambs. *Reche* 1086. Here the reference is to the post-Roman earthwork of Devil's Dyke.

Read Lancs. *Revet* 1202. 'Headland frequented by female roe-deer or by she-goats'. OE *rǣge* + *hēafod*.

Reading Readg. *Readingum c.*900, *Reddinges* 1086 (DB). '(Settlement of) the family or followers of a man called *Rēad(a)*'. OE pers. name + *-ingas*.

Reagill Cumbria. *Reuegile* 1176. 'Ravine frequented by foxes'. OScand. *refr* + *gil.*

Reanaclogheen (*Réidh na bCloichín*) Waterford. 'Clearing of the little stones'.

Reanascreena (*Rae na Scríne*) Cork. 'Level place of the shrine'.

Rear (*Rae*) Tipperary. 'Level place'.

Rearsby Leics. *Redresbi* 1086 (DB). 'Farmstead or village of a man called Hreitharr'. OScand. pers. name + *bý.*

Recess (*Sraith Salach*) Galway. 'Dirty river-meadow'.

Reculver Kent. *Regulbium c.*425, *Roculf* 1086 (DB). An ancient Celtic name meaning 'great headland'.

Redberth Pemb. *Ridebard* 1361. 'Ford with bushes'. Welsh *rhyd* + *perth.*

Redbourn, Redbourne, 'reedy stream', OE *hrēod* + *burna*: **Redbourn** Herts. *Reodburne c.*1060, *Redborne* 1086 (DB). **Redbourne** N. Lincs. *Radburne* 1086 (DB).

Redbridge Gtr. London. borough named from an old bridge across the River RODING first recorded as *Red Bridge* in 1746.

Redcar Red. & Cleve. *Redker c.*1170. 'Red or reedy marsh'. OE *rēad* or *hrēod* + OScand. *kjarr.*

Redcliff Bay N. Som. *Radeclive c.*1180. 'Red cliff or bank'. OE *rēad* + *clif.*

Reddish Stockp. *Rediche* 1212. 'Reedy ditch'. OE *hrēod* + *dīc.*

Redditch Worcs. *La Rededich* 1247. 'Red or reedy ditch'. OE *rēad* or *hrēod* + *dīc.*

Rede Suffolk. *Reoda* 1086 (DB). Probably '(place at) the reed-bed'. OE *hrēod.*

Redenhall Norfolk. *Redanahalla* 1086 (DB). Probably 'nook of land where reeds grow'. OE **hrēoden* + *halh.*

Redgorton Perth. *Rothgortanan c.*1250. 'Fort by the little enclosure'. Gaelic *ràth* + *gortain.*

Redgrave Suffolk. *Redgrafe* 11th cent. 'Reedy pit' from OE *hrēod* + *græf*, or 'red grove' from OE *rēad* + *grāf.*

Redhill N. Som. *Ragiol* 1086 (DB). Possibly 'hill at roe-deer edge'. OE *rā* + *ecg* + *hyll.*

Redhill Surrey. *Redehelde* 1301. 'Red slope'. OE *rēad* + *helde.*

Redhills Cavan. *Redhills* 1837. The village is apparently named from the reddish soil of the hills here.

Redisham Suffolk. *Redesham* 1086 (DB). Probably 'homestead or village of a man called **Rēad*'. OE pers. name + *hām.*

Redland Brist. *Thriddeland* 1209. Probably 'third part of an estate'. OE *thridda* + *land.*

Redlingfield Suffolk. *Radinghefelda* [*sic*] 1086 (DB), *Radlingefeld* 1166. 'Open land of the family or followers of a man called Rǣdel or **Rǣdla*'. OE pers. name + *-inga-* + *feld.*

Redlynch Somerset. *Redlisc* 1086 (DB). 'Reedy marsh'. OE *hrēod* + **lisc.*

Redlynch Wilts. *Radelynche* 1282. 'The red ledge of higher ground'. OE *rēad* + *hlinc.*

Redmarley D'Abitot Glos. *Reodemǣreleage* 963, *Ridmerlege* 1086 (DB), *Rudmarleye Dabetot* 1324. 'Woodland clearing with a reedy pond'. OE *hrēod* + *mere* + *lēah.* Manorial affix from the *d'Abitot* family, here in 1086.

Redmarshall Stock. on T. *Redmereshill c.*1208. 'Hill of the red pool' (from the red clay soil here). OE *rēad* + *mere* + *hyll.*

Redmile Leics. *Redmelde* 1086 (DB). '(Place with) red soil'. OE *rēad* + **mylde.*

Redmire N. Yorks. *Ridemare* 1086 (DB). 'Reedy pool'. OE *hrēod* + *mere.*

Redruth Cornwall. *Ridruth* 1259. 'Red ford'. OCornish *rid* + *rudh.*

Redwick S. Glos. *Hreodwican* 955–9, *Redeuuiche* 1086 (DB). 'Dairy farm among the reeds'. OE *hrēod* + *wīc.*

Redworth Darltn. *Redwortha* 1183.
'Enclosure where reeds grow'. OE *hrēod* +
worth.

Ree, Lough (*Loch Rí*) Longford,
Roscommon, Westmeath. 'King's lake'.

Reed Herts. *Retth* 1086 (DB). Probably OE
*rȳ(h)th 'a rough piece of ground'.

Reedham Norfolk. *Redham* 1044–7, *Redeham*
1086 (DB). 'Homestead or enclosure where
reeds grow'. OE *hrēod* + *hām* or *hamm*.

Reedness E. R. Yorks. *Rednesse* c.1170.
'Reedy headland'. OE *hrēod* + *næss*.

Reen (*Rinn*) Kerry. 'Point'.

Reens (*Roighne*) Cork. 'Choicest part'.

Reepham, 'manor held or run by a reeve', OE
rēfa + *hām*: **Reepham** Lincs. *Refam* 1086 (DB).
Reepham Norfolk. *Refham* 1086 (DB).

Reeth N. Yorks. *Rie* 1086 (DB), *Rithe* 1184.
Probably '(place at) the stream'. OE *rīth*.

Reigate Surrey. *Reigata* c.1170. 'Gate for
female roe-deer'. OE *ræge* + *geat*.

Reighton N. Yorks. *Rictone* 1086 (DB).
'Farmstead by the straight ridge'. OE **ric* + *tūn*.

Reiss Highland. 'Rising ground', from a word
related to OScand. *rísa* 'to rise'.

Relaghbeg (*Reidhleach Beag*) Cavan. 'Small
level place'.

Relick (*Reilig*) Westmeath. 'Graveyard'.

Remenham Wokhm. *Rameham* 1086 (DB).
Probably 'homestead or enclosure by the river-
bank'. OE *reoma* (genitive *-n*) + *hām* or *hamm*.

Rempstone Notts. *Rampestune* 1086 (DB).
Probably 'farmstead of a man called **Hrempi*'.
OE pers. name + *tūn*.

Renanirree (*Rae na nDoirí*) Cork. 'Level
place of the oak-trees'.

Rendcomb Glos. *Rindecumbe* 1086 (DB).
'Valley of a stream called *Hrinde*'. OE river-
name (meaning 'the pusher') + *cumb*.

Rendham Suffolk. *Rimdham, Rindham* 1086
(DB). 'Cleared homestead', or 'homestead by a
hill'. OE *rȳmed* or **rind(e)* + *hām*.

Rendlesham Suffolk. *Rendlæsham* c.730,
Rendlesham 1086 (DB). Possibly 'homestead of a

man called **Rendel*', OE pers. name + *hām*.
Alternatively the first element could be an
OE word **rendel* 'little shore'.

Renfrew Renf. *Reinfry* c.1128, *Renfriu* c.1150.
'Point of the current'. OWelsh *rhyn* + *frwd*. The
River Gryfe enters the Clyde here.

Renhold Beds. *Ranhale* 1220. Probably 'nook
of land frequented by roe-deer'. OE *rā* (genitive
plural *rāna*) + *halh*.

Renishaw Derbys. *Reynalddeschawe* 1281.
'Reynold's small wood or copse'. ME pers. name
or surname + *shawe* (OE *sceaga*).

Rennington Northum. *Reiningtun* 12th cent.
'Estate associated with a man called **Regna*'.
OE pers. name + *-ing-* + *tūn*.

Renton W. Dunb. The town was founded in
1782 and named after Cecilia *Renton*, daughter-
in-law of Jean Telfer, sister of the novelist Tobias
Smollett (1721–74), who was born nearby.

Renwick Cumbria. *Rauenwich* 1178.
Probably 'dwelling or farm of a man called
Hrafn or **Hræfn*'. OScand. or OE pers. name +
OE *wīc*.

Repps Norfolk. *Repes* 1086 (DB), *Repples* 1191.
Possibly 'the strips of land'. OE **reopul*.
Alternatively perhaps from OScand. *hreppr*
'district, community'.

Repton Derbys. *Hrypadun* 730–40,
Rapendune 1086 (DB). 'Hill of the tribe called
Hrype'. Old tribal name (*see* RIPON) + OE *dūn*.

Reston, North & Reston, South Lincs.
Ristone 1086 (DB). 'Farmstead near or among
brushwood'. OE *hrīs* + *tūn*.

Restormel Cornwall. *Rostormel* 1310.
Possibly 'moor at the bare hill'. Cornish **ros* +
tor + **moyl*.

Retford, East & Retford, West Notts.
Redforde 1086 (DB). 'Red ford'. OE *rēad* + *ford*.

Rettendon Essex. *Rettendun* c.1000,
Ratenduna 1086 (DB). Probably 'hill infested
with rats'. OE **rætten* + *dūn*.

Revesby Lincs. *Resuesbi* 1086 (DB).
'Farmstead or village of a man called Refr'.
OScand. pers. name + *bý*.

Rewe Devon. *Rewe* 1086 (DB). 'Row of
houses'. OE *ræw*.

Reydon Suffolk. *Rienduna* 1086 (DB). 'Hill where rye is grown'. OE *rygen* + *dūn*.

Reymerston Norfolk. *Raimerestuna* 1086 (DB). 'Farmstead of a man called Raimar'. OGerman pers. name + OE *tūn*.

Reynalton Pemb. *Reynaldeston* 1394. 'Reynald's farm'. OGerman pers. name + OE *tūn*.

Rhaeadr. *See* RHAYADER.

Rhaglan. *See* RAGLAN.

Rhayader (*Rhaeadr* or *Rhaeadr Gwy*) Powys. *Raidergoe* 1191, *Rayadyr Gwy* 14th cent., *Raiaderguy* 1543. '(Place by the) waterfall'. Welsh *rhaeadr*. The Welsh name adds the name of the River Wye (*Gwy*) (*see* HAY-ON-WYE).

Rhisga. *See* RISCA.

Rhiwabon. *See* RUABON.

Rhobell Fawr (mountain) Gwyd. 'Big saddle'. Welsh *rhobell* + *mawr*.

Rhode (*Ród*) Offaly. 'Route'.

Rhodes Minnis Kent. *Rode* 1327. '(Place at) the clearing', from OE *rodu*, with the addition of (*ge*)*mǽnnes* 'common land' (recorded in *Menessegate* 1327, 'gate to the common land').

Rhodesia Notts. A transferred name from the African countries known as Northern and Southern *Rhodesia* (now Zambia and Zimbabwe), so called after Cecil *Rhodes* (1853–1902).

Rhondda Rhon. *Rotheni* 12th cent. '(Place on the) River Rhondda'. The river-name means 'noisy one' (Welsh *rhoddni*).

Rhoose (*Y Rhws*) Vale Glam. *Rhos* 1533. '(Place on the) moor'. Welsh *rhos*.

Rhos-on-sea (*Llandrillo-yn-Rhos*) Conwy. The Welsh name means 'St Trillo's church in Rhos' (Welsh *llan* + *yn*), Rhos being a cantref (Welsh *rhos* 'moor').

Rhosllanerchrugog Wrex. *Rose lane aghregog* 1544. 'Moor of the heather glade'. Welsh *rhos* + *llannerch* + *grugog*.

Rhostryfan Gwyd. 'Moor by (Moel) Tryfan'. *Rhos Tryfan* 1827. Welsh *rhos*. The nearby hill *Moel Tryfan* has a name meaning 'bare hill with a distinctive summit' (Welsh *moel* + *tryfan*).

Rhostyllen Wrex. *Rhos Stellan* 1546. 'Moor on a ledge'. Welsh *rhos* + *ystyllen*.

Rhuddlan Denb. *Roelend* 1086 (DB), *Ruthelan* 1191. 'Red bank'. Welsh *rhudd* + *glan*.

Rhum. *See* RUM.

Rhuthun. *See* RUTHIN.

Rhws, Y. *See* RHOOSE.

Rhydaman. *See* AMMANFORD.

Rhyl Denb. *Ryhull* 1301, *Yrhill* 1578, *Rhyll* 1660. '(Place by) the hill'. Welsh *yr* + ME *hull*.

Rhymney (*Rhymni*) Cphy. *Remni* 1101, *Rempny* 1296, *Rymney* 1541. '(Place on the) River Rhymney'. The river-name means 'auger' (Welsh *rhwmp* + adjectival ending *-ni*), describing its boring action.

Rhynd Perth. 'Promontory'. Gaelic *rinn*.

Rhynie Aber. *Rhynyn* c.1230. 'Promontory'. Gaelic *rinn*.

Rhynie Highland. *Rathne* 1529. Probably 'little fort'. Gaelic *ràthan*.

Ribbesford Worcs. *Ribbedford* 1023, *Ribeford* 1086 (DB). 'Ford by a bed of ribwort or hound's-tongue'. OE *ribbe* + *bedd* + *ford*.

Ribble Valley (district) Lancs. A modern name, from the River Ribble (*see* next entry). **South Ribble** is named from the same river.

Ribbleton Lancs. *Ribleton* 1201. 'Farmstead on the River Ribble'. OE river-name (from **ripel* 'tearing' or 'boundary') + *tūn*.

Ribchester Lancs. *Ribelcastre* 1086 (DB). 'Roman fort on the River Ribble'. OE river-name (*see* previous name) + *ceaster*.

Ribigill Highland. *Regeboll* (no date). 'Farm on a ridge'. OScand. *hryggr* + *ból*.

Ribston, Great & Ribston, Little
N. Yorks. *Ripestan* 1086. Probably 'rock or stone by which ribwort or hound's-tongue grows'. OE *ribbe* + *stān*.

Riby Lincs. *Ribi* 1086 (DB). 'Farmstead where rye is grown'. OE *ryge* + OScand. *bý*.

Riccall N. Yorks. *Richale* 1086 (DB). 'Nook of land of a man called **Rīca*'. OE pers. name + *halh*.

Rich Hill Armagh. The name comes from Edward *Richardson*, who built Richhill Castle here in the 17th cent.

Richards Castle Herefs. *Castellum Ricardi* c.1180–6, *Richardescastel* 1349. From OFrench, ME *castel*, alluding to the Frenchman *Richard Scrope* whose son Osbern held the castle here in 1086 (DB).

Richborough Kent. *Rutupiae* c.150, *Ratteburg* 1197. 'Stronghold called **Repta*'. Reduced form of ancient Celtic name (probably 'muddy waters or estuary') + OE *burh*.

Richmond Gtr. London. *Richemount* 1502. Earlier called *West Sheen* (*see* SHEEN), renamed by Henry VII from the following place.

Richmond N. Yorks. *Richemund* c.1110. 'Strong hill'. OFrench *riche* + *mont*.

Richmondshire (district) N. Yorks. *shira de Richmond* 1174, *Richemundesir* 1198. 'District centred on RICHMOND (N. Yorks)'. OE *scīr*.

Rickinghall Inferior & Rickinghall Superior Suffolk. *Rikinghale* 10th cent., *Rikingahala* 1086 (DB). 'Nook of the family or followers of a man called **Rīca*'. OE pers. name + *-inga-* + *halh*.

Rickling Essex. *Richelinga* 1086 (DB). '(Settlement of) the family or followers of a man called **Rīcela* or a woman called Rīcola'. OE pers. name + *-ingas*.

Rickmansworth Herts. *Prichemareworde* [sic] 1086 (DB), *Richemaresworthe* c.1180. 'Enclosure of a man called **Rīcmǣr*'. OE pers. name + *worth*.

Riddlecombe Devon. *Ridelcoma* 1086 (DB). Possibly 'valley of a man called **Riddela*'. OE pers. name + *cumb*.

Riddlesden Brad. *Redelesden* 1086 (DB). 'Valley of a man called **Rēdel or Hrēthel*'. OE pers. name + *denu*.

Ridge Herts. *La Rigge* 1248. 'The ridge'. OE *hrycg*.

Ridgewell Essex. *Rideuuella* 1086 (DB). Probably 'spring or stream where reeds grow'. OE *hrēod* + *wella*.

Ridgmont Beds. *Rugemund* 1227. 'Red hill'. OFrench *rouge* + *mont*.

Riding Mill Northum. *Ryding* 1262. OE **ryding* 'a clearing'.

Riding, East, Riding, North, & Riding, West (ancient tripartite division of Yorkshire). *Estreding, Nortreding, Westreding* 1086 (DB). From OScand. *thrithjungr* 'a third part', the initial *th-* having coalesced with the final consonant of OE *ēast, north*, and *west* to give *Riding*.

Ridlington, 'estate associated with a man called **Rēdel or Hrēthel*', OE pers. name + *-ing-* + *tūn*: **Ridlington** Norfolk. *Ridlinketuna* 1086 (DB). **Ridlington** Rutland. *Redlinctune* 1086 (DB).

Ridware, Hamstall & Ridware, Pipe Staffs. *Rideware* 1004, 1086 (DB), *Hamstal Ridewar* 1242, *Pipe Ridware* 14th cent. Probably '(settlement of) the dwellers at the ford'. Celtic **rĭd* + OE *-ware*. Distinguishing affixes are from OE *hām-stall* 'homestead' and from the *Pipe* family, holders of the manor in the 13th cent.

Rievaulx N. Yorks. *Rievalle* 1157. 'Valley of the River Rye'. Celtic river-name (probably meaning 'stream') + OFrench *val(s)*.

Rillington N. Yorks. *Redlintone* 1086 (DB). 'Estate associated with a man called **Rēdel or Hrēthel*'. OE pers. name + *-ing-* + *tūn*.

Rimington Lancs. *Renitone* [sic] 1086 (DB), *Rimingtona* 1182–5. 'Farmstead on the boundary stream'. OE *rima* + *-ing* + *tūn*.

Rimpton Somerset. *Rimtune* 938, *Rintone* 1086 (DB). 'Farmstead on the boundary'. OE *rima* + *tūn*.

Rimswell E. R. Yorks. *Rimeswelle* 1086 (DB). Possibly 'spring or stream of a man called Hrímr or **Rymi*'. OScand. or OE pers. name + OE *wella*.

Rinaston (*Tre-Einar*) Pemb. *Villa Reyneri* c.1308, *Reynerhiston* 1315. 'Reyner's farm'. OGerman pers. name + OE *tūn* (Welsh *tref*).

Rinawade (*Rinn an Bháid*) Dublin. 'Point of the boat'.

Rindown (*Rinn Dúin*) Roscommon. 'Point of the fort'.

Rine (*Rinn*) Clare. 'Point'.

Rineanna (*Rinn Eanaigh*) Clare. 'Point of the marsh'.

Rineen (*Rinnín*) Clare. 'Little point'.

Ring (*Rinn*) Waterford. 'Point'.

Ringarogy (*Rinn Gearróige*) Cork. 'Point of the small portion'.

Ringaskiddy (*Rinn an Scídigh*) Cork. 'Scídioch's headland'.

Ringcurran (*Rinn Chorráin*) Cork. 'Point of the sickle'.

Ringfad (*Rinn Fhada*) Down. 'Long point'.

Ringland Norfolk. *Remingaland* 1086 (DB). Probably 'cultivated land of the family or followers of a man called *Rȳmi'. OE pers. name + *-inga-* + *land*.

Ringmer E. Sussex. *Ryngemere* 1276. 'Circular pool', or 'pool near a circular feature'. OE *hring* + *mere*.

Ringmore Devon. near Bigbury. *Reimore* [*sic*] 1086 (DB), *Redmore* 1242. 'Reedy moor'. OE *hrēod* +*mōr*.

Ringrone (*Rinn Róin*) Cork. 'Seal point'.

Ringsend (*An Rinn*) Dublin. 'The point'. English *end* was added to the Irish name to mean 'place at the end of the promontory'.

Ringsfield Suffolk. *Ringesfelda* 1086 (DB). 'Open land of a man called *Hring, or near a circular feature'. OE pers. name or OE *hring* + *feld*.

Ringshall Suffolk. *Ringhesehla, Ringeshala* 1086 (DB). Etymology uncertain. Possibly 'nook of land of a man called *Hring, or near a circular feature', OE pers. name or OE *hring* + *halh*. Alternatively perhaps 'circular shelter (for animals)', OE *hring* + *(ge)sell*. **Ringshall Stocks** (recorded thus from 1837) is from ModE *stock* 'tree-stump'.

Ringstead, 'circular enclosure, or place near a circular feature', OE *hring* + *stede*: **Ringstead** Norfolk. *Ringstyde c.*1050, *Rincsteda* 1086 (DB). **Ringstead** Northants. *Ringstede* 12th cent.

Ringwood Hants. *Rimucwuda* 955, *Rincvede* 1086 (DB). Probably 'wood on a boundary'. OE *rimuc* + *wudu*.

Ringwould Kent. *Roedligwealda* 861, *Redelingewalde* 1254. Probably 'woodland of the family or followers of a man called *Rēdel or Hrēthel'. OE pers. name + *-inga-* + *weald*.

Rinmore (*Rinn Mhór*) Donegal. 'Large point'.

Rinn (*Rinn*) Leitrim, Longford. 'Point'.

Rinnafaghla (*Rinn na Fochla*) Donegal. 'Point of the cave'.

Rinneen (*Rinnín*) Galway, Kerry. 'Little point'.

Rinns of Galloway (peninsula) Dumf. *Le Rynnys* 1460. 'Headlands of Galloway'. Gaelic *rinn*. See GALLOWAY.

Rinns of Islay (peninsula) Arg. 'Divisions of Islay'. Gaelic *rann*. The name refers to the three districts into which ISLAY was formerly divided.

Rinville (*Rinn Mhíl*) Galway. 'Point of the animal'.

Rinvyle (*Rinn Mhaoile*) Galway. 'Promontory of the hornless cow'.

Ripe E. Sussex. *Ripe* 1086 (DB). '(Place at) the edge or strip of land'. OE *rip.

Ripley, usually 'strip-shaped woodland clearing', OE *ripel* + *lēah*: **Ripley** Derbys. *Ripelei* 1086 (DB). **Ripley** Hants. *Riple* 1086 (DB). **Ripley** Surrey. *Ripelia* 1204. However the following may have a different origin: **Ripley** N. Yorks. *Ripeleia* 1086 (DB). Probably 'woodland clearing of the tribe called *Hrype'. Old tribal name (*see* RIPON) + OE *lēah*.

Riplingham E. R. Yorks. *Ripingham* [*sic*] 1086 (DB), *Ripplingeham* 1180. 'Homestead associated with a man called *Rip(p)el', OE pers. name + *-ing-* + *hām*, or 'homestead by the strip of land', OE *ripel* + *-ing* + *hām*.

Ripon N. Yorks. *Hrypis c.*715, *Ripum* 1086 (DB). '(Place in the territory of) the tribe called *Hrype'. Old tribal name (origin and meaning obscure) in a dative plural form *Hrypum*.

Rippingale Lincs. *Repinga* 806, *Repinghale* 1086 (DB). 'Nook of the family or followers of a man called *Hrepa'. OE pers. name + *-inga-* + *halh*.

Ripple, '(place at) the strip of land', OE *ripel*: **Ripple** Kent. *Ryple* 1087. **Ripple** Worcs. *Rippell* 708, *Rippel* 1086 (DB).

Ripponden Calder. *Ryburnedene* 1307. 'Valley of the River Ryburn'. OE river-name ('fierce or reedy stream') + *denu*.

Ripton, Abbot's & Ripton, King's Cambs. *Riptone c.*960, *Riptune* 1086 (DB). 'Farmstead by a strip of land'. OE *rip + tūn*. Affixes from early possession by the Abbot of Ramsey and the King.

Risborough, Monks & Risborough, Princes Bucks. *Hrisanbyrge* 903, *Riseberge* 1086 (DB). 'Hill(s) where brushwood grows'. OE **hrīsen + beorg*. Manorial affixes from early possession by the monks of Christchurch, Canterbury and by the Black Prince.

Risbury Herefs. *Riseberia* 1086 (DB). 'Fortress among the brushwood'. OE *hrīs + burh* (dative *byrig*).

Risby, 'farmstead or village among the brushwood', OScand. *hrís + bý*: **Risby** N. Lincs. *Risebi* 1086 (DB). **Risby** Suffolk. *Risebi* 1086 (DB).

Risca (*Rhisga*) Cphy. *Risca* 1330. '(Place of) bark'. Welsh *rhisg*. The reference may be to a house built of bark-covered logs or to bark used as shingles on walls.

Rise E. R. Yorks. *Risun* 1086 (DB). '(Place among) the brushwood'. OE *hrīs* in a dative plural form *hrīsum*.

Riseley, 'brushwood clearing', OE *hrīs + lēah*: **Riseley** Beds. *Riselai* 1086 (DB). **Riseley** Wokhm. *Rysle* 1300.

Rishangles Suffolk. *Risangra* 1086 (DB). 'Wooded slope where brushwood grows'. OE *hrīs + hangra*.

Rishton Lancs. *Riston* c.1205. 'Farmstead where rushes grow'. OE *risc + tūn*.

Rishworth Calder. *Risseworde* 12th cent. 'Enclosure where rushes grow'. OE *risc + worth*.

Rising, Castle Norfolk. *Risinga* 1086 (DB), *Castel Risinge* 1254. Probably '(settlement of) the family or followers of a man called **Risa*'. OE pers. name + *-ingas*, with the later addition of *castel* from the Norman castle here. Alternatively 'dwellers in the brushwood', OE *hrīs + -ingas*.

Risley, 'brushwood clearing', OE *hrīs + lēah*: **Risley** Derbys. *Riselei* 1086 (DB). **Risley** Warrtn. *Ryselegh* 1284.

Risplith N. Yorks. *Respleth* 1490. 'Slope overgrown with brushwood'. OE **hrispe + hlith*.

Rissington, Great & Rissington, Little Glos. *Rise(n)dune* 1086 (DB). 'Hill where brushwood grows'. OE **hrīsen + dūn*.

Riston, Long E. R. Yorks. *Ristune* 1086 (DB). 'Farmstead near or among brushwood'. OE *hrīs + tūn*.

Rivenhall End Essex. *Reuenhala* 1068, *Ruenhale* 1086 (DB). Probably 'rough or rugged nook of land'. OE *hrēof* (dative *-an*) + *halh*.

Riverhead Kent. *Reddride* 1278. 'Landing-place for cattle'. OE *hrīther + hȳth*.

Rivington Lancs. *Revington* 1202. 'Farmstead at the rough or rugged hill'. OE *hrēof + -ing + tūn*.

Roade Northants. *Rode* 1086 (DB). '(Place at) the clearing'. OE *rodu*.

Roberton S. Lan. *Villa Roberti* c.1155, *Robertstun* 1229. 'Robert's farm'. OE *tūn*.

Robertsbridge E. Sussex. *Pons Roberti* 1176, *Robartesbregge* 1445. Named from *Robert* de St Martin, founder of the 12th-cent. abbey here. The earliest form has Latin *pons* 'bridge'.

Robeston Wathen Pemb. *Villa Roberti* 1282, *Roberdeston* 1357, *Robertson Wathen* 1545. 'Robert's farm by Llangwathen'. OE *tūn*. *Llangwathen* is 'Gwaiddan's grove' (Welsh *llwyn*, 'grove', replaced by *llan*, 'church').

Robin Hood's Bay N. Yorks. *Robin Hoode Baye* 1532. A late name which probably arose from the popular ballads about this legendary outlaw.

Roborough Devon. *Raweberge* 1086 (DB). 'Rough hill'. OE *rūh + beorg*.

Roby Knows. *Rabil* [*sic*] 1086 (DB), *Rabi* 1185. 'Farmstead or village at a boundary'. OScand. *rá + bý*.

Rocester Staffs. *Rowcestre* 1086 (DB). OE *ceaster* 'Roman fort', possibly with *rūh* 'rough' or an unidentified pers. name.

Rochdale Rochdl. *Recedham* 1086 (DB), *Rachedal* c.1195. 'Valley of the River Roch'. River-name + OScand. *dalr*. The river-name itself is a back-formation from the old name *Recedham* 'homestead with a hall', OE *reced + hām*.

Roche Cornwall. *La Roche* 1233. 'The rock'. OFrench *roche*.

Rochester Medway. *Hrofaescaestir* 731, *Rovecestre* 1086 (DB). Probably 'Roman town or fort called *Hrofi*'. Ancient Celtic name (reduced from *Durobrivis* 4th cent. 'the walled town with the bridges') + OE *ceaster*.

Rochester Northum. *Rucestr* 1242. Possibly 'rough earthwork or fort'. OE *rūh + ceaster*.

Rochford, 'ford of the hunting-dog', OE *ræcc* + *ford*: **Rochford** Essex. *Rochefort* 1086 (DB). **Rochford** Worcs. *Recesford* 1086 (DB).

Rock Northum. *Rok* 1242. Probably ME *rokke* 'a rock, a peak'.

Rock Worcs. *Ak* 1224, *Roke* 1259. '(Place at) the oak-tree'. OE *āc*, with *R-* from ME *atter* 'at the'.

Rock of Dunamase (*Dún Masg*) Laois. 'Rock of the fort of Masg'.

Rock, The Tyrone. *The Rock* c.1835. The reference is to a rocky outcrop nearby.

Rockall (island) W. Isles. *Rocol* 1606. Possibly 'barren place in the stormy sea'. OScand. **rok* + *kollr*.

Rockbeare Devon. *Rochebere* 1086 (DB). 'Grove frequented by rooks'. OE *hrōc* + *bearu*.

Rockbourne Hants. *Rocheborne* 1086 (DB). 'Stream frequented by rooks'. OE *hrōc* + *burna*.

Rockcliffe Cumbria. *Rodcliua* 1185. 'Red cliff or bank'. OScand. *rauthr* + OE *clif*.

Rockcorry (*Buíochar*) Monaghan. *Rockorry* 1778. The village lies on rocky soil and was founded in the second half of the 18th cent. by Thomas Charles Stewart *Corry*. The Irish name means 'yellow land'.

Rockhampton S. Glos. *Rochemtune* 1086 (DB). 'Homestead by the rock, or frequented by rooks'. OE *rocc* or *hrōc* + *hām-tūn*.

Rockingham Northants. *Rochingeham* 1086 (DB). 'Homestead of the family or followers of a man called *Hrōc(a)'. OE pers. name + *-inga-* + *hām*.

Rockland, 'grove frequented by rooks', OScand. *hrókr* + *lundr*: **Rockland All Saints & Rockland St Peter** Norfolk. *Rokelund* 1086 (DB). Distinguishing affixes from the church dedications. **Rockland St Mary** Norfolk. *Rokelund* 1086 (DB). Affix from the church dedication.

Rockley Wilts. *Rochelie* 1086 (DB). 'Wood or clearing frequented by rooks'. OE *hrōc* + *lēah*.

Rodbourne, 'reedy stream', OE *hrēod* + *burna*: **Rodbourne** Wilts., near Malmesbury. *Reodburna* 701. **Rodbourne Cheney** Swindn. *Redborne* 1086 (DB), *Rodburne Chanu* 1304. Manorial affix from Ralph *le Chanu*, here in 1242.

Rodd Herefs. *La Rode* 1220. 'The clearing'. OE *rodu*.

Roddam Northum. *Rodun* 1201. '(Place at) the clearings'. OE *rodu* in a dative plural form *rodum*.

Rode, 'the clearing', OE *rodu*: **Rode Heath** Ches. *Rode* 1086 (DB), *Rodeheze* 1280. With the later addition of ME *hethe* 'heath'. **Rode, North** Ches. *Rodo* 1086 (DB).

Roden Tel. & Wrek. *Hrodene* late 10th cent., *Rodene* 1242. Named from the River Roden, a Celtic river-name probably meaning 'swift river'.

Roding, Abbess, Roding, Aythorpe, Roding, Beauchamp, Roding, High, Roding, Leaden, Roding, Margaret, & Roding, White Essex. *Rodinges* c.1050, 1086 (DB), *Roinges Abbatisse* 1237, *Roeng Aytrop* 1248, *Roynges Beuchamp* 1238, *High Roinges* 1224, *Ledeineroing* 1248, *Roinges Sancte Margaret* c.1245, *White Roeng* 1248. '(Settlement of) the family or followers of a man called *Hrōth(a)'. OE pers. name + *-ingas*. The affixes *Abbess*, *Aythorpe*, and *Beauchamp* are all manorial, from early possession by the Abbess of Barking, by a man called *Aitrop*, and by the *de Beauchamp* family respectively. *High* Roding refers to its situation, *Leaden* (OE *lēaden*) to the lead roof of the church, *Margaret* to the dedication of the church, and *White* to the colour of the church walls. The river-name **Roding** is a back-formation from the place name; the earlier name for the river was *Hyle*, *see* ILFORD.

Rodington Tel. & Wrek. *Rodintone* 1086 (DB). 'Farmstead on the River Roden'. Celtic river-name (*see* RODEN) + OE *tūn*, possibly with medial connective *-ing-*.

Rodley Glos. *Rodele* 1086 (DB). 'Clearing amongst the reeds'. OE *hrēod* + *lēah*.

Rodley Leeds. *Rothelaye* 1246. Probably 'woodland clearing of a man called Hrōthwulf or *Róthulfr'. OE or OScand. pers. name + OE *lēah*.

Rodmarton Glos. *Redmertone* 1086 (DB). Probably 'farmstead by a reedy pool'. OE *hrēod* + *mere* + *tūn*.

Rodmell E. Sussex. *Redmelle* [*sic*] 1086 (DB), *Radmelde* 1202. Probably '(place with) red soil'. OE *rēad* + **mylde*.

Rodmersham Kent. *Rodmaeresham c.*1100.
'Homestead or village of a man called
*Hrōthmǣr'. OE pers. name + *hām*.

Rodney Stoke Somerset. *See* STOKE.

Rodsley Derbys. *Redeslei* 1086 (DB). OE *lēah*
'woodland clearing', possibly with OE *hrēod*
'reed-bed'.

Roe (*Rua*) (river) Derry. 'Red one'.

Roehampton Gtr. London. *Rokehampton*
1350. 'Home farm frequented by rooks'. OE *hrōc*
+ *hām-tūn*.

Roffey W. Sussex. *La Rogheye* 1281. Probably
'the fence or enclosure for roe-deer', OE *rāh-
hege*, but 'the rough enclosure' if the first
element is OE *rūh*.

Rogart Highland. *Rothegorth c.*1230. Gaelic
ro-ghort 'big field'.

Rogate W. Sussex. *Ragata* 12th cent. 'Gate for
roe-deer'. OE *rā* + *geat*.

Rogerstone Newpt. *Rogerston* 1535. 'Roger's
farm'. OGerman pers. name + OE *tūn*.

Rogeston Pemb. *Villa Rogeri* 14th cent.,
Rogeriston 1370. 'Roger's farm'. OGerman pers.
name + OE *tūn*.

Rohallion Perth. 'Fort of the Caledonians'.
Gaelic *ràth*.

Roker Sundld. First recorded as *Roca (Battery
& Point)* in 1768, origin uncertain, but possibly a
transferred name from Cape *Roca* on the
Portuguese coast.

Rollesby Norfolk. *Rollesby* 1044–7,
Rotholfuesby 1086 (DB). 'Farmstead or village of
a man called Hrólfr'. OScand. pers. name + *bý*.

Rolleston Leics. *Rovestone* [*sic*] 1086 (DB).
Rolveston 1199. 'Farmstead of a man called
Hrōthwulf or Hrólfr'. OE or OScand. pers. name
+ OE *tūn*.

Rolleston Notts. *Roldestun* 1086 (DB).
'Farmstead of a man called Hróaldr'. OScand.
pers. name + OE *tūn*.

Rolleston Staffs. *Rothulfeston* 941, *Rolvestune*
1086 (DB). Identical in origin with ROLLESTON
(Leics.).

Rollright, Great & Rollright, Little
Oxon. *Rollandri* [*sic*] 1086 (DB), *Rollendricht*
1091. Probably 'groove (= gorge) at *Rodland

('wheel precinct', referring to the ancient
Rollright Stones)'. Celtic *rod + *land + *rïcc.

Rolvenden Kent. *Rovindene* [*sic*] 1086 (DB),
Ruluinden 1185. 'Woodland pasture
associated with a man called Hrōthwulf'. OE
pers. name + *-ing-* + *denn*.

Romaldkirk Durham. *Rumoldesc(h)erce*
1086 (DB). 'Church dedicated to St Rūmwald'.
OE pers. name + OScand. *kirkja*.

Romanby N. Yorks. *Romundrebi* 1086 (DB).
'Farmstead or village of a man called
Róthmundr'. OScand. pers. name + *bý*.

Romansleigh Devon. *Liega* 1086 (DB),
Reymundesle 1228. 'The wood or clearing'. OE
lēah, with the later addition of the Celtic saint's
name *Rumon* from the dedication of the church.

Romford Gtr. London. *Romfort* 1177.
Probably 'wide or spacious ford'. OE *rūm + ford*.
The river-name **Rom** is a back-formation from
the place name.

Romiley Stockp. *Rumelie* 1086 (DB).
'Spacious woodland clearing'. OE *rūm + lēah*.

Romney, Old & Romney, New Kent.
Rumenea 895, 11th cent., *Romenel* 1086 (DB).
Originally a river-name, from OE *ēa* 'river' and
an uncertain first element, perhaps *Rūmen* 'the
broad one' which may be an old name for
Romney Marsh. Settlers here are referred to
as *Merscware* as early as the 8th cent., 'the
marsh dwellers' from OE *mersc + -ware*.

Romsey Hants. *Rummæsig c.*970, *Romesy*
1086 (DB). 'Island, or dry ground in marsh, of a
man called *Rūm'. OE pers. name + *ēg*.

Romsley, probably 'wood or clearing of the
raven', OE *hræfn, hremn + lēah*: **Romsley**
Shrops. *Hremesleage* 1002, *Rameslege* 1086 (DB).
Romsley Worcs. *Romesle* 1270.

Rona (island) W. Isles. 'Rough island'.
OScand. *hraun + ey*.

Ronaldsay, North & Ronaldsay, South
(islands) Orkn. *See* NORTH RONALDSAY, SOUTH
RONALDSAY.

Rookhope Durham. *Rochop* 1242. 'Small
enclosed valley frequented by rooks'. OE *hrōc +
hop*.

Rookley I. of Wight. *Roclee* 1202. 'Wood or
clearing frequented by rooks'. OE *hrōc + lēah*.

Roos E. R. Yorks. *Rosse* 1086 (DB). Celtic **ros* 'moor, heath, promontory'.

Roose Cumbria. *Rosse* 1086 (DB). Identical in origin with the previous name.

Ropley Hants. *Roppele* 1172. 'Woodland clearing of a man called **Hroppa'. OE pers. name + *lēah*.

Ropsley Lincs. *Ropeslai* 1086 (DB). Probably 'woodland clearing of a man called **Hropp'. OE pers. name + *lēah*.

Rorrington Shrops. *Roritune* 1086 (DB). 'Estate associated with a man called **Hrōr'. OE pers. name + *-ing-* + *tūn*.

Rosbeg (*Ros Beag*) Donegal. 'Little point'.

Rosbercon (*Ros Ó mBearchón*) Kilkenny. 'Uí Bhearchon's grove'.

Roscommon (*Ros Comáin*) Roscommon. 'Comán's grove'.

Roscor (*Ros Corr*) Fermanagh. 'Pointed promontory'.

Roscrea (*Ros Cré*) Tipperary. 'Cré's point'.

Rose Ash Devon. *Aissa* 1086 (DB), *Assherowes* 1281. '(Place at) the ash-tree'. OE *æsc*, with later manorial affix from possession by a man called *Ralph* in the late 12th cent.

Roseacre Lancs. *Raysacre* 1283. 'Cultivated land with a cairn'. OScand. *hreysi* + *akr*.

Roseberry Topping Red. & Cleve. *Othenesberg* 1119 (15th-cent. copy). 'Hill or mound associated with the heathen god Odin (the Scandinavian equivalent of Woden)'. OScand. god-name + *berg*, with later addition of ModE dialect *topping* 'hill'. Initial *R-* from the final letter of the word *under* in the name of nearby NEWTOLN UNDER ROSEBERRY (earlier *Newton under Ouesbergh*).

Rosedale Abbey N. Yorks. *Russedal c.*1140. 'Valley of the horses, or of a man called Rossi'. OScand. *hross* or pers. name + *dalr*.

Roseden Northum. *Russeden* 1242. 'Valley where rushes grow'. OE **rysc* + *denu*.

Roseland Cornwall. *See* ST JUST.

Rosemarket Pemb. *Rosmarche c.*1230. 'Rhos market'. The medieval market was named from the cantref of *Rhos* (Welsh *rhos*, 'moor').

Rosemarkie Highland. *Rosmarkensis* 1128. 'Point of land by the horse stream'. Gaelic *ros* + *marc* + *-nidh*.

Rosenallis (*Ros Fhionnghlaise*) Laois. 'Grove of the bright stream'.

Rosgill Cumbria. *Rossegile* late 12th cent. 'Ravine of the horses'. OScand. *hross* + *gil*.

Rosguil (*Ros Goill*) Donegal. 'Goll's promontory'.

Roslea (*Ros Liath*) Fermanagh. 'Grey grove'.

Rosley Cumbria. *Rosseleye* 1272. 'Woodland clearing where horses are kept'. OScand. *hross* + OE *lēah*.

Rosliston Derbys. *Redlauestun* 1086 (DB). Probably 'farmstead of a man called Hrōthlāf'. OE pers. name + *tūn*.

Rosmuck (*Ros Muc*) Galway. 'Headland of pigs'.

Rosnakill (*Ros na Cille*) Donegal. 'Grove of the church'.

Ross, 'hill-spur, moor, heathy upland', Celtic **ros*: **Ross** Northum. *Rosse* 1208–10. **Ross on Wye** Herefs. *Rosse* 1086 (DB). Wye is an ancient pre-English river-name of unknown origin and meaning.

Ross (*Ros*) Meath. 'Grove'.

Ross (the historic county). 'Moorland or promontory'. Gaelic *ros*.

Ross Carbery (*Ros Ó gCairbre*) Cork. 'Uí Chairbre's point'.

Ross, Lough (*Loch Rois*) Armagh, Galway, Monaghan. 'Lake of the grove'.

Rossadrehid (*Ros an Droichid*) Tipperary. 'Grove of the bridge'.

Rosscahill (*Ros Chathail*) Galway. 'Cahill's grove'.

Rossendale (district) Lancs. *Rocendal* 1242, *Rossendale* 1292. Probably 'moor valley' (that of the River Irwell for which *see* IRLAM). Celtic **ros* (with suffix) + OE *dæl* or OScand. *dalr*.

Rosses, The (*Na Rosa*) (district) Donegal. (*do na*) *Rosaibh* 1603. 'The promontories'.

Rossett (*Yr Orsedd*) Wrex. *Rhossedh c.*1700. '(Place by) the hill'. Welsh *yr* + *gorsedd*.

Rossinan (*Ros Fhionáin*) Kilkenny. 'Fionán's grove'.

Rossington Donc. *Rosington c.*1190. Probably 'farmstead at the moor'. Celtic **ros* + OE *-ing-* + *tūn*.

Rossinver (*Ros Inbhir*) Leitrim. 'Grove of the estuary'.

Rosslare (*Ros Láir*) Wexford. 'Middle promontory'.

Rossmore (*Ros Mór*) Cork, Galway. 'Big promontory'.

Rostherne Ches. *Rodestorne* 1086 (DB). 'Thorn-tree of a man called Rauthr'. OScand. pers. name + OE *thorn* or *thyrne*.

Roston Derbys. *Roschintone* 1086 (DB). Possibly 'estate associated with a man called *Hrōthsige'. OE pers. name + *-ing-* + *tūn*.

Rostrevor (*Ros Treabhair*) Down. *Rose Trevor* 1613. 'Trevor's wood'. Edward *Trevor*, a Welshman, commanded the English garrison at Newry at the end of the Nine Years War (1594–1603) and later acquired land here.

Rosturk (*Ros Toirc*) Mayo. 'Grove of the boar'.

Rosyth Fife. *Rossyth c.*1170, *Westir Rossith* 1363. Meaning uncertain.

Rothbury Northum. *Routhebiria. c.*1125. 'Stronghold of a man called *Hrōtha'. OE pers. name + *burh* (dative *byrig*). Alternatively the first element may be OScand. *rauthr* 'red' (from the colour of the bed-rock in the area).

Rotherby Leics. *Redebi* 1086 (DB), *Retherby* 1206. 'Farmstead or village of a man called Hreitharr'. OScand. pers. name + *bý*.

Rotherfield, 'open land where cattle graze', OE *hrȳther* + *feld*: **Rotherfield** E. Sussex. *Hrytheranfeld c.*880, *Reredfelle* [*sic*] 1086 (DB). **Rotherfield Greys & Rotherfield Peppard** Oxon. *Redrefeld* 1086 (DB), *Retherfeld Grey* 1313, *Ruderefeld Pippard* 1255. Manorial affixes from early possession by the *de Gray* and *Pipard* families.

Rotherham Rothm. *Rodreham* 1086 (DB). 'Homestead or village on the River Rother'. Celtic river-name (meaning 'chief river') + OE *hām*.

Rotherhithe Gtr. London. *Rederheia c.*1105, *Retherhith* 1127. 'Landing-place for cattle'. OE *hrȳther* + *hȳth*.

Rothersthorpe Northants. *Torp* 1086 (DB), *Retherestorp* 1231. 'Outlying farm or hamlet of the counsellor or advocate'. OScand. *thorp* with the later addition of OE *rǣdere*.

Rotherwick Hants. *Hrytheruuica c.*1100. 'Cattle farm'. OE *hrȳther* + *wīc*.

Rothes Moray. *Rothes* 1238. '(Place by the) fort'. Gaelic *ràth*. *See also* GLENROTHES.

Rothesay Arg. (Bute). *Rothersay* 1321. 'Rother's island'. OScand. *ey*. The pers. name is that of *Roderick*, the son of Reginald, to whom Bute was granted in the 13th cent.

Rothley, probably 'wood with clearings', OE **roth* + *lēah*: **Rothley** Leics. *Rodolei* 1086 (DB). **Rothley** Northum. *Rotheley* 1233.

Rothwell, 'spring or stream by the clearing(s)', OE **roth* + *wella*: **Rothwell** Leeds. *Rodewelle* 1086 (DB). **Rothwell** Lincs. *Rodowelle* 1086 (DB). **Rothwell** Northants. *Rodewelle* 1086 (DB).

Rotsea E. R. Yorks. *Rotesse* 1086 (DB). 'Pool with scum on it', or 'pool of a man called Rōta'. OE *hrot* or pers. name + *sǣ*.

Rottingdean Bright. & Hove. *Rotingedene* 1086 (DB). Probably 'valley of the family or followers of a man called Rōta'. OE pers. name + *-inga-* + *denu*.

Rottington Cumbria. *Rotingtona c.*1125. 'Estate associated with a man called Rōt(a)'. OE pers. name + *-ing-* + *tūn*.

Rougham, probably 'homestead or village on rough ground', OE *rūh* (here used as a noun) + *hām*: **Rougham** Norfolk. *Ruhham* 1086 (DB). **Rougham Green** Suffolk. *Rucham c.*950, *Ruhham* 1086 (DB).

Roughton, 'farmstead on rough ground', OE *rūh* 'rough' (here used as a noun) + *tūn*: **Roughton** Lincs. *Rocstune* [*sic*] 1086 (DB), *Ructuna c.*1115. **Roughton** Norfolk. *Rugutune* 1086 (DB), *Ruhton* 1202. **Roughton** Shrops. *Roughton* 1316.

Roughty (*An Ruachtach*) (river) Kerry. 'The destructive one'.

Roundhay Leeds. *La Rundehaia c.*1180. 'Round enclosure'. OFrench *rond* + OE *hæg*.

Roundstone (*Cloch na Rón*) Galway. 'Stone of the seals'. Irish *rón* gave English *round*.

Roundway Wilts. *Rindweiam* 1149. Probably 'cleared way'. OE *rȳmed* + *weg*.

Rounton, East & Rounton, West N. Yorks. *Rontun, Runtune* 1086 (DB), *Rungtune* c.1130. 'Farmstead enclosed with poles, or near a causeway made with poles'. OE *hrung* + *tūn*.

Rous Lench Worcs. *See* LENCH.

Rousay (island) Orkn. *Hrolfsey* c.1260. 'Hrólfr's island'. OScand. pers. name + *ey*.

Rousdon Devon. *Done* 1086 (DB), *Rawesdon* 1285. '(Place at) the hill'. OE *dūn*, with manorial affix from the family of *Ralph* here from the 12th cent.

Routh E. R. Yorks. *Rutha* 1086 (DB). Possibly 'rough shaly ground'. OScand. *hrúthr*.

Rowde Wilts. *Rode* 1086 (DB). Probably '(place at) the reed-bed'. OE *hrēod*.

Rower, The (*An Robhar*) Kilkenny. 'The flood'.

Rowhedge Essex. *Rouhegy* 1346. 'Rough hedge or enclosure'. OE *rūh* + *hecg*.

Rowland Derbys. *Ralunt* 1086 (DB). 'Boundary grove', or 'grove frequented by roe-deer'. OScand. *rá* + *lundr*.

Rowlands Castle Hants. *Rolokescastel* c.1315, *Roulandes Castell* 1369. Originally 'castle of a man called Rolok', from ME *castel* and an OFrench pers. name, but later apparently changed through association with the heroic knight *Roland* of medieval legend.

Rowlands Gill Gatesd. A recent name, recorded thus from 1863, from the surname *Rowland* and ModE *gill* (OScand. *gil*) 'ravine'.

Rowley, 'rough wood or clearing', OE *rūh* + *lēah*: **Rowley** E. R. Yorks. *Rulee* 12th cent. **Rowley Hill** Kirkl. *Ruleia* c.1180. **Rowley Regis** Sandw. *Roelea* 1173. Affix is Latin *regis* 'of the king' from early possession by the Crown.

Rowsham Bucks. *Rollesham* 1170. 'Homestead or village of a man called Hrōthwulf'. OE pers. name + *hām*.

Rowsley Derbys. *Reuslege* [*sic*] 1086 (DB), *Rolvesle* 1204. 'Woodland clearing of a man called Hrōthwulf or Hrólfr'. OE or OScand. pers. name + OE *lēah*.

Rowston Lincs. *Rouestune* 1086 (DB). Possibly 'farmstead of a man called Hrólfr'. OScand. pers. name + OE *tūn*. Alternatively the first element may be OE *hrōf* 'roof', perhaps in some topographical sense.

Rowton Ches. *Rowa Christletona* 12th cent., *Roweton* 13th cent. 'The rough part of CHRISTLETON'. OE *rūh* + *tūn*.

Rowton Shrops. near High Ercall. *Routone* [*sic*] 1086 (DB), *Rowelton* 1195. Possibly 'farmstead by a rough hill'. OE *rūh* + *hyll* + *tūn*. Alternatively the first element may be an OE **rūhel* 'small rough place'.

Roxburgh Sc. Bord. *Rocisburc* 1127. '*Hrōc's fortress'. OE pers. name + *burh*.

Roxby N. Lincs. *Roxebi* 1086 (DB). 'Farmstead or village of a man called Hrókr'. OScand. pers. name + *bý*.

Roxby N. Yorks. near Hinderwell. *Roscebi* 1086 (DB). 'Farmstead or village of a man called Rauthr'. OScand. pers. name + *bý*.

Roxton Beds. *Rochesdone* 1086 (DB). Probably 'hill of a man called *Hrōc'. OE pers. name + *dūn*.

Roxwell Essex. *Rokeswelle* 1291. Probably 'spring or stream of a man called *Hrōc'. OE pers. name + *wella*.

Royal Leamington Spa Warwicks. *See* LEAMINGTON.

Roydon Essex. *Ruindune* 1086 (DB). 'Hill where rye is grown'. OE *rygen* + *dūn*.

Roydon Norfolk. near Diss. *Rygedune* c.1035, *Regadona* 1086 (DB). 'Rye hill'. OE *ryge* + *dūn*.

Royston Barns. *Rorestone* 1086 (DB). 'Farmstead of a man called *Hrōr or *Róarr'. OE or OScand. pers. name + OE *tūn*.

Royston Herts. *Roiston* 1286. Originally called *Crux Roaisie* 1184, 'cross (Latin *crux*) of a woman called Rohesia', later simply *Roeys* to which *toun* 'village' was added in the 13th cent.

Royton Oldham. *Ritton* 1226. 'Farmstead where rye is grown'. OE *ryge* + *tūn*.

Ruabon (*Rhiwabon*) Wrex. *Rywuabon* 1291, *Rhiwvabon* 1394. 'Mabon's hill'. Welsh *rhiw*. The identity of *Mabon* is uncertain.

Ruan, named from the patron saint of the parish, St Ruan or Rumon: **Ruan Lanihorne**

Cornwall. *Lanryhorn* 1270, 'parish of *Sanctus Rumonus* of *Lanyhorn*' 1327. The second part of the name means 'church-site of a man called *Rihoarn', Cornish *lann* + pers. name. **Ruan Minor** Cornwall. 'Church of *Sanctus Rumonus Parvus*' 1277. Affix is Latin *minor* 'smaller' (earlier *parvus* 'small') to distinguish it from the adjacent parish of Ruan Major.

Ruardean Glos. *Rwirdin* 1086 (DB). 'Rye or hill enclosure'. OE *ryge* or Celtic **riu* + OE *worthign*.

Rubane (*Rú Bán*) Down. *Rowbane* c.1615. 'White clearing'.

Rubery Birm. *Robery Hills* 1650. 'Rough mound or hill'. OE *rūh* + *beorg*.

Ruckcroft Cumbria. *Rucroft* 1211. 'Rye enclosure'. OScand. *rugr* + OE *croft*.

Ruckinge Kent. *Hroching* 786, *Rochinges* 1086 (DB). 'Place characterized by rooks', or 'place associated with a man called *Hrōc'. OE *hrōc* or pers. name + *-ing*.

Ruckland Lincs. *Rocheland* [*sic*] 1086 (DB), *Roclund* 12th cent. 'Grove frequented by rooks'. OScand. *hrókr* + *lundr*.

Ruddington Notts. *Roddintone* 1086 (DB). 'Estate associated with a man called Rudda'. OE pers. name + *-ing-* + *tūn*.

Rudford Glos. *Rudeford* 1086 (DB). 'Ford among the reeds'. OE *hrēod* + *ford*.

Rudge Shrops. *Rigge* 1086 (DB). '(Place at) the ridge'. OE *hrycg*.

Rudgwick W. Sussex. *Regwic* 1210. 'Dwelling or farm on a ridge'. OE *hrycg* + *wīc*.

Rudham, East & Rudham, West Norfolk. *Rudeham* 1086 (DB). 'Homestead or village of a man called Rudda'. OE pers. name + *hām*.

Rudston E. R. Yorks. *Rodestan* 1086 (DB). '(Place at) the rood-stone or stone cross'. OE **rōd-stān*.

Rudyard Staffs. *Rudegeard* 1002, *Rudierd* 1086 (DB). Probably 'yard or enclosure where rue is grown'. OE *rūde* + *geard*.

Rue Point (*Rubha*) Antrim. 'Clearing'.

Rufford Lancs. *Ruchford* 1212. 'Rough ford'. OE *rūh* + *ford*.

Rufforth York. *Ruford* 1086 (DB). Identical in origin with the previous name.

Rugby Warwicks. *Rocheberie* 1086 (DB), *Rokebi* 1200. 'Fortified place of a man called *Hrōca'. OE pers. name + *burh* (dative *byrig*) replaced by OScand. *bý* 'village'. Alternatively 'fort frequented by rooks', with OE *hrōc* as first element.

Rugeley Staffs. *Rugelie* 1086 (DB). 'Woodland clearing on or near a ridge'. OE *hrycg* + *lēah*.

Ruishton Somerset. *Risctun* 9th cent. 'Farmstead where rushes grow'. OE **rysc* + *tūn*.

Ruislip Gtr. London. *Rislepe* 1086 (DB). Probably 'leaping-place (across the river) where rushes grow'. OE **rysc* + *hlȳp*.

Rum (island) Highland. *Ruim* 677. Origin obscure, probably pre-Celtic.

Rumburgh Suffolk. *Romburch* c.1050, *Ramburc* [*sic*] 1086 (DB). 'Wide stronghold, or stronghold built of tree-trunks'. OE *rūm* or **hruna* + *burh*.

Runcorn Halton. *Rumcofan* c.1000. 'Wide bay or creek'. OE *rūm* + *cofa*.

Runcton W. Sussex. *Rochintone* 1086 (DB). 'Estate associated with a man called *Hrōc(a)'. OE pers. name + *-ing-* + *tūn*.

Runcton, North, Runcton, South, & Runcton Holme Norfolk. *Runghetuna* 1086 (DB), *Rungeton Holm* 1276. 'Farmstead enclosed with poles, or near a causeway made with poles'. OE *hrung* + *tūn*. Holme is probably OScand. *holmr* 'island, raised ground in marsh'.

Runfold Surrey. *Hrunigfeall* 974. 'Place where trees have been felled'. OE **hruna* + **(ge)feall*.

Runhall Norfolk. *Runhal* 1086 (DB). Possibly 'nook of land where there are tree-trunks'. OE **hruna* + *halh*.

Runham Norfolk. *Ronham* 1086 (DB). Possibly 'homestead or enclosure of a man called *Rūna', OE pers. name + *hām* or *hamm*. Alternatively the first element may be OE **hruna* 'tree-trunk', perhaps with reference to its use as a footbridge.

Runnington Somerset. *Runetone* 1086 (DB). Probably 'farmstead of a man called *Rūna'. OE pers. name (genitive *-n*) + *tūn*.

Runnymede Surrey. *Ronimede* 1215. 'Meadow at the island where councils are held'. OE *rūn* + *ēg* + *mǣd*.

Runswick N. Yorks. *Reneswike* 1273. Possibly 'creek of a man called Reinn'. OScand. pers. name + *vík*.

Runton, East & Runton, West Norfolk. *Runetune* 1086 (DB). 'Farmstead of a man called *Rūna or *Rúni'. OE or OScand. pers. name + OE *tūn*.

Runwell Essex. *Runewelle c.*940, *Runewella* 1086 (DB). 'Council spring', i.e. probably 'wishing-well'. OE *rūn* 'a secret, a mystery, a council' + *wella*.

Ruscombe Wokhm. *Rothescamp* 1091. 'Enclosed land of a man called *Rōt'. OE pers. name + *camp*.

Rush (*Ros*) Dublin. 'Promontory'.

Rushall Norfolk. *Riuessala* 1086 (DB). OE *halh* 'nook of land' with an uncertain first element, probably a pers. name.

Rushall Wilts. *Rusteselve* [*sic*] 1086 (DB), *Rusteshala* 1160. Probably 'nook of land of a man called *Rust'. OE pers. name + *halh*.

Rushbrooke Suffolk. *Ryssebroc c.*950, *Ryscebroc* 1086 (DB). 'Brook where rushes grow'. OE *rysc* + *brōc*.

Rushbury Shrops. *Riseberie* 1086 (DB). 'Fortified house or manor among the rushes'. OE *rysc* + *burh* (dative *byrig*).

Rushcliffe (district) Notts. *Riseclive* 1086 (DB). 'Slope where brushwood grows'. A revival of an old wapentake-name. OE *hrīs* + *clif*.

Rushden, 'valley where rushes grow', OE *ryscen* + *denu*: **Rushden** Herts. *Risendene* 1086 (DB). **Rushden** Northants. *Risedene* 1086 (DB).

Rushford Norfolk. *Rissewrth c.*1060, *Rusceuuorda* 1086 (DB). 'Enclosure where rushes grow'. OE *rysc* + *worth*.

Rushmere, 'pool where rushes grow', OE *rysc* + *mere*: **Rushmere** Suffolk. *Ryscemara* 1086 (DB). **Rushmere St Andrew** Suffolk. *Ryscemara* 1086 (DB). Affix from the dedication of the church.

Rushmoor (district) Hants. *Rishe more* 1567. An adoption of a local name also used of a small valley called Rushmoor Bottom. OE *rysc* + *mōr*.

Rushock Worcs. *Russococ* [*sic*] 1086 (DB), *Rossoc* 1166. Probably 'rushy place, rush-bed'. OE *ryscuc*.

Rusholme Manch. *Russum* 1235. '(Place at) the rushes'. OE *rysc* in a dative plural form *ryscum*.

Rushton, 'farmstead where rushes grow', OE *rysc* + *tūn*: **Rushton** Ches. *Rusitone* 1086 (DB). **Rushton** Northants. *Risetone* 1086 (DB). **Rushton Spencer** Staffs. *Risetone* 1086 (DB). Manorial affix from early possession by the *Spencer* family.

Rushwick Worcs. *Russewyk* 1275. 'Dairy farm among the rushes'. OE *rysc* + *wīc*.

Ruskey (*Rúscaigh*) Roscommon. 'Morass'.

Ruskington Lincs. *Rischintone* 1086 (DB). 'Farmstead where rushes grow'. OE *ryscen* (with Scand. *-sk-*) + *tūn*.

Rusland Cumbria. *Rolesland* 1336. 'Cultivated land of a man called Hrólfr or Hróaldr'. OScand. pers. name + *land*.

Rusper W. Sussex. *Rusparre* 1219. '(Place at) the rough spar or beam of wood'. OE *rūh* + *spearr*.

Rustington W. Sussex. *Rustinton* 1180. 'Estate associated with a man called *Rust(a)'. OE pers. name + *-ing-* + *tūn*.

Ruston N. Yorks. *Rostune* 1086 (DB). Possibly 'farmstead distinguished by its roof-beams or rafters'. OE *hrōst* + *tūn*.

Ruston Parva E. R. Yorks. *Roreston* 1086 (DB). 'Farmstead of a man called *Hrōr or *Róarr'. OE or OScand. pers. name + OE *tūn*. Affix is Latin *parva* 'little' to distinguish this place from LONG RISTON.

Ruston, East & Ruston, Sco Norfolk. *Ristuna* 1086 (DB), *Estriston* 1361, *Scouriston c.*1280. 'Farmstead near or among brushwood'. OE *hrīs* + *tūn*. Distinguishing affixes are OE *ēast* and OScand. *skógr* 'wood'.

Ruswarp N. Yorks. *Risewarp c.*1146. Possibly 'silted land overgrown with brushwood'. OE *hrīs* + *wearp*.

Rutherglen Glas. *Ruthirglen c.*1160. Gaelic *gleann* 'valley'. First element obscure.

Ruthin (*Rhuthun*) Denb. *Ruthin* 1253. 'Red fort'. Welsh *rhudd* + *din*.

Ruthven Aber. Ang., Highland. Gaelic *Ruadhainn* 'red place'.

Ruthwell Dumf. *Rovell* 1287. Possibly 'spring by a cross'. OE *rōd* + *wella*.

Rutland (the county). *Roteland* c.1060, 1086 (DB). 'Estate of a man called Rōta'. OE pers. name + *land*.

Rutland Island Donegal. *Rutland* 1837. The original Irish name of the island was *Inis Mhic an Doirn*, 'island of the son of Dorn'. It was renamed as now in 1785 in honour of Charles Manners, 4th Duke of *Rutland*, appointed Lord Lieutenant of Ireland the previous year.

Ruyton-XI-Towns Shrops. *Ruitone* 1086 (DB). 'Farmstead where rye is grown'. OE *ryge* + *tūn*. The parish was formerly composed of eleven townships.

Ryal Northum. *Ryhill* 1242. 'Hill where rye is grown'. OE *ryge* + *hyll*.

Ryarsh Kent. *Riesce* [sic] 1086 (DB), *Rierssh* 1242. 'Ploughed field used for rye'. OE *ryge* + *ersc*.

Ryburgh, Great & Ryburgh, Little Norfolk. *Reie(n)burh* 1086 (DB). Probably 'old encampment used for growing rye'. OE *ryge(n)* + *burh*.

Rydal Cumbria. *Ridale* 1240. 'Valley where rye is grown'. OE *ryge* + OScand. *dalr*.

Ryde I. of Wight. *La Ride* 1257. 'The small stream'. OE *rīth*.

Rye E. Sussex. *Ria* 1130. '(Place at) the island or dry ground in marsh'. OE *īeg*, with initial *R-* from ME *atter* 'at the'.

Rye (river) N. Yorks. *See* RIEVAULX.

Ryedale (district) N. Yorks. *Rydale* c.1167. 'Valley of the River Rye'. A revival of an old wapentake-name. Celtic river-name (*see* RIEVAULX) + OScand. *dalr*.

Ryehill, Ryhill, 'hill where rye is grown', OE *ryge* + *hyll*: **Ryehill** E. R. Yorks. *Ryel* c.1155. **Ryhill** Wakefd. *Rihella* 1086 (DB).

Ryhall Rutland. *Righale* c.1050, *Riehale* 1086 (DB). 'Land in a river-bend where rye is grown'. OE *ryge* + *halh*.

Ryhope Sundld. *Reofhoppas* c.1040, *Refhop* 1183. 'Rough valley'. OE *hrēof* + *hop* (originally in a plural form).

Ryknild or Icknield Street. *See* ICKNIELD WAY.

Ryle, Great & Ryle, Little Northum. *Rihull* 1212. 'Hill where rye is grown'. OE *ryge* + *hyll*.

Rylstone N. Yorks. *Rilestun* 1086 (DB). OE *tūn* 'farmstead' with an uncertain first element, possibly OE *rynel* 'a small stream'.

Ryme Intrinseca Dorset. *Rima* 1160. '(Place at) the edge or border'. OE *rima*. Latin affix means 'inner', distinguishing this place from the former manor of Ryme Extrinseca ('outer').

Ryston Norfolk. *Ristuna* 1086 (DB). 'Farmstead near or among brushwood'. OE *hrīs* + *tūn*.

Ryther N. Yorks. *Ridre* 1086 (DB). '(Place at) the clearing'. OE **ryther*.

Ryton, usually 'farm where rye is grown', OE *ryge* + *tūn*: **Ryton** Gatesd. *Riton* c.1150. **Ryton** Shrops. *Ruitone* 1086 (DB). **Ryton, Great** Shrops. *Ruiton* 1209. **Ryton-on-Dunsmore** Warwicks. *Ruyton* c.1045, *Rietone* 1086 (DB). Dunsmore (*Dunesmore* 12th cent.) is an old district-name meaning 'moor of a man called Dunn', OE pers. name + *mōr*.
 However the following has a different origin: **Ryton** N. Yorks. *Ritun* 1086 (DB). 'Farmstead on the River Rye'. Celtic river-name (*see* RIEVAULX) + OE *tūn*.

Sabden Lancs. *Sapeden c.*1140. 'Valley where fir-trees grow'. OE *sæppe* + *denu*.

Sacombe Herts. *Sueuecampe* 1086 (DB). 'Enclosed land of a man called *Swǣfa*'. OE pers. name + *camp*.

Sacriston Durham. *Segrysteynhogh* 1312. 'Hill-spur of the sacristan or sexton (of Durham Abbey)'. ME *secrestein* + *how* (OE *hōh*).

Sadberge Darltn. *Satberga c.*1150. Possibly OScand. *set-berg* 'a flat-topped hill'. Alternatively OScand. *sate* 'flat piece of ground' + *berg* 'hill'.

Saddington Leics. *Sadintone, Setintone* 1086 (DB), *Sadington* 1230. OE *tūn* 'farmstead, estate', possibly with a reduced form of the OE pers. name *Sǣgēat*, or an unrecorded OE pers. name **Sada*, with medial *-ing-* 'associated with'.

Saddle Bow Norfolk. *Sadelboge* 1198. 'The saddle-bow', referring to some arched or curved feature. OE *sadol* + *boga*.

Saddleworth Oldham. *Sadelwrth* late 12th cent. 'Enclosure on a saddle-shaped ridge'. OE *sadol* + *worth*.

Saffron Walden Essex. *See* WALDEN.

Sageston Pemb. *Sagerston* 1362. 'Sager's farm'. OE *tūn*.

Saggart (*Teach Sagart*) Dublin. 'Sagra's house'.

Saham Toney Norfolk. *Saham* 1086 (DB), *Saham Tony* 1498. 'Homestead by the pool'. OE *sǣ* + *hām*. Manorial affix from the *de Toni* family, here in the late 12th cent.

Saighton Ches. *Saltone* 1086 (DB). 'Farmstead where willow-trees grow'. OE *salh* + *tūn*.

Sain Nicolas. *See* ST NICHOLAS (Vale Glam.).

St Agnes Cornwall. 'Parish of *Sancta Agnes*' 1327. From the dedication of the church to St Agnes.

St Albans Herts. *Sancte Albanes stow* 1007, *Villa Sancti Albani* 1086 (DB). 'Holy place of St Alban', from the saint martyred here in AD 209. The early spellings contain OE *stōw* 'holy place' and Latin *villa* 'town'. *See also* WATLING STREET.

Saint Andras. *See* ST ANDREWS MAJOR.

St Andrews Fife. *Sancti Andree c.*1158. '(Place with the shrine of) St Andrew'. The relics of St Andrew are said to have been brought here in the 8th cent. Also known in the 12th and 13th centuries as *Kilrimont*, probably 'church of the royal hill', from Gaelic *cill* + *rígmonad*.

St Andrews Major (*Saint Andras*) Vale Glam. *Sancti Andree* 1254. '(Church of) St Andrew'. *Major* distinguishes the place from St Andrews Minor.

St Asaph (*Llanelwy*) Denb. *Sancto Asaph* 1291. '(Church of) St Asaph'. The Welsh name means 'church by the River Elwy' (Welsh *llan*), the river-name meaning 'driving one'.

St Austell Cornwall. 'Church of *Austol*' *c.*1150. From the dedication of the parish church to St Austol.

St Bees Cumbria. 'Church of *Sancta Bega*' *c.*1135. Named from St Bega, a female saint.

St Blazey Cornwall. 'Chapel of *Sanctus Blasius*' 1440. From the dedication of the church to St Blaise.

St Breock Cornwall. 'Church of *Sanctus Briocus*' 1259. From the patron saint of the church, St Brioc.

St Breward Cornwall. 'Church of *Sanctus Brewveredus*' *c.*1190. From the patron saint of the church, St Breward.

St Briavels Glos. '(Castle of) *Sanctus Briauel*' 1130. Named from the Welsh saint Briavel.

St Brides Pemb. *Sancta Brigida* 1242.
'(Church of) St Bridget'.

St Budeaux Plym. *Seynt Bodokkys* 1520.
From the dedication of the church to the Celtic
saint Budoc. Earlier *Bucheside* 1086 (DB). 'St
Budoc's hide of land', OE *hīd*.

St Buryan Cornwall. 'Church of *Sancta
Beriana*' *c.*939, *Sancta Berriona* 1086 (DB).
Named from St Beryan, a female saint.

St Clears (*Sanclêr*) Carm. *Sancto Claro* 1291,
Seint Cler 1331. '(Church of) St Clear'.

St Cleer Cornwall. 'Church of *Sanctus Clarus*'
1212. From the patron saint of the church,
St Cleer.

**St Columb Major & St Columb
Minor** Cornwall. 'Church of *Sancta Columba*'
*c.*1240. Named from St Columba, a female saint.

St Cross South Elmham Suffolk. *See*
ELMHAM.

St David's (*Tyddewi*) Pemb. *Ty Dewi* 1586.
'(Church of) St David'. The Welsh name means
'house of Dewi' (Welsh *tŷ* + *Dewi*, this being one
of the Welsh forms of *David*.)

St Devereux Herefs. 'Church of *Sanctus
Dubricius*' 1291. Named from St Dyfrig, a male
Welsh saint.

St Dogmaels (*Llandudoch*) Pemb. *Sancti
Dogmaelis* *c.*1208. '(Church of) St Dogmael'.
The Welsh name apparently means 'St Tydoch's
church' (Welsh *llan*), although no such saint is
known.

St Dogwells (*Llantydewi*) Pemb. *Ecclesia
St. Dogmaelis* 1215. '(Church of) St Dogmael'.
Compare ST DOGMAELS. The Welsh name
apparently means 'church of the house of St
David' (Welsh *llan* + *tŷ* + *Dewi*), perhaps by
association with ST DAVID'S.

St Edmundsbury (district) Suffolk.
A modern adoption of an early name of BURY ST
EDMUNDS, preserved in the diocese-name of
St Edmundsbury & Ipswich.

St Endellion Cornwall. 'Church of *Sancta
Endelienta*' 1260. Named from St Endilient, a
female saint.

St Florence Pemb. *Sanctus Florencius* 1248.
'(Church of) St Florence'.

St George's (*Llan Sain Siôr*) Vale Glam.
Sancti Georgii 1254. '(Church of) St George'. The
Welsh name translates the English (Welsh *llan*,
'church').

St Germans Cornwall. 'Church of *Sanctus
Germanus*' 1086 (DB). From the patron saint of
the church.

St Giles in the Wood Devon. *Stow St Giles*
1330. 'Holy place of St Giles', from OE *stōw* and
the patron saint of the church.

St Giles on the Heath Devon. 'Chapel of
Sanctus Egidius' 1202, *St Gylses en le Hethe* 1585.
From the dedication of the church to St Giles
(Latin *Egidius*), with affix 'on the heath' from
ME *hethe*.

St Helens I. of Wight. *Sancta Elena* 12th cent.
From the dedication of the church to St Helena.

St Helens St Hel. Named from the chapel of
St Helen, first recorded in 1552.

St Helier Channel Islands. Capital town of
JERSEY named from St Helier, a 6th-cent.
Belgian monk who is said to have lived as a
hermit in a cave above the town for 15 years.

St Ippollitts Herts. 'Church of *S. Ypollitus*'
1283. From the dedication of the 11th-cent.
church here to St Hippolytus.

St Ishmael's Pemb. *Ecclesia Sancti Ismael*
1291. '(Church of) St Ishmael'.

St Issey Cornwall. *Sanctus Ydi* 1195. From the
dedication of the church to St Idi.

St Ive Cornwall. *Sanctus Yvo* 1201. From the
dedication of the church to St Ivo, *see* next
name.

St Ives Cambs. *Sancto Ivo de Slepe* 1110.
Named from St Ivo whose relics were found
here in the 10th cent. The earlier name *Slepe* is
from OE **slæp* 'slippery place'.

St Ives Cornwall. *Sancta Ya* 1284. From the
dedication of the church to St Ya, a female saint.

St Ives Dorset. *Iuez* 1167. 'Place overgrown
with ivy'. OE **īfet*. *Saint* was only added in
recent times on the analogy of the previous
names.

St John's Point (*Rinn Eoin*) Down.
Ballihione al. Rinchione 1549. 'Eoin's point'.

Saint Johnston (*Baile Seangean*) Donegal.
*St Johnston c.*1655. The village is named after *St
Johnstoun*, the former name of PERTH, Scotland.

St Just Cornwall. 'Church of *Sanctus Justus*' 1291. From the dedication of the church to St Just. Often called **St Just in Penwith**, this last (*Penwid* 1186, probably 'end-district') being an old Cornish name for the Land's End peninsula.

St Just in Roseland Cornwall. 'Church of *Sanctus Justus*' c.1070. Identical in origin with the previous name. Roseland (*Roslande* 1259) is an old district name meaning 'promontory land', Cornish *ros* + OE *land*.

St Keverne Cornwall. *Sanctus Achebrannus* 1086 (DB). From the patron saint of the church, St Aghevran.

St Kew Cornwall. *Sanctus Cheus* 1086 (DB). From the patron saint of the church, St Kew. **St Kew Highway**, *Highway* 1699, is a hamlet so called from its situation on the main road to Wadebridge.

St Keyne Cornwall. 'Church of *Sancta Keyna*' 1291. Named from St Keyn, a female saint.

St Kilda (island) W. Isles. *Hert* 1370; *St. Kilda* 17th cent. Both the earlier and later names are obscure in origin.

St Lawrence, 'place with a church dedicated to St Lawrence': **St Lawrence** I. of Wight. 'Church of *Sanctus Laurencius*' 1255. Originally also called *Wathe* (*Wathe* 1311, *St Laurence de Wathe* 1340), '(place at) the ford' from OE *wæd*. **St Lawrence** Kent. 'Church of *Sanctus Laurencius*' 1253. **St Lawrence** Pemb. *Rectoria Sancti Laurentii* 1535.

St Leonards E. Sussex. *Villa de Sancto Leonardo juxta Hasting* ('manor of St Leonard near to Hastings') 1288. Named from the dedication of the original church here, destroyed by the sea in the 15th cent.

St Leonards, Chapel Lincs. Named from 'the chapel of St Leonard', first recorded in 1257.

St Levan Cornwall. 'Parish of *Sanctus Silvanus*' 1327. From the dedication of the church to St Selevan.

St Lythans (*Llwyneliddon*) Vale Glam. *Luin Elidon* 1160, *Llan Leiddan* 1590. 'Eliddon's grove'. Welsh *llwyn* + pers. name. Original Welsh *llwyn* was later taken to be *llan*, 'church', suggesting a dedication to a St Lythan.

St Margaret's Island Pemb. *Little Caldey alias St Margaretts Island* 1718. 'St Margaret's

island'. The island was originally known as *Little Caldy*. *See* CALDY ISLAND.

St Margaret's at Cliffe Kent. *Sancta Margarita* 1086 (DB), *Cliue c.*1100. 'Church dedicated to St Margaret at the place called Cliffe', from OE *clif* 'cliff'.

St Mary Bourne Hants. *See* BOURNE.

St Mary Cray Gtr. London. *See* CRAY.

St Mary's Bay Kent. named from St Mary in the Marsh (*Seyntemariecherche* 1240), with affix from OE *mersc*.

St Mary's Hoo Medway. *See* HOO.

St Mawes Cornwall. 'Town of *Sanctus Maudetus*' 1284. From the patron saint of the church, St Maudyth.

St Mawgan Cornwall. 'Church of *Sanctus Mauchanus*' 1257. From the dedication of the church to St Mawgan, *see also* MAWGAN. **Mawgan Porth** is from Cornish *porth* 'cove'.

St Mellion Cornwall. 'Church of *Sanctus Melanus*' 1259. From the patron saint of the church, St Melaine.

St Merryn Cornwall. *Sancta Marina* 1259. Apparently named from St Marina (a female saint), but possibly from St Merin (a male).

St Mewan Cornwall. 'Church of *Sanctus Mawanus*' 1291. From the dedication of the church to St Mewan.

St Michael Caerhays Cornwall. 'Chapel of *Sanctus Michael de Karihaes*' 1259. From the dedication of the church to St Michael the Archangel. The distinguishing affix is from the manor of Caerhays, a name obscure in origin and meaning.

St Michael Penkevil Cornwall. 'Church of *Sanctus Michael de Penkevel*' 1261. From the same dedication as previous name. The affix is from the manor of Penkevil, a Cornish name probably meaning 'horse headland'.

St Michael's on Wyre Lancs. *Michelescherche* 1086 (DB). 'Church of St Michael', with OE *cirice* in the early spelling. Wyre is a Celtic river-name meaning 'winding one'.

St Monance Fife. *Sanct Monanis* 1565. '(Church of) St Monans'.

Saint Mullin's (*Tigh Moling*) Kilkenny. 'Moling's house'.

St Neot Cornwall. *Sanctus Neotus* 1086 (DB). From the patron saint of the church, St Neot.

St Neots Cambs. *S' Neod* 12th cent. Also named from St Neot, whose relics were brought here from Cornwall in the 10th cent.

St Nicholas (*Tremarchog*) Pemb. *Villa Camerarii* 1287, *Tremarchoc* 1551, *Saynt Nycolas* 1554. '(Church of) St Nicholas'. The Welsh name means 'knight's farm' (Welsh *tref* + *marchog*), the first form above being the Latin equivalent.

St Nicholas (*Sain Nicolas*) Vale Glam. *Sancto Nicholao* 1153. '(Church of) St Nicholas'.

St Nicholas at Wade Kent. *Villa Sancti Nicholai* 1254, *St Nicholas at Wade* 1458. Named from the dedication of the church. Wade is from OE *wæd* 'a ford, a crossing-place' (leading to the Isle of Thanet).

St Osyth Essex. *Seynte Osithe* 1046. From the dedication of a priory here to St Ōsgȳth, a 7th-cent. princess. Its early name *Cice* 1086 (DB) is from OE **cicc* 'a bend'.

St Paul's Cray Gtr. London. *See* CRAY.

St Peter Port Channel Islands. Capital town of GUERNSEY named from the dedication of its 12th-cent. church and from *port* 'town with a harbour'.

St Peter's Kent. *Borgha sancti Petri* ('borough of St Peter') 1254. Named from the dedication of the church.

St Petrox Pemb. *Ecclesia de Sancto Petroco* 1291, *St Petrocks* 1603. '(Church of) St Petroc'.

St Quivox S. Ayr. '(Church of) St Caemhan'. The saint's name is represented in the form *Mochaemhoc*, with the Gaelic affectionate prefix *mo* 'my', and the diminutive suffix *-oc*.

St Tudy Cornwall. *Hecglostudic* 1086 (DB). From the dedication of the church to St Tudy, with Cornish *eglos* 'church' in the early spelling.

St Weonards Herefs. *Lann Sant Guainerth* *c.*1130. 'Church of St Gwennarth', with OWelsh **lann* 'church' in the early spelling.

St Winnow Cornwall. *San Winnuc* 1086 (DB). From the dedication of the church to St Winnoc.

Saintbury Glos. *Svineberie* [*sic*] 1086 (DB), *Seinesberia* 1186. 'Fortified place of a man called Sǣwine'. OE pers. name + *burh* (dative *byrig*).

Saintfield (*Tamhnaigh Naomh*) Down. *Tawnaghnym* 1605. 'Field of saints'.

Salcombe, 'salt valley', OE *sealt* + *cumb*: **Salcombe** Devon. *Saltecumbe* 1244. **Salcombe Regis** Devon. *Sealtcumbe* *c.*1060, *Selcome* 1086 (DB). Affix is Latin *regis* 'of the king'.

Salcott Essex. *Saltcot* *c.*1200. 'Building where salt is made or stored'. OE *sealt* + *cot*.

Sale Traffd. *Sale* *c.*1205. '(Place at) the sallow-tree'. OE *salh* in a dative form *sale*.

Saleby Lincs. *Salebi* 1086 (DB). Possibly 'farmstead or village of a man called Sali'. OScand. pers. name + *bȳ*. Alternatively the first element may be OE *salh* 'sallow, willow'.

Saleen (*Sáilín*) Cork. 'Little (arm of the) sea'.

Salehurst E. Sussex. *Salhert* [*sic*] 1086 (DB), *Salhirst* *c.*1210. 'Wooded hill where sallow-trees grow'. OE *salh* + *hyrst*.

Salesbury Lancs. *Sale(s)byry* 1246. 'Fortified place or manor where sallow-trees grow'. OE *salh* + *burh* (dative *byrig*).

Salford Beds. *Saleford* 1086 (DB). 'Ford where sallow-trees grow'. OE *salh* + *ford*.

Salford Oxon. *Saltford* 777, *Salford* 1086 (DB). 'Ford over which salt is carried'. OE *salt* + *ford*.

Salford Salford. *Salford* 1086 (DB). Identical in origin with SALFORD Beds.

Salford Priors & Abbots
Salford Warwicks. *Saltford* 714. *Salford* 1086 (DB). Identical in origin with SALFORD Oxon. Affixes from early possession by Kenilworth Priory and Evesham Abbey.

Salfords Surrey. *Salford* 1279. 'Ford where sallow-trees grow'. OE *salh* + *ford*.

Salhouse Norfolk. *Salhus* 1291. Probably 'the sallow-trees'. OE *salh* in a plural form.

Saling, Great Essex. *Salinges* 1086 (DB). Possibly '(settlement of) the dwellers among the sallow-trees', OE *salh* + *-ingas*. Alternatively 'places where sallow-trees grow', from the plural of an OE word **sal(h)ing*.

Salisbury Wilts. *Searobyrg c.*900, *Sarisberie* 1086 (DB). 'Stronghold at *Sorvio*'. OE *burh* (dative *byrig*) added to reduced form of original Celtic name *Sorviodunum* (obscure word + Celtic **dūno-* 'fort'), this referring in the 4th-cent. Antonine Itinerary to the Iron Age hill-fort at **Old Sarum**. The form *Sarum*, found from the 14th cent., arises from a misreading of the abbreviated spelling *Sar'* for *Sarisberie* and the like in medieval documents.

Salkeld, Great & Salkeld, Little
Cumbria. *Salchild c.*1110. Possibly 'sallow-tree wood'. OE *salh* + **hylte*. However the second element is perhaps rather an OE **hilde* 'slope' (a variant of *helde*).

Sallahig (*Saileáin*) Kerry. 'Place of willows'.

Salle Norfolk. *Salla* 1086 (DB). 'Wood or clearing where sallow-trees grow'. OE *salh* + *lēah*.

Sallins (*Na Solláin*) Kildare. 'The willow groves'.

Sallynoggin (*An Naigín*) Dublin. 'The Noggin'. Sally- may be English *sally* 'willow'.

Salmonby Lincs. *Salmundebi* 1086 (DB). 'Farmstead or village of a man called Salmundr'. OScand. pers. name + *bý*.

Salop (the county). *See* SHROPSHIRE.

Salperton Glos. *Salpretune* 1086 (DB). Probably 'farmstead on a salt-way'. OE *salt* + *here-pœth* + *tūn*.

Salt Staffs. *Selte* 1086 (DB). OE **selte* 'a salt-pit, a salt-works'.

Saltaire Brad. a modern name created for the new town founded in the valley of the River *Aire* (for which *see* AIRMYN) for the manufacture of alpaca by Sir Titus *Salt* in 1850.

Saltash Cornwall. *Esse* 1201, *Saltehasche* 1302. Originally '(place at) the ash-tree', OE *œsc*. Later addition *salt* from the production of salt here.

Saltburn-by-the-Sea Red. & Cleve. *Salteburnam c.*1185. '(Place by) the salt stream'. OE *salt* + *burna*.

Saltby Leics. *Saltebi* 1086 (DB). Possibly 'farmstead at the salty spring', or 'where salt is made'. OScand. *saltr* (adjective) or *salt* (noun) + *bý*.

Saltcoats N. Ayr. *Saltcoates* 1548. 'Buildings where salt is made or stored'. OE *salt* + *cot*.

Saltee Islands (*Oileán an tSalainn*) Wexford. 'Salt island'.

Salter Lancs. *Salterge c.*1150. 'Shieling where salt is found or kept'. OE *salt* + OScand. *erg*.

Salterforth Lancs. *Salterford* 13th cent. 'Ford used by salt-merchants'. OE *saltere* + *ford*.

Salterton, Budleigh Devon. *Saltre* 1210, *Salterne* 1405. OE *salt-œrn* 'building where salt is made or sold'. Affix from nearby Budleigh.

Salterton, Woodbury Devon. *Salterton* 1306. 'Farmstead of the salt-workers or salt-merchants'. OE *saltere* + *tūn*. Affix from nearby WOODBURY.

Saltfleet Lincs. *Salfluet* 1086 (DB). '(Place by) the salt stream'. OE *salt* + *flēot*.

Saltfleetby All Saints, Saltfleetby St Clements, & Saltfleetby St Peter Lincs. *Salflatebi* 1086 (DB). 'Farmstead or village by the salt stream'. OE *salt* + *flēot* + OScand. *bý*.

Saltford B. & NE. Som. *Sanford* [*sic*] 1086 (DB), *Salford* 1229, *Saltford* 1291. Probably 'salt-water ford', OE *salt* + *ford*, but the first element may originally have been *salh* 'sallow, willow'.

Salthouse Norfolk. *Salthus* 1086 (DB). 'Building for storing salt'. OE *salt* + *hūs*.

Saltmarshe E. R. Yorks. *Saltemersc* 1086 (DB). 'Salty or brackish marsh'. OE *salt* + *mersc*.

Saltney Ches./Flin. *Salteney c.*1230. 'Salty island (of dry ground in marsh)'. OE *salt* (dative *-an*) + *ēg*.

Salton N. Yorks. *Saletun* 1086 (DB). 'Farmstead where sallow-trees grow'. OE *salh* + *tūn*.

Saltwood Kent. *Sealtwuda c.*1000, *Salteode* 1086 (DB). 'Wood where salt is made or stored'. OE *sealt* + *wudu*.

Salvington W. Sussex. *Saluinton* 1249. 'Estate associated with a man called Sǣlāf or Sǣwulf'. OE pers. name + *-ing-* + *tūn*.

Salwarpe Worcs. *Salouuarpe* 817, *Salewarpe* 1086 (DB). Probably 'dark-coloured silted land'. OE *salu* + *wearp*.

Sambourne Warwicks. *Samburne* 714, *Sandburne* 1086 (DB). 'Sandy stream'. OE *sand* + *burna*.

Sambrook Tel. & Wrek. *Semebre* [*sic*] 1086 (DB), *Sambroc* 1256. Probably 'sandy brook'. OE *sand* + *brōc*.

Samlesbury Lancs. *Samelesbure* 1188. Probably 'stronghold near a shelf or ledge of land'. OE *scamol* + *burh* (dative *byrig*).

Sampford, 'sandy ford', OE *sand* + *ford*: **Sampford Arundel** Somerset. *Sanford* 1086 (DB), *Samford Arundel* 1240. Manorial affix from Roger *Arundel*, tenant in 1086. **Sampford Brett** Somerset. *Sanford* 1086 (DB), *Saunford Bret* 1306. Manorial affix from the *Bret* family, here in the 12th cent. **Sampford Courtenay** Devon. *Sanfort* 1086 (DB), *Saunforde Curtenay* 1262. Manorial affix from the *Curtenay* family, here in the 13th cent. **Sampford Peverell** Devon. *Sanforda* 1086 (DB), *Saunford Peverel* 1275. Manorial affix from the *Peverel* family, here in the 12th cent. **Sampford Spiney** Devon. *Sanforda* 1086 (DB), *Saundford Spyneye* 1304. Manorial affix from the *Spiney* family, here in the 13th cent.

Sanclêr. *See* ST CLEARS.

Sancreed Cornwall. 'Church of *Sanctus Sancretus*' 1235. From the patron saint of the church, St Sancred.

Sancton E. R. Yorks. *Santun* 1086 (DB). 'Farmstead with sandy soil'. OE *sand* + *tūn*.

Sand Hutton N. Yorks. *See* HUTTON.

Sandall, Kirk Donc. *Sandala* 1086 (DB), *Kirke Sandale* 1261. 'Sandy nook of land'. OE *sand* + *halh*. Affix is OScand. *kirkja* 'church'.

Sanday (island) Orkn. 'Sandy island'. OScand. *sand* + *ey*.

Sandbach Ches. *Sanbec* 1086 (DB). 'Sandy valley stream'. OE *sand* + *bæce*.

Sandbanks Poole. a modern name not recorded before *c.*1800, and self-explanatory.

Sanderstead Gtr. London. *Sondenstede* *c.*880, *Sandestede* 1086 (DB). 'Sandy homestead'. OE *sand* (noun) + *hǣm-styde*, or **sanden* (adjective) + *stede*.

Sandford, 'sandy ford', OE *sand* + *ford*: **Sandford** Cumbria. *Saunford c.*1210. **Sandford** Devon. *Sandforda* 930. **Sandford** Shrops. *Sanford* 1086 (DB). **Sandford on Thames** Oxon. *Sandforda* 1050, *Sanford* 1086 (DB). **Sandford Orcas** Dorset. *Sanford* 1086 (DB), *Sandford Horscoys* 1372. Manorial affix from the *Orescuils* family, here from the 12th cent. **Sandford St Martin** Oxon. *Sanford* 1086 (DB). Affix from the dedication of the church.

Sandgate Kent. *Sandgate* 1256. 'Gap or gate leading to the sandy shore'. OE *sand* + *geat*.

Sandhoe Northum. *Sandho* 1225. 'Sandy hill-spur'. OE *sand* + *hōh*.

Sandholes (*Clais an Ghainimh*) Tyrone. *Sandholes* 1834. '(Place by) sand pits'.

Sandhurst, 'sandy wooded hill', OE *sand* + *hyrst*: **Sandhurst** Brack. For. *Sandherst* 1175. **Sandhurst** Glos. *Sanher* [*sic*] 1086 (DB), *Sandhurst* 12th cent. **Sandhurst** Kent. *Sandhyrste c.*1100.

Sandhutton N. Yorks. *Hotune* 1086 (DB), *Sandhoton* 12th cent. 'Farmstead on a sandy spur of land'. OE *sand* + *hōh* + *tūn*.

Sandiacre Derbys. *Sandiacre* 1086 (DB). 'Sandy cultivated land'. OE *sandig* + *æcer*.

Sandon, 'sandy hill', OE *sand* + *dūn*: **Sandon** Essex. *Sandun* 1199. **Sandon** Herts. *Sandune c.*1000, *Sandone* 1086 (DB). **Sandon** Staffs. *Sandone* 1086 (DB).

Sandown I. of Wight. *Sande* [*sic*] 1086 (DB), *Sandham* 1271. 'Sandy enclosure or river-meadow'. OE *sand* + *hamm*.

Sandown Park Surrey. *Sandone* 1235. 'Sandy hill or down'. OE *sand* + *dūn*.

Sandridge Herts. *Sandrige* 1086 (DB). 'Sandy ridge'. OE *sand* + *hrycg*. The same name occurs in Devon.

Sandringham Norfolk. *Santdersincham* 1086 (DB). 'The sandy part of DERSINGHAM'. The affix is OE *sand*.

Sandtoft N. Lincs. *Sandtofte* 1157. 'Sandy building plot or curtilage'. OScand. *sandr* + *toft*.

Sandwell Sandw. *Saundwell* 13th cent. 'Sandy spring'. OE *sand* + *wella*.

Sandwich Kent. *Sandwicæ c.*710–20, *Sanduice* 1086 (DB). 'Sandy harbour or trading centre'. OE *sand* + *wīc*.

Sandwick Shet. *Sandvik* 1360. 'Sandy inlet'. OScand. *sand* + *vík*.

Sandwith Cumbria. *Sandwath* 1260. 'Sandy ford'. OScand. *sandr* + *vath*.

Sandy Beds. *Sandeie* 1086 (DB). 'Sandy island'. OE *sand* + *ēg*.

Sankey, Great Warrtn. *Sonchi c.*1180. Named from the River Sankey, a pre-English river-name of uncertain origin and meaning.

Sanquhar Dumf. *Sanchar* 1150. '(Place by) the old fort'. Gaelic *sean* + *cathair*.

Santon, 'farmstead with sandy soil', OE *sand* + *tūn*: **Santon Bridge** Cumbria. *Santon c.*1235. **Santon, Low** N. Lincs. *Santone* 1086 (DB).

Santry (*Seantrabh*) Dublin. 'Old dwelling'.

Sapcote Leics. *Scepecote* 1086 (DB). 'Shed(s) or shelter(s) for sheep'. OE *scēap* + *cot*.

Sapey, Lower & Sapey, Upper Herefs. *Sapian* 781, *Sapie* 1086 (DB). Probably from OE *sæpig* 'sappy, juicy', originally applied to Sapey Brook.

Sapiston Suffolk. *Sapestuna* 1086 (DB). OE *tūn* 'farmstead' with an uncertain first element, possibly a pers. name.

Sapley Cambs. *Sappele* 1227. 'Wood or clearing where fir-trees grow'. OE *sæppe* + *lēah*.

Sapperton, 'farmstead of the soap-makers or soap-merchants', OE **sāpere* + *tūn*: **Sapperton** Glos. *Sapertun c.*1075, *Sapletorne* [*sic*] 1086 (DB). **Sapperton** Lincs. *Sapretone* 1086 (DB).

Sarclet Highland. Possibly OScand. *saur* 'mud', or *sauthr* 'sheep' + *klettr* 'cliff'.

Sark (island) Channel Islands. *Serc c.*1040, *Serk* 1294. A development of an earlier *Sargia* (rationalized as Roman name *Cesarea* in the 4th-cent. *Maritime Itinerary*), of unknown origin and meaning.

Sarn Powys. '(Place by the) causeway'. Welsh *sarn*.

Sarnesfield Herefs. *Sarnesfelde* 1086 (DB). 'Open land by a road'. OWelsh *sarn* + OE *feld*.

Sarratt Herts. *Syreth c.*1085. Possibly 'dry or barren place'. OE **sīeret*.

Sarre Kent. *Serræ c.*763. Obscure in origin and meaning, but possibly an old pre-English name for the River Wantsum.

Sarsden Oxon. *Secedene* [*sic*] 1086 (DB), *Cerchesdena c.*1180. Probably 'valley of the church'. OE *cirice* + *denu*.

Sarum, Old Wilts. *See* SALISBURY.

Satley Durham. *Sateley* 1228. Possibly 'woodland clearing with stables or folds'. OE *set* (Old Northumbrian **seatu*) + *lēah*. Alternatively the first element may be OE *sǣt* 'lurking-place, trap (for animals)'.

Satterleigh Devon. *Saterlei* 1086 (DB). OE *lēah* 'woodland clearing' with either OE *sǣtere* 'robber' or the OE antecedent of the ME plant-name *seter-* 'hellebore'.

Satterthwaite Cumbria. *Saterthwayt* 1336. Probably 'clearing by a hut or shieling'. OScand. *sǣtr* + *thveit*.

Saughall, Great & Saughall, Little Ches. *Salhale* 1086 (DB). 'Nook of land where sallow-trees grow'. OE *salh* + *halh*.

Saul (*Sabhall*) Down. *Sabul Patricii c.*670. 'Barn'.

Saul Glos. *Salle* 12th cent. 'Wood or clearing where sallow-trees grow'. OE *salh* + *lēah*.

Saundby Notts. *Sandebi* 1086 (DB). Probably 'farmstead or village of a man called Sandi'. OScand. pers. name + *bý*. Alternatively the first element may be OScand. *sandr* 'sand'.

Saundersfoot Pemb. *Saunders foot* 1602. 'Saunders' foot'. A hill or cliff here must have been owned by one *Saunders*.

Saunderton Bucks. *Santesdune* 1086 (DB), *Santredon* 1196. OE *dūn* 'hill' with an uncertain first element.

Saval (*Sabhall*) Down. 'Barn'.

Savernake Forest Wilts. *Safernoc* 934. Probably 'district of a river called *Sabrina* or *Severn*', from a derivative of an ancient pre-English river-name (perhaps applied originally to the River Bedwyn) of unknown origin and meaning.

Sawbridgeworth Herts. *Sabrixteworde* 1086 (DB). 'Enclosure of a man called Sǣbeorht'. OE pers. name + *worth*.

Sawdon N. Yorks. *Saldene* early 13th cent. 'Valley where sallow-trees grow'. OE *salh* + *denu*.

Sawel (*Samhail*) (mountain) Tyrone. 'Likeness'. *Samhail Phite Meadhbea c.*1680. The full name is *Samhail Phite Méabha*, 'likeness to Maeve's vulva', referring to a hollow on the side of the mountain.

Sawley Derbys. *Salle* [*sic*] 1086 (DB), *Sallawa* 1166. 'Hill or mound where sallow-trees grow'. OE *salh* + *hlāw*.

Sawley Lancs. *Sotleie* [*sic*] 1086 (DB), *Sallaia* 1147. 'Woodland clearing where sallow-trees grow'. OE *salh* + *lēah*.

Sawley N. Yorks. *Sallege c.*1030. Identical in origin with the previous name.

Sawrey, Far & Sawrey, Near Cumbria. *Sourer* 1336. 'The sour or muddy grounds'. OScand. *saurr* in a plural form *saurar*.

Sawston Cambs. *Salsingetune* 970, *Salsiton* 1086 (DB). 'Farmstead of the family or followers of a man called *Salse'. OE pers. name + *-inga-* + *tūn*.

Sawtry Cambs. *Saltreiam* 974, *Saltrede* 1086 (DB). 'Salty stream'. OE *salt* + *rīth*.

Saxby, probably 'farmstead or village of a man called Saksi', OScand. pers. name + *bý*, but alternatively the first element may be OScand. *Saksar* 'Saxons': **Saxby** Leics. *Saxebi* 1086 (DB). **Saxby** Lincs. *Sassebi* 1086 (DB). **Saxby All Saints** N. Lincs. *Saxebi* 1086 (DB). Affix from the dedication of the church.

Saxelbye Leics. *Saxelbie* 1086 (DB). Probably 'farmstead or village of a man called Saxulfr'. OScand. pers. name + *bý*.

Saxham, Great & Saxham, Little Suffolk. *Saxham* 1086 (DB). Probably 'homestead of the Saxons'. OE *Seaxe* + *hām*.

Saxilby Lincs. *Saxebi* [*sic*] 1086 (DB), *Saxlabi c.*1115. Identical in origin with SAXELBY.

Saxlingham, 'homestead of the family or followers of a man called *Seaxel or Seaxhelm', OE pers. name + *-inga-* + *hām*: **Saxlingham** Norfolk. *Saxelingaham* 1086 (DB). **Saxlingham Nethergate** Norfolk. *Seaxlingaham* 1046, *Saiselingaham* 1086 (DB). The affix, first found in the 13th cent., means 'lower gate or street'.

Saxmundham Suffolk. *Sasmundeham* 1086 (DB). 'Homestead of a man called *Seaxmund'. OE pers. name + *hām*.

Saxondale Notts. *Saxeden* 1086 (DB), *Saxendala c.*1130. 'Valley of the Saxons'. OE *Seaxe* + *dæl* (replacing *denu*).

Saxtead Suffolk. *Saxteda* 1086 (DB). Probably 'place of a man called Seaxa'. OE pers. name + *stede*.

Saxthorpe Norfolk. *Saxthorp* 1086 (DB). 'Outlying farmstead or hamlet of a man called Saksi'. OScand. pers. name + *thorp*.

Saxton N. Yorks. *Saxtun* 1086 (DB). 'Farmstead of the Saxons, or of a man called Saksi'. OE *Seaxe* or OScand. pers. name + OE *tūn*.

Scackleton N. Yorks. *Scacheldene* 1086 (DB). Probably 'valley by a point of land'. OE *scacol* (with Scand. *sk-*) + *denu*.

Scaddy (*Sceadaigh*) Down. *Scaddy* 1886. 'Bare place'.

Scaftworth Notts. *Scafteorde* 1086 (DB). Probably 'enclosure of a man called *Sceafta or Skapti', OE or OScand. pers. name + OE *worth*. Alternatively 'enclosure made with poles', OE *sceaft* (with Scand. *sk-*) + *worth*.

Scagglethorpe N. Yorks. *Scachetorp* 1086 (DB). Probably 'outlying farmstead or hamlet of a man called Skakull or Skakli'. OScand. pers. name + *thorp*.

Scalby, 'farmstead or village of a man called Skalli', OScand. pers. name + *bý*: **Scalby** E. R. Yorks, *Scalleby* 1230. **Scalby** N. Yorks. *Scallebi* 1086 (DB).

Scaldwell Northants. *Scaldewelle* 1086 (DB). 'Shallow spring or stream'. OE **scald* (with Scand. *sk-*) + *wella*.

Scaleby Cumbria. *Schaleby c.*1235. 'Farmstead or village near the huts or shielings'. OScand. *skáli* + *bý*.

Scales, a common name in the North, 'the temporary huts or sheds', OScand. *skáli*; examples include: **Scales** Cumbria, near Aldingham. *Scales* 1269. **Scales** Cumbria, near Threlkeld. *Skales* 1323. **Scales** Lancs., near Newton. *Skalys* 1501.

Scalford Leics. *Scaldeford* 1086 (DB). 'Shallow ford'. OE **scald* (with Scand. *sk-*) + *ford*.

Scalloway Shet. *Scalwa, Schalvay* 1507. 'Bay by the hut'. OScand. *skáli* + *vágr*.

Scalpay (island) Highland (Skye). 'Boat-shaped island'. OScand. *skalpr* + *ey*.

Scamblesby Lincs. *Scamelesbi* 1086 (DB). Probably 'farmstead or village of a man called Skammel or Skammlaus'. OScand. pers. name or byname + *bý*.

Scampston N. Yorks. *Scameston* 1086 (DB). 'Farmstead of a man called Skammr or Skammel'. OScand. pers. name + OE *tūn*.

Scampton Lincs. *Scantone* 1086 (DB). 'Short farmstead, or farmstead of a man called Skammi'. OScand. *skammr* or pers. name + OE *tūn*.

Scapa Flow Orkn. *Scalpay* 1579. 'Boat-shaped isthmus'. OScand. *skalpr* + *eith*, with Scots *flow*. The 'isthmus' is the stretch of land south of Kirkwall on the eastern side of Scapa Bay.

Scarborough N. Yorks. *Escardeburg* c.1160. Probably 'stronghold of a man called Skarthi'. OScand. pers. name + OE *burh*.

Scarcliffe Derbys. *Scardeclif* 1086 (DB). 'Cliff or steep slope with a gap'. OE *sceard* (with Scand. *sk-*) + *clif*.

Scarcroft Leeds. *Scardecroft* 1166. Probably 'enclosure in a gap'. OE *sceard* (with Scand. *sk-*) + *croft*.

Scardaun (*Scardán*) Mayo. 'Tiny waterfall'.

Scargill Durham. *Scracreghil* 1086 (DB). Probably 'ravine frequented by mergansers or similar seabirds'. OScand. *skraki* + *gil*.

Scarisbrick Lancs. *Scharisbrec* c.1200. Possibly 'hill-side or slope by a hollow'. OScand. *skar* + *brekka*.

Scarle, North (Lincs.) **& Scarle, South** (Notts.). *Scornelei* 1086 (DB). 'Mud or dung clearing'. OE *scearn* (with Scand. *sk-*) + *lēah*.

Scarnageeragh (*Scarbh na gCaorach*) Monaghan. 'Shallow ford of the sheep'.

Scarning Norfolk. *Scerninga* 1086 (DB). Probably 'the muddy or dirty place'. OE *scearn* (with Scand. *sk-*) + *-ing*.

Scarriff (*Scairbh*) Clare, Cork, Galway. 'Shallow ford'.

Scarrington Notts. *Scarintone* 1086 (DB). Possibly 'muddy or dirty farmstead'. OE **scearnig* (with Scand. *sk-*) + *tūn*. Alternatively

the first element may be OE **scearning* 'dirty place'.

Scartaglen (*Scairteach an Ghlinne*) Kerry. 'Thicket of the valley'.

Scarteen (*Scairtín*) Kerry. 'Little thicket'.

Scartho NE. Lincs. *Scarhou* 1086 (DB). 'Mound near a gap, or one frequented by cormorants'. OScand. *skarth* or *skarfr* + *haugr*.

Scarva (*Scarbhach*) Down. *the Skarvagh* 1618. 'Place of the shallow ford'.

Scattery Island (*Inis Cathaigh*) Clare. 'Cathach's island'.

Scawby N. Lincs. *Scallebi* 1086 (DB). Probably 'farmstead or village of a man called Skalli'. OScand. pers. name + *bý*.

Scawton N. Yorks. *Scaltun* 1086 (DB). 'Farmstead in a hollow or with a shieling'. OScand. *skál* or *skáli* + OE *tūn*.

Schiehallion (mountain) Perth. 'Fairy hill of the Caledonians'. Gaelic *sidh*.

Scholar Green Ches. *Scholehalc* late 13th cent. 'Nook of land with a hut or shieling'. OScand. *skáli* + OE *halh*.

Scholes, a common name in the North, 'the temporary huts or sheds', OScand. *skáli*; examples include: **Scholes** Kirkl., near Holmfirth. *Scholes* 1284. **Scholes** Leeds. *Skales* 1258.

Scilly Isles *Sully* 12th cent. An ancient pre-English name first recorded in the 1st cent. AD, of unknown origin and meaning.

Sco Ruston Norfolk. *See* RUSTON.

Scole Norfolk. *Escales* 1191. 'The temporary huts or sheds'. OScand. *skáli*.

Scone Perth. *Sgoinde* 1020. Origin obscure.

Scopwick Lincs. *Scapeuic* 1086 (DB). 'Sheep farm'. OE *scēap* (with Scand. *sk-*) + *wīc*.

Scorborough E. R. Yorks. *Scogerbud* 1086 (DB). 'Booth or shelter in a wood'. OScand. *skógr* + *búth*.

Scorton, 'farmstead near a ditch or ravine', OScand. *skor* + OE *tūn*: **Scorton** Lancs. *Scourton* c.1550. **Scorton** N. Yorks. *Scortone* 1086 (DB).

Scotby Cumbria. *Scoteby* 1130. 'Farmstead or village of the Scots'. OScand. *Skottar* (genitive *Skotta*) + *bý*.

Scotch Corner N. Yorks. *Scotch Corner* 1860. Road junction on the Great North Road so called because the main road to SW Scotland via Carlisle branches off here.

Scotch Street Armagh. *Scotch Street* 1835. The village, with its single street, was founded by Scottish settlers. The equivalent Irish name is *Sráid na hAlbanach*, 'street of the Scotsmen'.

Scotforth Lancs. *Scozforde* 1086 (DB). 'Ford used by the Scot or Scots'. OE *Scot* (genitive singular *Scottes*, genitive plural *Scotta*) + *ford*.

Scothern Lincs. *Scotstorne, Scotorne* 1086 (DB). 'Thorn-tree of the Scot'. OE *Scot* (genitive singular *Scottes*) + *thorn*.

Scotland. 'Land of the Scots'. The original 'Scots' were the Celtic 5th- to 6th-cent. settlers from Ireland. *Scotia* became the name of the whole country in the 9th cent. after the merger of the Scottish and Pictish kingdoms. The meaning of the tribal name is uncertain.

Scotshouse Monaghan. *Scotshouse* 1835. 'Scott's house'. William *Scott* was granted an estate here in the 17th cent. The present village was founded in the early 19th cent.

Scotter Lincs. *Scottere* 1061-6, *Scot(e)re* 1086 (DB). Possibly 'sand- or gravel-bank of the Scots'. OScand. *Skottar* (genitive *Skotta*) + *eyrr*.

Scotterthorpe Lincs. *Scaltorp* 1086 (DB). Probably 'outlying farmstead or hamlet of a man called Skalli'. OScand. pers. name + *thorp*.

Scotton, 'farmstead of the Scots', OE *Scot* (genitive plural *Scotta*) + *tūn*: **Scotton** Lincs. *Scottun* 1060-6. *Scotone* 1086 (DB). **Scotton** N. Yorks., near Knaresborough. *Scotone* 1086 (DB). **Scotton** N. Yorks., near Richmond. *Scottune* 1086 (DB).

Scottow Norfolk. *Scoteho* 1044-7, *Scotohou* 1086 (DB). 'Hill-spur of the Scots'. OE *Scot* (genitive plural *Scotta*) + *hōh*.

Scoulton Norfolk. *Sculetuna* 1086 (DB). 'Farmstead of a man called Skúli'. OScand. pers. name + OE *tūn*.

Scrabo (*Screabach*) Down. *Scraboc c.*1275. 'Rough land'.

Scrabster Highland. *Skarabolstad* 1201. 'Homestead of the young seagull'. OScand. *skári* + *bólstathr*.

Scrafton, West N. Yorks. *Scraftun* 1086 (DB). 'Farmstead near a hollow'. OE *scræf* (with Scand. *sk-*) + *tūn*.

Scraghan (*Scrathán*) Westmeath. 'Sward'.

Scrahanagnave (*Screathan na gCnámh*) Kerry. 'Scree of the bones'.

Scrainwood Northum. *Scravenwod* 1242. 'Wood frequented by shrewmice or villains'. OE *scrēawa* (genitive plural *-ena* and with Scand. *sk-*) + *wudu*.

Scramoge (*Scramóg*) Roscommon. Meaning uncertain.

Scrane End Lincs. *Screinga* 12th cent. Possibly '(settlement of) the family or followers of a man called *Scīrhēah*' from an OE pers. name + *-ingas*, or 'structure(s) made of poles' from an OE *scræging* (with Scand. *sk-*).

Scraptoft Leics. *Scrapetoft* 1043, *Scrapentot* [sic] 1086 (DB). Probably 'homestead of a man called Skrápi'. OScand. pers. name + *toft*. Or from OScand. *skrap* 'scraps' (referring ro poor produce).

Scratby Norfolk. *Scroutebi c.*1020, *Scroteby* 1086 (DB). 'Farmstead or village of a man called Skrauti'. OScand. pers. name + *bý*.

Scrayingham N. Yorks. *Screngham, Escraingham* 1086 (DB). 'Homestead of the family or followers of a man called *Scīrhēah*' from an OE pers. name + *-inga-* + *hām*, or 'homestead with a structure of poles' from an OE *scræging* (with Scand. *sk-*) + *hām*.

Scredington Lincs. *Scredinctun* 1086 (DB). Possibly 'estate associated with a man called Scīrhēard'. OE pers. name (with Scand. *sk-*) + *-ing-* + *tūn*. Alternatively perhaps from an OE *scrēading* (with Scand. *sk-*) 'shred, thin strip of land' + *tūn*.

Screeb (*Scríb*) Galway. 'Furrow'.

Screen (*An Scrín*) Wexford. 'The shrine'.

Screggan (*An Screagán*) Offaly. 'The rough place'.

Scremby Lincs. *Screnbi* 1086 (DB). 'Farmstead or village of a man called *Skræma*'. OScand. pers. name + *bý*.

Scremerston Northum. *Schermereton* 1196. 'Estate of the Skermer family'. ME surname + OE *tūn*.

Scribbagh (*Scriobach*) Fermanagh. 'Rough land'.

Scriven N. Yorks. *Scrauing(h)e* 1086 (DB). 'Hollow place with pits'. OE **screfen* (with Scand. *sk-*).

Scrooby Notts. *Scrobi* 1086 (DB). Possibly 'farmstead or village of a man called Skropi'. OScand. pers. name + *bý*.

Scropton Derbys. *Scrotun* [*sic*] 1086 (DB), *Scropton* c.1141. Probably 'farmstead of a man called Skropi'. OScand. pers. name + OE *tūn*.

Scruton N. Yorks. *Scurvetone* 1086 (DB). 'Farmstead of a man called Skurfa'. OScand. pers. name + OE *tūn*.

Sculthorpe Norfolk. *Sculatorpa* 1086 (DB). 'Outlying farmstead or hamlet of a man called Skúli'. OScand. pers. name + *thorp*.

Scunthorpe N. Lincs. *Escumetorp* 1086 (DB). 'Outlying farmstead or hamlet of a man called Skúma'. OScand. pers. name + *thorp*.

Scur, Lough (*Loch an Scuir*) Leitrim. 'Lake of the camp'.

Sea Palling Norfolk. *See* PALLING.

Seaborough Dorset. *Seveberge* 1086 (DB). 'Seven hills or barrows'. OE *seofon + beorg*.

Seacombe Wirral. *Secumbe* c.1277. 'Valley by the sea'. OE *sǣ + cumb*.

Seacroft Leeds. *Sacroft* 1086 (DB). 'Enclosure by a pool or marsh'. OE *sǣ + croft*.

Seadavog (*Suí Dabhóg*) Donegal. 'Davóg's seat'.

Seaford E. Sussex. *Saforde* 12th cent. 'Ford by the sea'. OE *sǣ + ford*. The River Ouse flowed into the sea here until the 16th cent.

Seaforde Down. *Seaforde* 1837. 'Forde's seat'. Irish *suí* + English surname. The village arose in the 18th cent. and takes its name from the local *Forde* family.

Seaforth W. Isles. (Lewis). 'Inlet of the sea'. OScand. *sǽr + fjǫrthr*.

Seaforth Sefton. a recent name, from *Seaforth House* in Litherland which was named after Lord Seaforth in the early 19th cent.

Seagrave Leics. *Se(t)grave, Satgrave* 1086 (DB). OE *grǣf(a)* 'grove, coppiced wood' with an uncertain first element, probably OE *sēath* 'a pit, a pool'.

Seagry, Lower & Seagry, Upper Wilts. *Segrete* 1086 (DB). Possibly 'stream where sedge grows'. OE *secg + rīth*.

Seaham Durham. *Sæham* c.1040. 'Homestead or village by the sea'. OE *sǣ + hām*.

Seal Kent. *La Sela* 1086 (DB). Probably 'the hall or dwelling', OE *sele*. Alternatively 'the sallow-tree copse', OE **sele*.

Sealand Flin. *Sealand* 1726. 'Sea land'. The village is on land reclaimed from the sea north of the River Dee.

Seamer, 'lake or marsh pool', OE *sǣ + mere*: **Seamer** N. Yorks., near Scarborough. *Semær* 1086 (DB). **Seamer** N. Yorks., near Stokesley. *Semer* 1086 (DB).

Seapatrick (*Suí Phádraig*) Down. *Soyge-Patrick* 1546. 'Patrick's seat'. The church here is dedicated to St Patrick.

Searby Lincs. *Seurebi* 1086 (DB). 'Farmstead or village of the seafarers, or of a man called Sæfari'. OScand. **sæfari* or pers. name + *bý*.

Seasalter Kent. *Seseltre* 1086 (DB). 'Salt-works on the sea'. OE *sǣ + sealt-ærn*.

Seascale Cumbria. *Sescales* c.1165. 'Shieling(s) or hut(s) by the sea'. OScand. *sǽr + skáli*.

Seathwaite Cumbria. near Borrowdale. *Seuethwayt* 1292. 'Sedge clearing'. OScand. *sef + thveit*.

Seathwaite Cumbria. near Ulpha. *Seathwhot* 1592. 'Lake clearing'. OScand. *sǽr + thveit*.

Seaton, usually 'farmstead by the sea or by an inland pool', OE *sǣ + tūn*; examples include: **Seaton** Devon. *Setton* 1238. **Seaton** Cumbria, near Workington. *Setona* c.1174. **Seaton Carew** Hartlepl. *Setona* late 12th cent., *Seton Carrowe* 1345. Manorial affix from the *Carou* family, here in the 12th cent. **Seaton Delaval** Northum. *Seton de la Val* 1270. Manorial affix from the *de la Val* family, here in the 13th cent. **Seaton Ross** E. R. Yorks. *Seton* 1086 (DB), *Seaton Rosse* 1618. Manorial affix from the *Ross* family who held the vill from the 12th cent.
However the following has a different origin: **Seaton** Rutland. *Segentone, Seieton* 1086 (DB).

Probably 'farmstead of a man called *Sæga'. OE pers. name + *tūn*.

Seavington St Michael & Seavington St Mary Somerset. *Seofenempton c.*1025, *Sevenehantune* 1086 (DB). 'Village of seven homesteads'. OE *seofon* + *hām-tūn*. Distinguishing affixes from the dedications of the churches here.

Sebergham Cumbria. *Saburgham* 1223. Probably 'homestead of a woman called Sæburh'. OE pers. name + *hām*.

Seckington Warwicks. *Seccandun* late 9th cent., *Sechintone* 1086 (DB). 'Hill of a man called Secca'. OE pers. name (genitive *-an*) + *dūn*.

Sedbergh Cumbria. *Sedberge* 1086 (DB). OScand. *set-berg* 'a flat-topped hill'.

Sedgeberrow Worcs. *Segcgesbearuue* 777, *Seggesbarue* 1086 (DB). 'Grove of a man called *Secg'. OE pers. name + *bearu*.

Sedgebrook Lincs. *Sechebroc* 1086 (DB). 'Brook where sedge grows'. OE *secg* + *brōc*.

Sedgefield Durham. *Ceddesfeld c.*1040, *Seggesfeld c.*1180. 'Open land of a man called Cedd or *Secg'. OE pers. name + *feld*.

Sedgeford Norfolk. *Secesforda* 1086 (DB). Possibly 'ford of a man called *Secci'. OE pers. name + *ford*.

Sedgehill Wilts. *Seghulle* early 12th cent. 'Hill of a man called Secga, or near where sedge grows'. OE pers. name or OE *secg* + *hyll*.

Sedgemoor (district) Somerset. *Seggemore* 1263. 'Marshy ground where sedge grows'. OE *secg* + *mōr*.

Sedgley Dudley. *Secgesleage* 985, *Segleslei* [*sic*] 1086 (DB). 'Woodland clearing of a man called *Secg'. OE pers. name + *lēah*.

Sedgwick Cumbria. *Sigghiswic c.*1185. 'Dwelling or farm of a man called *Sicg'. OE pers. name + *wīc*.

Sedlescombe E. Sussex. *Selescome* [*sic*] 1086 (DB), *Sedelescumbe c.*1210. Probably 'valley with a house or dwelling'. OE *sedl* + *cumb*.

Seefin (*Suí Finn*) Cork, Down, Mayo, Waterford. 'Fionn's seat'.

Seend Wilts. *Sinda* 1190. Probably 'sandy place'. OE *sende*.

Seer Green Bucks. *La Sere* 1223. 'The dry or barren place'. OE *sēar*.

Seething Norfolk. *Sithinges* 1086 (DB). Probably '(settlement of) the family or followers of a man called *Sīth(a)'. OE pers. name + *-ingas*.

Sefton Sefton. *Sextone* [*sic*] 1086 (DB), *Sefftun c.*1220. 'Farmstead where rushes grow'. OScand. *sef* + OE *tūn*.

Seghill Northum. *Syghal* 1198. Possibly 'nook of land by a stream called *Sige'. OE river-name (meaning 'slow-moving') + *halh*.

Seighford Staffs. *Cesteforde* 1086 (DB). Possibly 'ford over which there was a dispute'. OE *ceast* 'strife' + *ford*.

Selattyn Shrops. *Sulatun* 1254. 'Settlement with ploughs or ploughlands', or 'settlement characterized by gullies'. OE *sulh* (genitive plural *sula*) + *tūn*.

Selborne Hants. *Seleborne* 903, *Selesburne* 1086 (DB). 'Stream by (a copse of) sallow-trees'. OE *sealh* or *sele* + *burna*.

Selby N. Yorks. *Seleby c.*1030, *Salebi* 1086 (DB). 'Farmstead or village near (a copse of) sallow-trees'. OE *sele* or OScand. *selja* + OScand. *bý*.

Selham W. Sussex. *Seleham* 1086 (DB). 'Homestead by a copse of sallow-trees'. OE *sele* + *hām*.

Selkirk Sc. Bord. *Selechirche c.*1120, *Seleschirche c.*1190, *Selkirk* 1306. 'Church by a hall'. OE *sele* + *cirice*.

Sellack Herefs. *Lann Suluc c.*1130. 'Church of Suluc (a pet-form of Suliau)'. From the dedication of the church (OWelsh *lann*).

Sellafield Cumbria. *Sellofeld* 1576. Probably 'land by the willow-tree mound'. OScand. *selja* + *haugr* + OE *feld*.

Sellindge Kent. *Sedlinges* 1086 (DB). Probably '(settlement of) those sharing a house or building', from OE *sedl* + *-ingas*. Alternatively 'place at a house or building', OE *sedl* + *-ing*.

Selling Kent. *Setlinges* 1086 (DB). Identical in origin with the previous name.

Selloo (*Suí Lú*) Monaghan. 'Lú's seat'.

Selly Oak Birm. *Escelie [sic]* 1086 (DB), *Selvele* 1204. 'Woodland clearing on a shelf or ledge'. OE *scelf* + *lēah*, with the later addition *oak*.

Selmeston E. Sussex. *Sielmestone* 1086 (DB). 'Farmstead of a man called Sigehelm'. OE pers. name + *tūn*.

Selsdon Gtr. London. *Selesdune c.*880. Possibly 'hill of a man called Seli'. OE pers. name + *dūn*. Alternatively the first element could be OE *sele* 'dwelling, hall' or OE **sele* 'sallow-tree copse'.

Selsey W. Sussex. *Seolesiae c.*715, *Seleisie* 1086 (DB). 'Island of the seal'. OE *seolh* + *ēg*. **Selsey Bill**, not recorded until the 18th cent., is from OE *bile* 'a bill, a beak' used topographically for 'promontory'.

Selsley Glos. not recorded before the late 18th cent., origin uncertain without earlier spellings.

Selston Notts. *Salestune* 1086 (DB). Probably 'farmstead of a man called *Sele or *Seli'. OE pers. name + *tūn*.

Selwood Somerset. *Seluudu c.*894. 'Wood where sallow-trees grow'. OE *sealh* + *wudu*.

Selworthy Somerset. *Seleuurde* 1086 (DB). 'Enclosure by a copse of sallow-trees'. OE **sele* + *worth*(*ig*).

Semer Suffolk. *Seamera* 1086 (DB). 'Lake or marsh pool'. OE *sǣ* + *mere*.

Semington Wilts. *Semneton* 13th cent. 'Farmstead on the stream called *Semnet*'. Pre-English river-name (of uncertain meaning) + OE *tūn*.

Semley Wilts. *Semeleage* 955. 'Woodland clearing on the River Sem'. Pre-English river-name (of uncertain meaning) + OE *lēah*.

Sempringham Lincs. *Sempingaham* 852, *Sepingeham [sic]* 1086 (DB), *Sempringham* 1150. Probably 'homestead of the family or followers of a man called Sempa'. OE pers. name + *-inga-* + *hām*.

Send Surrey. *Sendan* 960–2, *Sande* 1086 (DB). 'Sandy place'. OE **sende*.

Sennen Cornwall. 'Parish of *Sancta Senana*' 1327. From the female patron saint of the church.

Seskilgreen (*Seisíoch Chill Ghrianna*) Tyrone. 'Plot of the church of the sun'.

Seskin (*Seisceann*) Tipperary. 'Marsh'.

Seskinore (*Seisceann Odhar*) Tyrone. *Shaskannoure* 1613. 'Brownish marsh'.

Seskinryan (*Seisceann Ríain*) Carlow. 'Rían's marsh'.

Sessay N. Yorks. *Sezai* 1086 (DB). 'Island or well-watered land where sedge grows'. OE *secg* + *ēg*.

Setchey Norfolk. *Seche [sic]* 1086 (DB), *Sechithe* 13th cent. Possibly 'landing-place of a man called *Secci'. OE pers. name + *hȳth*.

Settle N. Yorks. *Setel* 1086 (DB). 'The house or dwelling'. OE *setl*.

Settrington N. Yorks. *Sendriton [sic]* 1086 (DB), *Seteringetune c.*1090. Possibly 'estate associated with a robber or with a man called *Sǣtere'. OE *sǣtere* or OE pers. name + *-ing-* + *tūn*.

Seven (river) N. Yorks. *See* SINNINGTON.

Seven Sisters Neat. The name was originally that of a colliery begun in 1872 and named after the pit owner's seven daughters.

Sevenhampton, 'village of seven homesteads', OE *seofon* + *hām-tūn*: **Sevenhampton** Glos. *Sevenhamtone* 1086 (DB). **Sevenhampton** Swindn. *Sevamentone* 1086 (DB).

Sevenoaks Kent. *Seouenaca c.*1100. '(Place by) seven oak-trees'. OE *seofon* + *āc*.

Severn (*Hafren*) (river) Powys–Shrops.–Somerset. *Sabrina* 2nd cent. AD, *Sauerna* 1086 (DB). An ancient pre-Celtic river-name of doubtful etymology.

Severn Stoke Worcs. *See* STOKE.

Sevington Kent. *Seivetone* 1086 (DB). 'Farmstead of a woman called Sǣgifu'. OE pers. name + *tūn*.

Sewerby E. R. Yorks. *Siuuardbi* 1086 (DB). 'Farmstead or village of a man called Sigvarthr'. OScand. pers. name + *bý*.

Sewstern Leics. *Sewesten [sic]* 1086 (DB), *Seustern c.*1125. Probably 'thorn-tree of a man called Sǣwīg'. OE pers. name + *thyrne*.

Sezincote Glos. *Ch*(*i*)*esnecote* 1086 (DB). 'Gravelly cottage(s)'. OE **cisen* + *cot*.

Shabbington Bucks. *Sobintone* 1086 (DB). 'Estate associated with a man called *Sc(e)obba*. OE pers. name + -*ing*- + *tūn*.

Shackerstone Leics. *Sacrestone* 1086 (DB), *Schakereston c.* 1260. 'The robber's farmstead'. OE *scēacere* + *tūn*.

Shadforth Durham. *Shaldeford* 1183. 'Shallow ford'. OE *sc(e)ald* + *ford*.

Shadingfield Suffolk. *Scadenafella* 1086 (DB). Possibly 'open land by the boundary valley'. OE *scēad* + *denu* + *feld*.

Shadoxhurst Kent. *Schettokesherst* 1239. OE *hyrst* 'wooded hill' with an uncertain first element, possibly a pers. name or an earlier place name.

Shaftesbury Dorset. *Sceaftesburi* 877, *Sceftesberie* 1086 (DB). 'Fortified place of a man called *Sceaft*, or on a shaft-shaped hill'. OE pers. name or OE *sceaft* + *burh* (dative *byrig*).

Shafton Barns. *Sceptun* [*sic*] 1086 (DB), *Scaftona c.*1160. 'Farmstead marked by a pole, or made with poles'. OE *sceaft* + *tūn*.

Shalbourne Wilts. *Scealdeburnan* 955, *Scaldeburne* 1086 (DB). 'Shallow stream'. OE *scҽald* + *burna*.

Shalcombe I. of Wight. *Eseldecome* 1086 (DB). 'Shallow valley'. OE *scҽald* + *cumb*.

Shalden Hants. *Scealdeden* 1046, *Seldene* 1086 (DB). 'Shallow valley'. OE *scҽald* + *denu*.

Shaldon Devon. not recorded before the early 17th cent., probably from OE *dūn* 'hill' but first element uncertain without early spellings.

Shalfleet I. of Wight. *Scealdan fleote* 838, *Seldeflet* 1086 (DB). 'Shallow stream or creek'. OE *scҽald* + *flēot*.

Shalford, 'shallow ford', OE *scҽald* + *ford*: **Shalford** Essex. *Scaldefort* 1086 (DB). **Shalford** Surrey. *Scaldefor* 1086 (DB).

Shalstone Bucks. *Celdestone* 1086 (DB). Possibly 'farmstead at the shallow place or stream'. OE *scҽald* (used as a noun) + *tūn*.

Shanagarry (*An Seangharraí*) Cork. 'The old garden'.

Shanagolden (*Seanghualainn*) Limerick. 'Old shoulder'.

Shanahoe (*Seanchua*) Laois. 'Old hollow'.

Shanbally (*An Seanbhaile*) Cork. 'The old townland'.

Shanballymore (*An Seanbhaile Mór*) Cork. 'The big old homestead'.

Shanco (*Seanchua*) Monaghan. 'Old hollow'.

Shandwick Highland. 'Sandy bay'. OScand. *sand* + *vík*.

Shane's Castle Antrim. *Castel Edaindaubchairrgi* 1490. The townland is named after *Shane* O'Neill, ruler of Lower Clandeboy from 1595 to 1617 and grandfather of Rose O'Neill, whose husband gave the name of RANDALSTOWN. The castle's original Irish name was *Éadan Dúcharraige*, 'brow of black rock'.

Shangton Leics. *Sanctone* 1086 (DB), *Schanketon* 1206. 'Farmstead or village at the shank of land or narrow ridge'. OE *scanca* + *tūn*.

Shankill (*Seanchill*) Antrim, Dublin. 'Old church'.

Shanklin I. of Wight. *Sencliz* 1086 (DB). 'Bank by the drinking cup' (referring to the waterfall here). OE *scenc* + *hlinc*.

Shanlaragh (*Seanlárach*) Cork. 'Old ruins'.

Shannon (*Sionainn*) (river) Clare–Limerick. *Senos c.*150, *Sinand* n.d. 'Old goddess'.

Shanonagh (*Sean dDomhnach*) Westmeath. 'Old church'.

Shanragh (*Seanráth*) Laois. 'Old fort'.

Shanrahan (*Seanraithean*) Tipperary. 'Old ferny place'.

Shantonagh (*Seantonnach*) Monaghan. 'Old quagmire'.

Shanvaus (*An Seanmhás*) Leitrim. 'The old plain'.

Shap Cumbria. *Hep c.*1190. 'The heap of stones' (referring to an ancient stone circle). OE *hēap*.

Shapinsay (island) Orkn. *Hjalpandisay c.*1250. 'Hjalpand's(?) island'. OScand. pers. name + *ey*.

Shapwick, 'sheep farm', OE *scēap* + *wīc*: **Shapwick** Dorset. *Scapeuuic* 1086 (DB). **Shapwick** Somerset. *Sapwic* 725, *Sapeswich* 1086 (DB).

Sharavogue (*Searbhóg*) Offaly. 'Bitter place'.

Shardlow Derbys. *Serdelau* 1086 (DB). 'Mound with a notch or indentation'. OE *sceard* + *hlāw*.

Shareshill Staffs. *Servesed* [*sic*] 1086 (DB), *Sarueshull* 1213, *Sarushulf* 1298. Possibly 'hill (or shelf of land) of a man called *Scearf'. OE pers. name + *hyll* (alternating with OE *scylf* in the early forms). Alternatively the first element may be a metathesized form of OE *scræf* 'cave, hut, hovel'.

Sharlston Wakefd. *Scharueston* c.1180. Possibly 'farmstead of a man called *Scearf'. OE pers. name + *tūn*. Alternatively the first element may be a metathesized form of OE *scræf* 'cave, hut, hovel'.

Sharnbrook Beds. *Scernebroc* 1086 (DB). 'Dirty or muddy brook'. OE *scearn* + *brōc*.

Sharnford Leics. *Scearnforda* 1004, *Scerneforde* 1086 (DB). 'Dirty or muddy ford'. OE *scearn* ⊦ *ford*.

Sharow N. Yorks. *Sharou* c.1130. 'Boundary hill-spur'. OE *scearu* + *hōh*.

Sharpenhoe Beds. *Scarpeho* 1197. 'Sharp or steep hill-spur'. OE *scearp* (dative *-an*) + *hōh*.

Sharperton Northum. *Scharberton* 1242. Possibly 'farmstead by the steep hill'. OE *scearp* + *beorg* + *tūn*.

Sharpness Glos. *Nesse* 1086 (DB), *Schobbenasse* 1368. 'Headland of a man called *Scobba'. OE pers. name + *næss*.

Sharrington Norfolk. *Scarnetuna* 1086 (DB). 'Dirty or muddy farmstead'. OE *scearn* (dative *-an*) + *tūn*.

Shaugh Prior Devon. *Scage* 1086 (DB). 'Small wood or copse'. OE *sc(e)aga*. Affix from its early possession by Plympton Priory.

Shavington Ches. *Santune* [*sic*] 1086 (DB), *Shawynton* 1260. 'Estate associated with a man called Scēafa'. OE pers. name + *-ing-* + *tūn*.

Shaw, 'small wood or copse', OE *sc(e)aga*: **Shaw** Oldham. *Shaghe* 1555. **Shaw** Swindn. *Schawe* 1256. **Shaw** W. Berks. *Essages* [*sic*] 1086 (DB), *Shage* 1167.

Shawbury Shrops. *Sawesberie* 1086 (DB), *Shawberia* c.1165. 'Manor house by a small wood or copse'. OE *sc(e)aga* + *burh* (dative *byrig*).

Shawell Leics. *Sawelle* [*sic*] 1086 (DB), *Schadewelle* 1224. 'Boundary spring or stream'. OE *scēath* + *wella*.

Shawford Hants. *Scaldeforda* 1208. 'Shallow ford'. OE **sceald* + *ford*.

Sheaf (river) Derbys.–Yorkshire. *See* SHEFFIELD.

Shearsby Leics. *Svevesbi* 1086 (DB). Possibly 'farmstead or village of a man called Swǣf or Skeifr'. OE or OScand. pers. name + OScand. *bý*.

Shebbear Devon. *Sceftbeara* 1050–73, *Sepesberie* 1086 (DB). 'Grove where shafts or poles are got'. OE *sceaft* + *bearu*.

Shebdon Staffs. *Schebbedon* 1267. 'Hill of a man called *Sceobba'. OE pers. name + *dūn*.

Shedfield Hants. *Scidafelda* 956. 'Open land where planks of wood are got, or where they are used as foot-bridges'. OE *scīd* + *feld*.

Sheean (*An Sián*) Mayo. 'The fairy mound'.

Sheefin (*Suí Finn*) Westmeath. 'Finn's seat'.

Sheefry Hills (*Cnoic Shiofra*) Mayo. 'Siofra's hills'.

Sheelin, Lough (*Loch Síleann*) Cavan, Meath, Westmeath. 'Sileann's lake'.

Sheen, 'the sheds or shelters', OE **scēo* in a plural form **scēon*: **Sheen** Staffs. *Sceon* 1002, 1086 (DB). **Sheen, East & Sheen, North**, Gtr. London. *Sceon* c.950. *West Sheen* (*West Shene* 1258) was the earlier name for RICHMOND.

Sheepstor Devon. *Sitelestorra* 1168. Possibly 'craggy hill thought to resemble a bar or bolt', OE *scytels* + *torr*. Alternatively the first element may be an OE **scitels* 'dung'.

Sheepwash Devon. *Schepewast* 1166. 'Place where sheep are dipped'. OE *scēap-wæsce*.

Sheepy Magna & Sheepy Parva Leics. *Scepehe* 1086 (DB). 'Island, or dry ground in marsh, where sheep graze'. OE *scēap* + *ēg*. Distinguishing affixes are Latin *magna* 'great' and *parva* 'little'.

Sheering Essex. *Sceringa* 1086 (DB). Probably '(settlement of) the family or followers of a man called *Scear(a)'. OE pers. name + *-ingas*.

Sheerness Kent. *Scerhnesse* 1203. Probably 'bright headland'. OE *scīr* + *næss*. Alternatively the first element may be OE *scear* 'plough-share' alluding to the shape of the headland.

Sheet Hants. *Syeta c.*1210. 'Projecting piece of land, corner or nook'. OE **scīete.*

Sheffield Sheff. *Scafeld* 1086 (DB). 'Open land by the River Sheaf'. OE river-name (from *scēath* 'boundary') + *feld.*

Sheffield Bottom W. Berks. *Sewelle* [*sic*] 1086 (DB), *Scheaffelda* 1167. 'Open land with a shelter'. OE **scēo* + *feld.*

Sheffield Green E. Sussex. *Sifelle* [*sic*] 1086 (DB), *Shipfeud* 1272. 'Open land where sheep graze'. OE *scēap* + *feld.*

Shefford, 'ford used for sheep', OE *scēap* or **scīep* + *ford*: **Shefford** Beds. *Sepford* 1220. **Shefford, East & Shefford, Great** W. Berks. *Siford* [*sic*] 1086 (DB), *Schipforda* 1167.

Shehy Mountains (*Cnoic na Seithe*) Cork. 'Hills of the hides'.

Sheinton Shrops. *Sc(h)entune* 1086 (DB), *Sheinton* 1222. Probably 'farmstead on a stream called **Scenc*'. OE *scenc* 'drinking cup' + *tūn* (*see* SHENTON).

Sheldon Birm. *Scheldon* 1189. 'Shelf hill'. OE *scelf* + *dūn.*

Sheldon Derbys. *Scelhadun* 1086 (DB). Probably 'heathy hill with a shelf or ledge'. OE *scelf* + *hǣth* + *dūn.*

Sheldon Devon. *Sildene* 1086 (DB). 'Shelf valley'. OE *scylf* + *denu.*

Sheldwich Kent. *Scilduuic* 784. 'Dwelling or farm with a shelter'. OE *sceld* + *wīc.*

Shelf Calder. *Scelf* 1086 (DB). 'Shelf of level ground'. OE *scelf.*

Shelfanger Norfolk. *Sceluangra* 1086 (DB). 'Wood on a shelf of level ground'. OE *scelf* + *hangra.*

Shelfield, 'hill with a shelf or plateau', OE *scelf* + *hyll*: **Shelfield** Warwicks. *Shelfhulle* 1221. **Shelfield** Wsall. *Scelfeld* [*sic*] 1086 (DB), *Schelfhul* 1271.

Shelford, 'ford at a shallow place', OE **sceldu* + *ford*: **Shelford** Notts. *Scelforde* 1086 (DB). **Shelford, Great & Shelford, Little** Cambs. *Scelford c.*1050, *Escelforde* 1086 (DB).

Shelley Kirkl. *Scelneleie* [*sic*] 1086 (DB), *Shelfleie* 1198. 'Woodland clearing on a shelf of level ground'. OE *scelf* + *lēah.*

Shellingford Oxon. *Scaringaford* 931, *Serengeford* 1086 (DB). 'Ford of the family or followers of a man called **Scear*'. OE pers. name + *-inga-* + *ford.*

Shellow Bowells Essex. *Scelga* 1086 (DB), *Scheuele Boueles* 1303. 'Winding river' (with reference to the River Roding). OE **Sceolge* from *sceolh* 'twisted'. Manorial affix from the *de Bueles* family, here in the 13th cent.

Shelton, 'farmstead on a shelf of level ground', OE *scelf* + *tūn*: **Shelton** Beds. *Eseltone* 1086 (DB). **Shelton** Norfolk. *Sceltuna* 1086 (DB). **Shelton** Notts. *Sceltun* 1086 (DB). **Shelton** Shrops. *Saltone* [*sic*] 1086 (DB), *Shelfton* 1221.

Shelve Shrops. *Schelfe* 1180. 'Shelf of level ground'. OE *scelf.*

Shelwick Herefs. *Scelwiche* 1086 (DB). 'Dwelling or farm with a shelter'. OE *sceld* + *wīc.*

Shenfield Essex. *Scenefelda* 1086 (DB). 'Beautiful open land'. OE *scēne* + *feld.*

Shenington Oxon. *Senendone* 1086 (DB). 'Beautiful hill'. OE *scēne* (dative *-an*) + *dūn.*

Shenley, 'bright or beautiful woodland glade', OE *scēne* + *lēah*: **Shenley** Herts. *Scenlai* 1086 (DB). **Shenley** Milt. K. *Senelai* 1086 (DB).

Shenstone Staffs. *Seneste* [*sic*] 1086 (DB), *Scenstan* 11th cent. 'Beautiful stone'. OE *scēne* + *stān.*

Shenton Leics. *Scenctune* 1002, *Scentone* 1086 (DB). 'Farmstead on the River Sence'. OE river-name (from *scenc* 'drinking cup') + *tūn.*

Shepherdswell (or *Sibertswold*) Kent. *Swythbrihteswealde* 944, *Sibertsuuald, -uualt* 1086 (DB). 'Forest of a man called Swīthbeorht'. OE pers. name + *weald.*

Shepley Kirkl. *Scipelei* 1086 (DB). 'Clearing where sheep are kept'. OE *scēap* + *lēah.*

Shepperton Surrey. *Scepertune* 959, *Scepertone* 1086 (DB). Probably 'farmstead of the shepherd(s)'. OE *scēap-hirde* + *tūn.*

Sheppey, Isle of Kent. *Scepeig* 696–716, *Scape* 1086 (DB). 'Island where sheep are kept'. OE *scēap* + *ēg.*

Shepreth Cambs. *Esceprid* 1086 (DB). 'Sheep stream'. OE *scēap* + *rīth*.

Shepshed Leics. *Scepe(s)hefde* 1086 (DB). 'Sheep headland'. OE *scēap* + *hēafod*.

Shepton, 'sheep farm', OE *scēap* + *tūn*: **Shepton Beauchamp** Somerset. *Sceptone* 1086 (DB), *Septon Belli campi* 1266. Manorial affix from the *de Beauchamp* (Latin *de Bello campo*) family, here in the 13th cent. **Shepton Mallet** Somerset. *Sepetone* 1086 (DB), *Scheopton Malet* 1228. Manorial affix from the *Malet* family, here in the 12th cent. **Shepton Montague** Somerset. *Sceptone* 1086 (DB), *Schuptone Montagu* 1285. Manorial affix from Drogo *de Montacute*, tenant in 1086.

Shepway (district) Kent. 'Trackway for sheep'. A revival of a lathe-name, preserved in a district of MAIDSTONE. OE *scēap* + *weg*.

Sheraton Durham. *Scurufatun* c.1040. Probably 'farmstead of a man called Skurfa'. OScand. pers. name + OE *tūn*.

Sherborne, Sherbourne, Sherburn, '(place at) the bright or clear stream', OE *scīr* + *burna*: **Sherborne** Dorset. *Scireburnan* 864, *Scireburne* 1086 (DB). **Sherborne** Glos. *Scireburne* 1086 (DB). **Sherborne St John & Monk Sherborne** Hants. *Sireburne*, *Sireborne* 1086 (DB), *Shireburna Johannis* 1167, *Schireburne Monachorum* c.1270. Distinguishing affixes from possession by Robert *de Sancto Johanne* in the 13th cent., and from the priory here (Latin *monachorum* 'of the monks'). **Sherbourne** Warwicks. *Scireburne* 1086 (DB). **Sherburn** Durham. *Scireburne* c.1170. **Sherburn** N. Yorks. *Scire(s)burne 1086 (db)*. **Sherburn in Elmet** *N. Yorks.* Scirburnan *c.900,* Scireburne *1086 (db). For the old district name Elmet, see* BARWICK.

Shercock (*Searcóg*) Cavan. *Skarkeoge al. Sharcocke* 1629. Meaning uncertain.

Shere Surrey. *Essira* 1086 (DB). '(Place at) the bright or clear stream'. OE *scīr*.

Shereford Norfolk. *Sciraford* 1086 (DB). 'Bright or clear ford'. OE *scīr* + *ford*.

Sherfield, probably 'bright open land', i.e. having sparse vegetation, OE *scīr* + *feld*: **Sherfield English** Hants. *Sirefelle* 1086 (DB). Manorial affix from *le Engleis* family, here in the 14th cent. **Sherfield on Loddon** Hants. *Schirefeld* 1179. Loddon is a Celtic river-name (possibly meaning 'muddy stream').

Sherford Devon. *Scireford* c.1050, *Sirefort* 1086 (DB). 'Bright or clear ford'. OE *scīr* + *ford*.

Sheriff Hutton N. Yorks. *See* HUTTON.

Sheriffhales Shrops. *Halas* 1086 (DB), *Shiruehales* 1301. 'Nooks of land', OE *halh* in a plural form. Affix is OE *scīr-rēfa* from possession by the *Sheriff* of Shropshire in 1086.

Sheringham Norfolk. *Silingeham* [*sic*] 1086 (DB), *Scheringham* 1242. 'Homestead of the family or followers of a man called Scīra'. OE pers. name + *-inga-* + *hām*.

Sherington Milt. K. *Serintone* 1086 (DB). 'Estate associated with a man called Scīra'. OE pers. name + *-ing-* + *tūn*.

Sherkin Island (*Inis Arcáin*) Cork. 'Arcán's island'.

Shernborne Norfolk. *Scernebrune* 1086 (DB). 'Dirty or muddy stream'. OE *scearn* + *burna*.

Sherrington Wilts. *Scearntune* 968, *Scarentone* 1086 (DB). 'Dirty or muddy farmstead'. OE *scearn* + *tūn*.

Sherston Wilts. *Scoranstan* 896, *Sorestone* 1086 (DB). 'Stone or rock on a steep slope'. OE **scora* + *stān*.

Sherwood Forest Notts. *Scirwuda* 955. 'Wood belonging to the shire'. OE *scīr* + *wudu*.

Sheskinatawy (*Seisceann an tSamhaid*) Donegal. 'Swamp of the sorrel'.

Sheskinshule (*Seisceann Siúil*) Tyrone. 'Moving marsh'.

Shetland (the unitary authority). *Haltland* c.1100, *Shetland* 1289. Various suggested interpretations, but probably reshaping of pre-Norse name.

Shevington Wigan. *Shefinton* c.1225. Probably 'farmstead near a hill called *Shevin*'. Celtic **ceṽn* 'ridge' + OE *tūn*.

Sheviock Cornwall. *Savioch* 1086 (DB). Possibly 'strawberry place'. Cornish *sevi* + suffix *-ek*.

Shiant Islands W. Isles. 'Hallowed islands'. Gaelic *seunta*.

Shields, 'temporary sheds or huts (used by fishermen)', ME *schele*: **Shields, North** N. Tyne. *Chelis* 1268. **Shields, South** S. Tyne. *Scheles* 1235, *Suthshelis* 1313.

Shifnal Shrops. *Scuffanhalch* 12th cent. Probably 'nook or hollow of a man called *Scuffa'. OE pers. name (genitive -*n*) + *halh*.

Shilbottle Northum. *Schipplingbothill* 1242. Probably 'dwelling of the men of SHIPLEY'. Place name + OE -*inga*- + *bōtl*.

Shildon Durham. *Sciluedon* 1214. 'Shelf hill'. OE *scylfe* + *dūn*.

Shillelagh (*Síol Éalaigh*) Wicklow. *Sil nElathig* 11th cent. *Shilleylle* 1549. 'Descendants of Éalach'.

Shillingford, probably 'ford of the family or followers of a man called *Sciell(a)', OE pers. name + -*inga*- + *ford*, but an OE stream-name *Scielling* 'the noisy one' may be an alternative first element: **Shillingford** Devon. *Sellingeford* 1179. **Shillingford** Oxon. *Sillingeforda* 1156. **Shillingford St George** Devon. *Selingeforde* 1086 (DB). Affix from the dedication of the church.

Shillingstone Dorset. *Akeford Skelling* 1220, *Shillyngeston* 1444. 'Schelin's estate (in OKEFORD)'. OE *tūn* with the name of the tenant in 1086 (DB).

Shillington Beds. *Scytlingedune* 1060, *Sethlindone* 1086 (DB). Possibly 'hill of the family or followers of a man called *Scyttel'. OE pers. name + -*inga*- + *dūn*. Alternatively the first element could be an OE hill-name *Scyteling*, from OE *scytel* 'a bar or bolt' + suffix -*ing*.

Shilton, 'farmstead on a shelf or ledge', OE *scylf*(*e*) + *tūn*: **Shilton** Oxon. *Scylftune* 1044. **Shilton** Warwicks. *Scelftone* 1086 (DB). **Shilton, Earl** Leics. *Sceltone* 1086 (DB), *Erle Shilton* 1576. Affix from its early possession by the Earls of Leicester.

Shimpling, probably '(settlement of) the family or followers of a man called *Scimpel', OE pers. name + -*ingas*: **Shimpling** Norfolk. *Simplingham* c.1035, *Simplinga* 1086 (DB). Sometimes with OE *hām* 'homestead' in early spellings. **Shimpling** Suffolk. *Simplinga* 1086 (DB).

Shincliffe Durham. *Scinneclif* c.1123. 'Cliff or bank haunted by a phantom or demon'. OE *scinna* + *clif*.

Shiney Row Sundld. Recorded thus in 1821, from *row* 'row of houses'. First part of name obscure.

Shinfield Wokhm. *Selingefelle* [*sic*] 1086 (DB), *Schiningefeld* 1167. 'Open land of the family or followers of a man called *Scīene'. OE pers. name + -*inga*- + *feld*.

Shinglis (*Seanlios*) Westmeath. 'Old fort'.

Shinrone (*Suí an Róin*) Offaly. 'Seat of the seal'.

Shipbourne Kent. *Sciburna* [*sic*] c.1100, *Scipburn* 1198. 'Sheep stream'. OE *scēap* + *burna*.

Shipdham Norfolk. *Scipdham* 1086 (DB). 'Homestead with a flock of sheep'. OE *scīpde* + *hām*.

Shiplake Oxon. *Siplac* 1163. 'Sheep stream'. OE *scēap* + *lacu*.

Shipley, 'clearing or pasture where sheep are kept', OE *scēap* or *scīep* + *lēah*; examples include: **Shipley** Brad. *Scipeleia* 1086 (DB). **Shipley** Northum. *Schepley* 1236. **Shipley** Shrops. *Sciplei* 1086 (DB). **Shipley** W. Sussex. *Scapeleia* 1073, *Sepelei* 1086 (DB).

Shipmeadow Suffolk. *Scipmedu* 1086 (DB). 'Meadow for sheep'. OE *scēap* + *mǣd* (dative *mǣdwe*).

Shippon Oxon. *Sipene* 1086 (DB). OE *scypen* 'a cattle-shed'.

Shipston on Stour Warwicks. *Scepuuæisctune* c.770, *Scepwestun* 1086 (DB). 'Farmstead by the sheep-wash'. OE *scēap-wæsce* + *tūn*. Stour is a Celtic or OE river-name probably meaning 'the strong one'.

Shipton, usually 'sheep farm', OE *scēap* or *scīep* + *tūn*; examples include: **Shipton** Shrops. *Scipetune* 1086 (DB). **Shipton Bellinger** Hants. *Sceptone* 1086 (DB), *Shupton Berenger* 14th cent. Manorial affix from the *Berenger* family, here in the 13th cent. **Shipton Gorge** Dorset. *Sepetone* 1086 (DB), *Shipton Gorges* 1594. Manorial addition from the *de Gorges* family, here in the 13th cent. **Shipton Moyne** Glos. *Sciptone* 1086 (DB), *Schipton Moine* 1287. Manorial affix from the *Moygne* family, here from the 13th cent. **Shipton under Wychwood** Oxon. *Sciptone* 1086 (DB). Wychwood is an old forest-name, *see* ASCOTT UNDER WYCHWOOD.
   However the following has a different origin: **Shipton** N. Yorks. *Hipton* 1086 (DB). 'Farmstead where rose-hips grow'. OE *hēope* + *tūn*.

Shirburn Oxon. *Scireburne* 1086 (DB). 'Bright or clear stream'. OE *scīr* + *burna*.

Shirebrook Derbys. *Scirebroc* 1202. 'Boundary brook' or 'bright brook'. OE *scīr* (as noun or adjective) + *brōc*.

Shireoaks Notts. *Shirakes* 12th cent. 'Oak-trees on the (county) boundary'. OE *scīr* + *āc*. In the similar name **Shire Oak** (Wsall.) the meaning is rather 'oak-tree marking the site of the shire assembly'.

Shirland Derbys. *Sirelunt* 1086 (DB). Probably 'bright or sparsely wooded grove'. OE *scīr* + OScand. *lundr*.

Shirley, 'bright woodland clearing' or perhaps sometimes 'shire clearing', OE *scīr* + *lēah*: **Shirley** Derbys. *Sirelei* 1086 (DB). **Shirley** Gtr. London. *Shirleye* 1314. **Shirley** Solhll. *Syrley* c.1240. **Shirley** Sotn. *Sirelei* 1086 (DB).

Shirrell Heath Hants. *Scirhiltæ* 826. Probably 'bright or sparsely wooded grove'. OE *scīr* + **hylte*, with the addition of *Heath* from the 17th cent.

Shirwell Devon. *Sirewelle* 1086 (DB). 'Bright or clear spring or stream'. OE *scīr* + *wella*.

Shobdon Herefs. *Scepedune* [sic] 1086 (DB), *Scobbedun* 1242. 'Hill of a man called **Sceobba*'. OE pers. name + *dūn*.

Shobrooke Devon. *Sceocabroc* 938, *Sotebroca* 1086 (DB). 'Brook haunted by an evil spirit'. OE *sceocca* + *brōc*.

Shocklach Ches. *Socheliche* 1086 (DB). 'Boggy stream haunted by an evil spirit'. OE *sceocca* + **læcc*.

Shoeburyness, North Shoebury Sthend. *Sceobyrig* early 10th cent., *Soberia* 1086 (DB), *Shoberynesse* 16th cent. 'Fortress providing shelter'. OE **scēo* + *burh* (dative *byrig*), with the later addition of *næss* 'promontory'.

Sholden Kent. *Shoueldune* 1176. Possibly 'hill thought to resemble a shovel'. OE *scofl* + *dūn*.

Sholing Sotn. *Sholling* 1251. Possibly 'uneven or sloping place'. OE *sceolh* + *-ing*.

Shoreditch Gtr. London. *Soredich* c.1148, *Schoredich* 1236. Possibly 'ditch leading to the river-bank (of the Thames)'. OE **scora* + *dīc*. An alternative recent suggestion is that the first element may rather be a Brittonic word **skor* 'fort, rampart' with reference to the Roman wall of the City.

Shoreham, 'homestead by a steep bank or slope', OE **scora* + *hām*: **Shoreham** Kent. *Scorham* 822. **Shoreham by Sea** W. Sussex. *Sorham* 1073, *Soreham* 1086 (DB).

Shorncote Glos. *Schernecote* 1086 (DB). 'Cottage(s) in a dirty or muddy place'. OE *scearn* + *cot*.

Shorne Kent. *Scorene* c.1100. 'Steep place'. OE **scoren*.

Shorwell I. of Wight. *Sorewelle* 1086 (DB). 'Spring or stream by a steep slope'. OE **scora* + *wella*.

Shotesham Norfolk. *Shotesham* 1044–7, *Scotesham* 1086 (DB). 'Homestead of a man called Scot'. OE pers. name + *hām*.

Shotley, possibly 'woodland clearing frequented by pigeons, or where shooting takes place, or on a steep slope', OE **sceote* or *sceot* or **scēot* + *lēah*: **Shotley** Northum. *Schotley* 1242. **Shotley** Suffolk *Scoteleia* 1086 (DB). **Shotley Bridge** Durham, *Shotleybrigg* 1613, is named from SHOTLEY Northum., with reference to the important bridge here over the River Derwent spanning the two counties of Northumberland and Durham.

Shottermill Surrey. *Shottover* 1537, *Schotouermyll* 1607. Probably 'mill of a family called Shotover'. ME surname + *myln*.

Shottery Warwicks. *Scotta rith* 699–709. Probably 'stream of the Scots'. OE *Scot* (genitive plural *-a*) + *rīth*. Alternatively the first element may be OE *sc(e)ota* 'trout'.

Shottesbrooke Winds. & Maid. *Sotesbroc* 1086 (DB), *Schottesbroc, Schotebroch* 1187. Probably 'brook of a man called Scot'. OE pers. name + *brōc*. Alternatively the first element may be OE *sc(e)ota* 'trout'.

Shotteswell Warwicks. *Soteswell* c.1140. Probably 'spring or stream of a man called Scot'. OE pers. name + *wella*.

Shottisham Suffolk. *Scotesham* 1086 (DB). 'Homestead of a man called Scot or **Scēot*'. OE pers. name + *hām*.

Shottle Derbys. *Sothelle* [sic] 1086 (DB), *Schethell* 1191. Probably 'hill with a steep slope'. OE **scēot* + *hyll*.

Shotton Durham. near Peterlee. *Sceottun* c.1040. Probably 'farmstead on or near a steep slope'. OE *scēot* + *tūn*.

Shotton Flin. *Schotton* 1283. Identical in origin with previous name.

Shotton Northum. near Mindrum. *Scotadun* c.1050. 'Hill of the Scots'. OE *Scot* (genitive plural -*a*) + *dūn*.

Shotts N. Lan. *Bertrum Schottis* 1552. '(Place by) the slopes'. OE *scēot*. *Bertrum* in the 16th-cent. spelling is probably a pers. name or surname.

Shotwick Ches. *Sotowiche* 1086 (DB), *Shotowica* c.1100. 'Dwelling or farm at a steep promontory'. OE *scēot* + *hōh* + *wīc*.

Shouldham Norfolk. *Sculdeham* 1086 (DB). Probably 'homestead owing a debt or obligation'. OE *sculd* + *hām*.

Shoulton Worcs. *Scolegeton* c.1220. Possibly 'farmstead at the stream called **Sceolge*'. OE river-name (meaning 'winding') + *tūn*.

Shraheens (*Na Sraithíní*) Mayo. 'The little river-meadow'.

Shrawardine Shrops. *Saleurdine* [*sic*] 1086 (DB), *Scrawardin* 1166. Possibly 'enclosure near a hollow or hovel'. OE *scræf* + *worthign*.

Shrawley Worcs. *Scræfleh* 804. 'Woodland clearing near a hollow or hovel'. OE *scræf* + *lēah*.

Shrewsbury Shrops. *Scrobbesbyrig* 11th cent., *Sciropesberie* 1086 (DB). 'Fortified place of the scrubland region'. Old district name (from an OE **scrobb*) + OE *burh* (dative *byrig*).

Shrewton Wilts. *Wintreburne* 1086 (DB), *Winterbourne Syreveton* 1232. 'The sheriff's manor on the stream formerly called *Wintreburne*'. OE *scīr-rēfa* + *tūn*. The 'winter stream' (OE *winter* + *burna*) is now called **Till** (a back-formation from TILSHEAD).

Shrigley, Pott Ches. *Schriggeleg* 1285, *Potte Shryggelegh* 1354. Two distinct names. Pott is from ME *potte* 'a deep hole'. Shrigley is 'woodland clearing frequented by mistle-thrushes', OE *scrīc* + *lēah*.

Shrivenham Oxon. *Scrifenanhamme* c.950, *Seriveham* 1086 (DB). 'River-meadow allotted by decree (to the church)'. OE *scrifen* + *hamm*.

Shronowen (*Srón Abhann*) Kerry. 'Snout of the river'.

Shropham Norfolk. *Screpham* 1086 (DB). OE *hām* 'homestead' with an uncertain first element, possibly a pers. name.

Shropshire (the county). *Sciropescire* 1086 (DB). Shortened form of old spelling for SHREWSBURY + OE *scīr* 'district'. The alternative form **Salop** is a revival of an old Norman contracted spelling of the name Shropshire.

Shroton Dorset. *See* IWERNE COURTNEY.

Shrough (*Sruth*) Tipperary. 'Stream'.

Shroughan (*Sruthán*) Wicklow. 'Stream'.

Shrove (*An tSrúibh*) Donegal. 'The snout'.

Shrue (*Sidh Ruadh*) Westmeath. 'Red fairy hill'.

Shrule (*Sruthair*) Mayo. 'Stream'.

Shuckburgh Warwicks. *Socheberge* 1086 (DB). 'Hill or mound haunted by an evil spirit'. OE *scucca* + *beorg*.

Shucknall Herefs. *Shokenhulle* 1377. 'Hill haunted by an evil spirit'. OE *scucca* (genitive -*n*) + *hyll*.

Shudy Camps Cambs. *See* CAMPS.

Shuna (island) Arg. Unidentified first element + OScand. *ey* 'island'.

Shurdington Glos. *Surditona* c.1150. Possibly 'estate associated with a man called **Scyrda*'. OE pers. name + -*ing*- + *tūn*.

Shurton Somerset. *Shureveton* 1219. 'The sheriff's manor'. OE *scīr-rēfa* + *tūn*.

Shustoke Warwicks. *Scotescote* [*sic*] 1086 (DB), *Schutestok* 1247. Probably 'outlying farmstead or hamlet of a man called **Scēot*'. OE pers. name + *stoc* (replacing *cot* 'cottage' of the DB form).

Shute Devon. *Schieta* c.1200. 'The corner or angle of land'. OE **scīete*.

Shutford Oxon. *Schiteford* c.1160. Probably 'ford of a man called **Scytta*'. OE pers. name + *ford*.

Shutlanger Northants. *Shitelhanger* 1162. 'Wooded slope where shuttles or wooden bars are obtained'. OE *scytel* + *hangra*.

Shuttington Warwicks. *Cetitone* [*sic*] 1086 (DB), *Schetintuna* c.1160. 'Estate associated with

a man called *Scytta'. OE pers. name + -ing-
+ tūn.

Sibbertoft Northants. *Sibertod* [sic] 1086
(DB), *Sibertoft* 1198. 'Homestead of a man called
Sigebeorht or Sigbjǫrn'. OE or OScand. pers.
name + OScand. toft.

Sibdon Carwood Shrops. *Sibetune* 1086
(DB), *Sipton Carswood* 1672. 'Farmstead of a
man called Sibba'. OE pers. name + tūn. Affix
from a nearby place Carwood, *Carwod* 1307,
probably 'rock wood', from OE carr + wudu.

Sibertswold Kent. See SHEPHERDSWELL.

Sibford Ferris & Sibford Gower Oxon.
Sibeford 1086 (DB), *Sibbard Ferreys* early 18th
cent., *Sibbeford Goyer* 1220. 'Ford of a man
called Sibba'. OE pers. name + ford. Manorial
affixes from early possession by the de Ferrers
and Guher families.

Sible Hedingham Essex. See HEDINGHAM.

Sibsey Lincs. *Sibolci* 1086 (DB). 'Island of a
man called Sigebald'. OE pers. name + ēg.

Sibson Cambs. *Sibestune* 1086 (DB).
'Farmstead of a man called Sibbi'. OScand. pers.
name + OE tūn.

Sibson Leics. *Sibetesdone* 1086 (DB). Possibly
'hill of a man called Sigeberht'. OE pers. name +
dūn.

Sibthorpe Notts. *Sibetorp* 1086 (DB).
'Outlying farm of a man called Sibba or Sibbi'.
OE or OScand. pers. name + OScand. thorp.

Sibton Suffolk. *Sibbetuna* 1086 (DB).
'Farmstead of a man called Sibba'. OE pers.
name + tūn.

Sicklinghall N. Yorks. *Sidingale* [sic] 1086
(DB), *Sicclinhala* c.1150. 'Nook of land of the
family or followers of a man called *Sicel'. OE
pers. name + -inga- + halh.

Sidbury Devon. *Sideberia* 1086 (DB).
'Fortified place by the River Sid'. OE river-name
(from sīd 'broad') + burh (dative byrig).

Sidbury Shrops. *Sudberie* 1086 (DB).
'Southern fortified place or manor'. OE sūth +
burh (dative byrig).

Sidcup Gtr. London. *Cetecopp* 1254. Probably
'seat-shaped or flat-topped hill'. OE *set-copp.

Siddington Ches. *Sudendune* 1086 (DB).
'(Place) south of the hill'. OE sūthan + dūn.

Siddington Glos. *Sudintone* 1086 (DB).
'(Land) south in the township'. OE sūth + in +
tūn.

Sidestrand Norfolk. *Sistran* [sic] 1086 (DB),
Sidestrande late 12th cent. 'Broad shore'. OE
sīd + strand.

Sidford Devon. *Sideford* 1238. 'Ford on the
River Sid'. OE river-name (see SIDBURY) + ford.

Sidlaw Hills Ang. *Seedlaws* 1799. Uncertain
first element + OE hlāw 'hill'.

Sidlesham W. Sussex. *Sideleham* 683.
'Homestead of a man called *Sidel'. OE pers.
name + hām.

Sidmouth Devon. *Sedemuda* 1086 (DB).
'Mouth of the River Sid'. OE river-name (see
SIDBURY) + mūtha.

Siefton Shrops. *Sireton* [sic] 1086 (DB),
Siveton 1257. Possibly 'farmstead of a woman
called Sigegifu'. OE pers. name + tūn.
Alternatively the first element may be an old
stream-name Siven of unknown origin.

Sigglesthorne E. R. Yorks. *Siglestorne* 1086
(DB). 'Thorn-tree of a man called Sigulfr'.
OScand. pers. name + thorn.

Silchester Hants. *Silcestre* 1086 (DB). Possibly
'Roman station by a willow copse'. OE *siele +
ceaster. Alternatively the first element may be a
reduced form of Calleva 'place in the woods',
the original Celtic name of this Roman city first
recorded in the 2nd cent. AD.

Sileby Leics. *Siglebi* 1086 (DB). 'Farmstead or
village of a man called Sigulfr'. OScand. pers.
name + bý.

Silecroft Cumbria. *Selecroft* 1211. 'Enclosure
where willows grow'. OScand. selja + OE croft.

Sili. See SULLY.

Silk Willoughby Lincs. See WILLOUGHBY.

Silkstone Barns. *Silchestone* 1086 (DB).
'Farmstead of a man called Sigelāc'. OE pers.
name + tūn.

Silksworth Sundld. *Sylceswurthe* c.1040.
'Enclosure of a man called Sigelāc'. OE pers.
name + worth.

Silloth Cumbria. *Selathe* 1292. 'Barn(s) by the
sea'. OScand. sǽr or OE sǽ + OScand. hlatha.

Silpho N. Yorks. *Silfhou c.*1160. 'Flat-topped hill-spur'. OE *scylfe* + *hōh*.

Silsden Brad. *Siglesdene* 1086 (DB). 'Valley of a man called Sigulfr'. OScand. pers. name + OE *denu*.

Silsoe Beds. *Siuuilessou* 1086 (DB). 'Hill-spur of a man called *Sifel'. OE pers. name + *hōh*.

Silton, Nether & Silton, Over N. Yorks. *Silftune* 1086 (DB). 'Farmstead by a shelf or ledge', or 'farmstead of a man called Sylfa'. OE *scylfe* or OScand. pers. name + OE *tūn*.

Silverdale Lancs. *Selredal* [sic] 1199, *Celverdale* 1292. 'Silver-coloured valley' (from the grey limestone here). OE *seolfor* + *dæl*.

Silverstone Northants. *Sulueston* 942, *Silvestone* 1086 (DB). Probably 'farmstead of a man called Sǣwulf or Sigewulf'. OE pers. name + *tūn*.

Silverton Devon. *Sulfretone* 1086 (DB). Probably 'farmstead near the gully ford'. OE *sulh* + *ford* + *tūn*.

Simonburn Northum. *Simundeburn* 1229. 'Stream of a man called Sigemund'. OE pers. name + *burna*.

Simonstone Lancs. *Simondestan* 1278. 'Boundary stone of a man called Sigemund'. OE pers. name + *stān*.

Simpson Milt. K. *Siwinestone* 1086 (DB). 'Farmstead of a man called Sigewine'. OE pers. name + *tūn*.

Sinderby N. Yorks. *Senerebi* 1086 (DB). 'Southern farmstead or village', or 'farmstead or village of a man called Sindri'. OScand. *syndri* or pers. name + *bý*.

Sindlesham Wokhm. *Sindlesham* 1220. Possibly 'homestead or enclosure of a man called *Synnel'. OE pers. name + *hām* or *hamm*.

Sineirl (*Suí an Iarla*) Antrim. 'Seat of the earl'.

Singleborough Bucks. *Sincleberia* 1086 (DB). 'Shingle hill'. OE *singel* + *beorg*.

Singleton Lancs. *Singletun* 1086 (DB). 'Farmstead with shingled roof'. OE *scingol* + *tūn*.

Singleton W. Sussex. *Silletone* [sic] 1086 (DB), *Sengelton* 1185. Probably 'farmstead by a burnt clearing'. OE *sengel* + *tūn*.

Sinnington N. Yorks. *Siuenintun* 1086 (DB). 'Farmstead near the River Seven'. Pre-English river-name (of uncertain meaning) + OE *-ing-* + *tūn*.

Sinton, Leigh Worcs. *Sothyntone in Lega* 13th cent. '(Place) south in the village (of LEIGH)'. OE *sūth* + *in* + *tūn*.

Sion Mills (*Muileann an tSiáin*) Tyrone. *the 'Sion Mills'* 1843. 'Mill of the fairy mound'.

Sipson Gtr. London. *Sibwineston c.*1150. 'Farmstead of a man called Sibwine'. OE pers. name + *tūn*.

Sissinghurst Kent. *Saxingherste c.*1180. 'Wooded hill of the family or followers of a man called Seaxa'. OE pers. name + *-inga-* + *hyrst*.

Siston S. Glos. *Sistone* 1086 (DB). 'Farmstead of a man called *Sige'. OE pers. name + *tūn*.

Sithney Cornwall. *St Sythyn* 1230. From the patron saint of the parish church, St Sithny.

Sitlington Wakefd. *See* MIDDLESTOWN.

Sittingbourne Kent. *Sidingeburn* 1200. Probably 'stream of the dwellers on the slope'. OE *sīde* + *-inga-* + *burna*.

Six Mile Bottom Cambs., hamlet so named 1801, from its situation in a valley (OE *botm*) six miles from Newmarket.

Sixhills Lincs. *Sisse* [sic] 1086 (DB), *Sixlei* 1196. 'Six woodland clearings'. OE *six* + *lēah*.

Sixmilecross Tyrone. *na gCorrach* 1930. The village crossroads is roughly six Irish miles from Omagh. The Irish name of the village is *Na Coracha Móra*, 'the big round hills'.

Sixpenny Handley Dorset. *See* HANDLEY.

Sizewell Suffolk. *Syreswell* 1240. Probably 'spring or stream of a man called Sigehere'. OE pers. name + *wella*.

Skagh (*Sceach*) Cork, Limerick. 'Hawthorn'.

Skahard (*Sceach Ard*) Limerick. 'High hawthorn'.

Skaill Orkn. 'Hut'. OScand. *skáli*.

Skara Brae Orkn. Origin not known, unless linked with ON *skarfr* 'cormorant'.

Skea (*Sceach*) Fermanagh, Tyrone. 'Hawthorn'.

Skeabost Highland. (Skye). 'Skíthi's farm'.
OScand. pers. name + *bólstathr*.

Skeagh (*Sceach*) Antrim, Cavan, Donegal,
Laois, Tipperary. 'Hawthorn'.

Skeaghmore (*Sceath Mhór*) Westmeath. 'Big
hawthorn'.

Skeard (*Scéird*) Mayo. 'Bleak hill'.

Skeeby N. Yorks. *Schirebi* [*sic*] 1086 (DB),
Schittebi 1187. Probably 'farmstead or village of
a man called Skíthi'. OScand pers. name + *bý*.

Skeffington Leics. *Sciftintone* 1086 (DB).
Probably 'estate associated with a man
called *Sceaft*'. OE pers. name (with Scand. *sk-*)
+ -*ing-* + *tūn*.

Skeffling E. R. Yorks. *Sckeftling* 12th cent.
Probably '(settlement of) the family or followers
of a man called *Sceaftel*'. OE pers. name (with
Scand. *sk-*) + -*ingas*.

Skegby Notts. near Mansfield. *Schegebi* 1086
(DB). 'Farmstead or village of a man called
Skeggi, or on a beard-shaped promontory'.
OScand. pers. name or *skegg* + *bý*.

Skegness Lincs. *Sceggenesse* 12th cent.
'Beard-shaped promontory', or 'promontory of
a man called Skeggi'. OScand. *skegg* or pers.
name + *nes*.

Skegoniel (*Sceitheog an Iarla*) Antrim.
Balliskeighog-Inerla 1605. 'Thorn bush of the
earl'.

Skelbo Highland. 'Shell farm'. OScand. *skel* +
bólstathr (or *ból*).

Skelbrooke Donc. *Scalebre, Scalebro* 1086
(DB), *Scelebroch* 1160–75. 'Stream near the hut
or shieling'. OE **scēla* (with Scand. *sk-*) + *brōc*.
The river-name **Skell** is probably a back-
formation from the place name.

Skeldyke Lincs. *Skeldyke* 1281. 'The
boundary ditch or drainage channel'. OScand.
skial + *dík*.

Skellingthorpe Lincs. *Scheldinchope* 1086
(DB). 'Enclosure in marsh associated with a man
called *Sceld*' from OE pers. name + -*ing-* + *hop*,
or 'enclosure in marsh by a shield-shaped hill'
from OE *sceld* + -*ing* + *hop*. Initial *Sk-* through
Scand. influence.

Skellow Donc. *Scanhalle* [*sic*] 1086 (DB),
*Scalehale c.*1190. 'Nook of land with a shieling or

hut'. OE **scēla* (influenced by OScand. *skáli*) +
halh.

Skelmanthorpe Kirkl. *Scelmertorp* 1086
(DB). 'Outlying farmstead of a man called
Skjaldmarr'. OScand. pers. name + *thorp*.

Skelmersdale Lancs. *Schelmeresdele* 1086
(DB). 'Valley of a man called *Skjaldmarr*'.
OScand. pers. name + *dalr*.

Skelpick Highland. 'Shelly place'. Perhaps
Gaelic *sgealbach*.

Skelton, 'farmstead on a shelf or ledge', OE
scelf (with Scand. *sk-*) + *tūn*: **Skelton** Cumbria.
*Sheltone c.*1160. **Skelton** E. R. Yorks. *Schilton*
1086 (DB). **Skelton** N. Yorks., near Richmond.
Scelton 12th cent. **Skelton** Red. & Cleve.
Scheltun 1086 (DB). **Skelton** York. *Scheltun*
1086 (DB).

Skelwith Bridge Cumbria. *Schelwath* 1246.
'Ford by the waterfall'. OScand. *skjallr*
'resounding' + *vath*. The waterfall referred to is
Skelwith Force (from OScand. *fors* 'waterfall').
The addition *Bridge* is found from the 17th cent.

Skenarget (*Sceach Airgid*) Tyrone. 'Silver
hawthorn'.

Skendleby Lincs. *Scheueldebi* [*sic*] 1086 (DB),
*Schendelebia c.*1150. OScand. *bý* 'farmstead,
village' possibly added to a place name meaning
'beautiful slope' from OE *scēne* (with Scand. *sk-*)
+ *helde*.

Skerne E. R. Yorks. *Schirne* 1086 (DB). Named
from Skerne Beck, 'the clear or pure stream',
from OScand. *skírn*, or (perhaps more likely)
from OE *scīr* (dative -*an*) + *ēa* with initial *sk-*
through Scand. influence.

Skerray Highland. Possibly OScand. *sker*
'skerry, reef' + uncertain second element.

Skerries (*Ynysoedd y Moelrhoniaid*) (islands)
Angl. 'Reefs'. OScand. *sker*. The Welsh name
means 'islands of the seals' (Welsh *ynys*, plural
ynysoedd + *y* + *moelrhon*, plural *moelrhoniaid*).

Skerries, The (*Na Sceirí*) (islands) Antrim.
The Skerries 1837. 'The rocky islands'.

Skerton Lancs. *Schertune* 1086 (DB), *Skerton*
1201. 'Farmstead by the reefs or sandbanks'.
OScand. *sker* + *tún*.

Sketty Swan. 'Ceti's island'. Welsh *ynys*.

Skewsby N. Yorks. *Scoxebi* 1086 (DB). 'Farmstead or village in the wood'. OScand. *skógr* + *bý*.

Skeyton Norfolk. *Scegutuna* 1086 (DB). 'Farmstead of a man called Skeggi'. OScand. pers. name + *tūn*.

Skibbereen (*An Sciobhairín*) Cork. 'The place of the little boats'.

Skibo Highland. *Schythebol* 1275. 'Skíthi's farm'. OScand. pers. name + *bólstathr* (or *ból*).

Skidbrooke Lincs. *Schitebroc* 1086 (DB). 'Dirty brook'. OE *scite* (with Scand. *sk-*) + *brōc*.

Skidby E. R. Yorks. *Scyteby* 972, *Schitebi* 1086 (DB). 'Farmstead or village of a man called Skyti', or 'dirty farmstead or village'. OScand. pers. name or *skítr* + *bý*.

Skiddernagh (*Sciodarnach*) Mayo. 'Place of puddles'.

Skilgate Somerset. *Schiligata* 1086 (DB). Possibly 'stony or shaly gate or gap'. OE **scilig* + *geat*.

Skillington Lincs. *Schillintune* 1086 (DB). Possibly 'estate associated with a man called **Scill(a)'*, OE pers. name + *-ing-* + *tūn*. Alternatively the first element may rather be OE *scilling* 'shilling' alluding to a rent. Initial *Sk-* through Scand. influence.

Skinburness Cumbria. *Skyneburg* 1175, *Skynburneyse* 1298. 'Promontory near the demon-haunted stronghold'. OE *scinna* (with Scand. *sk-*) + *burh* + OScand. *nes*.

Skinningrove Red. & Cleve. *Scineregrive* *c.*1175. 'Pit used by skinners or tanners'. OScand. *skinnari* + *gryfja*.

Skipsea E. R. Yorks. *Skipse* 1160. 'Lake used for ships'. OScand. *skip* + *sǽr*.

Skipton, 'sheep farm', OE *scīp* (with Scand. *sk-*) + *tūn*: **Skipton** N. Yorks. *Scipton* 1086 (DB). **Skipton on Swale** N. Yorks. *Schipetune* 1086 (DB). Swale is an OE river-name probably meaning 'rushing water'.

Skipwith N. Yorks. *Schipewic* 1086 (DB). 'Sheep farm'. OE *scīp* (with Scand. *sk-*) + *wīc* (replaced by OScand. *vithr* 'wood').

Skirlaugh, North & Skirlaugh, South E. R. Yorks. *Schirelai* 1086 (DB). 'Bright woodland clearing'. OE *scīr* (with Scand. *sk-*) + *lēah*.

Skirmett Bucks. *La Skiremote c.*1307. 'Meeting-place of the shire-court'. OE *scīr* (with Scand *sk-*) + (*ge*)*mōt*.

Skirpenbeck E. R. Yorks. *Scarpenbec* 1086 (DB). Probably 'stream which sometimes dries up'. OScand. **skerpin* + *bekkr*.

Skirwith Cumbria. *Skirewit* 1205. 'Wood belonging to the shire'. OE *scīr* (with Scand. *sk-*) + OScand. *vithr*.

Skokholm (island) Pemb. *Scogholm* 1219, *Stokholm* 1275. 'Island in a channel'. OScand. *stokkr* + *holmr*. Expected *St-* has become *Sk-* under the influence of neighbouring SKOMER.

Skomer (island) Pemb. *Skalmey* 1324. 'Cloven island'. OScand. *skálm* + *ey*.

Skreen (*Scrín*) Meath, Sligo. 'Shrine'.

Skull (*Scoil*) Cork. 'School'.

Skye (island) Highland. *Scitis c.*150, *Scia c.*700, *Skith c.*1250, *Skye* 1266. Pre-Celtic name, later reinterpreted as 'winged (island)'. Gaelic *sgiath*. The island's great peninsulas thrust out north and south like wings.

Slad, Slade, '(place at) the valley', ME *slade* (OE *slæd*): **Slad** Glos. *The Slad* 1779. **Slade** Devon, near Sheldon. *Slade* 1249. **Slade Green** Gtr. London. *Slade Green* 1561. With ME *grene* 'village green'.

Slade (*An Slaod*) Wexford. 'The fall-away of ground'.

Slaggyford Northum. *Slaggiford* 13th cent. 'Muddy ford'. ME **slaggi* + *ford*.

Slaidburn Lancs. *Slateborne* 1086 (DB). 'Stream by the sheep-pasture'. OE **slæget* + *burna*.

Slaithwaite Kirkl. *Sladweit* 1178, *Sclagtwayt* 1277. Probably 'clearing where timber was felled'. OScand. *slag* + *thveit*.

Slaley Northum. *Slaveleia* 1166. 'Muddy woodland clearing'. OE **slæf* + *lēah*.

Slamannan Falk. *Slethmanin* 1250. 'Moor of Manau'. Gaelic *sliabh*.

Slane (*Baile Shláine*) Meath. 'Homestead of fullness'.

Slanebeg (*Sleamhain Bheag*) Westmeath. 'Small sleek place'.

Slanemore (*Sleamhain Mhór*) Westmeath. 'Big sleek place'.

Slaney (*Sleine*) (river) Carlow–Wexford–Wicklow. 'Healthy one'.

Slapton, 'farmstead by a slippery place' or 'muddy farmstead', OE **slæp* or **slæp* + *tūn*: **Slapton** Bucks. *Slapetone* 1086 (DB). **Slapton** Devon. *Sladone* [*sic*] 1086 (DB), *Slapton* 1244. **Slapton** Northants. *Slaptone* 1086 (DB).

Slaugham W. Sussex. *Slacham* c.1100. 'Homestead or enclosure where sloe or blackthorn grows'. OE *slāh* + *hām* or *hamm*.

Slaughter, Lower & Slaughter, Upper Glos. *Sclostre* 1086 (DB). From an OE **slōhtre* 'muddy place' or 'ravine, deep channel'.

Slaughtmanus (*Sleacht Mhánasa*) Derry. *Laghtmanus* 1613. 'Grave mound of Mánas'.

Slawston Leics. *Slagestone* 1086 (DB). Probably 'farmstead of a man called Slagr'. OScand. pers. name + OE *tūn*.

Slea Head (*Ceann Sléibhe*) Kerry. 'Head of the mountain'.

Sleaford Lincs. *Sliowaforda* 852, *Eslaforde* 1086 (DB). 'Ford over the River Slea'. OE river-name (meaning 'muddy stream') + *ford*.

Sleagill Cumbria. *Slegill* c.1190. 'Ravine of the trickling stream, or of a man called **Slefa*'. OScand. *slefa* or pers. name + *gil*.

Sleat (peninsula) Highland (Skye). *Slate* c.1400. 'Level place'. OScand. *slétta*.

Sleatygraigue (*Gráig Shléibhte*) Laois. 'Hamlet of the mountains'.

Sledmere E. R. Yorks. *Slidemare* 1086 (DB). 'Pool in the valley'. OE *slæd* + *mere*.

Sleekburn, East & Sleekburn, West Northum. *Sliceburne* c.1050. 'Smooth stream, or muddy stream'. OE **slicu* or **slīc* + *burna*.

Sleightholme Durham. *Slethholm* 1234. Probably 'level raised ground'. OScand. *sléttr* + *holmr*.

Sleights N. Yorks. *Sleghtes* c.1223. 'Smooth or level fields'. OScand. *slétta*.

Slemish (*Sliabh Mis*) (mountain) Antrim. *Slébh Mis* 771. 'Mis's mountain'.

Slievaduff (*Sliabh an Daimh*) (mountain) Kerry. 'Mountain of the ox'.

Slieve Anierin (*Sliabh an Iarainn*) (mountain) Leitrim. 'Iron mountain'.

Slieve Bernagh (*Sliabh Bearnach*) (mountain) Clare, Down. 'Gapped mountain'.

Slieve Binnian (*Sliabh Binneáin*) (mountain) Down. *Great Bennyng* c.1568. 'Mountain of the little peak'.

Slieve Bloom (*Sliabh Bladhma*) (mountain) Laois, Offaly. 'Blaze mountain'.

Slieve Commedagh (*Sliabh Coimhéideach*) (mountain) Down. *Slieve Kimedia* c.1830. 'Guarding mountain'.

Slieve Croob (*Sliabh Crúibe*) (mountain) Down. *Crooby Mountaine* c.1657. 'Mountain of the hoof'.

Slieve Donard (*Sliabh Dónairt*) (mountain) Down. *Sliabh-Domha(n)gaird* 1645. 'St Dónart's mountain'.

Slieve Gamph (*Sliabh Gamh*) (mountain) Mayo, Sligo. 'Mountain of storms'.

Slieve Gullion (*Sliabh gCuillinn*) (mountain) Armagh (*moninni*). *Sleibi Culinn* c.830. 'Mountain of the slope'.

Slieve League (*Sliabh Liag*) (mountain) Donegal. *Sleibh Liacc* c.830. 'Mountain of the pillar stones'.

Slieve Mish (*Sliabh Mis*) (mountain) Kerry. 'Mountain of Mis'.

Slieve na Calliagh (*Sliabh na Calliagh*) (mountain) Meath. 'Mountain of the hag'.

Slieve Snaght (*Sliabh Sneachta*) (mountain) Donegal. *Sléibhte Sneachta* 1260. 'Mountain of snow'.

Slieveardagh (*Sliabh Ardach*) (mountain) Kilkenny, Tipperary. 'Mountain of the high field'.

Slieveaughty (*Sliabh Eachtaí*) (mountain) Galway. 'Eachta's mountain'.

Slievecallan (*Sliabh Calláin*) (mountain) Clare. 'Callán's mountain'.

Slievefelim (*Sliabh Eibhlinne*) (mountain) Limerick. 'Eibhlinn's mountain'.

Slievemore (*Sliabh Mór*) (mountain)
Tyrone. 'Big mountain'.

Slievenamon (*Sliabh na mBan*) (mountain)
Tipperary. 'Mountain of the women'.

Slieveroe (*Sliabh Rua*) (mountain) Kilkenny.
'Red mountain'.

Slievetooey (*Sliabh Tuaidh*) (mountain)
Donegal. *Slieve-a-tory* 1835. 'Northern
mountain'.

Sligo (*Sligeach*) Sligo. 'Abounding in shells'.

Slimbridge Glos. *Heslinbruge* [*sic*] 1086 (DB),
*Slimbrugia c.*1153. 'Bridge or causeway over a
muddy place'. OE *slīm* + *brycg*.

Slindon, probably 'sloping hill', OE **slinu* +
dūn: **Slindon** Staffs. *Slindone* 1086 (DB).
Slindon W. Sussex. *Eslindone* 1086 (DB).

Slingsby N. Yorks. *Selungesbi* 1086 (DB).
'Farmstead or village of a man called Slengr'.
OScand. pers. name + *bý*.

Slipton Northants. *Sliptone* 1086 (DB).
Probably 'muddy farmstead'. OE *slipa* + *tūn*.

Sloley Norfolk. *Slaleia* 1086 (DB). 'Woodland
clearing where sloe or blackthorn grows'. OE
slāh + *lēah*.

Sloothby Lincs. *Slodebi* 1086 (DB). Probably
'farmstead or village of a man called Slóthi'.
OScand. pers. name + *bý*.

Slough Slough. *Slo* 1195. 'The slough or miry
place'. OE *slōh*.

Slyne Lancs. *Sline* 1086 (DB). From OE **slinu*
'a slope'.

Smaghran (*Smeachrán*) Roscommon. 'Stripe
of land'.

Smailholm Sc. Bord. *Smalham* 1246.
'Narrow village'. OE *smæl* + *hām*.

Small Hythe Kent. *Smalide* 13th cent.,
Smalhede 1289. 'Narrow landing-place or
harbour'. OE *smæl* + *hȳth*.

Smallburgh Norfolk. *Smaleberga* 1086 (DB).
'Hill or mound on the River *Smale*'. Old name
for the River Ant (from OE *smæl* 'narrow' and *ēa*
'river') + OE *beorg*.

Smalley Derbys. *Smælleage* 1009, *Smalei*
1086 (DB). 'Narrow woodland clearing'. OE
smæl + *lēah*.

Smardale Cumbria. *Smeredal* 1190. Probably
'valley where butter is produced'. OScand.
smjǫr or OE *smeoru* + OScand. *dalr*.

Smarden Kent. *Smeredaenne c.*1100.
'Woodland pasture where butter is produced'.
OE *smeoru* + *denn*.

Smeaton, 'farmstead of the smiths', OE *smith*
+ *tūn*: **Smeaton, Great** N. Yorks. *Smithatune*
966–72, *Smidetune* 1086 (DB). **Smeaton, Kirk
& Smeaton, Little** N. Yorks. *Smedetone* 1086
(DB). Distinguishing affix is OScand. *kirkja*
'church'.

Smeeth Kent. *Smitha* 1018. From OE
smiththe 'a smithy'.

Smethwick Sandw. *Smedeuuich* 1086 (DB).
'Dwelling or building of the smiths'. OE *smith*
(genitive plural **smeotha*) + *wīc*.

Smisby Derbys. *Smidesbi* 1086 (DB). 'The
smith's farmstead'. OScand. *smithr* + *bý*.

Smithborough Monaghan. *Smithsborough*
1778. A man named *Smith* established cattle
fairs here in the latter half of the 18th cent. The
Irish name is *Na Mullaí*, 'the hilltops'.

Snaefell (mountain) Isle of Man. 'Snow
mountain'. OScand. *snǽǽr* + *fjall*.

Snailwell Cambs. *Sneillewelle c.*1050,
Snelleuuelle 1086 (DB). 'Spring or stream
infested with snails'. OE *snægl* + *wella*.

Snainton N. Yorks. *Snechintune* 1086 (DB).
Possibly 'farmstead associated with a man
called **Snoc*'. OE pers. name + *-ing-* + *tūn*.

Snaith E. R. Yorks. *Snaith c.*1080, *Esneid* 1086
(DB). 'Piece of land cut off'. OScand. *sneith*.

Snape, from OE **snæp* 'boggy piece of land' or
OScand. *snap* 'poor pasture'; examples include:
Snape N. Yorks. *Snape* 1154. **Snape** Suffolk.
Snapes 1086 (DB).

Snaresbrook Gtr. London. not recorded
before 1599, from ME *broke* 'brook' with an
uncertain first element, possibly a ME surname
or ME *snare* 'a snare or trap for catching wild
animals and birds'.

Snarestone Leics. *Snarchetone* 1086 (DB),
Snarkeston 1188. 'Farmstead of a man called
*Snar(o)c'. OE pers. name + *tūn*.

Snarford Lincs. *Snerteforde* 1086 (DB). 'Ford
of a man called Snǫrtr'. OScand. pers. name +
OE *ford*.

Snargate Kent. *Snergathe c.*1197. Probably 'gate or gap where snares for animals are placed'. OE *sneare* + *geat*.

Snave Kent. *Snaues* 1182, *Snaue* 1240. Probably 'the spit(s) or strip(s) of land'. OE **snafa*.

Snead Powys. *Sned* 1201. 'Detached plot of land'. OE *snǣd*.

Sneaton N. Yorks. *Snetune* 1086 (DB). 'Farmstead of a man called Snær, or with a detached piece of land'. OScand. pers. name or OE *snǣd* + *tūn*.

Sneem (*An tSnaidhm*) Kerry. 'The knot'. Many roads and rivers meet here.

Snelland Lincs. *Sneleslunt* 1086 (DB). 'Grove of a man called Snjallr or Snell'. OScand. or OE pers. name + OScand. *lundr*.

Snelston Derbys. *Snellestune* 1086 (DB). 'Farmstead of a man called Snell'. OE pers. name + *tūn*.

Snettisham Norfolk. *Snetesham* 1086 (DB). 'Homestead of a man called **Snǣt* or Sneti'. OE pers. name + *hām*.

Snitter Northum. *Snitere* 1176. Possibly 'weather-beaten place' from a word related to ME *sniteren* 'to snow'.

Snitterby Lincs. *Snetrebi* 1086 (DB). Probably 'farmstead or village of a man called **Snytra*'. OE pers. name + OScand. *bý*.

Snitterfield Warwicks. *Snitefeld* 1086 (DB). 'Open land frequented by snipe'. OE *snīte* + *feld*.

Snodhill Herefs. *Snauthil* 1195. Probably 'snowy hill'. OE **snāwede* + *hyll*.

Snodland Kent. *Snoddingland* 838, *Esnoiland* [*sic*] 1086 (DB). 'Cultivated land associated with a man called **Snodd*'. OE pers. name + *-ing-* + *land*.

Snoreham Essex. *Snorham* 1238. Probably 'homestead or enclosure by a rough hill'. OE **snōr* + *hām* or *hamm*.

Snoring, Great & Snoring, Little Norfolk. *Snaringes* 1086 (DB). Probably '(settlement of) the family or followers of a man called **Snear*'. OE pers. name + *-ingas*.

Snowdon (*Yr Wyddfa*) (mountain) Gwyd. *Snawdune* 1095. 'Snow hill'. OE *snāw* + *dūn*. The Welsh name means 'the mound' (Welsh *yr* + *gwyddfa*), referring to the mountain's use as a burial place.

Snowdonia (*Eryri*) (district) Gwyd. 'District of Snowdon'. See SNOWDON. The Welsh name means 'highlands' but has long been perceived as 'district of eagles' (Welsh *eryr* 'eagle').

Snowshill Glos. *Snawesille* 1086 (DB). 'Hill where the snow lies long'. OE *snāw* + *hyll*.

Soar (river) Warwicks.–Leics.–Notts. See BARROW UPON SOAR.

Soay (island) W. Isles. 'Sheep island'. OScand. *sauthr* + *ey*.

Soberton Hants. *Sudbertune* 1086 (DB). 'Southern grange or outlying farm'. OE *sūth* + *bere-tūn*.

Sodbury, Chipping & Sodbury, Old S. Glos. *Soppanbyrig* 872–915, *Sopeberie* 1086 (DB), *Cheping Sobbyri* 1269, *Olde Sobbury* 1346. 'Fortified place of a man called **Soppa*'. OE pers. name + *burh* (dative *byrig*). Distinguishing affixes from OE *cēping* 'market' and *ald* 'old'.

Soham Cambs. *Sægham c.*1000, *Saham* 1086 (DB). 'Homestead by a swampy pool'. OE **sǣge* + *hām*.

Soham, Earl & Soham, Monk Suffolk. *Saham* 1086 (DB). 'Homestead by a pool'. OE *sǣ* or **sā* + *hām*. Distinguishing affixes from early possession by the Earl of Norfolk and the monks of Bury St Edmunds.

Soho Gtr. London. *So Ho* 1632. A hunting cry, from the early association of this area with hunting. Another example of this name, **Soho** Somerset (near Vobster), is identical in origin, though legend has it that it recalls the use of the phrase as the Duke of Monmouth's password in the 1685 Rising.

Soldierstown Antrim. *Soldierstown* 1780. The name refers to a former military station here.

Solent, The Hants. *Soluente* 731. An ancient pre-English name of uncertain origin and meaning.

Solfach. See SOLVA.

Solihull Solhll. *Solihull* 12th cent. 'Muddy hill'. OE **sylig, *solig* + *hyll*. Alternatively the first element may be OE **sulig* 'pig-stye'.

Sollers Hope Herefs. See HOPE.

Sollom Lancs. *Solayn c.*1200. 'Muddy enclosure' from OScand. *sol* + **hegn*, or 'sunny slope' from OScand. *sól* + *hlein*.

Solva (*Solfach*) Pemb. *Saleuuach c.*1200. '(Place on the) River Solfach'. The river-name may mean 'poor one' (Welsh *salw*, 'ugly').

Solway Firth Cumbria. Dumf. *Sulewad* 1218. 'Estuary of the pillar ford'. OScand. *súl* + *vath* + *fjǫrthr*. The 'pillar' is the Lochmaben Stone marking the ford. Alternatively the first element may be OScand. *súla* 'solan goose'.

Somborne, King's, Somborne, Little, & Somborne, Upper Hants. *Swinburnan* 909, *Sunburne* 1086 (DB). 'Pig stream'. OE *swīn* + *burna*. Affix *King's* from its having been a royal estate.

Somerby, probably 'farmstead or village of a man called Sumarlithi', OScand. pers. name or *bý*: **Somerby** Leics. *Sumerlidebie* 1086 (DB). **Somerby** Lincs., near Brigg. *Sumertebi* 1086 (DB).

Somercotes, 'huts or cottages used in summer', OE *sumor* + *cot*: **Somercotes** Derbys. *Somercotes* 1276. **Somercotes, North & Somercotes, South** Lincs. *Summercotes* 1086 (DB).

Somerford, 'ford usable in summer', OE *sumor* + *ford*: **Somerford, Great & Somerford, Little** Wilts. *Sumerford* 937, *Sumreford* 1086 (DB). **Somerford Keynes** Glos. *Sumerford* 685, *Somerford Keynes* 1291. Manorial affix from the *de Kaynes* family, here in the 13th cent.

Somerleyton Suffolk. *Sumerledetuna* 1086 (DB). 'Farmstead of a man called Sumarlithi'. OScand. pers. name + OE *tūn*.

Somersal Herbert Derbys. *Summersale* 1086 (DB), *Somersale Herbert c.*1300. Possibly 'nook of land of a man called *Sumor'. OE pers. name + *halh*. Alternatively 'nook used in summer', from OE *sumor*. Manorial affix from the *Fitzherbert* family, here in the 13th cent.

Somersby Lincs. *Summerdebi* 1086 (DB). Probably 'farmstead or village of a man called Sumarlithi'. OScand. pers. name + *bý*.

Somerset (the county). *Sumersæton* 12th cent. '(District of) the settlers around Somerton'. Reduced form of SOMERTON + OE *sǣte*.

Somersham Cambs. *Summeresham c.*1000, *Sumersham* 1086 (DB). Probably 'homestead of a man called *Sumor or *Sunmær'. OE pers. name + *hām*.

Somersham Suffolk. *Sumersham* 1086 (DB). 'Homestead of a man called *Sumor'. OE pers. name + *hām*.

Somerton, usually 'farmstead used in summer', OE *sumor* + *tūn*: **Somerton** Norfolk. *Sumertonne* 1044–7, *Somertuna* 1086 (DB). **Somerton** Oxon. *Sumertone* 1086 (DB). **Somerton** Somerset. *Sumortun* 901–24, *Summertone* 1086 (DB).
  However the following has a different origin: **Somerton** Suffolk. *Sumerledetuna* 1086 (DB). 'Farmstead of a man called Sumarlithi'. OScand. pers. name + OE *tūn*.

Sompting W. Sussex. *Suntinga* 956, *Sultinges* 1086 (DB). '(Settlement of) the dwellers at the marsh'. OE **sumpt* + *-ingas*.

Sonning Wokhm. *Soninges* 1086 (DB). '(Settlement of) the family or followers of a man called *Sunna'. OE pers. name + *-ingas*. The same tribal group gives name to SUNNINGHILL and SUNNINGWELL.

Sopley Hants. *Sopelie* 1086 (DB). Possibly 'woodland clearing of a man called *Soppa'. OE pers. name + *lēah*. Alternatively the first element may be an OE **soppa* 'marsh'.

Sopworth Wilts. *Sopeworde* 1086 (DB). 'Enclosure of a man called *Soppa'. OE pers. name + *worth*.

Sorn E. Ayr. '(Place with a) kiln'. Gaelic *sorn*.

Sotby Lincs. *Sotebi* 1086 (DB). 'Farmstead or village of a man called Sóti'. OScand. pers. name + *bý*.

Sotterley Suffolk. *Soterlega* 1086 (DB). OE *lēah* 'woodland clearing' with an uncertain first element, possibly a pers. name.

Soudley, Lower & Soudley, Upper Glos. *Suleie* 1221. 'South woodland clearing'. OE *sūth* + *lēah*.

Soulbury Bucks. *Soleberie* 1086 (DB). 'Stronghold by a gully'. OE *sulh* + *burh* (dative *byrig*).

Soulby, 'farmstead of a man called Súl, or one made of posts', OScand. pers. name or *súl* + *bý*: **Soulby** Cumbria, near Brough. *Sulebi c.*1 60. **Soulby** Cumbria, near Penrith. *Suleby* 12 35.

Souldern Oxon. *Sulethorne c.*1160. 'Thorntree in or near a gully'. OE *sulh* + *thorn*.

Souldrop Beds. *Sultrop* 1196. 'Outlying farmstead near a gully'. OE *sulh* + *throp*.

Sourton Devon. *Swurantune c.*970, *Surintone* 1086 (DB). 'Farmstead by a neck of land or col'. OE *swēora* + *tūn*.

South as affix. *See* main name, e.g. for **South Acre** (Norfolk) *see* ACRE.

South Ronaldsay (island) Orkn. *Rögnvalsey c.*1150. 'Southern Rögnvaldr's island'. OScand. pers. name + *ey*. *Rögnvaldr* was the brother of Sigurd, first Jarl of Orkney in the late 9th cent. The island is *South* in relation to NORTH RONALDSAY, where the pers. name is different.

South Uist. *See* UIST.

Southall Gtr. London. *Suhaull* 1198, *Sudhale* 1204. 'Southern nook(s) of land'. OE *sūth* + *halh, see* NORTHOLT.

Southam, 'southern homestead or land in a river-bend', OE *sūth* + *hām* or *hamm*: **Southam** Glos. *Suth-ham c.*991, *Surham* [sic] 1086 (DB). **Southam** Warwicks. *Suthham* 998, *Sucham* [sic] 1086 (DB).

Southampton Sotn. *Homtun* 825, *Suthhamtunam* 962, *Hantone* 1086 (DB). 'Estate on a promontory'. OE *hamm* + *tūn*, with prefix *sūth* to distinguish this place from NORTHAMPTON (which has a different origin).

Southborough Kent. 'Borough of *Suth*' 1270, *la South Burgh* 1450. 'Southern borough' (of TONBRIDGE). OE *sūth* + *burh*.

Southbourne Bmouth. like Northbourne and Westbourne, a name of recent creation for a suburb of BOURNEMOUTH.

Southburgh Norfolk. *Berc* [sic] 1086 (DB), *Suthberg* 1291. 'South hill'. OE *sūth* + *beorg*.

Southchurch Sthend. *Suthcyrcan* 1042–66, *Sudcerca* 1086 (DB). 'Southern church'. OE *sūth* + *cirice*.

Southease E. Sussex. *Sueise* 966, *Suesse* 1086 (DB). 'Southern land overgrown with brushwood'. OE *sūth* + **hǣs(e)*.

Southend on Sea Sthend. *Sowthende* 1481. 'The southern end (of PRITTLEWELL parish)'. ME *south* + *ende*.

Southery Norfolk. *Suthereye* 942, *Sutreia* 1086 (DB). 'Southerly island'. OE *sūtherra* + *ēg*.

Southfleet Kent. *Suthfleotes* 10th cent., *Sudfleta* 1086 (DB). 'Southern place at the stream'. OE *sūth* + *flēot*.

Southgate Gtr. London. *Suthgate* 1370. 'Southern gate' (to Enfield Chase). OE *sūth* + *geat*.

Southill Beds. *Sudgiuele* 1086 (DB). 'Southern settlement of the tribe called *Gifle*'. OE *sūth* + old tribal name, *see* NORTHILL.

Southleigh Devon. *Lege* 1086 (DB), *Suthlege* 1228. '(Place at) the wood or clearing'. OE *lēah* with *sūth* to distinguish it from NORTHLEIGH.

Southminster Essex. *Suthmynster c.*1000, *Sudmunstra* 1086 (DB). 'Southern church'. OE *sūth* + *mynster*.

Southoe Cambs. *Sutham* [stc] 1086 (DB), *Sudho* 1186. 'Southern hill-spur'. OE *sūth* + *hōh*.

Southolt Suffolk. *Sudholda* 1086 (DB). 'South wood'. OE *sūth* + *holt*.

Southorpe Peterb. *Sudtorp* 1086 (DB). 'Southern outlying farmstead'. OScand. *súthr* + *thorp*.

Southowram Calder. *Overe* 1086 (DB), *Sudhoueram c.*1275. '(Place at) the ridges'. OE **ofer, *ufer* in a dative plural form **uferum*, with *sūth* to distinguish it from NORTHOWRAM.

Southport Sefton. a modern artificial name, first bestowed on the place in 1798.

Southrepps Norfolk. *Sutrepes* 1086 (DB), *Sutrepples* 1209. Probably 'south strips of land'. OE *sūth* + **reopul*.

Southrey Lincs. *Sutreie* 1086 (DB). 'Southerly island'. OE *sūtherra* + *ēg*.

Southrop Glos. *Sudthropa c.*1140. 'Southern outlying farmstead'. OE *sūth* + *throp*.

Southsea Portsm. *Southsea Castle c.*1600. Self-explanatory. The present place grew up round the castle built by Henry VIII in 1540.

Southwaite Cumbria. near Hesket. *Thouthweyt* 1272. 'Clay clearing'. OE *thōh* + OScand. *thveit*.

Southwark Gtr. London. *Sudwerca* 1086 (DB). 'Southern defensive work or fort'. OE *sūth* + *weorc*. Earlier called *Suthriganaweorc* 10th cent., 'fort of the men of SURREY'.

Southwell Notts. *Suthwellan* 958, *Sudwelle* 1086 (DB). 'South spring'. OE *sūth* + *wella*.

Southwick Dumf. *Suchayt c.*1280. *Suthayk* 1289. In spite of the earlier spelling, probably OE *sūth* + *wīc* 'southern dwelling or farm'.

Southwick, 'southern dwelling or (dairy) farm', OE *sūth* + *wīc*: **Southwick** Hants. *Sudwic c.*1140. **Southwick** Northants. *Suthwycan c.*980. **Southwick** Sundld. *Suthewich* 12th cent. **Southwick** W. Sussex. *Sudewic* 1073. **Southwick** Wilts. *Sudwich* 1196.

Southwold Suffolk. *Sudwolda* 1086 (DB). 'South forest'. OE *sūth* + *wald*.

Southwood Norfolk. *Sudwda* 1086 (DB). 'Southern wood'. OE *sūth* + *wudu*.

Sow (river) Somerset. *See* MIDDLEZOY.

Sowe (river) Covtry. *See* WALSGRAVE.

Sowerby, 'farmstead on sour ground', OScand. *saurr* + *bý*; examples include: **Sowerby** N. Yorks., near Thirsk. *Sorebi* 1086 (DB). **Sowerby**, **Sowerby Bridge** Calder. *Sorebi* 1086 (DB), *Sourebybrigge* 15th cent. The bridge is across the River Calder. **Sowerby, Brough** Cumbria. *Sowreby* 1235, *Soureby by Burgh* 1314. It is near to BROUGH. **Sowerby, Temple** Cumbria. *Sourebi* 1179, *Templessoureby* 1292. Affix from its early possession by the Knights Templar.

Sowton Devon. *Southton* 1420. 'South farmstead or village'. OE *sūth* + *tūn*. Its earlier name was *Clis* 1086 (DB), one of several places named from the River Clyst, *see* CLYST HYDON.

Soyland Calder. *Soland* 1274. 'Boggy cultivated land'. OE **sōh* 'bog, swamp' + *land*.

Spalding Lincs. *Spallinge* 1086 (DB), *Spaldingis c.*1115. '(Settlement of) the dwellers in *Spald*'. Old district name (possibly from OE **spald* 'ditch or trench') + *-ingas*.

Spaldington E. R. Yorks. *Spellinton* 1086 (DB). 'Farmstead of the tribe called *Spaldingas*' (who give name to SPALDING). OE tribal name (genitive plural *-inga-*) + *tūn*.

Spaldwick Cambs. *Spalduice* 1086 (DB). Possibly 'dwelling or farm by a trickling stream or ditch'. OE *spāld* or **spald* + *wīc*.

Spalford Notts. *Spaldesforde* 1086 (DB). Possibly 'ford over a trickling stream or ditch'. OE *spāld* or **spald* + *ford*.

Sparham Norfolk. *Sparham* 1086 (DB). 'Homestead or enclosure made with spars or shafts'. OE **spearr* + *hām* or *hamm*.

Sparkford Somerset. *Sparkeforda* 1086 (DB). 'Brushwood ford'. OE **spearca* + *ford*.

Sparkwell Devon. *Spearcanwille c.*1070, *Sperchewelle* 1086 (DB). 'Spring or stream where brushwood grows'. OE **spearca* + *wella*.

Sparsholt, 'wood of the spear' (perhaps referring to a spear-trap for wild animals or to a wood where spear-shafts are obtained), OE *spere* + *holt*: **Sparsholt** Hants. *Speoresholte* 901. Alternatively the first element of this name may be OE **spearr* 'a spar, a rafter'. **Sparsholt** Oxon. *Speresholte* 963, *Spersolt* 1086 (DB).

Spaunton N. Yorks. *Spantun* 1086 (DB). 'Farmstead with shingled roof'. OScand. *spánn* + OE *tūn*.

Spaxton Somerset. *Spachestone* 1086 (DB). Probably 'farmstead of a man called Spakr'. OScand. pers. name + OE *tūn*.

Speen W. Berks. *Spene* 821, *Spone* 1086 (DB). Probably 'place where wood-chips are found'. OE **spēne*. This OE name represents an adaptation of the older Latin name *Spinis* 'at the thorn bushes' (recorded in the 4th cent.).

Speenoge (*Spíonóg*) Donegal. '(Place of) gooseberries'.

Speeton N. Yorks. *Specton* 1086 (DB). Possibly 'enclosure where meetings are held'. OE *spēc* + *tūn*.

Speke Lpool. *Spec* 1086 (DB). Possibly OE *spēc* 'branches, brushwood'.

Speldhurst Kent. *Speldhirst* 8th cent. 'Wooded hill where wood-chips are found'. OE *speld* + *hyrst*.

Spellbrook Herts. *Spelebrok* 1287. 'Speech brook', i.e. perhaps 'brook by which meetings are held'. OE *spell* + *brōc*.

Spelsbury Oxon. *Speolesbyrig* early 11th cent., *Spelesberie* 1086 (DB). 'Stronghold of a man called **Spēol*'. OE pers. name + *burh* (dative *byrig*).

Spelthorne (district) Surrey. *Spelethorne* 1086 (DB). 'Thorn-tree where speeches are made'. A revival of an old hundred-name, so the name refers to meetings of the Hundred. OE *spell* + *thorn*.

Spen, from ME *spenne* 'a fence, an enclosure': **Spen** Kirkl. *Spen* 1308. This gives name to the modern district of **Spenborough**. **Spen, High** Gatesd. *Le Spen* 1312.

Spennithorne N. Yorks. *Speningetorp* [*sic*] 1086 (DB), *Speningthorn* 1184. Probably 'thorn-tree by the fence or enclosure'. OE or OScand. **spenning* + *thorn*.

Spennymoor Durham. *Spenningmore* 1292. Probably 'moor with a fence or enclosure'. OE or OScand. **spenning* + *mōr*.

Spetchley Worcs. *Spæclea* 967, *Speclea* 1086 (DB). 'Woodland clearing where meetings are held'. OE *spēc* + *lēah*.

Spetisbury Dorset. *Spehtesberie* 1086 (DB). 'Pre-English earthwork frequented by the green woodpecker'. OE **speoht* + *burh* (dative *byrig*).

Spexhall Suffolk. *Specteshale* 1197. 'Nook of land frequented by the green woodpecker'. OE **speoht* + *halh*.

Spey. *See* GRANTOWN-ON-SPEY.

Spiddal (*Spidéal*) Galway, Meath. 'Hospital'.

Spilsby Lincs. *Spilesbi* 1086 (DB). 'Farmstead or village of a man called **Spillir*'. OScand. pers. name + *bý*.

Spindlestone Northum. *Spindlestan* 1187. 'Stone or rock thought to resemble a spindle'. OE *spinele* + *stān*.

Spink (*Spinc*) Laois. 'Pinnacle'.

Spital in the Street Lincs. *Hospitale* 1204, *Spitelenthestrete* 1322. 'The hospital or religious house on the Roman road (Ermine Street)'. ME *spitel* (from OFrench *hospitale*) with OE *strǣt*.

Spithead Hants. not recorded before the 17th cent., 'headland of the sand-spit or pointed sandbank'. OE *spitu* + *hēafod*.

Spithurst E. Sussex. *Splytherst* 1296. 'Wooded hill with an opening or gap'. OE *hyrst* with ME *split*.

Spittaltown (*Baile an Ospidéil*) Westmeath. 'Town of the hospital'.

Spixworth Norfolk. *Spikesuurda* 1086 (DB). Probably 'enclosure of a man called **Spic*'. OE pers. name + *worth*.

Spofforth N. Yorks. *Spoford* [*sic*] 1086 (DB), *Spotford* 1218. 'Ford by a small plot of ground'. OE **spot* + *ford*.

Spondon Derby. *Spondune* 1086 (DB). 'Hill where wood-chips or shingles are got'. OE *spōn* + *dūn*.

Sporle Norfolk. *Sparlea* 1086 (DB). Probably 'wood or clearing where spars or shafts are obtained'. OE **spearr* + *lēah*.

Spratton Northants. *Spretone* 1086 (DB). Probably 'farmstead made of poles'. OE *sprēot* + *tūn*.

Spreakley Surrey. *Sprakele* 1225. 'Wood or woodland clearing characterized by twigs or shoots'. OE *sprǣc* + *lēah*.

Spreyton Devon. *Spreitone* 1086 (DB). 'Farmstead amongst brushwood'. OE **sprǣg* + *tūn*.

Spridlington Lincs. *Spredelintone* 1086 (DB). Possibly 'estate associated with a man called **Sprēotel* or **Sprin(g)del*'. OE pers. name + -*ing*- + *tūn*.

Springthorpe Lincs. *Springetorp* 1086 (DB). 'Outlying farmstead by a spring'. OE *spring* + OScand. *thorp*.

Sproatley E. R. Yorks. *Sprotele* 1086 (DB). 'Woodland clearing where young shoots grow'. OE *sprota* + *lēah*.

Sproston Green Ches. *Sprostune* 1086 (DB). 'Farmstead of a man called Sprow'. OE pers. name + *tūn*.

Sprotbrough Donc. *Sproteburg* 1086 (DB). 'Stronghold of a man called **Sprota*', or 'stronghold overgrown with shoots'. OE pers. name or *sprota* + *burh*.

Sproughton Suffolk. *Sproeston* 1191. Probably 'farmstead of a man called Sprow'. OE pers. name + *tūn*.

Sprowston Norfolk. *Sprowestuna* 1086 (DB). 'Farmstead of a man called Sprow'. OE pers. name + *tūn*.

Sproxton, probably 'farmstead of a man called **Sprok*', OScand. pers. name + OE *tūn*: **Sproxton** Leics. *Sprotone* [*sic*] 1086 (DB), *Sproxcheston* c.1130. **Sproxton** N. Yorks. *Sprostune* 1086 (DB).

Spurn Head E. R. Yorks. *Ravenserespourne* 1399. From ME **spurn* 'a spur, a projecting piece of land', in the early spelling with the place name *Ravenser* which means 'sandbank of a man called Hrafn', OScand. pers. name + *eyrr*.

Spurstow Ches. *Spuretone* [*sic*] 1086 (DB), *Sporstow* c.1200. 'Meeting-place on a trackway, or by a spur of land'. OE *spor* or *spura* + *stōw*.

Srah (*An tSraith*) Mayo. 'The river-meadow'.

Srahmore (*An Srath Mór*) Mayo. 'The big river-meadow'.

Srue (*Sruth*) Galway. 'Stream'.

Sruh (*Sruth*) Waterford. 'Stream'.

Stackallan (*Stigh Colláin*) Meath. 'Collán's house'.

Stackpole Elidor. *See* CHERITON.

Stacks of Duncansby (rock group) Highland. 'Steep rocks off Duncansby Head'. OScand. *stakkr*. *See* DUNCANSBY HEAD.

Stadhampton Oxon. *Stodeham* c.1135. 'Enclosure or river-meadow where horses are kept'. OE *stōd* + *hamm*. The *-ton* is a late addition.

Staffa (island) Arg. 'Pillar island'. OScand. *stafr* + *ey*. The reference is to the columns of basaltic rock.

Staffield Cumbria. *Stafhole* c.1225. 'Round hill with a staff or post'. OScand. *stafr* + *hóll*.

Stafford Staffs. *Stæfford* mid-11th cent., *Stadford* 1086 (DB). 'Ford by a landing-place'. OE *stæth* + *ford*. **Staffordshire** (OE *scīr* 'district') is first referred to in the 11th cent.

Stafford, West Dorset. *Stanford* 1086 (DB). 'Stony ford'. OE *stān* + *ford*.

Staffordstown Antrim. The hamlet is named after Martha *Stafford*, daughter of Sir Francis Stafford, who in the early 17th cent. married Sir Henry O'Neill, son of Shane O'Neill of Shane's Castle.

Stagsden Beds. *Stachedene* 1086 (DB). 'Valley of stakes or boundary posts'. OE *staca* + *denu*.

Stainburn N. Yorks. *Stanburne* c.972, *Sta(i)nburne* 1086 (DB). 'Stony stream'. OE *stān* (replaced by OScand. *steinn*) + *burna*.

Stainby Lincs. *Stigandebi* 1086 (DB). 'Farmstead or village of a man called Stígandi'. OScand. pers. name + *bý*.

Staincross Barns. *Staincros* 1086 (DB). 'Stone cross'. OScand. *steinn* + *kros*.

Staindrop Durham. *Standropa* c.1040. Probably 'valley with stony ground'. OE *stǣner* + *hop*.

Staines Surrey. *Stane* 11th cent., *Stanes* 1086 (DB). '(Place at) the stone(s)'. OE *stān*.

Stainfield Lincs. near Bardney. *Steinfelde* 1086 (DB). 'Stony open land'. OE *stān* (replaced by OScand. *steinn*) + *feld*.

Stainfield Lincs. near Rippingale. *Stentvith* 1086 (DB). 'Stony clearing'. OScand. *steinn* + *thveit*.

Stainforth, 'stony ford', OE *stān* (replaced by OScand. *steinn*) + *ford*: **Stainforth** Donc. *Steinforde* 1086 (DB). **Stainforth** N. Yorks. *Stainforde* 1086 (DB).

Staining Lancs. *Staininghe* 1086 (DB). Possibly '(settlement of) the family or followers of a man called *Stān, or the dwellers in a stony district'. OE pers. name or *stān* (with Scand. influence) + *-ingas*. Alternatively 'stony place', OE *stāning.

Stainland Calder. *Stanland* 1086 (DB). 'Stony cultivated land'. OE *stān* (replaced by OScand. *steinn*) + *land*.

Stainley, 'stony woodland clearing', OE *stān* (replaced by OScand. *steinn*) + *lēah*: **Stainley, North** N. Yorks. *Stanleh* c.972, *Nordstanlaia* 1086 (DB). **Stainley, South** N. Yorks. *Stanlai* 1086 (DB), *Southstainlei* 1198.

Stainmore, North Cumbria. *Stanmoir* c.990. 'Rocky or stony moor'. OE *stān* (replaced by OScand. *steinn*) + *mōr*.

Stainsacre N. Yorks. *Stainsaker* 1090–6. 'Cultivated land of a man called Steinn'. OScand. pers. name + *akr*.

Stainton, a frequent name in the North, usually 'farmstead on stony ground', OE *stān* (replaced by OScand. *steinn*) + *tūn*; examples include: **Stainton** Donc. *Stantone* 1086 (DB). **Stainton** Durham. *Stantun* c.1040. **Stainton, Market** Lincs. *Staintone* 1086 (DB), *Steynton Market* 1286. Affix is ME *merket*. **Stainton le Vale** Lincs. *Staintone* 1086 (DB). Affix means 'in the valley' from ME *vale*.

However the following has a different origin: **Stainton, Great & Stainton, Little** Darltn. *Staninctona* 1091. 'Farmstead at the stony place'. OE *stāning (influenced by OScand. *steinn*) + *tūn*.

Staintondale N. Yorks. *Steintun* 1086 (DB), *Staynton Dale* 1562. 'Farmstead on stony ground'. OE *stān* (replaced by OScand. *steinn*) + *tūn*, with the later addition of *dale* 'valley'.

Stair S. Ayr. '(Place by) stepping stones'. Gaelic *stair*.

Staithes N. Yorks. *Setonstathes* 1415. 'The landing-places'. OE *stæth*. Seaton ('sea farmstead') is a nearby place.

Stalbridge Dorset. *Stapulbrige* 860–6, *Stapplebrige* 1086 (DB). 'Bridge built on posts or piles'. OE *stapol* + *brycg*.

Stalham Norfolk. *Stalham* 1044–6, 1086 (DB). 'Homestead or enclosure by the fishing pool'. OE *stall* + *hām* or *hamm*.

Stalisfield Green Kent. *Stanefelle* [*sic*] 1086 (DB), *Stealesfelde* c.1100. Probably 'open land with a stall or stable'. OE *steall* + *feld*.

Stalling Busk N. Yorks. *Stalunesbusc* 1218. 'The stallion's bush'. ME *stalun* + OScand. *buski*.

Stallingborough NE. Lincs. *Stalingeburg* 1086 (DB). Possibly 'stronghold of the family or followers of a man called *Stalla*'. OE pers. name + *-inga-* + *burh*. Or the first element may be an *-ing* derivative of OE *stall* 'cattle stall, fishing pool' or *stalu* 'stem, post'.

Stalmine Lancs. *Stalmine* 1086 (DB). Probably 'mouth of the stream or pool'. OE *stæll* + OScand. *mynni*.

Stalybridge Tamesd. *Stauelegh* 13th cent., *Stalybridge* 1687. 'Bridge at the wood where staves are got'. OE *stæf* + *lēah*, with the later addition of *bridge*.

Stambourne Essex. *Stanburna* 1086 (DB). 'Stony stream'. OE *stān* + *burna*.

Stambridge, Great Essex. *Stanbruge* 1086 (DB). 'Stone bridge'. OE *stān* + *brycg*.

Stamford, 'stone ford' or 'stony ford', OE *stān* + *ford*: **Stamford** Lincs. *Steanford* 10th cent., *Stanford* 1086 (DB). **Stamford Bridge** E. R. Yorks. *Stanford brycg* c.1075. The ford here was replaced by a bridge at an early date.

Stamford Hill Gtr. London. *Saundfordhull* 1294. 'Hill by a sandy ford'. OE *sand* + *ford* + *hyll*. *Stam-* only from 1675.

Stamfordham Northum. *Stanfordham* 1188. 'Homestead or village at the stone ford'. OE *stān* + *ford* + *hām*.

Stamullen (*Steach Maoilín*) Meath. 'Maoilín's house'.

Stanbridge Beds. *Stanbrugge* 1165. 'Stone bridge'. OE *stān* + *brycg*.

Standen Kent. *Stankyndenn* 1334. 'The stony woodland pasture'. OE *stāniht* (dative *-an*) + *denn*.

Standish Wigan. *Stanesdis* 1178. 'Stony pasture or enclosure'. OE *stān* + *edisc*.

Standlake Oxon. *Stanlache* c.1155. 'Stony stream or channel'. OE *stān* + *lacu*.

Standon 'stony hill', OE *stān* + *dūn*: **Standon** Herts. *Standune* 944–6, *Standone* 1086 (DB). **Standon** Staffs. *Stantone* [*sic*] 1086 (DB), *Standon* 1190.

Stanfield Norfolk. *Stanfelda* 1086 (DB). 'Stony open land'. OE *stān* + *feld*.

Stanford, found in various counties, 'stone ford' or 'stony ford', OE *stān* + *ford*; examples include: **Stanford** Kent. *Stanford* 1035. **Stanford Bishop** Herefs. *Stanford* 1086 (DB). Affix from its early possession by the Bishop of Hereford. **Stanford Dingley** W. Berks. *Stanworde* 1086 (DB), *Staneford Deanly* 1535. Manorial affix from the *Dyngley* family, here in the 15th cent. **Stanford in the Vale** Oxon. *Stanford* 1086 (DB), *Stanford in le Vale* 1496. Affix from its situation in the VALE OF WHITE HORSE. **Stanford le Hope** Thurr. *Staunford* 1267, *Stanford in the Hope* 1361. Affix means 'in the bay' from ME *hope*. **Stanford Rivers** Essex. *Stanford* 1068, *Stanfort* 1086 (DB), *Stanford Ryueres* 1289. Manorial affix from the *Rivers* family, here in the 13th cent.

Stanghow Red. & Cleve. *Stanghou* 1280. 'Mound or tumulus marked by a post or pole'. OScand. *stong* + *haugr*.

Stanhoe Norfolk. *Stanhou* 1086 (DB). 'Stony hill-spur'. OE *stān* + *hōh*.

Stanhope Durham. *Stanhopa* c.1170. 'Stony valley'. OE *stān* + *hop*.

Stanion Northants. *Stanere* [*sic*] 1086 (DB), *Stanerna* 1162. 'Stone house(s) or building(s)'. OE *stān* + *ærn*.

Stanley, 'stony woodland clearing', OE *stān* + *lēah*; examples include: **Stanley** Derbys.

Stanlei 1086 (DB). **Stanley** Durham. *Stanley* 1228. **Stanley** Staffs. *Stanlega* 1130. **Stanley** Wakefd. *Stanlei* 1086 (DB). **Stanley, King's & Stanley, Leonard** Glos. *Stanlege* 1086 (DB), *Kingestanleg* 1220. *Stanllegh Leonardi* 1285. Affixes from ownership by the Crown and from the former dedication of the church to St Leonard.

Stanmer Bright. & Hove. *Stanmere* 765, 1086 (DB). 'Stony pool'. OE *stān* + *mere*.

Stanmore Gtr. London. *Stanmere* 1086 (DB). Identical in origin with the previous name.

Stanney Ches. *Stanei* 1086 (DB). 'Stone or rock island'. OE *stān* + *ēg*.

Stanningfield Suffolk. *Stanfelda* 1086 (DB), *Stanefeld* 1197. 'Stony open land'. OE *stān* (perhaps alternating with **stānen* 'stony') + *feld*.

Stanningley Leeds. *Stannyngley* 1562. Probably 'stony woodland clearing'. OE **stāning* 'stony place' or **stānen* 'stony' + *lēah*.

Stannington Northum. *Stanigton* 1242. 'Farmstead at the stony place'. OE **stāning* + *tūn*.

Stansfield Suffolk. *Stanesfelda* 1086 (DB). Probably 'open land of a man called **Stān*'. OE pers. name + *feld*.

Stanstead, Stansted, 'stony place', OE *stān* + *stede*: **Stanstead** Suffolk. *Stanesteda* 1086 (DB). **Stanstead Abbots** Herts. *Stanestede* 1086 (DB), *Stanstede Abbatis de Wautham* 1247. Affix from early possession by the Abbot of Waltham. **Stansted** Kent. *Stansted* 1231. **Stansted Mountfitchet** Essex. *Stanesteda* 1086 (DB), *Stansted Mounfichet* c.1290. Manorial affix from the *Muntfichet* family, here in the 12th cent.

Stanton, a common name, usually 'farmstead on stony ground', occasionally 'farmstead near standing stones', OE *stān* + *tūn*; examples include: **Stanton** Glos. *Stantone* 1086 (DB). **Stanton** Suffolk. *Stantuna* 1086 (DB). **Stanton Drew** B. & NE. Som. *Stantone* 1086 (DB), *Stanton Drogonis* 1253. Manorial affix from possession by one *Drogo* or *Drew* in 1225. **Stanton Harcourt** Oxon. *Stantone* 1086 (DB), *Stantone Harecurt* c.1275. Manorial affix from the *de Harecurt* family, here in the 12th cent. **Stanton, Long** Cambs. *Stantune* 1086 (DB), *Long Stanton* 1282. Affix refers to the length of the village. **Stanton St Quintin** Wilts. *Stantone* 1086 (DB), *Staunton St Quintin*

1283. Manorial affix from the *de Sancto Quintino* family, here in the 13th cent. **Stanton upon Hine Heath** Shrops. *Stantune* 1086 (DB), *Staunton super Hyne Heth* 1327. Affix means 'heath of the household servants', ME *hine* + *hethe*.

Stanwardine Shrops. *Staurdine* 1086 (DB), *Stanwardin* 1194. 'Enclosure made of stones, or on stony ground'. OE *stān* + *worthign*.

Stanway, 'stony road', OE *stān* + *weg*; examples include: **Stanway** Essex. *Stanwægun* c.1000, *Stanwega* 1086 (DB). **Stanway** Glos. *Stanwege* 12th cent.

Stanwell Surrey. *Stanwelle* 1086 (DB). 'Stony spring or stream'. OE *stān* + *wella*.

Stanwick Northants. *Stanwigga* 10th cent., *Stanwige* 1086 (DB). 'The rocking- or logan-stone'. OE *stān* + *wigga*.

Stapeley Ches. *Steple* [sic] 1086 (DB), *Stapeleg* 1260. Probably 'wood or clearing where posts are got'. OE *stapol* + *lēah*.

Stapenhill Staffs. *Stapenhille* 1086 (DB). '(Place at) the steep hill'. OE *stēap* (dative -*an*) + *hyll*.

Staple, '(place at) the pillar of wood or stone', OE *stapol*: **Staple** Kent. *Stapel* 1240. **Staple Fitzpaine** Somerset. *Staple* 1086 (DB). Manorial affix from the *Fitzpaine* family, here in the 14th cent.

Stapleford, 'ford marked by a post', OE *stapol* + *ford*; examples include: **Stapleford** Cambs. *Stapelforda* 956, *Stapleforde* 1086 (DB). **Stapleford** Notts. *Stapleford* 1086 (DB). **Stapleford Abbots & Stapleford Tawney** Essex. *Staplefort* 1086 (DB), *Staplford Abbatis Sancti Edmundi* 1255, *Stapilford Thany* 1291. Distinguishing affixes from early possession by the Abbot of Bury St Edmunds and by the *de Tany* family.

Staplegrove Somerset. *Stapilgrove* 1327. 'Grove where posts are got'. OE *stapol* + *grāf*.

Staplehurst Kent. *Stapelherst* 1226. 'Wooded hill where posts are got'. OE *stapol* + *hyrst*.

Stapleton, usually 'farmstead by a post or built on posts', OE *stapol* + *tūn*; examples include: **Stapleton** Brist. *Stapleton* 1215. **Stapleton** Leics. *Stapletone* 1086 (DB). However the following probably mean 'farmstead by a steep slope', OE *stēpel* + *tūn*:

Stapleton Herefs. *Stepeltone* 1286.
Stapleton Shrops. *Stepleton* 12th cent.

Staploe Beds. *Stapelho* 1203. 'Hill-spur marked by, or shaped like, a post'. OE *stapol + hōh*.

Starbeck N. Yorks. not recorded before 1817, probably 'sedge brook', from OScand. *star + bekkr*.

Starbotton N. Yorks. *Stamphotne [sic]* 1086 (DB), *Stauerboten* 12th cent. 'Valley where stakes are got'. OE **stæfer* (replacing OScand. *stafn* in the first form) + OScand. *botn*.

Starcross Devon. first recorded as *Star Crosse* 1689, probably referring to a cross or cross-roads where starlings gather, from dialect *stare* (OE *stær*) and *cross*.

Starston Norfolk. *Sterestuna* 1086 (DB). 'Farmstead or village of a man called Styrr'. OScand. pers. name + OE *tūn*.

Start Point Devon. *La Sterte* 1310. OE *steort* 'tongue of land, promontory'.

Startforth Durham. *Stretford c.*1050, *Stradford* 1086 (DB). 'Ford on a Roman road'. OE *strǣt + ford*.

Startley Wilts. *Stercanlei* 688, *Sterkele* 1249. 'The dense wood'. OE *stearc* 'stiff, hard' + *lēah*.

Stathe Somerset. *Stathe* 1233. 'The landing-place'. OE *stæth*.

Stathern Leics. *Stachedirne* 1086 (DB). 'Stake thorn-tree', i.e. 'thorn-tree marking a boundary'. OE *staca + thyrne*.

Staughton, Great (Cambs.) & **Staughton, Little** (Beds.). *Stoctun c.*1000, *Tochestone, Estone [sic]* 1086 (DB). 'Farmstead at an outlying hamlet'. OE *stoc + tūn*.

Staunton, usually 'farmstead on stony ground, or one near a standing stone', OE *stān + tūn*; examples include: **Staunton** Glos., near Hartpury. *Stantun* 972, 1086 (DB). **Staunton on Arrow** Herefs. *Stantun* 958, *Stantune* 1086 (DB). Arrow is a Celtic or pre-Celtic river-name (probably meaning 'swift').
   However the following has a different origin: **Staunton on Wye** Herefs. *Standune* 1086 (DB). 'Stony hill'. OE *stān + dūn*. For the river-name, *see* ROSS ON WYE.

Staveley, 'wood or clearing where staves are got', OE *stæf + lēah*: **Staveley** Cumbria. *Staueleya c.*1200. **Staveley** Derbys. *Stavelie*

1086 (DB). **Staveley** N. Yorks. *Stanlei [sic]* 1086 (DB), *Staflea* 1167. **Staveley-in-Cartmel** Cumbria. *Stavelay* 1282. *See* CARTMEL.

Staverton, usually 'farmstead made of or marked by stakes', OE *stæfer + tūn*: **Staverton** Glos. *Staruenton* 1086 (DB), *Staverton* 1248. **Staverton** Northants. *Stæfertun* 944, *Stavertone* 1086 (DB). **Staverton** Wilts. *Stavretone* 1086 (DB).
   However the following has a different origin: **Staverton** Devon. *Stofordtune c.*1070, *Stovretona* 1086 (DB). 'Farmstead by a stony ford.' OE *stān + ford + tūn*.

Stawell Somerset. *Stawelle* 1086 (DB). 'Stony spring or stream'. OE *stān + wella*.

Staxton N. Yorks. *Stacstone* 1086 (DB). 'Farmstead or village of a man called *Stakkr'. OScand. pers. name + OE *tūn*.

Stearsby N. Yorks. *Stirsbi* 1086 (DB). 'Farmstead or village of a man called Styrr'. OScand. pers. name + *bý*.

Stebbing Essex. *Stibinga* 1086 (DB). '(Settlement of) the family or followers of a man called *Stybba, or the dwellers among the tree-stumps'. OE pers. name or *stybb + -ingas*.

Stedham W. Sussex. *Steddanham* 960, *Stedeham* 1086 (DB). 'Homestead or enclosure of the stallion, or of a man called *Stedda'. OE *stēda* or pers. name + *hām* or *hamm*.

Steep Hants. *Stepe* 12th cent. 'The steep place'. OE **stīepe*.

Steeping, Great & Steeping, Little Lincs. *Stepinge* 1086 (DB). Probably '(settlement of) the family or followers of a man called Stēapa'. OE pers. name + *-ingas*.

Steeple, 'steep place', OE *stēpel*: **Steeple** Dorset. *Stiple* 1086 (DB). **Steeple** Essex. *Stepla* 1086 (DB).

Steeple as affix. *See* main name, e.g. for **Steeple Ashton** (Wilts.) *see* ASHTON.

Steeton Brad. *Stiuetune* 1086 (DB). Probably 'farmstead built of or amongst tree-stumps'. OE **styfic + tūn*.

Stella Gatesd. *Stellinglei c.*1145. 'Woodland clearing or pasture at the place for catching fish'. OE **stelling + lēah*.

Stelling Minnis Kent. *Steallinge c.*1090, *Stellinges* 1086 (DB). Possibly '(settlement of)

the family or followers of a man called *Stealla'.
OE pers. name + -ingas. Alternatively perhaps
from an OE *stealling 'a stall for cattle'. Affix is
from OE mænnes 'common land'.

Stemster Highland. Stambustar 1557.
'Farmstead with stones'. OScand. steinn +
bólstathr.

Stenhousemuir Falk. Stan house c.1200,
Stenhous 1601. 'Moorland by the stone house'.
OE stān + hūs, with the later addition of Scottish
muir 'moor'.

Stenness Shet. Steinsness c.970. 'Stone
headland'. OScand. steinn + nes.

Stenton E. Loth. Steinton 1150. 'Stone
village'. OE stān + tūn.

Stepaside Pemb. Stepaside 1694. The village
would have arisen by a wayside cottage or
modest alehouse inviting the traveller to 'step
aside' for rest and refreshment.

Stepney Gtr. London. Stybbanhythe c.1000,
Stibenhede 1086 (DB). Possibly 'landing-place of
a man called *Stybba'. OE pers. name (genitive
-n) + hȳth. Alternatively the first element may be
an OE *stybba (genitive -n) 'stump, pile' with
reference to the construction of the landing-
place.

Steppingley Beds. Stepigelai 1086 (DB).
'Woodland clearing of the family or followers of
a man called Stēapa'. OE pers. name + -inga- +
lēah.

Sternfield Suffolk. Sternesfelda 1086 (DB).
Possibly 'open land of a man called *Sterne'. OE
pers. name + feld.

Stert Wilts. Sterte 1086 (DB). 'Projecting piece
of land'. OE steort.

Stetchworth Cambs. Steuicheswrthe c.1050,
Stiuicesuuorde 1086 (DB). 'Enclosure amongst
the tree-stumps, or of a man called *Styfic'. OE
*styfic or pers. name + worth.

Stevenage Herts. Stithenæce c.1060,
Stigenace 1086 (DB). Probably '(place at) the
strong oak-tree'. OE stīth (dative -an) + āc
(dative āec).

Stevenston N. Ayr. Stevenstoun 1246.
'Steven's farm'. OE tūn.

Steventon, 'estate associated with a man
called *Stīf(a)', OE pers. name + -ing- + tūn, or
possibly 'farmstead at the tree-stump place', OE
*styf(ic)ing + tūn: **Steventon** Hants. Stivetune

1086 (DB). **Steventon** Oxon. Stivetune
1086 (DB).

Stevington Beds. Stiuentone 1086 (DB).
Probably identical in origin with the previous
two names.

Stewartby Beds. a recent name for a 'new
village' built in 1935 for employees in the brick-
making industry, named from Halley Stewart,
chairman of a local brick company.

Stewarton E. Ayr. Stewartoun 1201.
'Steward's estate'. OE tūn. The reference is to
Walter, Seneschal (High Steward) to King David
I, who owned the estate in the 12th cent.

Stewartstown Tyrone. Stewartoune c.1655.
Sir Andrew Stewart was granted land here in
1698 on which he built a castle. The Irish name
of Stewartstown is An Chraobh, 'the branch'.

Stewkley Bucks. Stiuelai [sic] 1086 (DB),
Stiuecelea 1182. 'Woodland clearing with tree-
stumps'. OE *styfic + lēah.

Stewton Lincs. Stivetone 1086. Probably
'farmstead built of or amongst tree-stumps'. OE
*styfic + tūn.

Steyning W. Sussex. Stæningum c.880,
Staninges 1086 (DB). Possibly '(settlement of)
the family or followers of a man called *Stān, or
the dwellers at the stony place'. OE pers. name
or OE stān or *stæne + -ingas. Alternatively
'stony places', from the plural of an OE word
*stāning or *stæning.

Steynton Pemb. Steinton 1291. 'Farm built of
stone'. OScand. steinn + OE tūn.

Stibbard Norfolk. Stabyrda [sic] 1086 (DB),
Stiberde 1202. 'Bank beside a path, a road-side'.
OE stīg + *byrde.

Stibbington Cambs. Stebintune 1086 (DB).
Probably 'estate associated with a man called
*Stybba'. OE pers. name + -ing- + tūn.
Alternatively the first element may be an OE
*stybbing 'tree-stump clearing'.

Stichill Sc. Bord. Stichele c.1170. OE *sticele
'steep place'.

Stickford Lincs. Stichesforde 1086 (DB),
Stikeford 1185. Possibly 'ford by the long strip of
land', OE sticca + ford. Alternatively perhaps
'sticky (i.e. muddy) ford', from the OE adjective
sticce.

Sticklepath Devon. Stikelepethe 1280. 'Steep
path'. OE sticol + pæth.

Stickney Lincs. *Stichenai* 1086 (DB). Possibly 'long strip of land between streams'. OE *sticca* (genitive *-n*) + *ēg*. Alternatively perhaps 'sticky (i.e muddy) raised ground in the marsh', from the OE adjective *Sticce* (dative *-an*) + *ēg*.

Stiffkey Norfolk. *Stiuekai* 1086 (DB). 'Island, or dry ground in marsh, with tree-stumps'. OE **styfic* + *ēg*.

Stifford, North Thurr. *Stithforde* c.1090, *Stiforda* 1086 (DB). Probably 'ford where lamb's cress or nettles grow'. OE *stīthe* + *ford*.

Stillingfleet N. Yorks. *Steflingefled* 1086 (DB). 'Stretch of river belonging to the family or followers of a man called **Stȳfel*'. OE pers. name + *-inga-* + *flēot*.

Stillington, probably 'estate associated with a man called **Stȳfel*', OE pers. name + *-ing-* + *tūn*: **Stillington** N. Yorks. *Stiuelinctun* 1086 (DB). **Stillington** Stock. on T. *Stillingtune* c.1190.

Stillorgan (*Stigh Lorgan*) Dublin. 'Lorcan's house'.

Stilton Cambs. *Stichiltone* 1086 (DB). 'Farmstead or village at a stile or steep ascent'. OE *stigel* + *tūn*.

Stinchcombe Glos. *Stintescombe* c.1155. 'Valley frequented by the sandpiper or dunlin'. OE **stint* + *cumb*.

Stinsford Dorset. *Stincteford* 1086 (DB). 'Ford frequented by the sandpiper or dunlin'. OE **stint* + *ford*.

Stirchley Tel. & Wrek. *Styrcleage* 1002. 'Clearing or pasture for young bullocks'. OE *styrc* + *lēah*.

Stirling Aber. *Strevelin* 1124, *Sterling* c.1470. Meaning uncertain. The name may have originally been that of the river, now the Forth, on which Stirling stands.

Stisted Essex. *Stistede* 1086 (DB), *Stidsted* 1198. Probably 'place where lamb's cress or nettles grow'. OE *stīthe* + *stede*.

Stithians Cornwall. *Sancta Stethyana* 1268. From the dedication of the church to St Stithian, a female saint.

Stittenham, High N. Yorks. *Stidnun* [sic] 1086 (DB), *Stiklum* c.1260. Probably '(place at) the steep ascents'. OE **sticel(e)* in a dative plural form **sticelum*.

Stivichall Covtry. *Stivichall* c.1144. 'Nook of land with tree-stumps'. OE **styfic* + *halh*.

Stixwould Lincs. *Stigeswalde* 1086 (DB). 'Forest of a man called Stígr'. OScand. pers. name + OE *wald*.

Stob Dubh (mountain) Arg. 'Black stump'. Gaelic *stob* + *dubh*.

Stoborough Dorset. *Stanberge* 1086 (DB). 'Stony hill or barrow'. OE *stān* + *beorg*.

Stock Essex. *Herewardestoc* 1234, *Stocke* 1337. 'Outlying farmstead or hamlet of a man called Hereweard'. OE pers. name + *stoc*. The longer name remained in use until the 17th cent.

Stock Green & Stock Wood Worcs. *La Stokke* 1271. 'The tree-stump'. OE *stocc*.

Stockbridge Hants. *Stocbrugge* 1221. 'Bridge made of logs or tree-trunks'. OE *stocc* + *brycg*. The same name occurs in other counties, *see* following names.

Stockbridge Edin. 'Bridge of tree-trunks'. OE *stocc* + *brycg*.

Stockbriggs S. Lan. 'Bridge of tree-trunks'. OE *stocc* + *brycg*.

Stockbury Kent. *Stochingeberge* 1086 (DB). 'Woodland pasture of the dwellers at the outlying farmstead'. OE *stoc* + *-inga-* + *bǣr*.

Stockerston Leics. *Stoctone* [sic] 1086 (DB), *Stocfaston* c.1130. 'Stronghold built of logs'. OE *stocc* + *fæsten*.

Stocking Pelham Herts. *See* PELHAM.

Stockingford Warwicks. *Stoccingford* 1157. 'Ford by the tree-stump clearing'. OE **stoccing* + *ford*.

Stockland, 'cultivated land of the outlying farmstead', OE *stoc* + *land*: **Stockland** Devon. *Stocland* 998. **Stockland Bristol**. Somerset. *Stocheland* 1086 (DB). Manorial affix from its possession by the chamber of Bristol.

Stockleigh English & Stockleigh Pomeroy Devon. *Stochelie* 1086 (DB), *Stokeley Engles* 1268, *Stokelegh Pomeray* 1261. Probably 'woodland clearing with tree-stumps'. OE *stocc* + *lēah*. Manorial affixes from early possession by the *Engles* and *de Pomerei* families.

Stockport Stockp. *Stokeport* c.1170. 'Market-place at an outlying hamlet'. OE *stoc* + *port*.

Stocksbridge Sheff. not on record before 1841, probably identical in origin with STOCKBRIDGE.

Stocksfield Northum. *Stokesfeld* 1242. 'Open land belonging to an outlying hamlet'. OE *stoc* + *feld*.

Stockton, a name found in several counties, probably 'farmstead at an outlying hamlet', from OE *stoc* + *tūn*, although some may be 'farmstead built of logs', OE *stocc* + *tūn*; examples include: **Stockton** Warwicks. *Stocton* 1249. **Stockton Heath** Warrtn. *Stocton* c.1200, *Stoaken Heath* 1682. **Stockton on Tees** Stock. on T. *Stocton* 1196. Tees is a Celtic or pre-Celtic river-name, possibly meaning 'surging river'.

Stockwell Gtr. London. *Stokewell* 1197. 'Spring or stream by a tree-stump'. OE *stocc* + *wella*.

Stockwith, East (Lincs.) & **Stockwith, West** (Notts.). *Stochithe* 12th cent. 'Landing-place made of logs or tree-stumps'. OE *stocc* + *hŷth*.

Stodmarsh Kent. *Stodmerch* 675. 'Marsh where a herd of horses is kept'. OE *stōd* + *mersc*.

Stoford, 'stony ford', OE *stān* + *ford*: **Stoford** Somerset, near Yeovil. *Stafford* 1225. **Stoford** Wilts. *Stanford* 943.

Stogumber Somerset. *Stoke Gunner* 1225. 'Outlying farmstead or hamlet'. OE *stoc*, with pers. name or surname *Gumer* of an early owner.

Stogursey Somerset. *Stoche* 1086 (DB), *Stok Curcy* 1212. 'Outlying farmstead or hamlet'. OE *stoc*, with manorial affix from the *de Curci* family, here in the 12th cent.

Stoke, a very common name, from OE *stoc* 'outlying farmstead or hamlet, secondary settlement'; examples include: **Stoke Bruerne** Northants. *Stoche* 1086 (DB), *Stokbruer* 1254. Manorial affix from the *Briwere* family, here in the 13th cent. **Stoke-by-Nayland** Suffolk. *Stoke* c.950, *Stokes* 1086 (DB), *Stokeneylond* 1272. See NAYLAND. **Stoke Climsland** Cornwall. *Stoke* 1266, *Stok in Clymeslond* 1302. Affix is from OE *land* 'estate' with an obscure first element. **Stoke D'Abernon** Surrey. *Stoche* 1086 (DB), *Stokes de Abernun* 1253. Manorial affix from the *de Abernun* family, here in the 12th cent. **Stoke Ferry** Norfolk. *Stoches* 1086 (DB), *Stokeferie*

1248. Affix is OScand. *ferja*, referring to a ferry over the River Wissey. **Stoke Fleming** Devon. *Stoc* 1086 (DB), *Stokeflemeng* 1270. Manorial affix from the family of *le Flemeng*, here in the 13th cent. **Stoke Gabriel** Devon. *Stoke* 1307, *Stokegabriel* 1309. Affix from the dedication of the church to St Gabriel. **Stoke Gifford** S. Glos. *Stoche* 1086 (DB), *Stokes Giffard* 1243. Manorial affix from the *Gifard* family, here from the 11th to the 14th cent. **Stoke, Limpley** Wilts. *Hangyndestok* 1263, *Stoke* 1333, *Lymply Stoke* 1585. Originally with OE *hangende* 'hanging' from its situation below a steep hillside. The later affix is to be associated with the dedication of a former chapel here called '*Our Lady of Limpley's Chapel*' in 1578. **Stoke Mandeville** Bucks. *Stoches* 1086 (DB), *Stoke Mandeville* 1284. Manorial affix from the *Mandeville* family who held the manor in the 13th cent. **Stoke on Trent** Stoke. *Stoche* 1086 (DB). For the river-name, *see* TRENTHAM. **Stoke Poges** Bucks. *Stoches* 1086 (DB), *Stokepogeis* 1292. Manorial affix from the family of *le Pugeis*, here in the 13th cent. **Stoke, Rodney** Somerset. *Stoches* 1086 (DB). Manorial affix from the *de Rodeney* family, here in the early 14th cent. **Stoke St Gregory** Somerset. *Stokes* 1225. Affix from the dedication of the church. **Stoke, Severn** Worcs. *Stoc* 972, *Stoche* 1086 (DB), *Savernestok* 1212. Affix from its situation on the River SEVERN. **Stoke sub Hamdon** Somerset. *Stoca* 1086 (DB), *Stokes under Hamden* 1248. Affix refers to Hamdon Hill, *Hamedone* c.1100, possibly 'hill of the enclosures' from OE *hamm* + *dūn*.

Stoke Newington Gtr. London. *See* NEWINGTON.

Stokeham Notts. *Estoches* 1086 (DB), *Stokum* 1242. '(Place at) the outlying farmsteads'. OE *stoc* in a dative plural form *stocum*.

Stokeinteignhead Devon. *Stoches* 1086 (DB), *Stokes in Tynhide* 1279. Originally 'outlying farmstead or hamlet', from OE *stoc*. Affix means 'in the district consisting of ten hides of land', from OE *tēn* + *hīd*.

Stokenchurch Bucks. *Stockenechurch* c.1200. 'Church made of logs'. OE *stoccen* + *cirice*.

Stokenham Devon. *Stokes* 1242, *Stok in Hamme* 1276. 'Outlying farmstead or hamlet in the area of cultivated ground'. OE *stoc* + *in* + *hamm*.

Stokesay Shrops. *Stoches* 1086 (DB), *Stok Say* 1256. Originally 'outlying farmstead or hamlet', from OE *stoc*. Manorial affix from the *de Say* family, here in the 12th cent.

Stokesby Norfolk. *Stokesbei* 1086 (DB). 'Village with an outlying farmstead or pasture'. OE *stoc* + OScand. *bý*.

Stokesley N. Yorks. *Stocheslage* 1086 (DB). 'Woodland clearing belonging to an outlying farmstead or hamlet'. OE *stoc* + *lēah*.

Ston Easton Somerset. *See* EASTON.

Stondon, 'stony hill', OE *stān* + *dūn*: **Stondon Massey** Essex. *Staundune* 1062, *Standon de Marcy* 1238. Manorial affix from the *de Marci* family, here in the 13th cent. **Stondon, Upper** Beds. *Standone* 1086 (DB).

Stone, 'place at the stone or stones', OE *stān*; examples include: **Stone** Bucks. *Stanes* 1086 (DB). **Stone** Glos. *Stane* 1204. **Stone** Kent, near Dartford. *Stane* 10th cent., *Estanes* 1086 (DB). **Stone** Staffs. *Stanes* 1187.

Stone Street Suffolk. Recorded thus from 1837, named from its situation on the Roman road so called, ModE *stone* + *street* (OE *strēt*).

Stonegrave N. Yorks. *Staningagrave* 757–8, *Stainegrif* 1086 (DB). 'Quarry of the people living by the stone or rock'. OE *stān* + *-inga-* + *græf* (influenced by OScand. *gryfja*).

Stonehaven Aber. *Stanehyve* 1587, *Steanhyve* 1629. 'Stone landing-place'. OE *stān* + *hȳth*.

Stonehenge Wilts. *Stanenges* c.1130. 'Stone gallows' (from a fancied resemblance of the monument to such). OE *stān* + *hengen*.

Stonehouse Glos. *Stanhus* 1086 (DB). 'The stone-built house'. OE *stān* + *hūs*.

Stoneleigh Warwicks. *Stanlei* 1086 (DB). 'Stony woodland clearing'. OE *stān* + *lēah*.

Stonely Cambs. *Stanlegh* 1260. Identical in origin with the previous name.

Stonesby Leics. *Stovenebi* 1086 (DB), *Stouenesbia* c.1130. Probably 'farmstead or village by a tree-stump', or 'at the tree-stump place'. OE or OScand. *stofn* + OScand. *bý*. Alternatively the first element may be an OScand. pers. name *Stofn*.

Stonesfield Oxon. *Stuntesfeld* 1086 (DB). 'Open land of a man called *Stunt'. OE pers. name + *feld*.

Stonham Aspal, Earl Stonham, & Little Stonham Suffolk. *Stonham* c.1040, *Stanham* 1086 (DB). 'Homestead by a stone or with stony ground'. OE *stān* + *hām*. *Aspal* and *Earl* are manorial affixes from the *de Aspale* family and from *Earl* Roger Bigod, here in the 13th cent.

Stonnall Staffs. *Stanahala* 1143. 'Stony nook of land'. OE *stān* + *halh*.

Stonor Oxon. *Stanora* late 10th cent. 'Stony hill-slope'. OE *stān* + *ōra*.

Stonton Wyville Leics. *Stantone* 1086 (DB), *Staunton Wyvile* 1265. 'Farmstead on stony ground'. OE *stān* + *tūn*. Manorial affix from the *de Wivill* family, here in the 13th cent.

Stony Stratford Milt. K. *See* STRATFORD.

Stoodleigh Devon. *Stodlei* 1086 (DB). 'Woodland clearing where a herd of horses is kept'. OE *stōd* + *lēah*.

Stopham W. Sussex. *Stopeham* 1086 (DB). 'Homestead or river-meadow of a man called *Stoppa, or by a hollow'. OE pers. name or OE *stoppa* + *hām* or *hamm*.

Stopsley Luton. *Stoppelee* 1198. 'Woodland clearing of a man called *Stoppa, or by a hollow'. OE pers. name or OE *stoppa* + *lēah*.

Storeton Wirral. *Stortone* 1086 (DB). 'Large farmstead', or 'farmstead near a young wood'. OScand. *stórr* or *storth* + *tún* or OE *tūn*.

Stormont Down. *Storm Mount* 1834. Probably 'storm mount', originally the name of an estate here.

Stornoway W. Isles. (Lewis). *Stornochway* 1511. 'Steering bay'. OScand. *stjórn* + *vágr*. The name seems to imply that ships had to manoeuvre carefully when entering or leaving the harbour.

Storridge Herefs. *Storugge* 13th cent. 'Stony ridge'. OE *stān* + *hrycg*.

Storrington W. Sussex. *Storgetune* 1086 (DB). Probably 'farmstead or village visited by storks'. OE *storc* + *tūn*.

Stortford, Bishop's Herts. *Storteford* 1086 (DB), *Bysshops Stortford* 1587. 'Ford by the tongues of land'. OE *steort* + *ford*. Affix from its early possession by the Bishop of London. The river-name **Stort** is a back-formation from the place name.

Storth Cumbria. *Storthes* 1349. 'Brushwood, plantation'. OScand. *storth*.

Stotfold Beds. *Stodfald* 1007, *Stotfalt* 1086 (DB). 'The stud-fold or enclosure for horses'. OE *stōd-fald*.

Stottesdon Shrops. *Stodesdone* 1086 (DB). Probably 'hill where a herd of horses is kept'. OE *stōd* + *dūn*.

Stoughton, 'farmstead at an outlying hamlet', OE *stoc* + *tūn*: **Stoughton** Leics. *Stoctone* 1086 (DB). **Stoughton** Surrey. *Stoctune* 12th cent. **Stoughton** W. Sussex. *Estone* [*sic*] 1086 (DB), *Stoctona* 1121.

Stoulton Worcs. *Stoltun* 840, 1086 (DB). 'Farmstead or village with a seat (used at meetings of the Hundred)'. OE *stōl* + *tūn*.

Stour Provost Dorset. *Stur* 1086 (DB), *Sture Preauus* 1270. 'Estate on the River Stour held by the Norman Abbey of Préaux' (during the 12th and 13th cents.), *see* STOUR, EAST & WEST.

Stour, East & Stour, West Dorset. *Sture* 1086 (DB). Named from the Dorset/Hants. River Stour, *see* STOURBRIDGE.

Stourbridge Dudley. *Sturbrug* 1255. 'Bridge over the River Stour'. Celtic or OE river-name (probably meaning 'the strong one') + OE *brycg*. There are no less than five major rivers in England called Stour, *see* following names.

Stourmouth Kent. *Sturmutha* late 11th cent. 'Mouth of the River Stour'. Celtic or OE river-name (*see* STOURBRIDGE) + OE *mūtha*.

Stourpaine Dorset. *Sture* 1086 (DB), *Stures Paen* 1243. 'Estate on the River Stour held by the *Payn* family' (here in the 13th cent.), *see* STOUR, EAST & WEST.

Stourport-on-Severn Worcs. *Stourport* c.1775. 'Port at the confluence of the Rivers Stour and Severn'. A recent name.

Stourton, 'farmstead or village on one of the rivers called Stour', Celtic or OE river-name (probably meaning 'the strong one') + OE *tūn*; **Stourton** Staffs. *Sturton* 1227. **Stourton** Warwicks. *Sturton* 1206. **Stourton** Wilts. *Stortone* 1086 (DB).

Stourton Caundle Dorset. *See* CAUNDLE.

Stoven Suffolk. *Stouone* 1086 (DB). '(Place at) the tree-stump(s)'. OE or OScand. *stofn*.

Stow, Stowe, 'assembly-place, holy place', OE *stōw*; examples include: **Stow** Lincs. *Stou* 1086 (DB). **Stow Bardolf** Norfolk. *Stou* 1086 (DB). Manorial affix from the *Bardulf* family, here in the 13th cent. **Stow Bedon** Norfolk. *Stou* 1086 (DB), *Stouwebidun* 1287. Manorial affix from the *de Bidun* family, here in the 13th cent. **Stow cum Quy** Cambs. *Stoua* 1086, *Stowe cum Quey* 1316. Quy is *Coeia* in 1086 (DB), 'cow island' from OE *cū* + *ēg*. Latin *cum* is 'with'. **Stowe** Staffs. *Stowe* 1242. **Stow Maries** Essex. *Stowe* 1222, *Stowe Mareys* 1420. Manorial affix from the *Mareys* family, here in the 13th cent. **Stow on the Wold** Glos. *Eduuardesstou* 1086 (DB), *Stoua* 1213, *Stowe on the Olde* 1574. Originally 'St Edward's holy place'. Later affix is from OE *wald* 'high ground cleared of forest'. **Stow, West** Suffolk. *Stowa* 1086 (DB), *Westowe* 1254.

Stowell, 'stony spring or stream', OE *stān* + *wella*: **Stowell** Somerset. *Stanwelle* 1086 (DB). **Stowell, West** Wilts. *Stawelle* 1176.

Stowey, Nether & Stowey, Over Somerset. *Stawei* 1086 (DB). 'Stone way'. OE *stān* + *weg*.

Stowford, 'stony ford', OE *stān* + *ford*; examples include: **Stowford** Devon, near Portgate. *Staford* 1086 (DB). **Stowford, East** Devon. *Staveford* 1086 (DB).

Stowlangtoft Suffolk. *Stou* 1086 (DB), *Stowelangetot* 13th cent. OE *stōw* 'place of assembly or holy place', with manorial affix from the *de Langetot* family, here in the 13th cent.

Stowmarket Suffolk. *Stou* 1086 (DB), *Stowmarket* 1268. OE *stōw* as in previous name, here with later addition referring to the important market here.

Stowting Kent. *Stuting* 1044, *Stotinges* 1086 (DB). Probably 'place characterized by a lumpy hillock', OE **stūt* + *-ing*. Alternatively 'place associated with a man called **Stūt*', with an OE pers. name as first element.

Strabane (*An Srath Bán*) Tyrone. *Sraith Bán* c.1616. 'The white riverside land'.

Strachan Aber. *Stratheyhan* 1153. Gaelic *srath* 'valley' + unknown element.

Stradbally (*Sráidbhaile*) Kerry, Laois, Waterford. 'Street town'.

Stradbroke Suffolk. *Statebroc* [*sic*] 1086 (DB), *Stradebroc* 1168. Possibly 'brook by a paved road'. OE *strǣt* + *brōc*.

Stradishall Suffolk. *Stratesella* 1086 (DB). 'Shelter (for animals) by a paved road'. OE *strǣt* + *(ge)sell*.

Stradone (*Sraith an Domhain*) Cavan. *Shraghadoone* 1629. Possibly 'river-meadow of the valley floor'.

Stradsett Norfolk. *Strateseta* 1086 (DB). 'Dwelling or fold on a Roman road'. OE *strǣt* + *(ge)set*.

Stragglethorpe Lincs. *Stragerthorp* 1242. OScand. *thorp* 'outlying farmstead or hamlet' with an uncertain first element, possibly the pers. name or surname of some early owner.

Stragolan (*Srath Gabhláin*) Fermanagh. *Stragolan* 1751. 'River-meadow of the fork'.

Strahart (*Sraith Airt*) Wexford. 'Art's river-meadow'.

Straid (*Sráid*) Antrim, Donegal, Mayo. 'Street'.

Straidarran (*Sráidbhaile Uí Áráin*) Derry. *Baile Uí Hárán* c.1680. 'Street town of the Uí Áráin'.

Straiton Midloth. *Stratone* 1296. 'Village on a Roman road'. OE *strǣt* + *tūn*.

Stralongford (*Srath Longfoirt*) Tyrone. *Shralonghert* 1609. 'River-meadow of the fortress'.

Stramshall Staffs. *Stagrigesholle* [*sic*] 1086 (DB), *Strangricheshull* 1227. Possibly 'hill of a man called *Strangrīc*'. OE pers. name + *hyll*.

Stranagalwilly (*Srath na Gallbhuaile*) Tyrone. *Shraghnagallnilly* 1661. 'River-meadow of the stone milking place'.

Strangeways Manch. *Strangwas* 1322. Probably '(place by) a stream with a strong current'. OE *strang* + *(ge)wæsc*.

Strangford Down. *Strangfiord* 1205. 'Strong sea inlet'. OScand. *strangr* + *fjǫrthr*. The village is named after Strangford Lough, a long sea inlet, whose Irish name is *Loch Cuan*, 'sea inlet of bays'. The Irish name of the village is *Baile Loch Cuan*, 'town on Strangford Loch'.

Stranmillis (*Sruthán Milis*) Antrim. (*bun*) *a' tSrutháin Mhilis* 1644. 'Sweet stream'.

Stranocum (*Sraith Nócam*) Antrim. *Stronokum* c.1659. Irish *srath*, 'river-meadow'. The second element is of obscure origin.

Stranorlar (*Srath an Urláir*) Donegal. *Stranhurland* 1590. 'River-meadow of the valley floor'.

Stranraer Dumf. *Stranrever* 1320. 'Broad point'. Gaelic *sròn* + *reamhar*. The 'broad point' is presumably that of the RINNS OF GALLOWAY nearby.

Stratfield, 'open land by a Roman road', OE *strǣt* + *feld*: **Stratfield Mortimer** W. Berks. *Stradfeld* 1086 (DB), *Stratfeld Mortimer* 1275. Manorial affix from the *de Mortemer* family, here from 1086. **Stratfield Saye & Stratfield Turgis** Hants. *Stratfeld* c.1060, *Stradfelle* 1086 (DB), *Stratfeld Say* 1277, *Stratfeld Turgys* 1289. Manorial affixes from the *de Say* and *Turgis* families, here in the 13th cent.

Stratford, 'ford on a Roman road', OE *strǣt* + *ford*; examples include: **Stratford** Gtr. London. *Strætforda* 1067. **Stratford, Fenny** Milt. K. *Fenni Stratford* 1252. Affix is OE *fennig* 'marshy'. **Stratford St Andrew** Suffolk. *Straffort* 1086 (DB). Affix from the dedication of the church. **Stratford St Mary** Suffolk. *Strætford* 962–91, *Stratford* 1086 (DB). Affix from the dedication of the church. **Stratford, Stony** Milt. K. *Stani Stratford* 1202. Affix is OE *stānig* 'stony'. **Stratford Tony** Wilts. *Stretford* 672, *Stradford* 1086 (DB), *Stratford Touny* 14th cent. Manorial affix from the *de Touny* family, here in the 13th cent. **Stratford upon Avon** Warwicks. *Stretfordæ* c.700, *Stradforde* 1086 (DB). Avon is a Celtic river-name meaning simply 'river'. **Stratford, Water** Bucks. *Stradford* 1086 (DB), *West Watrestretford* 1383. Affix is OE *wæter* 'river, stream'.

Stratford-on-Slaney (*Áth na Sráide*) Wicklow. 'Ford of the street'. The Irish name translates the English surname of Edward *Stratford*, 2nd Earl of Aldborough, who founded the village on the River *Slaney* in the late 18th cent.

Strath Beag Highland. 'Little valley'. Gaelic *srath* + *beag*.

Strathallan Perth. *Strathalun* 1187. 'Valley of the Allan Water'. Gaelic *srath*. The river has a pre-Celtic name meaning 'flowing water'.

Strathaven S. Lan. *Strathouren* c.1190. 'Valley of the Avon Water'. Gaelic *srath*. The

Celtic river-name means simply 'river', like the English Avons.

Strathblane Stir. *Strachblachan c.*1200, *Strachblahane* 1335. Perhaps 'valley of little flowers'. Gaelic *srath* + *bláthan*.

Strathbogie (district) Aber. *Strabolgin* 1187, *Strabolgy* 1335. Perhaps 'valley of the River Bogie'. Gaelic *srath*. The river-name means 'river with bag-shaped pools' (OGaelic *bolg*).

Strathclyde (district) N. Lan. 'Valley of the River Clyde'. Gaelic *srath*. For the river-name, see CLYDEBANK.

Strathdon Aber. 'Valley of the River Don'. Gaelic *srath*. The river-name means 'goddess'.

Strathearn Perth. *Sraithh'erni c.*950, *Stradearn* 1185. 'Valley of the River Earn'. Gaelic *srath*. See EARN.

Strathkinness Fife. *Stradkines* 1144. Gaelic *srath* 'valley' + obscure second element.

Strathmiglo Fife. *Scradimigglock c.*1200, *Stramygloke* 1294. 'Valley of the River Miglo'. Gaelic *srath*. *Miglo*, 'bog-loch' (Welsh *mign* + *llwch*), was the name of the marshy lake that now forms the upper reach of the River Eden.

Strathmore (valley) Ang., Perth. 'Great valley'. Gaelic *srath* + *mór*. The long valley separates the Scottish Highlands, to the north, from the Central Lowlands, to the south.

Strathnaver Highland. *Strathnauir* 1268. 'Valley of the River Naver'. Gaelic *srath*. The pre-Celtic river-name means 'river'.

Strathpeffer Highland. *Strathpefir* 1350. 'Valley of the River Peffery'. Gaelic *srath*. The river-name means 'radiant one' (Pictish *pevr*).

Strathwhillan N. Ayr. (Arran). *Terequhilane* 1445. Possibly 'land of Cuilean', or 'holly land'. Gaelic *tìr* 'land' + pers. name or *cuileann* 'holly'.

Strathy Highland. '(Place in) the valley'. Gaelic *srath*.

Stratton, usually 'farmstead or village on a Roman road', OE *stræt* + *tūn*; examples include: **Stratton Audley** Oxon. *Stratone* 1086 (DB), *Stratton Audeley* 1318. Manorial affix from the *de Alditheleg* family, here in the 13th cent. **Stratton, East & Stratton, West** Hants. *Strattone* 903, *Stratune* 1086 (DB). **Stratton on the Fosse** Somerset. *Stratone* 1086 (DB). On the great Roman road called FOSSE WAY. **Stratton St Margaret** Swindn. *Stratone*

1086 (DB). Affix from the dedication of the church. **Stratton St Michael & Long Stratton** Norfolk. *Stratuna*, *Stretuna* 1086 (DB). Distinguishing affixes from the dedication of the church and from the length of the village. **Stratton Strawless** Norfolk. *Stratuna* 1086 (DB), *Stratton Streles* 1269. The affix means literally 'lacking in straw', from OE *strēaw* + *-lēas*.

However the following has a different origin: **Stratton** Cornwall. *Strætneat c.*880, *Stratone* 1086 (DB). 'Valley of the River Neet'. Cornish *stras* + Celtic river-name (of uncertain meaning), with the later addition of OE *tūn* 'village'.

Streamstown (*Baile an tSrutháin*) Galway. 'Town of the stream'.

Streat E. Sussex. *Estrat* 1086 (DB). '(Place on or near) a Roman road'. OE *stræt*.

Streatham Gtr. London. *Estreham* [sic] 1086 (DB), *Stratham* 1175, *Streteham* 1247. 'Homestead or village on a Roman road'. OE *stræt* + *hām*.

Streatley, 'woodland clearing by a Roman road', OE *stræt* + *lēah*: **Streatley** Beds. *Strætlea c.*1053, *Stradlei* 1086 (DB). **Streatley** W. Berks. *Stretlea c.*690, *Estralei* 1086 (DB).

Street, '(place on or near) a Roman road or other paved highway', OE *stræt*; for example **Street** (Somerset, near Glastonbury): *Stret* 725.

Street (*Sráid*) Westmeath. 'Street'.

Streethay Staffs. *Stretheye* 1262. 'Enclosure on a Roman road'. OE *stræt* + *hæg*.

Streetly, 'woodland clearing by a Roman road', OE *stræt* + *lēah*: **Streetly** Wsall. *Strætlea* 957. **Streetly End** Cambs. *Stradleia* 1086.

Strefford Shrops. *Straford* 1086 (DB). 'Ford on a Roman road'. OE *stræt* + *ford*.

Strensall York. *Strenshale* 1086 (DB). From OE *halh* 'nook of land' with *strēon* (genitive *-es*) 'wealth, treasure, offspring, procreation', perhaps referring to fertile land or good fishing.

Strensham Worcs. *Strengesho* 972, *Strenchesham c.*1086. 'Homestead or enclosure of a man called *Strenge*. OE pers. name + *hām* or *hamm* (alternating with *hōh* 'promontory' in the early spelling).

Strete Devon. *Streta* 1194. '(Place on) the Roman road'. OE *stræt*.

Stretford Traffd. *Stretford* 1212. 'Ford on a Roman road'. OE *strǣt* + *ford*.

Strethall Essex. *Strathala* 1086 (DB). 'Nook of land by a Roman road'. OE *strǣt* + *halh*.

Stretham Cambs. *Stratham* c.970, *Stradham* 1086 (DB). 'Homestead or village on a Roman road'. OE *strǣt* + *hām*.

Strettington W. Sussex. *Stratone* 1086 (DB), *Estretementona* 12th cent. 'Farmstead or village of the dwellers by the Roman road'. OE *strǣt* + *hǣme* + *tūn*.

Stretton, 'farmstead or village on a Roman road', OE *strǣt* + *tūn*; examples include: **Stretton** Derbys. *Strǣttune* c.1002, *Stratune* 1086 (DB). **Stretton, All & Stretton, Church** Shrops. *Stratun(e)* 1086 (DB), *Aluredestretton, Chirich Stretton* 1262. Distinguishing affixes from the name of an early owner called *Alfred* and from OE *cirice* 'church'. **Stretton Grandison** Herefs. *Stratune* 1086 (DB), *Stretton Graundison* 1350. Manorial affix from the *Grandison* family, here in the 14th cent. **Stretton-on-Dunsmore** Warwicks. *Stratone* 1086 (DB). For Dunsmore, *see* RYTON. **Stretton Sugwas** Herefs. *Stratone* 1086 (DB), *Strattone by Sugwas* 1334. Sugwas is 'alluvial land frequented by sparrows', or 'marshy alluvial land', OE **sugge* or **sugga* + **wæsse*.

Strickland, Great & Strickland, Little Cumbria. *Stircland* late 12th cent. 'Cultivated land where young bullocks are kept'. OE *stirc* + *land*.

Stringston Somerset. *Strengestune* 1086 (DB). 'Farmstead or village of a man called **Strenge*'. OE pers. name + *tūn*.

Strixton Northants. *Strixton* 12th cent. 'Farmstead or village of a man called *Stríkr*'. OScand. pers. name + *tūn*.

Stroan (*Sruthán*) Antrim, Cavan, Kilkenny. 'Stream'.

Strokestown (*Béal na mBuillí*) Roscommon. 'Ford-mouth of the strokes'. The name appears to refer to a battle here.

Stroma (island) Highland. *Straumsey* 1150. 'Island in the current'. OScand. *straumr* + *ey*. Stroma lies in the current of the Pentland Firth. *Compare* STROMNESS.

Stromness Orkn. (North Ronaldsay). *Straumsness* 1150. 'Headland of the current'.

OScand. *straumr* + *nes*. The name alludes to the strong current off the headland.

Strone Arg. 'Headland'. Gaelic *sròn*.

Stronsay (island) Orkn. *Strjónsay* 13th cent. OScand. **strjón* 'wealth' + *ey*; with reference to the herring fishing.

Strood, Stroud, 'marshy land overgrown with brushwood', OE *strōd*; examples include: **Strood** Medway. *Strod* 889. **Stroud** Glos. (*La*) *Strode* 1200. **Stroud Green** Gtr. London, near Highbury. *Strode* 1407.

Struan Perth. '(Place by a) little stream'. Gaelic *sruthan*.

Strubby Lincs. near Alford. *Strobi* 1086 (DB). Probably 'farmstead or village of a man called Strútr'. OScand. pers. name + *bý*.

Struell (*Sruthail*) Down. (*Capella de*) *Strohull* 1306. 'Stream'.

Strumble Head Pemb. *Strumble head*. Probably 'stormy headland'. ModE *storm* + *bill*. *Head* was added when the sense of 'headland' in the original was forgotten.

Strumpshaw Norfolk. *Stromessaga* 1086 (DB). 'Tree-stump wood or copse'. OE **strump* + *sceaga*.

Stubbington Hants. *Stubitone* 1086 (DB). 'Estate associated with a man called **Stubba*', OE pers. name + *-ing-* + *tūn*, or 'farmstead at the tree-stump clearing', OE **stubbing* + *tūn*.

Stubbins Lancs. *Stubbyng* 1563. OE **stubbing* 'place with tree-stumps, a clearing'.

Stubbs, Walden N. Yorks. *Istop* [*sic*] 1086 (DB), *Stubbis* c.1180, *Stubbes Walding* 1280. 'The tree-stumps'. OE *stubb*. Manorial affix from an early owner called *Walding*, here in the 12th cent.

Stubton Lincs. *Stubetune* 1086 (DB). 'Farmstead or village of a man called **Stubba*, or where there are tree-stumps'. OE pers. name or *stubb* + *tūn*.

Studdal, East Kent. *Stodwalde* 1240. 'Forest where a herd of horses is kept'. OE *stōd* + *weald*.

Studham Beds. *Stodham* 1053–66, *Estodham* 1086 (DB). 'Homestead or enclosure where a herd of horses is kept'. OE *stōd* + *hām* or *hamm*.

Studland Dorset. *Stollant* 1086 (DB). 'Cultivated land where a herd of horses is kept'. OE *stōd* + *land*.

Studley, 'woodland clearing or pasture where a herd of horses is kept', OE *stōd* + *lēah*: **Studley** Oxon. *Stodleya c.*1185. **Studley** Warwicks. *Stodlei* 1086 (DB). **Studley** Wilts. *Stodleia* 12th cent. **Studley Roger** N. Yorks. *Stodlege c.*1030, *Stollai* 1086 (DB), *Stodelay Roger* 1228. Manorial affix from early possession by *Roger* de Mowbray or by Archbishop *Roger* of York.

Stukeley, Great & Stukeley, Little Cambs. *Stivecle* 1086 (DB). 'Woodland clearing with tree-stumps'. OE **styfic* + *lēah.*

Stuntney Cambs. *Stuntenei* 1086 (DB). 'Steep island'. OE *stunt* (dative *-an*) + *ēg.*

Sturmer Essex. *Sturmere c.*1000, 1086 (DB). 'Pool on the River Stour'. Celtic or OE river-name (probably 'the strong one') + OE *mere.*

Sturminster Marshall Dorset. *Sture minster* 9th cent., *Sturminstre* 1086 (DB), *Sturministra Marescal* 1268. 'Church on the River Stour'. Celtic or OE river-name (*see* STOUR, EAST & WEST) + OE *mynster*. Manorial affix from the *Mareschal* family, here in the 13th cent.

Sturminster Newton Dorset. *Nywetone, at Stoure* 968, *Newentone* 1086 (DB), *Sturminstr Nyweton* 1291. Originally two separate names for places on opposite sides of the river, 'new farmstead or village' from OE *nīwe* + *tūn*, and 'church on the River Stour' (*see* previous name).

Sturry Kent. *Sturgeh* 678, *Esturai* 1086 (DB). 'District by the River Stour'. Celtic or OE river-name (*see* STOURMOUTH) + OE **gē.*

Sturton, 'village on a Roman road', OE *strǣt* + *tūn*: **Sturton by Stow** Lincs. *Stratone* 1086 (DB). *See* STOW. **Sturton, Great** Lincs. *Stratone* 1086 (DB). **Sturton le Steeple** Notts. *Estretone* 1086 (DB). Affix *le Steeple*, first found in the 18th cent., refers to the tower of the church.

Stuston Suffolk. *Stutestuna* 1086 (DB). Probably 'farmstead or village of a man called **Stūt*'. OE pers. name + *tūn.*

Stutton N. Yorks. *Stouetun* 1086 (DB). 'Farmstead of a man called Stúfr', or 'one build of or amongst tree-stumps'. OScand. pers. name or **stúfr* + OE *tūn.*

Stutton Suffolk. *Stuttuna* 1086 (DB). Probably 'farmstead or estate on a stumpy hillock'. OE **stūt* + *tūn.*

Styal Ches. *Styhale c.*1200. 'Nook of land with a pigsty, or by a path'. OE *stigu* or *stīg* + *halh.*

Styrrup Notts. *Estirape* 1086 (DB), *Stirap* 1197. from OE *stīg-rāp* 'a stirrup', probably named from the hill to the east of the village, perhaps thought by early settlers to resemble a stirrup in shape.

Suckley Worcs. *Suchelei* 1086 (DB). 'Woodland clearing frequented by sparrows'. OE **succa* + *lēah.*

Sudborough Northants. *Sutburg* 1086 (DB). 'Southern fortification'. OE *sūth* + *burh.*

Sudbourne Suffolk. *Sutborne c.*1050, *Sutburna* 1086 (DB). 'Southern stream'. OE *sūth* + *burna.*

Sudbrooke Lincs. near Lincoln. *Sutbroc* 1086 (DB). 'Southern brook'. OE *sūth* + *brōc.*

Sudbury, 'southern fortification', OE *sūth* + *burh* (dative *byrig*): **Sudbury** Derbys. *Sudberie* 1086 (DB). **Sudbury** Gtr. London. *Suthbery* 1282. Here the second element may mean 'manor'. **Sudbury** Suffolk. *Suthbyrig c.*995, *Sutberia* 1086 (DB).

Suffield Norfolk. *Sudfelda* 1086 (DB). 'Southern open land'. OE *sūth* + *feld.*

Suffolk (the county). *Suthfolchi* 895, *Sudfulc* 1086 (DB). '(Territory of) the southern people (of the East Angles)'. OE *sūth* + *folc, see* NORFOLK.

Sugnall Staffs. *Sotehelle* [sic] 1086 (DB), *Sugenhulle* 1222. Probably 'hill frequented by sparrows'. OE **sugge* (genitive plural *-ena*) + *hyll*. Alternatively 'hill of a man called Sucga', from an OE pers. name (genitive *-n*).

Suir (*Siúr*) (river) Tipperary–Waterford. 'Sister'.

Sulgrave Northants. *Sulgrave* 1086 (DB). 'Grove near a gully or narrow valley'. OE *sulh* + *grāf.*

Sulham W. Berks. *Soleham* 1086 (DB). Probably 'homestead by a gully'. OE *sulh* + *hām.*

Sulhamstead W. Berks. *Silamested* 1198. 'Homestead by a gully or narrow valley'. OE *sulh* (genitive *sylh*) + *hām-stede.*

Sullington W. Sussex. *Sillinctune* 959, *Sillintone* 1086 (DB). Possibly 'farmstead or village by a willow copse'. OE **sieling* + *tūn*. Alternatively the first element may be OE **sielling* 'gift', denoting land given as a gift.

Sullom Voe Shet. *Sollom* 1507. OScand. *sólheimr* 'sunny farm' + *vágr* 'bay'.

Sully (*Sili*) Vale Glam. *Sulie* 1193, *Sulye* 1254. Origin uncertain. The name may be manorial.

Sumburgh Shet. *Svinaborgh* 1491. 'South fort'. OScand. *sunn* + *borg*.

Summerseat Bury. *Sumersett* 1556. 'Hut or shieling used in summer'. OScand. *sumarr* + *sǽtr*.

Sunbury Surrey. *Sunnanbyrg* 960, *Sunneberie* 1086 (DB). 'Stronghold of a man called **Sunna*'. OE pers. name + *burh* (dative *byrig*).

Sunderland, usually 'detached estate', OE *sundor-land*: **Sunderland** Cumbria. *Sonderland* 1278. **Sunderland** Sundld. *Sunderland c.*1168.
   However the following has a different origin: **Sunderland, North** Northum. *Sutherlannland* 12th cent. 'Southern cultivated land'. OE *sūtherra* + *land*.

Sundon Luton. *Sunnandune c.*1050, *Sonedone* 1086 (DB). 'Hill of a man called **Sunna*'. OE pers. name + *dūn*.

Sundridge Kent. *Sondresse* 1086 (DB). 'Separate or detached ploughed field'. OE *sundor* + *ersc*.

Sunningdale Winds. & Maid., a parish-name formed in the 19th cent. from the following name.

Sunninghill Winds. & Maid. *Sunningehull* 1190. 'Hill of the family or followers of a man called **Sunna*'. OE pers. name + *-inga-* + *hyll*. The same tribal group gives name to SONNING and SUNNINGWELL.

Sunningwell Oxon. *Sunningauuille* 9th cent., *Soningeuuel* 1086 (DB). 'Spring or stream of the family or followers of a man called **Sunna*'. OE pers. name + *-inga-* + *wella*.

Surbiton Gtr. London. *Suberton* 1179. 'Southern grange or outlying farm'. OE *sūth* + *bere-tūn*. See NORBITON.

Surfleet Lincs. *Sverefelt* [*sic*] 1086 (DB), *Surfliet* 1167. 'Sour creek or stream'. OE *sūr* + *flēot*.

Surlingham Norfolk. *Sutherlingaham* 1086 (DB). Probably 'homestead of the family or followers of a man called **Herela*', from OE pers. name + *-inga-* + *hām*, with *sūth* 'south'. Alternatively 'homestead of the people living to the south (of Norwich)', OE *sūther* + *-linga-* + *hām*.

Surrey (the county). *Suthrige* 722, *Sudrie* 1086 (DB). 'Southerly district' (relative to MIDDLESEX). OE *sūther* + **gē*.

Sussex (the county). *Suth Seaxe* late 9th cent., *Sudsexe* 1086 (DB). '(Territory of) the South Saxons'. OE *sūth* + *Seaxe*.

Sustead Norfolk. *Sutstede* 1086 (DB). 'Southern place'. OE *sūth* + *stede*.

Sutcombe Devon. *Sutecome* 1086 (DB). 'Valley of a man called **Sutta*'. OE pers. name + *cumb*.

Sutherland (the historic county). *Suthernelande c.*1250. 'Southern territory'. OScand. *suthr* + *land*. The region was a 'southern territory' to the Vikings who settled in Orkney and Shetland.

Sutterton Lincs. *Suterton* 1200. Probably 'farmstead or village of the shoe-maker(s)'. OE *sūtere* + *tūn*.

Sutton, a very common place name, 'south farmstead or village', i.e. one to the south of another settlement, OE *sūth* + *tūn*: examples include: **Sutton** Cambs. *Sudtone* 1086 (DB). **Sutton** Gtr. London. *Sudtone* 1086 (DB). **Sutton** Suffolk. *Suthtuna* 1086 (DB). **Sutton at Hone** Kent. *Sudtone* 1086 (DB), *Sutton atte hone* 1240. Affix is 'at the (boundary) stone' from OE *hān*. **Sutton Bonington** Notts. *Sudtone, Bonniton* 1086 (DB). Originally two separate manors, Bonington being 'estate associated with a man called Buna', OE pers. name + *-ing-* + *tūn*. **Sutton Bridge** Lincs. named from Long Sutton. **Sutton Coldfield** Birm. *Sutone* 1086 (DB), *Sutton in Colefeud* 1269. Coldfield is 'open land where charcoal is produced', OE *col* + *feld*. **Sutton Courtenay** Oxon. *Suthtun c.*870, *Sudtone* 1086 (DB), *Suttone Curteney* 1284. Manorial affix from the *Curtenai* family, here in the 12th cent. **Sutton, Full** E. R. Yorks. *Sudtone* 1086 (DB), *Fulesutton* 1234. Affix is OE *fūl* 'dirty'. **Sutton, Great & Sutton, Little** Ches. *Sudtone* 1086 (DB). **Sutton, Guilden** Ches. *Sudtone* 1086 (DB), *Guldenesutton c.*1200. Affix is OE *gylden* 'splendid, wealthy'. **Sutton in Ashfield** Notts. *Sutone* 1086 (DB), *Sutton in Essefeld* 1276.

Affix is an old district name, *see* ASHFIELD.
Sutton, Kings Northants. *Sudtone* 1086 (DB),
Kinges Sutton 1294. The manor was held by
King William in 1086. **Sutton, Long** Lincs.
Sudtone 1086 (DB). Affix refers to the length of
the village. **Sutton, Long** Somerset. *Sudton*
878, *Sutune* 1086 (DB), *Langesutton* 1312. Affix
from the length of the village. **Sutton on Hull**
K. upon Hull. *Sudtone* 1086 (DB), *Sutune iuxta
Hul* 1172. *See* KINGSTON upon Hull. **Sutton on
Sea** Lincs. *Sudtone* 1086 (DB). **Sutton on
Trent** Notts. *Sudtone* 1086 (DB). **Sutton
Valence** Kent. *Suthtun* 814, *Sudtone* 1086
(DB), *Sutton Valence* 1316. Manorial affix from
the *Valence* family, here in the 13th cent.

Sutton Hoo Suffolk. *See* HOO.

Swaby Lincs. *Suabi* 1086 (DB). 'Farmstead or
village of a man called Sváfi'. OScand. pers.
name + *bý*.

Swadlincote Derbys. *Sivardingescotes* 1086
(DB). Probably 'cottage(s) of a man called
*Sweartling or *Svartlingr'. OE or OScand. pers.
name + OE *cot*.

Swaffham, 'homestead of the Swabians', OE
Swǣfe + *hām*: **Swaffham** Norfolk. *Suafham*
1086 (DB). **Swaffham Bulbeck &
Swaffham Prior** Cambs. *Suafham* c.1050,
1086 (DB), *Swafham Bolebek* 1218, *Swafham
Prior* 1261. Manorial affixes from early
possession by the *de Bolebech* family and by the
Prior of Ely.

Swafield Norfolk. *Suafelda* [sic] 1086 (DB),
Suathefeld c.1150. Possibly 'open land
characterized by swathes or tracks'. OE *swæth* +
feld.

Swainby N. Yorks. near Whorlton. *Swaneby*
1314. Probably 'farmstead or village of the
young men'. OScand. *sveinn* + *bý*.

Swainsthorpe Norfolk. *Sueinestorp* 1086
(DB). 'Outlying farmstead or hamlet of a man
called Sveinn'. OScand. pers. name + *thorp*.

Swalcliffe Oxon. *Sualewclive* c.1166. 'Cliff or
slope frequented by swallows'. OE *swealwe* +
clif.

Swale (river), two examples probably
identical in origin, if from OE *su(e)alwe*
'rushing water': i) Kent *see* SWALECLIFFE,
ii) N. Yorks., *see* BROMPTON on Swale.

Swalecliffe Kent. *Swalewanclife* 949,
Soaneclive [sic] 1086 (DB). 'Swallow's riverbank',
OE *swealwe* + *clif*, or 'bank by (the estuary of)

the River SWALE' (the latter once a more
extensive river before coastal changes took
place).

Swallow Lincs. *Sualun* 1086 (DB). *Sualwa*
c.1115. '(Place at) the swallow-hole (where the
local stream disappears underground)'. OE
swalwe.

Swallowcliffe Wilts. *Swealewanclif* 940,
Svaloclive 1086 (DB). 'Swallow's cliff or slope'.
OE *swealwe* + *clif*.

Swallowfield Wokhm. *Sualefelle* 1086 (DB).
'Open land on the rushing stream'. OE *swealwe*
+ *feld*.

Swanage Dorset. *Swanawic* late 9th cent.,
Swanwic 1086 (DB). 'Farm of the herdsmen, or
farm where swans are reared'. OE *swān* or *swan*
+ *wīc*.

Swanbourne Bucks. *Suanaburna* 792,
Sueneborne 1086 (DB). 'Stream frequented by
swans'. OE *swan* + *burna*.

Swanland E. R. Yorks. *Suenelund* 1189.
'Grove of a man called Svanr or Sveinn'.
OScand. pers. name + *lundr*.

Swanley Kent. near Farningham. *Swanleg*
1203. Probably 'woodland clearing of the
herdsmen'. OE *swān* + *lēah*.

Swanmore Hants. *Suanemere* 1205. 'Pool
frequented by swans'. OE *swan* + *mere*.

Swannington, probably 'estate associated
with a man called *Swan', OE pers. name + *-ing-
+ *tūn*: **Swannington** Leics. *Suaninton* late
12th cent. **Swannington** Norfolk.
Sueningatuna 1086 (DB).

Swanscombe Kent. *Suanescamp* 677,
Svinescamp 1086 (DB). 'Enclosed land of the
herdsman'. OE *swān* + *camp*.

Swansea (*Abertawe*) Swan. *Sweynesse* c.1165,
Sueinesea 1190, *Swanesey* 1322. 'Sveinn's
island'. OScand. pers. name + *ey*. The Welsh
name means 'mouth of the Tawe' (Welsh *aber*),
the Celtic river-name perhaps meaning simply
'water'.

Swanton, 'farmstead or village of the
herdsmen', OE *swān* + *tūn*: **Swanton Abbot**
Norfolk. *Swanetonne* 1044–7, *Suanetuna* 1086
(DB). Affix from early possession by Holme
Abbey. **Swanton Morley** Norfolk.
Suanetuna 1086 (DB). Manorial affix from the *de
Morle* family, here in the 14th cent. **Swanton**

Novers Norfolk. *Suanetone* 1047–70, *Suanetunam* 1086 (DB), *Swanton Nowers* 1361. Manorial affix from the *de Nuiers* family, here in the 13th cent.

Swanwick, '(dairy) farm of the herdsmen', OE *swān* + *wīc*: **Swanwick** Derbys. *Swanwyk* late 13th cent. **Swanwick** Hants. *Suanewic* 1185.

Swarby Lincs. *Svarrebi* 1086 (DB). 'Farmstead or village of a man called Svarri'. OScand. pers. name + *bý*.

Swardeston Norfolk. *Suerdestuna* 1086 (DB). 'Farmstead or village of a man called *Sweord'. OE pers. name + *tūn*.

Swarkestone Derbys. *Suerchestune* 1086 (DB). Probably 'farmstead or village of a man called Swerkir'. OScand. pers. name + OE *tūn*.

Swarland Northum. *Swarland* 1242. 'Heavy cultivated land'. OE *swǣr* + *land*.

Swaton Lincs. *Svavetone* 1086 (DB). 'Farmstead or village of a man called *Swāfa or Sváfi'. OE or OScand. pers. name + OE *tūn*.

Swatragh (*Suaitreach*) Derry. *Ballitotry* 1613. '(Billeted) soldier'. The full form of the Irish name is *Baile an tSuaitrigh*, 'homestead of the (billeted) soldier'.

Swavesey Cambs. *Suauesye* 1086 (DB). 'Landing-place of a man called *Swǣf'. OE pers. name ('the Swabian') + *hȳth*.

Sway Hants. *Sueia* 1086 (DB). Possibly an OE river-name meaning 'noisy stream', or from OE *swæth* 'swathe, track'.

Swayfield Lincs. *Suafeld* [*sic*] 1086 (DB), *Suathefeld* 1206. Possibly 'open land characterized by swathes or tracks'. OE *swæth* + *feld*.

Swaythling Sotn. *Swǣthelinge* 909. Probably an old name of the stream here, uncertain in origin and meaning but possibly an OE **swǣtheling* meaning 'misty stream'.

Swefling Suffolk. *Sueflinga* 1086 (DB), *Sueftlinges* c.1150. Probably '(settlement of) the family or followers of a man called *Swiftel'. OE pers. name + *-ingas*.

Swell, Lower & Swell, Upper Glos. *Swelle* 706, *Svelle* 1086 (DB). OE **swelle* 'rising ground or hill'.

Swepstone Leics. *Scopestone* [*sic*] 1086 (DB), *Swepeston* c.1125. 'Farmstead or village of a man called *Sweppi'. OE pers. name + *tūn*.

Swerford Oxon. *Surford* 1086 (DB), *Swereford* 1194, 'Ford by a neck of land or col'. OE *swēora* + *ford*.

Swettenham Ches. *Suetenham* late 12th cent. 'Homestead or enclosure of a man called Swēta'. OE pers. name (genitive *-n*) + *hām* or *hamm*.

Swilland Suffolk. *Suinlanda* 1086 (DB). 'Tract of land where pigs are kept, swine pasture'. OE *swīn* + *land*.

Swillington Leeds. *Suillintune* 1086 (DB). Possibly 'farmstead near the pig hill (or clearing)'. OE *swīn* + *hyll* (or *lēah*) + *-ing-* + *tūn*.

Swilly (*Súileach*) (river) Donegal. *Suileach* 1258. 'Seeing one'. The name probably has supernatural connotations.

Swimbridge Devon. *Birige* 1086 (DB), *Svimbrige* 1225. '(Place at) the bridge held by a man called Sǣwine'. OE pers. name (of the tenant in 1086) + *brycg*.

Swinbrook Oxon. *Svinbroc* 1086 (DB). 'Pig brook'. OE *swīn* + *brōc*.

Swinderby Lincs. *Sunderby, Suindrebi* 1086 (DB). Possibly 'southern farmstead or village', OScand, *sundri* + *bý*. Alternatively 'animal farmstead where pigs are kept'. OScand. *svin* + *djúr* + *bý*.

Swindon, 'hill where pigs are kept', OE *swīn* + *dūn*: **Swindon** Glos. *Svindone* 1086 (DB). **Swindon** Staffs. *Swineduna* 1167. **Swindon** Swindn. *Svindune* 1086 (DB).

Swine E. R. Yorks. *Suuine* 1086 (DB). '(Place at) the creek or channel'. OE **swin*.

Swinefleet E. R. Yorks. *Swyneflet* c.1195. Probably 'stretch of river where pigs are kept'. OE *swīn* + *flēot*.

Swineshead Beds. *Suineshefet* 1086 (DB). 'Pig's headland'. OE *swīn* + *hēafod*.

Swineshead Lincs. *Suinesheabde* [*sic*] 786–96. 'Source of the creek'. OE **swin* + *hēafod*.

Swinford, 'pig ford', OE *swīn* + *ford*: **Swinford** Leics. *Svineford* 1086 (DB). **Swinford** Oxon. *Swynford* 931.

Sywell

Swingfield Minnis Kent. *Suinafeld* c.1100. 'Open land where pigs are kept'. OE *swīn* + *feld*. Affix is OE *mǣnnes* 'common land'.

Swinhoe Northum. *Swinhou* 1242. 'Swine headland'. OE *swīn* + *hōh*.

Swinhope Lincs. *Suinhope* 1086 (DB). 'Secluded valley where pigs are kept'. OE *swīn* + *hop*.

Swinithwaite N. Yorks. *Swiningethwait* 1202. 'Place cleared by burning'. OScand. *svithningr* + *thveit*.

Swinscoe Staffs. *Swyneskow* 1248. 'Swine wood'. OScand. *svín* + *skógr*.

Swinstead Lincs. *Suinhamstede* 1086 (DB). 'Homestead where pigs are kept'. OE *swīn* + *hām-stede*.

Swinton, 'pig farm', OE *swīn* + *tūn*: **Swinton** N. Yorks., near Masham, *Suinton* 1086 (DB). **Swinton** Rothm. *Suintone* 1086 (DB). **Swinton** Salford. *Suinton* 1258.

Swinton Sc. Bord. *Suineston* c.1098. 'Sveinn's farm'. OScand. pers. name + OE *tūn*.

Swithland Leics. *Swithelunde* 1224. 'Grove by the burnt clearing'. OScand. *svitha* + *lundr*.

Swords (*Sord*) Dublin. 'Sward'.

Swyncombe Oxon. *Svinecumbe* 1086 (DB). 'Valley where pigs are kept'. OE *swīn* + *cumb*.

Swynnerton Staffs. *Sulvertone* [sic] 1086 (DB), *Suinuerton* 1242. 'Farmstead by the pig ford'. OE *swīn* + *ford* + *tūn*.

Swyre Dorset. *Suere* 1086 (DB). OE *swēora* 'a neck of land, a col'.

Syde Glos. *Side* 1086 (DB). 'Long hill-slope'. OE *sīde*.

Sydenham Down. The name was adopted from SYDENHAM Gtr. London.

Sydenham Gtr. London. *Chipeham* 1206. Probably 'homestead or enclosure of a man called *Cippa*'. OE pers. name (genitive -*n*) + *hām* or *hamm*.

Sydenham Oxon. *Sidreham* [sic] 1086 (DB), *Sidenham* 1216. 'Broad or extensive enclosure'. OE *sīd* (dative -*an*) + *hamm*.

Sydenham Damerel Devon. *Sidelham* [sic] 1086 (DB), *Sydenham Albemarlie* 1297. Identical in origin with the previous name. Manorial affix from the *de Albemarle* family, here in the 13th cent.

Syderstone Norfolk. *Cidesterna* 1086 (DB). Possibly 'broad or extensive property', OE *sīd* + **sterne*. Alternatively 'pool of a man called **Siduhere*', OE pers. name + OScand. *tjǫrn*.

Sydling St Nicholas Dorset. *Sidelyng* 934, *Sidelince* 1086 (DB). 'Broad ridge'. OE *sīd* + *hlinc*. Affix from the dedication of the church.

Sydmonton Hants. *Sidemanestone* 1086 (DB). 'Farmstead or village of a man called *Sidumann*'. OE pers. name + *tūn*.

Syerston Notts. *Sirestune* 1086 (DB). 'Farmstead or village of a man called *Sigehere*'. OE pers. name + *tūn*.

Sykehouse Donc. *Sikehouse* 1404. 'House(s) by the stream'. OScand. *sík* + *hús*.

Symington S. Ayr. *Symondstona* 1293. 'Simon's farm'. OE *tūn*. The place is said to have been named after *Simon Lockhart* (c.1150).

Symondsbury Dorset. *Simonedesberge* 1086 (DB). 'Hill or barrow of a man called *Sigemund*'. OE pers. name + *beorg*.

Syresham Northants. *Sigresham* 1086 (DB). 'Homestead or enclosure of a man called *Sigehere*'. OE pers. name + *hām* or *hamm*.

Syston Leics. *Sitestone* 1086 (DB), *Sithestun* 1201. Possibly 'farmstead of a man called *Sigehæth*'. OE pers. name + *tūn*.

Syston Lincs. *Sidestan* 1086 (DB). 'Broad stone'. OE *sīd* + *stān*.

Sywell Northants. *Siwella* 1086 (DB). 'Seven springs'. OE *seofon* + *wella*.

Tabley, Over Ches. *Stabelei* [*sic*] 1086 (DB), *Thabbelewe* 12th cent. 'Woodland clearing of a man called Tæbba'. OE pers. name + *lēah*.

Tachbrook, Bishop's Warwicks. *Tæcelesbroc* 1033, *Taschebroc* 1086 (DB). 'The boundary brook'. OE **tæcels* + *brōc*. Manorial affix from its early possession by the Bishop of Chester.

Tacker, Lough (*Loch Tacair*) Cavan. 'Artificial lake'.

Tackley Oxon. *Tachelie* 1086 (DB). 'Woodland clearing of a man called **Tæcca*, or where young sheep are kept'. OE pers. name or OE **tacca* + *lēah*.

Tacolneston Norfolk. *Tacoluestuna* 1086 (DB). 'Farmstead or village of a man called Tātwulf'. OE pers. name + *tūn*.

Tacumshin, Lough (*Loch Theach Cuimsin*) Wexford. 'Lake of Cuimsin's house'.

Tadcaster N. Yorks. *Tatecastre* 1086 (DB). 'Roman town of a man called Tāta or **Tāda*'. OE pers. name + *cæster*.

Taddington Derbys. *Tadintune* 1086 (DB). 'Estate associated with a man called **Tāda*'. OE pers. name + *-ing-* + *tūn*.

Tadley Hants. *Tadanleage* 909. Probably 'woodland clearing of a man called **Tāda*'. OE pers. name + *lēah*.

Tadlow Cambs. *Tadeslaue* c.1080, *Tadelai* 1086 (DB). 'Tumulus of a man called **Tāda*'. OE pers. name + *hlāw*.

Tadmarton Oxon. *Tademærtun* 956, *Tademertone* 1086 (DB). Probably 'farmstead by a pool frequented by toads'. OE **tāde* + *mere* + *tūn*.

Tadworth Surrey. *Thæddeuurthe* 1062, *Tadeorde* 1086 (DB). Possibly 'enclosure of a man called **Thædda*'. OE pers. name + *worth*.

Taf (anglicized to **Taff**). *See* LLANDAFF.

Tafolog (mountain) Powys. 'Place of dock (plants)'. Welsh *tafol* + *-og*.

Tagheen (*Teach Chaoin*) Mayo. 'Pleasant house'.

Taghmon (*Teach Munna*) Westmeath, Wexford. 'Muna's house'.

Taghnafearagh (*Teach na bFiarach*) Westmeath. 'House of the lea fields'.

Taghshinny (*Teach Siní*) Longford. 'Sineach's house'.

Tagoat (*Teach Gót*) Wexford. 'Gót's house'.

Tai-bach Powys. 'Little houses'. Welsh *tŷ* (plural *tai*) + *bach*.

Tain Highland. *Tene* 1226, *Thayn* 1257. '(Place of the) river'. The name may be pre-Celtic, but the sense 'stream' seems likely.

Takeley Essex. *Tacheleia* 1086 (DB). 'Woodland clearing of a man called **Tæcca*, or where young sheep are kept'. OE pers. name or OE **tacca* + *lēah*.

Tal-y-llyn Conwy. 'End of the lake'. Welsh *tâl* + *y* + *llyn*.

Talacharn. *See* LAUGHARNE.

Talacre Flin. 'End of the arable land'. Welsh *tâl* + *acer*.

Talaton Devon. *Taletone* 1086 (DB). 'Farmstead or village on the River Tale'. OE river-name (meaning 'the swift one') + *tūn*.

Talbenny Pemb. *Talbenny* 1291. '(Place at the) end'. Welsh *tâl* + an obscure element.

Talbot Village Bmouth. model village built in the 1860s and named after the local landowners, the two *Talbot* sisters.

Talgarth Powys. 'End of the ridge'. Welsh *tâl* + *garth*.

Talisker Highland. (Skye). Possibly 'shelf rock'. OScand. *hjalli* + *sker*.

Talke Staffs. *Talc* 1086 (DB). A Celtic name for the ridge here, from Brittonic **talcen* 'forehead; end, front'.

Talkin Cumbria. *Talcan* c.1195. Probably a Celtic name meaning 'white brow (of a hill)', or identical with previous name.

Talla Bheith Forest Perth. *An t-All a' Bheithe* (no date). 'Crag of birch-trees'. Gaelic *all* + *beithe*.

Tallaght (*Tamhlacht*) Dublin. 'Plague burial place'.

Tallentire Cumbria. *Talentir c.*1160. Probably a Celtic name meaning 'end of the land'.

Tallington Lincs. *Talintune* 1086 (DB). Probably 'estate associated with a man called **Tæl* or **Tala*'. OE pers. name + *-ing-* + *tūn*.

Tallow (*Tulach*) Waterford. 'Hill'. The full Irish name is *Tulach an Iarainn*, 'hill of the iron', referring to former iron-ore workings here.

Talmine Highland. Uncertain, but perhaps 'level land'. Gaelic *talamh* + *min*.

Tal-sarn Cergn. 'End of the causeway'. Welsh *tâl* + *y* + *sarn*.

Talwrn Wrex. '(Place by the) field'. Welsh *talwrn*.

Tal-y-bont Cergn. 'End of the bridge'. Welsh *tâl* + *y* + *pont*. Compare BRIDGEND.

Talysarn Gwyd. *Talysarn* 1795. 'End of the causeway'. Welsh *tâl* + *y* + *sarn*.

Tamar (river) Cornwall–Devon, gives name to TAMERTON, *see* THAMES.

Tame (river), three examples. *See* THAMES.

Tamerton, 'farmstead or village on the River Tamar', Celtic or pre-Celtic river-name (for which *see* THAMES) + OE *tūn*: **Tamerton Foliot** Plym. *Tambretone* 1086 (DB), *Tamereton Foliot* 1262. Manorial affix from the *Foliot* family, here from the 13th cent.
Tamerton, North Cornwall. *Tamerton* 1180.

Tameside, named from the Lancs.–Ches. River **Tame**, an ancient river-name for the possible meaning of which *see* THAMES.

Tamlaght (*Tamhlacht*) Derry. 'Plague burial place'.

Tamlaght Finlagan (*Tamhlacht Fionnlugháin*) Derry. 'Fionnlugh's burial place'.

Tamlaght O'Crilly (*Tamlacht Úi Chroiligh*) Derry. 'O'Crilly's burial place'.

Tamnaghbane (*Tamhnach Bán*) Armagh. 'White cultivated spot'.

Tamney (*Tamhnaigh*) Donegal. *Tamenagh* c.1655. 'Cultivated spot'.

Tamnyrankin (*Tamhnaigh*) Derry. 'Rankin's cultivated spot'.

Tamworth Staffs. *Tamouuorthig* 781, *Tamuuorde* 1086 (DB). 'Enclosure on the River Tame'. Celtic or pre-Celtic river-name (for which *see* THAMES) + OE *worthig* (replaced by *worth*).

Tandoo Point Dumf. Perhaps 'black buttock'. Gaelic *ton* + *dubh*.

Tandragee (*Tóin re Gaoith*) Armagh. *Tóin re Gaoith* 1642. 'Backside to the wind'.

Tandridge Surrey. *Tenhric c.*965, *Tenrige* 1086 (DB). OE *hrycg* 'ridge, hill' with an uncertain first element, possibly an OE **tended* 'lighted' (perhaps with reference to fires lit on the ridge during meetings of Tandridge Hundred).

Tanfield Durham. *Tamefeld* 1175. 'Open land on the River Team'. Celtic or pre-Celtic river-name (for which *see* THAMES) + OE *feld*.

Tanfield, East & Tanfield, West N. Yorks. *Tanefeld* 1086 (DB). Possibly 'open land where shoots or osiers grow'. OE *tān* + *feld*.

Tang (*An Teanga*) Westmeath. 'The tongue'.

Tangley Hants. *Tangelea* 1175. Possibly 'woodland clearing at the spits of land'. OE *tang* + *lēah*.

Tangmere W. Sussex. *Tangmere* 680, *Tangemere* 1086 (DB). Possibly 'the adjacent pools'. OE *getang* + *mere*.

Tankersley Barns. *Tancresleia* 1086 (DB). 'Woodland clearing of a man called Tancred or T(h)ancrad'. OGerman pers. name + OE *lēah*.

Tankerton Kent. *Tangerton* 1240. Probably 'farmstead or estate of a man called Tancred or T(h)ancrad'. OGerman pers. name + OE *tūn*.

Tannington Suffolk. *Tatintuna* 1086 (DB). 'Estate associated with a man called Tāta'. OE pers. name + *-ing-* + *tūn*.

Tansley Derbys. *Taneslege* 1086 (DB). Probably 'woodland clearing of a man called *Tān*', OE pers. name + *lēah*. Alternatively the first element may be OE *tān* 'branch' (perhaps used of 'a valley branching off from the main valley').

Tansor Northants. *Tanesovre* 1086 (DB). Probably 'promontory or bank of a man called *Tān*', OE pers. name + *ofer* or *ōfer*. Alternatively the first element may be OE *tān* 'branch' as in previous name.

Tanton N. Yorks. *Tametun* 1086 (DB). 'Farmstead or village on the River Tame'. Celtic or pre-Celtic river-name (for which *see* THAMES) + OE *tūn*.

Tanworth-in-Arden Warwicks. *Tanewrthe* 1201. Possibly 'enclosure made with branches'. OE *tān* + *worth*. For Arden, *see* HENLEY-IN-ARDEN.

Taplow Bucks. *Thapeslau* 1086 (DB). 'Tumulus of a man called *Tæppa*'. OE pers. name + *hlāw*.

Tara (*Teamhair*) Meath. 'Conspicuous place'.

Tarbat Ness Highland. 'Headland by a portage'. Gaelic *tairbeart* + OScand. *nes*. *See* TARBERT, TARBET.

Tarbert Arg. '(Place by a) portage'. Gaelic *tairbeart*.

Tarbert (*Tairbeart*) Kerry. 'Isthmus'.

Tarbet Arg. '(Place by a) portage'. Gaelic *tairbeart*.

Tarbock Green Knows. *Torboc* 1086 (DB), *Thornebrooke* c.1240. Probably 'thorn-tree brook'. OE *thorn* + *brōc*.

Tardebigge Worcs. *Tærdebicgan* c.1000, *Terdeberie* [sic] 1086 (DB). Possibly a Celtic name from Brittonic *tarth* 'spring' + *pīg* 'point, peak'.

Tarelton (*Tír Eiltín*) Cork. 'Eiltín's country'.

Tarleton Lancs. *Tarelton, Tarleton* 13th cent. Possibly 'farmstead or estate of a man called Tharaldr'. OScand. pers. name + OE *tūn*.

Tarlton Glos. *Torentune* [sic] 1086 (DB), *Torleton* 1204. Probably 'farmstead at the thorn-tree clearing'. OE *thorn* + *lēah* + *tūn*.

Tarporley Ches. *Torpelei* 1086 (DB), *Thorperlegh* 1281. Possibly 'woodland clearing of the peasants or cottagers'. OE *thorpere* + *lēah*.

Tarrant, Celtic river-name possibly meaning 'the trespasser', i.e. 'river liable to floods', and giving name to the following: **Tarrant Crawford** Dorset. *Tarente* 1086 (DB), *Little Craweford* 1280. Distinguishing affix from *Craveford* 1086 (DB), 'ford frequented by crows', OE *crāwe* + *ford*. **Tarrant Gunville** Dorset. *Tarente* 1086 (DB), *Tarente Gundevill* 1233. Manorial affix from the *Gundeville* family, here in the 12th cent. **Tarrant Hinton** Dorset. *Terente* 9th cent., *Tarente* 1086 (DB), *Tarente Hyneton* 1280. 'Estate on the River Tarrant belonging to a religious community (Shaftesbury Abbey)'. OE *hīwan* + *tūn*. **Tarrant Keyneston** Dorset. *Tarente* 1086 (DB), *Tarente Kahaines* 1225. 'Estate held by the *Cahaignes* family', here from the 12th cent. **Tarrant Launceston** Dorset. *Tarente* 1086 (DB), *Tarente Loueweniston* 1280. 'Estate held by a man called *Lēofwine* or a family called *Lowin*'. **Tarrant Monkton** Dorset. *Tarente* 1086 (DB), *Tarent Moneketon* 1280. 'Estate belonging to the monks (of Tewkesbury Abbey)'. OE *munuc* + *tūn*. **Tarrant Rawston** Dorset. *Tarente* 1086 (DB). *Tarrant Rawston alias Antyocke* 1535. 'Ralph's estate', earlier 'estate held by the *Antioch* family'. **Tarrant Rushton** Dorset. *Tarente* 1086 (DB), *Tarente Russeus* 1280. 'Estate held by the *de Rusceaus* family', here in the 13th cent.

Tarring Neville E. Sussex. *Toringes* 1086 (DB), *Thoring Nevell* 1339. Possibly '(settlement of) the family or followers of a man called *Teorra*', OE pers. name + *-ingas*. Alternatively 'dwellers at the rocky hill', OE *torr* + *-ingas*. Manorial affix from the *de Neville* family, here in the 13th cent.

Tarring, West W. Sussex. *Teorringas* 946, *Terringes* 1086 (DB). '(Settlement of) the family or followers of a man called *Teorra*'. OE pers. name + *-ingas*.

Tarrington Herefs. *Tatintune* 1086 (DB). 'Estate associated with a man called Tāta'. OE pers. name + -*ing*- + *tūn*.

Tarvie Highland. 'Place of bulls'. Gaelic *tarbhaidh*.

Tarvin Ches. *Terve* [*sic*] 1086 (DB), *Teruen* 1185. Celtic river-name meaning 'boundary (stream)', the old name of the River Gowy.

Tasburgh Norfolk. *Taseburc* 1086 (DB). Probably 'stronghold of a man called *Tæsa'. OE pers. name + *burh*. The river-name **Tas** is a back-formation from the place name.

Tasley Shrops. *Tasseleya c.*1143. OE *lēah* 'wood, clearing' with an uncertain first element.

Tatenhill Staffs. *Tatenhyll* 942. 'Hill of a man called Tāta'. OE pers. name (genitive -*n*) + *hyll*.

Tatham Lancs. *Tathaim* 1086 (DB). 'Homestead of a man called Tāta'. OE pers. name + *hām*.

Tathwell Lincs. *Tadewelle* 1086 (DB). 'Spring or stream frequented by toads'. OE *tāde* + *wella*.

Tatsfield Surrey. *Tatelefelle* 1086 (DB). 'Open land of a man called Tātel'. OE pers. name + *feld*.

Tattenhall Ches. *Tatenale* 1086 (DB). 'Nook of land of a man called Tāta'. OE pers. name (genitive -*n*) + *halh*.

Tatterford Norfolk. *Taterforda* 1086 (DB). 'Ford of a man called Tāthere'. OE pers. name + *ford*.

Tattersett Norfolk. *Tatessete* 1086 (DB). 'Fold of a man called Tāthere'. OE pers. name + *set*.

Tattershall Lincs. *Tateshale* 1086 (DB). 'Nook of land of a man called Tāthere'. OE pers. name + *halh*.

Tattingstone Suffolk. *Tatistuna* [*sic*] 1086 (DB), *Tatingeston* 1219. Possibly 'farmstead of a man called *Tāting', OE pers. name + *tūn*. Alternatively 'farmstead at the place associated with a man called Tāta', OE pers. name + -*ing* + *tūn*.

Tatworth Somerset. *Tattewurthe* 1254. 'Enclosure of a man called Tāta'. OE pers. name + *worth*.

Taughmaconnell (*Teach Mhic Conaill*) Roscommon. 'Mac Conaill's house'.

Taunton Somerset. *Tantun* 737, *Tantone* 1086 (DB). 'Farmstead or village on the River Tone'. Celtic river-name (meaning 'fire', i.e. 'sparkling stream') + OE *tūn*.

Taunton Deane (district) Somerset. 'Valley by TAUNTON', *see* previous name. A revival of an old hundred-name, preserved in the Vale of Taunton Deane. OE *denu*.

Taur (*Teamhair*) Cork. 'Conspicuous place'.

Taverham Norfolk. *Taverham* 1086 (DB). 'Red-lead homestead or enclosure', perhaps referring to a red-painted building or to soil colour. OE *tēafor* + *hām* or *hamm*.

Tavistock Devon. *Tauistoce* 981, *Tavestoc* 1086 (DB). 'Outlying farmstead or hamlet by the River Tavy'. Celtic or pre-Celtic river-name (possibly 'dark' or simply 'stream') + OE *stoc*.

Tavy, Mary & Tavy, Peter Devon. *Tavi*, *Tawi* 1086 (DB), *Tavymarie* 1412, *Petri Tavy* 1270. Named from the River Tavy, *see* previous name. Distinguishing affixes from the dedications of their churches.

Tawin Island (*Tamhain*) Galway. 'Stump'.

Tawnaghlahan (*Tamhnach Lahan*) Donegal. 'Broad cultivated spot'.

Tawnyeely (*Tamhnaigh Aelaigh*) Leitrim. 'Cultivated spot of lime'.

Tawnyinah (*Tamhnaigh an Eich*) Mayo. 'Clearing of the horse'.

Tawnylea (*Tamhnaí Liatha*) Sligo. 'Grey cultivated spot'.

Tawstock Devon. *Tauestoca* 1086 (DB). 'Outlying farmstead or hamlet on the River Taw'. Celtic river-name (*see* next name) + OE *stoc*.

Tawton, 'farmstead or village on the River Taw', Celtic river-name (possibly 'strong or silent stream') + OE *tūn*: **Tawton, Bishop's** Devon. *Tautona* 1086 (DB), *Tautone Episcopi* 1284. Affix from its possession by the Bishop (Latin *episcopus*) of Exeter in 1086. **Tawton, North & Tawton, South** Devon. *Tawetone*, *Tavetone* 1086 (DB).

Taxal Derbys. *Tackeshale c.*1251. Possibly 'nook of land held on lease'. ME *tak* + OE *halh*.

Tay. *See* TAYPORT.

Tay, Lough (*Loch Té*) Wicklow. 'Lake of Té'. Té is a goddess.

Taynton Glos. *Tetinton* 1086 (DB). 'Estate associated with a man called Tǣta'. OE pers. name + *-ing-* + *tūn*.

Tayport Fife. 'Port on the River Tay'. If Celtic, the river-name perhaps means 'silent one'; but it may be pre-Celtic.

Tealby Lincs. *Tavelesbi* 1086 (DB). Possibly an ancient tribal name OE **Tāflas* or **Tǣflas* (representing the East Germanic *Taifali*) to which OScand. *bý* 'farmstead, village' was later added.

Team (river) Durham, gives name to TANFIELD, *see* THAMES.

Tean, Lower & Tean, Upper Staffs. *Tene* 1086 (DB). Named from the River Tean, a Celtic river-name probably meaning simply 'stream'.

Tearaght Island (*An Tiaracht*) Kerry. Perhaps 'the horse's hindquarters'.

Tebay Cumbria. *Tibeia* c.1160. 'Island of a man called Tiba'. OE pers. name + *ēg*.

Tebworth Beds. *Teobbanwyrthe* 962. 'Enclosure of a man called Teobba'. OE pers. name (genitive -*n*) + *worth*.

Tedavnet (*Tigh Damhnata*) Monaghan. *Thechdamnad* 1306. 'Damhnat's house'.

Tedburn St Mary & Venny Tedburn Devon. *Teteborne* 1086 (DB), *Fennytetteburne* 1299, *Sct. Marytedborne* 1577. '(Place at) the stream of a woman called Tette or a man called **Tetta*'. OE pers. name + *burna*. Distinguishing affixes from the dedication of the church and from OE *fennig* 'marshy'.

Teddington Glos. *Teottingtun* 780, *Teotintune* 1086 (DB). 'Estate associated with a man called **Teotta*'. OE pers. name + *-ing-* + *tūn*.

Teddington Gtr. London. *Tudintun* 969. 'Estate associated with a man called Tuda'. OE pers. name + *-ing-* + *tūn*.

Tedstone Delamere & Tedstone Wafre Herefs. *Tedesthorne* 1086 (DB), *Teddesthorn la Mare*, *Teddesthorne Wafre* 1249. Probably 'thorn-tree of a man called **Tēod*'. OE pers. name + *thorn*. Manorial affixes from early possession by the *de la Mare* and *le Wafre* families.

Teelin (*Teileann*) Donegal. 'Dish'. The reference is to the shape of the small bay nearby.

Teermaclane (*Tír Mhic Calláin*) Clare. 'Mac Callán's country'.

Teermore (*Tír Mór*) Westmeath. 'Large district'.

Teernacreeve (*Tír dá Craebh*) Westmeath. 'District of two sacred trees'.

Tees (river) Cumbria–Durham. *See* STOCKTON ON TEES.

Teesdale (district) Durham. *Tesedale* 12th cent. 'Valley of the River Tees'. Celtic or pre-Celtic river-name (*see* STOCKTON ON TEES) + OE *dæl* or OScand. *dalr*.

Teeton Northants. *Teche* [*sic*] 1086 (DB), *Teacne* 1195. '(Hill with) a beacon'. OE **tǣcne*.

Teevurcher (*Taobh Urchair*) Meath. 'Side of the cast'.

Teffont Evias & Teffont Magna Wilts. *Tefunte* 860, *Tefonte* 1086 (DB), *Teffunt Ewyas* 1275. 'Boundary spring'. OE **tēo* + **funta*. Manorial affix from possession by the barons of *Ewyas* in the 13th cent.; the other distinguishing affix is Latin *magna* 'great'.

Teigh Rutland. *Tie* 1086 (DB). OE *tēag* 'a small enclosure'.

Teignbridge (district) Devon. *Teingnebrige* 1187. 'Bridge over the River Teign'. A revival of an old hundred-name. Celtic river-name (*see* next name) + OE *brycg*.

Teigngrace Devon. *Taigne* 1086 (DB). *Teyngegras* 1331. Named from the River Teign, a Celtic river-name meaning 'the sweeper'. Manorial affix from the *Gras* family, here in the 14th cent.

Teignmouth Devon. *Tengemutha* 1044. 'Mouth of the River Teign'. Celtic river-name (*see* previous name) + OE *mūtha*.

Telford Tel. & Wrek. a modern name commemorating the engineer Thomas *Telford* (1757–1834), famous for his bridges, roads, and canals.

Tellisford Somerset. *Tefleford* 1001, *Tablesford* 1086 (DB). Possibly 'ford of the table or level place'. OE *tefl* (genitive -*es*) + *ford*. Or the first element may be an OE pers. name such as **Thēofol*.

Telscombe E. Sussex. *Titelescumbe* 966. 'Valley of a man called *Titel'. OE pers. name + *cumb*.

Teme (river) Shrops.–Worcs.–Herefs., gives name to TENBURY WELLS, *see* THAMES.

Temple as affix. *See* main name, e.g. for **Temple Cloud** B. & NE. Som., *see* CLOUD.

Temple, The Down. *The Temple* 1858. The name was originally that of a public house used as a meeting-place for a freemasons' lodge.

Templeboy (*Teampall Baoith*) Sligo. 'Baoth's church'.

Templebrendan (*Teampall Brendáin*) Mayo. 'Brendan's church'.

Templecombe Somerset. *Come* 1086 (DB), *Cumbe Templer* 1291. '(Place at) the valley', OE *cumb*, later with affix from its early possession by the Knights Templars.

Templecronan (*Teampall Crónáin*) Clare. 'Cronan's church'.

Templederry (*Teampall Doire*) Tipperary. 'Church of the oak wood'.

Templeetney (*Teampall Eithne*) Tipperary. 'Eithne's church'.

Templefinn (*Teampall Fionn*) Down. 'White church'.

Templeglentan (*Teampall an Ghleanntáin*) Limerick. 'Church of the valley'.

Templemartin (*Teampall Mártan*) Cork, Wexford. 'Mártan's church'.

Templemichael (*Teampall Mhichíl*) Cork. 'Michael's church'.

Templemore (*An Teampall Mór*) Tipperary. 'The big church'.

Templemoyle (*Teampall Maol*) Down. 'Bald church'.

Templenakilla (*Teampall na Cille*) Kerry. 'Church of the church'.

Templenoe (*Teampall Nua*) Kerry. 'New church'.

Templeogue (*Teach Mealóg*) Dublin. 'Mealóg's house'.

Templeoran (*Teampall Fhuaráin*) Westmeath. 'Church of the spring'.

Templeorum (*Teampall Fhothram*) Kilkenny. 'Church of Fothram'.

Templepatrick (*Teampall Phádraig*) Antrim, Mayo. 'Patrick's church'.

Templeshambo (*Teampall Seanbhoth*) Wexford. 'Church of the old huts'.

Templeton Devon. *Templum* 1206, *Templeton* 1334. 'Manor held by the Knights Templar'. ME *temple* + OE *tūn*.

Templetouhy (*Teampall Tuaithe*) Tipperary. 'Church of the territory'.

Tempo (*An tIompú Deiseal*) Fermanagh. *Tempodessell* 1622. 'The right-hand turn'.

Tenbury Wells Worcs. *Tamedeberie* 1086 (DB). 'Stronghold on the River Teme'. Celtic or pre-Celtic river-name (for which *see* THAMES) + OE *burh*. The addition *Wells* referring to the spa here is only recent.

Tenby (*Dinbych-y-pysgod*) Pemb. *Dinbych* c.1275, *Tynby* 1369. 'Little fort'. Welsh *din* + *bych*. Compare DENBIGH. The Welsh name means 'little fort of the fish' (Welsh *din* + *bych* + *y* + *pysgod*), referring to the town's importance as a fishing port.

Tendring Essex. *Tendringa* 1086 (DB). Possibly 'place where tinder or fuel is gathered'. OE *tynder* + *-ing*.

Tenterden Kent. *Tentwardene* 1179. 'Woodland pasture of the Thanet dwellers'. THANET + OE *-ware* + *denn*.

Terenure (*Tír an Iúir*) Dublin. 'Territory of the yew'.

Terling Essex. *Terlinges* 1017–35, *Terlingas* 1086 (DB). '(Settlement of) the family or followers of a man called Tyrhtel'. OE pers. name + *-ingas*. The river-name **Ter** is a back-formation from the place name.

Termon (*An Tearmann*) Cavan, Donegal. 'The sanctuary land'.

Termonfeckin (*Tearmann Feichín*) Louth. 'Feichín's sanctuary'.

Termonmaguirk (*Tearmann*) Tyrone. 'Maguirk's sanctuary'.

Tern (river) Shrops. *See* EATON UPON TERN.

Terrington N. Yorks. *Teurinctune* 1086 (DB). Possibly 'estate associated with a man called *Teofer'. OE pers. name + *-ing-* + *tūn*.

Terrington St Clement & Terrington St John Norfolk. *Tilinghetuna* 1086 (DB). *Terintona* 1121. 'Farmstead of the family or followers of a man called *Tir(a)'. OE pers. name + *-inga-* + *tūn*. Distinguishing affixes from the dedications of the churches.

Terryglass (*Tír dhá Ghlas*) Tipperary. 'Territory of two streams'.

Test Valley (district) Hants. A modern name, from the River Test (for which *see* TESTWOOD).

Teston Kent. *Terstan* 10th cent., *Testan* 1086 (DB). 'Stone with a gap or hole'. OE *tær* + *stān*.

Testwood Hants. *Lesteorde* [*sic*] 1086 (DB), *Terstewode* c.1185. 'Wood by the River Test'. Celtic river-name (probably meaning 'rapidly flowing one') + OE *wudu*.

Tetbury Glos. *Tettanbyrg* c.900, *Teteberie* 1086 (DB). 'Fortified place or manor house of a woman called Tette or a man called *Tetta'. OE pers. name + *burh* (dative *byrig*).

Tetchill Shrops. *Tetneshul* 13th cent. 'Hill of a man called *Tetīn'. OE pers. name + *hyll*.

Tetcott Devon. *Tetecote* 1086 (DB). 'Cottage of a woman called Tette or of a man called *Tetta'. OE pers. name + *cot*.

Tetford Lincs. *Tedforde* 1086 (DB). 'People's or public ford'. OE *thēod* + *ford*.

Tetney Lincs. *Tatenai* 1086 (DB). 'Island of a man called Tæta or a woman called Tæte'. OE pers. name (genitive *-n*) + *ēg*.

Tetsworth Oxon. *Tetleswrthe* c.1150. 'Enclosure of a man called *Tætel'. OE pers. name + *worth*.

Tettenhall Wolverh. *Teotanheale* early 10th cent., *Totenhale* 1086 (DB). 'Nook of land of a man called *Tēotta'. OE pers. name (genitive *-n*) + *halh*.

Teversal Notts. *Tevreshalt* 1086 (DB). 'Shelter of the painter or sorcerer, or of a man called *Teofer'. OE *tīefrere* or pers. name + *hald*.

Teversham Cambs. *Teuersham* 1086 (DB). 'Homestead of the painter or sorcerer, or of a man called *Teofer'. OE *tīefrere* or pers. name + *hām*.

Tevrin (*Teamhrín*) Westmeath. 'Small conspicuous hill'.

Tew, Great, Tew, Little, & Tew, Duns Oxon. *Tiwan* 1004, *Tewe, Teowe* 1086 (DB), *Donestiua* c.1210. Possibly an OE *tīewe* 'row or ridge'. Manorial affix from early possession by a man called *Dunn*.

Tewin Herts. *Tiwingum* 944–6, *Teuuinge* 1086 (DB). Possibly '(settlement of) the family or followers of a man called *Tīwa'. OE pers. name + *-ingas*.

Tewkesbury Glos. *Teodekesberie* 1086 (DB). 'Fortified place of a man called *Tēodec'. OE pers. name + *burh* (dative *byrig*).

Tey, Great & Tey, Marks Essex. *Tygan* c.950, *Teia* 1086 (DB), *Merkys Teye* 1475. From OE *tīege* 'enclosure'. Manorial affix from possession by the *de Merck* family.

Teynham Kent. *Tenham* 798, *Therham* [*sic*] 1086 (DB). Probably 'homestead of a man called *Tēna'. OE pers. name + *hām*.

Thakeham W. Sussex. *Taceham* 1086 (DB). 'Homestead with a thatched roof'. OE *thaca* + *hām*.

Thame Oxon. *Tame* c.1000, 1086 (DB). Named from the River Thame, a Celtic or pre-Celtic river-name, *see* THAMES.

Thames (river) Glos.–London. *Tamesis* 51 BC. An ancient river-name, possibly from a Celtic root *tam-* 'dark' or rather from a pre-Celtic root *tā-* 'melt, flow turbidly'. The river-names **Tame** (three examples, N. Yorks., Warwicks.–Staffs., Lancs.–Ches.), **Team** (Durham), **Teme** (Shrops.–Worcs.–Herefs.), **Thame** (Bucks.–Oxon.), and (with a different ending) **Tamar** (Cornwall–Devon) are from the same root and probably have a similar meaning.

Thames Ditton Surrey. *See* DITTON.

Thamesmead Gtr. London, a recent name for a new town (1967) by the Thames.

Thanet Kent. *Tanatus* 3rd cent., *Tanet* 1086 (DB). A Celtic name possibly meaning 'bright island', perhaps with reference to a beacon.

Thanington Kent. *Taningtune* 833. Possibly 'estate associated with a man called *Tān'. OE pers. name + *-ing-* + *tūn*.

Tharston Norfolk. *Therstuna* 1086 (DB). 'Farmstead or village of a man called Therir'. OScand. pers. name + OE *tūn*.

Thatcham W. Berks. *Thæcham c.*954, *Taceham* 1086 (DB). 'Thatched homestead, or river-meadow where thatching materials are got'. OE **thæcce* + *hām* or *hamm*.

Thatto Heath St Hel. *Thetwall* 12th cent. 'Spring or stream with a water-pipe'. OE *thēote* + *wella*.

Thaxted Essex. *Tachesteda* 1086 (DB). 'Place where thatching materials are got'. OE *thæc* + *stede*.

Theakston N. Yorks. *Eston* [*sic*] 1086 (DB), *Thekeston* 1157. Probably 'farmstead or village of a man called *Thēodec'. OE pers. name + *tūn*.

Thealby N. Lincs. *Tedulfbi* 1086 (DB). 'Farmstead or village of a man called Thjóthulfr'. OScand. pers. name + *bý*.

Theale, 'the planks' (referring to a bridge or building), OE *thel* in a plural form *thelu*: **Theale** Somerset. *Thela* 1176. **Theale** W. Berks. *Teile* 1208.

Thearne E. R. Yorks. *Thoren* 1297. 'The thorn-tree'. OE *thorn*.

Theberton Suffolk. *Thewardetuna* [*sic*] 1086 (DB), *Tiberton* 1176. 'Farmstead or village of a man called Thēodbeorht'. OE pers. name + *tūn*.

Thedden Grange Hants. *Tedena* [*sic*] 1168, *Thetdene* 1234. 'Valley with a water-pipe'. OE *thēote* + *denu*.

Theddingworth Leics. *Tedingeswarde* [*sic*] 1086 (DB), *Thedingewrth* 1207. Probably 'enclosure of the family or followers of a man called *Thēoda'. OE pers. name + *-inga-* + *worth*.

Theddlethorpe All Saints & Theddlethorpe St Helen Lincs. *Tedlagestorp* 1086 (DB). Possibly 'outlying farmstead or hamlet of a man called *Thēodlāc'. OE pers. name + OScand. *thorp*. Distinguishing affixes from the dedications of the churches.

Thelbridge Devon. *Talebrige* 1086 (DB). 'Plank bridge'. OE *thel* + *brycg*.

Thelnetham Suffolk. *Thelueteham* 1086 (DB). Possibly 'enclosure frequented by swans near a plank bridge'. OE *thel* + *elfitu* + *hamm*.

Thelwall Warrtn. *Thelwæle* 923. 'Pool by a plank bridge'. OE *thel* + *wēl*.

Themelthorpe Norfolk. *Timeltorp* 1203. Possibly 'outlying farmstead or hamlet of a man

called *Thȳmel or *Thymill'. OE or OScand. pers. name + *thorp*. Alternatively the first element may be OE *thȳmel* 'a thimble' alluding to the small size of the settlement.

Thenford Northants. *Taneford* 1086 (DB). 'Ford of the thegns or retainers'. OE *thegn* + *ford*.

Therfield Herts. *Therefeld* 1060, *Derevelde* 1086 (DB). 'Dry open land'. OE *thyrre* + *feld*.

Thetford, 'people's or public ford', OE *thēod* + *ford*: **Thetford** Norfolk. *Theodford* late 9th cent., *Tedfort* 1086 (DB). **Thetford, Little** Cambs. *Thiutforda c.*972, *Liteltedford* 1086 (DB). Affix is OE *lȳtel*.

Theydon Bois Essex. *Thecdene* 1062, *Teidana* 1086 (DB), *Teidon Boys* 1257. Probably 'valley where thatching materials are got'. OE *thæc* + *denu*. Manorial affix from the *de Bosco* or *de Boys* family, here in the 12th cent.

Thimbleby, 'farmstead or village of a man called *Thymill or *Thymli', OScand. pers. name + *bý*: **Thimbleby** Lincs. *Stimblebi* [*sic*] 1086 (DB), *Timblebi c.*1115. **Thimbleby** N. Yorks. *Timbelbi* 1086 (DB).

Thingwall Wirral. *Tinguelle* 1086 (DB). 'Field where an assembly meets'. OScand. *thing-vǫllr*.

Thirkleby N. Yorks. *Turchilebi* 1086 (DB). 'Farmstead or village of a man called Thorkell'. OScand. pers. name + *bý*.

Thirlby N. Yorks. *Trillebia* 1187. 'Farmstead or village of a man called *Thrylli, or of the thralls'. OScand. pers. name or *thrǽll* + *bý*.

Thirn N. Yorks. *Thirne* 1086 (DB). 'The thorn-tree'. OE *thyrne*.

Thirsk N. Yorks. *Tresch* 1086 (DB). OScand. **thresk* 'a marsh'.

Thirston, East Northum. *Thrasfriston* 1242. Possibly 'farmstead or village of a man called *Thræsfrith'. OE pers. name + *tūn*.

Thistleton, 'farmstead or village where thistles grow', OE *thistel* + *tūn*: **Thistleton** Lancs. *Thistilton* 1212. **Thistleton** Rutland. *Tisteltune* 1086 (DB).

Thixendale N. Yorks. *Sixtendale* 1086 (DB). 'Valley of a man called Sigsteinn'. OScand. pers. name + *dalr*.

Thockrington Northum. *Thokerinton* 1223.
'Farmstead or estate associated with a man
called *Thocer'. OE pers. name + -*ing*- + *tūn*.

Tholthorpe N. Yorks. *Turulfestorp* 1086 (DB).
'Outlying farmstead or hamlet of a man called
Thórulfr'. OScand. pers. name + *thorp*.

Thomastown Kilkenny. 'Thomas's town'.
The Irish name of Thomastown is *Baile Mhic
Andáin*, 'Fitzanthony's homestead'. *Thomas*
Fitzanthony had a castle here in the 13th cent.

Thompson Norfolk. *Tomestuna* 1086 (DB).
'Farmstead or village of a man called Tumi'.
OScand. pers. name + OE *tūn*.

Thompson's Bridge Fermanagh. A man
named *Thompson* owned a bridge over the
Sillees River here.

Thong Kent. *Thuange* c.1200. 'Narrow strip of
land'. OE *thwang*.

Thonoge (*Tonóg*) (river) Tipperary. 'Little
bottom'.

Thoralby N. Yorks. *Turoldesbi* 1086 (DB).
Probably 'farmstead or village of a man called
Thóraldr'. OScand. pers. name + *bý*.

Thoresby Notts. *Thuresby* 958, *Turesbi* 1086
(DB). 'Farmstead or village of a man called
Thúrir'. OScand. pers. name + *bý*.

Thoresby, North Lincs. *Toresbi* 1086 (DB).
'Farmstead or village of a man called Thórir'.
OScand. pers. name + *bý*.

Thoresby, South Lincs. *Toresbi* 1086 (DB).
Identical in origin with the previous name.

Thoresway Lincs. *Toreswe* 1086 (DB),
Thoresweie 1187. Possibly 'way or road of a man
called Thórir', OScand. pers. name + OE *weg*.
Alternatively the DB form may represent
OScand. **Thórswǣ* 'shrine dedicated to the
heathen god Thor' in which the second element
wǣ was later confused with OE *weg*.

Thorganby, 'farmstead or village of a man
called Thorgrímr', OScand. pers. name + *bý*:
Thorganby Lincs. *Turgrimbi* 1086 (DB).
Thorganby N. Yorks. *Turgisbi* [sic] 1086 (DB),
Turgrimebi 1192.

Thorington Suffolk. *Torentuna* 1086 (DB).
'Thorn-tree enclosure or farmstead'. OE *thorn*
or *thyrne* + *tūn*.

Thorlby N. Yorks. *Toreilderebi* 1086 (DB).
'Farmstead or village of a man called Thóraldr,

or of a woman called Thórhildr'. OScand. pers.
name + *bý*.

Thorley, 'thorn-tree wood or clearing', OE
thorn + *lēah*: **Thorley** Herts. *Torlei* 1086 (DB).
Thorley I. of Wight. *Torlei* 1086 (DB).

Thormanby N. Yorks. *Tormozbi* 1086 (DB).
'Farmstead or village of a man called
Thormóthr'. OScand. pers. name + *bý*.

Thornaby on Tees Stock. on T. *Tormozbi*
1086 (DB). Identical in origin with the previous
name.

Thornage Norfolk. *Tornedis* 1086 (DB).
'Thorn-tree enclosure or pasture'. OE *thorn* +
edisc.

Thornborough, 'hill where thorn-trees
grow', OE *thorn* + *beorg*: **Thornborough**
Bucks. *Torneberge* 1086 (DB). **Thornborough**
N. Yorks. *Thornbergh* 1198.

Thornbury, 'fortified place where thorn-trees
grow', OE *thorn* + *burh* (dative *byrig*):
Thornbury Devon. *Torneberie* 1086 (DB).
Thornbury Herefs. *Thornbyrig* c.1000,
Torneberie 1086 (DB). **Thornbury** S. Glos.
Turneberie 1086 (DB).

Thornby Northants. *Torneberie* 1086 (DB),
Thirnebi c.1160. 'Farmstead or village where
thorn-trees grow'. OScand. *thyrnir* + *bý*
(replacing OE *burh* 'stronghold').

Thorncombe, 'valley where thorn-trees
grow', OE *thorn* + *cumb*: **Thorncombe** Dorset,
near Blandford. *Tornecome* 1086 (DB).
Thorncombe Dorset, near Holditch.
Tornecoma 1086 (DB).

Thorndon Suffolk. *Tornduna* 1086 (DB). 'Hill
where thorn-trees grow'. OE *thorn* + *dūn*.

Thorne, '(place at) the thorn-tree', OE *thorn*:
Thorne Donc. *Torne* 1086 (DB). **Thorne St
Margaret** Somerset. *Torne* 1086 (DB), *Thorn
St Margaret* 1251. Affix from the dedication of
the church.

Thorner Leeds. *Tornoure* 1086 (DB). 'Ridge or
bank where thorn-trees grow'. OE *thorn* + **ofer*.

Thorney, usually 'thorn-tree island', OE
thorn + *ēg*; examples include: **Thorney** Peterb.
Thornige c.960, *Torny* 1086 (DB). **Thorney,
West** W. Sussex. *Thorneg* 11th cent., *Tornei*
1086 (DB).

However the following has a different origin:
Thorney Notts. *Torneshaie* 1086 (DB). 'Thorn-tree enclosure'. OE *thorn* + *haga*.

Thornfalcon Somerset. *Torne* 1086 (DB), *Thorn fagun* 1265. '(Place at) the thorn-tree', OE *thorn*. Manorial affix from early possession by a family called *Fagun*.

Thornford Dorset. *Thornford* 951, *Torneford* 1086 (DB). 'Ford where thorn-trees grow'. OE *thorn* + *ford*.

Thorngumbald E. R. Yorks. *Torne* 1086 (DB), *Thoren Gumbaud* 1297. '(Place at) the thorn-tree'. OE *thorn*. Manorial affix from the *Gumbald* family, here in the 13th cent.

Thornham, 'homestead or village where thorn-trees grow', OE *thorn* + *hām*: **Thornham** Norfolk. *Tornham* 1086 (DB). **Thornham Magna & Thornham Parva** Suffolk. *Thornham* 1086 (DB). Affixes are Latin *magna* 'great' and *parva* 'little'.

Thornhaugh Peterb. *Thornhawe* 1189. 'Thorn-tree enclosure'. OE *thorn* + *haga*.

Thornhill, 'hill where thorn-trees grow', OE *thorn* + *hyll*; examples include: **Thornhill** Derbys. *Tornhull* 1200. **Thornhill** Kirkl. *Tornil* 1086 (DB).

Thornley Durham. near Crook. *Thornley* 1382. 'Thorn-tree wood or clearing'. OE *thorn* + *lēah*.

Thornley Durham. near Peterlee. *Thornhlawa* 1071–80. 'Thorn-tree hill or mound'. OE *thorn* + *hlāw*.

Thornthwaite, 'thorn-tree clearing', OScand. *thorn* + *thveit*: **Thornthwaite** Cumbria, near Keswick. *Thornthwayt* 1254. **Thornthwaite** N. Yorks. *Tornthueit* 1230.

Thornton, a common name, 'thorn-tree enclosure or farmstead', OE *thorn* + *tūn*; examples include: **Thornton** Brad. *Torentone* 1086 (DB). **Thornton** Lancs. *Torentun* 1086 (DB). **Thornton, Bishop** N. Yorks. *Thorntune* c.1030, *Torentone* 1086 (DB). Affix refers to its early possession by the Archbishop of York. **Thornton, Childer** Ches. *Thorinthun* c.1210, *Childrethornton* 1288. Affix means 'of the young men' (from OE *cild*, genitive plural *cildra*), with reference to the young monks of St Werburgh's Abbey in Chester. **Thornton le Dale** N. Yorks. *Torentune* 1086 (DB). Affix is OScand. *dalr* 'valley'. **Thornton Heath** Gtr. London. *Thorneton Hethe* 1511. With OE *hǣth* 'heath'.

Thoroton Notts. *Toruertune* 1086 (DB). 'Farmstead or village of a man called Thorfrøthr'. OScand. pers. name + OE *tūn*.

Thorp, Thorpe, a common name, from OScand. *thorp* 'outlying farmstead or hamlet, dependent secondary settlement', except for a few South Country instances which are from OE *throp* of similar meaning; examples include: **Thorp Arch** Leeds. *Torp* 1086 (DB), *Thorp de Arches* 1272. Manorial affix from the *de Arches* family, here in the 11th cent. **Thorpe** Norfolk. *Torpe* 1254. **Thorpe** Surrey. *Thorp* 672–4, *Torp* 1086 (DB). **Thorpe-le-Soken** Essex. *Torp* 12th cent., *Thorpe in ye Sooken* 1612. Affix is from OE *sōcn* 'district with special jurisdiction'. **Thorpe Morieux** Suffolk. *Thorp* 962–91, *Torp(a)* 1086 (DB), *Thorp Morieux* 1330. Manorial affix from the *de Murious* family, here in 1201. **Thorpe on the Hill** Lincs. *Torp* 1086 (DB). **Thorpe St Andrew** Norfolk. *Torp* 1086 (DB). Affix from the dedication of the church. **Thorpe Salvin** Rothm. *Torp* 1086 (DB), *Thorpe Saluayn* 1255. Manorial affix from the *Salvain* family, here in the 13th cent. **Thorpe Willoughby** N. Yorks. *Torp* 1086 (DB), *Thorp Wyleby* 1276. Manorial affix from the *de Willeby* family, here in the 13th cent.

Thorpeness Suffolk. *Torp* 1086 (DB). From OScand. *thorp* 'outlying farmstead or hamlet, dependent secondary settlement', with the later addition of *ness* (OE *nœss*, OScand. *nes*) 'headland, promontory'.

Thorrington Essex. *Torinduna* [sic] 1086 (DB), *Torritona* 1202. Probably 'thorn-tree enclosure or farmstead'. OE *thorn* or *thyrne* + *tūn*.

Thorverton Devon. *Toruerton* 1182. Probably 'farmstead by the thorn-tree ford'. OE *thorn* + *ford* + *tūn*.

Thrandeston Suffolk. *Thrandeston* c.1035, *Thrandestuna* 1086 (DB). 'Farmstead or village of a man called Thrándr'. OScand. pers. name + OE *tūn*.

Thrapston Northants. *Trapestone* 1086 (DB). 'Farmstead or village of a man called *Thræpst'. OE pers. name + *tūn*.

Threapwood Ches. *Threpewood* 1548. 'Disputed wood'. OE *thrēap* + *wudu*.

Threave Dumf. 'Farm'. Gaelic *treabh*.

Three Bridges W. Sussex. *Le three bridges* 1613. Self-explanatory.

Three Rivers (district) Herts. A modern name, referring to the Rivers Chess (*see* CHESHAM), Gade (*see* GADDESDEN), and Colne (*see* COLNEY, LONDON).

Threekingham Lincs. *Trichingeham* 1086 (DB). 'Homestead of the tribe called Tricingas'. Old tribal name (origin obscure) + OE *hām*.

Threepwood Sc. Bord. *Trepewode c.*1230. 'Disputed woodland'. OE *thrēap + wudu*.

Threlkeld Cumbria. *Trellekell* 1197. 'Spring of the thralls or serfs'. OScand. *thræll + kelda*.

Threshfield N. Yorks. *Freschefelt* [*sic*] 1086 (DB), *Threskefeld* 12th cent. 'Open land where corn is threshed'. OE **thresc + feld*.

Thrigby Norfolk. *Trikebei* 1086 (DB). 'Farmstead or village of a man called **Thrykki*'. OScand. pers. name + *bý*.

Thringstone Leics. *Trangesbi* [*sic*] 1086 (DB), *Trengeston c.*1200. 'Farmstead or village of a man called **Thræingr*'. OScand. pers. name + OE *tūn* (replacing OScand. *bý*).

Thrintoft N. Yorks. *Tirnetofte* 1086 (DB). 'Thorn-tree homestead'. OScand. *thyrnir + toft*. Alternatively the first element may be the OScand. man's name *Thyrnir*.

Thriplow Cambs. *Tripelan* [*sic*] *c.*1050, *Trepeslau* 1086 (DB). 'Hill or tumulus of a man called **Tryppa*'. OE pers. name + *hlāw*.

Throcking Herts. *Trochinge* 1086 (DB). Possibly 'place where beams are used or obtained'. OE *throc + -ing*.

Throckley Newc. upon T. *Trokelawa* 1177. Possibly 'hill where beams are obtained'. OE *throc + hlāw*.

Throckmorton Worcs. *Trochemerton* 1176. Possibly 'farmstead by a pool with a beam bridge'. OE *throc + mere + tūn*.

Throphill Northum. *Trophil* 1166. 'Hamlet hill'. OE *throp + hyll*.

Thropton Northum. *Tropton* 1177. 'Estate with an outlying farmstead or hamlet', OE *throp + tūn*.

Throston, High Hartlepl. *Thoreston c.*1300. 'Farmstead or village of a man called Thórir or Thórr'. OScand. pers. name + OE *tūn*.

Throwleigh Devon. *Trule* 1086 (DB). 'Woodland clearing with or near a conduit'. OE *thrūh + lēah*.

Throwley Kent. *Trevelai* 1086 (DB). Identical in origin with the previous name.

Thrumpton Notts. near Ruddington. *Turmodestun* 1086 (DB). 'Farmstead or village of a man called Thormóthr'. OScand. pers. name + OE *tūn*.

Thrupp, 'outlying farmstead or hamlet', OE *throp*: **Thrupp** Glos. *Trop* 1261. **Thrupp** Oxon. *Trop* 1086 (DB).

Thrushelton Devon. *Tresetone* 1086 (DB). 'Farmstead or village frequented by thrushes'. OE **thryscele + tūn*.

Thruxton, 'estate or manor of a man called Thorkell', OScand. pers. name + OE *tūn*: **Thruxton** Hants. *Turkilleston* 1167. **Thruxton** Herefs. *Torchestone* [*sic*] 1086 (DB), *Turkelestona* 1160–70.

Thrybergh Rothm. *Triberge* 1086 (DB). 'Three hills'. OE *thrī + beorg*.

Thundersley Essex. *Thunreslea* 1086 (DB). 'Sacred grove of the heathen god Thunor'. OE god-name + *lēah*.

Thurcaston Leics. *Turchitelestone* 1086 (DB). 'Farmstead or village of a man called Thorketill'. OScand. pers. name + OE *tūn*.

Thurcroft Rothm. *Thurscroft* 1319. 'Enclosure of a man called Thórir'. OScand. pers. name + OE *croft*.

Thurgarton, 'farmstead or village of a man called Thorgeirr', OScand. pers. name + OE *tūn*: **Thurgarton** Norfolk. *Thurgartun* 1044–7, *Turgartuna* 1086 (DB). **Thurgarton** Notts. *Turgarstune* 1086 (DB).

Thurgoland Barns. *Turgesland* 1086 (DB). 'Cultivated land of a man called Thorgeirr'. OScand. pers. name + *land*.

Thurlaston, 'farmstead or village of a man called Thorleifr', OScand. pers. name + OE *tūn*: **Thurlaston** Leics. *Turlauestona* 1166. **Thurlaston** Warwicks. *Torlauestone* 1086 (DB).

Thurlby, 'farmstead or village of a man called Thórulfr', OScand. pers. name + *bý*: **Thurlby** Lincs., near Bourne. *Turolvebi* 1086 (DB). **Thurlby** Lincs., near Lincoln. *Turolue(s)bi* 1086 (DB).

Thurleigh Beds. *La Lega* 1086 (DB), *Thyrleye* 1372. '(Place at) the wood or clearing'. From OE *lēah* with the remains of the OE definite article *thǣre* (dative).

Thurles (*Durlas*) Tipperary. 'Strong fort'.

Thurlestone Devon. *Thyrelanstane* 847, *Torlestan* 1086 (DB). 'Stone or rock with a hole in it'. OE *thyrel* + *stān*.

Thurlow, Great & Thurlow, Little Suffolk. *Tridlauua* 1086 (DB). Probably 'burial mound of the warriors'. OE *thrȳth* + *hlāw*.

Thurloxton Somerset. *Turlakeston* 1195. 'Farmstead or village of a man called Thorlákr'. OScand. pers. name + OE *tūn*.

Thurlstone Barns. *Turulfestune* 1086 (DB). 'Farmstead or village of a man called Thórulfr'. OScand. pers. name + OE *tūn*.

Thurlton Norfolk. *Thuruertuna* 1086 (DB). 'Farmstead or village of a man called Thorfrithr'. OScand. pers. name + OE *tūn*.

Thurmaston Leics. *Turmodestone* 1086 (DB). 'Farmstead or village of a man called Thormóthr'. OScand. pers. name + OE *tūn*.

Thurnby Leics. *Turneby* 1156. 'Farmstead or village of a man called Thyrne, or where thorn-bushes grow'. OScand. pers. name or *thyrnir* + *bý*.

Thurnham Lancs. *Tiernun* [*sic*] 1086 (DB), *Thurnum* 1160. '(Place at) the thorn-trees'. OE *thyrne* in a dative plural form *thyrnum*.

Thurning, 'place where thorn-trees grow', OE *thyrne* + *-ing*: **Thurning** Norfolk. *Turninga* 1086 (DB). **Thurning** Northants. *Torninge* 1086 (DB).

Thurnscoe Barns. *Ternusche* 1086 (DB), *Thirnescoh c.*1090. 'Thorn-tree wood'. OScand. *thyrnir* + *skógr*.

Thurrock, Little & Thurrock, West Thurr. *Thurroce* 1040–2, *Thurrucca* 1086 (DB). 'Place where filthy water collects'. OE *thurruc*. Originally applied to a stretch of marshland west of Tilbury. *See also* GRAYS.

Thursby Cumbria. *Thoresby c.*1165. 'Farmstead or village of a man called Thórir'. OScand. pers. name + *bý*.

Thursford Norfolk. *Turesfort* 1086 (DB). 'Ford associated with a giant or demon'. OE *thyrs* + *ford*.

Thursley Surrey. *Thoresle* 1292. Probably 'sacred grove of the heathen god Thunor'. OE god-name + *lēah*.

Thurso Highland. *Thorsa* 1152. '(Place on the) River Thurso'. The Celtic river-name probably means 'bull river', from the old name of nearby Dunnet Head, known to the Romans as *Tarvedunum*, 'bull fort'.

Thurstaston Wirral. *Turstanetone* 1086 (DB). 'Farmstead or village of a man called Thorsteinn'. OScand. pers. name + *tún* or OE *tūn*.

Thurston Suffolk. near Bury. *Thurstuna* 1086 (DB). 'Farmstead or village of a man called Thóri'. OScand. pers. name + OE *tūn*.

Thurstonfield Cumbria. *Turstanfeld c.*1210. 'Open land of a man called Thorsteinn'. OScand. pers. name + OE *feld*.

Thurstonland Kirkl. *Tostenland* 1086 (DB). 'Cultivated land of a man called Thorsteinn'. OScand. pers. name + *land*.

Thurton Norfolk. *Tortuna* 1086 (DB). Probably 'thorn-tree enclosure or farmstead'. OE *thorn* or *thyrne* + *tūn*.

Thurvaston Derbys. *Turverdestune* 1086 (DB). 'Farmstead or village of a man called Thorfrithr'. OScand. pers. name + OE *tūn*.

Thuxton Norfolk. *Turstanestuna* 1086 (DB). 'Farmstead or village of a man called Thorsteinn'. OScand. pers. name + OE *tūn*.

Thwaite, 'the clearing, meadow, or paddock', OScand. *thveit*; examples include: **Thwaite** Suffolk. *Theyt* 1228. **Thwaite St Mary** Norfolk. *Thweit* 1254. Affix from the dedication of the church.

Thwing E. R. Yorks. *Tuuenc* 1086 (DB). 'Narrow strip of land'. OScand. *thvengr* or OE *thweng*.

Tibberton, 'farmstead or village of a man called Tīdbeorht', OE pers. name + *tūn*: **Tibberton** Glos. *Tebriston* 1086 (DB). **Tibberton** Tel. & Wrek. *Tetbristone* 1086 (DB). **Tibberton** Worcs. *Tidbrihtincgtun* 978–92, *Tidbertun* 1086 (DB).

Tibenham Norfolk. *Tybenham* 1044–7, *Tibenham* 1086 (DB). 'Homestead of a man called Tibba'. OE pers. name (genitive *-n*) + *hām*.

Tibohine (*Tigh Baoithín*) Mayo. 'Baoithín's church'.

Tibshelf Derbys. *Tibecel* 1086 (DB). *Tibbeshelf* 1179. 'Shelf or ledge of a man called Tibba'. OE pers. name + *scelf*.

Tibthorpe E. R. Yorks. *Tibetorp* 1086 (DB). 'Outlying farmstead or hamlet of a man called Tibbi or Tibba'. OScand. or OE pers. name + OScand. *thorp*.

Ticehurst E. Sussex. *Tycheherst* 1248. 'Wooded hill where young goats are kept'. OE *ticce*(*n*) + *hyrst*.

Tichborne Hants. *Ticceburna* c.909. 'Stream frequented by young goats'. OE *ticce*(*n*) + *burna*.

Tickencote Rutland. *Tichecote* 1086 (DB). 'Shed for young goats'. OE *ticcen* + *cot*.

Tickenham N. Som. *Ticaham* 1086 (DB). 'Homestead or enclosure of a man called Tica, or where young goats are kept'. OE pers. name (genitive *-n*) or *ticcen* + *hām* or *hamm*.

Tickhill Donc. *Tikehill* 12th cent. 'Hill of a man called Tica, or where young goats are kept'. OE pers. name or *ticce*(*n*) + *hyll*.

Ticknall Derbys. *Ticenheale* c.1002. 'Nook of land where young goats are kept'. OE *ticcen* + *halh*.

Tickton E. R. Yorks. *Tichetone* 1086 (DB). 'Farmstead of a man called Tica, or where young goats are kept'. OE pers. name or *ticce*(*n*) + *tūn*.

Ticloy (*Tigh Cloiche*) Antrim. 'House of stone'.

Tidcombe Wilts. *Titicome* 1086 (DB). Probably 'valley of a man called *Titta'. OE pers. name + *cumb*.

Tiddington Oxon. *Titendone* 1086 (DB). 'Hill of a man called *Tytta'. OE pers. name (genitive *-n*) + *dūn*.

Tiddington Warwicks. *Tidinctune* 969. 'Estate associated with a man called Tīda'. OE pers. name + *-ing-* + *tūn*.

Tideford Cornwall. *Tutiford* 1201. 'Ford over the River Tiddy'. Celtic river-name (probably meaning 'the soaker') + OE *ford*.

Tidenham Glos. *Dyddanhamme* 956, *Tedeneham* 1086 (DB). 'Enclosure of a man

called *Dydda'. OE pers. name (genitive *-n*) + *hamm*.

Tideswell Derbys. *Tidesuuelle* 1086 (DB). 'Spring or stream of a man called Tīdi'. OE pers. name + *wella*.

Tidmarsh W. Berks. *Tedmerse* 1196. 'People's or common marsh'. OE *thēod* + *mersc*.

Tidmington Warwicks. *Tidelminctune* 977, *Tidelmintun* 1086 (DB). 'Estate associated with a man called Tīdhelm'. OE pers. name + *-ing-* + *tūn*.

Tidworth, North (Wilts.) & **Tidworth, South** (Hants.). *Tudanwyrthe* c.990, *Tode*(*w*)*orde* 1086 (DB). 'Enclosure of a man called Tuda'. OE pers. name + *worth*.

Tiermaclane (*Tír Mhic Calláin*) Clare. 'Mac Callán's country'.

Tievemore (*Taobh Mór*) Donegal. 'Large side'.

Tiffield Northants. *Tifelde* 1086 (DB). Possibly 'open land with or near a meeting-place'. OE *tīg + feld*.

Tilbrook Cambs. *Tilebroc* 1086 (DB). Probably 'brook of a man called Tila'. OE pers. name + *brōc*. Alternatively the first element may be the OE adjective *til* 'useful' from which the pers. name derives.

Tilbury Thurr. *Tilaburg* 731, *Tiliberia* 1086 (DB). Probably 'stronghold of a man called Tila'. OE pers. name + *burh* (dative *byrig*). Alternatively the first element may be a lost stream-name *Tila* 'the useful one'.

Tilehurst Readg. *Tigelherst* 1167. 'Wooded hill where tiles are made'. OE *tigel + hyrst*.

Tilford Surrey. *Tileford* c.1140. 'Useful ford, or ford of a man called Tila'. OE *til* or pers. name + *ford*.

Tillicoultry Clac. *Tulycultri* 1195, *Tullicultre* c.1199. Possibly 'hill at the back of the settlement'. Gaelic *tulach + cùl + treabh*. Tillicoultry is on the edge of the Ochill Hills.

Tillingham Essex. *Tillingaham* c.1000, *Tillingham* 1086 (DB). 'Homestead of the family or followers of a man called Tilli'. OE pers. name + *-inga-* + *hām*.

Tillington Herefs. *Tullinton* c.1180. 'Estate associated with a man called Tulla or *Tylla'. OE pers. name + *-ing-* + *tūn*.

Tillington W. Sussex. *Tullingtun* 960. 'Estate associated with a man called Tulla'. OE pers. name + *-ing-* + *tūn*.

Tilly Whim Caves Dorset. *Tilly Whim* 1811. From the surname *Tilly* and *whim* 'a windlass' (used for the lowering of stone to boats from the quarry here, unworked since 1812).

Tilmanstone Kent. *Tilemanestone* 1086 (DB). 'Farmstead or village of a man called Tilmann'. OE pers. name + *tūn*.

Tilney All Saints & Tilney St Lawrence Norfolk. *Tilnea* 1170. 'Useful island, or island of a man called Tila'. OE *til* (dative *-an*) or pers. name (genitive *-n*) + *ēg*. Distinguishing affixes from the dedications of the churches.

Tilshead Wilts. *Tidulfhide* 1086 (DB). 'Hide of land of a man called Tīdwulf'. OE pers. name + *hīd, see* SHREWTON.

Tilstock Shrops. *Tildestok* 1211. 'Outlying farmstead or hamlet of a woman called Tīdhild'. OE pers. name + *stoc*.

Tilston Ches. *Tilleston* 1086 (DB). 'Stone of a man called Tilli or Tilla'. OE pers. name + *stān*.

Tilstone Fearnall Ches. *Tidulstane* 1086 (DB), *Tilston Farnhale* 1427. 'Stone of a man called Tīdwulf'. OE pers. name + *stān*. Affix is the name of a local place now lost, 'fern nook', OE *fearn* + *halh*.

Tilsworth Beds. *Pileworde* [*sic*] 1086 (DB), *Thuleswrthe* 1202. Probably 'enclosure of a man called *Thýfel* or *Tyfel*'. OE pers. name + *worth*.

Tilton Leics. *Tiletone* 1086 (DB). 'Farmstead or village of a man called Tila'. OE pers. name + *tūn*.

Timahoe (*Tigh Mochua*) Kildare, Laois. 'Mochua's house'.

Timberland Lincs. *Timberlunt* 1086 (DB). 'Grove where timber is obtained'. OE *timber* or OScand. *timbær* + *lundr*.

Timberscombe Somerset. *Timbrecumbe* 1086 (DB). 'Valley where timber is obtained'. OE *timber* + *cumb*.

Timble N. Yorks. *Timmel* *c.*972, *Timble* 1086 (DB). Possibly from an OE **tymbel* 'a fall of earth' or 'a tumbling stream'.

Timoleague (*Tigh Molaige*) Cork. 'Molaga's house'.

Timolin (*Tigh Moling*) Kildare. 'Moling's house'.

Timperley Traffd. *Timperleie* 1211-25. OE *lēah* 'wood, clearing', possibly with an OE **timper* 'spar, stretcher'.

Timsbury B. & NE. Som. *Timesberua* [*sic*] 1086 (DB), *Timberesberwe* 1233. 'Grove where timber is obtained'. OE *timber* + *bearu*.

Timsbury Hants. *Timbreberie* 1086 (DB). 'Timber or wooden fort or manor'. OE *timber* + *burh* (dative *byrig*).

Timworth Green Suffolk. *Timeworda* 1086 (DB). 'Enclosure of a man called **Tima*'. OE pers. name + *worth*.

Tinahely (*Tigh na hÉille*) Wicklow. 'House of the latchet'.

Tincleton Dorset. *Tincladene* 1086 (DB). 'Valley of the small farms'. OE **fynincel* + *denu*.

Tindale Cumbria. *Tindale* late 12th cent. 'Valley of the River South Tyne'. Celtic or pre-Celtic river-name (meaning simply 'river') + OScand. *dalr*.

Tingewick Bucks. *Tedinwiche* 1086 (DB). Probably 'dwelling or (dairy) farm at the place associated with a man called Tīda or Tēoda'. OE pers. name + *-ing* + *wīc*.

Tingrith Beds. *Tingrei* [*sic*] 1086 (DB), *Tingrith* *c.*1215. 'Stream by which assemblies are held'. OE *thing* + *rīth*.

Tinnakilla (*Tigh na Coille*) Wexford. 'House of the wood'.

Tinode (*Tigh an Fhóid*) Westmeath. 'House of the sod'.

Tinsley Sheff. *Tineslauue* 1086 (DB). 'Mound of a man called **Tynni*'. OE pers. name + *hlāw*.

Tintagel Cornwall. *Tintagol* *c.*1137. Probably 'fort by the neck of land'. Cornish **din* + **tagell*.

Tintern (*Tyndyrn*) Mon. *Dindyrn* *c.*1150. 'King's fortress'. Welsh *din* + *teyrn*.

Tintinhull Somerset. *Tintanhulle* 10th cent., *Tintenella* 1086 (DB). OE *hyll* 'hill' with an obscure first part, possibly a Celtic name from **dīn* 'fort' or **tīn* 'rump' + **tnou* 'valley'.

Tintwistle Derbys. *Tengestvisie* 1086 (DB). Probably 'river-fork of the prince'. OE *thengel* + *twisla*.

Tinwald Dumf. *Tynwalde c.*1280. 'Place of assembly'. OScand. *thing-vǫllr*.

Tinwell Rutland. *Tedinwelle* [*sic*] 1086 (DB), *Tinningewelle* mid-13th cent. Possibly 'spring or stream of the family or followers of a man called *Tȳni*'. OE pers. name + *-inga-* + *wella*.

Tipperary (*Tiobraid Árann*) Tipperary. 'Ára's well'.

Tipperty Aber. *Typpertay* 1501. 'Well place'. OGaelic *tiobartach*.

Tiptoe Hants. First on record in 1555, probably a manorial name from the *Typetot* family mentioned in this area from the 13th cent.

Tipton Sandw. *Tibintone* 1086 (DB). 'Estate associated with a man called Tibba'. OE pers. name + *-ing-* + *tūn*.

Tiptree Essex. *Tipentrie* 12th cent. Probably 'tree of a man called *Tippa*'. OE pers. name + *trēow*.

Tirawley (*Tír Amhalghaidh*) Mayo. 'Amhalghaidh's territory'.

Tirboy (*Tír Buí*) Galway. 'Yellow territory'.

Tiree (island) Arg. *Tir Iath c.*850, *Tiryad* 1343. 'Eth's land'. Gaelic *tìr*. *Eth* has not been identified and is probably pre-Celtic.

Tireragh (*Tír Fhiachrach*) Sligo. 'Fiachra's territory'.

Tirkane (*Tír Chiana*) Derry. *Tír Chiana c.*1740. Possibly 'Cian's territory'.

Tirley Glos. *Trineleie* 1086 (DB). 'Circular woodland clearing'. OE *trind* + *lēah*.

Tirnamona (*Tír na Móna*) Monaghan. 'Territory of the bog'.

Tisbury Wilts. *Tyssesburg c.*800. *Tisseberie* 1086 (DB). 'Stronghold of a man called *Tyssi*'. OE pers. name + *burh* (dative *byrig*).

Tissington Derbys. *Tizinctun* 1086 (DB). 'Estate associated with a man called Tīdsige'. OE pers. name + *-ing-* + *tūn*.

Tisted, East & Tisted, West Hants. *Ticcesstede* 932, *Tistede* 1086 (DB). 'Place where young goats are kept'. OE *ticce(n)* + *stede*.

Titchfield Hants. *Ticefelle* 1086 (DB). 'Open land where young goats are kept'. OE *ticce(n)* + *feld*.

Titchmarsh Northants. *Tuteanmersc* 973, *Ticemerse* 1086 (DB). 'Marsh of a man called Tyccea'. OE pers. name + *mersc*.

Titchwell Norfolk. *Ticeswelle c.*1035, *Tigeuuella* 1086 (DB). 'Spring or stream frequented by young goats'. OE *ticce(n)* + *wella*.

Titley Herefs. *Titelege* 1086 (DB). Probably 'woodland clearing of a man called *Titta*'. OE pers. name + *lēah*.

Titlington Northum. *Titlingtona* 12th cent. Probably 'estate associated with a man called *Titel*'. OE pers. name + *-ing-* + *tūn*.

Titsey Surrey. *Tydices eg* 10th cent., *Ticesei* 1086 (DB). 'Well-watered land of a man called *Tydic*'. OE pers. name + *ēg*.

Tittensor Staffs. *Titesovre* [*sic*] 1086 (DB), *Titneshovere* 1236. 'Ridge of a man called *Titten*'. OE pers. name + *ǒfer*.

Tittleshall Norfolk. *Titeshala* [*sic*] 1086 (DB), *Titleshal* 1200. 'Nook of land of a man called *Tyttel*'. OE pers. name + *halh*.

Tiverton Ches. *Tevreton* 1086 (DB). 'Red-lead farmstead', perhaps denoting a red-painted building or where red-lead was available. OE *tēafor* + *tūn*.

Tiverton Devon. *Twyfyrde* 880–5, *Tovretona* 1086 (DB). 'Farmstead or village at the double ford'. OE *twī-fyrde* + *tūn*.

Tivetshall St Margaret & Tivetshall St Mary Norfolk. *Teueteshala* 1086 (DB). OE *halh* 'nook of land', possibly with an old form of dialect *tewhit* 'lapwing'. Distinguishing affixes from the dedications of the churches.

Tixall Staffs. *Ticheshale* 1086 (DB). 'Nook of land where young goats are kept'. OE *ticcen* (genitive *ticnes*) + *halh*.

Tixover Rutland. *Tichesovre* 1086 (DB). 'Promontory where young goats are pastured'. OE *ticcen* + *ǒfer*.

Toames (*Tuaim*) Cork. 'Mound'.

Tobarmacdugh (*Tobar Mhic Dhuach*) Galway. 'Mac Duach's well'.

Tober (*Tobar*) Cavan. 'Well'.

Toberavilla (*Tobar an Bhile*) Kerry. 'Well of the sacred tree'.

Toberbreedy (*Tobar Bhríde*) Cork. 'Bríd's well'.

Toberbunny (*Tobar Bainne*) Dublin. 'Well of milk'.

Tobercurry (*Tobar an Choire*) Sligo. 'Well of the cauldron'.

Toberdoney (*Tobar Domhnaigh*) Antrim, Louth, etc. 'Sunday's well'.

Tobermogue (*Tobar Mhodóg*) Westmeath. 'Modóg's well'.

Tobermore (*Tobar Mór*) Derry. *Tobarmore* 1613. 'Big well'.

Tobermory Arg. (Mull). *Tibbermore* 1540. 'St Mary's well'. Gaelic *tiobar* + *Moire*.

Tockenham Wilts. *Tockenham* 854, *Tocheham* 1086 (DB). 'Homestead or enclosure of a man called *Toca*'. OE pers. name + *hām* or *hamm*.

Tockholes Black. w. Darw. *Tocholis* c.1200. 'Hollows of a man called *Toca or Tók(i)*'. OE or OScand. pers. name + *hol*.

Tockington S. Glos. *Tochintune* 1086 (DB). 'Estate associated with a man called *Toca*'. OE pers. name + *-ing-* + *tūn*.

Tockwith N. Yorks. *Tocvi* 1086 (DB). 'Wood of a man called Tóki'. OScand. pers. name + *vithr*.

Todber Dorset. *Todeberie* 1086 (DB). Possibly 'hill or grove of a man called Tota'. OE pers. name + *beorg* or *bearu*. Alternatively the first element may be OE *tōte* 'look-out place'.

Toddington Beds. *Totingedone* 1086 (DB). 'Hill of the family or followers of a man called Tuda'. OE pers. name + *-inga-* + *dūn*.

Toddington Glos. *Todintun* 1086 (DB). 'Estate associated with a man called Tuda'. OE pers. name + *-ing-* + *tūn*.

Todenham Glos. *Todanhom* 804, *Teodeham* 1086 (DB). 'Enclosed valley of a man called Tēoda'. OE pers. name (genitive *-n*) + *hamm*.

Todmorden Calder. *Tottemerden* 1246. 'Boundary valley of a man called Totta'. OE pers. name + *mǣre* + *denu*.

Todwick Rothm. *Tatewic* 1086 (DB). 'Dwelling or (dairy) farm of a man called Tāta'. OE pers. name + *wīc*.

Toem (*Tuaim*) Tipperary. 'Mound'.

Toft, Tofts, from OScand. *toft* 'curtilage or homestead'; examples include: **Toft** Cambs. *Tofth* 1086 (DB). **Toft Monks** Norfolk. *Toft* 1086 (DB). Affix from its possession by the Norman Abbey of Préaux in the 12th cent.

Tofts, West Norfolk. *Stofftam* [*sic*] 1086 (DB), *Toftes* 1199.

Toftrees Norfolk. *Toftes* 1086 (DB). Identical in origin with the previous name.

Toghblane (*Teach Bhláin*) Down. 'Bláin's house'.

Togher (*Tóchar*) Cork, Louth, Offaly. 'Causeway'.

Togherbane (*Tóchar Bán*) Kerry. 'White causeway'.

Togherbeg (*Tóchar Beag*) Galway, Wicklow. 'Little causeway'.

Togston Northum. *Toggesdena* 1130. 'Valley of a man called *Tocg*'. OE pers. name + *denu*.

Tolland Somerset. *Taa land* late 11th cent., *Talanda* 1086 (DB). OE *land* 'cultivated land', possibly with OE *tā* 'toe' used figuratively for 'narrow strip of land between streams'.

Tollard Royal Wilts. *Tollard* 1086 (DB), *Tollard Ryall* 1535. 'Hollow hill'. Celtic **tull* + *ardd*. Affix from lands here held c.1200 by King John.

Toller Fratrum & Toller Porcorum
Dorset. *Tolre* 1086 (DB), *Tolre Fratrum, Tolre Porcorum* 1340. Named from the River *Toller*, now the River Hooke. *Toller* is an old Celtic river-name meaning 'hollow stream'. The humorously contrasting affixes are Latin *fratrum* 'of the brethren' (with reference to early possession by the Knights Hospitaller) and *porcorum* 'of the pigs' (with reference to its herds of swine).

Tollerton N. Yorks. *Toletun* 972. *Tolentun* 1086 (DB). 'Farmstead or village of the tax-gatherers'. OE *tolnere* + *tūn*.

Tollerton Notts. *Troclauestune* 1086 (DB). 'Farmstead or village of a man called Thorleifr'. OScand. pers. name + OE *tūn*.

Tollesbury Essex. *Tolesberia* 1086 (DB). 'Stronghold of a man called *Toll*'. OE pers. name + *burh* (dative *byrig*).

Tolleshunt D'Arcy & Tolleshunt Major Essex. *Tollesfuntan* c.1000, *Toleshunta* 1086 (DB). 'Spring of a man called *Toll*'. OE pers. name + **funta*. Manorial affixes from possession by the *Darcy* family (here in the 15th cent.) and by a tenant called *Malger* in 1086.

Tolpuddle Dorset. *Pidele* 1086 (DB), *Tollepidele* 1210. 'Estate on the River Piddle belonging to a woman called Tóla'. OE river-name (*see* PIDDLEHINTON) with OScand. pers. name. The Tóla in question is recorded as giving her lands including Tolpuddle to Abbotsbury Abbey before 1066.

Tolsta Head W. Isles. (Lewis). 'Thørlfr's farm'. OScand. pers. name + *stathr*, with English *head* 'headland'.

Tolworth Gtr. London. *Taleorde* 1086 (DB). 'Enclosure of a man called *Tala'. OE pers. name + *worth*.

Tomhaggard (*Teach Moshagard*) Wexford. 'Moshagra's house'.

Tomich Highland. 'Place of knolls'. Gaelic *tomach*.

Tomintoul Moray. 'Barn knoll'. Gaelic *toman t-sabail*.

Ton (*Tóin*) Monaghan. 'Bottom'.

Tonbridge Kent. *Tonebrige* 1086 (DB). Probably 'bridge belonging to the estate or manor'. OE *tūn* + *brycg*.

Tonduff (*Tóin Dubh*) Donegal. 'Black bottom'.

Tone (river) Somerset. *See* TAUNTON.

Tong, 'tongs-shaped feature', e.g. 'fork of a river', OE *tang*, **twang* (apparently replacing a related form **tweonga* of similar meaning): **Tong** Brad. *Tuinc* [*sic*] 1086 (DB), *Tange* 1176. **Tong** Shrops. *Tweongan* 10th cent., *Tvange* 1086 (DB).

Tonge Leics. *Tunge* 1086 (DB). 'The tongue of land'. OScand. *tunga*.

Tongham Surrey. *Tuangham* 1189. 'Homestead or enclosure by the fork of a river'. OE *tang* or **twang* + *hām* or *hamm*.

Tongue Highland. *Toung* 1542. '(Place on) the tongue of land'. OScand. *tunga*.

Tonragee (*Tóin re Gaoith*) Mayo. 'Backside to the wind'.

Tonypandy Rhon. 'Grassland of the fulling mill'. Welsh *ton* + *y* + *pandy*.

Toom (*Tuaim*) Tipperary. 'Mound'.

Toomard (*Tuaim Ard*) Galway. 'High mound'.

Toombeola (*Tuaim Beola*) Galway. 'Beola's mound'.

Toome (*Tuaim*) Antrim. 'Burial mound'. (*for*) *Fertais Tuamma* c.900. The earlier name was *Fearsaid Thuama*, 'sandbank ford of Toome'.

Toomebridge (*Droichead Thuama*) Antrim. 'Bridge of the mound'.

Toomyvara (*Tuaim Uí Mheára*) Tipperary. 'Ó Meára's burial mound'.

Toor (*Tuar*) Limerick. 'Bleach green'.

Tooraree (*Tuar an Fraoigh*) Limerick. 'Bleach green of the heather'.

Tooreen (*An Tuairín*) Mayo. 'The little bleach green'.

Tooreennafersha (*Tuairín na Feirste*) Kerry. 'Little bleach green of the sandbank'.

Tooreennahone (*Tuairín na hÓn*) Kerry. 'Little bleach green of the hole'.

Toorevagh (*Tuar Riabhach*) Westmeath. 'Grey bleach green'.

Toormore (*An Tuar Mór*) Cork. 'The big bleach green'.

Toot Baldon Oxon. *See* BALDON.

Tooting Bec & Tooting Graveney Gtr. London. *Totinge* 672–4, *Totinges* 1086 (DB), *Totinge de Bek* 1255, *Thoting Gravenel* 1272. Possibly '(settlement of) the family or followers of a man called Tōta', from OE pers. name + *-ingas*. Alternatively 'people of the look-out place', from OE **tōt* + *-ingas*. Distinguishing affixes from early possession by the Norman Abbey of Bec-Hellouin and by the *de Gravenel* family.

Topcliffe N. Yorks. *Topeclive* 1086 (DB). 'The highest part of the bank or cliff'. OE *topp* + *clif*.

Topcroft Norfolk. *Topecroft* 1086 (DB). 'Enclosure of a man called Tópi'. OScand. pers. name + OE *croft*.

Toppesfield Essex. *Topesfelda* 1086 (DB). 'Open land on or by a hill-top'. OE *topp* + *feld*.

Topsham Devon. *Toppeshamme* 937, *Topeshant* [sic] 1086 (DB). 'Promontory called **Topp* (the top or summit)'. OE *topp* + *hamm*.

Torbay Torbay. *Torrebay* 1401. 'Bay near TORRE'. ME *bay*.

Torbryan Devon. *Torre* 1086 (DB), *Torre Briane* 1238. 'The rocky hill'. OE *torr*, with manorial affix from the *de Brione* family, here in the 13th cent.

Torfaen (the unitary authority). 'Stone gap'. Welsh *tor* + *maen*.

Torksey Lincs. *Turecesieg c*.900, *Torchesey* 1086 (DB). Probably 'island (of dry ground in marsh) of a man called **Turoc*'. OE pers. name + *ēg*.

Tormarton S. Glos. *Tormentone* [sic] 1086 (DB), *Tormarton* 1166. Probably 'farmstead by the pool where thorn-trees grow'. OE *thorn* + *mere* + *tūn*.

Torpenhow Cumbria. *Torpennoc* 1163. 'Ridge of the hill with a rocky peak'. OE *torr* + Celtic **penn* + OE *hōh*.

Torpoint Cornwall. *Tor-point* 1746, earlier *The Torr* 1617. ME *torr* 'crag' + *point*.

Torquay Torbay. *Torrekay* 1591. 'Quay near Torre', *see* TORRE. ME *key*.

Torrance E. Dunb. 'Little hills'. Gaelic *torran*.

Torre Devon. *Torre* 1086 (DB). 'The rocky hill'. OE *torr*.

Torridge (district) Devon. A modern adoption of the Celtic river-name Torridge (for which *see* TORRINGTON, BLACK).

Torridon Highland. *Torvirlane* 1464. 'Place of portage'. Gaelic *tairbheartan*.

Torrington, Black, Torrington, Great, & Torrington, Little Devon. *Tori(n)tona* 1086 (DB), *Blaketorrintun* 1219. 'Farmstead or village on the River Torridge'. Celtic river-name (meaning 'turbulent stream') + OE *tūn*. Affix is OE *blæc* 'dark-coloured' (referring to the river here).

Torrington, East & Torrington, West Lincs. *Terintone* 1086 (DB). 'Estate associated with a man called **Tir(a)*'. OE pers. name + *-ing-* + *tūn*.

Torrisholme Lancs. *Toredholme* 1086 (DB). 'Island of a man called Thóraldr'. OScand. pers. name + *holmr*.

Tortington W. Sussex. *Tortinton* 1086 (DB). 'Estate associated with a man called **Torhta*'. OE pers. name + *-ing-* + *tūn*.

Tortworth S. Glos. *Torteword* 1086 (DB). 'Enclosure of a man called **Torhta*'. OE pers. name + *worth*.

Torver Cumbria. *Thoruergh* 1190-9. 'Turf-roofed shed or shieling', or 'shieling of a man called **Thorfi*'. OScand. *torf* or pers. name + *erg*.

Torworth Notts. *Turdeworde* 1086 (DB). Probably 'enclosure of a man called Thórthr'. OScand. pers. name + OE *worth*.

Tory Island (*Toraigh*) Donegal. (*fasughadh*) *Toraighe* 611. 'Abounding in rocky heights'.

Toseland Cambs. *Toleslund* 1086 (DB). 'Grove of a man called Tóli'. OScand. pers. name + *lundr*.

Tosson, Great Northum. *Tossan* 1205. Possibly 'look-out stone'. OE **tōt* + *stān*.

Tostock Suffolk. *Totestoc* 1086 (DB). 'Outlying farmstead or hamlet by the look-out place'. OE **tōt* + *stoc*.

Totham, Great & Totham, Little Essex. *Totham c*.950, *Tot(e)ham* 1086 (DB). 'Homestead or village by the look-out place'. OE **tōt* + *hām*.

Totland I. of Wight. *Totland* 1608. 'Cultivated land or estate with a look-out place'. OE **tōt* + *land*.

Totley Sheff. *Totingelei* 1086 (DB). 'Woodland clearing of the family or followers of a man called Tota'. OE pers. name + *inga-* + *lēah*.

Totnes Devon. *Totanæs c*.1000, *Toteneis* 1086 (DB). 'Promontory of a man called Totta'. OE pers. name + *næss*.

Tottenham Gtr. London. *Toteham* 1086 (DB), *Totenham* 1189. 'Homestead or village of a man called Totta'. OE pers. name (genitive *-n*) + *hām*.

Tottenhill Norfolk. *Tottenhella* 1086 (DB). 'Hill of a man called Totta'. OE pers. name (genitive *-n*) + *hyll*.

Totteridge Gtr. London. *Taderege* 12th cent., *Taterige* 1230. 'Ridge of a man called Tāta'. OE pers. name + *hrycg*.

Totternhoe Beds. *Totenehou* 1086 (DB). 'Hill-spur with a look-out house'. OE *tōt + ærn + hōh*.

Tottington Bury. *Totinton* 1212. Probably 'estate associated with a man called Tota'. OE pers. name + *-ing- + tūn*. Alternatively the first element may be OE *tōt* 'look-out hill'.

Totton Hants. *Totintone* 1086 (DB). Identical in origin with the previous name.

Tour (*Tuar*) Limerick. 'Bleach green'.

Tourlestrane (*Tuar Loistreáin*) Sligo. 'Loistreán's bleach green'.

Tourmakeady (*Tuar Mhic Éadaigh*) Mayo. 'Mac Éadaigh's bleach green'.

Tournafulla (*Tuar na Fola*) Limerick. 'Bleach green of the blood'.

Tow Law Durham. *Tollawe* 1423. Possibly 'look-out hill or mound'. OE *tōt + hlāw*.

Towcester Northants. *Tofeceaster* early 10th cent., *Tovecestre* 1086 (DB). 'Roman fort on the River Tove'. OE river-name (meaning 'slow') + *ceaster*.

Towednack Cornwall. 'Parish of *Sanctus Tewennocus*' 1327. From the dedication of the church to St Winwaloe (of which name *To-Winnoc* is a pet-form).

Tower Hamlets Gtr. London. term used at least as early as the 17th cent. for 'hamlets under the jurisdiction of the Tower of London'.

Towersey Oxon. *Eie* 1086 (DB), *Turrisey* 1240. 'The island held by the *de Turs* family' (here from the 13th cent.). OE *ēg* with later manorial affix.

Towthorpe York. *Touetorp* 1086 (DB). 'Outlying farmstead or hamlet of a man called Tófi'. OScand. pers. name + *thorp*.

Towton N. Yorks. *Touetun* 1086 (DB). 'Farmstead or village of a man called Tófi'. OScand. pers. name + *tūn*.

Towy. See TYWI VALLEY.

Towyn. See TYWYN.

Toxteth Lpool. *Stochestede* [*sic*] 1086 (DB), *Tokestath* 1212. 'Landing-place of a man called Tóki or *Tōk*'. OScand. pers. name + *stoth*.

Toye (*Tuaith*) Down. 'Territory'.

Toynton All Saints & Toynton St Peter Lincs. *Totintun(e)* 1086 (DB). 'Estate associated with a man called Tota'. OE pers. name + *tōt* 'look-out hill'. Affixes from the dedications of the churches.

Toynton, High Lincs. *Tedin-*, *Todintune* 1086 (DB). 'Estate associated with a man called Tēoda'. OE pers. name + *-ing- + tūn*.

Trafford Park Traffd. *Stratford* 1206. 'Ford on a Roman road'. OE *strǣt + ford*.

Trafford, Bridge & Trafford, Mickle Ches. *Trosford*, *Traford* 1086 (DB). 'Trough ford'. OE *trog + ford*. Distinguishing affixes are OE *brycg* and OScand. *mikill* 'great'.

Trafrask (*Trá Phraisce*) Cork. 'Strand of kale'.

Tralee (*Trá Lí*) Kerry. 'Strand of the (river) Lí'.

Trallwng, Y. See WELSHPOOL.

Tramore (*Trá Mhór*) Waterford. 'Big strand'.

Tranarossan (*Trá na Rosán*) Donegal. 'Strand of the Rosses'.

Tranent E. Loth. *Tranent* 1210. 'Farm on the streams'. Welsh *tref + nant* (plural *neint*). Compare TREFNANT.

Tranmere Wirral. *Tranemul* late 12th cent. 'Sandbank frequented by cranes'. OScand. *trani + melr*.

Tranwell Northum. *Trennewell* 1268. 'Spring or stream frequented by cranes'. OScand. *trani* + OE *wella*.

Trasna Island (*Trasna*) Fermanagh. *Trassna* 1610. 'Island across'.

Trawden Lancs. *Trochdene* 1296. 'Trough-shaped valley'. OE *trog + denu*.

Trawmore (*Trá Mhór*) Mayo. 'Big strand'.

Trawsfynydd Gwyd. *Trausvenith* 1292. '(Place by a route) across the mountain'. Welsh *traws + mynydd*.

Tre-Einar. See RINASTON.

Tre-Gwyr. See GOWERTON.

Treales Lancs. *Treueles* 1086 (DB). 'Farmstead of a court'. Welsh *tref* + *llys*.

Treamlod. *See* AMBLESTON.

Trearddur Angl. *Tre Iarthur* 1609, *Treyarddur* 1691. 'Iarddur's farm' (Welsh *tref*). *Iarddur* owned one of the largest farms in northwest Anglesey in medieval times.

Treborough Somerset. *Traberge* 1086 (DB). 'Hill or mound growing with trees'. OE *trēow* + *beorg*.

Tredegar Blae. The town arose in the 19th cent. on land owned by Baron Tredegar, whose title came from *Tredegar*, 'Tegyr's farm' (Welsh *tref*, 'farm'), near Newport. *Compare* NEW TREDEGAR.

Tredington Warwicks. *Tredingctun* 757, *Tredinctun* 1086 (DB). 'Estate associated with a man called Tyrdda'. OE pers. name + *-ing-* + *tūn*.

Treeton Rothm. *Tretone* 1086 (DB). Probably 'farmstead built with posts'. OE *trēow* + *tūn*.

Trefaldwyn. *See* MONTGOMERY.

Trefdraeth. *See* NEWPORT (Pemb.).

Trefeglwys Powys. 'Farm by the church'. Welsh *tref* + *eglwys*.

Treffynnon. *See* HOLYWELL.

Trefgarn Pemb. *Traueger* 1324, *Trefgarn* 1368. 'Farm by the rock'. Welsh *tref* + *carn*.

Trefnant Denb. *Trevenant* 1661. 'Farm on the stream'. Welsh *tref* + *nant*.

Treforys. *See* MORRISTON.

Trefriw Conwy. *Treffruu* 1254. 'Farm on the hill'. Welsh *tref* + *rhiw*.

Trefwrdan. *See* JORDANSTON.

Trefyclo. *See* KNIGHTON.

Trefynwy. *See* MONMOUTH.

Tregaron Cergn. *tre garon* 1566, *Caron alias Tre Garon* 1763. 'Village on the River Caron'. Welsh *tref*. The river is named from St *Caron*, to whom the church is dedicated.

Tregatwg. *See* CADOXTON.

Tregetin. *See* KEESTON.

Treglemais Pemb. *Trefclemens* 1326. 'Clement's farm'. Welsh *tref*.

Tregony Cornwall. *Trefhrigoni* 1049, *Treligani* 1086 (DB). Possibly 'farm of a man called *Rigni*. Cornish *tre* + pers. name.

Trelales. *See* LALESTON.

Treletert. *See* LETTERSTON.

Tremadog Gwyd. 'Village of Madocks'. Welsh *tref*. The name is that of William Alexander *Madocks*, who gave the name of nearby PORTHMADOG.

Tremaine Cornwall. *Tremen* c.1230. 'Farm of the stone'. Cornish *tre* + *men*.

Tremarchog. *See* ST NICHOLAS (Pemb.).

Trematon Cornwall. *Tremetone* 1086 (DB). Cornish *tre* 'farm' + unknown word or pers. name + OE *tūn* 'estate'.

Treneglos Cornwall. *Treneglos* 1269. 'Farm of the church'. Cornish *tre* + *an* 'the' + *eglos*.

Trent Dorset. *Trente* 1086 (DB). Originally the name of the stream here, a Celtic river-name possibly meaning 'the trespasser', i.e. 'river liable to floods'.

Trentham Stoke. *Trenham* 1086 (DB). 'Homestead or river-meadow on the River Trent'. Celtic river-name (identical with previous name) + OE *hām* or *hamm*.

Trentishoe Devon. *Trendesholt* [sic] 1086 (DB), *Trenlesho* 1203. 'Round hill-spur'. OE *trendel* + *hōh*.

Treopert. *See* GRANSTON.

Treorchy (*Treorci*) Rhon. 'Village on the River Orci'. Welsh *tref*. The meaning of the river-name is unknown.

Tresco I. of Scilly. *Trescau* 1305. 'Farm of elder-trees'. Cornish *tre* + *scaw*.

Tresimwn. *See* BONVILSTON.

Treswell Notts. *Tireswelle* 1086 (DB). 'Spring or stream of a man called *Tīr*'. OE pers. name + *wella*.

Tretire Herefs. *Rythir* 1212. 'Long ford'. Welsh *rhyd* + *hir*.

Trevose Head Cornwall. first recorded in the 17th cent., named from a farm called Trevose which is *Trenfos* 1302, 'farm by the bank or dyke' (perhaps referring to an earlier fort), from Cornish *tre* + *an* 'the' + *fos*.

Trewalter. *See* WALTERSTON.

Trewhitt, High Northum. *Tirwit* 1150–62. Possibly 'river-bend where resinous wood is obtained'. OScand. *tyri* + OE **wiht*.

Trewyddel. *See* MOYLGROVE.

Treyford W. Sussex. *Treverde* 1086 (DB). 'Ford marked by a tree or with a tree-trunk bridge'. OE *trēow* + *ford*.

Trien (*An Trian*) Mayo. 'The third portion'.

Trillick (*Trileac*) Tyrone. (*airchindeach*) *Trelecc* 814. 'Megalithic tomb'.

Trim (*Baile Átha Troim*) Meath. 'Town of the ford of the elder-tree'.

Trimdon Durham. *Tremeldona* 1196. 'Hill with a wooden preaching cross'. OE *trēo-mǣl* + *dūn*.

Trimingham Norfolk. *Trimingeham* 1185. 'Homestead of the family or followers of a man called *Trymma'. OE pers. name + *-inga-* + *hām*.

Trimley Suffolk. *Tremelaia* 1086 (DB). 'Woodland clearing of a man called *Trymma'. OE pers. name + *lēah*.

Trimpley Worcs. *Trinpelei* 1086 (DB). 'Woodland clearing of a man called *Trympa'. OE pers. name + *lēah*.

Tring Herts. *Treunge* [*sic*] 1086 (DB), *Trehangr* 1199. Possibly 'wooded slope'. OE *trēow* + *hangra*.

Tritlington Northum. *Turthlyngton* c.1170. 'Estate associated with a man called Tyrhtel'. OE pers. name + *-ing-* + *tūn*.

Tromman (*Tromán*) Meath. 'Little place where elders grow'.

Trommaun (*Tromán*) Roscommon. 'Little place where elders grow'.

Troon S. Ayr. *Le Trone* 1371, *Le Trune* 1464. '(Place by) the headland'. Welsh *trwyn*.

Trossachs (hills) Stir. 'Transverse hills'. The name is said to be based on Welsh *traws*, 'across', referring to the hills that divide Loch Katrine from Loch Achray.

Trostan (*Trostán*) (mountain) Antrim. *Trostan* 1780. 'Pole'.

Troston Suffolk. *Trostingtun* c.1000, *Trostuna* 1086 (DB). 'Estate associated with a man called *Trost(a)'. OE pers. name + *-ing-* + *tūn*.

Trottiscliffe Kent. *Trottes clyva* 788, *Totesclive* [*sic*] 1086 (DB). 'Cliff or hill-slope of a man called *Trott'. OE pers. name + *clif*.

Trotton W. Sussex. *Traitone* [*sic*] 1086 (DB), *Tratinton* 12th cent. Possibly 'estate associated with a man called *Trætt'. OE pers. name + *-ing-* + *tūn*. Alternatively the first element may be an OE **trǣding* 'path, stepping stones'.

Troutbeck, '(place on) the trout stream', OE *truht* + OScand. *bekkr*: **Troutbeck** Cumbria, near Ambleside. *Trutebek* 1272. **Troutbeck** Cumbria, near Penruddock. *Troutbek* 1332.

Trowbridge Wilts. *Straburg* [*sic*] 1086 (DB), *Trobrigge* 1184. 'Tree-trunk bridge'. OE *trēow* + *brycg*.

Trowell Notts. *Trowalle* 1086 (DB). 'Tree-stream', perhaps referring to a tree-trunk used as a bridge. OE *trēow* + *wella*.

Trowse Newton Norfolk. *Treus*, *Newotuna* 1086 (DB). Originally two separate names, 'wooden house' from OE *trēow* + *hūs*, and 'new farmstead' from OE *nīwe* + *tūn*.

Trull Somerset. *Trendle* 1225. 'Circular feature'. OE *trendel*.

Trumpington Cambs. *Trumpintune* c.1050, *Trumpintone* 1086 (DB). 'Estate associated with a man called *Trump(a)'. OE pers. name + *-ing-* + *tūn*.

Trunch Norfolk. *Trunchet* 1086 (DB). Probably 'wood on an upland or plateau'. Celtic **trūm* + **cę̄d*.

Truro Cornwall. *Triueru* c.1173. Possibly a Cornish name meaning '(place of) great water turbulence'.

Trusham Devon. *Trisma* 1086 (DB). Probably a Celtic name, 'place of thorns', from Primitive Cornish **drȳs* 'thornbush' + suffix.

Trusley Derbys. *Trusselai* 1166. Possibly 'brushwood clearing'. OE *trūs* + *lēah*.

Trusthorpe Lincs. *Dreuistorp* [*sic*] 1086 (DB), *Struttorp* 1196. Probably 'outlying farmstead or hamlet of a man called Strutr'. OScand. pers. name + *thorp*.

Trym (river) Glos. *See* WESTBURY ON TRYM.

Trysull Staffs. *Treslei* 1086 (DB). Named from the River Trysull, probably a Celtic river-name meaning 'tumultuous one'.

Tuam (*Tuaim*) Galway. 'Mound'.

Tuamgraney (*Tuaim Gréine*) Clare. 'Grian's mound'.

Tubber (*An Tobar*) Galway. 'The well'.

Tubbrid (*Tiobraid*) Kilkenny, Tipperary. 'Well'.

Tubney Oxon. *Tobenie* 1086 (DB). 'Island, or dry ground in marsh, of a man called *Tubba'. OE pers. name (genitive -*n*) + *ēg*.

Tuddenham, 'homestead or village of a man called Tud(d)a', OE pers. name (genitive -*n*) + *hām*: **Tuddenham** Suffolk, near Ipswich. *Tude(n)ham, Toddenham* 1086 (DB). **Tuddenham** Suffolk, near Lackford. *Tode(n)ham, Totenham* 1086 (DB). **Tuddenham, East & Tuddenham, North** Norfolk. *East Tudenham, Nord Tudenham* 1086 (DB).

Tudeley Kent. *Tivedele* 1086 (DB). Possibly 'wood or clearing overgrown with ivy'. OE *īfede* + *lēah*, with initial *T-* from the OE preposition *æt* 'at'.

Tudhoe Durham. *Tudhou* 1243. 'Hill-spur of a man called Tud(d)a'. OE pers. name + *hōh*.

Tuesley Surrey. *Tiwesle* 1086 (DB). Possibly 'wood or clearing dedicated to the heathen god Tīw', from OE god-name + *lēah*. Alternatively the first element may be an OE man's name *Tīwhere*.

Tuffley Glos. *Tuffelege* 1086 (DB). 'Woodland clearing of a man called Tuffa'. OE pers. name + *lēah*.

Tufnell Park Gtr. London. named from one William *Tufnell*, lord of the manor of Barnsbury in 1753.

Tufton Hants. *Tochiton* 1086 (DB). 'Estate associated with a man called *Toca or Tucca'. OE pers. name + *-ing-* + *tūn*.

Tugby Leics. *Tochebi* 1086 (DB), *Tokebi* 1176. 'Farmstead or village of a man called Tóki'. OScand. pers. name + *bý*.

Tugford Shrops. *Dodefort* [*sic*] 1086 (DB), *Tuggeford* c.1138. 'Ford of a man called *Tucga'. OE pers. name + *ford*.

Tulla (*Tulach*) Clare. 'Hill'.

Tullagh (*Tulach*) Donegal. 'Hill'.

Tullaghan (*Tulachán*) Leitrim, Sligo. 'Little hill'.

Tullaghanmore (*Tulachán Mór*) Westmeath. 'Big hillock'.

Tullagher (*Tulachar*) Kilkenny. 'Hilly place'.

Tullaghoge (*Tulach Óg*) Tyrone. (*oe*) *Telach Occ* 913. 'Hillock of the youths'.

Tullaherin (*Tulach Thirim*) Kilkenny. 'Dry hill'.

Tullakeel (*Tulach Chaoil*) Kerry. 'Narrow hill'.

Tullamore (*Tulach Mhór*) Kerry, Offaly. 'Big hill'.

Tullibody Clac. *Tullibotheny* 1195. 'Hill of the hut'. Gaelic *tulach* + *both*.

Tullow (*An Tulach*) Carlow. 'The hill'.

Tullroan (*Tulach Ruáin*) Kilkenny. 'Ruán's hill'.

Tully (*Tulach*) Galway. 'Hill'.

Tullyallen (*Tulaigh Álainn*) Louth. 'Beautiful hill'.

Tullyco (*Tulaigh Chuach*) Cavan. 'Cuckoo hill'.

Tullycrine (*Tulach an Chrainn*) Limerick. 'Hill of the tree'.

Tullylish (*Tulach Lis*) Down. 'Hill of the fort'.

Tullyrone (*Tulaigh Uí Ruáin*) Armagh. *Tulliroan* 1611. 'Ó Ruáin's hill'.

Tullyvin (*Tulaigh Bhinn*) Cavan. *Tullyvin* 1835. Possibly 'hill of the peak'.

Tulrahan (*Tulach Shrutháin*) Mayo. 'Hill of the stream'.

Tulse Hill Gtr. London. named from the *Tulse* family, recorded as living here in the 17th cent.

Tumby Lincs. *Tunbi* 1086 (DB). Possibly 'farmstead or village of a man called Tumi'.

OScand. pers. name + *bý*. Alternatively the first element may be OScand. *tún* 'fence, enclosure'.

Tummel (river) Pert. *Abhain Teimheil* (no date). '(River of) darkness'. OGaelic **temel*. The name alludes to the river's heavily wooded gorges.

Tunbridge Wells Kent. named from TONBRIDGE, *Wells* referring to the medicinal springs here discovered in the 17th cent. The affix *Royal* was bestowed on the town by Edward VII.

Tunnyduff (*Tonnach Dhubh*) Cavan. 'Black quagmire'.

Tunstall, found in various counties, from OE **tūn-stall* 'the site of a farm, a farmstead'; examples include: **Tunstall** Kent. *Tunestelle* 1086 (DB). **Tunstall** Stoke. *Tunstal* 1212. **Tunstall** Suffolk. *Tunestal* 1086 (DB).

Tunstead Norfolk. *Tunstede* 1044–7, *Tunesteda* 1086 (DB). OE *tūn-stede* 'farmstead'.

Tunworth Hants. *Tuneworde* 1086 (DB). Probably 'enclosure of a man called Tunna'. OE pers. name + *worth*.

Tuosist (*Tuath Ó Síosta*) Kerry. 'Territory of the Uí Síosta'.

Tupsley Herefs. *Topeslage* 1086 (DB). Possibly 'ram's woodland clearing'. ME *tup* (or a byname formed from this word) + OE *lēah*.

Tur Langton Leics. *See* LANGTON.

Ture (*An tIúr*) Donegal. 'The yew-tree'.

Turkdean Glos. *Turcandene* 716–43, *Turchedene* (DB). 'Valley of a river called *Turce*'. Lost Celtic river-name (probably meaning 'boar') + OE *denu*.

Turlough (*Turlach*) Mayo. 'Fen'.

Turloughmore (*An Turlach Mór*) Galway. 'The large fen'.

Turnastone Herefs. *Thurneistun* 1242. 'Estate of a family called *de Turnei*'. OE *tūn*, with the name of a family recorded in the area in the 12th cent.

Turners Puddle Dorset. *Pidele* 1086 (DB), *Tonerespydele* 1268. 'Estate on the River Piddle held by the Toner family'. OE river-name (**pidele* 'a marsh or fen') with name of family here from 1086.

Turnhouse Edin. Obscure.

Turnworth Dorset. *Torneworde* 1086 (DB). 'Thorn-bush enclosure'. OE *thyrne* + *worth*.

Turreagh (*Tor Riabhach*) Antrim. *Torreagh* 1659. 'Grey crag'.

Turriff Aber. *Turrech* 1273, *Turreth* c.1300, *Turreff* c.1500. Obscure.

Turton Bottoms Black. w. Darw. *Turton* 1212. 'Farmstead or village of a man called Thóri or Thórr'. OScand. pers. name + OE *tūn*. Affix is from OE *botm* 'valley bottom'.

Turvey Beds. *Torueie* 1086 (DB). 'Island with good turf or grass'. OE *turf* + *ēg*.

Turville Bucks. *Thyrefeld* 796. 'Dry open land'. OE *thyre* + *feld*.

Turweston Bucks. *Turvestone* 1086 (DB). Possibly 'farmstead or village of a man called Thorfrøthr or *Thorfastr'. OScand. pers. name + OE *tūn*.

Tutbury Staffs. *Toteberia* 1086 (DB), *Stutesberia* c.1150. 'Stronghold of a man called Tutta or *Stūt'. OE pers. name + *burh* (dative *byrig*).

Tutnall Worcs. *Tothehel* [*sic*] 1086 (DB), *Tottenhull* 1262. 'Hill of a man called Totta'. OE pers. name (genitive -*n*) + *hyll*.

Tuttington Norfolk. *Totingtonne* 1044–7, *Tutincghetuna* 1086 (DB). 'Estate associated with a man called Tutta'. OE pers. name + -*ing*- + *tūn*.

Tuxford Notts. *Tuxfarne* [*sic*] 1086 (DB), *Tukesford* 12th cent. Possibly 'ford of a man called *Tuk', OScand. pers. name + OE *ford*, but the first element may be an early form of *tusk* 'tussock, tuft of rushes'.

Tweedmouth Northum. *Tuedemue* 1208–10. 'Mouth of the River Tweed'. Celtic or pre-Celtic river-name (possibly meaning 'powerful one') + OE *mūtha*.

Twelve Pins, The (*Beanna Beola*) (mountains) Galway. 'Twelve peaks'. English *pin* is Irish *binn*, 'peak'. The Irish name means 'peaks of Beola'.

Twemlow Green Ches. *Twamlawe* 12th cent. '(Place at) the two tumuli'. OE *twēgen* (dative *twǣm*) + *hlāw*.

Twickenham Gtr. London. *Tuicanhom* 704, *Twykenham* 1279. Probably 'river-bend land of a man called *Twicca'. OE pers. name (genitive *-n*) + *hamm*. Alternatively the first element may be OE **twicce* 'river-fork'.

Twigworth Glos. *Tuiggewrthe* 1216. Possibly 'enclosure of a man called Twicga', OE pers. name + *worth*. Alternatively 'enclosure made with twigs', from OE *twigge*.

Twineham W. Sussex. *Tuineam* late 11th cent. '(Place) between the streams'. OE *betwēonan* + *ēa* (dative plural *ēam*).

Twinstead Essex. *Tumesteda* [*sic*] 1086 (DB), *Twinstede* 1203. Probably 'double homestead'. OE *twinn* + *stede*.

Twitchen Devon. *Twechon* 1442. OE *twicen(e)* 'cross-roads'.

Twomileborris (*Buiríos Léith*) Tipperary. Twomileborris is two Irish miles from Leighmore (Leamakevoge), whose Irish name is *Liath Mór Mo-Chaomhóg*, 'big grey place of my little Kevin'. Hence its Irish name, meaning 'borough of Liath'.

Twycross Leics. *Tvicros* 1086 (DB). '(Place with) two crosses'. OE *twī-* + *cros*.

Twyford, found in various counties, 'double ford'. OE **twī-ford* or **twī-fyrde*; examples include: **Twyford** Hants. *Tuifyrde* c.970, *Tviforde* 1086 (DB). **Twyford** Wokhm. *Tuiford* 1170.

Twyning Glos. *Bituinæum* 814, *Tveninge* 1086 (DB). Originally '(place) between the rivers' from OE *betwēonan* + *ēa* (dative *ēam*), later 'settlement of those living there' with the addition of OE *-ingas*.

Twywell Northants. *Twiwel* 1013, *Tuiwella* 1086 (DB). 'Double spring or stream'. OE *twī-* + *wella*.

Tydd St Giles (Cambs.) **& Tydd St Mary** (Lincs.). *Tid, Tite* 1086 (DB). Possibly OE *titt* 'teat' in transferred sense 'small hill' (perhaps with reference to an original saltern or salthill here). Distinguishing affixes from the church dedications. **Tydd Gote** (*Tyddegot* 1316) is from ME *gote* 'sluice'.

Tyddewi. See ST DAVID's.

Tyfarnham (*Teach Farannáin*) Westmeath. 'Farannán's house'.

Tyldesley Wigan. *Tildesleia* c.1210. 'Woodland clearing of a man called Tilwald'. OE pers. name + *lēah*.

Tyler Hill Kent. *Tylerhelde* 1304. 'Slope of the tile-makers'. OE **tiglere* + *helde*.

Tynagh (*Tíne*) Galway. 'Watercourse'.

Tynan (*Tuínéan*) Armagh. (*airchindeach*) *Tuidnidha* 1072. 'Watercourse'.

Ty-nant Conwy. 'House in the valley'. Welsh *tŷ* + *nant*.

Tyndyrn. See TINTERN.

Tyne (river) Northum.–Durham. See TYNEMOUTH.

Tynedale (district) Northum. *Tindala* 1158. 'Valley of the River Tyne'. Pre-Celtic river-name (*see* TYNEMOUTH) + OScand. *dalr*.

Tyneham Dorset. *Tigeham* 1086 (DB). Probably 'goat's enclosure'. OE **tige* (genitive *-an*) + *hamm*.

Tynemouth N. Tyne. *Tinanmuthe* 792. 'Mouth of the River Tyne'. Pre-Celtic river-name (meaning simply 'flowing one, river') + OE *mūtha*. The unitary authorities **North Tyneside** and **South Tyneside** are named from the same river.

Tyrella (*Teach Riala*) Down. (*o*) *Thigh Riaghla* c.1630. 'Riail's church'.

Tyringham Milt. K. *Telingham* [*sic*] 1086 (DB), *Tringeham* 1130. 'Homestead or river-meadow of the family or followers of a man called **Tīr(a)*'. OE pers. name + *-inga-* + *hām* or *hamm*.

Tyrone (*Tir Eoghain*) (the county). (*tighernuis*) *Thíre hEoghain* c.1500. 'Territory of Eoghan'.

Tyrrellspass (*Bealach an Tirialaigh*) Westmeath. In 1597 Captain Richard *Tyrrell* and Piers Lacy ambushed and defeated an English force on the road here. The Irish name means 'Tyrrell's road'.

Tysoe Warwicks. *Tiheshoche* 1086 (DB). Probably 'hill-spur of the heathen god Tīw'. OE god-name + *hōh*.

Tytherington, 'estate associated with a man called **Tydre*', OE pers. name + *-ing-* + *tūn*, or 'stock-breeding farmstead', with an OE **tȳd(d)ring* as first element: **Tytherington** Ches. *Tidderington* c.1245. **Tytherington** S. Glos. *Tidrentune* 1086 (DB). **Tytherington** Wilts. *Tuderinton* 1242.

Tytherleigh Devon. *Tiderlege* 12th cent. Probably 'young wood or clearing', referring to new growth. OE *tīedre* + *lēah*.

Tytherley, East & Tytherley, West Hants. *Tiderlei* 1086 (DB). Identical in origin with the previous name.

Tytherton Lucas & East Tytherton Wilts. *Tedrintone* 1086 (DB). Identical in origin with TYTHERINGTON. Manorial affix from the *Lucas* family, here in the 13th cent.

Tywardreath Cornwall. *Tiwardrai* 1086 (DB). 'House on the beach'. Cornish *ti* + *war* + *treth*.

Tywi Valley Carm. 'Valley of the River Tywi'. The Tywi (Towy) has a Celtic name that may mean 'the dark one'.

Tywyn Gwyd. *Thewyn* 1254, *Tewyn* 1291, *Towyn, or Tywyn* 1868. '(Place by the) seashore'. Welsh *tywyn*.

Uamh Bheag (mountain) Stir. '(Mountain with) the little cave'. Gaelic *uamh* + *beag* (feminine *bheag*).

Ubbeston Green Suffolk. *Upbestuna* 1086 (DB). 'Farmstead or village of a man called Ubbi'. OScand. pers. name + OE *tūn*.

Ubley B. & NE. Som. *Hubbanlege* late 10th cent., *Tumbeli* [sic] 1086 (DB). 'Woodland clearing of a man called Ubba'. OE pers. name + *lēah*.

Uckerby N. Yorks. *Ukerby* 1198. 'Farmstead or village of a man called *Úkyrri or *Útkári'. OScand. pers. name + *bý*.

Uckfield E. Sussex. *Uckefeld* 1220. 'Open land of a man called Ucca'. OE pers. name + *feld*.

Uckington Glos. *Hochinton* 1086 (DB). 'Estate associated with a man called Ucca'. OE pers. name + *-ing-* + *tūn*.

Uddingston S. Lan. *Odistoun* 1296, *Odingstoune* 1475. OE *tūn* 'farmstead, estate' with an uncertain pers. name, possibly OScand. *Oddr*.

Udimore E. Sussex. *Dodimere* [sic] 1086 (DB), *Odumer* 12th cent. Possibly 'wooded pond'. OE *wudig* + *mere*.

Uffculme Devon. *Offecoma* 1086 (DB). 'Estate on the Culm river held by a man called Uffa'. OE pers. name + Celtic river-name (*see* CULLOMPTON).

Uffington, 'estate associated with a man called Uffa', OE pers. name + *-ing-* + *tūn*: **Uffington** Lincs. *Offintone* 1086 (DB). **Uffington** Shrops. *Ofitone* 1086 (DB).

Uffington Oxon. *Uffentune* 10th cent., *Offentone* 1086 (DB). 'Estate of a man called Uffa'. OE pers. name (genitive *-n*) + *tūn*.

Ufford, 'enclosure of a man called Uffa', OE pers. name + *worth*: **Ufford** Peterb. *Uffawyrtha* 948. **Ufford** Suffolk. *Uffeworda* 1086 (DB).

Ufton Warwicks. *Hulhtune* 1043, *Ulchetone* 1086 (DB). Possibly 'farmstead with a shed or hut'. OE **huluc* + *tūn*.

Ufton Nervet W. Berks. *Offetune* 1086 (DB), *Offeton Nernut* 1284 'Farmstead or village of a man called Uffa'. OE pers. name + *tūn*. Manorial affix from the *Neyrnut* family, here in the 13th cent.

Ugborough Devon. *Ulgeberge* [sic] 1086 (DB), *Uggeberge* 1200. 'Hill or mound of a man called *Ugga'. OE pers. name + *beorg*.

Uggeshall Suffolk. *Uggiceheala* 1086 (DB). 'Nook of land of a man called *Uggeca'. OE pers. name + *halh*.

Ugglebarnby N. Yorks. *Ugleberdesbi* 1086 (DB). 'Farmstead or village of a man called *Uglubárthr'. OScand. pers. name + *bý*.

Ugley Essex. *Uggele* c.1041, *Ugghelea* 1086 (DB). 'Woodland clearing of a man called *Ugga'. OE pers. name + *lēah*.

Ugthorpe N. Yorks. *Ughetorp* 1086 (DB). 'Outlying farmstead or hamlet of a man called Uggi'. OScand. pers. name + *thorp*.

Uig Highland (Skye). *Wig* 1512. '(Place by a) bay'. Gaelic *ùig* (from OScand. *vík*).

Uig W. Isles (Lewis). *Vye* 1549. '(Place by a) bay'. Gaelic *ùig* (from OScand. *vík*).

Uisge Labhair (river) Highland. 'Noisy water'. Gaelic *uisge* + *labhar*.

Uist (island) W. Isles. *Iuist* 1282, *Ouiste* 1373. 'Inner abode'. OScand. *í* + *vist*. This is a Scandinavian reinterpretation of a pre-Celtic name.

Ulbster Highland. *Ulbister* 1538. 'Wolf's dwelling'. OScand. *ulfr* + *bólstathr*.

Ulceby, 'farmstead or village of a man called Ulfr', OScand. pers. name + *bý*: **Ulceby** Lincs. *Ulesbi* 1086 (DB). **Ulceby** N. Lincs. *Ulvesbi* 1086 (DB).

Ulcombe Kent. *Ulancumbe* 946, *Olecumbe* 1086 (DB). 'Valley of the owl, or of a man called *Ūla'. OE *ūle* or OE pers. name + *cumb*.

Uldale Cumbria. *Ulvesdal* 1216. 'Valley of a man called Ulfr, or one frequented by wolves'. OScand. pers. name or *ulfr* + *dalr*.

Uley Glos. *Euuelege* 1086 (DB). 'Yew-tree wood or clearing'. OE *īw* + *lēah*.

Ulgham Northum. *Wlacam* 1139, *Ulweham* 1242. 'Valley or nook frequented by owls'. OE *ūle* + *hwamm*.

Ullapool Highland. *Ullabill* 1610. 'Wolf's farm'. OScand. *ulfr* + *boeli*.

Ullard (*Ulaidh Ard*) Kilkenny. 'High penitential station'.

Ullauns (*Ulán*) Kilkenny. 'Penitential station'.

Ullenhall Warwicks. *Holehale* 1086 (DB). 'Nook of land of a man called *Ulla'. OE pers. name (genitive -*n*) + *halh*.

Ulleskelf N. Yorks. *Oleschel* [sic] 1086 (DB), *Ulfskelf* 1170-7. 'Shelf or ledge of a man called Ulfr'. OScand. pers. name + *skjalf* or OE *scelf* (with Scand. *sk*-).

Ullesthorpe Leics. *Ulestorp* 1086 (DB). 'Outlying farmstead or hamlet of a man called Ulfr'. OScand. pers. name + *thorp*.

Ulley Rothm. *Ollei* 1086 (DB). Probably 'woodland clearing frequented by owls'. OE *ūle* + *lēah*.

Ullingswick Herefs. *Ullingwic* 1086 (DB). 'Dwelling or (dairy) farm associated with a man called *Ulla'. OE pers. name + -*ing*- + *wīc*.

Ullock Cumbria, near Mockerkin. *Uluelaik* 1279. 'Place where wolves play'. OScand. *ulfr* + *leikr*.

Ullswater Cumbria. *Ulueswater* c.1230. 'Lake of a man called Ulfr'. OScand. pers. name + OE *wæter*.

Ulpha Cumbria, near Coniston. *Wolfhou* 1279. 'Hill frequented by wolves'. OScand. *ulfr* + *haugr*.

Ulrome E. R. Yorks. *Ulfram* 1086 (DB). Probably 'homestead or village of a man called Wulfhere or a woman called Wulfwaru'. OE pers. name + *hām*.

Ulster (*Ulaidh*) (the province). 'Land of the Ulstermen'. *Ouolountoi* c.150. OScand. possessive -*s* + Irish *tír*. The meaning of the tribal name is obscure.

Ulverston Cumbria. *Ulurestun* 1086 (DB). 'Farmstead or village of a man called Wulfhere or Ulfarr'. OE or OScand. pers. name + OE *tūn*.

Umberleigh Devon. *Umberlei* 1086 (DB), *Wumberlegh* 1270. Possibly 'woodland clearing by the meadow stream'. OE *winn* + *burna* + *lēah*.

Ummamore (*Iomaidh*) Westmeath. 'Contention'.

Ummeraboy (*Iomaire Buí*) Cork. 'Yellow ridge'.

Ummeracam (*Iomaire Cam*) Armagh. 'Crooked ridge'.

Umrycam (*Iomaire Cam*) Derry, Donegal. 'Crooked ridge'.

Underbarrow Cumbria. *Underbarroe* 1517. '(Place) under the hill'. OE *under* + *beorg*.

Underriver Kent. *Underevere* 1279, *Sub le Ryver* 1477. '(Place) under the hill-brow'. OE *under* + *yfer*.

Underwood Notts. *Underwode* 1287. '(Place) within or near a wood'. OE *under* + *wudu*.

Unst (island) Shet. *Ornyst* c.1200. Ancient pre-Scandinavian name, origin obscure.

Unstone Derbys. *Onestune* 1086 (DB). Probably 'farmstead or village of a man called *Ōn'. OE pers. name + *tūn*.

Unthank Cumbria, near Gamblesby. *Unthank* 1254. Either '(land held) without consent, i.e. a squatter's holding', or a derogatory name for 'unthankful land, i.e. land difficult to cultivate'. OE *unthanc*. There are other examples of the name in Derbys., Durham, Northumberland and Cumbria.

Up as affix. *See* main name, e.g. for **Up Cerne** (Dorset) *see* CERNE.

Upavon Wilts. *Oppavrene* 1086 (DB). '(Settlement) higher up the River Avon'. OE *upp* + Celtic river-name (meaning simply 'river').

Upchurch Kent. *Upcyrcean c.*1100. 'Church standing high up'. OE *upp* + *cirice*.

Upham Hants. *Upham* 1201. 'Upper homestead or enclosure'. OE *upp* + *hām* or *hamm*.

Uphill N. Som. *Opopille* 1086 (DB). '(Place) above the creek'. OE *uppan* + *pyll*.

Upleadon Glos. *Ledene* 1086 (DB), *Upleden* 1253. '(Settlement) higher up the River Leadon'. OE *upp* + Celtic river-name (*see* HIGHLEADON).

Upleatham Red. & Cleve. *Upelider* 1086 (DB), *Uplithum c.*1150. '(Place at) the upper slopes'. OE *upp* + *hlith*, or OScand. *upp*(*i*) + *hlíth*, in a dative plural form.

Uplowman Devon. *Oppaluma* 1086 (DB). '(Settlement) higher up the River Loman'. OE *upp* + Celtic river-name (probably meaning 'elm river').

Uplyme Devon. *Lim* 1086 (DB), *Uplim* 1238. '(Settlement) higher up the River Lim'. OE *upp* + Celtic river-name (meaning simply 'stream').

Upminster Gtr. London. *Upmynstre* 1062, *Upmunstra* 1086 (DB). 'Higher minster or church'. OE *upp* + *mynster*.

Upney Gtr. London. *Upney* 1456. '(Settlement) upon the island (of dry ground in marsh)'. OE *uppan* + *ēg*.

Upottery Devon. *Upoteri* 1005, *Otri* 1086 (DB). '(Settlement) higher up the River Otter'. OE *upp* + river-name (*see* OTTERY).

Upper as affix. *See* main name, e.g. for **Upper Arley** (Worcs.) *see* ARLEY.

Upperland (*Áth an Phoirt Leathain*) Derry. *Aghfortlany* 1613. 'Ford of the broad (river) bank'. The English name arose from a garbled pronunciation of the Irish.

Upperthong Kirkl. *Thwong* 1274, *Overthong* 1286. 'Narrow strip of land'. OE *thwang*, with *uferra* 'upper' to distinguish it from NETHERTHONG.

Uppingham Rutland. *Yppingeham* 1067. 'Homestead or village of the hill-dwellers'. OE *yppe* + *-inga-* + *hām*.

Uppington Shrops. *Opetone* 1086 (DB), *Oppinton* 1195. Possibly 'estate associated with a man called *Uppa*'. OE pers. name + *-ing-* + *tūn*.

Upsall N. Yorks., near Thirsk. *Upsale* 1086 (DB). 'Higher dwelling(s)'. OScand. *upp* + *salr*.

Upton, a common name, usually 'higher farmstead or village', OE *upp* + *tūn*; examples include: **Upton** Dorset, near Lytchett. *Upton* 1463. **Upton** Slough. *Opetone* 1086 (DB). **Upton** Wirral. *Optone* 1086 (DB). **Upton Cheyney** S. Glos. *Vppeton* 1190, *Upton Chaune* 1325. Manorial affix from the *Cheyney* family. **Upton Grey** Hants. *Upton* 1202, *Upton Grey* 1281. Manorial affix from the *de Grey* family, here in the 13th cent. **Upton Hellions** Devon. *Uppetone Hyliun* 1270. Manorial affix from the *de Helihun* family, here in the 13th cent. **Upton Pyne** Devon. *Opetone* 1264, *Uppetone Pyn* 1283. Manorial affix from the *de Pyn* family, here in the 13th cent. **Upton St Leonards** Glos. *Optune* 1086 (DB), *Upton Sancti Leonardi* 1287. Affix from the dedication of the church. **Upton Scudamore** Wilts. *Uptune c.*990, *Upton* 1086 (DB), *Upton Squydemor* 1275. Manorial affix from the *de Skydemore* family, here in the 13th cent. **Upton Snodsbury** Worcs. *Snoddesbyri* 972, *Snodesbyrie* 1086 (DB), *Upton juxta Snodebure* 1280. Originally two separate names, the earlier being 'stronghold of a man called *Snodd*', OE pers. name + *burh* (dative *byrig*). **Upton upon Severn** Worcs. *Uptune* 897, *Uptun* 1086 (DB).
   However the following has a different origin: **Upton Lovell** Wilts. *Ubbantun* 957, *Ubbedon Lovell* 1476. 'Farmstead or village of a man called Ubba'. OE pers. name + *tūn*, with manorial affix from the *Lovell* family, here in the 15th cent.

Upwaltham W. Sussex. *Waltham* 1086 (DB), *Up Waltham* 1371. 'Homestead or village in a forest'. OE *w*(*e*)*ald* + *hām*, with *upp* 'higher'.

Upwell Cambs./Norfolk. *Wellan* 963, *Upwell* 1221. '(Place) higher up the stream'. OE *wella* with affix *upp* to distinguish it from OUTWELL.

Upwey Dorset. *Wai*(*e*) 1086 (DB), *Uppeweie* 1241. '(Settlement) higher up the River Wey'. Pre-English river-name (*see* WEYMOUTH) with OE *upp*.

Upwood Cambs. *Upwude* 974, *Upehude* [*sic*] 1086 (DB). 'Higher wood'. OE *upp* + *wudu*.

Urbalreagh (*Earball Riabhach*) Derry. ' ;rey end piece'.

Urchfont Wilts. *Ierchesfonte* 1086 (DB). 'Spring of a man called *Eohrīc'. OE pers. name + **funta*.

Urishay Herefs. *Haya Hurri* 1242. 'Enclosure of a man called Wulfrīc'. OE pers. name (Norman form *Urri*) + *hæg*.

Urlingford (*Áth na nUrlainn*) Kilkenny. 'Ford of the shafts'.

Urmston Traffd. *Wermeston* 1194. 'Farmstead or village of a man called *Wyrm or Urm'. OE or OScand. pers. name + OE *tūn*.

Urquhart Fife. *Petnaurcha* 1128. 'Portion of the cast'. Pictish **pett* + OGaelic *aurchoir*. The reference is perhaps to a method of allocating land.

Urquhart Highland. *Airchardan c.*1100. '(Place by a) thicket'. British *air cardden*.

Urrin (*Urrainn*) (river) Wexford. 'Division'.

Urswick, Great & Urswick, Little Cumbria. *Ursewica c.*1150. 'Dwelling or (dairy) farm by the bison lake'. OE *ūr* + *sǣ* + *wīc*.

Ushaw Moor Durham. *Vlueshahe* late 12th cent. Probably 'small wood frequented by wolves'. OE *wulf* (with loss of *w-* due to Scand. influence) + *scaga*. Alternatively the first element may be the OScand. pers. name *Ulfr*.

Usk (*Brynbuga*) Mon. *Uscha* 1100, *Uisc* 1150. '(Place on the) River Usk'. The Celtic river-name may mean 'abounding in fish' or alternatively simply 'water'. The Welsh name means 'Buga's hill' (Welsh *bryn*).

Usselby Lincs. *Osoluebi c.*1115. 'Farmstead or village of a man called Ōswulf'. OE pers. name + OScand. *bý*.

Utterby Lincs. *Uthterby* 1150–60. Probably 'farmstead or village of a man called Ūhtrēd or Ūhthere'. OE pers. name + OScand. *bý*.

Uttlesford (district) Essex. *Udelesforda* 1086 (DB). 'Ford of a man called *Udel'. A revival of an old hundred-name, otherwise preserved in Uttlesford Bridge (in WENDENS AMBO). OE pers. name + *brycg*.

Uttoxeter Staffs. *Wotocheshede* 1086 (DB). Probably 'heath of a man called *Wuttuc'. OE pers. name + **hǣddre*.

Uxbridge Gtr. London. *Wixebrug c.*1145. 'Bridge of the tribe called the *Wixan*'. OE tribal name + *brycg*.

Uzmaston Pemb. *Villa Osmundi* 1230. 'Osmund's farm'. OE pers. name + *tūn*.

Vale Royal Ches. *Vallis Regalis* 1277, *Valroyal* 1357. 'Royal valley', the site of a monastery founded by Edward I. OFrench *val* + *roial* (in Latin in the first spelling).

Vale of White Horse (district) Oxon. *The vale of Whithors* 1368. Named from the pre-historic figure of a horse cut into the chalk on White Horse Hill.

Valencia (*Dairbhre*) Kerry. 'Place of oaks'. The English name represents Irish *Béal Inse*, 'estuary of the island'.

Valley (*Y Fali*) Angl. 'The valley'. The name refers to the valley-like cutting from which rubble was extracted to build the Stanley Embankment from mainland Anglesey to Holy Island in the 1820s.

Vange Essex. *Fengge* 963, *Phenge* 1086 (DB). 'Fen or marsh district'. OE *fenn* + **gē*.

Vaternish (peninsula) Highland (Skye). *Watternes* 1501. 'Water promontory'. OScand. *vatn* + *nes*, probably influenced by English *water*.

Vauxhall Gtr. London. *Faukeshale* 1279. 'Hall or manor of a man called *Falkes*'. OFrench pers. name + OE *hall*.

Velly Devon. *See* CLOVELLY.

Venn Ottery Devon. *See* OTTERY.

Venny Tedburn Devon. *See* TEDBURN.

Ventnor I. of Wight. '(Farm of) *Vintner*' 1617. Probably a manorial name from a family called *le Vyntener*. The old name of Ventnor was *Holeweia c.*1200, 'the hollow way, or the way in a hollow', from OE *hol* + *weg*.

Ventry (*Fionntrá*) Kerry. 'White strand'.

Vernham Dean Hants. *Ferneham* 1210. 'Homestead or enclosure where ferns grow'. OE *fearn* + *hām* or *hamm*, with the later addition of *denu* 'valley'.

Verwood Dorset. *Beuboys* 1288, *Fairwod* 1329. 'Beautiful wood'. OE *fæger* + *wudu* (alternating with OFrench *beu* + *bois* in the earliest spelling).

Veryan Cornwall. *Sanctus Symphorianus* 1281. From the dedication of the church to St Symphorian.

Vexford, Lower Somerset. *Fescheforde* 1086 (DB). Possibly 'fresh-water ford'. OE *fersc* + *ford*.

Victoria Gtr. London. District centred on Victoria Station, opened in 1860, named after Queen *Victoria* (1819–1901). There are similar names commemorating the queen in Manchester, Cornwall, and several other counties.

Victoria Bridge Tyrone. A bridge was built here in 1842 and named after Queen Victoria.

Vigo Village Kent, modern name apparently commemorating the capture of the Spanish port of Vigo by the British fleet in 1719.

Vilanstown (*Baile na Bhileanach*) Westmeath. *Villenston c.*1590. 'Farmstead of the villeins'.

Vine's Cross E. Sussex, a late name from *cross* 'cross-roads', to be associated with a local family called *Vyne* recorded from the 16th cent.

Virginia Cavan. *Virginia* 1611. The town, founded in 1611 and originally named *Lough Ramor*, was soon after renamed from the newly established colony of Virginia in America. Its Irish name is *Achad an Iúir*, 'field of the yew-tree'.

Virginia Water Surrey, first recorded in 1749, originally a fanciful name, commemorating the American colony of Virginia, for the artificial lake created here in 1748 by the Duke of Cumberland.

Virginstow Devon. *Virginestowe c.*1180. 'Holy place of (St Bridget) the Virgin'. Saint's name (from the dedication of the church) + OE *stōw*.

Vobster Somerset. *Fobbestor* 1234. 'Rocky hill of a man called *Fobb'. OE pers. name + *torr*.

Voe Shet. '(Place by a) little bay'. OScand. *vágr*.

Vowchurch Herefs. *Fowchirche* 1291. 'Coloured church'. OE *fāh* + *cirice*.

Vyrnwy, Lake (reservoir) Powys. 'Lake of the River Vyrnwy'. The Welsh form of the river-name is *Efyrnwy*, possibly a personal name based on *ebur* 'yew-tree' (as well as 'cow-parsnip') (Welsh *efwr*).

Wackerfield Durham. *Wacarfeld c.*1040. Probably 'open land of the watchful ones, or where osiers grow'. OE *wacor* (genitive plural *wacra*) or **wācor* + *feld*.

Wacton Norfolk. *Waketuna* 1086 (DB). 'Farmstead or village of a man called *Waca'. OE pers. name + *tūn*.

Wadborough Worcs. *Wadbeorgas* 972, *Wadberge* 1086 (DB). 'Hills where woad grows'. OE *wād* + *beorg*.

Waddesdon Bucks. *Votesdone* 1086 (DB). 'Hill of a man called *Weott'. OE pers. name + *dūn*.

Waddingham Lincs. *Wadingeham* 1086 (DB). 'Homestead of the family or followers of a man called Wada'. OE pers. name + *-inga-* + *hām*.

Waddington, 'estate associated with a man called Wada', OE pers. name + *-ing-* + *tūn*: **Waddington** Lancs. *Widitun [sic]* 1086 (DB), *Wadingtun c.*1231. **Waddington** Lincs. *Wadintune* 1086 (DB).

Waddon Gtr. London. *Waddone c.*1115. 'Hill where woad grows'. OE *wād* + *dūn*.

Wadebridge Cornwall. *Wade* 1358, *Wadebrygge* 1478. OE *wæd* 'a ford' with the later addition of *brycg* from the 15th cent. when a bridge was built.

Wadenhoe Northants. *Wadenho* 1086 (DB). 'Hill-spur of a man called Wada'. OE pers. name (genitive -n) + *hōh*.

Wadhurst E. Sussex. *Wadehurst* 1253. 'Wooded hill of a man called Wada'. OE pers. name + *hyrst*.

Wadshelf Derbys. *Wadescel* 1086 (DB). 'Shelf or level ground of a man called Wada'. OE pers. name + *scelf*.

Wadworth Donc. *Wadewrde* 1086 (DB). 'Enclosure of a man called Wada'. OE pers. name + *worth*.

Wainfleet All Saints Lincs. *Wenflet* 1086 (DB). 'Creek or stream that can be crossed by a wagon'. OE *wægn* + *flēot*. Affix from the dedication of the church.

Wainscott Medway. *Wainscot* 1819. 'The shed for carts or wagons'. ME *wain* + *cot*.

Waitby Cumbria. *Watebi c.*1170. 'Wet farmstead'. OScand. *vátr* + *bý*.

Wakefield Wakefd. *Wachefeld* 1086 (DB). 'Open land where wakes or festivals take place'. OE **wacu* + *feld*.

Wakering, Great & Wakering, Little Essex. *Wacheringa* 1086 (DB). '(Settlement of) the family or followers of a man called Wacer'. OE pers. name + *-ingas*.

Wakerley Northants. *Wacherlei* 1086 (DB). 'Woodland clearing of the watchful ones, or where osiers grow'. OE *wacor* (genitive plural *wacra*) or **wācor* + *lēah*.

Wakes Colne Essex. *See* COLNE.

Walberswick Suffolk. *Walberdeswike* 1199. 'Dwelling or (dairy) farm of a man called Walbert'. OGerman pers. name + OE *wīc*.

Walberton W. Sussex. *Walburgetone* 1086 (DB). 'Farmstead or village of a woman called Wealdburh or Waldburg'. OE or OGerman pers. name + *tūn*.

Walcot, Walcote, Walcott, 'cottage(s) of the Britons', OE *walh* (genitive plural *wala*) + *cot*; examples include: **Walcot** Lincs., near Folkingham. *Walecote* 1086 (DB). **Walcote** Leics. *Walecote* 1086 (DB). **Walcott** Norfolk. *Walecota* 1086 (DB).

Walden, 'valley of the Britons', OE *walh* (genitive plural *wala*) + *denu*: **Walden** N. Yorks. *Waldene* 1270. **Walden, King's &**

Walden, Paul's Herts. *Waledene* 888, *Waldene* 1086 (DB). Distinguishing affixes from possession by the King in 1086 and by St Paul's in London from 1544. **Walden, Saffron** Essex. *Wealadene* c.1000, *Waledana* 1086 (DB), *Saffornewalden* 1582. Affix (ME *safron*) refers to the cultivation of the saffron plant here.

Walden Stubbs N. Yorks. *See* STUBBS.

Walderslade Medway. *Waldeslade* 1190. 'Valley in a forest'. OE *weald* + *slæd*.

Walderton W. Sussex. *Walderton* 1168. 'Farmstead or village of a man called Wealdhere'. OE pers. name + *tūn*.

Waldingfield, Great & Waldingfield, Little Suffolk. *Wealdingafeld* c.995, *Waldingefelda* 1086 (DB). 'Open land of the forest dwellers'. OE *weald* + *-inga-* + *feld*.

Walditch Dorset. *Waldic* 1086 (DB). 'Ditch with a wall or embankment'. OE *weall* or *walu* + *dīc*.

Waldridge Durham. *Walrigge* 1297. Probably 'ridge with or by a wall'. OE *wall* + *hrycg*.

Waldringfield Suffolk. *Waldringfeld* c.950, *Waldringafelda* 1086 (DB). 'Open land of the family or followers of a man called Waldhere'. OE pers. name + *-inga-* + *feld*.

Waldron E. Sussex. *Waldrene* 1086 (DB). 'House in the forest'. OE *weald* + *ærn*.

Wales (*Cymru*). '(Land of the) foreigners'. OE *walh* (plural *walas*). The name was used by the Anglo-Saxons of the 'alien' Celts. The Welsh name has a Celtic meaning 'compatriot'.

Wales Rothm. *Wales* 1086 (DB). '(Settlement of) the Britons'. OE *walh* (plural *walas*). This place name is thus identical in origin with WALES, the name of the principality.

Walesby, 'farmstead or village of a man called Valr', OScand. pers. name + *bý*: **Walesby** Lincs. *Walesbi* 1086 (DB). **Walesby** Notts. *Walesbi* 1086 (DB).

Walford Herefs. near Ross. *Walecford* 1086 (DB). 'Briton ford'. OE *walh* + *ford*.

Walford Herefs. near Wigmore. *Waliforde* 1086 (DB). Probably 'ford near a spring'. OE *wælla* + *ford*.

Walford Shrops. *Waleford* 1086 (DB). Identical in origin with the previous name.

Walgherton Ches. *Walcretune* 1086 (DB). 'Farmstead or village of a man called Walhhere'. OE pers. name + *tūn*.

Walgrave Northants. *Waldgrave* 1086 (DB). 'Grove belonging to OLD'. OE *grāf*.

Walkden Salford. *Walkeden* 1325. Possibly 'valley of a man called *Walca'. OE pers. name + *denu*. Alternatively the first element may be a lost OE river-name **W(e)alce* 'rolling one'.

Walker Newc. upon T. *Waucre* 1242. 'Marsh by the (Roman) wall'. OE *wall* + OScand. *kjarr*.

Walkerburn Sc. Bord. '(Place by) the fullers' stream'. OE *walcere* + *burna*.

Walkeringham Notts. *Wacheringeham* 1086 (DB). Probably 'homestead of the family or followers of a man called Walhhere'. OE pers. name + *-inga-* + *hām*.

Walkern Herts. *Walchra* [*sic*] 1086 (DB), *Walkern* 1222. 'Building for fulling cloth'. OE **walc* + *ærn*.

Walkhampton Devon. *Walchentone* 1084, *Wachetona* [*sic*] 1086 (DB). Probably 'farmstead of the dwellers on a stream called *Wealce'. OE river-name (meaning 'the rolling one') + *hǣme* + *tūn*.

Walkington E. R. Yorks. *Walchinton* 1086 (DB). 'Estate associated with a man called *Walca'. OE pers. name + *-ing-* + *tūn*.

Wall, usually '(place at) the wall', from OE *wall*: **Wall** Northum. *Wal* 1166. With reference to the Roman Wall. **Wall** Staffs. *Walla* 1167. With reference to the Roman town here.

However the following have a different origin:
Wall, East & Wall under Heywood Shrops. *Well* 1200, *Estwalle* 1255, *Walle sub Eywode* 1235. '(Place at) the spring or stream' from OE (Mercian) *wælla*. Affix means 'within or near the enclosed wood', OE *hæg* or *hege* + *wudu*.

Wallasey Wirral. *Walea* 1086 (DB), *Waleyesegh* 1351. 'The island of Waley (Britons' island)'. OE *walh* (genitive plural *wala*) + *ēg* with the later addition of a second explanatory *ēg*.

Wallingford Oxon. *Welingaforda* c.895, *Walingeford* 1086 (DB). 'Ford of the family or followers of a man called Wealh'. OE pers. name + *-inga-* + *ford*.

Wallington Gtr. London. *Waletona c.*1080, *Waletone* 1086 (DB). 'Farmstead or village of the Britons'. OE *walh* (genitive plural *wala*) + *tūn*.

Wallington Hants. *Waletune* 1233. Probably identical in origin with the previous name.

Wallington Herts. *Wallingtone* [*sic*] 1086 (DB), *Wandelingetona* 1280. 'Farmstead of the family or followers of a man called *Wændel*'. OE pers. name + *-inga-* + *tūn*.

Wallop, Middle, Wallop, Nether, & Wallop, Over Hants. *Wallope* 1086 (DB). Possibly 'valley with a spring or stream'. OE *wella, wælla* + *hop*. Alternatively the first element may be OE *weall* 'a wall' or *walu* 'a ridge, an embankment'.

Wallsend N. Tyne. *Wallesende c.*1085. 'End of the (Roman) wall'. OE *wall* + *ende*.

Walmer Kent. *Walemere* 1087. 'Pool of the Britons'. OE *walh* (genitive plural *wala*) + *mere*.

Walmersley Bury. *Walmeresley* 1262. Possibly 'woodland clearing of a man called Waldmǣr or Walhmǣr'. OE pers. name + *lēah*.

Walmley Birm. *Warmelegh* 1232. 'Warm wood or clearing'. OE *wearm* + *lēah*.

Walney, Isle of Cumbria. *Wagneia* 1127. Probably 'killer-whale island', from OScand. *vǫgn* + *ey*. Alternatively the first element may be OE **wagen* 'quaking sands'.

Walpole Suffolk. *Walepola* 1086 (DB). 'Pool of the Britons'. OE *walh* (genitive plural *wala*) + *pōl*.

Walpole St Andrew & Walpole St Peter Norfolk. *Walepol c.*1050, *Walpola* 1086 (DB). Perhaps 'pool by the (Roman) bank', OE *wall* + *pōl*, but possibly identical in origin with previous name. Affixes from the dedications of the churches.

Walsall Wsall. *Waleshale* 1163. 'Nook of land or valley of a man called Walh'. OE pers. name + *halh*. Alternatively the first element could be OE *walh* 'Welshman'.

Walsden Calder. *Walseden* 1235. Possibly 'valley of a man called *Walsa'. OE pers. name + *denu*.

Walsgrave on Sowe Covtry. *Sowa* 1086 (DB), *Woldegrove* 1411. 'Grove in or near a forest'. OE *wald* + *grāf*. Originally named from the River Sowe itself, a pre-English river-name of unknown meaning.

Walsham, 'homestead or village of a man called Walh', OE pers. name + *hām*: **Walsham le Willows** Suffolk. *Wal(e)sam* 1086 (DB). Affix means 'among the willow-trees'. **Walsham, North** Norfolk. *Northwalsham* 1044–7, *Walsam* 1086 (DB). **Walsham, South** Norfolk. *Suthwalsham* 1044–7, *Walesham* 1086 (DB).

Walsingham, Great & Walsingham, Little Norfolk. *Walsingaham c.*1035, 1086 (DB). 'Homestead of the family or followers of a man called Wæls'. OE pers. name + *-inga-* + *hām*.

Walsoken Cambs. *Walsocne* 974, *Walsoca* 1086 (DB). 'Jurisdictional district near the (Roman) bank'. OE *wall* + *sōcn*.

Walterston (*Trewalter*) Vale Glam. *Waltervilla c.*1102, *Walterston* 1540. 'Walter's farm'. OE *tūn* (Welsh *tref*). The Welsh name translates the English.

Walterstone Herefs. *Walterestun* 1249. 'Walter's manor or estate'. OE *tūn*. From its possession by *Walter* de Lacy in the late 11th cent.

Waltham, found in various counties, 'homestead or village in a forest' (probably indicating a royal hunting estate), OE *w(e)ald* + *hām*; examples include: **Waltham** NE. Lincs. *Waltham* 1086 (DB). **Waltham Abbey** Essex. *Waltham* 1086 (DB). Affix from the medieval Abbey of Holy Cross here. **Waltham, Bishops** Hants. *Waltham* 904, 1086 (DB). Affix from its early possession by the Bishop of Winchester. **Waltham Cross** Herts. *Walthamcros* 1365. 'Cross near WALTHAM [ABBEY]'. Named from the 'Eleanor Cross' set up here by Edward I in memory of Queen Eleanor in 1290. **Waltham, Great & Waltham, Little** Essex. *Waltham, Waldham* 1086 (DB), *parua Waltam* 1197, *Waltham Magna* 1238. **Waltham on the Wolds** Leics. *Waltham* 1086 (DB). *See* WOLDS. **Waltham St Lawrence & White Waltham** Winds. & Maid. *Wealtham* 940, *Waltham* 1086 (DB), *Waltamia Sancti Laurencii* 1225, *Wytewaltham* 1243. Distinguishing affixes from the dedication of the church and from OE *hwīt* 'white' referring to chalky soil.

Waltham Forest Gtr. London. new London borough recalling old *foresta de Wautham* 1261 (so named from WALTHAM ABBEY).

Walthamstow Gtr. London. *Wilcumestowe c.*1075, *Wilcumestou* 1086 (DB). 'Holy place where guests are welcome', or 'holy place of a woman called Wilcume'. OE *wilcuma* or pers. name + *stōw*.

Walton, a common name, often 'farmstead or village of the Britons', from OE *walh* (genitive plural *wala*) + *tūn*; examples include: **Walton** Derbys. *Waletune* 1086 (DB). **Walton** Suffolk. *Waletuna* 1086 (DB). **Walton, Higher** Warrtn. *Waletona* 1154–60. **Walton-le-Dale** Lancs. *Waletune* 1086 (DB), *Walton in La Dale* 1304. Affix means 'in the valley' from OScand. *dalr*. **Walton-on-Thames** Surrey. *Waletona* 1086 (DB). **Walton on the Naze** Essex. *Walentonie* 11th cent., *Walton at the Naase* 1545. Affix means 'on the promontory', from OE *næss*. **Walton upon Trent** Derbys. *Waletune* 942, 1086 (DB).

However several Waltons have a different origin; examples include: **Walton in Gordano** N. Som. *Waltona* 1086 (DB). 'Farmstead in a forest or with a wall'. OE *w(e)ald* or *w(e)all* + *tūn*. For the affix, *see* CLAPTON. **Walton on the Hill** Surrey. *Waltone* 1086 (DB). Identical in origin with the previous name. **Walton, West** Norfolk. *Waltuna* 1086 (DB). 'Farmstead or village by the (Roman) bank'. OE *w(e)all* + *tūn*. **Walton, Wood** Cambs. *Waltune* 1086 (DB), *Wodewalton* 1300. Probably 'farmstead in a forest'. OE *w(e)ald* + *tūn*, with the later addition of ME *wode* 'woodland'.

Walworth, 'enclosure of the Britons', OE *walh* (genitive plural *wala*) + *worth*: **Walworth** Darltn. *Walewrth* 1207. **Walworth** Gtr. London. *Wealawyrth* 1001, *Waleorde* 1086 (DB).

Wambrook Somerset. *Awambruth* [*sic*] 802–39 (17th-cent. copy), *Wambroc* 1215–20. Probably 'brook in a hollow'. OE *wamb* 'womb, belly' (here used figuratively in a topographical sense) + *brōc*.

Wanborough Surrey. *Weneberge* 1086 (DB). Probably 'hill or mound of a man called *Wenna'. OE pers. name + *beorg*.

Wanborough Swindn. *Wænbeorgon* 854, *Wemberge* 1086 (DB). '(Place at) the tumour-shaped mounds'. OE *wenn* + *beorg*.

Wandsworth Gtr. London. *Wendleswurthe* 11th cent., *Wandelesorde* 1086 (DB). 'Enclosure of a man called *Wændel'. OE pers. name + *worth*. The river-name **Wandle** is a back-formation from the place name.

Wangford Suffolk. near Southwold. *Wankeforda* 1086 (DB), *Wangeford* 1238.

Possibly 'ford by the open ground'. OE *wang* + *ford*.

Wanlip Leics. *Anlepe* 1086 (DB). 'The lonely or solitary place'. OE *ānlīepe*.

Wansbeck (district) Northum. A modern adoption of the ancient river-name Wansbeck (*Wenspic* 1137), of uncertain origin and meaning (later associated with OScand. *bekkr* 'stream').

Wansdyke (ancient embankment, now district-name in B. & NE. Som.). *Wodnes dic* 903. 'Dyke associated with the heathen war-god Wōden'. OE god-name + *dīc*.

Wansford Peterb. *Wylmesforda* 972. 'Ford by a spring or whirlpool'. OE *wylm*, *wælm* + *ford*.

Wansford E. R. Yorks. *Wandesford* 1176. 'Ford of a man called *Wand or *Wandel'. OE pers. name + *ford*.

Wanstead Gtr. London. *Wænstede* c.1055, *Wenesteda* 1086 (DB). 'Place by a tumour-shaped mound or where wagons are kept'. OE *wænn* or *wæn* + *stede*.

Wanstrow Somerset. *Wandestreu* 1086 (DB). 'Tree of a man called *Wand or *Wandel'. OE pers. name + *trēow*.

Wantage Oxon. *Waneting* c.880, *Wanetinz* 1086 (DB). '(Place at) the fluctuating stream'. Derivative of OE *wanian* 'to decrease' + *-ing*.

Wapley S. Glos. *Wapelei* 1086 (DB). 'Woodland clearing by the spring'. OE *wapol* + *lēah*.

Wappenbury Warwicks. *Wapeberie* 1086 (DB). 'Stronghold of a man called *Wæppa'. OE pers. name (genitive *-n*) + *burh* (dative *byrig*).

Wappenham Northants. *Wapeham* 1086 (DB). 'Homestead or enclosure of a man called *Wæppa'. OE pers. name (genitive *-n*) + *hām* or *hamm*.

Wapping Gtr. London. *Wapping* c.1220. Possibly '(settlement of) the family or followers of a man called *Wæppa', or '*Wæppa's place'. OE pers. name + *-ingas* or *-ing*. Alternatively 'marshy or miry place', from a derivative of a word related to OE *wapol* 'pool, marsh, spring'.

Warbleton E. Sussex. *Warborgetone* 1086 (DB). 'Farmstead or village of a woman called Wǣrburh'. OE pers. name + *tūn*.

Warblington Hants. *Warblitetone* 1086 (DB), *Werblinton* 1186. 'Farmstead or estate associated with a woman called Wǣrblīth'. OE pers. name + *-ing-* + *tūn*.

Warborough Oxon. *Wardeberg* 1200. 'Watch or look-out hill'. OE *weard* + *beorg*.

Warboys Cambs. *Weardebusc* 974. Probably 'bush of a man called *Wearda'. OE pers. name + *busc* (with influence from OFrench *bois* 'wood'). Alternatively the first element may be OE *weard* 'watch, protection'.

Warbstow Cornwall. 'Chapel of *Sancta Werburga'* 1282, *Warberstowe* 1309. 'Holy place of St Wǣrburh'. OE female saint's name + *stōw*.

Warburton Traffd. *Wareburgetune* 1086 (DB). 'Farmstead or village of a woman called Wǣrburh'. OE pers. name + *tūn*.

Warcop Cumbria. *Wardecop* 1197, *Warthecopp* 1199–1225. Probably 'look-out ridge'. OE *weard* (influenced by OScand. *vǫrthr*) + *copp*.

Warden, 'watch or look-out hill', OE *weard* + *dūn*: **Warden** Kent. *Wardon* 1207. **Warden** Northum. *Waredun* c.1175. **Warden, Chipping** Northants. *Waredone* 1086 (DB), *Chepyng Wardoun* 1389. Affix is OE *cēping* 'market'. **Warden, Old** Beds. *Wardone* 1086 (DB), *Old Wardon* 1495. Affix is OE *eald* 'old'.

Wardington Oxon. *Wardinton* c.1180. Probably 'estate associated with a man called *Wearda or Wǣrheard'. OE pers. name + *-ing-* + *tūn*.

Wardlaw Highland. *Wardelaue* 1210. 'Watch hill'. OE *weard* + *hlāw*.

Wardle, 'watch or look-out hill', OE *weard* + *hyll*: **Wardle** Ches. *Warhelle* [*sic*] 1086 (DB), *Wardle* 1184. **Wardle** Rochdl. *Wardhul* c.1193.

Wardley Rutland. *Werlea* 1067. Probably 'wood or clearing near a weir', OE *wer* + *lēah*. Alternatively the first element may be OE *weard* 'watch, protection' if the *-d-* spelling (found from 1241) is original and not intrusive.

Wardlow Derbys. *Wardelawe* 1258. 'Watch or look-out hill'. OE *weard* + *hlāw*.

Ware Herts. *Waras* 1086 (DB). 'The weirs'. OE *wær*.

Wareham Dorset. *Werham* late 9th cent., *Warham* 1086 (DB). 'Homestead or river-

meadow by a weir'. OE *wer, wær + hām* or *hamm*.

Warehorne Kent. *Werahorna* 830, *Werahorne* 1086 (DB). Probably 'guarding horn of land, look-out promontory'. OE **horna* with a word *wer- related to OE *werian* 'to guard or defend'.

Warenford Northum. *Warneford* 1256. 'Ford over Warren Burn'. Celtic river-name (probably 'alder river' with the later addition of OE *burna* 'stream') + OE *ford*.

Waresley Cambs. *Wederesle* [*sic*] 1086 (DB), *Wereslea* 1169. Probably 'woodland clearing of a man called *Wether or *Wǣr'. OE pers. name + *lēah*.

Warfield Brack. For. *Warwelt* [*sic*] 1086 (DB), *Warefeld* 1171. 'Open land by a weir'. OE *wer, wær + feld*.

Wargrave Wokhm. *Weregrave* 1086 (DB). 'Pit or trench by the weirs'. OE *wer* + *græf*.

Warham Norfolk. *Warham* 1086 (DB). 'Homestead or village by a weir'. OE *wær* + *hām*.

Waringsford Down. *Waringsford* 1743. The village is named from a former house and estate recorded in 1744 as 'the seat of Henry Waring Esq.'.

Waringstown Down. The village is named from William *Warren* or *Waring*, who built a house here in 1666.

Wark, from OE (*ge*)*weorc* 'fortification': **Wark** Northum., near Bellingham. *Werke* 1279. **Wark** Northum., near Cornhill. *Werch* 1158.

Warkleigh Devon. *Warocle* 1100–3, *Wauerkelegh* 1242. Possibly 'spider wood or clearing'. OE **wæferce* + *lēah*.

Warkton Northants. *Wurcingtun* 946, *Werchintone* 1086 (DB). 'Estate associated with a man called *Weorc(a)'. OE pers. name + *-ing-* + *tūn*.

Warkworth Northum. *Werceworthe* c.1050. 'Enclosure of a man called *Weorca'. OE pers. name + *worth*.

Warlaby N. Yorks. *Werlegesbi* 1086 (DB). 'Farmstead or village of the traitor or troth-breaker'. OE *wērloga* + OScand. *bý*.

Warley, Great & Warley, Little Essex. *Werle* c.1045, *Wareleia* 1086 (DB). 'Wood or clearing near a weir or subject to a covenant'. OE *wer* or *wǣr* + *lēah*.

Warlingham Surrey. *Warlyngham* 1144. 'Homestead of the family or followers of a man called *Wǣrla*. OE pers. name + *-inga-* + *hām*.

Warmfield Wakefd. *Warnesfeld* 1086 (DB). 'Open land frequented by wrens, or where stallions are kept'. OE *wærna* or *wǣrna* + *feld*.

Warmingham Ches. *Warmincham* 1259. 'Homestead of the family or followers of a man called Wǣrma or Wǣrmund', or 'homestead at Wǣrma's or Wǣrmund's place'. OE pers. name + *-inga-* or *-ing* + *hām*.

Warmington Northants. *Wyrmingtun* c.980, *Wermintone* 1086 (DB). 'Estate associated with a man called *Wyrma*'. OE pers. name + *-ing-* + *tūn*.

Warmington Warwicks. *Warmintone* 1086 (DB). 'Estate associated with a man called Wǣrma or Wǣrmund'. OE pers. name + *-ing-* + *tūn*.

Warminster Wilts. *Worgemynster* c.912, *Guerminstre* 1086 (DB). 'Church on the River Were'. OE river-name (meaning 'winding') + *mynster*.

Warmsworth Donc. *Wermesford* [*sic*] 1086 (DB), *Wermesworth* c.1105. 'Enclosure of a man called *Wǣrmi* or Wǣrmund'. OE pers. name + *worth*.

Warmwell Dorset. *Warmewelle* 1086 (DB). 'Warm spring'. OE *wearm* + *wella*.

Warnborough, North & Warnborough, South Hants. *Weargeburnan* 973–4, *Wergeborne* 1086 (DB). Possibly 'stream where criminals were drowned', OE *wearg* + *burna*, though *wearg* may have an earlier sense 'wolf', hence perhaps 'stream frequented by wolves'.

Warnford Hants. *Wernæford* c.1053, *Warneford* 1086 (DB). 'Ford frequented by wrens or one used by stallions'. OE *wærna* or *wǣrna* + *ford*. Alternatively the first element may be an OE man's name *Wǣrna*.

Warnham W. Sussex. *Werneham* 1166. 'Homestead or enclosure of a man called *Wǣrna*, or where stallions are kept'. OE pers. name or *wǣrna* + *hām* or *hamm*.

Warninglid W. Sussex. *Warthynglithe* 13th cent. Probably 'hill-slope associated with a man called Wearda'. OE pers. name + *-ing-* + *hlith*.

Warrenpoint Down. *Warings's Point* 1744. 'Warren's point'. It is uncertain whether the *Warren* (or *Waring*) here was related to the family who gave the name to WARINGSTOWN.

Warrington Milt. K. *Wardintone* c.1175. Probably 'estate associated with a man called *Wearda* or *Wǣrheard*'. OE pers. name + *-ing-* + *tūn*.

Warrington Warrtn. *Walintune* [*sic*] 1086 (DB), *Werington* 1246. 'Farmstead or village by the weir or river-dam'. OE *wering* + *tūn*.

Warsash Hants. *Weresasse* 1272. 'Ash-tree by the weir, or of a man called Wǣr'. OE *wer* or pers. name + *æsc*.

Warslow Staffs. *Wereslei* [*sic*] 1086 (DB), *Werselow* 1300. Possibly 'hill with a watch-tower'. OE *weard-seld* + *hlāw*.

Warsop Notts. *Wareshope* 1086 (DB). Probably 'enclosed valley of a man called *Wǣr*'. OE pers. name + *hop*.

Warter E. R. Yorks. *Wartre* 1086 (DB). 'The gallows for criminals'. OE *wearg-trēow*.

Warthill N. Yorks. *Wardhilla* 1086 (DB). 'Watch or look-out hill'. OE *weard* + *hyll*.

Wartling E. Sussex. *Werlinges* [*sic*] 1086 (DB), *Wertlingis* 12th cent. '(Settlement of) the family or followers of a man called *Wyrtel*'. OE pers. name + *-ingas*.

Wartnaby Leics. *Worcnodebie* 1086 (DB). Probably 'farmstead or village of a man called *Wærcnōth*'. OE pers. name + OScand. *bý*.

Warton, 'watch or look-out farmstead', OE *weard* + *tūn*: **Warton** Lancs., near Carnforth. *Wartun* 1086 (DB). **Warton** Lancs., near Kirkham. *Wartun* 1086 (DB). **Warton** Northum. *Wartun* 1236.

Warton Warwicks. *Wavertune* c.1155. 'Farmstead by a swaying tree or near marshy ground'. OE *wæfre* + *tūn*.

Warwick Cumbria. *Warthwic* 1131. 'Dwelling or farm on the bank'. OE *waroth* + *wīc*.

Warwick Warwicks. *Wærincwicum* 1001, *Warwic* 1086 (DB). 'Dwellings by the weir or river-dam'. OE *wæring* + *wīc*. **Warwickshire** (OE *scīr* 'district') is first referred to in the 11th cent.

Wasdale Head Cumbria. *Wastedale* 1279, *Wascedaleheved* 1334. 'Valley of the water or lake'. OScand. *vatn* (genitive *-s*) + *dalr*, with the later addition of ME *heved* 'head, upper end'.

Wash, The Lincs. /Norfolk. *The Wasshes* c.1545. OE *wæsc* 'sandbank washed by the sea', originally used of two stretches fordable at low water.

Washaway Cornwall. Recorded thus from 1699, probably from ModE *washway* 'hollow road'.

Washbourne Devon. *Waseborne* 1086 (DB). Probably 'stream used for washing (sheep or clothes)'. OE *wæsce* + *burna*. Alternatively the first element may be OE *(ge)wæsc* 'flood'.

Washbourne, Great & Washbourne, Little Glos. *Uassanburnan* 780, *Waseborne* 1086 (DB). 'Stream with alluvial land'. OE **wæsse* + *burna*.

Washbrook Suffolk. *Wasebroc* 1198. 'Brook used for washing (sheep or clothes)', or 'brook liable to flood'. OE *wæsce* or *(ge)wæsc* + *brōc*.

Washfield Devon. *Wasfelte* 1086 (DB). Probably 'open land near a place used for washing (sheep or clothes)'. OE *wæsce* + *feld*.

Washford Somerset. *Wecetford* c.960. Probably 'ford over the stream called **Wæcet*'. Celtic name (perhaps the original name of Washford Water) + OE *ford*, *see* WATCHET.

Washford Pyne Devon. *Wasforde* 1086 (DB). Probably 'ford at the place for washing (sheep or clothes)'. OE *wæsce* + *ford*. Manorial affix from the *de Pinu* family, here in the 13th cent.

Washingborough Lincs. *Washingeburg* 1086 (DB), *Wasingburg* c.1157. Probably 'stronghold of the family or followers of a man called **Wassa*'. OE pers. name + *-inga-* + *burh*.

Washington Sundld. *Wassyngtona* 1183. 'Estate associated with a man called **Hwæssa*'. OE pers. name + *-ing-* + *tūn*.

Washington W. Sussex. *Wessingatun* 946–55, *Wasingetune* 1086 (DB). 'Estate of the family or followers of a man called **Wassa*'. OE pers. name + *-inga-* + *tūn*.

Wasing W. Berks. *Walsince* 1086 (DB). Perhaps originally the name of the stream here, with obscure element + OE *-ing*.

Wasperton Warwicks. *Waspertune* 1043, *Wasmertone* [sic] 1086 (DB). Probably 'pear orchard by alluvial land'. OE **wæsse* + *peru* + *tūn*.

Wass N. Yorks. *Wasse* 1541. Probably 'the fords'. OScand. *vath* in a plural form.

Watchet Somerset. *Wæcet* 962, *Wacet* 1086 (DB). Probably 'under the wood' from Celtic **cĕd* 'wood' with prefix **wo-*, perhaps originally applied to the stream here, *see* WASHFORD.

Watchfield Oxon. *Wæclesfeld* 931, *Wachenesfeld* 1086 (DB). 'Open land of a man called **Wæcel* or **Wæccīn*'. OE pers. name + *feld*.

Watendlath Cumbria. *Wattendlane* late 12th cent. Probably 'lane to the end of the lake'. OScand. *vatn* + *endi* with OE *lane* (replaced by OScand. *hlatha* 'barn').

Water Orton Warwicks. *See* ORTON.

Water Stratford Bucks. *See* STRATFORD.

Waterbeach Cambs. *Vtbech* 1086 (DB), *Waterbech* 1237. OE *bece* 'stream, valley' or *bæc* (locative **bece*) 'low ridge', with the addition of OE *ūt* 'outer' and *wæter* 'water' to distinguish it from LANDBEACH.

Waterden Norfolk. *Waterdenna* 1086 (DB). 'Valley with a stream or lake'. OE *wæter* + *denu*.

Wateresk (*Uachtar Easc*) Down. 'Upper channel'.

Waterfall Staffs. *Waterfal* 1201. OE *wæter-gefall* 'place where a stream disappears into the ground'.

Waterford (*Port Láirge*) Waterford. *Vadrefiord* (no date). 'Wether inlet'. OScand. *vethr* + *fjǫrthr*. Wethers were loaded on to boats here to be taken to other ports. The Irish name means 'bank of the haunch'.

Waterhouses Staffs. On record from the late 16th cent., 'the houses on the stream', ModE *water* + *house*.

Wateringbury Kent. *Uuotryngebyri* 964–95, *Otringeberge* 1086 (DB). Possibly 'stronghold of the family or followers of a man called Ōhthere'. OE pers. name + *-inga-* + *burh* (dative *byrig*).

Waterloo Sefton. named from the *Royal Waterloo Hotel* (founded in 1815 and so called after the famous battle of that year). Similar names commemorating the battle are found in other counties, for example **Waterloo** Gtr. London, **Waterlooville** Hants.

Watermillock Cumbria. *Wethermeloc* early 13th cent. 'Little bare hill where wether-sheep graze'. Celtic **m̨el* with diminutive suffix, to which OE *wether* has been added.

Waterperry Oxon. *Perie* 1086 (DB), *Waterperi c.*1190. '(Place at) the pear-tree(s)'. OE *pyrige* with the later addition of ME *water* to distinguish it from WOODPERRY.

Waterston Pemb. *Walterystone* 1407. 'Walter's farm'. OGerman pers. name + OE *tūn*.

Watford, 'ford used when hunting', OE *wāth* + *ford*: **Watford** Herts. *Watford c.*945. **Watford** Northants. *Watford* 1086 (DB).

Wath, 'the ford', OScand. *vath*; examples include: **Wath** N. Yorks., near Ripon. *Wat* 1086 (DB). **Wath upon Dearne** Rothm. *Wade* 1086 (DB). For the river-name, *see* BOLTON UPON DEARNE.

Watling Street (Roman road from Dover to Wroxeter). *Wæclinga stræt* late 9th cent. 'Roman road associated with the family or followers of a man called **Wæcel*'. OE pers. name + *-inga-* + *strǣt*. An early name for ST ALBANS is *Wæclingaceaster c.*900, 'Roman fort of **Wæcel's* people', and the road-name was no doubt applied to the stretch of road between St Albans and London before it was extended to the whole length. The name Watling Street was later transferred to several other Roman roads.

Watlington Norfolk. *Watlingetun* 11th cent. Possibly 'farmstead of the family or followers of a man called **Hwætel* or **Wacol*'. OE pers. name + *-inga-* + *tūn*.

Watlington Oxon. *Wæclinctune* 887, *Watelintone* 1086 (DB). Probably 'estate associated with a man called **Wæcel*'. OE pers. name + *-ing-* + *tūn*.

Watnall Notts. *Watenot* [*sic*] 1086 (DB), *Watenho* 1202. Possibly 'hill-spur of a man called **Wata*'. OE pers. name (genitive *-n*) + *hōh*.

Watten Highland. *Watne* 1230. '(Place by a) lake'. OScand. *vatn*.

Wattisfield Suffolk. *Watlesfelda* 1086 (DB). 'Open land of a man called **Wacol* or **Hwætel*'. OE pers. name + *feld*.

Wattisham Suffolk. *Wecesham* 1086 (DB). 'Homestead or village of a man called **Wæcci*'. OE pers. name + *hām*.

Watton E. R. Yorks. *Uetadun* 731, *Wattune* 1086 (DB), 'Wet hill, or hill of the wet places'. OE *wǣt* (as adjective or noun) + *dūn*.

Watton Norfolk. *Wadetuna* 1086 (DB). 'Farmstead of a man called Wada'. OE pers. name + *tūn*.

Watton at Stone Herts. *Wattun* 969, *Wodtone* 1086 (DB), *Watton atte Stone* 1311. 'Farmstead where woad is grown'. OE *wād* + *tūn*. Affix 'at the stone' (OE *stān*) from an old stone here.

Wattstown Rhon. Edmund *Watts* was a coal-mine owner here in the 19th cent.

Waun, Y. *See* CHIRK.

Wavendon Milt. K. *Wafandun* 969, *Wavendone* 1086 (DB). 'Hill of a man called **Wafa*'. OE pers. name (genitive *-n*) + *dūn*.

Waveney (district) Suffolk, named from the River Waveney which is *Wahenhe* 1275, 'the river by a quagmire', OE **wagen* + *ēa*.

Waverley (district) Surrey. *Wauerleia* 1147. Probably 'woodland clearing by the marshy ground'. A revival of an old local name, otherwise preserved in Waverley Abbey (near FARNHAM). OE *wǣfre* + *lēah*.

Waverton Ches. *Wavretone* 1086 (DB). 'Farmstead by a swaying tree'. OE *wǣfre* + *tūn*.

Waverton Cumbria. *Wauerton* 1183. 'Farmstead by the River Waver'. OE river-name (from *wǣfre* 'winding') + *tūn*.

Wawne E. R. Yorks. *Wagene* 1086 (DB). Possibly from an OE **wagen* 'quaking bog, quagmire' or a Celtic name from **waun* 'marsh'.

Waxham Norfolk. *Waxtonesham* 1044-7, *Wactanesham*, *Wacstenesham* 1086 (DB). Probably 'homestead of a man called **Wægstān*', OE pers. name + *hām*. Alternatively 'homestead by the stone where watch was kept', OE *wacu* + *stān* + *hām*.

Waxholme E. R. Yorks. *Waxham* 1086 (DB). 'Homestead where wax (from bees) is produced'. OE *weax* + *hām*.

Wayford Somerset. *Waiford* 1206. 'Ford on a way or road'. OE *weg* + *ford*.

Weald, 'the woodland or forest', OE *weald*: **Weald Bassett, North** Essex. *Walda* 1086 (DB), *Welde Basset* 1291. Manorial affix from the *Basset* family, here in the 13th cent. **Weald,**

South Essex. *Welde* 1062, *Welda* 1086 (DB).
Weald, The Kent–Hants. *Waldum* (a dative
plural form) 1185.

Wealden (district) E. Sussex. A modern
adjectival derivative of Weald, *see* THE WEALD in
previous entry.

Wealdstone Gtr. London. a late name,
'boundary stone of HARROW WEALD'.

Wear (river) Durham. *See*
MONKWEARMOUTH.

Weare, '(place at) the weir or fishing
enclosure', OE *wer*: **Weare** Somerset. *Werre*
1086 (DB). **Weare Giffard** Devon. *Were* 1086
(DB), *Weregiffarde* 1328. Manorial affix from the
Giffard family, here in the 13th cent.

Weasenham Norfolk. *Wesenham* 1086 (DB).
Possibly 'homestead or village of a man called
Weosa. OE pers. name (genitive -*n*) + *hām*.

Weaverham Ches. *Wivreham* 1086 (DB).
River-name (from OE *wēfer* 'winding' or from
ancestor of Welsh *gwefr* 'amber-coloured') + OE
hām.

Weaverthorpe N. Yorks. *Wifretorp* 1086
(DB). 'Outlying farmstead or hamlet of a man
called Víthfari'. OScand. pers. name + *thorp*.

Weddington Warwicks. *Watitune* 1086 (DB).
Possibly 'estate associated with a man called
Hwæt. OE pers. name + -*ing*- + *tūn*.

Wedmore Somerset. *Wethmor* late 9th cent.,
Wedmore 1086 (DB). Possibly 'marsh used for
hunting'. OE *wǣthe* + *mōr*.

Wednesbury Sandw. *Wadnesberie* 1086
(DB). 'Stronghold associated with the heathen
god Wōden'. OE god-name + *burh* (dative
byrig).

Wednesfield Wolverh. *Wodnesfeld* 996,
Wodnesfelde 1086 (DB). 'Open land of the
heathen god Wōden'. OE god-name + *feld*.

Weedon, 'hill with a heathen temple', OE
wēoh + *dūn*: **Weedon** Bucks. *Weodune* 1066.
Weedon Bec Northants. *Weodun* 944,
Wedone 1086 (DB), *Wedon Beke* 1379. Manorial
affix from its possession by the Norman Abbey
of Bec-Hellouin in the 12th cent. **Weedon
Lois** Northants. *Wedone* 1086 (DB), *Leyes
Weedon* 1475. The affix is possibly from a 'well
of St Loys or Lewis' in the parish, but it may be
manorial.

Weeford Staffs. *Weforde* 1086 (DB). Probably
'ford by a heathen temple'. OE *wēoh* + *ford*.

Week, Weeke, 'the dwelling, the specialized
farm or trading settlement', OE *wīc*; examples
include: **Week St Mary** Cornwall. *Wich* 1086
(DB), *Seintemarywyk* 1321. Affix from the
dedication of the church. **Weeke** Hants. *Wike*
1248.

Weekley Northants. *Wiclea* 956, *Wiclei* 1086
(DB). Probably 'wood or clearing near an earlier
Romano-British settlement'. OE *wīc* + *lēah*.

Weeley Essex. *Wilgelea* 11th cent., *Wileia*
1086 (DB). 'Wood or clearing where willow-trees
grow'. OE *wilig* + *lēah*.

Weeting Norfolk. *Watinge* c.1050, *Wetinge*
1086 (DB). 'Wet or damp place'. OE *wēt* + -*ing*.

Weeton, 'farmstead where willow-trees
grow', OE *wīthig* + *tūn*: **Weeton** E. R. Yorks.
Wideton 1086 (DB). **Weeton** Lancs. *Widetun*
1086 (DB). **Weeton** N. Yorks. *Widitun*
1086 (DB).

Weighton, Little E. R. Yorks. *Widetone* 1086
(DB). Identical in origin with the previous
names.

Weighton, Market E. R. Yorks. *Wicstun*
1086 (DB). 'Farmstead by an earlier Romano-
British settlement'. OE *wīc* + *tūn*. The affix
Market first occurs in the early 19th cent.

Welborne, Welbourn, Welburn, 'stream
fed by a spring', OE *wella* + *burna*; examples
include: **Welborne** Norfolk. *Walebruna* 1086
(DB). **Welbourn** Lincs. *Wellebrune* 1086 (DB).
Welburn N. Yorks., near Bulmer. *Wellebrune*
1086 (DB).

Welbury N. Yorks. *Welleberge* 1086 (DB). 'Hill
with a spring'. OE *wella* + *beorg*.

Welby Lincs. *Wellebi* 1086 (DB). 'Farmstead
or village by a spring or stream'. OE *wella* +
OScand. *bý*.

Welcombe Devon. *Walcome* 1086 (DB).
'Valley with a spring or stream'. OE *wella* +
cumb.

Weldon, Great & Weldon, Little
Northants. *Weledone* 1086 (DB). 'Hill with a
spring or by a stream'. OE *wella* + *dūn*.

Welford Northants. *Wellesford* 1086 (DB).
'Ford by the spring or over the stream'. OE *wella*
+ *ford*.

Welford W. Berks. *Weligforda* 949, *Waliford* 1086 (DB). 'Ford where willow-trees grow'. OE *welig* + *ford*.

Welford on Avon Warwicks. *Welleford* 1086 (DB), *Welneford* 1177. 'Ford by the springs'. OE *wella* (genitive plural *-ena*) + *ford*. On the River AVON.

Welham Leics. *Weleham* 1086 (DB). Possibly 'homestead by the stream', OE *wella* + *hām*. Alternatively 'homestead of a man called **Wēola*', from an OE pers. name.

Welham Notts. *Wellun* 1086 (DB). '(Place at) the springs'. OE *wella* in a dative plural form *wellum*.

Welhamgreen Herts. *Wethyngham* c.1315, *Whelamgrene* 1467. 'Willow-tree enclosure'. OE **wīthign* + *hamm*, later with ME *grene* 'village green'.

Well, '(place at) the spring or stream', OE *wella*; examples include: **Well** Lincs. *Welle* 1086 (DB). **Well** N. Yorks. *Welle* 1086 (DB).

Welland (river) Northants.–Lincs. *Weolud* 921. A Celtic or pre-Celtic river-name of uncertain meaning.

Wellesbourne Hastings & Wellesbourne Mountford Warwicks. *Welesburnan* 840, *Waleborne* 1086 (DB). Possibly 'stream with a pool'. OE *wēl* + *burna*. Distinguishing affixes from possession by the *de Hastanges* family (from the 14th cent.) and the *de Munford* family (from the 12th cent.).

Welling Gtr. London. *Wellyngs* 1362. Possibly a manorial name from the *Willing* family, here in 1301. Alternatively perhaps '(place at) the springs', from an OE **w(i)elling*.

Wellingborough Northants. *Wedlingeberie* 1086 (DB), *Wendlingburch* 1178. 'Stronghold of the family or followers of a man called **Wændel*'. OE pers. name + *-inga-* + *burh*.

Wellingham Norfolk. *Walnccham* [*sic*] 1086 (DB), *Uuelingheham* c.1190. 'Homestead of the dwellers by a spring or stream'. OE *wella* + *-inga-* + *hām*.

Wellingore Lincs. *Wellingoure* 1086 (DB). Probably 'promontory of the dwellers by a spring or stream'. OE *wella* + *-inga-* + **ofer*.

Wellington, probably 'estate associated with a man called **Wēola*', OE pers. name + *-ing-* + *tūn*: **Wellington** Herefs. *Weolintun* early 11th cent., *Walintone* 1086 (DB). **Wellington** Somerset. *Weolingtun* 904, *Walintone* 1086 (DB). **Wellington** Tel. & Wrek. *Walitone* 1086 (DB), *Welintun* c.1145.

Wellow B. & NE. Som. *Weleuue* 1084. Originally the name of the stream here, a Celtic river-name possibly meaning 'winding'.

Wellow I. of Wight. *Welig* c.880, *Welige* 1086 (DB). '(Place at) the willow-tree'. OE *welig*.

Wellow Notts. *Welhag* 1207. 'Enclosure near a spring or stream'. OE *wella* + *haga*.

Wellow, East & Wellow, West Hants. *Welewe* c.880, *Weleve* 1086 (DB). Identical in origin with WELLOW (B. & NE. Som.).

Wells, 'the springs', OE *wella* in a plural form: **Wells** Somerset. *Willan* c.1050, *Welle* 1086 (DB). **Wells-next-the-Sea** Norfolk. *Guelle* [*sic*] 1086 (DB), *Wellis* 1291.

Wellsborough Leics. *Wethelesberge* 1181, *Weulesbergh* 1285. 'Hill of the wheel', i.e. 'hill with a circular shape or feature'. OE *hweowol* + *beorg*.

Welnetham, Great & Welnetham, Little Suffolk. *Hvelfiham* [*sic*] 1086 (DB), *Weluetham* 1170. Possibly 'enclosure frequented by swans near a water-wheel or other circular feature'. OE *hwēol* + *elfitu* + *hamm*.

Welney Norfolk. *Wellenhe* c.14th cent. 'River called *Welle*'. The earlier name of Old Croft River (from OE *wella* 'stream') + OE *ēa*.

Welsh Bicknor Herefs. *See* BICKNOR.

Welsh Frankton Shrops. *See* FRANKTON.

Welshampton Shrops. *Hantone* 1086 (DB), *Welch Hampton* 1649. Originally 'high farmstead or estate', from OE *hēah* (dative *hēan*) + *tūn*. The later affix 'Welsh' refers to its proximity to a detached portion of Flintshire.

Welshpool (*Y Trallwng*) Powys. *Pola* 1253, *Walshe Pole* 1477. '(Place by the) Welsh pool'. OE *welisc* + *pōl*. The name emphasizes the town's position on the Welsh side of the border. The Welsh name means 'the very wet swamp' (Welsh *y* + *tra* + *llwng*), referring to the same pool.

Welton, usually 'farmstead by a spring or stream', OE *wella* + *tūn*; examples include: **Welton** E. R. Yorks. *Welleton* 1086 (DB). **Welton le Marsh** Lincs. *Waletune* 1086 (DB).

Affix means 'in the marshland'. **Welton le Wold** Lincs. *Welletune* 1086 (DB). Affix means 'on the wold(s)' with loss of preposition, *see* WOLDS.

Welwick E. R. Yorks. *Welwic* 1086 (DB). 'Dwelling or (dairy) farm near a spring or stream'. OE *wella + wīc*.

Welwyn Herts. *Welingum c.*945, *Welge* 1086 (DB). '(Place at) the willow-trees'. OE *welig* in a dative plural form *weligum*.

Wem Shrops. *Weme* 1086 (DB). 'Dirty or muddy place'. OE **wemm*.

Wembdon Somerset. *Wadmendune* 1086 (DB). Probably 'hill of the huntsmen'. OE **wǣthe-mann + dūn*.

Wembley Gtr. London. *Wembalea* 825. 'Woodland clearing of a man called **Wemba*'. OE pers. name + *lēah*.

Wembury Devon. *Wenbiria* late 12th cent. OE *burh* (dative *byrig*) 'fortified place, stronghold' with an uncertain first element, possibly OE *wenn*, *wænn* 'tumour-shaped mound'.

Wembworthy Devon. *Mameorda* [sic] 1086 (DB), *Wemeworth* 1207. 'Enclosure of a man called **Wemba*'. OE pers. name + *worth*.

Wendens Ambo Essex. *Wendena* 1086 (DB). Probably 'winding valley'. OE **wende + denu*. Affix is Latin *ambo* 'both', referring to the union of the two parishes called Wenden in 1662.

Wendlebury Oxon. *Wandesberie* [sic] 1086 (DB), *Wendelberi c.*1175. 'Stronghold of a man called **Wændla*'. OE pers. name + *burh* (dative *byrig*).

Wendling Norfolk. *Wenlinga* 1086 (DB). Possibly '(settlement of) the family or followers of a man called **Wændel*'. OE pers. name + *-ingas*.

Wendover Bucks. *Wændofran c.*970, *Wendoure* 1086 (DB). Originally the name of the stream here, a Celtic river-name meaning 'white waters'.

Wendron Cornwall. 'Church of *Sancta Wendrona*' 1291. Named from the patron saint of the church, a female saint of whom nothing is known.

Wendy Cambs. *Wandei* 1086 (DB). 'Island at a river-bend'. OE **wende + ēg*.

Wenham, Great & Wenham, Little Suffolk. *Wenham* 1086 (DB). Possibly 'homestead or enclosure with pastureland'. OE **wynn + hām* or *hamm*.

Wenhaston Suffolk. *Wenadestuna* 1086 (DB). 'Farmstead or village of a man called **Wynhæth*'. OE pers. name + *tūn*.

Wenlock, Little & Wenlock, Much Shrops. *Wininicas* 675–90, *Wenlocan, Winlocan* 9th cent., *Wenloch* 1086 (DB). Probably OE *loca* 'enclosed place' (hence perhaps 'monastery') added to the first part of old Celtic name *Wininicas* (obscure, but possibly 'white area' or the like referring to limestone of Wenlock Edge, from Celtic **wïnn* 'white'). Affix is OE *mycel* 'great'.

Wennington Cambs. *Weningtone c.*960. 'Estate associated with a man called **Wenna*'. OE pers. name + *-ing- + tūn*.

Wennington Gtr. London. *Winintune* 1042–4. 'Estate associated with a man called Wynna'. OE pers. name + *-ing- + tūn*.

Wennington Lancs. *Wennigetun* 1086 (DB). 'Farmstead on the River Wenning'. OE river-name (meaning 'dark stream') + *tūn*.

Wensley Derbys. *Wodnesleie* 1086 (DB). 'Sacred grove of the heathen god Wōden'. OE god-name + *lēah*.

Wensley N. Yorks. *Wendreslaga* 1086 (DB), *Wendesle* 1203. 'Woodland clearing of a man called **Wændel*'. OE pers. name + *lēah*.
Wensleydale is *Wandesleydale c.*1150, from OScand. *dalr* 'valley'.

Wentbridge Wakefd. *Wentbrig* 1302. 'Bridge across the River Went'. Pre-English river-name of uncertain origin + OE *brycg*.

Wentnor Shrops. *Wantenoure* 1086 (DB), *Wontenoure c.*1200, *Wentenour* 1251. 'Flat-topped ridge of a man called **Wonta* or **Wenta*'. OE pers. name (genitive -*n*) + **ofer*.

Wentworth, probably 'enclosure of a man called Wintra', OE pers. name + *worth*:
Wentworth Cambs. *Winteuuorde* 1086 (DB).
Wentworth Rothm. *Wintreuuorde* 1086 (DB).

Wenvoe (*Gwenfô*) Vale Glam. *Wnfa* 1153, *Wenvo c.*1262. Meaning uncertain.

Weobley Herefs. *Wibelai* 1086 (DB). 'Woodland clearing of a man called **Wiobba*'. OE pers. name + *lēah*.

Were (river) Wilts. *See* WARMINSTER.

Wereham Norfolk. *Wigreham* 1086 (DB). Probably 'homestead on a stream called *Wigor* (the winding one)'. Celtic river-name (an old name for the River Wissey) + OE *hām*.

Werrington Peterb. *Witheringtun* 972, *Widerintone* 1086 (DB). 'Estate associated with a man called Wither'. OE pers. name + *-ing-* + *tūn*.

Werrington Cornwall. *Ulvredintone* 1086 (DB). 'Estate associated with a man called Wulfrǣd'. OE pers. name + *-ing-* + *tūn*.

Wervin Ches. *Wivrevene* 1086 (DB). 'Cattle fen, or quaking fen'. OE *weorf* or **wifer* + *fenn*.

Wesham Lancs. *Westhusum* 1189. '(Place at) the westerly houses'. OE *west* + *hūs* in a dative plural form.

Wessex (ancient Anglo-Saxon kingdom centred on Winchester). *West Seaxe* late 9th cent. '(Territory of) the West Saxons'. OE *west* + *Seaxe*.

Wessington Derbys. *Wistanestune* 1086 (DB). 'Farmstead or village of a man called Wīgstān'. OE pers. name + *tūn*.

West as affix. *See* main name, e.g. for **West Acre** (Norfolk) *see* ACRE.

West End Hants. *Westend* 1607. Possibly from OE *wēsten* 'waste-land'.

West Kilbride N. Ayr. 'Western (place by) St Brigid's church'. *West* distinguishes this place from EAST KILBRIDE, although there are almost thirty miles between them.

West Lothian. *See* MIDLOTHIAN.

West Williamston. *See* EAST WILLIAMSTON.

Westbere Kent. *Westbere* 1212. 'Westerly cattle shed or swine pasture'. OE *west* + *bÿre* or *bær*.

Westborough Lincs. *Westburg* 1086 (DB). 'Westerly stronghold or manor'. OE *west* + *burh*.

Westbourne W. Sussex. *Burne* 1086 (DB), *Westbourne* 1305. '(Place at) the stream'. OE *burna*, with the later addition of *west* to distinguish it from EASTBOURNE.

Westbury, 'westerly stronghold or fortified place', OE *west* + *burh* (dative *byrig*); examples include: **Westbury** Shrops. *Wesberie* 1086 (DB). **Westbury** Wilts. *Westberie* 1086 (DB).

Westbury on Severn Glos. *Wesberie* 1086 (DB). **Westbury on Trym** Brist. *Westbyrig* 791–6, *Hvesberie* [*sic*] 1086 (DB). Trym is an OE river-name (probably 'strong one').

Westby Lancs. *Westbi* 1086 (DB). 'Westerly farmstead or village'. OScand. *vestr* + *bý*.

Westcliffe on Sea Sthend. a self-explanatory name of recent origin.

Westcote, Westcott, 'westerly cottage(s)', OE *west* + *cot*; examples include: **Westcote** Glos. *Westcote* 1315. **Westcott** Surrey. *Westcote* 1086 (DB).

Westdean E. Sussex. *Dene* 1086 (DB), *Westdene* 1291. 'West valley'. OE *west* + *denu*. 'West' in relation to EASTDEAN.

Westerdale N. Yorks. *Westerdale* c.1165. 'More westerly valley'. OScand. *vestari* + *dalr*.

Westerfield Suffolk. *Westrefelda* 1086 (DB). '(More) westerly open land'. OE **wester* or *westerra* + *feld*.

Westergate W. Sussex. *Westgate* 1230. 'More westerly gate or gap'. OE *westerra* + *geat*. *See* EASTERGATE.

Westerham Kent. *Westarham* 871–89, *Oistreham* [*sic*] 1086 (DB). 'Westerly homestead'. OE **wester* + *hām*.

Westerleigh S. Glos. *Westerlega* 1176. 'More westerly woodland clearing'. OE *westerra* + *lēah*.

Western Isles (the unitary authority). The Outer Hebrides, the most westerly island chain in Scotland. The Gaelic name of the islands is *Na h-Eileanan an Iar*, 'The Islands of the West', or for the unitary authority *Eilean Siar*, 'Western Isles'.

Westfield, 'westerly open land', OE *west* + *feld*: **Westfield** E. Sussex. *Westewelle* [*sic*] 1086 (DB), *Westefelde* c.1115. **Westfield** Norfolk. *Westfelda* 1086 (DB).

Westgate on Sea Kent. *Westgata* 1168. 'Westerly gate or gap'. OE *west* + *geat*.

Westhall Suffolk. *Westhala* 1139. 'Westerly nook of land'. OE *west* + *halh*.

Westham E. Sussex. *Westham* 1222. 'Westerly promontory'. OE *west* + *hamm*.

Westhampnett W. Sussex. *Hentone* 1086 (DB), *Westhamptonette* 1279. Originally 'high farmstead', from OE *hēah* (dative *hēan* + *tūn*), with the later addition of *west* and OFrench *-ette* 'little'.

Westhead Lancs. *Westhefd c.*1190. 'Westerly headland or ridge'. OE *west + hēafod*.

Westhide Herefs. *Hide* 1086 (DB), *Westhyde* 1242. 'Westerly hide of land'. OE *west + hīd*.

Westhope Shrops. *Weshope* 1086 (DB). 'Westerly enclosed valley'. OE *west + hop*.

Westthorpe Suffolk. *Westtorp* 1086 (DB). 'Westerly outlying farmstead or hamlet'. OScand. *vestr + thorp*.

Westhoughton Bolton. *Halcton c.*1210, *Westhalcton c.*1240. 'Westerly farmstead in a nook of land'. OE *west + halh + tūn*.

Westleigh Devon. *Weslega* 1086 (DB). 'Westerly wood or clearing'. OE *west + lēah*.

Westleton Suffolk. *Westledestuna* 1086 (DB). 'Farmstead or village of a man called Vestlithi'. OScand. pers. name + OE *tūn*.

Westley, 'westerly wood or clearing', OE *west + lēah*: **Westley** Suffolk. *Westlea* 1086 (DB). **Westley Waterless** Cambs. *Westle c.*1045, *Weslai* 1086 (DB), *Westle Waterles* 1285. Affix means 'wet clearings', from OE *wæter + lēas* (the plural of *lēah*).

Westlinton Cumbria. *Westlevington c.*1200. 'Farmstead by the River Lyne'. Celtic river-name (possibly 'the smooth one') + OE *tūn*, with *west* to distinguish it from KIRKLINTON.

Westmeath (*An Iarmhí*) (the county). 'Western Meath'. The county was created out of the province of MEATH in 1542.

Westmeston E. Sussex. *Westmæstun c.*765, *Wesmestun* 1086 (DB). 'Most westerly farmstead'. OE *westmest + tūn*.

Westmill Herts. *Westmele* 1086 (DB). 'Westerly mill'. OE *west + myln*.

Westminster Gtr. London. *Westmynster c.*975. 'West monastery', i.e. to the west of London. OE *west + mynster*.

Westmorland (historical county). *Westmoringaland c.*1150. 'District of the people living west of the moors' (alluding to the North Yorkshire Pennines). OE *west + mōr + -inga- + land*.

Westnewton Northum. *Niwetona* 12th cent. 'New farmstead'. OE *nīwe + tūn*, with *west* to distinguish it from KIRKNEWTON.

Weston, a very common place name, 'west farmstead or village', i.e. one to the west of another settlement, OE *west + tūn*; examples include: **Weston** Herts. *Westone* 1086 (DB). **Weston** Lincs. *Westune* 1086 (DB). **Weston, Buckhorn** Dorset. *Westone* 1086 (DB), *Boukeresweston* 1275. Manorial affix from some medieval owner called *Bouker*. **Weston, Edith** Rutland. *Westona* 1113, *Weston Edith* 1275. Affix probably from its possession by Queen *Ēadgȳth*, wife of Edward the Confessor, in 1086. **Weston, Hail** Cambs. *Heilweston* 1199. Affix is the original Celtic name (meaning 'dirty stream') of the River Kym. **Weston in Gordano** N. Som. *Weston* 1086 (DB), *Weston in Gordene* 1343. Affix is an old district name, *see* CLAPTON. **Weston Rhyn** Shrops. *Westone* 1086 (DB), *Weston Ryn* 1302. Affix refers to nearby Rhyn, from Welsh *rhyn* 'peak, hill'. **Weston Subedge** Glos. *Westone* 1086 (DB), *Weston sub Egge* 1255. Affix means 'under the edge or escarpment' (of the Cotswolds) from Latin *sub* and OE *ecg*. **Weston super Mare** N. Som. *Weston c.*1230, *Weston super Mare* 1349. The Latin affix means 'on the sea'. **Weston Turville** Bucks. *Westone* 1086 (DB), *Westone Turvile* 1302. Manorial affix from the *de Turvile* family, here in the 12th cent. **Weston under Penyard** Herefs. *Westune* 1086 (DB). Affix from nearby Penyard (Hill), *Peniard* 1228, identical with PENNARD Somerset.

Westoning Beds. *Westone* 1086 (DB), *Westone Ynge* 1365. 'West farmstead', OE *west + tūn*, with manorial affix from its possession by the *Ing* family in the 14th cent.

Westonzoyland Somerset. *Sowi* 1086 (DB), *Westsowi c.*1245. 'The westerly manor of the estate called *Sowi*', thus distinguished from MIDDLEZOY and with the later addition of *land* 'estate'.

Westow N. Yorks. *Wiuestou* 12th cent. 'Holy place of the women, or of a woman called *Wīfe*'. OE *wīf* or pers. name + *stōw*.

Westray (island) Orkn. *Vesturey c.*1260. 'Western island'. OScand. *vestr + ey*. Westray is the westernmost of the islands in northern Orkney.

Westward Cumbria. *Le Westwarde* 1354. 'Western division (of a forest)'. ME *west + warde*.

Westward Ho! Devon, modern name commemorating the novel of this name by Charles Kingsley (published in 1855) largely set in this locality.

Westwell, 'westerly spring or stream', OE *west* + *wella*: **Westwell** Kent. *Welle* 1086 (DB), *Westwell* 1226. **Westwell Leacon** (recorded thus from 1798) is from Kentish dialect *leacon* 'wet common land'. **Westwell** Oxon. *Westwelle* 1086 (DB).

Westwick, 'westerly dwelling or (dairy) farm', OE *west* + *wīc*: **Westwick** Cambs. *Westuuiche* 1086 (DB). **Westwick** Durham. *Westewic* 1091. **Westwick** Norfolk. *Westuuic* 1086 (DB).

Westwood Wilts. *Westwuda* 987, *Westwode* 1086 (DB). 'Westerly wood'. OE *west* + *wudu*.

Wetheral Cumbria. *Wetherhala* c.1100. 'Nook of land where wether-sheep are kept'. OE *wether* + *halh*.

Wetherby Leeds. *Wedrebi* 1086 (DB). 'Wether-sheep farmstead'. OScand. *vethr* + *bý*.

Wetherden Suffolk. *Wederdena* 1086 (DB). 'Valley where wether-sheep are kept'. OE *wether* + *denu*.

Wetheringsett Suffolk. *Weddreringesete* c.1035, *Wederingaseta* 1086 (DB). Probably 'fold of the people of WETHERDEN'. Reduced form of previous name + *-inga-* + *set*.

Wethersfield Essex. *Witheresfelda* 1086 (DB). 'Open land of a man called Wihthere or *Wether'. OE pers. name + *feld*.

Wettenhall Ches. *Watenhale* 1086 (DB). 'Wet nook of land'. OE *wēt* (dative *-an*) + *halh*.

Wetton Staffs. *Wettindun* 1252. Possibly 'wet hill'. OE *wēt* (dative *-an*) + *dūn*.

Wetwang E.R. Yorks. *Wetuuangha* 1086 (DB). Probably 'field for the trial of a legal action'. OScand. *vætt-vangr*.

Wetwood Staffs. *Wetewode* 1291. 'Wet wood'. OE *wēt* + *wudu*.

Wexcombe Wilts. *Wexcumbe* 1167. 'Valley where wax (from bees) is found'. OE *weax* + *cumb*.

Wexford (*Loch Garman*) Wexford. *Weysford* 15th cent. 'Inlet by the sandbank'. OIrish *escir* + OScand. *fjǫrthr*. The Irish name of Wexford means 'lake of the (river) Garma'.

Weybourne Norfolk. *Wabrune* 1086 (DB). OE *burna* 'spring, stream' with an uncertain first element, possibly an OE or pre-English name of the river from a root *war- 'water', or an OE *wagu 'quagmire', or OE *wær* 'weir, river-dam'.

Weybread Suffolk. *Weibrada* 1086 (DB). OE *brǣdu* 'broad stretch of land' with OE *weg* 'road' or a lost pre-English river-name *Wey*.

Weybridge Surrey. *Webruge* 1086. 'Bridge over the River Wey'. Ancient pre-English river-name of unknown origin and meaning + OE *brycg*.

Weyhill Hants. *La Wou* c.1270. Possibly 'the hill-spur climbed by a road'. OE *weg* + *hōh*, with the later addition of *hill*. Alternatively the original name may be from OE *wēoh* '(heathen) temple'.

Weymouth Dorset. *Waimouthe* 934. 'Mouth of the River Wey'. Ancient pre-English river-name of unknown origin and meaning + OE *mūtha*.

Whaddon, usually 'hill where wheat is grown', OE *hwǣte* + *dūn*; examples include: **Whaddon** Bucks. *Hwǣtædun* 966–75, *Wadone* 1086 (DB). **Whaddon** Cambs. *Wadone* 1086 (DB).
  However the following has a different second element: **Whaddon** Wilts., near Salisbury. *Watedene* 1086 (DB). 'Valley where wheat is grown'. OE *hwǣte* + *denu*.

Whale Cumbria. *Vwal* 1178. OScand. *hvál* 'an isolated round hill'.

Whaley Derbys. near Bolsover. *Walley* 1230. Possibly 'woodland clearing by a spring or stream'. OE *wælla* + *lēah*.

Whaley Bridge Derbys. *Weile* c.1250. 'Woodland clearing by a road'. OE *weg* + *lēah*.

Whalley Lancs. *Hwælleage* 11th cent., *Wallei* 1086 (DB). 'Woodland clearing on or near a round hill'. OE *hwæl* + *lēah*.

Whalsay (island) Shet. *Hvalsey* c.1250. 'Whale-shaped island'. OScand. *hvalr* + *ey*.

Whalton Northum. *Walton* 1203. Possibly 'farmstead by a round hill'. OE *hwæl* + *tūn*.

Wham N. Yorks. *Quane* [sic] 13th cent., *Wham* 1771. 'The marshy valley or hollow'. OE *hwamm*.

Whaplode Lincs. *Cappelad* 810, *Copelade* 1086 (DB). 'Watercourse or channel where eelpouts are found'. OE **cwappa* + *lād*.

Wharfe N. Yorks. *Warf* 1224. OScand. *hvarf* or *hverfi* 'a bend or corner'.

Wharfe (river) Yorkshire. *See* BURLEY IN WHARFEDALE.

Wharles Lancs. *Quarlous* 1249. Probably 'hills or mounds near a stone circle'. OE *hwerfel* + *hlāw*.

Wharncliffe Side Sheff. *Querncliffe* 1406, *Wharnetliffe Side* 1634. 'Cliff where querns or millstones are obtained'. OE *cweorn* + *clif*, with the later addition of ME *side* 'hill-side'.

Wharram Percy & Wharram le Street N. Yorks. *Warran* 1086 (DB), *Wharrom Percy* 1291, *Warrum in the Strete* 1333. Possibly '(place at) the kettles or cauldrons' (perhaps used in some topographical sense). OE *hwer* in a dative plural form. Distinguishing affixes from early possession by the *de Percy* family and from proximity to an ancient road ('on the street' from OE *strǣt*).

Wharton Ches. *Wanetune* [*sic*] 1086 (DB), *Waverton* 1216. Probably 'farmstead by a swaying tree or near marshy ground'. OE *wæfre* + *tūn*.

Whashton N. Yorks. *Whassingetun* c.1160. Probably 'estate associated with a man called **Hwæssa*'. OE pers. name + *-ing-* + *tūn*.

Whatcombe Dorset. *Watecumbe* 1288. 'Wet valley', or 'valley where wheat is grown'. OE *wǣt* or *hwǣte* + *cumb*.

Whatcote Warwicks. *Quatercote* [*sic*] 1086 (DB), *Whatcote* 1206. 'Cottage(s) near which wheat is grown'. OE *hwǣte* + *cot*.

Whatfield Suffolk. *Watefelda* 1086 (DB). Probably 'open land where wheat is grown'. OE *hwǣte* + *feld*.

Whatley Somerset. near Chard. *Watelege* 1086 (DB). 'Woodland clearing where wheat is grown'. OE *hwǣte* + *lēah*.

Whatlington E. Sussex. *Watlingetone* 1086 (DB). Probably 'farmstead of the family or followers of a man called **Hwætel*'. OE pers. name + *-inga-* + *tūn*.

Whatstandwell Derbys. *Wattestanwell ford* 1390. Named from a certain *Wat* or *Walter* *Stonewell* who had a house near the ford in 1390.

Whatton, probably 'farmstead where wheat is grown', OE *hwǣte* + *tūn*: **Whatton** Notts. *Watone* 1086 (DB). **Whatton, Long** Leics. *Watton* 1190. Affix (recorded from the 14th cent.) refers to the length of the village.

Wheatacre Norfolk. *Hwateaker* 1086 (DB). 'Cultivated land used for wheat'. OE *hwǣte* + *æcer*.

Wheathampstead Herts. *Wathemestede* c.960, *Watamestede* 1086 (DB). 'Homestead where wheat is grown'. OE *hwǣte* + *hām-stede*.

Wheatley, usually 'clearing where wheat is grown', OE *hwǣte* + *lēah*; examples include: **Wheatley** Oxon. *Hwatelega* 1163. **Wheatley Lane** Lancs. *Watelei* 1086 (DB). **Wheatley, North & Wheatley, South** Notts. *Wateleie* 1086 (DB).
   However the following has a different origin: **Wheatley Hill** Durham. *Wuatlaue* 1180. 'Hill where wheat is grown'. OE *hwǣte* + *hlāw*, with later addition (from 16th cent.) of *hill*.

Wheaton Aston Staffs. *See* ASTON.

Wheddon Cross Somerset. *Wheteden* 1243. 'Wheat valley'. OE *hwǣte* + *denu*.

Wheelock Ches. *Hoiloch* 1086 (DB). Named from the River Wheelock, a Celtic river-name meaning 'winding'.

Wheelton Lancs. *Weltona* c.1160. 'Farmstead with a water-wheel or near some other circular feature'. OE *hwēol* + *tūn*.

Whenby N. Yorks. *Quennebi* 1086 (DB). 'Farmstead or village of the women'. OScand. *kona* (genitive plural *kvenna*) + *bý*.

Whepstead Suffolk. *Wepstede* 942–51, *Huepestede* 1086 (DB). Probably 'place where brushwood grows'. OE **hwip(p)e* + *stede*.

Wherstead Suffolk. *Weruesteda* 1086 (DB). 'Place by a wharf or shore'. OE *hwearf* + *stede*.

Wherwell Hants. *Hwerwyl* 955, *Hwerwillom* c.1121. 'Cauldron stream(s)'. OE *hwer* + *wella*, referring to eddies in the River Test.

Wheston, Whetstone, from OE *hwet-stān* 'a whetstone', probably referring to standing stones or to places where stone suitable for whetstones was found: **Wheston** Derbys. *Whetstan* 1251. **Whetstone** Gtr. London.

Wheston 1417. **Whetstone** Leics. *Westham* [*sic*] 1086 (DB), *Whetestan* 12th cent.

Whicham Cumbria. *Witingham* 1086 (DB). 'Homestead of the family or followers of a man called Hwīta', or 'homestead at Hwīta's place'. OE pers. name + *-inga-* or *-ing* + *hām*.

Whichford Warwicks. *Wicford* 1086 (DB), *Wicheforda* c.1130. Probably 'ford of the tribe called the Hwicce'. OE tribal name + *ford*.

Whickham Gatesd. *Quicham* 1196. 'Homestead or enclosure with a quickset hedge'. OE *cwic* + *hām* or *hamm*.

Whiddy Island (*Faoide*) Cork. 'Bad weather island'.

Whilton Northants. *Woltone* 1086 (DB). 'Farmstead with a water-wheel, or on a round hill'. OE *hwēol* + *tūn*.

Whimple Devon. *Winple* 1086 (DB). Originally the name of the stream here, a Celtic name meaning 'white pool or stream'.

Whinburgh Norfolk. *Wineberga* 1086 (DB). 'Hill where gorse grows'. OScand. **hvin* + OE *beorg*.

Whippingham I. of Wight. *Wippingeham* 735, *Wipingeham* 1086 (DB). 'Homestead of the family or followers of a man called **Wippa*'. OE pers. name + *-inga-* + *hām*.

Whipsnade Beds. *Wibsnede* 1202. 'Detached plot of a man called **Wibba*'. OE pers. name + *snǣd*.

Whissendine Rutland. *Wichingedene* 1086 (DB), *Wissenden* 1203. Possibly 'valley of the family or followers of a man called **Hwicce*', OE pers. name + *-inga-* + *denu*. Alternatively 'valley of the tribe called the Hwicce', OE tribal name (genitive plural *Hwiccena*) + *denu*.

Whissonsett Norfolk. *Witcingkeseta* 1086 (DB), *Wichingseta* 1191. Possibly 'Fold of the people who dwell at or come from a place called *Wic* ("the specialized farm or trading settlement")'. OE *Wic* + *-inga-* + (*ge*)*set* 'fold for animals', *see* WITCHINGHAM, GREAT Norfolk.

Whistley Green Wokhm. *Wisclea* 968, *Wiselei* 1086 (DB). 'Marshy-meadow clearing'. OE *wisc* + *lēah*.

Whiston Knows. *Quistan* 1190. 'The white stone'. OE *hwīt* + *stān*.

Whiston Northants. *Hwiccingtune* 974, *Wicentone* 1086 (DB). 'Farmstead of the tribe called the Hwicce'. OE tribal name (genitive plural *-na*) + *tūn*.

Whiston Rothm. *Witestan* 1086 (DB). 'The white stone'. OE *hwīt* + *stān*.

Whiston Staffs. *Witestun* c.1002, *Witestone* 1086 (DB). 'Farmstead of a man called **Witi*'. OE pers. name + *tūn*.

Whitacre, Nether & Whitacre, Over Warwicks. *Witacre* 1086 (DB). 'White cultivated land'. OE *hwīt* + *æcer*.

Whitbeck Cumbria. *Witebec* c.1160. 'White stream'. OScand. *hvítr* + *bekkr*.

Whitburn S. Tyne. *Whiteberne* 1183. 'White barn'. OE *hwīt* + *bere-ærn*.

Whitburn W. Loth. *Whiteburne* 1296. '(Place by) the white stream'. OE *hwīt* + *burna*.

Whitby Ches. *Witeberia*, *Witebia* c.1100, *Whiteby* 1241. 'White (i.e. stone-built) manor-house or village'. OE *hwīt* + *burh* (dative *byrig*) replaced by OScand. *bý*.

Whitby N. Yorks. *Witeby* 1086 (DB). 'White farmstead or village, or of a man called Hvíti'. OScand. *hvítr* or pers. name + *bý*.

Whitchurch, 'white church', i.e. probably 'stone-built church', OE *hwīt* + *cirice*; examples include: **Whitchurch** B. & NE. Som. *Hwitecirce* 1065. **Whitchurch** Hants. *Hwitancyrice* 909. **Whitchurch** Shrops. *Album Monasterium* 1199 (here the name is rendered in Latin), *Whytchyrche* 13th cent. Called *Westune* 'west farmstead' in 1086 (DB).
Whitchurch Canonicorum Dorset. *Witcerce* 1086 (DB), *Whitchurch Canonicorum* 1262. Latin affix means 'of the canons', referring to early possession by the canons of Salisbury. The dedication of the church here to St Candida (St Wite) may be derived from the place name rather than vice versa.

Whitcott Keysett Shrops. *Hodecote Keyset* 1284. 'Cottage(s) of a man called Hoda'. OE pers. name + *cot*. Manorial affix from the surname *Kesyat*.

White as affix. *See* main name, e.g. for **White Notley** (Essex) *see* NOTLEY.

White City Gtr. London. so called from the white-stuccoed walls of the stadium and exhibition centre built c.1908.

White Ladies Aston Worcs. *See* ASTON.

Whiteabbey Antrim. *White-Abbey*, in Irish *Mainistir Fhionn c.*1700. The name refers to a 13th-cent. abbey of the Premonstratensian Order, popularly known as the 'White Canons'. The English name may translate original Irish *Mainistir Fhionn*, 'white monastery'.

Whitechapel Gtr. London. *Whitechapele* 1340. 'The white chapel', i.e. probably 'stone-built chapel'. OE *hwīt* + ME *chapele*.

Whitecross Armagh. *Whitecross c.*1835. The reference is to whitewashed houses at a crossroads here.

Whitefield Bury. *Whitefeld* 1292. 'White open land'. OE *hwīt* + *feld*.

Whitegate Ches. *Whytegate* 1540. 'The white gate'. OE *hwīt* + *geat*, referring to the outer gate of Vale Royal Abbey.

Whitehaven Cumbria. *Qwithofhavene c.*1135. 'Harbour near the white headland'. OScand. *hvítr* + *hofuth* + *hafn*.

Whitehead Antrim. *the White-Head* 1683. The town takes its name from the headland *White Head*, itself named in contrast to nearby *Black Head*.

Whitehill Fermanagh. *Whitehill* 1837. 'White hill'.

Whitehouse Antrim. 'White house'. The name was originally that of a fortified house built *c.*1574.

Whiteley Village Surrey. Named after William *Whiteley* (of Whiteley's Stores in London) who had this model village with almshouses built in the early 1900s.

Whiteparish Wilts. *La Whytechyrch* 1278, *Whyteparosshe* 1289. Originally 'the white church', later replaced by 'parish'. OE *hwīt* + *cirice* and ME *paroche*.

Whiterock Antrim. *White Rock c.*1830. 'White rock'. The name relates to a nearby limestone quarry.

Whitestaunton Somerset. *Stantune* 1086 (DB), *Whitestaunton* 1337. 'Farmstead on stony ground', OE *stān* + *tūn*, with later affix referring to the limestone quarries here.

Whitestone Devon. *Hwitastane c.*1100, *Witestan* 1086 (DB). '(Place at) the white stone'. OE *hwīt* + *stān*.

Whitewell Antrim. *Whitewell* 1858. 'White well'.

Whitfield, 'white open land', OE *hwīt* + *feld*: **Whitfield** Kent. *Whytefeld* 1228. **Whitfield** Northants. *Witefelle* 1086 (DB). **Whitfield** Northum. *Witefeld* 1254.

Whitford Devon. *Witefort* 1086 (DB). 'White ford'. OE *hwīt* + *ford*.

Whitgift E. R. Yorks. *Witegift c.*1070. Probably 'dowry land of a man called Hvítr'. OScand. pers. name + *gipt*.

Whitgreave Staffs. *Witegraue* 1193. 'White grove or copse'. OE *hwīt* + *grǣfe*.

Whithorn Dumf. *Candida Casa c.*730, *æt Hwitan Ærne c.*890. 'White building', i.e. 'stone-built church'. OE *hwīt* + *ærn*.

Whitland (*Hendy-gwyn*) Carm. *Alba Domus* 1191, *Alba Landa* 1214, *Whitland* 1309. 'White land'. OE *hwīt* + *land*. The Welsh name means 'old white house' (Welsh *hen* + *tŷ* + *gwyn*).

Whitley, 'white wood or clearing', OE *hwīt* + *lēah*; examples include: **Whitley** Readg. *Witelei* 1086 (DB). **Whitley Bay** N. Tyne. *Wyteleya* 12th cent. **Whitley, Higher & Whitley, Lower** Ches. *Witelei* 1086 (DB).

Whitmore Staffs. *Witemore* 1086 (DB). 'White moor or marsh'. OE *hwīt* + *mōr*.

Whitnash Warwicks. *Witenas* 1086 (DB). '(Place at) the white ash-tree'. OE *hwīt* (dative *-an*) + *æsc*.

Whitney Herefs. *Witenie* 1086 (DB). 'White island'. OE *hwīt* (dative *-an*) + *ēg*. Alternatively the first element may be the OE pers. name *Hwīta* (genitive *-n*).

Whitsbury Hants. *Wiccheberia c.*1130. 'Fortified place where wych-elms grow'. OE *wice* + *burh* (dative *byrig*).

Whitstable Kent. *Witestaple*, *Witenestaple* 1086 (DB). 'White post', or 'post of the councillors'. OE *hwīt* (dative *-an*) or *wita* (genitive plural *-ena*) + *stapol*.

Whitstone Cornwall. *Witestan* 1086 (DB). 'White stone'. OE *hwīt* + *stān*.

Whittingham Northum. *Hwitincham c.*1050. 'Homestead of the family or followers of a man called Hwīta', or 'homestead at Hwīta's place'. OE pers. name + *-inga-* or *-ing* + *hām*.

Whittingslow Shrops. *Witecheslawe* 1086 (DB). 'Tumulus of a man called Hwittuc'. OE pers. name + *hlāw*.

Whittington, 'estate associated with a man called Hwīta', OE pers. name + *-ing-* + *tūn*; examples include: **Whittington** Shrops. *Wititone* 1086 (DB). **Whittington** Staffs. *Hwituntune* 925. **Whittington, Great** Northum. *Witynton* 1233.

Whittle-le-Woods Lancs. *Witul* c.1160. 'White hill'. OE *hwīt* + *hyll*. Later affix means 'in the woodland'.

Whittlebury Northants. *Witlanbyrig* c.930. 'Stronghold of a man called *Witla'. OE pers. name + *burh* (dative *byrig*).

Whittlesey Cambs. *Witlesig* 972, *Witesie* 1086 (DB). 'Island of a man called *Wittel'. OE pers. name + *ēg*.

Whittlesford Cambs. *Witelesforde* 1086 (DB). 'Ford of a man called *Wittel'. OE pers. name + *ford*.

Whitton, usually 'white farmstead' or 'farmstead of a man called Hwīta', OE *hwīt* or pers. name + *tūn*, for example: **Whitton** Stock. on T. *Wittune* 1208–10. **Whitton** Suffolk. *Widituna* [sic] 1086 (DB), *Witton* 1212.
    However the following has a different origin: **Whitton** N. Lincs. *Witenai* 1086 (DB). 'Hwīta's island of land'. OE masculine pers. name (genitive *-n*) + *ēg*.

Whittonstall Northum. *Quictunstal* 1242. 'Farmstead with a quickset hedge'. OE *cwic* + *tūn-stall*.

Whitwell, 'white spring or stream', OE *hwīt* + *wella*; examples include: **Whitwell** Derbys. *Hwitewylle* c.1002, *Witewelle* 1086 (DB). **Whitwell** I. of Wight. *Quitewell* 1212.

Whitwick Leics. *Witewic* 1086 (DB). 'White dwelling or (dairy) farm', or 'farm of a man called Hwīta'. OE *hwīt* or pers. name + *wīc*.

Whitwood Wakefd. *Witewde* 1086 (DB). 'White wood' (referring to colour of tree-bark or blossom). OE *hwīt* + *wudu*.

Whitworth Lancs. *Whiteworth* 13th cent. 'White enclosure'. OE *hwīt* + *worth*.

Whixall Shrops. *Witehala* [sic] 1086 (DB), *Whitekeshal* 1241. 'Nook of land of a man called Hwittuc'. OE pers. name + *halh*.

Whixley N. Yorks. *Cucheslage* 1086 (DB). 'Woodland clearing of a man called *Cwic'. OE pers. name + *lēah*.

Whorlton Durham. *Queorningtun* c.1040. Probably 'farmstead at the mill-stone place or the mill stream'. OE *cweorn* + *-ing* + *tūn*.

Whorlton N. Yorks. *Wirveltune* 1086 (DB). 'Farmstead near the round-topped hill'. OE *hwerfel* + *tūn*.

Whyle Herefs. *Hvilech* 1086 (DB), *Whilai*, *Wihale* 12th cent. Possibly from OE *lēah* 'wood, clearing', first element uncertain.

Whyteleafe Surrey. named from *White Leaf Field* 1839, so called from the aspens that grew there.

Wibtoft Leics. /Warwicks. *Wibbetofte* 1002, *Wibetot* 1086 (DB). 'Homestead of a man called Wibba or Vibbi'. OE or OScand. pers. name + *toft*.

Wichenford Worcs. *Wiceneford* 11th cent. 'Ford by the wych-elms'. OE *wice* (genitive plural *-ena*) + *ford*.

Wichling Kent. *Winchelesmere* 1086 (DB), *Winchelinge* 1220–4. Probably '(settlement of) the family or followers of a man called *Wincel', OE pers. name + *-ingas*. The alternative name used in the 11th and 12th centuries means '*Wincel's pool or boundary', second element OE *mere* or *mǣre*.

Wick, 'the dwelling, the specialized farm or trading settlement', OE *wīc*; examples include: **Wick** S. Glos. near Kingswood. *Wike* 1189. **Wick** Worcs., near Pershore. *Wiche* 1086 (DB). **Wick St Lawrence** N. Som. *Wike* 1225. Affix from the dedication of the church.

Wick Highland. *Vik* 1140, *Weke* 1455. '(Place by) the bay'. OScand. *vík*.

Wicken, 'the dwellings, the specialized farm or trading settlement', OE *wīc* in the dative plural form *wīcum* or a ME plural form *wiken*: **Wicken** Cambs. *Wicha* 1086 (DB), *Wiken* c.1200. **Wicken** Northants. *Wiche* 1086 (DB), *Wicne* c.1235. **Wicken Bonhunt** Essex. *Wica* 1086 (DB), *Wykes Bonhunte* 1238. Originally two separate names, Bonhunt (*Banhunta* 1086 (DB), possibly being 'place where people were summoned for hunting', OE *bann* + *hunte*.

Wickenby Lincs. *Wichingebi* 1086 (DB). 'Farmstead or village of a man called Vikingr, or of the vikings'. OScand. pers. name or *vikingr* + *bý*.

Wickersley Rothm. *Wicresleia* 1086 (DB). 'Woodland clearing of a man called Víkarr'. OScand. pers. name + OE *lēah*.

Wickford Essex. *Wicforda c.*975, *Wicfort* 1086 (DB). Probably 'ford by an earlier Romano-British settlement'. OE *wīc* + *ford*.

Wickham, usually 'homestead associated with a *vicus*, i.e. an earlier Romano-British settlement', OE **wīc-hām*; examples include: **Wickham** Hants. *Wicham* 925–41, *Wicheham* 1086 (DB). **Wickham** W. Berks. *Wicham* 1167. **Wickham Bishops** Essex. *Wicham* 1086 (DB), *Wykham Bishops* 1313. Affix from its possession by the Bishop of London. **Wickham Market** Suffolk. *Wikham* 1086 (DB). Affix from the important medieval market here. **Wickham Skeith** Suffolk. *Wichamm* 1086 (DB), *Wicham Skeyth* 1368. Affix is OScand. *skeith* 'a racecourse'. **Wickham, West** Gtr. London. *Wic hammes gemǣru* 973, *Wicheham* 1086 (DB). The early form contains OE *gemǣre* 'boundary' and may show confusion with *hamm* 'enclosure'.

Wickhambreaux Kent. *Wicham* 948, *Wicheham* 1086 (DB), *Wykham Breuhuse* 1270. Identical in origin with the previous names. Manorial affix from the *de Brayhuse* family, here in the 13th cent.

Wickhambrook Suffolk. *Wicham* 1086 (DB), *Wichambrok* 1254. Identical in origin with the previous names, but with the later addition of OE *brōc* 'brook'.

Wickhamford Worcs. *Wicwona* 709, *Wiquene* 1086 (DB), *Wikewaneford* 1221. 'Ford at the place called *Wicwon*'. OE *ford*, see CHILDSWICKHAM.

Wickhampton Norfolk. *Wichamtuna* 1086 (DB). Probably 'homestead with or near a dairy farm'. OE *wīc* + *hām-tūn*.

Wicklewood Norfolk. *Wikelewuda* 1086 (DB). Probably 'wood at **Wiclēah* (wych-elm clearing)'. OE *wice* + *lēah* + *wudu*.

Wicklow (*Cill Mhántáin*) Wicklow. *Wykynoelo* (no date). 'Vikings' meadow'. OScand. *víkingr* + *ló*. The Irish name means 'Mantán's church'.

Wickmere Norfolk. *Wicmera* 1086 (DB). 'Pool by a dwelling or (dairy) farm'. OE *wīc* + *mere*.

Wickwar S. Glos. *Wichen* 1086 (DB), *Wykewarre* 13th cent. Originally 'the dwellings or specialized farm', from OE *wīc* in a plural form. Later manorial affix from the family of *la Warre*, here from the early 13th cent.

Widcombe, North & Widcombe, South B. & NE. Som. *Widecomb* 1303. 'Wide valley', or 'willow-tree valley'. OE *wīd* or *wīthig* + *cumb*.

Widdington Essex. *Widintuna* 1086 (DB). 'Farmstead or village where willow-trees grow'. OE **withign* + *tūn*.

Widdrington Northum. *Vuderintuna c.*1160. 'Estate associated with a man called **Widuhere*'. OE pers. name + *-ing-* + *tūn*.

Wide Open N. Tyne. Recorded thus from 1863, a mining village so named from its exposed situation.

Widecombe in the Moor Devon. *Widecumba* 12th cent. Probably 'valley where willow-trees grow'. OE *wīthig* + *cumb*. Affix refers to its situation on DARTMOOR.

Widford, 'ford where willow-trees grow', OE *wīthig* + *ford*: **Widford** Essex. *Witford* 1202. **Widford** Herts. *Wideford* 1086 (DB).

Widmerpool Notts. *Wimarspol* 1086 (DB). 'Wide lake (or willow-tree lake) pool'. OE *wīd* or *wīthig* + *mere* + *pōl*.

Widnes Halton. *Wydnes c.*1200. 'Wide promontory'. OE *wīd* + *næss*.

Wield Hants. *Walde* 1086 (DB), *Welde* 1256. 'The woodland or forest'. OE *weald*.

Wigan Wigan. *Wigan* 1199. A Celtic name, 'little settlement', from a diminutive of Brittonic **wīg* 'homestead, settlement' (later Welsh *gwig* 'wood').

Wigborough, Great Essex. *Wicgheberga* 1086 (DB). 'Hill of a man called Wicga'. OE *pers. name* + *beorg*.

Wiggenhall Norfolk. *Wigrehala* 1086 (DB), *Wiggenhal* 1196. 'Nook of land of a man called Wicga'. OE pers. name (genitive *-n*) + *halh*.

Wigginton, 'farmstead of, or associated with, a man called Wicga', OE pers. name (genitive *-n* or + *-ing-*) + *tūn*: **Wigginton** Herts. *Wigentone* 1086 (DB). **Wigginton** Oxon. *Wigentone* 1086 (DB). **Wigginton** Staffs. *Wigetone* 1086 (DB). **Wigginton** York. *Wichintun* 1086 (DB).

Wigglesworth N. Yorks. *Winchelesuuorde*
1086 (DB). 'Enclosure of a man called Wincel'.
OE *pers. name* + *worth*.

Wiggonby Cumbria. *Wygayneby* 1278.
'Farmstead or village of a man called Wigan'.
Celtic pers. name + OScand. *bý*.

Wighill N. Yorks. *Duas Wicheles* 1086 (DB),
Wikale 1219. 'Nook of land with a dairy farm or
by an earlier Romano-British settlement'. OE
wīc + *halh* (in the plural form in 1086, with Latin
duas 'two').

Wight, Isle of. *Vectis c.*150, *Wit* 1086 (DB).
A Celtic name possibly meaning 'place of the
division', referring to its situation between the
two arms of the Solent.

Wighton Norfolk. *Wistune* 1086 (DB).
'Dwelling place, farmstead with a dwelling'. OE
wīc-tūn.

Wigmore Herefs. *Wigemore* 1086 (DB).
Probably 'quaking marsh'. OE *wicga* + *mōr*.

Wigmore Medway. *Wydemere* 1275. 'Broad
pool'. OE *wīd* + *mere*.

Wigsley Notts. *Wigesleie* 1086 (DB).
'Woodland clearing of a man called *Wicg'. OE
pers. name + *lēah*.

Wigsthorpe Northants. *Wykingethorp* 1232.
'Outlying farmstead or hamlet of a man called
Vikingr'. OScand. pers. name + *thorp*.

Wigston Magna Leics. *Wichingestone* 1086
(DB). 'Farmstead or estate of a man called
*Wicing or Vikingr'. OE or OScand. pers. name +
OE *tūn*. Affix is Latin *magna* 'great' to
distinguish this place from **Wigston Parva**
(*Wicestan* 1086 (DB)) which has a quite different
origin, either 'rocking-stone' from OE *wigga* +
stān, or 'stone of a man called Wicga' from OE
pers. name + *stān*.

Wigtoft Lincs. *Wiketoft* 1187. Probably
'homestead by a (former) creek'. OScand. *vík* +
toft.

Wigton Cumbria. *Wiggeton* 1163. 'Farmstead
or village of a man called Wicga'. OE pers. name
+ *tūn*.

Wigtown Dumf. *Wigeton* 1266. 'Dwelling
place'. OE *wīc-tūn*.

Wigwig Shrops. *Wigewic* 1086 (DB). Probably
'dairy farm of a man called Wicga'. OE pers.
name + *wīc*.

Wike, 'the dwelling, the specialized farm or
trading settlement', OE *wīc*; for example **Wike**
Leeds. *Wich* 1086 (DB).

Wilbarston Northants. *Wilbertestone* 1086
(DB). 'Farmstead or village of a man called
Wilbeorht'. OE pers. name + *tūn*.

Wilberfoss E. R. Yorks. *Wilburcfosa* 1148.
'Ditch of a woman called Wilburh'. OE pers.
name + *foss*.

**Wilbraham, Great & Wilbraham,
Little** Cambs. *Wilburgeham c.*975,
Wiborgham 1086 (DB). 'Homestead or village of
a woman called Wilburh'. OE pers. name +
hām.

Wilburton Cambs. *Wilburhtun* 970,
Wilbertone 1086 (DB). 'Farmstead or village of a
woman called Wilburh'. OE pers. name + *tūn*.

Wilby Norfolk. *Willebeih* 1086 (DB).
'Farmstead by the willow-trees', or possibly
'circle of willow-trees'. OE *wilig* + OScand. *bý*
or OE *bēag*.

Wilby Northants. *Willabyg c.*1067, *Wilebi* 1086
(DB). 'Farmstead of a man called Willa or Villi'.
OE or OScand. pers. name + *bý*.

Wilby Suffolk. *Wilebey* 1086 (DB). Probably
'circle of willow-trees'. OE *wilig* + *bēag*.

Wilcot Wilts. *Wilcotum* 940, *Wilcote* 1086
(DB). 'Cottages by the stream or spring'. OE
wiella + *cot* (dative plural *-um*).

Wildboarclough Ches. *Wildeborclogh* 1357.
'Deep valley frequented by wild boars'. OE
wilde-bār + *clōh*.

Wilden Beds. *Wildene* 1086 (DB). Possibly
'willow-tree valley'. OE *wilig* + *denu*.

Wilden Worcs. *Wineladuna* [sic] 1182,
Wiveldon 1299. Probably 'hill of a man called
*Wifela'. OE pers. name + *dūn*.

Wildsworth Lincs. *Winelesworth* [sic] 1199,
Wyveleswurth 1280. Probably 'enclosure of a
man called *Wifel'. OE pers. name + *worth*.

Wilford Nott. *Wilesford* 1086 (DB). Probably
'willow-tree ford'. OE *wilig* + *ford*.

Wilkesley Ches. *Wiuelesde* [sic] 1086 (DB),
Wivelescle 1230. 'The beetle's, or *Wifel's, claw
of land'. OE *wifel* or OE pers. name (from the
same word) + *clēa*.

Willand Devon. *Willelanda* 1086 (DB). 'Waste land', or 'cultivated land reverted to waste'. OE *wilde* + *land*.

Willaston, 'farmstead or village of a man called Wīglāf', OE pers. name + *tūn*: **Willaston** Ches., near Hooton. *Wilaveston* 1086 (DB). **Willaston** Ches., near Nantwich. *Wilavestune* 1086 (DB).

Willen Milt. K. *Wily* 1189. '(Place at) the willow-trees'. OE **wilig* in a dative plural form **wilgum*.

Willenhall Covtry. *Wilenhala* 12th cent. 'Nook or small valley where willow-trees grow'. OE **wiligen* + *halh*.

Willenhall Wsall. *Willanhalch* 732, *Winenhale* [sic] 1086 (DB). 'Nook or small valley of a man called Willa'. OE pers. name (genitive *-n*) + *halh*.

Willerby, 'farmstead or village of a man called Wilheard', OE pers. name + OScand. *bý*: **Willerby** E. R. Yorks. *Wilgardi* [sic] 1086 (DB), *Willardebi* 1196. **Willerby** N. Yorks. *Willerdebi* 1125–30.

Willersey Glos. *Willerseye* 709, *Willersei* 1086 (DB). 'Island of a man called Wilhere or Wilheard'. OE pers. name + *ēg*.

Willersley Herefs. *Willaveslege* 1086 (DB). Probably 'woodland clearing of a man called Wīglāf'. OE pers. name + *lēah*.

Willesborough Kent. *Wifelesberg* 863. 'The beetle's, or **Wifel's, hill'. OE *wifel* or pers. name + *beorg*.

Willesden Gtr. London. *Willesdone* 939, *Wellesdone* 1086 (DB). 'Hill with a spring or by a stream'. OE *wiell* + *dūn*.

Willett Somerset. *Willet* 1086 (DB). Named from the River Willett, an old river-name of uncertain origin and meaning.

Willey, usually 'willow-tree wood or clearing', OE **wilig* + *lēah*; examples include: **Willey** Shrops. *Wilit* [sic] 1086 (DB), *Wilileg* 1199. **Willey** Warwicks. *Welei* 1086 (DB).
    However the following has a different origin: **Willey** Surrey. *Weoleage* 909. 'Sacred grove with a heathen temple'. OE *wēoh* + *lēah*.

Williamscot Oxon. *Williamescote* 1166. 'Cottage(s) of a man called Willelm or William'. OGerman pers. name + OE *cot*.

Willingale Doe & Willingale Spain Essex. *Willinghehala* 1086 (DB), *Willingeshale Doe* 1270, *Wylinghehale Spayne* 1269. 'Nook of land of the family or followers of a man called Willa'. OE pers. name + *-inga-* + *halh*. Manorial affixes from the *de Ou* family, here in the 12th cent., and from the *de Ispania* family, here from 1086.

Willingdon E. Sussex. *Willendone* 1086 (DB). Probably 'hill of a man called Willa'. OE pers. name (genitive *-n*) + *dūn*.

Willingham, sometimes 'homestead of the family or followers of a man called **Wifel', OE pers. name + *-inga-* + *hām*: **Willingham** Cambs., near Cambridge. *Vuivlingeham c.*1050, *Wivelingham* 1086 (DB). **Willingham** Lincs. *Wilingeham* 1086 (DB). **Willingham, North** Lincs. *Wiuilingeham* 1086 (DB).
    However the following are 'homestead of the family or followers of a man called Willa', OE pers. name + *-inga-* + *hām*: **Willingham, Cherry** Lincs. *Wilingeham* 1086 (DB), *Chyry Wylynham* 1386. Affix is ME *chiri* 'cherry-tree'. **Willingham, South** Lincs. *Ulingeham* 1086 (DB).

Willington Beds. *Welitone* 1086 (DB). 'Willow-tree farmstead'. OE **wilign* + *tūn*.

Willington Derbys. *Willetune* 1086 (DB). Probably identical in origin with the previous name.

Willington Durham. *Wyvelintun c.*1190. 'Estate associated with a man called **Wifel'. OE pers. name + *-ing-* + *tūn*.

Willington N. Tyne. *Wiflintun c.*1085. Identical in origin with the previous name.

Willington Warwicks. *Ullavintone* 1086 (DB). 'Estate associated with a man called Wulflāf'. OE pers. name + *-ing-* + *tūn*.

Willington Corner Ches. *Winfletone* 1086 (DB). 'Farmstead or village of a woman called Wynflǣd'. OE pers. name + *tūn*. Affix, first used in the 19th cent., refers to a corner of Delamere Forest.

Willisham Tye Suffolk. *Willauesham c.*1040. 'Homestead or enclosure of a man called Wīglāf'. OE pers. name + *hām* or *hamm*. *Tye* is dialect *tye* (from OE *tēag*) 'a large common pasture', as in other Suffolk names.

Willitoft E. R. Yorks. *Wilgetot* 1086 (DB). 'Willow-tree homestead'. OE **wilig* + OScand. *toft*.

Williton Somerset. *Willettun* 904, *Willetone* 1086 (DB). 'Farmstead or village on the River Willett'. Old river-name (*see* WILLETT) + OE *tūn*.

Willoughby, usually 'farmstead by the willow-trees', OE *wilig* + OScand. *bý*, although some may be 'circle of willow-trees', OE *wilig* + *bēag*; examples include: **Willoughby** Lincs., near Alford. *Wilgebi* 1086 (DB). **Willoughby** Warwicks. *Wiliabyg* 956, *Wilebei* 1086 (DB). **Willoughby, Silk** Lincs. *Wilgebi* 1086 (DB). Affix is a reduced form of a nearby place *Silkebi* 1212, 'farmstead of a man called Silki, or near a gully', OScand. pers. name or OE *sīoluc* + OScand. *bý*.

Willoughton Lincs. *Wilchetone* 1086 (DB). 'Farmstead or village where willow-trees grow'. OE *wilig* + *tūn*.

Wilmcote Warwicks. *Wilmundigcotan* 1016, *Wilmecote* 1086 (DB). 'Cottage(s) associated with a man called Wilmund'. OE pers. name + *-ing-* + *cot*.

Wilmington Devon. *Wilelmitone* 1086 (DB). 'Estate associated with a man called Wilhelm'. OE pers. name + *-ing-* + *tūn*.

Wilmington E. Sussex. *Wilminte* [*sic*] 1086 (DB), *Wilminton* 1189. 'Estate associated with a man called Wīghelm or Wilhelm'. OE pers. name + *-ing-* + *tūn*.

Wilmington Kent. near Dartford. *Wilmintuna* 1089. 'Estate associated with a man called Wīghelm'. OE pers. name + *-ing-* + *tūn*.

Wilmslow Ches. *Wilmesloe* c.1250. 'Mound of a man called Wīghelm'. OE pers. name + *hlāw*.

Wilnecote Staffs. *Wilmundecote* 1086 (DB). 'Cottage(s) of a man called Wilmund'. OE pers. name + *cot*.

Wilpshire Lancs. *Wlypschyre* 1246. OE *scīr* 'district, estate' with first element probably Celtic *wlīb* 'wet' (the ancestor of OWelsh *gulip*).

Wilsford, 'the beetle's, or *Wifel's, ford', OE *wifel* or pers. name + *ford*: **Wilsford** Lincs. *Wivelesforde* 1086 (DB). **Wilsford** Wilts., near Pewsey. *Wifelesford* 892, *Wivlesford* 1086 (DB). **Wilsford** Wilts., near Salisbury. *Wiflesford* 1086 (DB).

Wilsill N. Yorks. *Wifeleshealh* c.1030, *Wifleshale* 1086 (DB). '*Wifel's or Vífill's nook', or 'beetle's nook'. OE or OScand. pers. name, or OE *wifel*, + *halh*.

Wilson Leics. *Wyvelestona* 12th cent. 'Farmstead of a man called *Wifel or Vífill'. OE or OScand pers. name + *tūn*.

Wilstone Herts. *Wivelestorn* 1220. 'The beetle's, or *Wifel's, thorn-tree'. OE *wifel* or pers. name + *thorn*.

Wilton, usually 'farmstead or village where willow-trees grow', OE *wilig* + *tūn*; examples include: **Wilton** Norfolk. *Wiltuna* 1086 (DB). **Wilton** Red. & Cleve. *Wiltune* 1086 (DB). **Wilton, Bishop** E. R. Yorks. *Wiltone* 1086 (DB). Affix from its early possession by the Archbishops of York.
  However the following have a different origin: **Wilton** Wilts., near Burbage. *Wulton* 1227. Probably 'farmstead near a spring or stream'. OE *wiella* + *tūn*. **Wilton** Wilts., near Salisbury. *Uuiltun* 838, *Wiltune* 1086 (DB). 'Farmstead or village on the River Wylye'. Pre-English river-name (*see* WYLYE) + OE *tūn*.

Wiltshire (the county). *Wiltunscir* 870, *Wiltescire* 1086 (DB). 'Shire centred on WILTON (near Salisbury)'. OE *scīr*.

Wimbish Essex. *Winebisc, Wimbisc* c.1040, *Wimbeis* 1086 (DB). Possibly 'bushy copse of a man called Wine'. OE pers. name + *(ge)bysce*.

Wimbledon Gtr. London. *Wunemannedune* c.950. Probably 'hill of a man called *Wynnmann'. OE pers. name + *dūn*.

Wimblington Cambs. *Wimblingetune* c.975. Probably 'estate associated with a man called Wynnbald'. OE pers. name + *-ing-* + *tūn*.

Wimborne Minster Dorset. *Winburnan* late 9th cent., *Winburne* 1086 (DB), *Wymburneminstre* 1236. Originally the name of the river here (now called Allen), 'meadow stream' from OE *winn* + *burna*. Affix is OE *mynster* 'monastery (church)'.

Wimborne St Giles Dorset. *Winburne* 1086 (DB), *Vpwymburn Sancti Egidij* 1268, *Upwymbourne St Giles* 1399. Like WIMBORNE MINSTER, named from the river here. The early affix is OE *upp* 'higher up (the river)'. *St Giles* (Latin *Egidius*) is from the dedication of the church.

Wimbotsham Norfolk. *Winebotesham* 1086 (DB). 'Homestead of a man called Winebald'. OE pers. name + *hām*.

Wimpole Cambs. *Winepole* 1086 (DB). 'Pool of a man called Wina'. OE pers. name + *pōl*.

Wimpstone Warwicks. *Wylmestone* 1313.
'Farmstead of a man called Wilhelm or
Wīghelm'. OE pers. name + *tūn*.

Wincanton Somerset. *Wincaletone* 1086
(DB). 'Farmstead on (an arm of) the River Cale'.
Celtic river-name (of uncertain origin, but
prefixed by **wïnn* 'white') + OE *tūn*.

Winch, East & Winch, West Norfolk.
Estwinic, Wesuuinic 1086 (DB). 'Farmstead with
meadow-land'. OE **winn* + *wīc*.

Wincham Ches. *Wimundisham* 1086 (DB).
'Homestead of a man called Wīgmund'. OE
pers. name + *hām*.

Winchcombe Glos. *Wincelcumbe c.*810, 1086
(DB). 'Valley with a bend in it'. OE **wincel* +
cumb.

Winchelsea E. Sussex. *Winceleseia* 1130.
'Island by a river-bend'. OE **wincel* + *ēg*.

Winchendon Bucks. *Wincandone* 1004,
Wichendone 1086 (DB). Possibly 'hill with a
winch (for pulling up carts)', or 'at a bend or
corner'. OE *wince* (genitive *-an*) + *dūn*.

Winchester Hants. *Ouenta c.*150,
*Uintancæstir c.*730, *Wincestre* 1086 (DB).
'Roman town called *Venta*'. Pre-Celtic name
(possibly 'favoured or chief place') + OE *ceaster*.

Winchfield Hants. *Winchelefeld* 1229. 'Open
land by a nook or corner'. OE **wincel* + *feld*.

Winchmore Hill Gtr. London. *Wynsemerhull*
1319. Possibly 'Wynsige's boundary hill'. OE
pers. name + *mǣre* + *hyll*. Or the second
element may be OE *mere* 'pool'.

Wincle Ches. *Winchul c.*1190. 'Hill of a man
called **Wineca*', or 'hill by a bend or with a
winch'. OE pers. name or *wince* + *hyll*.

Windermere Cumbria. *Winandermere* 12th
cent. 'Lake of a man called Vinandr'. OScand.
pers. name (genitive *-ar*) + OE *mere*.

Winderton Warwicks. *Winterton* 1166.
'Farmstead used in winter'. OE *winter* + *tūn*.

Windlesham Surrey. *Windesham* 1178,
Windlesham 1227. Probably 'homestead of a
man called **Windel*'. OE pers. name + *hām*.
Alternatively the first element may be OE
*windels 'a windlass'.

Windley Derbys. *Winleg* 12th cent. 'Meadow
clearing'. OE **winn* + *lēah*.

Windrush Glos. *Wenric* 1086 (DB). Named
from the River Windrush, a Celtic river-name
possibly meaning 'white fen'.

Windsor Winds. & Maid. *Windlesoran c.*1060,
Windesores 1086 (DB). 'Bank or slope with a
windlass'. OE **windels* + *ōra*.

Winestead E. R. Yorks. *Wifestede* 1086 (DB).
'Homestead of the women, or of a woman called
*Wīfe'. OE *wīf* or pers. name + *stede*.

Winfarthing Norfolk. *Wineferthinc* 1086
(DB). 'Quarter of an estate belonging to a man
called Wina'. OE pers. name + *feorthung*.

Winford N. Som. *Wunfrod c.*1000, *Wenfrod*
1086 (DB). A Celtic river-name meaning 'white
or bright stream'. Celtic **wïnn* + **frud*.

Winforton Herefs. *Widferdestune* [sic] 1086
(DB), *Wynfreton* 1265. Probably 'farmstead or
estate of a man called Winefrith'. OE pers.
name + *tūn*.

Winfrith Newburgh Dorset. *Winfrode*
1086 (DB). Identical in origin with WINFORD.
Manorial affix from the *Newburgh* family, here
from the 12th cent.

Wing Bucks. *Weowungum* 966–75, *Witehunge*
[sic] 1086 (DB). Possibly '(settlement of) the
family or followers of a man called **Wiwa*', OE
pers. name + *-ingas* (dative plural *-ingum*).
Alternatively '(settlement of) the dwellers at, or
devotees of, a heathen temple', OE *wīg, wēoh* +
-ingas.

Wing Rutland. *Wenge* 12th cent. 'The field'.
OScand. *vengi*.

Wingate(s), 'wind-swept gap(s) or pass(es)',
OE **wind-geat*: examples include: **Wingate**
Durham. *Windegatum* 1071–80. **Wingates**
Northum. *Wyndegates* 1208.

Wingerworth Derbys. *Wingreurde* 1086
(DB). 'Enclosure of a man called **Winegār*'. OE
pers. name + *worth*.

Wingfield Beds. *Winfeld c.*1200. 'Open land
used for pasture, or at a bend or corner'. OE
*winn or *wince* + *feld*.

Wingfield Suffolk. *Wingefeld c.*1035,
Wighefelda 1086 (DB). Probably 'open land of
the family or followers of a man called **Wīga*'.
OE pers. name + *-inga-* + *feld*. Alternatively the
first element may be OE *wīg* 'heathen temple'.

Wingfield, North & Wingfield, South
Derbys. *Wynnefeld* 1002, *Winnefelt, Winefeld*

1086 (DB). 'Open land used for pasture'. OE *winn + feld.

Wingham Kent. *Uuigincggaham* 834, *Wingheham* 1086 (DB). 'Homestead of the family or followers of a man called *Wīga'. OE pers. name + -inga- + hām. Alternatively the first element may be OE wīg 'heathen temple'.

Wingrave Bucks. *Withungrave* [sic] 1086 (DB), *Wiungraua* 1163. Possibly 'grove of the family or followers of a man called *Wiwa', OE pers. name + -inga- + grāf. Alternatively 'grove of the dwellers at, or devotees of, a heathen temple', OE wīg, wēoh + -inga- + grāf.

Winkburn Notts. *Wicheburne* [sic] 1086 (DB), *Winkeburna* c.1150. 'Stream of a man called *Wineca, or with bends in it'. OE pers. name or *wincel (with Scand. -k-) + burna.

Winkfield Brack. For. *Winecanfeld* 942, *Wenesfelle* [sic] 1086 (DB). 'Open land of a man called *Wīneca'. OE pers. name + feld.

Winkleigh Devon. *Wincheleia* 1086 (DB). 'Woodland clearing of a man called *Wineca'. OE pers. name + lēah.

Winksley N. Yorks. *Wincheslaie* 1086 (DB). 'Woodland clearing of a man called Winuc'. OE pers. name + lēah.

Winmarleigh Lancs. *Wynemerislega* 1212. 'Woodland clearing of a man called Winemǣr'. OE pers. name + lēah.

Winnersh Wokhm. *Wenesse* 1190. 'Ploughed field by meadow'. OE *winn + ersc.

Winscales Cumbria. *Wyndscales* 1227. 'Temporary huts or sheds in a windy place'. OE wind + OScand. skáli.

Winscombe N. Som. *Winescumbe* c.965, *Winescome* 1086 (DB). 'Valley of a man called Wine'. OE pers. name + cumb.

Winsford, 'ford of a man called Wine', OE pers. name + ford: **Winsford** Ches. *Wyneford* c.1334. **Winsford** Somerset. *Winesford* 1086 (DB).

Winsham Somerset. *Winesham* 1046, 1086 (DB). 'Homestead or enclosure of a man called Wine'. OE pers. name + hām or hamm.

Winshill Staffs. *Wineshylle* 1002. 'Hill of a man called Wine'. OE pers. name + hyll.

Winskill Cumbria. *Wyndscales* 1292. 'Temporary huts or sheds in a windy place'. OE wind + OScand. skáli.

Winslade Hants. *Winesflot* 1086 (DB). 'Spring or channel of a man called Wine'. OE pers. name + flōde.

Winsley Wilts. *Winesleg* 1242. 'Woodland clearing of a man called Wine'. OE pers. name + lēah.

Winslow Bucks. *Wineshlauu* 795, *Weneslai* [sic] 1086 (DB). 'Mound of a man called Wine'. OE pers. name + hlāw.

Winson Glos. *Winestune* 1086 (DB). 'Farmstead or village of a man called Wine'. OE pers. name + tūn.

Winspit Dorset. First recorded in 1786, possibly 'stone pit with a winch', from OE wince.

Winster Cumbria. *Winster* 13th cent. Named from the River Winster, a Celtic river-name meaning 'white stream' or an OScand. river-name meaning 'the left one'.

Winster Derbys. *Winsterne* 1086 (DB). 'Thorn-tree of a man called Wine'. OE pers. name + thyrne.

Winston, 'farmstead or village of a man called Wine', OE pers. name + tūn: **Winston** Durham. *Winestona* 1091. **Winston** Suffolk. *Winestuna* 1086 (DB).

Winstone Glos. *Winestan* 1086 (DB). '(Boundary) stone of a man called Wynna'. OE pers. name + stān.

Winterborne, Winterbourne, originally a river-name 'winter stream, i.e. stream flowing most strongly in winter', OE winter + burna; examples of places named from their situation on the various rivers so called include:
Winterborne Came Dorset. *Wintreburne* 1086 (DB), *Winterburn Caam* 1280. Affix from its early possession by the Norman Abbey of Caen. **Winterborne Clenston** Dorset. *Wintreburne* 1086 (DB), *Wynterburn Clencheston* 1303. Affix means 'estate (tūn) of the Clench family'. **Winterborne Stickland** Dorset. *Winterburne* 1086 (DB), *Winterburn Stikellane* 1203. Affix means '(with a) steep lane', from OE sticol + lane. **Winterborne Whitechurch** Dorset. *Wintreburne* 1086 (DB), *Wynterborn Wytecherch* 1268. Affix means '(with a) white, i.e. stone-built, church', OE hwīt + cirice. **Winterborne Zelston** Dorset. *Wintreborne* 1086 (DB), *Wynterbourn Selyston*

1350. Affix means 'estate (*tūn*) of the *de Seles* family'. **Winterbourne** S. Glos. *Wintreborne* 1086 (DB). **Winterbourne Abbas** Dorset. *Wintreburne* 1086 (DB), *Wynterburn Abbatis* 1244. Latin affix 'of the abbot' refers to early possession by the Abbey of Cerne.

Winterbourne Bassett & Winterbourne Monkton Wilts. *Wintreburne* 950, 1086 (DB), *Winterburn Basset* 1242, *Moneke Wynterburn* 1251. Distinguishing affixes from early possession by the *Basset* family and by the monks of Glastonbury Abbey. **Winterbourne Dauntsey, Winterbourne Earls & Winterbourne Gunner** Wilts. *Wintreburne* 1086 (DB), *Wynterburne Dauntesie* 1268, *Winterburne Earls* 1250, *Winterburn Gonor* 1267. Distinguishing affixes from early possession by the *Danteseye* family, by the Earls of Salisbury, and by a lady called *Gunnora* (here in 1249).

Winteringham N. Lincs. *Wintringeham* 1086 (DB). 'Homestead of the family or followers of a man called Wintra'. OE pers. name + *-inga-* + *hām*.

Wintersett Wakefd. *Wintersete c.*1125. 'Fold used in winter'. OE *winter* + *set*.

Winterslow Wilts. *Wintreslev* 1086 (DB). 'Mound or tumulus of a man called Winter'. OE pers. name + *hlǣw*.

Winterton N. Lincs. *Wintringatun c.*1067, *Wintrintune* 1086 (DB). 'Farmstead of the family or followers of a man called Wintra'. OE pers. name + *-inga-* + *tūn*.

Winterton-on-Sea Norfolk. *Wintertonne* 1044–7, *Wintretuna* 1086 (DB). 'Farmstead used in winter'. OE *winter* + *tūn*.

Winthorpe Lincs. *Winetorp* 12th cent. 'Outlying farmstead or hamlet of a man called Wina'. OE pers. name + OScand. *thorp*.

Winthorpe Notts. *Wimuntorp* 1086 (DB). 'Outlying farmstead or hamlet of a man called Wīgmund or Vigmundr'. OE or OScand. pers. name + *thorp*.

Winton Cumbria. *Wyntuna c.*1094. Probably 'pasture farmstead'. OE **winn* + *tūn*.

Winton Bmouth. a suburb named from the Earl of Eglinton, created also Earl of *Winton* in 1859, a kinsman of the Talbot sisters (*see* TALBOT VILLAGE).

Wintringham N. Yorks. *Wentrigham* 1086 (DB), *Wintringham* 1169. 'Homestead of the

family or followers of a man called Wintra'. OE pers. name + *-inga-* + *hām*.

Winwick, 'dwelling or (dairy) farm of a man called Wina', OE pers. name + *wīc*: **Winwick** Cambs. *Wineuuiche* 1086 (DB). **Winwick** Northants. *Winewican* 1043, *Winewiche* 1086 (DB).

Winwick Warrtn. *Winequic* 1170, *Winewich* 1204. 'Dwelling or (dairy) farm of a man called **Wineca*'. OE pers. name + *wīc*.

Wirksworth Derbys. *Wyrcesuuyrthe* 835, *Werchesworde* 1086 (DB). 'Enclosure of a man called **Weorc*'. OE pers. name + *worth*.

Wirral Ches. *Wirhealum, Wirheale* early 10th cent. '(Place at) the nook(s) where bog-myrtle grows'. OE *wīr* + *halh* (dative plural *halum*).

Wirswall Ches. *Wireswelle* 1086 (DB). 'Spring or stream of a man called Wīghere'. OE pers. name + *wella*.

Wisbech Cambs. *Wisbece* 1086 (DB). Possibly 'marshy-meadow valley or ridge'. OE *wisc* or **wisse* + *bece* or *bæc* (locative **bece*). Alternatively the first element may be the River Wissey, itself an OE name meaning 'marshy stream'.

Wisborough Green W. Sussex. *Wisebregh* 1227. 'Marshy-meadow hill'. OE *wisc* + *beorg*.

Wiseton Notts. *Wisetone* 1086 (DB). 'Farmstead of a man called Wīsa', or 'marshy-meadow farmstead'. OE pers. name or *wisc* + *tūn*.

Wishaw N. Lan. *Witscaga* 1086 (DB). OE *sceaga* 'wood' with doubtful first element, possibly OE **wiht* 'bend'.

Wishaw Warwicks. *Witscaga* 1086 (DB). Probably identical with previous name.

Wishford, Great Wilts. *Wicheford* 1086 (DB). 'Ford where wych-elms grow'. OE *wice* + *ford*.

Wiske (river) N. Yorks. *See* APPLETON WISKE.

Wisley Surrey. *Wiselei* 1086 (DB). 'Marshy-meadow clearing'. OE *wisc* + *lēah*.

Wispington Lincs. *Wispinctune* 1086 (DB). Possibly 'farmstead at the place where brushwood grows'. OE **wisp* + *-ing* + *tūn*.

Wissett Suffolk. *Wisseta* 1086 (DB), *Witseta* 1165. Possibly 'fold of a man called Witta'. OE pers. name + *set*.

Wissington Suffolk. *Wiswythetun* c.1000. 'Farmstead or estate of a woman called Wīgswīth'. OE pers. name + *tūn*.

Wistanstow Shrops. *Wistanestou* 1086 (DB). 'Holy place of St Wīgstān'. OE saint's name (Wīgstān was a Mercian prince murdered in 849 or 850) + *stōw*.

Wistanswick Shrops. *Wistaneswick* 1274. 'Dwelling or (dairy) farm of a man called Wīgstān'. OE pers. name + *wīc*.

Wistaston Ches. *Wistanestune* 1086 (DB). 'Farmstead or village of a man called Wīgstān'. OE pers. name + *tūn*.

Wiston W. Sussex. *Wistanestun* 1086 (DB). 'Farmstead or village of a man called Wīgstān or Winestān'. OE pers. name + *tūn*.

Wiston (*Cas-wis*) Pemb. *Castellum Wiz* 1146, *Wistune* 1319. 'Wizo's farmstead'. OE *tūn*. *Wizo* came from Flanders to set up a castle here in the 12th cent. Its Welsh name was *Castell Gwis*, 'Wizo's castle', which gave the present Welsh name.

Wiston S. Lan. *Wicestun* c.1155. OE *tūn* 'farmstead' with uncertain first element.

Wistow, 'the dwelling place', OE *wīc-stōw*: **Wistow** Cambs. *Wicstoue* 974, *Wistov* 1086 (DB). **Wistow** N. Yorks. *Wicstow* c.1030.

Wistow Leics. *Wistanestov* 1086 (DB). 'Holy place of St Wīgstān', from OE *stōw*, thus identical with WISTANSTOW Shrops.

Wiswell Lancs. *Wisewell* 1207. Possibly 'spring or stream near a marshy meadow'. OE *wisc* or **wisse* + *wella*.

Witcham Cambs. *Wichamme* 970, *Wiceham* 1086 (DB). 'Promontory where wych-elms grow'. OE *wice* + *hamm*.

Witchampton Dorset. *Wichemetune* 1086 (DB). Probably 'farmstead of the dwellers at a village associated with an earlier Romano-British settlement'. OE *wīc* + *hǣme* + *tūn*.

Witchford Cambs. *Wiceford* 1086 (DB). 'Ford where wych-elms grow'. OE *wice* + *ford*.

Witchingham, Great Norfolk. *Wicinghaham* 1086 (DB), *Wichingeham* 1130. Possibly 'homestead or village of the people who dwell at or come from a place called *Wic*

("the specialized form or trading settlement")'. OE *wīc* + *-inga-* +*hām*, *see* WHISSONETT Norfolk.

Witcombe, Great Glos. *Wydecomb* 1220. 'Wide valley'. OE *wīd* + *cumb*.

Witham Essex. *Witham* late 9th cent., 1086 (DB). Probably 'homestead near a river-bend'. OE **wiht* + *hām*.

Witham Friary Somerset. *Witeham* 1086 (DB). 'Homestead of a councillor, or of a man called Witta'. OE *wita* or pers. name + *hām*. The affix *Friary*, found from the 16th cent., alludes loosely to the Carthusian monastery founded here in 1178.

Witham on the Hill Lincs. *Witham* 1086 (DB). Possibly 'homestead in a bend'. OE **wiht* + *hām*.

Witham, North & Witham, South Lincs. *Widme* 1086 (DB). Named from the River Witham, a Celtic or pre-Celtic river-name of uncertain origin.

Witheridge Devon. *Wiriga* [sic] 1086 (DB), *Wytherigge* 1256. 'Willow-tree ridge', or 'ridge where wether-sheep are kept'. OE *wīthig* or *wether* + *hrycg*.

Witherley Leics. *Wytherdele* c.1204. 'Woodland clearing of a woman called Wīgthrȳth'. OE pers. name + *lēah*.

Withern Lincs. *Widerne* 1086 (DB). Probably 'house in the wood'. OE *widu, wudu* + *ærn*.

Withernsea E. R. Yorks. *Widfornessei* 1086 (DB). Possibly 'lake at the place near the thorn-tree'. OE *with* + *thorn* + *sǣ*.

Withernwick E. R. Yorks. *Widforneuuic* 1086 (DB). Possibly 'dairy farm of the place near the thorn-tree'. OE *with* + *thorn* + *wīc*.

Withersdale Street Suffolk. *Weresdel* [sic] 1086 (DB), *Wideresdala* 1184. 'Valley where wether-sheep are kept'. OE *wether* + *dæl*, with the later addition of *street* in the sense 'hamlet'.

Withersfield Suffolk. *Wedresfelda* 1086 (DB). 'Open land where wether-sheep are kept.' OE *wether* + *feld*.

Witherslack Cumbria. *Witherslake* c.1190. 'Valley of the wood, or of the willow-tree'. OScand. *vithr* (genitive *vithar*) or *víth* (genitive *víthjar*) + *slakki*.

Withiel Cornwall. *Widie* 1086 (DB). 'Wooded place'. Cornish *guydh* 'trees' with suffix **-yel*.

Withiel Florey Somerset. *Withiglea* 737, *Wythele Flory* 1305. 'Wood or clearing where willow-trees grow'. OE *wīthig* + *lēah*. Manorial affix from the *de Flury* family, here in the 13th cent.

Withington, usually 'farmstead with a willow copse', OE **wīthign* + *tūn*; examples include: **Withington** Ches. *Widinton* 1185. **Withington** Herefs. *Widingtune* 1086 (DB).
However the following has a different origin: **Withington** Glos. *Wudiandun* 737, *Widindune* 1086 (DB). 'Hill of a man called Widia'. OE pers. name (genitive -*n*) + *dūn*.

Withnell Lancs. *Withinhull* c.1160. 'Hill where willow-trees grow'. OE **wīthigen* + *hyll*.

Withybrook Warwicks. *Wythibroc* 12th cent. 'Willow-tree brook'. OE *wīthig* + *brōc*.

Withycombe, 'valley where willow-trees grow', OE *wīthig* + *cumb*: **Withycombe** Somerset. *Widicumbe* 1086 (DB).

Withycombe Raleigh Devon. *Widecome* 1086 (DB), *Widecombe Ralegh* 1465. Manorial affix from the *de Ralegh* family, here in the early 14th cent.

Withyham E. Sussex. *Withiham* 1230. 'Willow-tree enclosure or promontory'. OE *wīthig* + *hamm*.

Withypool Somerset. *Widepolle* 1086 (DB). 'Willow-tree pool'. OE *wīthig* + *pōl*.

Witley Surrey. *Witlei* 1086 (DB). 'Woodland clearing in a bend, or of a man called Witta'. OE **wiht* or pers. name + *lēah*.

Witley, Great & Witley, Little Worcs. *Wittlæg* 964, *Witlege* 1086 (DB). 'Woodland clearing in a bend'. OE **wiht* + *lēah*.

Witnesham Suffolk. *Witdesham* [sic] 1086 (DB), *Witnesham* 1254. Possibly 'homestead of a man called **Wittīn*'. OE pers. name + *hām*.

Witney Oxon. *Wyttanige* 969, *Witenie* 1086 (DB). 'Island, or dry ground in marsh, of a man called Witta'. OE pers. name (genitive -*n*) + *ēg*.

Wittenham, Little & Wittenham, Long Oxon. *Wittanham* c.865, *Witeham* 1086 (DB). 'River-bend land of a man called Witta'. OE pers. name (genitive -*n*) + *hamm*.

Witter (*Uachtar*) Down. 'Upper place'.

Wittering Peterb. *Witheringaeige* 972, *Witheringham* 1086 (DB), *Witeringa* 1167. '(Island or homestead of) the family or followers of a man called Wither'. OE pers. name + -*ingas* (in genitive plural with *ēg* and *hām* in the early forms).

Wittering, East & Wittering, West W. Sussex. *Wihttringes* 683, *Westringes* [sic] 1086 (DB). '(Settlement of) the family or followers of a man called Wihthere'. OE pers. name + -*ingas*.

Wittersham Kent. *Wihtriceshamme* 1032. 'Promontory of a man called Wihtrīc'. OE pers. name + *hamm*.

Witton, a common name, usually 'farmstead in or by a wood', OE *wudu* or *widu* + *tūn*; for example: **Witton Bridge** Norfolk. *Widituna* 1086 (DB). **Witton, East & Witton, West** N. Yorks. *Witun* 1086 (DB). **Witton Gilbert** Durham. *Wyton* 1195. Manorial affix from its possession by *Gilbert* de la Ley in the 12th cent. **Witton-le-Wear** Durham. *Wudutun* c.1040. On the River Wear, for which *see* MONKWEARMOUTH.
However some Wittons have a different origin: **Witton** Birm. *Witone* 1086 (DB). Possibly 'farmstead by an earlier Romano-British settlement'. OE *wīc* + *tūn*. **Witton** Ches. *Witune* 1086 (DB). 'Estate with a salt-works'. OE *wīc* + *tūn*.

Wiveliscombe Somerset. *Wifelescumb* 854, *Wivelescome* 1086 (DB). 'Valley of a man called **Wifel*'. OE pers. name + *cumb*, but *see* next name.

Wivelsfield E. Sussex. *Wifelesfeld* c.765. Possibly 'open land of a man called **Wifel*', OE pers. name + *feld*. Alternatively the first element in this and the previous name may be the noun *wifel* 'weevil' denoting 'weevil-infested land'.

Wivenhoe Essex. *Wiunhov* 1086 (DB). 'Hill-spur of a woman called **Wīfe*'. OE pers. name (genitive -*n*) + *hōh*.

Wiveton Norfolk. *Wiuetuna, Wiuentona* 1086 (DB), *Wyveton* 1226. 'Farmstead or village of a woman called **Wīfe*'. OE pers. name (genitive -*n*) + *tūn*. Alternatively perhaps 'farmstead of the women', from genitive plural *wīfa* of OE *wīf*.

Wix Essex. *Wica* 1086 (DB). 'The dwellings or specialized farm'. OE *wīc* in a ME plural form *wikes*.

Wixford Warwicks. *Wihtlachesforde* 962, *Witelavesford* [sic] 1086 (DB). 'Ford of a man called Wihtlāc'. OE pers. name + *ford*.

Wixoe Suffolk. *Wlteskeou* [*sic*] 1086 (DB), *Widekeshoo* 1205. 'Hill-spur of a man called Widuc'. OE pers. name + *hōh*.

Woburn Beds. *Woburne* 1086 (DB). '(Place at) the crooked or winding stream'. OE *wōh* + *burna*. It gives name to nearby **Woburn Sands** Bucks.

Wokefield Park W. Berks. *Weonfelda c.*950, *Hocfelle* 1086 (DB). Probably 'open land of a man called *Weohha'. OE pers. name + *feld*.

Woking Surrey. *Wocchingas c.*712, *Wochinges* 1086 (DB). '(Settlement of) the family or followers of a man called *Wocc(a)'. OE pers. name + *-ingas*.

Wokingham Wokhm. *Wokingeham* 1146. 'Homestead of the family or followers of a man called *Wocc(a)'. OE pers. name + *-inga-* + *hām*.

Woldingham Surrey. *Wallingeham* [*sic*] 1086 (DB), *Waldingham* 1204. Probably 'homestead of the forest dwellers'. OE *weald* + *-inga-* + *hām*. Alternatively 'homestead of the family or followers of a man called *Wealda', with an OE pers. name as first element.

Wolds, The (upland districts, i. Leics., ii. Lincs., iii. N. Yorks. & E. R. Yorks.), from OE *wald* 'high forest-land, later cleared', *see* FOSTON ON THE WOLDS, GARTON ON THE WOLDS, etc.

Wolferton Norfolk. *Wulferton* 1166. 'Farmstead or village of a man called Wulfhere'. OE pers. name + *tūn*.

Wolford, Great & Wolford, Little Warwicks. *Wolwarde* 1086 (DB). 'Place protected against wolves'. OE *wulf* + *weard*.

Wolfsdale Pemb. *Wolvedale* 1312. 'Valley frequented by wolves'. OE *wulf* + *dæl*.

Wollaston Northants. *Wilavestone* 1086 (DB), *Wullaueston* 1190. 'Farmstead or village of a man called Wīglāf or Wulflāf'. OE pers. name + *tūn*.

Wollaston Shrops. *Willavestune* 1086 (DB). 'Farmstead or village of a man called Wīglāf'. OE pers. name + *tūn*.

Wollerton Shrops. *Ulvretone* 1086 (DB), *Wluruntona c.*1135. 'Farmstead or village of a woman called Wulfrūn'. OE pers. name + *tūn*.

Wolsingham Durham. *Wlsingham c.*1150. Possibly 'homestead of the family or followers of a man called Wulfsige', OE pers. name + *-inga-* +

hām. Alternatively 'homestead at *Wulsing* (Wulfsige's place)', OE pers. name + *-ing* + *ham*.

Wolstanton Staffs. *Wlstanetone* 1086 (DB). 'Farmstead or village of a man called Wulfstān'. OE pers. name + *tūn*.

Wolston Warwicks. *Ulvricetone* 1086 (DB). 'Farmstead or village of a man called Wulfrīc'. OE pers. name + *tūn*.

Wolvercote Oxon. *Ulfgarcote* 1086 (DB). 'Cottage associated with a man called Wulfgār'. OE pers. name + *-ing-* + *cot*.

Wolverhampton Wolverh. *Heantune* 985, *Wolvrenehamptonia c.*1080. Originally 'high farmstead', from OE *hēah* (dative *hēan*) + *tūn*, later with the addition of the OE pers. name *Wulfrūn*, the lady to whom the manor was given in 985.

Wolverley Worcs. *Wulfferdinleh* 866, *Ulwardelei* 1086 (DB). 'Woodland clearing associated with a man called Wulfweard'. OE pers. name + *-ing-* + *lēah*.

Wolverton Milt. K. *Wluerintone* 1086 (DB). 'Estate associated with a man called Wulfhere'. OE pers. name + *-ing-* + *tūn*.

Wolvey Warwicks. *Ulveia* 1086 (DB). Probably 'enclosure protected against wolves'. OE *wulf* + *hege* or *hæg*.

Wolviston Stock. on T. *Oluestona* 1091. 'Farmstead or village of a man called Wulf'. OE pers. name + *tūn*.

Wombleton N. Yorks. *Winbeltun* 1086 (DB). 'Farmstead or village of a man called Wynbald or Winebald'. OE pers. name + *tūn*.

Wombourne Staffs. *Wamburne* 1086 (DB). '(Place at) the crooked or winding stream'. OE *wōh* (dative *wōn*) + *burna*.

Wombwell Barns. *Wanbuelle* 1086 (DB). 'Spring or stream in a hollow, or of a man called *Wamba'. OE *wamb* or pers. name + *wella*.

Womenswold Kent. *Wimlincgawald* 824. Possibly 'forest of the family or followers of a man called *Wīmel'. OE pers. name + *-inga-* + *weald*.

Wonersh Surrey. *Woghenhers* 1199. 'Crooked ploughed field'. OE *wōh* (dative *wōn*) + *ersc*.

Wonson Devon. *Woneston* 1244. Possibly '(place at) the crooked stone'. OE *wōh* (dative *wōn*) + *stān*.

Wonston Dorset. *Wolmerston* 1280. 'Farmstead or estate of a man called Wulfmǣr'. OE pers. name + *tūn*.

Wonston, South Hants. *Wynsigestune* 901, *Wenesistune* 1086 (DB). 'Farmstead or estate of a man called Wynsige'. OE pers. name + *tūn*.

Wooburn Bucks. *Waburna* c.1075, *Waborne* 1086 (DB). Probably 'stream with its banks walled up, or with a dam'. OE *wǣg* + *burna*. Alternatively the first element may be OE *wōh* 'crooked, winding'.

Wood as affix. *See* main name, e.g. for **Wood Dalling** (Norfolk) *see* DALLING.

Wood Green Gtr. London. *Wodegrene* 1502. 'Village green in or near woodland'. ME *wode* + *grene*.

Woodale N. Yorks. *Wulvedale* 1223. 'Valley frequented by wolves'. OE *wulf* + *dæl*.

Woodbastwick Norfolk. *Bastwik* 1044–7, *Bastuuic* 1086 (DB), *Wodbastwyk* 1253. 'Farm where bast (the bark of the lime-tree used for rope-making) is got'. OE *bæst* + *wīc*, with the later addition of ME *wode* 'wood'.

Woodborough Notts. *Udeburg* 1086 (DB). 'Fortified place by the wood'. OE *wudu* + *burh*.

Woodborough Wilts. *Wideberghe* 1208. 'Wooded hill'. OE *wudu* + *beorg*.

Woodbridge Suffolk. *Oddebruge* c.1050, *Wudebrige* 1086 (DB). 'Wooden bridge', or 'bridge near the wood'. OE *wudu* + *brycg*.

Woodburn, East Northum. *Wodeburn* 1265. 'Stream in the wood'. OE *wudu* + *burna*.

Woodbury Devon. *Wodeberia* 1086 (DB). 'Fortified place by the wood'. OE *wudu* + *burh* (dative *byrig*).

Woodbury Salterton Devon. *See* SALTERTON.

Woodchester Glos. *Uuduceastir* 716–43, *Widecestre* 1086 (DB). 'Roman camp in the wood'. OE *wudu* + *ceaster*.

Woodchurch Kent. *Wuducirce* c.1100. 'Wooden church', or 'church by the wood'. OE *wudu* + *cirice*.

Woodcote, Woodcott, 'cottage(s) in or by a wood', OE *wudu* + *cot*; examples include: **Woodcote** Gtr. London. *Wudecot* 1200.

Woodcote Oxon. *Wdecote* 1109. **Woodcott** Hants., near Litchfield. *Odecote* 1086 (DB).

Woodditton Cambs. *Dictune* 1022, *Wodeditone* 1227. 'Farmstead by a ditch or dyke', OE *dīc* + *tūn*, with affix *wode* 'wood' to distinguish this place from FEN DITTON.

Woodeaton Oxon. *Etone* 1086 (DB), *Wudeetun* 1185. 'Farmstead by a river'. OE *ēa* + *tūn*, with the later affix *wudu* 'wood'.

Woodend, '(place at) the end of the wood', ME *wode* + *ende*; for example **Woodend** Northants. *Wodende* 1316.

Woodford, 'ford in or by a wood', OE *wudu* + *ford*; examples include: **Woodford** Gtr. London. *Wudeford* 1062, *Wdefort* 1086 (DB). **Woodford** Northants., near Thrapston. *Wodeford* 1086 (DB). **Woodford** Stockp. *Wideforde* 1248. **Woodford** Wilts. *Wuduforda* 972. **Woodford Halse** Northants. *Wodeford* 1086 (DB). Affix from the manor of HALSE.

Woodhall, 'hall in or by a wood', OE *wudu* + *h(e)all*; for example **Woodhall Spa** Lincs. *Wudehalle* 12th cent. (with affix since early 19th cent. from its reputation as a watering-place).

Woodham, usually 'homestead or village in or by a wood', OE *wudu* + *hām*; examples include: **Woodham** Surrey. *Wodeham* 672–4. **Woodham Ferrers, Woodham Mortimer, & Woodham Walter** Essex. *Wudaham* c.975, *Udeham, Odeham, Wdeham* 1086 (DB), *Wodeham Ferrers* 1230, *Wodeham Mortimer* 1255, *Wodeham Walter* 1238. Manorial affixes from early possession of estates here by the *de Ferrers, Mortimer,* and *Fitzwalter* families.

Woodhay, East (Hants.) **& Woodhay, West** (W. Berks.) *Wideheia* c.1150. 'Wide enclosure'. OE *wīd* (dative *-an*) + *hæg*.

Woodhorn Northum. *Wudehorn* 1178. 'Wooded horn of land or promontory'. OE *wudu* + *horn*.

Woodhouse, 'house(s) in or near a wood', OE *wudu* + *hūs*; examples include: **Woodhouse** Leics. *Wodehuses* c.1220. **Woodhouse Eaves** has affix from OE *efes* 'edge of a wood' with reference to CHARNWOOD FOREST. **Woodhouse** Sheff. *Wdehus* 1200.

Woodhurst Cambs. *Wdeherst* 1209. Originally OE *hyrst* 'wooded hill' (probably used here as the name of a district) with the later addition of ME *wode* 'wood'.

Woodland(s), 'cultivated land in or near a wood', OE *wudu* + *land*; examples include: **Woodland** Devon. *Wodelonde* 1328. **Woodlands** Dorset. *Wodelande* 1244.

Woodleigh Devon. *Wuduleage* c.1010, *Odelie* 1086 (DB). 'Clearing in a wood'. OE *wudu* + *lēah*.

Woodlesford Leeds. *Wridelesford* 12th cent. 'Ford near a thicket'. OE **wrīdels* + *ford*.

Woodmancote, Woodmancott, 'cottage(s) of the woodsmen or foresters', OE *wudu-mann* + *cot*; examples include: **Woodmancote** Glos., near Rendcomb. *Wodemancote* 12th cent. **Woodmancott** Hants. *Woedemancote* 903, *Udemanecote* 1086 (DB).

Woodmansey E. R. Yorks. *Wodemanse* 1289. 'Pool of the woodsman or forester'. OE *wudu-mann* + *sǣ*.

Woodmansterne Surrey. *Odemerestor* [sic] 1086 (DB), *Wudemaresthorne* c.1190. 'Thorn-tree by the boundary of the wood'. OE *wudu* + *mǣre* + *thorn*.

Woodnesborough Kent. *Wanesberge* [sic] 1086 (DB), *Wodnesbeorge* c.1100. 'Hill or mound associated with the heathen god Wōden'. OE god-name + *beorg*.

Woodnewton Northants. *Niwetone* 1086 (DB), *Wodeneuton* 1255. 'New farmstead (in woodland)'. OE *nīwe* + *tūn* with the later addition of ME *wode*.

Woodperry Oxon. *Perie* 1086 (DB), *Wodeperi* 1242. '(Place at) the pear-tree(s)', from OE *pyrige*, with the later addition of ME *wode* 'wood' to distinguish it from WATERPERRY.

Woodplumpton Lancs. *Pluntun* 1086 (DB), *Wodeplumpton* 1327. 'Farmstead (in woodland) where plum-trees grow'. OE *plūme* + *tūn* with the later addition of ME *wode*.

Woodrising Norfolk. *Risinga* 1086 (DB), *Woderisingg* 1291. Probably, like CASTLE RISING, OE pers. name **Risa* or OE *hrīs* 'brushwood' + *-ingas*, with the later addition of ME *wode* 'wood'.

Woodsetts Rothm. *Wudesete* c.1220. 'Folds (for animals) in the wood'. OE *wudu* + *set*.

Woodsford Dorset. *Wardesford* 1086 (DB). 'Ford of a man called **Weard*'. OE pers. name + *ford*.

Woodstock Oxon. *Wudestoce* c.1000, *Wodestoch* 1086 (DB). 'Settlement in woodland'. OE *wudu* + *stoc*.

Woodston Peterb. *Widestun, Wudestun* 973, *Wodestun* 1086 (DB). OE *tūn* 'farmstead, estate' with an uncertain first element, possibly OE *widu, wudu* 'wood' with an analogical late OE genitive in *-es*, or a contracted form of OE *wudu-efes* 'edge or border of a wood'.

Woodthorpe, 'outlying farmstead or hamlet in woodland', OE *wudu* + OScand. *thorp*: **Woodthorpe** Derbys. *Wodethorpe* 1258. **Woodthorpe** Lincs. *Wdetorp* 12th cent.

Woodton Norfolk. *Wodetuna* 1086 (DB). 'Farmstead in or near a wood'. OE *wudu* + *tūn*.

Woodville Derbys. a modern name dating from 1845, until then known as *Wooden-Box* from the wooden hut here for collecting toll at the turnpike.

Woodyates, East & Woodyates, West Dorset. *Wdegeate* 9th cent., *Odiete* 1086 (DB). '(Place at) the gate or gap in the wood'. OE *wudu* + *geat*.

Woofferton Shrops. *Wulfreton* 1221. 'Farmstead of a man called Wulfhere or Wulffrith'. OE pers. name + *tūn*.

Wookey Somerset. *Woky* 1065. '(Place at) the trap or snare for animals'. OE *wōcig*, probably originally with reference to Wookey Hole (*Wokyhole* 1065, with OE *hol* 'ravine').

Wool Dorset. *Welle* 1086 (DB). '(Place at) the spring or springs'. OE *wiella*.

Woolacombe Devon. *Wellecome* 1086 (DB). 'Valley with a spring or stream'. OE *wiella* + *cumb*.

Woolaston Glos. *Odelaweston* 1086 (DB). 'Farmstead of a man called Wulflāf'. OE pers. name + *tūn*.

Woolbeding W. Sussex. *Welbedlinge* 1086 (DB). Probably 'place associated with a man called Wulfbeald'. OE pers. name + *-ing*.

Wooler Northum. *Wulloure* 1187. Probably 'spring promontory'. OE *wella* + **ofer*.

Woolfardisworthy Devon. near Crediton. *Ulfaldeshodes* [sic] 1086 (DB), *Wolfardesworthi* 1264. 'Enclosure of a man called Wulfheard'. OE pers. name + *worthig*.

Woolhampton W. Berks. *Ollavintone* 1086 (DB). 'Estate associated with a man called Wulflāf'. OE pers. name + *-ing-* + *tūn*.

Woolhope Herefs. *Hope* 1086 (DB), *Wulvivehop* 1234. '(Place at) the valley'. OE *hop*, with later addition from a woman called *Wulfgifu* who gave the manor to Hereford Cathedral in the 11th cent.

Woolland Dorset. *Wonlonde* 934, *Winlande* 1086 (DB). 'Estate with pasture or meadow'. OE **wynn* + *land*.

Woolley, usually 'wood or clearing frequented by wolves', OE *wulf* + *lēah*; examples include: **Woolley** Cambs. *Ciluelai* [*sic*] 1086 (DB), *Wulueleia* 1158. **Woolley** Wakefd. *Wiluelai* 1086 (DB).

Woolpit Suffolk. *Wlpit* 10th cent., *Wlfpeta* 1086 (DB). 'Pit for trapping wolves'. OE *wulf-pytt*.

Woolscott Warwicks. *Wulscote c.*1235. Probably 'cottage(s) of a man called Wulfsige'. OE pers. name + *cot*.

Woolstaston Shrops. *Ulestanestune* 1086 (DB). 'Farmstead of a man called Wulfstān'. OE pers. name + *tūn*.

Woolsthorpe Lincs. near Grantham. *Ulestanestorp* 1086 (DB). 'Outlying farmstead or hamlet of a man called Wulfstān'. OE pers. name + OScand. *thorp*.

Woolston Devon. *Ulsistone* 1086 (DB). 'Farmstead of a man called Wulfsige'. OE pers. name + *tūn*.

Woolston Sotn. *Olvestune* 1086 (DB). 'Farmstead of a man called Wulf'. OE pers. name + *tūn*.

Woolston Warrtn. *Ulfitona* 1142. 'Farmstead of a man called Wulfsige'. OE pers. name + *tūn*.

Woolstone Oxon. *Olvricestone* 1086 (DB). 'Estate of a man called Wulfrīc'. OE pers. name + *tūn*. The *Wulfrīc* in question is named in 10th-cent. charters.

Woolstone, Great & Woolstone, Little Milt. K. *Wlsiestone* 1086 (DB). 'Farmstead of a man called Wulfsige'. OE pers. name + *tūn*.

Woolton Lpool. *Uluentune* 1086 (DB). 'Farmstead of a man called *Wulfa'. OE pers. name (genitive *-n*) + *tūn*.

Woolverstone Suffolk. *Uluerestuna* 1086 (DB). 'Farmstead of a man called Wulfhere'. OE pers. name + *tūn*.

Woolverton Somerset. *Wulfrinton* 1196. Probably 'estate associated with a man called Wulfhere'. OE pers. name + *-ing-* + *tūn*.

Woolwich Gtr. London. *Uuluuich* 918, *Hulviz* [*sic*] 1086 (DB). 'Trading settlement or harbour for wool'. OE *wull* + *wīc*.

Woore Shrops. *Wavre* 1086 (DB). Possibly '(place by) the swaying tree'. OE *wæfre*.

Wootton, a common name, 'farmstead in or near a wood', OE *wudu* + *tūn*; examples include: **Wootton** Northants. *Witone* 1086 (DB). **Wootton** Oxon., near Abingdon. *Wuttune* 985. **Wootton Bassett** Wilts. *Wdetun* 680, *Wodetone* 1086 (DB), *Wotton Basset* 1272. Manorial affix from the *Basset* family, here in the 13th cent. **Wootton Fitzpaine** Dorset. *Wodetone* 1086 (DB), *Wotton Fitz Payn* 1392. Manorial affix from the *Fitz Payn* family, here in the 14th cent. **Wootton, Glanvilles** Dorset. *Widetone* 1086 (DB), *Wotton Glaunuill* 1288. Manorial affix from the *Glanville* family, here in the 13th cent. **Wootton, Leek** Warwicks. *Wottona* 1122, *Lecwotton* 1285. Affix presumably OE *lēac* 'leek' from their cultivation here. **Wootton, North & Wootton, South** Norfolk. *Wdetuna* 1086 (DB), *Nordwitton* 1166, *Sudwutton* 1182. **Wootton Rivers** Wilts. *Wdutun* 804, *Otone* 1086 (DB), *Wotton Ryvers* 1332. Manorial affix from the *de Rivere* family, here in the 13th cent. **Wootton St Lawrence** Hants. *Wudatune* 990, *Odetone* 1086 (DB). Affix from the dedication of the church. **Wootton Wawen** Warwicks. *Uuidutuun* 716–37, *Wotone* 1086 (DB), *Wageneswitona c.*1142. Affix from its possession in the 11th cent. by a man called *Wagen* (OScand. *Vagn*).

Worbarrow Bay Dorset. *Wyrebarowe* 1462, *Worthbarrow baye* 1575. 'Hill where watch is kept'. OE **wierde* + *beorg*, with the later addition of *bay*. The reference is to **Worbarrow Tout** (first recorded 1841, from OE *tōte* 'look-out hill').

Worcester Worcs. *Weogorna civitas* 691, *Wigranceastre* 717, *Wirecestre* 1086 (DB). 'Roman town of the Weogora tribe'. Pre-English folk-name (possibly from a Celtic river-name meaning 'winding river') + OE *ceaster*.
Worcestershire (OE *scīr* 'district') is first referred to in the 11th cent.

Wordsley Dudley. *Wuluardeslea* 12th cent. 'Woodland clearing of a man called Wulfweard'. OE pers. name + *lēah*.

Worfield Shrops. *Wrfeld* 1086 (DB). 'Open land on the River Worfe'. OE river-name (meaning 'winding') + *feld*.

Worgret Dorset. *Vergroh, Weregrote* 1086 (DB), *Wergerod* 1202. 'Gallows for criminals'. OE *wearg-rōd*.

Workington Cumbria. *Wirkynton c.*1125. 'Estate associated with a man called *Weorc'. OE pers. name + *-ing-* + *tūn*.

Worksop Notts. *Werchesope* 1086 (DB). Possibly 'enclosure or valley of a man called *Weorc'. OE pers. name + *hop*. Alternatively the first element may be OE *weorc* 'a work, a building'.

Worlaby N. Lincs. *Uluricebi* 1086 (DB). 'Farmstead or village of a man called Wulfrīc'. OE pers. name + OScand. *bý*.

Worldham, East & Worldham, West Hants. *Werildeham* 1086 (DB). Probably 'homestead of a woman called *Wǣrhild'. OE pers. name + *hām*.

World's End Hants. Recorded thus from 1759, a whimsical name for a place considered rather remote or inaccessible. The same name occurs in Gtr. London, W. Berks., W. Sussex, and other counties.

Worle N. Som. *Worle* 1086 (DB). Probably 'wood or clearing frequented by wood-grouse'. OE *wōr* + *lēah*.

Worleston Ches. *Werblestune* [*sic*] 1086 (DB), *Weruelestona c.*1100. Possibly 'farmstead at the clearing for cattle'. OE *weorf* + *lēah* (genitive *lēas*) + *tūn*.

Worlingham Suffolk. *Werlingaham* 1086 (DB). Probably 'homestead of the family or followers of a man called *Wērel'. OE pers. name + *-inga-* + *hām*.

Worlington Devon. *Ulvredintone* 1086 (DB). 'Estate associated with a man called Wulfrēd'. OE pers. name + *-ing-* + *tūn*.

Worlington Suffolk. *Wirilintona* 1086 (DB), *Wridelingeton* 1201. 'Farmstead by the winding stream'. OE *wride* + *wella* + *-ing-* + *tūn*.

Worlingworth Suffolk. *Wilrincgawertha c.*1035, *Wyrlingwortha* 1086 (DB). 'Enclosure of the family or followers of a man called Wilhere'. OE pers. name + *-inga-* + *worth*.

Wormbridge Herefs. *Wermebrig* 1207. 'Bridge on Worm Brook'. Celtic river-name (meaning 'dark stream') + OE *brycg*.

Wormegay Norfolk. *Wermegai* 1086 (DB). 'Island of the family or followers of a man called *Wyrma'. OE pers. name + *-inga-* + *ēg*.

Wormhill Derbys. *Wruenele* [*sic*] 1086 (DB), *Wermehull c.*1105. Probably 'hill of a man called *Wyrma'. OE pers. name + *hyll*.

Wormingford Essex. *Widemondefort* 1086 (DB). 'Ford of a man called *Withermund'. OE pers. name + *ford*.

Worminghall Bucks. *Wermelle* [*sic*] 1086 (DB), *Wirmenhale c.*1218. Probably 'nook of land of a man called *Wyrma'. OE pers. name (genitive *-n*) + *halh*.

Wormington Glos. *Wermetune* 1086 (DB). 'Estate associated with a man called *Wyrma'. OE pers. name + *-ing-* + *tūn*.

Worminster Somerset. *Wormester* 946. 'Rocky hill of the snake or dragon'. OE *wyrm* + *torr*.

Wormleighton Warwicks. *Wilmanlehttune* 956, *Wimelestone* 1086 (DB). 'Herb garden of a man called *Wilma'. OE pers. name + *lēac-tūn*.

Wormley Herts. *Wrmeleia c.*1060, *Wermelai* 1086 (DB). 'Woodland clearing infested with snakes'. OE *wyrm* + *lēah*.

Worms Head (*Penrhyn Gwyr*) Swan. *Wormyshede* 15th cent. 'Snake's head'. OE *wyrm* + *hēafod*. *Compare* GREAT ORMES HEAD. The Welsh name means 'Gower promontory' (Welsh *penrhyn*). *See* GOWER.

Wormshill Kent. *Godeselle* [*sic*] 1086 (DB), *Wotnesell c.*1225, *Worneshelle* 1254. Possibly 'hill of the heathen god Wōden', OE god-name + *hyll*. Alternatively 'shelter for a herd of pigs', OE *weorn* + *(ge)sell*.

Wormsley Herefs. *Wermeslai* 1086 (DB). 'Woodland clearing of a man called *Wyrm', or 'one infested with snakes'. OE pers. name or *wyrm* + *lēah*.

Wormwood Scrubs Gtr. London. *Wormeholte c.*1195, *Wormholtwode* 1437. 'Snake-infested wood or thicket'. OE *wyrm* + *holt* (also *wudu*), with later addition of *scrub* 'low stunted tree, brushwood'.

Worplesdon Surrey. *Werpesdune* 1086 (DB).
'Hill with a path or track'. OE **werpels* + *dūn*.

Worrall Sheff. *Wihale* 1086 (DB), *Wirhal* 1218.
Probably 'nook of land where bog-myrtle
grows'. OE *wīr* + *halh*.

Worsall, High & Worsall, Low   N. Yorks.
Wirceshel, Wercheshale 1086 (DB). 'Nook of land
of a man called *Weorc'. OE pers. name + *halh*.

Worsbrough Barns. *Wircesburg* 1086 (DB).
'Stronghold of a man called *Wyrc'. OE pers.
name + *burh*.

Worsley Salford. *Werkesleia* 1196,
Wyrkitheley 1246. Possibly 'woodland clearing
of a woman called *Weorcgӯth or of a man
called *Weorchæth'. OE pers. name + *lēah*.

Worstead Norfolk. *Wrthestede* 1044–7,
Wrdestedam 1086 (DB). 'Site of an enclosure,
a farmstead'. OE *worth* + *stede*.

Worsthorne Lancs. *Worthesthorn* 1202.
'Thorn-tree of a man called *Weorth'. OE pers.
name + *thorn*.

Worston Lancs. *Wrtheston* 1212. 'Farmstead
of a man called *Weorth'. OE pers. name + *tūn*.

Worth, 'the enclosure, the enclosed
settlement', OE *worth*; examples include:
Worth Kent. *Wurth* 1226. **Worth** W. Sussex.
Orde 1086 (DB). **Worth Matravers** Dorset.
Wirde 1086 (DB), *Worth Matrauers* 1664.
Manorial affix from the *Mautravers* family, here
from the 14th cent.

Wortham Suffolk. *Wrtham* c.950, *Wortham*
1086 (DB). 'Homestead with an enclosure'. OE
worth + *hām*.

Worthen Shrops. *Wrdine* 1086 (DB). 'The
enclosure'. OE *worthign*.

Worthing Norfolk. *Worthing* 1282. Probably
identical in origin with the previous name.

Worthing W. Sussex. *Ordinges* 1086 (DB).
Probably '(settlement of) the family or followers
of a man called *Weorth'. OE pers. name + *-ingas*.

Worthington Leics. *Werditone* 1086 (DB),
Wrthington c. 1130. Probably 'estate associated
with a man called *Weorth'. OE pers. name + *-
ing-* + *tūn*.

**Worthy, Headbourne, Worthy, Kings,
& Worthy, Martyr**   Hants. *Worthige* 825,
Ordie 1086 (DB), *Hydeburne Worthy* c.1270,
Chinges Ordia 1157, *Wordia le Martre* 1243. 'The

enclosure(s)'. OE *worthig*. The affix
Headbourne is an old stream-name ('stream by
the hides of land', OE *hīd* + *burna*), the other
affixes being from early possession by the king
and the *le Martre* family.

Worting Hants. *Wyrtingas* 960, *Wortinges*
1086 (DB). Possibly 'the herb gardens', from the
plural of an OE word **wyrting*.

Wortley Barns. *Wirtleie* 1086 (DB).
'Woodland clearing used for growing
vegetables'. OE *wyrt* + *lēah*.

Worton Wilts. *Wrton* 1173. 'The vegetable
garden'. OE *wyrt-tūn*.

Worton, Nether & Worton, Over   Oxon.
Ortune 1050–2, *Hortone* 1086 (DB). 'Farmstead
by a bank or slope'. OE *ōra* + *tūn*.

Wotherton Shrops. *Udevertune* 1086 (DB).
'Farmstead by the woodland ford'. OE *wudu* +
ford + *tūn*.

Wotter Devon. *Wodetorre* 1263. 'Wooded
rocky hill', or 'rocky hill by a wood'. OE *wudu* +
torr.

Wotton, 'farmstead in or near a wood', OE
wudu + *tūn*; examples include: **Wotton**
Surrey. *Wodetones* 1086 (DB). **Wotton under
Edge** Glos. *Wudutune* 940, *Vutune* 1086 (DB).
Affix refers to its situation under the Cotswold
escarpment. **Wotton Underwood** Bucks.
Wudotun 848, *Oltone* [sic] 1086 (DB). Affix
means 'within the wood'.

Woughton on the Green   Milt. K.
Ulchetone [sic] 1086 (DB), *Wocheton* 1163.
Probably 'farmstead of a man called *Wēoca'.
OE pers. name + *tūn*.

Wouldham Kent. *Uuldaham* 811, *Oldeham*
1086 (DB). 'Homestead of a man called *Wulda'.
OE pers. name + *hām*.

Wrabness Essex. *Wrabenasa* 1086 (DB).
Possibly 'headland of a man called *Wrabba'.
OE pers. name + *næss*.

Wrafton Devon. *Wratheton* 1238. Possibly
'farmstead built on piles or timber supports'. OE
wrathu + *tūn*.

Wragby, 'farmstead or village of a man called
Wraggi', OScand. pers. name + *bý*: **Wragby**
Lincs. *Waragebi* 1086 (DB). **Wragby** Wakefd.
Wraggebi 1160–70.

Wramplingham Norfolk. *Wranplincham*
1086 (DB). OE *hām* 'homestead' with an

uncertain first element, probably a pers. name or tribal name.

Wrangle Lincs. *Werangle* 1086 (DB). 'Crooked stream or other feature'. OE **wrengel* or OScand. **vrengill.*

Wrantage Somerset. *Wrentis* 1199. 'Pasture for stallions'. OE **wræna* + **etisc.*

Wrath, Cape. *See* CAPE WRATH.

Wratting, 'place where crosswort or hellebore grows', OE *wrætt* + *-ing*: **Wratting, Great** & **Wratting, Little** Suffolk. *Wratinga* 1086 (DB). **Wratting, West** Cambs. *Wreattinge* 974, *Waratinge* 1086 (DB).

Wrawby N. Lincs. *Waragebi* 1086 (DB). Probably 'farmstead or village of a man called **Wraghi*'. OScand. pers. name + *bý.*

Wraxall, 'nook or hollow frequented by the buzzard or other bird of prey', OE **wrocc* + *halh*: **Wraxall** Dorset. *Brocheshale* [sic] 1086 (DB), *Wrokeshal* 1196. **Wraxall** N. Som. *Werocosale* 1086 (DB). **Wraxall, North** Wilts. *Werocheshalle* 1086 (DB). **Wraxall, South** Wilts. *Wroxhal* 1227.

Wray, Wrea, Wreay, 'secluded nook or corner of land', OScand. *vrá*; examples include: **Wray** Lancs. *Wra* 1227. **Wray, High** Cumbria. *Wraye* c.1535. **Wrea Green** Lancs. *Wra* 1201. **Wreay** Cumbria, near Soulby. *Wra* 1487.

Wraysbury Winds. & Maid. *Wirecesberie* [sic] 1086 (DB), *Wiredesbur* 1195. 'Stronghold of a man called Wīgrēd'. OE pers. name + *burh* (dative *byrig*).

Wreake (river) Leics. *See* FRISBY ON THE WREAKE.

Wrecclesham Surrey. *Wrecclesham* 1225. Probably 'homestead or enclosure of a man called **Wrecel*'. OE pers. name + *hām* or *hamm.*

Wrecsam. *See* WREXHAM.

Wrekin, The Shrops. *See* WROXETER.

Wrelton N. Yorks. *Wereltun* 1086 (DB). Possibly 'farmstead by the hill where criminals are hanged'. OE *wearg* + *hyll* + *tūn.*

Wrenbury Ches. *Wareneberie* 1086 (DB). 'Stronghold of the wren, or of a man called **Wrenna*'. OE *wrenna* or pers. name + *burh* (dative *byrig*).

Wreningham Norfolk. *Wreningham* c.1060, *Urnincham* [sic] 1086 (DB). Probably 'homestead of the family or followers of a man called **Wrenna*'. OE pers. name + *-inga-* + *hām.*

Wrentham Suffolk. *Wretham* [sic] 1086 (DB), *Wrentham* 1228. Probably 'homestead of a man called **Wrenta*'. OE pers. name + *hām.*

Wressle E. R. Yorks. *Weresa* [sic] 1086 (DB), *Wresel* 1183. From OE **wrǣsel* 'something twisted', perhaps referring to broken ground or a winding river.

Wrestlingworth Beds. *Wrastlingewrd* c.1150. Probably 'enclosure of the family or followers of a man called **Wrǣstel*'. OE pers. name + *-inga-* + *worth.*

Wretham, East Norfolk. *Wretham* 1086 (DB). 'Homestead where crosswort or hellebore grows'. OE *wrætt* + *hām.*

Wretton Norfolk. *Wretton* 1198. 'Farmstead where crosswort or hellebore grows'. OE *wrætt* + *tūn.*

Wrexham (*Wrecsam*) Wrex. *Wristlesham* 1161, *Gwregsam* 1291. 'Wryhtel's water-meadow'. OE pers. name + *hamm.*

Wribbenhall Worcs. *Gurberhale* [sic] 1086 (DB), *Wrubbenhale* c.1160. 'Nook of land of a man called **Wrybba*'. OE pers. name (genitive *-n*) + *halh.*

Wrightington Lancs. *Wrichtington* 1202. Possibly 'farmstead or village of the wrights or carpenters'. OE *wyrhta* (genitive plural *-ena*) + *tūn.*

Wrinehill Staffs. *Wrinehull* 1225. OE *hyll* 'hill' with an uncertain first element, probably an old district-name *Wryme* or *Wrine* (OE **Wrīma*) perhaps meaning 'river-bend'.

Wrington N. Som. *Wringtone* 926, *Weritone* [sic] 1086 (DB). 'Farmstead on the river called *Wring*'. Earlier or alternative name (possibly meaning 'winding stream') for the River Yeo + OE *tūn.*

Writtle Essex. *Writele* 1066–76, *Writelam* 1086 (DB). Originally a river-name from OE **writol* 'babbling', an earlier name for the River Wid.

Wrockwardine Tel. & Wrek. *Recordine* 1086 (DB), *Wrocwurthin* 1196. 'Enclosure by the hill called the Wrekin'. Celtic name (*see* WROXETER) + OE *worthign.*

Wroot N. Lincs. *Wroth* 1157. 'Snout-like spur of land'. OE *wrōt*.

Wrotham Kent. *Uurotaham* 788, *Broteham* [*sic*] 1086 (DB). 'Homestead of a man called *Wrōta*'. OE pers. name + *hām*.

Wroughton Swindn. *Wervetone* 1086 (DB). 'Farmstead on the river called *Worf*'. Celtic river-name meaning 'winding stream' (an old name of the River Ray) + OE *tūn*.

Wroxall I. of Wight. *Wroccesheale* 1038–44, *Warochesselle* 1086 (DB). 'Nook or hollow frequented by the buzzard or other bird of prey'. OE *wrocc* + *halh*.

Wroxeter Shrops. *Ouirokōnion* c.150, *Rochecestre* 1086 (DB). 'Roman fort at or near *Uriconio* or the Wrekin'. Ancient Celtic name (possibly 'town of Virico') + OE *ceaster*.

Wroxham Norfolk. *Vrochesham* 1086 (DB). 'Homestead or enclosure of the buzzard, or of a man called *Wrocc*'. OE *wrocc* or pers. name + *hām* or *hamm*.

Wroxton Oxon. *Werochestan* 1086 (DB). Probably 'stone of the buzzards or other birds of prey'. OE *wrocc* + *stān*.

Wyaston Derbys. *Widerdestune* [*sic*] 1086 (DB), *Wyardestone* 1244. 'Farmstead of a man called Wīgheard'. OE pers. name + *tūn*.

Wyberton Lincs. *Wibertune* 1086 (DB). Possibly 'farmstead of a man called Wīgbeorht or of a woman called Wīgburh'. OE pers. name + *tūn*.

Wyboston Beds. *Wiboldestone* 1086 (DB). 'Farmstead of a man called Wīgbeald'. OE pers. name + *tūn*.

Wybunbury Ches. *Wimeberie* 1086 (DB), *Wybbunberi* c.1210. 'Stronghold or manor-house of a man called Wīgbeorn'. OE pers. name + *burh* (dative *byrig*).

Wychavon (district) Worcs. A modern name, combining an early name of DROITWICH with the river-name AVON.

Wychbold Worcs. *Uuicbold* 692, *Wicelbold* 1086 (DB). 'Dwelling near the trading settlement'. OE *wīc* + *bold*.

Wychwood (old forest) Oxon. *See* ASCOTT UNDER WYCHWOOD.

Wycliffe Durham. *Witcliue* 1086 (DB). Probably 'cliff by the river-bend'. OE *wiht* + *clif*.

Wycomb Leics. *Wiche* 1086 (DB), *Wicham* 1207. 'Homestead associated with a *vicus*, i.e. an earlier Romano-British settlement'. OE *wīc-hām*.

Wycombe (district) Bucks. A modern name, with reference to HIGH WYCOMBE, *see* next name.

Wycombe, High & Wycombe, West Bucks. *Wicumun* c.970, *Wicumbe* 1086 (DB). '(Place) at the dwellings or settlements'. OE *wīc* in a dative plural form *wīcum*. The river-name **Wye** here is probably only a back-formation from the place name.

Wyddfa, Yr. *See* SNOWDON.

Wyddgrug, Yr. *See* MOLD.

Wyddial Herts. *Widihale* 1086 (DB). 'Nook of land where willow-trees grow'. OE *wīthig* + *halh*.

Wye (river) Wales, Herefs. *See* ROSS ON WYE, HAY-ON-WYE. The river-name **Wye** in Derbyshire is identical in origin.

Wye Kent. *Uuiæ* 839, *Wi* 1086 (DB). '(Place at) the heathen temple'. OE *wīg*.

Wyke, 'the dwelling, the specialized farm or trading settlement', OE *wīc*; examples include: **Wyke** Surrey. *Wucha* 1086 (DB). **Wyke Regis** Dorset. *Wike* 984, *Kingeswik* 1242. Anciently a royal manor, hence the Latin affix *Regis* 'of the king'.

Wykeham N. Yorks. *Wicham* 1086 (DB). Probably 'homestead associated with an earlier Romano-British settlement'. OE *wīc-hām*.

Wylam Northum. *Wylum* 12th cent. Probably '(place at) the fish-traps'. OE *wīl* or *wīle* in a dative plural form *wīlum*.

Wylye Wilts. *Wilig* 901, *Wili* 1086 (DB). Named from the River Wylye, a pre-English river-name possibly meaning 'tricky stream', i.e. one liable to flood.

Wymering Portsm. *Wimeringes* 1086 (DB). '(Settlement of) the family or followers of a man called Wīgmǣr'. OE pers. name + *-ingas*.

Wymeswold Leics. *Wimundeswald* 1086 (DB). 'Forest land of a man called Wīgmund'. OE pers. name + *wald*.

Wymington Beds. *Wimentone* 1086 (DB). Probably 'farmstead of a man called Wīdmund'. OE pers. name + *tūn*.

Wymondham, 'homestead of a man called Wīgmund', OE pers. name + *hām*:
Wymondham Leics. *Wimundesham* 1086 (DB). **Wymondham** Norfolk. *Wimundham* 1086 (DB).

Wymondley, Great & Wymondley, Little Herts. *Wilmundeslea* 11th cent., *Wimundeslai* 1086 (DB). 'Woodland clearing of a man called Wilmund'. OE pers. name + *lēah*.

Wynford Eagle Dorset. *Wenfrot* 1086 (DB), *Wynfrod Egle* 1288. A Celtic river-name meaning 'white or bright stream', Celtic **wïnn* + **frud*. Manorial affix from the *del Egle* family, here in the 13th cent.

Wyre (district) Lancs. A modern adoption of the Lancs. river-name Wyre, perhaps of Celtic origin and meaning 'winding river', as for next name.

Wyre Forest Worcs., Shrops. *Wyre* c.1080. Possibly from a Celtic river-name meaning 'winding river'.

Wyre Piddle Worcs. *See* PIDDLE.

Wyrley, Great & Wyrley, Little Staffs. *Wereleia* 1086 (DB). 'Woodland clearing where bog-myrtle grows'. OE *wīr* + *lēah*.

Wysall Notts. *Wisoc* 1086 (DB), *Wisho* 1199. Possibly 'hill-spur of the heathen temple'. OE *wīg* + *hōh*.

Wytham Oxon. *Wihtham* c.957, *Winteham* [*sic*] 1086 (DB). 'Homestead or village in a river-bend'. OE **wiht* + *hām*.

Wytheford, Great Shrops. *Wicford* [*sic*] 1086 (DB), *Widiford* 1195. 'Ford where willow-trees grow'. OE *wīthig* + *ford*.

Wyton Cambs. *Witune* 1086 (DB), *Wictun* 1253. 'Dwelling-place, farmstead with a dwelling'. OE *wīc-tūn*.

Wyverstone Suffolk. *Wiuerthestune* 1086 (DB). 'Farmstead of a man called Wīgferth'. OE pers. name + *tūn*.

Wyvis Forest Highland. 'Forest by Ben Wyvis'. The mountain name means 'high hill', Gaelic *beinn* + *uais* (locative of *uas*).

Yaddlethorpe N. Lincs. *Iadulfestorp* 1086 (DB). 'Outlying farmstead or hamlet of a man called Ēadwulf'. OE pers. name + OScand. *thorp*.

Yafforth N. Yorks. *Iaforde* 1086 (DB). Possibly 'ford over the river (Wiske)'. OE *ēa* + *ford*.

Yalding Kent. *Hallinges* 1086 (DB), *Ealding* 1207. '(Settlement of) the family or followers of a man called Ealda'. OE pers. name + *-ingas*.

Yanworth Glos. *Janeworth c.*1050, *Teneurde* [*sic*] 1086 (DB). 'Enclosure used for lambs, or of a man called *Gæna*'. OE *ēan* or pers. name + *worth*.

Yapham E. R. Yorks. *Iapun* 1086 (DB). '(Place at) the steep slopes'. OE *gēap* in a dative plural form *gēapum*.

Yapton W. Sussex. *Abbiton c.*1187. 'Estate associated with a man called *Eabba*'. OE pers. name + *-ing-* + *tūn*.

Yarburgh Lincs. *Gereburg* 1086 (DB), *Jerdeburc* 12th cent. 'The earthwork, the fortification built of earth'. OE *eorth-burh*.

Yarcombe Devon. *Ercecombe* 10th cent., *Erticoma* 1086 (DB). 'Valley of the River Yarty'. Old river-name of uncertain origin + OE *cumb*.

Yardley, 'wood or clearing where rods or spars are obtained', OE *gyrd* + *lēah*: **Yardley** Birm. *Gyrdleah* 972, *Gerlei* 1086 (DB). **Yardley Gobion** Northants. *Gerdeslai* 1086 (DB), *Yerdele Gobioun* 1353. Manorial affix from the *Gubyun* family, here in the 13th cent. **Yardley Hastings** Northants. *Gerdelai* 1086 (DB), *Yerdele Hastings* 1316. Manorial affix from the *de Hastinges* family, here in the 13th cent.

Yarkhill Herefs. *Geardcylle* 811, *Archel* 1086 (DB). 'Kiln with a yard or enclosure'. OE *geard* + *cyln*.

Yarlet Staffs. *Erlide* 1086 (DB). Possibly 'gravelly slope, or eagle slope'. OE *ēar* or *earn* + *hlith*.

Yarlington Somerset. *Gerlincgetuna* 1086 (DB). 'Farmstead of the family or followers of a man called *Gerla*'. OE pers. name + *-inga-* + *tūn*.

Yarm Stock. on T. *Iarun* 1086 (DB). '(Place at) the fish-weirs'. OE *gear* in a dative plural form *gearum*.

Yarmouth I. of Wight. *Ermud* 1086 (DB), *Ernemuth* 1223. 'Gravelly or muddy estuary'. OE *ēaren* + *mūtha*. The river-name **Yar** is a back-formation from the place name.

Yarmouth, Great Norfolk. *Gernemwa* 1086 (DB). '(Place at) the mouth of the River Yare'. Celtic river-name (probably 'babbling stream') + OE *mūtha*.

Yarnfield Staffs. *Ernefeld* 1266. 'Open land frequented by eagles'. OE *earn* + *feld*.

Yarnscombe Devon. *Hernescome* 1086 (DB). 'Valley of the eagle'. OE *earn* + *cumb*.

Yarnton Oxon. *Ærdintune* 1005, *Hardintone* 1086 (DB). 'Estate associated with a man called *Earda*'. OE pers. name + *-ing-* + *tūn*.

Yarpole Herefs. *Iarpol* 1086 (DB), 'Pool with a weir or dam for catching fish'. OE *gear* + *pōl*.

Yarrow Sc. Bord. *Gierwa c.*1120. '(Place by) Yarrow Water'. The river-name means 'rough one' (Welsh *garw*).

Yarty (river) Devon. *See* YARCOMBE.

Yarwell Northants. *Jarwelle* 1166. 'Spring by the dam(s) for catching fish'. OE *gear* + *wella*.

Yate S. Glos. *Geate* 779, *Giete* 1086 (DB). '(Place at) the gate or gap'. OE *geat*.

Yateley Hants. *Yatele* 1248. 'Woodland clearing with or near a gate or gap'. OE *geat* + *lēah*.

Yatesbury Wilts. *Etesberie* 1086 (DB). 'Stronghold of a man called Gēat'. OE pers. name + *burh* (dative *byrig*).

Yattendon W. Berks. *Etingedene* 1086 (DB). Probably 'valley of the family or followers of a man called Gēat or Ēata'. OE pers. name + *-inga-* + *denu*.

Yatton N. Som. *Iatune* 1086 (DB). 'Farmstead by a river'. OE *ēa* + *tūn*.

Yatton Herefs. *Getune* 1086 (DB). 'Farmstead near a gate or gap'. OE *geat* + *tūn*.

Yatton Keynell Wilts. *Getone* 1086 (DB), *Yatton Kaynel* 1289. Identical in origin with the previous name. Manorial affix from the *Caynel* family, here in the 13th cent.

Yaverland I. of Wight. *Ewerelande* 683, *Evreland* 1086 (DB). 'Cultivated land where boars are kept'. OE *eofor* + *land*.

Yaxham Norfolk. *Jachesham* 1086 (DB). 'Homestead or enclosure of the cuckoo or of a man called *Gēac*'. OE *gēac* or pers. name + *hām* or *hamm*.

Yaxley, 'wood or clearing of the cuckoo', OE *gēac* + *lēah*: **Yaxley** Cambs. *Geaceslea* 963–84, *Iacheslei* 1086 (DB). **Yaxley** Suffolk. *Jacheslea* 1086 (DB).

Yazor Herefs. *Iavesovre* 1086 (DB), *Iagesoure* 1242. 'Ridge of a man called Iago'. Welsh pers. name + OE **ofer*.

Yeading Gtr. London. *Geddinges* 716–57, *Yeddings* 1325. '(Settlement of) the family or followers of a man called Geddi'. OE pers. name + *-ingas*.

Yeadon Leeds. *Iadun* 1086 (DB). Probably 'steep hill'. OE **gēh* + *dūn*.

Yealand Conyers & Yealand Redmayne Lancs. *Jalant* 1086 (DB), *Yeland Coygners* 1301, *Yeland Redman* 1341. 'High cultivated land'. OE *hēah* + *land*. Distinguishing affixes from early possession by the *de Conyers* and *Redeman* families.

Yealmpton Devon. *Elintona* 1086 (DB). 'Farmstead on the River Yealm'. Celtic or pre-Celtic river-name (of doubtful meaning) + OE *tūn*.

Yearsley N. Yorks. *Eureslage* 1086 (DB). 'Wood or clearing of the wild boar, or of a man called **Eofor*'. OE *eofor* or pers. name + *lēah*.

Yeaton Shrops. *Aitone* 1086 (DB), *Eton* 1327. 'Farmstead or estate on the river' (probably denoting a settlement which performed a special local function in relation to the river). OE *ēa* + *tūn*.

Yeaveley Derbys. *Gheveli* 1086 (DB). 'Woodland clearing of a man called **Geofa*'. OE pers. name + *lēah*.

Yedingham N. Yorks. *Edingham* 1170–5. 'Homestead of the family or followers of a man called Ēada'. OE pers. name + *-inga-* + *hām*.

Yelden Beds. *Giveldene* 1086 (DB). 'Valley of the tribe called *Gifle*'. Old tribal name (for which *see* NORTHILL) + OE *denu*.

Yeldham, Great & Yeldham, Little Essex. *Geldeham* 1086 (DB). Probably 'homestead liable to pay a certain tax'. OE *gield* + *hām*.

Yelford Oxon. *Aieleforde* 1086 (DB). 'Ford of a man called **Ægel*'. OE pers. name + *ford*.

Yell (island) Shet. *Yala* c.1250. Ancient pre-Scandinavian name, origin obscure.

Yelling Cambs. *Gillinge* 974, *Gellinge* 1086 (DB). Probably '(settlement of) the family or followers of a man called **Giella*'. OE pers. name + *-ingas*.

Yelvertoft Northants. *Gelvrecote* [sic] 1086 (DB), *Gelvertoft* 12th cent. Possibly 'homestead of a man called **Geldfrith*'. OE pers. name + OScand. *toft* (replacing OE *cot* 'cottage'). Alternatively the first part of the name may be 'ford at a pool', from an OE **gēol* + *ford*.

Yelverton Devon. *Elleford* 1291, *Elverton* 1765. Originally 'elder-tree ford', from OE *ellen* + *ford*, with the later addition of *toun* 'village'.

Yeoford Devon. *Ioweford* 1242. 'Ford on the River Yeo'. OE river-name (possibly 'yew stream') + *ford*.

Yeolmbridge Cornwall. *Yambrigge* 13th cent. Possibly 'bridge by the river-meadow'. OE *ēa* + *hamm* + *brycg*.

Yeovil Somerset. *Gifle* c.880, *Givele* 1086 (DB). '(Place on) the River *Gifl*'. Celtic river-name (meaning 'forked river'), an earlier name for the River Yeo (the form of which has been influenced by OE *ēa* 'river').

Yeovilton Somerset. *Geveltone* 1086 (DB). 'Farmstead on the River *Gifl*'. Celtic river-name (*see* YEOVIL) + OE *tūn*.

Yerbeston Pemb. *Jerbardeston* 14th cent. 'Gerbard's farm'. OGerman pers. name + OE *tūn*.

Yetlington Northum. *Yettlinton* 1187. Possibly 'estate associated with a man called *Gēatela*'. OE pers. name + *-ing- + tūn*.

Yetminster Dorset. *Etiminstre* 1086 (DB). 'Church of a man called Ēata'. OE pers. name + *mynster*.

Yettington Devon. *Yethemeton* 1242. 'Farmstead of the dwellers by the gate or gap'. OE *geat + hǣme + tūn*.

Yiewsley Gtr. London. *Wiuesleg* 1235. Probably 'woodland clearing of a man called *Wife*'. OE pers. name + *lēah*.

Ynys Byr. *See* CALDY ISLAND.

Ynys Cantwr (island) Pemb. 'Precentor's island'. Welsh *ynys + cantwr*. The island was formerly owned by the precentor of St David's.

Ynys Deullyn (island) Pemb. 'Island in the two lakes'. Welsh *ynys + dau + llyn*.

Ynys Dewi. *See* RAMSEY ISLAND.

Ynys Enlli. *See* BARDSEY ISLAND.

Ynys Gybi. *See* HOLY ISLAND.

Ynys Seiriol. *See* PUFFIN ISLAND.

Ynys-ddu Cphy. 'Black island'. Welsh *ynys + du*.

Ynysoedd y Moelrhoniaid. *See* SKERRIES.

Yockenthwaite N. Yorks. *Yoghannesthweit* 1241. 'Clearing of a man called Eogan'. OIrish pers. name + OScand. *thveit*.

Yockleton Shrops. *Ioclehuile* 1086 (DB). Possibly 'hill by a small manor or estate'. OE *geocled + hyll* (later replaced by *tūn*).

Yokefleet E. R. Yorks. *Iucufled* 1086 (DB). Probably 'creek or stream of a man called Jókell'. OScand. pers. name + *flēot*.

Yoker Glas. *Pont Yochyrr* 1505. '(Place on) low-lying ground'. Gaelic *euchair*.

York *Ebórakon c.*150, *Eboracum, Euruic* 1086 (DB). An ancient Celtic name meaning 'estate of a man called Eburos' or (more probably) 'yew-tree estate'. **Yorkshire** (OE *scīr* 'district') is first referred to in the 11th cent.

York Town Surrey, so named in the early 19th cent. after Frederick, Duke of York, who founded Sandhurst College in 1812.

Yorkletts Kent. *Yoclete* 1254. 'The small manor or estate'. OE *geocled*.

Yorton Shrops. *Iartune* 1086 (DB). 'Farmstead with a yard'. OE *geard + tūn*.

Youghal (*Eochaill*) Cork, Tipperary. 'Yew wood'.

Youlgreave Derbys. *Giolgrave* 1086 (DB). 'Yellow grove or pit'. OE *geolu + grǣfe* or *grǣf*.

Youlthorpe E. R. Yorks. *Aiulftorp* 1086 (DB), *Joletorp* 1166. 'Outlying farmstead or hamlet of a man called Eyjulfr or Jól(i)'. OScand. pers. name + *thorp*.

Youlton N. Yorks. *Ioletun* 1086 (DB). 'Farmstead of a man called Jóli'. OScand. pers. name + OE *tūn*.

Yoxall Staffs. *Iocheshale* 1086 (DB). Possibly 'nook comprising a yoke or measure of land'. OE *geoc + halh*.

Yoxford Suffolk. *Gokesford* 1086 (DB). 'Ford wide enough for a yoke of oxen'. OE *geoc + ford*. The river-name **Yox** is a back-formation from the place name.

Ysceifiog Flin. *Schiviau* 1086 (DB). '(Place) abounding in elder-trees'. Welsh *ysgaw*.

Ystalyfera Neat. *Ynys Tal y Veran* 1582, *Tir Ynystalverran* 1604, *Staleyfera* 1729, *Ystal-y-fera* 1831. Possibly 'river-meadow at the end of the short share (of land)'. Welsh *ynys + tâl + y + ber + rhan*.

Ystrad Mynach Cphy. *Ystrad manach* 1635, *Ystrad y Mynach* 1833. 'Valley of the monk'. Welsh *ystrad + mynach*.

Ystradgynlais Powys. *Stradgenles* 1372, *Estradgynles* 1493. 'Valley of Cynlais'. Welsh *ystrad*. The identity of Cynlais is unknown.

Ystumllwynarth. *See* OYSTERMOUTH.

Zeal, probably 'the hall or dwelling', OE *sele*, alternatively 'the sallow-tree copse', OE **sele*: **Zeal Monachorum** Devon. *Sele Monacorum* 1275, *Monekenesele* 1346. Affixes Latin *monachorum* and ME *monekene* mean 'of the monks' and refer to early possession by Buckfast Abbey. **Zeal, South** Devon. *La Sele* 1168.

Zeals Wilts. *Seale* 956, *Sele* 1086 (DB), *Seles* 1176. 'The sallow-tree(s)'. OE *sealh*.

Zelah Cornwall. *Sele* 1311. Identical in origin with ZEAL.

Zennor Cornwall. 'Church of *Sanctus Sinar*' *c*.1170. From the female patron saint of the church.

Zetland. *See* SHETLAND.

Glossary of Some Common Elements in British Place Names

In this list, OE stands for Old English, ME for Middle English, OFrench for Old French, and OScand. for Old Scandinavian. The OE letters 'thorn' and 'eth' have been rendered *th* throughout. The OE letter *æ* ('ash') represents a sound between *a* and *e*. Elements with an asterisk are postulated or hypothetical forms, that is they are words not recorded in independent use or only found in use at a later date.

á *OScand.* river
aber *Pictish, Welsh* river-mouth, estuary, confluence
āc *OE* oak-tree
*ācen *OE* made of oak, growing with oak-trees
achadh *Gaelic, Irish* field
ād *OE* funeral pyre, beacon
æcer *OE* plot of cultivated or arable land
ǣl, ēl *OE* eel
æppel *OE* apple-tree
ærn *OE* house, building, dwelling
æsc *OE* ash-tree
æspe *OE* aspen-tree, white poplar
ǣwell *OE* river-source, copious spring
ǣwelm *OE* river-source, copious spring
a(i)rd *Gaelic,* **aird** *Irish,* **ardd** *Welsh* height, point, promontory
akr *OScand.* plot of cultivated or arable land
ald, eald *OE* old, long used
alor *OE* alder-tree
*anger *OE* grassland, pasture
ānstīg, anstig *OE* single track, steep track
apuldor *OE* apple-tree
askr *OScand.* ash-tree
àth *Gaelic,* **áth** *Irish* ford
austr *OScand.* east
bæc *OE* back, ridge
bæce, bece *OE* stream in a valley
bærnet *OE* land cleared by burning
bæth *OE* bath, artificial pool
*bagga *OE* bag, *probably* badger
baile *Gaelic, Irish* farmstead, homestead, village, townland
banke *OScand.* bank, hill-slope
bār *OE* boar
bēacen *OE* beacon, signal

beag *Gaelic, Irish* small
béal *Irish* mouth
bēam *OE* tree-trunk, beam of timber
bēan *OE* bean
bearu *OE* grove, wood
beau, bel *OFrench* fine, beautiful
bēce *OE* beech-tree
beinn *Gaelic,* **beann, binn** *Irish* peak, mountain
bekkr *OScand.* stream
bēl *OE* fire, funeral pyre, beacon
*bel *OE* glade, dry ground in marsh
bēo *OE* bee
beofor *OE* beaver
beonet *OE* bent-grass, coarse grass
beorc *OE* birch-tree
beorg, berg *OE* rounded hill, mound, tumulus
*bēos *OE* bent-grass, rough grass
bere *OE* barley
bere-ærn *OE* barn, store-house for barley
bere-tūn, bær-tūn *OE* barley farm, corn farm, *later* outlying grange or demesne farm
bere-wīc *OE* barley farm, *later* outlying grange or dependent farm
berg *OScand.* hill
*bic, *bica *OE* pointed hill or ridge
bīcere *OE* bee-keeper
bigging *ME* a building
*billing *OE* hill, ridge
birce *OE* birch-tree
blæc *OE* black, dark-coloured
blaen *Welsh* end, river-source, upland
blá(r) *OScand.* dark, cheerless, exposed
bōc *OE* (i) beech-tree, (ii) book, charter
bōc-land *OE* land granted by charter

boga *OE* bow, arch, bend
*boi(a) *OE* boy, servant
ból *OScand.* farm
bólstathr *OScand.* dwelling, homestead, farm
bóndi *OScand.* peasant landowner
bóthar *Irish* road
botm, *bothm *OE*, **botn** *OScand.* broad river valley, valley bottom
bōthl, bōtl, bold *OE* special house or building
box *OE* box-tree
brād *OE* broad, spacious
brēc, bræc *OE* land broken up for cultivation
bred *OE* board, plank
*breh *Celtic* hill
brekka *OScand.* hill-slope
brēmel, brembel *OE* bramble, blackberry bush
brend *ME* burnt, cleared by burning
brēr *OE* brier, wild rose
Brettas *OE* Britons
brōc *OE* brook, stream, *often used of muddy streams*
brocc *OE* badger
brōm *OE* thorny bush, broom
brunnr *OScand.* spring, stream
brycg *OE* bridge, *sometimes* causeway
bryn *Welsh* hill
bucc *OE* buck, stag
bucca, bucc *OE* buck (male deer or he-goat)
bula *OE* bull
būr *OE* cottage, dwelling
(ge)būr *OE* peasant holding land for rent or services
burh (*dative* **byrig**) *OE* fortified place, stronghold, *variously applied to Iron-Age hill-forts, Roman and Anglo-Saxon fortifications, and fortified houses, later to manors or manor houses and to towns or boroughs*
burh-tūn, byrh-tūn *OE* farmstead near or belonging to a stronghold or manor house, fortified farmstead
burna *OE* stream, *often used of clear streams*
bury *ME* (*from dative* byrig *of OE* burh) manor, manor house
butere *OE* butter
búth, bōth *OScand.* booth, temporary shelter
bý *OScand.* farmstead, village, settlement

byden *OE* vessel, tub, hollow
byge *OE* river-bend
byht *OE* river-bend
*cā *OE* jackdaw
*cadeir *Celtic* chair, lofty place
caer *Welsh* fort, stronghold
cærse *OE* cress, water-cress
caiseal *Irish* stone fort
caisleán *Irish* castle
cāl *OE* cabbage
calc, cealc *OE* chalk, limestone
cald, ceald *OE* cold
calf, cealf *OE* calf
calu *OE* bald, bare, lacking vegetation, *often bare hill*
camb *OE* hill-crest, ridge
camp *OE* enclosed piece of land, *originally open uncultivated land on the edge of a Romano-British settlement*
caol *Gaelic, Irish* narrow (place), strait
carraig *Gaelic, Irish*, **carreg** *Welsh* rock, stone
cat(t) *OE* cat, wild-cat
ceann *Gaelic, Irish* head, headland, end
cēap *OE* trade, market
ceaster, cæster *OE* Roman station or walled town, old fortification or earthwork
*cę̄d *Celtic* forest, wood
ceorl *OE* freeman, peasant
cēping, cīeping *OE* market
cēse, *cīese *OE* cheese
cild *OE* child, youth, younger son, young nobleman
cill *Gaelic, Irish* church, churchyard
cirice *OE* church
cisel, ceosol *OE* gravel, shingle
clǣfre *OE* clover
clǣg *OE* clay, clayey soil
clif *OE* cliff, steep slope, river-bank
cloch *Irish* stone
*clōh *OE* ravine, deep valley
*clopp(a) *OE* lump, hill
cluain *Gaelic, Irish* meadow
cniht *OE* youth, servant, retainer
cnoc *Gaelic, Irish* (round) hill
cnoll *OE* hill-top, *later* hillock
cocc *OE* (i) heap, hillock, (ii) cock of wild bird, woodcock
coed *Welsh* trees, wood
cofa *OE* chamber, cave, cove
coill *Irish*, **coille** *Gaelic* wood

col *OE* coal, charcoal
cōl *OE* cool
copp *OE* hill-top, summit
cot *OE* cottage, hut, shelter
cran, corn *OE* crane, *also probably* heron *or similar bird*
craobh *Gaelic, Irish* tree
crāwe *OE* crow
croft *OE* enclosure, small enclosed field
cros *OE* cross
crug *Welsh* hill, mound, cairn
*crūg *Celtic* hill, mound, tumulus
cū *OE* cow
cumb *OE* coomb, valley, *used particularly of relatively short or broad valleys*
*Cumbre *OE* the Cymry, the Cumbrian Britons
cwēn *OE* queen
cwene *OE* woman
cweorn *OE* quern, hand-mill
cwm *Welsh* valley
cyln *OE* kiln
cyne- *OE* royal
cyning *OE* king
cȳta *OE* kite
dæl *OE* pit, hollow, valley
dalr *OScand.* valley
denn *OE* woodland pasture, especially for swine
denu *OE* valley, *used particularly of relatively long, narrow valleys*
dēop *OE* deep
dēor *OE* animal, *also* deer
derne, dierne *OE* hidden, overgrown with vegetation
dīc *OE* ditch, dyke, *also* embankment
din *Welsh* fortress, stronghold
*djúr, dýr *OScand.* animal, *also* deer
doire *Gaelic, Irish* oak grove
dôl *Welsh* meadow
domhnach *Irish* church
dræg *OE* a portage or slope used for dragging down loads, *also* a dray, a sledge
druim *Gaelic*, **droim** *Irish*, **trum** *Welsh* ridge
dūn *OE* hill, *used particularly of low hills with fairly level and extensive summits*
dùn *Gaelic*, **dún** *Irish* fortress, castle
ēa *OE* river
earn *OE* eagle
ēast *OE* east, eastern
ecg *OE* edge, escarpment

edisc *OE* enclosure, enclosed park
ēg, īeg *OE* island, land partly surrounded by water, dry ground in marsh, well-watered land, promontory
*eglēs *Celtic* Romano-British Christian church
eik *OScand.* oak-tree
elfitu *OE* swan
elle(r)n *OE* elder-tree
elm *OE* elm-tree
ende *OE*, **endi** *OScand.* end, district of an estate
ened *OE* duck
Engle *OE* the Angles, *later* the English
eofor *OE* wild boar
eorl *OE* nobleman
eowestre *OE* sheep-fold
eowu *OE* ewe
erg, ǽrgi *OScand.* shieling, hill pasture, summer pasture
ersc *OE* ploughed land
eski *OScand.* place growing with ash-trees
ey *OScand.* island
fæger *OE* fair, pleasant
fæsten *OE* stronghold
fāg, fāh *OE* variegated, multi-coloured
fald *OE* fold, enclosure for animals
fearn *OE* fern, bracken
feld *OE* open country, tract of land cleared of trees
fenn *OE* fen, marshy ground
fennig *OE* marshy, muddy
ferja *OScand.* ferry
fjǫrthr *OScand.* sea inlet, firth
fleax *OE* flax
flēot *OE* estuary, inlet, creek, *also* stream
fola *OE*, **foli** *OScand.* foal
folc *OE* folk, tribe, people
ford *OE* ford, river-crossing
fox *OE* fox
fugol *OE* (wild) fowl, bird
fūl *OE* foul, dirty
*funta *OE* spring (*a loan-word from Latin* fons, fontis *and therefore possibly used of a spring characterized by Roman building work*)
fyrhth(e) *OE* woodland, *often* sparse woodland or scrub
fyrs *OE* furze
gærs, græs *OE*, **gres** *OScand.* grass

gāra *OE* triangular plot of ground, point of land

garthr *OScand.* enclosure

gāt *OE*, **geit** *OScand.* goat

gata *OScand.* road, street

*gē *OE* district, region

gēac *OE* cuckoo

geard *OE* yard, enclosure

geat *OE* gate, gap, pass

gil *OScand.* deep narrow valley, ravine

glan *Welsh* river-bank, hillock

glas *Gaelic*, *Irish* grey-green

gleann *Gaelic*, *Irish*, **glyn** *Welsh* valley

gōs *OE* goose

græf, *grafa *OE* pit, trench

grǣfe, grāf(a) *OE* grove, copse, coppiced wood

grēne *OE* (i) green-coloured, (ii) grassy place, village green

grēot *OE* gravel

gwaun *Welsh* moorland

hæcc *OE* hatch, hatch-gate, flood-gate

hæg, gehæg *OE* enclosure

hǣme *OE* inhabitants, dwellers

*hǣr *OE* rock, heap of stones

*hǣs, *hǣse *OE* (land overgrown with) brushwood

hæsel *OE* hazel-tree

hǣth *OE* heath, heather, uncultivated land overgrown with heather

hafoc *OE* hawk

haga *OE*, **hagi** *OScand.* hedged enclosure

hagu-thorn *OE* hawthorn

halh *OE* nook or corner of land, *often used of land in a hollow or river-bend, or of dry ground in marsh*

hālig *OE* holy

hām *OE* homestead, village, manor, estate

hamm *OE* enclosure, land hemmed in by water or marsh or higher ground, land in a river-bend, river-meadow, promontory

*hamol, *hamel *OE* crooked, flat-topped

hām-stede, hǣm-styde *OE* homestead, site of a dwelling

hām-tūn *OE* home farm or settlement, enclosure in which a homestead stands

hana *OE* cock (of wild bird)

hangra *OE* sloping wood, wood on a slope

hār *OE* hoar, grey, *also* boundary

hara *OE* hare

*hāth *OE* heath, heather, uncultivated land

haugr *OScand.* hill, mound, tumulus

hēafod *OE* head, headland, end of a ridge, river-source

hēah *OE* high, *also* chief

hearg *OE* heathen shrine or temple

hecg(e) *OE* hedge

hēg, hīeg *OE* hay

helde, hielde *OE* slope

hengest *OE* stallion

henn *OE* hen (of wild bird)

heorot *OE* hart, stag

here *OE* army

hīd *OE* hide of land, amount of land for the support of one free family and its dependants (*usually about 120 acres*)

hind *OE* hind, doe

hīwan (*genitive* **hīgna**) *OE* household, members of a family or religious community

hīwisc *OE* household, amount of land for the support of a family

hlāw, hlǣw *OE* tumulus, mound, hill

hlinc *OE* bank, ledge, terrace

hlith *OE*, **hlíth** *OScand.* hill-slope

hlōse *OE* pigsty

hōc *OE* hook or corner of land, land in a bend

hōh *OE* heel of land, projecting hill-spur

hol *OE* (i) *as noun* a hole or hollow, (ii) *as adjective* hollow, deep

holegn *OE* holly

holmr *OScand.* island, promontory, raised ground in marsh, river-meadow

holt *OE* wood, thicket, *often used of a single-species wood*

hop *OE* small enclosed valley, enclosure in marsh or moor

horn *OE*, *OScand.*, ***horna** *OE* horn, horn-shaped hill or piece of land

hors *OE* horse

horu *OE* filth, dirt

hræfn *OE*, **hrafn** *OScand.* raven

hramsa *OE* wild garlic

hrēod *OE* reed, rush

hrīs *OE*, **hrís** *OScand.* brushwood, shrubs

hrīther *OE* ox, cattle

hrōc *OE*, **hrókr** *OScand.* rook

hrycg *OE*, **hryggr** *OScand.* ridge

hund *OE* dog

hunig *OE* honey

hunta *OE* huntsman
hūs *OE*, **hús** *OScand.* house
*hvin *OScand.* gorse
hwǣte *OE* wheat
hwēol *OE* wheel, circular feature
hwerfel, hwyrfel *OE*, **hvirfill** *OScand.* circle, circular feature
hwīt *OE*, **hvítr** *OScand.* white
hyll *OE* hill
hyrst *OE* wooded hill
hȳth *OE* landing-place or harbour, inland port
inbhir *Gaelic* river-mouth, confluence
-ing *OE suffix* place or stream characterized by, place belonging to
-ing- *OE connective particle implying* associated with *or* called after
-inga- *OE genitive (possessive) case of* **-ingas**
-ingas *OE plural suffix* people of, family or followers of, dwellers at
inis *Gaelic, Irish,* **ynys** *Welsh* island
īw, ēow, *ig *OE* yew-tree
kaldr *OScand.* cold
kalfr *OScand.* calf
karl *OScand.* freeman, peasant
kelda *OScand.* spring
kirkja *OScand.* church
kjarr *OScand.* marsh overgrown with brushwood
konungr, kunung *OScand.* king
krók *OScand.* bend, land in a river-bend
kross *OScand.* cross
lacu *OE* small stream, side-channel
lād *OE* water-course, **gelād** *OE* (difficult) river-crossing
*lǣcc, *lǣce *OE* stream, bog
lǣs *OE* pasture, meadow-land
lamb *OE* lamb
land *OE, OScand.* tract of land, estate, cultivated land
*lann *Cornish, OWelsh,* **llan** *Welsh* churchyard, church-site, church
lang *OE*, **langr** *OScand.* long
lāwerce *OE* lark
lēac *OE* leek, garlic
lēah *OE* wood, woodland clearing or glade, *later* pasture, meadow
leitir *Irish* hill-side
līn *OE*, **lín** *OScand.* flax
lind *OE* lime-tree
linn *Gaelic, Irish* pool

lios *Irish* ring fort
llech *Welsh* stone, rock
llyn *Welsh* lake
loc, loca *OE* lock, fold, enclosure
loch *Gaelic, Irish* lake
lundr *OScand.* small wood or grove
lȳtel *OE* little
ma *Welsh* plain, low-lying ground
mǣd *OE* meadow
mǣgden *OE* maiden
mǣl, mēl *OE* cross, crucifix
maen *Welsh* stone
mǣne, gemǣne *OE* held in common, communal
mǣre, gemǣre *OE* boundary
maes *Welsh* field
mainistir *Irish* monastery
*mapel, *mapul, mapuldor *OE* maple-tree
marr *OScand.* fen, marsh
meall *Gaelic* round hill, mountain
mearc *OE* boundary
melr *OScand.* sandbank, sand-hill
mēos *OE* moss
mere, mær(e) *OE* pond, pool, lake
mersc *OE* marsh, marshland
micel, mycel *OE*, **mikill** *OScand.* great
middel *OE* middle
*mönïth *Celtic,* **mynydd** *Welsh* mountain, hill
mont *OFrench, ME* mount, hill
mōr *OE*, **mór** *OScand.* moor, marshy ground, barren upland
mos *OE*, **mosi** *OScand.* moss, marsh, bog
munuc *OE* monk
mūs *OE*, **mús** *OScand.* mouse
mūtha *OE* river-mouth, estuary
myln *OE* mill
mynster *OE* monastery, church of a monastery, minster or large church
mýrr *OScand.* mire, bog, swampy ground
mȳthe, gemȳthe *OE* confluence of rivers
nant *Welsh* stream, valley
nǣss, ness *OE*, **nes** *OScand.* promontory, headland
nīwe, nēowe *OE* new
north *OE* north, northern
ōfer *OE* bank, margin, shore
*ofer, *ufer *OE* flat-topped ridge, hill, promontory
ōra *OE* (i) shore, hill-slope, flat-topped ridge, (ii) ore, iron-ore

oter *OE* otter

oxa *OE* ox

pæth *OE* path, track

pen *Welsh* head, top, end

*penn *Celtic* head, end, hill

penn *OE* pen, fold, enclosure for animals

peru *OE* pear

*pett *Pictish* share, portion (of land)

pirige, pyrige *OE* pear-tree

plega *OE* play, games, sport

plŭme *OE* plum, plum-tree

pōl, *pull *OE* pool, pond, creek

pont *Welsh* bridge

port *OE* (i) harbour, (ii) town, market town, market, (iii) gate

porth *Cornish, Welsh* harbour

prēost *OE* priest

pres, prys *Welsh* brushwood, thicket

pyll *OE* tidal creek, pool, stream

pytt *OE* pit, hollow

rá *OScand.* (i) roe, roebuck, (ii) boundary

ráth *Irish* ring fort

rauthr *OScand.* red

rēad *OE* red

rēfa *OE* reeve, bailiff

rhiw *Welsh* hill

*ric *OE* raised straight strip of land, narrow ridge

risc, *rysc *OE* rush

rīth, rīthig *OE* small stream

rodu *OE* clearing

*ros *Celtic*, ros *Gaelic, Irish*, rhos *Welsh* promontory; moorland; wood

rŭh *OE* rough

ryge *OE*, rugr *OScand.* rye

sǽ *OE* sǽr *OScand.* sea, inland lake

sǽte *OE plural* dwellers, settlers

sǽtr *OScand.* shieling, hill pasture, summer pasture

salh, sealh *OE* sallow, willow

salt *OE* (i) salt, (ii) salty (*adjective*)

sand *OE*, sandr *OScand.* sand

saurr *OScand.* mud, dirt, sour ground

sceaga *OE* small wood, copse

scēap, scēp, scīp, *scīep *OE* sheep

scēat *OE* corner or angle of land, projecting wood or piece of land

*scēla *OE* temporary hut or shelter

scelf, scielf, scylfe *OE*, skjalf *OScand.* shelf of level ground, ledge

scēne, scīene *OE* bright, beautiful

scīr *OE* (i) shire, district, (ii) bright, clear

scucca *OE* evil spirit, demon

Seaxe *OE* the Saxons

sele *OE* dwelling, house, hall

*sele, *siele *OE* sallow or willow copse

seofon *OE* seven

set, geset *OE* dwelling, stable, fold

sīc *OE* small stream

sīd *OE* large, extensive

sīde *OE*, side *ME* hill-side, land alongside a river or wood

skáli *OScand.* temporary hut or shed

skógr *OScand.* wood

slæd *OE* valley

*slæp *OE* slippery muddy place

slakki *OScand.* shallow valley

sliabh *Gaelic, Irish* mountain, moor

slōh *OE* slough, mire

sol *OE* muddy place

srath *Gaelic* (wide) valley

stæth *OE*, stǫth *OScand.* landing-place

stall, steall *OE* stall for animals, fishing pool

stān *OE*, steinn *OScand.* stone, rock, boundary stone

stapol *OE* post, pillar of wood or stone

stede *OE* enclosed pasture, place, site

steort *OE* tail or tongue of land

stīg *OE*, stígr *OScand* upland path, narrow road

stoc *OE* place, outlying farmstead or hamlet, secondary or dependent settlement

stocc *OE*, stokkr *OScand.* tree-trunk, stump, log

stōd *OE* stud, herd of horses

stōw *OE* place, assembly-place, holy place

strǽt *OE* Roman road, paved road

strōd *OE* marshy land overgrown with brushwood

sumor *OE*, sumarr *OScand.* summer

sūth *OE* south, southern

swan *OE* swan

swān *OE* herdsman, peasant

swīn *OE* svín *OScand.* swine, pig

tâl *Welsh* end

teach *Irish* house

teampall *Irish*, teampull *Gaelic* church

thing *OE*, *OScand.* assembly, meeting

thorn *OE*, *OScand.* thorn-tree

thorp *OScand.* secondary settlement, dependent outlying farmstead or hamlet

throp *OE* dependent outlying farmstead, hamlet

Thunor *OE heathen Saxon god* (*corresponding to the Scandinavian* Thor)

thveit *OScand.* clearing, meadow, paddock

thyrne *OE*, **thyrnir** *OScand.* thorn-tree

thyrs *OE* giant, demon

ticce(n) *OE* kid, young goat

tir *Gaelic*, **tír** *Irish* land, territory

Tīw, Tig *OE heathen Germanic god*

tobar *Gaelic, Irish* well

tóchar *Irish* causeway

toft, topt *OScand.* site of a house or building, curtilage, homestead

torr *OE* rock, rocky hill

*tōt(e) *OE* look-out place

toun *ME* (*from OE* tūn) village, manor, estate

tre *Cornish*, **tre(f)** *Welsh* farmstead, homestead, hamlet, village

trēo(w) *OE* tree, post, beam

tūn *OE* enclosure, farmstead, village, manor, estate

tún *OScand.* enclosure, farmstead

twī- *OE* double

twisla *OE* fork of a river, junction of streams

upp *OE* higher up

vágr *OScand.* bay, creek

vath *OScand.* ford

vík *OScand.* bay, creek, inlet

vithr *OScand.* wood

vrá, rá *OScand.* nook or corner of land

wād *OE* woad

wæd, gewæd *OE* ford, crossing-place

*wæsse *OE* riverside land liable to flood

wæter *OE* water, river, lake

wald, weald *OE* woodland, forest, high forest-land later cleared

walh, wealh *OE* Briton, Welshman

wall, weall *OE* wall, bank

-ware *OE plural* dwellers

weard *OE* watch, ward, protection

weg *OE* way, track, road

wella, well(e), wiella, wiell(e), wælla, wæll(e) *OE* spring, stream

weorc, geweorc *OE* building, fortification

wer, wær *OE* weir, river-dam, fishing-enclosure in a river

west *OE* west, western

wēt, wæt *OE* wet, damp

wether *OE* wether, ram

wīc *OE* earlier Romano-British settlement; dwelling, specialized farm or building, dairy farm; trading or industrial settlement, harbour

wice *OE* wych-elm

*wīc-hām *OE* homestead associated with an earlier Romano-British settlement

wīd *OE* wide, spacious

wīg, wēoh *OE* heathen shrine or temple

*wilig, welig *OE* willow-tree

wisc, *wisse *OE* marshy meadow

wīthig *OE* withy, willow-tree

Wōden *OE heathen Germanic god* (*corresponding to the Scandinavian* Othin)

wōh *OE* twisted, crooked

worth, worthig, worthign *OE* enclosure, enclosed farmstead or settlement

wudu, *earlier* **widu** *OE* wood, forest, *also* timber

wulf *OE*, **ulfr** *OScand.* wolf

wyrm, wurm *OE* reptile, snake, *also* dragon

yfer *OE* edge or brow of a hill

ystrad *Welsh* valley, valley-floor

ynys *Welsh* island

Select Bibliography for Further Reading

Other Dictionaries and Works of Reference

Census of Ireland 1851, general alphabetical index to the townlands and towns, parishes and baronies of Ireland (Dublin 1861, reprinted Baltimore 1984).

Charles, B. G., *Non-Celtic Place Names in Wales* (London 1938).

Ekwall, E., *The Concise Oxford Dictionary of English Place Names*, 4th edition (Oxford 1960).

Field, J., *English Field-Names: a Dictionary* (Newton Abbot 1972, reprinted Gloucester 1989).

Forster, K., *A Pronouncing Dictionary of English Place Names* (London 1981).

Gelling, M., Nicolaisen, W. F. H., and Richards, M., *The Names of Towns and Cities in Britain* (London 1970, revised reprint 1986).

Hogan, E. I., *Onomasticon Goedelicum, an index, with identifications, to the Gaelic names of places and tribes* (Dublin 1910).

Joyce, P. W., *The Origin and History of Irish Names of Places*, 3 volumes (Dublin 1869–1913).

—— *Irish Local Names Explained* (Dublin 1870), reprinted as *A Pocket Guide* (Belfast 1984).

McKay, P., *A Dictionary of Ulster Place Names*, 2nd edition (Belfast 2007).

Mills, A. D., *A Dictionary of English Place Names* (Oxford 1991, 1998).

Owen, H. W., *The Place Names of Wales* (Cardiff 1998, 2000).

—— and Morgan, R., *Dictionary of the Place Names of Wales* (Llandysul 2007).

Padel, O. J., *Cornish Place Name Elements*, EPNS 56/57 (Cambridge 1985).

Parsons, D. and Styles, T., with Hough, C., *The Vocabulary of English Place Names (Á-Box), (Brace-Cæster), (Ceafor-Cockpit)* (Nottingham 1997, 2000, 2004)

Pointon, G. E. (ed.), *BBC Pronouncing Dictionary of British Names*, 2nd edition (Oxford 1983).

Room, A., *A Concise Dictionary of Modern Place Names in Great Britain and Ireland* (Oxford 1983).

—— *A Dictionary of Irish Place Names* (Belfast 1994).

Smith, A. H., *English Place Name Elements*, EPNS 25, 26 (Cambridge 1956).

Spittal, J. and Field, J., *A Reader's Guide to the Place Names of the United Kingdom* (Stamford 1990).

tSuirbhéireacht Ordanáis, An [The Ordnance Survey of Ireland] *Gasaitéar na hÉireann/Gazetteer of Ireland* (Dublin 1989); *Liostaí Logainmneacha*/Lists of Irish forms of names: *Counties Limerick, Louth, Waterford* (Dublin 1991), *County Kilkenny* (Dublin 1993), *Offaly* (Dublin 1994), *County Monaghan* (Dublin 1996).

Watts, V. E., *The Cambridge Dictionary of English Place Names* (Cambridge 2004).

County Surveys and Monographs

Armstrong, A. M., Mawer, A., Stenton, F. M., and Dickins, B., *The Place Names of Cumberland*, EPNS 20–2 (Cambridge 1950–2).

Beckensall, S., *Place Names and Field Names of Northumberland* (Stroud 2006).

Broderick, G., *Place Names of the Isle of Man*, volumes 1–7 (Tubingen 1994–2005).

—— *A Dictionary of Manx Place Names* (Nottingham 2006).

Cameron, K., *The Place Names of Derbyshire*, EPNS 27–9 (Cambridge 1959).

—— *The Place Names of Lincolnshire*, EPNS 58, 64/65, 66, 71 (Nottingham 1985, 1991, 1992, 1996), (with Field. J. and Insley, J.), EPNS 73, 77 (Nottingham 1997, 2001), (with Insley, J. and Cameron, J.), EPNS 85

(Nottingham 2010) (several volumes to follow).

—— (with contributions by Insley, J.), *A Dictionary of Lincolnshire Place Names* (Nottingham 1998).

Charles, B. G., *The Place Names of Pembrokeshire*, volumes I and II (Aberystwyth 1992).

Coates, R., *The Place Names of Hampshire* (London 1989).

—— *The Ancient and Modern Names of the Channel Islands* (Stamford 1991).

Coplestone-Crow, B., *Herefordshire Place Names*, British Archaeological Reports, British Series 214 (Oxford 1989).

Cox, B., *The Place Names of Rutland*, EPNS 67/68/69 (Nottingham 1994).

—— *The Place Names of Leicestershire*, EPNS 75, 78, 81, 84 (Nottingham 1998, 2002, 2004, 2009) (some volumes to follow).

—— *A Dictionary of Leicestershire and Rutland Place-Names* (Nottingham 2005).

Cox, R., *The Gaelic Place Names of Carloway, Isle of Lewis* (Dublin 2002).

Cullen, Paul, *The Place Names of the Lathes of St Augustine and Shipway, Kent* (unpublished DPhil. thesis, University of Sussex, 1997).

Dodgson, J. McN., *The Place Names of Cheshire*, EPNS 44–8, 54 (Cambridge 1970–81), (with Rumble, A. R.), *The Place Names of Cheshire*, EPNS 74 (Nottingham 1997).

Duignan, W. H., *Notes on Staffordshire Place Names* (London 1902).

Ekwall, E., *The Place Names of Lancashire* (Manchester 1922).

Fägersten, A., *The Place Names of Dorset* (Uppsala 1933).

Field, J., *Place Names of Greater London* (London 1980).

Fraser, I. A., *The Place Names of Arran* (Glasgow 1999).

Gelling, M., *The Place Names of Berkshire*, EPNS 49–51 (Cambridge 1973–6).

—— *The Place Names of Oxfordshire*, EPNS 23–4 (Cambridge 1953–4).

—— with Foxall, H. D. G., *The Place Names of Shropshire*, EPNS 62/3, 70, 76, 80, 82 (Nottingham 1990, 1995, 2001, 2004, 2006) (some volumes to follow).

Glover, J., *The Place Names of Kent* (London 1976).

Gover, J. E. B., Mawer, A., and Stenton, F. M., *The Place Names of Devon*, EPNS 8, 9 (Cambridge 1931–2).

—— *The Place Names of Hertfordshire*, EPNS 15 (Cambridge 1938).

—— *The Place Names of Northamptonshire*, EPNS 10 (Cambridge 1933).

—— *The Place Names of Nottinghamshire*, EPNS 17 (Cambridge 1940).

—— *The Place Names of Wiltshire*, EPNS 16 (Cambridge 1939).

—— with Bonner, A., *The Place Names of Surrey*, EPNS 11 (Cambridge 1934).

—— with Houghton, F. T. S., *The Place Names of Warwickshire*, EPNS 13 (Cambridge 1936).

—— with Madge, S. J., *The Place Names of Middlesex (apart from the City of London)*, EPNS 18 (Cambridge 1942).

Horovitz, D., *The Place Names of Staffordshire* (Brewood 2005).

Hughes, A. J. and Hannan, R. J., *Place Names of Northern Ireland*, volume 2, County Down II: *The Ards Peninsula* (Belfast 1992).

Jones, G. T. and Roberts, T., *The Place Names of Anglesey* (Bangor 1996).

Kökeritz, H., *The Place Names of the Isle of Wight* (Uppsala 1940).

McAleer, P., *Townland Names of Co. Tyrone* (c.1936, reprinted Draperstown 1988).

Macdonald, A., *The Place Names of West Lothian* (Edinburgh 1941).

MacGabhann, F., *Place Names of Northern Ireland*, volume 7, County Antrim II: *Ballycastle and North-East Antrim* (Belfast 1997).

McKay, P., *Place Names of Northern Ireland*, volume 4, County Antrim I: *The Baronies of Toome* (Belfast 1995).

—— *Place Names of Northern Irela~* volume 8, County Fermana~ *and District: the Parish~* (Belfast 2004).

—— and Muh~ *Names a~*

Mawer, A., *7~ Northumber~* 1920).

Owen, O. ... *Flints* ... *Padel, O.* ... *Place Na* ...

—— and Stenton, F. M., *The Place Names of Bedfordshire and Huntingdonshire*, EPNS 3 (Cambridge 1926).

—— *The Place Names of Buckinghamshire*, EPNS 2 (Cambridge 1925).

—— with Gover, J. E. B., *The Place Names of Sussex*, EPNS 6, 7 (Cambridge 1929–30).

—— and Houghton, F. T. S., *The Place Names of Worcestershire*, EPNS 4 (Cambridge 1927).

Mills, A. D., *The Place Names of Dorset*, EPNS 52, 53, 59/60, 86/87 (Cambridge 1977–89, Nottingham 2010) (one volume to follow).

—— *Dorset Place Names: their Origins and Meanings* (Wimborne 1986, revised and reprinted Newbury 1998).

—— *The Place Names of the Isle of Wight* (Stamford 1996).

—— *A Dictionary of London Place Names* (Oxford 2001 second edition, 2010).

—— *Discover Dorset Place Names* (Wimborne 2008).

Mills, D., *The Place Names of Lancashire* (London 1976).

Moore Munn, Alfred, *Notes on the Place Names of the Parishes and Townlands of the County of Londonderry* (Cookstown 1925, reprinted Draperstown 1985).

Morgan, R., *A Study of Radnorshire Place Names* (Llanrwst 1998).

—— and Powell, R. F. P., *A Study of Breconshire Place Names* (Llanrwst 1999).

—— *A Study of Montgomeryshire Place Names* (Llanrwst 2001).

—— *Place Names of Gwent* (Llanrwst 2005).

Muhr, K., *Place Names of Northern Ireland*, volume 6, County Down IV: *North-West Down/Iveagh* (Belfast 1996).

Oakden, J. P., *The Place Names of Staffordshire*, EPNS 55 (Cambridge 1984) (several volumes to follow).

Ó Mainnín, M. B., *Place Names of Northern Ireland*, volume 3, County Down III: *The Mournes* (Belfast 1993).

Muraíle, N., *Mayo Places: their Names and Origins* (Dublin 1985).

H. W., *The Place Names of East shire* (Cardiff 1994).

L., *A Popular Dictionary of Cornish nes* (Penzance 1988).

Pierce, G. O., *The Place Names of Dinas Powys Hundred* (Cardiff 1968).

—— *Place Names in Glamorgan* (Cardiff 2002).

Power, P., *Place Names of the Decies* (London 1907, reprinted Cork 1952).

Price, L., *The Place Names of County Wicklow*, parts 1–7 (Dublin 1945–67).

Reaney, P. H., *The Place Names of Cambridgeshire and the Isle of Ely*, EPNS 19 (Cambridge 1943).

—— *The Place Names of Essex*, EPNS 12 (Cambridge 1935).

Sandred, K. I. (and Lindström, B.), *The Place Names of Norfolk*, EPNS 61, 72, 79 (Nottingham 1989, 1996, 2002) (several volumes to follow).

Skeat, W. W., *The Place Names of Suffolk* (Cambridge 1913).

Smith, A. H., *The Place Names of Gloucestershire*, EPNS 38–40 (Cambridge 1964–5).

—— *The Place Names of the East Riding of Yorkshire and York*, EPNS 14 (Cambridge 1937).

—— *The Place Names of the North Riding of Yorkshire*, EPNS 5 (Cambridge 1928).

—— *The Place Names of the West Riding of Yorkshire*, EPNS 30–7 (Cambridge 1961–3).

—— *The Place Names of Westmorland*, EPNS 42, 43 (Cambridge 1967).

Stewart, J., *Shetland Place Names* (Lerwick 1987).

Taylor, S. and Márkus, G., *The Place names of Fife*, volume 1, *West Fife between Leven and Forth*, volume 2, *Central Fife between the rivers Leven and Eden*, volume 3, *St Andrews and the East Neuk* (Donington 2006–2010).

Toner, G. and Ó Mainnín, M. B., *Place Names of Northern Ireland*, volume 1, County Down I: *Newry and South-West Down* (Belfast 1992).

Toner, G., *Place Names of Northern Ireland*, volume 5, County Derry I: *The Moyola Valley* (Belfast 1996).

Turner, A. G. C., *The Place Names of North Somerset* (unpublished typescript, 1950).

Wallenberg, J. K., *Kentish Place Names* (Uppsala 1931).

—— *The Place Names of Kent* (Uppsala 1934).

Walsh, P., *The Place Names of Westmeath* (Dublin 1957).

Watts, V., *A Dictionary of County Durham Place Names* (Nottingham 2002).

—— (edited by Paul Cavill) *The Place Names of County Durham* EPNS 83 (Nottingham 2007) (several volumes to follow).

Whaley, D., *A Dictionary of Lake District Place Names* (Nottingham 2006).

Various Studies on the Interpretation and Significance of Place Names

Cameron, K., *English Place Names*, new edition (London 1996).

—— (ed.), *Place Name Evidence for the Anglo-Saxon Invasion and Scandinavian Settlements*, EPNS (Nottingham 1977) (collection of eight essays).

—— 'The Significance of English Place Names', *Proceedings of the British Academy* 62 (Oxford 1976).

Carroll, J. and Parsons, D., *Anglo-Saxon Mint-Names 1. Axbridge-Hythe* (Nottingham 2007).

Cavill, P., Harding, S. E., and Jesch, J., *Wirral and its Viking Heritage* (Nottingham 2000).

Cavill, P and Broderick, G. (eds.), *Language Contact in the Place Names of Britain and Ireland* (Nottingham 2007).

Coates, R. and Breeze, A., with a contribution by Horovitz, D., *Celtic Voices, English Places: Studies of the Celtic Impact on Place Names in England* (Stamford 2000).

Copley, G. J., *Archaeology and Place Names in the Fifth and Sixth Centuries*, British Archaeological Reports, British Series 147 (Oxford 1986).

—— *Early Place Names of the Anglian Regions of England*, British Archaeological Reports (Oxford 1988).

—— *English Place Names and their Origins* (Newton Abbot 1968).

Dorward, D., *Scotland's Place Names* (Edinburgh 1995).

Ekwall, E., *English Place Names in -ing*, 2nd edition (Lund 1962).

—— *English River-Names* (Oxford 1928).

—— *Old English wic in Place Names* (Uppsala 1964).

Fellows-Jensen, G., *Scandinavian Settlement Names in the East Midlands* (Copenhagen 1978).

—— *Scandinavian Settlement Names in the North-West* (Copenhagen 1985).

—— *Scandinavian Settlement Names in Yorkshire* (Copenhagen 1972).

Field, J., *A History of English Field-Names* (London 1993).

Flanagan, D. and Flanagan, L., *Irish Place Names* (Dublin 1994).

Gammeltoft, P. and Jørgensen, B. (eds.), *Names Through the Looking-Glass: Festschrift in Honour of Gillian Fellows-Jensen* (Copenhagen 2006).

Gelling, M., *Place Names in the Landscape* (London 1984).

—— *Signposts to the Past*, 3rd edition (Chichester 1997).

—— *The West Midlands in the Early Middle Ages* (Leicester 1992).

—— and Cole, A., *The Landscape of Place Names* (Stamford 2000).

Higham, N. (ed.), *Britons in Anglo-Saxon England* (Woodbridge 2007).

Hooke, D., *The Anglo-Saxon Landscape: the Kingdom of the Hwicce* (Manchester 1985).

—— (ed.), *Anglo-Saxon Settlements* (Oxford 1988).

Hough, C., 'Women in English place names', in *'Lastworda betst'. Essays in Memory of Christine E. Fell with her Unpublished Writings*, edited by C. Hough and K. A. Lowe (Donington 2002).

Jackson, K. H., *Language and History in Early Britain* (Edinburgh 1953).

Lawson, Z. (ed.), *Aspects of Lancashire History: Essays in Memory of Mary Higham* (Preston 2007).

Mackay, G., *Scottish Place Names* (New Lanark 2000).

Mawer, A. and Stenton, F. M. (eds.), *Introduction to the Survey of English Place Names*, EPNS 1 (Cambridge 1924).

Muhr, K., *Celebrating Ulster's Townlands* (Belfast 1999).

Nicolaisen, W. F. H., *Scottish Place Names: Their Study and Significance*, new edition (Edinburgh 2001).

Padel, O. J. and Parsons, D. N. (eds.), *A Commodity of Good Names: Essays in*

Honour of Margaret Gelling (Donington 2008).

Reaney, P. H., *The Origins of English Place Names* (London 1960).

Rivet, A. L. F. and Smith, C., *The Place Names of Roman Britain* (London 1979).

Ross, D., *Scottish Place Names* (Edinburgh 2001).

Rumble, A. R. and Mills, A. D. (eds.), *Names, Places and People: An Onomastic Miscellany in Memory of John McNeal Dodgson* (Stamford 1997).

Sandred, K. I., *English Place Names in -stead* (Uppsala 1963).

Scott, M., *Scottish Place Names* (Edinburgh 2008).

Taylor, S., *The Uses of Place Names* (Edinburgh 1998).

Wainwright, F. T., *Archaeology and Place Names and History* (London 1962).

Watson, W. J., *The History of the Celtic place names of Scotland* (Edinburgh and London 1926).

—— *Scottish Place Name Papers* (London 2002).

Dictionary of British Place Names: Appendix of Web Links

(⊕) SEE WEB LINKS

This is a web-linked dictionary. To access the websites listed below, go to the dictionary's web page at www.oup.com/uk/reference/resources/britishplacenames,click on web links in the Resources section and click straight through to the relevant websites.

Ainmean-Àite na h-Alba ~ Gaelic Place Names of Scotland
National Gazetteer of Gaelic Place Names.

Archif Melville Richards (Place Name Database at the University of Bangor, North Wales)
Comprehensive archive of Welsh place names created by the late Professor Melville Richards.

British History Online (Institute of Historical Research)
Access to important printed source materials for place name research.

The British Library
Information on their collections, help with research.

Bunachar Logainmneacha na hÉireannFiontar: The Placenames Database of Ireland
Comprehensive list of Irish place names.

The Domesday Book Online
Information about Domesday places and landowners, and about the origins of place names.

The English Place Name Society
Information about membership of the Society, which publishes the results of the Survey of English Place-Names and an annual journal.

Institute for Name-Studies (at the University of Nottingham)
Information about the English Place Name Society, the ongoing Survey of English Place Names, and another major project, The Vocabulary of English Place Names.

Landscapes of Governance: Assembly Sites in England 5th–11th Centuries
UCL Institute of Archaeology: national research project bringing archaeology, place names and written sources together.

Langscape:The Language of Landscape Project: Reading the Anglo-Saxon Countryside
Makes accessible a rich body of material relating to the English countryside of a thousand years ago and more (based on a corpus of Anglo-Saxon boundary clauses).

'Localhistoryonline'
Lists local history societies and courses (with web links to these), also publishes *Local History Magazine*.

The National Archives at Kew
Information about the range of archives held here and in other record repositories, research guides to local history.

Northern Ireland Place Name Project at Queen's University Belfast
Centre for the study of Gaelic place names in the UK and for research into the origin and meaning of the place names of Northern Ireland. Information about their database, gazetteer and archive.

Ordnance Survey: Great Britain's national mapping agency
Information about maps and the mapping of place names, introductions to the language types of British place names.

'People's Place Names: Your Names for Places in Great Britain'
Research project run by Cardiff University to gather information about vernacular and informal place names.

'Perceptions of Place: English Place Name Study and Regional Variety'

Project funded by Arts and Humanities Research Council.

Prosopography of Anglo-Saxon England

Database providing information relating to all the recorded inhabitants of England from the late 6th to the late 11th century.

The Romano-British Settlements

Information about Roman sites and place names in Britain.

Royal Historical Society

Publishes 'Bibliography of British and Irish History'.

The Scottish Place Name Society ~ Comann Ainmean-Àlte na h-Alba

Information about membership of the Society.

The Shetland Place Names Project

Information on Shetland place names accessible through a database linked to digital maps.

Society for Name Studies in Britain and Ireland

Information about name studies for members of the Society and other interested individuals. The Society publishes an annual journal, *Nomina*.

Ulster Place Name Society

Information about membership of the Society, which publishes a journal.

The Welsh Language Board

Information about Welsh place names and their standardization.

Oxford Paperback Reference

The Kings of Queens of Britain
John Cannon and Anne Hargreaves

A detailed, fully-illustrated history ranging from mythical and pre-conquest rulers to the present House of Windsor, featuring regional maps and genealogies.

A Dictionary of World History

Over 4,000 entries on everything from prehistory to recent changes in world affairs. An excellent overview of world history.

A Dictionary of British History
Edited by John Cannon

An invaluable source of information covering the history of Britain over the past two millennia. Over 3,000 entries written by more than 100 specialist contributors.

Review of the parent volume
'the range is impressive . . . truly (almost) all of human life is here'

Kenneth Morgan, Observer

Oxford Paperback Reference

The Concise Oxford Dictionary of English Etymology
T. F. Hoad

A wealth of information about our language and its history, this reference source provides over 17,000 entries on word origins.

'A model of its kind'

Daily Telegraph

A Dictionary of Euphemisms
R. W. Holder

This hugely entertaining collection draws together euphemisms from all aspects of life: work, sexuality, age, money, and politics.

Review of the previous edition
'This ingenious collection is not only very funny but extremely instructive too'

Iris Murdoch

The Oxford Dictionary of Slang
John Ayto

Containing over 10,000 words and phrases, this is the ideal reference for those interested in the more quirky and unofficial words used in the English language.

'hours of happy browsing for language lovers'

Observer

OXFORD

Oxford Paperback Reference

The Concise Oxford Companion to English Literature
Margaret Drabble and Jenny Stringer

Based on the best-selling *Oxford Companion to English Literature*, this is an indispensable guide to all aspects of English literature.

Review of the parent volume
'a magisterial and monumental achievement'

Literary Review

The Concise Oxford Companion to Irish Literature
Robert Welch

From the ogam alphabet developed in the 4th century to Roddy Doyle, this is a comprehensive guide to writers, works, topics, folklore, and historical and cultural events.

Review of the parent volume
'Heroic volume ... It surpasses previous exercises of similar nature in the richness of its detail and the ecumenism of its approach.'

Times Literary Supplement

A Dictionary of Shakespeare
Stanley Wells

Compiled by one of the best-known international authorities on the playwright's works, this dictionary offers up-to-date information on all aspects of Shakespeare, both in his own time and in later ages.

Oxford Paperback Reference

The Concise Oxford Dictionary of Quotations
FIFTH EDITION
Edited by Elizabeth Knowles

Now based on the highly acclaimed sixth edition of *The Oxford Dictionary of Quotations*, this new edition maintains its extensive coverage of literary and historical quotations, and contains completely up-to-date material. A fascinating read and an essential reference tool.

The Oxford Dictionary of Political Quotations
Edited by Antony Jay

This lively and illuminating dictionary from the writer of 'Yes Minister' presents a vintage crop of over 4,000 political quotations. Ranging from the pivotal and momentous to the rhetorical, the sincere, the bemused, the tongue-in-cheek, and the downright rude, examples include memorable words from the old hands as well as from contemporary politicians.

'funny, striking, thought-provoking and incisive...will appeal to those browsing through it as least as much as to those who wish to use it as a work of reference'
Observer

Oxford Dictionary of Modern Quotations
Edited by Elizabeth Knowles

The answers to all your quotation questions lie in this delightful collection of over 5,000 of the twentieth century's most famous quotations.

'Hard to sum up a book so useful, wayward and enjoyable'   *Spectator*

The Oxford Dictionary of Literary Quotations
Edited by Peter Kemp

Containing 4,000 of the most memorized and cited literary quotations, this dictionary is an excellent reference work as well as an enjoyable read.

The Oxford Dictionary of Humorous Quotations
Edited by Ned Sherrin

From the sharply witty to the downright hilarious, this sparkling collection will appeal to all senses of humour.

Oxford Paperback Reference

The Oxford Dictionary of Art & Artists
Ian Chilvers

Based on the highly praised *Oxford Dictionary of Art*, over 2,500 up-to-date entries on painting, sculpture, and the graphic arts.

'the best and most inclusive single volume available, immensely useful and very well written'

Marina Vaizey, *Sunday Times*

The Concise Oxford Dictionary of Art Terms
Michael Clarke

Written by the Director of the National Gallery of Scotland, over 1,800 terms cover periods, styles, materials, techniques, and foreign terms.

A Dictionary of Architecture and Landscape Architecture
James Stevens Curl

Over 6,000 entries and 250 illustrations cover all periods of Western architectural history.

'splendid . . . you can't have a more concise, entertaining, and informative guide to the words of architecture.'

Architectural Review

'excellent, and amazing value for money . . . by far the best thing of its kind.'

Professor David Walker

Oxford Paperback Reference

The Concise Oxford Dictionary of World Religions
Edited by John Bowker

Over 8,200 entries containing unrivalled coverage of all the major world religions, past and present.

'covers a vast range of topics ... is both comprehensive and reliable'

The Times

The Oxford Dictionary of Saints
David Farmer

From the famous to the obscure, over 1,400 saints are covered in this acclaimed dictionary.

'an essential reference work'

Daily Telegraph

The Concise Oxford Dictionary of the Christian Church
E. A. Livingstone

This indispensable guide contains over 5,000 entries and provides full coverage of theology, denominations, the church calendar, and the Bible.

'opens up the whole of Christian history, now with a wider vision than ever'

Robert Runcie, former Archbishop of Canterbury

Oxford Paperback Reference

Concise Medical Dictionary

Over 12,000 clear entries covering all the major medical and surgical specialities make this one of our best-selling dictionaries.

'"No home should be without one" certainly applies to this splendid medical dictionary'

Journal of the Institute of Health Education

'An extraordinary bargain'

New Scientist

'Excellent layout and jargon-free style'

Nursing Times

A Dictionary of Nursing

Comprehensive coverage of the ever-expanding vocabulary of the nursing professions. Features over 10,000 entries written by medical and nursing specialists.

An A–Z of Medicinal Drugs

Over 4,000 entries cover the full range of over-the-counter and prescription medicines available today. An ideal reference source for both the patient and the medical professional.